Spectral, Convolution and Numerical Techniques in Circuit Theory

Fuad Badrieh

Spectral, Convolution and Numerical Techniques in Circuit Theory

Springer

Fuad Badrieh
Micron Technology
Boise, ID, USA

ISBN 978-3-030-10057-5 ISBN 978-3-319-71437-0 (eBook)
https://doi.org/10.1007/978-3-319-71437-0

Printed on acid-free paper

This Springer imprint is published by the registered company Springer International Publishing AG part of Springer Nature.
The registered company address is: Gewerbestrasse 11, 6330 Cham, Switzerland

I dedicate this book to my wife Nell and our children Ahmad, Salwa, Laila, Mahdi and Munir, and to my mother and late father.

Preface

This text deals with applied techniques for solving circuit problems. The techniques span spectral, convolution, and numerical ones. The applications span *RLC* circuits, distributed ones (such as transmission and diffusion lines), ones with feedback, multi-port networks, and even some rudimentary transistors. In pretty much all of the problems, we apply analytic techniques which mainly follow the divide-and-conquer principle. Be it number of harmonics (both in space and time), convolution time step, or finite difference time step, the applied methods give the user a good answer quickly and a better one if one uses more harmonics or finer time step. In all cases, predicted answer is compared to SPICE simulations, and in pretty much all cases, we see excellent match.

The backbone of the text gathers from various fields of engineering and mathematics. From electrical engineering, it builds on topics such as signals and systems, basic circuit theory, multi-port networks, feedback, and transistor modeling. From electromagnetics, it builds on electrostatics, diffusion lines, and transmission lines. From mathematics, it builds on differential and partial differential equations, functional mathematics, numerical analysis, and complex analysis. And finally, from computer science, it builds on algorithms, coding, SPICE, and graphics.

The book is intended for the student, researcher, and practitioner. In school it could be used for an upper undergraduate or a starter graduate class, which would typically follow a class in signals and systems and one or two classes in circuits. In industry, it can be used in any circuit design involving *RLC* elements, signal integrity, power delivery, analog design, feedback, and RF modeling. The book can also be used as a sample of scientific computation and visualization.

Boise, ID, USA — Fuad Badrieh

Acknowledgments

I'd like to acknowledge the support from Micron Technology for allowing me to work on this project and for their understanding. I'd like to pass special thanks and acknowledgment for my supervisor Brent Keeth (Senior Fellow), for supporting me throughout the 4 years it took to finish this work. Thanks are also due to Tony Huff and Emily Berriochoa for positive feedback and encouragement; Adam El-Mansouri for helpful discussions; and John Wilford (Director) and John Schreck (VP) for having an open mind about proceeding with this work. Outside of Micron I'd like to thank Professor Ata Zadehgol (University of Idaho) and Professor Vishal Saxena (also University of Idaho) for valuable discussions and for initial book review for the publisher. Due thanks are for Dr. D. Ferry who was my supervisor in graduate work and who set a high example for learning and publishing. From the publisher side, I would like to acknowledge Charles (Chuck) Glaser for seeing the potential behind this work, and Murugesan Tamilselvan and Monica Paul for an excellent job in editing and producing this work.

Contents

1 Introduction

1.1 Scope of Book

Consider the *RLC* circuit shown in Fig. 1.1. It is a multi-branch circuit, with linear elements *RLC* and linear laws which are derived from Maxwell's equations. The circuit is stimulated by an input source. This system is ubiquitous in the field of electrical engineering, but when broken down to numbers and equations it could easily represent other fields of engineering, such as mechanical, aerospace, chemical, ..., etc. Really we could swap in mechanical springs and pulleys, or pipes and valves, or loads and beams—any system that is distributed and linear. The system has inner components, is stimulated, and we want to measure some outcome (output voltage here). How can we do that?

The most common starting methods are

- **Analytic differential equations**: We could set up KVL (Kirchoff Voltage Law) and KCL (Kirchoff Current Law), derive the integro-differential equations, apply suitable initial conditions, and solve for the various currents and voltages. Analytics could also involve series expansion in terms of polynomials.
- **Numerical finite difference**: We could discretize the analytic differential equations, setup a linear system, solve it (or invert matrices), and arrive at i and v, for each branch, for all time.
- **Software**: Of course we could throw it at SPICE and click the simulate button!

While the above methods work, they suffer for an inherent limitation: every time the stimulus changes we have to re-solve the *whole* system—every branch, every branch current, and every node voltage! Is there nothing unique about the system that can be reutilized? Do we have to start from zero every time?

It turns out, there is something unique about the system, and that uniqueness can be harnessed for the second round of analysis. The idea here is to "characterize the system" and then use that information to facilitate the solution for any later stimulus. To achieve this we fall back on two, very powerful methods:

- **Spectral methods**: Here we characterize the system in the *frequency* domain, and derive the corresponding *transfer function*. This function ties output to input for every frequency point. That is, if the input is a sine/cosine at frequency ω_0, the transfer function gives the corresponding output at that same frequency. But what if the input is not a sine/cosine? The beauty here lies in the fact that we can represent pretty much any input in terms of sines/cosines, and using *superposition* we are guaranteed a solution!
- **Convolution methods**: In essence we apply a rather arbitrary input, be it impulse, step, ramp

F. Badrieh, *Spectral, Convolution and Numerical Techniques in Circuit Theory*,
https://doi.org/10.1007/978-3-319-71437-0_1

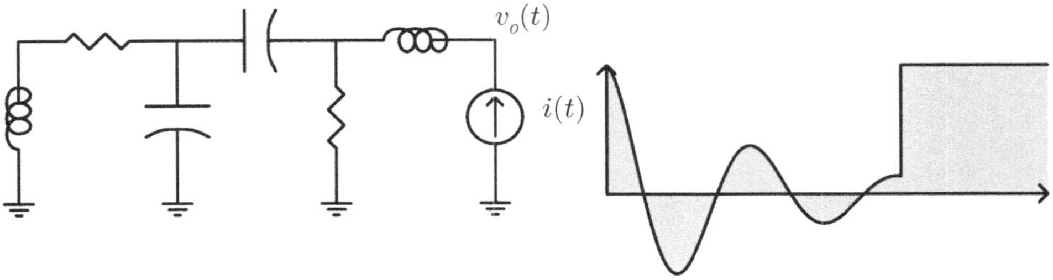

Fig. 1.1 Arbitrary linear system with arbitrary stimulus

(or even others), figure the solution for that particular stimulus, then convolve that solution with any other input stimulus. To achieve this we would need the relation between the used impulse and the arbitrary stimulus.

It is these last two methods that form the bone of this text; but, and for completeness we do touch on a few other methods, just so reference, and to put things in perspective. As such, the remaining of this chapter will give a quick survey on the methods used, and the remaining of the text will drill into them!

Another advantage of the last two methods (spectral and convolution) is that sometimes the system is characterized rather arbitrary (or by brute force), via measurement or through a simulation tool, and the inner workings (branches and nodes) are not readily available for the conventional methods, such as analytic or numerical differential equations. The beauty of spectral/convolution methods is that they can take this single measurement/simulation output and completely determine the system workings for *any* other stimulus. This is a very powerful concept and will take time to fully absorb, but after all that's what the rest of the book is here for!

1.2 Steady State Methods

This method assumes that the stimulus is in the form of $e^{j\omega t}$. It replaces R, C, and L with their frequency dependent impedances, applies KVL and KCL, ends up with algebraic equations tying input to output, and solves for output. This method is good and sufficient for *steady state response*, but does not capture the *transient* part of the solution. Also, typically it is assumed that storage elements are initialized to zero state; i.e., initial conditions are zero.

1.3 Analytic Differential Equations Methods

Here the various relations tying currents to voltages are used in the time domain, and KVL and KCL applied to them yield a set of differential equations tying input to output. The differential equations are then solved for and result in final currents and voltages. Initial conditions can be taken into account, and solutions are comprehensive in that they include both transient and steady state components. Another variant of this method is polynomial series expansion: here the solution is assumed as a series, the series is plugged into KVL/KCL, and relations arise which dictate the series coefficients.

1.4 Numerical Differential Equations Methods

Here KVL and KCL differential equations are discretized (in time), differentiation and integration converted to algebraic relations, and currents and voltages solved for next time step based on prior time step solution. This method is very powerful, especially with power of hardware, and covers transient and steady state solutions altogether. Quite often, and for a large system, the discretization ends up with a linear system, of the form $Ax = b$, and then matrices are factored or inverted to give the solution at each time step.

1.5 Transform/Frequency Techniques

Similar to the steady state, elements are assigned frequency dependent impedances, but inputs are not limited to $e^{j\omega t}$. Inputs can assume any form, such as unit steps, impulses, or negative exponentials. By knowing the response to $e^{j\omega t}$ and by decomposing any input in terms of $e^{j\omega t}$, one is able to predict the output for any input. Hence, solution captures both transient and steady state parts. Also, initial conditions can be folded in the flow.

1.6 Convolution Methods

Here, response to an impulse or step input (or for that matter a ramp, quadratic, ...) is figured, and that response is convolved with any other input to find response to that input. That is, by knowing response to an impulse (or step, or ramp ...) we are able to predict output for all responses. This method applies in the time domain, and can take care of initial conditions.

1.7 Superposition

A complicated circuit can also be tackled using superposition. For example, if there are multiple stimuli to the circuit, we can enable them one at a time, find the corresponding solution, and *then* add all solutions in the end. If the input has multiple terminals, we can stimulate one at a time, and in the end generate a multi-port model which can be fed as-is into many SPICE-compatible tools.

1.8 Linearization

Many nonlinear effects can be linearized locally. As such, we can attempt to model complex nonlinear systems using simple linear models, such as *RLC* elements. We can even attempt "binned" linear models, for each operating point range.

1.9 Continuous Media

Spectral and convolution techniques also find wide application in *continuous media*, such as transmission and diffusion lines. For those cases we are able to derive exact solutions for voltages and currents which can be used to gauge the accuracy of the finite-segment representation.

1.10 Summary

This chapter gave a top-level introduction to the scope of the book and the various methods covered. The two primary methods are the spectral and convolution ones; but for completeness we touch on other methods, such as steady state, analytic differential equations, numerical techniques, and such. Also the chapter touched on some important concepts throughout the text, such as superposition, linearization, and continuous media.

2 Steady State Solutions to Circuit Problems

2.1 Introduction

Steady state AC theory assumes that input is of the form $e^{j\omega_0 t}$, replaces R, C, and L with their frequency dependent impedances, and applies KVL and KCL to end up with a set of algebraic equations, from which all currents and voltages are derived. Notice that the ω_0 is a particular angular frequency; the answer (be it current or voltage) would apply only for that frequency! This method is illustrated via a few examples.

2.2 Series *RC* Driven by Sine Function

Consider the series RC network shown in Fig. 2.1. We want to find out branch current and output voltage. Rather than dealing with $\sin\omega_0 t$ let's for now assume that input is $e^{j\omega_0 t}$. The impedance of the network is given by

$$Z(\omega_0) = R + \frac{1}{j\omega_0 C} = \frac{1 + j\omega_0 RC}{j\omega_0 C} \tag{2.1}$$

Current is given by

$$I(\omega_0) = \frac{V(\omega_0)}{Z(\omega_0)} = \frac{j\omega_0 C}{1 + j\omega_0 RC} \tag{2.2}$$

Output voltage is given by current times impedance of cap

$$V_o(\omega_0) = I(\omega_0)\frac{1}{j\omega_0 C} = \frac{1}{1 + j\omega_0 RC} \tag{2.3}$$

So that in the time domain

$$v_o(t) = \frac{1}{1 + j\omega_0 RC} e^{j\omega_0 t} \tag{2.4}$$

That is, if our input is $e^{j\omega_0 t}$ then our output is

$$e^{j\omega_0 t} \rightarrow \frac{1}{1 + j\omega_0 RC} e^{j\omega_0 t} \tag{2.5}$$

Similarly

$$e^{-j\omega_0 t} \rightarrow \frac{1}{1 - j\omega_0 RC} e^{-j\omega_0 t} \tag{2.6}$$

Our real input is a sine which is given by

$$\sin\omega_0 t = \frac{e^{j\omega_0 t} - e^{-j\omega_0 t}}{2j} \tag{2.7}$$

Then our real output would be

$$v_o(t) = \frac{1}{2j}\left[\frac{1}{1 + j\omega_0 RC} e^{j\omega_0 t} - \frac{1}{1 - j\omega_0 RC} e^{-j\omega_0 t}\right] \tag{2.8}$$

F. Badrieh, *Spectral, Convolution and Numerical Techniques in Circuit Theory*,
https://doi.org/10.1007/978-3-319-71437-0_2

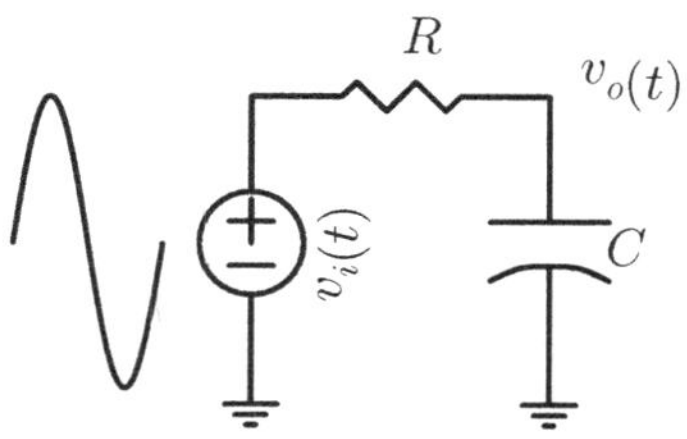

Fig. 2.1 Series RC driven by sine function

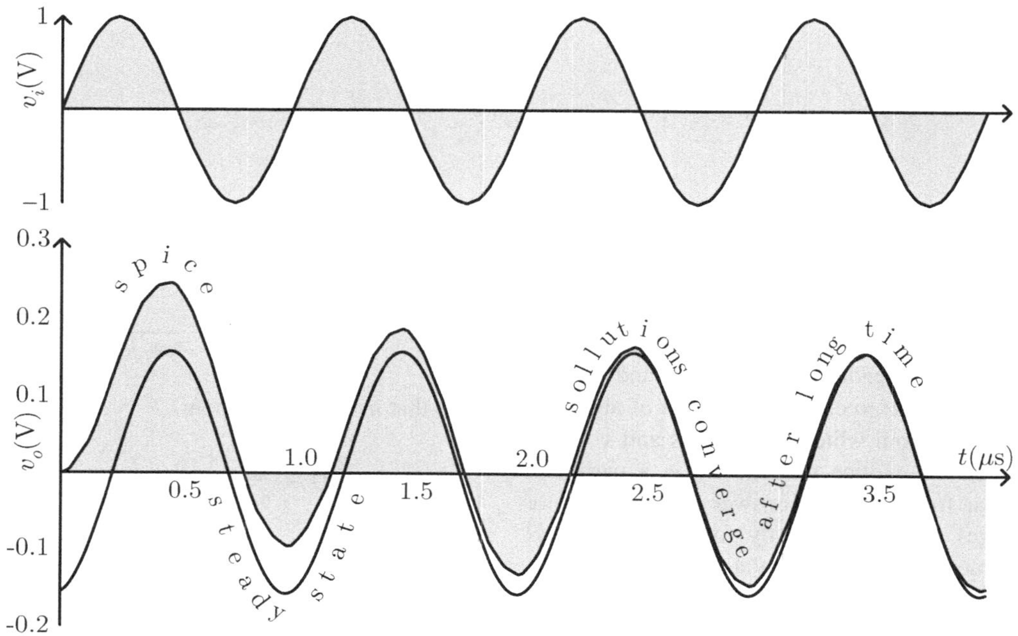

Fig. 2.2 Response of series RC network to sine input: exact and steady state results. Case of $R = 1\ \Omega$, $C = 1\ \mu$F, and $\omega_0 = 2\pi \times 10^6$

Now we will need to do a series of simplifications.

$$
\begin{aligned}
v_o(t) &= \frac{1}{2j}\left[\frac{\cos\omega_0 t + j\sin\omega_0 t}{1 + j\omega_0 RC} - \frac{\cos\omega_0 t - j\sin\omega_0 t}{1 - j\omega_0 RC}\right] \\
&= \frac{1}{2j}\frac{[1 - j\omega_0 RC][\cos\omega_0 t + j\sin\omega_0 t] - [1 + j\omega_0 RC][\cos\omega_0 t - j\sin\omega_0 t]}{1 + \omega_0^2 R^2 C^2} \\
&= \frac{1}{2j}\frac{2j\sin\omega_0 t - 2j\omega_0 RC\cos\omega_0 t}{1 + \omega_0^2 R^2 C^2} \\
&= \boxed{\frac{\sin\omega_0 t - \omega_0 RC\cos\omega_0 t}{1 + \omega_0^2 R^2 C^2}}
\end{aligned}
\tag{2.9}
$$

Results are plotted in Fig. 2.2 along with SPICE results. Notice that sure enough, when things settle down, both results match. That is, steady state results match exact results, only after transient components have died.

2.3 Series *RC* Driven by Cosine Input

Again consider the series *RC* network shown in Fig. 2.1 but this time input voltage is a cosine function (as opposed to a sine one). We want to find out branch current and output voltage. From the prior section we know that a complex exponential input gave the following response:

$$e^{j\omega_0 t} \rightarrow \frac{1}{1+j\omega_0 RC} e^{j\omega_0 t}, \quad \text{and} \quad e^{-j\omega_0 t} \rightarrow \frac{1}{1-j\omega_0 RC} e^{-j\omega_0 t} \tag{2.10}$$

Using the definition of the cosine function

$$\cos \omega_0 t = \frac{e^{j\omega_0 t} + e^{-j\omega_0 t}}{2} \tag{2.11}$$

we can tell that the response would be

$$\begin{aligned} v(t) &= \frac{1}{2}\frac{e^{j\omega_0 t}}{1+j\omega_0 RC} + \frac{1}{2}\frac{e^{-j\omega_0 t}}{1-j\omega_0 RC} \\ &= \frac{1}{2}\frac{[1-j\omega_0 RC][\cos \omega_0 t + j \sin \omega_0 t] + [1+j\omega_0 RC][\cos \omega_0 t - j \sin \omega_0 t]}{1+\omega_0^2 R^2 C^2} \\ &= \boxed{\frac{\cos \omega_0 t + \omega_0 RC \sin \omega_0 t}{1+\omega_0^2 R^2 C^2}} \end{aligned} \tag{2.12}$$

These results, along with SPICE results, are shown in Fig. 2.3. Notice that once things settle down, again steady state solution matches exact one.

2.4 Series *RC* Driven by Periodic Pulse Input (with 0 DC Average)

Consider the series *RC* network in Fig. 2.4 which is driven by a periodic pulse (with zero average). How can we use steady state to predict steady state output? We are accustomed to input being a sine/cosine (prior two sections), but here we have a square input. As will be shown later in the text, it turns out we can represent an arbitrary periodic shape via a combination of sines/cosines. In particular, we have

$$v_i(t) = A_0 + \sum_{n=1}^{\infty} A_n \cos \omega_n t + \sum_{n=1}^{\infty} B_n \sin \omega_n t \tag{2.13}$$

where

$$\omega_n = n\omega_0, \quad \text{and} \quad \omega_0 = \frac{2\pi}{T} \tag{2.14}$$

In this case, the B_n drops out and we end up with

$$A_0 = 0.0, \quad A_n = \frac{2}{\pi n} \sin \frac{n\pi}{2} \tag{2.15}$$

so that

$$\boxed{v_i(x) = \sum_{n=1}^{\infty} \frac{2}{\pi n} \sin \frac{n\pi}{2} \cos \omega_n t} \tag{2.16}$$

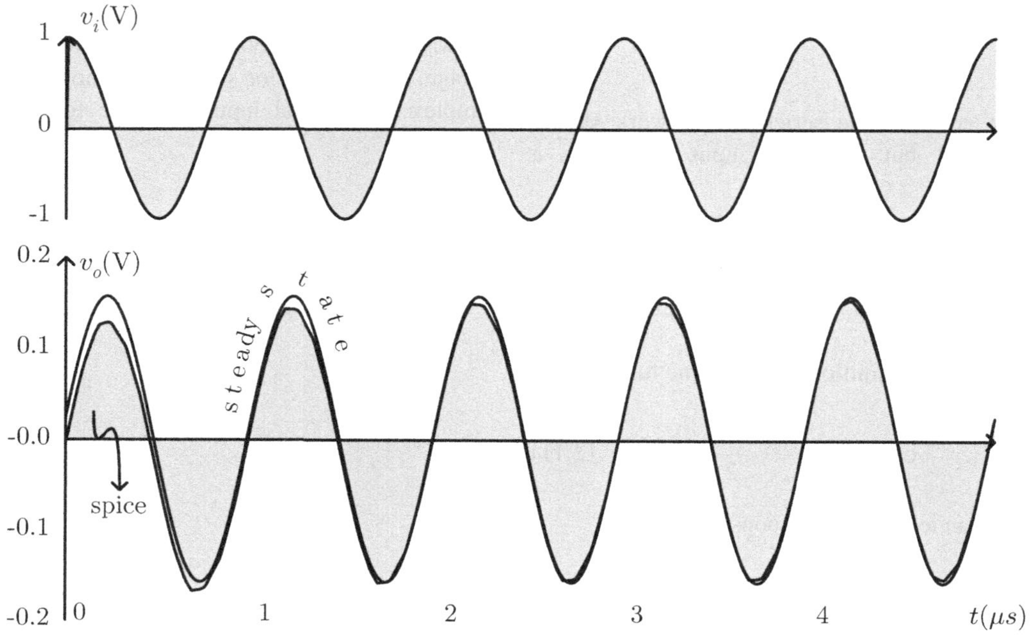

Fig. 2.3 Response of series RC network to cosine input: exact and steady state results. Case of $R = 1\,\Omega$, $C = 1\,\mu$F, and $\omega_0 = 2\pi \times 10^6$

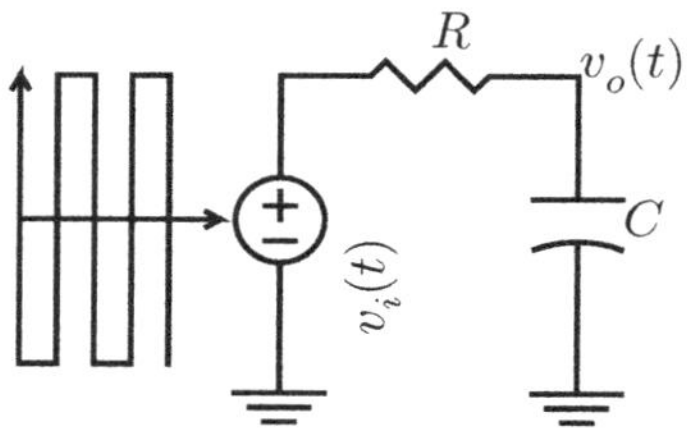

Fig. 2.4 Series RC driven by periodic pulse (with 0 DC average)

We can validate this by plotting this expansion, as shown in Fig. 2.5. Recall from the prior section that if input was $\cos \omega_n t$ the output was

$$\cos \omega_0 t \rightarrow \frac{\cos \omega_0 t + \omega_0 RC \sin \omega_0 t}{1 + \omega_0^2 R^2 C^2} \tag{2.17}$$

While our input is not a single cosine, it is a series thereof. By superposition we then have

$$\boxed{v_o(t) = \sum_{n=1}^{\infty} \left[\frac{2}{\pi n} \sin \frac{n\pi}{2} \right] \left[\frac{\cos \omega_n t + \omega_n RC \sin \omega_n t}{1 + \omega_n^2 R^2 C^2} \right]} \tag{2.18}$$

Results along those of SPICE are shown in Fig. 2.6 Notice the excellent match!

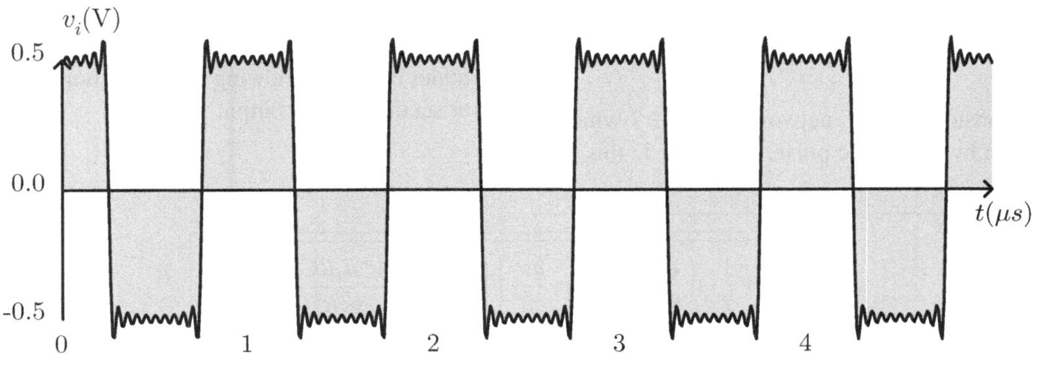

Fig. 2.5 Periodic pulse as series expansion of sines/cosine

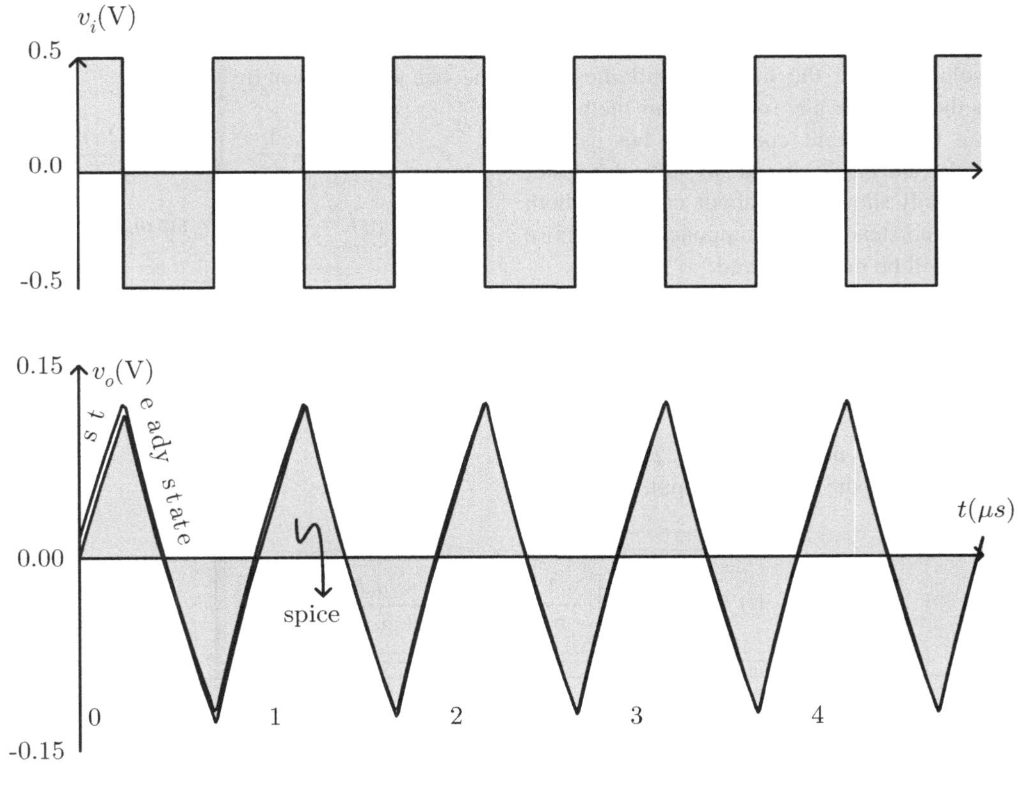

Fig. 2.6 Periodic pulse response of series *RC* circuit: pulse with zero average. Case of $R = 1\,\Omega$, $C = 1\,\mu F$, and $\omega_0 = 2\pi \times 10^6$

2.5 Series *RC* Driven by Periodic Pulse Input (with DC Offset)

Consider the *RC* network in Fig. 2.7 which is driven by a periodic pulse, of period T; this time the pulse has a net DC average (0.5). We again expand this as a Fourier series, but this time add an adder of 1/2. Following what was done in the prior section we get output voltage

$$v_o(t) = \frac{1}{2} + \sum_{n=1}^{\infty} \left[\frac{2}{\pi n} \sin \frac{n\pi}{2}\right] \left[\frac{\cos \omega_n t + \omega_n RC \sin \omega_n t}{1 + \omega_n^2 R^2 C^2}\right] \tag{2.19}$$

These results alongside those of SPICE are shown in Fig. 2.8. Notice that once things settle down, our results match very well to those of SPICE. That is, our steady state solution matches total solution once the transient part dies off. This is the price we pay for using this method—we lose the transient component, but capture accurately (and easily) the steady state part. Later chapters will show ways about capturing both transient and steady state components, but those methods will be more involved.

2.6 Series *RC* Driven by Periodic Triangular Input

Consider the *RC* network in Fig. 2.9 which is driven by a periodic triangular input, of period T. We can decompose the triangular signal (as shown in later chapters) in terms of a DC, sine, and cosine signals. The DC is merely the average, the cosine vanishes since the signal is odd, and the sine term is given by

$$b_n = -\frac{1}{n\pi}, \qquad \omega_n = \frac{2\pi n}{T}$$
$$v_i(t) = \frac{1}{2} + \sum_n b_n \sin \omega_n t \tag{2.20}$$

This is shown in Fig. 2.10. Since we already know the response to a single sine, we can use superposition to find the response due to a series of sines:

$$v_o(t) = \frac{1}{2} + \sum_{n=1}^{\infty} \left[-\frac{1}{\pi n}\right] \left[\frac{\sin \omega_n t - \omega_n RC \cos \omega_n t}{1 + \omega_n^2 R^2 C^2}\right] \tag{2.21}$$

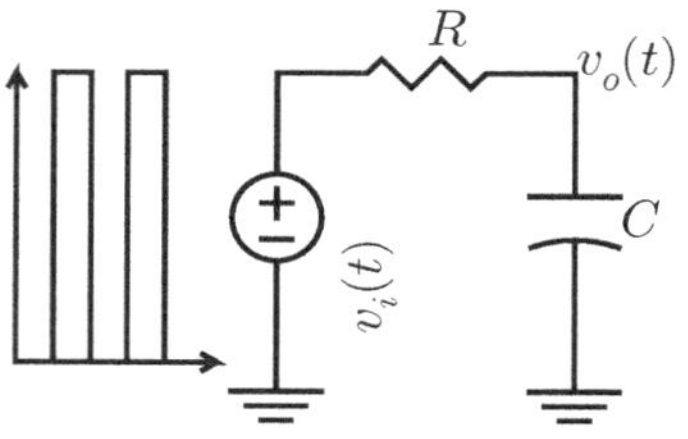

Fig. 2.7 Series *RC* driven by periodic pulse

These results as well as those of SPICE are shown in Fig. 2.11. Notice that after a while, the steady state solution converges to the exact solution.

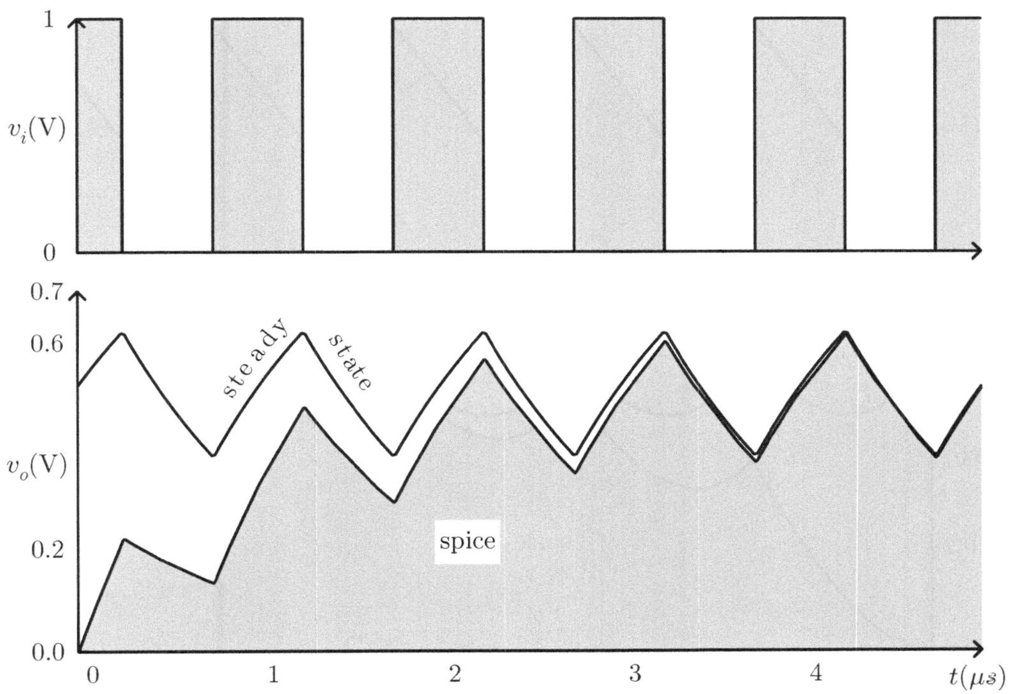

Fig. 2.8 Periodic pulse response of series RC circuit. Case of $R = 1\,\Omega$, $C = 1\,\mu$F, and $\omega_0 = 2\pi \times 10^6$

Fig. 2.9 Series RC driven by periodic triangle

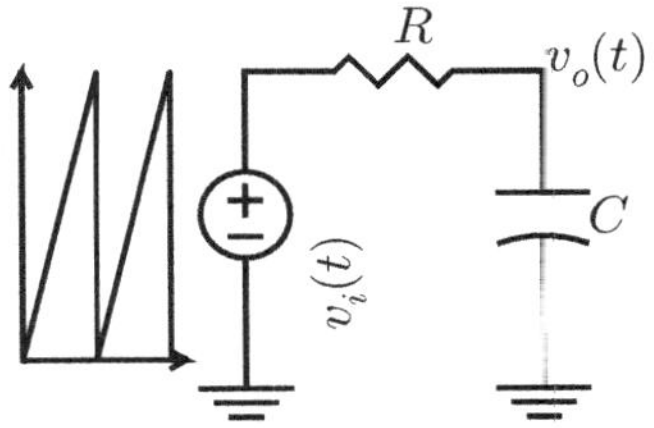

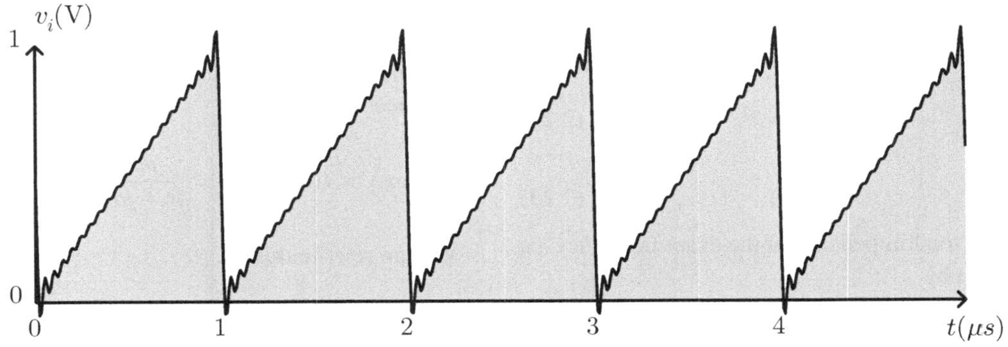

Fig. 2.10 Series RC driven by periodic triangular input (shown as a series expansion)

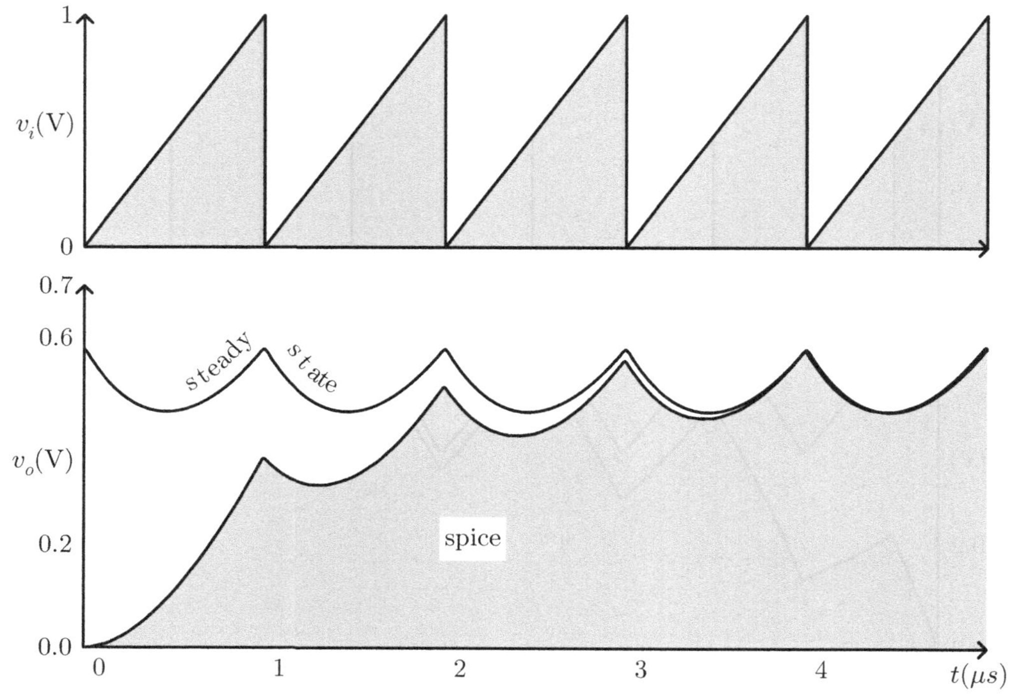

Fig. 2.11 Series RC driven by periodic triangular input: output voltage response. Case of $R = 1\,\Omega$, $C = 1\,\mu$F and $\omega_0 = 2\pi \times 10^6$

2.7 Series *RL* Driven by Sine Input

Consider the series RL shown in Fig. 2.12; it is driven by a sine input (of period ω_0) and we want to find output voltage. Rather than starting with the sine we will use the complex exponential $e^{j\omega_0 t}$ and figure the sine results backwards. So let's set our input as

$$v_i(t) = e^{j\omega_0 t} \tag{2.22}$$

or

$$V_i(\omega_0) = 1 \tag{2.23}$$

The total impedance at the exact frequency ω_0 is given by

$$Z_(\omega_0) = R + j\omega_0 L \tag{2.24}$$

Current at that frequency is

$$I(\omega_0) = \frac{1}{R + j\omega_0 L} \tag{2.25}$$

Fig. 2.12 Series RL driven by sine input

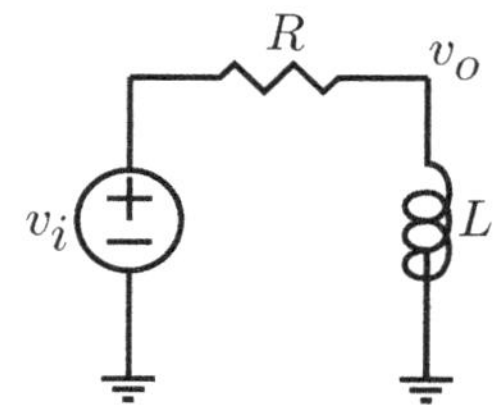

Output voltage is then current times output impedance

$$V_o(\omega_0) = I(\omega_0) j\omega_0 L = \frac{j\omega_0 L}{R + j\omega_0 L} \tag{2.26}$$

We can rewrite this as

$$V_o(\omega_0) = \frac{j\omega_0 L(R - j\omega_0 L)}{R^2 + \omega_0^2 L^2} = \frac{\omega_0^2 L^2 + j\omega_0 RL}{R^2 + \omega_0^2 L^2} \tag{2.27}$$

This transfer function is shown in Fig. 2.13. Notice at DC and low frequency, the output voltage is zero and phase shift is 90°; that is,

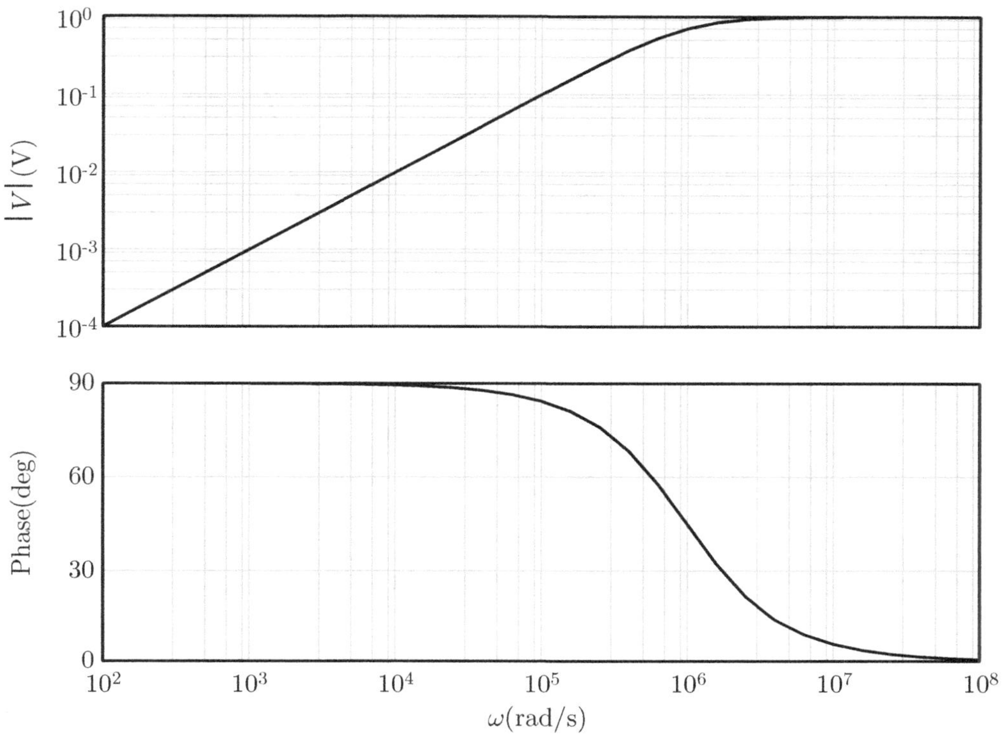

Fig. 2.13 Series *RL* transfer function. Case of $R = 1\,\Omega$ and $L = 1\,\mu$H

since the impedance of the inductor is small, most (if not all) of the input voltage falls on the resistor. But at high frequency, and as the inductive impedance grows, more and more of input voltage resides across the inductor, and the phase shift goes to zero. At really high frequency output voltage assumes input one. Now we go back into the time domain; we simply multiply $V_o(\omega_0)$ by $e^{j\omega_0 t}$:

$$v_o(t) = \frac{\omega_0^2 L^2 + j\omega_0 RL}{R^2 + \omega_0^2 L^2} e^{j\omega_0 t} \tag{2.28}$$

Expand the complex exponential

$$v_o(t) = \frac{\left[\omega_0^2 L^2 + j\omega_0 RL\right]\left[\cos\omega_0 t + j\sin\omega_0 t\right]}{R^2 + \omega_0^2 L^2} \tag{2.29}$$

Collect terms

$$v_o(t) = \frac{\left[\omega_0^2 L^2 \cos\omega_0 t - \omega_0 RL\sin\omega_0 t\right] + j\left[\omega_0^2 L^2 \sin\omega_0 t + \omega_0 RL\cos\omega_0 t\right]}{R^2 + \omega_0^2 L^2} \tag{2.30}$$

Recalling that

$$e^{j\omega_0 t} = \cos\omega_0 t + j\sin\omega_0 t \tag{2.31}$$

we conclude that for a sine input, output voltage would be

$$\boxed{v_o(t) = \frac{\omega_0^2 L^2 \sin\omega_0 t + \omega_0 RL\cos\omega_0 t}{R^2 + \omega_0^2 L^2}} \tag{2.32}$$

Results and comparison to SPICE for case of $\omega_0 = 2\pi \times 10^6$ Hz are shown in Fig. 2.14. Notice

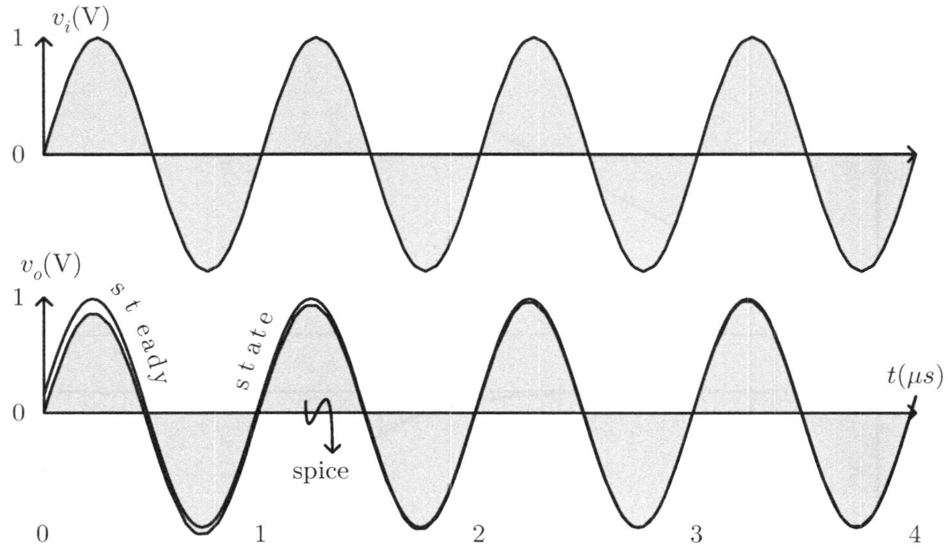

Fig. 2.14 Series *RL* response to sine input: $\omega_0 = 2\pi \times 10^6$ Hz ($R = 1\,\Omega$ and $L = 1\,\mu$H)

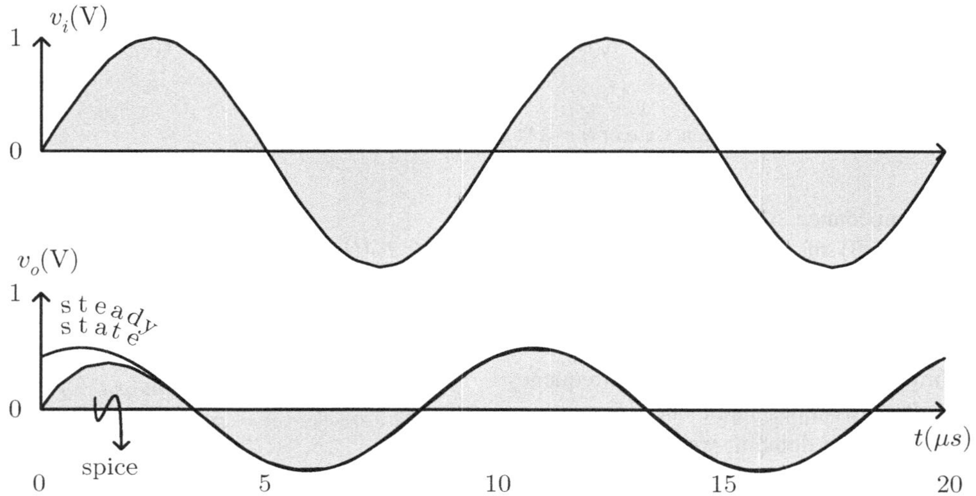

Fig. 2.15 Series *RL* response to sine input: $\omega_0 = 2\pi \times 10^5$ Hz ($R = 1\,\Omega$ and $L = 1\,\mu$H)

that for this case, most of the voltage (of not all of it) falls across the inductor, because the frequency is high. If we were to cut down frequency to $\omega_0 = 2\pi \times 10^5$ Hz then we get more of a distribution as shown in Fig. 2.15. In fact as we reduce the frequency, the voltage across the inductor continues diminishing as shown in Fig. 2.16.

2.8 Alternative Flow for Steady State

There is an alternative method to steady state, which is best explained via an example. Consider again the series *RL* network, driven by a sine

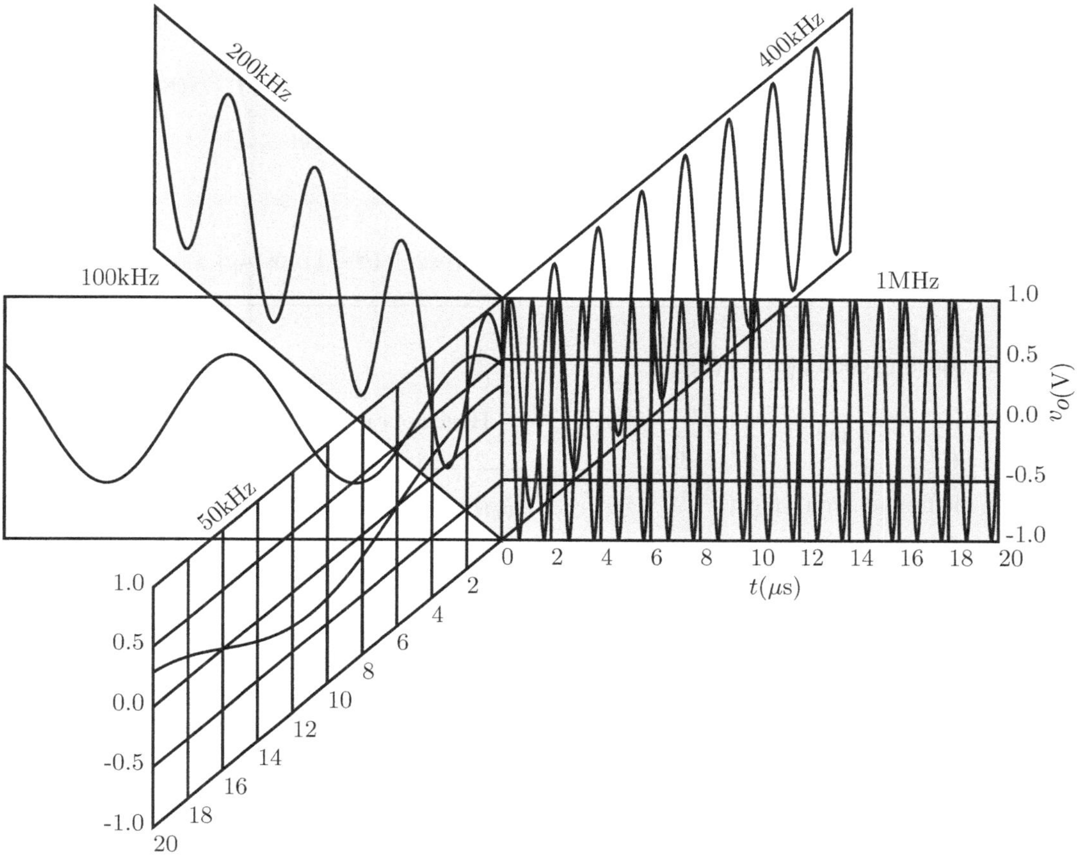

Fig. 2.16 Series RL response to sine input with reducing frequency (counterclockwise) ($R = 1\,\Omega$ and $L = 1\,\mu\text{H}$)

function. From the prior section we know that the voltage across the inductor is given by

$$v_o(t) = \frac{\omega_0^2 L^2 \sin\omega_0 t + \omega_0 RL\cos\omega_0 t}{R^2 + \omega_0^2 L^2} \tag{2.33}$$

This was obtained by assuming a complex excitation, using impedance function, figuring total response then taking imaginary part thereof. How about we try something different. The main assumption here is that voltage across either of R or L must assume the form of

$$v(t) = A\cos\omega_0 t + B\sin\omega_0 t \tag{2.34}$$

We can rationalize this by noting that after things settle down, the only "driven" oscillation is that of the input signal; and if that varies at ω_0 then are we too off assuming that the result must also oscillate at ω_0, albeit having sine and cosine components? Let's try the following. Assume voltage across resistor is

$$v_R(t) = a\cos\omega_0 t + b\sin\omega_0 t \tag{2.35}$$

Assume further that voltage across the inductor is

$$v_L(t) = c\cos\omega_0 t + d\sin\omega_0 t \tag{2.36}$$

We have four unknowns: a, b, c, and d. We know that sum of voltages across R and L must add up to input stimulus; then we would have

$$(a+c)\cos\omega_0 t + (b+d)\sin\omega_0 t = \sin\omega_0 t \tag{2.37}$$

Equating coefficients we get

$$c = -a; \qquad d = 1 - b \tag{2.38}$$

So we are left off with two coefficients; we need more equations to solve for them. We know by KCL, resistor current must equal inductor one; that is

$$\frac{v_R(t)}{R} = \frac{1}{L}\int v_L(t)dt \tag{2.39}$$

Plugging in we get

$$\frac{a}{R}\cos\omega_0 t + \frac{b}{R}\sin\omega_0 t = \frac{1}{\omega_0 L}\left[-a\sin\omega_0 t + (b-1)\cos\omega_0 t\right] \tag{2.40}$$

Again equating coefficients we get

$$\frac{b}{R} = -\frac{a}{\omega_0 L}, \qquad \text{and} \qquad \frac{a}{R} = \frac{b-1}{\omega_0 L} \tag{2.41}$$

From the first relation we get

$$b = -\frac{aR}{\omega_0 L} \tag{2.42}$$

Put into the second relation and get

$$\frac{a}{R} = \frac{b-1}{\omega_0 L} = \frac{-\frac{aR}{\omega_0 L} - 1}{\omega_0 L} = \frac{-aR - \omega_0 L}{\omega_0^2 L^2} \tag{2.43}$$

Rearrange to get

$$a\left[\frac{1}{R} + \frac{R}{\omega_0^2 L^2}\right] = -\frac{1}{\omega_0 L}$$

$$a\frac{\omega_0^2 L^2 + R^2}{R\omega_0^2 L^2} = -\frac{1}{\omega_0 L}$$

$$a = -\frac{R\omega_0 L}{R^2 + \omega_0^2 L^2} \tag{2.44}$$

This implies that the coefficient multiplying the cosine term in the inductor voltage is $\frac{R\omega_0 L}{R^2+\omega_0^2 L^2}$, which we can check right away (Eq. (2.33)) to be correct. Next we solve for b

$$b = -\frac{aR}{\omega_0 L} = \frac{R}{\omega_0 L}\frac{R\omega_0 L}{R^2 + \omega_0^2 L^2} = \frac{R^2}{R^2 + \omega_0^2 L^2} \tag{2.45}$$

The coefficient multiplying the sine term in the inductor voltage is

$$1 - b = 1 - \frac{R^2}{R^2 + \omega_0^2 L^2} = \frac{\omega_0^2 L^2}{R^2 + \omega_0^2 L^2} \tag{2.46}$$

Hence our inductor voltage is

$$\boxed{v_L(t) = \frac{R\omega_0 L}{R^2 + \omega_0^2 L^2}\cos\omega_0 t + \frac{\omega_0^2 L^2}{R^2 + \omega_0^2 L^2}\sin\omega_0 t} \tag{2.47}$$

in complete agreement with Eq. (2.33).

2.9 Series *LC* Driven by Sine Input

Consider the LC network in Fig. 2.17 which is driven by a sine, of period T; we wish to find output voltage across the cap. Again instead of using a sine we use a complex exponential. The total impedance is

$$Z(\omega_0) = j\omega_0 L + \frac{1}{j\omega_0 C} = \frac{1 - \omega_0^2 LC}{j\omega_0 C} \tag{2.48}$$

Total current is inverse of this

$$I(\omega_0) = \frac{j\omega_0 C}{1 - \omega_0^2 LC} \tag{2.49}$$

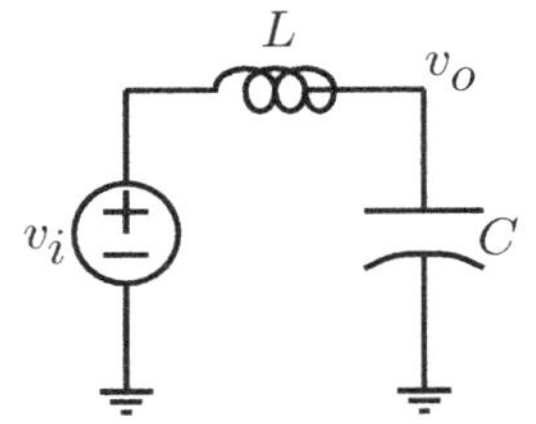

Fig. 2.17 Series LC driven by sine input

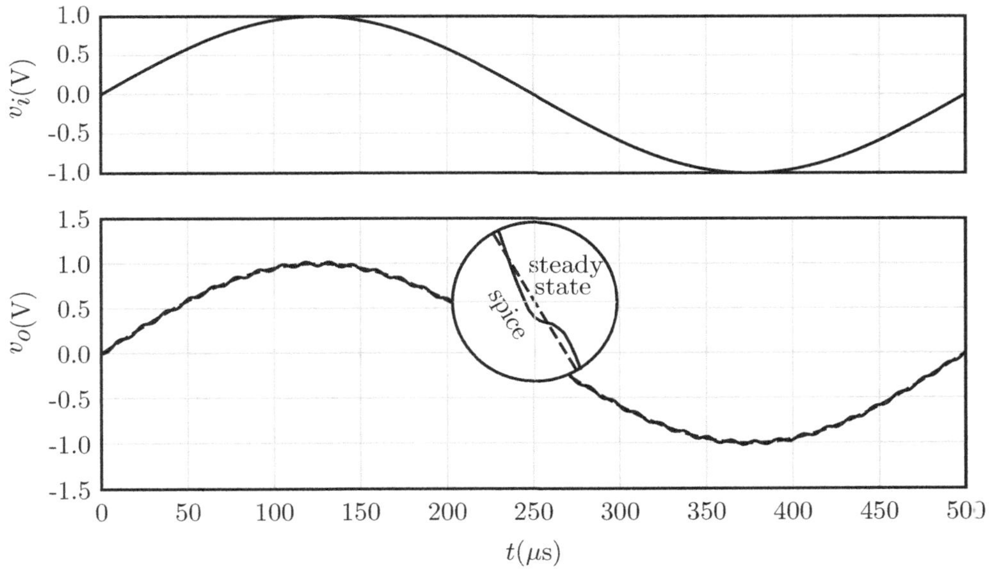

Fig. 2.18 Series *LC* response to sine input with frequency 2 MHz. Case of $C = 2\,\text{nF}$ and $L = 2\,\text{nH}$

Output voltage is total current times cap impedance

$$V_o(\omega_0) = \frac{1}{1 - \omega_0^2 LC} \tag{2.50}$$

In time we then have

$$v_o(t) = \frac{e^{j\omega_0 t}}{1 - \omega_0^2 LC} \tag{2.51}$$

For a sine input, we simply take the imaginary part of this

$$\boxed{v_o(t) = \frac{\sin \omega_0 t}{1 - \omega_0^2 LC}} \tag{2.52}$$

This, along with SPICE results, are plotted in Fig. 2.18 for the case of input frequency of 2 MHz. Notice the good match, but notice that the SPICE ones seem to have another signal (of higher frequency) riding atop them. Let's try another input frequency, 5 MHz; this gives Fig. 2.19. Again we see same issue that SPICE results seem to include another frequency signature. Let's try yet another input frequency, 20 MHz; this gives Fig. 2.20. Again we see same issue. Looks like no matter what we do, we are unable to match the high-f signature that SPICE is showing. As we will find out later, this high-f signature is related to an *LC* "resonance," and that comprises part of the transient response, which appears to be undying here! Since our method does not capture transient behavior, but instead steady state one, we will have to come back to this problem later in the text!

2.10 Two-Branch *RLC* Network Driven by Sine Input

Consider the 2-branch *RLC* network shown in Fig. 2.21. It is driven by a sine function of angular frequency ω_0. We want to find the two branch currents. Assume that driving function is e^{st}:

$$v_i(t) = e^{st}, \qquad s = j\omega_0 \tag{2.53}$$

and currents are

$$i_1(t) = I_1 e^{st}, \qquad \text{and} \qquad i_2(t) = I_2 e^{st} \tag{2.54}$$

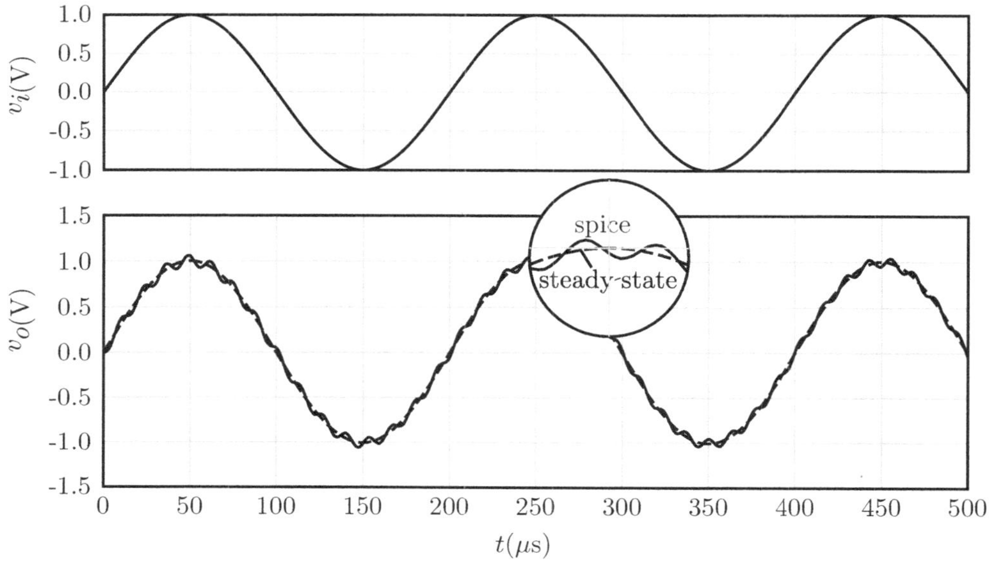

Fig. 2.19 Series LC response to sine input with frequency 5 MHz. Case of $C = 2$ nF and $L = 2$ nH

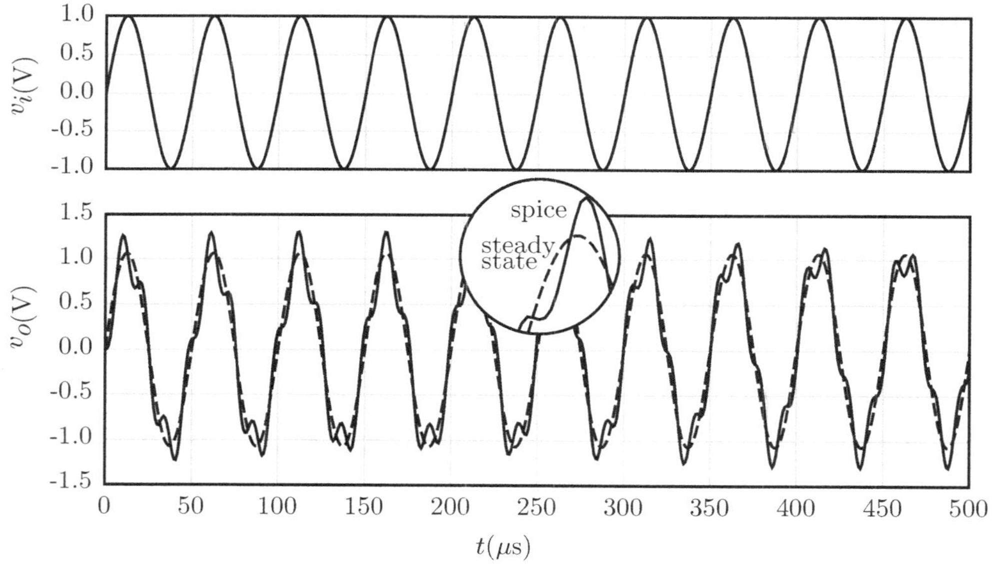

Fig. 2.20 Series LC response to sine input with frequency 20 MHz. Case of $C = 2$ nF and $L = 2$ nH

Fig. 2.21 Two-branch RLC network driven by sine input

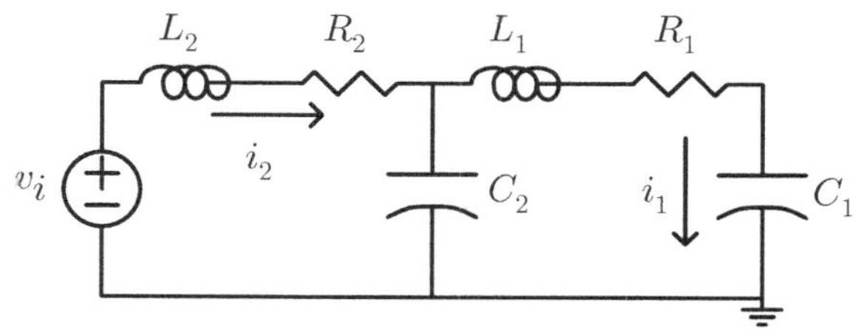

Doing KVL around the right loop (and dividing by e^{st}) we get

$$I_1\left[\frac{1}{sC_1}+\frac{1}{sC_2}+R_1+sL_1\right]-I_2\frac{1}{sC_2}=0 \tag{2.55}$$

Doing KVL around the left loop we get

$$I_2\left[\frac{1}{sC_2}+R_2+sL_2\right]-I_1\frac{1}{sC_1}=\boxed{1} \tag{2.56}$$

Notice the 1 at the RHS of the second equation (as opposed to 0 in the first) which is due to the driving source. In matrix form we have

$$\begin{bmatrix}\frac{1}{sC_1}+\frac{1}{sC_2}+R_1+sL_1 & -\frac{1}{sC_2}\\ -\frac{1}{sC_1} & \frac{1}{sC_2}+R_2+sL_2\end{bmatrix}\begin{bmatrix}I_1\\ I_2\end{bmatrix} = \begin{bmatrix}0\\ 1\end{bmatrix} \tag{2.57}$$

This is a system of 2×2 equations which we can solve, even symbolically. However, to simplify things let's assume some *RLC* numbers and use numerical (rather than symbolical) matrices; for example assume

$$R_1=1,\quad C_1=1\,\mu,\quad L_1=1\,\mu,\quad R_2=2,\quad C_2=2\,\mu,\quad L_2=2\,\mu \tag{2.58}$$

where resistors are in Ω, caps in Farads, and inductors in Henrys. Furthermore, assume input has frequency

$$\omega_0=2\pi\times 5\times 10^5\text{rad/s} \tag{2.59}$$

Then our complex matrix reduces to

$$\begin{bmatrix}1+2.66j & 0+0.16j\\ 0+0.32j & 2+6.12j\end{bmatrix}\begin{bmatrix}I_1\\ I_2\end{bmatrix}=\begin{bmatrix}0\\ 1\end{bmatrix} \tag{2.60}$$

This can be solved and results are

$$\begin{bmatrix}I_1\\ I_2\end{bmatrix}=\begin{bmatrix}-0.0054+0.0068j\\ 0.0485-0.1478j\end{bmatrix} \tag{2.61}$$

That is, if $v_i(t)=e^{j\omega_0 t}$ then

$$i_1(t)=(-0.0054+0.0068j)e^{j\omega_0 t},\qquad i_2(t)=(+0.0485-0.1478j)e^{j\omega_0 t} \tag{2.62}$$

But our input is not $e^{j\omega_0 t}$—rather it is $\sin(\omega_0 t)$! But the sine is simply the imaginary part of the complex exponential; hence by superposition we simply take the imaginary part of our solution.

$$i_1(t)=\Im\left[(-0.0054+0.0068j)e^{j\omega_0 t}\right],\qquad i_2(t)=\Im\left[(+0.0485-0.1478j)e^{j\omega_0 t}\right] \tag{2.63}$$

which gives (with $\omega_0=2\pi\times 5\times 10^5$)

$$i_1(t)=-0.0054\sin(\omega_0 t)+0.0068\cos(\omega_0 t),\quad i_2(t)=+0.0485\sin(\omega_0 t)-0.1478\cos(\omega_0 t), \tag{2.64}$$

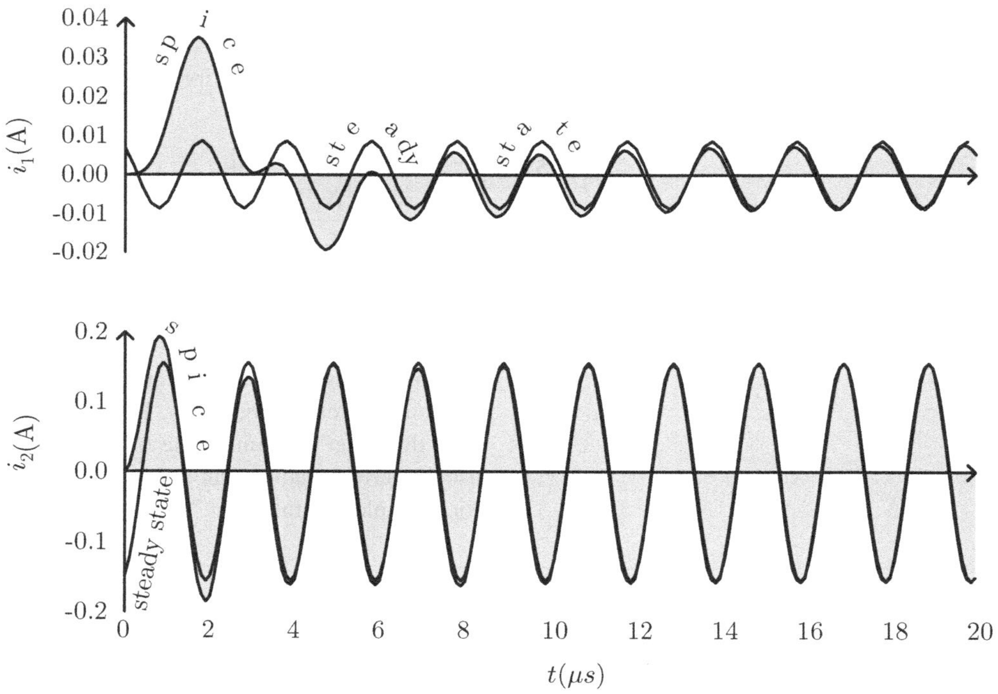

Fig. 2.22 Two-branch *RLC* network currents as a function of sine input. Case of $R_1 = 1\,\Omega$, $C_1 = 1\,\mu$F, $L_1 = 1\,\mu$H, $R_2 = 2\,\Omega$, $C_2 = 2\,\mu$F and $L_2 = 2\,\mu$H

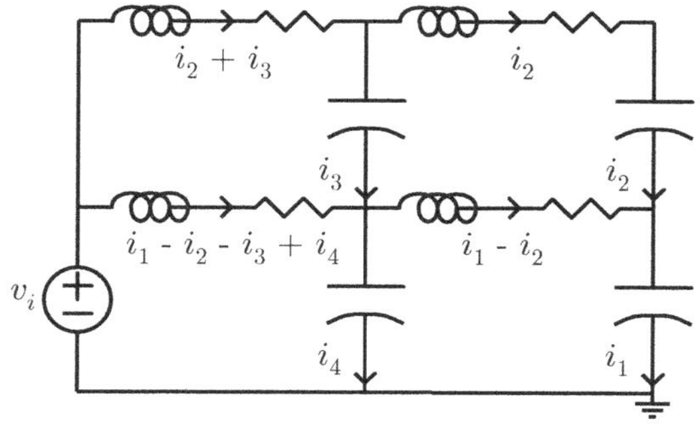

Fig. 2.23 Four-branch *RLC* network driven by sine input

To recap: assume input in the form of complex exponential. Find response, then use superposition to figure response due to real input, which in this case was a sine function (which relates to the complex exponential as being the imaginary part thereof). We examine our solution and comparison to SPICE in Fig. 2.22, Notice that sure enough when things have settled down, our results match exactly those of SPICE.

2.11 Four-Branch *RLC* Network Driven by Sine Input

Consider the 4-branch *RLC* network shown in Fig. 2.23. It is driven by a sine function of angular frequency ω_0. We want to find the four branch currents. Doing KVL around lower right, upper right, upper left, and lower left we get a set of 4×4 equations:

$$\begin{bmatrix} \frac{1}{sC}+R+sL & -R-sL & 0 & -\frac{1}{sC} \\ -R-sL & \frac{1}{sC}+2R+2sL & -\frac{1}{sC} & 0 \\ -R-sL & 2R+2sL & \frac{1}{sC}+2R+2sL & -R-sL \\ R+sL & -R-sL & -R-sL & \frac{1}{sC}+R+sL \end{bmatrix} \begin{bmatrix} I_1 \\ I_2 \\ I_3 \\ I_4 \end{bmatrix} = \begin{bmatrix} 0 \\ 0 \\ 0 \\ 1 \end{bmatrix} \quad (2.65)$$

For the particular case of

$$R = 1\,\Omega, \quad C = 1\,\mu\text{F}, \quad L = 1\,\mu\text{H}, \quad \text{and} \quad \omega_0 = 2\pi \times 10^6 \quad (2.66)$$

the matrix reduces to

$$\begin{bmatrix} 1+6.12j & -1-6.28j & 0+0.000j & 0+0.159j \\ -1-6.28j & 2+12.41j & 0+0.159j & 0+0.000j \\ -1-6.28j & 2+12.57j & 2+12.407j & -1-6.283j \\ 1+6.28j & -1-6.28j & -1-6.283j & 1+6.124j \end{bmatrix} \begin{bmatrix} I_1 \\ I_2 \\ I_3 \\ I_4 \end{bmatrix} = \begin{bmatrix} 0 \\ 0 \\ 0 \\ 1 \end{bmatrix} \quad (2.67)$$

Solving this system gives us

$$\begin{bmatrix} I_1 \\ I_2 \\ I_3 \\ I_4 \end{bmatrix} = \begin{bmatrix} -0.008+0.023j \\ -0.005+0.014j \\ +0.034-0.182j \\ +0.065-0.355j \end{bmatrix} \quad (2.68)$$

We multiply this by $e^{j\omega_0 t}$ and then take the imaginary part. Finally we get

$$\begin{aligned} i_1(t) &= -0.008\sin(\omega_0 t) + 0.023\cos(\omega_0 t) \\ i_2(t) &= -0.005\sin(\omega_0 t) + 0.014\cos(\omega_0 t) \\ i_3(t) &= +0.034\sin(\omega_0 t) - 0.182\cos(\omega_0 t) \\ i_4(t) &= +0.065\sin(\omega_0 t) - 0.355\cos(\omega_0 t) \end{aligned} \quad (2.69)$$

Figure 2.24 shows our results and comparison to SPICE. When things settle down, we get very good match.

2.12 Nine-Branch *RLC* Network Driven by Sine Input

Consider the 9-branch *RLC* network shown in Fig. 2.25. It is driven by a sine function of angular frequency ω_0. We want to find the various nine-branch currents. We assign the currents as shown in the figure. In this case we end up with 18 currents. We would need a set of 18 equations. Nine of those are obtained by doing KVL around the 9 small loops. For example, doing KVL around bottom right loop gives

$$I_1\frac{1}{sC} + (R+sL)I_{12} - I_2\frac{1}{sC} = 0 \quad (2.70)$$

Doing KVL around the next loop above gives

$$I_4\frac{1}{sC} + (R+sL)I_{11} - I_5\frac{1}{sC} - (R+sL)I_{12} = 0 \quad (2.71)$$

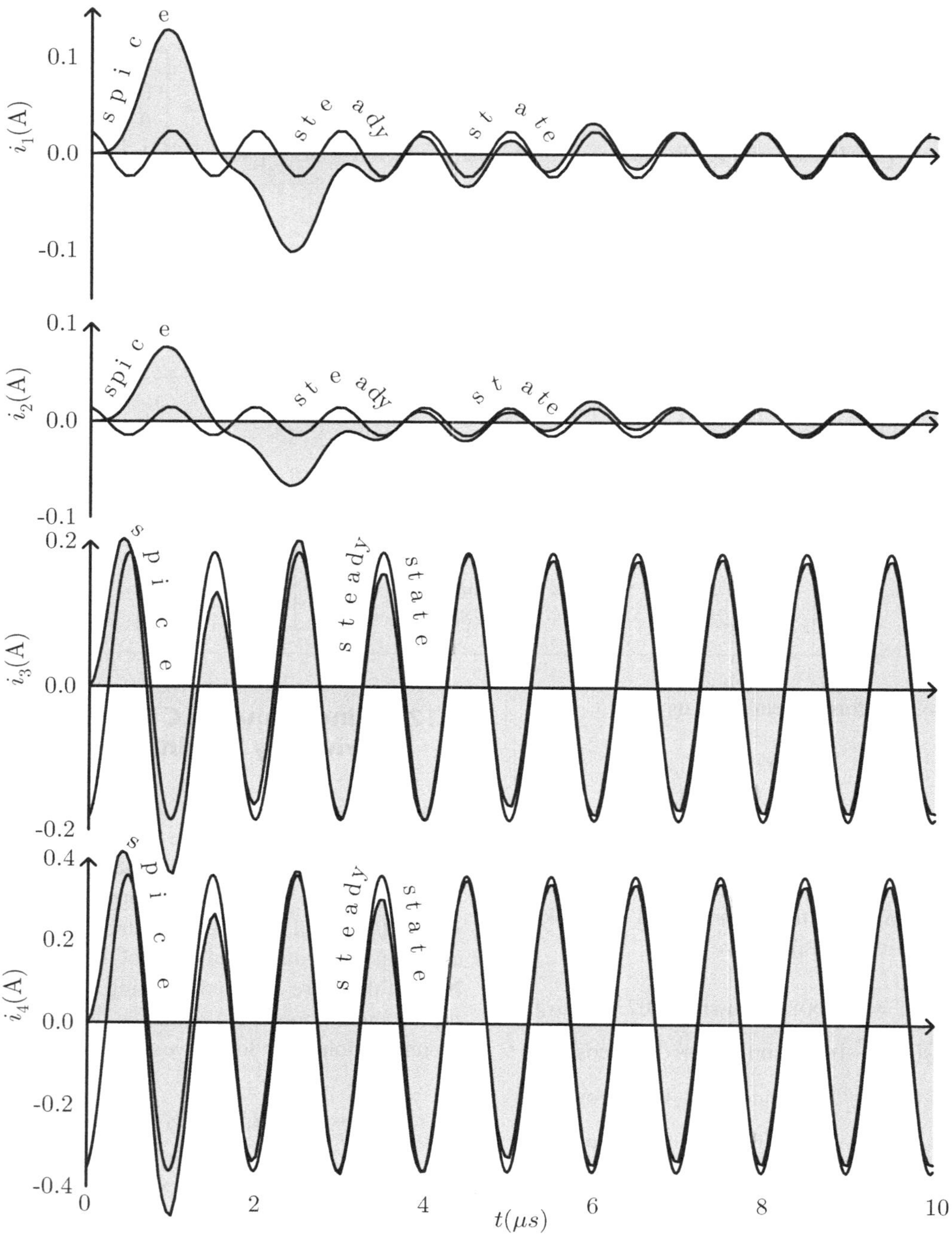

Fig. 2.24 Four-branch *RLC* network currents for a sine input

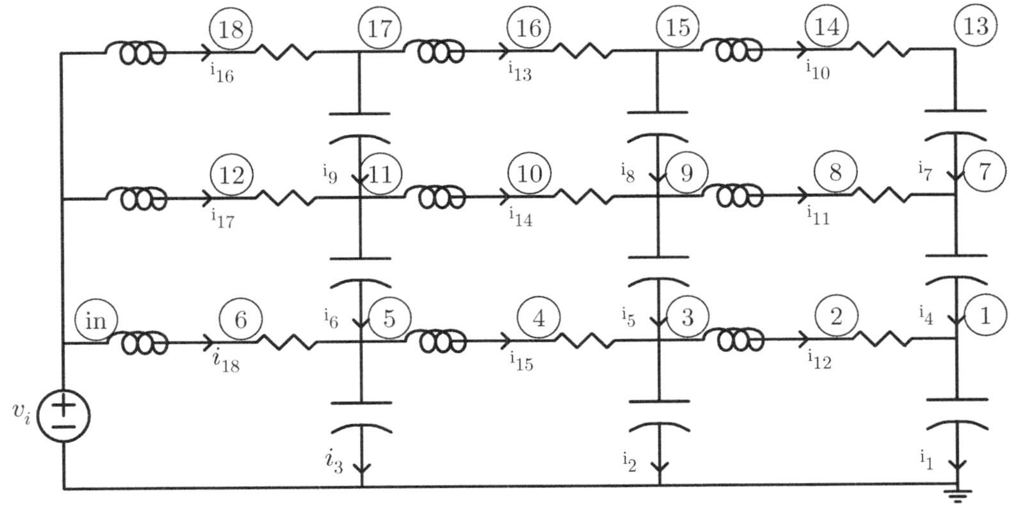

Fig. 2.25 Nine-branch *RLC* network driven by sine input

and so forth. Next we get the remaining 9 equations by doing KCL at the nodes 1,3,5,7,9, 11,13,15, and 17. For example, at node 1 we have

$$I_1 - I_4 - I_{12} = 0 \tag{2.72}$$

Similarly at node 7 we get

$$I_4 - I_7 - I_{11} = 0 \tag{2.73}$$

and so forth. The full set of equation follows. Let $Z = R + sL$; then

$$I_1\frac{1}{sC} - I_2\frac{1}{sC} + I_{12}Z = 0$$

$$I_4\frac{1}{sC} - I_5\frac{1}{sC} + I_{11}Z - I_{12}Z = 0$$

$$I_7\frac{1}{sC} - I_8\frac{1}{sC} + I_{10}Z - I_{11}Z = 0$$

$$I_2\frac{1}{sC} - I_3\frac{1}{sC} + I_{15}Z = 0$$

$$I_5\frac{1}{sC} - I_6\frac{1}{sC} + I_{14}Z - I_{15}Z = 0$$

$$I_8\frac{1}{sC} - I_9\frac{1}{sC} + I_{13}Z - I_{14}Z = 0$$

$$I_3\frac{1}{sC} + I_{18}Z = \boxed{1}$$

$$I_6\frac{1}{sC} + I_{17}Z - I_{18}Z = 0$$

$$I_9\frac{1}{sC} + I_{16}Z - I_{17}Z = 0$$

$$I_1 - I_4 - I_{12} = 0$$

$$I_4 - I_7 - I_{11} = 0$$

$$I_7 - I_{10} = 0$$

$$I_2 + I_{12} - I_5 - I_{15} = 0$$

$$I_5 + I_{11} - I_8 - I_{14} = 0$$

$$I_8 + I_{10} - I_{13} = 0$$

$$I_9 + I_{13} - I_{16} = 0$$

$$I_6 + I_{14} - I_9 - I_{17} = 0$$

$$I_3 + I_{15} - I_6 - I_{18} = 0$$

Again we end up with 18 equation which assume the form

$$ZI = V \tag{2.74}$$

where Z is an 18×18 matrix, I is the unknown current vector of size 18×1, and V is the forced voltage, which is a column 18×1 all zeroes, except the entry corresponding to the KVL loop which covers the bottom left loop; there V has the value unity. For the particular case of

$$R = 1\,\Omega, \qquad C = 1\,\mu\text{F}, \qquad L = 1\,\mu\text{H}, \qquad \text{and} \qquad \omega_0 = 2\pi \times 10^6 \tag{2.75}$$

the numerical matrix (set to single digit format, in order to fit) is shown in Fig. 2.26. Solving for this linear system, multiplying by $e^{j\omega_0 t}$ then taking the imaginary part gives the solution

$$\begin{aligned}
i_1(t) &= -0.671 \sin \omega_0 t + 0.542 \cos \omega_0 t & i_{10}(t) &= -0.299 \sin \omega_0 t + 0.242 \cos \omega_0 t\\
i_2(t) &= +0.300 \sin \omega_0 t - 0.931 \cos \omega_0 t & i_{11}(t) &= -0.239 \sin \omega_0 t + 0.193 \cos \omega_0 t\\
i_3(t) &= +1.298 \sin \omega_0 t - 0.479 \cos \omega_0 t & i_{12}(t) &= -0.132 \sin \omega_0 t + 0.107 \cos \omega_0 t\\
i_4(t) &= -0.539 \sin \omega_0 t + 0.435 \cos \omega_0 t & i_{13}(t) &= -0.160 \sin \omega_0 t - 0.179 \cos \omega_0 t\\
i_5(t) &= +0.246 \sin \omega_0 t - 0.753 \cos \omega_0 t & i_{14}(t) &= -0.133 \sin \omega_0 t - 0.139 \cos \omega_0 t\\
i_6(t) &= +0.994 \sin \omega_0 t - 0.254 \cos \omega_0 t & i_{15}(t) &= -0.079 \sin \omega_0 t - 0.071 \cos \omega_0 t\\
i_7(t) &= -0.299 \sin \omega_0 t + 0.242 \cos \omega_0 t & i_{16}(t) &= +0.377 \sin \omega_0 t - 0.283 \cos \omega_0 t\\
i_8(t) &= +0.139 \sin \omega_0 t - 0.421 \cos \omega_0 t & i_{17}(t) &= +0.324 \sin \omega_0 t - 0.289 \cos \omega_0 t\\
i_9(t) &= +0.537 \sin \omega_0 t - 0.104 \cos \omega_0 t & i_{18}(t) &= +0.225 \sin \omega_0 t - 0.295 \cos \omega_0 t
\end{aligned}$$

Figure 2.27 shows comparison between our results and SPICE ones.

2.13 Summary

This chapter deals with steady state solution for *RLC* circuits. All stimuli are assumed to be of the form $e^{j\omega t}$ and so are all responses. Each of the *RLC* elements is replaced with its frequency dependent impedance, followed by applications of KCL and KVL, and this results in a set of algebraic equations. The solution of the algebraic solutions gives the coefficients of the sought responses. The steady state methods correctly and accurately capture the system solution after long enough time has lapsed. It, however, does not capture the transient part of the response, which typically happens from time zero up to some other time after which things have settled down. We applied the method to various examples, ranging from simple ones to a nine-branch circuit, and for each case we compared our results to SPICE, quite often with excellent agreement. The only case we had some issues was the ideal *LC* tank, as the transient component of this circuit never dies out! The chapter also touched on linear system solvers which take the form of $Ax = b$, which A is a matrix, x is the sought solution, and b is the input.

A =

0-0.3i	0+0.3i	0+0.0i	0+0.0i	0+0.0i	0+0.0i	0+0.0i	0+0.0i	0+0.0i	0+0i	0+0i	1+3i	0+0i	0+0i	0+0i	0+0i	0+0i	0+0i
0+0.0i	0+0.0i	0+0.0i	0-0.3i	0+0.3i	0+0.0i	0+0.0i	0+0.0i	0+0.0i	0+0i	1+3i	-1-3i	0+0i	0+0i	0+0i	0+0i	0+0i	0+0i
0+0.0i	0+0.0i	0+0.0i	0+0.0i	0+0.0i	0+0.0i	0-0.3i	0+0.3i	0+0.0i	1+3i	-1-3i	0+0i	0+0i	0+0i	0+0i	0+0i	0+0i	0+0i
0+0.0i	0-0.3i	0+0.3i	0+0.0i	0+0.0i	0+0.0i	0+0.0i	0+0.0i	0+0.0i	0+0i	0+0i	0+0i	0+0i	0+0i	1+3i	0+0i	0+0i	0+0i
0+0.0i	0+0.0i	0+0.0i	0+0.0i	0-0.3i	0+0.3i	0+0.0i	0+0.0i	0+0.0i	0+0i	0+0i	0+0i	0+0i	1+3i	-1-3i	0+0i	0+0i	0+0i
0+0.0i	0+0.0i	0+0.0i	0+0.0i	0+0.0i	0+0.0i	0+0.0i	0-0.3i	0+0.3i	0+0i	0+0i	0+0i	1+3i	-1-3i	0+0i	0+0i	0+0i	0+0i
0+0.0i	0+0.0i	0-0.3i	0+0.0i	0+0.0i	0+0.0i	0+0.0i	0+0.0i	0+0.0i	0+0i	0+0i	0+0i	0+0i	0+0i	0+0i	0+0i	0+0i	1+3i
0+0.0i	0+0.0i	0+0.0i	0+0.0i	0+0.0i	0-0.3i	0+0.0i	0+0.0i	0+0.0i	0+0i	0+0i	0+0i	0+0i	0+0i	0+0i	0+0i	1+3i	-1-3i
0+0.0i	0+0.0i	0+0.0i	0+0.0i	0+0.0i	0+0.0i	0+0.0i	0+0.0i	0-0.3i	0+0i	0+0i	0+0i	0+0i	0+0i	0+0i	1+3i	-1-3i	0+0i
1+0.0i	0+0.0i	0+0.0i	-1+0.0i	0+0.0i	0+0.0i	0+0.0i	0+0.0i	0+0.0i	0+0i	0+0i	-1+0i	0+0i	0+0i	0+0i	0+0i	0+0i	0+0i
0+0.0i	0+0.0i	0+0.0i	1+0.0i	0+0.0i	0+0.0i	-1+0.0i	0+0.0i	0+0.0i	0+0i	-1+0i	0+0i	0+0i	0+0i	0+0i	0+0i	0+0i	0+0i
0+0.0i	0+0.0i	0+0.0i	0+0.0i	0+0.0i	0+0.0i	1+0.0i	0+0.0i	0+0.0i	-1+0i	0+0i	0+0i	0+0i	0+0i	0+0i	0+0i	0+0i	0+0i
0+0.0i	1+0.0i	0+0.0i	0+0.0i	1+0.0i	0+0.0i	0+0.0i	0+0.0i	0+0.0i	0+0i	0+0i	1+0i	0+0i	0+0i	-1+0i	0+0i	0+0i	0+0i
0+0.0i	0+0.0i	0+0.0i	0+0.0i	1+0.0i	0+0.0i	0+0.0i	1+0.0i	0+0.0i	0+0i	1+0i	0+0i	0+0i	-1+0i	0+0i	0+0i	0+0i	0+0i
0+0.0i	0+0.0i	0+0.0i	0+0.0i	0+0.0i	0+0.0i	0+0.0i	1+0.0i	0+0.0i	1+0i	0+0i	0+0i	-1+0i	0+0i	0+0i	0+0i	0+0i	0+0i
0+0.0i	0+0.0i	0+0.0i	0+0.0i	0+0.0i	0+0.0i	0+0.0i	0+0.0i	1+0.0i	0+0i	0+0i	0+0i	1+0i	0+0i	0+0i	-1+0i	0+0i	0+0i
0+0.0i	0+0.0i	0+0.0i	0+0.0i	0+0.0i	1+0.0i	0+0.0i	0+0.0i	-1+0.0i	0+0i	0+0i	0+0i	0+0i	1+0i	0+0i	0+0i	-1+0i	0+0i
0+0.0i	0+0.0i	1+0.0i	0+0.0i	0+0.0i	-1+0.0i	0+0.0i	0+0.0i	0+0.0i	0+0i	0+0i	0+0i	0+0i	0+0i	1+0i	0+0i	0+0i	-1+0i

x	= b
-0.67+0.54i	0
0.30-0.93i	0
1.30-0.48i	0
-0.54+0.44i	0
0.25-0.75i	0
0.99-0.25i	0
-0.30+0.24i	1
0.14-0.42i	0
0.54-0.10i	0
-0.30+0.24i	0
-0.24+0.19i	0
-0.13+0.11i	0
-0.16-0.18i	0
-0.13-0.14i	0
-0.08-0.07i	0
0.38-0.28i	0
0.32-0.29i	0
0.23-0.30i	0

Fig. 2.26 Numerical values for matrix A corresponding to KVL and KCL applied to Fig. 2.25; also forced voltage and current solution

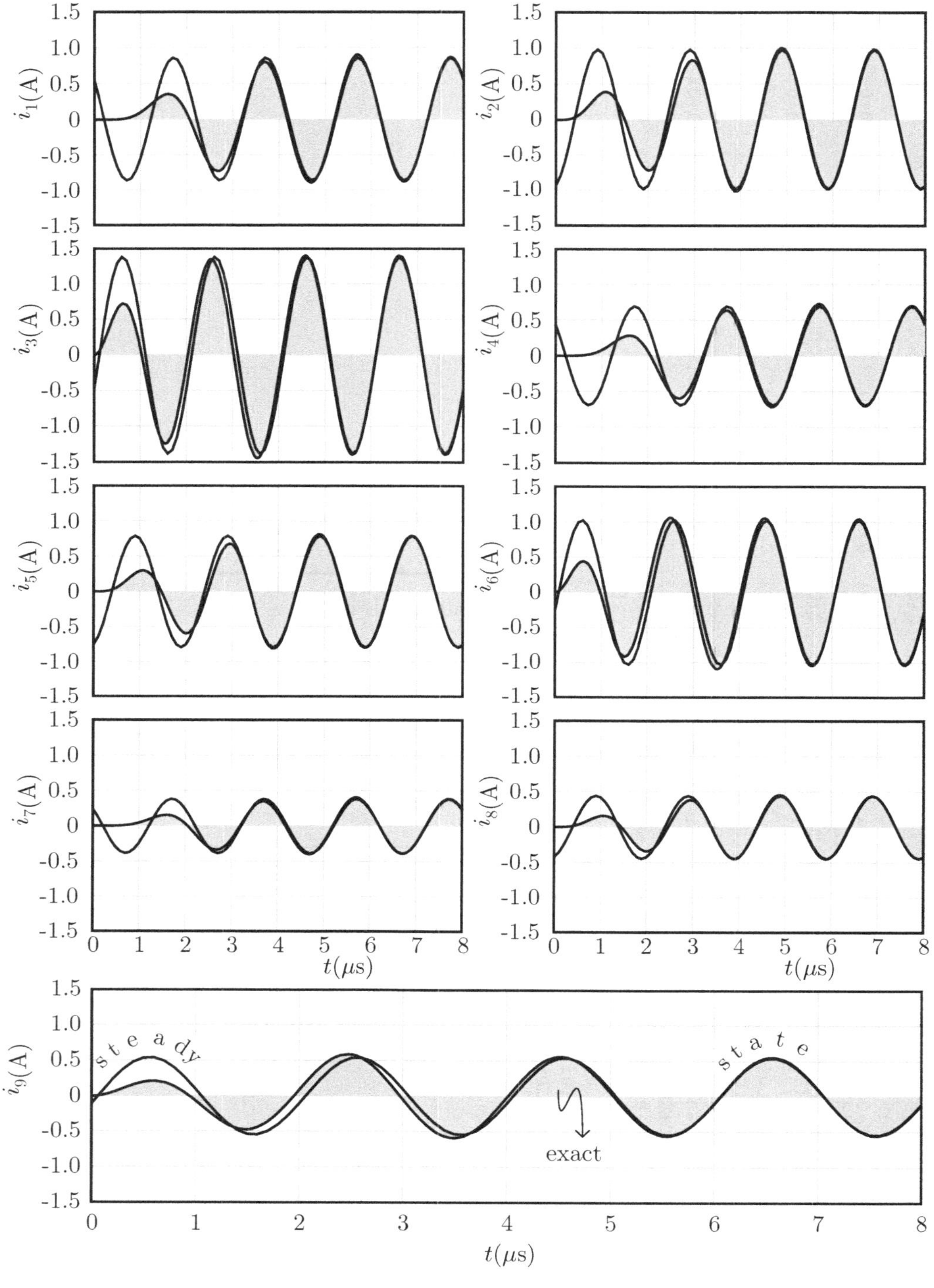

Fig. 2.27 Nine-branch *RLC* currents due to sine input

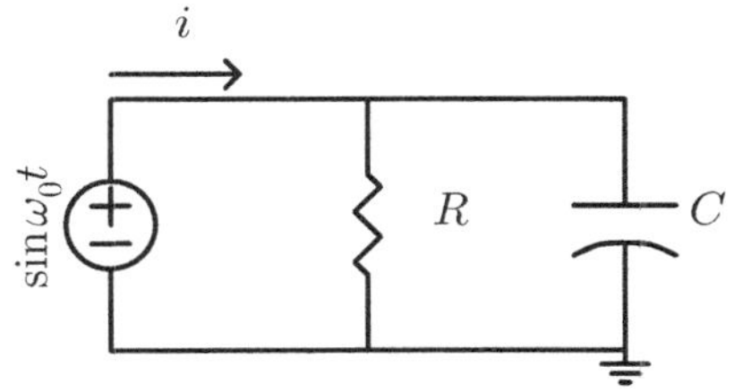

Fig. 2.28 Parallel *RC* driven by sine

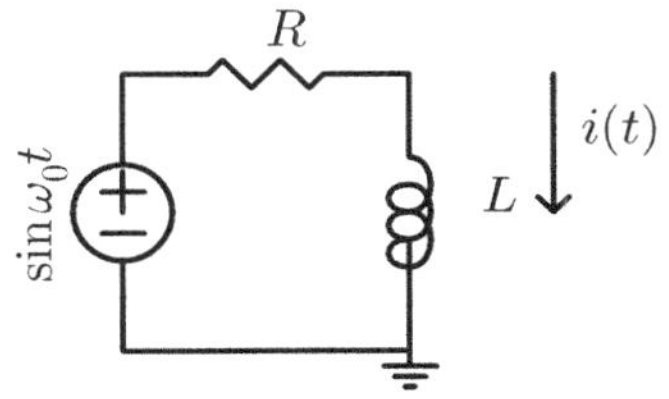

Fig. 2.29 Series *RL* driven by sine

2.14 Problems

1. A parallel *RC* circuit (Fig. 2.28) is driven by an input sine function $v_i(t) = \sin\omega_0 t$; find input current.
Answer:

$$i(t) = \frac{1}{R}\sin\omega_0 t + \omega_0 C\cos\omega_0 t$$

2. A series *RL* circuit (Fig. 2.29) is driven by an input sine function $v_i(t) = \sin\omega_0 t$; find input current.
Answer:

$$i(t) = \frac{1}{R^2 + \omega_0^2 L^2}\left[R\sin\omega_0 t - \omega_0 L\cos\omega_0 t\right]$$

3. For the case of $R = 1$, $L = 0.1$, and $\omega_0 = 2\pi$, write SPICE file and simulate prior Problem 2. Compare to steady state solution. See sample solution in Fig. 2.30.

4. Repeat Problem 2 assuming input voltage is $\cos\omega_0 t$.
Answer:

$$i(t) = \frac{1}{R^2 + \omega_0^2 L^2}\left[R\cos\omega_0 t + \omega_0 L\sin\omega_0 t\right]$$

5. The circuit in Fig. 2.31 is driven by a sine function $\sin\omega_0 t$; find steady state input current. Also, assume $R_1 = 1$, $R_2 = 2$, $C = 0.1$, and $\omega_0 = 2\pi$; write SPICE input deck and compare results to steady state one; see Fig. 2.32.
Answer:

$$i(t) = \frac{R_1 + R_2}{R_1 R_2}\frac{1}{\frac{1}{R_2^2C^2} + \omega_0^2} \times \left[\left(\frac{1}{(R_1 + R_2)R_2C^2} + \omega_0^2\right)\sin\omega_0 t + \left(\frac{-\omega_0}{(R_1 + R_2)C} + \frac{\omega_0}{R_2 C}\right)\cos\omega_0 t\right]$$

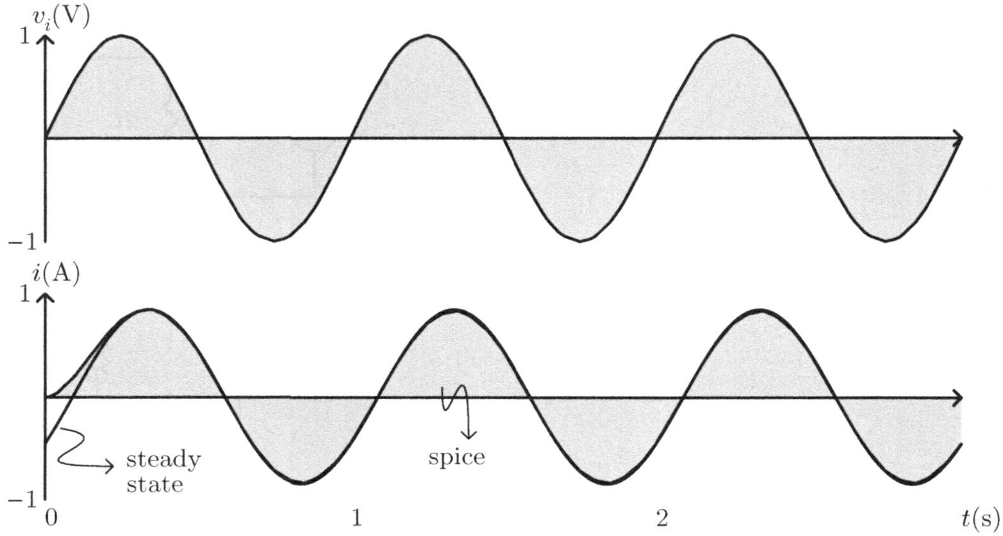

Fig. 2.30 Answer to Problem 3

Fig. 2.31 *RC* network driven by sine voltage

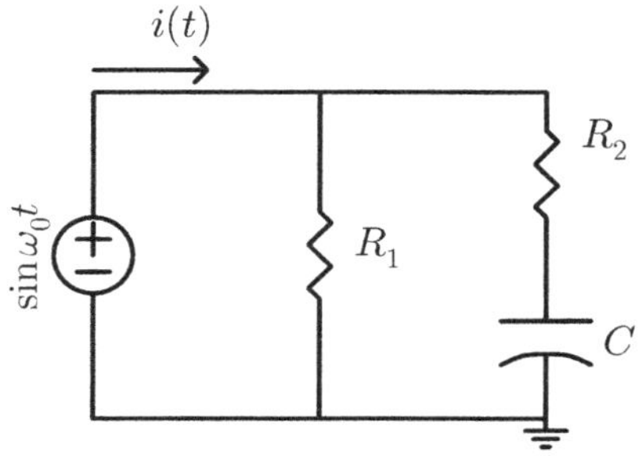

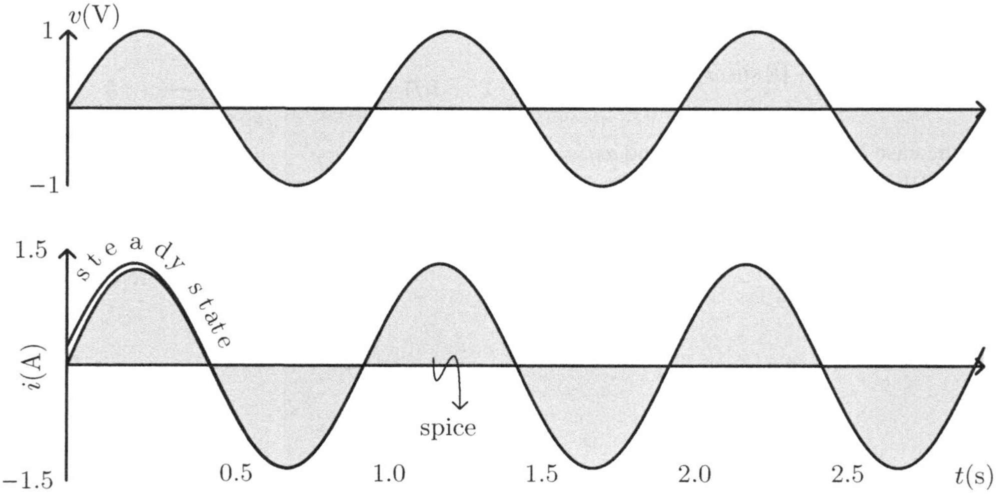

Fig. 2.32 Answer of second part of Problem 5

3 Differential Equation Solution to Circuit Problems

3.1 Introduction

We saw in the prior section that steady state techniques can give us the steady state solution, but lack that part of the solution which transports the system from the zero state to the steady state; that is, it lacks the transient solution. Differential equation techniques don't suffer from this limitation. If done correctly this method yields the general solution—transient part and steady state one. The field of differential equations is very rich and classic; we won't (even attempt to) develop the theory from scratch but rather apply what most of us have learned in basic differential equations classes to the problems we face, as the need arises. This is best illustrated by a few detailed examples.

3.2 Series *RC* Driven by Sine Function

Again consider the simple *RC* network shown in Fig. 3.1 which is driven by a causal sine function; we state causal in the sense it was zero for negative time, and was applied starting at time zero. Doing KVL around the loop we have

$$\frac{1}{C}\int_0^t i(t) + Ri(t) = \sin \omega_0 t \tag{3.1}$$

Let's think a bit about the solution. On the right side we have a $\sin \omega_0 t$, so surely the solution would have a sine too. If we assume

$$i_p(t) \sim \sin \omega_0 t \qquad \text{(not complete)} \tag{3.2}$$

we arrive at the contradiction once we integrate this current. When integrating we get a cosine term, so on the left we have a sine and a cosine, but on the right a sine only; hence, a sine-only won't do. By the same reasoning we cannot have a cosine only. How about a combination?

$$i_p(t) = A \sin \omega_0 t + B \cos \omega_0 t \tag{3.3}$$

If we plug this back into the differential equation we get

$$\frac{1}{C\omega_0}\left[-A \cos \omega_0 t + B \sin \omega_0 t\right] + R\left[A \sin \omega_0 t + B \cos \omega_0 t\right] = \sin \omega_0 t \tag{3.4}$$

Notice that we are not paying attention to initial conditions; that is, it is most likely that this solution will NOT be complete, and hence

F. Badrieh, *Spectral, Convolution and Numerical Techniques in Circuit Theory*,
https://doi.org/10.1007/978-3-319-71437-0_3

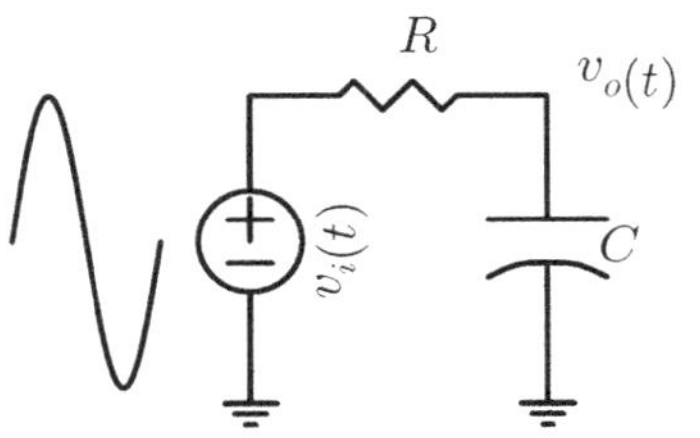

Fig. 3.1 Series RC driven by sine function

by definition this solution is typically coined the "particular solution" (as opposed to total solution). We will come back to initial conditions (and ramifications on total solution) shortly. Collect terms

$$\left[\frac{B}{C\omega_0} + RA\right]\sin\omega_0 t + \left[\frac{-A}{C\omega_0} + RB\right]\cos\omega_0 t = \sin\omega_0 t \tag{3.5}$$

This can be true only if

$$\frac{B}{C\omega_0} + RA = 1, \qquad \text{and} \qquad \frac{-A}{C\omega_0} + RB = 0 \tag{3.6}$$

From the second equation we get

$$B = \frac{A}{RC\omega_0} \tag{3.7}$$

Putting this into the first equation we get

$$\frac{A}{RC^2\omega_0^2} + RA = 1, \qquad A\left[\frac{1}{RC^2\omega_0^2} + R\right] = 1,$$

$$A\frac{1 + R^2C^2\omega_0^2}{RC^2\omega_0^2} = 1 \tag{3.8}$$

which gives for A

$$\boxed{A = \frac{RC^2\omega_0^2}{1 + R^2C^2\omega_0^2}} \tag{3.9}$$

Knowing A we get B

$$\boxed{B = \frac{C\omega_0}{1 + R^2C^2\omega_0^2}} \tag{3.10}$$

Our current is then

$$\boxed{i_p(t) = \frac{RC^2\omega_0^2}{1 + R^2C^2\omega_0^2}\sin\omega_0 t + \frac{C\omega_0}{1 + R^2C^2\omega_0^2}\cos\omega_0 t} \tag{3.11}$$

This looks good so far, but there is one problem (as anticipated). The current at time zero comes out nonzero!!

$$i(0) = \frac{C\omega_0}{1 + R^2C^2\omega_0^2} \neq 0 \tag{3.12}$$

So this solution does NOT satisfy the initial conditions. Towards that goal, let's reexamine the differential equation setting the forcing side to zero

$$\frac{1}{C}\int_0^t i_h(t)dt + Ri_h(t) = 0 \tag{3.13}$$

The solution to this is simple

$$\boxed{i_h(t) = Ae^{-t/RC}} \tag{3.14}$$

where A is a constant to be determined. This solution is coined "homogeneous solution," which is that solution due to a zero driving function. In other words, it would deal with initial conditions and discharge of cap. Our total solution now becomes

$$i(t) = \frac{RC^2\omega_0^2}{1 + R^2C^2\omega_0^2}\sin\omega_0 t + \frac{C\omega_0}{1 + R^2C^2\omega_0^2}\cos\omega_0 t + A^{-t/RC} \tag{3.15}$$

We can find the value of A by again looking at the initial condition:

$$i(0) = \frac{C\omega_0}{1 + R^2C^2\omega_0^2} + A = 0$$

$$\Rightarrow A = -\frac{C\omega_0}{1 + R^2C^2\omega_0^2} \tag{3.16}$$

So finally our total solution is

$$i(t)=\frac{C\omega_0}{1+R^2C^2\omega_0^2}\left[RC\omega_0\sin\omega_0 t+\cos\omega_0 t-e^{-t/RC}\right] \tag{3.17}$$

Let's us confirm this really is the solution!
First voltage across the cap:

$$\begin{aligned}\frac{1}{C}\int_0^i(t)dt &= \frac{\omega_0}{1+R^2C^2\omega_0^2}\left[-RC\cos\omega_0 t+\frac{1}{\omega_0}\sin\omega_0 t+RCe^{-t/RC}\right]_0^t\\ &= \frac{\omega_0}{1+R^2C^2\omega_0^2}\left[-RC\cos\omega_0 t+RC+\frac{1}{\omega_0}\sin\omega_0 t+RCe^{-t/RC}-RC\right]\\ &= \frac{\omega_0}{1+R^2C^2\omega_0^2}\left[-RC\cos\omega_0 t+\frac{1}{\omega_0}\sin\omega_0 t+RCe^{-t/RC}\right]\end{aligned} \tag{3.18}$$

Next voltage across the resistor:

$$v_r(t) = \frac{RC\omega_0}{1+R^2C^2\omega_0^2}\left[RC\omega_0\sin\omega_0 t+\cos\omega_0 t-e^{-t/RC}\right] \tag{3.19}$$

If we add these two voltages, we see that sure enough, the cosine and exponential terms cancel out and we end up only with the sine term. So our answer is

$$v_o = \frac{\omega_0}{1+R^2C^2\omega_0^2}\left[-RC\cos\omega_0 t+\frac{1}{\omega_0}\sin\omega_0 t+RCe^{-t/RC}\right] \tag{3.20}$$

These results, and comparison to SPICE ones, are shown in Fig. 3.2. We observe perfect match; that is, we are able to predict both transient and steady state results. Let's reflect on what was done. We *assumed* a particular solution of the same form of the driving function. Then we assumed a *homogeneous* solution which is the response due to zero driving function (i.e., to deal with initial conditions). Then we combined both and set constants to ensure correct initial conditions.

3.3 Series *LC* Driven by Unit Step Function

Consider the *LC* circuit shown in Fig. 3.3. A unit step input is applied, and we want to find branch current as well as output voltage (across the cap). Doing KVL around the loop we get

$$\frac{1}{C}\int_0^t i(t)dt+L\frac{di(t)}{dt}=1,\qquad t>0 \tag{3.21}$$

Differentiate with respect to time and get

$$\frac{1}{C}i(t)+L\frac{d^2i(t)}{dt^2}=0,\qquad \text{or} \tag{3.22}$$

$$\frac{d^2i(t)}{dt^2} = -\frac{1}{LC}i(t) \tag{3.23}$$

Notice that the particular solution here is zero, and that the homogeneous solution becomes the general solution:

$$i_p(t)=0;\qquad i(t)=i_h(t) \tag{3.24}$$

The general solution to this is

$$i(t)=A\sin\omega_0 t+B\cos\omega_0 t,\qquad \omega_0^2=\frac{1}{LC} \tag{3.25}$$

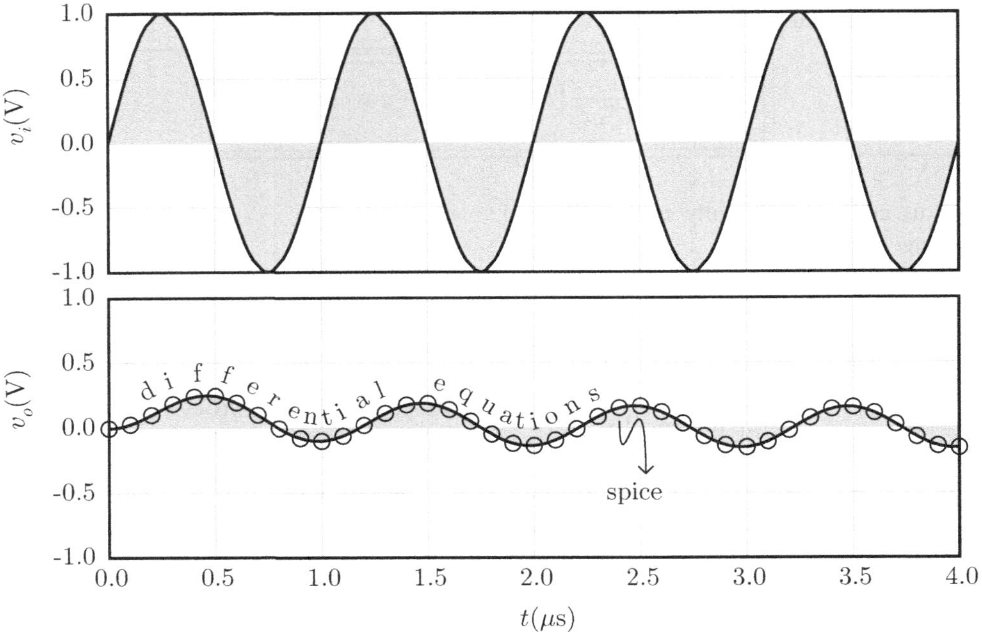

Fig. 3.2 Response of series *RC* network to sine input: SPICE and theoretical results. Case of $R = 1\ \Omega$, $C = 1\ \mu F$, and $\omega_0 = 2\pi \times 10^6$

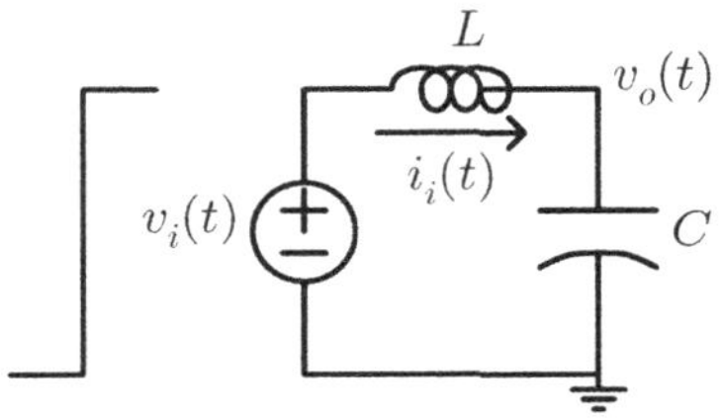

Fig. 3.3 Series *LC* driven by unit step function

We evaluate *A* and *B* by applying the initial conditions. We know that immediately after applying input voltage, branch current is zero

$$i(0) = 0, \qquad \text{which gives} \tag{3.26}$$

$$B = 0, \qquad i(t) = A \sin \omega_0 t \tag{3.27}$$

The initial current is zero because of the inductor; the inductor has high impedance at high frequency, and right at the moment of applying the input step we have a high frequency event! We also know that immediately after applying voltage, all the voltage (one here) will be across the inductor; that is

$$L\frac{di}{dt} = 1, \qquad \text{so that} \tag{3.28}$$

$$LA\omega_0 \cos \omega_0 t = 1\Big|_{t=0} \tag{3.29}$$

which gives

$$A = \frac{1}{L\omega_0} = \sqrt{\frac{C}{L}}, \qquad \text{so that finally} \tag{3.30}$$

$$\boxed{i(t) = \sqrt{\frac{C}{L}} \sin \omega_0 t, \qquad \omega_0 = \frac{1}{\sqrt{LC}}} \tag{3.31}$$

Notice that current decreases with inductance, which makes sense. It also increases with cap, which also makes sense. Both *L* and *C* can be viewed as impedances, and the bigger (smaller) impedance the smaller (larger) current. Notice also that average current is zero. The output voltage across the cap is simply the integral of current (divided by the cap):

$$v_o(t) = \frac{1}{C}\int_0^t i(t)dt = -\frac{1}{C}\sqrt{\frac{C}{L}}\frac{1}{\omega_0}\cos\omega_0 t\bigg|_0^t$$
$$= -\frac{1}{\sqrt{LC}}\sqrt{LC}\cos\omega_0 t\bigg|_0^t \qquad (3.32)$$

$$\boxed{v_o(t) = 1 - \cos\omega_0 t} \qquad (3.33)$$

Even though input voltage is constant (after time zero), output voltage oscillates, due to the oscillation of current. While the average current comes out zero, the average output voltage comes out 1. That is, since average input voltage is 1, that of output also comes out 1 (albeit latter oscillates). Figure 3.4 compares our results to SPICE; we notice perfect agreement.

3.4 Series *LC* Driven by Impulse Input

Consider again the *LC* circuit shown in Fig. 3.5. Rather than unit input, we apply an impulse one. While the input blows up, the area under it does not; in fact the area under the impulse is set to one. We want to find current and output voltage.

Again doing KVL around the loop and differentiating we get

$$\frac{1}{C}i(t)dt + L\frac{d^2i(t)}{dt^2} = 0, \qquad t > 0 \qquad (3.34)$$

which again gives

$$\frac{d^2i(t)}{dt^2} = -\frac{1}{LC}i(t) \qquad (3.35)$$

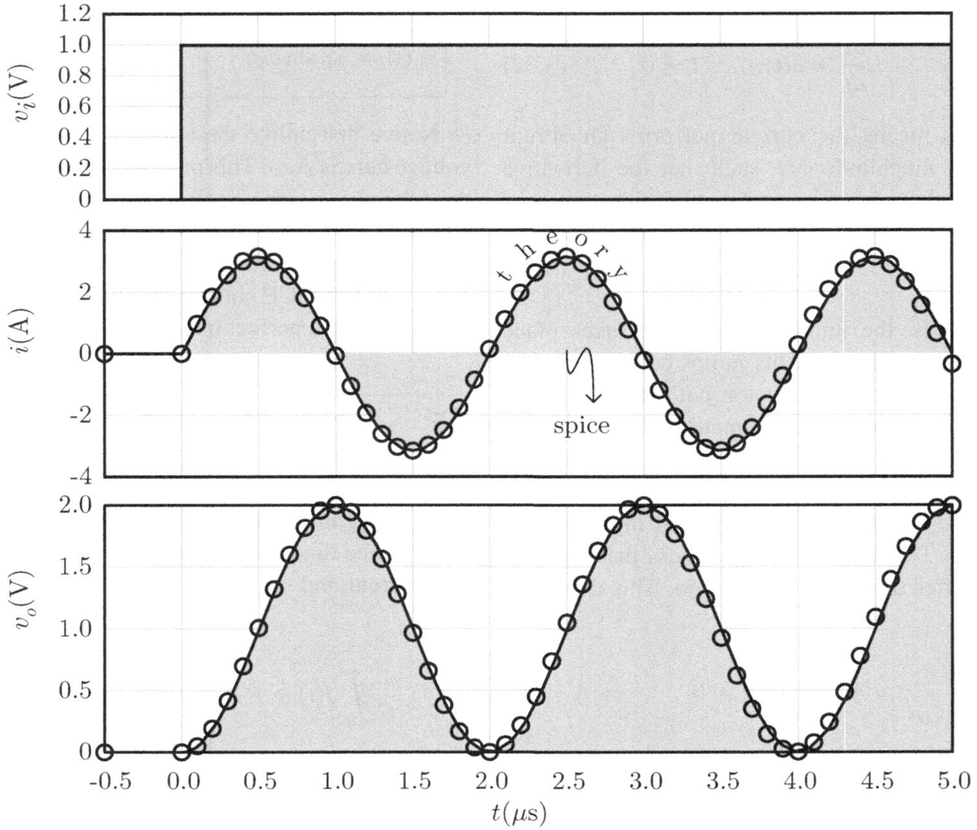

Fig. 3.4 Unit step response of series *LC* and comparison to SPICE simulations. Case of $L = 0.1\,\mu$H and $C = 1\,\mu$F

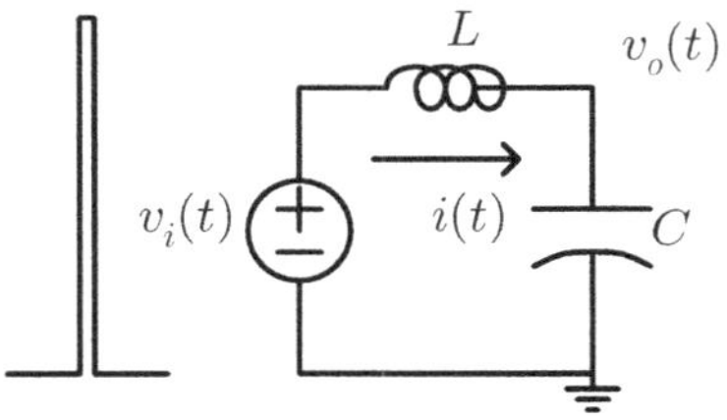

Fig. 3.5 Series *LC* driven by impulse function

The solution to this assumes the form

$$i(t) = A\sin\omega_0 t + B\cos\omega_0 t, \qquad \omega_0 = \frac{1}{\sqrt{LC}} \tag{3.36}$$

To get A and B we have to evaluate the initial conditions. Right at time zero, while the impulse is on, the current will NOT be zero. The impulse voltage lands on the inductor, hence we know the initial voltage across the inductor is an impulse too; then we use

$$L\frac{di}{dt} = \delta(t), \qquad t = 0 \tag{3.37}$$

This means the current performs an abrupt step of magnitude $1/L$ such that the derivative gives a delta function; this then implies that

$$B = \frac{1}{L} \tag{3.38}$$

That is, the jump in the cosine takes place between 0 and $1/L$. This jumps causes the delta function across the inductor, with strength (area) of 1 (definition of delta function). To find A we have to consider the derivative of current right after time zero. We know that after time zero, the impulse input had passed by, and input voltage is zero. This means voltage across the inductor, right after time zero, is zero too. This then says that

$$L\frac{di}{dt}\bigg|_{t=0+} = A\omega_0\cos\omega_0 t = 0, \qquad \Rightarrow A = 0 \tag{3.39}$$

We finally arrive at total current

$$\boxed{i(t) = \frac{1}{L}\cos\omega_0 t, \qquad \omega_0 = \frac{1}{\sqrt{LC}}} \tag{3.40}$$

That is, current is zero for negative time. It jumps abruptly at time zero to $1/L$, forming a delta voltage across the inductor. That is, at time zero we have a delta voltage, as prescribed. After time zero, the rate of change of current starts at zero, since the voltage across the network (really across the inductor) is zero. From then onwards, it oscillates in the form of a cosine function, with frequency $1/\sqrt{LC}$. Notice that current is inversely proportional to inductance, as expected. Inductance is an impedance, and the larger the impedance the smaller the current. Notice furthermore that current is independent of capacitance; since input is an impulse, the inductor totally dominates the impedance. Lastly notice that average current is zero, similar to the unit step input case. The voltage across the cap is related to the integral of this current

$$v_o(t) = \frac{1}{C}\int_0^t i(t)dt = \frac{1}{LC}\frac{1}{\omega_0}\sin\omega_0 t\bigg|_0^t \tag{3.41}$$

$$\boxed{v_o(t) = \omega_0\sin\omega_0 t, \qquad \omega_0 = \frac{1}{\sqrt{LC}}} \tag{3.42}$$

Notice that unlike the unit step case, average voltage here is zero. This makes sense, since after the delta passes, input voltage averages to zero (in fact it is zero), and it makes sense that average output voltage is zero too. Our derived results are compared to SPICE simulation ones, as shown in Fig. 3.6. Notice perfect match.

3.5 Series *LC* Driven by Sine Input

Consider again the *LC* circuit shown in Fig. 3.7. Input is sine function (frequency ω_1), we want to find current and output voltage. Doing KVL we get

$$\frac{1}{C}\int_0^t i(t)dt + L\frac{di}{dt} = \sin\omega_1 t \tag{3.43}$$

Differentiate once and get

$$\frac{1}{C}i(t)dt + L\frac{d^2i}{dt^2} = \omega_1\cos\omega_1 t \tag{3.44}$$

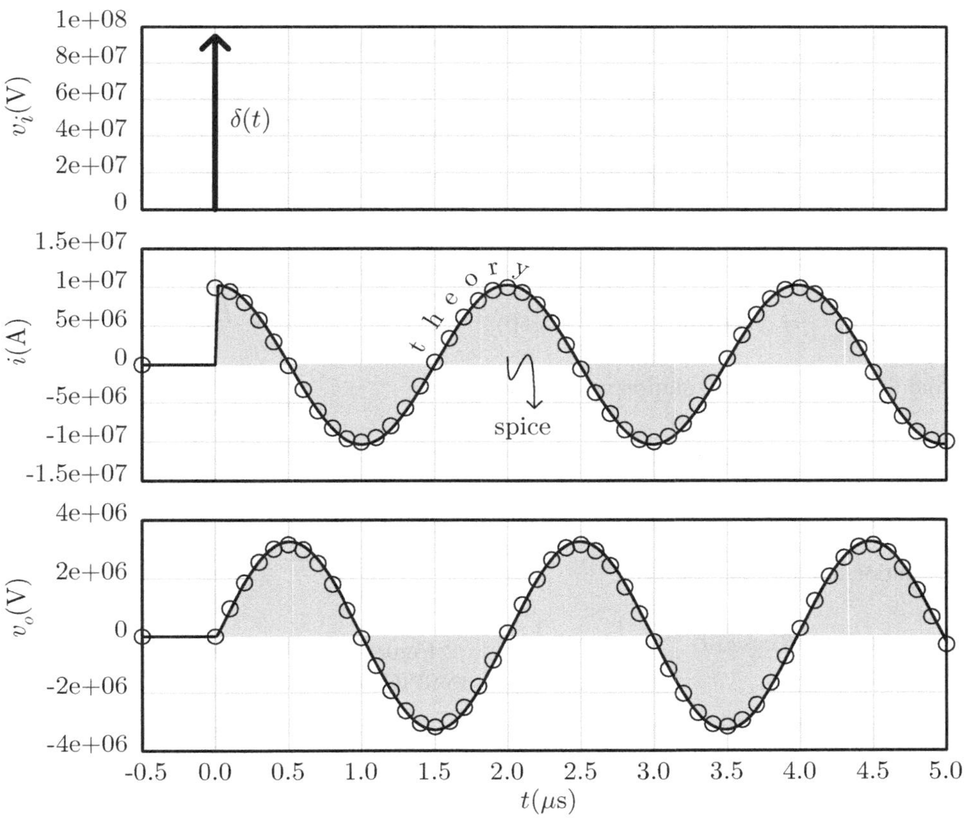

Fig. 3.6 Impulse response of series *LC* and comparison to SPICE simulations. Case of $L = 0.1\,\mu$H and $C = 1\,\mu$F

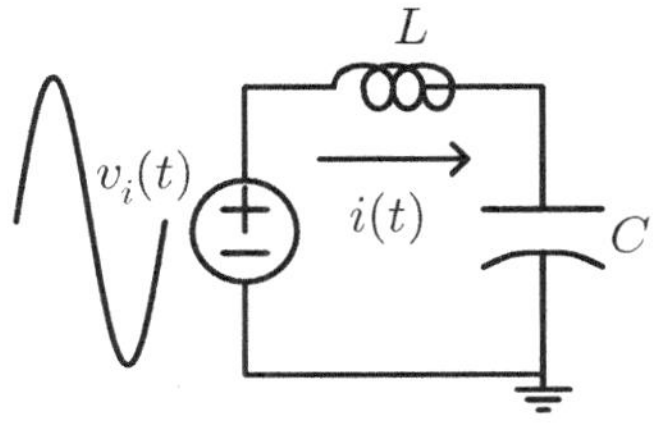

Fig. 3.7 Series *LC* driven by sine input

To solve the particular solution we assume

$$i(t) = A\sin\omega_1 t + B\cos\omega_1 t \tag{3.45}$$

Plugging in we get

$$\frac{1}{C}\left[A\sin\omega_1 t + B\cos\omega_1 t\right] - L\omega_1^2\left[A\sin\omega_1 t + B\cos\omega_1 t\right] = \omega_1\cos\omega_1 t \tag{3.46}$$

Equating coefficient we get

$$\frac{A}{C} - AL\omega_1^2 = 0, \qquad A\left[\frac{1}{C} - L\omega_1^2\right] = 0,$$

$$\Rightarrow A = 0 \tag{3.47}$$

Next

$$B\left[\frac{1}{C} - L\omega_1^2\right] = \omega_1, \qquad B\frac{1 - LC\omega_1^2}{C} = \omega_1,$$

$$\Rightarrow B = \frac{C\omega_1}{1 - LC\omega_1^2} \tag{3.48}$$

Then our particular solution is

$$i_p(t) = \frac{C\omega_1}{1 - LC\omega_1^2}\cos\omega_1 t \tag{3.49}$$

We can verify this is true as follows:

$$\frac{1}{C}i(t)+L\frac{d^2i}{dt^2} = \frac{\omega_1}{1-LC\omega_1^2}\cos\omega_1 t$$
$$-L\frac{C\omega_1}{1-LC\omega_1^2}\omega_1^2\cos\omega_1 t$$
$$=\omega_1\cos\omega_1 t\frac{1-LC\omega_1^2}{1-LC\omega_1^2}$$
$$=\omega_1\cos\omega_1 t \tag{3.50}$$

To find the homogeneous solution we need to solve

$$\frac{1}{C}i(t)+L\frac{d^2i}{dt^2}=0, \quad \text{or} \quad \frac{d^2i}{dt^2}=-\frac{1}{LC}i(t) \tag{3.51}$$

The solution assumes the form

$$i_h(t)=A\sin\omega_0 t+B\cos\omega_0 t, \quad \omega_0=\frac{1}{\sqrt{LC}} \tag{3.52}$$

Our total solution is then

$$i(t)=\frac{C\omega_1}{1-LC\omega_1^2}\cos\omega_1 t+A\sin\omega_0 t+B\cos\omega_0 t \tag{3.53}$$

$$i(t)=\frac{1}{L}\frac{\omega_1}{\omega_0^2-\omega_1^2}\cos\omega_1 t+A\sin\omega_0 t+B\cos\omega_0 t \tag{3.54}$$

To find A and B we apply initial conditions. We know at time zero, initial current is zero; then

$$i(0)=\frac{1}{L}\frac{\omega_1}{\omega_0^2-\omega_1^2}+B=0, \quad B=-\frac{1}{L}\frac{\omega_1}{\omega_0^2-\omega_1^2} \tag{3.55}$$

Our current now becomes

$$i(t)=\frac{1}{L}\frac{\omega_1}{\omega_0^2-\omega_1^2}[\cos\omega_1 t-\cos\omega_0 t]+A\sin\omega_0 t \tag{3.56}$$

The second initial condition is that derivative of current is zero; this gives

$$\left.\frac{di}{dt}\right|_{t=0}=0\Rightarrow A=0 \tag{3.57}$$

Finally we get complete solution

$$\boxed{i(t)=\frac{1}{L}\frac{\omega_1}{\omega_0^2-\omega_1^2}[\cos\omega_1 t-\cos\omega_0 t]} \tag{3.58}$$

Output voltage across the cap is the integral of this current (divided by C):

$$v_o(t)=\frac{1}{C}\int_0^t i(t)dt=\frac{1}{LC}\frac{\omega_1}{\omega_0^2-\omega_1^2}$$
$$\left[\frac{1}{\omega_1}\sin\omega_1 t\Big|_0^t-\frac{1}{\omega_0}\sin\omega_0 t\Big|_0^t\right] \tag{3.59}$$

$$\boxed{v_o(t)=\frac{\omega_0^2\omega_1}{\omega_0^2-\omega_1^2}\left[\frac{1}{\omega_1}\sin\omega_1 t-\frac{1}{\omega_0}\sin\omega_0 t\right]} \tag{3.60}$$

Figure 3.8 shows our results and comparison to SPICE.

3.6 Two-Branch *RC* Network Driven by Unit Step Input

Consider the two-branch RC network shown in Fig. 3.9. We want to find the branch currents as a function of time. Doing KVL around right loop we get

$$\frac{1}{C_1}\int_0^t i_1(t)dt+R_1 i_1=\frac{1}{C_2}\int_0^t i_2(t)dt \tag{3.61}$$

Differentiate both sides with respect to time and get

$$\frac{1}{C_1}i_1(t)+R_1\frac{di_1}{dt}=\frac{1}{C_2}i_2(t), \quad \text{or} \tag{3.62}$$

$$\boxed{i_2(t)=\frac{C_2}{C_1}i_1(t)+R_1C_2\frac{di_1(t)}{dt}} \tag{3.63}$$

We will later need the derivative of $i_2(t)$ as well:

$$\frac{di_2(t)}{dt}=\frac{C_2}{C_1}\frac{di_1(t)}{dt}+R_1C_2\frac{d^2i_1(t)}{dt^2} \tag{3.64}$$

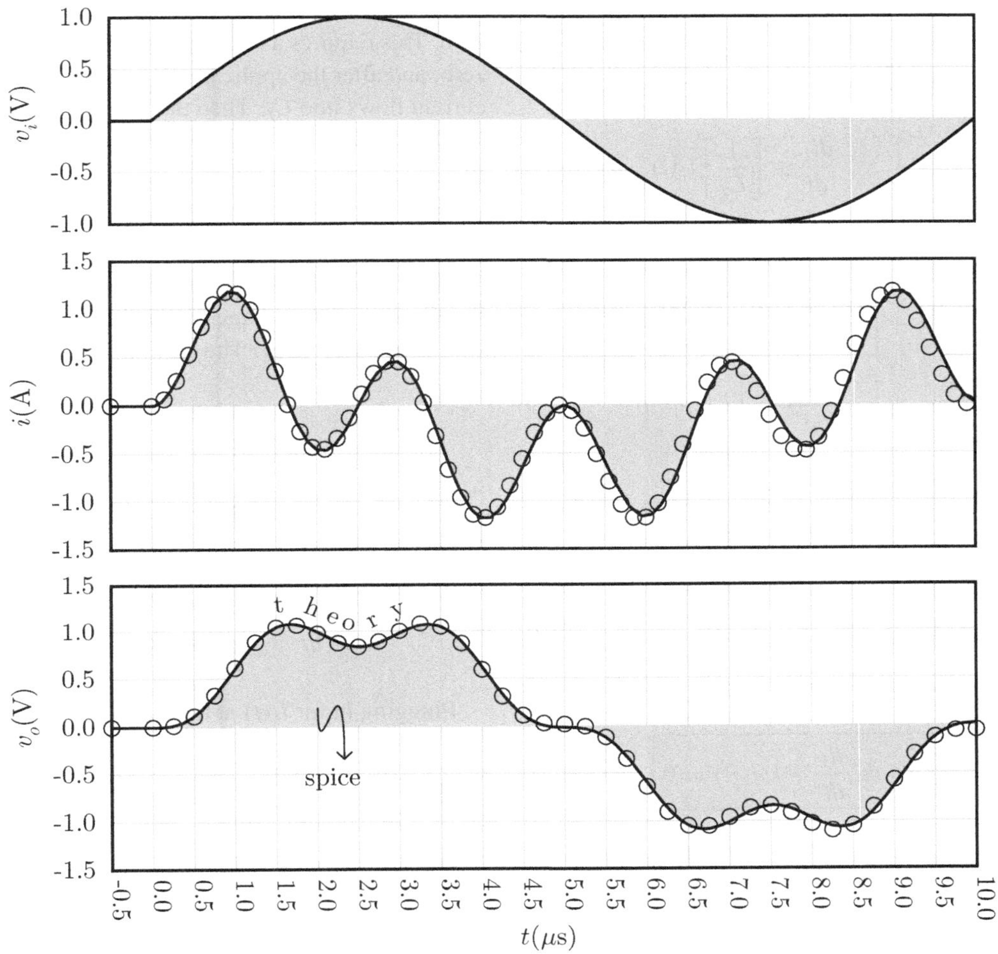

Fig. 3.8 Sine response of series LC and comparison to SPICE simulations. Case of $L = 0.1\,\mu$H, $C = 1\,\mu$F, and $\omega_0 = 2\pi \times 10^5$ rad/s

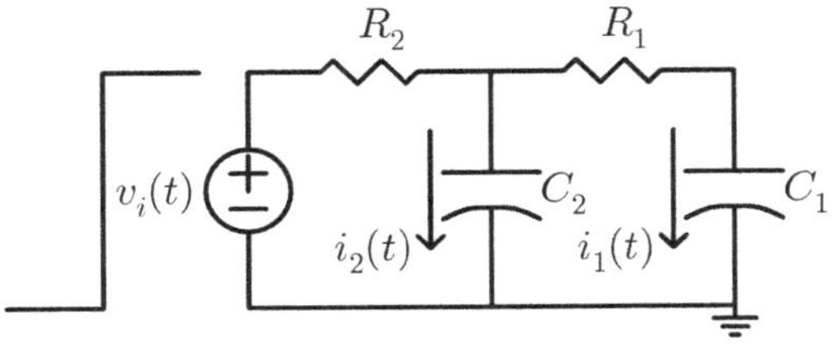

Fig. 3.9 Two-branch RC network driven by step input

Doing KVL around left loop we get

$$\boxed{\frac{1}{C_2}i_2(t) + R_2\frac{di_2}{dt} + R_2\frac{di_1}{dt} = 0} \qquad (3.65)$$

Plugging in for i_2 and its derivative we get

$$\frac{1}{C_2}\left[\frac{C_2}{C_1}i_1(t) + R_1C_2\frac{di_1(t)}{dt}\right] + R_2 \left[\frac{C_2}{C_1}\frac{di_1(t)}{dt} + R_1C_2\frac{d^2i_1(t)}{dt^2}\right] + R_2\frac{di_1}{dt} = 0 \qquad (3.66)$$

Collect terms

$$[R_1R_2C_2]\frac{d^2i_1}{dt^2} + \left[R_1 + R_2 + R_2\frac{C_2}{C_1}\right]\frac{di_1}{dt} + \left[\frac{1}{C_1}\right]i_1(t) = 0 \qquad (3.67)$$

Simplify:

$$[R_1R_2C_2]\frac{d^2i_1}{dt^2}+\left[\frac{(R_1+R_2)C_1+R_2C_2}{C_1}\right]\frac{di_1}{dt}+\left[\frac{1}{C_1}\right]i_1(t)=0 \quad (3.68)$$

Divide by $R_1R_2C_2$

$$\frac{d^2i_1}{dt^2}+\left[\frac{(R_1+R_2)C_1+R_2C_2}{R_1R_2C_1C_2}\right]\frac{di_1}{dt}+\left[\frac{1}{R_1R_2C_1C_2}\right]i_1(t)=0 \quad (3.69)$$

To save space, define

$$k_1=\frac{(R_1+R_2)C_1+R_2C_2}{R_1R_2C_1C_2}, \quad \text{and} \quad k_2=\frac{1}{R_1R_2C_1C_2} \quad (3.70)$$

Then we have

$$\boxed{\frac{d^2i_1}{dt^2}+k_1\frac{di_1}{dt}+k_2i_1(t)=0} \quad (3.71)$$

Assume solution of the form e^{st} and plug back; we get

$$e^{st}\left[s^2+k_1s+k_2\right]=0 \quad (3.72)$$

In order for this to hold for all time we must have:

$$\boxed{s^2+k_1s+k_2=0} \quad (3.73)$$

The two solutions as

$$s_{1,2}=-\frac{k_1}{2}\pm\sqrt{(k_1/2)^2-k_2} \quad (3.74)$$

Then the general solution becomes

$$\boxed{i_1(t)=Ae^{s_1t}+Be^{s_2t}} \quad (3.75)$$

The zero initial condition is $i_1(0)=0$ which implies

$$B=-A \quad (3.76)$$

Then

$$i_1(t)=A\left[e^{s_1t}-e^{s_2t}\right] \quad (3.77)$$

We will also need the derivative of $i_1(t)$ at time zero. This requires a bit of thinking! Right at time zero, and after the application of the unit step, all current flows into C_2. Then the voltage across C_2 would follow something like

$$v_{C_2}(t)=1-e^{-t/R_2C_2}, \qquad t \quad \text{small} \quad (3.78)$$

Notice that this voltage is independent of the right branch. The voltage across the right branch would be this voltage. The current would be

$$i_1(t)=\frac{1-e^{-t/R_2C_2}}{R_1}, \qquad t \quad \text{small} \quad (3.79)$$

It follows that the derivative of $i_1(t)$ around time zero is

$$\left.\frac{di_1(t)}{dt}\right|_{t=0}=\frac{1}{R_1R_2C_2} \quad (3.80)$$

Plugging in for $i_1(t)$ at time zero we get

$$A\left[s_1-s_2\right]=\frac{1}{R_1R_2C_2}, \qquad \text{which sets} \quad (3.81)$$

$$A=\frac{1}{s_1-s_2}\frac{1}{R_1R_2C_2} \quad (3.82)$$

So finally our solution is

$$\boxed{i_1(t)=\frac{1}{s_1-s_2}\frac{1}{R_1R_2C_2}\left[e^{s_1t}-e^{s_2t}\right]} \quad (3.83)$$

$$\boxed{s_{1,2}=-\frac{k_1}{2}\pm\sqrt{(k_1/2)^2-k_2}} \quad (3.84)$$

$$\boxed{k_1=\frac{(R_1+R_2)C_1+R_2C_2}{R_1R_2C_1C_2}, \quad \text{and } k_2=\frac{1}{R_1R_2C_1C_2}} \quad (3.85)$$

and recalling

$$i_2(t)=\frac{C_2}{C_1}i_1(t)+R_1C_2\frac{di_1(t)}{dt} \quad (3.86)$$

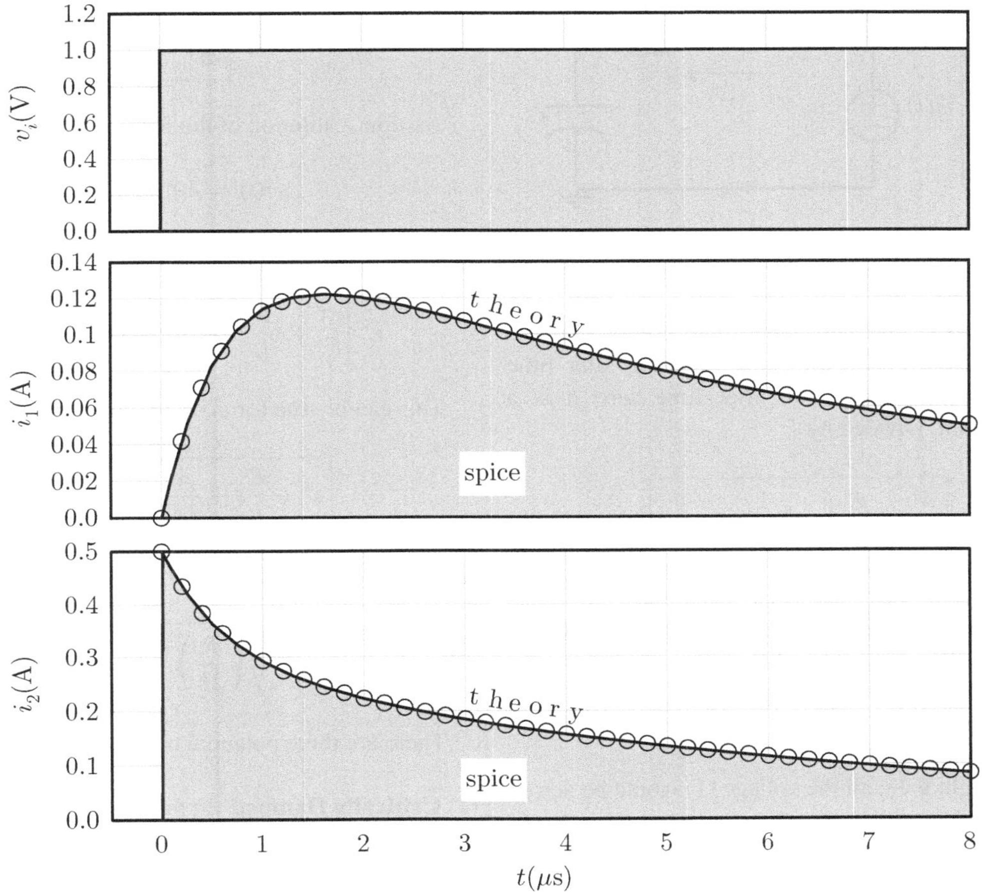

Fig. 3.10 Two-branch *RC* network response due to step input voltage. Case of $R_1 = 1\,\Omega$, $R_2 = 2\,\Omega$, $C_1 = 1\,\mu F$ and $C_2 = 2\,\mu F$

we get the second current

$$\boxed{i_2(t) = \frac{1}{s_1 - s_2}\frac{1}{R_1 R_2 C_2}\left\{\frac{C_2}{C_1}\left[e^{s_1 t} - e^{s_2 t}\right] + R_1 C_2\left[s_1 e^{s_1 t} - s_2 e^{s_2 t}\right]\right\}} \tag{3.87}$$

Figure 3.10 compares our results to SPICE; we observe excellent match.

3.7 Series *RLC* Driven by Step Input

Consider the classic series *RLC* circuit shown in Fig. 3.11. It is subjected to a unit step input, as shown in the figure. We'd like to find branch current and output voltage. Doing KVL around the circuit we get

$$\frac{1}{C}\int_0^t i(t)dt + L\frac{di}{dt} + Ri = u(t) \tag{3.88}$$

Differentiate with respect to time (to get rid of the integral) and we get

$$L\frac{d^2i}{dt^2} + R\frac{di}{dt} + \frac{1}{C}i(t) = 0 \tag{3.89}$$

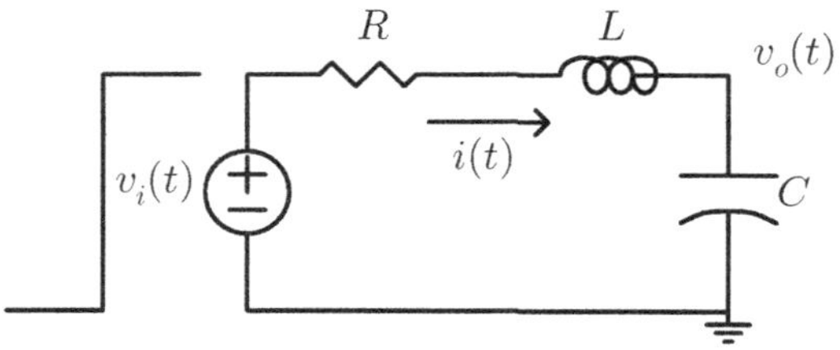

Fig. 3.11 Series *RLC* network driven by step input

Notice that here we are implying that the derivative of the unit step function, after time zero, is zero; that is, after time zero it is a constant. Divided by L

$$\boxed{\frac{d^2i}{dt^2} + \frac{R}{L}\frac{di}{dt} + \frac{1}{LC}i(t) = 0} \tag{3.90}$$

To solve this we need the initial conditions. Right when the step is applied, the inductor chokes the network and no current flow; hence

$$\boxed{i(0) = 0} \tag{3.91}$$

Right then, all the voltage (1) would be across the inductor; hence

$$L\left.\frac{di}{dt}\right|_0 = 1 \Rightarrow \boxed{\left.\frac{di}{dt}\right|_0 = \frac{1}{L}} \tag{3.92}$$

Assume a solution of the form

$$i(t) = Ae^{st} \tag{3.93}$$

Plug in and get

$$e^{st}\left[s^2 + s\frac{R}{L} + \frac{1}{LC}\right] = 0 \tag{3.94}$$

This can be true for all time only if

$$\boxed{s^2 + \frac{R}{L}s + \frac{1}{LC} = 0} \tag{3.95}$$

This quadratic equation gives two solutions

$$s = -\frac{R}{2L} \pm \sqrt{\left(\frac{R}{2L}\right)^2 - \frac{1}{LC}} \tag{3.96}$$

There are three potential outcomes:

(1) Critically Damped In this case

$$\boxed{\left(\frac{R}{2L}\right)^2 - \frac{1}{LC} = 0, \quad \text{or} \quad \frac{R}{2L} = \sqrt{\frac{1}{LC}}, \quad \text{or} \quad R = 2\sqrt{\frac{L}{C}}} \tag{3.97}$$

and we have a single solution

$$s = -\frac{R}{2L} \tag{3.98}$$

and

$$i(t) = Ae^{-\frac{R}{2L}t} \tag{3.99}$$

Notice that this forms a single solution, but since we have a differential equation of order 2, we would expect another (linearly independent) solution! Also, having a single solution would make it impossible to satisfy *two* initial conditions. Therefore we need another solution. Let's define

$$b = \frac{R}{L} \tag{3.100}$$

Then our differential equation, under the single-root case, becomes

$$i'' + bi' + \left(\frac{b}{2}\right)^2 i = 0 \tag{3.101}$$

We know that a solution of the form $e^{-bt/2}$ solves this. We can double check by direct substitution:

$$e^{-bt/2}\left[\left(\frac{b}{2}\right)^2 + b\left(-\frac{b}{2}\right) + \left(\frac{b}{2}\right)^2\right] = e^{-bt/2}\left[\frac{b^2}{2} - \frac{b^2}{2}\right] = 0 \tag{3.102}$$

As another solution let us try something like

$$i_2(t) = te^{-bt/2} \tag{3.103}$$

We can tell if this is reasonable if it leads us somewhere! This solution has

$$\frac{di_2}{dt} = -\frac{b}{2}te^{-bt/2} + e^{-bt/2} = \boxed{e^{-bt/2}\left[1 - \frac{b}{2}t\right]},$$

and (3.104)

$$\frac{di_2^2}{dt} = -\frac{b}{2}\left[-\frac{b}{2}te^{-bt/2} + e^{-bt/2}\right] - \frac{b}{2}e^{-bt/2}$$

$$= e^{-bt/2}\left[-\frac{b}{2} - \frac{b}{2} + \frac{b^2}{4}t\right] = \boxed{e^{-bt/2}\left[\frac{b^2}{4}t - b\right]} \tag{3.105}$$

Plug back into Eq. (3.101) and get

$$e^{-bt/2}\left[\left(\frac{b^2}{4}t - b\right) + b\left(1 - \frac{b}{2}t\right) + \frac{b^2}{4}t\right]$$

$$= e^{-bt/2}\left[\frac{b^2}{2}t - \frac{b^2}{2}t - b + b\right] = \boxed{0} \tag{3.106}$$

which means our second solution is valid! Then the general solution would be a linear combination of both solutions:

$$\boxed{i(t) = (A + Bt)e^{-\frac{R}{2L}t}} \tag{3.107}$$

Now we apply first initial condition and that is $i(0) = 0$; this gives $A = 0$ and we end up with

$$i(t) = Bte^{-\frac{R}{2L}t} \tag{3.108}$$

The second initial condition $di/dt(0) = 1/L$ gives

$$B\left[e^{-\frac{R}{2L}t} - \frac{R}{2L}te^{-\frac{R}{2L}t}\right]_{t=0} = \frac{1}{L}, \qquad \Rightarrow B = \frac{1}{L} \tag{3.109}$$

Hence finally we get

$$\boxed{i(t) = \frac{1}{L}te^{-\frac{R}{2L}t}} \tag{3.110}$$

The output voltage is

$$v_o(t) = \frac{1}{C}\int_0^t i(t)dt = \frac{1}{LC}\int_0^t te^{-\frac{R}{2L}t}dt$$

$$= \frac{1}{LC}\left[t\left(-\frac{2L}{R}\right)e^{-Rt/2L} + \frac{2L}{R}\int e^{-Rt/2L}dt\right]_0^t$$

$$= \frac{1}{LC}\left[-t\frac{2L}{R}e^{-Rt/2L} - \left(\frac{2L}{R}\right)^2 e^{-Rt/2L}\right]_0^t$$

$$= \frac{1}{LC}e^{-Rt/2L}\frac{2L}{R}\left[-t - \frac{2L}{R}\right]_0^t$$

$$= \frac{1}{LC}\frac{2L}{R}\left\{e^{-Rt/2L}\left[-t - \frac{2L}{R}\right] + \frac{2L}{R}\right\} \tag{3.111}$$

Now use the condition

$$\frac{1}{LC} = \left(\frac{R}{2L}\right)^2 \tag{3.112}$$

and get

$$v_o(t) = \left(\frac{R}{2L}\right)^2\frac{2L}{R}\left\{e^{-Rt/2L}\left[-t - \frac{2L}{R}\right] + \frac{2L}{R}\right\}$$

$$= \boxed{\frac{R}{2L}\left\{e^{-Rt/2L}\left[-t - \frac{2L}{R}\right] + \frac{2L}{R}\right\}} \tag{3.113}$$

Figure 3.12 shows comparison between our results and those of SPICE.

(2) Over Damped Here the response has too much dampening, and hence the name "over damped"; in other words, too much friction/loss. Let's recall the roots

$$s_{1,2} = -\frac{R}{2L} + \pm\sqrt{\left(\frac{R}{2L}\right)^2 - \frac{1}{LC}} \tag{3.114}$$

The over damped response is characterized by the relation

$$\left(\frac{R}{2L}\right)^2 - \frac{1}{LC} > 0 \quad \text{or} \quad R > 2\sqrt{\frac{L}{C}} \tag{3.115}$$

In this case the term under the square root comes out positive, such that both roots come out real. Let's define a couple of variables to simplify

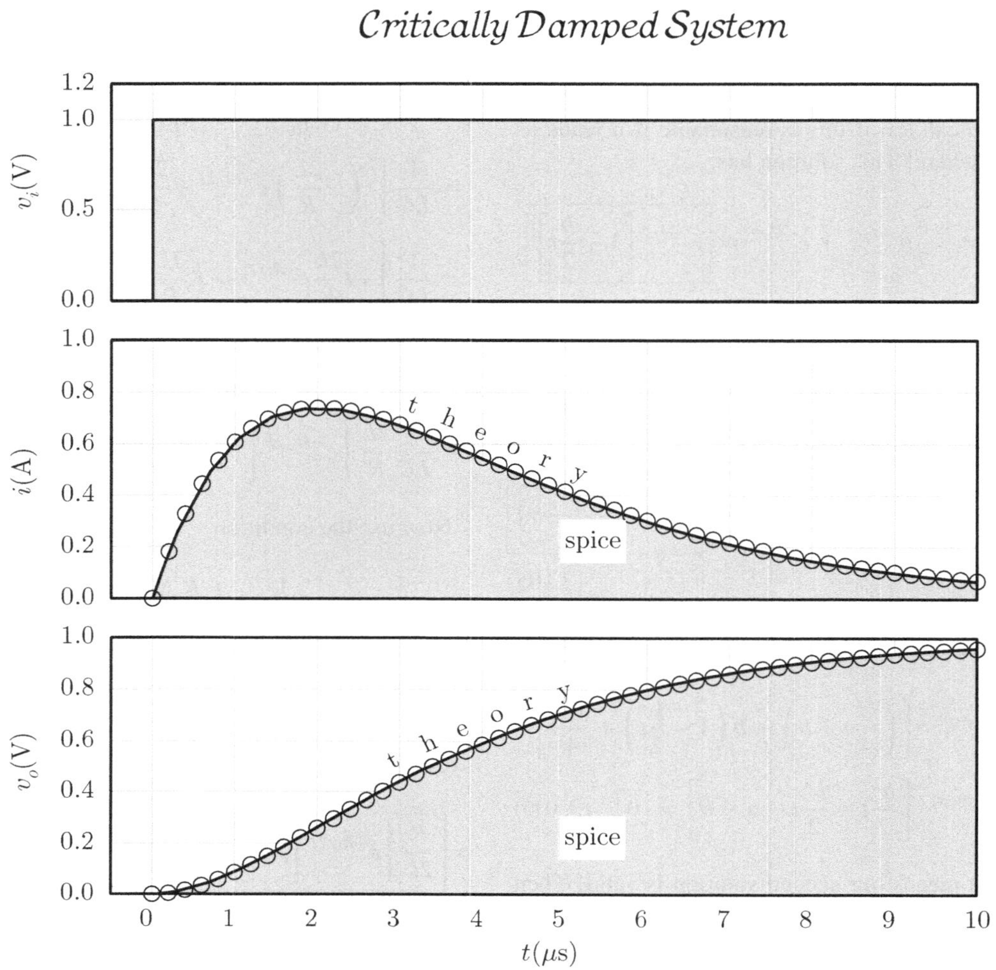

Fig. 3.12 Step response of series *RLC* circuit: critically damped case. Case of $R = 1\,\Omega$, $L = 1\,\mu\text{H}$, and $C = 4\,\mu\text{F}$

the algebra; let

$$\boxed{\alpha = \frac{R}{2L}, \qquad \text{and} \qquad \omega_{LC}^2 = \frac{1}{LC}} \tag{3.116}$$

Then the roots become

$$s_{1,2} = -\alpha \pm \sqrt{\alpha^2 - \omega_{LC}^2} \tag{3.117}$$

The general solution then becomes

$$\boxed{i(t) = Ae^{-\alpha t + \sqrt{\alpha^2 - \omega_{LC}^2}t} + Be^{-\alpha t - \sqrt{\alpha^2 - \omega_{LC}^2}t}} \tag{3.118}$$

To find A and B we plug in the initial condition. The first condition $i(0) = 0$ gives

$$A + B = 0 \Rightarrow B = -A \tag{3.119}$$

The second condition $di/dt(0) = 1/L$ yields

$$A\left[-\alpha + \sqrt{\alpha^2 - \omega_{LC}^2} + \alpha + \sqrt{\alpha^2 - \omega_{LC}^2}\right] = \frac{1}{L}, \tag{3.120}$$

$$\text{or } A = \frac{1}{2L}\frac{1}{\sqrt{\alpha^2 - \omega_{LC}^2}} \tag{3.121}$$

Hence our solution becomes

$$\boxed{i(t) = \frac{1}{2L} \frac{1}{\sqrt{\alpha^2 - \omega_{LC}^2}} \left[e^{-\alpha t + \sqrt{\alpha^2 - \omega_{LC}^2} t} - e^{-\alpha t - \sqrt{\alpha^2 - \omega_{LC}^2} t} \right]} \tag{3.122}$$

The output voltage across the cap is figured in turn by integrating this current

$$\boxed{v_o(t) = \frac{1}{2LC} \frac{1}{\sqrt{\alpha^2 - \omega_{LC}^2}} \left[\frac{e^{-\alpha t + \sqrt{\alpha^2 - \omega_{LC}^2} t} - 1}{-\alpha + \sqrt{\alpha^2 - \omega_{LC}^2}} - \frac{e^{-\alpha t - \sqrt{\alpha^2 - \omega_{LC}^2} t} - 1}{-\alpha - \sqrt{\alpha^2 - \omega_{LC}^2}} \right]} \tag{3.123}$$

Figure 3.13 shows our results and comparison to SPICE.

(3) Under Damped In this case, the damping is small, and we should expect to see oscillations. Quantitatively we have

$$\boxed{\left(\frac{R}{2L}\right)^2 - \frac{1}{LC} < 0 \quad \text{or} \quad R < 2\sqrt{\frac{L}{C}}} \tag{3.124}$$

Recall the root equation

$$s_{1,2} = -\frac{R}{2L} \pm \sqrt{\left(\frac{R}{2L}\right)^2 - \frac{1}{LC}} \tag{3.125}$$

This means the term under the root is negative, and we should expect to see imaginary numbers. Let's rearrange as follows:

$$s_{1,2} = -\frac{R}{2L} \pm \sqrt{- - \left[\left(\frac{R}{2L}\right)^2 - \frac{1}{LC}\right]} \tag{3.126}$$

Pull out the $\sqrt{-1}$ and get

$$s_{1,2} = -\frac{R}{2L} \pm j\sqrt{\frac{1}{LC} - \left(\frac{R}{2L}\right)^2} \tag{3.127}$$

Now the term under the root is positive. Again using our simplified symbols we get

$$s_{1,2} = -\alpha \pm j\sqrt{\omega_{LC}^2 - \alpha^2} \tag{3.128}$$

Let's do further symbolic simplification; let

$$\omega_0^2 = \omega_{LC}^2 - \alpha^2 \tag{3.129}$$

Then we finally get for the roots

$$\boxed{s_{1,2} = -\alpha \pm j\omega_0} \tag{3.130}$$

and our solution for the under damped case comes out

$$\boxed{i(t) = Ae^{-\alpha t + j\omega_0 t} + Be^{-\alpha t - j\omega_0 t}} \tag{3.131}$$

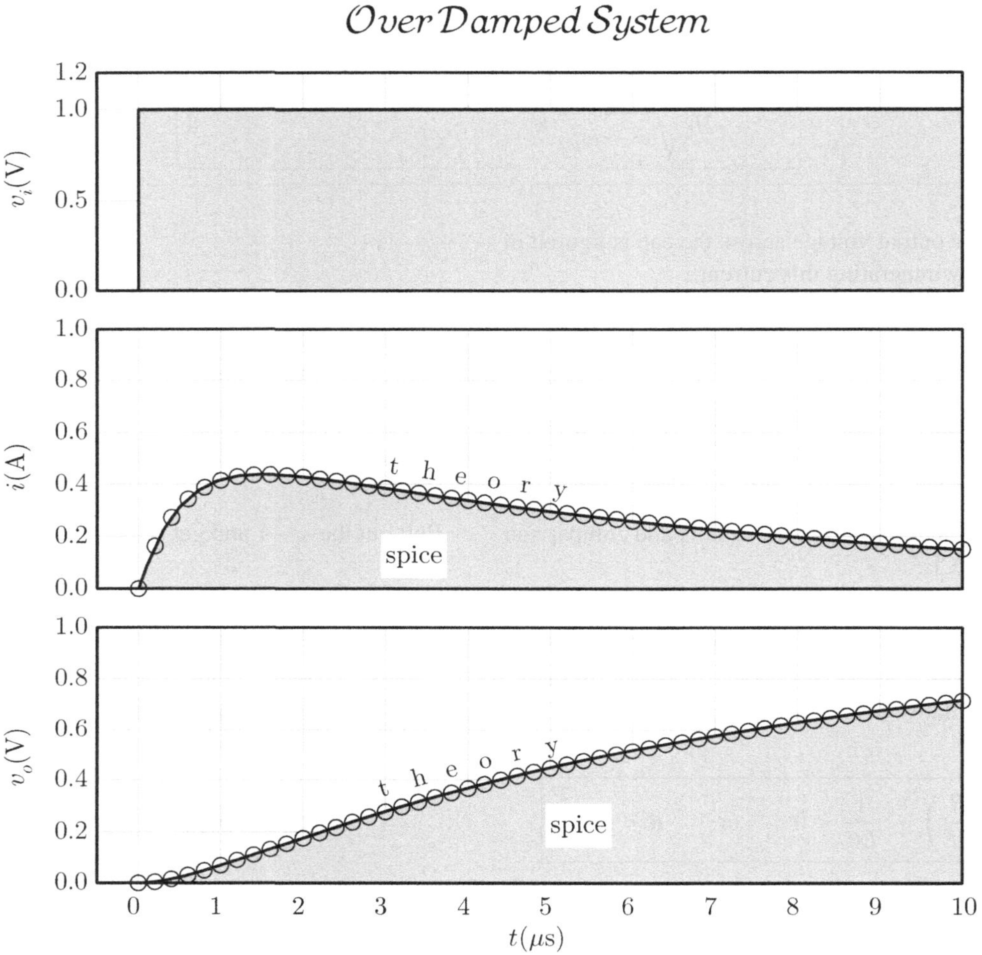

Fig. 3.13 Step response of series *RLC* circuit: over damped case. Case of $R = 2\,\Omega$, $L = 1\,\mu$H, and $C = 4\,\mu$F

To find A and B we again apply the initial conditions. The first IC $i(0) = 0$ gives

$$B = -A \tag{3.132}$$

The second condition $di/dt(0) = 1/L$ yields

$$A\left[-\alpha + j\omega_0 + \alpha + j\omega_0\right] = \frac{1}{L}, \quad \text{or} \tag{3.133}$$

$$A = \frac{1}{L}\frac{1}{2j\omega_0} \tag{3.134}$$

Then our solution becomes

$$i(t) = \frac{1}{L\omega_0}e^{-\alpha t}\frac{e^{j\omega_0 t} - e^{-j\omega_0 t}}{2j}, \quad \text{or} \tag{3.135}$$

$$\boxed{i(t) = \frac{1}{L\omega_0}e^{-\alpha t}\sin\omega_0 t} \tag{3.136}$$

To find output voltage we integrate current across the cap

$$v_o(t) = \frac{1}{LC\omega_0}\int_0^t e^{-\alpha t}\sin\omega_0 t\,dt = \frac{\omega_{LC}^2}{\omega_0} \times \int_0^t e^{-\alpha t}\sin\omega_0 t\,dt \tag{3.137}$$

To evaluate the integral we use integration by parts

$$\int e^{-\alpha t}\sin\omega_0 t\,dt = \frac{-1}{\omega_0}e^{-\alpha t}\cos\omega_0 t$$
$$-\frac{\alpha}{\omega_0}\int e^{-\alpha t}\cos\omega_0 t dt$$
$$= -\frac{1}{\omega_0}e^{-\alpha t}\cos\omega_0 t - \frac{\alpha}{\omega_0}$$
$$\left(\frac{1}{\omega_0}e^{-\alpha t}\sin\omega_0 t + \frac{\alpha}{\omega_0}\int e^{-\alpha t}\sin\omega_0 t dt\right)$$
$$= -\frac{1}{\omega_0}e^{-\alpha t}\cos\omega_0 t - \frac{\alpha}{\omega_0^2}e^{-\alpha t}\sin\omega_0 t$$
$$-\frac{\alpha^2}{\omega_0^2}\int e^{-\alpha t}\sin\omega_0 t dt \tag{3.138}$$

Collect terms and get

$$\int e^{-\alpha t}\sin\omega_0 t\,dt = \frac{\omega_0^2}{\alpha^2+\omega_0^2}$$
$$\left[-\frac{1}{\omega_0}e^{-\alpha t}\cos\omega_0 t - \frac{\alpha}{\omega_0^2}e^{-\alpha t}\sin\omega_0 t\right]$$
$$= -\frac{1}{\alpha^2+\omega_0^2}e^{-\alpha t}\left[\omega_0\cos\omega_0 t + \alpha\sin\omega_0 t\right]$$
$$= -\frac{1}{\omega_{LC}^2}e^{-\alpha t}\left[\omega_0\cos\omega_0 t + \alpha\sin\omega_0 t\right] \tag{3.139}$$

Including the time integration at zero and plugging back into Eq. (3.137) we get

$$v_o(t) = \frac{\omega_{LC}^2}{\omega_0}$$
$$\frac{1}{\omega_{LC}^2}\left[\omega_0 - e^{-\alpha t}\left(\omega_0\cos\omega_0 t + \alpha\sin\omega_0 t\right)\right]$$
$$= \boxed{1 - \frac{1}{\omega_0}e^{-\alpha t}\left[\omega_0\cos\omega_0 t + \alpha\sin\omega_0 t\right]} \tag{3.140}$$

Our results and comparison to SPICE ones are shown in Fig. 3.14. Notice now we get damped oscillations in the response. Let's try a few dissipation cases. As shown in Fig. 3.15 the smaller the R the more pronounced the oscillations.

3.8 Summary

In this section we introduced the method of solving circuit problems by differential equations. Most often we end up with a particular solution and a homogeneous one. The particular solution matches the driving function while the homogeneous one solves the un-driven case. When added together the constants are set to satisfy the initial conditions. The method of differential equations provides the total solution which is comprised of the transient one and the steady state one. We tried the method on a few example, and used various stimuli such as step, impulse, and sine. We wrapped the chapter with the very important case of series RCL network. This latter one was split into critically, over, and under damped cases. We gathered from the examples that LC interactions give oscillations while R components create dissipation. In all cases and throughout the problems we verified our answers with SPICE simulations.

3.9 Problems

1. A parallel RC network, as shown in Fig. 3.16, is driven by a sine input current $\sin\omega_0 t$; find output voltage.
   Answer:
   $$v = \frac{\omega_0 R^2 C}{1+\omega_0^2 R^2 C^2}\left[\frac{1}{\omega_0 RC}\sin\omega_0 t - \cos\omega_0 t + e^{-t/RC}\right]$$
2. Start with Problem 1 and assume $R = 1$, $C = 0.1$ and $\omega_0 = 2\pi$. Plot the solution and compare to SPICE results. Also, discern the transient and steady state solution from total solution. See sample solution in Fig. 3.17.
3. Start with Problem 1 and assume input current now is e^{-st}; Find output voltage analytically. Next assume $R = 1$, $C = 0.5$, and $s = 1$; plot results and compare to SPICE (see Fig. 3.18).
   Answer:
   $$v(t) = \frac{R}{1-sRC}\left[e^{-st} - e^{-t/RC}\right]$$

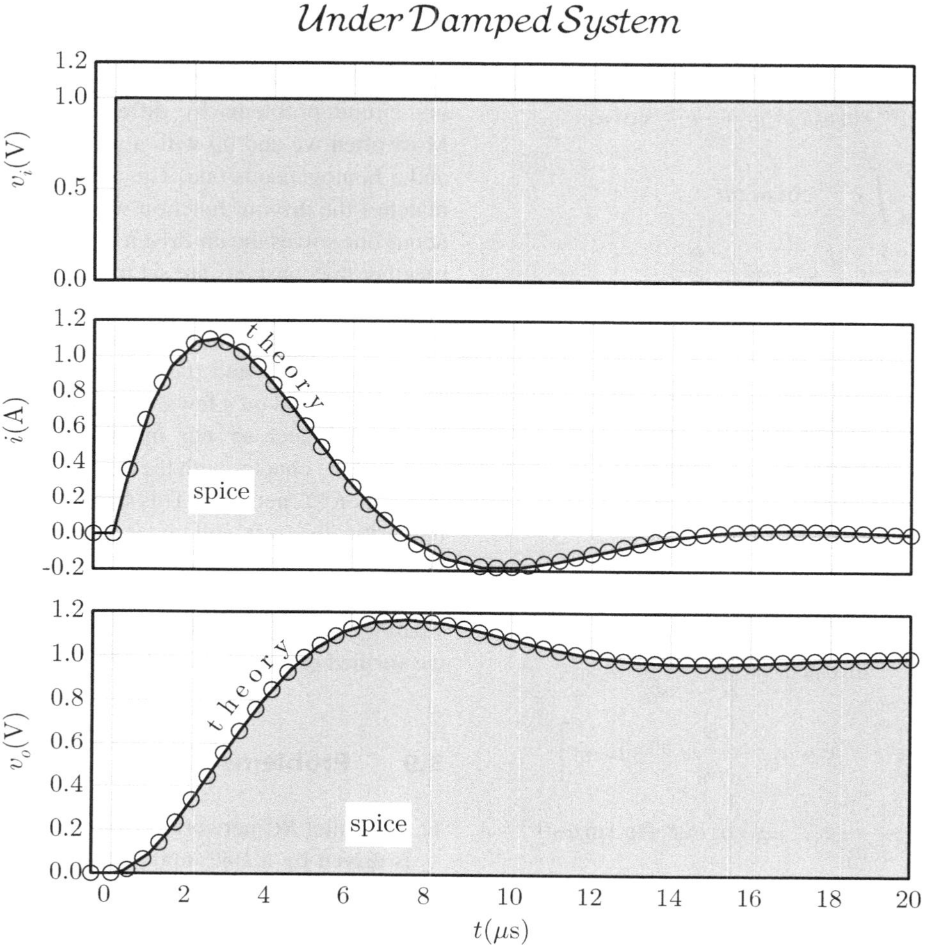

Fig. 3.14 Step response of series *RLC* circuit: under damped case. Case of $R = 0.5\,\Omega$, $L = 1\,\mu$H, and $C = 4\,\mu$F

4. Start with Problem 3. What happens if $s = \frac{1}{RC}$? Plot results and compare to SPICE for $R = 1$ and $C = 0.5$. See sample solution in Fig. 3.19.
   Answer:
$$v(t) = \frac{1}{C} t e^{-t/RC}$$
5. Consider the *RC* circuit in Fig. 3.20. Assume input current is e^{at}; find output voltage.
   Answer:
$$v_o(t) = R_2 \left[\frac{1 + aR_1 C}{1 + a(R_1 + R_2)C} \left(e^{at} - e^{-t/(R_1+R_2)C} \right) + \frac{R_1}{R_1 + R_2} e^{-t/(R_1+R_2)C} \right]$$
6. Take solution of Problem 5 and set $R_1 = 1$, $R_2 = 2$, $C = 0.5$, and $a = -1$; plot solution and compare to SPICE. Answer in Fig. 3.21.
7. Take solution of Problem 5 and replace a with $j\omega_0$. Then take imaginary part of solution to arrive at solution for case driven by $\sin \omega_0 t$.
   Answer:
$$v_o(t) = \frac{1}{1 + \omega_0^2 (R_1 + R_2)^2 C^2} \left\{ \left[R_2 + \omega_0^2 R_1 R_2 (R_1 + R_2) C^2 \right] \sin \omega_0 t + \omega_0 R_2^2 C \left(-\cos \omega_0 t + e^{-t/(R_1+R_2)C} \right) \right\}$$

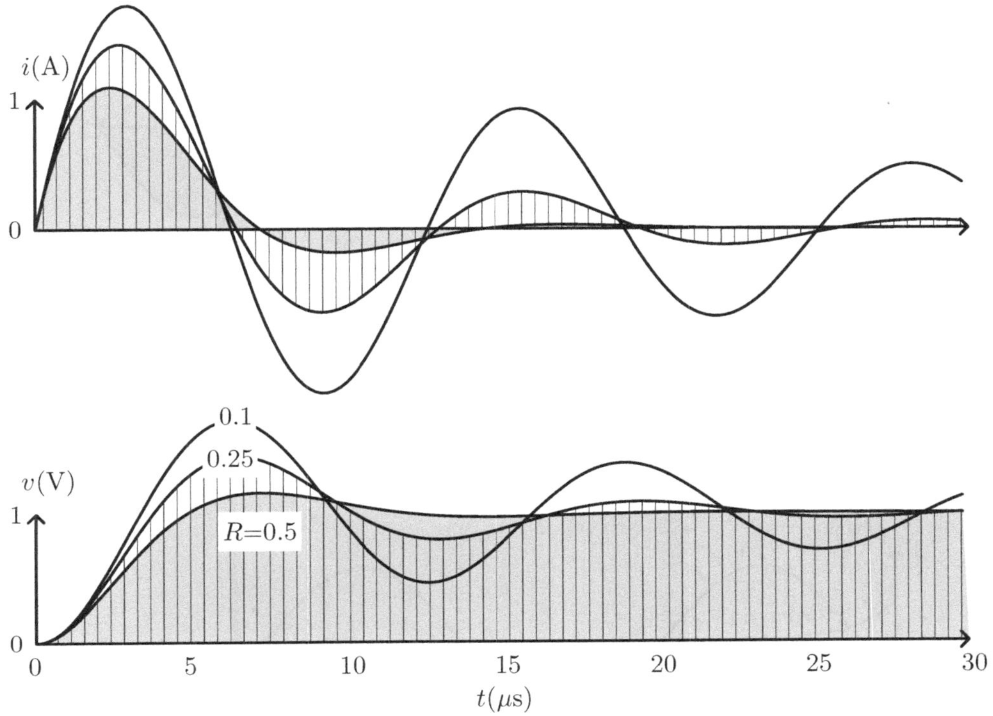

Fig. 3.15 Step response of series *RLC* circuit: under damped case. Case of $L = 1\,\mu H$, $C = 4\,\mu F$, and variable R

Fig. 3.16 Parallel *RC* driven by sine current

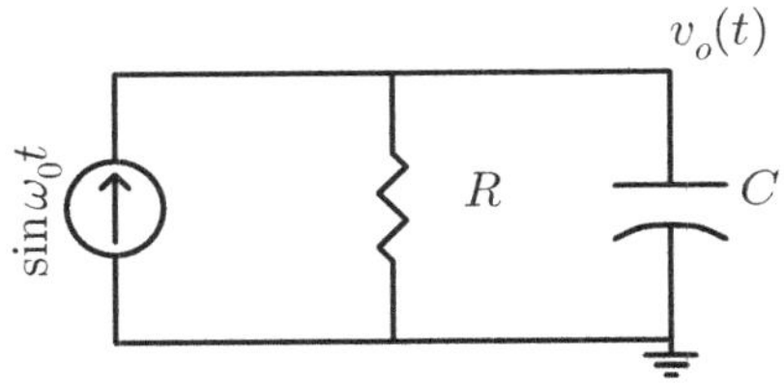

8. Plot solution of Problem 7 assuming $R_1 = 1$, $R_2 = 2$, $C = 0.1$, and $\omega_0 = 2\pi$, and compare to SPICE. Discern transient and steady state components of total solution. See sample run in Fig. 3.22.

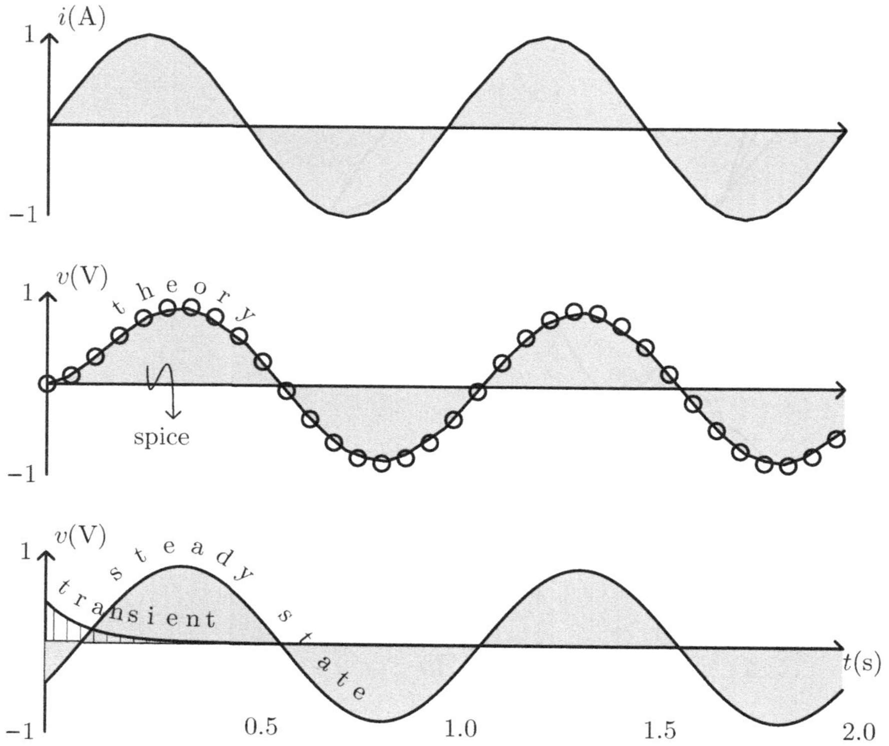

Fig. 3.17 Answer to Problem 2

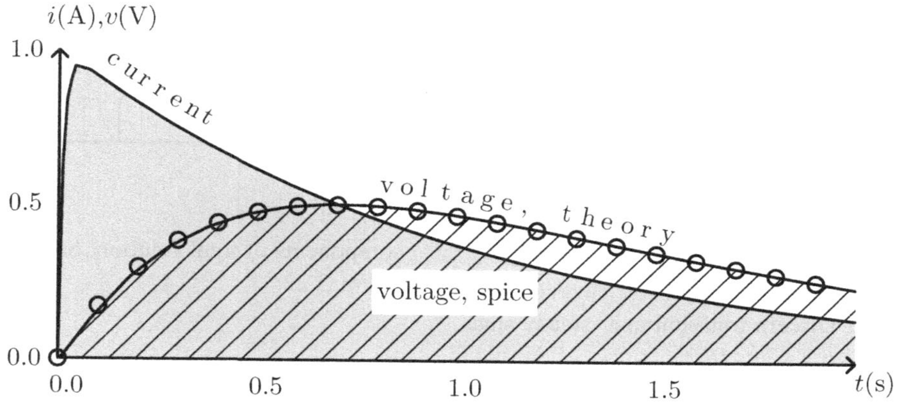

Fig. 3.18 Answer to Problem 3

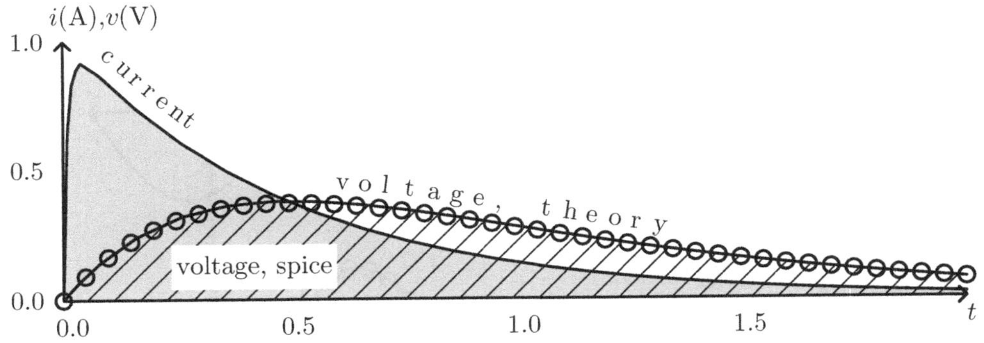

Fig. 3.19 Solution to Problem 4

Fig. 3.20 *RC* network driven by e^{at} current

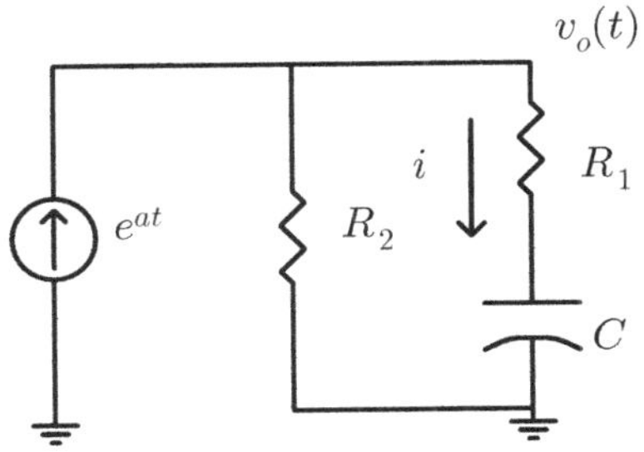

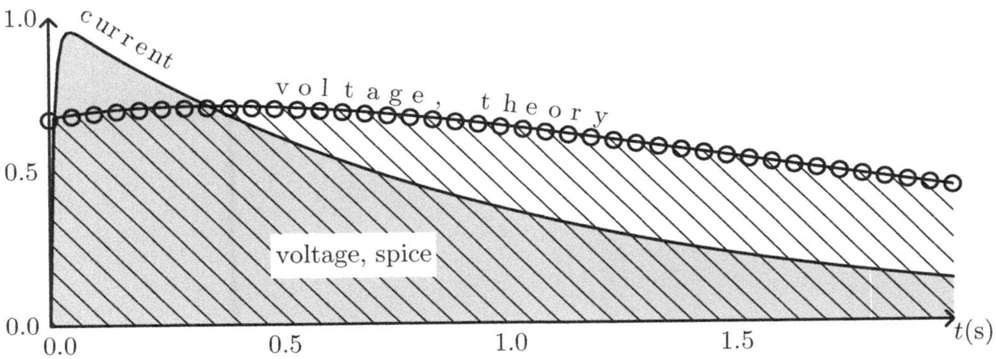

Fig. 3.21 Solution to Problem 6

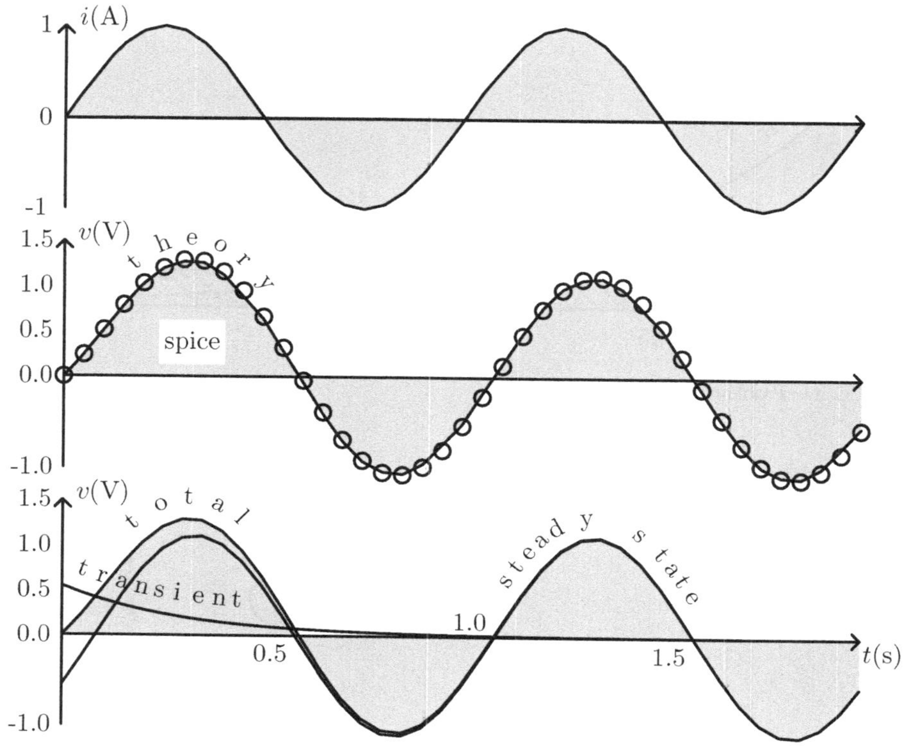

Fig. 3.22 Solution to Problem 8

4 Series Expansion Solution for Circuit Problems

4.1 Introduction

In some cases, we cannot find the exact solution of the differential equations governing a circuit, as was done in the last chapter. The technique shown here enables us to get an approximation to the solution, which is a better one with more expansion terms. It relies on assuming a polynomial form of the solution, and figuring the expansion coefficients based on the governing differential equation, and recursively. The series expansion method is really a very general one; the only drawback is that sometimes the convergence is really slow! For our purpose we won't care much about convergence efficiency; what we mostly care about is an approximate technique to solve the circuit when exact solutions are not available. Also, the expansion here will pave the way for another type of series expansion—the Fourier series, treated in detail in later chapters. Rather than spending much time on developing the theory from scratch, for our purpose it is best to demonstrate the method via a few examples.

4.2 Two-Branch *RC* Network Driven by Unit Step Input

As a first example of this method, consider again the two-branch *RC* network which is driven by a unit input voltage, as shown in Fig. 4.1. We want to find the branch currents as a function of time. Set the two unknown variables as current through right cap $i_1(t)$ and current through left cap $i_2(t)$. Doing KVL around the right loop we get

$$\frac{1}{C_1}\int_0^t i_1(t)dt + i_1R_1 - \frac{1}{C_2}\int_0^t i_2(t)dt = 0 \quad (4.1)$$

Differentiate once with respect to time and get

$$\boxed{\frac{1}{C_1}i_1(t) + R_1\frac{di_1}{dt} - \frac{1}{C_2}i_2(t) = 0} \quad (4.2)$$

Next do KVL around left loop:

$$\frac{1}{C_2}\int_0^t i_2(t)dt + R_2i_2 + R_2i_1 = 1 \quad (4.3)$$

F. Badrieh, *Spectral, Convolution and Numerical Techniques in Circuit Theory*,
https://doi.org/10.1007/978-3-319-71437-0_4

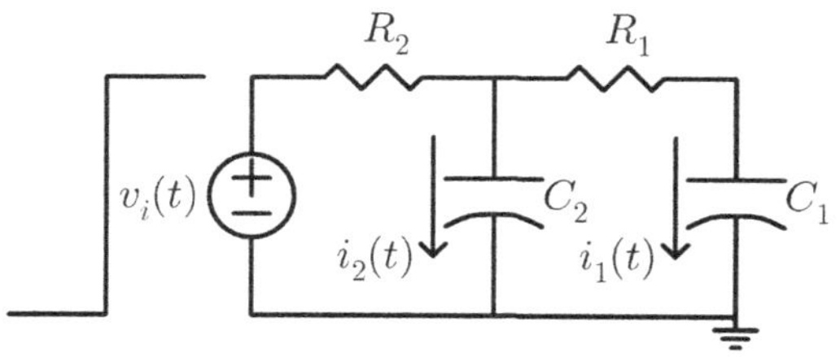

Fig. 4.1 Two-branch RC network driven by step input

Again differential with respect to time and get

$$\boxed{\frac{1}{C_2}i_2(t) + R_2\frac{di_2}{dt} + R_2\frac{di_1}{dt} = 0} \tag{4.4}$$

Assume that we can expand each of the currents as a polynomial series

$$\boxed{i_1(t) = a_0 + a_1t + a_2t^2 + a_3t^3 + \cdots, \quad \text{and} \quad i_2(t) = b_0 + b_1t + b_2t^2 + b_3t^3 + \cdots} \tag{4.5}$$

such that

$$\boxed{\frac{di_1(t)}{dt} = a_1 + 2a_2t + 3a_3t^2 + \cdots, \quad \text{and} \quad \frac{di_2(t)}{dt} = b_1 + 2b_2t + 3b_3t^2 + \cdots} \tag{4.6}$$

Plug into the first loop equation and get

$$\frac{1}{C_1}\left[a_0 + a_1t + a_2t^2 + \cdots\right] + R_1\left[a_1 + 2a_2t + 3a_3^2\right] - \frac{1}{C_2}\left[b_0 + b_1t + b_2t^2 + \cdots\right] = 0 \tag{4.7}$$

Doing the same for the second loop equation we get

$$\begin{aligned}&\frac{1}{C_2}\left[b_0 + b_1t + b_2t^2 + \cdots\right] + R_2\left[b_1 + 2b_2t + 3b_3t^2 + \cdots\right]\\&+R_2\left[a_1 + 2a_2t + 3a_3t^2 + \cdots\right] = 0\end{aligned} \tag{4.8}$$

Looking at last two equations, we see that they hold true for all time only if the coefficients of each power of t were identically zero. Assume we know the very first coefficients of each series: a_0 and b_0. From the first equation, and looking at coefficients of t^0 we arrive at

$$\frac{a_0}{C_1} + R_1a_1 - \frac{b_0}{C_2} = 0 \Rightarrow \boxed{a_1 = \frac{1}{R_1}\left[\frac{b_0}{C_2} - \frac{a_0}{C_1}\right]} \tag{4.9}$$

Doing the same on the second equation we get

$$\frac{b_0}{C_2} + R_2 b_1 + R_2 a_1 = 0$$

$$\Rightarrow \boxed{b_1 = -\frac{1}{R_2}\left[\frac{b_0}{C_2} + R_2 a_1\right]} \quad (4.10)$$

Back to the first equation, and examining t^1 coefficients we arrive at

$$a_2 = \frac{1}{2R_1}\left[\frac{b_1}{C_2} - \frac{a_1}{C_1}\right] \quad (4.11)$$

And from the second equation

$$b_2 = -\frac{1}{2R_2}\left[\frac{b_1}{C_2} + 2R_2 a_2\right] \quad (4.12)$$

and so forth. In general we arrive at the recursive relation

$$\boxed{a_n = \frac{1}{nR_1}\left[\frac{b_{n-1}}{C_2} - \frac{a_{n-1}}{C_1}\right], \text{ and } b_n = -\frac{1}{nR_2}\left[\frac{b_{n-1}}{C_2} + nR_2 a_n\right]} \quad (4.13)$$

From the initial conditions we do in fact know that

$$a_0 = 0, \qquad \text{and} \qquad b_0 = \frac{1}{R_2} \quad (4.14)$$

So we know all there needs to be known to carry on the series expansion. So our solution is

$$i_1(t) = \sum_n a_n t^n, \qquad i_2(t) = \sum_n b_n t^n \quad (4.15)$$

with the coefficients given in the last two equations. Figure 4.2 shows our series approximation and comparison to SPICE. We can see that the more expansion terms we use, the better the approximation.

4.3 Two-Branch *RL* Network Driven by Unit Step Input

Consider the 2-branch *RL* circuit shown in Fig. 4.3. It is subjected to a unit step input potential as shown in the figure. We want to find the two ranch currents. Assume we can expand

$$i_1(t) = a_0 + a_1 t + a_2 t^2 + a_3 t^3 + \cdots,$$
$$i_2(t) = b_0 + b_1 t + b_2 t^2 + b_3 t^3 + \cdots \quad (4.16)$$

with the corresponding first derivatives

$$\frac{di_1(t)}{dt} = a_1 + 2a_2 t + 3a_3 t^2 + \cdots,$$
$$\frac{di_2(t)}{dt} = b_1 + 2b_2 t + 3b_3 t^2 + \cdots \quad (4.17)$$

Then

$$i_1(t) - i_2(t) = (a_0 - b_0) + (a_1 - b_1)t + (a_2 - b_2)t^2 + (a_3 - b_3)t^3 + \cdots, \quad (4.18)$$

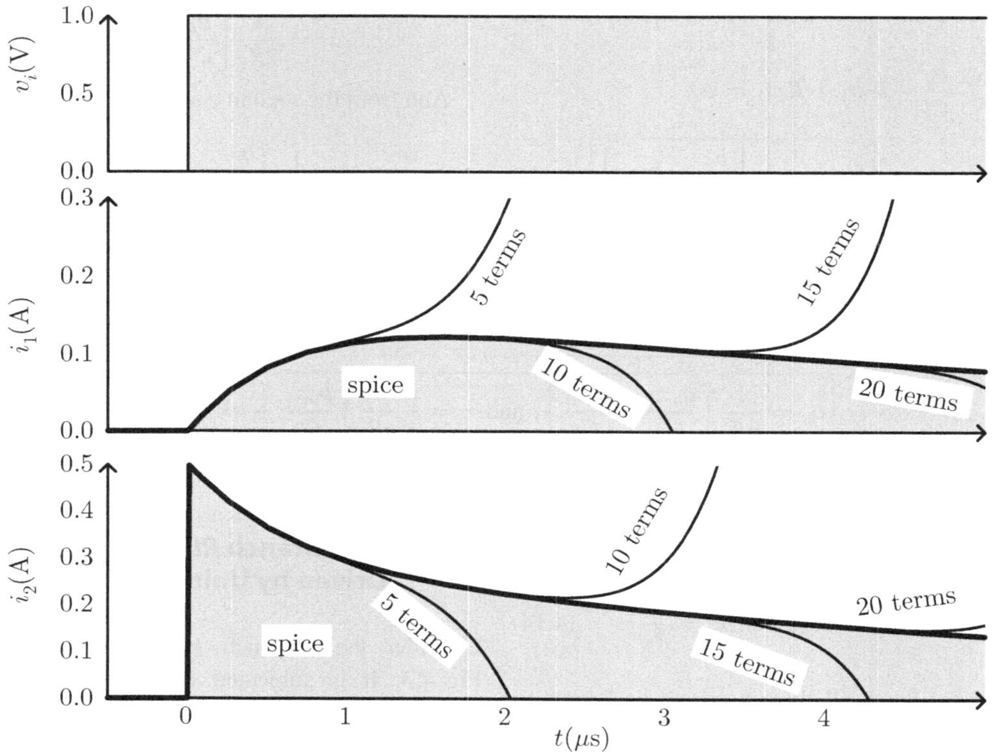

Fig. 4.2 Two-branch *RC* network response to step input voltage. Case of $R_1 = 1\,\Omega$, $R_2 = 2\,\Omega$, $C_1 = 1\,\mu$F and $C_2 = 2\,\mu$F

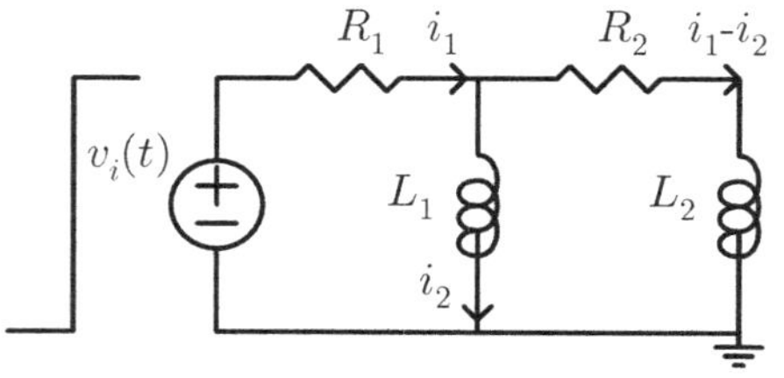

Fig. 4.3 Two-branch *RL* network driven by step input

and

$$\frac{di_1(t)}{dt} - \frac{di_2(t)}{dt} = (a_1 - b_1) + 2(a_2 - b_2)t + 3(a_3 - b_3)t^2 + \cdots \quad (4.19)$$

Doing KVL around left loop we get

$$\boxed{v_i(t) - i_1 R_1 - L_1 \frac{di_2(t)}{dt} = 0} \quad (4.20)$$

Plugging in the series expansion gives

$$1 - R_1(a_0 + a_1 t + a_2 t^2 + \cdots) - L_1(b_1 + 2b_2 t + 3b_3 t^2 + \cdots) = 0 \quad (4.21)$$

Collecting terms gives

$$(1 - R_1 a_0 - 1L_1 b_1) + (-R_1 a_1 - 2L_1 b_2)t + (-R_1 a_2 - 3L_1 b_3)t^2 + = 0 \quad (4.22)$$

This can be true only if each term in the parenthesis is identically zero; for example, assuming we know the a's we can get the b's as follows:

$$b_1 = \frac{1 - R_1 a_0}{1L_1}, \quad b_2 = \frac{-R_1 a_1}{2L_1}, \quad b_3 = \frac{-R_1 a_2}{3L_1}, \cdots \quad (4.23)$$

Doing KVL around the second loop we get

$$L_1 \frac{di_2}{dt} - R_2(i_1 - i_2) - L_2 \frac{d(i_1 - i_2)}{dt} = 0, \qquad \text{which simplifies to} \tag{4.24}$$

$$\boxed{(L_1 + L_2)\frac{di_2}{dt} - R_2(i_1 - i_2) - L_2 \frac{di_1}{dt} = 0} \tag{4.25}$$

Apply series expansion:

$$\begin{aligned}(L_1 + L_2)(b_1 + 2b_2 t + 3b_3 t^2 + \cdots) - R_2\left[(a_0 - b_0) + (a_1 - b_1)t + (a_2 - b_2)t^2 + \cdots\right] \\ - L_2(a_1 + 2a_2 t + 3a_3 t^2 + \cdots) = 0\end{aligned} \tag{4.26}$$

Collecting terms we get

$$\begin{aligned}[(L_1 + L_2)b_1 - L_2 a_1 - R_2(a_0 - b_0)] + [2(L_1 + L_2)b_2 - 2L_2 a_2 - R_2(a_1 - b_1)]t \\ + [3(L_1 + L_2)b_3 - 3L_2 a_3 - R_2(a_2 - b_2)]t^2 = 0\end{aligned} \tag{4.27}$$

Again this can be true only if each power of t is zero; for example, if we know the b's then we can get the a's as follows:

$$\begin{aligned}a_1 &= \frac{1(L_1 + L_2)b_1 - R_2(a_0 - b_0)}{1L_2}, \quad a_2 = \frac{2(L_1 + L_2)b_2 - R_2(a_1 - b_1)}{2L_2}, \\ a_3 &= \frac{3(L_1 + L_2)b_3 - R_2(a_2 - b_2)}{3L_2}, \cdots\end{aligned} \tag{4.28}$$

In general we then have

$$\boxed{b_1 = \frac{1 - R_1 a_0}{L_1}, \quad b_n = -\frac{R_1 a_{n-1}}{nL_1}, (n > 1), \qquad \text{and}} \tag{4.29}$$

$$\boxed{a_n = \frac{n(L_1 + L_2)b_n - R_2(a_{n-1} - b_{n-1})}{nL_2}, (n > 0)} \tag{4.30}$$

with the initial conditions

$$b_0 = a_0 = 0 \tag{4.31}$$

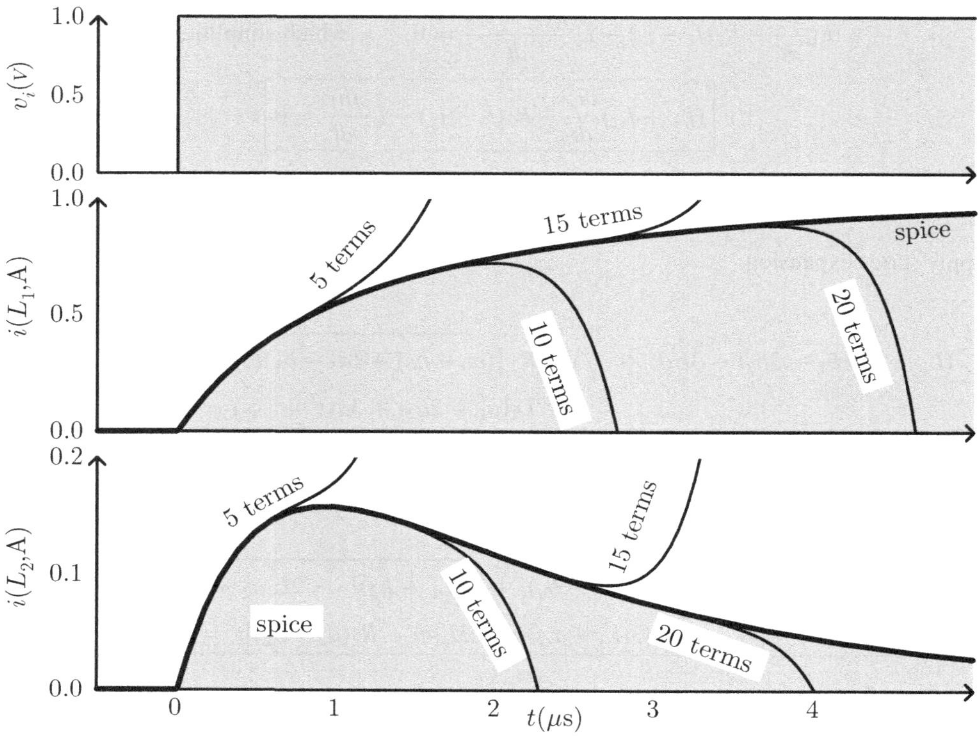

Fig. 4.4 Two-branch *RL* network driven by step input: inductor currents. Case of $R_1 = 1\,\Omega$, $R_2 = 2\,\Omega$, $L_1 = 1\,\mu$H, and $L_2 = 2\,\mu$H

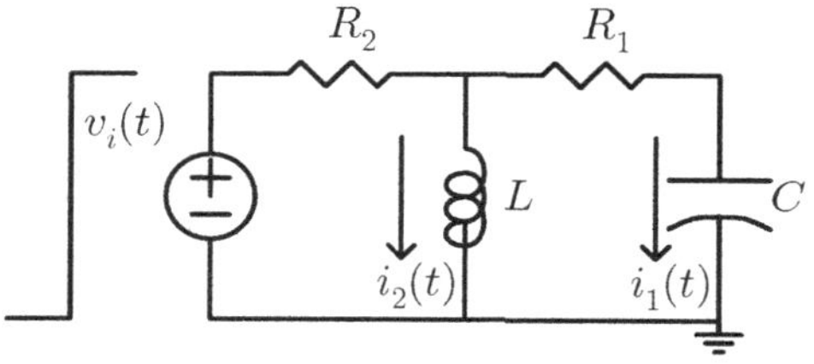

Fig. 4.5 Two-branch *RLC* network driven by step input

Notice that right after the step turns on we have zero current because the impedance of the inductors is infinite at high frequency. Only after things settle down would current start to conduct.

So our solution is

$$i_1(t) = \sum_n a_n t^n, \qquad i_2(t) = \sum_n b_n t^n \quad (4.32)$$

with the coefficients as just derived. Figure 4.4 shows inductor currents from SPICE and using our theory; very good match is observed.

4.4 Two-Branch *RLC* Network Driven by Unit Step Input

Consider next the two-branch *RLC* network shown in Fig. 4.5. It is driven by a unit step input voltage; we want to find the branch currents. Doing KVL around the outer loop we get

$$\frac{1}{C}\int_0^t i_1(t)dt + (R_1 + R_2)i_1 + R_2 i_2 = 1 \quad (4.33)$$

Differentiate with respect to time and get

$$\boxed{\frac{1}{C}i_1(t) + (R_1 + R_2)\frac{di_1}{dt} + R_2\frac{di_2}{dt} = 0} \quad (4.34)$$

Now do KVL around the left loop

$$\boxed{L\frac{di_2}{dt} + R_2 i_2 + R_2 i_1 = 1} \quad (4.35)$$

Notice that we do not differentiate this last equation. Now assume both currents can be expanded as a polynomial series

$$\boxed{i_1(t) = a_0 + a_1 t + a_2 t^2 + a_3 t^3 + \cdots \quad \text{and} \quad i_2(t) = b_0 + b_1 t + b_2 t^2 + b_3 t^3 + \cdots} \quad (4.36)$$

Plug back into Eq. (4.34) and get

$$\begin{aligned} &\frac{1}{C}\left[a_0 + a_1 t + a_2 t^2 + a_3 t^3 + \cdots\right] \\ &+ (R_1 + R_2)\left[a_1 + 2a_2 t + 3a_3 t^2 + \cdots\right] \\ &+ R_2\left[b_1 + 2b_2 t + 3b_3 t^2 + \cdots\right] = 0 \end{aligned} \quad (4.37)$$

Plug into Eq. (4.35) and get

$$\begin{aligned} &L\left[b_1 + 2b_2 t + 3b_3 t^2 + \cdots\right] \\ &+ R_2\left[a_0 + a_1 t + a_2 t^2 + a_3 t^3 + \cdots\right] \\ &+ R_2\left[b_0 + b_1 t + b_2 t^2 + b_3 t^3 + \cdots\right] = 1 \end{aligned} \quad (4.38)$$

Assume initial conditions such that

$$\boxed{a_0 = \frac{1}{R_1 + R_2}, \qquad b_0 = 0} \quad (4.39)$$

That is, right after the step turns on the inductor is open and the capacitor is short. All the current goes through the cap; and this happens to be unity divided by total resistance. From Eq. (4.38) we equate the t^0 terms and get

$$b_1 = \frac{1 - R_2(a_0 + b_0)}{L} \quad (4.40)$$

We put this back into Eq. (4.37) and get

$$a_1 = -\frac{\frac{a0}{C} + R_2 b_1}{R_1 + R_2} \quad (4.41)$$

Now we put this back into Eq. (4.38) and get b_2

$$b_2 = -\frac{R_2(a_1 + b_1)}{2L} \quad (4.42)$$

and similarly get

$$a_2 = -\frac{\frac{a1}{C} + 2R_2 b_2}{2(R_1 + R_2)} \quad (4.43)$$

and so forth. In general we then have

$$\boxed{b_0 = 0, \quad b_1 = \frac{1 - R_2(a_0 + b_0)}{L}, \quad b_n = -\frac{R_2(a_{n-1} + b_{n-1})}{nL}, \quad (n > 1), \text{and}} \quad (4.44)$$

$$\boxed{a_0 = \frac{1}{R_1 + R_2}, \quad a_n = -\frac{\frac{a_{n-1}}{C} + nR_2 b_n}{n(R_1 + R_2)}, \quad (n > 0)} \quad (4.45)$$

That is, we are able to recursively calculate the polynomial series expansion coefficients. Once those are known, then in theory we have our solution.

$$i_1(t) = \sum_n a_n t^n, \qquad i_2(t) = \sum_n b_n t^n \tag{4.46}$$

Figure 4.6 shows our results and comparison to SPICE ones. Notice that the more terms we include in the series expansion, the more accurate results we get.

4.5 Three-Branch *RC* Network Driven by Unit Step Input

Consider next the three-branch *RC* network shown in Fig. 4.7. It is driven by a unit step input voltage; we want to find the various branch currents. Doing KVL around right loop (and differentiating) we get

$$\frac{1}{C_1} i_1 + R_1 \frac{di_1}{dt} - \frac{1}{C_2} i_2 = 0 \tag{4.47}$$

Doing KVL around middle loop (and differentiating) we get

$$\frac{1}{C_2} i_2 + R_2 \left[\frac{di_1}{dt} + \frac{di_2}{dt} \right] - \frac{1}{C_3} i_3 = 0 \tag{4.48}$$

And finally doing KVL around left loop we get

$$\frac{1}{C_3} i_3 + R_3 \left[\frac{di_1}{dt} + \frac{di_2}{dt} + \frac{di_3}{dt} \right] = 0 \tag{4.49}$$

The initial conditions are

$$i_1(0) = 0, \qquad i_2(0) = 0, \qquad i_3(0) = \frac{1}{R_3} \tag{4.50}$$

Expand the currents as follows:

$$\begin{aligned} i_1(t) &= a_0 + a_1 t + a_2 t^2 + a_3 t^3 + \cdots, \\ i_2(t) &= b_0 + b_1 t + b_2 t^2 + b_3 t^3 + \cdots \\ i_3(t) &= c_0 + c_1 t + c_2 t^2 + c_3 t^3 + \cdots \end{aligned} \tag{4.51}$$

Plug into the first KVL equation (4.47) and get

$$\begin{aligned} &\frac{1}{C_1}\left[a_0 + a_1 t + a_2 t^2 + a_3 t^3 + \cdots\right] \\ &+ R_1 \left[a_1 + 2a_2 t + 3a_3 t^2 + \cdots\right] \\ &- \frac{1}{C_2}\left[b_0 + b_1 t + b_2 t^2 + b_3 t^3 + \cdots\right] = 0 \end{aligned} \tag{4.52}$$

Plug into the second KVL equation (4.48) and get

$$\begin{aligned} &\frac{1}{C_2}\left[b_0 + b_1 t + b_2 t^2 + b_3 t^3 + \cdots\right] \\ &+ R_2\big[(a_1 + b_1) + 2(a_2 + b_2)t \\ &+ 3(a_3 + b_3)t^2 + \cdots\big] \\ &- \frac{1}{C_3}\left[c_0 + c_1 t + c_2 t^2 + c_3 t^3 + \cdots\right] = 0 \end{aligned} \tag{4.53}$$

And the third Eq. (4.49) gives

$$\begin{aligned} &\frac{1}{C_3}\left[c_0 + c_1 t + c_2 t^2 + c_3 t^3 + \cdots\right] \\ &+R_3\left[(a_1 + b_1 + c_1) + 2(a_2 + b_2 + c_2)t \right. \\ &\left. +3(a_3 + b_3 + c_3)t^2 + \cdots\right] = 0 \end{aligned} \tag{4.54}$$

Recalling the initial conditions which translate to

$$a_0 = 0, \qquad b_0 = 0, \qquad c_0 = \frac{1}{R_3} \tag{4.55}$$

From the first (new) KVL equation (4.52) we get a relation for the new a_{n+1}:

$$\boxed{a_{n+1} = \frac{-\frac{a_n}{C_1} + \frac{b_n}{C_2}}{(n+1)R_1}} \tag{4.56}$$

From the second (new) KVL equation (4.53) we get a relation for the new b_{n+1}:

$$\boxed{b_{n+1} = \frac{-\frac{b_n}{C_2} + \frac{c_n}{C_3}}{(n+1)R_2} - a_{n+1}} \tag{4.57}$$

And finally from the third (new) KVL equation (4.54) we get a relation for the new c_{n+1}:

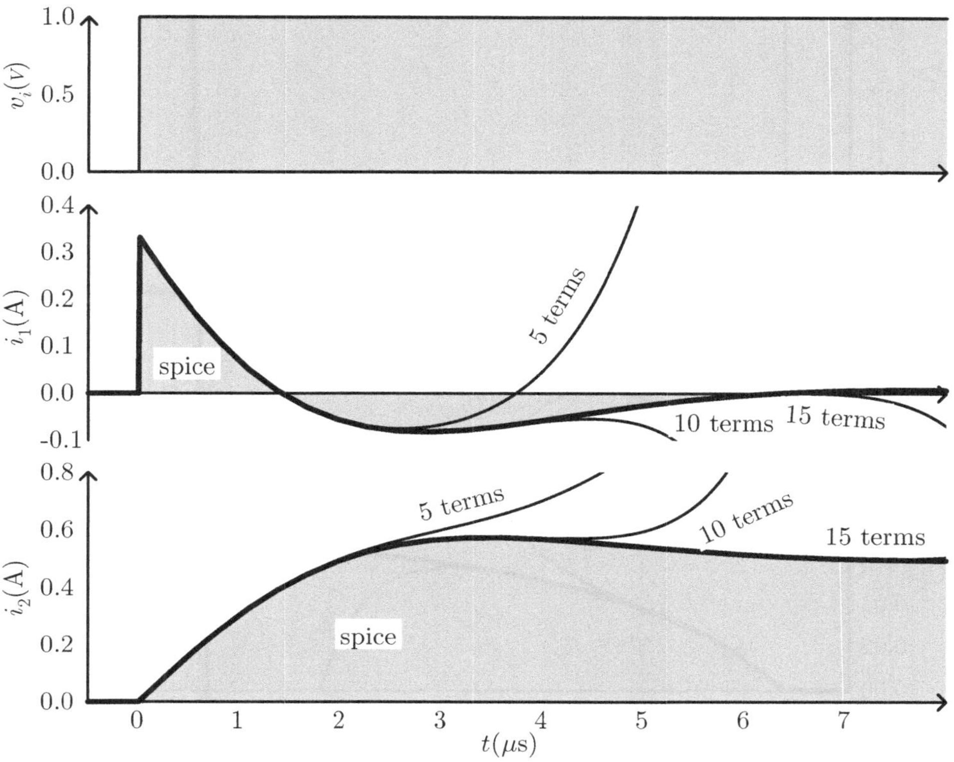

Fig. 4.6 Two-branch *RLC* network response to unit step input voltage. Case of $R_1 = 1\,\Omega$, $R_2 = 2\,\Omega$, $L = 1\,\mu$H and $C = 1\,\mu$F

Fig. 4.7 Three-branch *RC* network driven by step input

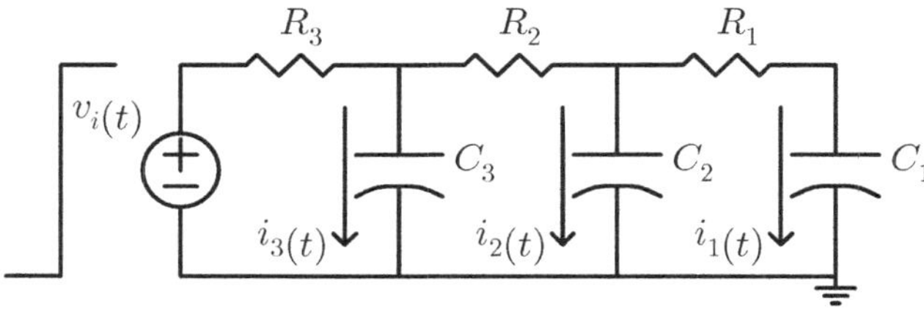

$$\boxed{c_{n+1} = \frac{-\frac{c_n}{C_3}}{(n+1)R_3} - (a_{n+1} + b_{n+1})} \tag{4.58}$$

Knowing the three recursion relations we know the three functions

$$i_1(t) = \sum_n a_n t^n, \quad i_2(t) = \sum_n b_n t^n,$$
$$i_3(t) = \sum_n c_n t^n \tag{4.59}$$

and these are plotted in Fig. 4.8 along with SPICE results. Again very good agreement.

4.6 Comparison to Fourier Series

It is worth pointing that this method gives the better approximation for initial time, and loses accuracy for larger time. The more terms we include, the more accurate the large time results become. The Fourier series expansion (to be covered later), on the other hand, gives relatively the same accuracy on both ends of the time spectrum. The more terms we include here, the better the overall accuracy becomes. That is, in the polynomial series expansion there is a difference in performance between the small and large time

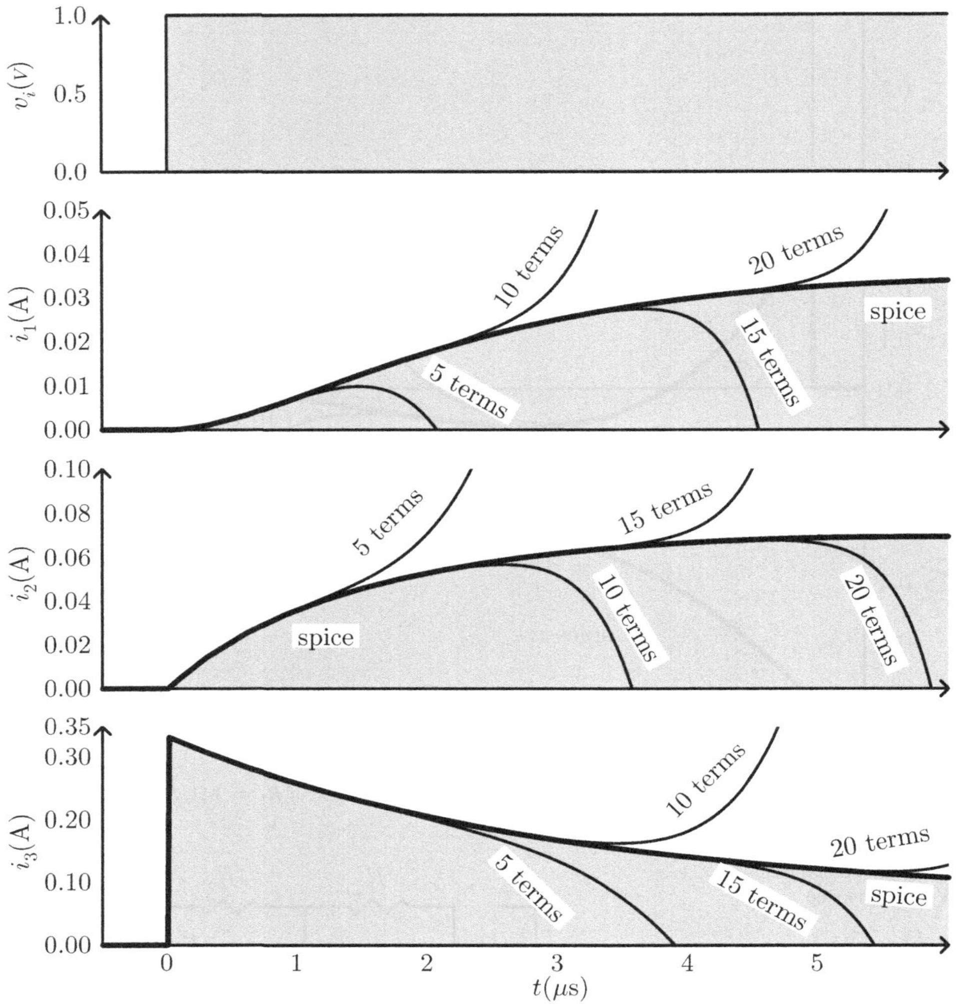

Fig. 4.8 Three-branch *RC* network response to unit step input. Case of $R_1 = 1\,\Omega$, $R_2 = 2\,\Omega$, $R_3 = 3\,\Omega$, $C_1 = 1\,\mu F$, $C_2 = 2\,\mu F$ and $C_3 = 3\,\mu F$

scale, but in the Fourier series (transform) there is not. They are really two different expansion methods, and one (the polynomial method) could be thought off as a "series" expansion while the other (Fourier) as a "parallel" expansion!

4.7 Summary

The method of polynomial series expansion is a pretty generic method for solving linear circuit problems. Each of the unknowns is assumed to be of the form of a polynomial time series, with coefficients which are determined based on the initial conditions and the governing differential equation. Typically a recursive relation is generated which gives the higher order power coefficients in terms of the lower ones. The method was demonstrated on a few examples, ending with a 3-branch *RC* network. In all cases including more series terms resulted in more accurate results. The method is of value for simple cases, but becomes difficult and less efficient for more complicated cases.

4.8 Problems

1. A parallel RC network, as shown in Fig. 4.9, is driven by a negative exponential function of the form e^{-st}; find output voltage using series expansion. Next, assume $R = 1$, $C = 0.5$, and $s = 4$; plot voltage and compare to SPICE; see sample solution in Fig. 4.10.
   Answer:

$$a_0 = 0, \quad a_1 = \frac{1}{C}, \quad a_n = \frac{1}{nC}\left[\frac{(-s)^{n-1}}{(n-1)!} - \frac{a_{n-1}}{R}\right]$$

2. Repeat Problem 1 but assume input is a sine function $i_i(t) = \sin \omega_0 t$. Next, assume $R = 1$, $C = 0.5$, and $\omega_0 = 2\pi$; plot results and compare to SPICE; see sample solution in Fig. 4.11.
   Answer:

$$a_0 = 0, \quad a_1 = 0, \quad a_n = -\frac{1}{nC}\frac{a_{n-1}}{R}, \quad (n \text{ odd})$$

$$a_n = \frac{1}{nC}\left[\frac{\omega_0^{n-1}(-1)^{n/2-1}}{(n-1)!} - \frac{a_{n-1}}{R}\right] \quad (n \text{ even})$$

3. The circuit in Fig. 4.12 is driven by a unit step current; find output voltage. Assume $R_1 = 0.5$, $R_2 = 1.0$, and $C = 0.5$; plot solution and compare to SPICE; see sample solution in Fig. 4.13.
   Answer:

$$i(t) = \sum a_n t^n \quad a_0 = \frac{R_2}{R_1 + R_2}$$

$$a_n = -\frac{1}{R_1 + R_2}\frac{a_{n-1}}{nC}$$

$$v_o = (1 - i(t))R_2$$

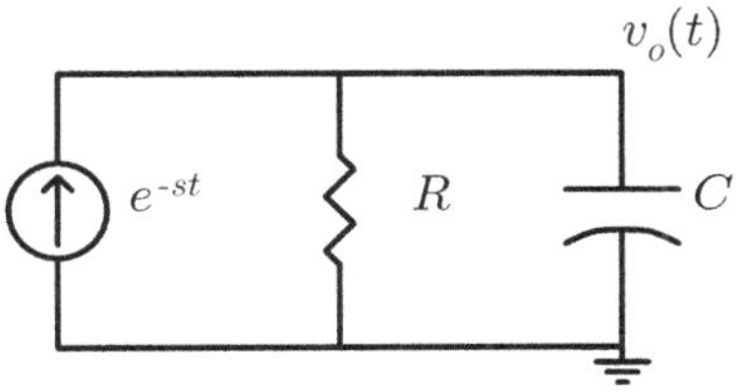

Fig. 4.9 Parallel RC driven by negative exponential current

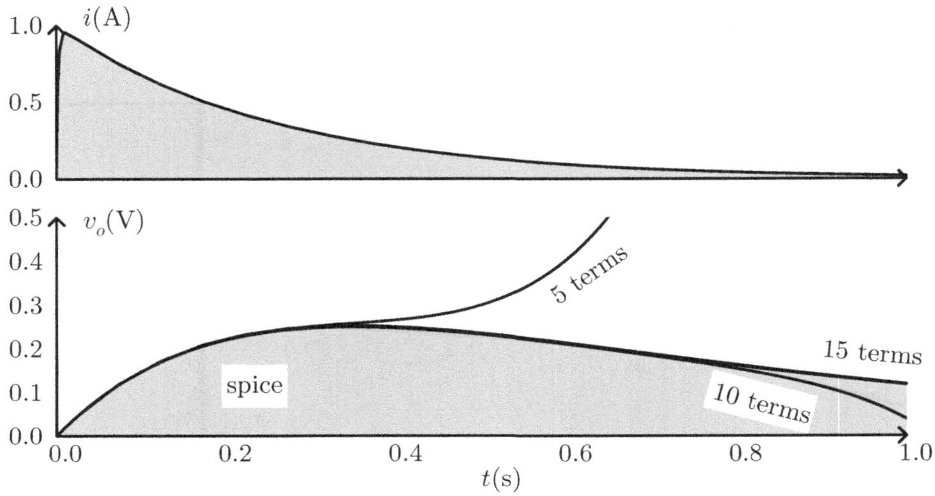

Fig. 4.10 Sample solution to Problem 1

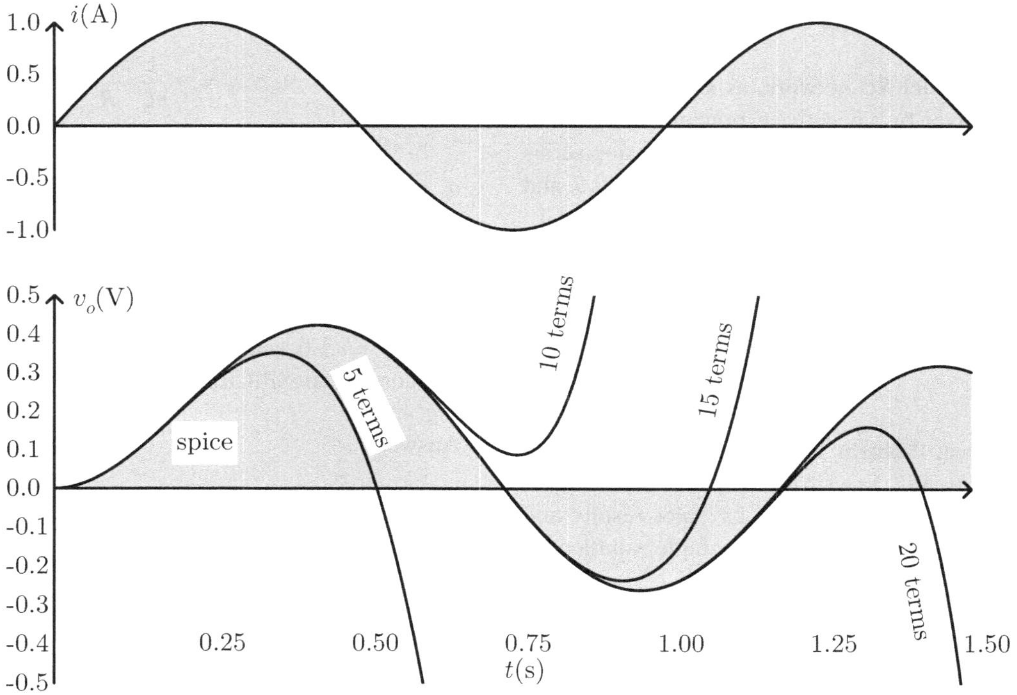

Fig. 4.11 Answer to Problem 2

Fig. 4.12 *RC* circuit for Problem 3

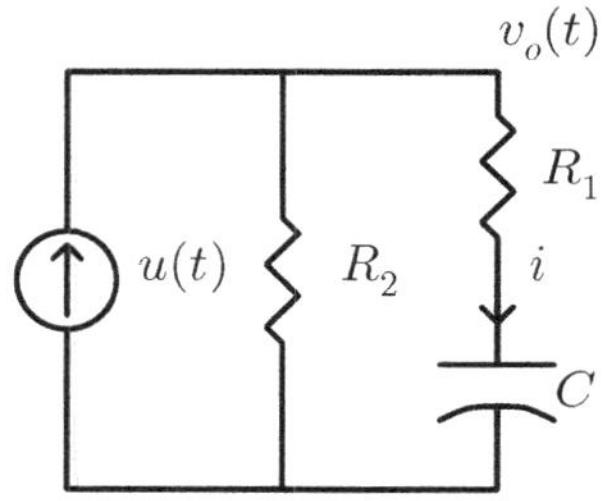

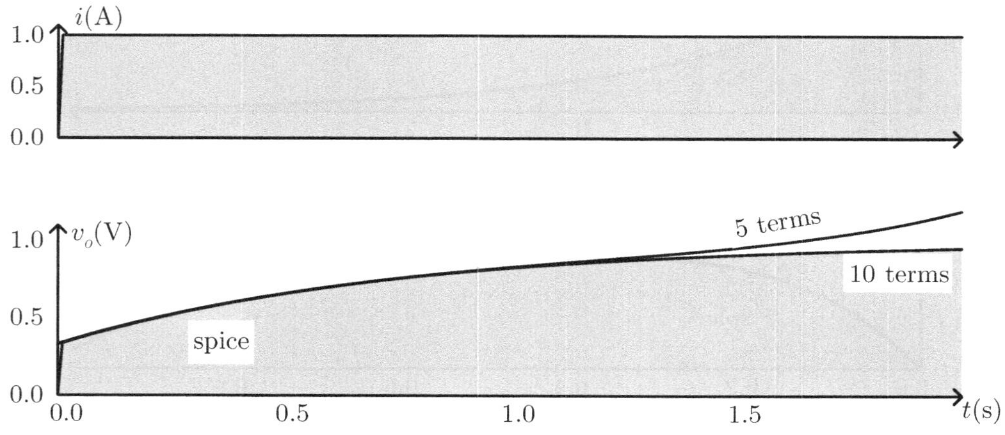

Fig. 4.13 Answer to Problem 3

4. The *RLC* circuit in Fig. 4.14 has the cap precharged to 1 V; find cap voltage for all time. Next, assume $R = 1$, $L = 0.5$, and $C = 0.1$; plot voltage and compare to SPICE. See sample solution in Fig. 4.15.

Answer:

$$v_o(t) = 1 + \frac{1}{C}\left[a_0 t + \frac{a_1}{2}t^2 + \frac{a_2}{3}t^3 + \frac{a_3}{4}t^4 + \cdots\right]$$

$$a_0 = 0, \quad a_1 = -\frac{1}{L},$$

$$a_n = -\frac{1}{nL}\left[\frac{1}{(n-1)C}a_{n-2} + Ra_{n-1}\right]$$

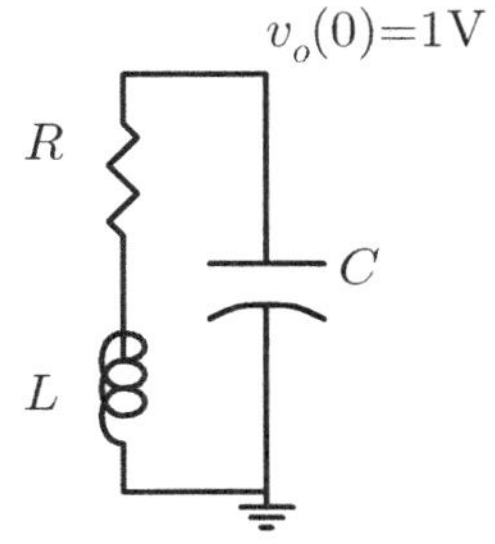

Fig. 4.14 *RLC* circuit precharged to 1 V

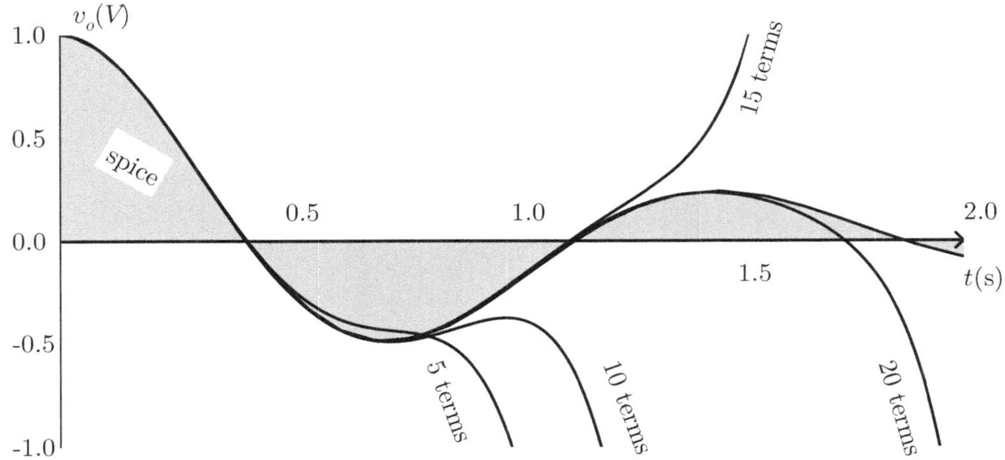

Fig. 4.15 Solution to Problem 4

5 Numerical Differential Equation Solution to Circuit Problems

5.1 Introduction

Circuit problems give rise to differential equations. Those equations can be solved directly when possible, or numerically. The area of numerical solutions to differential equations is a very advanced and developed one, and here we only shed some light on the most basic principles behind the simplest method. This work is not about numerical techniques, but to form a coherent picture on how circuit problems are solved, some introductory material about this topic maybe useful. Again rather than dig up the theory from the roots we demonstrate it via application to a few common examples.

5.2 Series *RC* Network Driven by Sine Input Voltage

Consider the series *RC* network shown in Fig. 5.1. On the input side we have a sine voltage (frequency ω_0) and our goal is to solve for current and output voltage. Doing KVL around the network gives

$$\frac{1}{C}\int_0^t i(\tau)d\tau + iR = \sin\omega_0 t \tag{5.1}$$

Assume we know current at time zero:

$$i_0 = i(0) \tag{5.2}$$

We want to find current i_1 at a small interval Δt after time zero.

$$i_1 = i(\Delta t) \tag{5.3}$$

The area between time zero and Δt forms a rectangle of width Δt and height which is the average of current at time 0 (i_0) and current at time Δt (i_1); then we have

$$\frac{\Delta t}{2C}\left[i_0 + i_1\right] + Ri_1 = v_1 \tag{5.4}$$

where $v_1 = v(\Delta t)$. Collecting terms gives

$$i_1\left[\frac{\Delta t}{2C} + R\right] = v_1 - \frac{\Delta t}{2C}i_0 \tag{5.5}$$

which can be solved as

$$\boxed{i_1 = \frac{1}{\frac{\Delta t}{2C} + R}\left[v_1 - \frac{\Delta t}{2C}i_0\right]} \tag{5.6}$$

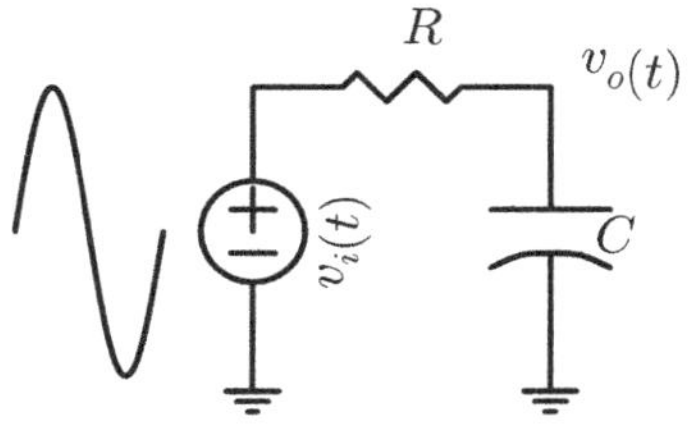

Fig. 5.1 Series *RC* driven by sine function

F. Badrieh, *Spectral, Convolution and Numerical Techniques in Circuit Theory*,
https://doi.org/10.1007/978-3-319-71437-0_5

Moving on to time $2\Delta t$ we have

$$i_2 = i(2\Delta t) \tag{5.7}$$

Putting back into the integral equation we have

$$\frac{\Delta t}{2C}\left[(i_0 + i_1) + (i_1 + i_2)\right] + Ri_2 = v_2 \tag{5.8}$$

Collecting terms and solving we get

$$i_2 = \frac{1}{\frac{\Delta t}{2C} + R}\left[v_2 - \frac{\Delta t}{2C}(i_0 + 2i_1)\right] \tag{5.9}$$

Similarly we get for $i_3 = i(3\Delta t)$

$$i_3 = \frac{1}{\frac{\Delta t}{2C} + R}\left[v_3 - \frac{\Delta t}{2C}(i_0 + 2i_1 + 2i_2)\right] \tag{5.10}$$

By now we see the pattern. We can form the n sequence as

$$\boxed{i_n = \frac{1}{\frac{\Delta t}{2C} + R}\left[v_n - \frac{\Delta t}{2C}\left(i_0 + 2\sum_{m=1}^{n-1} i_m\right)\right], \quad n > 1} \tag{5.11}$$

The output voltage is simply the integral of current divided by cap:

$$\boxed{v_{o,n} = \frac{1}{C}\sum_{m=0}^{n} i_m} \tag{5.12}$$

Figure 5.2 shows comparison between SPICE and numerical algorithm with 0.1 μs time discretization. Notice the excellent agreement.

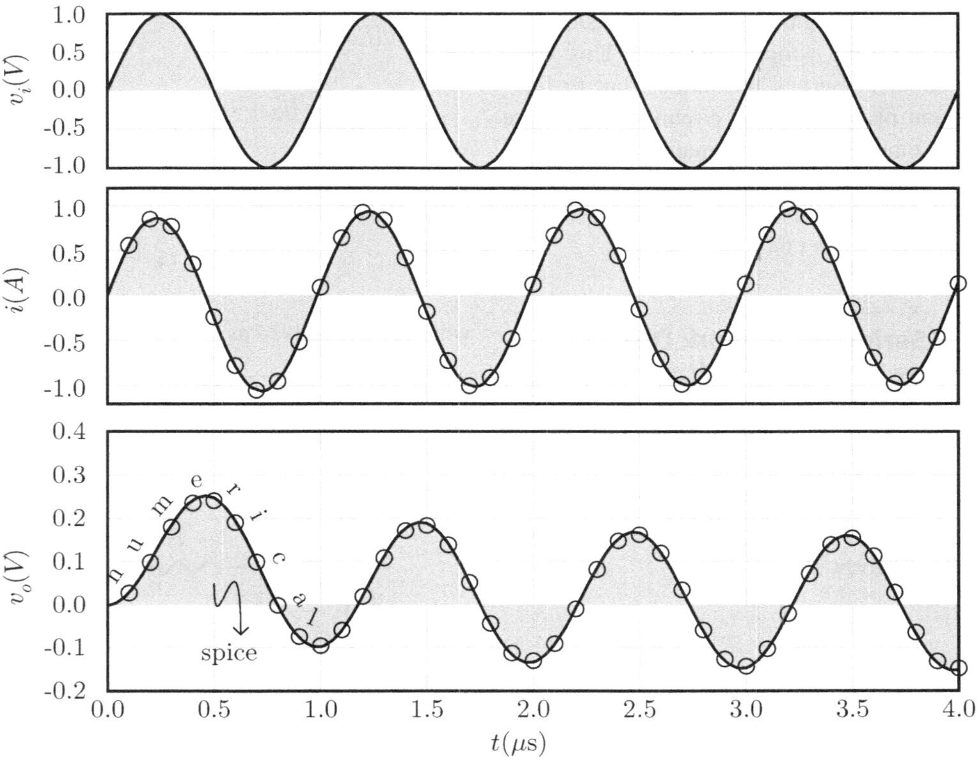

Fig. 5.2 Response of series RC network to sine input: SPICE and numerical solutions. Case of $R = 1\,\Omega$, $C = 1\,\mu F$, and $\omega_0 = 2\pi \times 10^6$

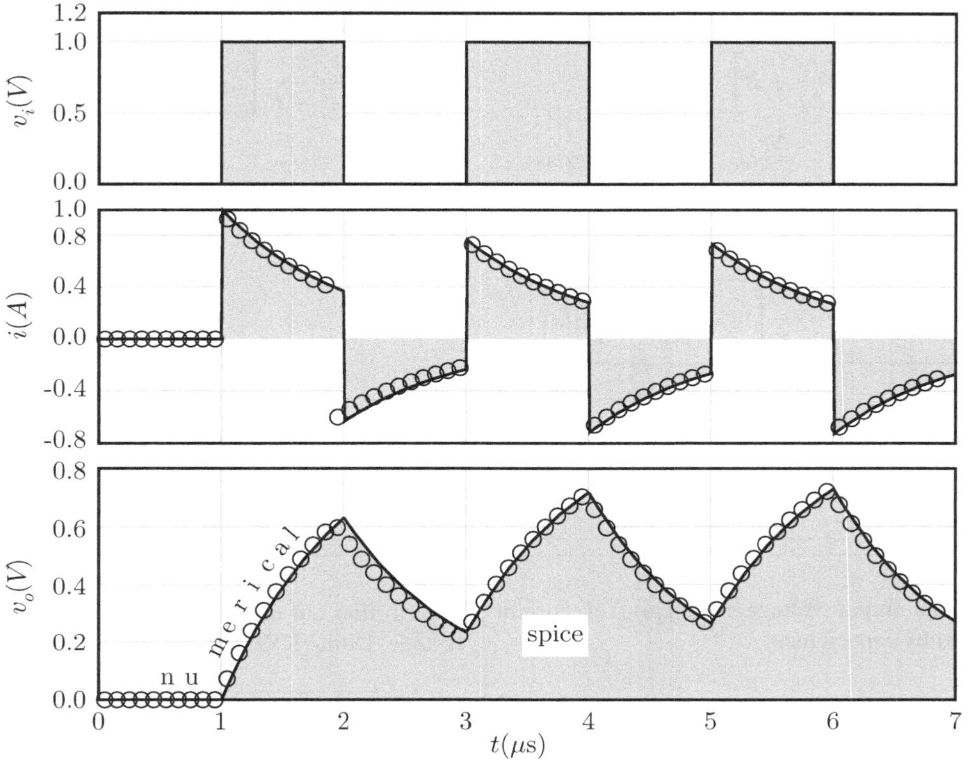

Fig. 5.3 Response of series *RC* network to square pulse input: SPICE and numerical solutions. Case of $R = 1\,\Omega$, $C = 1\,\mu F$, and $\omega_0 = 2\pi \times 10^6$

5.3 Series *RC* Network Driven by Square Pulse Input

Consider next the same *RC* network with a periodic pulse input. Input voltage is shown at the top of Fig. 5.3, followed by branch current and output voltage. This case had a time grid of $\Delta t = 0.02\,\mu s$ to try capture the abrupt transitions. As seen from the figure, the numerical method accurately captures both transient and steady state response.

5.4 Series *RLC* Network Driven by Periodic Pulse

Consider the series *RLC* network shown in Fig. 5.4. Input is a periodic pulse voltage and we want to find both current and output voltage. The governing integro/differential equation is

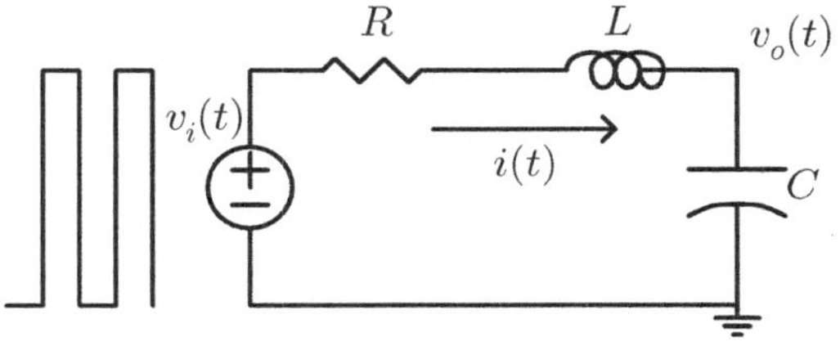

Fig. 5.4 Series *RLC* network

$$\boxed{\frac{1}{C}\int_0^t i(\tau)d\tau + L\frac{di(t)}{dt} + i(t)R = v_i(t)} \tag{5.13}$$

Assume that initial current i_0 is zero, and we want to find current i_1 after some time step Δt. Then we have

$$\frac{\Delta t}{2C}[i_0 + i_1] + \frac{L}{\Delta t}[i_1 - i_0] + Ri_1 = v_1 \tag{5.14}$$

Collect terms

$$i_1\left[\frac{\Delta t}{2C}+\frac{L}{\Delta t}+R\right] = v_1+\frac{L}{\Delta t}i_0-\frac{\Delta t}{2C}i_0, \quad \text{which solves as} \tag{5.15}$$

$$\boxed{i_1=\frac{1}{\frac{\Delta t}{2C}+\frac{L}{\Delta t}+R}\left[v_1+\frac{L}{\Delta t}i_0-\frac{\Delta t}{2C}i_0\right]} \tag{5.16}$$

After another Δt we have

$$i_2=\frac{1}{\frac{\Delta t}{2C}+\frac{L}{\Delta t}+R}\left[v_2+\frac{L}{\Delta t}i_1-\frac{\Delta t}{2C}(i_0+2i_1)\right] \tag{5.17}$$

The pattern appears and can be written as

$$\boxed{i_n=\frac{1}{\frac{\Delta t}{2C}+\frac{L}{\Delta t}+R}\left[v_n+\frac{L}{\Delta t}i_{n-1}-\frac{\Delta t}{2C}\left(i_0+2\sum_{m=1}^{n-1}i_m\right)\right], \quad n>1} \tag{5.18}$$

Again output voltage is integral of current divided by capacitance:

$$\boxed{v_{o,n}=\frac{1}{C}\sum_{m=0}^{n}i_m} \tag{5.19}$$

Figure 5.5 shows input voltage (top), branch current (middle), and output voltage (bottom). Shown are SPICE results and those of numerical algorithm. Here we used a time step of $\Delta t = 0.02\,\mu$s to capture the fast rise edge. Notice the excellent agreement, including initial and final values, as well as ramp rate and oscillatory effects.

5.5 Two-Branch *RC* Network Driven by Periodic Pulse

Consider the 2-branch *RC* network shown in Fig. 5.6. It is driven by a periodic pulse, and we want to find out the various currents and output voltage. Doing KVL around left loop we have

$$\boxed{\frac{1}{C_1}\int_0^t i_x(\tau)d\tau + R_1(i_x+i_y)=v_i(t)} \tag{5.20}$$

Doing KVL around outer loop we have

$$\boxed{\frac{1}{C_2}\int_0^t i_y(\tau)d\tau + R_2 i_y + R_1(i_x+i_y)=v_i(t)} \tag{5.21}$$

Assume that initial conditions are

$$i_x^0=i_x(0)=0; \qquad i_y^0=i_y(0)=0 \tag{5.22}$$

At time Δt we want to find the two currents

$$i_x^1=i_x(\Delta t), \qquad i_y^1=i_y(\Delta t) \tag{5.23}$$

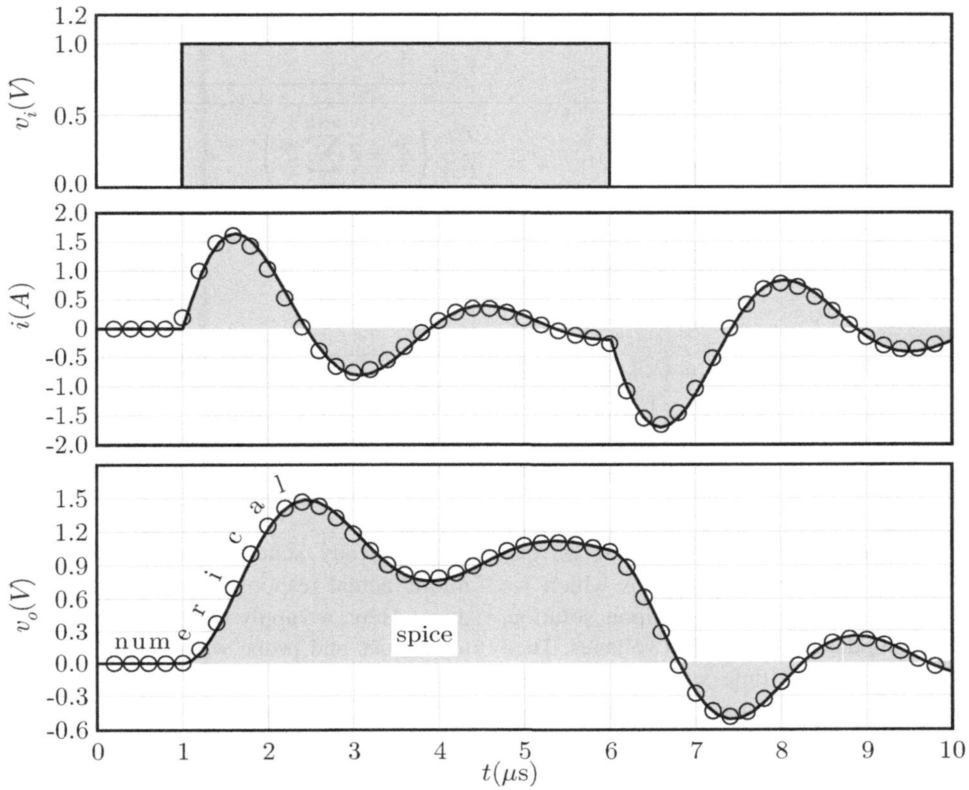

Fig. 5.5 Series *RLC* response to periodic pulse input. Case of $R = 0.2\,\Omega$, $C = 1\,\mu$F, and $L = 0.2\,\mu$H

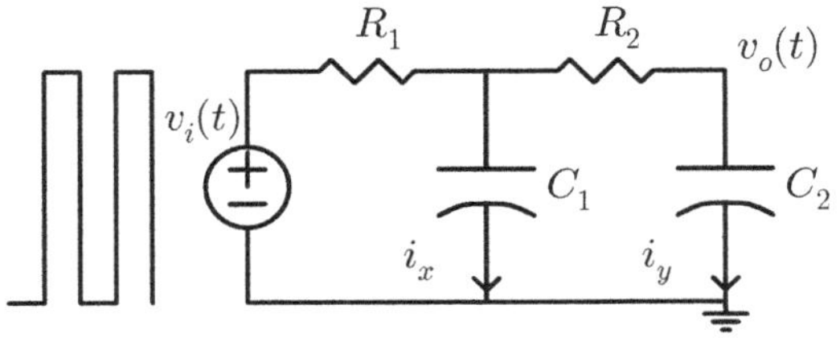

Fig. 5.6 Two branch *RC* network driven by periodic pulse

The discretized version of Eq. (5.20) gives

$$\boxed{\left[\frac{\Delta t}{2C_1} + R_1\right] i_x^1 + R_1 i_y^1 = v_1 - \frac{\Delta t}{2C_1} i_x^0} \tag{5.24}$$

The discretized version of Eq. (5.21) gives

$$\boxed{\left[\frac{\Delta t}{2C_2} + R_2 + R_1\right] i_y^1 + R_1 i_x^1 = v_1 - \frac{\Delta t}{2C_2} i_y^0} \tag{5.25}$$

At a generic time step $n\Delta t$ the discretized version of the first equation gives

$$\left[\frac{\Delta t}{2C_1}+R_1\right]i_x^n+R_1 i_y^n=v_n-\frac{\Delta t}{2C_1}\left(i_x^0+2\sum_{m=1}^{n-1}i_x^m\right),\quad n>1 \tag{5.26}$$

and the discretized version of the second equation gives

$$\left[\frac{\Delta t}{2C_2}+R_2+R_1\right]i_y^n+R_1 i_x^n=v_n-\frac{\Delta t}{2C_2}\left(i_y^0+2\sum_{m=1}^{n-1}i_y^m\right),\quad n>1 \tag{5.27}$$

That is, for each time interval $n\Delta t$ we end up with a 2×2 system of equations which we can solve using Cramer's rule. Upon solution, we have the updated currents and voltages. Then we move to the next time step, and so forth. Figure 5.7 shows input voltage, the two branch currents, and the output voltage. It shows SPICE simulations and numerical results. Very good match is observed.

5.6 Nine-Branch *RLC* Network Driven by Sine Input

Let's revisit the 9-branch *RLC* network which we dealt with in the steady state response chapter. It is reshown here in Fig. 5.8 for convenience. In the steady state chapter, we were able to mimic actual response after things have settled down. Here, we apply finite difference numerical techniques and probe whether we can duplicate the whole response!

Current assignment is shown in the figure; we have 18 such unknowns. Again we need 18 equations. As before, we end up with 9 KVL equations and 9 KCL ones. For example, going around the bottom right loop KVL gives

$$\frac{1}{C}\int_0^t i_1(t)dt-\frac{1}{C}\int_0^t i_2(t)dt+\left[R+L\frac{d}{dt}\right]i_{12}(t)=0 \tag{5.28}$$

Similarly, doing KVL on the loop just above gives

$$\frac{1}{C}\int_0^t i_4(t)dt-\frac{1}{C}\int_0^t i_5(t)dt+\left[R+L\frac{d}{dt}\right]i_{11}(t)-\left[R+L\frac{d}{dt}\right]i_{12}(t)=0 \tag{5.29}$$

The same is done for the rest of the loops, taking special consideration for the bottom left loop which has the source; for this loop KVL gives

$$\frac{1}{C}\int_0^t i_3(t)dt+\left[R+L\frac{d}{dt}\right]i_{18}(t)=v_i(t) \tag{5.30}$$

Having formed the 9 KVL equations, we move to the remaining 9 KCL ones. These simply state that at each node sum of currents equals zero. For example, at node $\boxed{1}$ we have

$$i_1-i_4-i_{12}=0 \tag{5.31}$$

and at node $\boxed{7}$ we have

$$i_4-i_7-i_{11}=0 \tag{5.32}$$

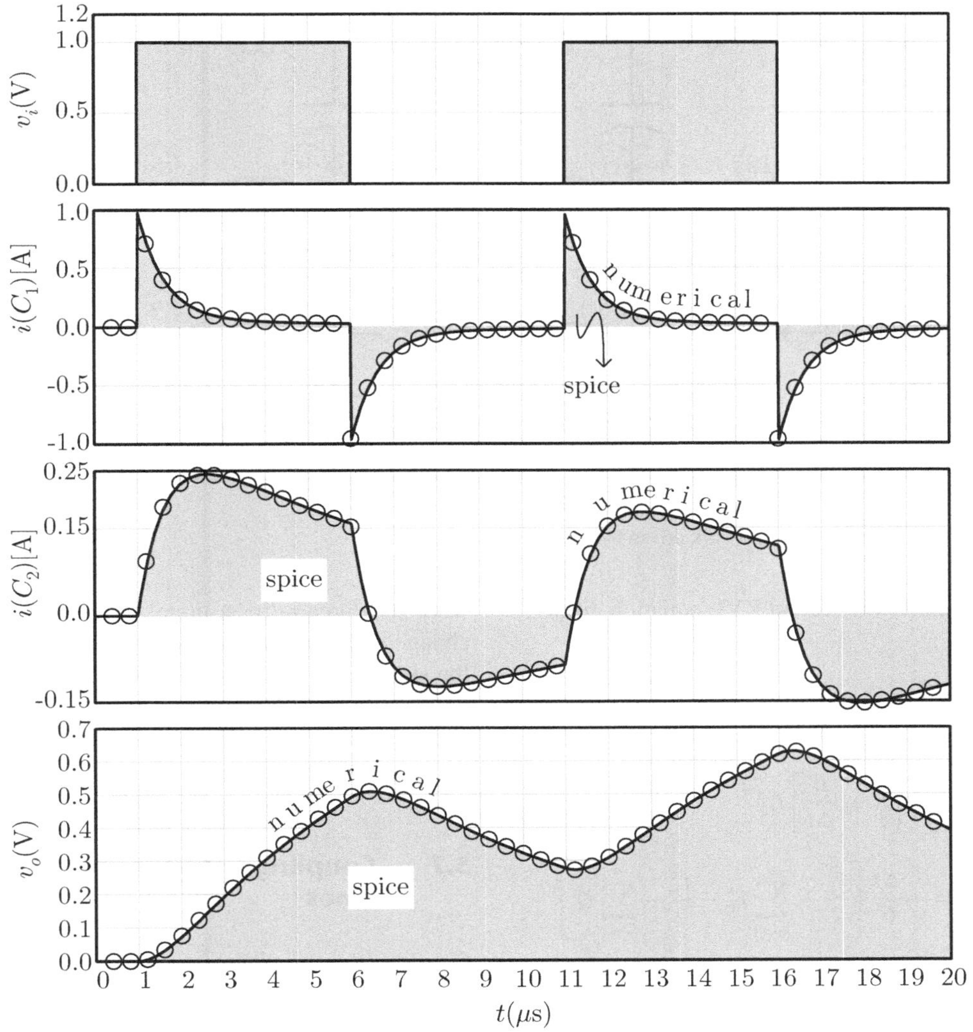

Fig. 5.7 Response of two branch *RC* network to a periodic pulse. Case of $R_1 = 1\,\Omega$, $R_2 = 1\,\Omega$, $C_1 = 1\,\mu F$, and $C_2 = 2\,\mu F$

and so on. Next we convert from time-continuous form to discretized one. For example, the first KVL equation above gives

$$\frac{\Delta t}{C}\left[\frac{i_1^0}{2} + \sum_{m=1}^{n} i_1^m + \frac{i_1^{n+1}}{2}\right] + R i_{12}^{n+1}$$
$$+\frac{L}{\Delta t}\left[i_{12}^{n+1} - i_{12}^n\right]$$
$$-\frac{\Delta t}{C}\left[\frac{i_2^0}{2} + \sum_{m=1}^{n} i_2^m + \frac{i_2^{n+1}}{2}\right] = 0 \quad (5.33)$$

where i^{n+1} is current at the next time step, and i^n is current at the current time step. Above equation can be rearranged to put all the $n+1$ terms on the left side and the other terms on the right side:

$$\frac{\Delta t}{2C} i_1^{n+1} - \frac{\Delta t}{2C} i_2^{n+1} + \left[R + \frac{L}{\Delta t}\right] i_{12}^{n+1}$$
$$= -\frac{\Delta t}{C}\left[\frac{i_1^0}{2} + \sum_{m=1}^{n} i_1^m - \frac{i_2^0}{2} - \sum_{m=1}^{n} i_2^m\right] + \frac{L}{\Delta t} i_{12}^n \quad (5.34)$$

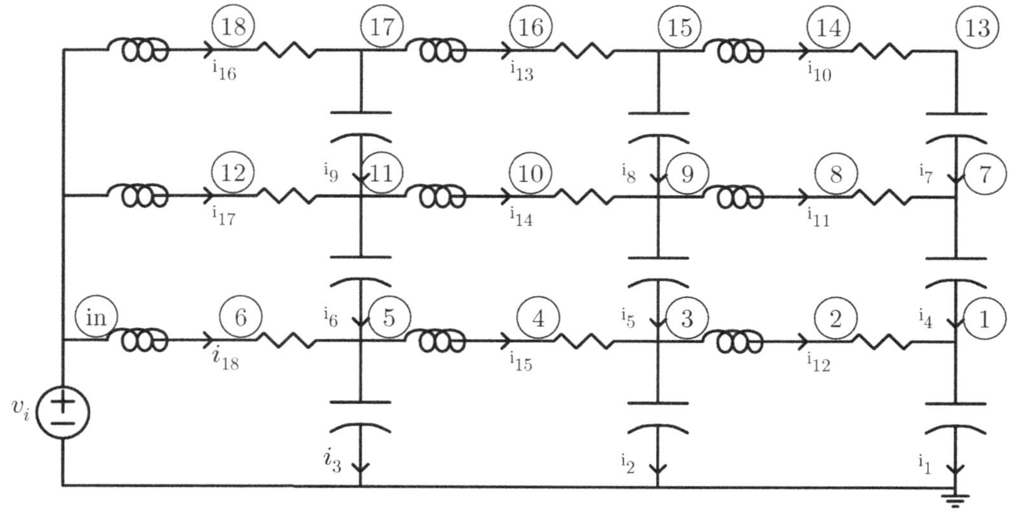

Fig. 5.8 Nine-branch *RLC* network driven by sine input

Similarly the second KVL equation gives

$$\frac{\Delta t}{2C} i_4^{n+1} - \frac{\Delta t}{2C} i_5^{n+1} + \left[R + \frac{L}{\Delta t}\right] i_{11}^{n+1} - \left[R + \frac{L}{\Delta t}\right] i_{12}^{n+1} = -\frac{\Delta t}{C}\left[\frac{i_4^0}{2} + \sum_{m=1}^{n} i_4^m - \frac{i_5^0}{2} - \sum_{m=1}^{n} i_5^m\right] + \frac{L}{\Delta t} i_{11}^n - \frac{L}{\Delta t} i_{12}^n \quad (5.35)$$

and so forth. Again we end up with a system of 18 × 18 equations with 18 unknowns (the currents). Starting with zero initial conditions we solve the linear system for the next step $(n + 1)$ current in terms of the prior step one. Then we repeat for $n + 2$ and so forth. If we choose our time increment small enough we should get convergence. Figure 5.9 shows our results and comparison to SPICE; notice that unlike the steady state methods, this method matches the transient (initial) part of the solution (of course and the rest of it).

While at it, and in contrast to the steady state solution, let's see what the response would look like for a step input. Notice that 99% of the work has already been done! All that needs to changed is swapping in the unit step input for the sine one; the rest of the setup, including the 18 × 18 matrix, remains the same. Figure 5.10 shows our results and comparison to SPICE; notice excellent agreement.

5.7 Coupling Between Two *RC* Lines

Consider the two *RC* lines shown in Fig. 5.11. Input is applied on one side, and we want to find voltage at the victim (node 2). Each line has its cap, and coupling happens through C_{12}. We can assign six currents, as shown in the figure. Notice that since node 1 is open we don't need to assign current through R_2 since it is zero. To solve for the six currents we need six equations. Four of those would be KVL, two KCL. Going around the lower left loop we have

$$\frac{1}{C_{11}}\int_0^t i_1(t)dt + i_2(t)R_1 = 0 \quad (5.36)$$

Going around the cap-only loop we have

$$\frac{1}{C_{11}}\int_0^t i_1(t)dt - \frac{1}{C_{12}}\int_0^t i_3(t)dt - \frac{1}{C_{22}}\int_0^t i_6(t)dt = 0 \quad (5.37)$$

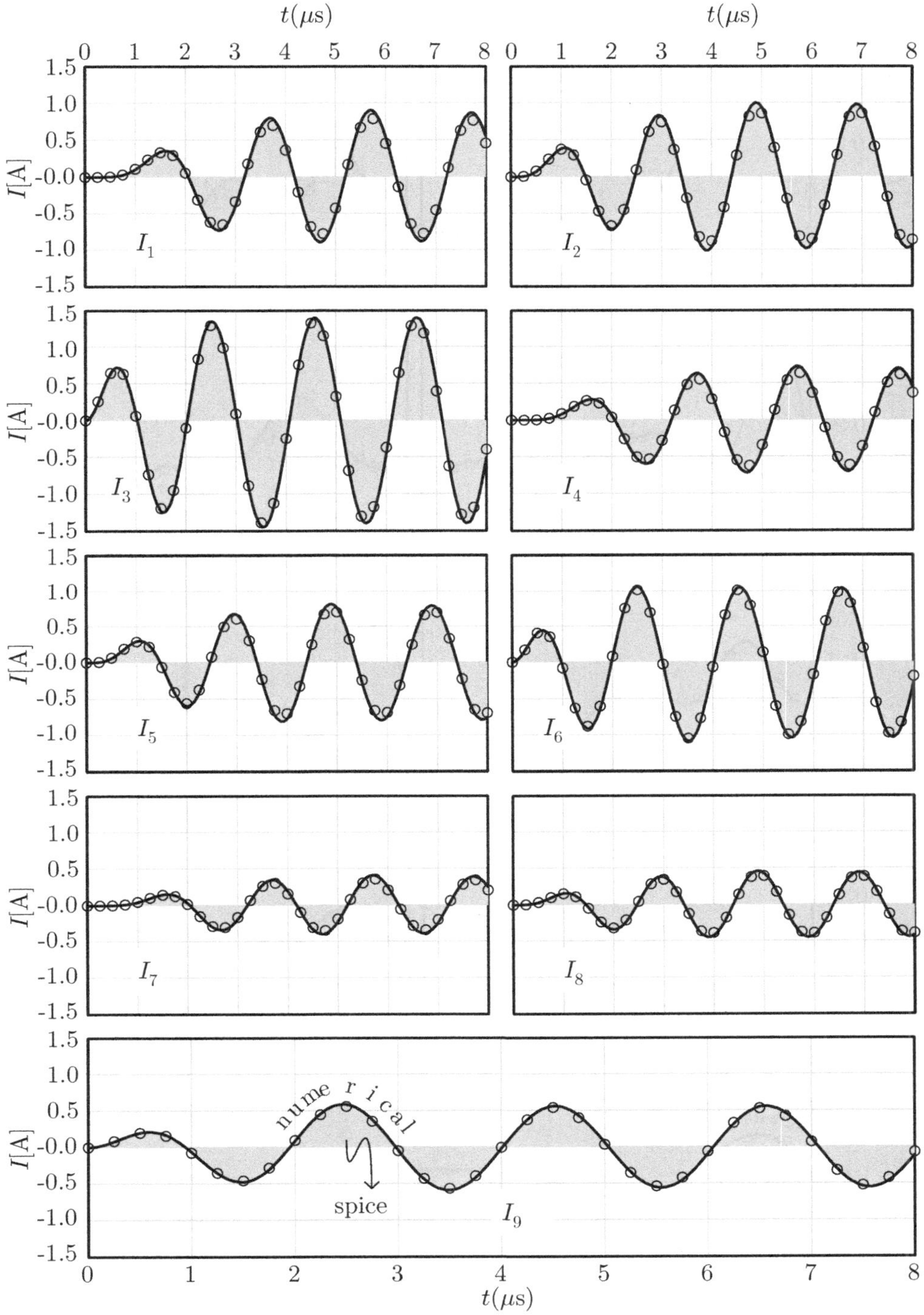

Fig. 5.9 Nine-branch *RLC* network response to sine input—solid SPICE; discrete theory. Case of $R = 1\ \Omega$, $C = 1\ \mu$F, $L = 1\ \mu$H, and $\omega_0 = 1\pi \times 10^6$

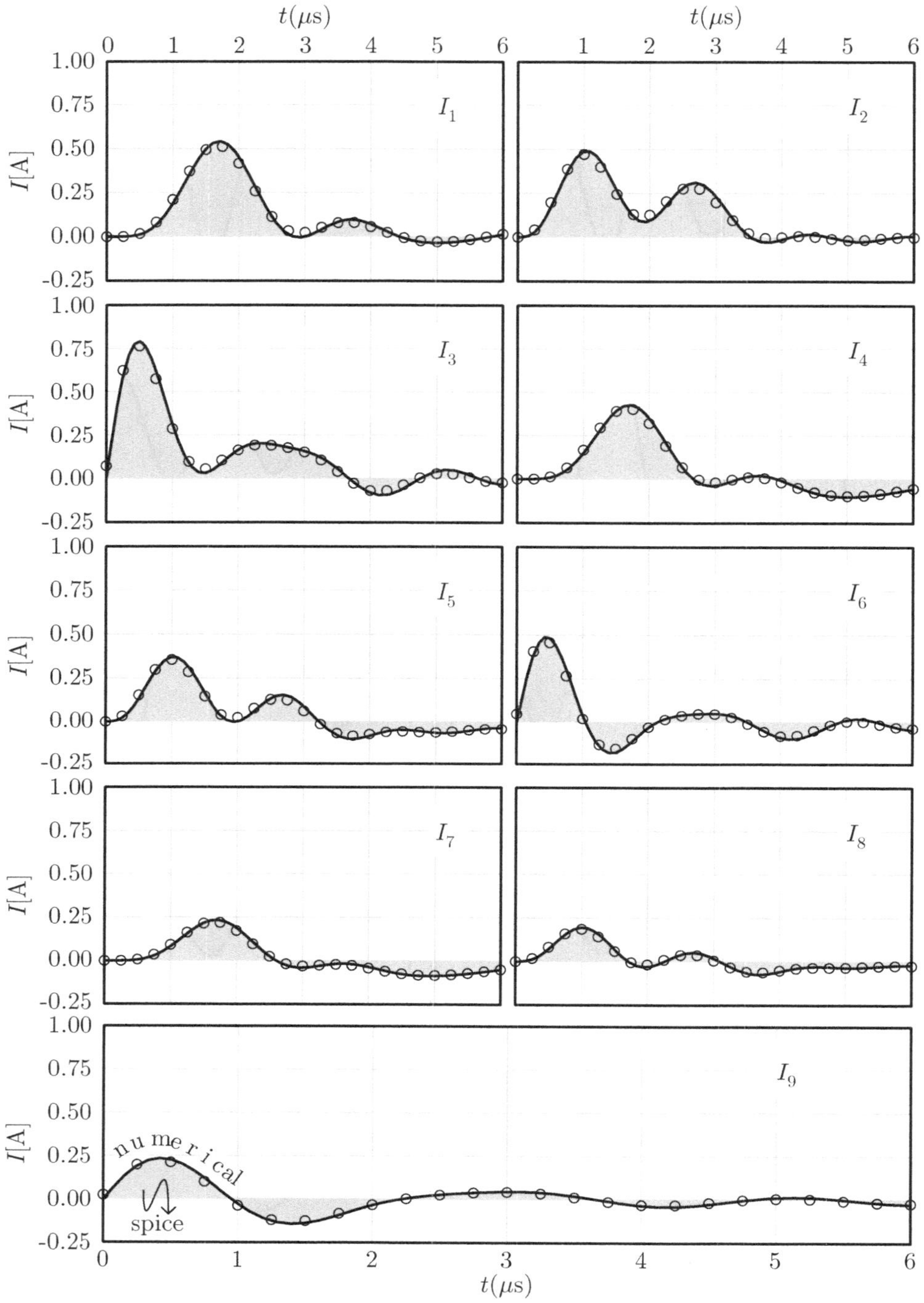

Fig. 5.10 Nine-branch *RLC* network response to unit step input—solid SPICE; discrete theory. Case of $R = 1\,\Omega$, $C = 1\,\mu\mathrm{F}$, and $L = 1\,\mu\mathrm{H}$

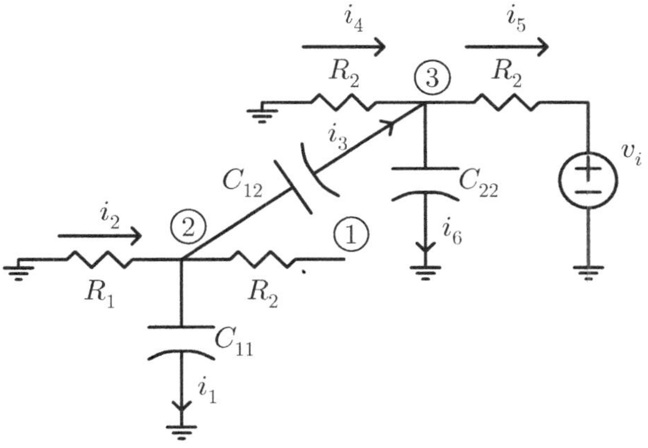

Fig. 5.11 Coupling between two *RC* lines

Going around the top right loop we have

$$\frac{1}{C_{22}}\int_0^t i_6(t)dt - i_5R_2 = v_i(t) = u(t) \quad (5.38)$$

Going around the top left loop we have

$$\frac{1}{C_{22}}\int_0^t i_6(t)dt + i_4R_2 = 0 \quad (5.39)$$

Next KCL. Adding currents at node 2 we get

$$i_1 + i_3 - i_2 = 0 \quad (5.40)$$

Adding currents at node 3 we get

$$i_5 + i_6 - i_4 - i_3 = 0 \quad (5.41)$$

Next we discretize these six equations. For example, the first differential equation becomes

$$\frac{\Delta t}{2C_{11}}i_1^{n+1} + R_1 i_2^{n+1} = -\frac{\Delta t}{C_{11}}\left[\frac{i_1^0}{2} + \sum_{m=1}^{n} i_1^m\right] \quad (5.42)$$

The second equation becomes

$$\frac{\Delta t}{2C_{11}}i_1^{n+1} - \frac{\Delta t}{2C_{12}}i_3^{n+1} - \frac{\Delta t}{2C_{22}}i_6^{n+1} =$$
$$-\frac{\Delta t}{C_{11}}\left[\frac{i_1^0}{2} + \sum_{m=1}^{n} i_1^m\right] + \frac{\Delta t}{C_{12}}\left[\frac{i_3^0}{2} + \sum_{m=1}^{n} i_3^m\right]$$
$$+\frac{\Delta t}{C_{22}}\left[\frac{i_6^0}{2} + \sum_{m=1}^{n} i_6^m\right] \quad (5.43)$$

and so forth. Again we end up with a 6×6 matrix, with six unknowns. For each time step we figure new current $(n+1)$ in terms of old current (n) by solving the linear system.

Notice that in order to proceed with the iteration, we need to know the initial conditions; that is what are the six currents right when the step input is applied? We can assume zero initial conditions, but we can expedite things a bit if we use the real initial conditions! Towards that end let's assume some *RC* numbers

$$R_1 = 1\,\Omega, R_2 = 2\,\Omega, C_{11} = 1\,\mu\text{F},$$
$$C_{12} = 0.5\,\mu\text{F}, \text{ and } C_{22} = 2\,\mu\text{F} \quad (5.44)$$

By inspection we notice that current through the left two resistors would start at zero:

$$i_2(0) = i_4(0) = 0 \quad (5.45)$$

Current along the top right resistor would simply be input voltage divided by R_2; for the case $R_2 = 2\,\Omega$ we have

$$i_5(0) = -\frac{1}{2} \quad (5.46)$$

To find i_6 and i_3 we need to figure equivalent cap of $C_{12} = 0.5\,\mu$F in series with $C_{11} = 1\,\mu$F; this comes out to $1/3\,\mu$F. Keeping in mind $C_{22} = 2\,\mu$F, now by impedance division we figure

$$i_6(0) = \frac{1}{2}\frac{3}{3+\frac{1}{2}} \quad (5.47)$$

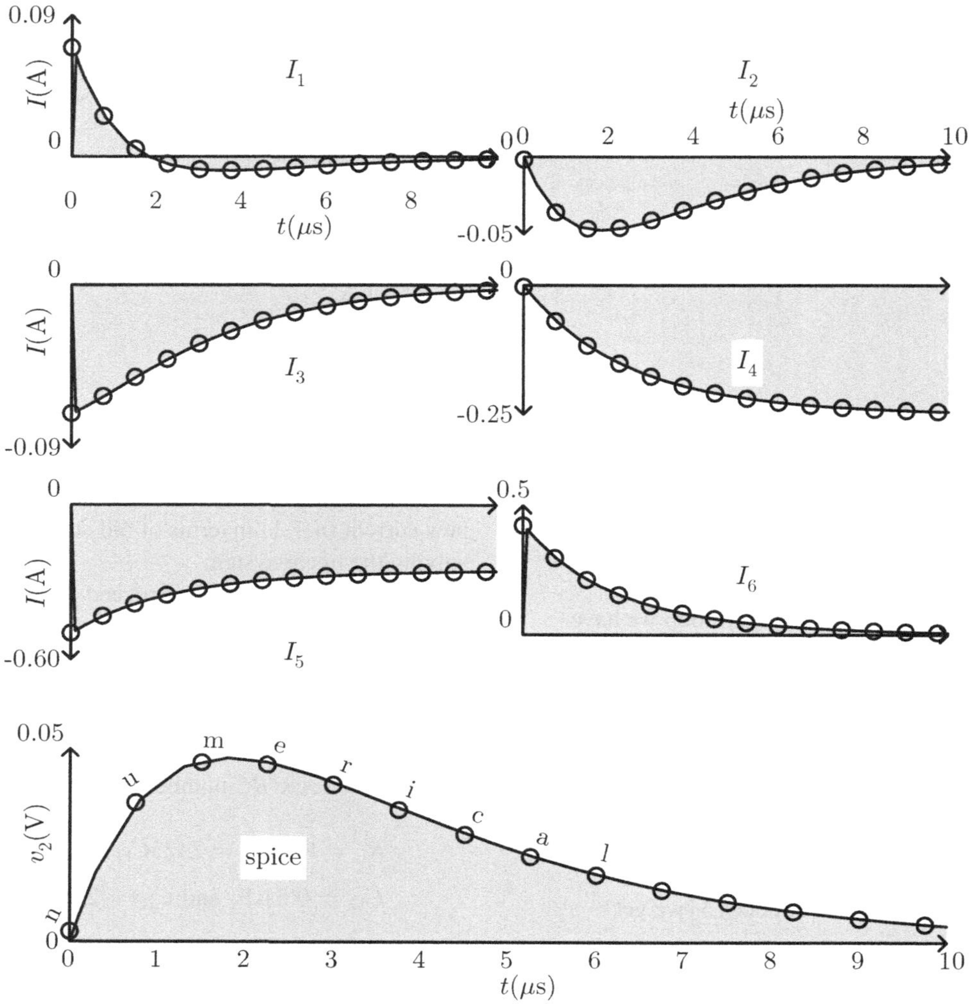

Fig. 5.12 Response of two *RC* lines due to step input on one line. Case of $R_1 = 1\,\Omega$, $R_2 = 2\,\Omega$, $C_{11} = 1\,\mu\text{F}$, $C_{12} = 0.5\,\mu\text{F}$, and $C_{22} = 2\,\mu\text{F}$

Similarly, we get i_3 and i_1 (which is negative of the former):

$$i_3(0) = -\frac{1}{2}\frac{\frac{1}{2}}{3+\frac{1}{2}}, \quad \text{and} \quad i_1(0) = +\frac{1}{2}\frac{\frac{1}{2}}{3+\frac{1}{2}} \tag{5.48}$$

Our results are tested against SPICE as shown in Fig. 5.12 in the form of branch currents and output voltage; excellent agreement is observed.

5.8 Miniature Power Plane

Consider the miniature power plane shown in Fig. 5.13. A unit step input is applied at the upper right corner; we would like to monitor how the various nodes charge in time. As shown in the figure we have 40 unknown currents; we would need 40 equations. We can get those as shown in Fig. 5.14.

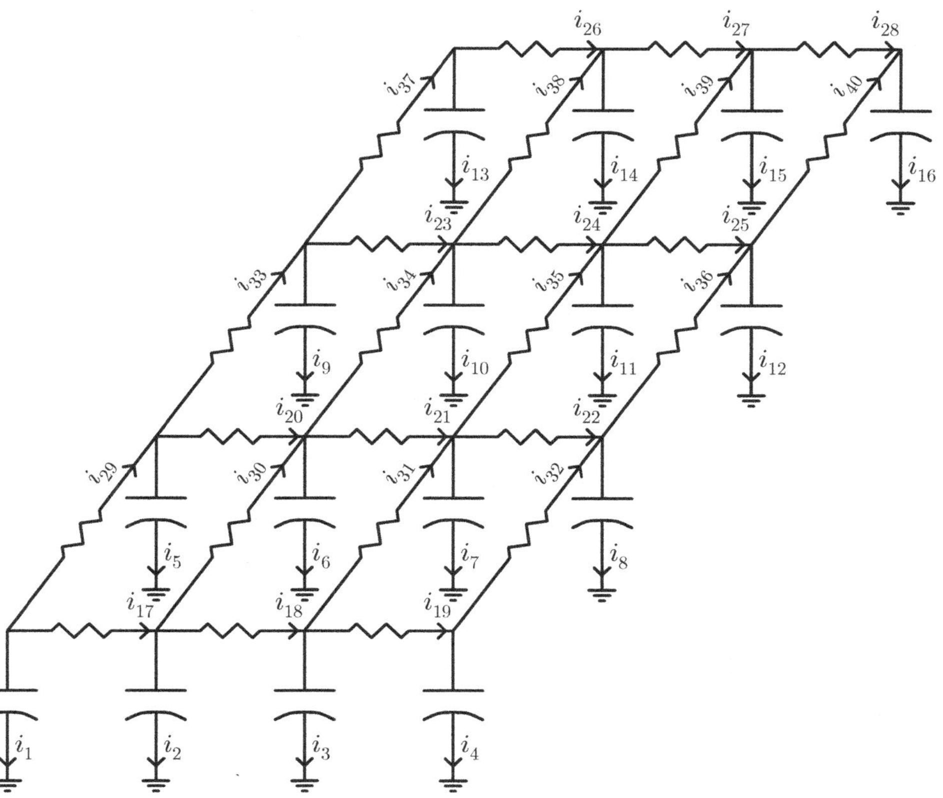

Fig. 5.13 Miniature *RC* power plane

- 12 KVL equations as shown in solid loops.
- 12 KVL equations as shown in dashed loops.
- 16 KCL equations as shown in circled numbers.

For example, doing KVL around the bottom right solid loop gives

$$\frac{1}{C}\int i_4(t)dt + i_{19}(t)R - \frac{1}{C}\int i_3(t)dt = 0 \quad (5.49)$$

and the same for the rest of the 11 solid loops. Then, as a second example, doing KVL around the bottom right dashed loop gives

$$\frac{1}{C}\int i_8(t)dt + i_{32}(t)R - \frac{1}{C}\int i_4(t)dt = 0 \quad (5.50)$$

and the same for the rest of the 11 dashed loops. This totals 24 equations. Next, as a third example doing KCL at node (1) gives

$$i_4(t) + i_{32}(t) - i_{19}(t) = 0 \quad (5.51)$$

and the same for the 15 remaining nodes. This would total 40 equations, as needed. Special attention warrants the initial conditions. When a step input is applied to node (13) we would have

$$i_{16}(0) = C\frac{dv(t)}{dt} = C\frac{1}{\Delta t} \quad (5.52)$$

Next the initial conditions for the two resistors that connect to node (13) are

$$i_{28}(0) = i_{40}(0) = -\frac{1}{R} \quad (5.53)$$

That is, since one node is a unity voltage and the other (the two caps) are at ground, the current is simply negative 1 divided by resistance. Finally the initial conditions on the two caps that tie to the two resistors just mentioned are

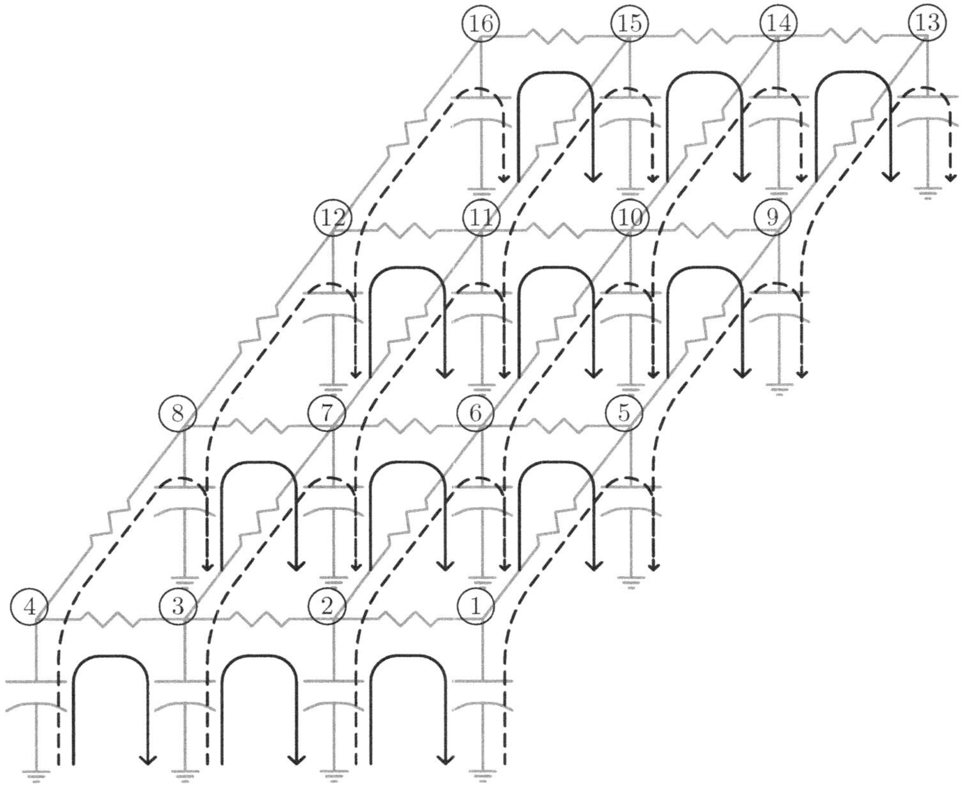

Fig. 5.14 Forty equations for power plane in Fig. 5.13

$$i_{12}(0) = i_{15}(0) = \frac{1}{R} \tag{5.54}$$

That is, their current is negative that of their corresponding resistors, by KCL. The rest of the caps and resistors have zero initial current. With initial conditions specified, we solve the 40×40 matrix at each increment time point and do the same for all time. The 40×40 system for time zero is shown in Fig. 5.15. The vector b has the initial conditions. Starting with those we solve for the next current vector I, knowing to change b to accommodate the new charge (integral of current). Then resolve for the next current I; and so forth. Notice that while b and I keep updating, the matrix A remains the same. When all said and done we get results in Fig. 5.16. The figure shows voltage across the caps using SPICE and simulations; notice excellent agreement.

5.9 Simple Power Delivery Network Problem

Consider the simplified PDN problem shown in Fig. 5.17. It is comprised of a power source (far left), some board/package *RLC* (middle), some die cap (right), and current demand. Assume current demand is a step input, but allow for the various currents to differ in step height. We would like to figure die voltage levels.

As shown in the figure we have eight unknown currents; we would need eight equations. We get those as follows:

- 4 KVL equations around the 4 *RLC* loops, and
- 4 KCL equations at nodes 1–4.

For example the first KVL equation is applied to the lower-most *RLC* branch as follows:

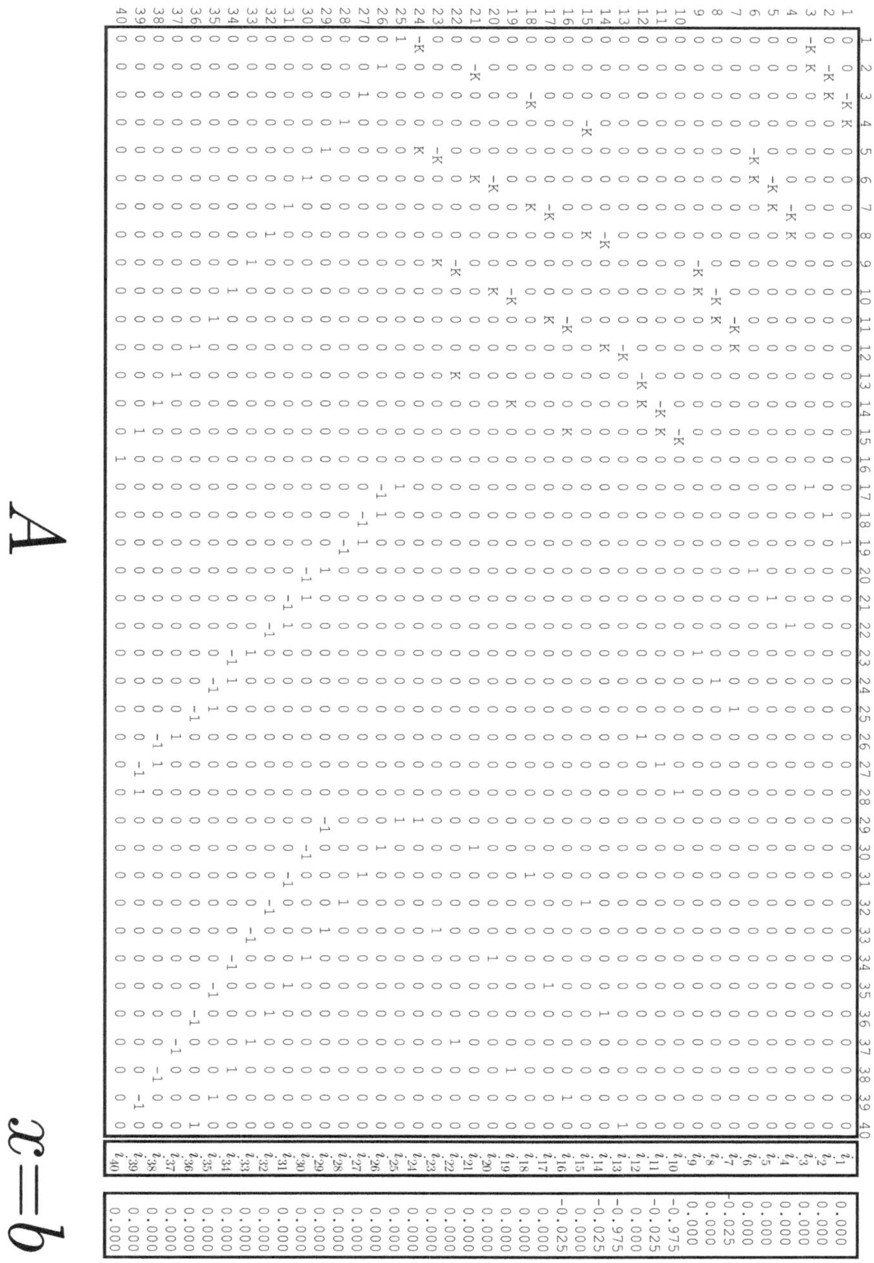

Fig. 5.15 40 × 40 system for power plane. The variable K equals $\frac{\Delta t}{2C}$. Resistance R was set to 1 Ω, capacitor C to 1 µF, while Δt was set to 0.05 µs

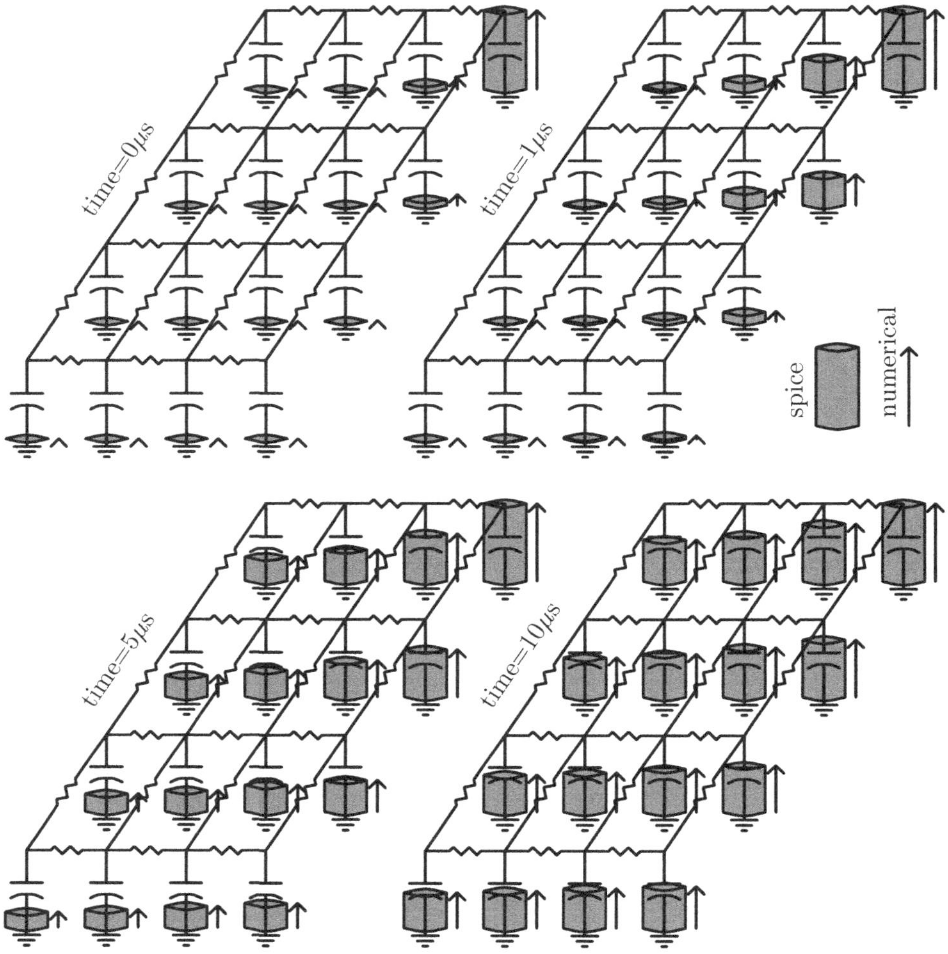

Fig. 5.16 Response of miniature *RC* power plane to unit step input at upper right corner: cylinders SPICE; arrows theory. Case of $R = 1\ \Omega$ and $C = 1\ \mu\text{F}$

$$\frac{1}{C}\int i_1(t)dt + L_1\frac{di_5}{dt} + K_{12}\frac{di_6}{dt} + K_{13}\frac{di_7}{dt} + K_{14}\frac{di_8}{dt} + Ri_5 = V_{CC} \tag{5.55}$$

The second KVL equation for the second *RLC* branch is

$$\frac{1}{C}\int i_2(t)dt + L_2\frac{di_6}{dt} + K_{12}\frac{di_5}{dt} + K_{23}\frac{di_7}{dt} + K_{24}\frac{di_8}{dt} + Ri_6 = V_{CC} \tag{5.56}$$

and so forth. On the other hand, applying the first KCL equation at node (1) gives

$$i_1 + I_1 - i_5 = 0 \tag{5.57}$$

and applying the same equation at node (2) gives

$$i_2 + I_2 - i_6 = 0 \tag{5.58}$$

and so forth. These set of eight equations are next discretized and solved for each time step. Figure 5.18 shows voltage levels at the four die sites for a sample run of

$$\begin{aligned} &R = 1, C = 1\mu, L = 1\mu, \Delta t = 0.05\mu, \\ &K_{12} = 0.50\mu, K_{13} = 0.33\mu, K_{14} = 0.25\mu, \\ &K_{23} = 0.50\mu, K_{24} = 0.33\mu, K_{34} = 0.50\mu \end{aligned} \tag{5.59}$$

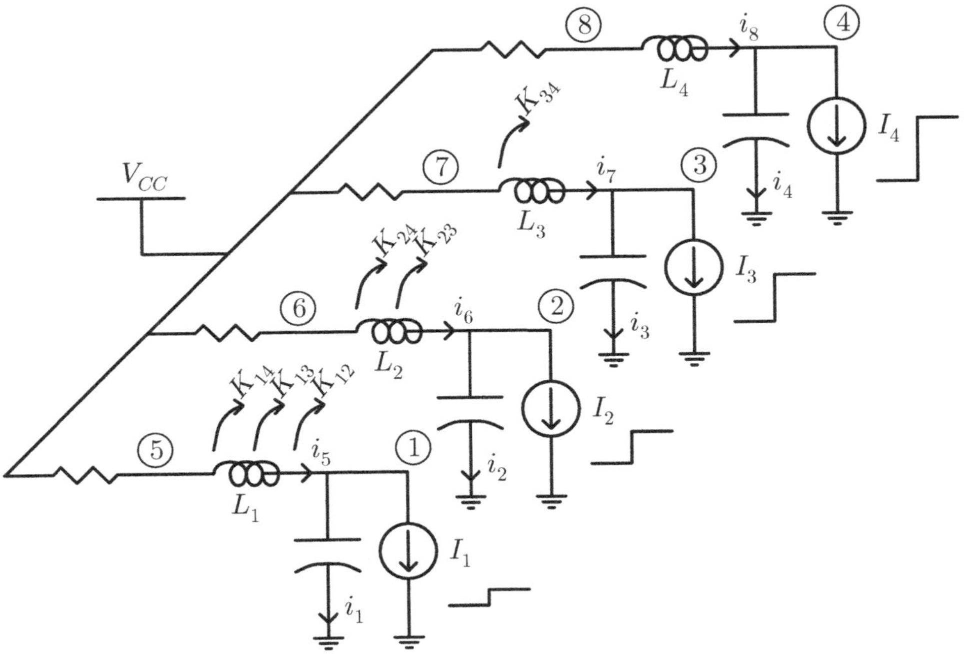

Fig. 5.17 Simplified PDN problem comprised of power source, PCB/PKG, die cap, and current demand

where resistance is in Ω, capacitance in F, and inductance in H. We can see from the figure that theory matches simulations very well.

5.10 Simple Wave Propagation Problem

Consider the transmission line shown in Fig. 5.19. A unit step input is applied at the input and we want to monitor voltage across the line, as a function of time. The transmission line has $L_0 = 0.1 \times 10^{-6}$ inductance per unit length, $C_0 = 0.1 \times 10^{-9}$ capacitance per unit length, and length 0.1 m. It has a characteristic impedance of $Z_0 = \sqrt{L_0/C_0} = 31.6\,\Omega$ which we use to set the input resistance tying the voltage source to the transmission line

$$R = Z_0 \tag{5.60}$$

That is, we set source impedance to characteristic impedance to eliminate any backwards waves impinging on the source from reflecting back onto the line! As shown in the figure we have 20 unknown currents which we need to solve for. We would need 20 equations. Ten of those come from KVL going around the 10 loops. For example, going around outermost loop we get

$$\frac{1}{C}\int i_{11}(t)dt + L\sum_{n=1}^{10}\frac{di_n(t)}{dt} + Ri_1(t) = 1 \tag{5.61}$$

Going around the next loop, starting with i_{12} and going around all the inductors on the left we arrive at

$$\frac{1}{C}\int i_{12}(t)dt + L\sum_{n=1}^{9}\frac{di_n(t)}{dt} + Ri_1(t) = 1 \tag{5.62}$$

and so on until the last loop on the left

$$\frac{1}{C}\int i_{20}(t)dt + L\frac{di_1(t)}{dt} + Ri_1(t) = 1 \tag{5.63}$$

This gives the first 10 equations. Next we do KCL at the 10 labeled nodes; for example at node (10) we get

$$-i_{10} + i_{11} = 0 \tag{5.64}$$

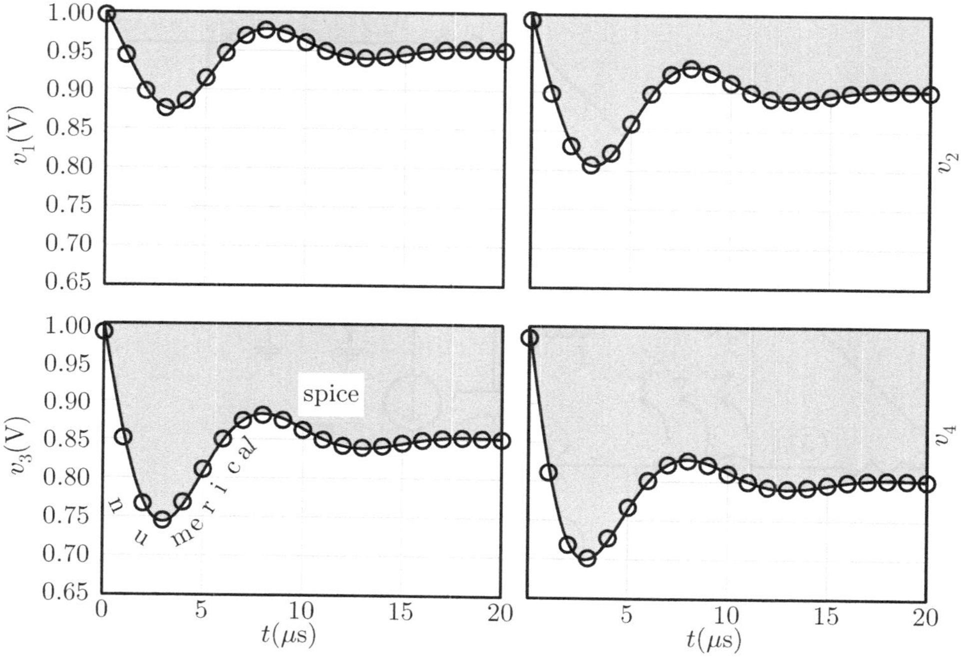

Fig. 5.18 Voltage at die side of simplified PDN problem

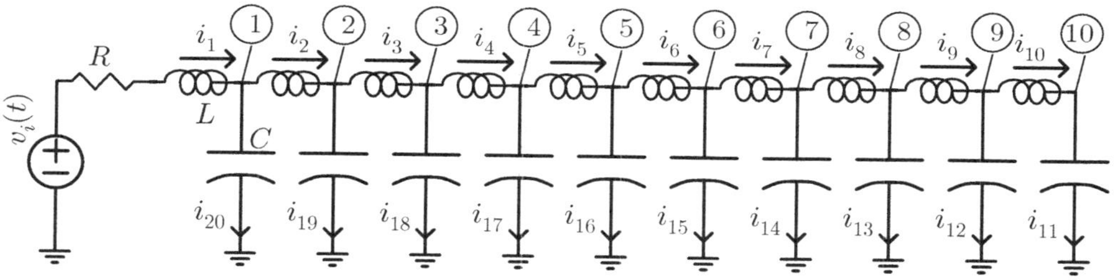

Fig. 5.19 Discretized transmission line for wave propagation

and at node (9) we have

$$-i_9 + i_{10} + i_{12} = 0 \tag{5.65}$$

and so on until we get to node (1) where we have

$$-i_1 + i_2 + i_{20} = 0 \tag{5.66}$$

This would have given the last 10 equations, for a total of 20 equations. Now we discretize as in prior sections and end up with system of equations

$$Ax = b \tag{5.67}$$

where x is the unknown current and b is the vector holding information about the prior iteration current. For each new time step, we update b and get new current x. Figure 5.20 shows response across the LC ladder, as a function of time. Notice how the wave flows from left towards right; it hits the termination and fully reflects (open termination); then it finally hits the input and never comes back (perfect termination). The speed of the wave is

$$v = \frac{1}{\sqrt{L_0 C_0}} = 316 \times 10^6 \text{ m/s} \tag{5.68}$$

The time it would take to travel from left to right is

$$t = \frac{\text{length}}{v} = \frac{0.1}{316 \times 10^6} = 316 \text{ ps} \tag{5.69}$$

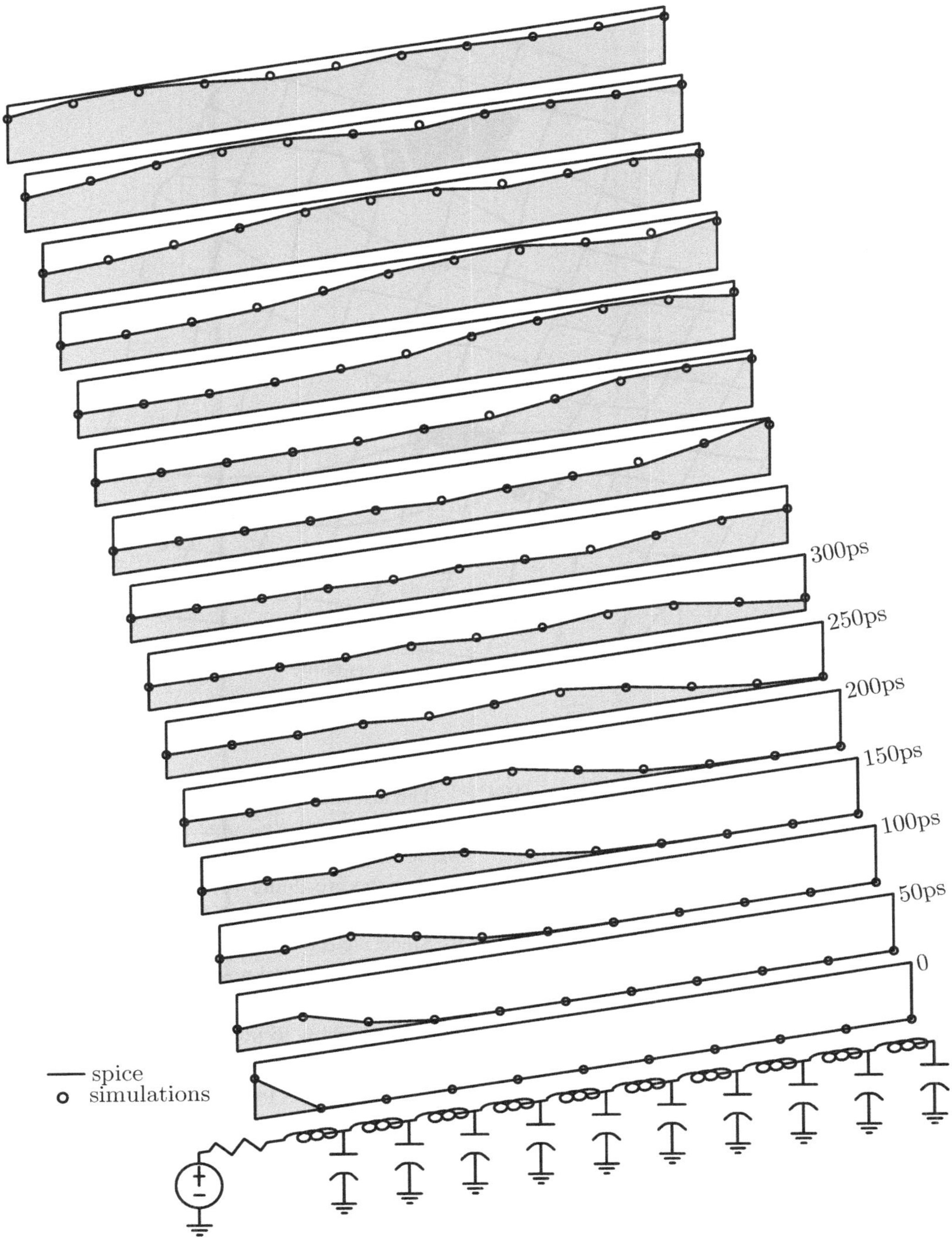

Fig. 5.20 Simple wave propagation. Case $L_0 = 0.1 \times 10^{-6}$ H/m, $C_0 = 0.1 \times 10^{-9}$ F/m, and length 0.1 m

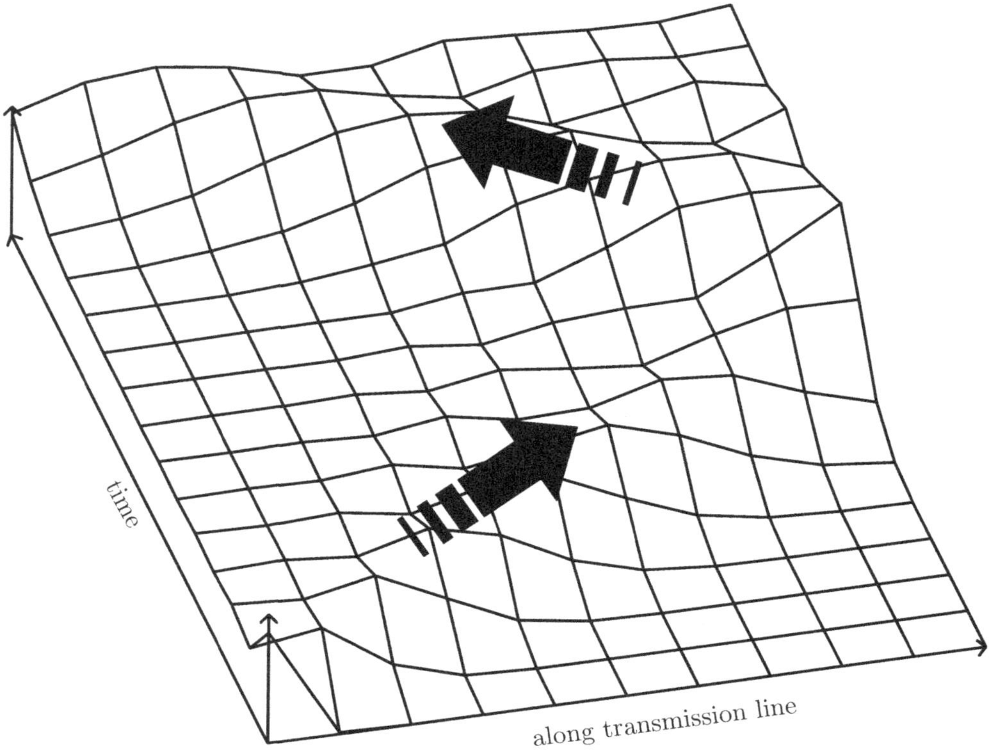

Fig. 5.21 Simple wave propagation

which can be confirmed in the simulations. We can replot this in a different way as shown in Fig. 5.21.

5.11 Some Notes on Numerical Methods

Between resistors, inductors, and capacitors we have three scenarios

- Simple scaling—this applies for resistors, where we simply scale currents.
- Time differentiation—this applies for inductors, where current value depends on current value and next value in time.
- Time integration—this applies for capacitors, where current value depends on current value and ALL prior values.

The last case warrants special attention. When integrating from, e.g., time zero to current point we follow the process shown in Fig. 5.22. Assuming we are doing KVL at time t_4, we need to account for all currents from zero time till t_4. Since we are trying to figure current at time t_4 that value is unknown, but putting it into an algebraic equation would allow us to do so. Question is, how to put i_4—do we fully include it or partially? As shown in the figure, when integrating current between t_3 and t_4 we cannot assume all the weight falls on i_4—instead we split it half/half with the proper current, i_3. Once i_4 is figured, and when subsequently working on i_5 we include it again, with a weight of half too. So, when all is done current at time t_4 would have been fully included in the integration: once when trying to figure it, and once when trying to figure the subsequent current, each time being included with a weight of half.

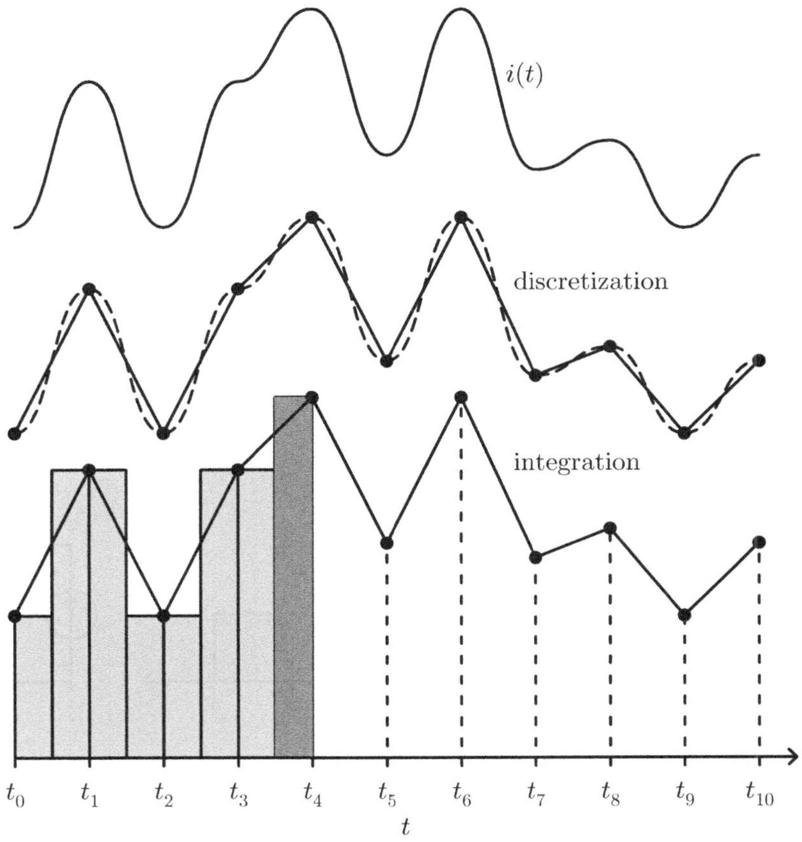

Fig. 5.22 Some note on numerical integration

5.12 Summary

This chapter dealt with the general topic of numerical methods applied to circuit problems. In particular we used finite difference to recast the governing differential equation into an algebraic form where the new solution at time $n\Delta t$ is related to the old solution at time $(n-1)\Delta t$. We showed how resistive currents scale, inductive ones tie new to old values, and capacitive currents relate to all prior values. This method is extremely general and works for all *RLC* circuits, even with mutual elements, as well feedback ones. We applied the method to examples with input stimuli ranging between unit steps, periodic pulse, sinusoids, and negative exponential. A few noteworthy examples were the power delivery one, the miniature power plane one, and the wave propagation one. In all cases we compare our results to SPICE, and with small enough time resolution we get excellent match. In some cases we end up with a linear system in the form of a matrix, in which case the matrix has to be solved for.

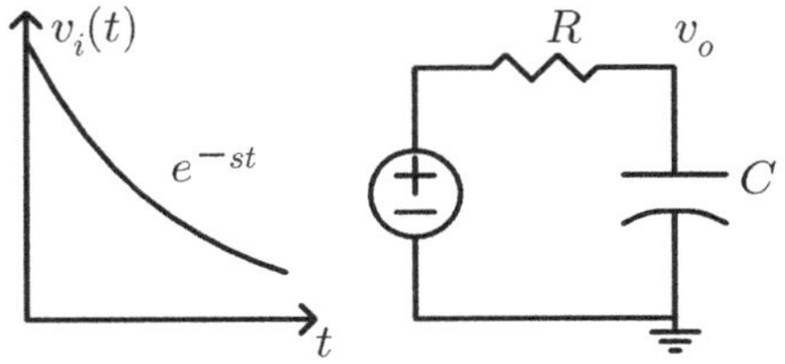

Fig. 5.23 *RC* network driven by negative exponential voltage

5.13 Problems

1. The *RC* network in Fig. 5.23 is driven by a negative exponential e^{-st}. Derive a numerical algorithm to find cap current and cap voltage

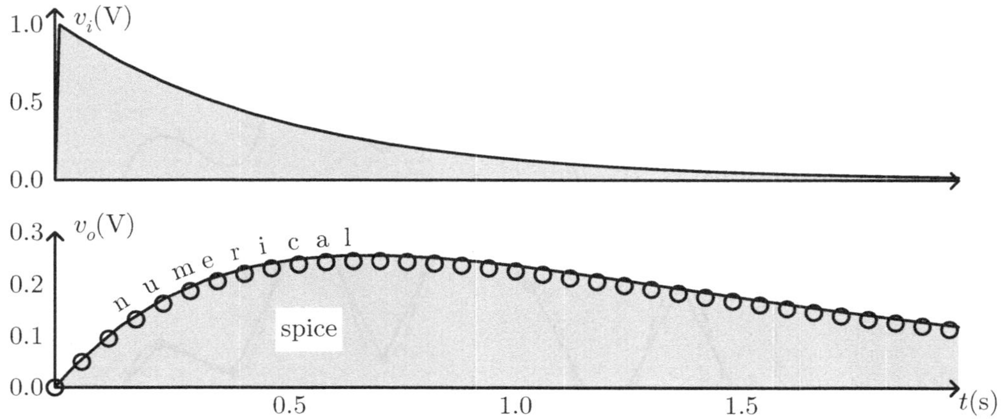

Fig. 5.24 Solution to Problem 1

Fig. 5.25 *RC* network driven by current source

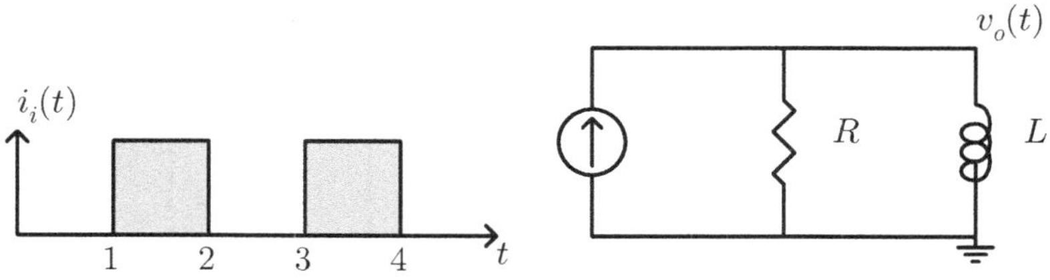

as a function of time for the case $R = 1$, $C = 1.0$, and $s = 2$. Compare to SPICE. Sample solution in Fig. 5.24.

2. The *RL* network in Fig. 5.25 is driven by a current source $i(t)$; derive an algorithmic expression for output voltage. Next, assume $R_1 = 1$, $L = 0.5$, and input current periodic with period 2 and 50% duty cycle; plot results and compare to SPICE. See sample solution in Fig. 5.26.
Answer:

$$v(0) = 0$$

$$v_n = \frac{1}{\frac{\Delta t}{2L} + 1/R}$$

$$\left[i_n - \frac{\Delta t}{2L}\left(v_0 + 2\sum_{m=1}^{n-1} v_m \right)\right], \quad n > 1$$

3. The *RLC* network in Fig. 5.27 is driven by a sine function offset by 1 V; derive an equation for output voltage assuming the sine has been running for a long time; i.e., 1 DC offset is really DC, and not a step. For case of $R = 1$, $L = 0.5$, $C = 0.1$, and $v_i(t) = 1 + \frac{1}{5}\sin \pi t$, plot results and compare to SPICE. See sample solution in Fig. 5.28.
Answer:

$$i_n = \frac{1}{\left[\frac{\Delta t}{2C} + R + \frac{L}{\Delta t}\right]}$$

$$\left\{ v_n - 1 - \frac{\Delta t}{2C}\left[i_0 + 2\sum_{m=1}^{n-1} i_m \right] + \frac{L}{\Delta t} i_{n-1} \right\}$$

$$v_{o,n} = 1 + \frac{\Delta t}{C}\sum_{m=1}^{n} i_m + R i_n$$

4. Repeat Problem 3 but this time insert an output current source of the form $1-e^{-t}$ as shown in Fig. 5.29; find output voltage. See sample solution in Fig. 5.30.

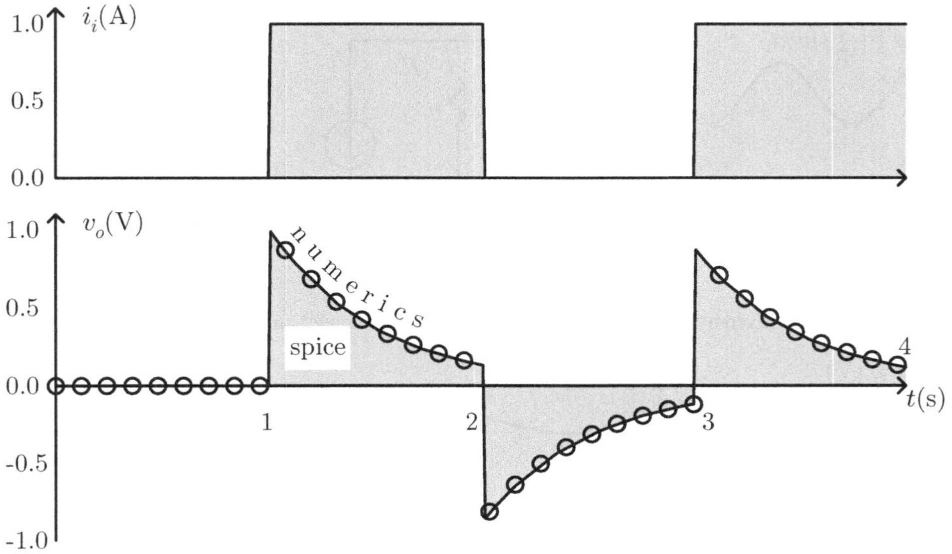

Fig. 5.26 Answer to Problem 2

Fig. 5.27 *RLC* network driven by sine voltage

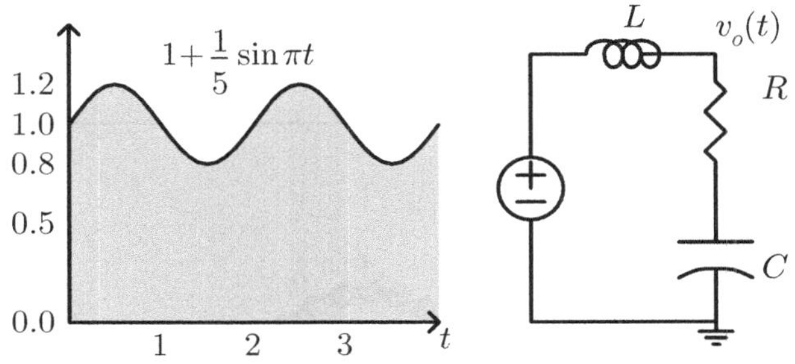

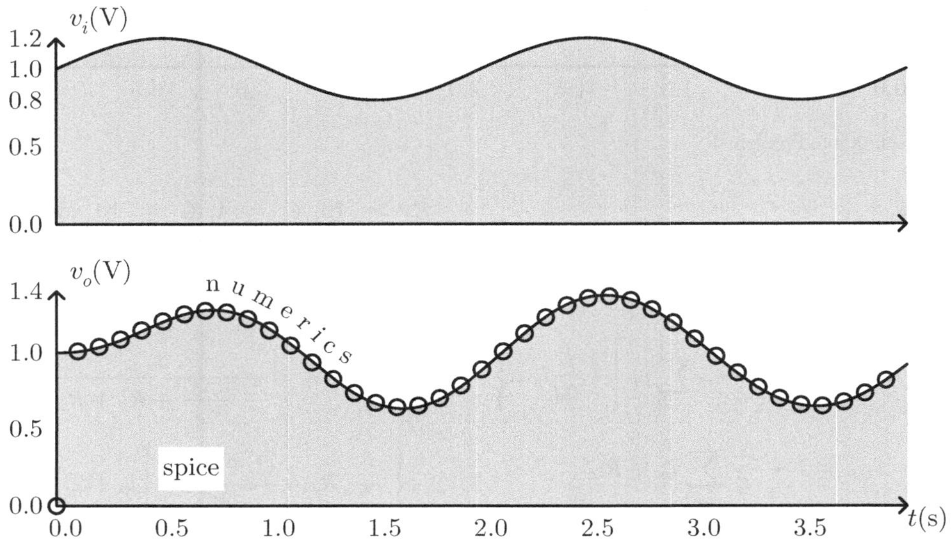

Fig. 5.28 Solution to Problem 3

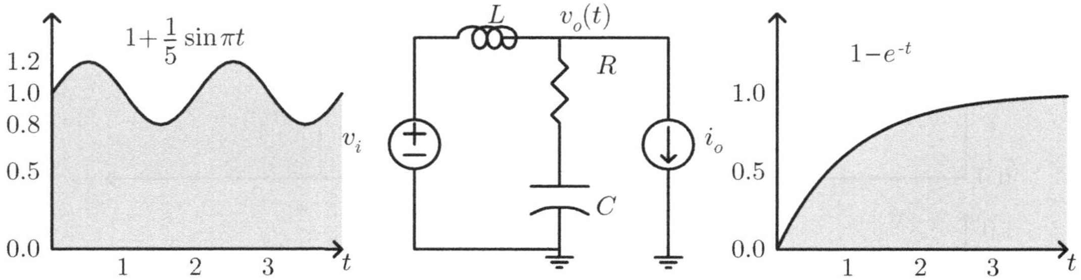

Fig. 5.29 *RLC* network driven by sine voltage input and negative exponential output current

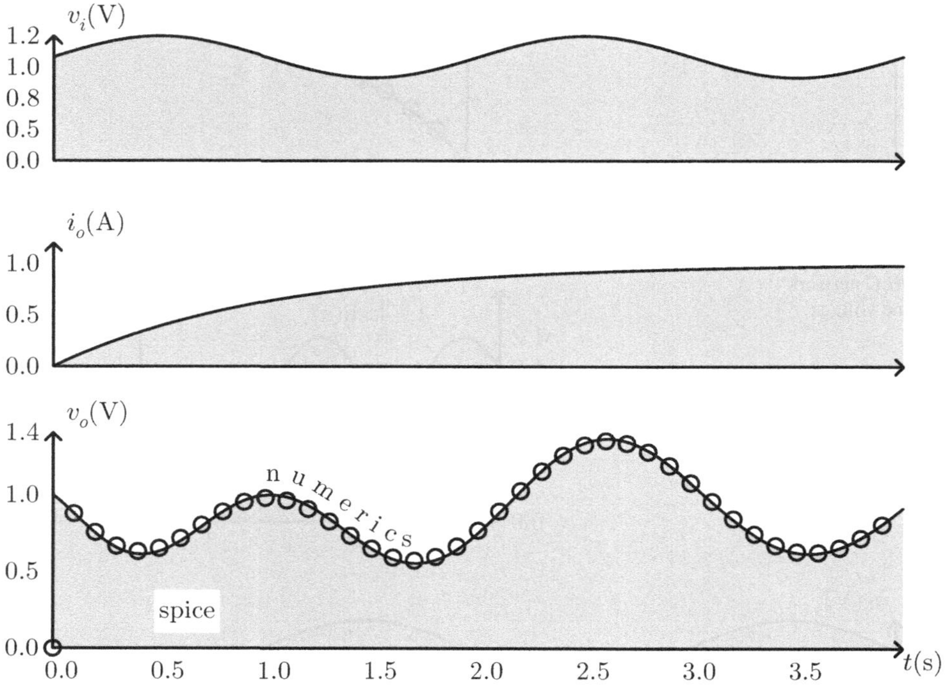

Fig. 5.30 Solution to Problem 4

Answer:

$$i_n = \frac{1}{\left[\frac{\Delta t}{2C} + R + \frac{L}{\Delta t}\right]}$$

$$\left\{v_n - 1 - Le^{-n\Delta t} - \frac{\Delta t}{2C}\left[i_0 + 2\sum_{m=1}^{n-1} i_m\right] + \frac{L}{\Delta t} i_{n-1}\right\}$$

$$v_{o,n} = 1 + \frac{\Delta t}{C}\sum_{m=1}^{n} i_m + Ri_n$$

5. Consider the feedback problem in Fig. 5.31. Derive an algorithmic expression for output voltage, given $i_o(t)$. Next, assume $R_1 = 1$, $R_2 = 10$, $C = 1$, $G = 10$, and $i_o(t) = u(t)$; plot results and compare to SPICE. See sample results in Fig. 5.32

Answer:

$$i_n = \frac{1}{\frac{(1+G)\Delta t}{2C} + R_1 + R_2}$$

$$\left\{-i_{o,n}R_2 - \frac{(1+G)\Delta t}{2C}\left[i_0 + 2\sum_{m=1}^{n-1} i_m\right]\right\}$$

$$v_{o,n} = i_n R_1 + \frac{\Delta t}{C}\sum_{m=0}^{n} i_m$$

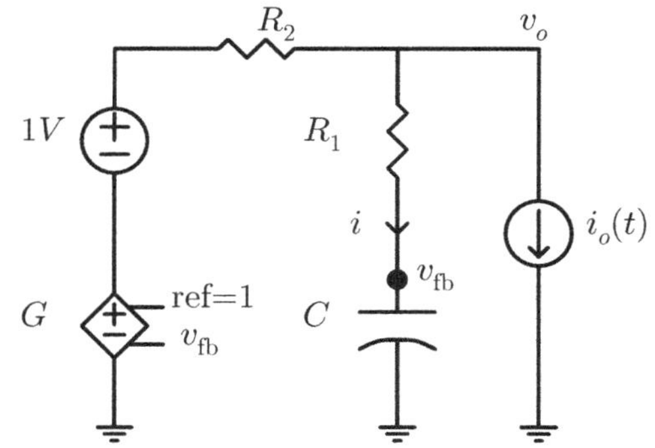

Fig. 5.31 *RC* problem with feedback

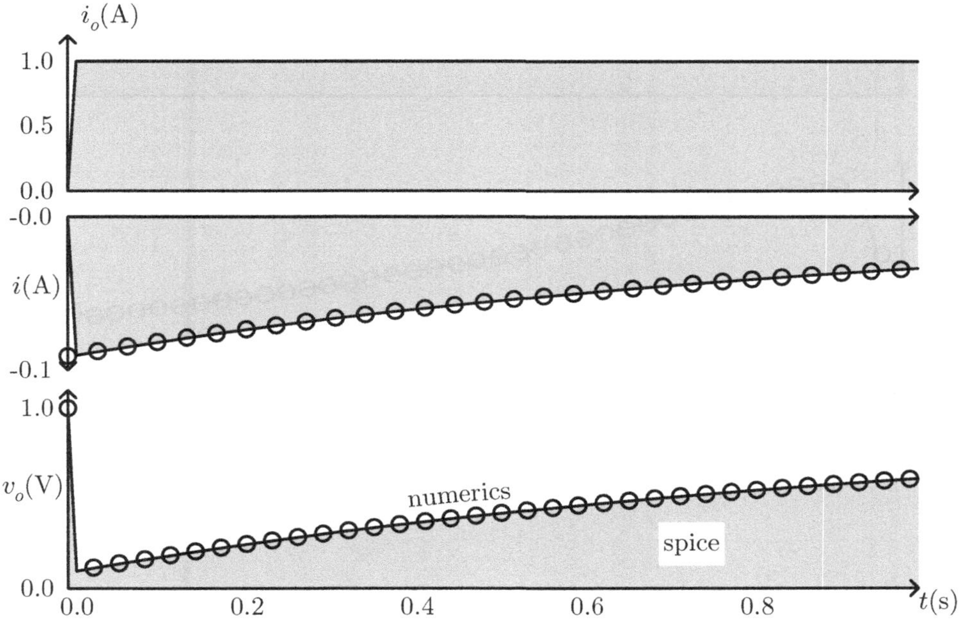

Fig. 5.32 Solution to Problem 5

6. Consider the *RC* circuit in Fig. 5.33; derive an algorithmic expression for current across C_1. Next, assume input voltage is $v_i(t) = u(t)$, $R_1 = 1$, $R_2 = 2\,\Omega$, $C_1 = 0.5$, and $C_2 = 0.1\,\text{F}$; plot results and compare to SPICE. See sample solution in Fig. 5.34.

Answer:

$$i_0 = 0; \quad i_1 = \frac{\Delta t v_0}{R_1 R_2 C_2}$$

$$i_{n+1} = i_n + \frac{\Delta t}{R_1 R_2 C_2}\left\{ v_n - \frac{\Delta t}{2C_1} \left[i_0 + 2\sum_{m=1}^{n-1} i_m + i_n \right] - i_n \left(R_1 + R_2 + R_2 \frac{C_2}{C_1} \right) \right\}$$

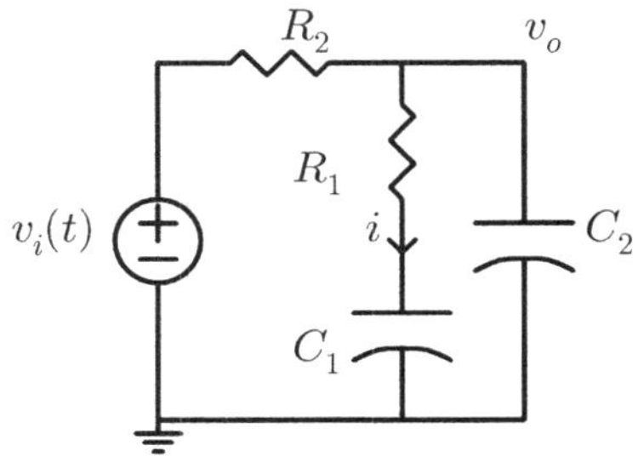

Fig. 5.33 *RC* network for Problem 6

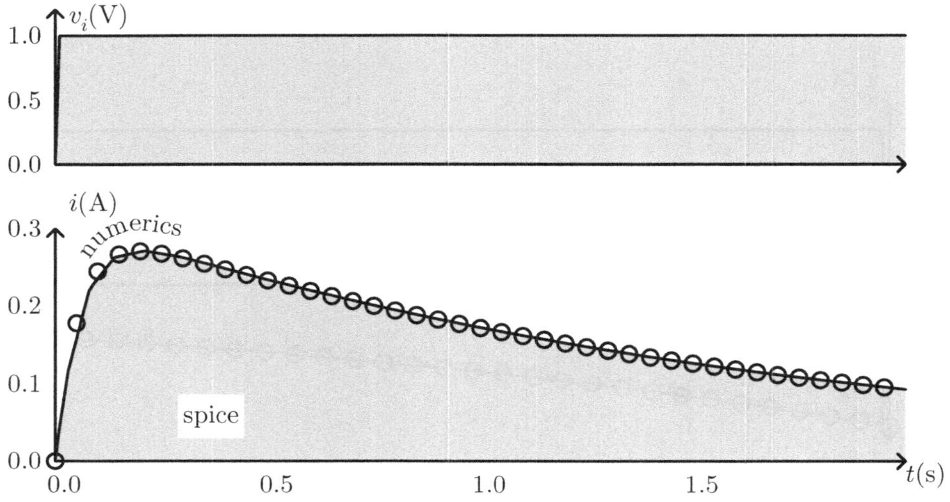

Fig. 5.34 Solution to Problem 6

6 Fourier Series and Periodic Functions

6.1 Introduction

So far we have dealt with various methods of solving circuit problems such as steady state, analytic differential equations, polynomial expansion, and numerical differential equations. Each of the methods had its pros and cons. For example, the steady state method misses the transient part of the solution. The polynomial expansion is not efficient. The numerical differential equation methods sheds little insight into the operation of the circuit, and has heavy reliance on computers. Finally the analytic differential equation method works only for simple driving functions, and cannot deal with cases where input is a mixture of squares, triangles, circles, etc. This now leads us to another class of methods—the spectral ones. The spectral methods rely on Fourier analysis, and towards that end we need to build the needed mathematical tools. This method cannot be overemphasized. It is generic and robust and time invested in it is worthy time! It all stars with Fourier series, which leads to Fourier transform, then into Laplace transform. In essence the spectral flow states the following: if we know the solution due to a sine/cosine input, then we know the solution to any input! How? And here comes the Fourier part of analysis; this states that pretty much any signal can be expressed as a summation of sines and cosines. All that is left then is linear theory and superposition, and the problem is solved! Let's then take a break from circuits and review and practice the Fourier series.

6.2 Fourier Series

The Fourier series states that if a function is periodic in T then we can reconstruct the function in terms of an infinite series of sines and cosines, also of period T and integral multiplication thereof:

$$f(t) = a_0 + \sum_{n=1}^{n=\infty} a_n \cos \omega_n t + \sum_{n=1}^{n=\infty} b_n \sin \omega_n t \tag{6.1}$$

where

$$\omega_n = \frac{2\pi n}{T} \tag{6.2}$$

$$a_0 = \frac{1}{T}\int_{-T/2.}^{T/2} f(t)dt \tag{6.3}$$

$$a_n = \frac{2}{T}\int_{-T/2.}^{T/2} f(t)\cos \omega_n t dt \tag{6.4}$$

$$b_n = \frac{2}{T}\int_{-T/2.}^{T/2} f(t)\sin \omega_n t dt \tag{6.5}$$

The above listing has been around—it seems—for ever, and rather than digging into proving it, let's assume it holds and start using it. Carrying on with the philosophy of this text,

F. Badrieh, *Spectral, Convolution and Numerical Techniques in Circuit Theory*,
https://doi.org/10.1007/978-3-319-71437-0_6

which is an application one, let's demonstrate the above flow by as many examples (and problems) as we can fit within a chapter!

6.3 Periodic Pulse

Our first example is periodic pulse of width 0.5, period 1.0, and min/max value of −1/ + 1 respectively. It is defined as

$$f(t) = \begin{cases} +1 & -\frac{1}{4} < t < \frac{1}{4} \\ -1 & +\frac{1}{4} < |t| < \frac{1}{2} \\ \text{Periodic in } T,\ T = 1 \end{cases} \tag{6.6}$$

and is shown in Fig. 6.1. Since the average of the signal is zero, then we conclude that the DC term is zero, $a_0 = 0$. Furthermore, since this function is even, then all the sine terms would be zero, $b_n = 0$. We are left with $n \neq 0$ cosine terms. In particular,

$$\begin{aligned} a_n &= \frac{2}{T}\int_{-T/2}^{T/2} f(t)\cos(\omega_n t)dt = \frac{2}{1}\int_{-1/2}^{1/2} f(t)\cos(\omega_n t)dt = 4\int_{0}^{1/2} f(t)\cos(\omega_n t)dt \\ &= 4\int_0^{1/4}\cos(\omega_n t)dt - 4\int_{1/4}^{1/2}\cos(\omega_n t)dt = \frac{4}{\omega_n}\left[\sin(\omega_n t)\Big|_0^{1/4} - \sin(\omega_n t)\Big|_{1/4}^{1/2}\right] \\ &= \frac{4}{2\pi n}\left[\sin\left(\frac{2\pi n}{4}\right) - \sin\left(\frac{2\pi n}{2}\right) + \sin\left(\frac{2\pi n}{4}\right)\right] \\ &= \frac{2}{\pi n}\left[2\sin\left(\frac{\pi n}{2}\right) - \sin(\pi n)\right] = \boxed{\frac{4}{\pi n}\sin\left(\frac{\pi n}{2}\right)} \end{aligned} \tag{6.7}$$

What this says is that we can approximate the pulse function with the following series:

$$f(t) = \sum_{n=1}^{\infty} \frac{4}{\pi n}\sin\left(\frac{\pi n}{2}\right)\cos(2\pi n t) \tag{6.8}$$

Notice that when n is even, the sine function evaluates to zero, and we can abstract the above series even shorter

$$f(t) = \sum_{n=1,3,\ldots}^{\infty} \frac{4}{\pi n}\sin\left(\frac{\pi n}{2}\right)\cos(2\pi n t) \tag{6.9}$$

Let's see how good this evaluates to! Figure 6.2 shows the original function along with the Fourier series, for different term count. Clearly as we add more terms we get better representation. Another way we can look at our reconstruction is shown in Fig. 6.3.

Having established the soundness of our approach let's analyze what we call the frequency spectrum of the signal. Recall we had found above that we can represent this periodic pulse signal in terms of cosines, of various frequencies, and the scaling factor of each cosine term is what we refer to as the spectrum a_n. Figure 6.4 shows this spectrum in terms of value (top) and absolute values (bottom). There are many pieces of information which can be deduced from such spectrum plots.

1. First, the strongest component is of course the first harmonic; in this case it has magnitude ~ 1.2, as compared to the pulse magnitude of 1; that is to first order, the first cosine function (first harmonic) has the same period as the original function and almost the same magnitude. This sounds like a reasonable starting point!
2. Since the starting function has 50/50 duty cycles, when even number of cosine periods fit inside the period, those cosines contribute nothing to the series; that is, when n is even, we get zero harmonics (since the time integration of the function against the cosine would have been zero).

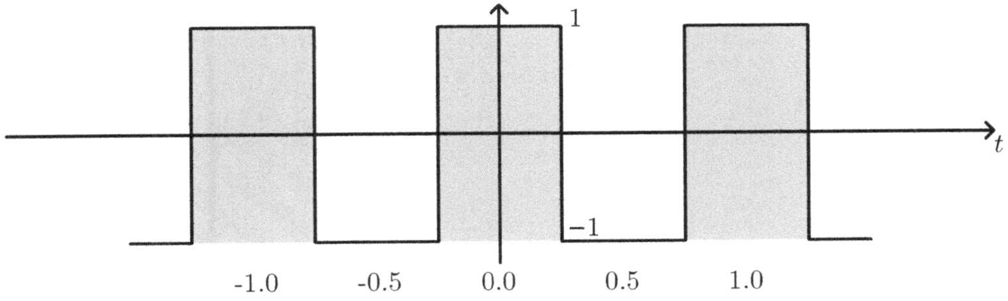

Fig. 6.1 Pulse function with zero average

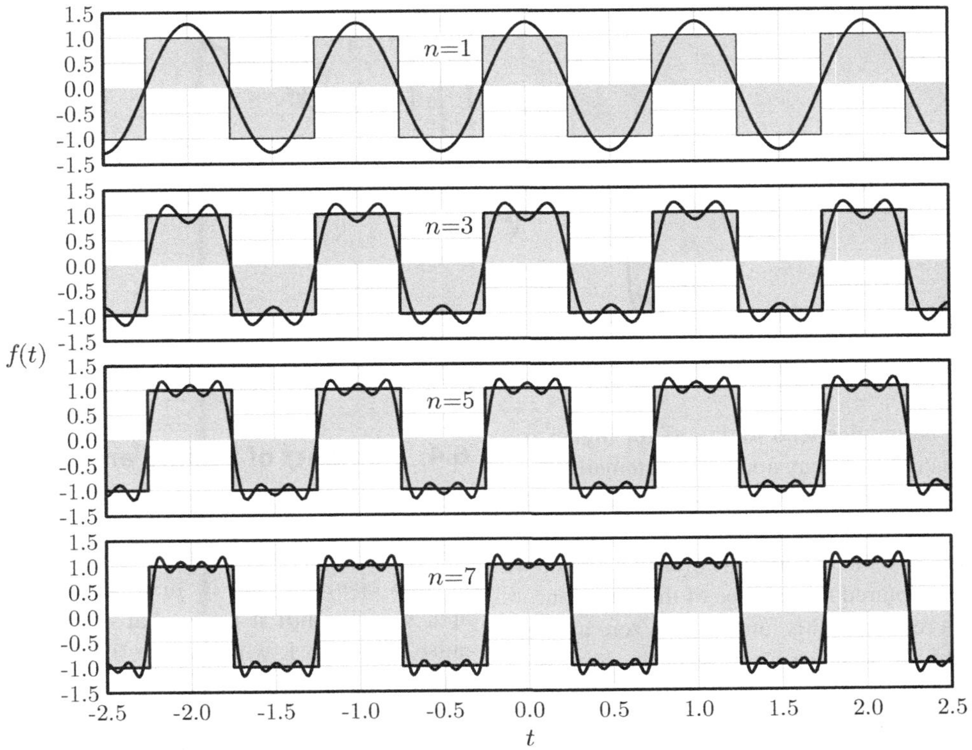

Fig. 6.2 Pulse function with zero average and approximation with Fourier series: From top to bottom we have $n = 1, 3, 5$, and 7

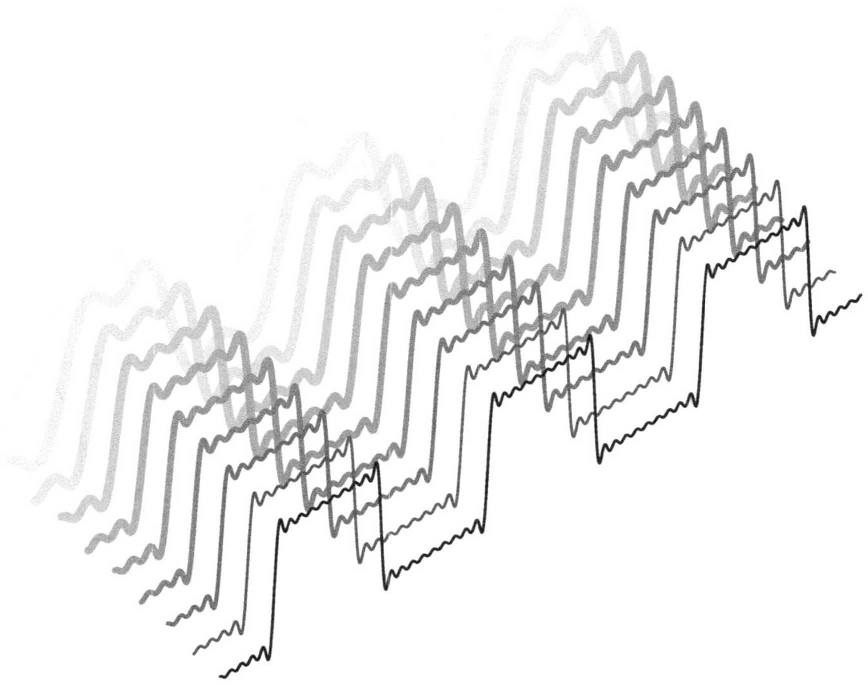

Fig. 6.3 Reconstruction of pulse using harmonics

3. The harmonics tend to die off for higher frequencies; they may not do so systematically in general, but here they die off systematically. That is, after some frequency we need not carry any more harmonics; at that point, we have captured the essence of this function.
4. As a reminder, this function has zero average, so the DC term (or $n = 0$ term) is zero.

In summary, the spectrum has all the information needed to reconstruct the starting signal. *Even though the spectrum itself looks nothing like the starting signal*, it holds the *key* to the harmonic series expansion. When this key is used (to calibrate each sine/cosine function), we are assured to regain the starting signal. To wrap it up, another fancy way of displaying signal evolution versus number of harmonics is shown in Fig. 6.5.

6.4 Impact of Scaling and DC Shift

Let's take the example from the prior section and do two operations on it: first, scale it by 1/2 and, second, shift it by so that it is a periodic pulse from 0 to 1, with average 0.5, as shown in Fig. 6.6. We then have

$$f(t) = \begin{cases} 1 & |t| < 0.25 \\ 0 & \text{else wise} \end{cases} \tag{6.10}$$

Before starting to evaluate Fourier components, let's try and see if we can salvage what we have done in the prior section. Both signals are pulse-like and have the same period. They differ only in magnitude and in average.

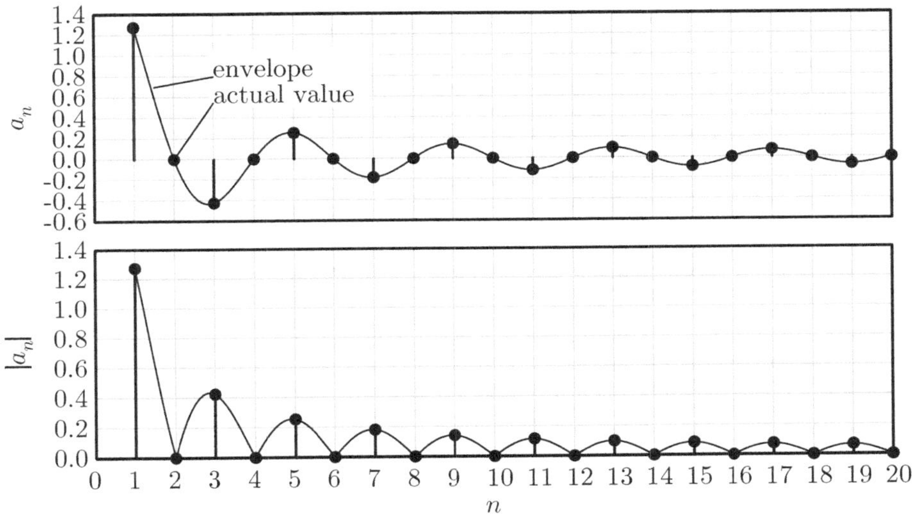

Fig. 6.4 Spectral content of periodic pulse (with zero average) with period 1

1. The prior section had peak-to-peak value of 2 while this one has 1. So, we will need to scale the Fourier spectrum by half.
2. The prior section had zero DC average, while this one has 0.5.

Knowing these differences, and building around them we arrive at the spectrum for the nonzero average pulse

$$b_n = 0 \quad \text{(even signal)}, \; a_0 = \frac{1}{2} \quad \text{DC average},$$

$$a_n = \frac{2}{\pi n} \sin\left(\frac{\pi n}{2}\right) \tag{6.11}$$

Having evaluated the Fourier series components, we simply plot the Fourier series and get results in Fig. 6.7. We can see clearly that the more harmonic components we add in, the better representation is the Fourier series. For reference, the spectral content is also shown in Fig. 6.7. Notice that this spectrum looks identical to that in prior section except that it is scaled by 0.5 and it has a DC component.

6.5 Impact of Shifting Pulse by Quarter Wave Length

Let's shift the periodic pulse (of 0.5 average) to the right by 0.25, as shown in Fig. 6.8 such that:

$$f(t) = \begin{cases} 1 & 0 < t < \frac{1}{2} \\ 0 & \text{else wise} \\ \text{period } 1 \end{cases} \tag{6.12}$$

Notice that, other than the DC average, this function is odd, in the sense that if it is mirrored about the y-axis, it flips sign. Since the function is odd we conclude that the cosine terms are zero (a_n=0), and only DC and sine terms are needed in the Fourier series expansion. The Fourier series coefficients are then

$$\begin{aligned} a_0 &= \frac{1}{2} \\ a_n &= 0 \text{ (odd signal)} \\ b_n &= \frac{1 - \cos n\pi}{n\pi} \end{aligned} \tag{6.13}$$

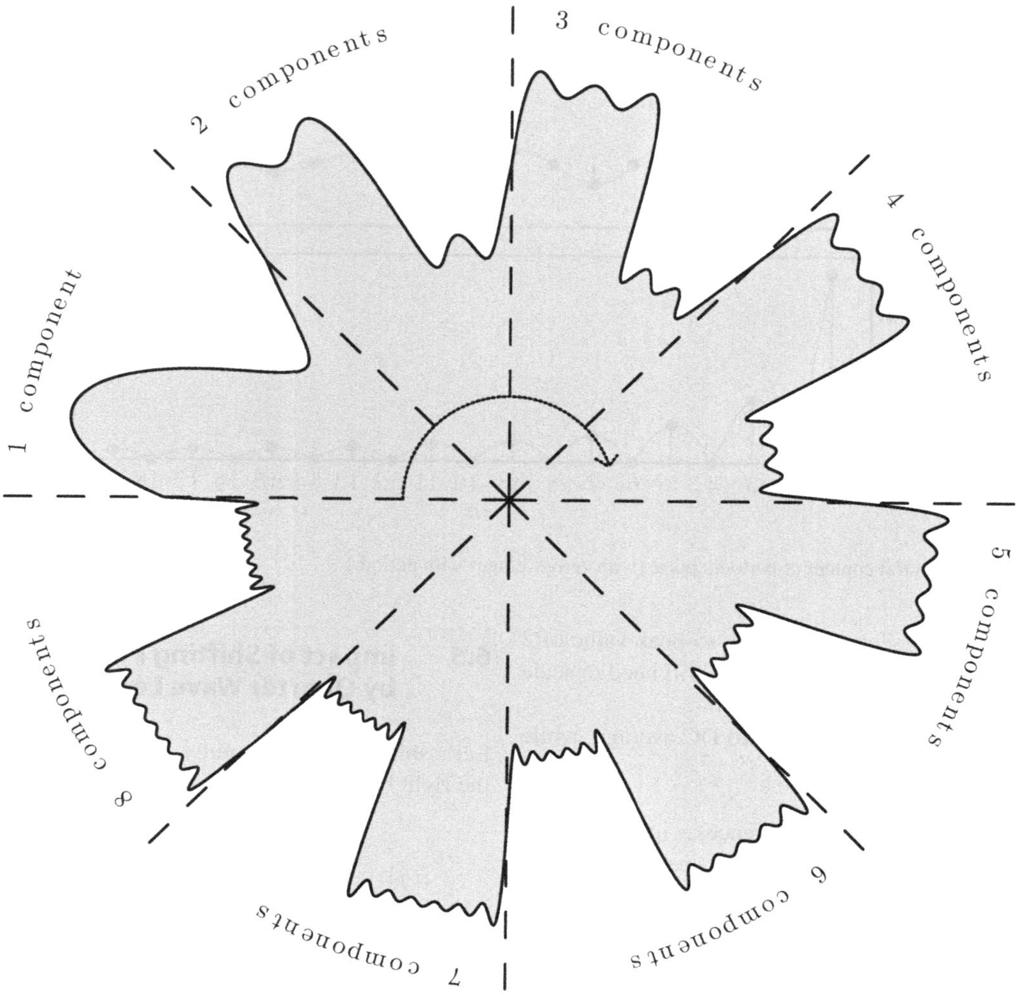

Fig. 6.5 Reconstruction of pulse using harmonics (fancy display)

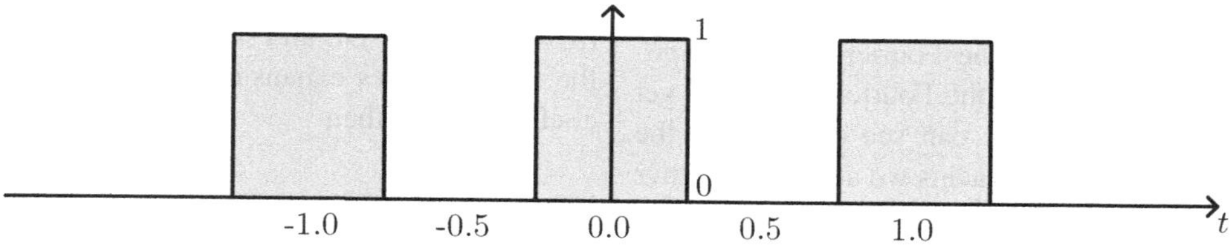

Fig. 6.6 Periodic pulse function with nonzero average

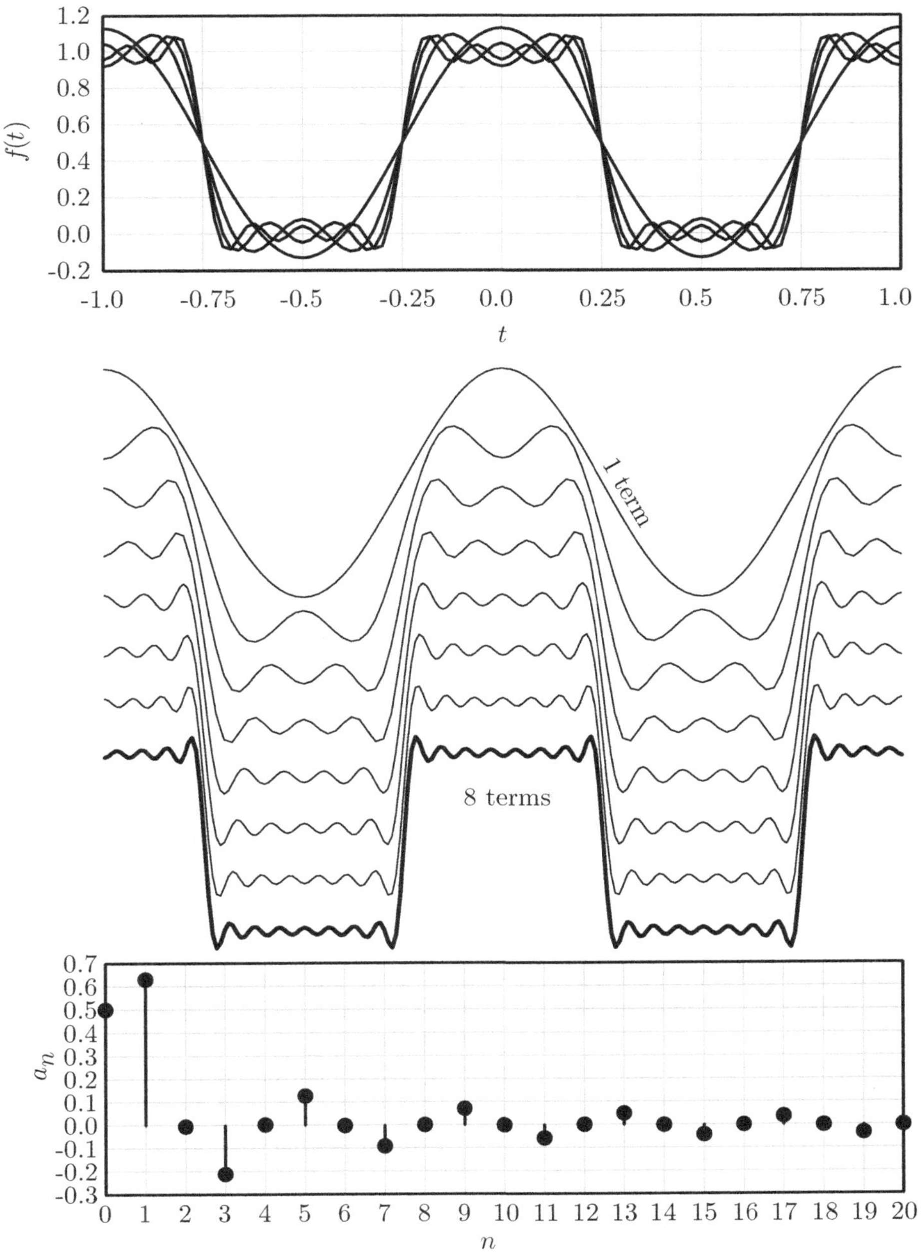

Fig. 6.7 Fourier series reconstruction and spectrum of shifted periodic pulse in Fig. 6.6

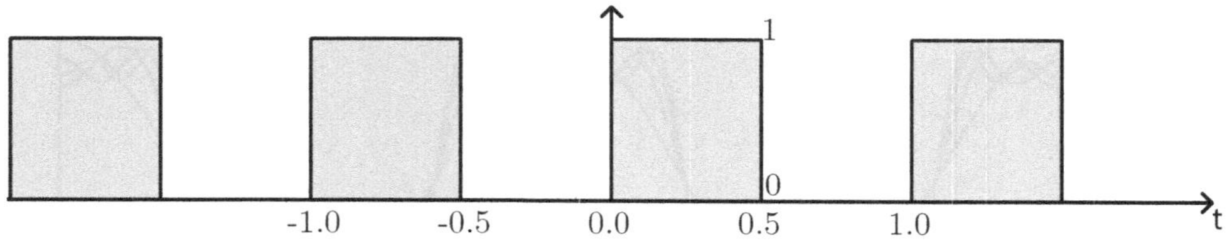

Fig. 6.8 Periodic pulse function with nonzero average, shifted by 1/4 wave length

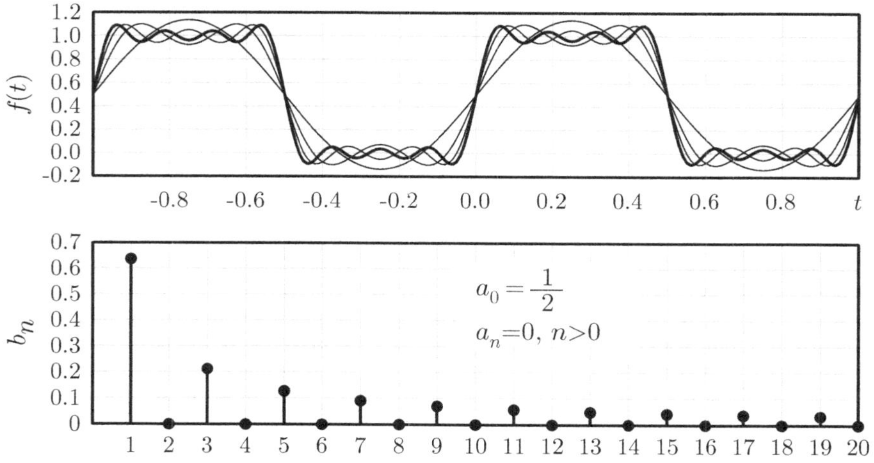

Fig. 6.9 Periodic pulse with average 0.5, shifted to the right by quarter wave length

If we plug these terms back into the Fourier series, and expand we get the reconstruction as shown in Fig. 6.9.

6.6 Impact of Shifting Pulse by Arbitrary Wave Length

Let's next shift the centered periodic pulse (of 0.5 average) to the right, this time by $1/5$, so that it is centered around 0.2, as shown in Fig. 6.10 such that:

$$f(t) = \begin{cases} 1 & -0.05 < t < 0.45 \\ 0 & \text{else} \\ \text{period } 1 \end{cases} \tag{6.14}$$

Now the function is neither odd nor even; hence it would include both cosine and sine coefficients. In particular:

$$a_0 = \frac{1}{2}$$

$$\begin{aligned} a_n &= \frac{2}{n\pi} \sin \frac{n\pi}{2} \cos \frac{2\pi n}{5} \\ b_n &= \frac{2}{n\pi} \sin \frac{n\pi}{2} \sin \frac{2\pi n}{5} \end{aligned} \tag{6.15}$$

If we plug those terms back into the Fourier series, and expand we get the reconstruction as shown in Fig. 6.11. Notice that now we have both A_n and B_n nonzero; but even then, the magnitude of the sum of the squares (as shown towards the bottom) still matches the prior two sections.

6.7 Impact of Pulse Width

Let's take the pulse function (centered, with pulse width 0.5 and DC average 0.5) and decrease its width (keeping the period the same), but increase the height such that total area under pulse remains the same; this is shown in Fig. 6.12. All

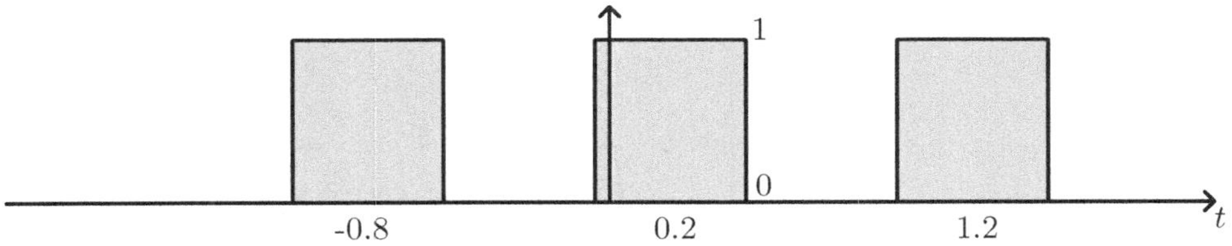

Fig. 6.10 Periodic pulse function with nonzero average, shifted by 1/5 wave length

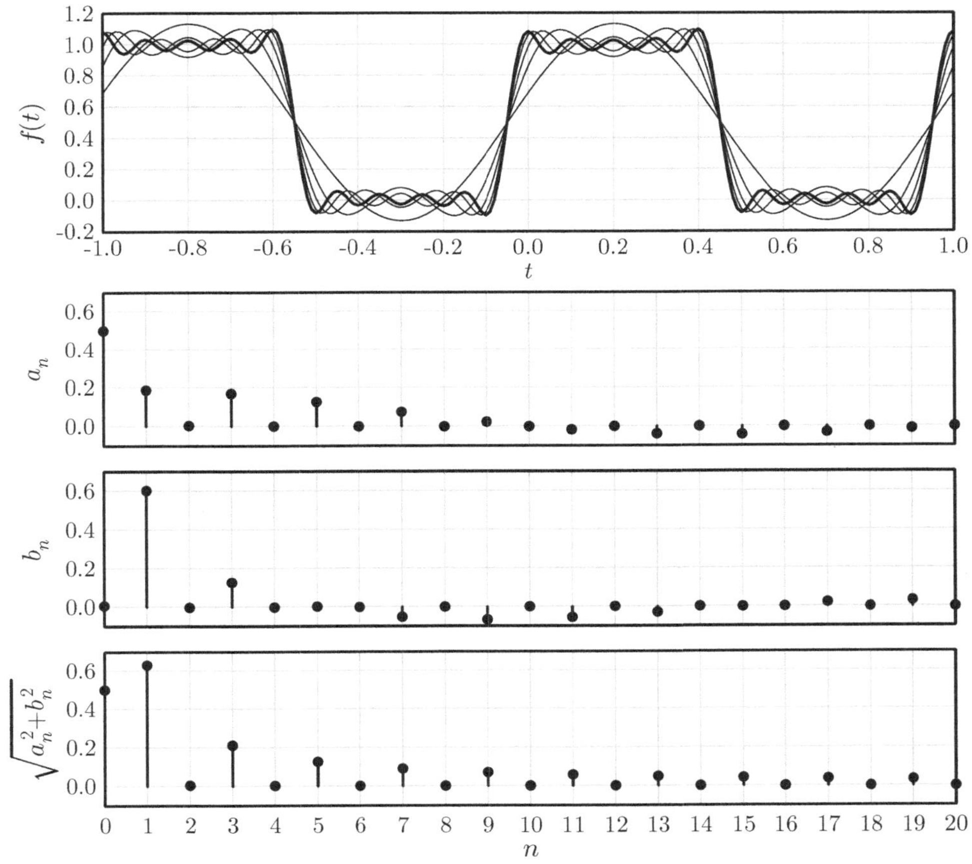

Fig. 6.11 Periodic pulse with average 0.5, shifted to the right by 1/5 wave length

Fig. 6.12 Periodic pulse with decreasing pulse width and increasing pulse height

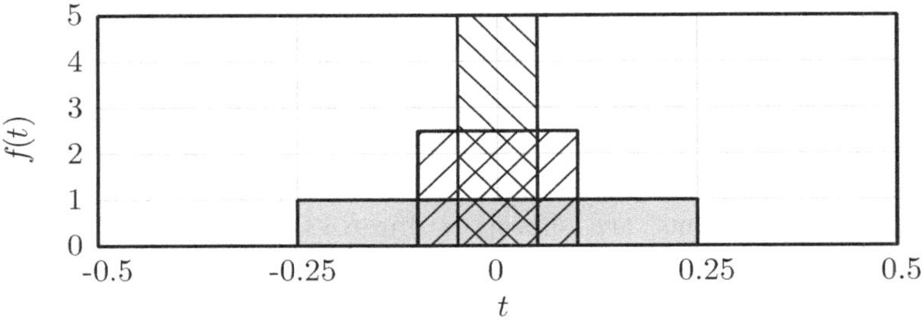

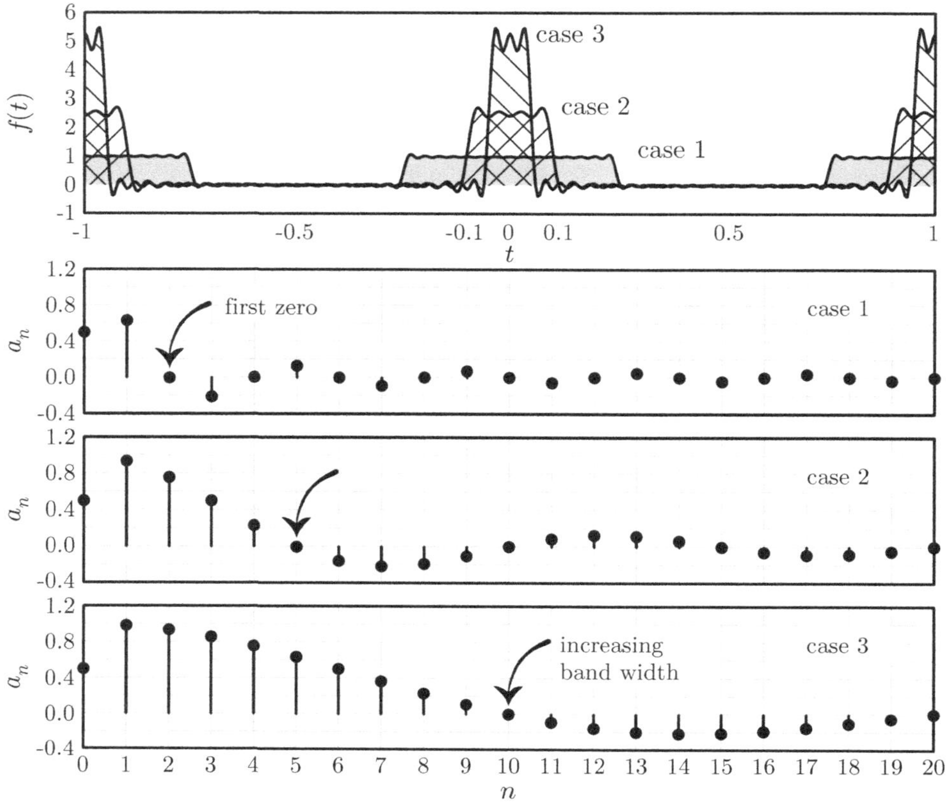

Fig. 6.13 Periodic pulse with different pulse width, but same period and area

three cases would have the same DC average, so that

$$a_0 = \frac{1}{2} \tag{6.16}$$

All would also have the same b_n coefficients, set to zero, since all cases are even. What is left is the a_n coefficients which are given by

$$a_{n,\text{case }1} = 1.0 \times \frac{2}{\pi n} \sin \frac{2\pi n}{4},$$
$$a_{n,\text{case }2} = 2.5 \times \frac{2}{\pi n} \sin \frac{2\pi n}{10},$$
$$a_{n,\text{case }3} = 5.0 \times \frac{2}{\pi n} \sin \frac{2\pi n}{20} \tag{6.17}$$

The corresponding time series as well as Fourier coefficients are shown in Fig. 6.13. Notice the very important observation: ***as the pulse width decreases, the spread in spectrum increases!!*** In fact if we continue to decrease the pulse width, and increase its height (such that its area remains half) we should expect the spectrum to converge to 1 (other than the DC value which remains 0.5). We could almost read out this observation by examining the trend in the three graphs.

6.8 Impact of Time Elongation

Let's again take the symmetric pulse and this time expand it in along the time-axis, again insuring that the DC value (average) remains the same; we'll try the three cases shown in Fig. 6.14.

All three cases are symmetric, so $b_n = 0$. For all three cases we have the same average, so that

$$a_0 = \frac{1}{2} \tag{6.18}$$

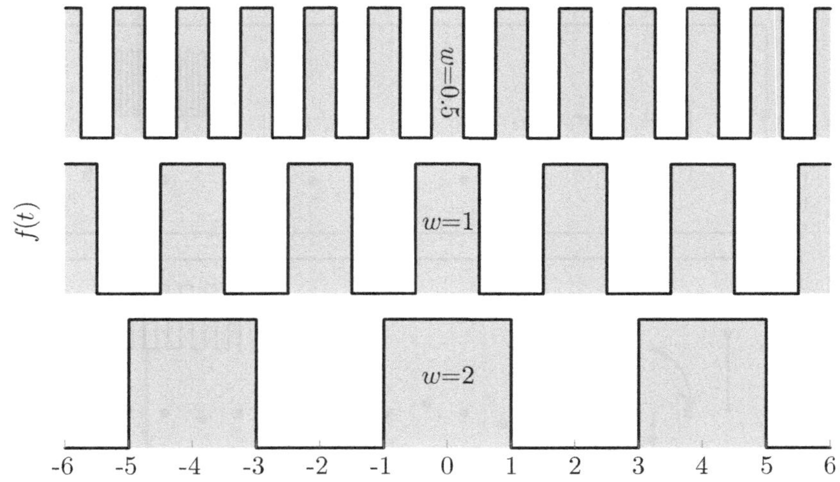

Fig. 6.14 Periodic pulse with three different periods, but with same DC average

Notice that while the pulse width is in fact increasing, the average remains the same, since the latter is the area divided by the period, and both are growing at the same rate. What remains is the cosine terms, and those are given by

$$a_n = \frac{2}{\pi n} \sin \frac{\pi n}{2} \tag{6.19}$$

for each of the three case! But how come they have the same FS coefficients, but look so different? The caveat is that the fundamental frequencies *are* different; in particular

$$\omega_{n,\text{case 1}} = \frac{2\pi n}{1}, \quad \omega_{n,\text{case 2}} = \frac{2\pi n}{2},$$
$$\omega_{n,\text{case 3}} = \frac{2\pi n}{4} \tag{6.20}$$

So while the integration limits (0.25, 0.5 and 1) differ, so does the denominator in the expression $\omega_n = \frac{2\pi n}{T}$. The spectrum content for the three cases is in turn shown in Fig. 6.15. Notice now the x-axis is ω_n rather than a_n. Notice also that all three cases look the same, but that the spectrum is "squeezed" for the cases with most time elongation. For example, other than the DC term being equal, the first harmonic coefficient (around 0.6) is also the same. What differs is the location of the harmonic: for the first case it is at 2π while it is 1π and $\frac{\pi}{2}$ for the second and third cases, respectively.

6.9 Impact of Period

Still adhering to the periodic pulse, let's examine the impact of the period. We'll start with the smaller period case, and incrementally increase it, still keeping the DC average the same. But unlike the prior section, where we increase pulse width, this time we leave pulse width the same, and instead increase pulse height. The three variants are shown in Fig. 6.16.

Again, since the signal is even, the sine terms cancel: $b_n = 0$. Also, per construction

$$a_0 = \frac{1}{2} \tag{6.21}$$

The remaining components evaluate to

$$a_{n,\text{case 1}} = 2\frac{\sin\frac{n\pi}{2}}{\pi n}, \qquad \omega_{n,\text{case 1}} = \frac{2\pi n}{1},$$
$$a_{n,\text{case 2}} = 4\frac{\sin\frac{n\pi}{4}}{\pi n}, \qquad \omega_{n,\text{case 2}} = \frac{2\pi n}{2},$$
$$a_{n,\text{case 3}} = 8\frac{\sin\frac{n\pi}{8}}{\pi n}, \qquad \omega_{n,\text{case 3}} = \frac{2\pi n}{4} \tag{6.22}$$

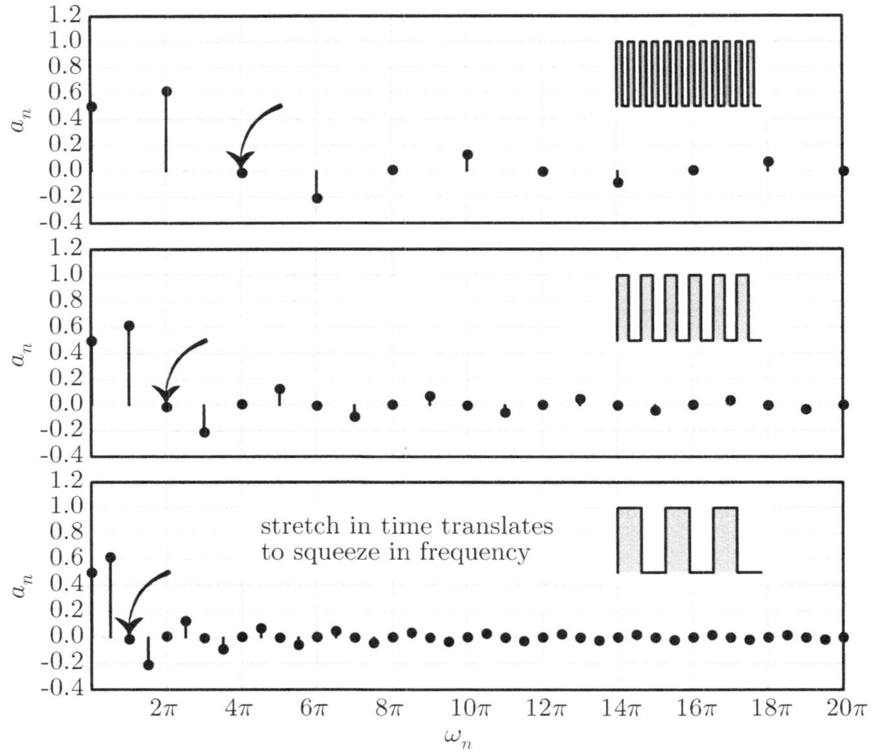

Fig. 6.15 Fourier spectrum of the three cases in Fig. 6.14

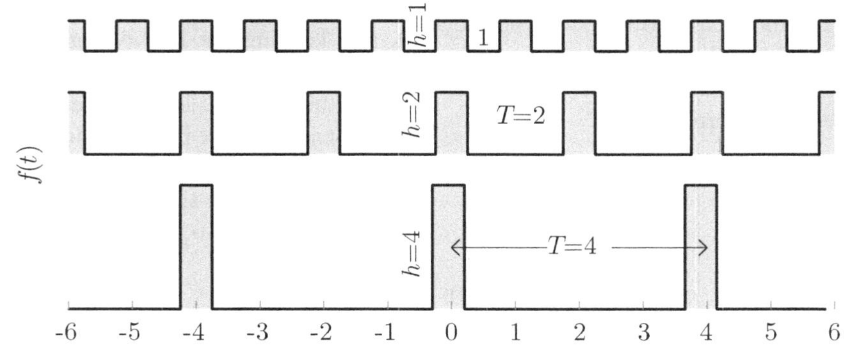

Fig. 6.16 Periodic pulse with three different periods, but with same DC average (via pulse height stretch)

The spectrum content for the three cases is in turn shown in Fig. 6.17. Notice at least three things

- Case with larger period has smaller fundamental frequency.
- Case with larger period has finer frequency increments.
- All three cases have same Fourier component at a given (shared) frequency.

We will find out later that what is happening here is that the function is becoming "aperiodic," and as it does it is "filling" the spectrum. In fact when the period becomes infinite, the spectrum is no longer discrete, but instead a continuous one

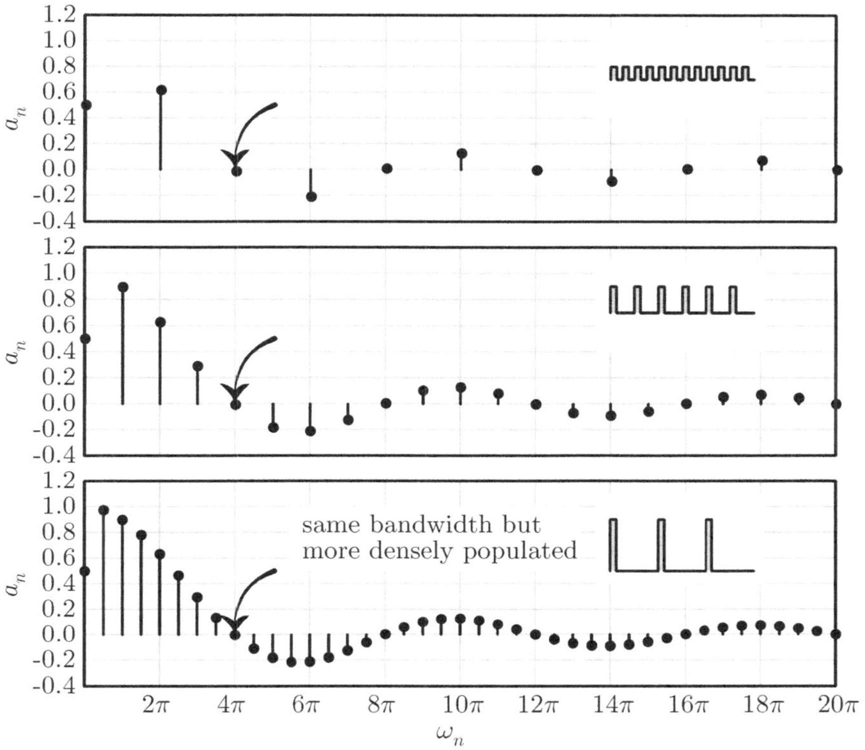

Fig. 6.17 Fourier spectrum of the three cases in Fig. 6.16

and its name is nothing other than the Fourier *transform*! More on this later.

6.10 Asymmetric Periodic Triangular Pulse

Enough of squares—let's try something different for a change! Consider the periodic triangular signal shown in Fig. 6.18; it has a width of τ and period T. Since this signal is neither even nor odd, we should expect it to have both sine and cosine Fourier components.

In particular $a_0 = \frac{\tau}{2}$,

$$a_n = \frac{2}{\tau}\left[-\frac{1-\cos\omega_n\tau}{\omega_n^2} + \tau\frac{\sin\omega_n\tau}{\omega_n}\right]$$

$$b_n = \frac{2}{\tau}\left[\frac{\sin\omega_n\tau}{\omega_n^2} - \tau\frac{\cos\omega_n\tau}{\omega_n}\right] \qquad (6.23)$$

Notice that we can calculate a_0 directly as the area under the curve, divided by the period; or by taking the limit $n \to 0$ of a_n and dividing by 2. The former case simply gives $\tau/2$ which is the area of the triangle, given by height (1 here) times half width (τ here), divided by period (1 here). The other method involving the limit must be done carefully. We will use the following approximations:

$$\cos x \sim 1 - \frac{x^2}{2}; \qquad \sin x \sim x \qquad (x \ll 1) \qquad (6.24)$$

Then the a_n terms approach

$$a_0 = \lim_{n\to 0}\frac{1}{2}\frac{2}{\tau}\left[\frac{-\omega_n^2\tau^2/2}{\omega_n^2} + \tau\frac{\omega_n\tau}{\omega_n}\right] = \frac{1}{\tau}\frac{\tau^2}{2} = \frac{\tau}{2} \qquad (6.25)$$

Figure 6.19 shows the Fourier spectrum. Figure 6.20 shows the time series reconstruction

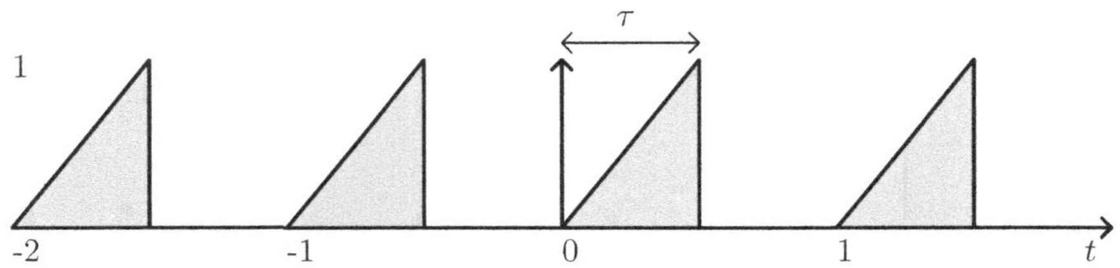

Fig. 6.18 Periodic triangular pulse

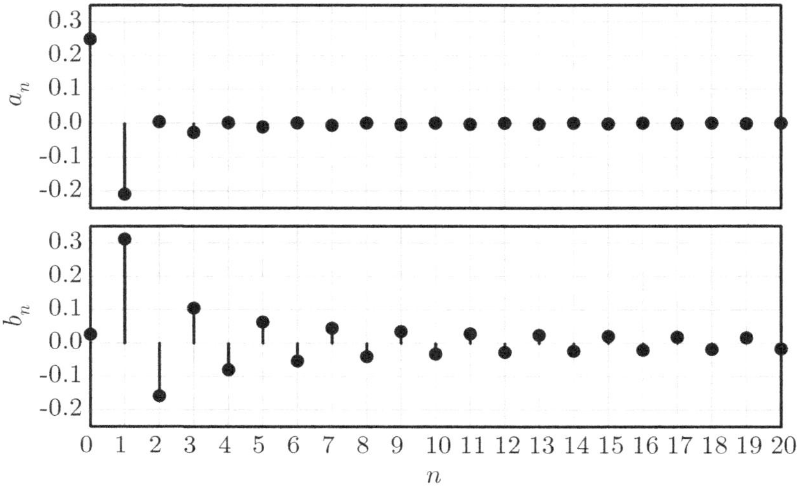

Fig. 6.19 Fourier spectrum of asymmetric triangular periodic pulse

using various number of Fourier components; as can be seen, with more Fourier components we get better signal reconstruction. Figure 6.21 shows the same thing. Figure 6.22 shows reconstruction for three τ values; again the time series resembles what we'd expected pretty well!

6.11 Symmetric Periodic Triangular Pulse

Consider next the symmetric version of the triangular pulse as shown in Fig. 6.23. Rather than doing the hard work to find the Fourier coefficients, we capitalize on last section's results. Since this new pulse is symmetric, we simply set the odd components to zero, and the even ones to *twice* the even part of last section. Remember, this pulse has twice the area per harmonic and hence the need for the two scaling! Then we would have

$$a_0 = \frac{1}{\tau}, \quad a_n = \frac{4}{\tau}\left[-\frac{1-\cos\omega_n\tau}{\omega_n^2} + \tau\frac{\sin\omega_n\tau}{\omega_n}\right] \tag{6.26}$$

Results of time series for different τ values shown in Fig. 6.24. Notice the beautiful transition between $\tau = 0.4$ and $\tau = 0.5$. Since the "cliff" is no longer needed the reconstruction is very sharp. In other words, since there is no longer much high-frequency spectrum in the signal, using the same number of reconstruction harmonics yields a much sharper reconstruction signal.

6.12 Periodic Parabolic Pulse

Consider next the periodic parabolic pulse shown in Fig. 6.25; it has a period 1 and pulse width 2τ. Since the pulse is even, b_n is zero, and by integration we arrive at

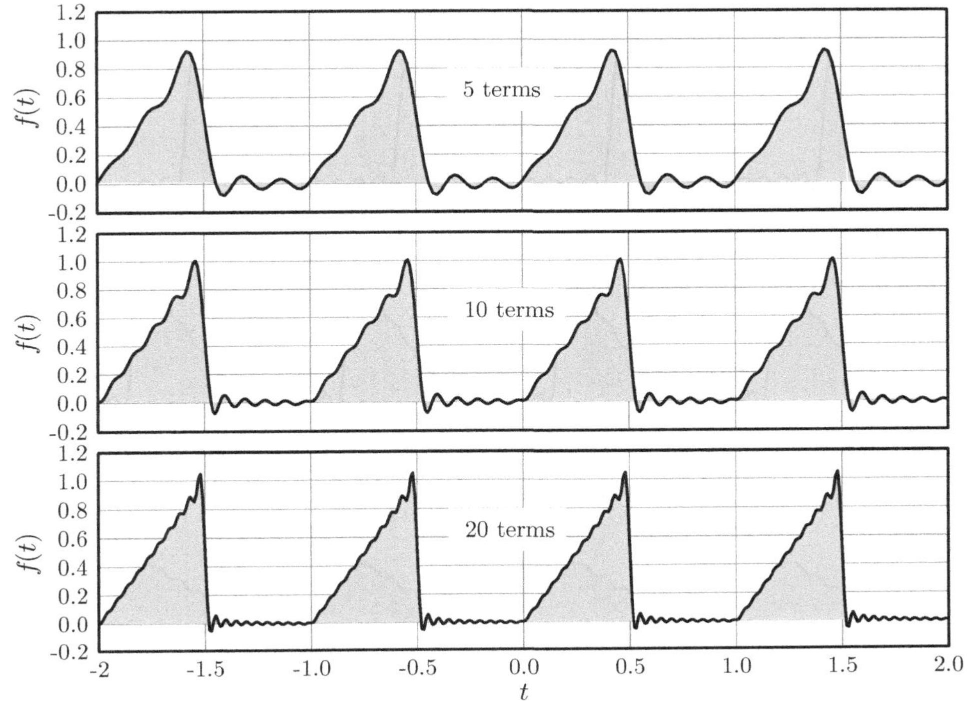

Fig. 6.20 Asymmetric periodic triangular pulse reconstruction

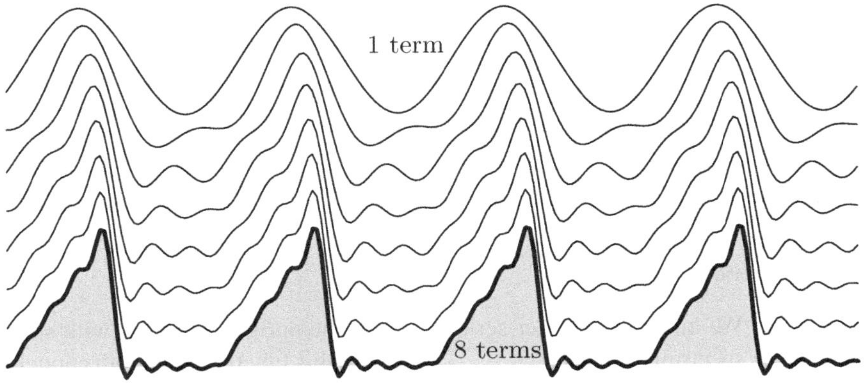

Fig. 6.21 Asymmetric periodic triangular pulse reconstruction

$$a_0 = \tau \frac{2}{3}, a_n = \frac{4}{\tau^2}$$
$$\left[\frac{-2 \sin \omega_n \tau}{\omega_n^3} + \frac{2\tau \cos \omega_n \tau}{\omega_n^2} + \frac{\tau^2 \sin \omega_n \tau}{\omega_n} \right] \tag{6.27}$$

Figure 6.26 shows time series for different τ values. Notice again the transition between $\tau = 0.4$ and $\tau = 0.5$. Because the cliff (or the high frequency content) is eliminated, using the same harmonics in the reconstruction yields a much better replica.

6.13 Periodic Inverse Parabolic Pulse

We can construct the inverse parabolic pulse shown in Fig. 6.27 by adding (negative) the parabolic one to the periodic pulse, again as

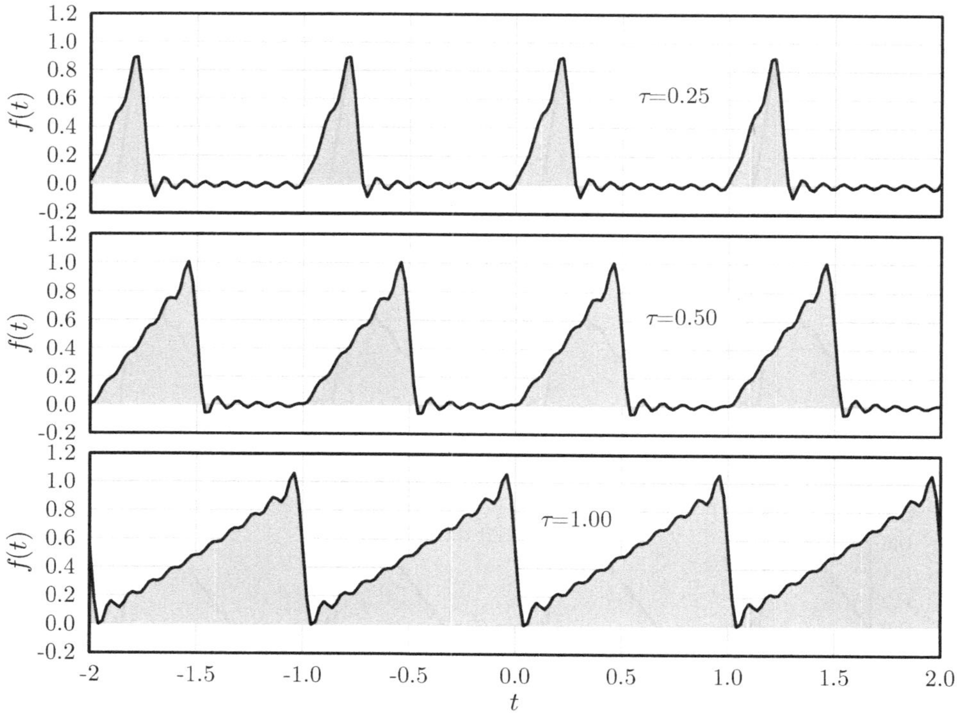

Fig. 6.22 Asymmetric periodic triangular pulse with different width

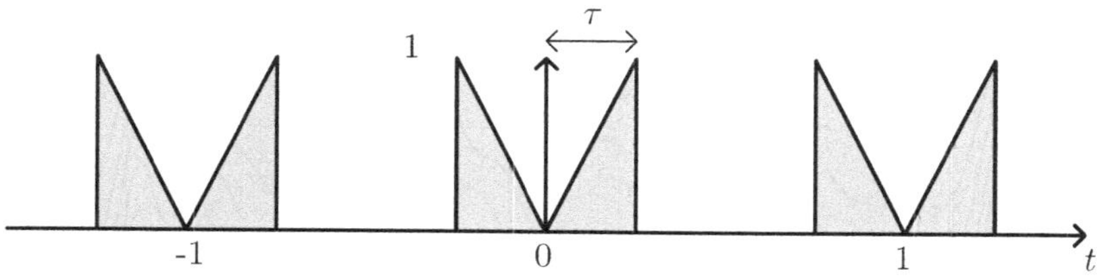

Fig. 6.23 Symmetric periodic triangular pulse

shown in the figure. We know the Fourier series for each, so our new spectrum becomes

$$a_0 = -\tau\frac{2}{3} + 2\tau$$

$$a_n = -\frac{4}{\tau^2}\left[\frac{-2\sin\omega_n\tau}{\omega_n^3} + \frac{2\tau\cos\omega_n\tau}{\omega_n^2} + \right.$$
$$\left. + \frac{\tau^2\sin\omega_n\tau}{\omega_n}\right] + 4\frac{\sin\omega_n\tau}{\omega_n} \qquad (6.28)$$

Figure 6.28 shows time series reconstruction for different τ values. Figure 6.29 shows the same thing in a different representation. Notice that unlike the upright parabola, with sharp discontinuities, this function is smooth enough that only a handful of harmonics are needed to produce good resolution.

6.14 The "Quadratic" Cosine Function

If we take the inverse parabolic function from last section, duplicate it, negate it, then shift it by half cycle; then finally add it to the original function, as shown in Fig. 6.30 we get something that resembles a cosine function. This is not a real cosine function because it only has 0 and

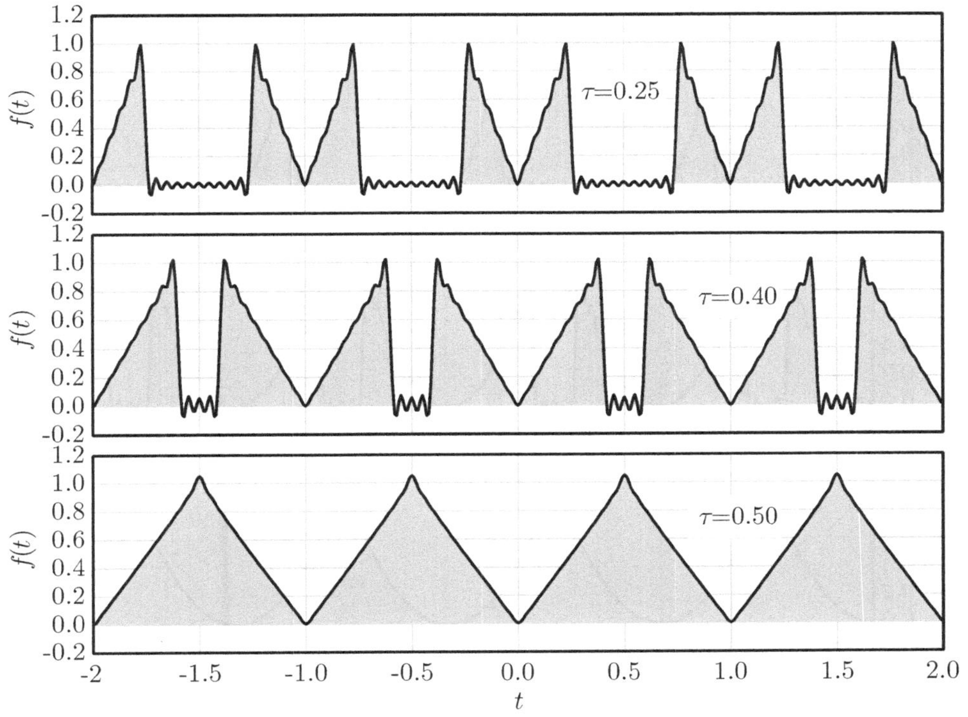

Fig. 6.24 Symmetric triangular pulse for different τ values: reconstruction using Fourier components

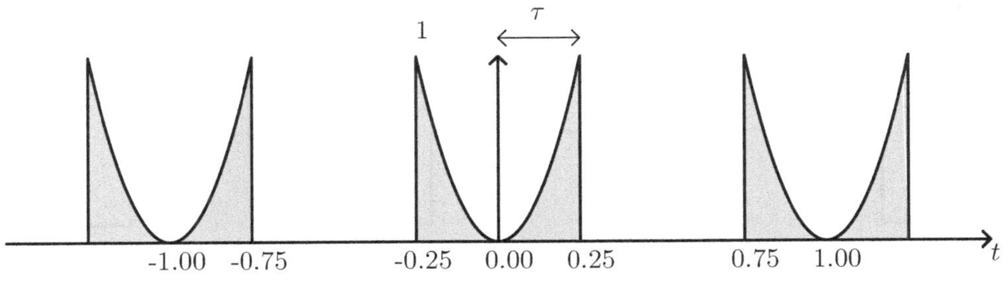

Fig. 6.25 Periodic parabolic pulse

2nd powers of x; nonetheless, we want to find its Fourier series. Let

$$a_n = -\frac{4}{\tau^2}\left[\frac{-2\sin\omega_n\tau}{\omega_n^3} + \frac{2\tau\cos\omega_n\tau}{\omega_n^2} + \frac{\tau^2\sin\omega_n\tau}{\omega_n}\right] + 4\frac{\sin\omega_n\tau}{\omega_n} \quad (6.29)$$

Then our Fourier series becomes

$$f(t) = \sum_n a_n\left[1 - \cos(\omega_n 2\tau)\right]\cos\omega_n t \quad (6.30)$$

The time series function is shown in Fig. 6.31. Also shown is a real cosine; as can be seen the resemblance is remarkable!! Remember the Taylor series expansion of the cosine:

$$\cos x = 1 - \frac{x^2}{2} + \frac{x^4}{4!} - \cdots \quad (6.31)$$

Apparently using only the first two terms in the series expansion gives good enough results!

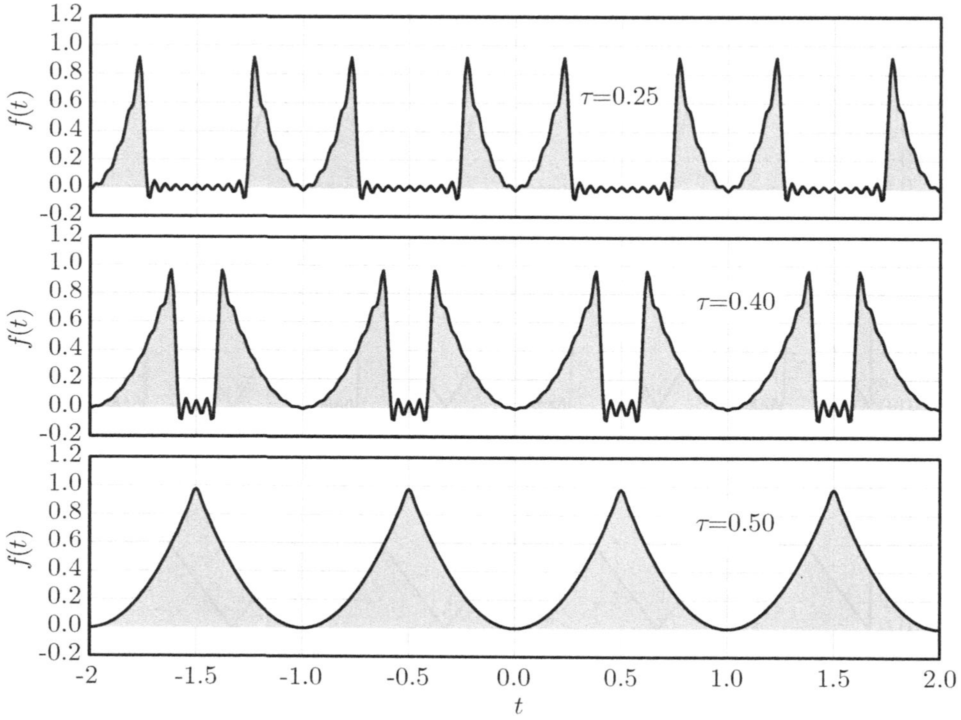

Fig. 6.26 Parabolic pulse for different τ values: reconstruction using Fourier components

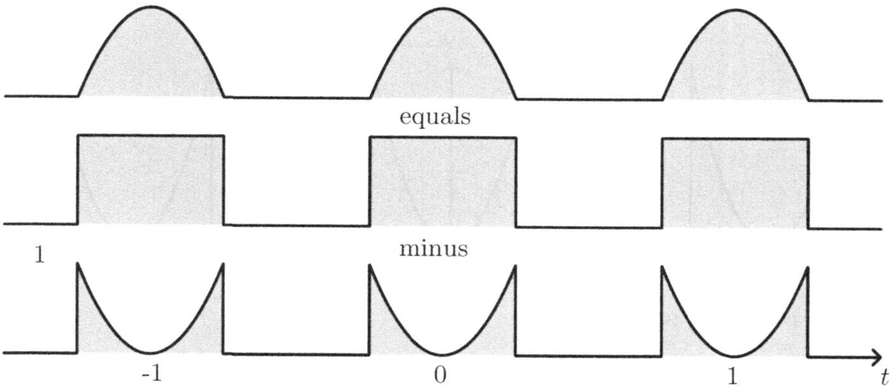

Fig. 6.27 Periodic inverse parabolic pulse and decomposition in terms of pulse and parabola

6.15 Fourier Series of Negative Exponential

Let us define the periodic negative exponential shown in Fig. 6.32 as

$$f(t) = e^{-at}, \qquad 0 < t < T$$
$$f(t) \quad \text{periodic in } T \tag{6.32}$$

This function is neither even nor odd; so we would expect both b_n and a_n. Also, its average is nonzero, so we would expect a_0 as well. By direct integration we get

$$a_0 = \frac{1}{aT}\left[1 - e^{-aT}\right],$$
$$a_n = \frac{2}{T}\frac{a\left[1 - e^{-aT}\cos\omega_n T\right]}{a^2 + \omega_n^2}, \qquad \text{and} \tag{6.33}$$

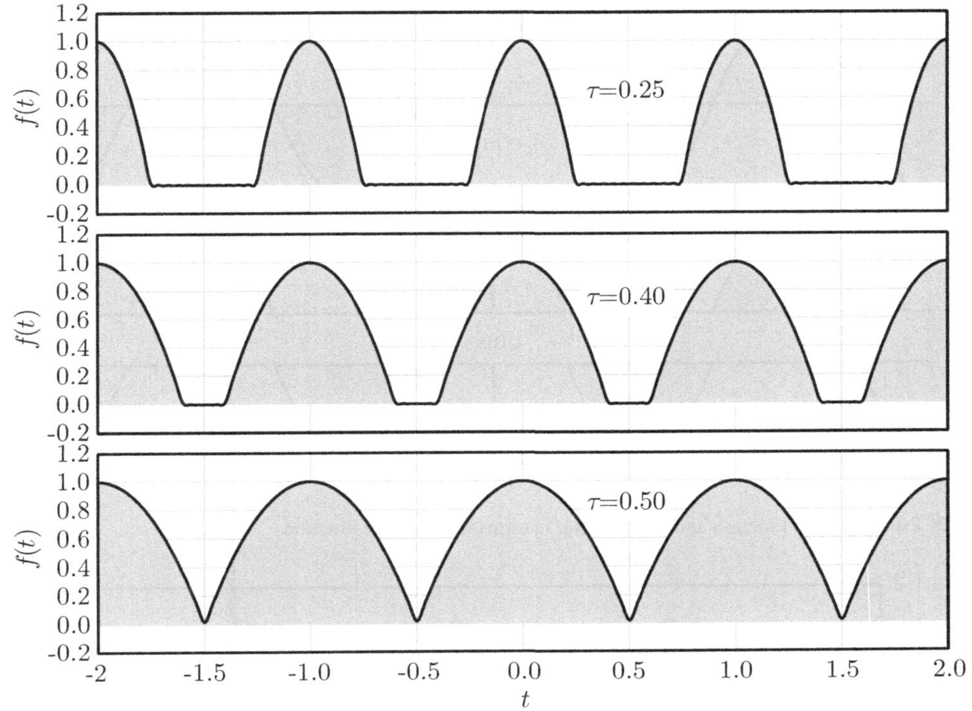

Fig. 6.28 Inverse parabolic pulse for different τ values: reconstruction using Fourier components

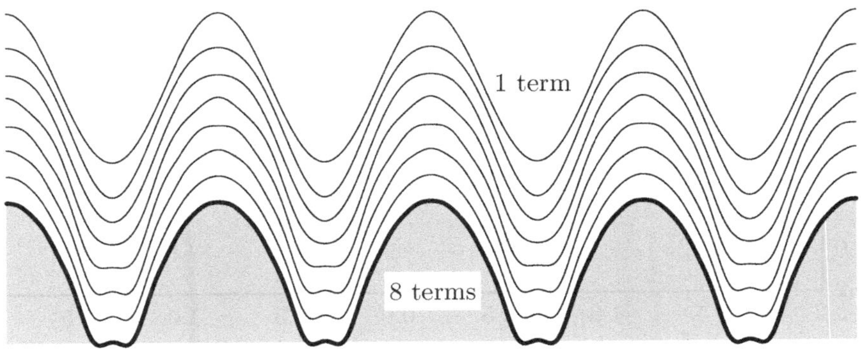

Fig. 6.29 Another way of conveying information shown in Fig. 6.28 (case $\tau = 0.4$)

$$b_n = \frac{2}{T}\frac{\omega_n\left[1 - e^{-aT}\cos\omega_n T\right]}{a^2 + \omega_n^2} \tag{6.34}$$

Then we would have

$$f(t) = a_0 + \sum_n a_n \cos\omega_n t + b_n \sin\omega_n t \tag{6.35}$$

Figure 6.33 shows the reconstruction using the Fourier components, for different a values. Figure 6.34 shows impact of adding harmonics for case of $a = -5$.

6.16 Fourier Series of Symmetric Negative Exponential

The symmetric negative exponential is shown in Fig. 6.35. The Fourier spectrum is very close to that of prior section, with some minor modifica-

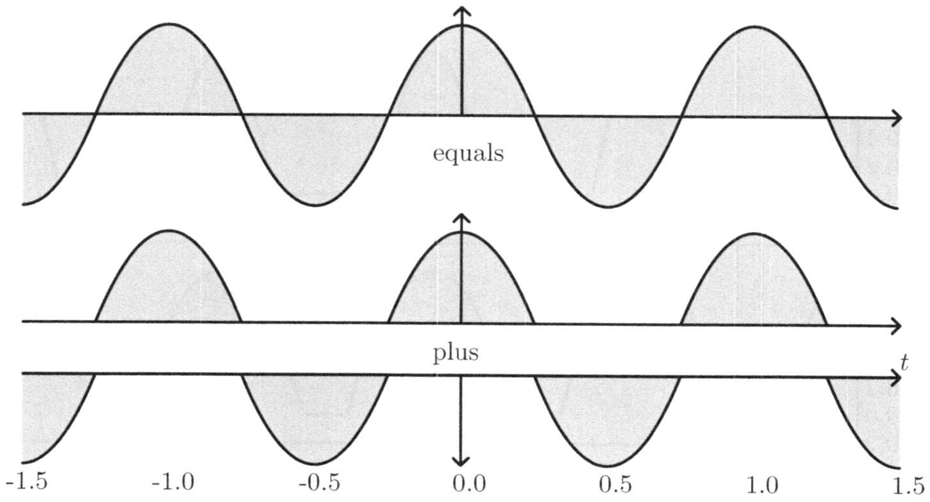

Fig. 6.30 Pseudo cosine function and composition in terms of parabolic function

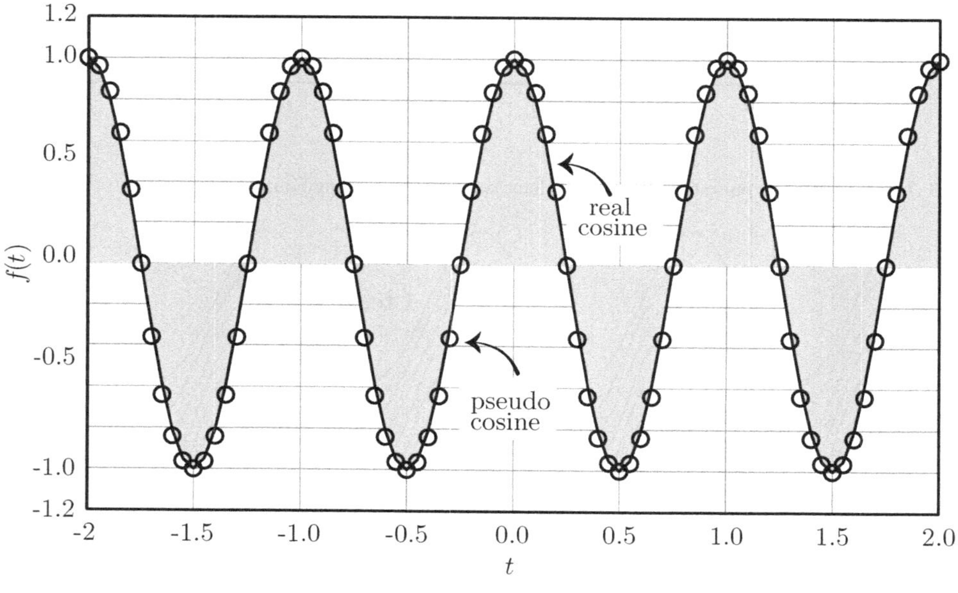

Fig. 6.31 Pseudo cosine function as a time series and a real cosine

Fig. 6.32 Periodic negative exponential

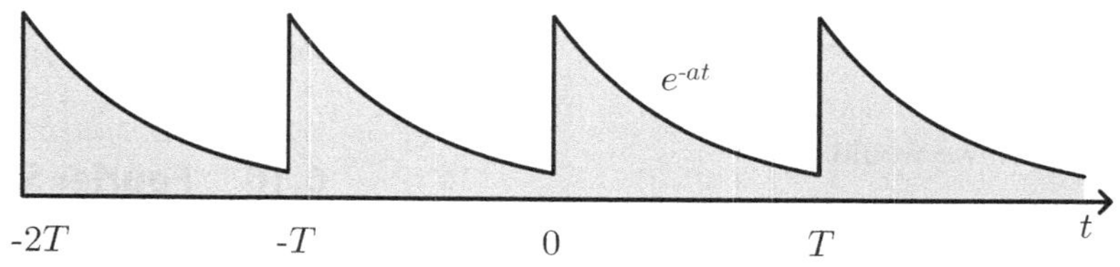

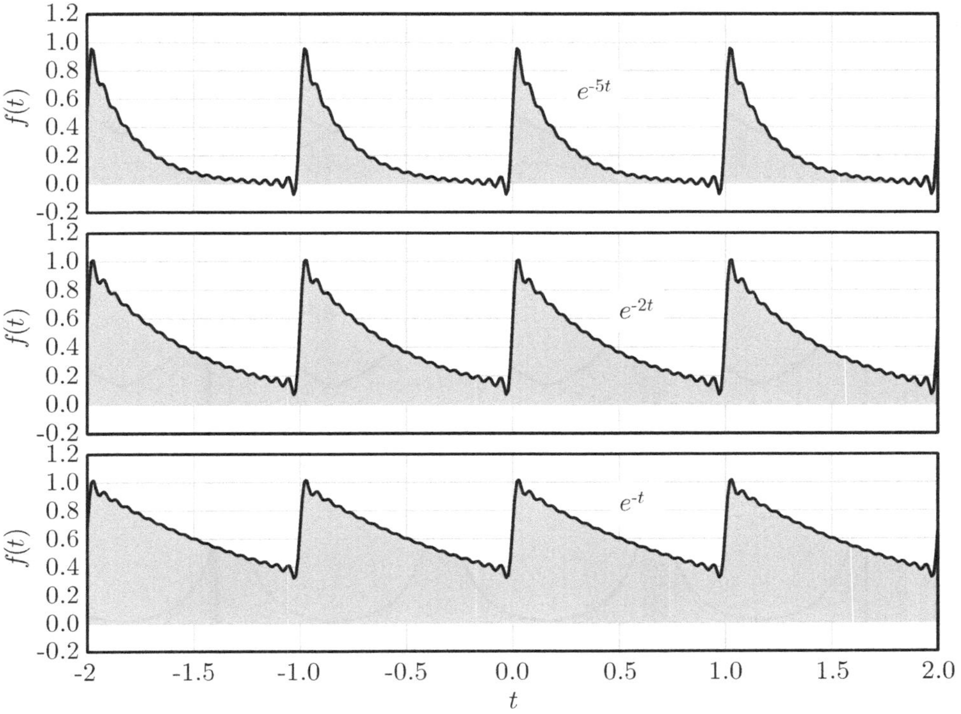

Fig. 6.33 Negative exponential and reconstruction using Fourier series, for different a values

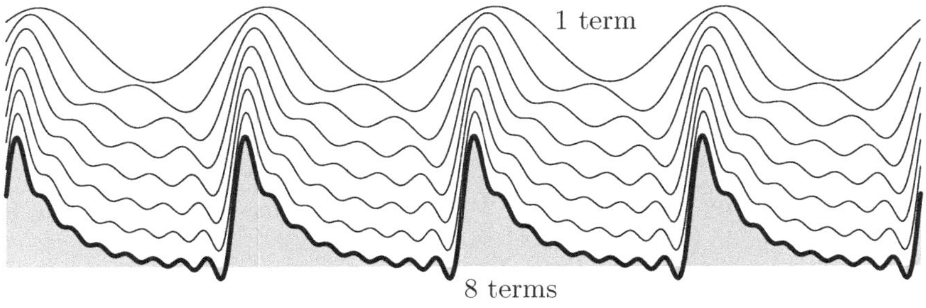

Fig. 6.34 Negative exponential and reconstruction using Fourier series (case $a = -5$)

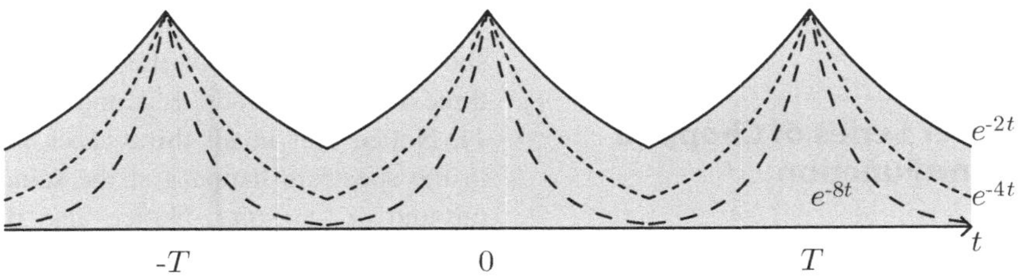

Fig. 6.35 Periodic symmetric negative exponential

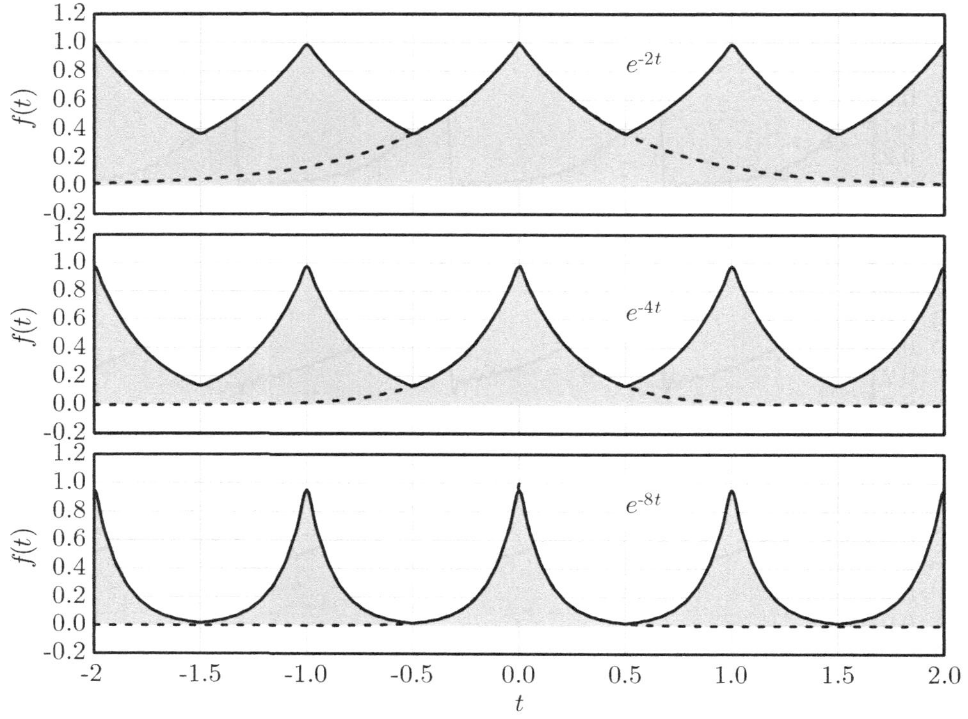

Fig. 6.36 Symmetric negative exponential function for different a values: original (dashed) and periodic version construction (solid)

tions (albeit the odd terms drop completely). In particular we have

$$a_n = \frac{4}{T}\frac{a\left[1 - e^{-aT/2}\cos\omega_n T/2\right]}{a^2 + \omega_n^2} \tag{6.36}$$

Figure 6.36 shows our results in the form of a time series, for different a values, and shows that for all three cases we get identical match to the original function. Figure 6.37 shows evolution of the time series versus number of harmonics.

6.17 Fourier Series of Chopped Cosine Function

The chopped cosine function (of frequency ω_0, or period T_0) is defined as that part of the cosine function with value above 0! It is shown as the solid line in Fig. 6.38. This particular case has a period of 1 such that $\omega_0 = 2\pi$. This function is even, so b_n=0. The DC and even harmonics are given by direct integration as:

$$a_n = \frac{2}{T}\left[\frac{\sin\left[(\omega_n - \omega_0)\frac{T_0}{4}\right]}{\omega_n - \omega_0} + \frac{\sin\left[(\omega_n + \omega_0)\frac{T_0}{4}\right]}{\omega_n + \omega_0}\right] \tag{6.37}$$

Notice that we have two periods—T and T_0; T is the period which we can set at will; T_0 is the period which defined the width of the cosine lobe (which is not varied here!). Figure 6.39 shows the spectrum of the chopped cosine for three different periods: 1, 2, and 4 (keeping $T_0 = 1$). Notice that in all three cases the first zero in the spectrum happens at the same frequency (around $\omega_n = 6\pi$). Notice also that the one with larger period has (a) lower fundamental, (b) denser spectrum, and (c) typically lower spectrum values. We'd expect lower spectrum values for larger T since we divide by T when figuring harmonics. What if we were to scale all three

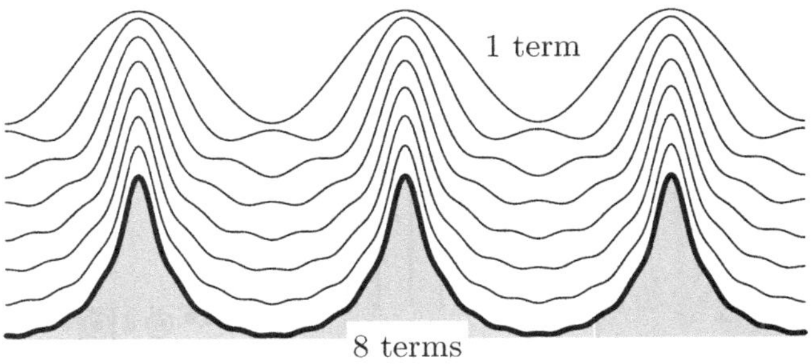

Fig. 6.37 Symmetric negative exponential function reconstruction using 1–8 harmonics

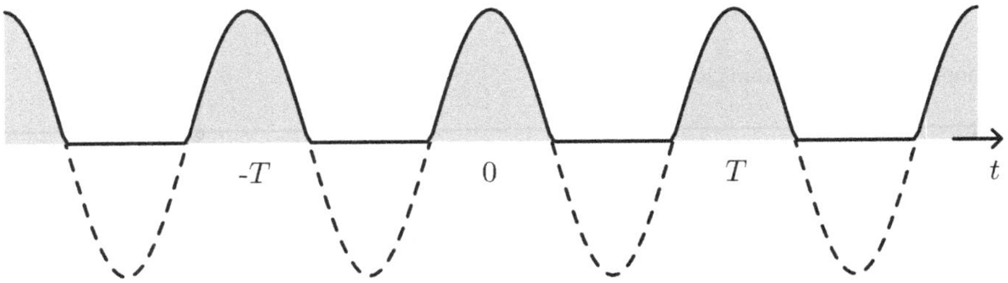

Fig. 6.38 Chopped cosine function

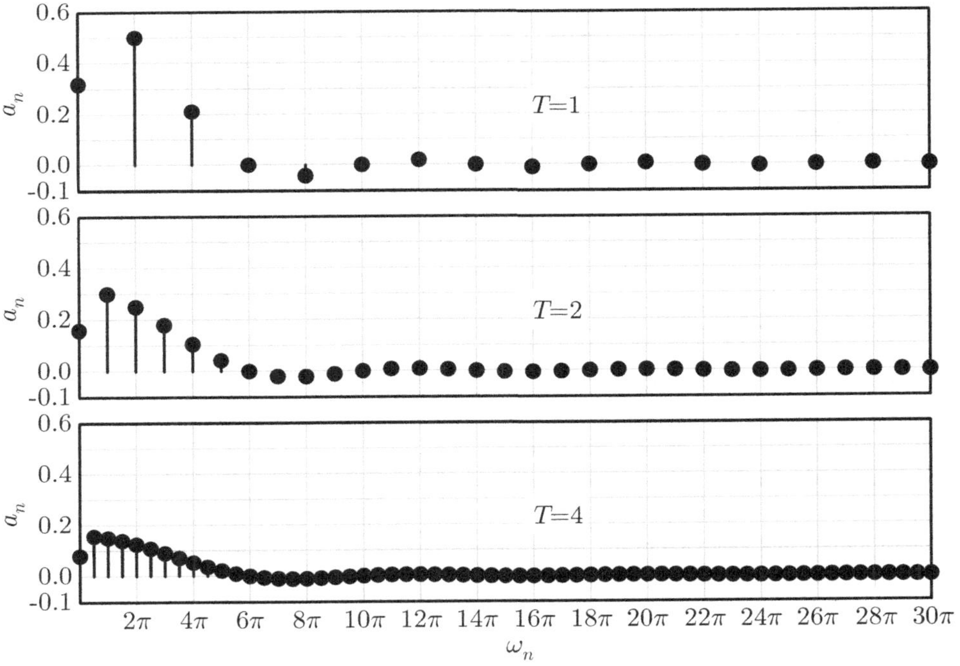

Fig. 6.39 Chopped cosine spectrum

cases by their corresponding periods? Result is shown in Fig. 6.40. Finally what if we were to introduce "negative" frequency, divide the a_n by two, and mirror them? We would see later how this can be done; but if we do that we get results in Fig. 6.41. Notice that in all three cases results are consistent and follow the same envelope.

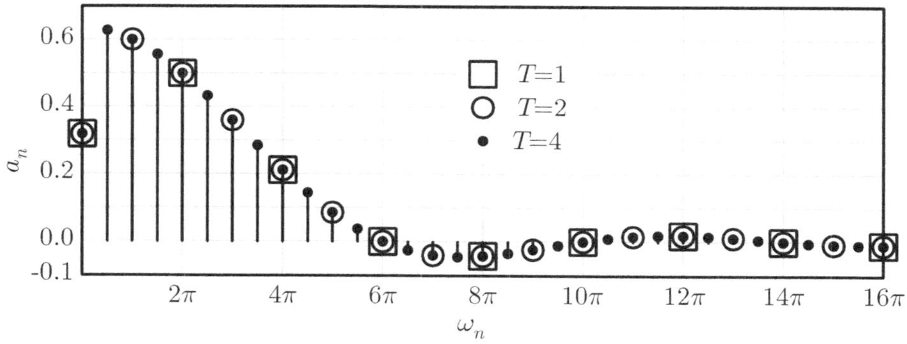

Fig. 6.40 Chopped cosine spectrum (scaled by T)

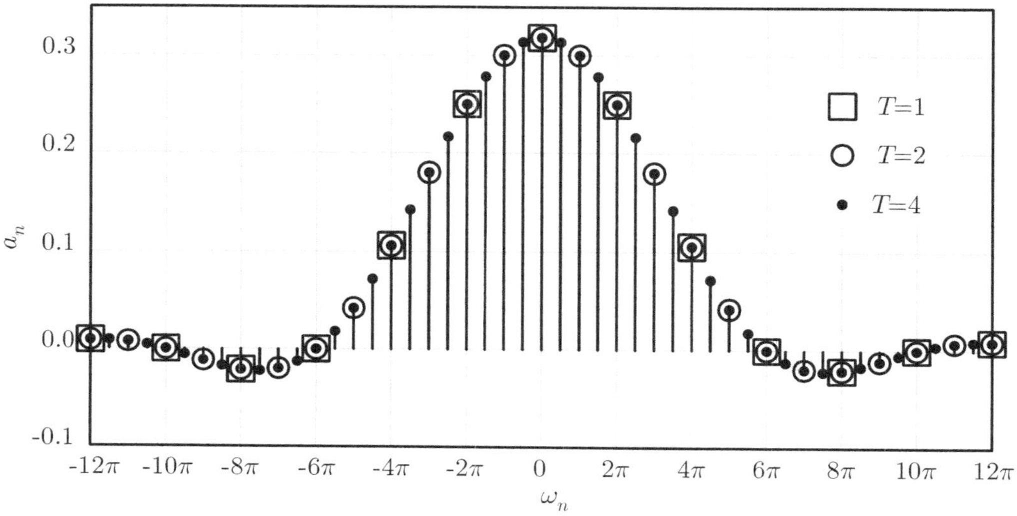

Fig. 6.41 Chopped cosine spectrum (scaled by T), divided by 2, and mirrored to accommodate negative frequency

Figure 6.42 shows sample reconstruction of the chopped cosine as a function of harmonics. Figure 6.43 (top side) shows reconstruction of time series and confirms our results. Now we can arbitrarily change the period of the positive pulse, but still keep the width of the pulse as half of the original period; this is shown in middle and bottom parts of the figure.

6.18 Fourier Series of Absolute Value of Cosine Function

The absolute cosine function (again of frequency ω_0 and period T_0) is shown in Fig. 6.44. Rather than do the integration work, we utilize the work of the prior section and simply set T to half T_0! Then we have

$$a_n = \frac{4}{T_0}\left[\frac{\sin(\omega_n - \omega_0)\frac{T_0}{4}}{\omega_n - \omega_0} + \frac{\sin(\omega_n + \omega_0)\frac{T_0}{4}}{\omega_n + \omega_0}\right] \tag{6.38}$$

Figure 6.45 shows that surprisingly using only a handful of harmonics we are able to duplicate the original function rather well!

6.19 Fourier Series of Cosine Squared

The cosine squared function (of frequency ω_0 and period T_0) is shown in Fig. 6.46. Notice that how it differs from the absolute cosine (also shown in

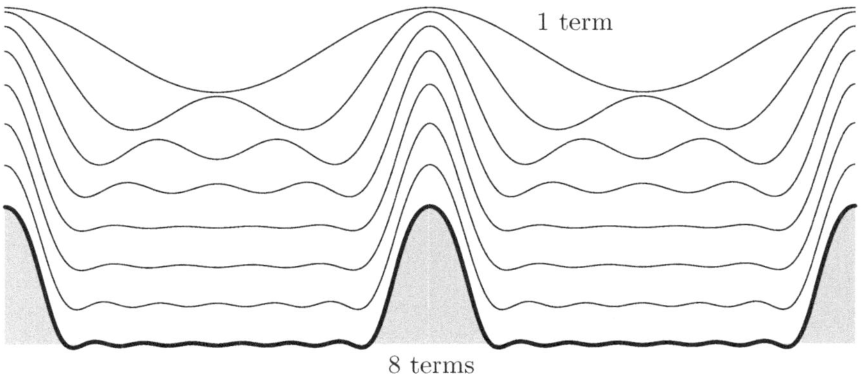

Fig. 6.42 Chopped cosine reconstruction using 1–8 harmonics

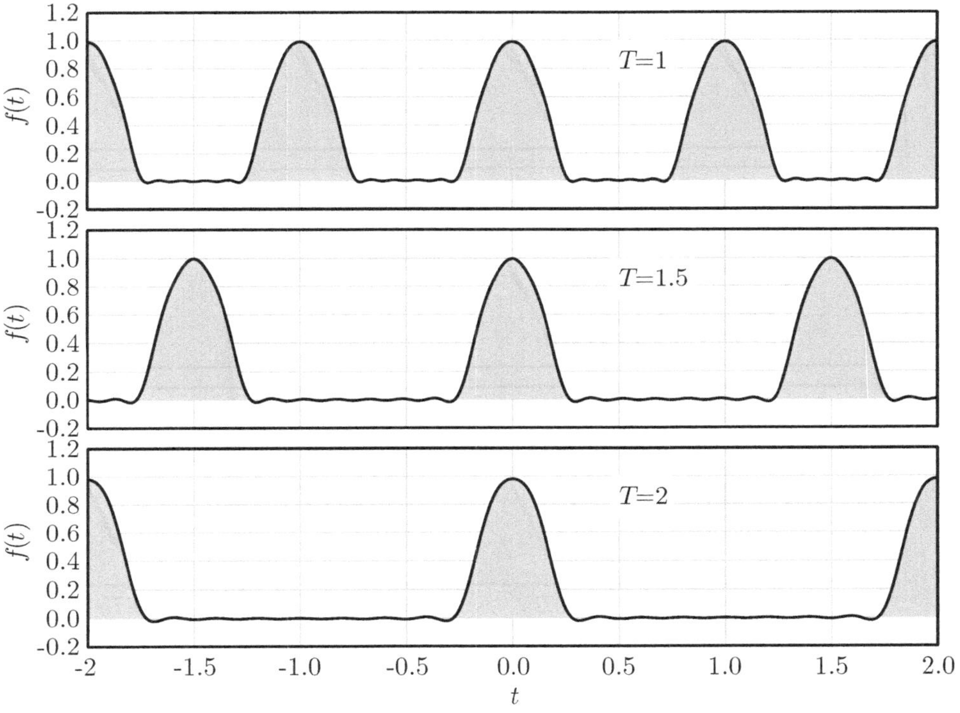

Fig. 6.43 Chopped cosine (top) and period alterations (middle and bottom)

figure), especially at the inflection points. We can easily get the spectrum by recalling that

$$\cos A \cos B = \frac{1}{2}\left[\cos(A+B) + \cos(A-B)\right] \tag{6.39}$$

In our case we get

$$\cos^2 \omega_0 t = \frac{1}{2}\left[1 + \cos 2\omega_0 t\right] \tag{6.40}$$

Notice that this already is in the form of a Fourier series; in particular we have

$$b_n = 0; \qquad a_0 = \frac{1}{2}; \qquad a_1 = 0; \qquad a_2 = \frac{1}{2}; \qquad a_{n>2} = 0 \tag{6.41}$$

Notice that we treated the function as being periodic in $T = 2\pi/\omega_0$, but we could have also

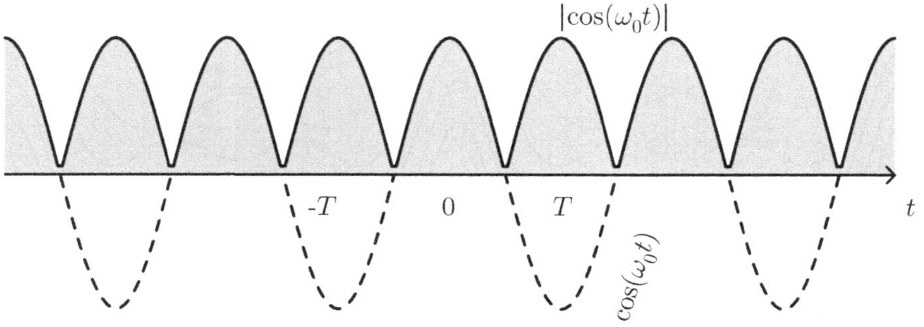

Fig. 6.44 Absolute cosine function

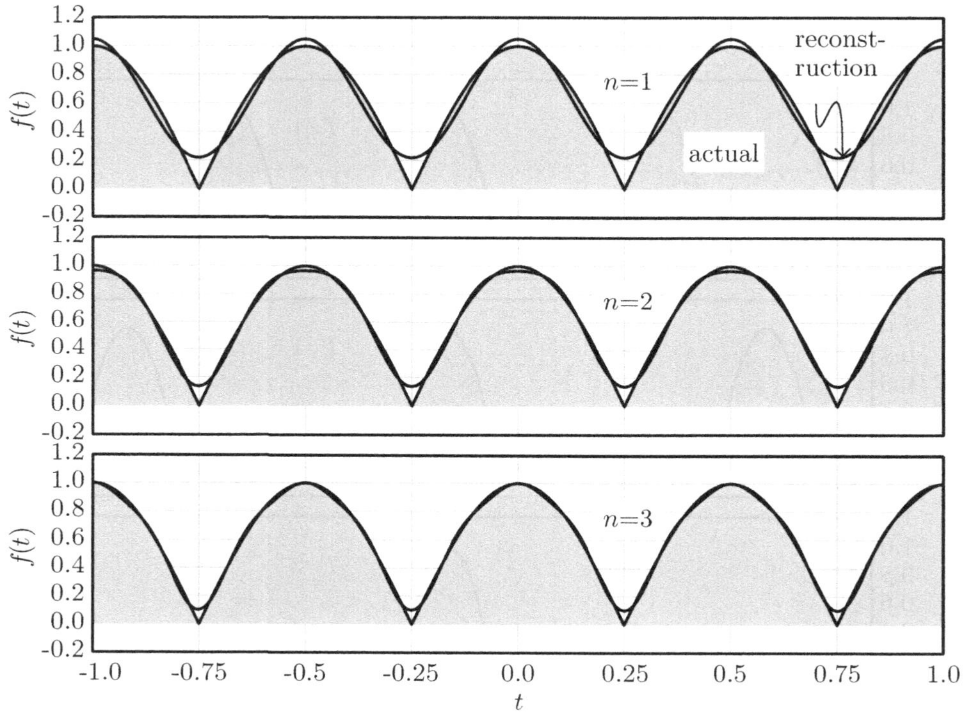

Fig. 6.45 Absolute cosine function reconstruction using 1, 2, and 3 harmonics

Fig. 6.46 Cosine squared function

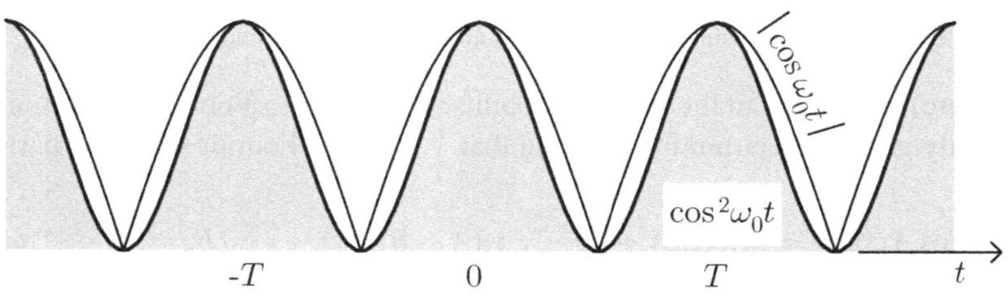

treated it as being periodic in half that period! Either way, results should come out the same. This Fourier spectrum for the case $\omega_0 = 2\pi$ is shown in Fig. 6.47. Notice that the spectrum of this signal is as simple as that: a DC component and one at 4π.

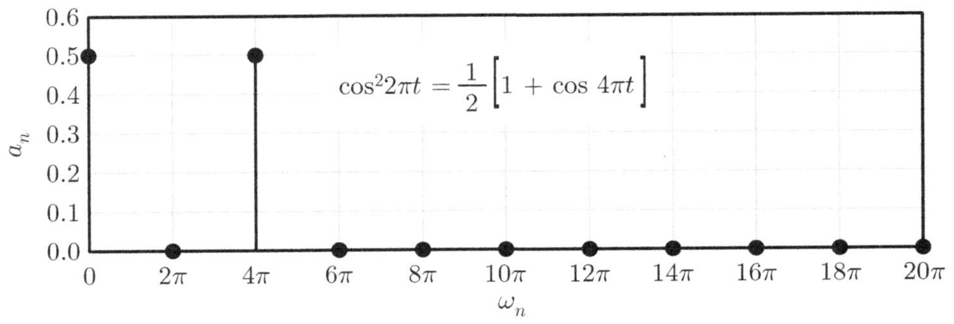

Fig. 6.47 Fourier spectrum of cosine squared (of frequency $\omega_0 = 2\pi$)

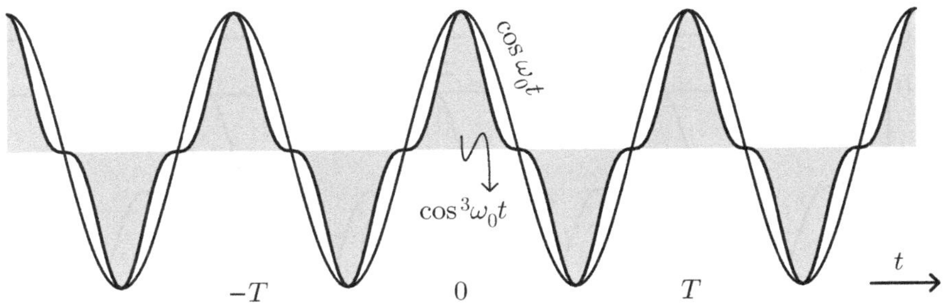

Fig. 6.48 Cosine cubed function

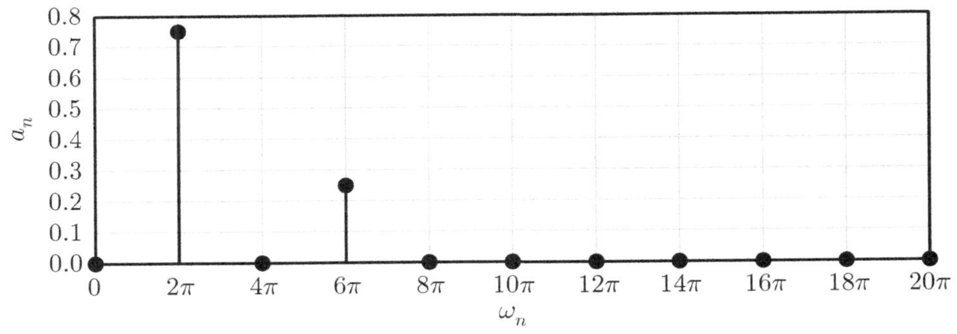

Fig. 6.49 Fourier spectrum of cosine cubed function; case of $\omega_0 = 2\pi$

6.20 Fourier Series of Cosine Cubed

Next is the cosine function, cubed (of frequency ω_0 and period T_0) as shown in Fig. 6.48. Again using trig identities we get

$$\cos^3 \omega_0 t = \frac{1}{2}\cos \omega_0 t + \frac{1}{4}\cos \omega_0 t + \frac{1}{4}\cos 3\omega_0 t \tag{6.42}$$

Notice that the average of this function is zero, so $a_0 = 0$; also since it is even then $b_n = 0$.

Then our series becomes

$$a_1 = \frac{3}{4}, \qquad a_3 = \frac{1}{4} \tag{6.43}$$

with all other components identically zero. Figure 6.49 shows the Fourier spectrum. Figure 6.50 shows our results and the constituents terms.

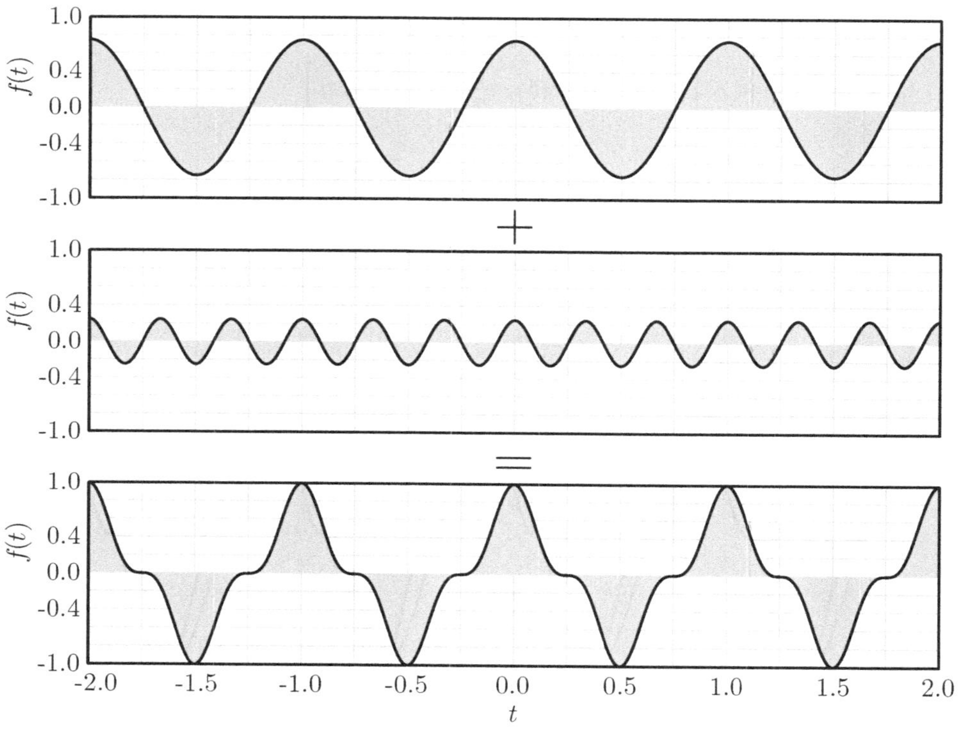

Fig. 6.50 Cosine cubes (bottom) and its constituents

6.21 Patching Signals

We already know that the Fourier series of the centered pulse (of width τ and period T) is

$$a_{n_1} = \frac{4}{T}\frac{\sin \omega_n \tau/2}{\omega_n} \tag{6.44}$$

We also know that the Fourier series of the chopped cosine (whose lobe is defined in terms of T_0) and is periodic in T is

$$a_{n_2} = \frac{2}{T}\left[\frac{\sin(\omega_n - \omega_0)\frac{T_0}{4}}{\omega_n - \omega_0} + \frac{\sin(\omega_n + \omega_0)\frac{T_0}{4}}{\omega_n + \omega_0}\right] \times \cos \omega_n \tau \tag{6.45}$$

(Notice that we added a $\cos \omega_n \tau$ since we want to shift this signal to the right by τ; more on this later in the book when time shifting properties are studied.) What happens if we add both spectrum?

$$a_n = a_{n_1} + a_{n_2} \tag{6.46}$$

The result is shown in the bottom of Fig. 6.51 and compared to the starting two signals. Sure enough, the time series of the new spectrum results in the sum of the individual signals!

6.22 Summary

This chapter is the first in studying spectral techniques which start with the basic Fourier series. From now on, and for the following few chapters the emphasis will be on harmonic functions (sines/cosines) and on how to use them to build other functions. The whole premise is that if we can decompose a signal in terms of sines/cosines (i.e., obtain the its spectrum) and if we know the system solution to sines/cosines, then we know the system solution to the new signal. It is natural to start the spectral analysis journey with periodic functions, and that's why this chapter started with the Fourier series of periodic functions. We examined a good amount of signals, ranging from the periodic pulse, to the triangular

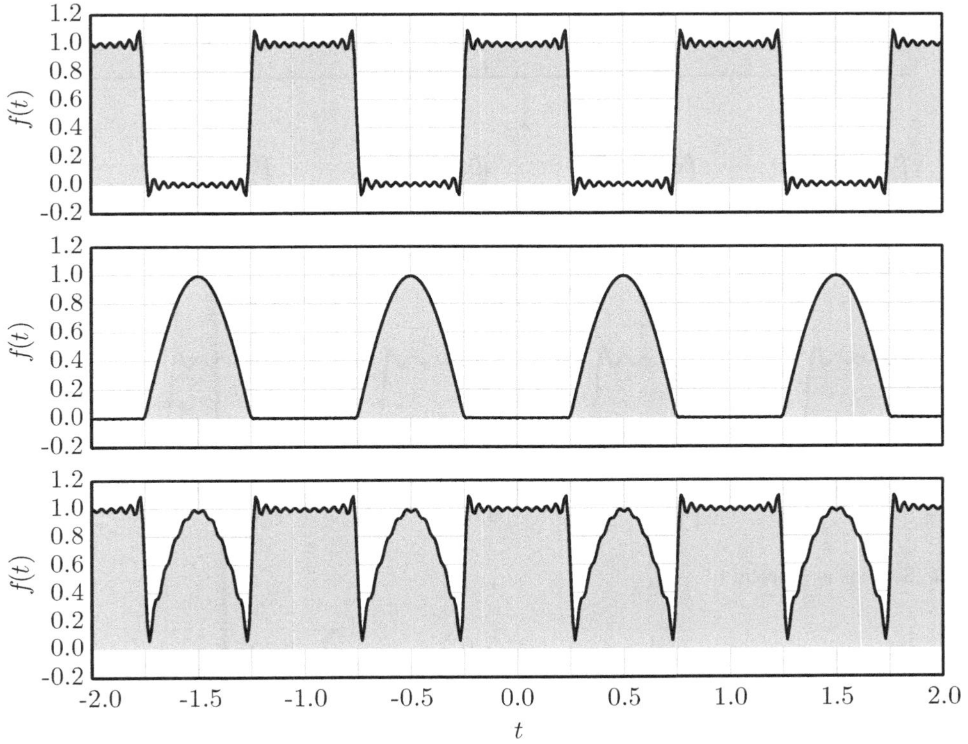

Fig. 6.51 Summation of pulse (top) and chopped signal (middle) as obtained by adding their frequency spectrum then going back to time series (bottom)

one and into the quadratic one. We studied the impact of changing pulse width, period, and both on signal spectrum. We also studied impact of signal shifting and signal stitching. We studied the meaning of spectrum plots, the even/odd properties, and the DC term. And we learned how to take the spectrum and use it to rebuild the time series. The next stop will be the complex Fourier series.

6.23 Problems

1. The periodic pulse of width 0.5 and period 1, as shown back in Fig. 6.8, has the Fourier coefficients shown back in Eq. (6.13). Plot the even components and odd ones, then the sum; see sample results in Fig. 6.52.
2. Fill in the steps to derive the Fourier coefficients in Eq. (6.23). For the case $\tau = 0.5$, plot the even, odd, and total solution. See sample solution in Fig. 6.53.
3. Find the Fourier series of the notched square shown in Fig. 6.54 without doing any explicit integration. Use only the general formula for a pulse of width 2τ and period T such that

$$a_0 = \frac{2\tau}{T}, \qquad \text{and}$$

$$a_n = \frac{4}{T}\frac{\sin \omega_n \tau}{\omega_n}$$

Hint: decompose the signal in terms of two squares as shown in the figure. See sample solution at the bottom of Fig. 6.54.

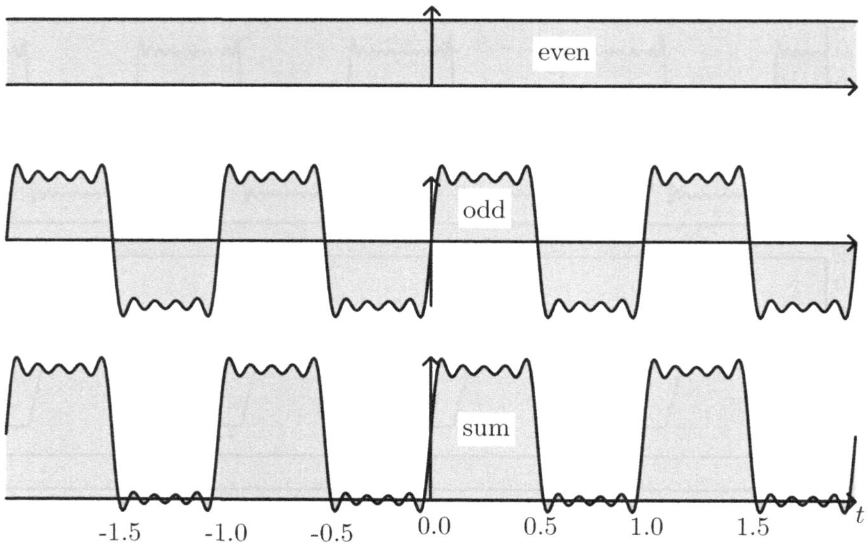

Fig. 6.52 Solution to Problem 1

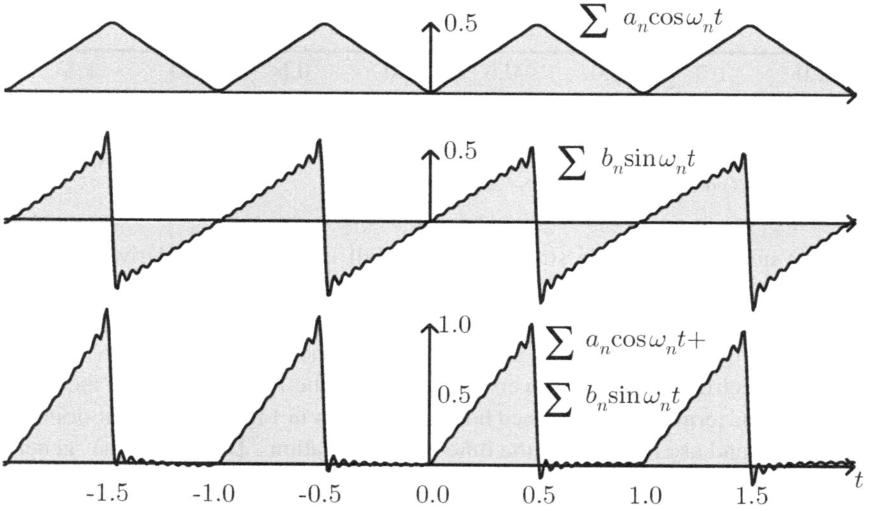

Fig. 6.53 Solution to Problem 2

4. Fill in the steps to generate Eq. (6.27), repeated below for convenience, for the Fourier coefficients of the parabolic pulse. For the case $T = 1$ and $\tau = 0.25$, plot each of the three terms separately, then the sum. See sample results in Fig. 6.55

$$a_0 = \tau\frac{2}{3},$$

$$a_n = \frac{4}{\tau^2}\left[\frac{-2\sin\omega_n\tau}{\omega_n^3} + \frac{2\tau\cos\omega_n\tau}{\omega_n^2} + \frac{\tau^2\sin\omega_n\tau}{\omega_n}\right]$$

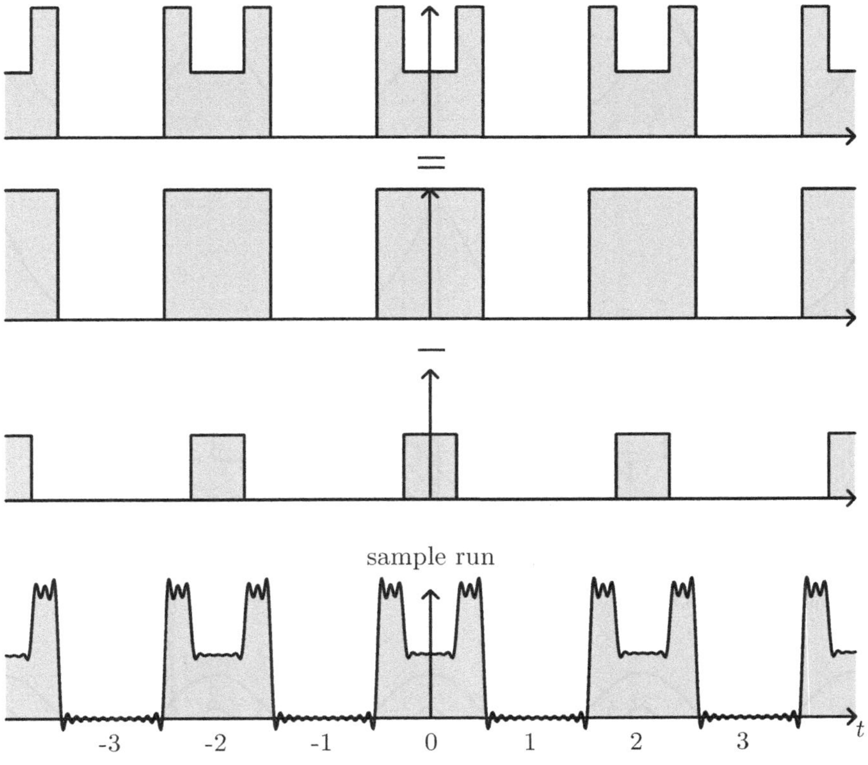

Fig. 6.54 Solution to Problem 3

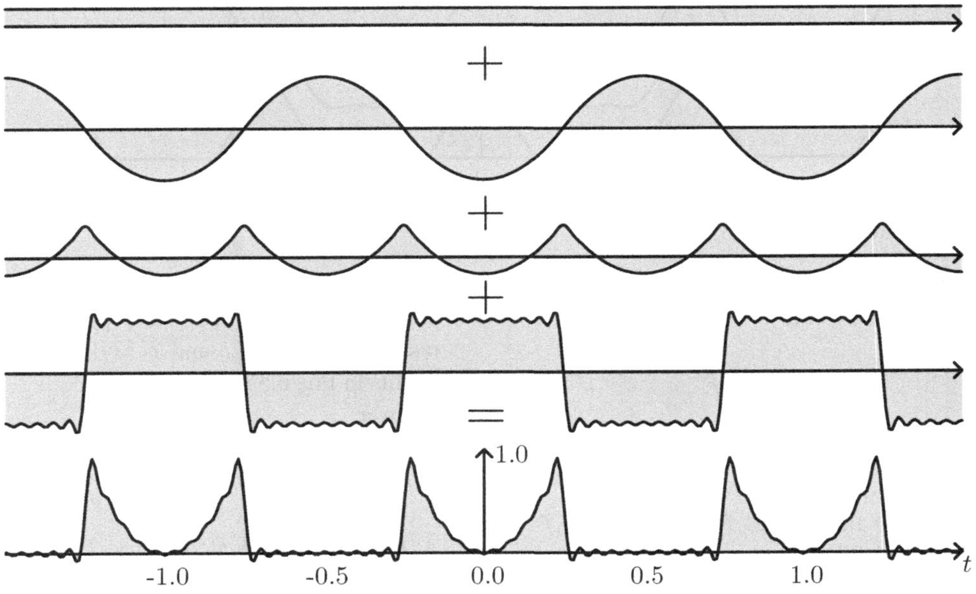

Fig. 6.55 Solution to Problem 4

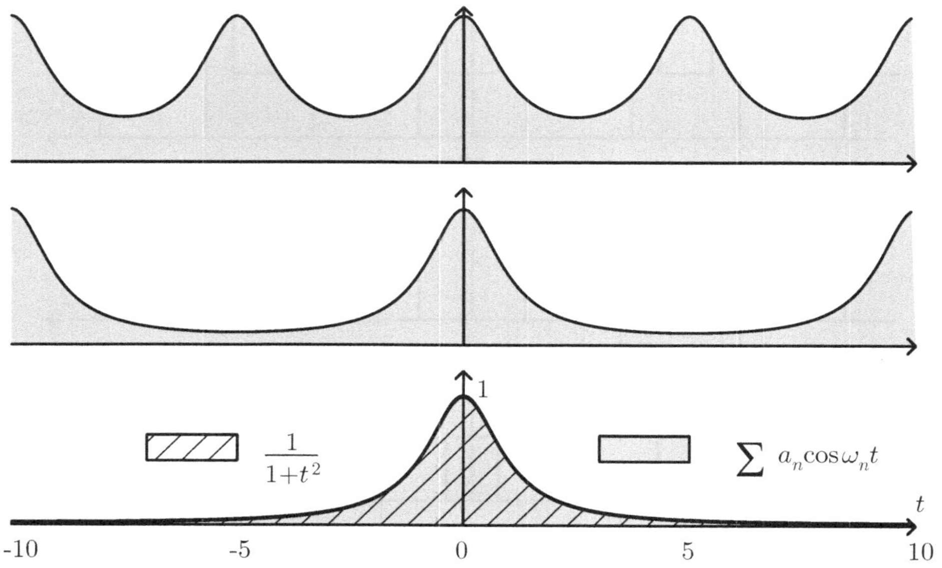

Fig. 6.56 Solution to Problem 5

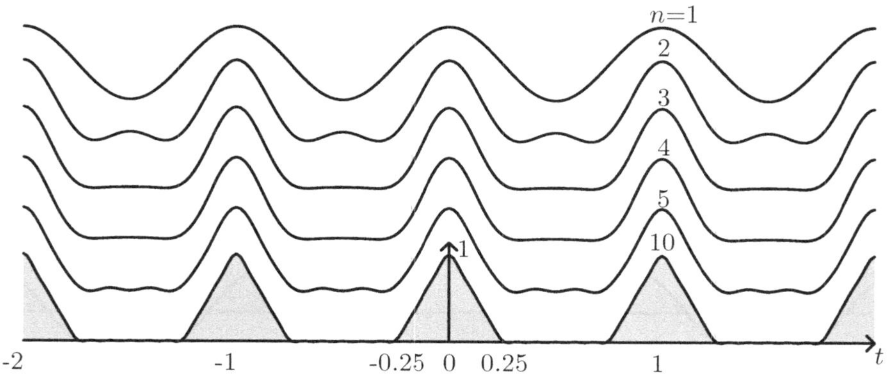

Fig. 6.57 Solution to Problem 6

5. A time function is periodic in T and has the Fourier coefficients

$$a_0 = \frac{\pi}{T}$$

$$a_n = \frac{2\pi}{T} e^{-\omega_n}$$

Plot the time series for three cases: $T = 5$, 10, and 20; compare last case to a time function with definition

$$f(t) = \frac{1}{1+t^2}$$

See sample results in Fig. 6.56.

6. Find Fourier series of symmetric, upright triangle of total width 2τ and period T. Plot results for $\tau = 0.25$ and $T = 1$. See sample results in Fig. 6.57.
Answer:

$$a_0 = \frac{\tau}{T}, \quad a_n = \frac{4}{\tau T} \frac{1 - \cos\omega_n \tau}{\omega_n^2}$$

7. A function is repeated every T and is defined as 1 throughout the period, as shown in Fig. 6.58. Show that the Fourier coefficients are

$$a_0 = 1, \qquad b_n = 0, \qquad a_n = 0 \tag{6.47}$$

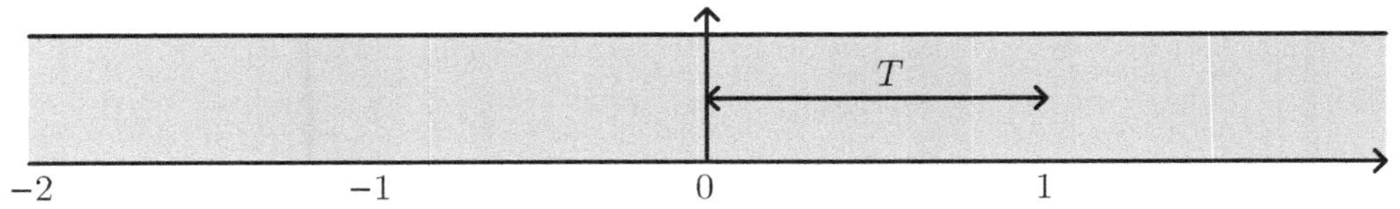

Fig. 6.58 Periodic function defined as 1 throughout (Problem 7)

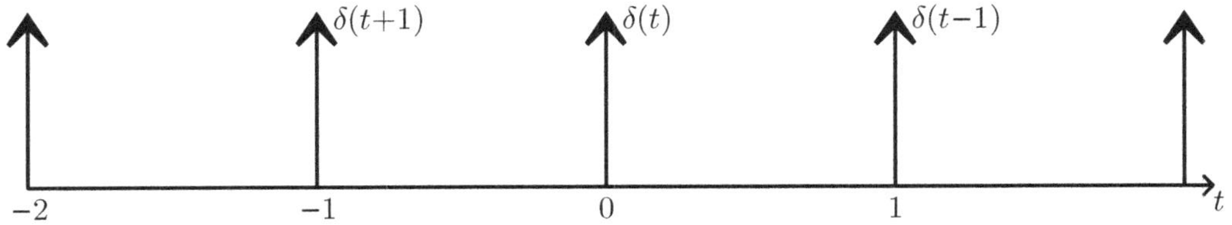

Fig. 6.59 Periodic delta function (Problem 8)

8. Show the steps in deriving the Fourier coefficients of the periodic delta function, of period T, as shown in Fig. 6.59. The delta function can be thought of as a very tall rectangle with very small width, but with unity area.

$$a_0 = \frac{1}{T}, \qquad a_n = \frac{2}{T}, \qquad b_n = 0$$

7 Complex Fourier Series

7.1 Introduction

In the prior chapter we chose as basis the sine and cosine functions, both of which are real. We ended up with something like this:

$$f(t) = a_0 + \sum_{n=1}^{\infty} a_n \cos \omega_n t + \sum_{n=1}^{\infty} b_n \sin \omega_n t \quad (7.1)$$

Notice that in this expansion we have to worry about two constants (or sequences): a_n and b_n. Wouldn't be nice if we were to worry about a single constant, say c_n? As will be shown below, we can do that, but in order to achieve that we'd need to resort into complex representation.

7.2 Orthogonality of Complex Harmonics

The main idea behind Fourier series is expanding a function in terms of a set of orthogonal functions. So far, the orthogonal functions have been the sine and cosine function. As we will show now, the complex exponential set is also orthogonal over its period. Start with the complex exponential periodic in T and defined as

$$C_n(t) = e^{j\omega_n t}, \qquad \omega_n = \frac{2\pi n}{T} \quad (7.2)$$

To show that this function is orthogonal we multiply it by another complex exponential, with different frequency and integrate over the period. The other complex exponential is defined by

$$C_m(t) = e^{j\omega_m t}, \qquad \omega_m = \frac{2\pi m}{T} \quad (7.3)$$

When we form the product we get

$$\begin{aligned} p_{nm}(t) &= e^{j\omega_n t} e^{j\omega_m t} \\ &= e^{j(\omega_n + \omega_m)t} \end{aligned} \quad (7.4)$$

Notice that the product is also periodic in T, independent of m and n; as such, when integrated over the period it would give zero

$$\int_T p_{nm}(t) = 0, \qquad m \neq -n \quad (7.5)$$

The only time the product gives a nonzero integral is when $m = -n$; then we have

$$\begin{aligned} p_{n,-n}(t) &= e^{j\omega_n t} e^{j\omega_{-n} t} \\ &= 1 \end{aligned} \quad (7.6)$$

and the integral over the period would give

$$\int_T p_{n,-n}(t) = \int_T 1 \cdot dt = T \quad (7.7)$$

F. Badrieh, *Spectral, Convolution and Numerical Techniques in Circuit Theory*,
https://doi.org/10.1007/978-3-319-71437-0_7

So what we have shown is that

1. The complex exponentials are orthogonal over a period; and
2. To form the orthogonality relationship, we need negative indices (on top of positive ones).

7.3 Derivation of the Complex Fourier Series

Having established that the complex exponentials form an orthogonal set (including negative indices) we now assume that we can expand an arbitrary periodic function (in T) in terms of those complex exponentials

$$\boxed{f(t) = \sum_{n=-\infty}^{\infty} c_n e^{j\omega_n t}, \qquad \omega_n = \frac{2\pi n}{T}} \tag{7.8}$$

To find the expansion terms c_n we multiply both sides by $e^{j\omega_m t}$ and integrate over time

$$\begin{aligned} \int_T f(t) e^{j\omega_m t} dt &= \int_T e^{j\omega_m t} \sum_{n=-\infty}^{\infty} c_n e^{j\omega_n t} dt, \qquad \omega_n = \frac{2\pi n}{T} \\ &= \sum_{n=-\infty}^{\infty} c_n \int_T e^{j(\omega_m + \omega_n)t} dt \end{aligned} \tag{7.9}$$

We already know that the right-hand side of the above equation will be zero except when $m = -n$, at which point we get

$$\int_T f(t) e^{-j\omega_n t} dt = T c_n \tag{7.10}$$

or

$$\boxed{c_n = \frac{1}{T} \int_T f(t) e^{-j\omega_n t} dt} \tag{7.11}$$

7.4 Relationship Between Complex and Real Fourier Series

It may appear that we will face some problems as a result of using *complex* series expansion to *real* functions; after all, if our starting function is real, its expansion must come out real. Also, we already know that the real Fourier series does work for real functions, so if things were to add up in the end, there must be a relation between the complex and real Fourier series; that is, we should be able to find a key mapping between the two. First, recall the real FS

$$f(t) = \sum_{n=0}^{\infty} a_n \cos \omega_n t + \sum_{n=1}^{\infty} b_n \sin \omega_n t \tag{7.12}$$

Next, recall the complex FS

$$f(t) = \sum_{n=-\infty}^{\infty} c_n e^{j\omega_n t} \tag{7.13}$$

Right from the start we notice the following differences:

1. The real FS has two expansion coefficients (a_n and b_n), while the complex FS has only one (c_n).
2. The real FS includes only positive n indices (including zero), while the complex FS includes both negative and positive n indices.
3. The real FS has real expansion coefficients, while the complex one has complex expansion coefficients.

Let's start by assuming that the complex FS expansion coefficients c_n is split into real and imaginary parts as follows:

$$c_n = c_n^r + jc_n^i \tag{7.14}$$

Put those back into the complex FS

$$\begin{aligned} f(t) &= \sum_{-n}^{n} c_n e^{j\omega_n t} \\ &= \sum_{-n}^{n} (c_n^r + jc_n^i)\,[\cos\omega_n t + j\sin\omega_n t] \\ &= \sum_{-n}^{n} (c_n^r \cos\omega_n t - c_n^i \sin\omega_n t) + j\left[c_n^r \sin\omega_n t + c_n^i \cos\omega_n t\right] \end{aligned} \tag{7.15}$$

As can be seen above, the right side has a real term and an imaginary one. But the left side has only a real term (the starting real function). In order for the equality to hold, we need to make sure that the imaginary part of the RHS goes to zero (overall). Let's take the n and $-n$ terms of the imaginary part

$$\text{Imag } n \text{ and } -n \text{ term} = c_n^r \sin\omega_n t + c_n^i \cos\omega_n t + c_{-n}^r \sin\omega_{-n} t + c_{-n}^i \cos\omega_{-n} t \tag{7.16}$$

Now use the even/odd properties of the cosine/sine functions and collect terms

$$\text{Imag } n \text{ and } -n \text{ term} = (c_n^i + c_{-n}^i)\cos\omega_n t + (c_n^r - c_{-n}^r)\sin\omega_n t \tag{7.17}$$

It is clear from the above that in order for the imaginary n and $-n$ expansion to go to zero we would need the following two conditions:

$$\begin{aligned} c_n^i &= -c_{-n}^i \\ c_n^r &= c_{-n}^r \end{aligned} \tag{7.18}$$

In other words (for a real signal $f(t)$) we need the **real** part of the complex exponential Fourier component to be *even* in frequency and the **imaginary** part to be *odd*. Now we go back to the real part of Eq. (7.15). Let us pick the n and $-n$ terms

$$\text{Real } n \text{ and } -n \text{ term} = c_n^r \cos\omega_n t - c_n^i \sin\omega_n t + c_{-n}^r \cos\omega_{-n} t - c_{-n}^i \sin\omega_{-n} t \tag{7.19}$$

Again use even/real properties of cosine/sine functions to get

$$\text{Real } n \text{ and } -n \text{ term} = (c_n^r + c_{-n}^r)\cos\omega_n t - (c_n^i - c_{-n}^i)\sin\omega_n t \tag{7.20}$$

Now using the relation in Eq. (7.18) we get

$$\text{Real } n \text{ and } -n \text{ term} = 2c_n^r \cos\omega_n t - 2c_n^i \sin\omega_n t \tag{7.21}$$

In other words, we can now drop the negative n indices and our complex FS gives us

$$f(t) = \sum_{n=0}^{\infty} 2c_n^r \cos\omega_n t - 2c_n^i \sin\omega_n t \tag{7.22}$$

Comparing this to the real FS given by

$$f(t) = \sum_{n=0}^{\infty} a_n \cos\omega_n t + b_n \sin\omega_n t \tag{7.23}$$

we immediately arrive at the relation tying the real FS expansion coefficients to complex ones

$$\begin{aligned} a_n &= 2c_n^r \qquad (n \neq 0) \\ b_n &= -2c_n^i \end{aligned} \tag{7.24}$$

For the case $n = 0$ both real and complex Fourier series give the same coefficient

$$a_0 = c_0 \tag{7.25}$$

since that is simply the area of the signal divided by the period. Let's get some practice applying the complex Fourier series, and in the process verify the above relation.

7.5 Centered Pulse

As a first example of how to use the complex Fourier series let's take the centered pulse of width τ and period T shown in Fig. 7.1. We already know from using the real Fourier series that this function can be expanded as

$$f(t) = a_0 + \sum_{n=1}^{\infty} a_n \cos\omega_n t \tag{7.26}$$

where

$$\boxed{a_n = \frac{4}{T}\frac{\sin\omega_n\tau/2}{\omega_n}}, \qquad \omega_n = \frac{2\pi n}{T} \tag{7.27}$$

Using the complex FS instead we have

$$f(t) = \sum_{n=-\infty}^{\infty} c_n e^{j\omega_n t} \tag{7.28}$$

where

$$\begin{aligned} c_n &= \frac{1}{T}\int_T f(t)e^{-j\omega_n t}dt = \frac{1}{T}\int_{-\tau/2}^{\tau/2} e^{-j\omega_n t}dt \\ &= \frac{1}{T}\frac{e^{-j\omega_n\tau/2} - e^{j\omega_n\tau/2}}{-j\omega_n} = \boxed{\frac{2}{T}\frac{\sin\omega_n\tau/2}{\omega_n}} \end{aligned} \tag{7.29}$$

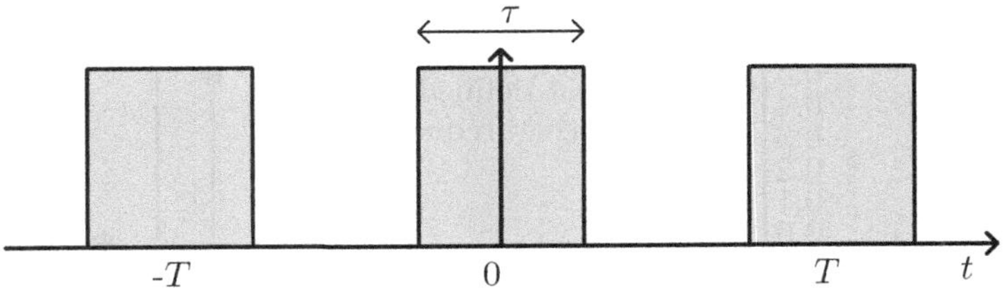

Fig. 7.1 Centered pulse

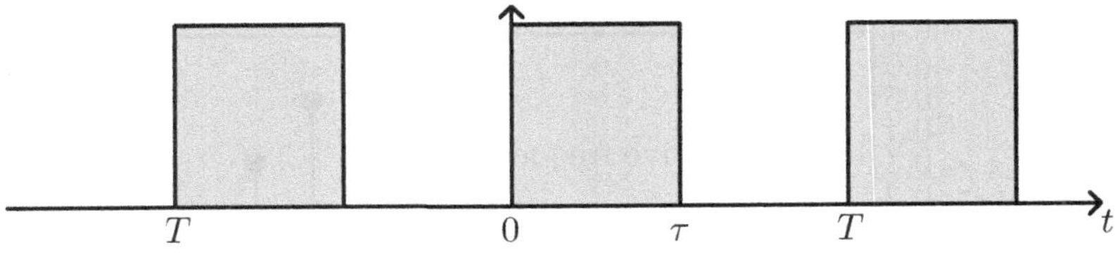

Fig. 7.2 Offset pulse

Comparing both real and complex FS we do in fact confirm results of Eq. (7.24); namely

$$a_n = \frac{4}{T}\frac{\sin \omega_n \tau/2}{\omega_n} = 2c_n^r, \quad \text{and} \tag{7.30}$$

$$b_n = 0 = -2c_n^i \tag{7.31}$$

7.6 Offset Pulse

Consider next the offset pulse of width τ and period T as shown in Fig. 7.2. Using the real Fourier series we already know that

$$f(t) = a_0 + \sum_n a_n \cos \omega_n t + \sum_n b_n \sin \omega_n t \tag{7.32}$$

where

$$a_n = \frac{2}{T}\frac{\sin \omega_n \tau}{\omega_n}, \qquad b_n = \frac{2}{T}\frac{1 - \cos \omega_n t}{\omega_n} \tag{7.33}$$

Now we use the complex FS

$$f(t) = \sum_{-n}^{n} c_n e^{j\omega_n t} \tag{7.34}$$

where

$$c_n = \frac{1}{T}\int_0^\tau f(t) e^{-j\omega_n t} dt = -\frac{1}{T}\frac{e^{-j\omega_n \tau} - e^{-j\omega_n 0}}{j\omega_n}$$
$$= \frac{1}{T}\frac{1 - e^{-j\omega_n \tau}}{j\omega_n} \tag{7.35}$$

Now let's compare Eq. (7.33) to (7.35). The real part of the latter is

$$c_n^r = \frac{1}{T}\frac{j \sin \omega_n \tau}{j\omega_n} = \frac{1}{T}\frac{\sin \omega_n \tau}{\omega_n} \tag{7.36}$$

which is exactly half a_n. Next the imaginary part of (7.35) is

$$c_n^i = -\frac{1}{T}\frac{1 - \cos \omega_n t}{\omega_n} \tag{7.37}$$

which again is exactly (negative) half that of b_n.

7.7 Negative Frequencies

As seen above, and would be used throughout the book, we are having to deal with *negative* frequencies. To illustrate the meaning of this consider again the centered pulse; it had a real FS representation of

$$f(t) = a_0 + \sum_{n=1}^{\infty} a_n \cos \omega_n t, \qquad a_n = \frac{4}{T}\frac{\sin \omega_n \tau}{\omega_n} \tag{7.38}$$

By the same notion the complex FS representation looked like this

$$f(t) = \sum_{n=-\infty}^{\infty} c_n e^{j\omega_n t}, \qquad c_n = \frac{2}{T}\frac{\sin \omega_n \tau}{\omega_n} \tag{7.39}$$

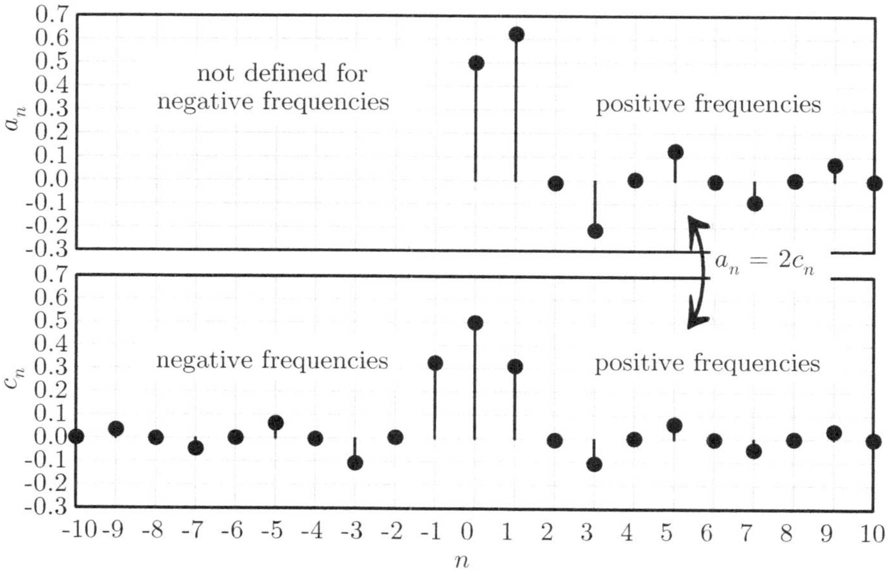

Fig. 7.3 Fourier spectrum of symmetric pulse function: top real FS; bottom complex one

If we plot both spectrums we get Fig. 7.3. Notice how the real series has a spectrum defined only for positive frequencies (positive n). The complex series, on the other hand, has spectrum defined for all frequencies (positive and negative). The two spectrum *are* related; specifically, for positive frequencies, the real spectrum simply equals *twice* that of the complex. Finally, notice that the zero frequency coefficient (a_0 and c_0) are the *same*; this is so because that coefficient really represents the *average* of the signal, and this does not change as a function of which FS is used!

7.8 Complex Fourier Series of Sines/Cosines

Let's find the complex Fourier series of the cosine function: $\cos \omega_0 t$. Rather than do the integration we already know that

$$\cos \omega_0 t = \frac{1}{2} e^{j\omega_0 t} + \frac{1}{2} e^{-j\omega_0 t} \tag{7.40}$$

That is, the cosine is simply the sum of two complex exponential! Since the complex FS is a series of complex exponentials, we already know that the series simply collapses to two terms: one at ω_0 and the other at $-\omega_0$ (with a factor of half). So the spectrum is *real* and looks like Fig. 7.4.

As for the sine function we know that

$$\sin \omega_0 t = \frac{1}{2j} e^{j\omega_0 t} - \frac{1}{2j} e^{-j\omega_0 t} = -\frac{j}{2} e^{j\omega_0 t} + \frac{j}{2} e^{-j\omega_0 t} \tag{7.41}$$

So this one is purely *imaginary* and looks like Fig. 7.5.

7.9 Single-Sided Negative Exponential

Let's find the complex Fourier series of the negative exponential

$$f(t) = e^{-at}, \qquad t > 0, \qquad \text{periodic in } T \tag{7.42}$$

as shown in Fig. 7.6. Plugging in the complex Fourier equations we get

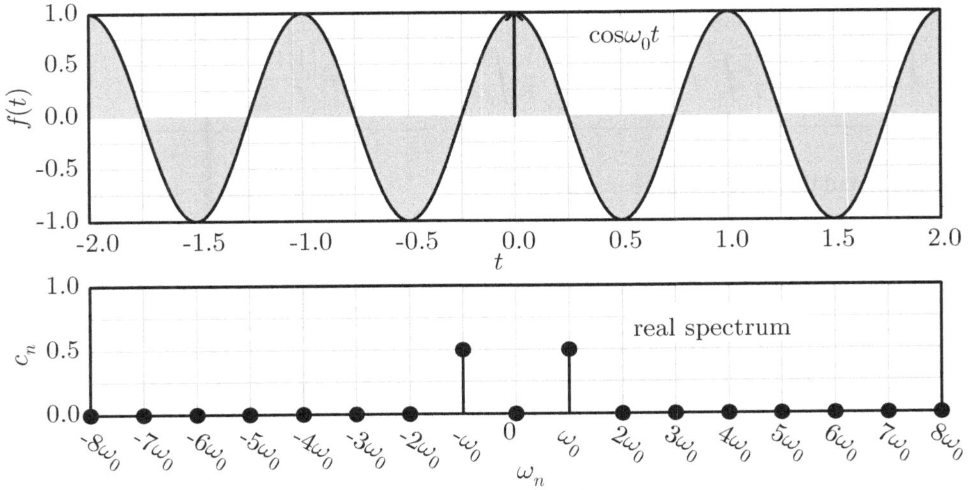

Fig. 7.4 Cosine and complex Fourier coefficients

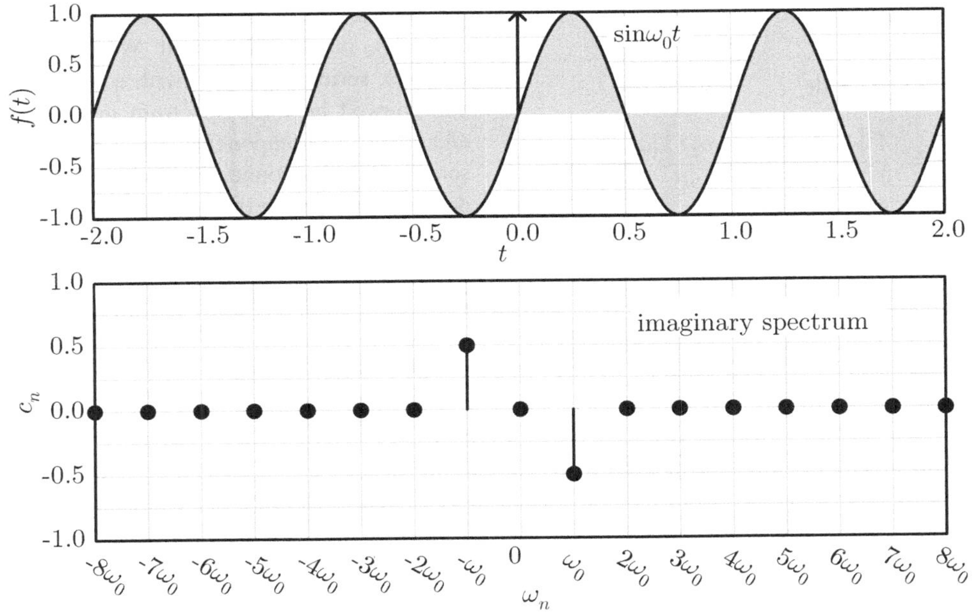

Fig. 7.5 Sine and complex Fourier coefficients

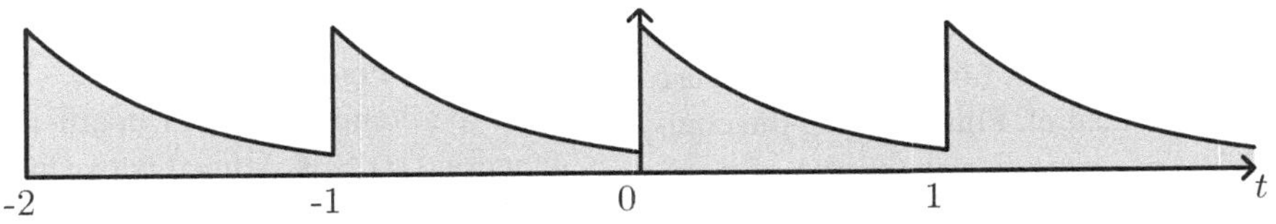

Fig. 7.6 Single-sided negative exponential with period $T = 1$

$$c_n = \frac{1}{T}\int_0^T f(t)e^{-j\omega_n t}dt = \frac{1}{T}\int_0^T e^{-t(a+j\omega_n)}dt = \frac{1}{T}\frac{1-e^{-(a+j\omega_n)T}}{a+j\omega_n} \tag{7.43}$$

We can expand this in terms of real and imaginary parts

$$\begin{aligned} C_n &= \frac{1}{T}\frac{\left[1-e^{-(a+j\omega_n)T}\right][a-j\omega_n]}{a^2+\omega_n^2} \\ &= \frac{1}{T}\frac{\left[1-e^{-aT}\cos\omega_n T + je^{-aT}\sin\omega_n T\right][a-j\omega_n]}{a^2+\omega_n^2} \end{aligned} \tag{7.44}$$

The sine term vanishes since

$$\sin\omega_n T = \sin\frac{2\pi n}{T}T = \sin 2\pi n \tag{7.45}$$

and we are left with

$$C_n = \frac{1}{T}\frac{\left[1-e^{-aT}\cos\omega_n T\right][a-j\omega_n]}{a^2+\omega_n^2} \tag{7.46}$$

So finally the time series is

$$f(t) = \sum_{n=-\infty}^{\infty}\left\{\frac{1}{T}\frac{\left[1-e^{-aT}\cos\omega_n T\right][a-j\omega_n]}{a^2+\omega_n^2}\right\}e^{j\omega_n t} \tag{7.47}$$

Figure 7.7 shows time series reconstruction.

7.10 Summary

The complex Fourier series builds on the real one. But rather than worrying about two constants (a_n and b_n) we now worry about a single one (c_n), albeit it being complex. Also rather than using sines/cosines as our basis functions we now use the complex exponential. Finally, using the complex Fourier series necessitates the use of negative frequencies (or negative indices). Other than that, the main idea remains the same. We went through the trouble of migrating to the complex plane paving the way for the Fourier transform and Laplace ones, which work naturally in the complex plane. In the chapter we established the relation between c_n on the one hand and a_n and b_n on the other hand. We illustrated the complex series expansion with a few examples, and showed how the spectrum now has a real and imaginary components. The complex Fourier series will be the foundation and starting point for deriving the Fourier transform, next.

7.11 Problems

1. The complex Fourier coefficients of the upright, symmetric triangle of total width τ and period T is

$$c_0 = \frac{\tau}{T}, \qquad c_n = \frac{2}{\tau T}\frac{1-\cos\omega_n\tau}{\omega_n^2}$$

Take the derivative of the above equation, with respect to ω_n, multiply by j, and plot the corresponding time series. How does this series compare to the original triangular function times the function $f(t) = t$? See sample solution in Fig. 7.8.

2. Find the complex Fourier coefficients of the function $f(t) = t$, defined between $-T/2$ and $T/2$. For the case $T = 1$ plot the time series; see sample solution in Fig. 7.9.

Answer: $c_n = 2j\dfrac{\cos\omega_n T/2}{\omega_n}$

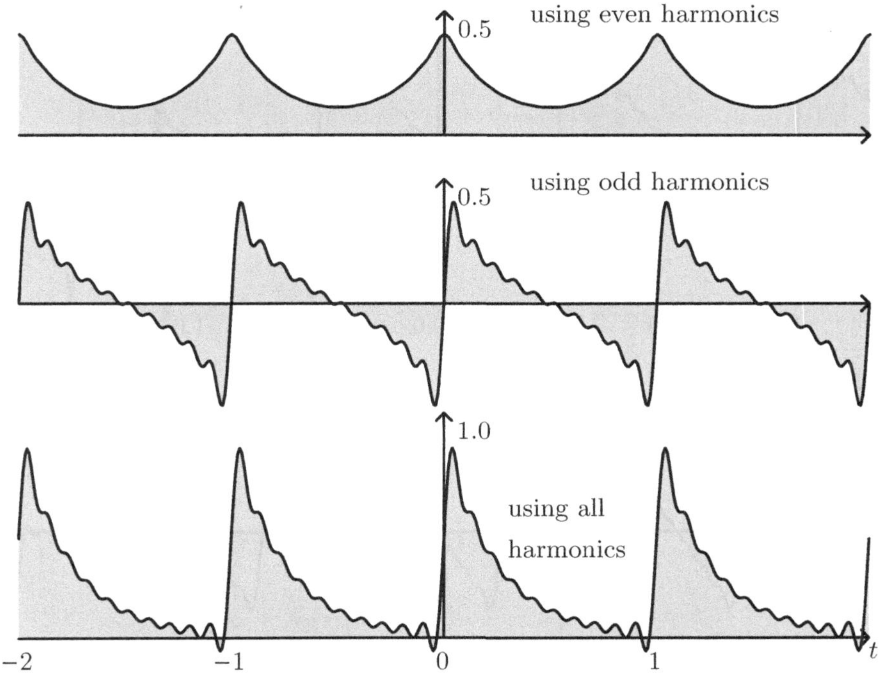

Fig. 7.7 Time series of single-sided negative exponential with $a = 4$ and $T = 1$

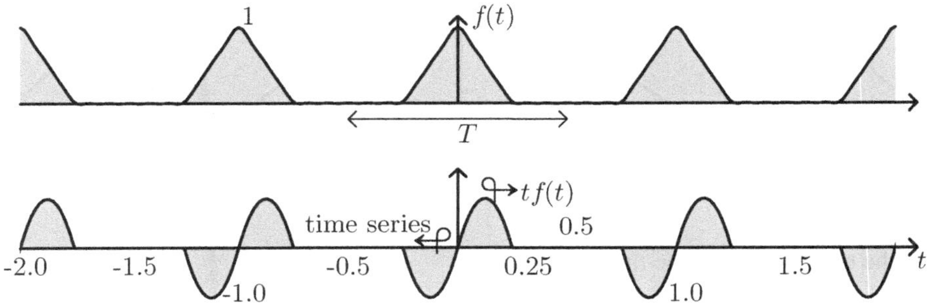

Fig. 7.8 Solution to Problem 1

3. Start with same assumptions as Problem 2, but now the function is defined only between $-T/4$ and $T/4$, and is zero otherwise. Find the complex Fourier coefficients and plot time series results for case $T = 1$. See sample results in Fig. 7.10.
   Answer:

$$c_n = \frac{2j}{T}\left[\frac{T}{4}\frac{\cos\omega_n T/4}{\omega_n} - \frac{\sin\omega_n T/4}{\omega_n^2}\right]$$

4. Find the complex Fourier coefficients for the function $f(t) = t^2$, defined between $-\frac{1}{2}$ and $\frac{1}{2}$. Plot resulting time series; see sample results in Fig. 7.11.
   Answer:

$$\tau = 0.5, \qquad c_0 = \frac{2\tau^3}{3}, \qquad c_n = 4\tau\frac{\cos\omega_n\tau}{\omega_n^2}$$

5. If the complex Fourier coefficients for the function $f(t) = e^{j\omega_0 t}$ is $c_n = \delta(\omega_n - \omega_0)$,

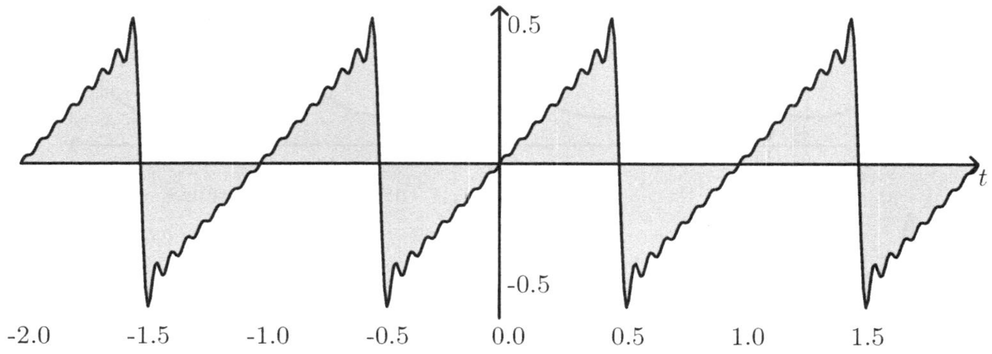

Fig. 7.9 Solution to Problem 2

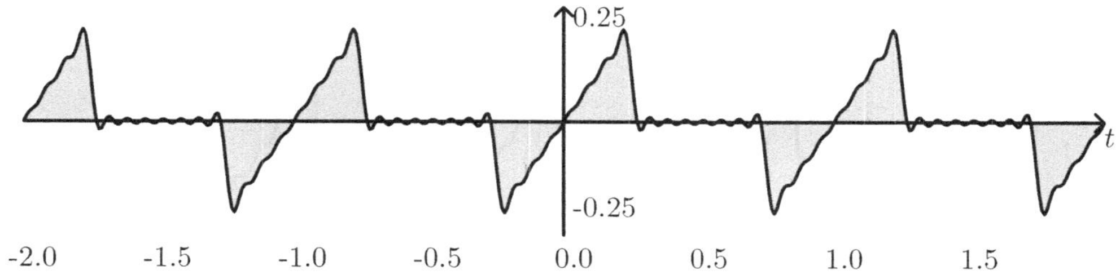

Fig. 7.10 Solution to Problem 3

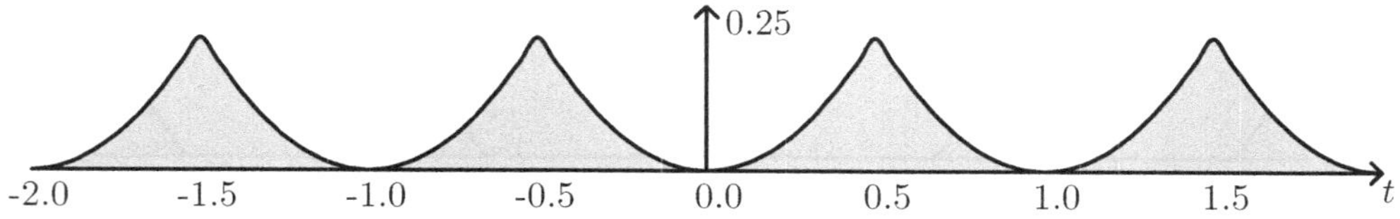

Fig. 7.11 Solution to Problem 4

what is the complex FS of the function $f(t) = e^{j(\omega_0 t+\phi)}$?

Answer:

$$c_n = e^{j\phi}\delta(\omega_n - \omega_0)$$

6. If the complex Fourier coefficients for the function $f(t) = \cos\omega_0 t$ is $c_n = \frac{1}{2}\left[\delta(\omega_n - \omega_0) + \delta(\omega_n + \omega_0)\right]$, what is the complex FS of the function $f(t) = \cos(\omega_0 t + \phi)$?

Answer:

$$c_n = \frac{1}{2}\left[e^{j\phi}\delta(\omega_n - \omega_0) + e^{-j\phi}\delta(\omega_n + \omega_0)\right]$$

7. Take the answer to Problem 6 and set $\phi = -\frac{\pi}{2}$; what would be the inverse transform?

Answer:

$$f(t) = \sin\omega_0 t$$

8 Fourier Transform

8.1 Introduction

The limitation of the Fourier series is the assumption that the starting function is periodic. While this covers a good class of functions, not all functions are periodic. For example, the delta function, unit step, or single pulse functions are not periodic. We can expand the concept of the Fourier series to non-periodic functions by using a simple trick of assuming a periodic function of period T, and taking the limit as $T \to \infty$!

8.2 Derivation of Fourier Transform

We start with the complex Fourier series and do some manipulations. A random function is represented as a linear combination of complex exponentials, with discrete frequencies ω_n.

$$\boxed{f(t) = \sum_{-\infty}^{\infty} c_n e^{j\omega_n t}, \qquad \omega_n = \frac{2\pi n}{T}} \tag{8.1}$$

The expansion coefficients are

$$\boxed{c_n = \frac{1}{T}\int_{-T/2}^{T/2} f(t) e^{-j\omega_n t} dt} \tag{8.2}$$

where T is the period. Combine both to get

$$f(t) = \sum_{n=-\infty}^{n=\infty} \left[\frac{1}{T}\int_{-T/2}^{T/2} f(t) e^{-j\omega_n t} dt\right] e^{j\omega_n t} \tag{8.3}$$

Since $\omega_n = \frac{2\pi n}{T}$ then

$$d\omega_n = \frac{2\pi}{T}, \qquad \text{or} \qquad \frac{1}{T} = \frac{d\omega_n}{2\pi} \tag{8.4}$$

Replace T by the above derived expression

$$f(t) = \sum_{n=-\infty}^{n=\infty} \left[\frac{d\omega_n}{2\pi}\int_{-T/2}^{T/2} f(t) e^{-j\omega_n t} dt\right] e^{j\omega_n t} \tag{8.5}$$

Rearrange:

$$f(t) = \frac{1}{2\pi}\sum_{n=-\infty}^{n=\infty} \left[\int_{-T/2}^{T/2} f(t) e^{-j\omega_n t} dt\right] e^{j\omega_n t} d\omega_n \tag{8.6}$$

In the limit as $T \to \infty$,

$$T \to \infty, \quad \omega_n \to \omega, \quad d\omega_n \to d\omega, \quad \text{and} \quad \sum \to \int \tag{8.7}$$

Then the Riemann summation formula converges to an integration

$$f(t) = \frac{1}{2\pi}\int_{-\infty}^{\infty} \left[\int_{-\infty}^{\infty} f(t) e^{-j\omega t} dt\right] e^{j\omega t} d\omega \tag{8.8}$$

F. Badrieh, *Spectral, Convolution and Numerical Techniques in Circuit Theory*,
https://doi.org/10.1007/978-3-319-71437-0_8

and hence the Fourier transform pair is defined

$$\boxed{f(t) = \frac{1}{2\pi}\int_{-\infty}^{\infty} F(\omega)e^{j\omega t}d\omega} \tag{8.9}$$

where

$$\boxed{F(\omega) = \int_{-\infty}^{\infty} f(t)e^{-j\omega t}dt} \tag{8.10}$$

Let's demonstrate the Fourier transform with some examples.

8.3 Fourier Transform of Pulse

The pulse function of width τ is defined as

$$f(t) = 1 \qquad -\frac{\tau}{2} < t < \frac{\tau}{2} \tag{8.11}$$

$$= 0 \qquad \text{else} \tag{8.12}$$

Notice that unlike before, this function is NOT periodic. The Fourier transform is then

$$F(\omega) = \int_{-\infty}^{\infty} f(t)e^{-j\omega t}dt = \int_{-\tau/2}^{\tau/2} e^{-j\omega t}dt$$

$$= \frac{-1}{j\omega}e^{-j\omega t}\Big|_{-\tau/2}^{\tau/2} = \boxed{\frac{2}{\omega}\sin\frac{\omega\tau}{2}} \tag{8.13}$$

Then, this states that

$$\boxed{f(t) = \frac{1}{2\pi}\int_{-\infty}^{\infty} \frac{2}{\omega}\sin\frac{\omega\tau}{2}e^{j\omega t}d\omega} \tag{8.14}$$

A plot of the FT is shown in Fig. 8.1. Since the Fourier transform came out even, only the multiplication by the real part of the complex exponential survives; hence

$$f(t) = \frac{1}{2\pi}\int_{-\infty}^{\infty} \frac{2}{\omega}\sin\frac{\omega\tau}{2}\cos(\omega t)d\omega \tag{8.15}$$

Furthermore, due to symmetry we can rewrite as

$$\boxed{f(t) = \frac{1}{\pi}\int_{0}^{\infty} \frac{2}{\omega}\sin\frac{\omega\tau}{2}\cos(\omega t)d\omega} \tag{8.16}$$

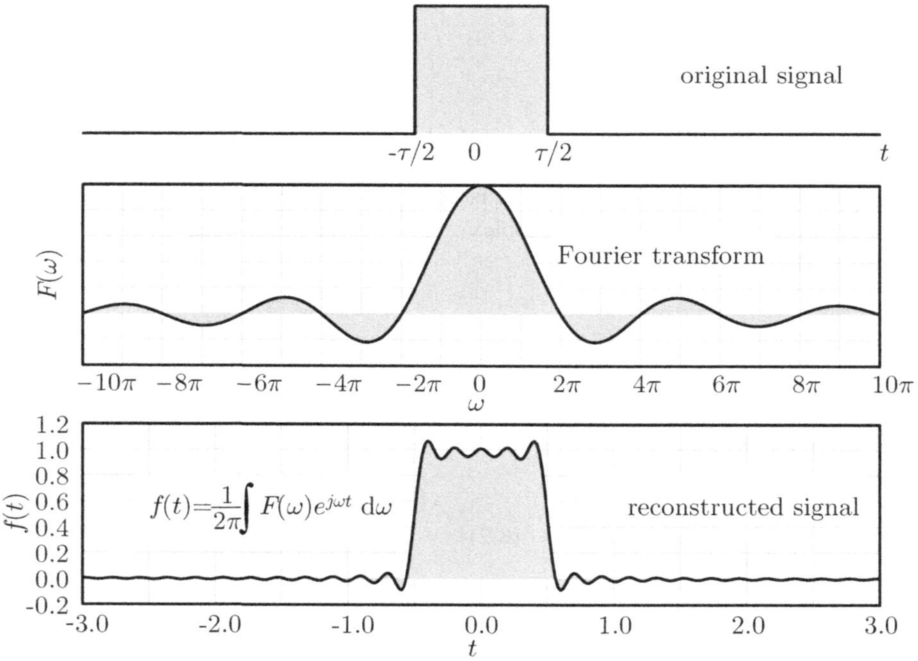

Fig. 8.1 Unit pulse, its Fourier transform, and its time series (inverse Fourier transform). Case of $\tau = 1$

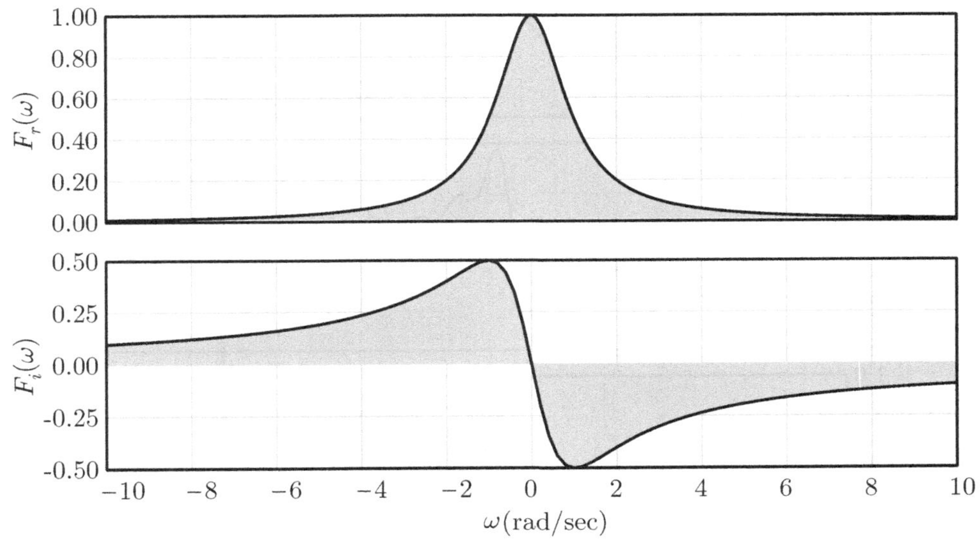

Fig. 8.2 Fourier transform of the single-sided e^{-t} function—both real and imaginary parts

Figure 8.1 shows a plot of the above equation, using some finite frequency steps.

8.4 Fourier Transform of Single-Sided e^{-at}

The single-sided negative exponential is defined as follows:

$$f(t) = 0, \qquad -\infty < t < 0 \tag{8.17}$$

$$= e^{-at}, \qquad 0 < t < \infty \tag{8.18}$$

The Fourier transform is derived as

$$\begin{aligned} F(\omega) &= \int_{-\infty}^{\infty} f(t)e^{-j\omega t}dt = \int_{-\infty}^{\infty} e^{-at}e^{-j\omega t}dt \\ &= \int_{0}^{\infty} e^{-t(a+j\omega)}dt = \frac{-1}{j\omega + a}e^{-t(a+j\omega)}\Big|_{0}^{\infty} \\ &= \boxed{\frac{1}{j\omega + a}} \end{aligned} \tag{8.19}$$

Unlike the pulse case, the FT here is complex: it has both a real part and an imaginary part. As such, we need to plot each of those to see how the FT behaves in the frequency domain; this is shown in Fig. 8.2. Notice that the real part is even while the imaginary one is odd.

We can verify this is correct either by doing complex integration, or numerically. Let's try the latter.

$$\begin{aligned} F(\omega) &= \frac{1}{a + j\omega} = \frac{a - j\omega}{a^2 + \omega^2} \\ &= \boxed{\frac{a}{a^2 + \omega^2} - \frac{j\omega}{a^2 + \omega^2}} \end{aligned} \tag{8.20}$$

To get back the time function we multiply the FT by the complex exponential $e^{j\omega t}$ and integrate over all frequency, starting from $-\infty$ and ending at ∞. Because the real part of the FT is even, when multiplied by the imaginary of the complex exponential, it forms an odd function (vs. frequency) which when integrated over frequency amounts to zero. Similarly, because the imaginary part of the FT is odd, when multiplied by the real part of the complex exponential, it also forms an odd function which again when integrated over frequency amounts to zero. Hence the only remaining terms are

$$\boxed{f(t) = \frac{1}{2\pi}\int_{-\infty}^{\infty} \frac{1}{a^2 + \omega^2}\left[a\cos\omega t + \omega\sin\omega t\right]d\omega} \tag{8.21}$$

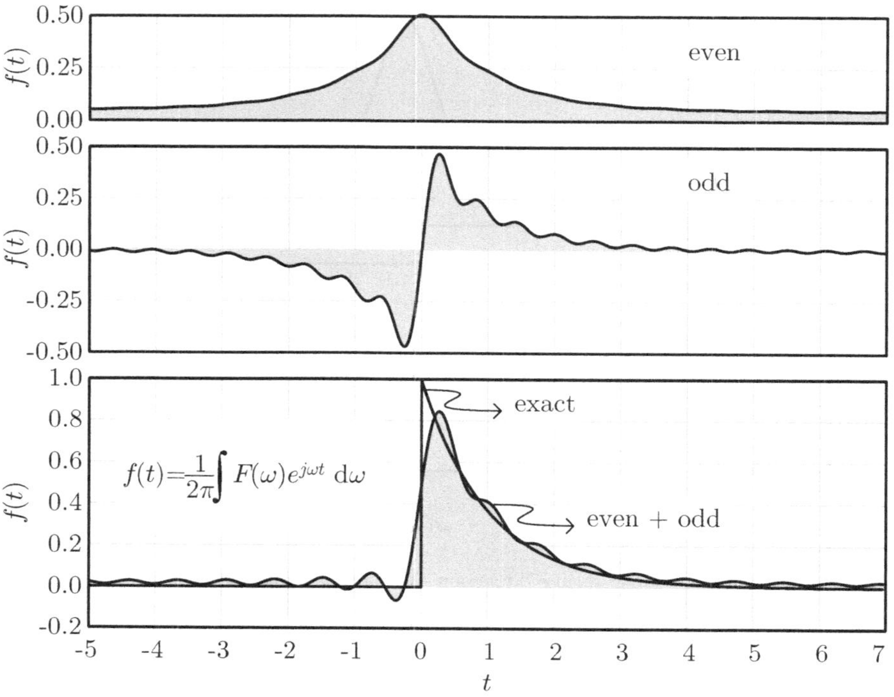

Fig. 8.3 Function e^{-t} and comparison to that from inverse FT

Figure 8.3 shows original function and that obtained from inverse FT using a finite frequency range. It is expected that as we encroach to higher frequencies the fit would get better.

8.5 Fourier Transform of $e^{-|at|}$

The symmetric negative exponential function is defined as

$$f(x) = \begin{cases} e^{at} & t < 0 \\ e^{-at} & t > 0 \end{cases} \tag{8.22}$$

From the prior section we know that

$$\int_0^{\infty} e^{-at} e^{-j\omega t} dt = \frac{1}{j\omega + a} \tag{8.23}$$

Similarly we arrive at

$$\int_{-\infty}^{0} e^{at} e^{-j\omega t} dt = -\frac{1}{j\omega - a} \tag{8.24}$$

The sum Fourier transform is then

$$\begin{aligned} \mathcal{F}\left[e^{-|at|}\right] &= \frac{1}{j\omega + a} - \frac{1}{j\omega - a} \\ &= \boxed{\frac{2a}{\omega^2 + a^2}} \end{aligned} \tag{8.25}$$

When we do the frequency integration, the imaginary part of the complex exponential drops out, and we end up with

$$\boxed{e^{-|at|} = \frac{2}{\pi}\int_0^{\infty} \frac{a}{a + \omega^2}\cos(\omega t) d\omega} \tag{8.26}$$

Let's try plotting this last result; this is shown in Fig. 8.4. The more frequency points we include in the numerical integration, and the higher in frequency we include, the better the approximation becomes.

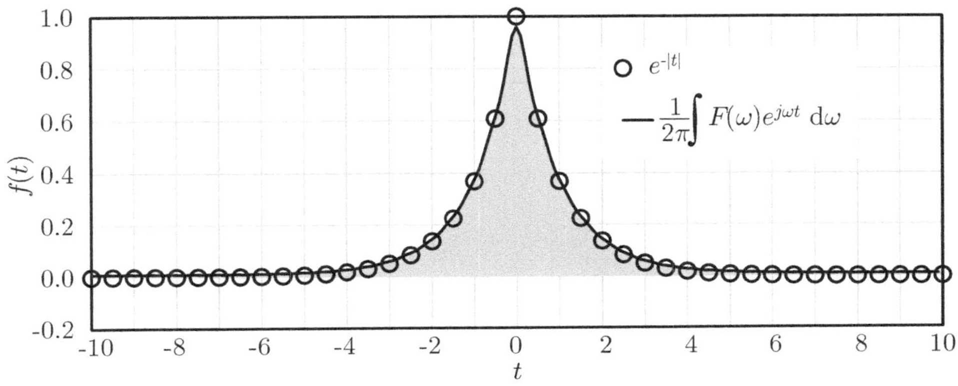

Fig. 8.4 Function $e^{-|at|}$ and comparison to that from inverse FT

8.6 Fourier Transform of DC Function

The DC function is defined as

$$f(t) = \text{const}, \qquad -\infty < t < \infty \tag{8.27}$$

To find the Fourier transform of this we proceed to direct integration:

$$\begin{aligned} F(\omega) &= \int_{-\infty}^{\infty} f(t)e^{-j\omega t}dt = \int_{-\infty}^{\infty} e^{-j\omega t}dt \\ &= -\frac{e^{-j\omega t}}{j\omega}\bigg|_{-\infty}^{\infty} \\ &= -\frac{1}{j\omega}\left[e^{-j\omega\infty} - e^{j\omega\infty}\right] \\ &= \boxed{2\frac{\sin(\omega\infty)}{\omega}} \end{aligned} \tag{8.28}$$

Now we will make connection with one of the most important concepts in this book—the delta function. It can (and will) be shown that the following limit exists:

$$\boxed{\lim_{T\to\infty} \frac{\sin\omega T}{\omega} \to \pi\delta(\omega)} \tag{8.29}$$

This is a sinc function whose value at frequency zero is T and oscillates with a period $2\pi/T$; the larger T, the higher the value at frequency zero, and the smaller the period (or larger the oscillating frequency). The integral, however, remains finite—and independent of T—and is evaluated (either analytically or numerically) to be exactly π! Based on this we conclude that

$$\boxed{F(\omega) = 2\pi\delta(\omega)} \tag{8.30}$$

That is, the DC function transforms to a single frequency—and that is the zero frequency. At that frequency it has infinite energy. The relation between the constant function and its FT is portrayed in Fig. 8.5.

8.6.1 Alternate Derivation

Another way we can derive the FT of the DC function is by taking the limit of the double-sided negative exponential:

$$\text{DC} = \lim_{a\to 0} e^{-a|t|} \tag{8.31}$$

We can verify this by observing the plot in Fig. 8.6. Recall from Eq. (8.25) that

$$e^{-a|t|} \to \frac{2a}{a^2+\omega^2} \tag{8.32}$$

What we want to show now is that this FT is a (2π) delta function as $a \to 0$. Recall the three conditions of the delta function

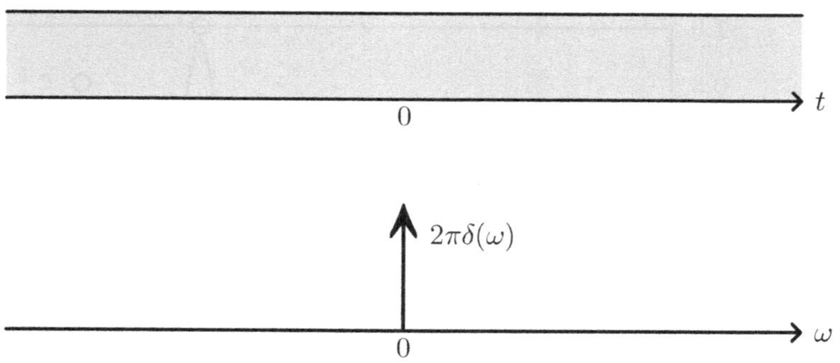

Fig. 8.5 DC function (top) and its FT (bottom)

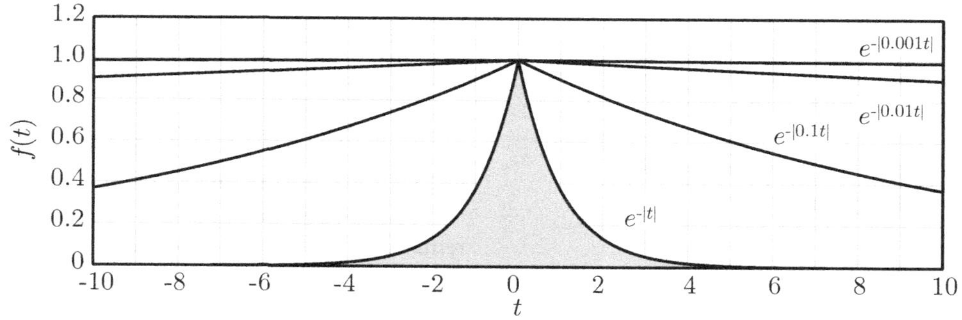

Fig. 8.6 DC function as $\lim_{a\to 0} e^{-a|t|}$

1. The function is zero everywhere where $\omega \neq 0$. For this condition we see that the limit in fact goes to zero

$$\lim_{a\to 0}\frac{2a}{a^2+\omega^2} = \frac{0}{0+\omega^2} = 0 \qquad (\omega \neq 0) \tag{8.33}$$

2. The second condition is that the function is infinite at $\omega = 0$. We can verify this is true by observing

$$\lim_{a\to 0}\left.\frac{2a}{a^2+\omega^2}\right|_{\omega=0} = \lim_{a\to 0}\frac{2a}{a^2} = \lim_{a\to 0}\frac{2}{a} = \infty \tag{8.34}$$

3. Finally we need to verify that the integral under this function is (2π) unity. We do the integration as follows:

$$\int_{-\infty}^{\infty}\frac{2a}{a^2+\omega^2}d\omega = \frac{2}{a}\int_{-\infty}^{\infty}\frac{1}{1+\frac{\omega^2}{a^2}}d\omega \tag{8.35}$$

Let $u = \frac{\omega}{a}$ and $du = \frac{d\omega}{a}$. Then the integral becomes

$$I = 2\int_{-\infty}^{\infty}\frac{1}{1+u^2}du \tag{8.36}$$

Let $u = \tan\theta$ and $du = \sec^2\theta$. Then the integral becomes

$$I = 2\int_{-\pi/2}^{\pi/2}\frac{\sec^2\theta}{\sec^2\theta}d\theta = 2\int_{-\pi/2}^{\pi/2}d\theta = 2\pi \tag{8.37}$$

That is, not only the integral is in fact 2π; it is independent of a.

So we have verified that this FT does in fact go a delta function (times 2π) in the limit $a \to 0$. In other words, we have shown that the FT of the DC function is $2\pi\delta(\omega)$. Figure 8.7 shows this FT function for different values of a and shows tentatively how this function becomes a delta one for small a.

8.7 Fourier Transform of Delta Function

The Fourier transform of the delta function can be derived by applying the FT formula:

$$F(\omega) = \int_{-\infty}^{\infty} f(t)e^{-j\omega t}dt = \int_{-\infty}^{\infty}\delta(t)e^{-j\omega t}dt = 1 \tag{8.38}$$

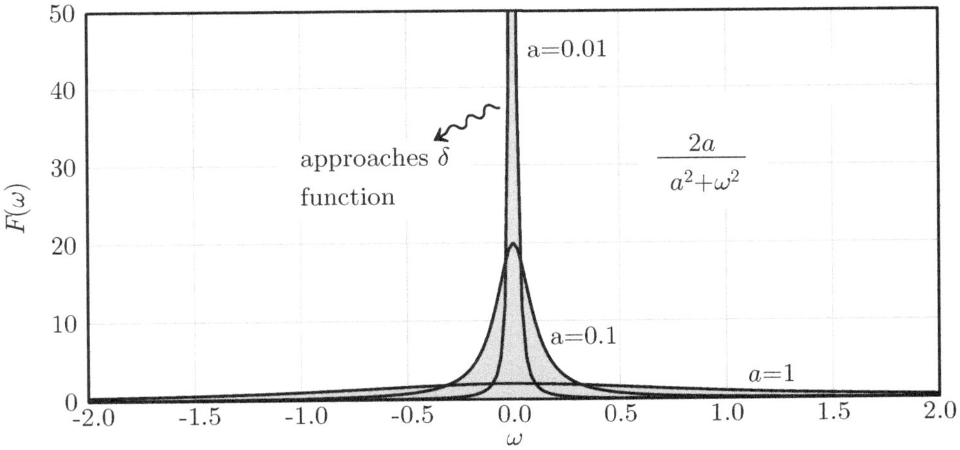

Fig. 8.7 Function and limit to delta function

That is,

$$\boxed{\mathcal{F}[\delta(t)] = F(\omega) = 1} \tag{8.39}$$

Let's verify this is true. We take the inverse transform and get

$$\begin{aligned} f(t) &= \frac{1}{2\pi}\int_{-\infty}^{\infty} F(\omega)e^{j\omega t}d\omega \\ &= \frac{1}{2\pi}\int_{-\infty}^{\infty} e^{j\omega t}d\omega = \lim_{\omega\to\infty}\frac{1}{2\pi}\frac{2\sin\omega t}{t} = \delta(t) \end{aligned} \tag{8.40}$$

where we have used the definition that $\lim_{\omega\to\infty}\frac{\sin\omega t}{t} = \pi\delta(t)$.

8.8 Fourier Transform of Unit Step Function

The unit step function is defined as

$$u(t) = \begin{cases} 0 & t < 0 \\ 1 & t > 0 \end{cases} \tag{8.41}$$

We plug in the FT formula and get

$$\begin{aligned} F(\omega) &= \int_{-\infty}^{\infty} f(t)e^{-j\omega t}dt = \int_{0}^{\infty} e^{-j\omega t}dt \\ &= -\frac{e^{-j\omega t}}{j\omega}\bigg|_0^{\infty} = \frac{1}{j\omega} - \lim_{t\to\infty}\frac{e^{-j\omega t}}{j\omega} \end{aligned} \tag{8.42}$$

The last term evaluates to

$$\lim_{t\to\infty}\frac{e^{-j\omega t}}{j\omega} = \lim_{t\to\infty}\left[\frac{\cos\omega t}{j\omega} - j\frac{\sin\omega t}{j\omega}\right] \tag{8.43}$$

We can identify—see Eq. (8.29)—the sine limit at

$$\lim_{t\to\infty}\frac{\sin\omega t}{\omega} = \pi\delta(\omega) \tag{8.44}$$

but we have to make a decision what to make of

$$\lim_{t\to\infty}\frac{\cos\omega t}{\omega} = ??? \tag{8.45}$$

While the cosine limit shares many properties with the sine one, such as oscillations, decaying for large ω and having large values towards $\omega \to 0$ it suffers from the following two symptoms:

1. The cosine limit is odd in frequency, making its integral over all frequencies vanish.
2. Around $\omega = 0$ the cosine limit does not aggregate to a self-contained pulse, of finite area such as the sine limit; instead, the cosine limit averages zero around zero frequency.

In the limit of large t, the cosine limit oscillates rapidly, and on average cancels out; that is, around any frequency integral, we can set t large enough, such that the average of the cosine limit goes to zero. This is in contrast to the sine limit, which would average to a nonzero value around $\omega = 0$, no matter how large t is set. As such, we make the decision to treat the cosine limit as

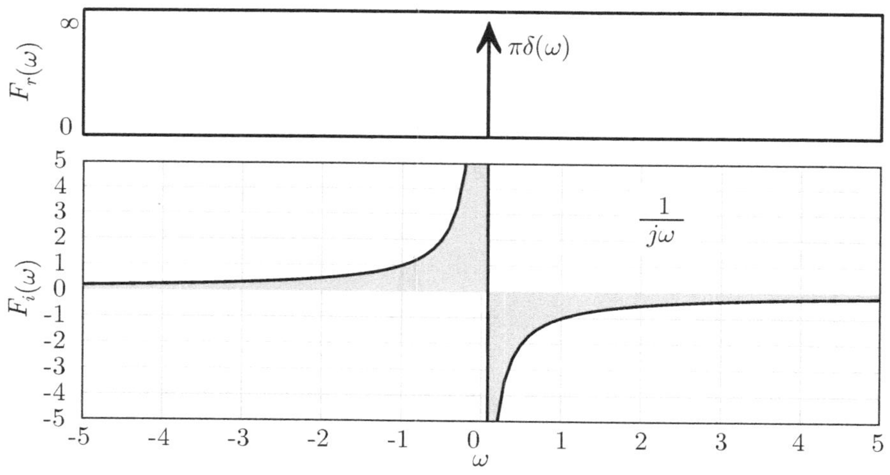

Fig. 8.8 Fourier transform of the unit step function

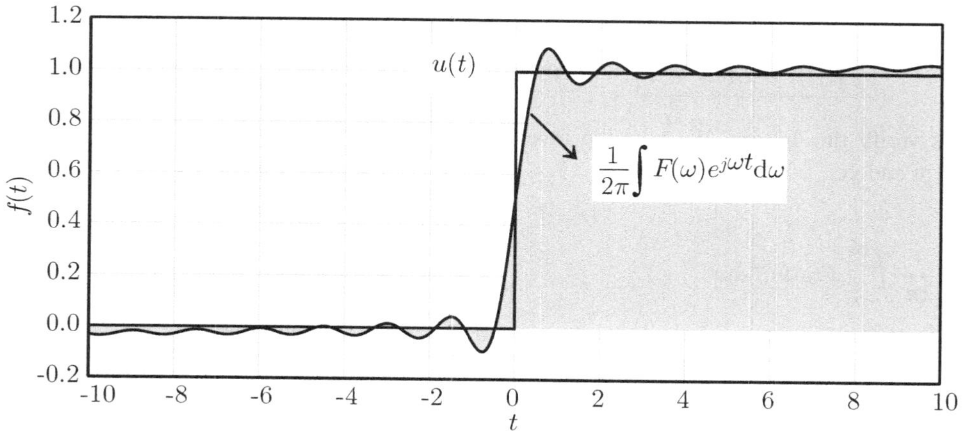

Fig. 8.9 Inverse transform yields unit step function

$$\boxed{\lim_{t\to\infty} \frac{\cos\omega t}{\omega} = 0} \tag{8.46}$$

Then our FT becomes

$$\boxed{\mathcal{F}[u(t)] = F(\omega) = \frac{1}{j\omega} + \pi\delta(\omega)} \tag{8.47}$$

The FT is shown in Fig. 8.8.

A plot of the inverse transform and comparison to original function is shown in Fig. 8.9.

8.9 Fourier Transform of Signum Function

The signum (or sign) function is defined as follows:

$$\text{sig}(t) = \begin{cases} 0 & t < -\frac{1}{2} \\ 1 & t > \frac{1}{2} \end{cases} \tag{8.48}$$

To find the FT we do time integration of this function against the complex exponential. Since this function is odd, the integration against the cosine part of the complex exponential vanishes. The sine integration part, in turn, is symmetric around time zero and the FT becomes

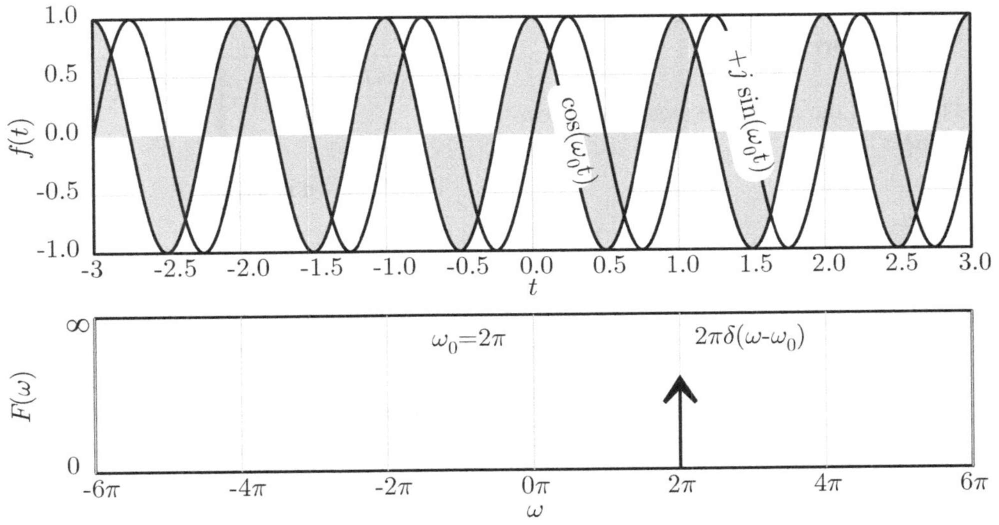

Fig. 8.10 Complex exponential and Fourier transform

$$
\begin{aligned}
F(\omega) &= \int_{-\infty}^{\infty} f(t)e^{-j\omega t}dt = -j\int_{0}^{\infty} \sin\omega t dt \\
&= j\frac{\cos\omega t}{\omega}\Big|_{0}^{\infty} = \lim_{t\to\infty}\left[j\frac{\cos\omega t}{\omega} + \frac{1}{j\omega}\right]
\end{aligned}
$$

The cosine limit vanishes (see Eq. (8.46)) and we arrive at

$$
\boxed{\mathcal{F}[\text{sig}(t)] = F(\omega) = \frac{1}{j\omega}} \tag{8.49}
$$

8.10 Fourier Transform of Complex Exponential

The complex exponential of frequency ω_0 is given by

$$
e^{j\omega_0 t} = \cos\omega_0 t + j\sin\omega_0 t \tag{8.50}
$$

The FT is obtained as follows:

$$
\begin{aligned}
F(\omega) &= \int_{-\infty}^{\infty} f(t)e^{-j\omega t}dt = \int_{-\infty}^{\infty} e^{j\omega_0 t}e^{-j\omega t}dt \\
&= \int_{-\infty}^{\infty} e^{-t(j\omega - j\omega_0)}dt
\end{aligned} \tag{8.51}
$$

The resultant integrand has both odd and even parts; the odd (sin) part vanishes and we end up with

$$
\begin{aligned}
F(\omega) &= 2\int_{0}^{\infty} \cos[t(\omega-\omega_0)]dt \\
&= 2\frac{\sin[t(\omega-\omega_0)]}{\omega-\omega_0}\Bigg|_{0}^{\infty} \\
&= 2\lim_{t\to\infty} \frac{\sin[t(\omega-\omega_0)]}{\omega-\omega_0}
\end{aligned}
$$

$$
\boxed{\mathcal{F}[e^{j\omega_0 t}] = 2\pi\delta(\omega-\omega_0)} \tag{8.52}
$$

This is shown in Fig. 8.10. We can verify this by forming the inverse FT

$$
\begin{aligned}
f(t) &= \frac{1}{2\pi}\int_{-\infty}^{\infty} F(\omega)e^{j\omega t}d\omega \\
&= \frac{1}{2\pi}\int_{-\infty}^{\infty} 2\pi\delta(\omega-\omega_0)e^{j\omega t}d\omega \\
&= e^{j\omega_0 t}
\end{aligned} \tag{8.53}
$$

Similarly we get

$$
\boxed{\mathcal{F}[e^{-j\omega_0 t}] = 2\pi\delta(\omega+\omega_0)} \tag{8.54}
$$

8.11 Fourier Transform of Cosine Function

To find the FT of the cosine we recall the following identities:

$$\cos \omega_0 t = \frac{1}{2}\left[e^{j\omega_0 t} + e^{-j\omega_0 t}\right], \tag{8.55}$$

$$e^{j\omega_0 t} \to 2\pi\delta(\omega-\omega_0), \text{ and } e^{-j\omega_0 t} \to 2\pi\delta(\omega+\omega_0) \tag{8.56}$$

Then we get

$$\boxed{\cos \omega_0 t \to \pi\left[\delta(\omega-\omega_0) + \delta(\omega+\omega_0)\right]} \tag{8.57}$$

This is shown in Fig. 8.11.

8.12 Fourier Transform of Sine Function

Similarly, to find the FT of the sine function we recall

$$\sin \omega_0 t = \frac{1}{2j}\left[e^{j\omega_0 t} - e^{-j\omega_0 t}\right] \tag{8.58}$$

Then we get

$$\boxed{\sin \omega_0 t \to \pi j\left[\delta(\omega+\omega_0) - \delta(\omega-\omega_0)\right]} \tag{8.59}$$

This is shown in Fig. 8.12.

8.13 Fourier Transform of Single-Sided Complex Exponential

The single-sided (or causal) complex exponential with frequency ω_0 is defined as

$$f(t) = \begin{cases} 0 & t < 0 \\ e^{j\omega_0 t} & t > 0 \end{cases} \tag{8.60}$$

Or equivalently

$$f(t) = u(t)e^{j\omega_0 t} \tag{8.61}$$

The Fourier transform is figured as follows:

$$\begin{aligned} F(\omega) &= \int_{-\infty}^{\infty} f(t)e^{-j\omega t}dt = \int_0^{\infty} e^{j\omega_0 t}e^{-j\omega t}dt = \int_0^{\infty} e^{-j(\omega-\omega_0)t}dt \\ &= -\frac{1}{j(\omega-\omega_0)}e^{-j(\omega-\omega_0)t}\Big|_0^{\infty} = \frac{1}{j(\omega-\omega_0)} - \frac{e^{-j(\omega-\omega_0)\infty}}{j(\omega-\omega_0)} \\ &= \boxed{\frac{1}{j(\omega-\omega_0)} + \pi\delta(\omega-\omega_0)} \end{aligned} \tag{8.62}$$

That is, the single-sided complex exponential has a real and an imaginary Fourier transform parts. The real part is a delta function while the imaginary one is an inverse frequency function. This is shown in Fig. 8.13.

8.14 Fourier Transform of Single-Sided Cosine Function

Here we find the Fourier transform of the single-sided cosine function defined by

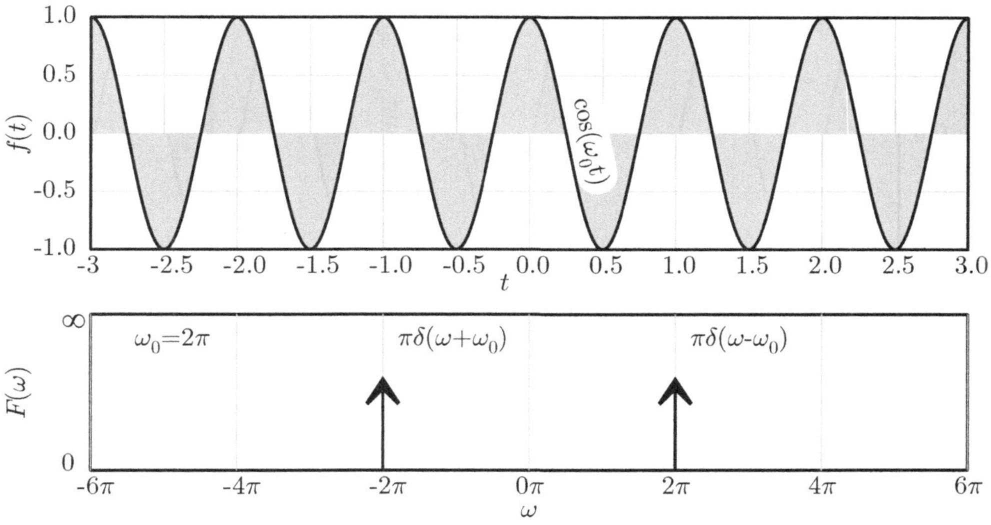

Fig. 8.11 Cosine function and Fourier transform

$$f(t) = \begin{cases} 0 & t < 0 \\ \cos(\omega_0 t) & t > 0 \end{cases} \tag{8.63}$$

We know that

$$\cos \omega_0 t = \frac{1}{2}\left[e^{j\omega_0 t} + e^{-j\omega_0 t}\right] \tag{8.64}$$

We also know from prior section that

$$u(t)e^{j\omega_0 t} \rightarrow \pi\delta(\omega - \omega_0) + \frac{1}{j(\omega - \omega_0)} \tag{8.65}$$

Then we would get

$$F(\omega) = \frac{\pi}{2}\left[\delta(\omega + \omega_0) + \delta(\omega - \omega_0)\right] + \frac{1}{2}\left[\frac{1}{j(\omega - \omega_0)} + \frac{1}{j(\omega + \omega_0)}\right]$$

$$= \boxed{\frac{\pi}{2}\left[\delta(\omega + \omega_0) + \delta(\omega - \omega_0)\right] + \frac{1}{j}\frac{\omega}{\omega^2 - \omega_0^2}} \tag{8.66}$$

Results are shown in Fig. 8.14.

8.14.1 Alternate Method

We can accomplish the same thing by recognizing that this function is the result of multiplying the cosine function times the unit step function

$$t(t) = u(t)\cos(\omega_0 t) \tag{8.67}$$

To find the FT we will use three ingredients: first the FT of the cosine function

$$\cos(\omega_0 t) \rightarrow \pi\left[\delta(\omega - \omega_0) + \delta(\omega + \omega_0)\right] \tag{8.68}$$

Second the FT of the unit step function

$$u(t) \rightarrow \pi\delta(\omega) + \frac{1}{j\omega} \tag{8.69}$$

And lastly the convolution theorem:

$$f_1(T) \cdot f_2(t) \rightarrow \frac{1}{2\pi}F_1(\omega) * F_2(\omega) \tag{8.70}$$

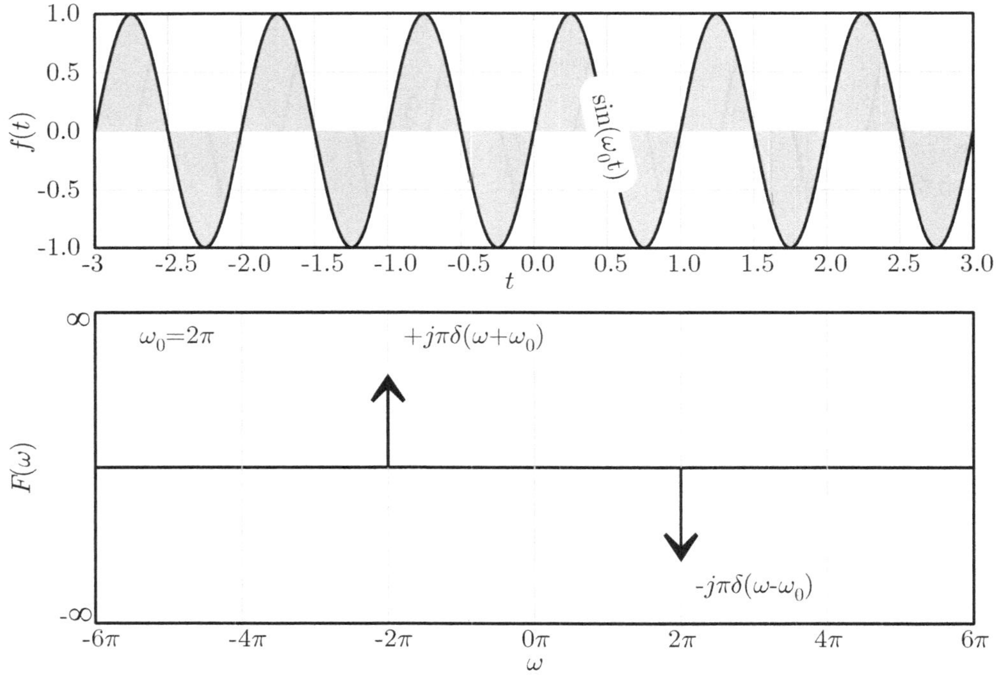

Fig. 8.12 Sine function and Fourier transform

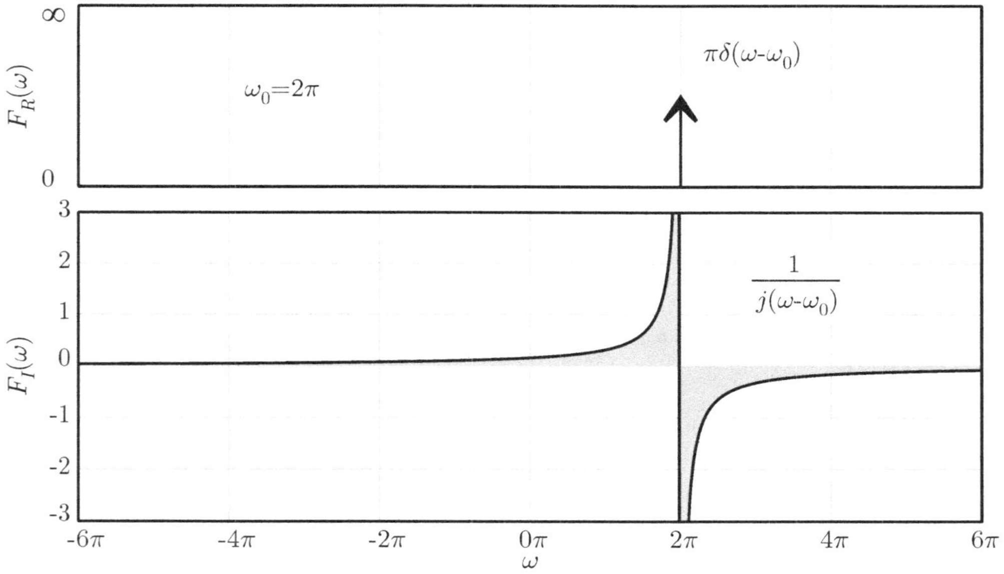

Fig. 8.13 Fourier transform of single-sided complex function

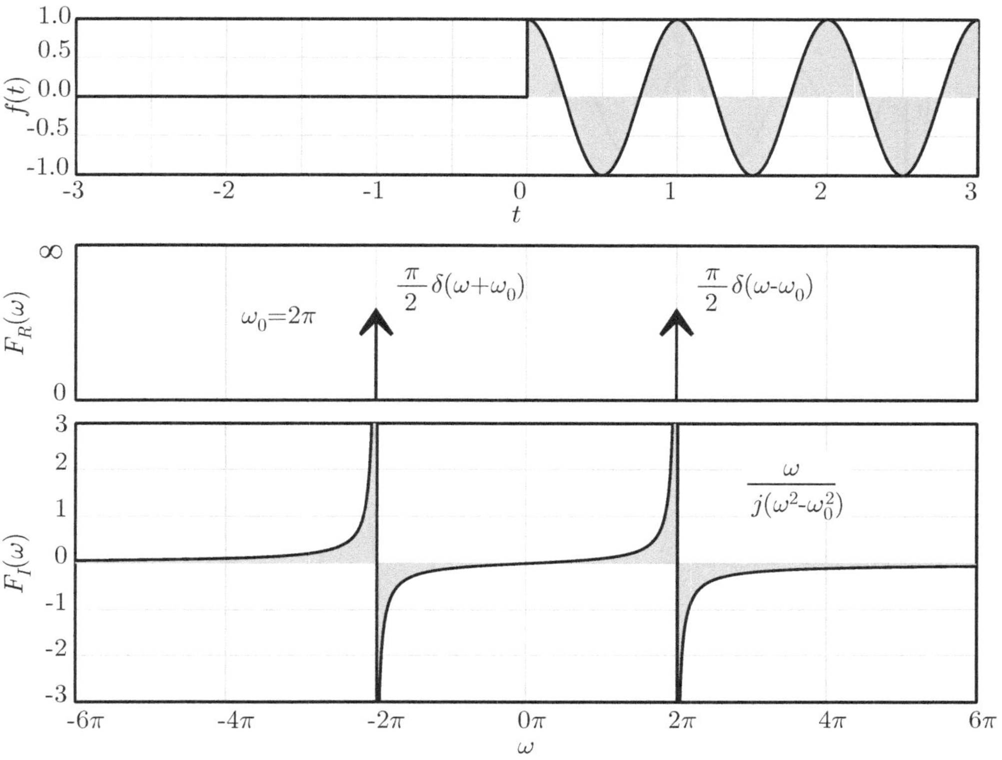

Fig. 8.14 Fourier transform of single-sided cosine function

Our FT then becomes

$$F(\omega) = \frac{1}{2\pi}\left\{\pi^2\left[\delta(\omega-\omega_0)+\delta(\omega+\omega_0)\right] + \frac{\pi}{j}\left[\frac{1}{\omega-\omega_0}+\frac{1}{\omega+\omega_0}\right]\right\}$$
$$= \frac{\pi}{2}\left[\delta(\omega-\omega_0)+\delta(\omega+\omega_0)\right] + \frac{1}{j}\frac{\omega}{\omega^2-\omega_0^2} \quad (8.71)$$

in agreement with Eq. (8.66). Notice that this FT is comprised of two components: The first corresponds to a cosine function (divided by 2) and the second corresponds to a cosine (again divided by 2), multiplied by the signum function; that is a cosine which flips sign for negative time. It just happens that when we add them we get identically zero for negative time, and the normal cosine function for positive time. This is shown in Fig. 8.15.

8.15 Fourier Transform of Single-Sided Sine Function

Similar to prior section, here we find the Fourier transform of the single-sided sine function defined by

$$f(t) = u(t)\sin(\omega_0 t) \quad (8.72)$$

Knowing the FT of the sine, unit step function, and using convolution theory we get

$$F(\omega) = \frac{1}{2\pi}\left\{\left[\frac{\pi}{j}\left[\delta(\omega-\omega_0)-\delta(\omega+\omega_0)\right]\right] * \left[\pi\delta(\omega)+\frac{1}{j\omega}\right]\right\}$$
$$= \frac{1}{2\pi}\left\{\frac{\pi^2}{j}\left[\delta(\omega-\omega_0)-\delta(\omega+\omega_0)\right] - \pi\left[\frac{1}{\omega-\omega_0}-\frac{1}{\omega+\omega_0}\right]\right\}$$

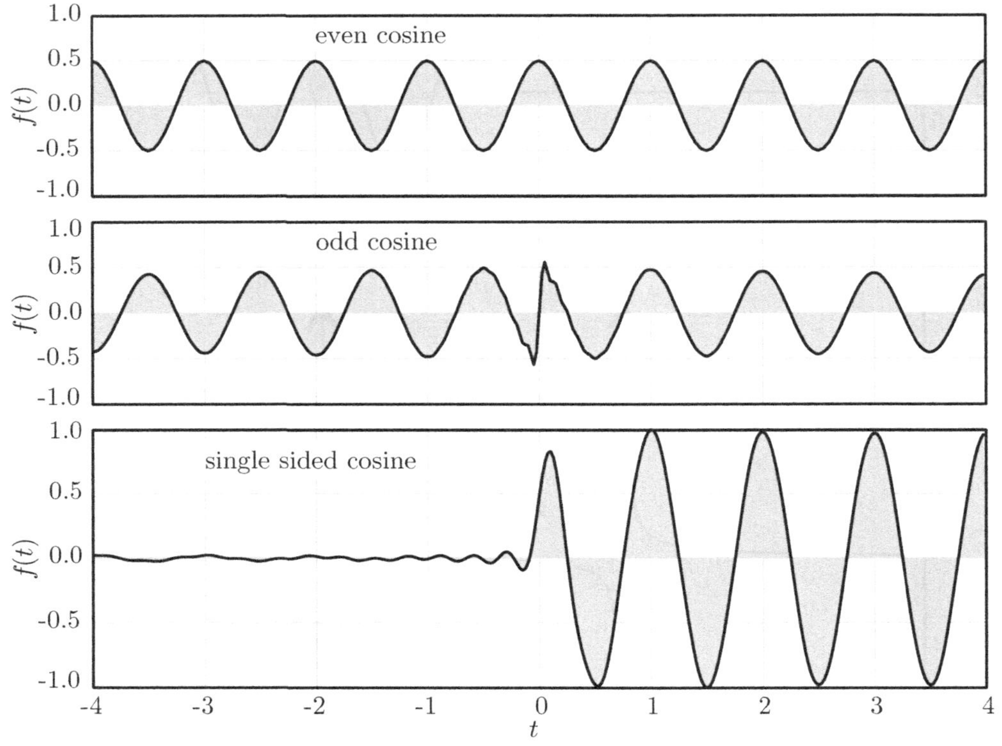

Fig. 8.15 Single-sided cosine function and reconstruction from inverse FT

$$= \boxed{\frac{\pi}{2j}\left[\delta(\omega-\omega_0)-\delta(\omega+\omega_0)\right]-\frac{\omega_0}{\omega^2-\omega_0^2}} \tag{8.73}$$

This transform is shown in Fig. 8.16. Again notice that this FT is comprised of two components. This first component corresponds to a sine function (divided by 2) and the second component corresponds to a sine function (divided by 2) which is flipped in sign for negative time. When both are added we are assured to regain the target function, which is zero for negative time and sine for positive time. Figure 8.17 confirms our flow and methods.

8.16 Summary

This is the third chapter along the journey of spectral analysis. We started with the real Fourier series, then complex one, and used this latter one to cross the bridge from the periodic world to the non-periodic one—the Fourier transform. The Fourier transform is extremely important and generic. In fact if we know the Fourier transform then we can get the Fourier series rather easily. Also, the Fourier transform will later pave the way for the Laplace transform (more on this later). The main premise in this chapter is that we can represent an aperiodic signal in terms of a summation (integral) of complex exponentials with various frequencies. Implicit in the chapter is that the starting signal is "finite" or "integrable" although we were able sometimes (such is the case of the unit step function or the continuous sines/cosines) to still arrive at the Fourier transform of "singular" functions. We derived the Fourier transform from the complex Fourier series by smoothly increasing the period to infinity and changing the summation to integration (over frequency). We illustrated the Fourier transform with ample examples, and problems. But this is just the first installment! More on the Fourier transform, including properties and further examples, follow in the next few chapters.

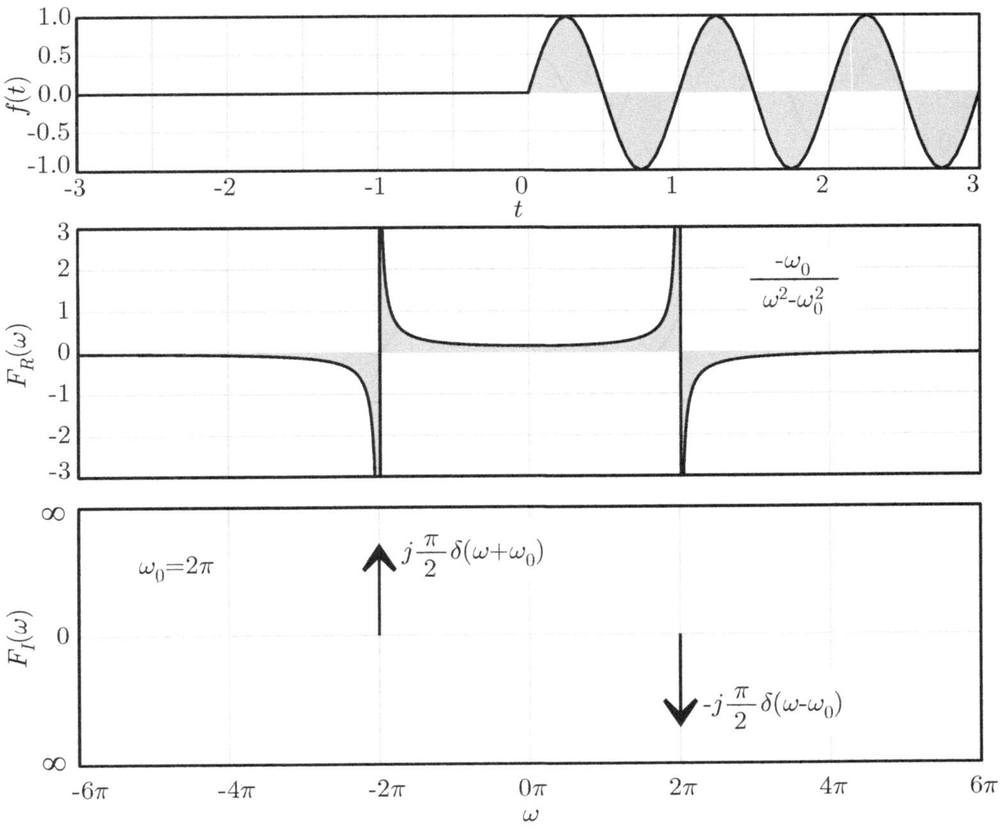

Fig. 8.16 Fourier transform of single-sided sine function

8.17 Problems

1. Find the Fourier transform of the shifted pulse, defined between 0 and τ. Plot the time series for $\tau = 0.5$. See sample solution in Fig. 8.18.
   Answer:
$$F(\omega) = \frac{1 - e^{-j\omega\tau}}{j\omega}$$
2. Start with solution to Problem 1. Split the Fourier transform into real part and imaginary one. For the case of $\tau = 0.5$ plot the time series for each case, then add both time series. See sample solution in Fig. 8.19.
   Answer:
$$\Re F(\omega) = \frac{\sin \omega\tau}{\omega}$$
$$\Im F(\omega) = \frac{1 - \cos \omega\tau}{j\omega}$$
3. The Fourier transform of the DC function was shown to be $2\pi\delta(\omega)$; prove this is correct by finding the inverse transform. Use the following fact (sampling property) about the delta function
$$\int_{-\infty}^{\infty} f(x)\delta(x - x_0) = f(x_0)$$
4. If an aperiodic function has the following Fourier *transform*
$$F(\omega) = \frac{1}{1 + \omega^2}$$
   what would the periodic version of this function, with period T, have as a Fourier *series*? Plot time series for both cases assuming $T = 10$. See sample results in Fig. 8.20.

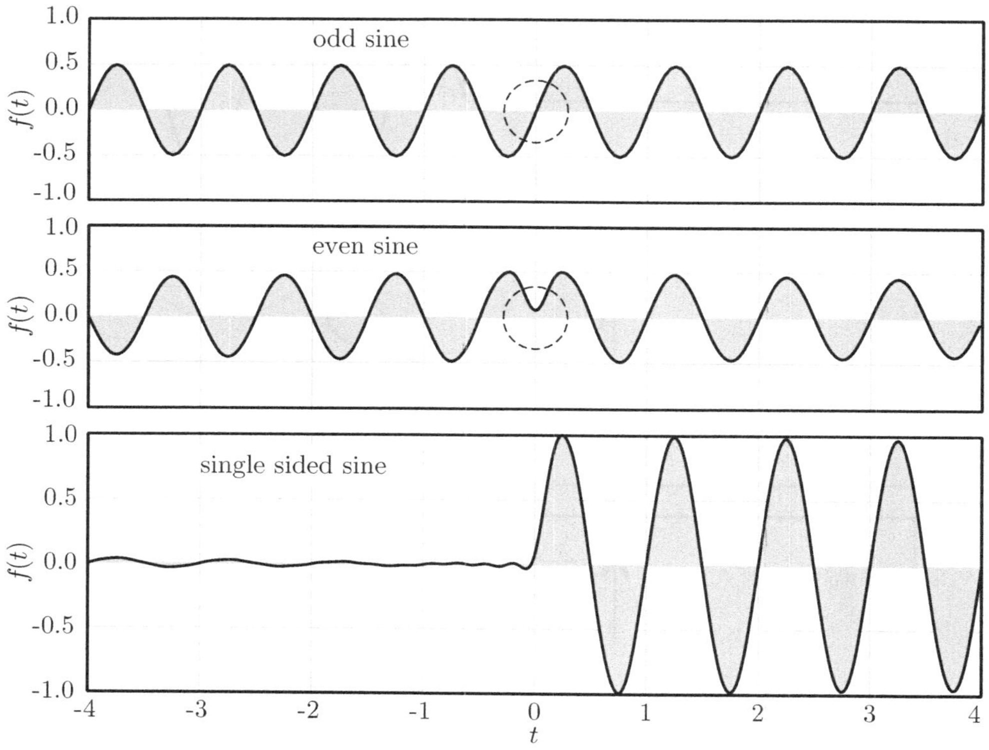

Fig. 8.17 Single-sided sine function and reconstruction from inverse FT

5. If a periodic function, with period T, has the following Fourier *series*

$$F(\omega_n) = \frac{-j\omega_n}{1+\omega_n^2}$$

what would the aperiodic version of this function have as a Fourier *transform*? Plot time series for both cases assuming $T = 10$. See sample solution in Fig. 8.21.

6. The Fourier transform of the signum function is

$$\text{signum}(t) \rightarrow \frac{1}{j\omega}$$

What this means is that the inverse transform of $\frac{1}{j\omega}$ is -0.5 for negative time and 0.5 for positive time. The inverse transform amounts to frequency integration; how is it that a singular function $\frac{1}{j\omega}$ which literally blows up, let alone not being continuous at zero be integrable? Hint: we are not integrating $\frac{1}{j\omega}$ alone; instead we are integrating $\frac{e^{j\omega t}}{j\omega}$. And due to symmetry only the sine part of the complex exponential remains so really we are integrating $\frac{\sin \omega t}{\omega}$. Plot this last term and convince yourself that it *is* integrable (in the frequency domain), for any t value! See sample solution in Fig. 8.22.

7. A function has the following Fourier transform

$$F(\omega) = \frac{1 - e^{-t_0(a+j\omega)}}{a + j\omega}$$

What is the function (in time domain)? Next, plot the time series of the first term of $F(\omega)$ then the time series for the second term, and finally the sum. Assume $a = 1$ and $t_0 = 2$. See sample solution in Fig. 8.23. Answer:

$$f(t) = e^{-at},\ (0 < t < t_0);\ 0,\ (\text{other wise})$$

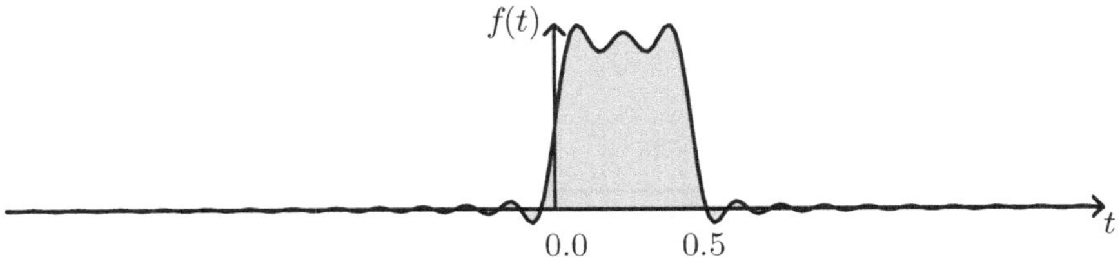

Fig. 8.18 Solution to Problem 1

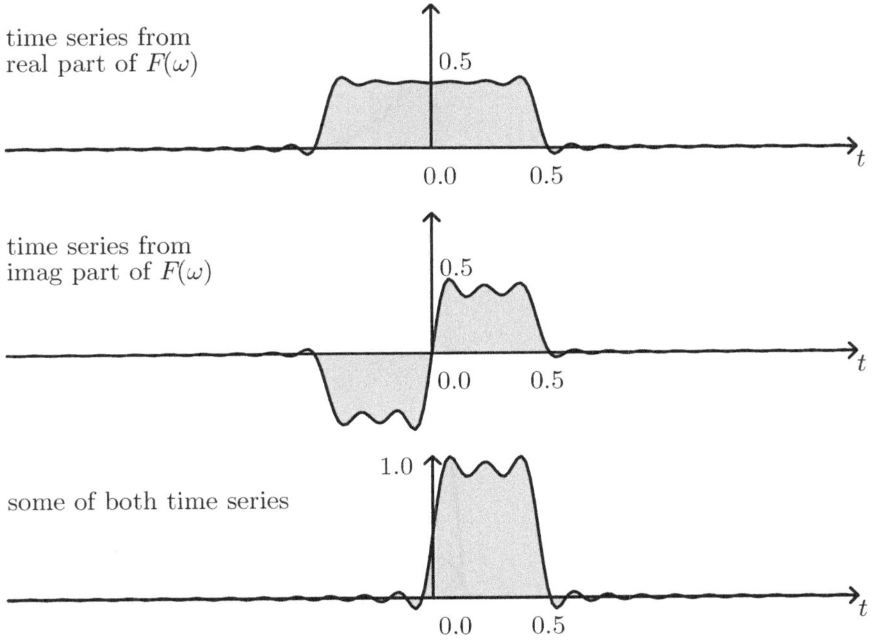

Fig. 8.19 Solution to Problem 2

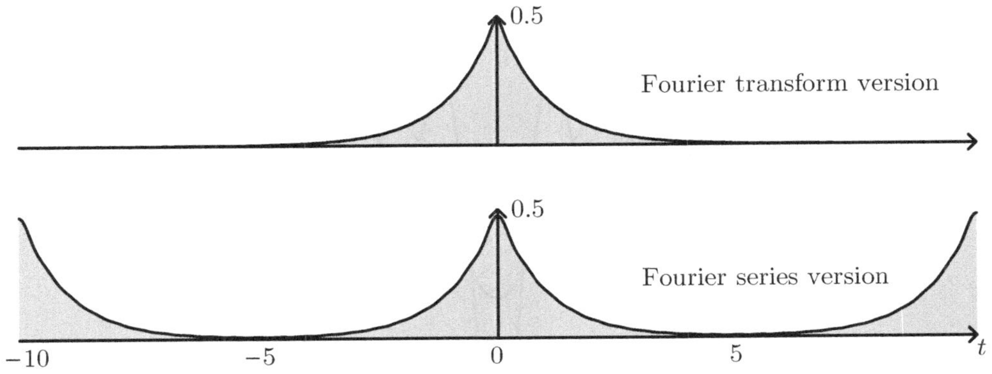

Fig. 8.20 Solution to Problem 4

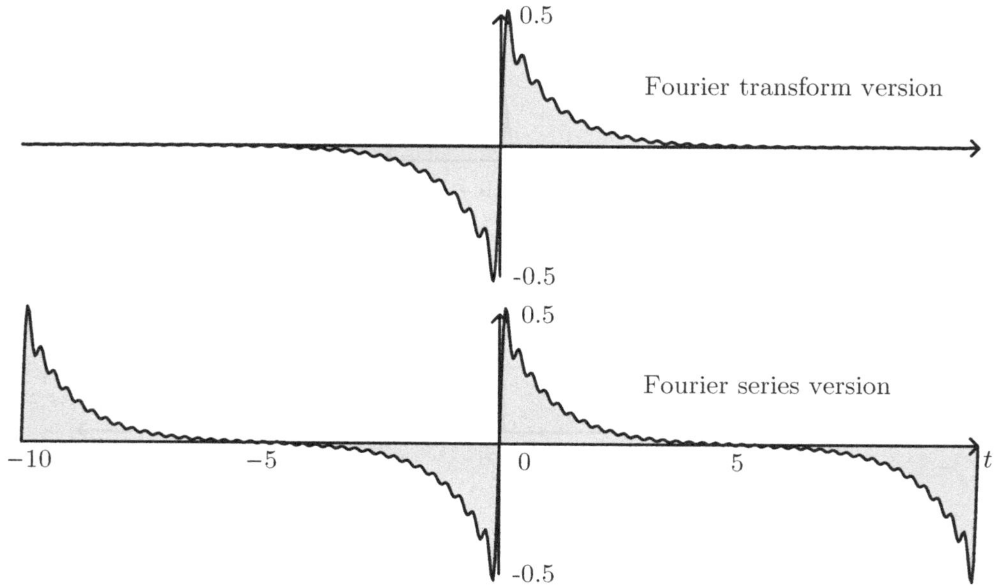

Fig. 8.21 Solution to Problem 5

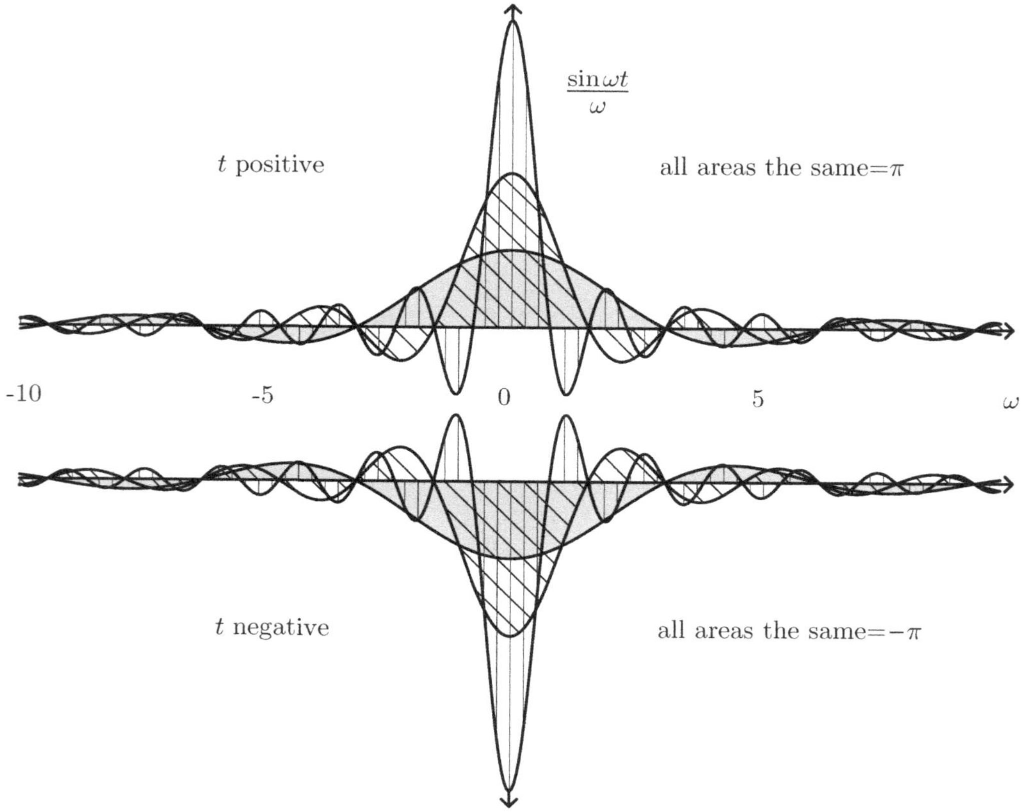

Fig. 8.22 Solution to Problem 6

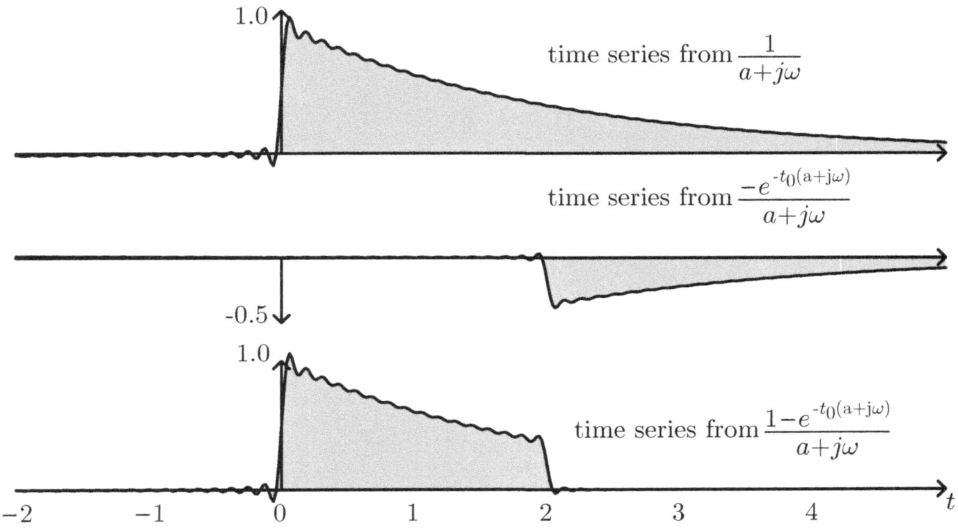

Fig. 8.23 Solution to Problem 7

9 Properties of the Fourier Transforms

9.1 Introduction

Having introduced in the last chapter the Fourier transform for dealing with aperiodic functions, and having shown some applications examples we may ask the question—how can we expedite deriving the Fourier transform of a given signal? Turns out, there is a set of very useful properties associated with the Fourier transform pair that should enable us to derive the Fourier transform of a new signal—relatively easily—*if* we know how to relate it to another one, with a priori known Fourier transform. In deriving those properties we also learn more about the machinery of the Fourier transform which makes the upcoming material doubly worth time invested on it!

9.2 Linearity of Fourier Transform

Basically this property states that the Fourier transform of the sum (or difference) of two functions is the sum (or difference) of the individual transforms.

$$\boxed{\text{Fourier transform of } [f(t)+g(t)] = F(\omega)+G(\omega)} \tag{9.1}$$

Another flavor of this linearity property is

$$\text{Fourier transform of } [af(t)] = aF(\omega) \tag{9.2}$$

where a is some scaling factor. The linearity property is an extremely handy one, and many complex functions can be broken down into summation of simpler ones. While this seems like a trivial property, it has tremendous use, as many signals can be decomposed in terms of more elementary ones.

9.2.1 First Example of Linearity

Consider for example the unit step function which has as FT

$$u(t) \rightarrow \pi\delta(\omega) + \frac{1}{j\omega} \tag{9.3}$$

Similarly, the mirrored unit step function has the FT

$$u(-t) \rightarrow \pi\delta(\omega) - \frac{1}{j\omega} \tag{9.4}$$

If we add the two step functions we would get a constant in the time domain (the DC function). In the frequency domain we get

$$\begin{aligned} u(t) + u(-t) \rightarrow \pi\delta(\omega) &+ \frac{1}{j\omega} + \pi\delta(\omega) - \frac{1}{j\omega} \\ &= 2\pi\delta(\omega) \end{aligned} \tag{9.5}$$

which is the correct answer; recall the DC function transforms to $2\pi\delta(\omega)$.

F. Badrieh, *Spectral, Convolution and Numerical Techniques in Circuit Theory*,
https://doi.org/10.1007/978-3-319-71437-0_9

9.2.2 Second Example of Linearity

As a second example of linearity consider the FT of the cosine function

$$\cos(\omega_0 t) \to \pi\delta(\omega - \omega_0) + \pi\delta(\omega + \omega_0) \quad (9.6)$$

On the other hand, the sine function has the FT

$$\sin(\omega_0 t) \to -j\pi\delta(\omega - \omega_0) + j\pi\delta(\omega + \omega_0) \quad (9.7)$$

If we add, in time, the cosine to the sine (with latter multiplied by j) we get

$$\cos\omega_0 t + j\sin\omega_0 t = e^{j\omega_0 t} \quad (9.8)$$

In the frequency domain the sum gives

$$\begin{aligned} &\cos\omega_0 t + j\sin\omega_0 t \to \pi\delta(\omega - \omega_0) \\ &+\pi\delta(\omega + \omega_0) + \pi\delta(\omega - \omega_0) \\ &-\pi\delta(\omega + \omega_0) = 2\pi\delta(\omega - \omega_0) \end{aligned} \quad (9.9)$$

again which is the correct FT for the complex exponential $e^{j\omega_0 t}$; recall

$$e^{j\omega_0 t} \to 2\pi\delta(\omega - \omega_0) \quad (9.10)$$

9.3 Time Scaling of Fourier Transform

If the time signals is scaled (multiplied) by a constant a, then the Fourier transform is scaled by the inverse of a and spread out in frequency as follows:

$$\boxed{\mathcal{F}[f(at)] = \frac{1}{|a|}F\left(\frac{\omega}{a}\right)} \quad (9.11)$$

Proof. Start with definition of FT

$$\mathcal{F}[f(t)] = \int_{-\infty}^{\infty} f(t)e^{-j\omega t}dt \quad (9.12)$$

Assume that a is positive, and scale time by a

$$\mathcal{F}[f(at)] = \int_{-\infty}^{\infty} f(at)e^{-j\omega t}dt \quad (9.13)$$

Let $u = at$ such that $t = u/a$ and $dt = du/a$; then

$$\begin{aligned} \mathcal{F}[f(at)] &= \frac{1}{a}\int_{-\infty}^{\infty} f(u)e^{-j\frac{\omega}{a}u}du \\ &= \frac{1}{a}F\left(\frac{\omega}{a}\right) \qquad (a \text{ positive}) \end{aligned} \quad (9.14)$$

Next assume that a is negative and use same variable substitution. The only difference now is that the limits of integration get reversed, so that we pick a negative sign

$$\begin{aligned} \mathcal{F}[f(at)] &= -\frac{1}{a}\int_{-\infty}^{\infty} f(u)e^{-j\frac{\omega}{a}u}du \\ &= -\frac{1}{a}F\left(\frac{\omega}{a}\right) \qquad (a \text{ negative}) \end{aligned} \quad (9.15)$$

Between these last two results we arrive at what we wanted to prove:

$$\mathcal{F}[f(at)] = \frac{1}{|a|}F\left(\frac{\omega}{a}\right) \quad (9.16)$$

9.3.1 Example of Time Scaling of Fourier Transform

Consider for example the single-sided negative exponential

$$u(t)e^{-t} \to \frac{1}{j\omega + 1} \quad (9.17)$$

Let us find the FT of the time scaled version

$$f(t) = u(t)e^{-2t} \quad (9.18)$$

Based on the property just proposed we should get

$$F(\omega) = \frac{1}{2}\frac{1}{j\frac{\omega}{2} + 1} = \frac{1}{2}\frac{2}{j\omega + 2} = \frac{1}{j\omega + 2} \quad (9.19)$$

which is the correct answer, recalling

$$u(t)e^{-at} \rightarrow \frac{1}{j\omega + a} \tag{9.20}$$

9.4 Reciprocity of the Fourier Transform

This states that

$$\boxed{\begin{array}{l}\text{if } f(t) \rightarrow \quad F(\omega)\\ \text{then } F(t) \rightarrow 2\pi f(-\omega)\end{array}} \tag{9.21}$$

This is a very important result, but not very easy to understand. The proof is as follows. Start with the FT definition

$$G(\omega) = \int_{-\infty}^{\infty} g(t)e^{-j\omega t}dt \tag{9.22}$$

Once we know the FT we can find its inverse

$$g(t) = \frac{1}{2\pi}\int_{-\infty}^{\infty} G(\omega)e^{j\omega t}d\omega \tag{9.23}$$

Now change from angular to linear frequency

$$g(t) = \int_{-\infty}^{\infty} G(2\pi f)e^{j2\pi ft}df \tag{9.24}$$

Change the sign of time

$$g(-t) = \int_{-\infty}^{\infty} G(2\pi f)e^{-j2\pi ft}df \tag{9.25}$$

Now let $t = f$ and $f = t$

$$g(-f) = \int_{-\infty}^{\infty} G(2\pi t)e^{-j2\pi ft}dt = \int_{-\infty}^{\infty} G(2\pi t)e^{-j\omega t}dt \tag{9.26}$$

Examining the right side we notice it is nothing other than the Fourier transform of $G(2\pi t)$! That is

$$\mathcal{F}[G(2\pi t)] = g(-f) \tag{9.27}$$

Now use the scaling property of the FT and get

$$\mathcal{F}\left[G\left(\frac{2\pi t}{2\pi}\right)\right] = 2\pi g(-f2\pi) \tag{9.28}$$

Or put in final format

$$\boxed{\mathcal{F}[G(t)] = 2\pi g(-\omega)} \tag{9.29}$$

which is what we set forth to achieve.

9.4.1 First Example of Reciprocity of the Fourier Transform

Consider the time delta function which has a FT of

$$\delta(t) \rightarrow 1 \tag{9.30}$$

By the reciprocity theorem this would imply that

$$1 \rightarrow 2\pi\delta(-\omega) = 2\pi\delta(\omega) \tag{9.31}$$

which we know to be true! (Notice that we made use of the fact that the delta function is even.)

9.4.2 Second Example of Reciprocity of the Fourier Transform

We know that the cosine function has the FT

$$\cos\omega_0 t \rightarrow \pi\delta(\omega - \omega_0) + \pi\delta(\omega + \omega_0) \tag{9.32}$$

By the reciprocity theorem this would imply that

$$\pi\delta(t - t_0) + \pi\delta(t + t_0) \rightarrow 2\pi\cos(\omega t_0)? \tag{9.33}$$

Let's check! We know that

$$\delta(t) \rightarrow 1 \tag{9.34}$$

and we know (as shown in the time shifting properties) that

$$\pi\delta(t - t_0) \rightarrow \pi e^{-j\omega t_0}, \text{ and } \pi\delta(t + t_0) \rightarrow \pi e^{j\omega t_0} \tag{9.35}$$

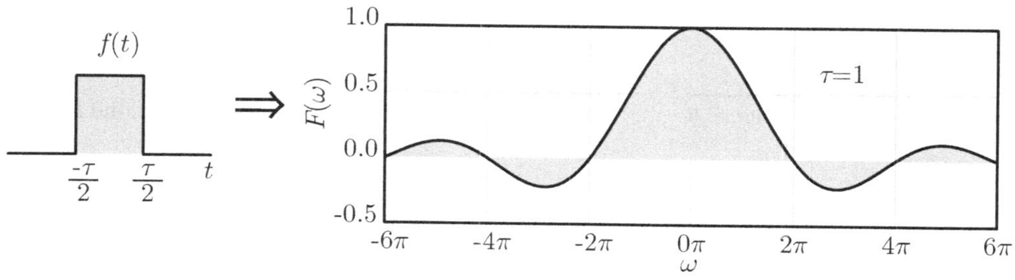

Fig. 9.1 Centered pulse and Fourier transform

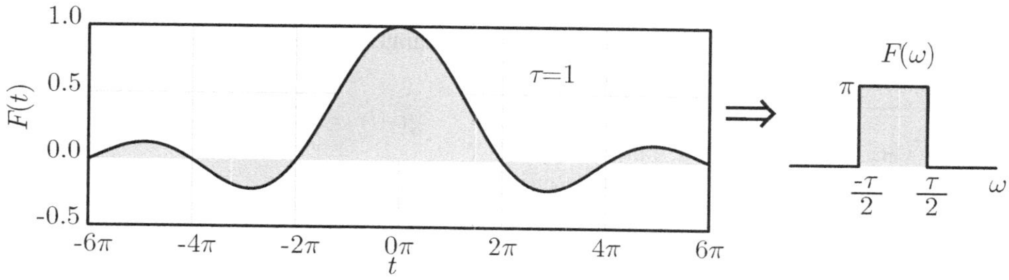

Fig. 9.2 Sinc function and Fourier transform

Adding these last two transforms we get

$$\pi e^{-j\omega t_0} + \pi e^{j\omega t_0} = \pi\,[2\cos(\omega t_0)] = 2\pi\cos(\omega t_0) \tag{9.36}$$

indeed agreeing with Eq. (9.33).

9.4.3 Third Example of Reciprocity of the Fourier Transform

We know that the centered pulse function, of width τ has the FT

$$[\text{centered pulse in time, of width } \tau] \rightarrow 2\frac{\sin\omega\tau/2}{\omega} \tag{9.37}$$

This is shown in Fig. 9.1. Using the reciprocity theorem this would imply that

$$\frac{\sin t\tau/2}{t} \rightarrow \pi\,[\text{centered pulse in frequency, with width } \tau] \tag{9.38}$$

In other words we should get the transform shown in Fig. 9.2. How can we confirm this last result? In other words how can we confirm that

$$\frac{\sin t\tau/2}{t} = \frac{1}{2\pi}\int_{-\tau/2}^{\tau/2} \pi e^{j\omega t} d\omega \tag{9.39}$$

We can carry on the integration as

$$\frac{1}{2\pi}\int_{-\tau/2}^{\tau/2} \pi e^{j\omega t} d\omega = \int_{0}^{\tau/2} \cos(\omega t) d\omega$$
$$= \frac{1}{t}\sin(\omega t)\Big|_{\omega=0}^{\omega=\tau/2} = \frac{1}{t}\sin(t\tau/2) \tag{9.40}$$

which gets us what we wanted. We can also carry on the frequency integration numerically (just as an exercise) and as shown in Fig. 9.3.

9.5 Time Shifting of Fourier Transform

If a signal shifts in time by t_0 then the Fourier transform gets multiplied by $e^{-j\omega t_0}$

$$\boxed{\text{Fourier transform of } [f(t-t_0)] = e^{-j\omega t_0} F(\omega)} \tag{9.41}$$

To prove this we do the following. Assume that

$$f(t) \rightarrow F(\omega) \tag{9.42}$$

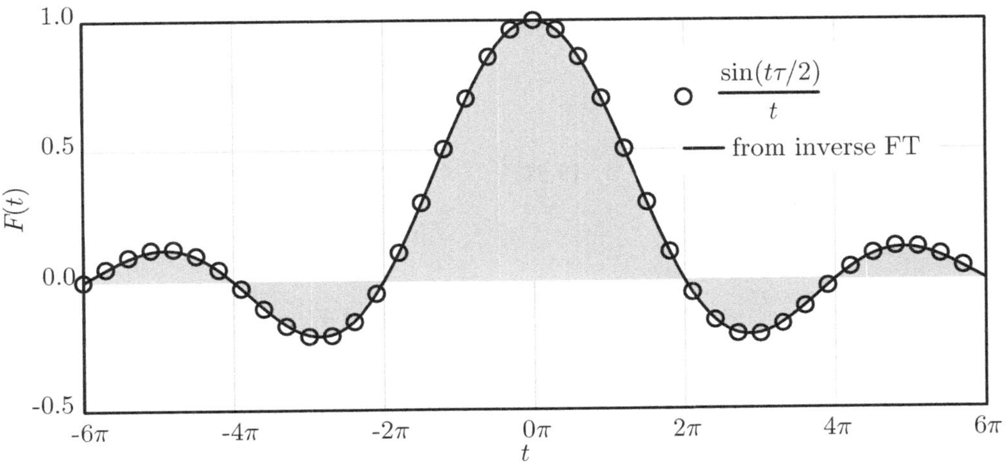

Fig. 9.3 Time sinc function and that from inverse FT of pulse function in frequency domain

The Fourier transform of the shifted signal is

$$\mathcal{F}[f(t-t_0)] = \int_{-\infty}^{\infty} f(t-t_0)e^{-j\omega t}dt \quad (9.43)$$

Do the following variable substitution:

$$u = t - t_0; \qquad du = dt; \qquad t = u + t_0 \quad (9.44)$$

so that

$$\begin{aligned} \mathcal{F}[f(t-t_0)] &= \int_{-\infty}^{\infty} f(u)e^{-j\omega(u+t_0)}du \\ &= e^{-j\omega t_0}\int_{-\infty}^{\infty} f(u)e^{-j\omega u}du \\ &= e^{-j\omega t_0}F(\omega) \end{aligned} \quad (9.45)$$

Notice that multiplication by $e^{-j\omega t_0}$ does not change the amplitude of the Fourier transform, since $|e^{-j\theta}| = 1$. The multiplication of the Fourier transform by $e^{-j\omega t_0}$ simply amounts to shifting the sinusoids in time (while doing inverse transform) by the t_0. To show this more clearly take

$$\begin{aligned} e^{-j\omega t_0}e^{j\omega t} &= e^{j\omega(t-t_0)} \\ &= \cos\omega(t-t_0) + j\sin\omega(t-t_0) \end{aligned} \quad (9.46)$$

So we can see both sine and cosine functions in the inverse Fourier transform get shifted by in time by t_0. This makes sense, since by simply shifting the original signal in time, we are not altering its frequency spectrum—instead we are simply delaying/or pushing it in time.

9.5.1 First Example of Time Shifting of Fourier Transform: The Offset Pulse

As an application example of this theory consider finding the FT of the shifted pulse of width τ and time center at $\tau/2$; that is

$$f(t) = \begin{cases} 1 & 0 < t < \tau \\ 0 & \text{else wise} \end{cases} \quad (9.47)$$

Rather than doing the time integration manually we can use the time-shift property. Recall that the FT of the pulse function, centered at time zero is

$$\mathcal{F}[\text{centered pulse}] = 2\frac{\sin\omega\tau/2}{\omega} \quad (9.48)$$

By using the time-shifting property, and recognizing that the shifted pulse is the centered pulse shifted to the right by $t = \tau/2$, we get the FT of the desired function as

$$\mathcal{F}[f(t)] = \mathcal{F}[\text{pulse function}] \cdot e^{-j\omega\tau/2} \quad (9.49)$$

$$= 2\frac{\sin \omega\tau/2}{\omega} e^{-j\omega\tau/2} \quad (9.50)$$

Let's process this further using

$$\sin \omega = \frac{1}{2j}\left[e^{j\omega} - e^{-j\omega}\right] \quad (9.51)$$

Then we have

$$\mathcal{F}[f(t)] = \frac{e^{j\omega\tau/2} - e^{-j\omega\tau/2}}{j\omega} e^{-j\omega\tau/2} = \frac{1 - e^{-j\omega\tau}}{j\omega} \quad (9.52)$$

How do we know this is true? Let's do the integration directly!

$$\mathcal{F}[f(t)] = \int_0^\tau e^{-j\omega t} dt = -\frac{1}{j\omega} e^{-j\omega t}\Big|_0^\tau = \frac{1 - e^{-j\omega\tau}}{j\omega} \quad (9.53)$$

in agreement with Eq. (9.52). We can summarize our findings as shown in Fig. 9.4.

9.5.2 Second Example of Time Shifting of Fourier Transform: Deriving Pulse FT from Unit Step One

As a second example of the time shifting property we will next show how we can derive the FT of the (centered) pulse knowing the FT of the unit step one! First we recognize that the centered pulse of width τ can be written in terms of the unit step function as

$$\text{pulse function} = u(t+\tau/2) - u(t-\tau/2) \quad (9.54)$$

This is shown in Fig. 9.5. We can now derive the FT of the pulse knowing those of the (shifted) unit step functions. First recall

$$u(t) \rightarrow \pi\delta(\omega) + \frac{1}{j\omega} \quad (9.55)$$

Then

$$u(t - \tau/2) \rightarrow e^{-j\omega\tau/2}\left[\pi\delta(\omega) + \frac{1}{j\omega}\right]$$
$$= \pi\delta(\omega) + \frac{e^{-j\omega\tau/2}}{j\omega} \quad (9.56)$$

and

$$u(t + \tau/2) \rightarrow \pi\delta(\omega) + \frac{e^{+j\omega\tau/2}}{j\omega} \quad (9.57)$$

Then the FT of the pulse is

$$\mathcal{F}[\text{pulse}] = \left[\pi\delta(\omega) + \frac{e^{+j\omega\tau/2}}{j\omega}\right]$$
$$-\left[\pi\delta(\omega) + \frac{e^{-j\omega\tau/2}}{j\omega}\right] = \frac{e^{+j\omega\tau/2} - e^{-j\omega\tau/2}}{j\omega}$$
$$= \boxed{\frac{2\sin \omega\tau/2}{\omega}} \quad (9.58)$$

as expected.

9.6 Frequency Shifting of Fourier Transform

If the Fourier transform shifts in frequency (to the right), then the inverse is multiplied by a complex exponential of that frequency.

$$\boxed{\text{Fourier transform of } \left[e^{j\omega_0 t} f(t)\right] = F(\omega - \omega_0)} \quad (9.59)$$

To prove this assume that

$$f(t) \rightarrow F(\omega) \quad (9.60)$$

Multiply $f(t)$ by $e^{j\omega_0 t}$ and find Fourier transform of result

$$\begin{aligned}\mathcal{F}[e^{j\omega_0 t} f(t)] &= \int_{-\infty}^{\infty} e^{j\omega_0 t} f(t) e^{-j\omega t} dt \\ &= \int_{-\infty}^{\infty} f(t) e^{-j\omega t + j\omega_0 t} dt \\ &= \int_{-\infty}^{\infty} f(t) e^{-jt(\omega-\omega_0)} dt \\ &= F(\omega - \omega_0)\end{aligned} \quad (9.61)$$

9.6.1 Example of Frequency Shifting of Fourier Transform

A sample application of this theory is the single-sided complex exponential

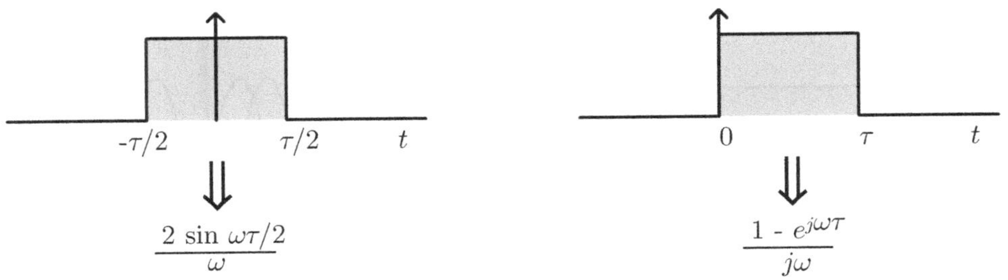

Fig. 9.4 Centered and offset pulse and corresponding Fourier transforms

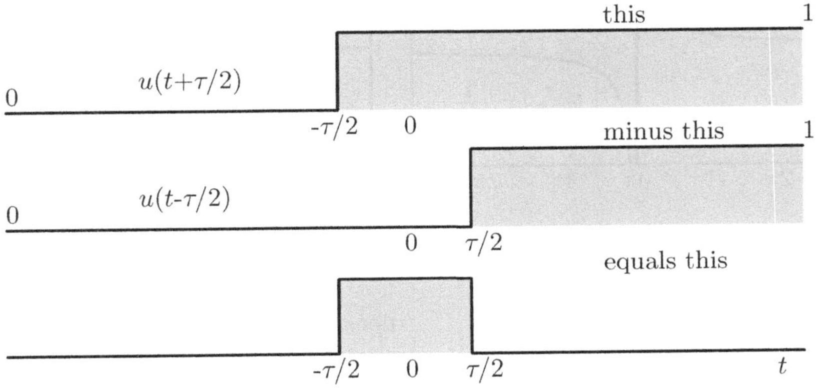

Fig. 9.5 Centered pulse composition in terms of shifted unit step functions

$$f(t) = \begin{cases} e^{j\omega_0 t} & t > 0 \\ 0 & \text{otherwise} \end{cases} \tag{9.62}$$

We know we can obtain this function by simply multiplying the normal complex exponential (defined for both negative and positive times) times the unit step function

$$f(t) = e^{j\omega_0 t} \times u(t) \tag{9.63}$$

We also know that the latter has the FT

$$u(t) \to \pi\delta(\omega) + \frac{1}{j\omega} \tag{9.64}$$

This then would imply that

$$\boxed{e^{j\omega_0 t} u(t) \to \pi\delta(\omega - \omega_0) + \frac{1}{j(\omega - \omega_0)}} \tag{9.65}$$

in agreement with results from prior chapter (Eq. (8.62)). These results are shown graphically in Fig. 9.6.

9.7 Time Differentiation of Fourier Transform

This property states that if we take the time derivative of the original function, the Fourier transform of the result equals the original Fourier transform times $j\omega$

$$\boxed{\text{Fourier transform of } \left[\frac{df(t)}{dt}\right] = j\omega F(\omega)} \tag{9.66}$$

We can prove this using two methods.

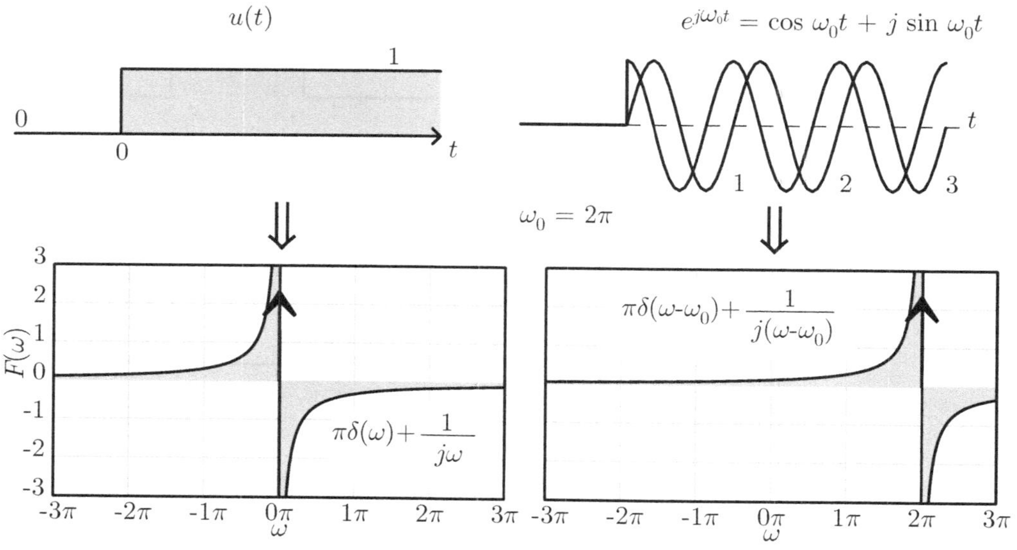

Fig. 9.6 Single-sided complex exponential and Fourier transform

First Proof Assume that

$$f(t) \to F(\omega) \tag{9.67}$$

Differentiate this signal and find Fourier transform of result

$$\mathcal{F}\left[\frac{df(t)}{dt}\right] = \int_{-\infty}^{\infty} \frac{df(t)}{dt} e^{-j\omega t} dt \tag{9.68}$$

Use integration by part such that

$$u = e^{-j\omega t}, \quad du = -j\omega e^{-j\omega t}, \quad dv = \frac{df(t)}{dt}, \quad v = f(t) \tag{9.69}$$

$$\mathcal{F}\left[\frac{df(t)}{dt}\right] = f(t)e^{-j\omega t}\Big|_{-\infty}^{\infty} + j\omega \int_{-\infty}^{\infty} f(t)e^{-j\omega t} dt = \boxed{j\omega F(\omega)} \tag{9.70}$$

where we have assumed that

$$f(t)e^{-j\omega t}\Big|_{-\infty}^{\infty} = 0, \tag{9.71}$$

meaning $f(t)$ dies off to zero at large times (positive or negative). Examples of such functions are the pulse one, or any time-contained signal. This would apply to any signal that is integrable, which means it is finite in time. For other cases, things get tricky because the very existence of the Fourier transform assumes the signal is integrable. Rather than dwell on the convergence issue, let's test the theory on some examples (after showing the second proof).

Second Proof This proof is easier to show. Start with the inverse Fourier transform

$$f(t) = \frac{1}{2\pi} \int_{-\infty}^{\infty} F(\omega) e^{j\omega t} d\omega \tag{9.72}$$

Take the time derivative of both sides

$$\begin{aligned} \frac{df(t)}{dt} &= \frac{d}{dt}\frac{1}{2\pi} \int_{-\infty}^{\infty} F(\omega) e^{j\omega t} d\omega \\ &= \frac{1}{2\pi} \int_{-\infty}^{\infty} F(\omega) \frac{de^{j\omega t}}{dt} d\omega \\ &= \frac{1}{2\pi} \int_{-\infty}^{\infty} F(\omega) j\omega e^{j\omega t} d\omega \\ &= \frac{1}{2\pi} \int_{-\infty}^{\infty} [j\omega F(\omega)] e^{j\omega t} d\omega \end{aligned} \tag{9.73}$$

Clearly the term in the bracket $[j\omega F(\omega)]$ would be the Fourier transform of the left side of the equation, $df(t)/dt$.

9.7.1 First Example of Time Differentiating Property: Signum and Delta Functions

We know that the FT of the signum function is

$$\mathcal{F}[\mathrm{sig}(t)] = \frac{1}{j\omega} \tag{9.74}$$

If we take the time derivative of this function we get nothing other than the delta function

$$\frac{d}{dt}\mathrm{sig}(t) = \delta(t) \tag{9.75}$$

By the time differentiation property, we would expect the Fourier transform of the derivative (the delta function) to be

$$\mathcal{F}[\delta(t)] = j\omega \times \frac{1}{j\omega} = 1 \tag{9.76}$$

which we know to be true! This is illustrated in Fig. 9.7.

9.7.2 Second Example of Time Differentiating Property: Sine and Cosine Functions

We know that the FT of the sine function is

$$\mathcal{F}[\sin\omega_0 t] = -j\pi\delta(\omega-\omega_0) + j\pi\delta(\omega+\omega_0) \tag{9.77}$$

If we take the time derivative of this function we get

$$\frac{d}{dt}\sin\omega_0 t = \omega_0\cos\omega_0 t \tag{9.78}$$

By the time differentiation property, we would expect the Fourier transform of the derivative (the scaled cosine function) to be

$$\begin{aligned}\mathcal{F}[\omega_0\cos\omega_0 t] &= j\omega \times \Big[-j\pi\delta(\omega-\omega_0)\\ &\quad + j\pi\delta(\omega+\omega_0)\Big]\\ &= \omega\pi\delta(\omega-\omega_0) - \omega\pi\delta(\omega+\omega_0)\\ &= \omega_0\pi\Big[\delta(\omega-\omega_0)\\ &\quad + \delta(\omega+\omega_0)\Big]\end{aligned} \tag{9.79}$$

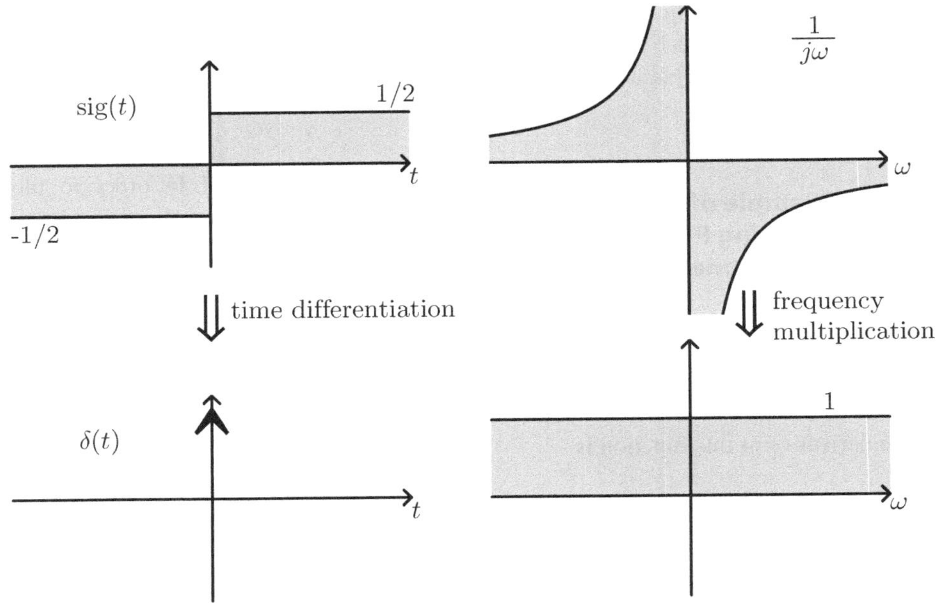

Fig. 9.7 Illustration of time differentiation property of the Fourier transform applied on the signum and delta functions

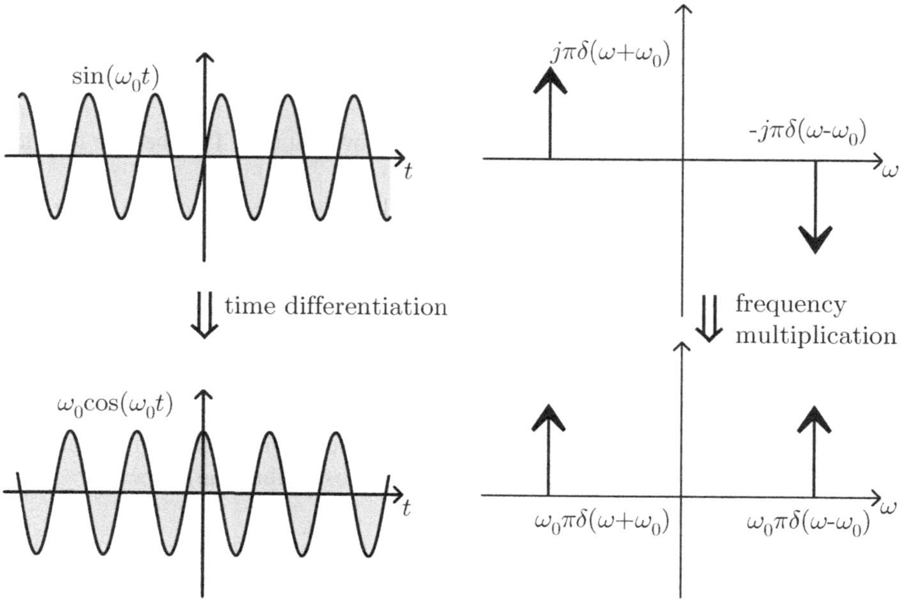

Fig. 9.8 Illustration of time differentiation property of the Fourier transform applied on the sine and cosine functions

where we have used the sampling property of the delta function. Divide both sides by ω_0 and arrive at

$$\cos \omega_0 t \rightarrow \pi \left[\delta(\omega - \omega_0) + \delta(\omega + \omega_0)\right] \quad (9.80)$$

which we know to be true. So we have shown that we are able to derive the FT of the cosine knowing the FT of the sine. This is illustrated in Fig. 9.8.

9.7.3 Third Example of Time Differentiating Property: Negative Exponential

Consider the single-sided negative exponential defined by

$$f(t) = u(t)e^{-at} \quad (9.81)$$

The time derivative of this function is

$$\frac{d}{dt}\left[u(t)e^{-at}\right] = \delta(t) - au(t)e^{-at} \quad (9.82)$$

Notice the presence of the delta function at the origin. It surfaces because we have a discontinuity there. The Fourier transform of the derivative follows

$$\delta(t) - au(t)e^{-at} \rightarrow 1 - \frac{a}{a + j\omega} = \frac{j\omega}{a + j\omega} \quad (9.83)$$

But this is nothing more than the Fourier transform of the original function times $j\omega$

$$\frac{j\omega}{a + j\omega} = \frac{1}{a + j\omega} \times j\omega \quad (9.84)$$

confirming the theory! In order to plot these FTs we need to expand them in terms of real and imaginary parts. For the function itself we expand

$$\frac{1}{a + j\omega} = \frac{a}{a^2 + \omega^2} - j\frac{\omega}{a^2 + \omega^2} \quad (9.85)$$

For the FT of the derivative of the function we have

$$\frac{j\omega}{a + j\omega} = \frac{\omega^2}{a^2 + \omega^2} + j\frac{a\omega}{a^2 + \omega^2} \quad (9.86)$$

Results are shown in Fig. 9.9.

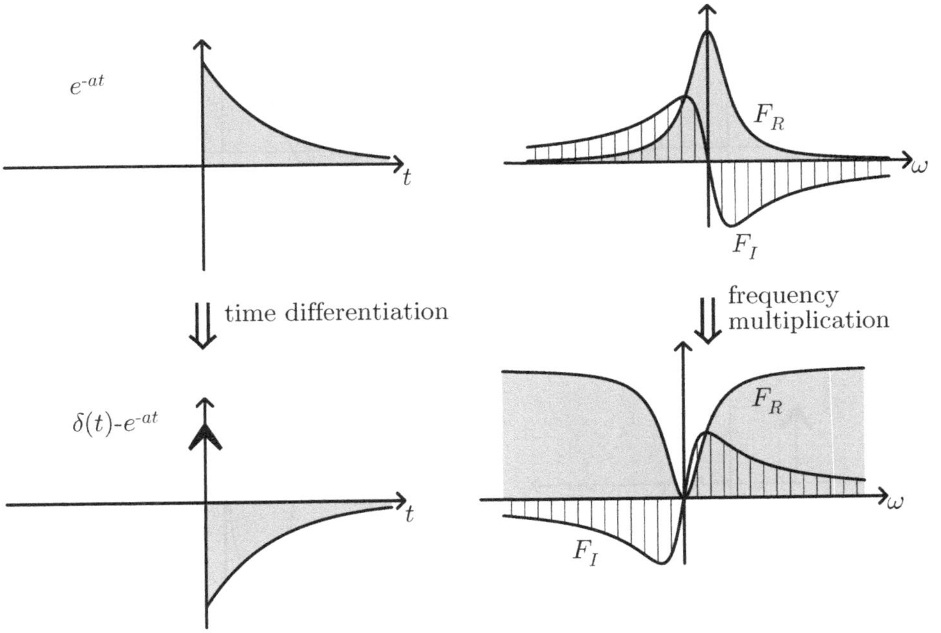

Fig. 9.9 Illustration of time differentiation property of the Fourier transform applied on the negative exponential

9.7.4 Fourth Example of Time Differentiating Property: Pulse Function

We know that a pulse of width 2τ has a FT of

$$\text{pulse of width } 2\tau \rightarrow 2\frac{\sin\omega\tau}{\omega} \tag{9.87}$$

If we take the time derivative of this we get two delta functions as shown in Fig. 9.10—one at $-\tau$ and the other at τ. Now if we find the FT of these two delta functions directly we get

$$\delta(t+\tau) - \delta(t-\tau) \rightarrow e^{j\omega\tau} - e^{-j\omega\tau} = 2j\sin\omega\tau \tag{9.88}$$

On the other hand if we find the FT using the frequency multiplication property we get

$$2\frac{\sin\omega\tau}{\omega} \times j\omega = 2j\sin\omega\tau \tag{9.89}$$

in exact agreement with Eq. (9.88).

9.8 Time Integration Property of Fourier Transform

This property states that the Fourier transform of the integral of $f(t)$ is the Fourier transform of $f(t)$, divided by $j\omega$ plus a scaled delta function:

$$\boxed{\text{Fourier transform of } \left[\int_{-\infty}^{t} f(\tau)d\tau\right] = \frac{1}{j\omega}F(\omega) + \pi F(0)\delta(\omega)} \tag{9.90}$$

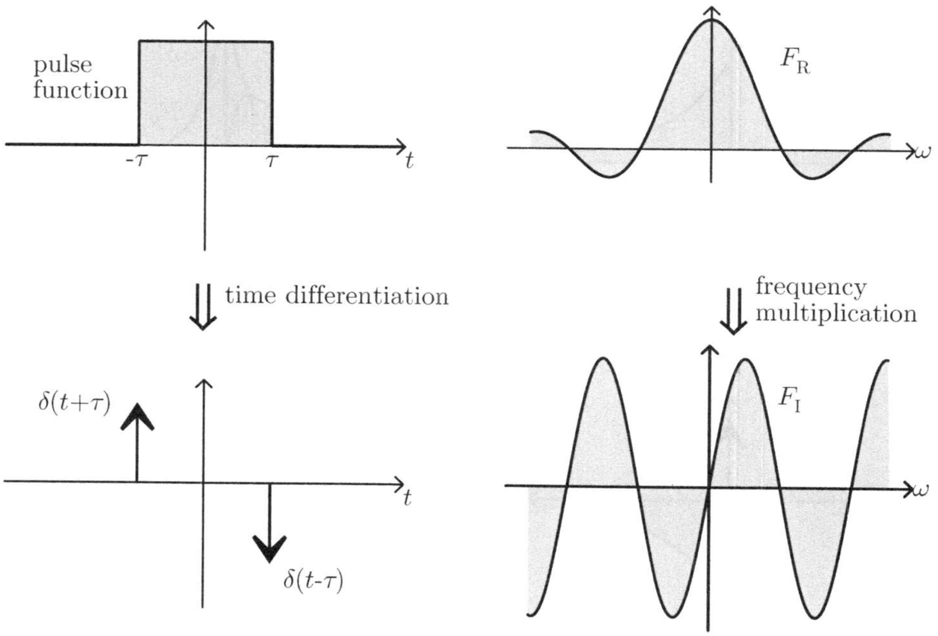

Fig. 9.10 Illustration of time differentiation property of the Fourier transform applied to the pulse function. Here $F_R = 2\frac{\sin\omega\tau}{\omega}$ and $F_I = 2j\sin\omega\tau$

Proof.

$$\mathcal{F}\left[\int_{-\infty}^{t} f(\tau)d\tau\right] = \int_{-\infty}^{\infty}\left[\int_{-\infty}^{t} f(\tau)d\tau\right]e^{-j\omega t}dt \tag{9.91}$$

$$\text{Let } u = \int_{-\infty}^{t} f(\tau)d\tau, \quad du = f(t),$$
$$dv = e^{-j\omega t}, \quad v = -\frac{1}{j\omega}e^{-j\omega t} \tag{9.92}$$

Use integration by parts

$$\begin{aligned}\mathcal{F}[\quad] &= -\int_{-\infty}^{t} f(\tau)d\tau\frac{e^{-j\omega t}}{j\omega}\Big|_{-\infty}^{\infty} \\ &\quad +\frac{1}{j\omega}\int_{-\infty}^{\infty} f(t)e^{-j\omega t}dt \\ &= -\int_{-\infty}^{\infty} f(\tau)d\tau\frac{e^{-j\omega\infty}}{j\omega} + \frac{1}{j\omega}F(\omega) \\ &= -F(0)\frac{e^{-j\omega\infty}}{j\omega} + \frac{1}{j\omega}F(\omega) \\ &= \pi F(0)\delta(\omega) + \frac{1}{j\omega}F(\omega)\end{aligned} \tag{9.93}$$

where we have used the fact that

$$\lim_{T\to\infty}\frac{\sin\omega T}{\omega} = \pi\delta(\omega) \tag{9.94}$$

9.8.1 First Example of Time Integration Property of Fourier Transform: Delta Function

We know that the delta function has the FT

$$\delta(t) \to 1 \tag{9.95}$$

If we integrate the delta function we get the unit step function

$$\int_{-\infty}^{t}\delta(\tau)d\tau = u(t) \tag{9.96}$$

This latter one has the FT

$$u(t) \to \pi\delta(\omega) + \frac{1}{j\omega} \tag{9.97}$$

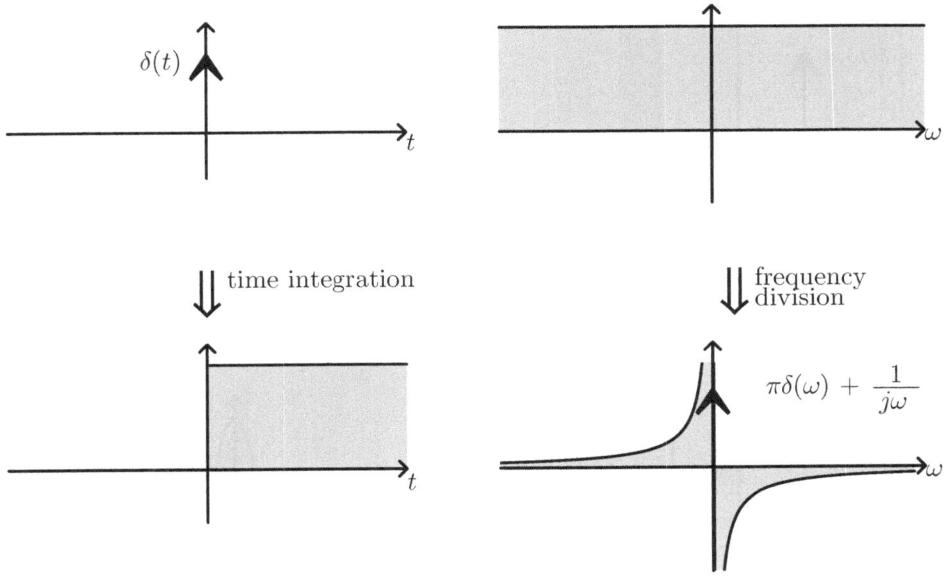

Fig. 9.11 Illustration of the time integration property applied on delta function

If we were to use the time integration property we would get

$$\int_{-\infty}^{t} \delta(\tau)d\tau \rightarrow \pi F(0)\delta(\omega) + \frac{F(\omega)}{j\omega}$$
$$= \pi\delta(\omega) + \frac{1}{j\omega} \qquad (9.98)$$

in perfect agreement with the expected result. Notice that we use the fact that $F(\omega) = 1$ which implies that $F(0) = 1$. The results of this section are illustrated in Fig. 9.11.

9.8.2 Second Example of Time Integration Property of Fourier Transform: Two Shifted Delta's

Two delta functions in the time domain centered at $-\tau$ and at τ, with latter having a negative sign transform to

$$\delta(t + \tau) - \delta(t - \tau) \rightarrow 2j \sin \omega\tau \qquad (9.99)$$

If we integrate the delta functions in time we get the pulse function

$$\int_{-\infty}^{t} \delta(u + \tau) - \delta(u - \tau)du$$
$$= [\text{pulse function of width}]\ 2\tau \qquad (9.100)$$

The FT of the integral we know to be

$$[\text{pulse function of width}]\ 2\tau \rightarrow 2\frac{\sin \omega\tau}{\omega} \qquad (9.101)$$

If we were to instead use the time integration property we would get

$$\int_{-\infty}^{t} \delta(u + \tau) - \delta(u - \tau)du \rightarrow \frac{F(\omega)}{j\omega}$$
$$= \frac{2j \sin \omega\tau}{j\omega} = \frac{2 \sin \omega\tau}{\omega} \qquad (9.102)$$

again in exact agreement with the expected output. Notice we used the fact that $F(0) = 0$. These results are illustrated in Fig. 9.12.

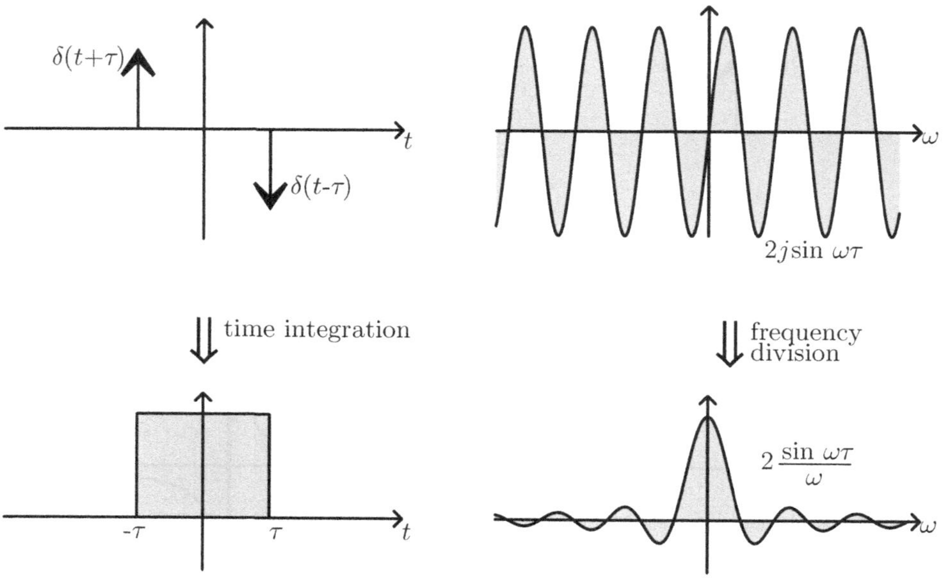

Fig. 9.12 Illustration of the time integration property applied to two shifted delta functions (whose integral is the pulse function)

9.8.3 Third Example of Time Integration Property of Fourier Transform: Checker Pulse Function

The checker pulse function (of pulse width τ) shown in top left part of Fig. 9.13 has the FT

$$\text{checker pulse} \rightarrow -2j\frac{1-\cos\omega\tau}{\omega} \tag{9.103}$$

If we were to integrate it in the time domain we'd get the function shown in lower left part of Fig. 9.13. By the integration property we'd expect the FT of the integral to be

$$\begin{aligned}\int_{-\infty}^{t} \text{checker pulse} &= \pi\mathcal{F}(0)\delta(\omega) + \frac{F(\omega)}{j\omega} \\ &= -2\frac{1-\cos\omega\tau}{\omega^2}\end{aligned} \tag{9.104}$$

where we have used the fact that here $F(0) = 0$. This is shown in lower right part of Fig. 9.13. We were able to verify that this result is accurate by doing numerical inverse Fourier transform, which did in fact reproduce the triangle-shape function.

9.9 Frequency Differentiation of Fourier Transform

Similar to the time differentiation property, the frequency differentiation property states that the frequency derivative of a transform function is the transform of the original function, multiplied by time.

$$\boxed{\text{Fourier transform of } [tf(t)] = j\frac{d}{d\omega}F(\omega)} \tag{9.105}$$

This can be proved as follows.

$$F(\omega) = \int_{-\infty}^{\infty} f(t)e^{-j\omega t}dt \tag{9.106}$$

$$\begin{aligned}j\frac{d}{d\omega}F(\omega) &= j\frac{d}{d\omega}\int_{-\infty}^{\infty} f(t)e^{-j\omega t}dt \\ &= j\int_{-\infty}^{\infty} f(t)\frac{d}{d\omega}\left(e^{-j\omega t}\right)dt \\ &= j\int_{-\infty}^{\infty} f(t)(-jt)e^{-j\omega t}dt \\ &= \int_{-\infty}^{\infty} [tf(t)]\,e^{-j\omega t}dt\end{aligned} \tag{9.107}$$

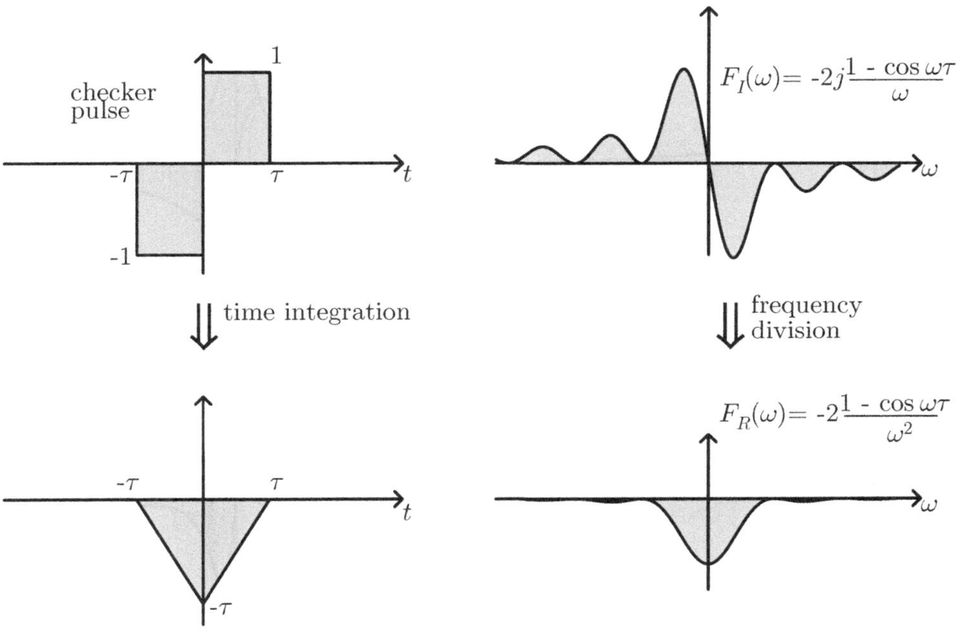

Fig. 9.13 Illustration of the time integration property applied on the checker pulse function

Clearly, the right side of the last equation is the FT of the function $tf(t)$. Another way about the proof is to evaluate the inverse FT of the frequency derivative of the FT:

$$g(t) = \frac{1}{2\pi} \int j \frac{dF(\omega)}{d\omega} e^{j\omega t} d\omega \tag{9.108}$$

Let

$$u = e^{j\omega t}; \quad du = jte^{j\omega t};$$
$$dv = \frac{dF(\omega)}{d\omega}; \text{and} \quad v = F(\omega) \tag{9.109}$$

Using integration by parts we get

$$g(t) = \frac{j}{2\pi} e^{j\omega t} F(\omega) \Big|_{-\infty}^{\infty} - \frac{j}{2\pi} \int F(\omega) jte^{j\omega t} d\omega$$
$$= t \frac{1}{2\pi} \int F(\omega) e^{j\omega t} d\omega = tf(t) \tag{9.110}$$

where we have assumed that $\lim_{\omega \to \infty} F(\omega) = 0$. That is, the spectrum of the function is band-limited.

9.9.1 First Example of Frequency Differentiation of Fourier Transform: Negative Exponential

We know that the single-sided negative exponential has the FT

$$u(t)e^{-at} \rightarrow \frac{1}{a + j\omega} \tag{9.111}$$

Based on the frequency differentiation property, we would expect then that the FT of the negative exponential times is equal to the frequency derivative of the original FT

$$tu(t)e^{-at} \rightarrow j \frac{d}{d\omega} \frac{1}{a + j\omega} = -j \frac{j}{(a + j\omega)^2}$$
$$= \boxed{\frac{1}{(a + j\omega)^2}} \tag{9.112}$$

We can decompose this into real and imaginary parts

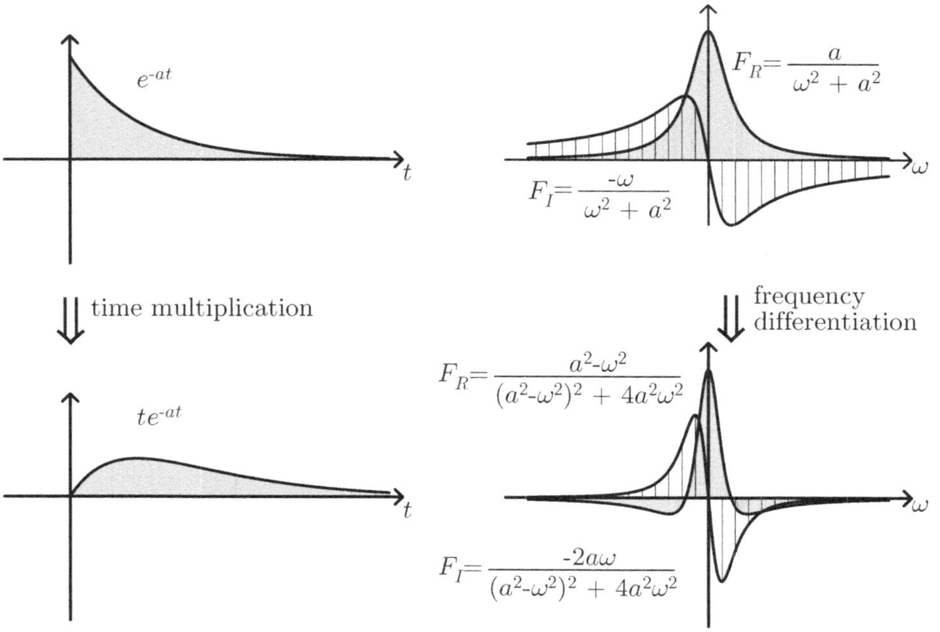

Fig. 9.14 Illustration of the frequency differentiation property applied on the negative exponential

$$\begin{aligned} \frac{1}{(a+j\omega)^2} &= \frac{1}{(a^2-\omega^2)+j(2a\omega)} \\ &= \frac{(a^2-\omega^2)-j(2a\omega)}{(a^2-\omega^2)^2+4a^2\omega^2} \end{aligned} \quad (9.113)$$

The real and imaginary parts of the original and differentiated Fourier transforms are shown in Fig. 9.14, along side the corresponding time signals.

9.9.2 Second Example of Frequency Differentiation of Fourier Transform: Symmetric Negative Exponential

We know (see Eq. (8.25)) that the FT of the symmetric negative exponential is

$$e^{-a|t|} \to \frac{2a}{a^2+\omega^2} \quad (9.114)$$

Based on the frequency differentiation property, if we multiply this function in the time domain by t then the FT of the results would be

$$te^{-a|t|} \to j\frac{d}{d\omega}\frac{2a}{a^2+\omega^2} = \boxed{-j\frac{4a\omega}{(a^2+\omega^2)^2}} \quad (9.115)$$

The time and frequency transform functions are shown in Fig. 9.15.

9.10 Frequency Integration of Fourier Transform

This theorem states that if

$$f(t) \to F(\omega), \text{ then} \quad (9.116)$$

$$\boxed{\pi f(0)\delta(t) - \frac{f(t)}{jt} \to \int_{-\infty}^{\omega} F(u)du} \quad (9.117)$$

For the special case where the function is zero at time zero, the theorem collapses to

$$\boxed{\frac{f(t)}{t} \to \frac{1}{j}\int_{-\infty}^{\omega} F(u)du, \text{ special case } f(0) = 0} \quad (9.118)$$

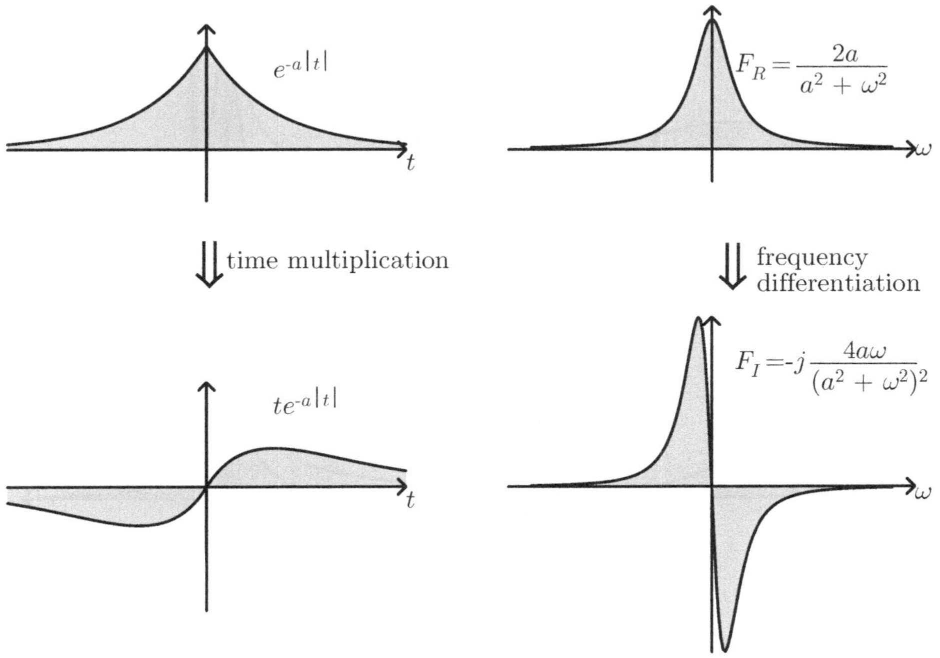

Fig. 9.15 Illustration of the frequency differentiation property applied on the symmetric negative exponential

Proof. Define $g(t)$ such that it is the inverse transform of the frequency integral of $F(\omega)$:

$$g(t) = \frac{1}{2\pi} \int_{-\infty}^{\infty} \left[\int_{-\infty}^{\omega} F(u)du \right] e^{j\omega t} d\omega \quad (9.119)$$

Let

$$\int_{-\infty}^{\omega} F(u)du = U; \quad dU = F(\omega);$$

$$dV = e^{j\omega t}; \quad V = \frac{1}{jt} e^{j\omega t} \quad (9.120)$$

Then using integration by parts we get

$$g(t) = \frac{1}{2\pi} \left[\int_{-\infty}^{\omega} F(u)du \frac{1}{jt} e^{j\omega t} \Big|_{-\infty}^{\infty} - \frac{1}{jt} \int_{-\infty}^{\infty} F(\omega) e^{j\omega t} d\omega \right] \quad (9.121)$$

$$g(t) = \pi f(0)\delta(t) - \frac{f(t)}{jt} \quad (9.122)$$

where we have used the fact that

$$\lim_{\omega \to \infty} \frac{\sin \omega t}{t} = \pi \delta(t) \quad (9.123)$$

9.10.1 First Example of Frequency Integration of Fourier Transform: Negative Exponential

We know that the negative exponential, times time, has the FT

$$u(t)te^{-at} \to \frac{1}{(a + j\omega)^2} \quad (9.124)$$

If we do the frequency integration (divided by j) we get

$$\frac{1}{j} \int_{-\infty}^{\omega} \frac{1}{(a + ju)^2} du = \frac{1}{a + ju} \Big|_{-\infty}^{\omega} = \frac{1}{a + j\omega} \quad (9.125)$$

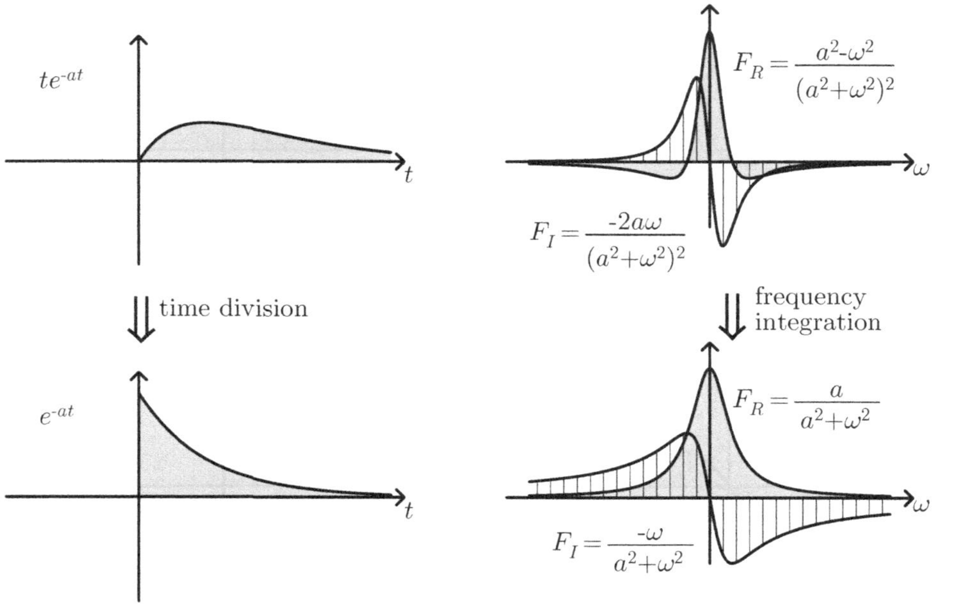

Fig. 9.16 Illustration of the frequency integration property applied on the negative exponential

But this is nothing other than the FT of

$$u(t)e^{-at} \rightarrow \frac{1}{a + j\omega} \tag{9.126}$$

Notice in this case it happened that $f(0) = 0$. So we have shown that frequency integration amounts to time division; this is shown in Fig. 9.16.

9.10.2 Second Example of Frequency Integration of Fourier Transform: Sine Function

We know that the sine function has a FT of

$$\sin \omega_0 t \rightarrow j\pi \, [\delta(\omega + \omega_0) - \delta(\omega - \omega_0)] \tag{9.127}$$

If we integrate in the frequency domain we get

$$\frac{1}{j} \int_{-\infty}^{\omega} (j\pi) \, [\delta(u + \omega_0) - \delta(u - \omega_0)] \, du$$
$$= \pi \times [\text{ pulse function if width } 2\ \omega_0] \tag{9.128}$$

But this is nothing other than the Fourier transform of the sinc function!

$$\frac{\sin \omega_0 t}{t} \rightarrow \pi \times [\text{ pulse function if width } 2\ \omega_0] \tag{9.129}$$

We can verify this by simply carrying on the integration

$$\frac{1}{2\pi} \int_{-\omega_0}^{\omega_0} \pi e^{j\omega t} d\omega = \frac{1}{2} \left[\frac{e^{j\omega t}}{jt} \right] \Bigg|_{-\omega_0}^{\omega_0}$$
$$= \frac{1}{2jt} \left[e^{j\omega_0 t} - e^{-j\omega_0 t} \right] = \frac{\sin \omega_0 t}{t} \tag{9.130}$$

The frequency integration property is illustrated in Fig. 9.17.

9.11 Time Convolution of Fourier Transform

This states that the Fourier transform of the convolution of two signals is the product of the individual Fourier transforms.

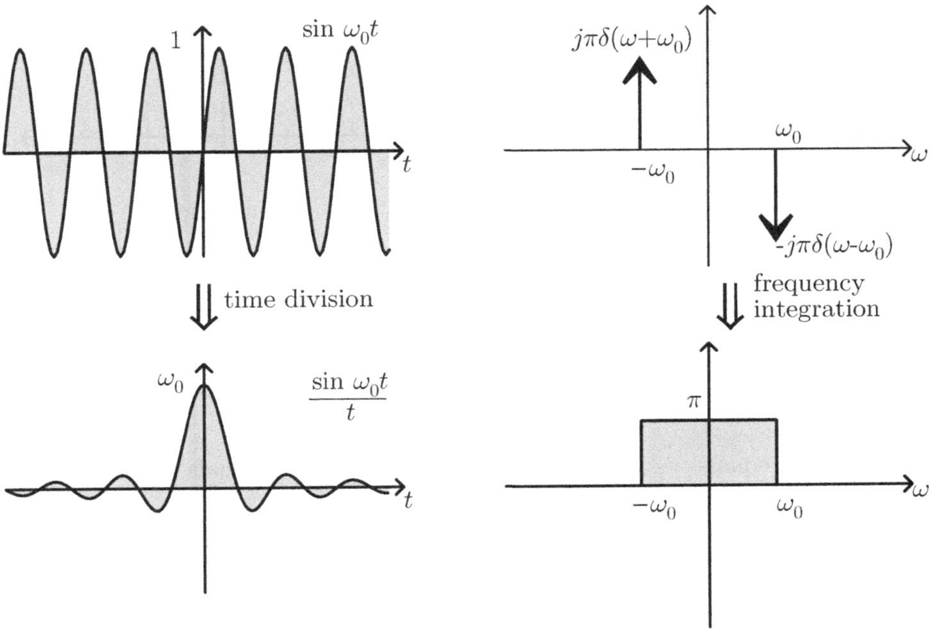

Fig. 9.17 Illustration of the frequency integration property applied to the sine function

$$\boxed{f(t) * g(t) \rightarrow F(\omega) \cdot G(\omega)} \tag{9.131}$$

To prove this we do the following. Assume two functions $f(t)$ and $g(t)$. The convolution of both functions is

$$f(t) * g(t) = \int_{-\infty}^{\infty} f(\tau)g(t-\tau)d\tau \tag{9.132}$$

The FT of the convolved function is

$$\begin{aligned}
&\mathcal{F}[f(t) * g(t)] \\
&= \int_{-\infty}^{\infty} \left[\int_{-\infty}^{\infty} f(\tau)g(t-\tau)d\tau\right] e^{-j\omega t}dt \\
&= \int_{-\infty}^{\infty} \left[\int_{-\infty}^{\infty} e^{-j\omega t} f(\tau)g(t-\tau)d\tau\right] dt \\
&= \int_{-\infty}^{\infty} \left[\int_{-\infty}^{\infty} e^{-j\omega t} f(\tau)g(t-\tau)dt\right] d\tau \\
&= \int_{-\infty}^{\infty} f(\tau) \left[\int_{-\infty}^{\infty} e^{-j\omega t} g(t-\tau)dt\right] d\tau \\
&= \int_{-\infty}^{\infty} f(\tau)\left[G(\omega)e^{-j\omega\tau}\right] d\tau \\
&= G(\omega) \int_{-\infty}^{\infty} f(\tau)e^{-j\omega\tau} d\tau \\
&= G(\omega)F(\omega)
\end{aligned} \tag{9.133}$$

where we have used the time shift property:

$$\begin{aligned}
\text{If} \quad & f(t) \rightarrow F(\omega) \\
\text{then} \quad & f(t-\tau) \rightarrow F(\omega)e^{-j\omega\tau}
\end{aligned} \tag{9.134}$$

9.11.1 First Example of Time Convolution of Fourier Transform: The Delta and DC Functions

We know that the delta function transforms to

$$\delta(t) \rightarrow 1 \tag{9.135}$$

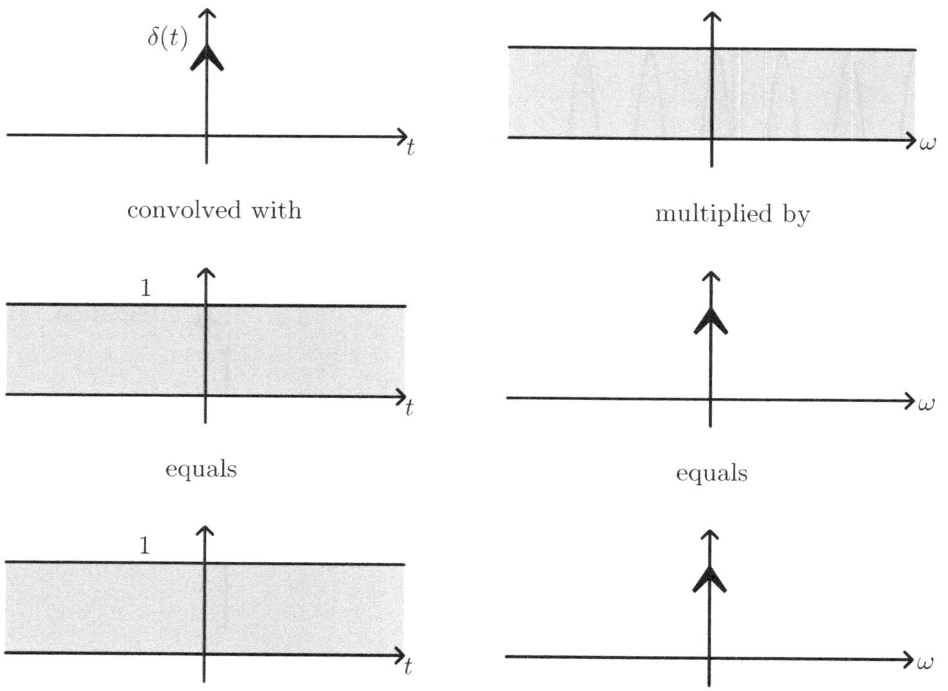

Fig. 9.18 Illustration of the time convolution property applied to the delta and DC functions

We also know that the DC function transforms to

$$1 \rightarrow 2\pi\delta(\omega) \tag{9.136}$$

If we convolve both functions in the time domain we get nothing other than the DC functions

$$\delta(t) * 1 = 1 \tag{9.137}$$

This would transform to

$$\delta(t) * 1 = 1 \rightarrow 2\pi\delta(\omega) \tag{9.138}$$

But this is nothing other than the product of the two Fourier transforms!

$$2\pi\delta(\omega) = 2\pi\delta(\omega) \times 1 \tag{9.139}$$

Hence we have verified the time convolution property! This is shown graphically in Fig. 9.18.

9.11.2 Second Example of Time Convolution of Fourier Transform: Two Deltas and a Pulse Function

Start with two delta functions, centered at $-t_0$ and t_0; their FT is

$$\delta(t + t_0) + \delta(t - t_0) \rightarrow e^{j\omega t_0} + e^{-j\omega t_0} = 2\cos\omega t_0 \tag{9.140}$$

We want to convolve with the pulse function, of width 2τ.

$$\text{pulse of width } 2\tau \rightarrow 2\frac{\sin\omega\tau}{\omega} \tag{9.141}$$

The convolution results in two pulses, each still of width 2τ, but they are centered at $-t_0$ and t_0. We would like to find the FT of the convolution result. Based on our theory we get

$$\boxed{[\delta(t + t_0) + \delta(t - t_0)] * [\text{pulse of width } 2\ \tau] \rightarrow 4\frac{\sin\omega\tau}{\omega}\cos\omega t_0} \tag{9.142}$$

We can simplify by noting that

$$\sin A \cos B = \frac{1}{2}\left[\sin(A+B) + \sin(A-B)\right] \tag{9.143}$$

Then we get

$$[\delta(t+t_0) + \delta(t-t_0)] * [\text{pulse of width } 2\,\tau] \rightarrow 2\frac{\sin[\omega(\tau+t_0)] + \sin[\omega(\tau-t_0)]}{\omega} \tag{9.144}$$

Flip a sign and get

$$\boxed{[\delta(t+t_0) + \delta(t-t_0)] * [\text{ pulse of width } 2]\tau \rightarrow 2\frac{\sin(\omega(t_0+\tau)) - \sin(\omega(t_0-\tau))}{\omega}} \tag{9.145}$$

But this is nothing more than the results we would have gotten by direct integration! Notice that since the convolution results are even, the FT ends up being

$$F(\omega) = 2\int_{t_0-\tau}^{t_0+\tau} \cos \omega t dt$$
$$= 2\frac{\sin(\omega(t_0+\tau)) - \sin(\omega(t_0-\tau))}{\omega} \tag{9.146}$$

which equals results predicted by the property! All these are illustrated in Fig. 9.19.

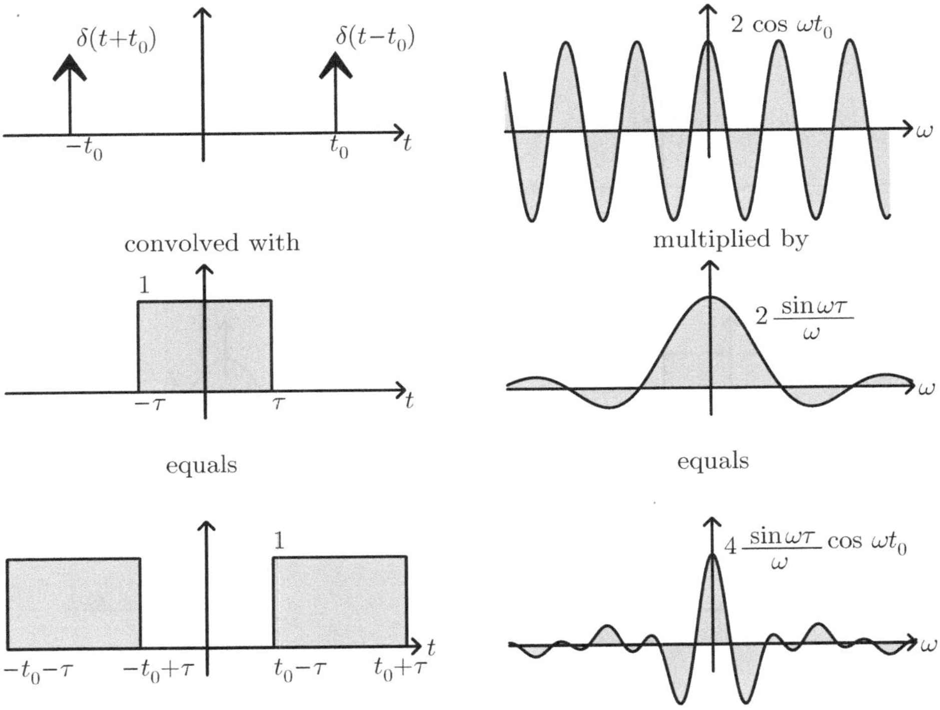

Fig. 9.19 Illustration of the time convolution property applied to two delta functions and a single pulse one

9.11.3 Third Example of Time Convolution of Fourier Transform: Two Opposite Deltas and Pulse Function

Consider two delta functions, located at τ and $-\tau$ with latter having negative sign:

$$\delta(t-\tau)-\delta(t+\tau) \tag{9.147}$$

The FT of this function is

$$\delta(t-\tau)-\delta(t+\tau) \rightarrow e^{-j\omega\tau}-e^{j\omega\tau}$$
$$= -2j\sin(\omega\tau) = \boxed{\frac{2}{j}\sin(\omega\tau)} \tag{9.148}$$

Next consider the pulse function if width 2τ; it has a FT of

$$\text{pulse of width } 2\,\tau \rightarrow 2\frac{\sin\omega\tau}{\omega} \tag{9.149}$$

If we convolve the delta couple with a pulse if width 2τ we would get a checker pulse as shown in Fig. 9.20. By direct integration that FT of the checker pulse is

$$\text{checker pulse} \rightarrow 2\frac{1-\cos(2\omega\tau)}{j\omega} \tag{9.150}$$

By the convolution property we also get

$$[\delta(t-\tau)-\delta(t+\tau)] * \text{pulse of width } 2\tau$$
$$\rightarrow \frac{2}{j}\sin(\omega\tau)\times 2\frac{\sin\omega\tau}{\omega}$$
$$= 2\frac{1-\cos(2\omega\tau)}{j\omega} \tag{9.151}$$

which matches the prior results; hence we have confirmed the convolution property on this example. Results illustrated in Fig. 9.20.

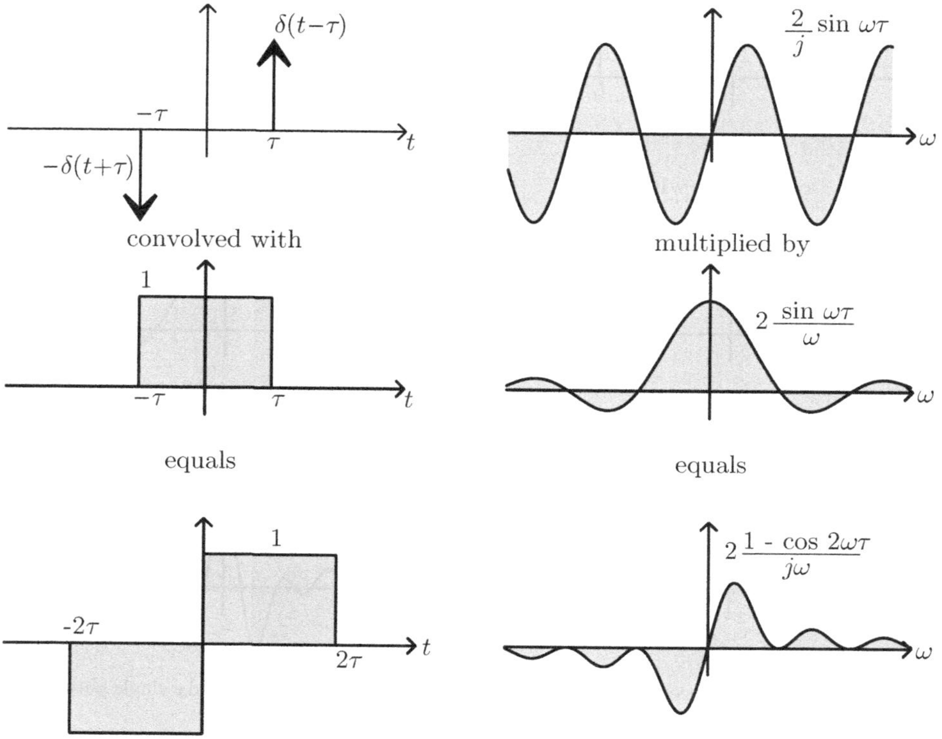

Fig. 9.20 Illustration of the time convolution property applied to two opposite delta functions and a single pulse one

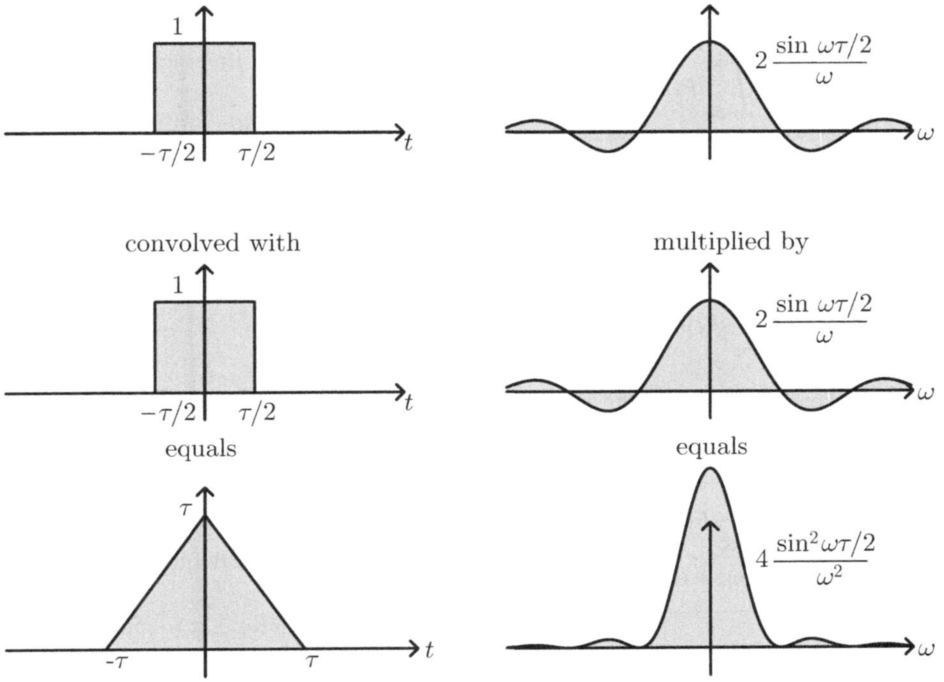

Fig. 9.21 Illustration of the time convolution property applied to pulse function convolved with itself

9.11.4 Fourth Example of Time Convolution of Fourier Transform: Two Pulse Functions

We know that the pulse function of width τ has a FT

$$\text{pulse of width } \tau \rightarrow 2\frac{\sin \omega\tau/2}{\omega} \tag{9.152}$$

If we convolve this function with itself we would get a shape that resembles a "hat" with width 2τ as shown in Fig. 9.21. We want to find the FT of this function. Based on the convolution property the FT of the convolution result is

$$\text{pulse of width } \tau * \text{pulse of width } \tau \rightarrow 4\frac{\sin^2 \omega\tau/2}{\omega^2} \tag{9.153}$$

This is illustrated in Fig. 9.21.

9.12 Frequency Convolution of the Fourier Transform

We are after the FT of the product of two functions in the time domain

$$\mathcal{F}[f(t) \cdot g(t)] \tag{9.154}$$

We proceed by the FT definition

$$\mathcal{F}[f(t)g(t)] = \int_{-\infty}^{\infty} f(t)g(t)e^{-j\omega t}dt \tag{9.155}$$

Expand $f(t)$ in terms of its inverse Fourier transform and get

$$\mathcal{F}[f(t)g(t)] = \int_{-\infty}^{\infty} \left[\frac{1}{2\pi}\int_{-\infty}^{\infty} F(u)e^{jut}du\right] g(t)e^{-j\omega t}dt \tag{9.156}$$

Rearrange the order of integration

$$\mathcal{F}[f(t)g(t)] = \frac{1}{2\pi}\int_{-\infty}^{\infty} F(u) \left[\int_{-\infty}^{\infty} e^{jut} g(t) e^{-j\omega t} dt\right] du$$
$$= \frac{1}{2\pi}\int_{-\infty}^{\infty} F(u) \left[\int_{-\infty}^{\infty} g(t) e^{-jt(\omega-u)} dt\right] du \quad (9.157)$$

But the bracketed term is nothing more than $G(\omega - u)$! Hence we have

$$\boxed{\begin{aligned}\mathcal{F}[f(t)g(t)] &= \frac{1}{2\pi}\int_{-\infty}^{\infty} F(u)G(\omega - u)du \\ &= \frac{1}{2\pi}F(\omega) * G(\omega)\end{aligned}} \quad (9.158)$$

9.12.1 First Example of Frequency Convolution of the Fourier Transform: Cosine Times Unit Step Function

We know that the cosine function has the FT

$$\cos\omega_0 t \rightarrow \pi\left[\delta(\omega-\omega_0) + \delta(\omega+\omega_0)\right] \quad (9.159)$$

We also know that the unit step function has the FT

$$u(t) \rightarrow \pi\delta(\omega) + \frac{1}{j\omega} \quad (9.160)$$

Based on the convolution property, the FT of the product is the convolution of the FTs; the convolution (divided by 2π) is computed as

$$\frac{1}{2}\left[\delta(\omega-\omega_0) + \delta(\omega+\omega_0)\right] * \left[\pi\delta(\omega) + \frac{1}{j\omega}\right]$$
$$= \frac{1}{2}\left[\pi\left[\delta(\omega-\omega_0) + \delta(\omega+\omega_0)\right] + \frac{1}{j(\omega-\omega_0)} + \frac{1}{j(\omega+\omega_0)}\right]$$
$$= \frac{\pi}{2}\left[\delta(\omega-\omega_0)+\delta(\omega+\omega_0)\right] + \frac{1}{j}\frac{\omega}{\omega^2-\omega_0^2} \quad (9.161)$$

But this agrees with results computed in prior chapter (Eq. (8.66)). So, we have confirmed the frequency convolution property. Results are illustrated in Fig. 9.22.

9.12.2 Second Example of Frequency Convolution of the Fourier Transform: Two Sine Functions

A sine function with frequency π has the FT

$$\sin\pi t \rightarrow j\pi\left[\delta(\omega+\pi) - \delta(\omega-\pi)\right] \quad (9.162)$$

Another sine with frequency 2π has the FT

$$\sin 2\pi t \rightarrow j\pi\left[\delta(\omega+2\pi) - \delta(\omega-2\pi)\right] \quad (9.163)$$

If we multiply them in the time domain we get

$$\sin\pi t \times \sin 2\pi t = \frac{1}{2}\left[\cos\pi t - \cos 3\pi t\right] \quad (9.164)$$

This would have the FT

$$\frac{1}{2}\left[\cos\pi t - \cos 3\pi t\right] \rightarrow \frac{\pi}{2}\left[\delta(\omega+1\pi) + \delta(\omega-1\pi) - \delta(\omega+3\pi) - \delta(\omega-3\pi)\right] \quad (9.165)$$

If on the other hand we use the frequency convolution property, and as shown in Fig. 9.23 we do in fact get the same answer!

9.13 Summary

This chapter dealt with the properties of the Fourier transform. Not only are these properties extremely useful, but throughout their derivation we delved deeply into the very definition of

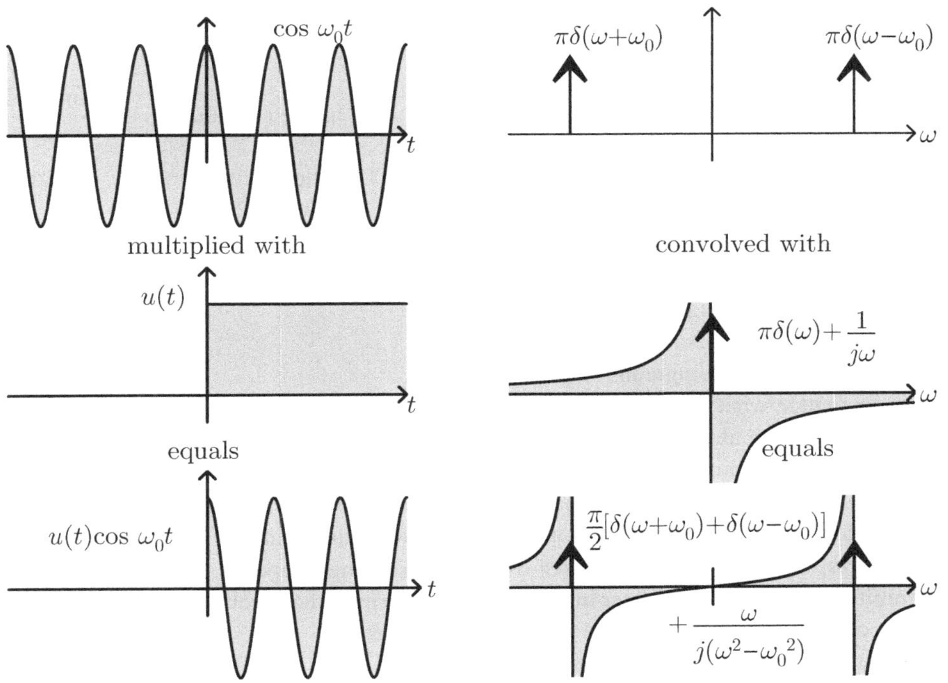

Fig. 9.22 Illustration of the frequency convolution property applied to cosine and unit step functions

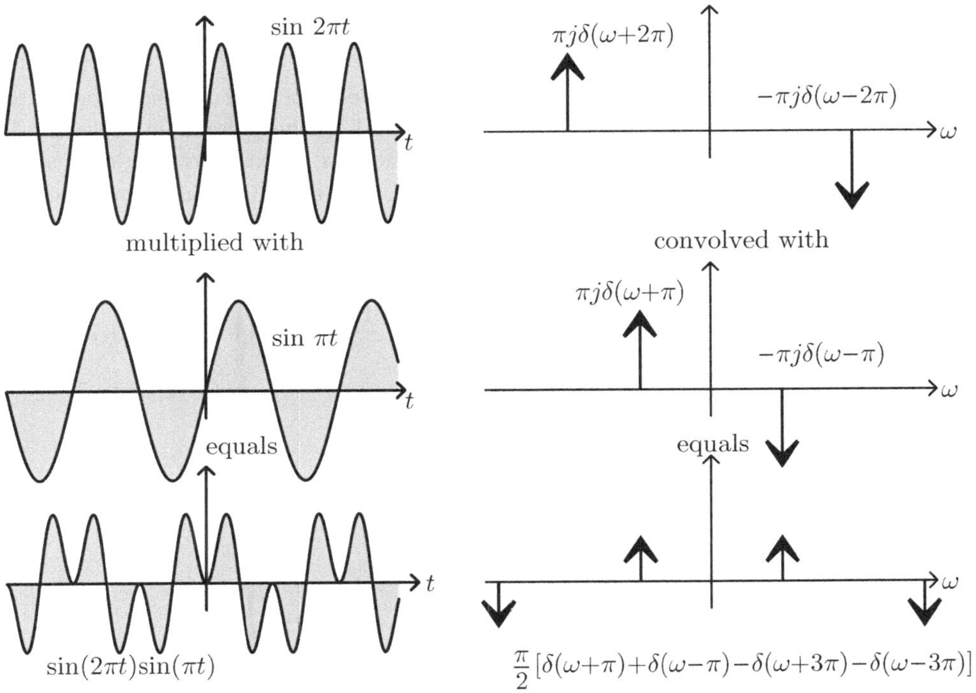

Fig. 9.23 Illustration of the frequency convolution property applied to two sine functions

the transform, and hopefully have gained much insight into its operation. The main chapter theme is that rather than doing the transform derivation from scratch every time we encounter a new function, why not tie the new function to something we already are familiar with, and then use the transform properties to arrive at the answer. It is often easier to add/subtract/scale rather than to integration. Another beneficial outcome of this chapter is that throughout the multitude of presented examples we saw over and over how everything ended up being consistent is the sense that no matter what way we followed to derive the FT of a particular function we always ended up with the same answer! Many of those properties will be used again once we encounter the Laplace transform, later. Also, we got some exposure to convolution and that on its own warrants multiple chapters, again later in the text. In the next chapter we will dig more into the Fourier transform, but now we have the advantage of being quipped with this new set of arsenal tools!

9.14 Problems

1. We know that a cosine function has the FT

$$\cos\omega_0 t \rightarrow \pi\left[\delta(\omega-\omega_0)+\delta(\omega+\omega_0)\right]$$

Use the scaling property to find the FT of $\cos 2\omega_0 t$. Hint: may need to use the following identity:

$$\delta\left(\frac{x}{2}\right)=2\delta(x)$$

2. We know that the pulse function of width τ has the FT

$$[\text{pulse of width } \tau] \rightarrow 2\frac{\sin\omega\tau/2}{\omega}$$

Use the scaling property to derive FT of pulse of width 2τ.

3. We know that the FT of the symmetric negative exponential is

$$e^{-a|t|} \rightarrow \frac{2a}{a^2+\omega^2}$$

Use the reciprocity property to find the FT of the function $f(t)=\frac{1}{1+t^2}$. Plot time series of this FT and compare to $f(t)$. See sample solution in Fig. 9.24.

4. Knowing that $\delta(t)\rightarrow 1$, and using the time shifting property, derive the Fourier transform of the function $\frac{1}{2}\left[\delta(t-t_0)+\delta(t+t_0)\right]$. Answer:

$$F(\omega)=\cos\omega t_0$$

5. Knowing the FT of shifted pulse of width τ

$$[\text{pulse of width } \tau] \rightarrow \frac{1-e^{-j\omega\tau}}{j\omega}$$

and using the time shifting property, derive the FT of the pulse train, of count 5, and period 2τ. Next, draw the time series for the case $\tau = 0.5$. See sample solution in Fig. 9.25. Answer:

$$\frac{1-e^{-j\omega\tau}}{j\omega}\left[1+e^{-j2\omega\tau}+e^{-j4\omega\tau}+ e^{-j6\omega\tau}+e^{-j8\omega\tau}\right]$$

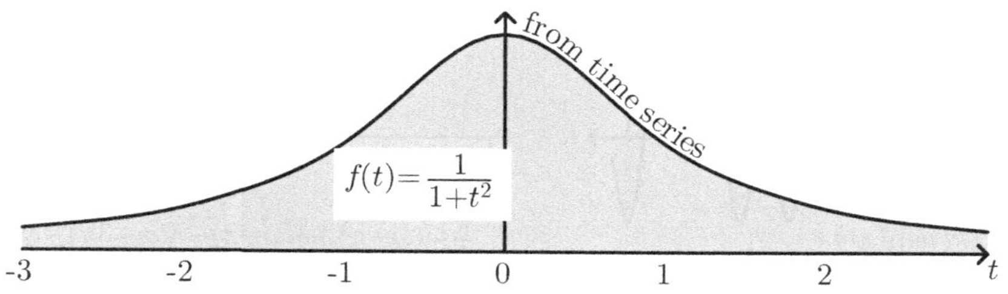

Fig. 9.24 Solution to Problem 3

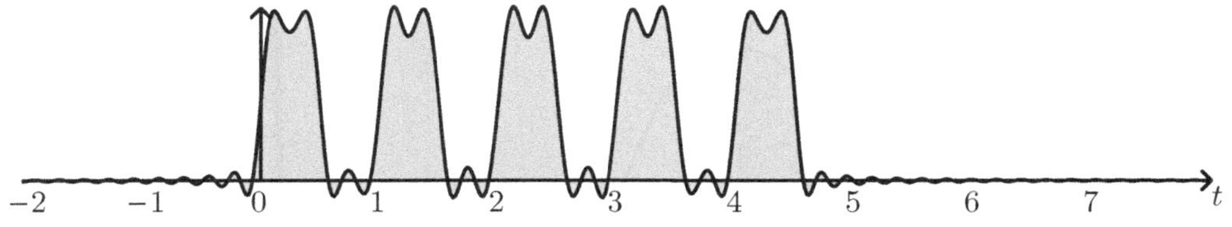

Fig. 9.25 Solution to Problem 5

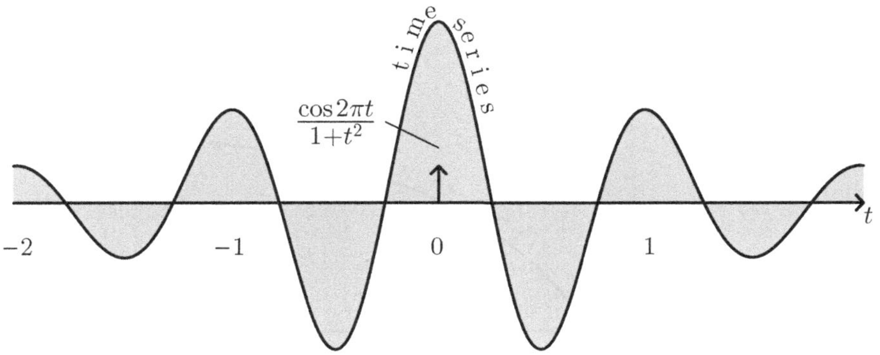

Fig. 9.26 Solution to Problem 6

6. Find the FT of the function

$$f(t) = \frac{\cos \omega_0 t}{1 + t^2}$$

using the frequency shifting property and the fact that

$$\frac{1}{1 + t^2} \rightarrow \pi e^{-|\omega|}$$

For the case $\omega_0 = 2\pi$ plot the $f(t)$ and its time series. See sample results in Fig. 9.26.
Answer:

$$F(\omega) = \frac{\pi}{2}\left[e^{-|\omega-\omega_0|} + e^{-|\omega+\omega_0|}\right]$$

7. Find the Fourier transform of the function

$$f(t) = \frac{-2t}{(1 + t^2)^2}$$

Then plot the time series and compare to original function. Hint: use fact that

$$\frac{1}{1 + t^2} \rightarrow \pi e^{-|\omega|}$$

See sample solution in Fig. 9.27.

Answer:

$$F(\omega) = \pi e^{-|\omega|} j\omega$$

8. Consider the function

$$f(t) = u(t)\left[1 - e^{-t}\right]$$

Find the Fourier transform of this function directly and find it using the time integration property, taking into account that $f(t) = \int_0^t e^{-\tau} d\tau$. Plot the time series and compare to the original function. See sample solution in Fig. 9.28.
Answer:

$$F(\omega) = \pi\delta(\omega) + \frac{1}{j\omega} + \frac{1}{1 + j\omega}$$

9. Start with the fact that the Fourier transform of the sinc function is

$$\frac{\sin \omega_0 t}{t} \rightarrow \pi[\text{pulse of width } 2\ \omega_0]$$

and use the frequency differentiation (or time multiplication) property to derive the FT of the sine function $\sin \omega_0 t$.

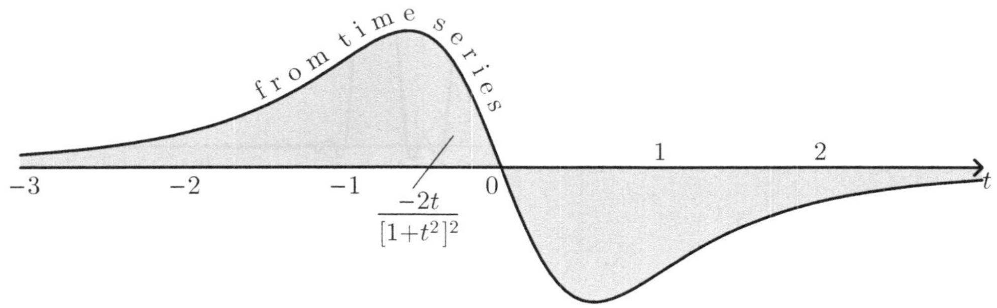

Fig. 9.27 Solution to Problem 7

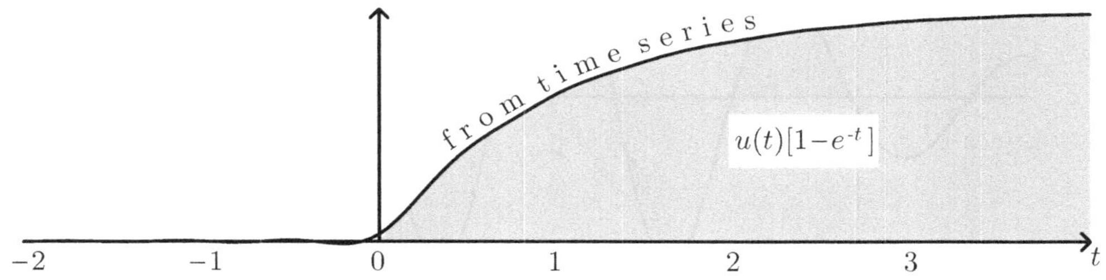

Fig. 9.28 Solution to Problem 8

10. Use the frequency integration property to derive the Fourier transform of the function

$$f(t) = \frac{1 - \cos \omega_0 t}{t}$$

For the case $\omega_0 = 2\pi$ plot the time series and compare to actual function. See sample results in Fig. 9.29.
Answer: the spectrum is an odd pulse function, in the frequency domain, (positive on the right and negative on the left) with pulse width ω_0 and height π/j.

11. What is the convolution of the single-sided negative exponential e^{-at} with itself? What is the Fourier transform of the resulting convolution? For the case $a = 1$, plot the time series results and compare to convolution. See sample results in Fig. 9.30.

Answer:

$$f(t) * f(t) = te^{-at}; \qquad F(\omega) = \frac{1}{(a + j\omega)^2}$$

12. Prove the following theory using the frequency convolution property:

$$\text{if} \quad f(t) \to F(\omega), \quad \text{then}$$

$$f(t) \cos \omega_0 t \to \frac{F(\omega + \omega_0) + F(\omega - \omega_0)}{2}$$

For the case $f(t) = e^{-0.5t}$ and $\omega_0 = 2\pi$ plot the time series and compare to original function. See sample results in Fig. 9.31.

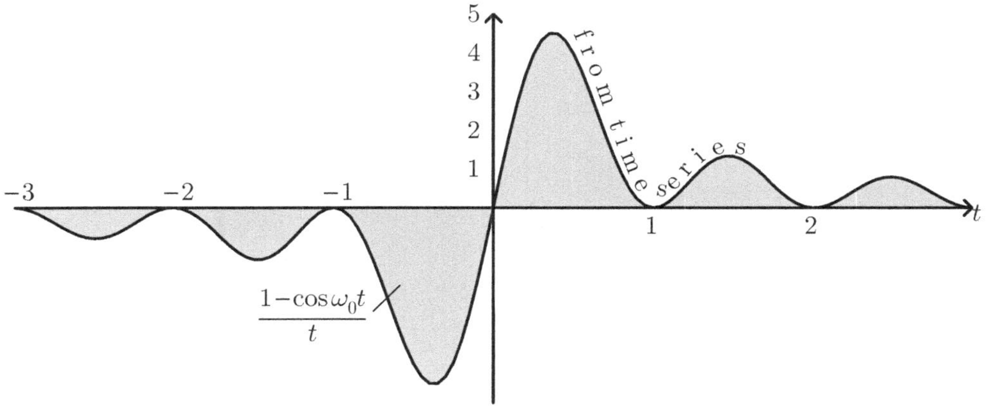

Fig. 9.29 Solution to Problem 10

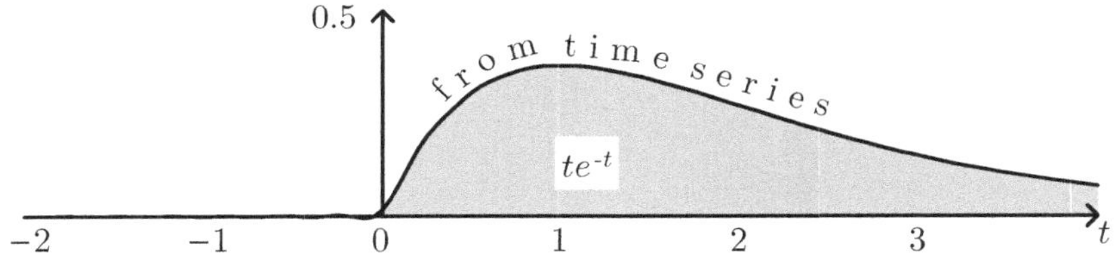

Fig. 9.30 Solution to Problem 11

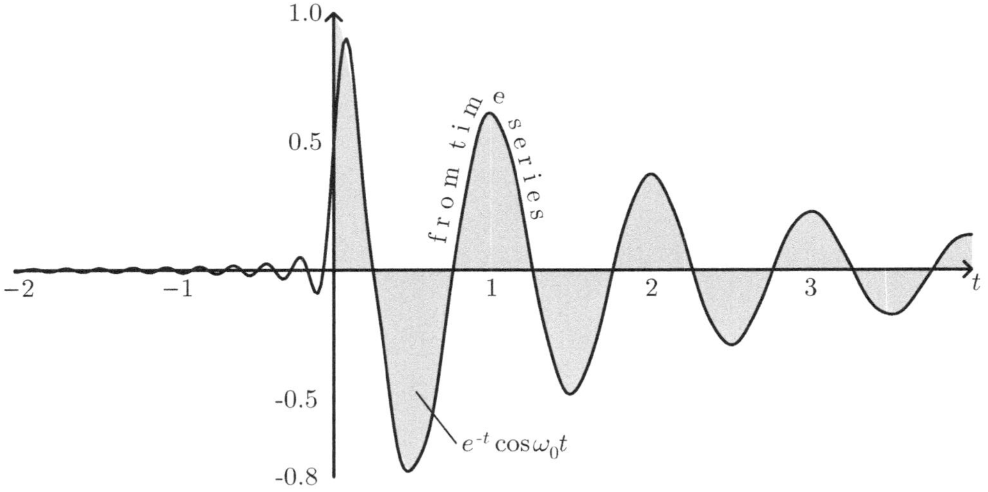

Fig. 9.31 Solution to Problem 12

10 Further Examples/Topics on Fourier Transform

10.1 Introduction

This chapter deals with more examples and practice of finding the FT of important/relevant functions. It also covers some more advanced topics. The key point here is plenty of practice and experience. While the prior chapters covered conventional signals, here we introduce some exotic ones. We also stress on signal cropping. Finally we recap by highlighting the flexibility of the Fourier transform. The organization of the chapter will be mostly one example after another!

10.2 Fourier Transform of Stair Signum Function

The stair signum function is defined as

$$f(t,\tau)=\begin{cases}\frac{1}{2} & t>\tau\\ \frac{-1}{2} & t<-\tau\\ 0 & -\tau<t<\tau\end{cases} \tag{10.1}$$

We recognize that this function can be written in terms of the unit step function as follows:

$$f(t,\tau)=\frac{1}{2}\left\{u(t-\tau)-u(-(t+\tau))\right\} \tag{10.2}$$

In terms of Fourier transform we know that

$$u(t)\rightarrow\pi\delta(\omega)+\frac{1}{j\omega} \tag{10.3}$$

Using the time shifting property, we conclude that

$$u(t-\tau)\rightarrow\left[\pi\delta(\omega)+\frac{1}{j\omega}\right]e^{-j\omega\tau},\quad \text{and} \tag{10.4}$$

$$u(-(t+\tau))\rightarrow\left[\pi\delta(\omega)-\frac{1}{j\omega}\right]e^{+j\omega\tau} \tag{10.5}$$

Subtracting and dividing by two give

$$F(\omega,\tau)=\frac{e^{-j\omega\tau}+e^{+j\omega\tau}}{2j\omega}=\boxed{\frac{\cos\omega\tau}{j\omega}} \tag{10.6}$$

Notice that for the special case of $\tau=0$ this collapses to the FT of the signum function

$$\text{signum function}\rightarrow\frac{1}{j\omega} \tag{10.7}$$

Results are shown in Fig. 10.1.

10.3 Fourier Transform of Odd e^{-at}

We'd like to find the FT of the odd negative exponential defined as

$$f(t)=\begin{cases}e^{-at} & t>0\\ -e^{at} & t<0\end{cases} \tag{10.8}$$

We know from before that the single-sided negative exponential has the FT

F. Badrieh, *Spectral, Convolution and Numerical Techniques in Circuit Theory*,
https://doi.org/10.1007/978-3-319-71437-0_10

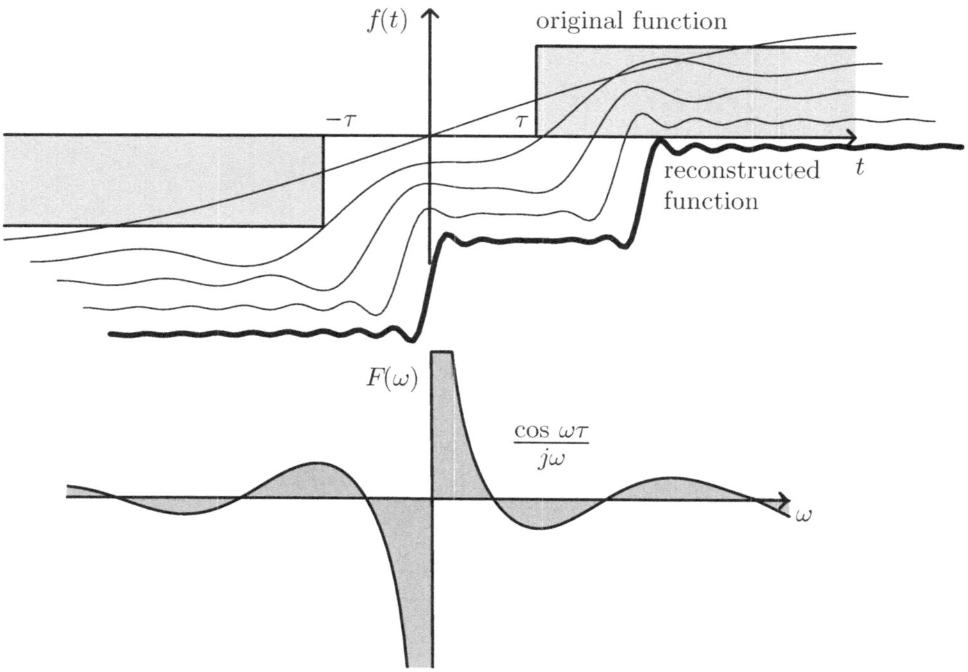

Fig. 10.1 Stair signum function and Fourier transform

$$u(t)e^{-at} \rightarrow \frac{1}{a+j\omega} \tag{10.9}$$

We can construct $f(t)$ by adding the single-sided negative exponential to its mirrored version, after multiplying latter by negative 1. That is, if we take the negative exponential, flip it about the y-axis, negate it, then add to the negative exponential, and we get the "odd" negative exponential. In the frequency domain, we accordingly flip the frequency sign, and negate as well, then add. Our combined FT then becomes

$$F(\omega) = \frac{1}{a+j\omega} - \frac{1}{a-j\omega} \tag{10.10}$$

We can simplify further

$$F(\omega) = \frac{a-j\omega-a-j\omega}{a^2+\omega^2} = \frac{-2j\omega}{a^2+\omega^2} \tag{10.11}$$

(Notice that this is nothing other than (twice) the odd part of the FT $\frac{1}{a+j\omega}$.) The derived results are shown in Fig. 10.2.

10.4 Fourier Transform of Cosine Times $e^{-a|t|}$

We know that the symmetric negative exponential has the FT

$$e^{-a|t|} \rightarrow \frac{2a}{a^2+\omega^2} \tag{10.12}$$

We also know that the cosine function has the FT

$$\cos \omega_0 t \rightarrow \pi\left[\delta(\omega+\omega_0)+\delta(\omega-\omega_0)\right] \tag{10.13}$$

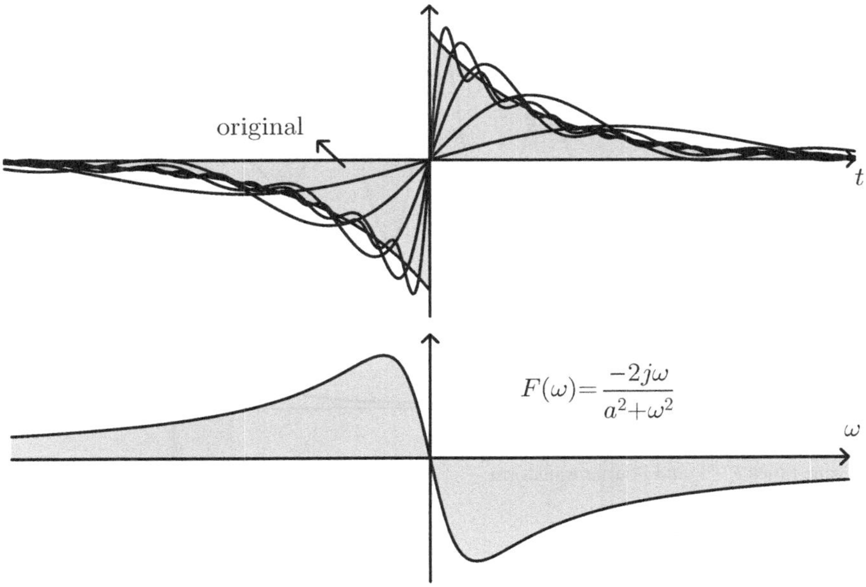

Fig. 10.2 Odd negative exponential and Fourier transform

If we multiply the two signals in the time domain we get

$$f(t) = e^{-a|t|} \cos \omega_0 t \tag{10.14}$$

Based on the frequency convolution property we would expect the product to have the FT

$$F(\omega) = \boxed{\frac{a}{a^2 + (\omega + \omega_0)^2} + \frac{a}{a^2 + (\omega - \omega_0)^2}} \tag{10.15}$$

Results are shown in Fig. 10.3. Let's take some limits. For example, as a approaches 0, the negative exponential approaches the DC function, and the time product approaches the ideal cosine function; as such the FT should approach $\pi\,[\delta(\omega + \omega_0) + \delta(\omega - \omega_0)]$. This in fact is the case as shown in Fig. 10.4. On the other hand if we make ω_0 smaller, the cosine approaches the DC, and we collapse to the FT of the negative symmetric exponential; this is shown in Fig. 10.5.

10.5 Fourier Transform of Cosine Times Single-Sided Negative Exponential

We know that the cosine function has the FT

$$\cos \omega_0 t \rightarrow \pi\,[\delta(\omega + \omega_0) + \delta(\omega - \omega_0)] \tag{10.16}$$

We also know that the single-sided negative exponential has the FT

$$u(t)e^{-at} \rightarrow \frac{1}{a + j\omega} \tag{10.17}$$

If we multiply in the time domain we get

$$f(t) = u(t)e^{-at} \cos \omega_0 t \tag{10.18}$$

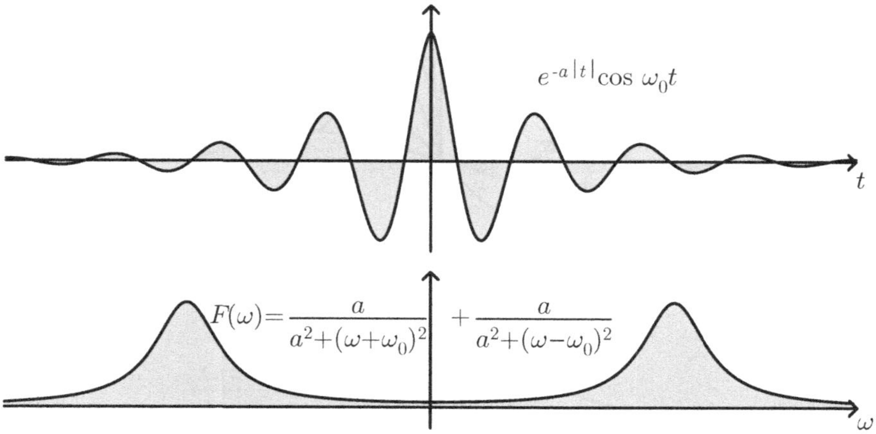

Fig. 10.3 Cosine times $e^{-a|t|}$ and Fourier transform

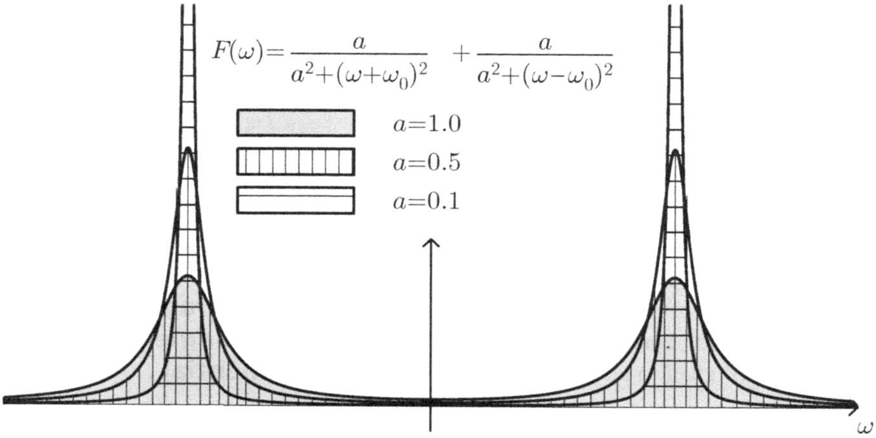

Fig. 10.4 Fourier transform of cosine times $e^{-a|t|}$, for different a values

Based on the frequency convolution property, the time product would have the FT

$$F(\omega) = \frac{1}{2}\left[\frac{1}{a + j(\omega + \omega_0)} + \frac{1}{a + j(\omega - \omega_0)}\right] \tag{10.19}$$

We can simplify further to extract the real and imaginary parts:

$$F(\omega) = \frac{1}{2}\left[\frac{a - j(\omega + \omega_0)}{a^2 + (\omega + \omega_0)^2} + \frac{a - j(\omega - \omega_0)}{a^2 + (\omega - \omega_0)^2}\right] \tag{10.20}$$

Results are shown in Fig. 10.6.

10.6 Fourier Transform of Cropped Cosine Function

The cropped cosine function of frequency ω_0 and width 2τ is defined as

$$f(t) = \begin{cases} \cos\omega_0 t & -\tau < t < \tau \\ 0 & \text{else wise} \end{cases} \tag{10.21}$$

It is the nominal cosine function (with period $2\pi/\omega_0$) cropped such that it is zero for $-\tau < t < \tau$. We quickly recognize that this function is the result of multiplying the cosine function times a pulse of width 2τ. We recall the convolution theorem which states

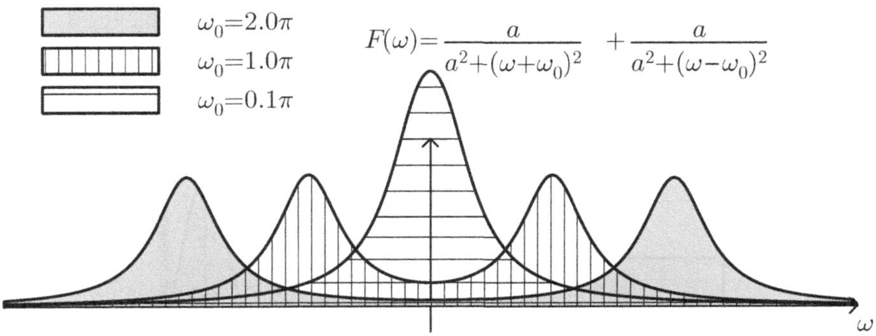

Fig. 10.5 Fourier transform of cosine times $e^{-a|t|}$, for differentiation ω_0 values

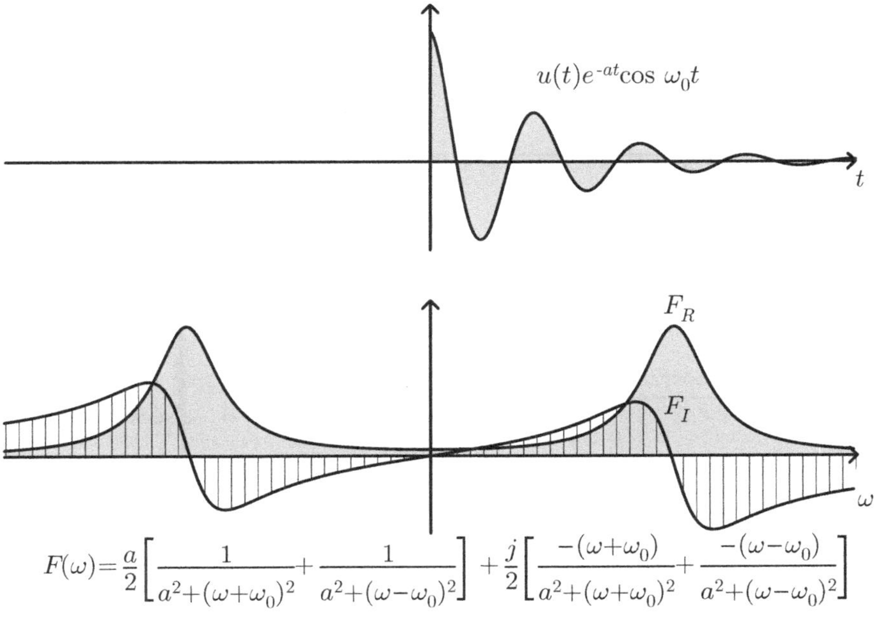

Fig. 10.6 Cosine times single-sided negative exponential

$$\mathcal{F}\left[f_1(t)\cdot f_2(t)\right]=\frac{1}{2\pi}F_1(\omega)*F_2(\omega) \quad (10.22)$$

We know the FT of the cosine as

$$\mathcal{F}[\cos\omega_0 t]=\pi\left[\delta(\omega-\omega_0)+\delta(\omega+\omega_0)\right] \quad (10.23)$$

and we know the FT of the pulse function of width 2τ

$$\mathcal{F}[\text{pulse of width } 2\tau]=2\frac{\sin\omega\tau}{\omega} \quad (10.24)$$

The convolution results are then

$$F(\omega)=\frac{\sin(\omega-\omega_0)\tau}{\omega-\omega_0}+\frac{\sin(\omega+\omega_0)\tau}{\omega+\omega_0} \quad (10.25)$$

Results are shown in Fig. 10.7.

10.7 Fourier Transform Cosine Squared

We can find the FT of the cosine squared using at least two methods. The first method is based on the fact that

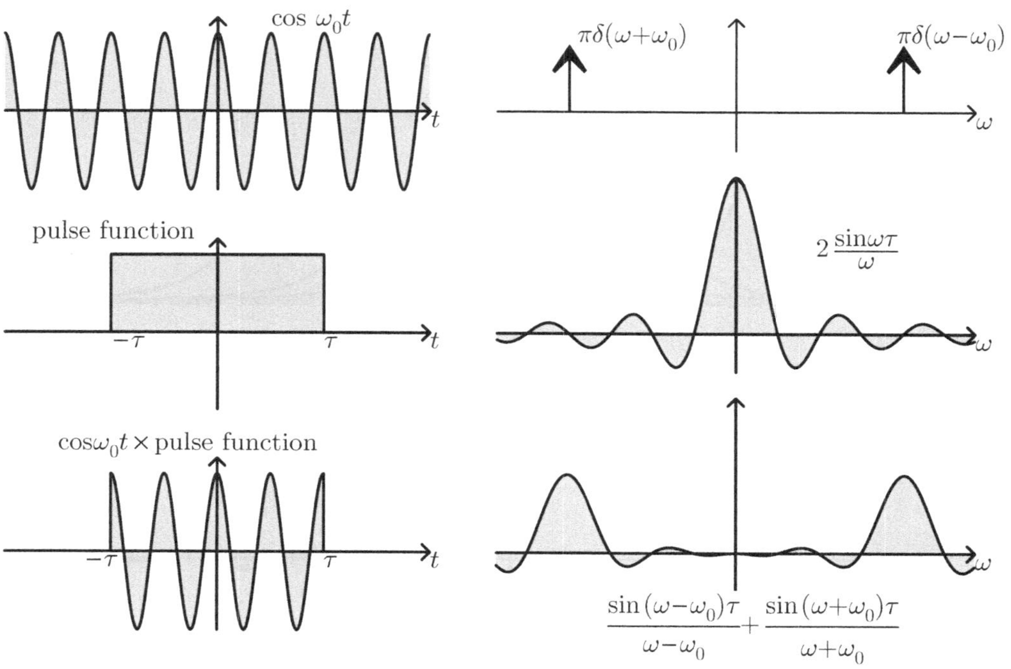

Fig. 10.7 Cropped cosine and Fourier transform

$$\cos^2(\omega_0 t) = \frac{1 + \cos(2\omega_0 t)}{2} \tag{10.26}$$

Clearly the FT would then be

$$\cos^2(\omega_0 t) \to \pi\delta(\omega) + \frac{\pi}{2}\left[\delta(\omega + 2\omega_0) + \delta(\omega - 2\omega_0)\right] \tag{10.27}$$

The second method is using frequency convolution, and this gives the same answer, as shown in Fig. 10.8. (Notice that convolution results have already been divided by 2π.)

10.8 Fourier Transform of (Cropped) *t*

The function $f(t) = t$ itself is problematic so far as finding a Fourier transform, since the signal grows without bound, and integrating such signal is not possible. However, we can find the FT of the *cropped* t function defined by

$$f(t) = \begin{cases} t & -t_0 < t < t_0 \\ 0 & \text{else wise} \end{cases} \tag{10.28}$$

We can derive the FT of this function by recognizing that it can be defined as the product of a pulse function and t

$$f(t) = \text{pulse of width } 2t_0 \times t \tag{10.29}$$

We also recall the frequency differentiation property

$$tf(t) \to j\frac{d}{d\omega}F(\omega) \tag{10.30}$$

Then our FT comes out

$$F(\omega) = j\frac{d}{d\omega}2\frac{\sin \omega t_0}{\omega} = \boxed{2j\left[-\frac{\sin \omega t_0}{\omega^2} + t_0\frac{\cos \omega t_0}{\omega}\right]} \tag{10.31}$$

Results are shown in Fig. 10.9.

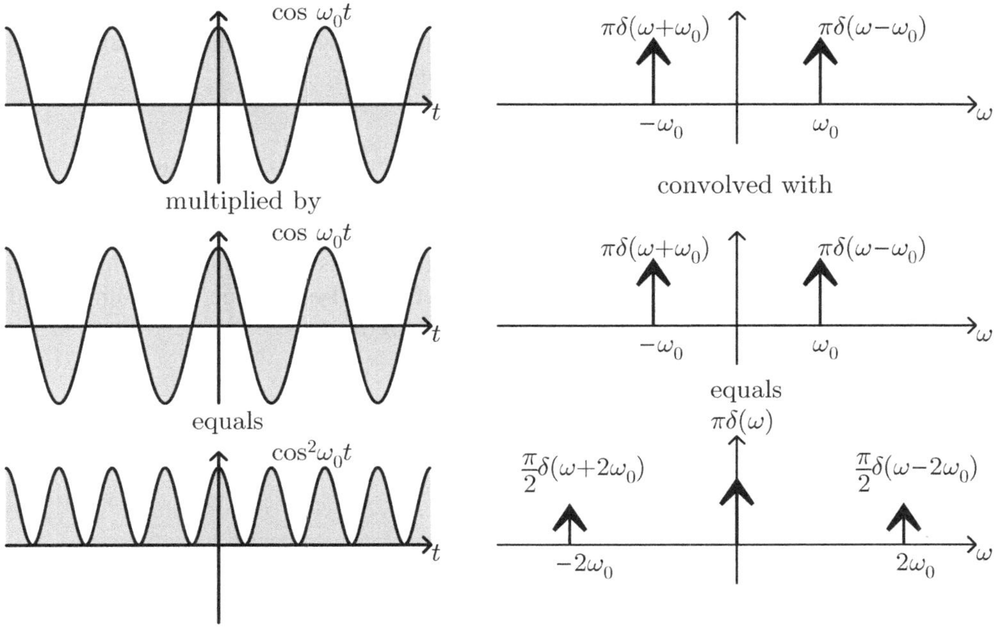

Fig. 10.8 Cosine squared and Fourier transform

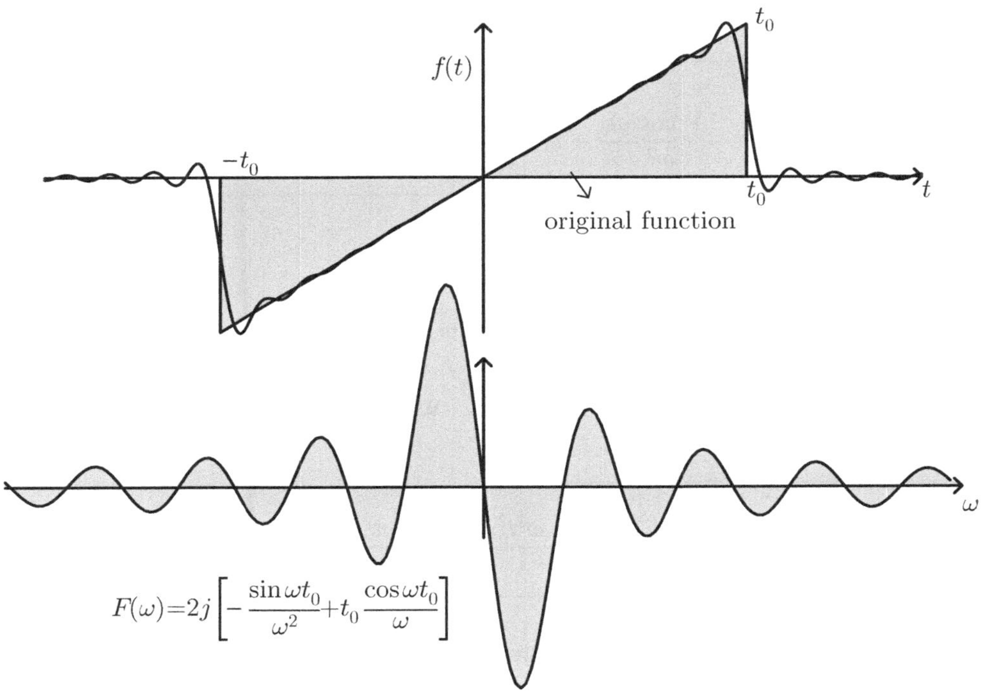

Fig. 10.9 Cropped t function and Fourier transform

10.9 Fourier Transform of (Cropped) Single-Sided *t*

The single-sided t function is defined as

$$f(t)=\begin{cases} t & 0<t<\infty \\ 0 & \text{else wise}\end{cases} \tag{10.32}$$

However, this function is non-integrable. A similar function is the cropped single-sided t defined by

$$f(t)=\begin{cases} t & 0<t<t_0 \\ 0 & \text{else wise}\end{cases} \tag{10.33}$$

We can find the FT of this by recognizing that this function is the result of multiplying t times the offset pulse function, of width t_0, and left edge at zero

$$f(t)=t\times \text{pulse function of width } t_0 \text{ and center at } \frac{t_0}{2} \tag{10.34}$$

We know the FT of the shifted pulse as

$$\text{pulse of width } t_0 \text{ and center} \frac{t_0}{2} \rightarrow \frac{1-e^{-j\omega t_0}}{j\omega} \tag{10.35}$$

Using the frequency differentiation property we get

$$\boxed{F(\omega)=-\frac{1-e^{-j\omega t_0}}{\omega^2}-\frac{t_0 e^{-j\omega t_0}}{j\omega}} \tag{10.36}$$

This can be decomposed into real and imaginary parts

$$F(\omega)=-\frac{1-\cos\omega t_0}{\omega^2}+t_0\frac{\sin\omega t_0}{\omega}+j\left[-\frac{\sin\omega t_0}{\omega^2}+t_0\frac{\cos\omega t_0}{\omega}\right] \tag{10.37}$$

Results are shown in Fig. 10.10. It is interesting to evaluate $F(0)$; since we have ω in the denominator, we have to be careful. The imaginary part goes to zero. The real part is evaluated after using the polynomial expansion of sine and cosine

$$\sin x \sim x-\frac{x^3}{3!}+\cdots, \quad \text{and} \quad \cos x \sim 1-\frac{x^2}{2}+\cdots \tag{10.38}$$

We then get

$$F(\omega\rightarrow 0)\sim -\frac{1-(1-\omega^2 t_0^2/2)}{\omega^2}+t_0\frac{\omega t_0}{\omega}=-\frac{t_0^2}{2}+t_0^2=\frac{t_0^2}{2} \tag{10.39}$$

which is nothing other than the area under the time curve!

10.10 Fourier Transform of (Cropped) Absolute Value of *t*

Again this function is non-integrable, but the cropped version is integrable. We define the cropped $|t|$ as

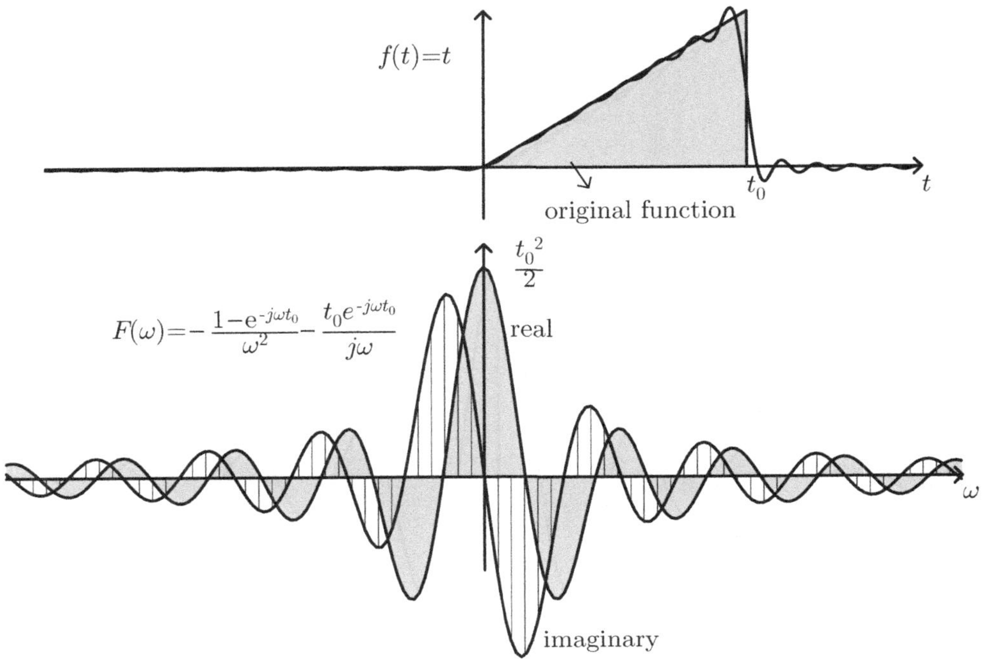

Fig. 10.10 Cropped single-sided t function and Fourier transform

$$f(t) = \begin{cases} |t| & -t_0 < t < t_0 \\ 0 & \text{else wise} \end{cases} \tag{10.40}$$

Rather than doing lengthy derivations, we can figure the FT by simply taking the even part of the FT of prior section, and multiply by two; hence

$$F(\omega) = 2\left[-\frac{1-\cos\omega t_0}{\omega^2} + t_0\frac{\sin\omega t_0}{\omega}\right] \tag{10.41}$$

Results are shown in Fig. 10.11. Notice the limit

$$F(0) = t_0^2 \tag{10.42}$$

which would be the area under the time curve.

10.11 Fourier Transform of (Cropped) t^2

Again the function $f(t) = t^2$ itself is problematic so far as finding a Fourier transform, since the signal grows without bound, and integrating such signal is not possible. However, we can find the FT of the *cropped* t^2 function defined by

$$f(t) = \begin{cases} t^2 & -t_0 < t < t_0 \\ 0 & \text{else wise} \end{cases} \tag{10.43}$$

We recognize that this function is simply t times the t function as derived in Eq. (10.31). Utilizing the frequency differentiation property again we get

$$F(\omega) = j\frac{d}{d\omega}\left\{2j\left[-\frac{\sin\omega t_0}{\omega^2} + t_0\frac{\cos\omega t_0}{\omega}\right]\right\}$$
$$= -2\left[2\frac{\sin\omega t_0}{\omega^3} - t_0\frac{\cos\omega t_0}{\omega^2} - t_0\frac{\cos\omega t_0}{\omega^2}\right.$$
$$\left.-t_0^2\frac{\sin\omega t_0}{\omega}\right] \tag{10.44}$$
$$= \boxed{-2\left[-t_0^2\frac{\sin\omega t_0}{\omega} - 2t_0\frac{\cos\omega t_0}{\omega^2} + 2\frac{\sin\omega t_0}{\omega^3}\right]}$$

Results are shown in Fig. 10.12. It's enlightening to examine the zero-frequency value of the FT. As seen in the above equation, we have

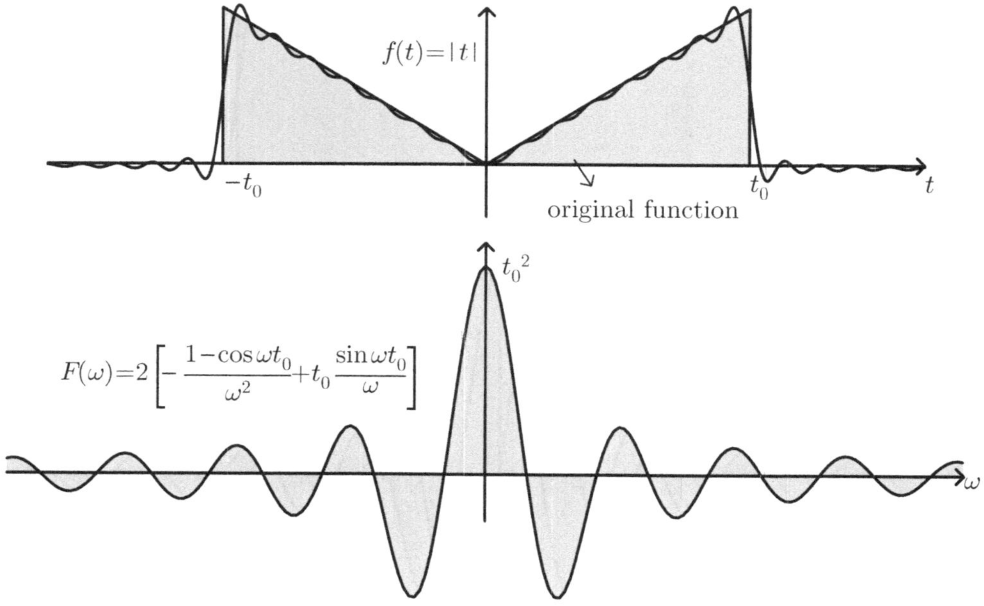

Fig. 10.11 Cropped $|t|$ function and Fourier transform

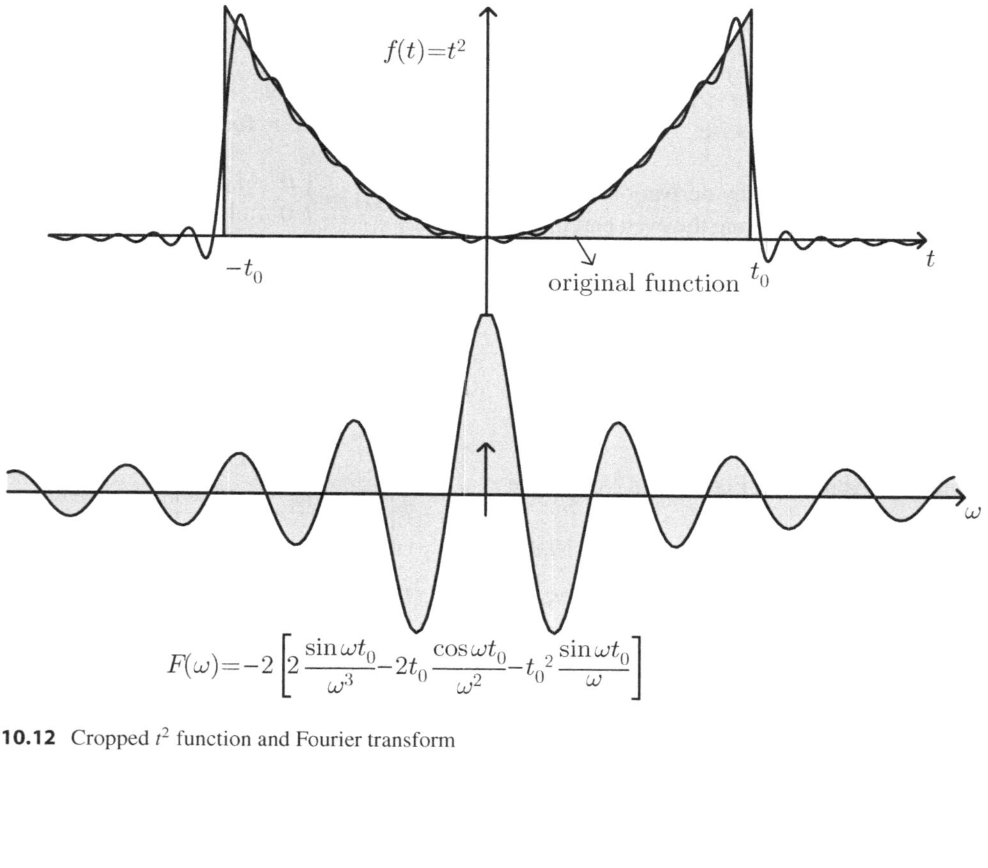

Fig. 10.12 Cropped t^2 function and Fourier transform

ω in the denominator, and hence evaluating the zero limit needs some attention. We would need to expand the sine and cosine as a polynomial series. We will use

$$\sin x \sim x - \frac{x^3}{3!} + \cdots, \text{ and } \cos x \sim 1 - \frac{x^2}{2} + \cdots \tag{10.45}$$

Notice that we will use at least the first two terms, since using 1 alone would give us incomplete results. We would get then

$$\begin{aligned} F(0) &= -2\left[\frac{2}{\omega^3}\left(\omega t_0 - \frac{\omega^3 t_0^3}{3!}\right)\right. \\ &\quad \left. -2\frac{t_0}{\omega^2}\left(1 - \frac{\omega^2 t_0^2}{2}\right) - \frac{t_0^2}{\omega}(\omega t_0)\right] \\ &= -2\left[2\frac{t_0}{\omega^2} - \frac{t_0^3}{3} - 2\frac{t_0}{\omega^2} + t_0^3 - t_0^3\right] \\ &= \boxed{\frac{2}{3}t_0^3} \end{aligned} \tag{10.46}$$

Notice that this is nothing other than the integral of the function over time; that is

$$2\int_0^{t_0} t^2 dt = \frac{2}{3}t_0^3 \tag{10.47}$$

10.12 Fourier Transform of (Cropped) t^4

The function of interest is defined as

$$f(t) = \begin{cases} t^4 & -t_0 < t < t_0 \\ 0 & \text{else wise} \end{cases} \tag{10.48}$$

We know the FT would be

$$f(t) \to \frac{d^4}{d\omega^4}\frac{2\sin\omega t_0}{\omega} \tag{10.49}$$

Notice that j times itself four times gives 1; hence the unity factor before the fourth derivative. From the prior section we know the second derivative of the FT

$$\frac{d^2}{d\omega^2}F(\omega) = -2t_0^2\frac{\sin\omega t_0}{\omega} - 4t_0\frac{\cos\omega t_0}{\omega^2} + 4\frac{\sin\omega t_0}{\omega^3} \tag{10.50}$$

The third derivative would be

$$\begin{aligned} \frac{d^3}{d\omega^3}F(\omega) &= -2t_0^3\frac{\cos\omega t_0}{\omega} + 2t_0^2\frac{\sin\omega t_0}{\omega^2} \\ &\quad + 4t_0^2\frac{\sin\omega t_0}{\omega^2} + 8t_0\frac{\cos\omega t_0}{\omega^3} + 4t_0\frac{\cos\omega t_0}{\omega^3} \\ &\quad - 12\frac{\sin\omega t_0}{\omega^4} \\ &= -2t_0^3\frac{\cos\omega t_0}{\omega} + 6t_0^2\frac{\sin\omega t_0}{\omega^2} + 12t_0\frac{\cos\omega t_0}{\omega^3} \\ &\quad - 12\frac{\sin\omega t_0}{\omega^4} \end{aligned} \tag{10.51}$$

The fourth derivative comes out

$$\begin{aligned} \frac{d^4}{d\omega^4}F(\omega) &= +2t_0^4\frac{\sin\omega t_0}{\omega} + 2t_0^3\frac{\cos\omega t_0}{\omega^2} \\ &\quad + 6t_0^3\frac{\cos\omega t_0}{\omega^2} - 12t_0^2\frac{\sin\omega t_0}{\omega^3} - 12t_0^2\frac{\sin\omega t_0}{\omega^3} \\ &\quad - 36t_0\frac{\cos\omega t_0}{\omega^4} - 12t_0\frac{\cos\omega t_0}{\omega^4} + 48\frac{\sin\omega t_0}{\omega^5} \\ &= +2t_0^4\frac{\sin\omega t_0}{\omega} + 8t_0^3\frac{\cos\omega t_0}{\omega^2} \\ &\quad - 24t_0^2\frac{\sin\omega t_0}{\omega^3} - 48t_0\frac{\cos\omega t_0}{\omega^4} + 48\frac{\sin\omega t_0}{\omega^5} \end{aligned} \tag{10.52}$$

Results are shown in Fig. 10.13.

10.13 Fourier Transform of $\frac{1}{1+t^2}$

Let's find the FT of the function

$$f(t) = \frac{1}{1+t^2} \tag{10.53}$$

In order to avoid the complex time integration as shown below

$$F(\omega) = \int_{-\infty}^{\infty}\frac{1}{1+t^2}e^{-j\omega t}dt \tag{10.54}$$

we recall a prior result (Eq. (8.25)), and that is

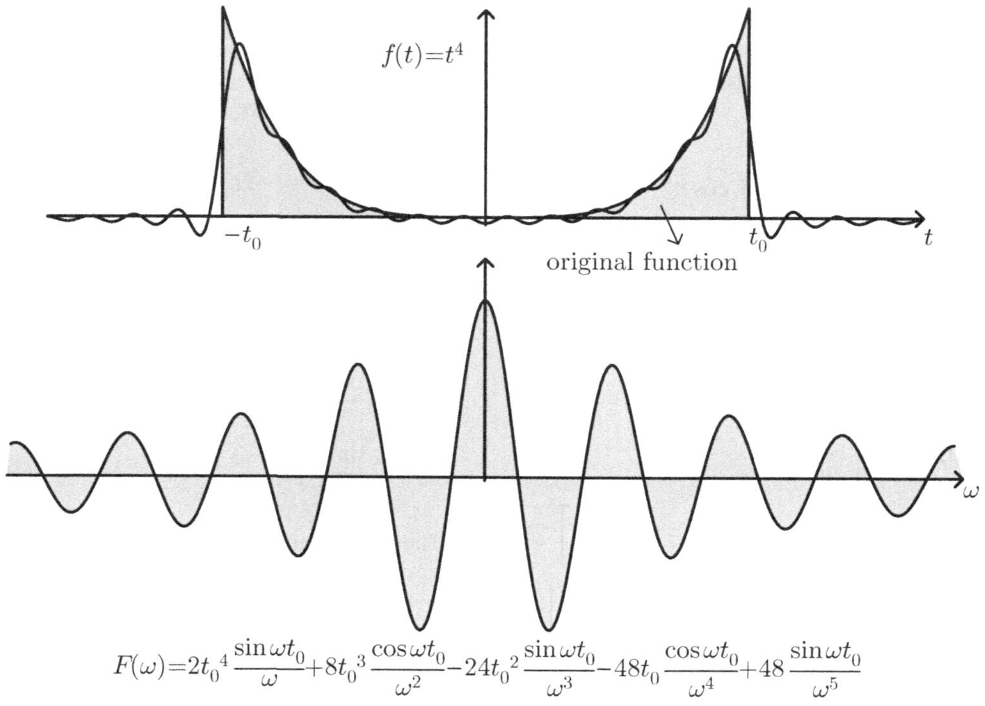

Fig. 10.13 Cropped t^4 function and Fourier transform

$$\mathcal{F}\left[e^{-|t|}\right] = \frac{2}{1+\omega^2} \tag{10.55}$$

and we recall the reciprocity property which stated that

$$\begin{aligned} &\text{if } f(t) \to F(\omega) \\ &\text{then } F(t) \to 2\pi f(-\omega) \end{aligned} \tag{10.56}$$

This then tells us that

$$\mathcal{F}\left[\frac{1}{1+t^2}\right] = \pi e^{-|\omega|} \tag{10.57}$$

Notice that the FT is real and symmetric, since our signal is real and symmetric too. Results are shown in Fig. 10.14.

10.14 Fourier Transform of $\frac{t}{1+t^2}$

Here we deal with the function

$$f(t) = \frac{t}{1+t^2} \tag{10.58}$$

Recall from the prior section that we had established that

$$\frac{1}{1+t^2} \to \pi e^{-|\omega|} \tag{10.59}$$

All we have to do now is multiply the time version of the above function by t and use the time multiplication property of the FT, repeated here for convenience:

$$\begin{aligned} &\text{if } f(t) \to F(\omega) \\ &\text{then } tf(t) \to j\frac{dF(\omega)}{d\omega} \end{aligned} \tag{10.60}$$

Based on this we conclude that

$$\mathcal{F}\left[\frac{t}{1+t^2}\right] = j\pi\frac{d\left(e^{-|\omega|}\right)}{d\omega} \tag{10.61}$$

Notice that the function $e^{-|\omega|}$ is even and that it's derivative equals itself for negative frequency and negative itself for positive frequency

$$\mathcal{F}\left[\frac{t}{1+t^2}\right] = \begin{cases} j\pi e^{-|\omega|} & \omega < 0 \\ -j\pi e^{-|\omega|} & \omega > 0 \end{cases} \tag{10.62}$$

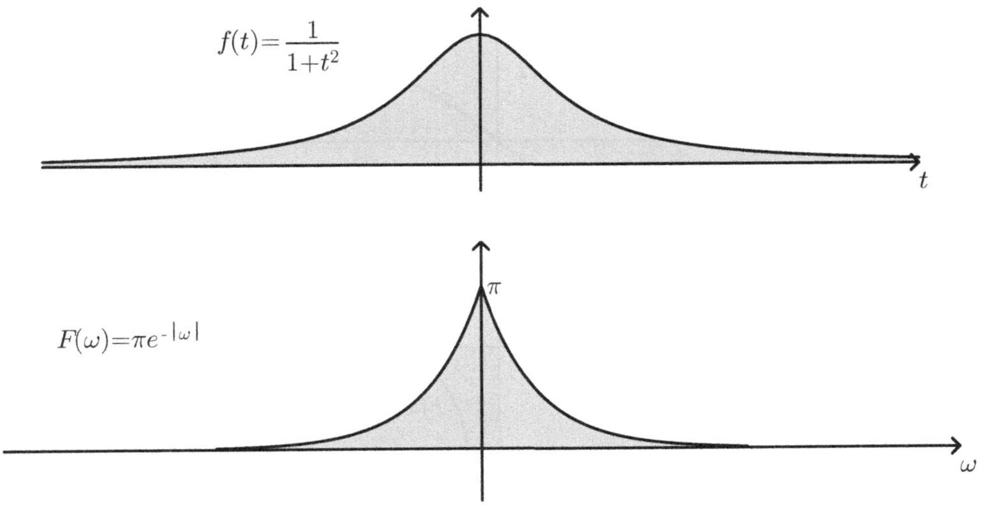

Fig. 10.14 Function $\frac{1}{1+t^2}$ and Fourier transform

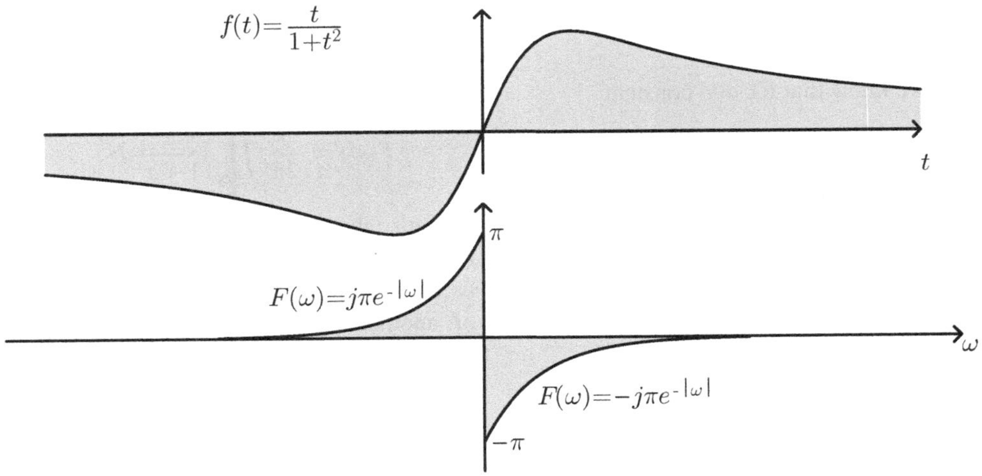

Fig. 10.15 Function $\frac{t}{1+t^2}$ and Fourier transform

Results are shown in Fig. 10.15.

10.15 Fourier Transform of Arctan(*t*)

To find the Fourier transform of the function

$$f(t) = \text{atan}(t) \tag{10.63}$$

we'd need to use a few tricks! First, recall that the arc tangent function is related to an integral involving $1/(1+t^2)$ as shown below

$$\int_{-\infty}^{t} \frac{1}{1+\tau^2} d\tau = \text{atan}(t) + C \tag{10.64}$$

To enforce that the arctan be $-\pi/2$ at $-\infty$ we set the constant C as follows:

$$\text{atan}(t) = -\frac{\pi}{2} + \int_{-\infty}^{t} \frac{1}{1+\tau^2} d\tau \tag{10.65}$$

Second, recall also (Eq. (10.57)) that we had already derived the FT of the integrand as

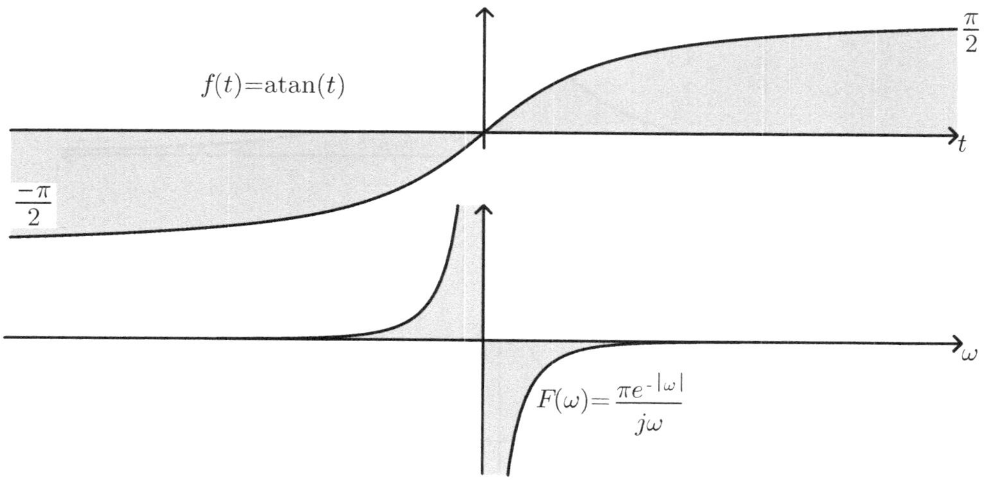

Fig. 10.16 Function atan(t) and Fourier transform

$$\frac{1}{1+t^2} \to \pi e^{-|\omega|} \quad (10.66)$$

Third we know that FT of a constant

$$-\frac{\pi}{2} \to -\pi^2\delta(\omega) \quad (10.67)$$

Finally recall the time integration property of the FT which states that

$$\text{if } f(t) \to F(\omega)$$
$$\text{then } \int f(t)dt \to \frac{F(\omega)}{j\omega} + \pi\delta(\omega) \times F(0) \quad (10.68)$$

Here, $F(0) = \pi$, and then the FT of the desired function becomes

$$F(\omega) = -\pi^2\delta(\omega) + \frac{\pi e^{-|\omega|}}{j\omega} + \pi^2\delta(\omega)$$
$$= \frac{\pi e^{-|\omega|}}{j\omega} \quad (10.69)$$

Results are shown in Fig. 10.16.

10.16 Fourier Transform of $\frac{1}{1+t^4}$

The goal here is to find the Fourier transform of the function

$$f(t) = \frac{1}{1+t^4} \quad (10.70)$$

which would be

$$F(\omega) = \frac{1}{2\pi}\int_{-\infty}^{\infty} \frac{e^{-j\omega t}}{1+t^4}dt \quad (10.71)$$

Let's take another look at this integral. By no means it is an easy one, and it cannot necessarily be hammered in terms of a derivative or integral of another function, which means we cannot necessarily use the various properties of the FT. It turns out, however, that this integral can be done using complex integration, and as will be shown in the chapter on complex analysis (Eqs. (A.262) and (A.263)), the transform comes out

$$F(\omega) = \begin{cases} \frac{\pi e^{\omega/\sqrt{2}}}{\sqrt{2}}\left[\cos\frac{\omega}{\sqrt{2}} - \sin\frac{\omega}{\sqrt{2}}\right] & \omega < 0 \\ \frac{\pi e^{-\omega/\sqrt{2}}}{\sqrt{2}}\left[\cos\frac{\omega}{\sqrt{2}} + \sin\frac{\omega}{\sqrt{2}}\right] & \omega > 0 \end{cases} \quad (10.72)$$

Results are shown in Fig. 10.17.

10.17 Fourier Transform of the Hat Function

The hat function is defined between $-\tau$ and τ

$$f(t) = \begin{cases} \tau + t & -\tau < t < 0 \\ \tau - t & 0 < t < \tau \\ 0 & \text{else wise} \end{cases} \quad (10.73)$$

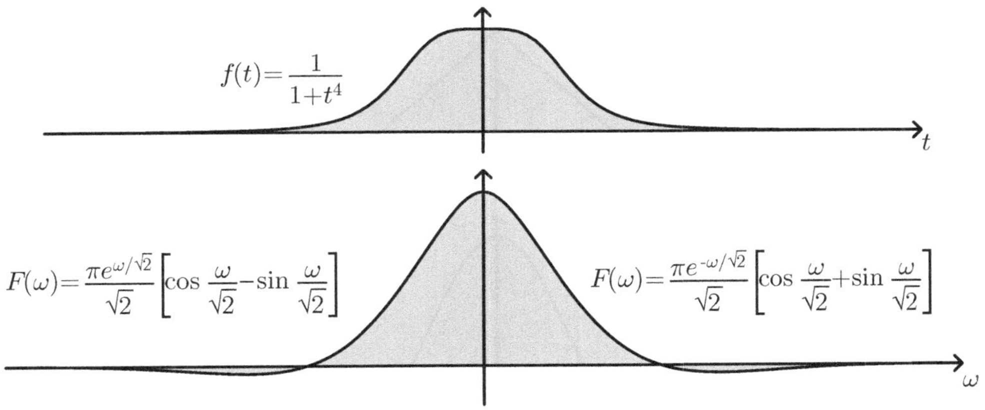

Fig. 10.17 Function $\frac{1}{1+t^4}$ and Fourier transform

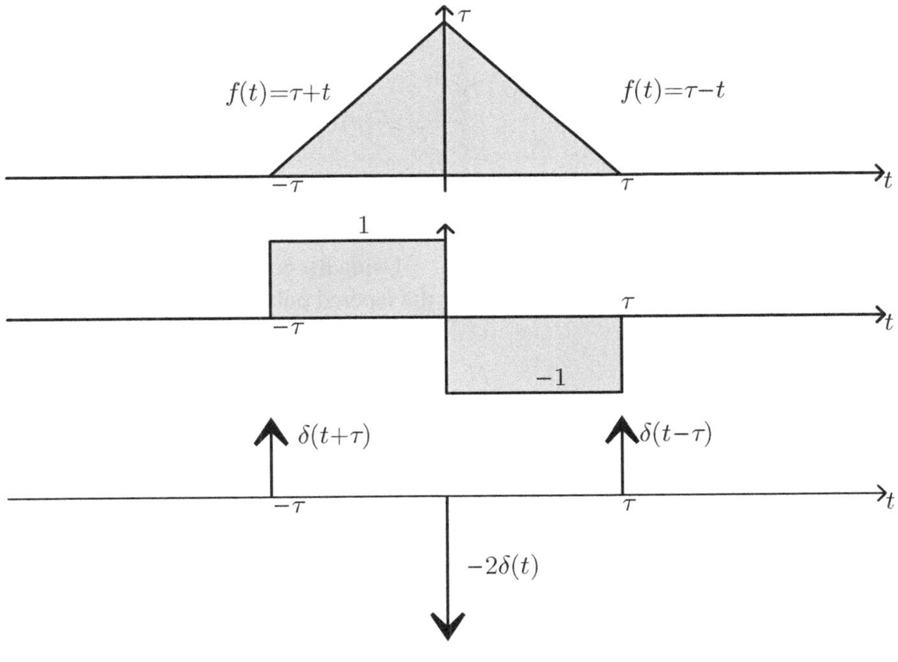

Fig. 10.18 Hat function and time derivatives

We already know from Chap. 9 (Properties of Fourier Transform) using the time integration property (or the time convolution one), that the FT of the hat function, and as was shown in Eq. 9.153, restated below (after expansion) is

$$F(\omega) = 2\frac{1 - \cos \omega\tau}{\omega^2} \tag{10.74}$$

Let's use yet another method to figure the FT—the time differentiation one. Recall

$$\frac{d}{dt}f(t) \to j\omega F(\omega), \qquad \text{and} \tag{10.75}$$

$$\frac{d^2 2}{dt^2}f(t) \to -\omega^2 F(\omega) \tag{10.76}$$

If we take the first derivative of the hat function we get the checker pulse as shown in Fig. 10.18. If we take the second derivative we get three impulses, as shown in the same figure

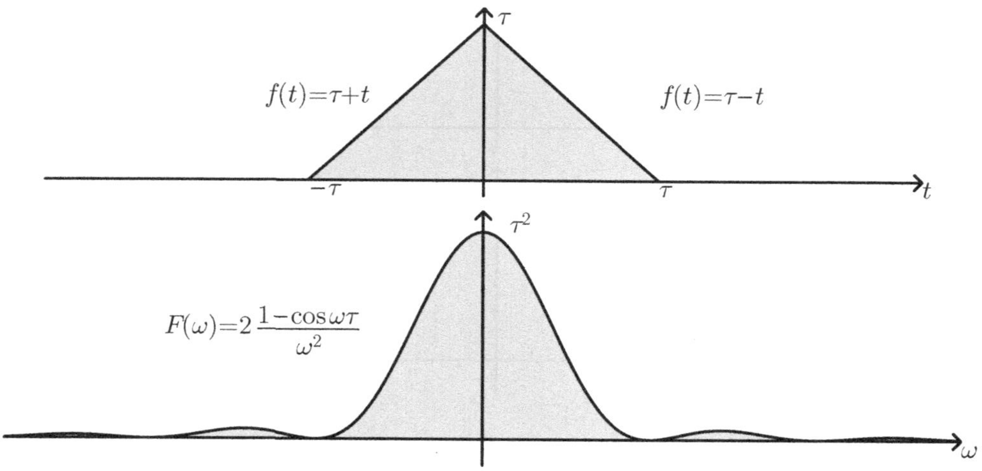

Fig. 10.19 Hat function and Fourier transform

$$\frac{d^2}{dt^2}f(t) = \delta(t+\tau) + \delta(t-\tau) - 2\delta(t) \quad (10.77)$$

Those transform to

$$\delta(t+\tau) + \delta(t-\tau) - 2\delta(t) \rightarrow e^{j\omega\tau} + e^{-j\omega\tau} - 2$$
$$= \boxed{2\left(-1+\cos\omega\tau\right)} \quad (10.78)$$

This would then imply that

$$\boxed{F(\omega) = 2\frac{1-\cos\omega\tau}{\omega^2}} \quad (10.79)$$

in agreement with Eq. 10.74; results in Fig. 10.19.

10.18 Fourier Transform of Tapered Pulse

Here we calculate the FT of the tapered pulse which has min width of $2t_0$, and max width of $2t_1$. Again instead of doing the time integration manually, we can use convolution. A minute worth of thought would show that this function is the convolution of two pulse functions: one with width $t_1 - t_0$ and height $1/(t_1 - t_0)$ and the other with width $(t_1 + t_0)$ and height 1, as shown in Fig. 10.20. These two pulse functions have the following FT transforms:

$$\mathcal{F}[p_1(t)] = \frac{2}{t_1 - t_0}\frac{\sin\omega[(t_1 - t_0)/2]}{\omega} \quad (10.80)$$

$$\mathcal{F}[p_2(t)] = 2\frac{\sin\omega[(t_0 + t_1)/2]}{\omega} \quad (10.81)$$

Using the convolution property, then the FT of the tapered pulse is the product of these two FTs

$$\mathcal{F}[f(t)] = \frac{4}{t_1 - t_0}\frac{\sin\omega[(t_1 - t_0)/2]}{\omega}\frac{\sin\omega[(t_0 + t_1)/2]}{\omega} \quad (10.82)$$

Results are shown in Fig. 10.21. Notice that first zero happens when

$$\omega[(t_0 + t_1)/2] = \pi, \quad \text{or when} \quad \omega = \frac{2\pi}{t_0 + t_1} \quad (10.83)$$

10.19 Fourier Transform of Asymmetric Triangular Checker Pulse

The asymmetric triangular checker pulse is shown in Fig. 10.22. It has a total width (total outer edge) of 2τ and peak 1. Rather than doing direct integration we will fall back on a few handy facts/properties. Take the first derivative, then the second one and end up with 2 delta

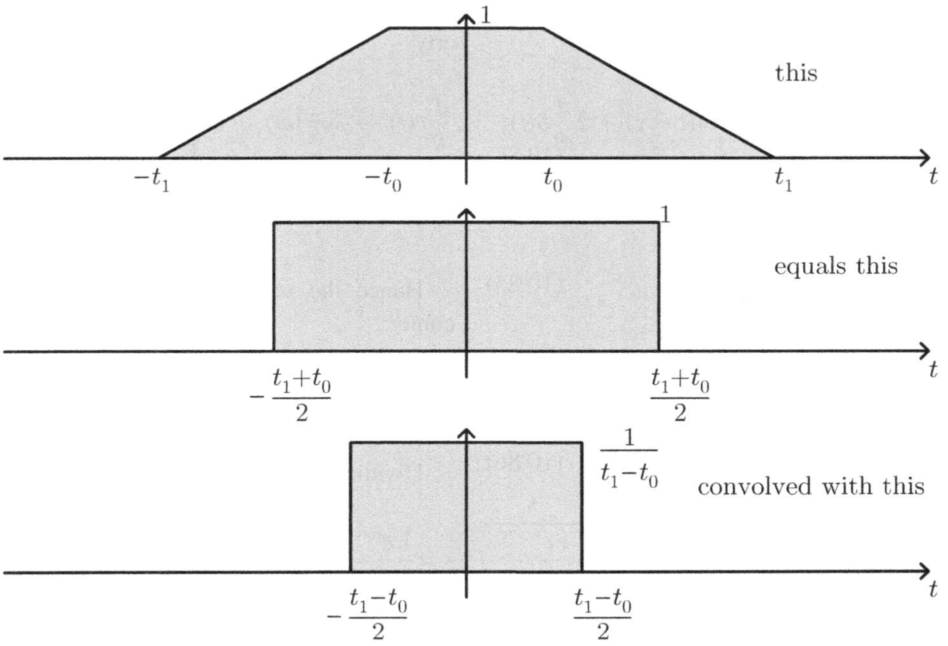

Fig. 10.20 Tapered pulse function as convolution of two pulses

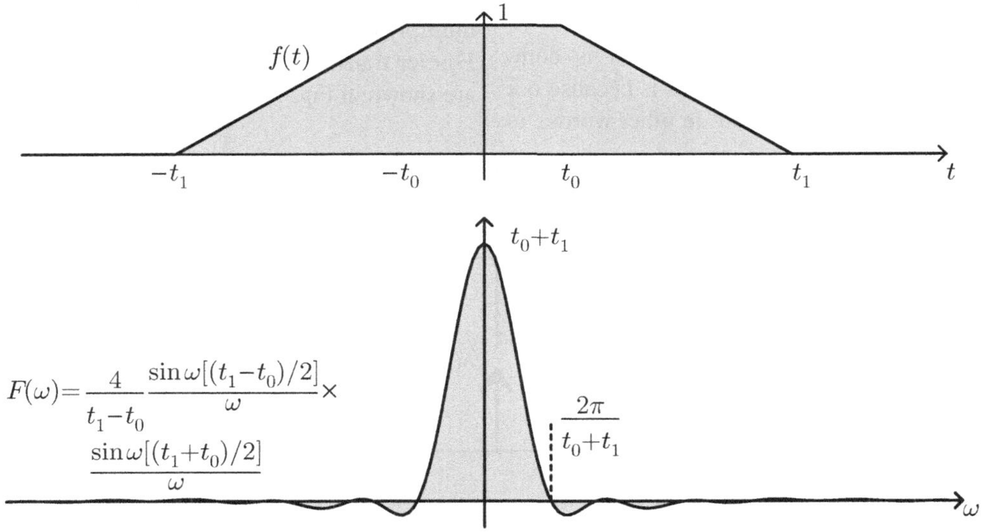

Fig. 10.21 Tapered pulse function and its Fourier transform

functions and a time derivative delta function as shown in the figure.

$$\frac{d^2f(t)}{dt^2} = -\frac{1}{\tau}\delta(t+\tau) + \frac{1}{\tau}\delta(t-\tau) + 2\frac{d}{dt}\delta(t) \tag{10.84}$$

We know that

$$\delta(t) \rightarrow 1 \tag{10.85}$$

and that

$$\delta(t+\tau) \rightarrow e^{j\omega\tau}, \quad \text{and} \quad \delta(t-\tau) \rightarrow e^{-j\omega\tau} \tag{10.86}$$

We also will use the time differentiation property

$$\frac{d}{dt}f(t) \rightarrow j\omega F(\omega), \quad \text{and} \quad \frac{d^2}{dt^2}f(t) \rightarrow -\omega^2 F(\omega) \tag{10.87}$$

which gives

$$\frac{d}{dt}\delta(t) \rightarrow j\omega \tag{10.88}$$

Hence the second derivative transform becomes

$$\frac{d^2f(t)}{dt^2} \rightarrow -\frac{1}{\tau}e^{j\omega\tau} + \frac{1}{\tau}e^{-j\omega\tau} + 2j\omega \tag{10.89}$$

Dividing by $-\omega^2$ gives us back the FT of $f(t)$

$$F(\omega)=\frac{1}{\tau}\frac{-e^{j\omega\tau}+e^{-j\omega\tau}}{-\omega^2}+\frac{2}{j\omega}=\frac{1}{\tau}\frac{e^{j\omega\tau}-e^{-j\omega\tau}}{\omega^2}+\frac{2}{j\omega} \tag{10.90}$$

$$\boxed{F(\omega) = 2\left[\frac{1}{j\omega} + \frac{j}{\tau}\frac{\sin\omega\tau}{\omega^2}\right]} \tag{10.91}$$

Notice that the Fourier transform is completely imaginary (and odd). This is because our time function is real and odd. In other words, to get an odd function out of the inverse transform we only use the sine component of $e^{j\omega t}$; and since that component is imaginary, we need to multiply by j to make the outcome real. Hence the Fourier transform is odd and imaginary. Results are shown in Fig. 10.23.

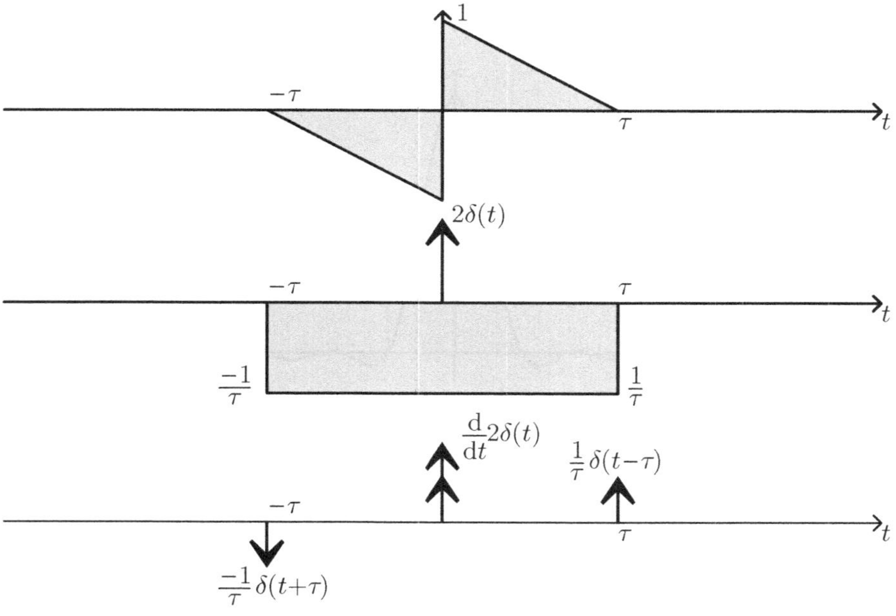

Fig. 10.22 Asymmetric triangular checker pulse and derivatives

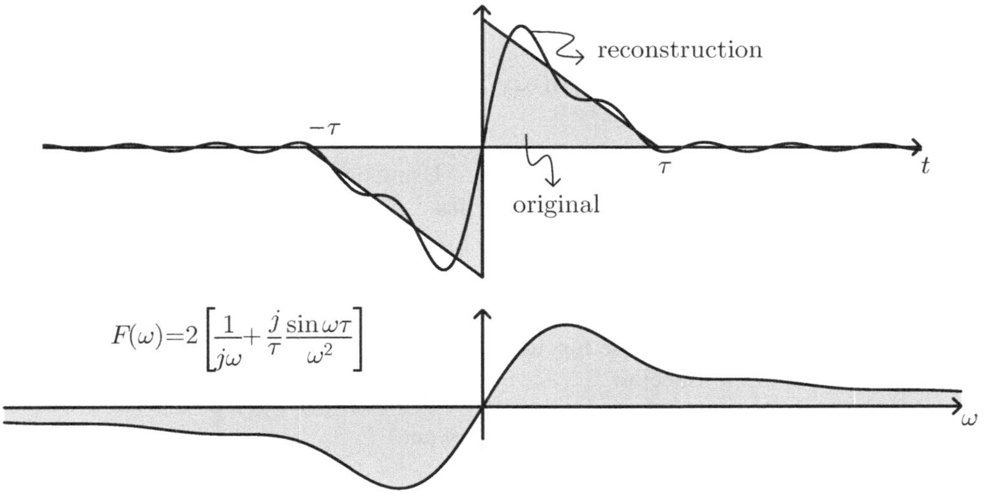

Fig. 10.23 Asymmetric triangular checker pulse and Fourier transform

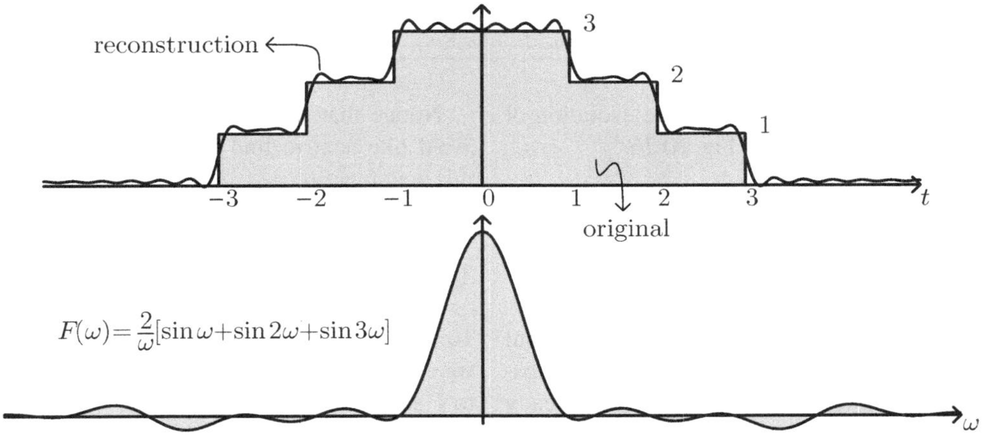

Fig. 10.24 Symmetric 3-step stair function and Fourier transform

10.20 Fourier Transform of Symmetric 3-Step Stair

The symmetric 3-step stair function is defined as

$$f(t) = \begin{cases} 3 & |t| < 1 \\ 2 & 1 < |t| < 2 \\ 1 & 2 < |t| < 3 \\ 0 & \text{else wise} \end{cases} \tag{10.92}$$

We quickly realize that this is the sum of three pulses: one of width 2, one of width 4, and one of width 6. We simply use linearity to find FT of aggregate function

$$\boxed{F(\omega) = \frac{2}{\omega}\,[\sin\omega + \sin 2\omega + \sin 3\omega]} \tag{10.93}$$

Results are shown in Fig. 10.24.

10.21 Fourier Transform of Truncated Pulse Train

Here we find the FT of a truncated pulse train. For demonstration purposes we choose a train of 9 pulses: one at the origin, four to the left, and four to the right. Each pulse has 0.5 width and spacing between pulses is 1; see Fig. 10.25 (top).

For reference

$$p(t) = \begin{cases} 1 & |x| < \frac{1}{4} \\ 0 & \text{elsewhere} \end{cases} \tag{10.94}$$

Since our pulse has width 0.5, its FT is

$$\mathcal{F}[p(t)] = 2\frac{\sin \omega/4}{\omega} \tag{10.95}$$

To get 9 pulses we recognize that the desired function is the convolution of the pulse function with 9 delta functions, as shown below

$$\begin{aligned} f(t) &= p(t) * [\delta(t+4) + \delta(t+3) + \delta(t+2) \\ &+ \delta(t+1) + \delta(t) + \delta(t-1) + \delta(t-2) \\ &+ \delta(t-3) + \delta(t-4)] \end{aligned} \tag{10.96}$$

Using time convolution property we conclude that

$$\begin{aligned} F(\omega) &= 2\frac{\sin \omega/4}{\omega} \cdot \left[e^{-j4\omega} + e^{-j3\omega} + e^{-j2\omega} + e^{-j1\omega} + \right. \\ &\left. + e^{j0\omega} + e^{j1\omega} + e^{j2\omega} + e^{j3\omega} + e^{j4\omega}\right] \end{aligned} \tag{10.97}$$

Combining negative and positive frequencies we get

$$\boxed{F(\omega) = 2\frac{\sin \omega/4}{\omega} \cdot [1 + 2\cos 1\omega + 2\cos 2\omega + 2\cos 3\omega + 2\cos 4\omega]} \tag{10.98}$$

See Fig. 10.25 (bottom). A time evolution of the time series is shown in Fig. 10.26.

10.22 The Ramped Unit Step Function

The unit step function is a very important input for many reasons, including full load application and sampling of all frequencies. We already know the FT of the ideal step input (with abrupt rise time) but many a times the rise time is more gradual. Here we look at the impact of rise time on the transfer function of the unit step.

The ideal unit step input has zero rise time. It is defined as

$$u_{\text{ideal}}(t) = \begin{cases} 0 & t < 0 \\ 1 & t > 0 \end{cases} \tag{10.99}$$

We define the ramped unit step as having a rise time τ. For simplicity we set the rise time symmetric such that

$$u(t) = \begin{cases} 0 & t < -\frac{\tau}{2} \\ \frac{1}{2} + \frac{t}{\tau} & -\frac{\tau}{2} < t < \frac{\tau}{2} \\ 1 & t > \frac{\tau}{2} \end{cases} \tag{10.100}$$

Notice that using this definition $u(0) = \frac{1}{2}$. We'd like next to find the Fourier transform of this new function.

10.22.1 The Hard Way

To find the FT of the ramped unit step function we split the function into two parts: one from $-\frac{\tau}{2}$ to $\frac{\tau}{2}$ and the other after $\frac{\tau}{2}$. That is, we decompose the function in terms of a right triangle about the y axis, and an ideal unit step function shifted to the right by $\frac{\tau}{2}$. We know the FT of the ideal step function

$$u_{\text{ideal}}(t) \rightarrow \pi\delta(\omega) + \frac{1}{j\omega} \tag{10.101}$$

The shifted ideal unit set function would have FT

$$u_{\text{ideal}}\left(t - \frac{\tau}{2}\right) \rightarrow \pi\delta(\omega) + \frac{e^{-j\omega\tau/2}}{j\omega} \tag{10.102}$$

Next we find the FT of the triangle function. We can furthermore decompose this function in terms of a pulse of magnitude $\frac{1}{2}$ and the same pulse multiplied by $\frac{t}{\tau}$; that is

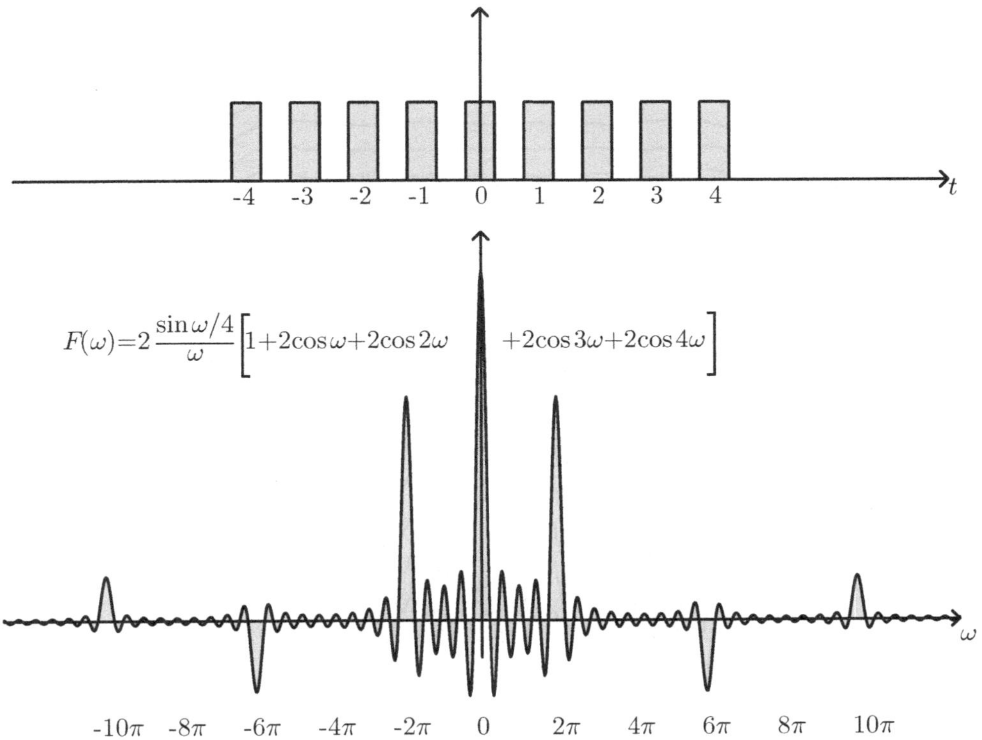

Fig. 10.25 Truncated pulse function and Fourier transform

$$\text{triangle}(t) = \frac{1}{2}\text{pulse}(t) + \frac{t}{\tau}\text{pulse}(t) \quad (10.103)$$

We know the pulse (of width τ) has a FT of

$$\text{pulse} \rightarrow 2\frac{\sin\omega\frac{\tau}{2}}{\omega} \quad (10.104)$$

Then the FT of the triangle, and utilizing the frequency differentiation property, becomes

$$\begin{aligned}
\text{triangle} &\rightarrow \frac{\sin\omega\frac{\tau}{2}}{\omega} + \frac{j}{\tau}\frac{d}{d\omega}\frac{2\sin\omega\frac{\tau}{2}}{\omega} \\
&= \frac{\sin\omega\frac{\tau}{2}}{\omega} + \frac{j}{\tau}\left[2\frac{\cos\omega\frac{\tau}{2}}{\omega} \times \frac{\tau}{2} - \frac{2\sin\omega\frac{\tau}{2}}{\omega^2}\right] \\
&= \frac{\sin\omega\frac{\tau}{2}}{\omega} + j\frac{\cos\omega\frac{\tau}{2}}{\omega} - \frac{2j}{\tau}\frac{\sin\omega\frac{\tau}{2}}{\omega^2} \\
&= -\frac{e^{-j\omega\tau/2}}{j\omega} - \frac{2j}{\tau}\frac{\sin\omega\frac{\tau}{2}}{\omega^2} \quad (10.105)
\end{aligned}$$

When we add this FT to that in Eq. (10.102) we finally get

$$u(t) \rightarrow \pi\delta(\omega) + \frac{2}{\tau}\frac{\sin\omega\frac{\tau}{2}}{j\omega^2} \quad (10.106)$$

10.22.2 The Easy Way

The easy way about this is to recognize that the ramped unit step function can be obtained as a result of convolution between a pulse function (of width τ and height $\frac{1}{\tau}$) and an ideal unit step function: that is

$$u(t) = \frac{1}{\tau}\text{pulse}(t) * u_{\text{ideal}}(t) \quad (10.107)$$

Now we simply use the time convolution property of the FT which states that the FT of the convolved functions is the product of the FT of the individual functions to get

$$u(t) \rightarrow \frac{1}{\tau}\frac{2\sin\omega\frac{\tau}{2}}{\omega} \cdot \left[\pi\delta(\omega) + \frac{1}{j\omega}\right]$$

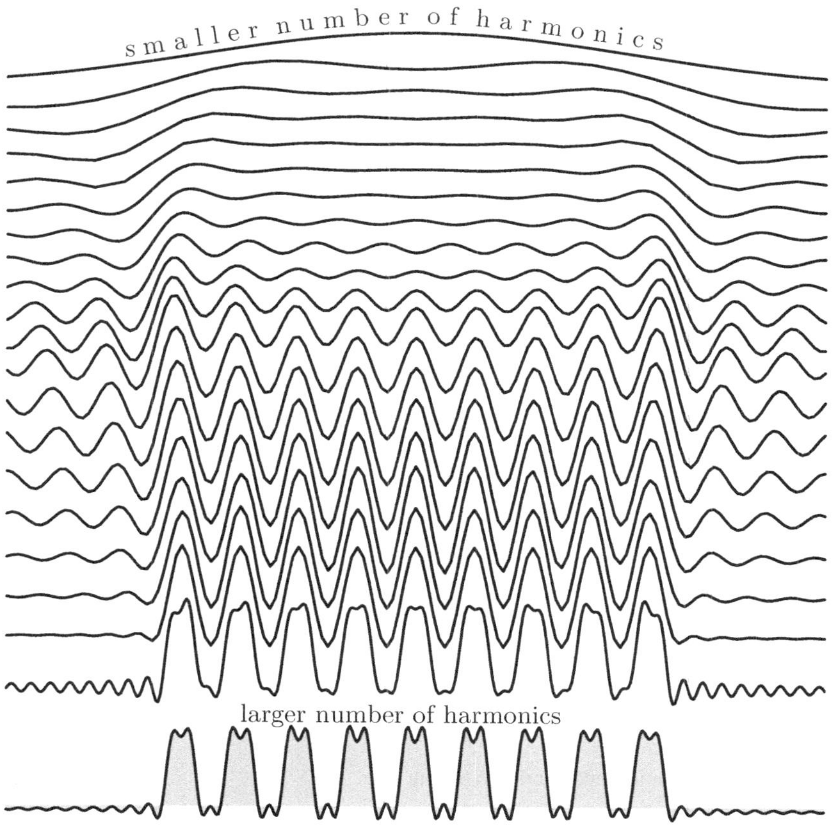

Fig. 10.26 Truncated pulse reconstruction as a function of number of harmonics

$$= \boxed{\pi\delta(\omega) + \frac{2}{\tau}\frac{\sin\omega\frac{\tau}{2}}{j\omega^2}} \tag{10.108}$$

in agreement with Eq. (10.106). Results are shown in Fig. 10.27. Notice that the low frequency content of the slanted step function is almost identical to that of the ideal one; but the high frequency content is smaller, indicating that we don't need a lot of high frequency harmonics, since the edge in the slanted step is smoother (as compared to the ideal one).

10.23 The Multi-Step Signum Function

We know the Fourier transform of the ideal signum function, as well as that with two steps (Eq. (10.6)). Here we will look at the impact of number of steps in the signum function, and compare to the smooth, slanted one. Figure 10.28 shows the signum function for various step count; the premise is that with larger step count, it should approach the slanted, smooth signum function. In the subsequent steps we calculate the Fourier transform as a function of step count.

1. Single step:

$$F(\omega) = \frac{1}{1}\left[\frac{1}{j\omega}\right] \tag{10.109}$$

2. Two steps:

$$F(\omega) = \frac{1}{2}\left[2\frac{\cos\omega\frac{\tau}{2}}{j\omega}\right] \tag{10.110}$$

3. Three steps:

$$F(\omega) = \frac{1}{3}\left[\frac{1}{j\omega} + 2\frac{\cos\omega\frac{\tau}{2}}{j\omega}\right] \tag{10.111}$$

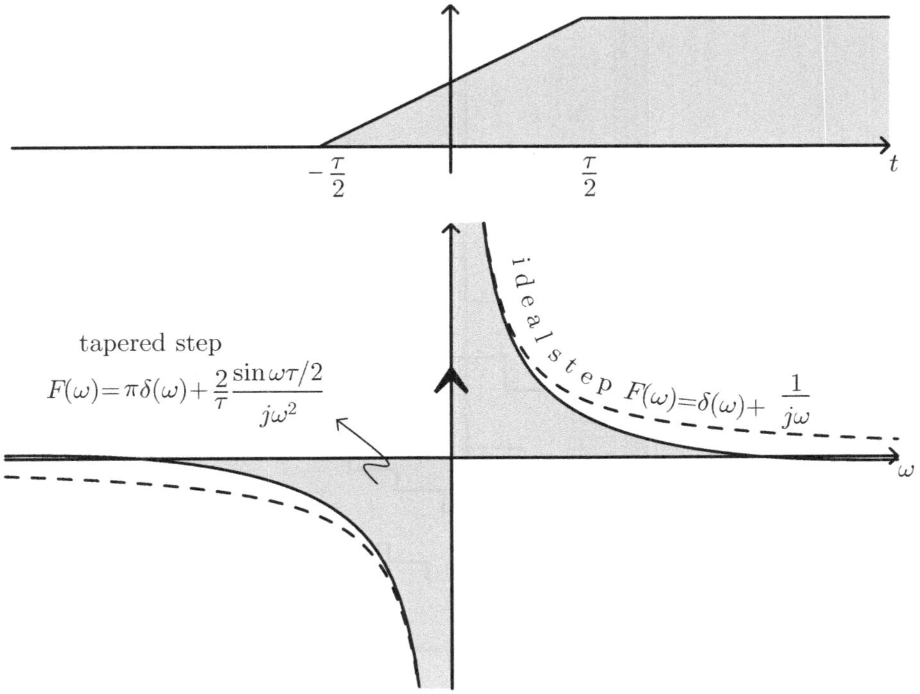

Fig. 10.27 Slanted step function and Fourier transform

4. Four steps:

$$F(\omega) = \frac{1}{4}\left[2\frac{\cos\omega\tau\frac{3}{6}}{j\omega} + 2\frac{\cos\omega\tau\frac{1}{6}}{j\omega}\right] \tag{10.112}$$

5. Five steps:

$$F(\omega) = \frac{1}{5}\left[\frac{1}{j\omega} + 2\frac{\cos\omega\tau\frac{4}{8}}{j\omega} + 2\frac{\cos\omega\tau\frac{2}{8}}{j\omega}\right] \tag{10.113}$$

6. Six steps:

$$F(\omega) = \frac{1}{6}\left[2\frac{\cos\omega\tau\frac{5}{10}}{j\omega} + 2\frac{\cos\omega\tau\frac{3}{10}}{j\omega} + 2\frac{\cos\omega\tau\frac{1}{10}}{j\omega}\right] \tag{10.114}$$

7. Seven steps:

$$F(\omega) = \frac{1}{7}\left[\frac{1}{j\omega} + 2\frac{\cos\omega\tau\frac{6}{12}}{j\omega} + 2\frac{\cos\omega\tau\frac{4}{12}}{j\omega} + 2\frac{\cos\omega\tau\frac{2}{12}}{j\omega}\right] \tag{10.115}$$

8. Eight steps:

$$F(\omega) = \frac{1}{8}\left[2\frac{\cos\omega\tau\frac{7}{14}}{j\omega} + 2\frac{\cos\omega\tau\frac{5}{14}}{j\omega} + 2\frac{\cos\omega\tau\frac{3}{14}}{j\omega} + 2\frac{\cos\omega\tau\frac{1}{14}}{j\omega}\right] \tag{10.116}$$

9. Nine steps:

$$F(\omega) = \frac{1}{9}\left[\frac{1}{j\omega} + 2\frac{\cos\omega\tau\frac{8}{16}}{j\omega} + 2\frac{\cos\omega\tau\frac{6}{16}}{j\omega} + 2\frac{\cos\omega\tau\frac{4}{16}}{j\omega} + 2\frac{\cos\omega\tau\frac{2}{16}}{j\omega}\right] \tag{10.117}$$

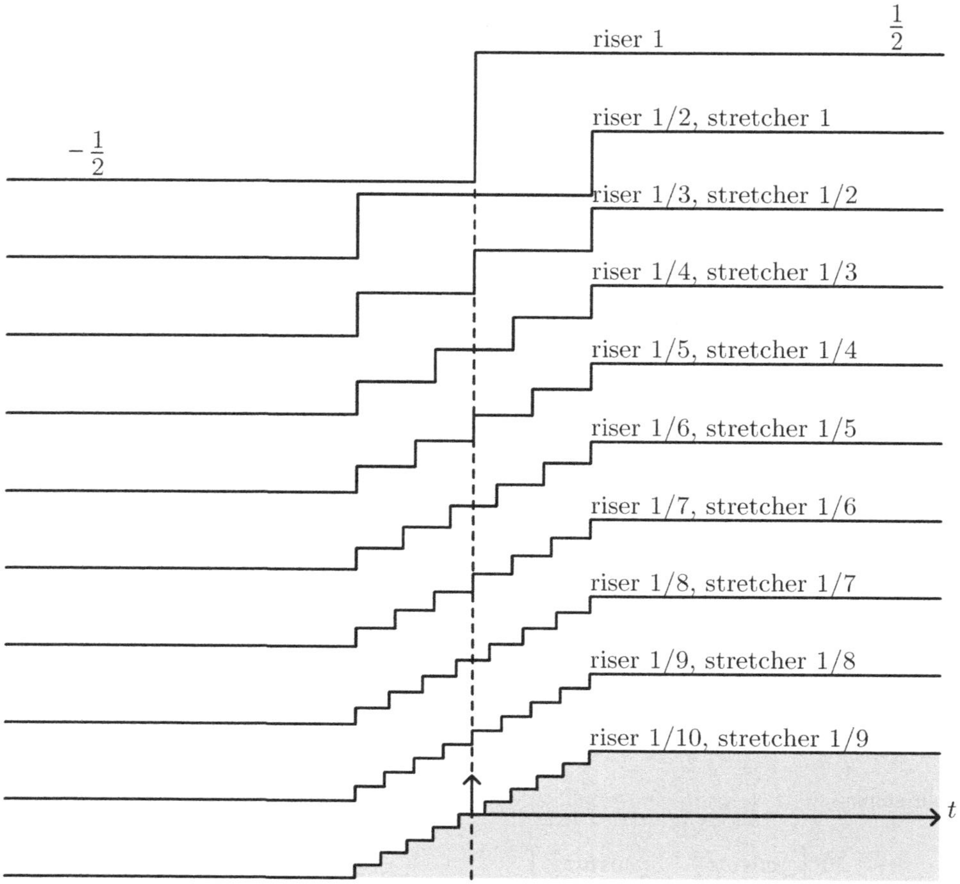

Fig. 10.28 Multi-step signum function with increasing step count

10. Ten steps:

$$F(\omega) = \frac{1}{10}\left[2\frac{\cos\omega\tau\frac{9}{18}}{j\omega} + 2\frac{\cos\omega\tau\frac{7}{18}}{j\omega} + 2\frac{\cos\omega\tau\frac{5}{18}}{j\omega} + 2\frac{\cos\omega\tau\frac{3}{18}}{j\omega} + 2\frac{\cos\omega\tau\frac{1}{18}}{j\omega}\right] \quad (10.118)$$

The various Fourier transforms are shown in Fig. 10.29. Also shown for reference is the slanted, smooth case. Notice that with increase in step count we approach the smooth case, as expected.

10.24 Flexibility of the Fourier Transform Applied to the Unit Step Function

We wrap this chapter by demonstrating the flexibility of the Fourier transform by deriving the unit step FT using seven different methods!!

Direct Integration

We can find the FT of the unit step function by direct integration:

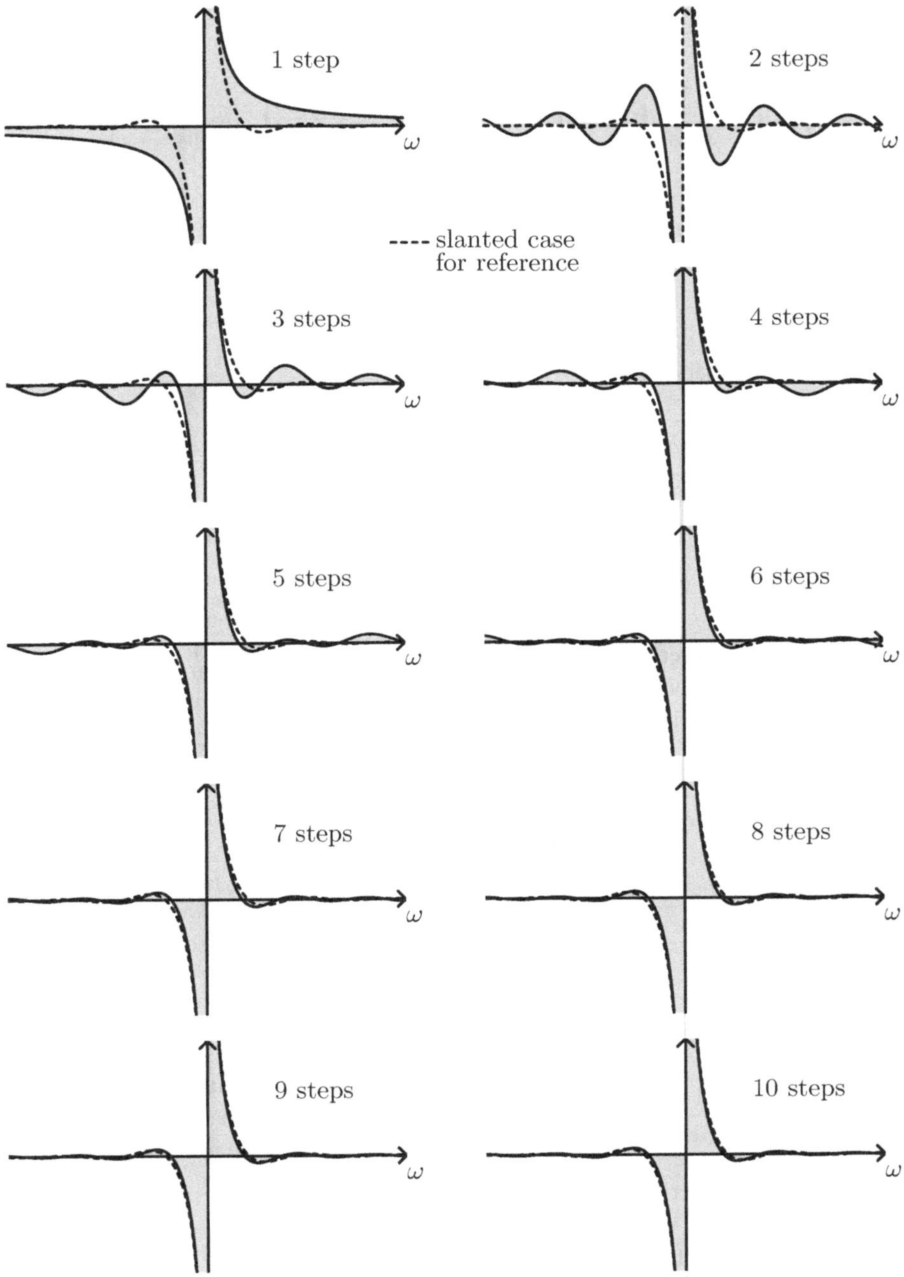

Fig. 10.29 Multi-step signum function Fourier transform, as a function of number of steps, and comparison to slanted version

$$\mathcal{F}[u(t)] = \int_{-\infty}^{\infty} u(t)e^{-j\omega t}dt = \int_{0}^{\infty} e^{-j\omega t}dt = -\frac{e^{-j\omega t}}{j\omega}\Big|_0^{\infty} = \frac{1}{j\omega} - \frac{e^{-j\omega\infty}}{j\omega}$$
$$= \frac{1}{j\omega} - \frac{\cos\omega\infty}{j\omega} + \frac{\sin\omega\infty}{\omega}$$

$$\boxed{\mathcal{F}[u(t)] = \pi\delta(\omega) + \frac{1}{j\omega}} \quad (10.119)$$

where we have used the fact that

$$\lim_{\tau\to\infty} \frac{\sin\omega\tau}{\omega} = \pi\delta(\omega) \quad (10.120)$$

and that on average

$$\lim_{\tau\to\infty} \frac{\cos\omega\tau}{\omega} = 0 \quad (10.121)$$

Using Signum and DC Function
We know that we can represent the unit step function as the sum of the signum function and half the DC function

$$u(t) = \frac{1}{2} + \text{signum}(t) \quad (10.122)$$

The 1/2 has the FT

$$\frac{1}{2} \to \pi\delta \quad (10.123)$$

and the signum function has the FT

$$\text{signum}(t) \to \frac{1}{j\omega} \quad (10.124)$$

Using superposition we arrive at

$$\boxed{u(t) \to \pi\delta(\omega) + \frac{1}{j\omega}} \quad (10.125)$$

Using the Single-Sided Decaying Exponential
We know that the transform function of the single-sided decaying function is

$$\mathcal{F}[e^{-at}u(t)] = \frac{1}{a+j\omega} \quad (10.126)$$

If we carry on the complex expansion we get

$$\frac{1}{a+j\omega} = \frac{a-j\omega}{a^2+\omega^2} \quad (10.127)$$

We can get the unit step function by taking the limit of a going to zero as follows:

$$u(t) = \lim_{a\to 0} e^{-at}u(t) \quad (10.128)$$

Hence the sought transform can be achieved by taking the same limit in frequency

$$F(\omega) = \lim_{a\to 0} \frac{a-j\omega}{a^2+\omega^2} = \lim_{a\to 0} \frac{a}{a^2+\omega^2} - j\frac{\omega}{a^2+\omega^2}$$
$$= \boxed{\pi\delta(\omega) + \frac{1}{j\omega}} \quad (10.129)$$

where we have used the following fact:

$$\lim_{a\to 0} \frac{a}{a^2+\omega^2} = \pi\delta(\omega) \quad (10.130)$$

Using the Time Integration Property of the Fourier Transform We know that we can get the unit step by time integrating the delta function

$$u(t) = \int_{-\infty}^{t} \delta(\tau)d\tau \quad (10.131)$$

We also know that the delta function transfers to

$$\delta(t) \to 1 \quad (10.132)$$

Finally we know the time integration property of the FT which states that

$$\int_{-\infty}^{t} f(\tau)d\tau \rightarrow \pi F(0)\delta(\omega) + \frac{F(\omega)}{j\omega} \quad (10.133)$$

Combining the above we arrive at

$$\boxed{F(\omega) = \pi\delta(\omega) + \frac{1}{j\omega}} \quad (10.134)$$

where we have used the fact that $F(0) = 1$

Using the Fourier Transform of the Pulse Function

We know that the pulse of width τ and left edge at 0 has the FT

$$\text{pulse of width } \tau \rightarrow \frac{1 - e^{-j\omega\tau}}{j\omega} \quad (10.135)$$

If we now let $\tau \rightarrow \infty$ we get

$$\boxed{F(\omega) = \pi\delta(\omega) + \frac{1}{j\omega}} \quad (10.136)$$

where we have used the fact that

$$\lim_{\tau\rightarrow\infty} \frac{e^{j\omega\tau}}{j\omega} = \pi\delta(\omega) \quad (10.137)$$

Using the Fourier Transform of the Single-Sided Cosine Function

We know that the FT of the single-sided cosine function, defined as zero for negative time, and cosine for positive time, and with frequency ω_0 is

$$\mathcal{F}[u(t)\cos\omega_0 t] = \frac{\pi}{2}[\delta(\omega - \omega_0) + \delta(\omega + \omega_0)] + \frac{\omega}{j(\omega^2 - \omega_0^2)} \quad (10.138)$$

See Eq. (8.66). We can achieve a unit step function out of the single-sided cosine function simply by letting $\omega_0 \rightarrow 0$; that is

$$\lim_{\omega_0\rightarrow 0} u(t)\cos\omega_0 t = u(t) \quad (10.139)$$

In accordance with this, the FT of the unit step function would be the limit of the FT of the single-sided cosine, as $\omega_0 \rightarrow 0$; that is

$$\mathcal{F}[u(t)] = \lim_{\omega_0\rightarrow 0} \frac{\pi}{2}[\delta(\omega - \omega_0) + \delta(\omega + \omega_0)] + \frac{\omega}{j(\omega^2 - \omega_0^2)}$$
$$= \frac{\pi}{2}[\delta(\omega) + \delta(\omega)] + \frac{\omega}{j\omega^2}$$

$$\boxed{F(\omega) = \pi\delta(\omega) + \frac{1}{j\omega}} \quad (10.140)$$

Using the Fourier Transform of the Slanted Unit Step

We know that the Fourier transform of the slanted unit step function (with rise time τ) is

$$\frac{2}{\tau}\frac{\sin\omega\frac{\tau}{2}}{j\omega^2} + \pi\delta(\omega) \quad (10.141)$$

See Eq. (10.106). We can achieve the ideal unit step function by taking the limit of the slanted step function as $\tau \rightarrow 0$; then the corresponding FT is

$$F(\omega) = \lim_{\tau\rightarrow 0}\left[\pi\delta(\omega) + \frac{2}{\tau}\frac{\sin\omega\frac{\tau}{2}}{j\omega^2}\right]$$
$$= \pi\delta(\omega) + \frac{2}{\tau} \times \left(\omega\frac{\tau}{2}\right)\frac{1}{j\omega^2}$$
$$= \boxed{\pi\delta(\omega) + \frac{1}{j\omega}} \quad (10.142)$$

The seven aforementioned methods are illustrated in Fig. 10.30.

10.25 Summary

In this chapter we examined further examples/topics on the Fourier transform. Beyond the conventional functions, such as the pulse and unit step (treated in last chapter) this chapter pushed the envelope by tackling some unconventional—yet still important and useful—

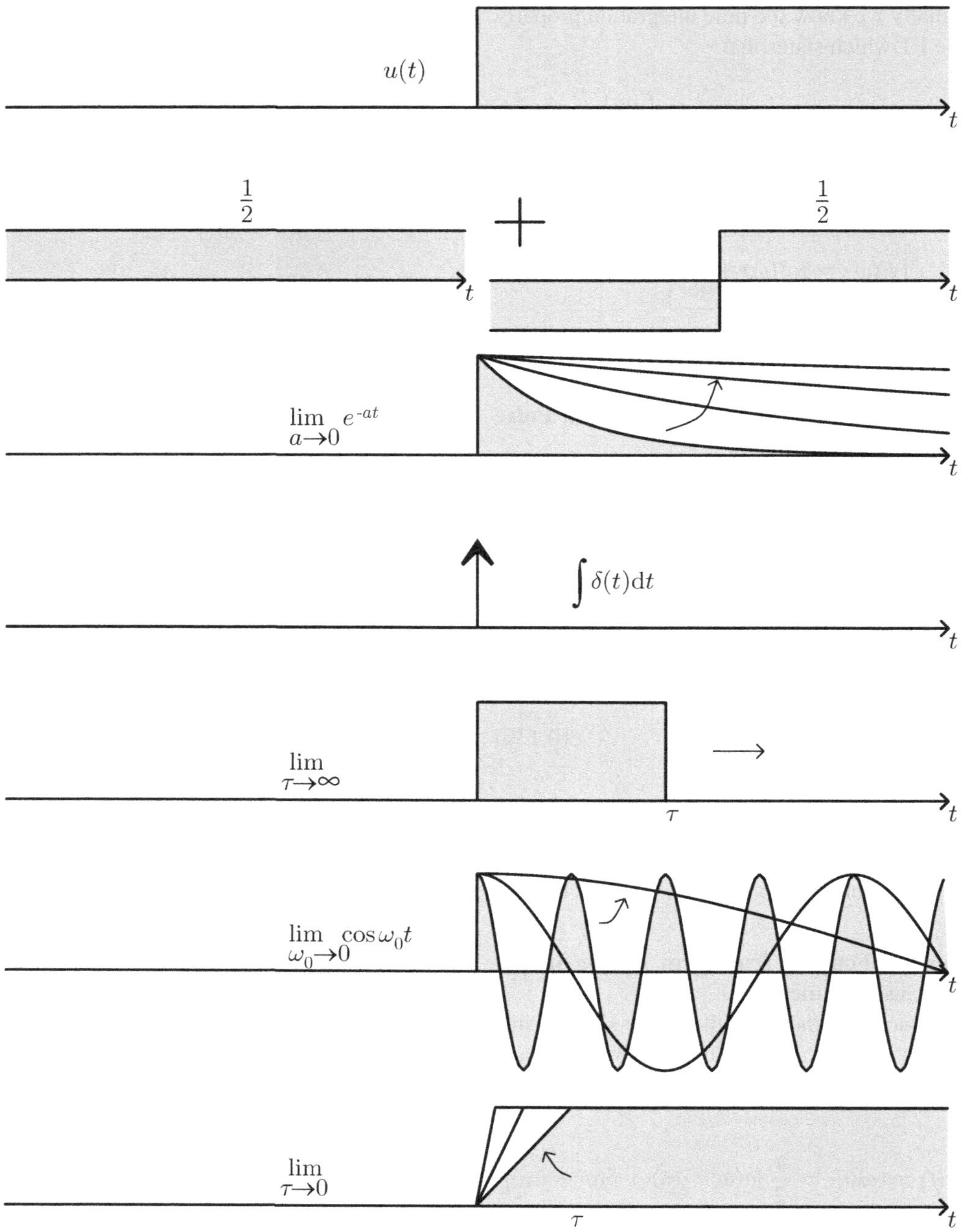

Fig. 10.30 Seven methods to derive the Fourier transform of the unit step function

functions. This includes stair signum function, odd negative exponential, product of cosine and negative exponential, cropped cosine, powers of t (cropped in time), variants of $\frac{t}{1+t^n}$, $\arctan(t)$, the hat function and variants thereof, multi-step stair function, truncated pulse train, and the slanted unit step function. For each case we derived the FT either explicitly or by using the properties developed in the last chapter. In some cases we built the solution via increments of constituent ones. We finally wrapped the chapter with a demonstration of the Fourier transform flexibility where we derived the FT of the unit step using at least seven methods. No matter how we look at it, and from any angle, the underlying theory works and delivers the same answer. It is assuring to see for oneself that all routes lead to the same destination. With this chapter we wrap the basic theory and application of the Fourier transform and we are ready to tackle two remaining advanced topics: Fourier transform of periodic functions (next chapter) and numerical and approximate techniques in finding the Fourier transform (next to following chapter).

10.26 Problems

1. Derive the Fourier transform of the stair-signum function using the time differentiation property. Start with the time signal, take the derivative, find the transform, and then divide by $j\omega$; see sample solution in Fig. 10.31.
2. Show that the two limits in Figs. 10.4 and 10.5 are correct.
3. Take the limit $a \rightarrow 0$ and $\omega_0 \rightarrow 0$ of Eq. (10.19) and ensure they agree with the single-sided cosine transform, and the negative exponential one.
4. For the function resulting from multiplying the cosine function of frequency ω_0 times the pulse function of width 2τ, the corresponding Fourier transform was derived in Eq. (10.25). Starting with that equation, what is the transform for the limit (a) $\tau \rightarrow \infty$ and (b) $\omega_C \rightarrow 0$?
5. The Fourier transform of $\cos^2 \omega_0 t$ has been shown in Fig. 10.8. Since both cosine and its square are even functions, the resulting FT is real (and even); confirm this is the case by examining the FT plots. Next, sketch the convolution steps leading to the bottom right plot in the figure.
6. The Fourier transform of the cropped t function defined between $-t_0$ and t_0 was derived in Eq. (10.31). Rederive the transform using the following steps. Start with the pulse function, of width $2t_0$; integrate it, then take out the average (DC) contribution; then subtract from it the stair-signum function to arrive at the desired function. Throughout the process, keep track of the applied steps, and apply in parallel the corresponding steps in the frequency domain. The above steps are partially captured in aiding Fig. 10.32.
7. Consider the Fourier transform of the cropped t^4 function, defined between $-t_0$ and t_0, and as derived in Eq. (10.52). Evaluate the function at zero frequency, assuming some t_0 (for example 3) and confirm that it matches the DC integral

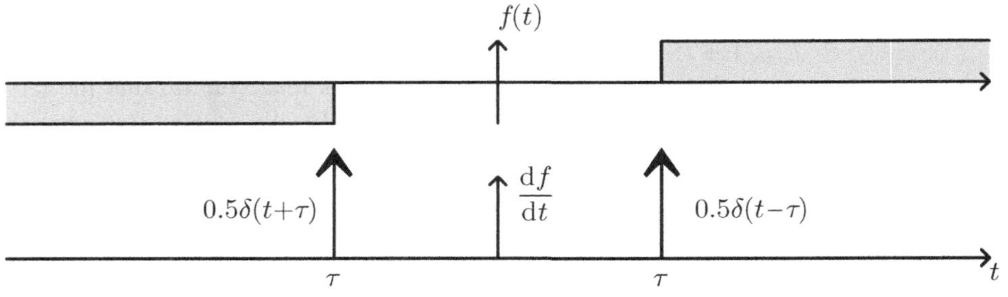

Fig. 10.31 Partial solution to Problem 1

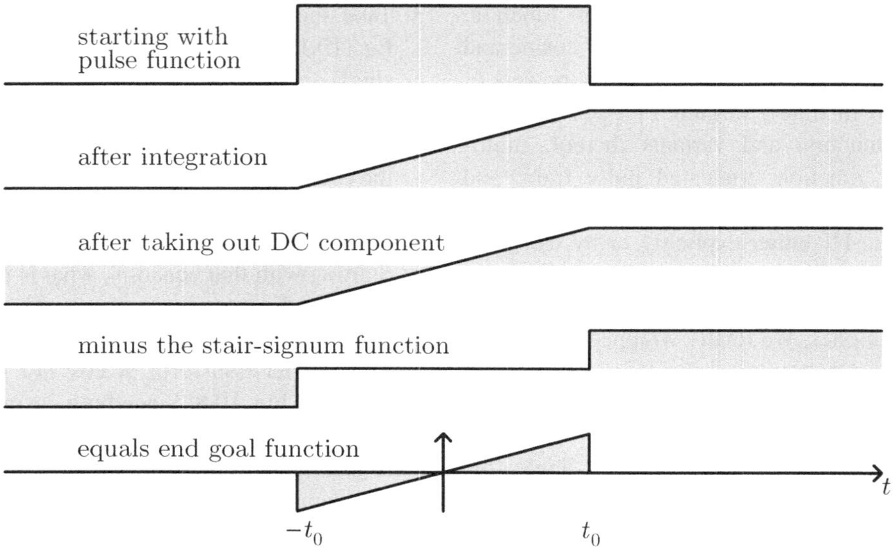

Fig. 10.32 Partial solution to Problem 6

$$F(0) = \int_{-t_0}^{t_0} t^4 dt = 2\frac{t_0^5}{5}$$

Hint: most calculators/programs would choke when dividing by zero; as a way around this evaluate the FT for very small (say 0.01) frequency, but not identically equal to zero!

8. It was shown in Eq. (10.72) that the Fourier transform of the function $1/(1+t^4)$ is (repeated here for convenience)

$$\frac{1}{1+t^4} \to \begin{cases} \frac{\pi e^{\omega/\sqrt{2}}}{\sqrt{2}}\left[\cos\frac{\omega}{\sqrt{2}} - \sin\frac{\omega}{\sqrt{2}}\right] & \omega < 0 \\ \frac{\pi e^{-\omega/\sqrt{2}}}{\sqrt{2}}\left[\cos\frac{\omega}{\sqrt{2}} + \sin\frac{\omega}{\sqrt{2}}\right] & \omega > 0 \end{cases}$$

What is the value at zero frequency? Compare this to the integral

$$F(0) = \int_{-\infty}^{\infty} \frac{1}{1+t^4} dt$$

To evaluate this integral numerically assume it dies off completely by $t = \pm 5$. Then use the fact that it is even which boils down to

$$F(0) \sim 2\int_0^5 \frac{1}{1+t^4} dt$$

Next use time increment of 0.1 and convert integral to summation

$$F(0) \sim 2 \times 0.1 \sum_{n=0}^{50} \frac{1}{1+(0.1n)^4}$$

Use a programmable calculator or write a script to evaluate this!

Answer:

$$F(0) = \frac{\pi}{\sqrt{2}}$$

9. The hat function of width 2τ and height τ was shown in Eq. (10.79) to have a FT

$$\text{Hat function of width } 2\,\tau \to 2\frac{1-\cos\omega\tau}{\omega^2}$$

What is the zero frequency value? How does that compare to the area under the triangle, valued at τ^2? Also, since the time function is real, confirm that the FT is real and even. Finally, at what frequency does the FT first hits the zero value?

10. Equation (10.82) showed the Fourier transform of the tapered pulse of bottom width $2t_1$, top width $2t_0$, and max height 1. We can derive the same results by taking into account that the tapered pulse can be obtained

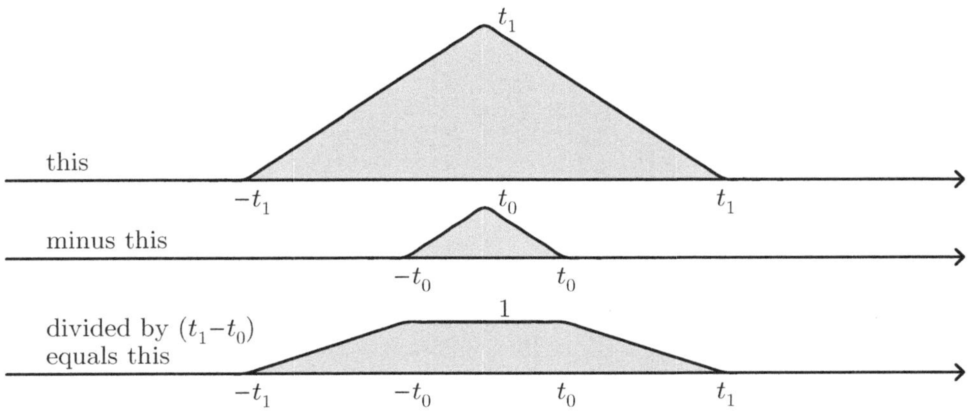

Fig. 10.33 Solution to Problem 10

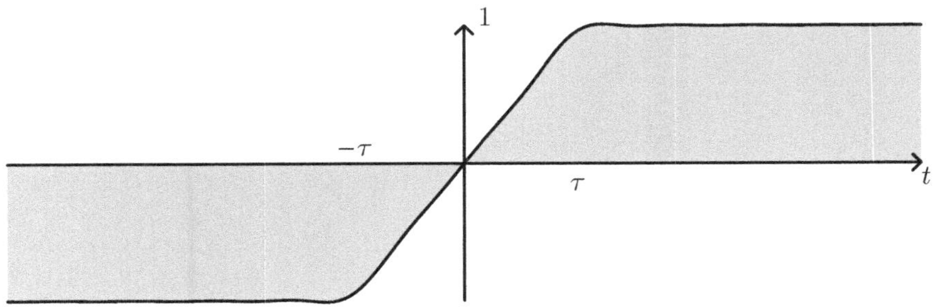

Fig. 10.34 Solution to Problem 11

by subtracting a symmetric triangular pulse of width $2t_0$ and max height t_0 from one of base width $2t_1$ and max height t_1. Use results in this chapter for each triangular FT, form the difference, and plot the time series for case $t_1 = 3$ and $t_0 = 1$. See solution in Fig. 10.33. Hint: since the answer comes out as the difference between cosine functions, while Eq. (10.82) had sine ones, you may need the following identity to match both answers:

$$\sin A \sin B = \frac{1}{2}\left[\cos(A - B) - \cos(A + B)\right]$$

11. What is the inverse Fourier transform of the function

$$F(\omega) = \frac{2}{j\tau}\frac{\sin \omega\tau}{\omega^2}$$

For the case τ=1, plot the time series. See sample solution in Fig. 10.34.

12. A pulse of width $2t_0$ is repeated three times: one at $t = 0$, another at $t = -t_1$, and another at $t = t_1$. Find its Fourier transform.
Answer:

$$F(\omega) = 2\frac{\sin \omega t_0}{\omega}\left[1 + 2\cos \omega t_1\right]$$

11 Fourier Transform of Periodic Signals

11.1 Introduction

By its definition, a Fourier transform is defined for aperiodic signals. However, the Fourier transform is flexible enough to work even for periodic signals as we'll show below. In the process we will get a better and deeper understanding for both Fourier series and Fourier transform. We will also get appreciation for the delta function, which will serve as a bridge between the two Fourier methods.

11.2 Relation Between Fourier Transform and Fourier Series

If the function is periodic, we know that it can be represented as a Fourier series. That is, we can represent the function as

$$f(t) = \sum_{n=-\infty}^{\infty} A_n e^{j\omega_n t}, \qquad \text{where} \tag{11.1}$$

$$A_n = \frac{1}{T} \int_{-\infty}^{\infty} f(t) e^{-j\omega_n t} dt \tag{11.2}$$

Now we wish to find the FT of this periodic function. Towards that end we need to recall the Fourier transform of the complex exponential (see Eq. (8.52))

$$\mathcal{F}\left[e^{j\omega_0 t}\right] = 2\pi\delta(\omega - \omega_0) \tag{11.3}$$

Using linearity we proceed to find the FT of the periodic function

$$\begin{aligned}\mathcal{F}[f(t)] &= \mathcal{F}\left[\sum_{n=-\infty}^{\infty} A_n e^{j\omega_n t}\right] \\ &= \sum_{n=-\infty}^{\infty} A_n \mathcal{F}\left[e^{j\omega_n t}\right] \\ &= 2\pi \sum_{n=-\infty}^{\infty} A_n \delta(\omega - \omega_n)\end{aligned} \tag{11.4}$$

$$\boxed{F_n(\omega) = 2\pi \sum_{n=-\infty}^{\infty} A_n \delta(\omega - \omega_n)} \tag{11.5}$$

This is a very powerful relation; let's summarize it verbally

- The FT of a periodic function is a sequence of delta (impulse) functions. It is not smooth; it is not continuous; it is very abrupt and happens in the form of a delta function sequence.
- The spacing between the delta functions is uniform. It is not linear, random, or some other relation. Only uniform intervals of frequencies between the delta functions.
- The spacing between the delta functions happens in integral increments of ω_0 such that $\omega_n = n\omega_0$. Furthermore, ω_0 is the fundamental frequency of the periodic function.

F. Badrieh, *Spectral, Convolution and Numerical Techniques in Circuit Theory*,
https://doi.org/10.1007/978-3-319-71437-0_11

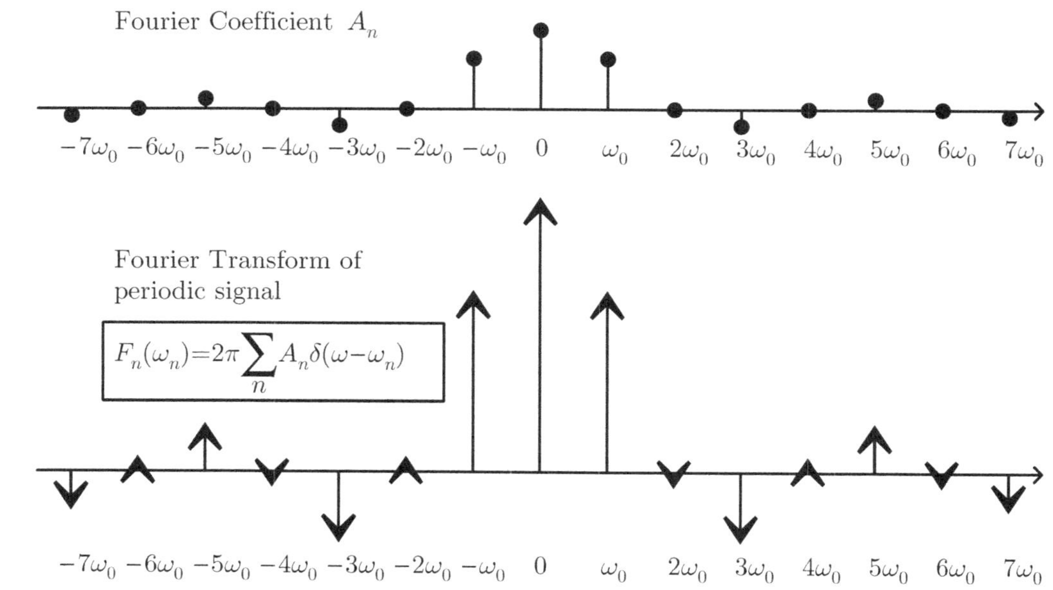

Fig. 11.1 Relation between Fourier coefficient and Fourier transform

- The *magnitude* of the various delta functions is nothing but the Fourier *series* coefficients (scaled by 2π).

That is, if we know the FS of a function, we know its FT, in accordance with Eq. (11.5). Let's get some practice on using Eq. (11.5) by finding the FT of the periodic pulse.

11.3 Fourier Transform of the Periodic Pulse

The *periodic* pulse has width 2τ and period T. Its FS coefficients (the Fourier series scaling terms) are

$$A_n = \frac{2}{T}\frac{\sin\omega_n\tau}{\omega_n} \tag{11.6}$$

Based on Eq. (11.5) we conclude that the FT of this *periodic* pulse is

$$\boxed{\mathcal{F}\,[\text{periodic pulse}] = 2\pi\sum_{n=-\infty}^{\infty}\frac{2}{T}\frac{\sin\omega_n\tau}{\omega_n}\delta(\omega-\omega_n)} \tag{11.7}$$

These results are illustrated in Fig. 11.1.

11.4 Relation Between Fourier Transform of Aperiodic Signal to That of Periodic One

We already established the relation between the Fourier transform of the periodic function and the Fourier coefficient of the same function as

$$F_n(\omega) = 2\pi\sum_{n=-\infty}^{\infty} A_n\delta(\omega-\omega_n) \tag{11.8}$$

That is the Fourier transform becomes a series of delta functions, spaced at $n\omega_0$ (where $\omega_0 = \frac{2\pi}{T}$, T is the period), and the delta functions are scaled by the Fourier coefficient at those frequencies. Now we set for deriving the relation between the Fourier transform of the periodic function and the Fourier transform of the aperiodic one! There is a quick and easy way about this. First recall that

$$A_n = \frac{1}{T}\int_{-\infty}^{\infty} f(t)e^{-j\omega t}dt, \qquad \text{and} \tag{11.9}$$

$$F(\omega) = \int_{-\infty}^{\infty} f(t)e^{-j\omega t}dt \tag{11.10}$$

A quick comparison between the above shows that

$$F(\omega_n) = TA_n(\omega_n), \qquad \text{or} \qquad A_n = \frac{1}{T}F(\omega_n) \tag{11.11}$$

That is, the Fourier transform (of the aperiodic signal) equals the period T times the Fourier coefficient. If we plug this back into Eq. (11.8) we quickly arrive at

$$F_n(\omega) = \frac{2\pi}{T} \sum_{n=-\infty}^{\infty} F(\omega_n)\delta(\omega - \omega_n) \tag{11.12}$$

But the ratio $2\pi/T$ is nothing more than the fundamental frequency:

$$\omega_0 = \frac{2\pi}{T} \tag{11.13}$$

As such we arrive at

$$\boxed{F_n(\omega) = \omega_0 \sum_{n=-\infty}^{\infty} F(\omega_n)\delta(\omega - \omega_n)} \tag{11.14}$$

This is yet another very powerful statement:

The Fourier transform of a periodic signal equals a series of delta functions, positioned at $n\omega_0$ and scaled by ω_0 times the value of the non-periodic Fourier transform, evaluated at $n\omega_0$.

11.5 The Periodic Pulse Again

Take the pulse again (of width 2τ); the non-periodic version has the FT

$$F(\omega) = 2\frac{\sin \omega\tau}{\tau} \tag{11.15}$$

Based on the above section, the periodic version would have the FT

$$F_n(\omega) = 2\omega_0 \sum_{n=-\infty}^{\infty} \frac{\sin \omega_n\tau}{\omega_n}\delta(\omega - \omega_n) \tag{11.16}$$

Notice that the ***FT of the non-periodic version is continuous, but that of the periodic one is non-continuous!!*** Both have the same overall shape, which is the sinc function; but one is continuous, and the other is a series of delta functions. These results are illustrated in Fig. 11.2.

11.6 Impact of Period

We know that the Fourier transform of the non-periodic signal has **no** sense of a period and that it is continuous. The Fourier transform of the periodic version, however, does have a sense of the period, and is discontinuous! How then does it depend on the period? As can be deduced by examining Eq. (11.14) we conclude that **increasing the period T does three things:**

1. Decrease the fundamental frequency $\omega_0 = \frac{2\pi}{T}$.
2. Increase the number of the delta functions per frequency range; that is, since ω_0 is smaller, we are able to fit more $n\omega_0$ per ω range.
3. The strength of the delta functions is reduced such that it is multiplied by ω_0; that is, the area under the delta function goes down.

This does not mean the spectrum dies off! While the strength of each delta function in the spectrum does go down, we end up with *more* of these delta functions. So we have more delta functions, but with smaller magnitude; but the *area* under the spectrum remains the same! Think of it as a distribution: we can either assign 10 delta functions, each with larger magnitude; or we can assign 100 delta functions, each with smaller magnitude. In the end, the content (area) in both distributions comes out the same. These conclusions are illustrated graphically (for the pulse case) in Fig. 11.3.

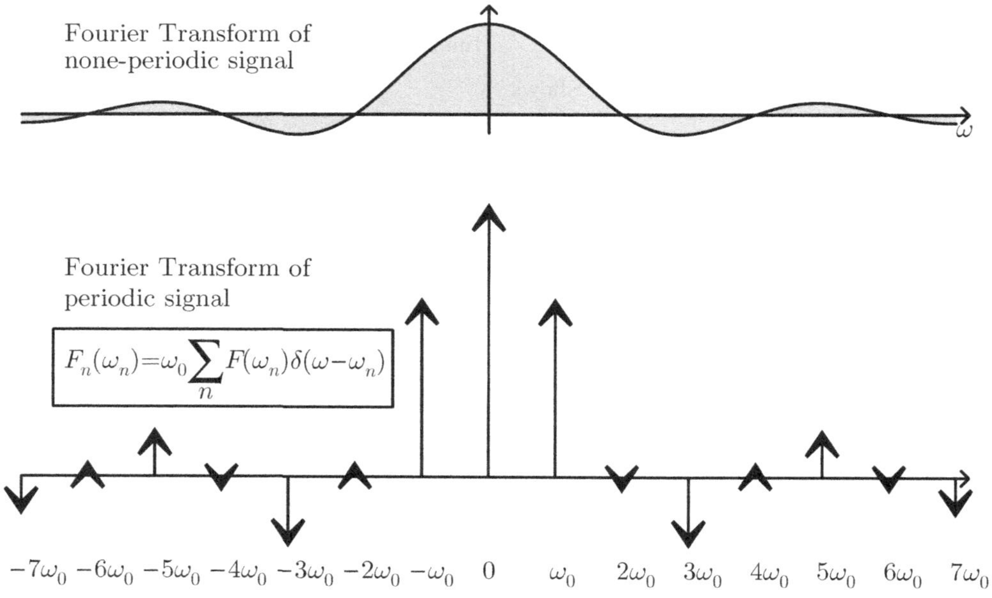

Fig. 11.2 Relation between Fourier transform of periodic and non-periodic signal

11.7 The Hat Function

We know that the (non-periodic) hat function, defined between $-\tau$ and τ, and of unity peak has the FT

$$F(\omega) = \frac{2}{\tau}\frac{1-\cos\omega\tau}{\omega^2} \tag{11.17}$$

(See Eq. (10.79) scaled by $1/\tau$). As per the above development, we now know the FT of the periodic version is

$$\boxed{F_n(\omega) = \frac{2}{\tau}\omega_0 \sum_{n=-\infty}^{\infty} \frac{1-\cos\omega_n\tau}{\omega_n^2}\delta(\omega-\omega_n)} \tag{11.18}$$

These results are shown in Fig. 11.4. Now if we increase the period we increase the resolution of the FT, but at the same time decrease the strength of the delta functions, as shown in Fig. 11.5.

11.8 The Inverted Parabola

The inverted parabola, defined between -1 and 1, and having the relation

$$f(t) = 1 - t^2 \tag{11.19}$$

has the (non-periodic) FT

$$F(\omega) = 2\left[2\frac{\sin\omega}{\omega^3} - 2\frac{\cos\omega}{\omega^2}\right] \tag{11.20}$$

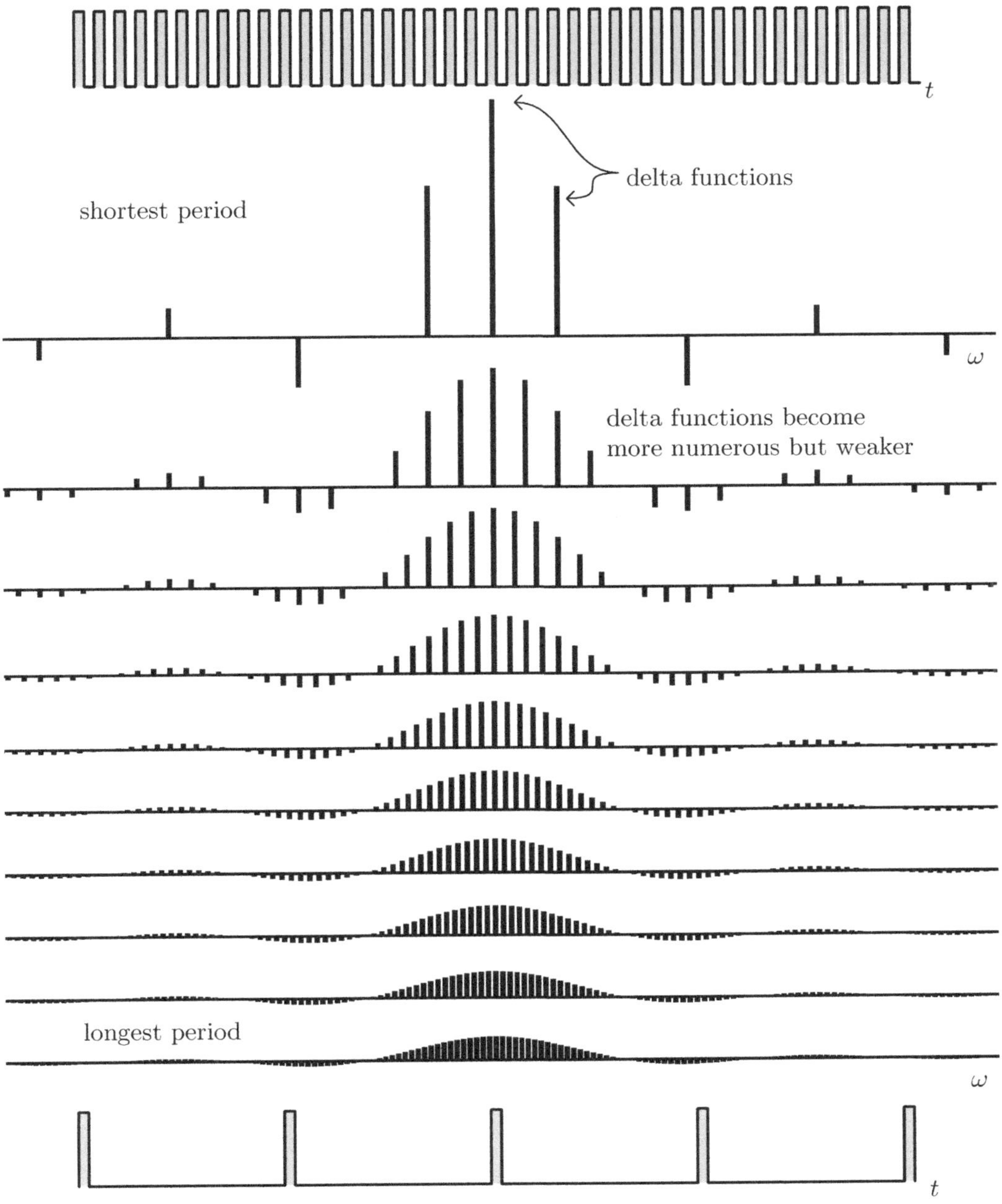

Fig. 11.3 Fourier transform of the periodic pulse as a function of period; larger period results in denser Fourier transform plots, but lower in magnitude

The corresponding Fourier transform of the periodic version is then

$$F_n(\omega) = 2\omega_0 \sum_{n=-\infty}^{\infty} \left[2\frac{\sin \omega_n}{\omega_n^3} - 2\frac{\cos \omega_n}{\omega_n^2} \right] \delta(\omega - \omega_n) \tag{11.21}$$

Notice that this essentially is Eq. (6.28). This is shown in Fig. 11.6.

11.9 The Delta Train Function

In analyzing the transition between Fourier transform and series we will need to rely on the delta train function. Recall the definition of the delta function

$$\delta(t) = \begin{cases} \infty & t = 0 \\ 0 & t \neq 0 \end{cases} \tag{11.22}$$

subject to the condition that

$$\int_{-\infty}^{\infty} \delta(t)dt = 1 \tag{11.23}$$

Now we define the delta train function as the delta function repeated every T time units

$$\delta_N(t) = \sum_{n=-\infty}^{n=\infty} \delta(t - nT) \tag{11.24}$$

The Fourier transform of the non-periodic delta function is simply

$$F(\omega) = 1 \tag{11.25}$$

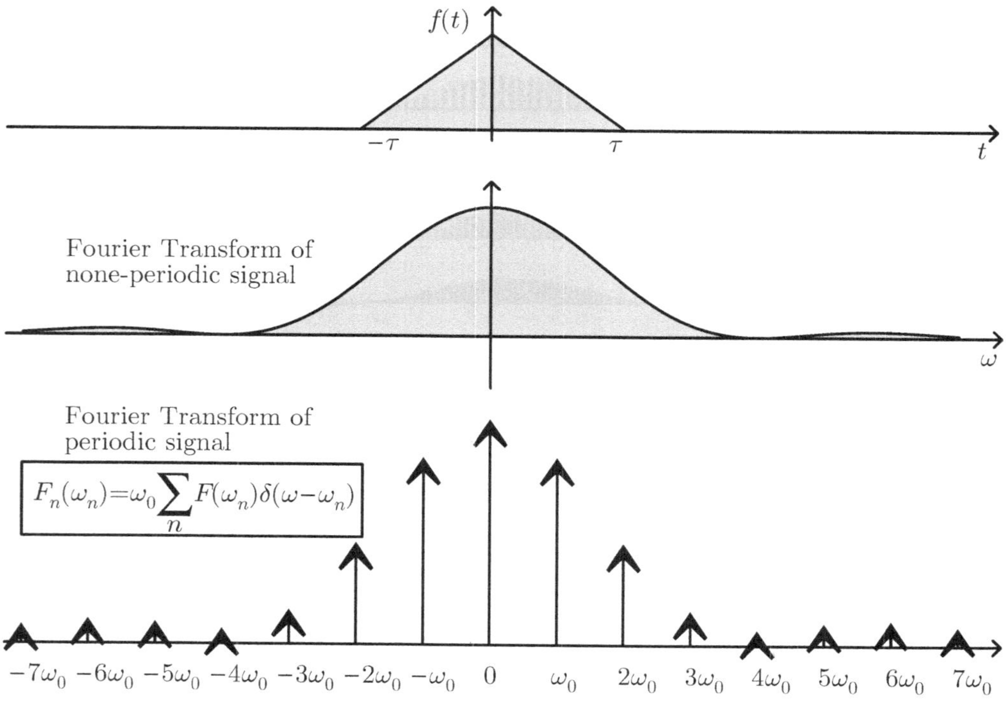

Fig. 11.4 Relation between Fourier transform of periodic and non-periodic signal (applied to hat function)

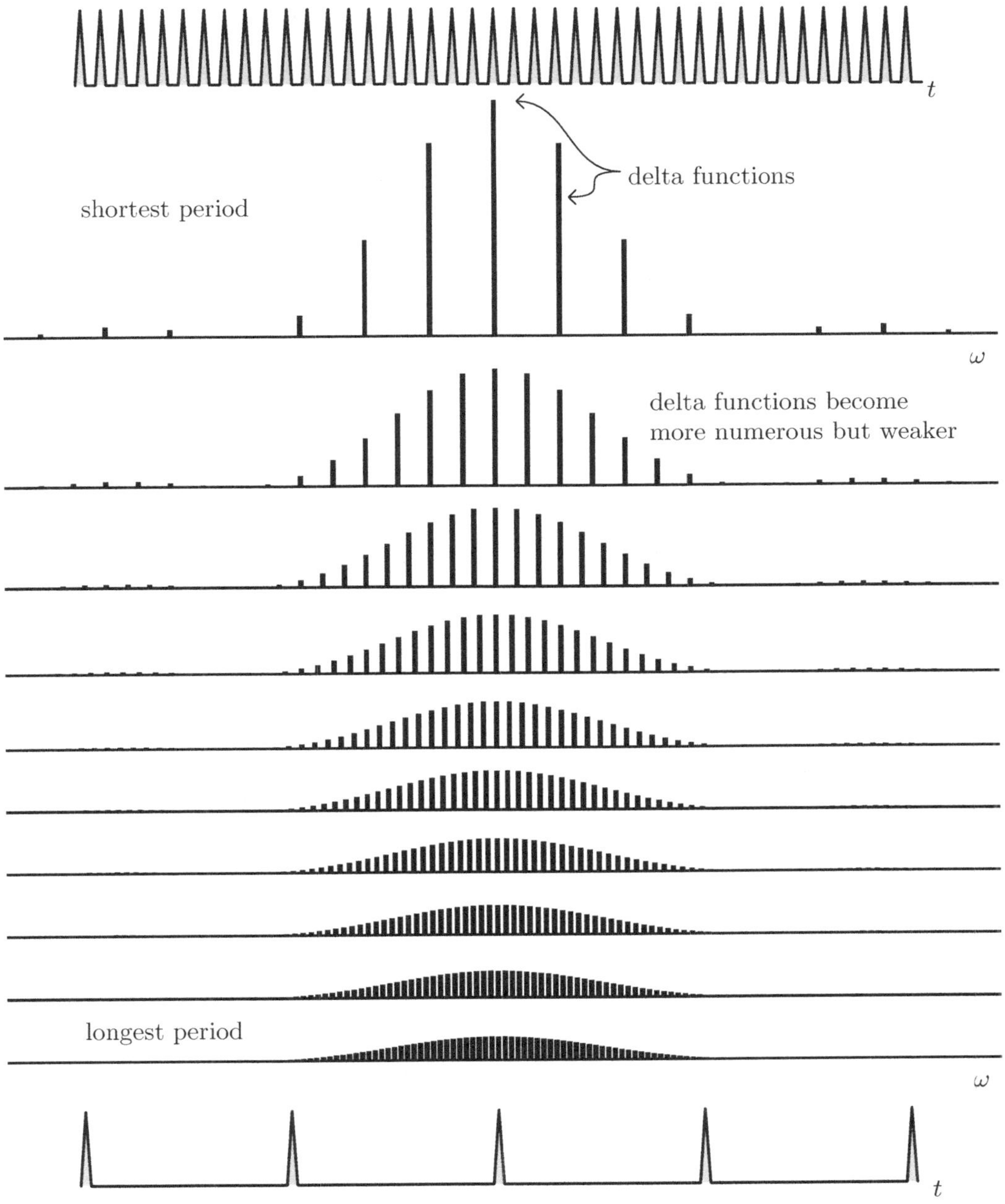

Fig. 11.5 Fourier transform of periodic hat function as a function of period

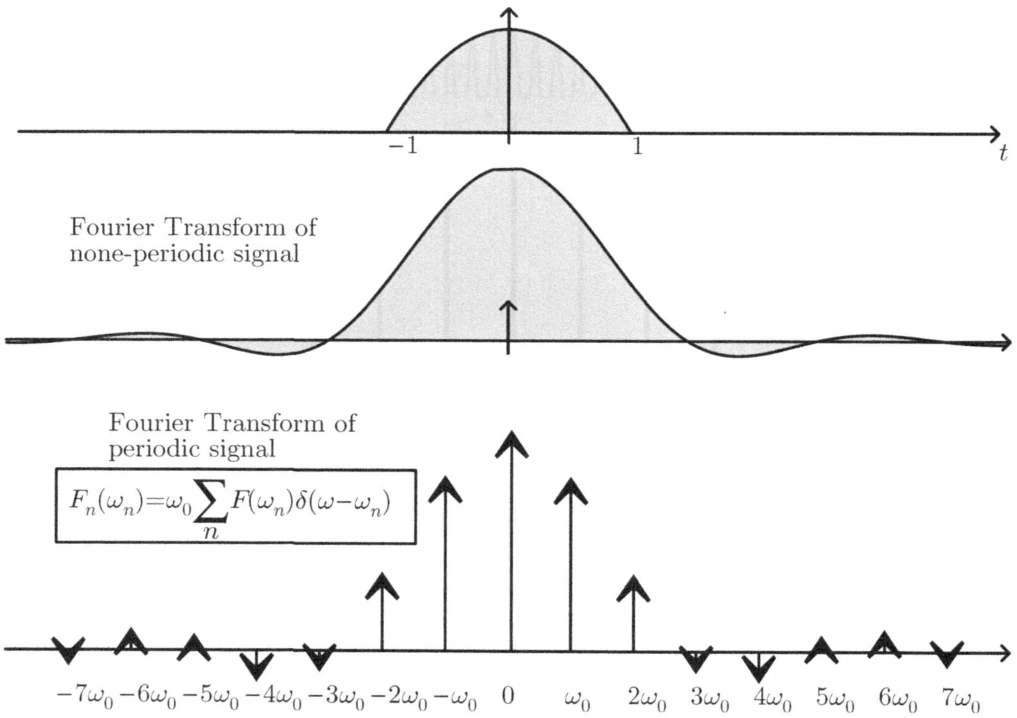

Fig. 11.6 Fourier transform of non-periodic and periodic inverted parabola function

The Fourier transform of the periodic one would then be

$$\boxed{F_n(\omega) = \omega_0 \sum_{n=-\infty}^{\infty} \delta(\omega - \omega_n)} \tag{11.26}$$

Figure 11.7 shows results and shows impact of increasing period; larger time period results in denser spectrum, but lower strength.

11.10 Summary

On the one hand we mastered the (complex) Fourier series of periodic functions; on the other hand we mastered the Fourier transform of aperiodic functions. In this chapter we studied the gray area of representing periodic signals via the Fourier transform! If we know the Fourier series, then we can use Eq. (11.5) to figure the Fourier transform of the same periodic signal. On the other hand, if we know the Fourier transform of the aperiodic signal, then we can use Eq. (11.14) again to figure the Fourier transform of the same periodic signal. Either way, we learned the fundamental difference between the Fourier transform of the single-timer version and the periodic one by the fact that the latter assumes the shape of a sequence of impulse functions while the former is a continuous, smooth one. The overall envelope in both cases remains the same; what differs is how the envelope is filled. In one case (the aperiodic one) the envelope is filled smoothly; in the other case (periodic one) it is "sampled" via a sequence of impulse functions. Sticking with the periodic case, we observed that increasing the period T resulted in a finer-sampled spectrum, but with similar area. As we push the period T to infinity, the sampled spectrum ought to become so densely sampled that for all purposes it becomes continuous—and hence we recover the Fourier transform of the non-periodic signal (since $T \to \infty$). We tried a few sample cases, ranging from the pulse, hat, and inverted parabola. Finally we studied the periodic delta function (delta train) and showed that its transform is—as expected—a delta train.

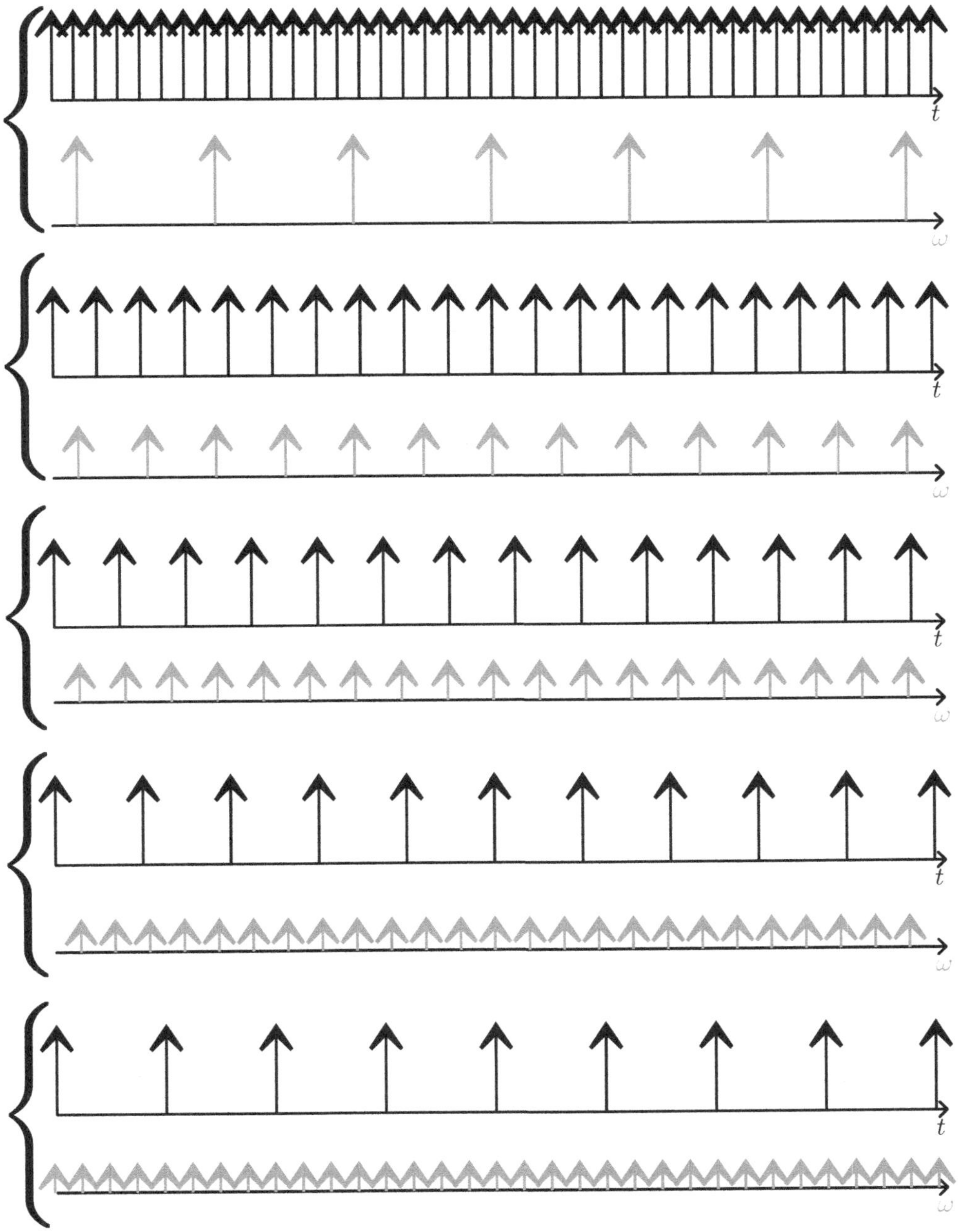

Fig. 11.7 Delta train in time and Fourier transform in frequency domain, for different period T

11.11 Problems

1. The Fourier transform of the absolute value function $|t|$, defined between $-\tau$ and τ, was derived in Eq. (10.41), repeated here for convenience:

$$F(\omega) = 2\frac{\cos\omega\tau - 1}{\omega^2} + 2\tau\frac{\sin\omega\tau}{\omega}$$

What is the Fourier transform of the periodic version of this, with period T? For the case $\tau = 1$, plot the time series for the aperiodic case, and for the periodic case with $T = 4$ and $T = 2$. See sample solution in Fig. 11.8.
Answer:

$$F(\omega_n) = \omega_0 \sum_{n=-\infty}^{\infty} \left[2\frac{\cos\omega_n\tau - 1}{\omega_n^2} + 2\tau\frac{\sin\omega_n\tau}{\omega_n}\right]\delta(\omega - \omega_n), \qquad \omega_0 = \frac{2\pi}{T}$$

2. The Fourier transform of the cropped quadratic function t^2, defined between $-\tau$ and τ, was derived in Eq. (10.44), repeated here for convenience:

$$F(\omega) = -4\frac{\sin\omega\tau}{\omega^3} + 4\tau\frac{\cos\omega\tau}{\omega^2} + 2\tau^2\frac{\sin\omega\tau}{\omega}$$

What is the Fourier transform of the periodic version of this, with period T? For the case $\tau = 1$, plot the time series for the aperiodic case, and for the periodic case with $T = 4$ and $T = 2$. See sample solution in Fig. 11.9.
Answer:

$$F(\omega_n) = \omega_0 \sum_{n=-\infty}^{\infty} \left[-4\frac{\sin\omega_n\tau}{\omega_n^3} + 4\tau\frac{\cos\omega_n\tau}{\omega_n^2} + 2\tau^2\frac{\sin\omega_n\tau}{\omega_n}\right]\delta(\omega - \omega_n), \qquad \omega_0 = \frac{2\pi}{T}$$

3. The pulse function $f(t)$ of width 2τ has the Fourier transform

$$F(\omega) = 2\frac{\sin\omega\tau}{\omega}$$

What is the Fourier transform of $f(t) + f(t - T) + f(t + T)$? How about $f(t) + f(t - T) + f(t + T) + f(t - 2T) + f(t + 2T)$? For the case $\tau = 0.5$ and $T = 2$, plot the time series and the Fourier transform. What is happening to the Fourier transform as the function becomes more "periodic"? What will happen when the function becomes totally periodic? See sample results in Fig. 11.10.

4. The single-cycled cosine function with frequency ω_0 has the Fourier transform

$$F(\omega) = \frac{\sin(\omega - \omega_0)\tau}{\omega - \omega_0} + \frac{\sin(\omega + \omega_0)\tau}{\omega + \omega_0}, \qquad \tau = \frac{\pi}{\omega_0}$$

What is the Fourier transform of the three-cycled cosine function? What is it for the 5- and 7-cycled cosine function? For the case $\omega_0 = 2\pi$ plot the time series for each case, and the corresponding Fourier transform. How does the latter look like as we add more cycles? What is the limit of infinite cycle count? See Fig. 11.11 for sample solution.

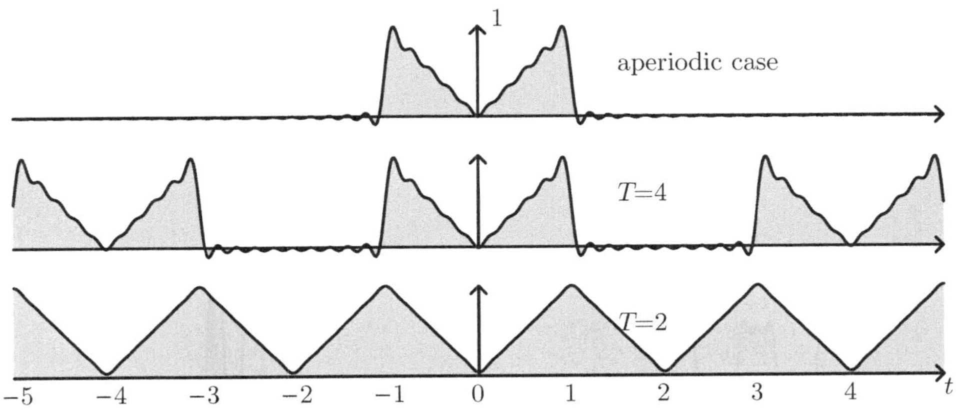

Fig. 11.8 Solution to Problem 1

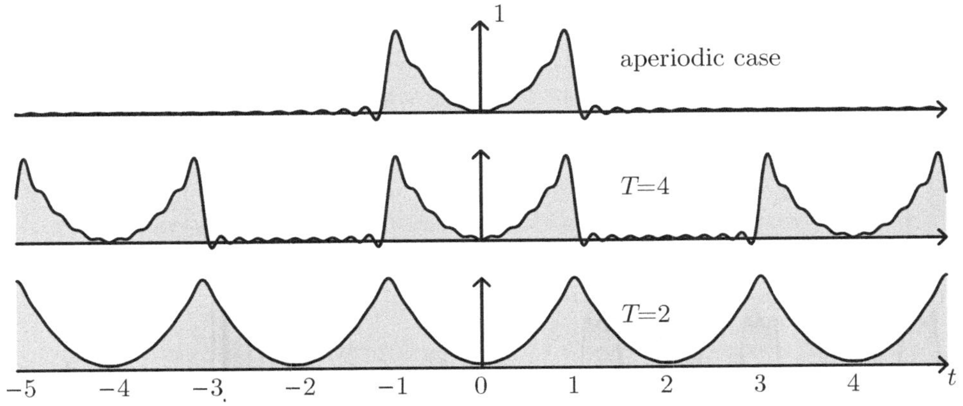

Fig. 11.9 Solution to Problem 2

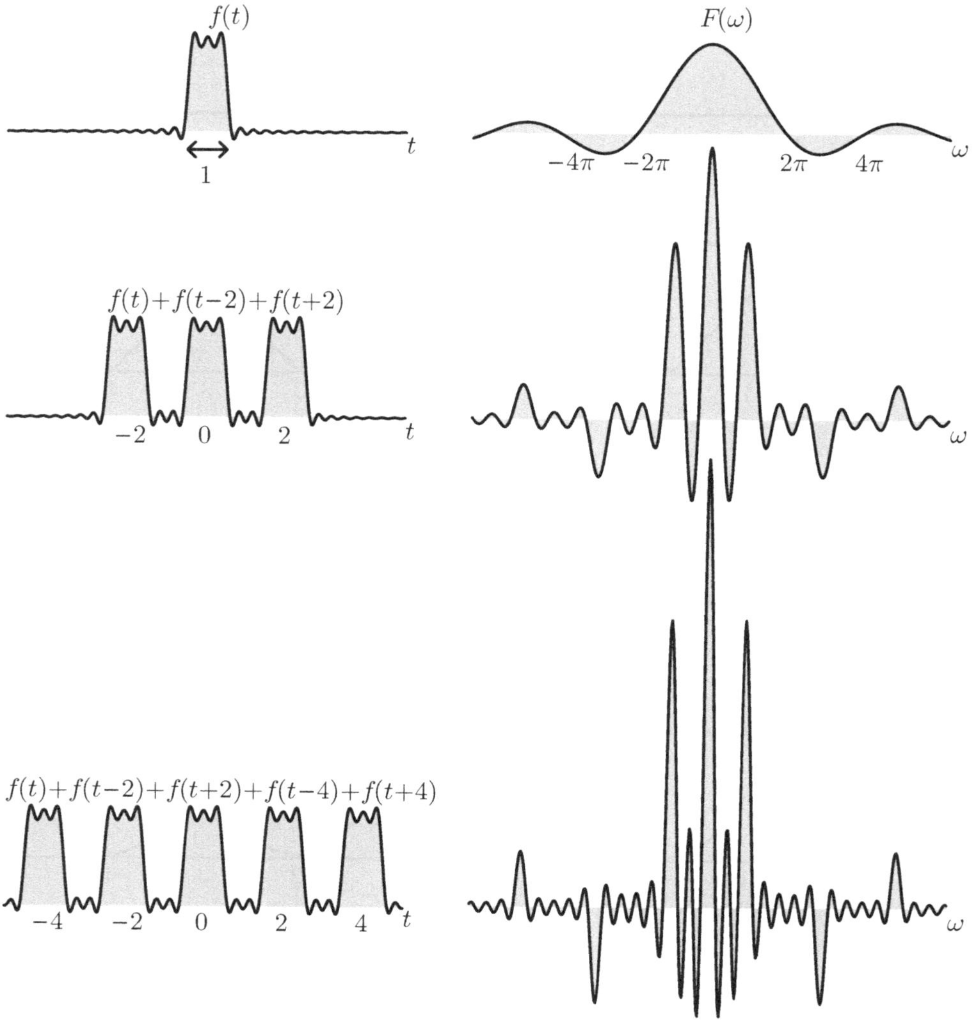

Fig. 11.10 Solution to Problem 3

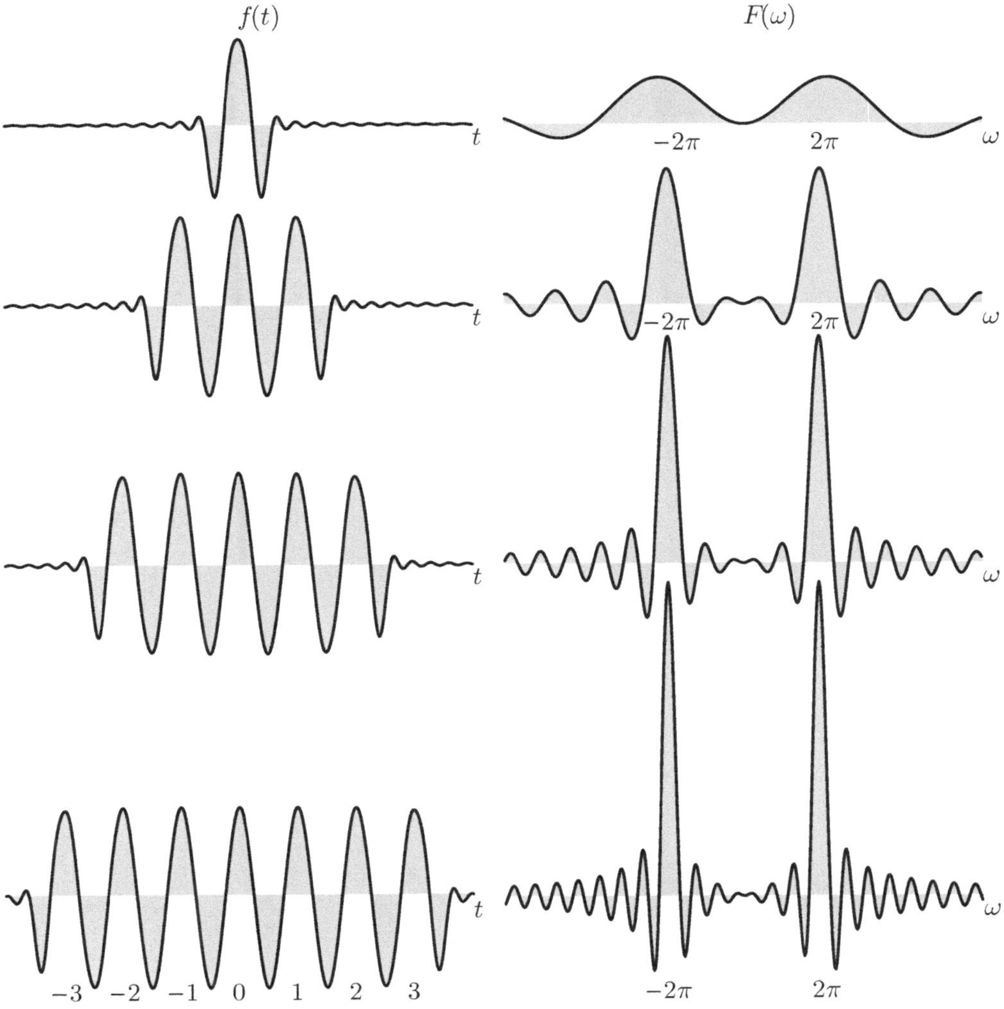

Fig. 11.11 Solution to Problem 4

12 Approximate and Numerical Techniques in Fourier Transform

12.1 Introduction

This is the last chapter covering the basics and theory of the Fourier series/Transform. It deals with approximate and numerical techniques. In many cases there are no closed-form solutions for the Fourier transform; that is, the Fourier integral cannot be carried on (at least easily). In those cases we can revert to approximate and numerical methods. These techniques include Taylor series, linear fit, impulse expansion, unit step expansion, and brute force numerical integration. As a way to demonstrate these methods we choose a sample function—in this case the half circle function, defined between $-1 < t < -1$ where

$$f(t) = \begin{cases} \sqrt{1-t^2} & |t| < 1 \\ 0 & \text{else where} \end{cases} \tag{12.1}$$

The Fourier integral would be

$$F(\omega) = \int_{-1}^{1} \sqrt{1-t^2} e^{j\omega t} dt \tag{12.2}$$

Even if this integral is solvable, the time function can still be used to demonstrate the various proposed methods.

12.2 Fourier Transform of Arbitrary Function Using Taylor Series

The premise behind the Taylor expansion is that the target function defined between $-t_0$ and t_0 can be expanded as

$$\begin{aligned} f(t) &= f(0) + \frac{df(0)}{dt} t + \frac{d^2 f(0)}{dt^2} \frac{t^2}{2} + \frac{d^3 f(0)}{dt^3} \frac{t^3}{3!} \\ &+ \frac{d^4 f(0)}{dt^4} \frac{t^4}{4!} + \cdots \end{aligned} \tag{12.3}$$

If we are able to expand the function in terms of a Taylor series, we are able to use the principle of superposition to find the resulting Fourier transform. In order to do that we would need to know the FT of the various terms in the polynomial expansion. In particular

F. Badrieh, *Spectral, Convolution and Numerical Techniques in Circuit Theory*,
https://doi.org/10.1007/978-3-319-71437-0_12

1. The constant term:

$$1 \to 2\frac{\sin \omega t_0}{\omega} \tag{12.4}$$

2. The linear term:

$$t \to 2j\left[t_0\frac{\cos \omega t_0}{\omega} - \frac{\sin \omega t_0}{\omega^2}\right] \tag{12.5}$$

3. The quadratic term:

$$t^2 \to -2\left[-t_0^2\frac{\sin \omega t_0}{\omega} - 2t_0\frac{\cos \omega t_0}{\omega^2} + 2\frac{\sin \omega t_0}{\omega^3}\right] \tag{12.6}$$

4. Order 3 term:

$$t^3 \to -2j\left[-t_0^3\frac{\cos \omega t_0}{\omega} + 3t_0^2\frac{\sin \omega t_0}{\omega^2} + 6t_0\frac{\cos \omega t_0}{\omega^3} - 6\frac{\sin \omega t_0}{\omega^4}\right] \tag{12.7}$$

5. Order 4 term:

$$t^4 \to 2\left[t_0^4\frac{\sin \omega t_0}{\omega} + 4t_0^3\frac{\cos \omega t_0}{\omega^2} - 12t_0^2\frac{\sin \omega t_0}{\omega^3} - 24t_0\frac{\cos \omega t_0}{\omega^4} + 24\frac{\sin \omega t_0}{\omega^5}\right] \tag{12.8}$$

6. Order n:

$$t^n \to \cdots \tag{12.9}$$

(Notice that we flipped the sign for the t_2 term, but not for the t_4 one; this is due to the fact that $j \times j = -1$ but $j^4 = 1$.) These functions are shown in Fig. 12.1. Notice that when the time function is even, the Fourier transform is both real and even. On the other hand, when the time function is odd, the Fourier transform is both imaginary and odd! Knowing these elementary transformations, we next move to determining the Taylor expansion. Recall

$$f(t) = \sqrt{1-t^2} \tag{12.10}$$

$$\frac{d}{dt}f(t) = \frac{1}{2}\frac{-2t}{(1-t^2)^{1/2}} = \frac{-t}{(1-t^2)^{1/2}} \tag{12.11}$$

$$\frac{d^2}{dt^2}f(t) = \frac{-1}{(1-t^2)^{1/2}} - \frac{t^2}{(1-t^2)^{3/2}} = \frac{t^2-1-t^2}{(1-t^2)^{3/2}} = \frac{-1}{(1-t^2)^{3/2}} \tag{12.12}$$

$$\frac{d^3}{dt^3}f(t) = \frac{3}{2}\frac{-2t}{(1-t^2)^{5/2}} = -3\frac{t}{(1-t^2)^{5/2}} \tag{12.13}$$

$$\frac{d^4}{dt^4}f(t) = \frac{-3}{(1-t^2)^{5/2}} + \frac{15}{2}\frac{-2t^2}{(1-t^2)^{7/2}} = \frac{-3(1-t^2)-15t^2}{(1-t^2)^{7/2}} = \frac{-3-12t^2}{(1-t^2)^{7/2}} \tag{12.14}$$

Our Taylor expansion, after evaluating these expressions at time zero, then becomes

$$\begin{aligned} f(t) &\sim 1-\tfrac{1}{2}t^2-\tfrac{3}{4!}t^4+\cdots \\ &= 1-\tfrac{1}{2}t^2-\tfrac{1}{8}t^4+\cdots \end{aligned} \tag{12.15}$$

Now we can use the FT's for the various t^n powers as derived at the beginning of the section. Let $t_0 = 1$. Then the desired Fourier transform becomes

$$F(\omega) = F_1(\omega) - \frac{1}{2}F_2(\omega) - \frac{1}{8}F_4(\omega) \tag{12.16}$$

Results are shown in Fig. 12.2. As can be observed we get a relatively good fit. It is expected of course the more Taylor series terms we carry, the better the approximation. So in summary, if we can decompose the function in terms of cropped powers of t^n, and knowing the FT of each power, we ought to be able to build the FT via superposition.

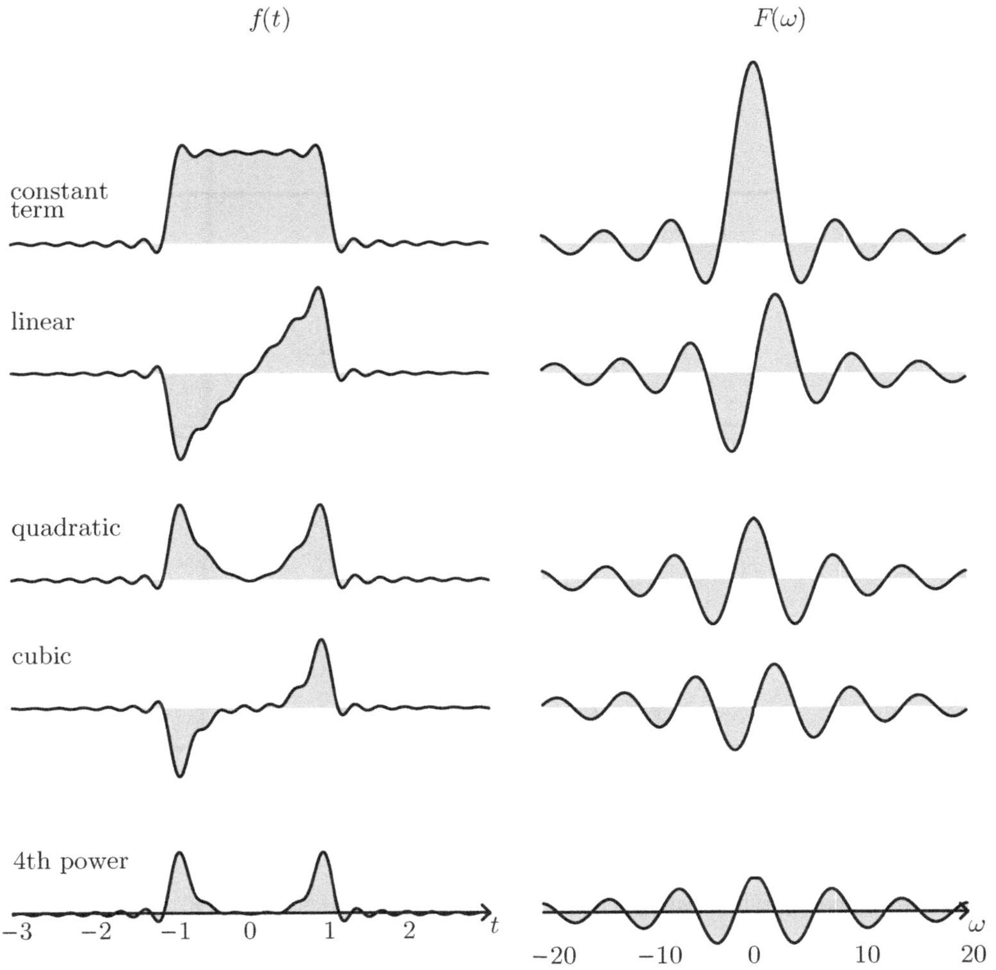

Fig. 12.1 Powers of t and their Fourier transform

12.3 Fourier Transform of Arbitrary Function Using Pulse Series

We can expand the half circle via summation of scaled, shifted pulse functions. The more pulses we use the better results we will get. For example, start with 5 pulses, each of width 0.4 (recall $5 \times 0.4 = 1$, which is the circle width); we would then have

$$f(t) \sim f(-0.8)p(t+0.8) + f(-0.4)p(t+0.4) + f(0.0)p(t-0.0) + f(0.4)p(t-0.4) + f(0.8)p(t-0.8) \tag{12.17}$$

where

$$p(t) = \begin{cases} 1 & -0.2 < t < 0.2 \\ 0 & \text{elsewhere} \end{cases} \tag{12.18}$$

This is shown in Fig. 12.3. Recall that

$$p(t) \to 2\frac{\sin \omega(0.2)}{\omega} \tag{12.19}$$

and recall the time shifting property

$$f(t-t_0) \to F(\omega)e^{-j\omega t_0} \tag{12.20}$$

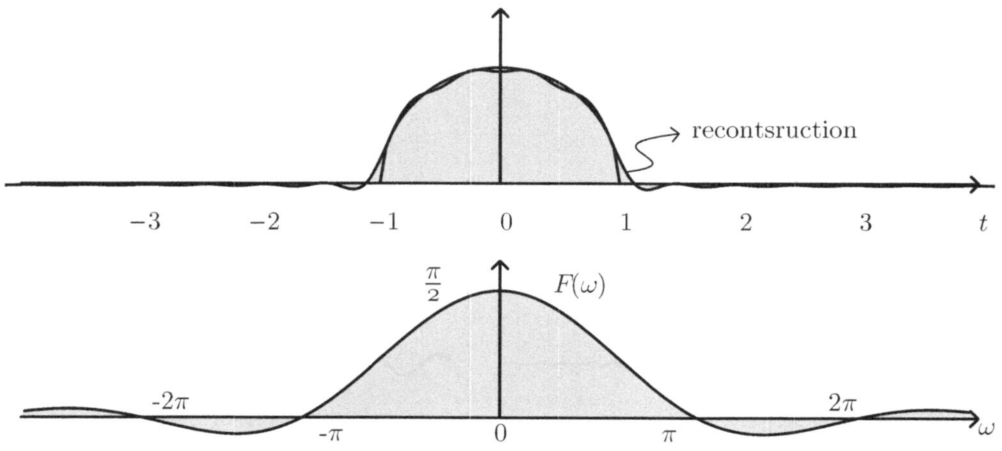

Fig. 12.2 Half circle function and Fourier transform based on Taylor series expansion

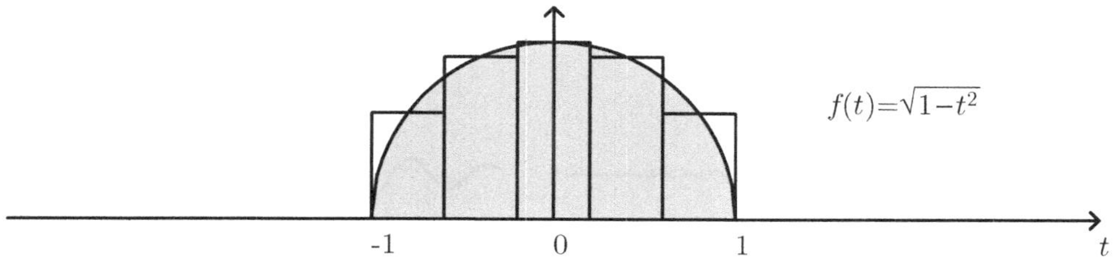

Fig. 12.3 Half circle expansion in terms of 5 pulses

Then our desired Fourier transform comes out

$$F(\omega) = 2\frac{\sin\omega(0.2)}{\omega}\left[f(-0.8)\times 2\cos\omega(0.8) + f(-0.4)\times 2\cos\omega(0.4) + f(0)\times 1\right] \tag{12.21}$$

Results are shown in Fig. 12.4. While the overall fit is close, we notice the fit really is resembling more the pulse sequence rather than the half circle! But what else should we expect? What we put in is what we get out? To remedy this let's put in a better approximation to start with!

Higher Resolution

We can refine our approximation by including 10 pulses, as shown in Fig. 12.5. Then we would have

$$\begin{aligned} f(t) \sim\, & f(-0.9)p(t+0.9) + f(-0.7)p(t+0.7) + f(-0.5)p(t+0.5) + f(-0.3)p(t+0.3) \\ & + f(-0.1)p(t+0.1) + f(+0.1)p(t-0.1) + f(+0.3)p(t-0.3) + f(+0.5)p(t-0.5) \\ & + f(+0.7)p(t-0.7) + f(+0.9)p(t-0.9) \end{aligned} \tag{12.22}$$

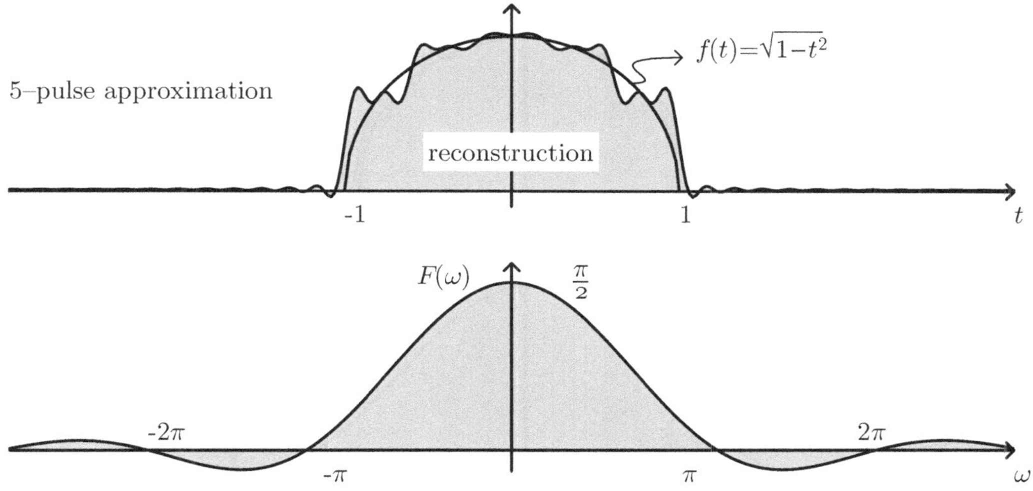

Fig. 12.4 Half circle and Fourier transform (5-pulse case)

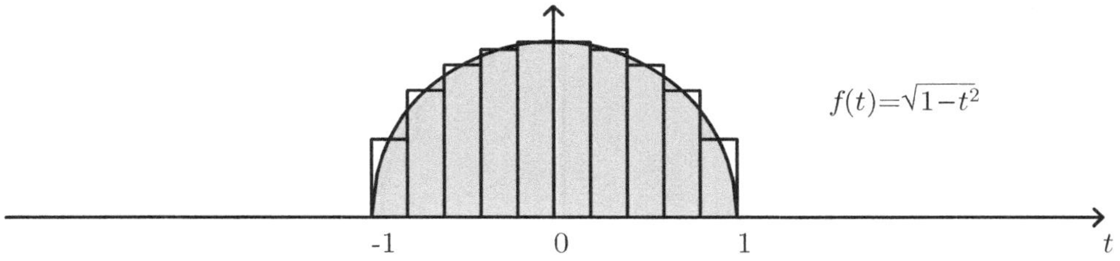

Fig. 12.5 Half circle expansion in terms of 10 pulses

This is shown in Fig. 12.5. The corresponding transfer function becomes

$$F(\omega) = 2\frac{\sin \omega(0.1)}{\omega}\Big[f(-0.9) \times 2\cos\omega(0.9) + f(-0.7) \times 2\cos\omega(0.7) + f(-0.5) \times 2\cos\omega(0.5) + f(-0.3) \times 2\cos\omega(0.3) + f(-0.1) \times 2\cos\omega(0.1)\Big] \tag{12.23}$$

Notice the sinc function now assumes pulse width of 0.2, as opposed 0.4 before. Results are shown in Fig. 12.6. Notice that with higher resolution (in terms of used pulse function count) we now get better fit.

12.4 Fourier Transform of Arbitrary Function Using Step Function Series

Instead of a series of pulses, we can choose a series of unit step functions, again scaled and shifted to represent an arbitrary function. For example, assume we discretize t into 10 segments, so that each segment width is 0.2. We can approximate the circle by

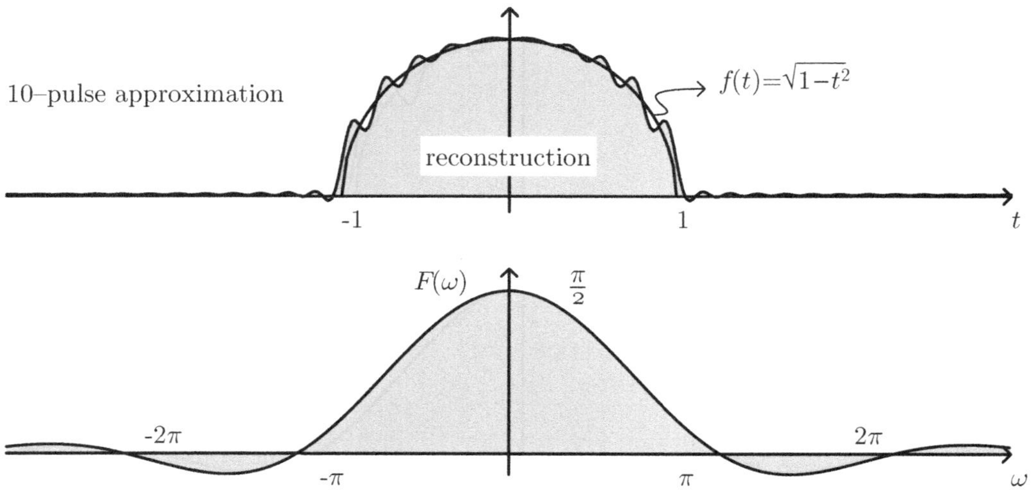

Fig. 12.6 Half circle and Fourier transform (10-pulse case)

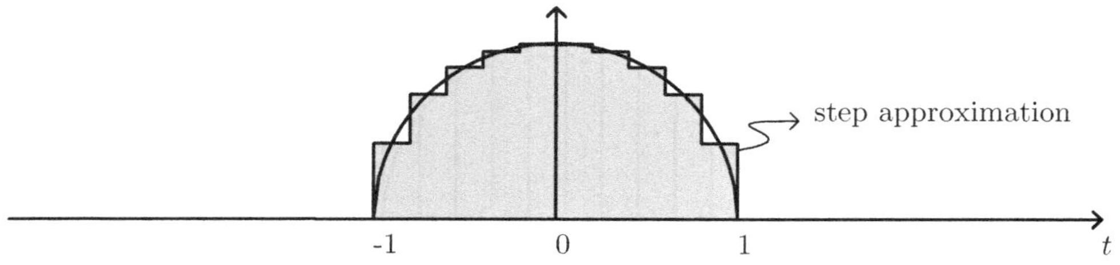

Fig. 12.7 Half circle decomposition in terms of unit step functions

$$
\begin{aligned}
f(t) \sim\ & u(t+1.0)\,[f(-0.9)-f(-1.0)] + u(t+0.8)\,[f(-0.7)-f(-0.9)] \\
& + u(t+0.6)\,[f(-0.5)-f(-0.7)] + u(t+0.4)\,[f(-0.3)-f(-0.5)] \\
& + u(t+0.2)\,[f(-0.1)-f(-0.3)] + u(t+0.0)\,[f(+0.1)-f(-0.1)] \\
& + u(t-0.2)\,[f(+0.3)-f(+0.1)] + u(t-0.4)\,[f(+0.5)-f(+0.3)] \\
& + u(t-0.6)\,[f(+0.7)-f(+0.5)] + u(t-0.8)\,[f(+0.9)-f(+0.7)] \\
& + u(t-1.0)\,[f(+1.0)-f(+0.9)]
\end{aligned}
\tag{12.24}
$$

Notice that the edge points require special treatment, since there are no points before $-t_0$ and no points after t_0. Other than those, for every anchor point, we take a unit step function at the point and multiply by derivative of the original function around that point (times Δt, which is 0.2 here). The derivative (times $\Delta t = 0.2$) at, e.g., -0.8 is evaluated by taking $f(-0.7)$ minus $f(-0.9)$. Results are shown in Fig. 12.7. Now that we have the original function in terms of unit step functions, and recalling

$$
u(t) \rightarrow \pi\delta(\omega) + \frac{1}{j\omega} \tag{12.25}
$$

and again using the time shifting property we end up with

$$F(\omega) \sim \frac{1}{j\omega}\Big[2j\sin\omega(1.0)[f(-0.9)-f(-1.0)] + 2j\sin\omega(0.8)[f(-0.7)-f(-0.9)]$$
$$+ 2j\sin\omega(0.6)[f(-0.5)-f(-0.7)] + 2j\sin\omega(0.4)[f(-0.3)-f(-0.5)]$$
$$+ 2j\sin\omega(0.2)[f(-0.1)-f(-0.3)]\Big] \quad (12.26)$$

where we have used

$$2j\sin x = e^{jx} - e^{-jx} \quad (12.27)$$

and the fact that the delta functions cancel out since the derivative for positive t is the negative of that at negative t. Results are shown in Fig. 12.8.

12.5 Fourier Transform of Arbitrary Function Using Time Differentiation Property

The premise here is to fit the target function with a series of linear segments, and take the time derivative (as many times as needed) until we get a sequence of delta functions. Since we know that FT of the delta function, and using the time differentiation property which states

$$\frac{df}{dt} \to j\omega F(\omega), \quad \text{and} \quad (12.28)$$

$$\frac{d^2f}{dt^2} \to -\omega^2 F(\omega), \quad \text{and} \quad (12.29)$$

we should be able to find an approximation for the FT of the target function. Again sticking with the half circle function, let's fit it with five linear segments as shown in Fig. 12.9.

Let

$$t_0 = -1.0,\ t_1 = -0.6,\ t_2 = -0.2,\ t_3 = +0.2,$$
$$t_4 = +0.6,\ t_5 = +1.0 \quad (12.30)$$

Further, let

$$f_0 = f(t_0),\ f_1 = f(t_1),\ f_2 = f(t_2),\ f_3 = f(t_3),$$
$$f_4 = f(t_4),\ f_5 = f(t_5) \quad (12.31)$$

Taking the first derivative, we end up with five rectangles, with magnitudes

$$\left.\frac{df}{dt}\right|_{t_0} = \frac{f(t_1)-f(t_0)}{\Delta t},\ \left.\frac{df}{dt}\right|_{t_1} = \frac{f(t_2)-f(t_1)}{\Delta t},\ \left.\frac{df}{dt}\right|_{t_2} = \frac{f(t_3)-f(t_2)}{\Delta t} \quad (12.32)$$

$$\left.\frac{df}{dt}\right|_{t_3} = \frac{f(t_4)-f(t_3)}{\Delta t}\ \left.\frac{df}{dt}\right|_{t_4} = \frac{f(t_5)-f(t_4)}{\Delta t} \quad (12.33)$$

Taking the second derivative we end up with 6 delta functions as

$$\frac{d^2f(t)}{dt^2} = \frac{f(t_1)-f(t_0)}{\Delta t}\delta(t-t_0) + \frac{[f(t_2)-f(t_1)]-[f(t_1)-f(t_0)]}{\Delta t}\delta(t-t_1)$$
$$+ \frac{[f(t_3)-f(t_2)]-[f(t_2)-f(t_1)]}{\Delta t}\delta(t-t_2) + \frac{[f(t_4)-f(t_3)]-[f(t_3)-f(t_2)]}{\Delta t}\delta(t-t_3)$$
$$+ \frac{[f(t_5)-f(t_4)]-[f(t_4)-f(t_3)]}{\Delta t}\delta(t-t_4) + \frac{0-[f(t_5)-f(t_4)]}{\Delta t}\delta(t-t_5) \quad (12.34)$$

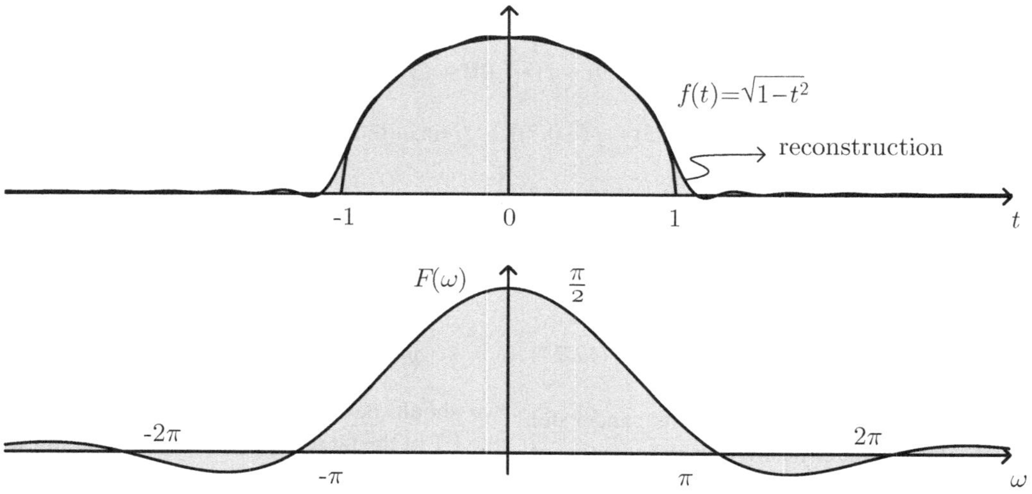

Fig. 12.8 Half circle and Fourier transform using 11 step functions

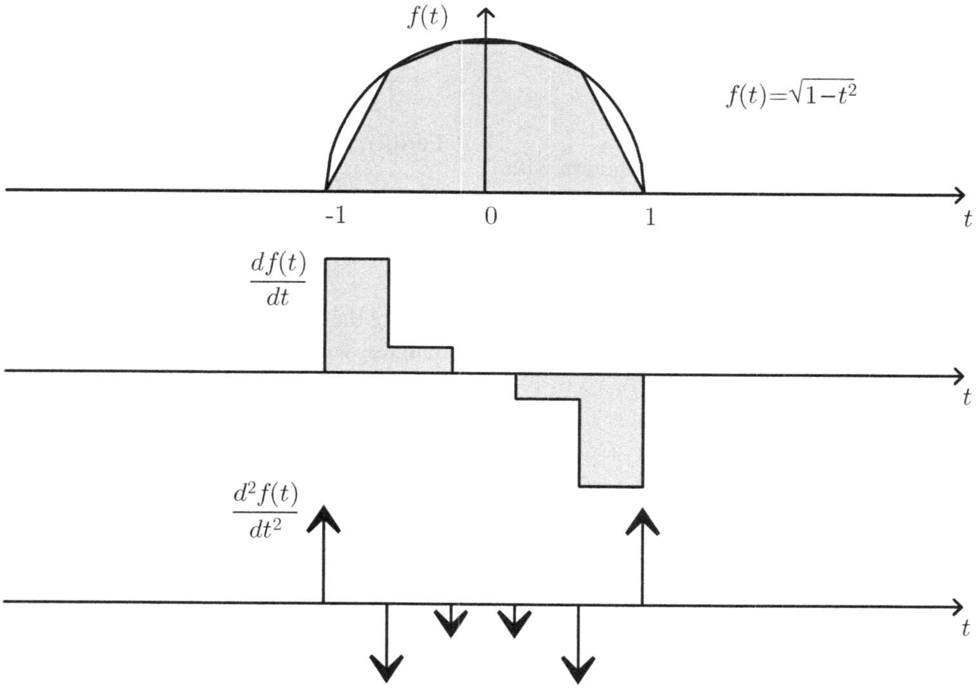

Fig. 12.9 Half circle decomposition in terms of (5) linear segments, and corresponding first and second derivatives

which simplifies to

$$
\begin{aligned}
\frac{d^2f(t)}{dt} &= \frac{f(t_1)-f(t_0)}{\Delta t}\delta(t-t_0) + \frac{f(t_2)-2f(t_1)+f(t_0)}{\Delta t}\delta(t-t_1) \\
&+ \frac{f(t_3)-2f(t_2)+f(t_1)}{\Delta t}\delta(t-t_2) + \frac{f(t_4)-2f(t_3)+f(t_2)}{\Delta t}\delta(t-t_3) \\
&+ \frac{f(t_5)-2f(t_4)+f(t_3)}{\Delta t}\delta(t-t_4) + \frac{-f(t_5)+f(t_4)}{\Delta t}\delta(t-t_5)
\end{aligned} \tag{12.35}
$$

Recalling that

$$\delta(t+t_0) \rightarrow e^{j\omega t_0} \tag{12.36}$$

and using the symmetry of the circle (and the time differentiation property) we end up with the Fourier transform

$$
\begin{aligned}
F(\omega) = -\frac{2}{\omega^2 \Delta t}\Big[& [f(t_1)-f(t_0)]\cos\omega t_0 + \left\{f(t_2)-2f(t_1)+f(t_0)\right\}\cos\omega t_1 \\
& + \left\{f(t_3)-2f(t_2)+f(t_1)]\right\}\cos\omega t_2 \Big]
\end{aligned} \tag{12.37}
$$

Results are shown in Fig. 12.10. Notice that we do in fact get what resembles the 5-segment linear model, which in its turn is supposed to resemble the circle. If we use instead more segments such as 10 and as shown in Fig. 12.11 we get better results, and as shown in Fig. 12.12.

12.6 Fourier Transform of Arbitrary Function Using Numerical Integration

If all fails, or as a verification technique we can always compute the Fourier transform numerically. Recall

$$F(\omega) = \int_{-\infty}^{\infty} f(t)e^{-j\omega t}dt \tag{12.38}$$

To find the FT at a given frequency ω, form the product

$$f(t)e^{-j\omega t} \tag{12.39}$$

which is a function of time, then integrate it over all time. In our case, time is restricted to

$$-1 < t < 1 \tag{12.40}$$

since the function is defined as zero outside that range. Hence for our case

$$F(\omega) = \int_{-1}^{1} f(t)e^{-j\omega t}dt \tag{12.41}$$

Plugging in for $f(t)$ we get

$$F(\omega) = \int_{-1}^{1} \sqrt{1-t^2}e^{-j\omega t}dt \tag{12.42}$$

Since our function $f(t)$ is even in time, when multiplied with the odd component of $e^{-j\omega t}$, which is $\sin\omega t$, and integrated over time, that component of the integral cancels out; that is it comes out to zero. Hence we are left with

$$F(\omega) = \int_{-1}^{1} \sqrt{1-t^2}\cos\omega t dt \tag{12.43}$$

Again for each ω form the product $\sqrt{1-t^2}\cos\omega t$, which would be a function of

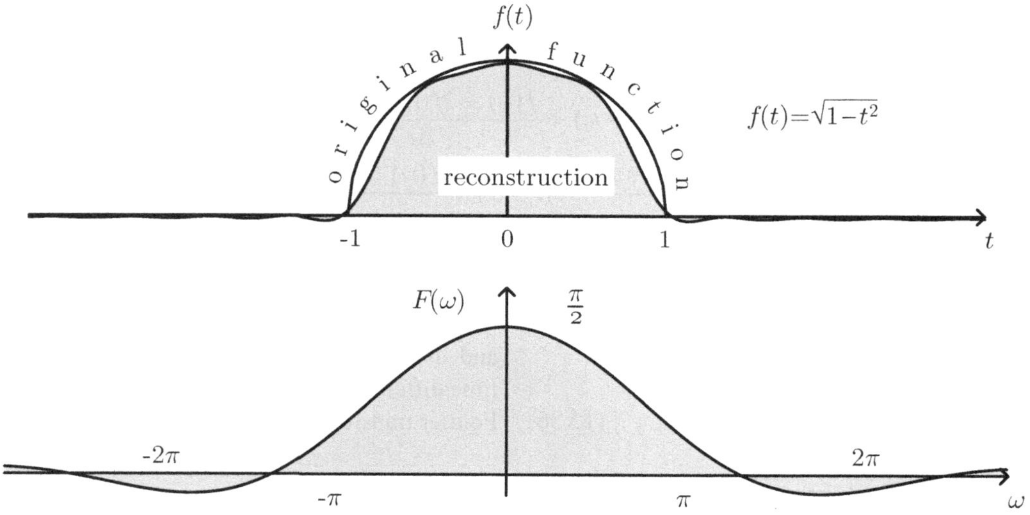

Fig. 12.10 Half circle Fourier transform (using 5 linear segments)

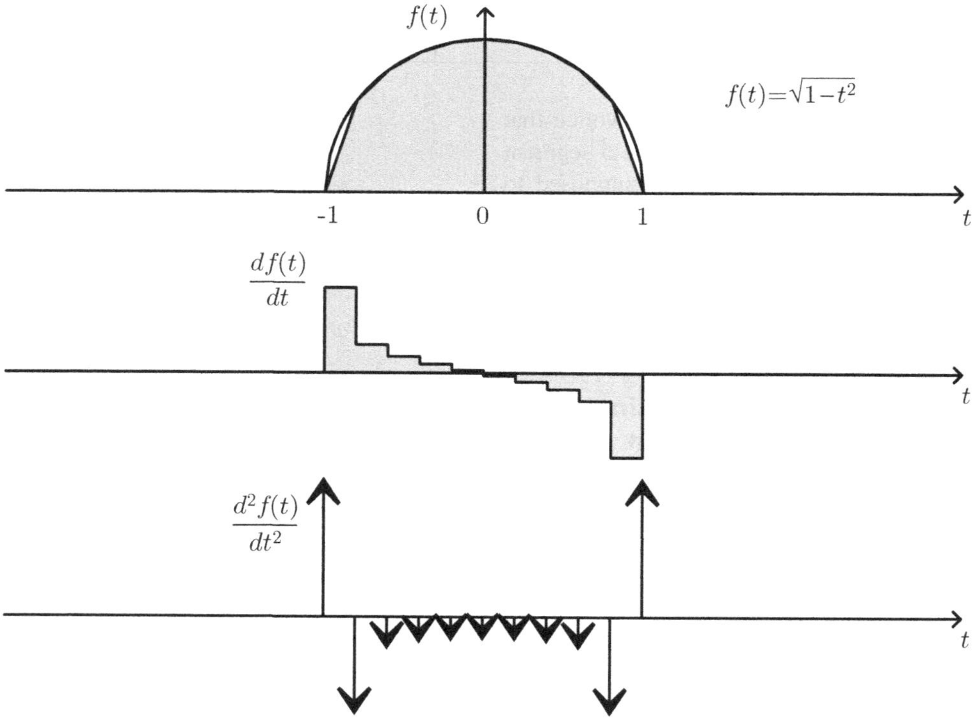

Fig. 12.11 Half circle decomposition in terms of (10) linear segments, and corresponding first and second derivatives

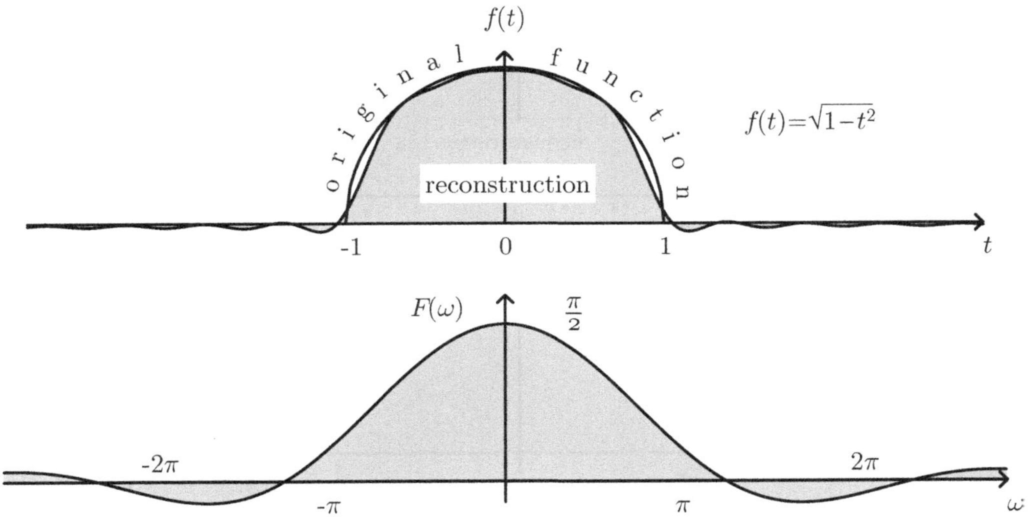

Fig. 12.12 Half circle Fourier transform (bottom) and time series (top) (using 10 linear segments)

time, and defined only within $-1 < t < 1$, and simply find the area underneath it. Then repeat for the next ω and so forth. The snippet below shows a complete C program that does the integration.

```
#include <stdio.h>
#include <math.h>
#include <stdlib.h>

float f(float t);

int main() {

float t, w, sum;
float dt, dw;
int i, j;

dt = 2./100.;
dw = 20./200.;
for(i=0;i<=200;i++){
w = -10. + (float) i * dw;
sum=0.0;
  for(j=0; j<=100; j++) {
    t=-1.+ (float) j * dt;
    sum=sum+f(t) * cos(w*t)*dt;
  }
  printf("%9.3f %9.3f\n", w, sum);
}
}

float f(float t) {
return sqrt(1. - t*t);
}
```

Figure 12.13 shows the results. Notice excellent agreement with prior methods. As mentioned above, since our starting function was even we ended up integrating only the cosine part of the complex exponential, and the Fourier transform came out real. Had our function turned out to be odd, then we will find out that the cosine integration would cancel out and we would end up only with the sine integration, in which case the result would be purely imaginary. In that case, we can still use real integration and simply multiply by j in the end. Lastly if our function was neither even nor odd, then we would need both the cosine and sine parts of the complex exponential. In that case we can do the cosine integration first, followed by the sine one, which gets multiplied by j, then simply add both. Main take here is that even though the Fourier integration—in the most generic sense—is complex, we can still do the numerical steps using real analysis, then patch things in the end. Or of course we can use a programming package than can deal with

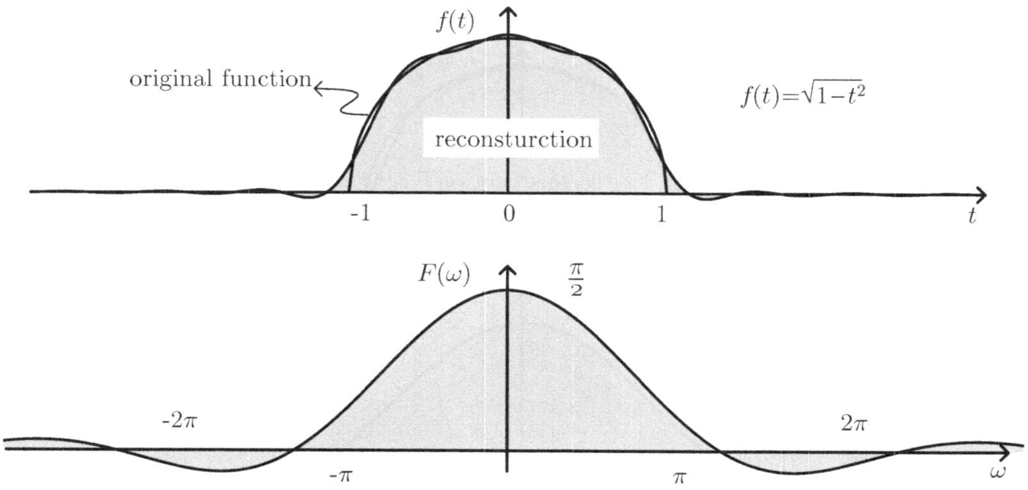

Fig. 12.13 Half circle Fourier transform using numerical integration

complex numbers from the start; in fact the C programming language is capable of just that! (So are Fortran, Python, R, Matlab, and the list goes on.)

12.7 Summary

This last chapter in the Fourier series/transform collection dealt with approximate and numerical methods for finding the Fourier transform. Of course if we know how to integrate the FT integral directly we won't need these methods; but more often is the case that the integral does not have a closed-form solution. In that case we need some methods to at least get an approximation to the integral. And preferably the methods are flexible enough to give us better accuracy as a function of some input knobs. Turns out there are such methods and they range from the Taylor series expansion, pulse sequence approximation, unit step sequence approximation, linear segment approximation, quadratic and higher order segment approximation (not covered here), and finally brute force numerical integration. In fact with little imaginations we can come up with even more such methods! For each of the presented methods we generated the corresponding Fourier transform and then plotted the time series and compared to the original function. In all cases we saw real evidence that the finer we tune in the input knobs the better approximation we get. The determining factors are always the same: time and frequency! We always end up with two plots: one for the function (or approximation thereof) versus time and the other for the Fourier transform versus frequency. Two faces of the same coin, and both hold the key information about the signal. If both are accurate, then they can be used interchangeably.

12.8 Problems

1. A function defined between -1 and 1 has the Taylor series

$$f(t) = 1 + t^2 + t^3$$

Find the Fourier transform and plot the time series. See sample solution in Fig. 12.14. Answer: Straightforward application of constant, quadratic, and order 3 term equations from Sect. 12.2.

2. Approximate the cosine function $f(t) = \cos t$ defined between $-\pi/2$ and $\pi/2$ up to order 4; then find the Fourier transform of the resulting

Taylor series and plot the time series. See sample solution in Fig. 12.15.

Answer:

$$f(t) = 1 - \frac{1}{2}t^2 + \frac{1}{24}t^4$$

3. Approximate the single-sine function $f(t) = \sin \pi t$, defined between -1 and 1 using 10 pulses, each of width 0.2. Find the corresponding Fourier transform, and plot the time series. Try three cases, and for each case assume $d\omega = 0.25$. For the first case integrate up to $\omega = 5$; for the second case integrate up to $\omega = 20$; and for the third case integrate up to $\omega = 40$. It is the convention that higher frequency integration limits result in more accurate time series; but is this the case here? Are we really after sharp pulse waveforms, or after the sine waveform? Does this suggest some filtering afterwards? See sample solution in Fig. 12.16.

Answer:

$$F(\omega) = -2j\frac{\sin(0.1\omega)}{\omega}\,[2\sin(0.1\pi)\sin(0.1\omega) + 2\sin(0.3\pi)\sin(0.3\omega) + 2\sin(0.5\pi)\sin(0.5\omega) + 2\sin(0.7\pi)\sin(0.7\omega) + 2\sin(0.9\pi)\sin(0.9\omega)]$$

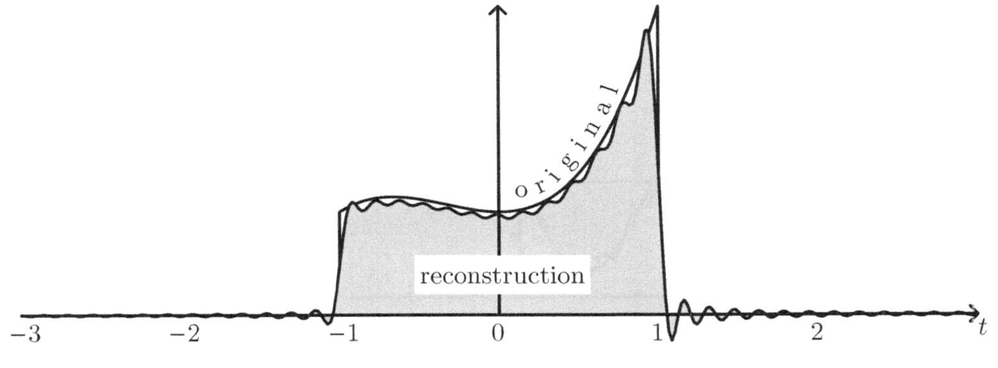

Fig. 12.14 Solution to Problem 1

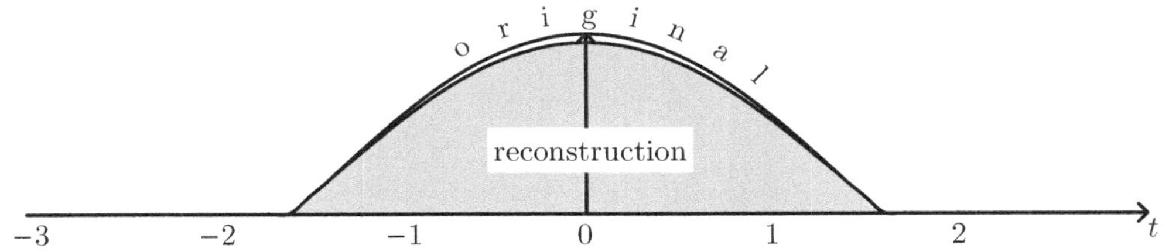

Fig. 12.15 Solution to Problem 2

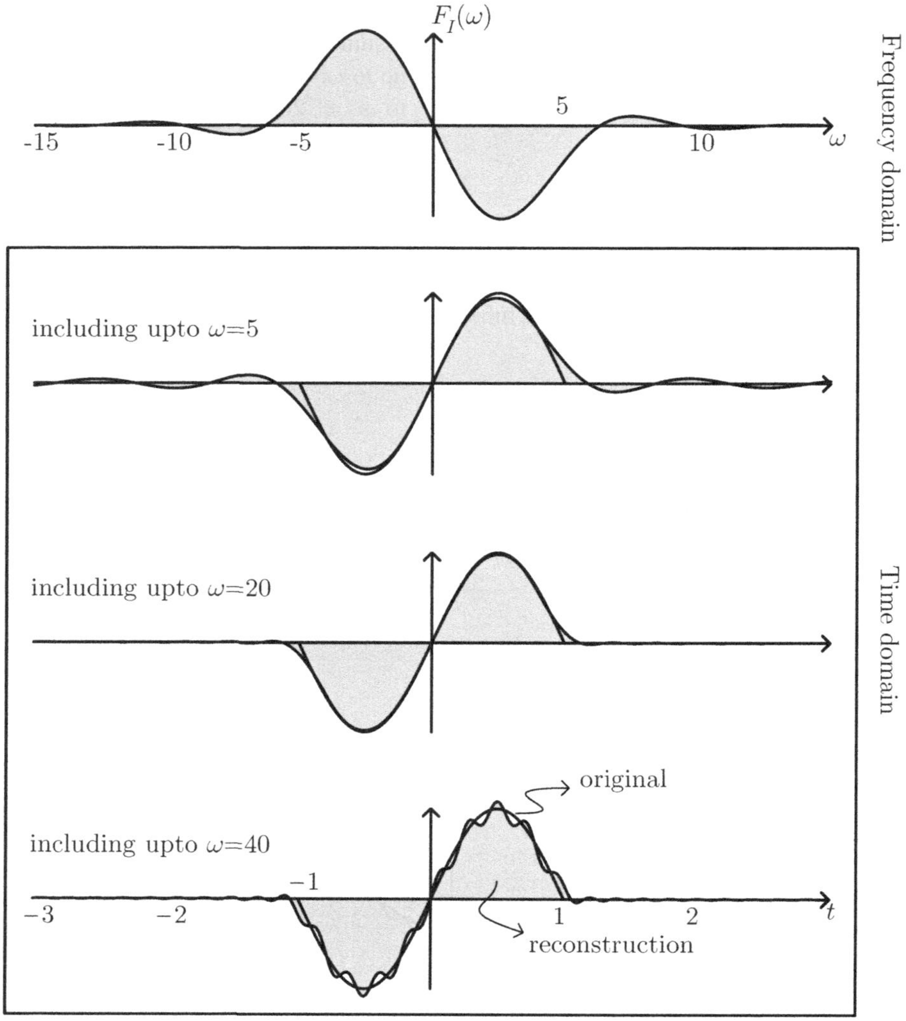

Fig. 12.16 Solution to Problem 3

4. A function $f(t)$ defined between -2 and 2 is given by

$$f(t) = e^{\sin \pi t}$$

Discretize the time domain into 40 segments, each of width $\Delta t = 1$, and approximate the function with 41 unit step functions. Next, find the Fourier transform of those unit step functions. Finally, plot the corresponding time series. See sample solution in Fig. 12.17.

5. Consider the cosine function $f(t) = \cos(t)$ defined between $-\pi/2$ and $\pi/2$, and zero otherwise. Approximate it with 8 linear segments. Take the first and then the second derivatives to end up with 9 delta functions. Find the Fourier transform of those delta functions, and then divide by $-\omega^2$ to get the Fourier transform of the original function. Plot the time series and compare to original function. See sample solution in Fig. 12.18.
Answer:

$$t_0, t_1, t_2, t_3, t_4, t_5 = \frac{\pi}{2}\left[\frac{-4}{4}, \frac{-3}{4}, \frac{-2}{4}, \frac{-1}{4}, \frac{0}{4}, \frac{1}{4}\right], \qquad \Delta t = \frac{\pi}{2}\frac{1}{4}$$

$$\begin{aligned} F(\omega) = &\Big\{2\left[f(t_1) - f(t_0)\right]\cos t_0\omega + 2\left[f(t_2) + f(t_0) - 2f(t_1)\right]\cos t_1\omega \\ &+ 2\left[f(t_3) + f(t_1) - 2f(t_2)\right]\cos t_2\omega + 2\left[f(t_4) + f(t_2) - 2f(t_3)\right]\cos t_3\omega \\ &= 2\left[f(t_5) + f(t_3) - 2f(t_4)\right]\cos t_4\omega\Big\} \times \frac{-1}{\omega^2 \Delta t} \end{aligned}$$

6. Consider the sine function $f(t) = \sin t$ defined between $-\pi/2$ and $\pi/2$, and zero otherwise. Approximate it with 8 linear segments. Take first then second derivatives until end up with delta function, then find Fourier transform of those. Then divide the FT of the delta functions from the first derivative by $j\omega$ and the FT of the delta functions from the second derivative by $-\omega^2$ to find the Fourier transform of the original function. Plot the time series and compare to original function. See sample results in Fig. 12.19.
Answer:

$$t_0, t_1, t_2, t_3, t_4 = \frac{\pi}{2}\left[\frac{-4}{4}, \frac{-3}{4}, \frac{-2}{4}, \frac{-1}{4}, \frac{0}{4}\right],$$

$$\Delta t = \frac{\pi}{2}\frac{1}{4}$$

$$\begin{aligned} F(\omega) = &\Big\{\left[f(t_1) - f(t_0)\right]\sin t_0\omega + \left[f(t_2) + f(t_0) - 2f(t_1)\right]\sin t_1\omega \\ &+ \left[f(t_3) + f(t_1) - 2f(t_2)\right]\sin t_2\omega \\ &+ \left[f(t_4) + f(t_2) - 2f(t_3)\right]\sin t_3\omega\Big\} \times \frac{2j}{\omega^2 \Delta t} \\ &-2\,\frac{\cos t_0\omega}{j\omega} \end{aligned}$$

7. Consider the function $f(t) = \sin(2\pi t^2)$ defined between -1 and 1, and zero otherwise. Write a computer program to find the Fourier transform between $\omega = -20$ and $\omega = 20$. Plot the time series and compare to original function. See sample solution in Fig. 12.20.

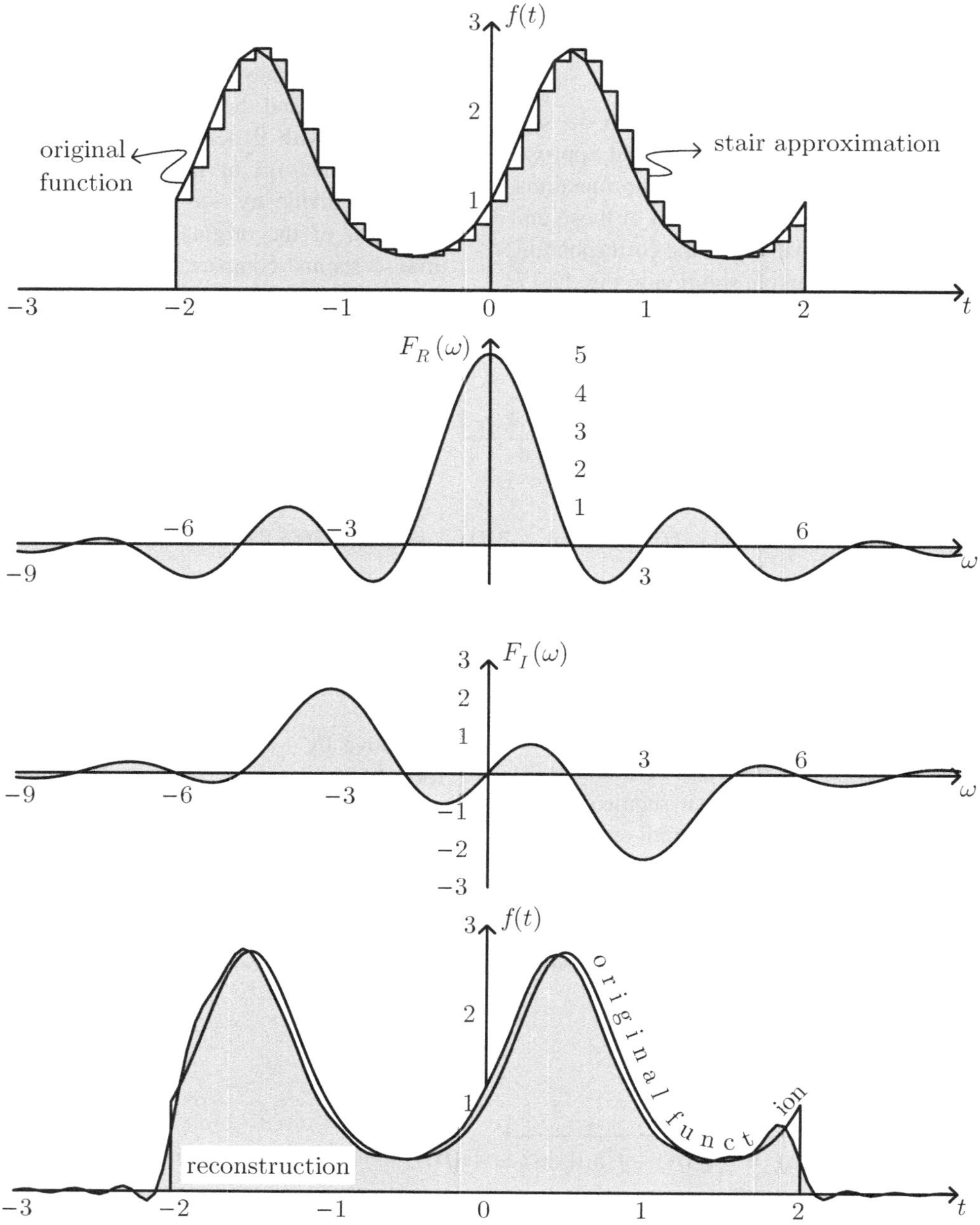

Fig. 12.17 Solution to Problem 4

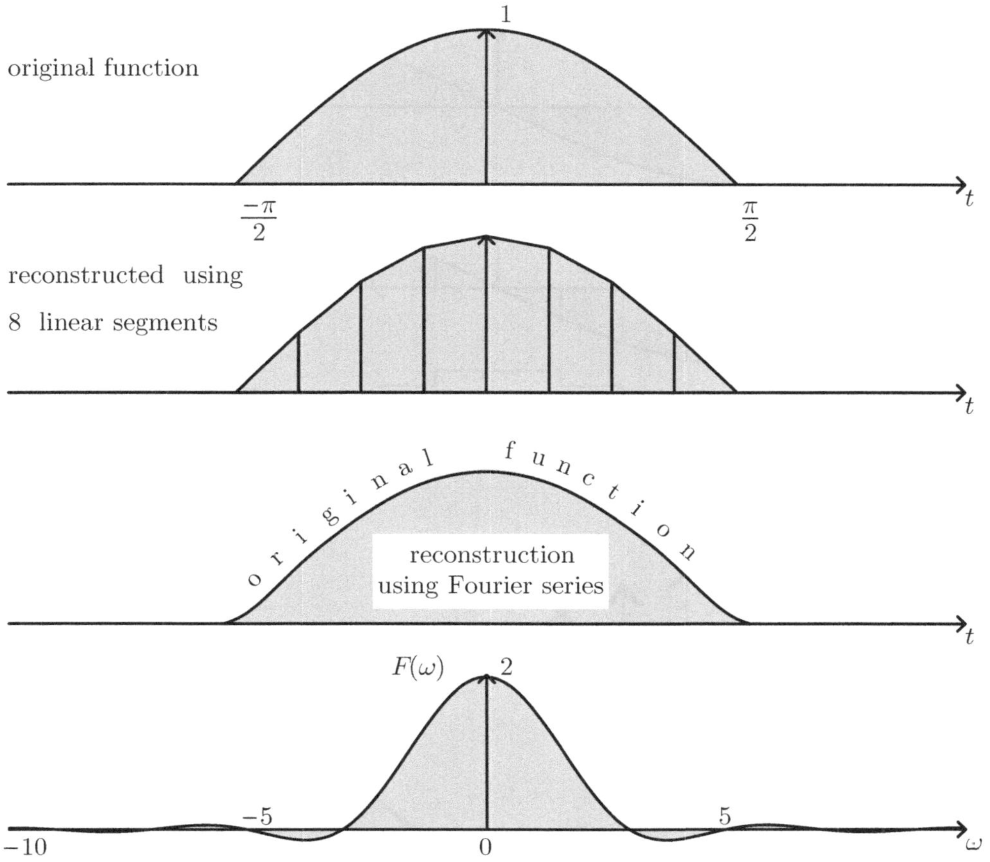

Fig. 12.18 Solution to Problem 5

Answer:

```
#include <stdio.h>
#include <math.h>
#include <stdlib.h>

float f(float t);

int main() {
float t, w, sum;
float dt, dw;
int i, j;
dt = 2./100.;
dw = 40./200.;
for(i=0;i<=200;i++){
w = -20. + (float) i * dw;
sum=0.0;
  for(j=0; j<=100; j++) {
    t=-1.+ (float) j * dt;
    sum=sum+f(t) * cos(w*t)*dt;
  }
  printf("%9.3f %9.3f\n", w, sum);
}
}

float f(float t) {
return sin(t*t*3.14*2.);
}
```

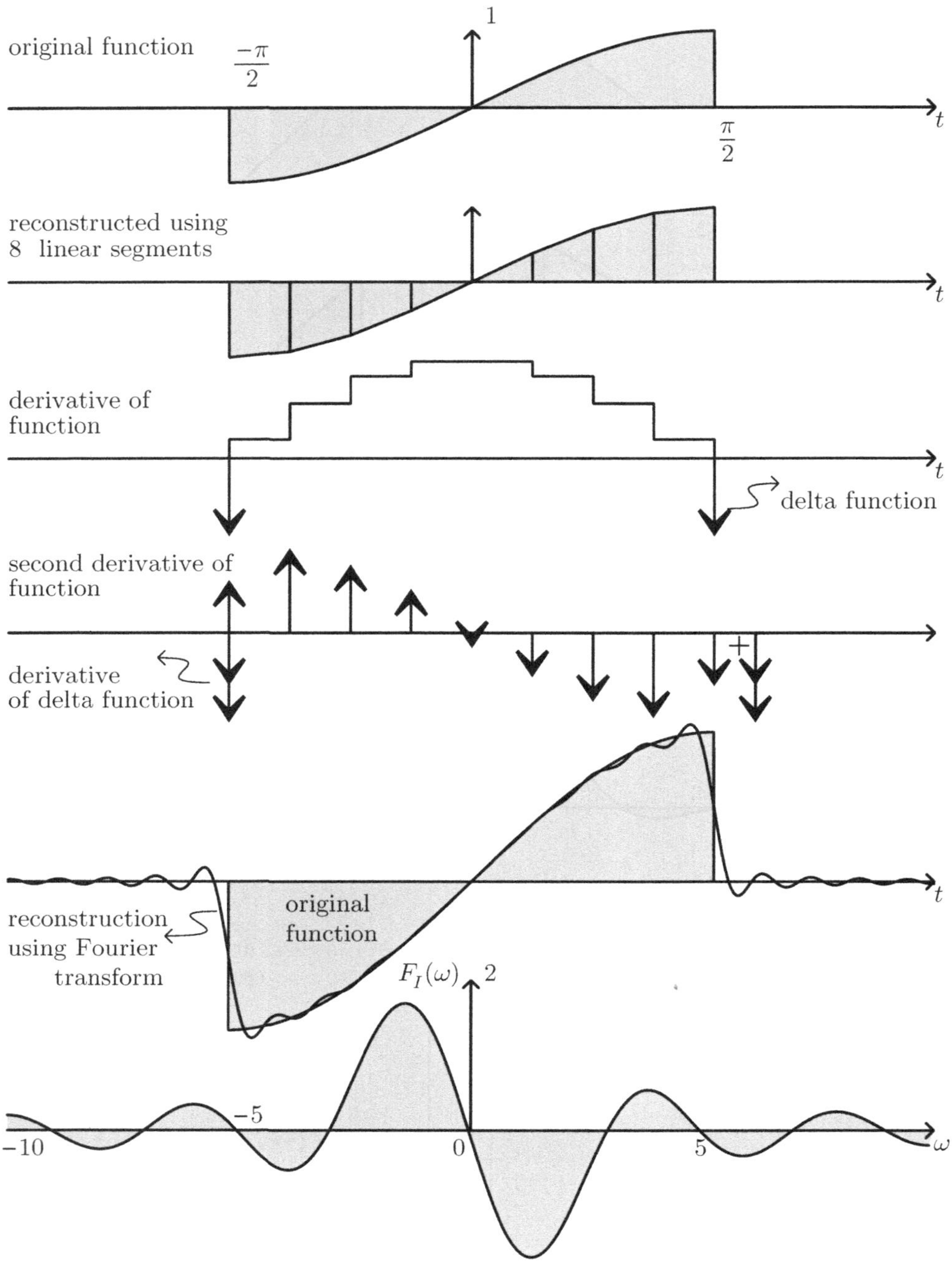

Fig. 12.19 Solution to Problem 6

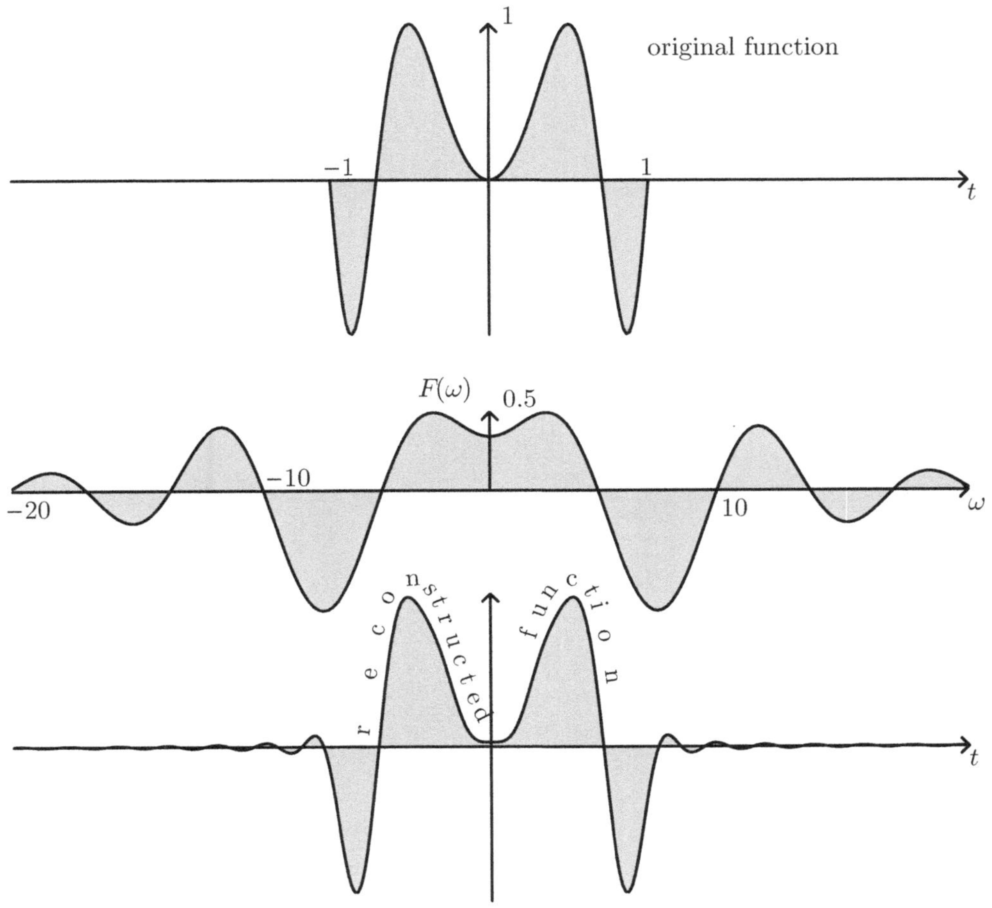

Fig. 12.20 Solution to Problem 7

13 Bandwidth

13.1 Introduction

After some seven chapters on various topics in the frequency domain, resembled by the Fourier series and transform by now we realize the importance and depth of the frequency concept. Assuming that is the case we may ask the question—over what frequency range should we be interested? The whole frequency spectrum? Positive, negative, low, high, band limited? Towards answering this question we turn to the concept of bandwidth. Bandwidth refers to the span of the frequency content of the signal. That is, over how wide of a frequency range does the Fourier transform extend. The meaning of bandwidth will be illustrated via a few examples.

13.2 The Time Delta Function and Infinite Frequency Bandwidth

The time delta function—we know by now—has the Fourier transform

$$\delta(t) \to 1 \tag{13.1}$$

In this case we say that the bandwidth is infinite! If we look at the FT (versus frequency) we see that it never decays (let alone go to zero) as a function of frequency. This says that in order to reconstruct a delta function using a summation of harmonics we need an infinite count thereof; or the bandwidth is infinite. This is shown in Fig. 13.1. The time delta function is one extreme that will for the most part be the worst signal so far as having the highest spectral content. In other words, for most practical applications the time delta function will be the highest in frequency content. (In reality the derivative of the time delta function, and higher derivatives will have even more spectral contents; but for most applications it is the normal delta function that gets used the most.)

13.3 The DC Time Function and Zero Frequency Bandwidth

On the other extreme there is the DC function which spans all time, but has zero bandwidth; recall

$$1 \to 2\pi\delta(\omega) \tag{13.2}$$

and this is illustrated in Fig. 13.2. What this states is that to reconstruct the DC function in the time domain we don't need to use a lot of harmonics; in fact we don't need to use *any* harmonic *other* than the zero frequency one! In other words, of all the complex exponentials of the form $e^{\omega t}$ we will only pick the one with $\omega = 0$, and pick it "strongly" in the form of a delta function in the frequency domain.

F. Badrieh, *Spectral, Convolution and Numerical Techniques in Circuit Theory*,
https://doi.org/10.1007/978-3-319-71437-0_13

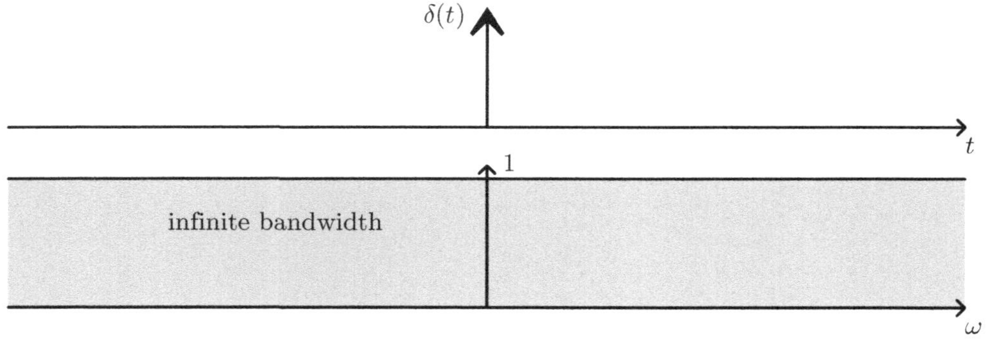

Fig. 13.1 Delta function and Fourier transform

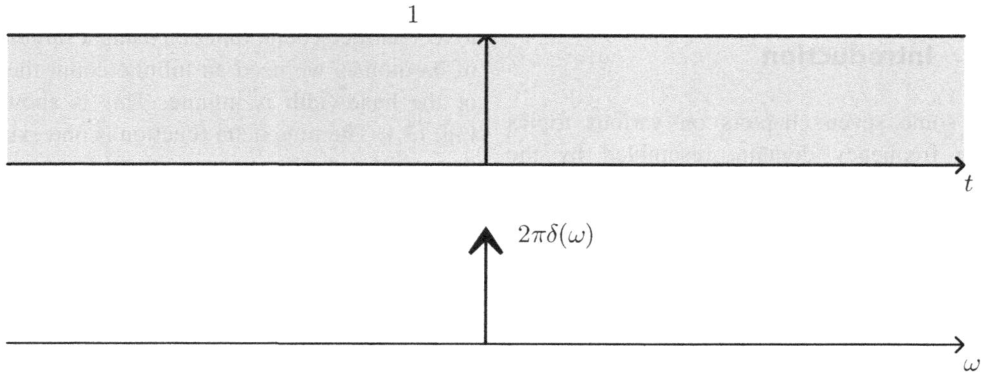

Fig. 13.2 DC function and Fourier transform

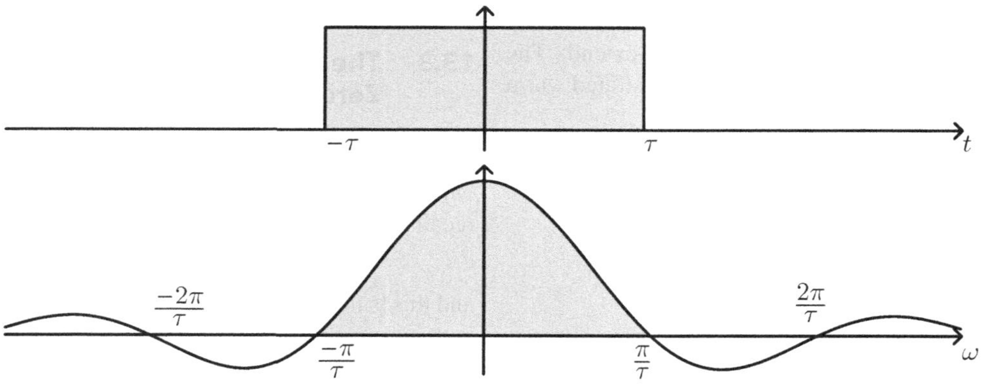

Fig. 13.3 Pulse function and Fourier transform

13.4 The Time Pulse Function with Finite Bandwidth

Between the two extremes (infinite bandwidth and zero bandwidth) we have for example the pulse function of width 2τ; it has the FT

$$\text{pulse of width } 2\tau \rightarrow 2\frac{\sin\omega\tau}{\omega} \tag{13.3}$$

Results are shown in Fig. 13.3. We can see that while the frequency spectrum extends to ∞, it does so in a decaying fashion. That is,

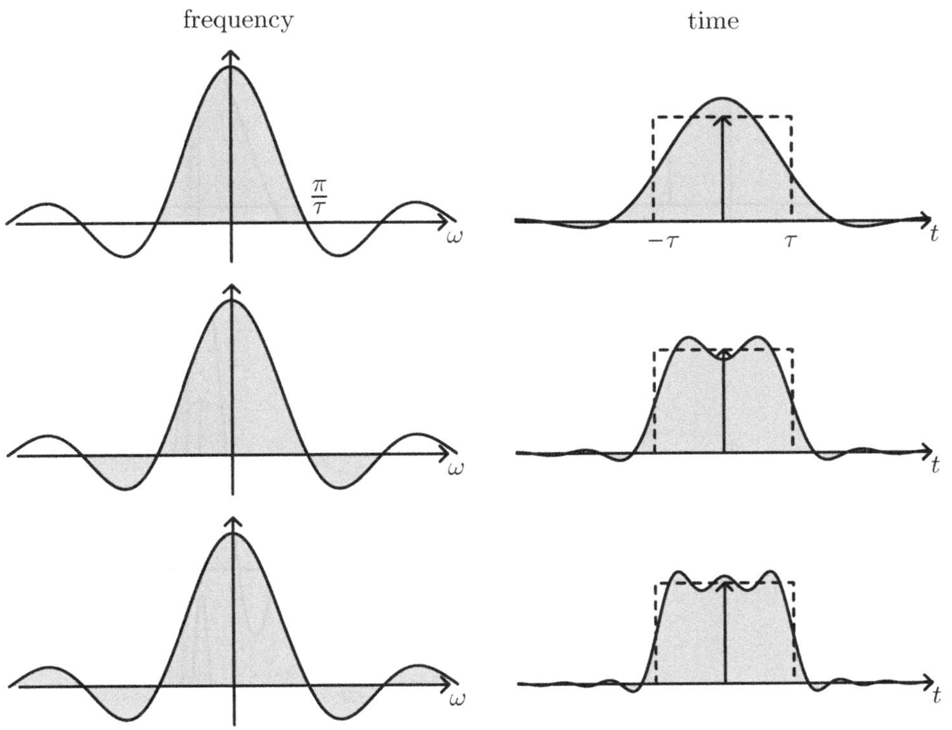

Fig. 13.4 Pulse function reconstruction as a function of BW inclusion

for all practical purposes the spectrum dies off at high frequency. Where should we mark the bandwidth (BW) tick? When the FT settles to zero? That would be pretty difficult, since technically the signal goes to 0 only at $\omega = \infty$. We can qualitatively state that within a few $\frac{\pi}{\tau}$'s the spectrum would have decayed enough such that any additional frequency contribution won't amount to much change during the reconstruction of the time series. For example, to zero order we can say the BW is where the FT first hits zero.

In this case we set the FT to zero

$$2\frac{\sin \omega\tau}{\omega} = 0 \tag{13.4}$$

This would imply that

$$\sin \omega\tau = 0 \tag{13.5}$$

which would solve for ω which we coin "BW".

$$\boxed{\mathrm{BW} \sim \frac{\pi}{\tau}} \tag{13.6}$$

That is, to first order, the frequency band between −BW and BW reasonably captures the pulse. This region has been shaded gray in the figure. Let's next test our thinking by building the time series of the pulse using only the spectrum between −BW and BW; results are shown at the top of Fig. 13.4. If we go up to 2 BW we get the middle figure; and finally if we include up to 3 BW we get the bottom of the figure. Clearly the more spectrum we include, the more accurate the reconstruction; but at least to zero order, including that between −BW and +BW gives us a rough estimate what the signal looks like in time.

Figure 13.5 shows the impact of pulse width on BW; as can be seen, wider pulses have narrower BWs. In fact if we make the pulse width infinite, the bandwidth would converge to zero, in agreement with the DC function.

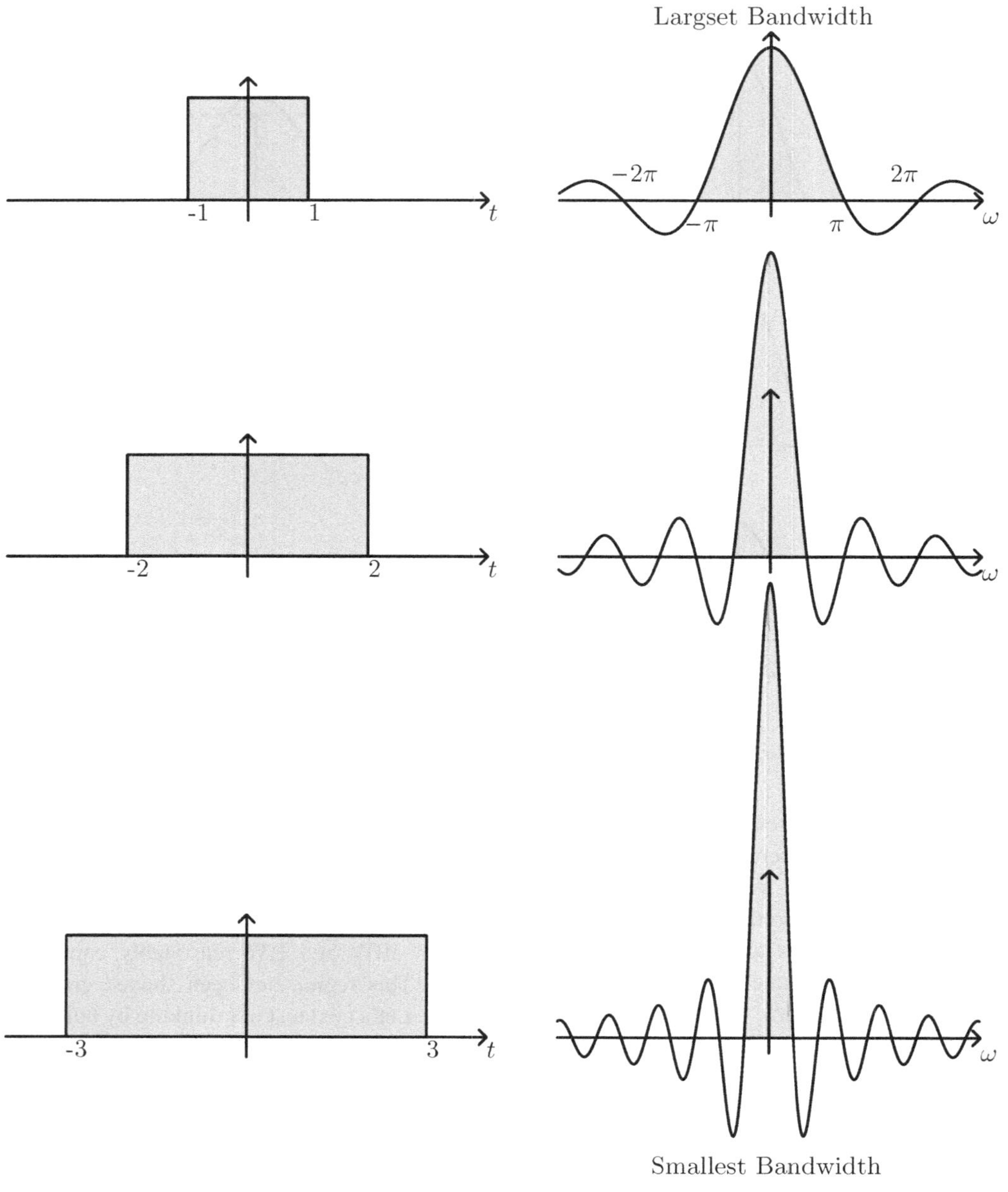

Fig. 13.5 Pulse function and Fourier transform for different pulse widths

13.5 The Tapered Pulse Function

A sibling of the ideal pulse is the tapered one. It has a bottom width of $2\tau_1$ and upper width $2\tau_2$. Its FT, and based on Eq. (10.82), is

$$\text{tapered pulse} \rightarrow \frac{2}{\tau_1 - \tau_2}\frac{1}{\omega^2}\left[\cos\omega\tau_2 - \cos\omega\tau_1\right] \tag{13.7}$$

Figure 13.6 shows various tapered pulses, with different rise times, starting with ideal pulse and ending with rise time with slope 1.5. We can see that as the pulse becomes more tapered, the first zero of the FT actually pushes out in frequency. This, however, does not mean that the tapered pulse has a higher bandwidth than the ideal pulse, since intuition tells us that the ideal pulse is the one with higher BW. Hence our definition of the BW coinciding with the first zero needs to be refined.

A way about comparing bandwidths is to look at both spectrums, identify the area under the curve (as shown in gray), and gauge which one has larger cumulative area (absolute-wise) at the given frequency, and the one that does so is deemed to have larger bandwidth. For example, we can see that the tapered pulse shown in case 3 of Fig. 13.6 has less localized area than the ideal pulse, and as such we refer to it as the one with smaller bandwidth. In other words, the ideal pulse (with wider span spectrum, in terms of gray area) is the one with higher bandwidth (even though it sustains its first zero before the tapered one). In conclusion

> The spectrum that extends further in frequency (in terms of cumulative, absolute area) is deemed to have larger bandwidth than the one more localized.

13.6 The Hat Pulse Function

The hat pulse function, defined between $-\tau$ and τ, with unity peak at time zero was shown (Eq. (10.74)) to have the transfer function

$$\text{hat function} \rightarrow 2\frac{1-\cos\omega\tau}{\omega^2} \tag{13.8}$$

We'd like to compare its bandwidth to the ideal pulse. We can already tell it would have a lower bandwidth, since the hat function is really the limit of the tapered pulse as $\tau_2 \rightarrow 0$. But just as a confirmation, let's compare the spectrum of the hat function to the ideal pulse; this is shown in Fig. 13.7.

Clearly as shown in the figure, the hat function has a *lower* bandwidth as compared to the ideal pulse. We deduce this by observing that the gray area under the curve fades away as a function of frequency faster than the pulse function! Let us test this claim. Let's build the time series of both the hat and pulse functions *using the same frequency range*! That is, for each case include the same number of harmonics; specifically, and for pulse width of 1 let's include from each spectrum only those frequencies up to 2π. Results are shown in Fig. 13.8. As can clearly be seen from the figure, the hat reconstruction has fully matured and is almost an exact duplicate of the original one; as for the pulse function, on the other hand, it is far from complete. What this suggests is that the BW of the pulse is *larger* than that of the hat; in other words, we need more frequency harmonics to reconstruct the pulse function as compared to the hat one!

13.7 The Upright, Centered Triangular Pulse

The upright, centered triangular pulse has a base width 2τ, and is at 90°; it is given by

$$f(t) = \frac{1}{2} + \frac{1}{2}\frac{t}{\tau}, \qquad -\tau < t < \tau \tag{13.9}$$

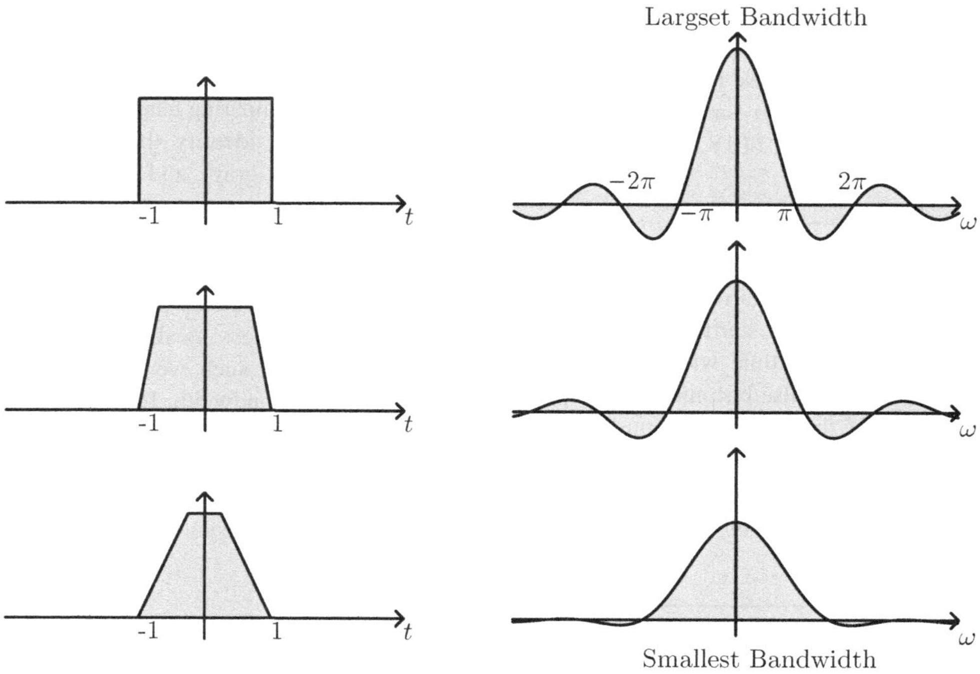

Fig. 13.6 Tapered pulse and Fourier transform, for different slew rate

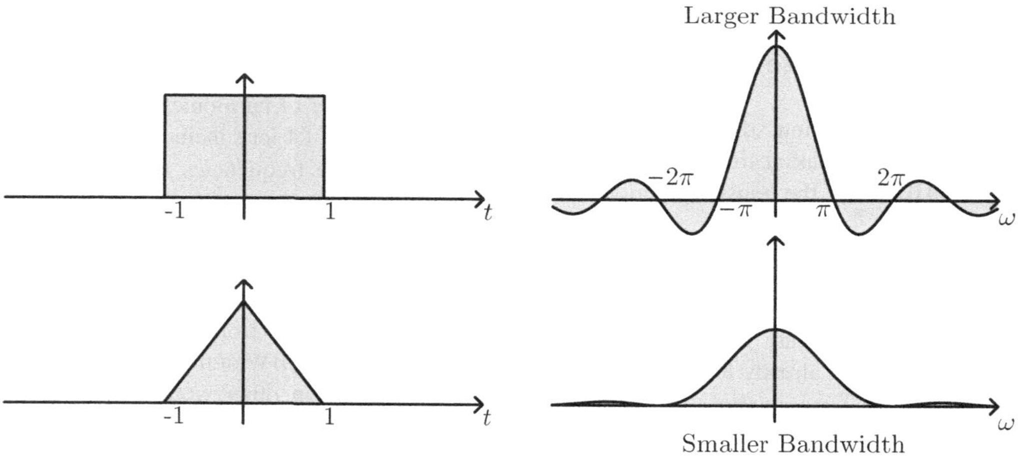

Fig. 13.7 Hat function and Fourier transform, and comparison to ideal pulse

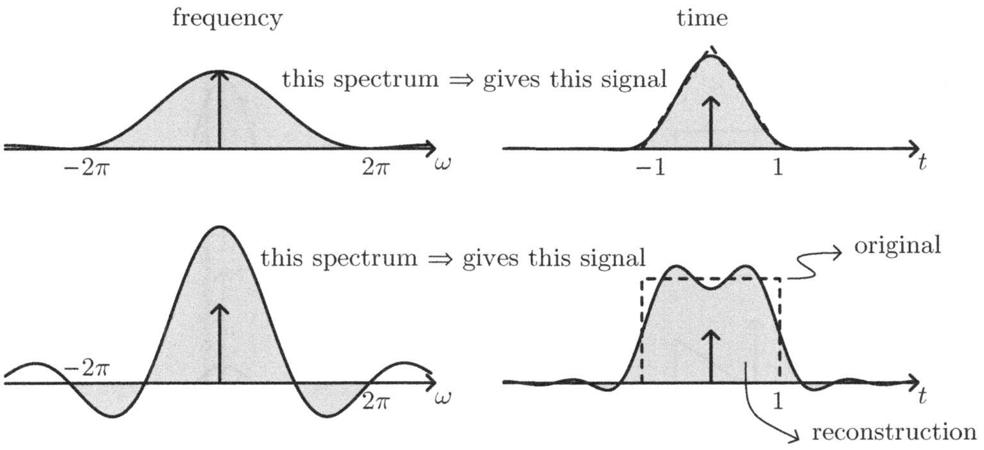

Fig. 13.8 Reconstruction of hat function and pulse one using same frequency range

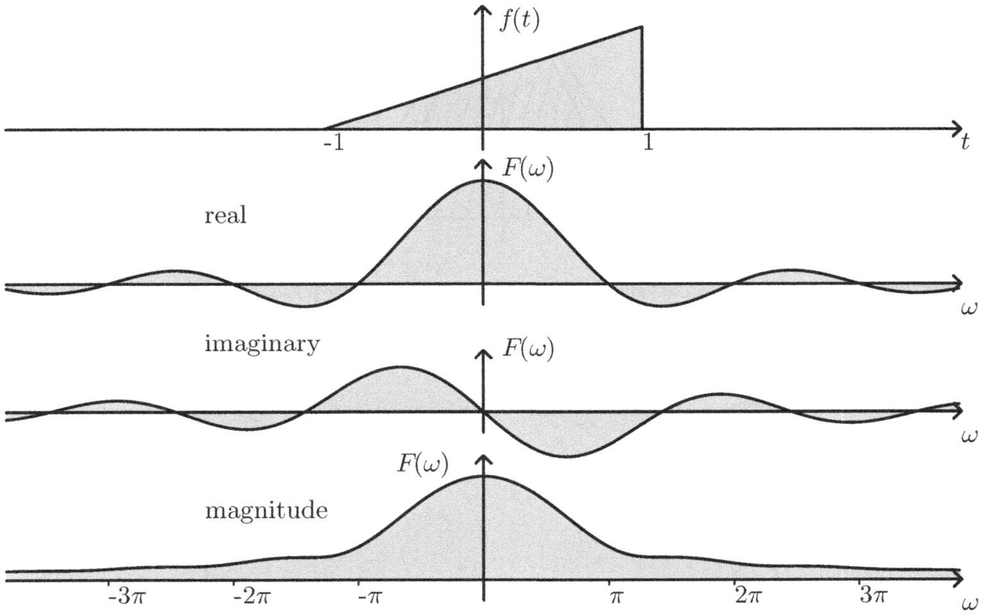

Fig. 13.9 Fourier transform of upright, centered triangular pulse

Its Fourier transform is

$$\text{upright, centered triangular pulse} \rightarrow \frac{\sin \omega\tau}{\omega} + j\left[\frac{\cos \omega\tau}{\omega} - \frac{1}{\omega^2\tau}\sin \omega\tau\right] \tag{13.10}$$

Since this function has both real and imaginary parts, it won't be straightforward comparing it to the ideal pulse, which has only real part. One way about this is to take the magnitude of the Fourier transform, as shown in Fig. 13.9.

Now we are ready to compare to the ideal pulse case. Figure 13.10 shows the results. On the right side we see the FT of both cases. It is not clear right away which one has larger bandwidth, but what is clear is that the DC value of the

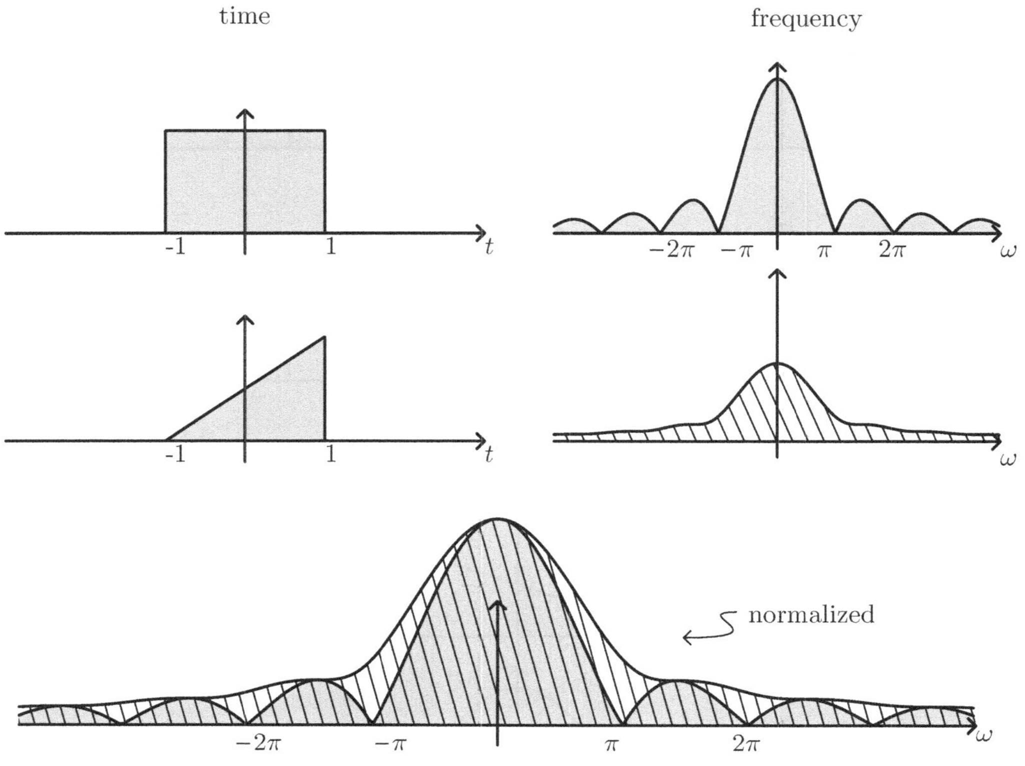

Fig. 13.10 Fourier transform (absolute value) of upright, centered triangular pulse, and comparison to ideal pulse

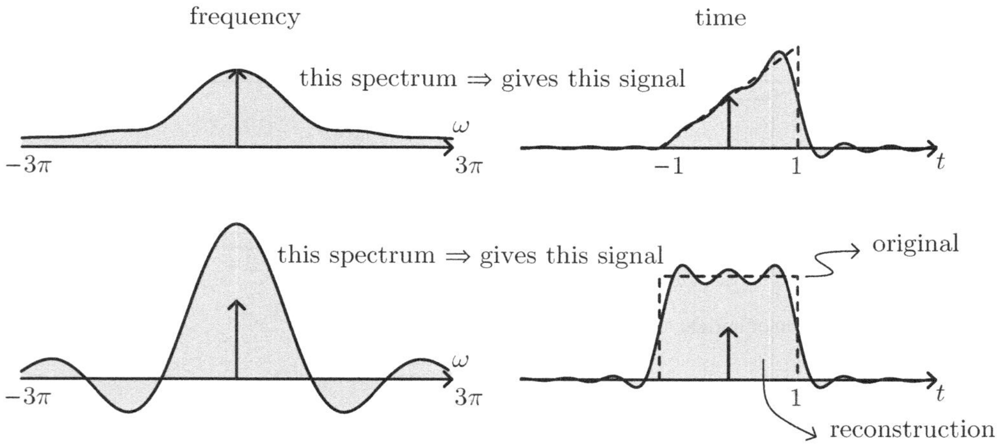

Fig. 13.11 Reconstruction of pulse and upright triangle using same frequency range

ideal pulse is 2× that of the triangular pulse. This makes sense since the triangle case has half the area (in time). To make the comparison on equal footing, we first scale up the triangular case and then overlay both traces as shown in the bottom of the figure. As seen from the figure, it looks like both spectrums have about the same bandwidth, especially having normalized the DC value!! To verify this last statement, let's build the time series for each case, including up to the same number of harmonics (3π here); this is shown in Fig. 13.11. As shown in the figure, the time

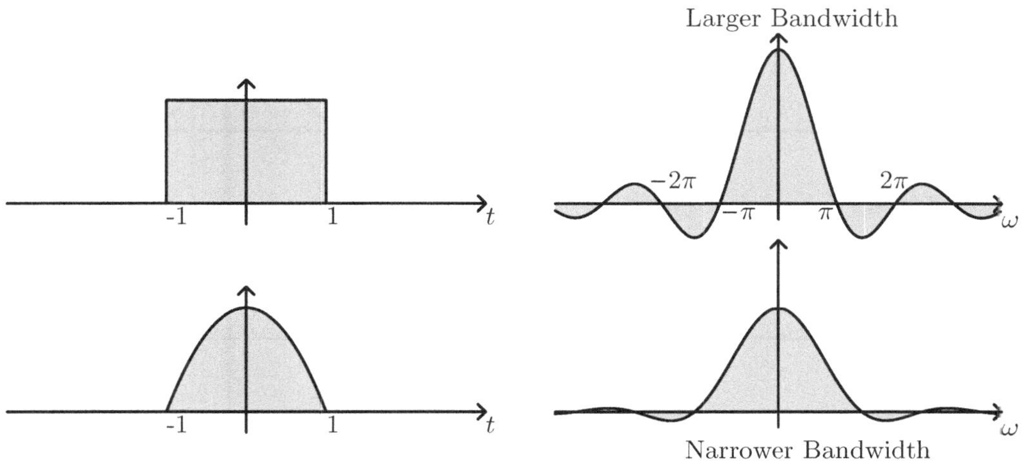

Fig. 13.12 Inverted parabola spectrum and comparison to ideal pulse

series shows about the same deviation (from the original wave forms) at the right edge; but at the left edge, we see that the triangular case has much better match. So, *absolute*-wise, the triangle case seems to have lower bandwidth (in the sense it was reconstructed to within a better accuracy—as compared to the ideal pulse—given a number of harmonics). But if we take the deviation in both cases, and divide by total area (in time), to get *relative* deviation we see that they both have the same deviation percentage; that is, they were both reconstructed to about the same resolution (again using the same number of harmonics). Hence we say that they both have the same bandwidth! In other words, in order to get the same relative deviation in signal reconstruction (compared to original signal), both signals end up using the same number of harmonics. This kind of makes sense, since the bottleneck really is the specific feature of the signal that has the highest frequency content, and that happened to be the right edge, which is shared amongst both signals.

13.8 The Inverted Parabola

The inverted parabola, of base width 2, and defined as

$$f(t) = 1 - t^2, \qquad -1 < t < 1 \qquad (13.11)$$

has the Fourier transform

$$F(\omega) = 4\left[\frac{\sin\omega}{\omega^3} - \frac{\cos\omega}{\omega^2}\right] \qquad (13.12)$$

Figure 13.12 shows the spectrum, and comparison to the ideal pulse. As can be seen, the bandwidth of the inverted parabola is narrower than the ideal pulse; this makes sense.

While the comparison made sense to the ideal pulse, it is not as easy when comparing to the hat function—which one has larger bandwidth? Figure 13.13 shows results and shows that both have about the same bandwidth; in fact both have the same limit dependence at high frequency (case of $\tau = 1$):

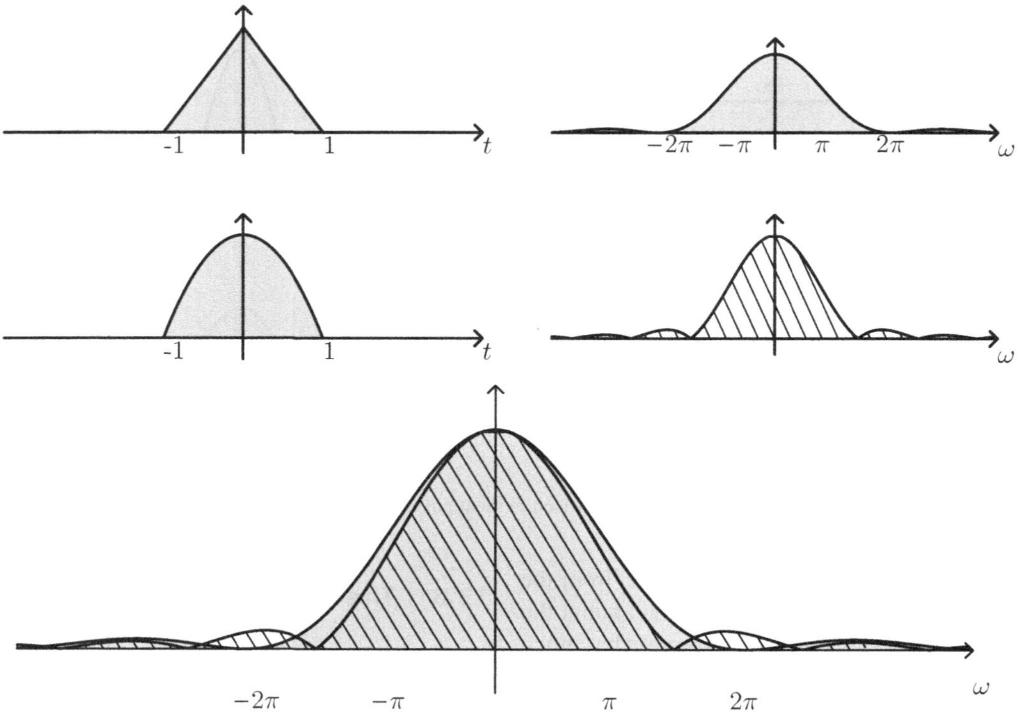

Fig. 13.13 Inverted parabola spectrum and comparison to hat function

$$\text{hat function } F(\omega) = 2\frac{1-\cos\omega}{\omega^2} \sim \frac{1}{\omega^2} \text{ at high frequency} \tag{13.13}$$

$$\text{inverted parabola } F(\omega) = 4\left[\frac{\sin\omega}{\omega^3} - \frac{\cos\omega}{\omega^2}\right] \sim \frac{1}{\omega^2} \text{ at high frequency} \tag{13.14}$$

13.9 The Ideal and Tapered Step Input

The ideal step input with infinite edge rate has the FT

$$u(t) \rightarrow \pi\delta(\omega) + \frac{1}{j\omega} \tag{13.15}$$

The real part is a delta function (sampling only the zero frequency range), and hence has zero bandwidth; but the imaginary part extends all the way to ∞, albeit decaying all the while. The slanted, tapered unit step function, on the other hand, has a rise time of 2τ, and it has the FT

$$\text{unit step with rise time } 2\tau \rightarrow \pi\delta(\omega) + \frac{1}{\tau}\frac{\sin\omega\tau}{j\omega^2} \tag{13.16}$$

Notice that when τ is pretty small, the sine term approaches

$$\lim_{\tau\rightarrow 0} \sin\omega\tau = \omega\tau \tag{13.17}$$

and the fraction term approaches

$$\lim_{\tau\rightarrow 0} \frac{1}{\tau}\frac{\sin\omega\tau}{j\omega^2} = \frac{1}{\tau}\frac{\omega\tau}{j\omega^2} = \frac{1}{j\omega} \tag{13.18}$$

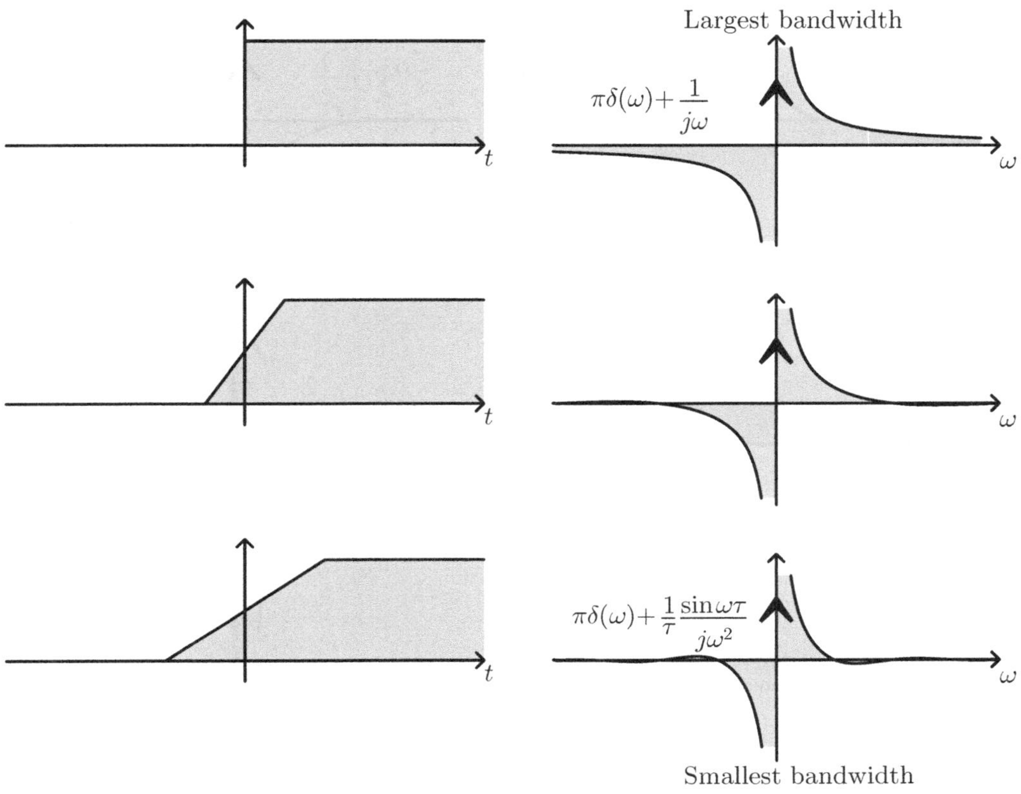

Fig. 13.14 Ideal and slanted steps, and corresponding frequency spectrum

which matches the FT of the ideal case! Figure 13.14 shows the ideal step, and the slanted one (with different ramp rates), and the corresponding frequency spectrum. As can be seen, the tapered step has lower bandwidth, and more so with elongated ramp rates.

13.10 Two Step Inputs with Same Rise Time, but Different Ramp Rate

Two step inputs share the same rise time of 2τ, but the step value of the second (max minus min) is $2\times$ the former—which one has higher bandwidth? Strictly speaking, both have the same bandwidth, which we define as that imaginary frequency at which the ratio of the remnant frequency spectrum to the total frequency spectrum drops below a certain threshold. But at a given frequency, including the BW and beyond, the case with $2\times$ would have higher energy content. This does not mean it has higher BW—only higher frequency content, which in fact would be the same for ALL frequency content. Results are shown in Fig. 13.15.

13.11 Summary

Every signal has frequency content; in some cases, like the impulse function, the content covers the whole frequency spectrum. In other cases, like the DC function, it samples only a single frequency (zero here). In between typically a signal has a finite bandwidth. Bandwidth is defined as that frequency region whose content is enough (after doing the inverse transform) to recreate the signal close enough. In this chapter we studied various examples, including different

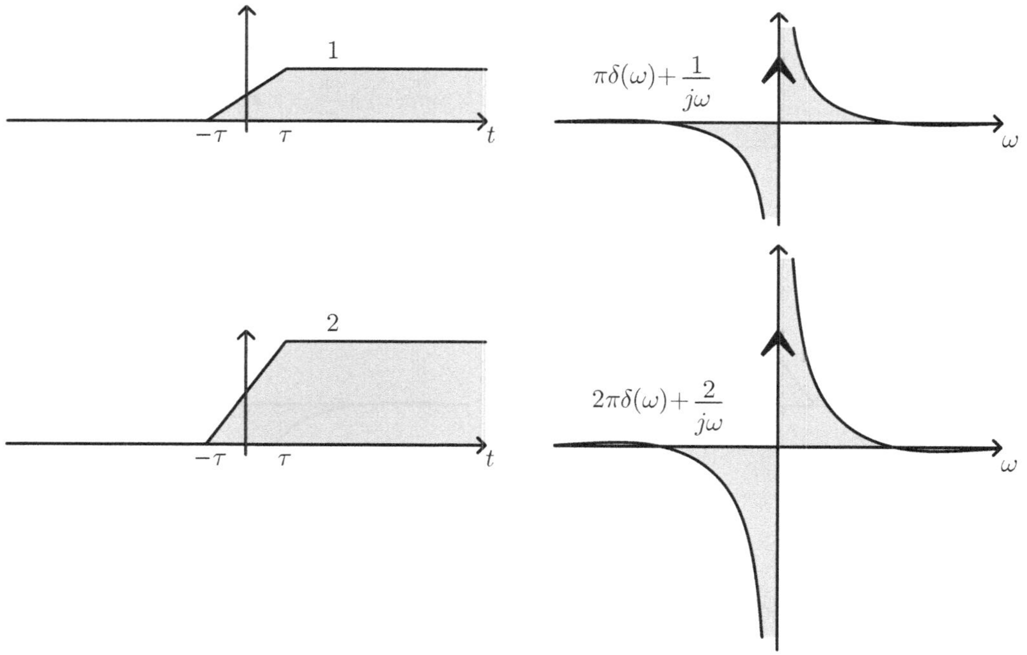

Fig. 13.15 Two steps with same rise time, but different ramp rate, have same bandwidth

pulse shapes to get a better feel of bandwidth extent. Typically two signals are compared for bandwidth, and the process outlined here is to overlay the two signals (after some normalization, if needed) and simply identify that signal whose frequency spectrum extends the furthest. Normalization is sometimes needed when doing comparison just to ensure we are comparing relative strength, as opposed to absolute one. We wrapped the chapter with examination of the ideal and slanted unit step function and confirmed the latter has narrower bandwidth.

13.12 Problems

1. Which signal has larger bandwidth—the single lobe pulse function of width 1, or the two-lobe, odd pulse, with each lobe also having same pulse width of 1? See sample results in Fig. 13.16 and fill in the details.
2. Which signal has larger bandwidth—the signum function, or the unit step one?
3. A pulse of width 1 and height 1, and another of width 1 and twice the height—which one has larger bandwidth?
4. Which has larger bandwidth—the single timer pulse function, or the mirrored one, of center-to-center separation T? See Fig. 13.17 for sample hints.
5. Which has larger bandwidth—the nominal cosine function, or the odd cosine function? See Fig. 13.18 for sample hints.
6. Which has higher bandwidth—the symmetric triangle with base width 4 and height 2, or the mirrored triangle with base width 2 and height 1, as shown in Fig. 13.19? The same figure shows hints for solution.

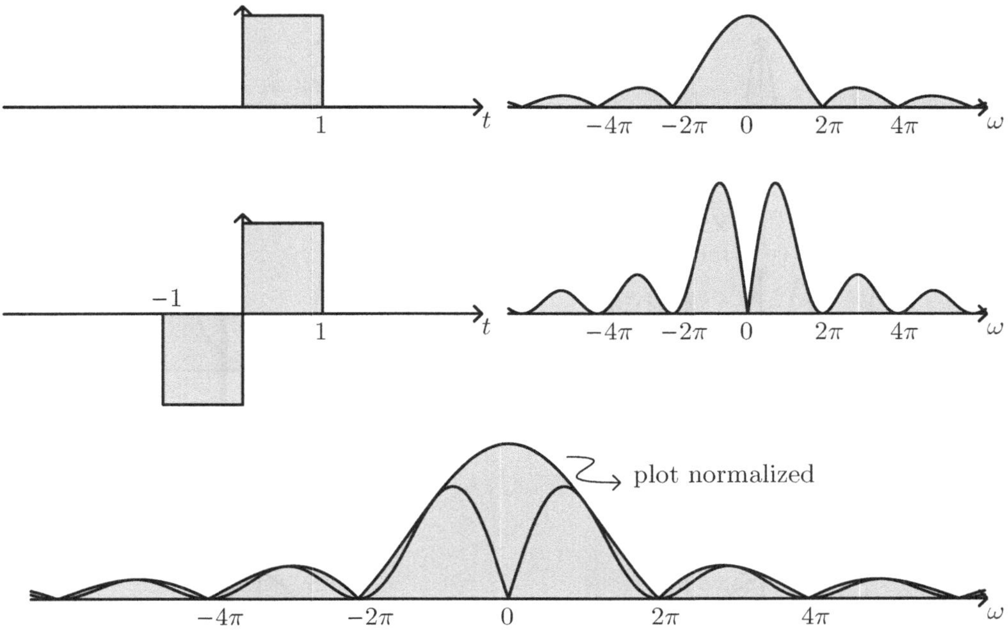

Fig. 13.16 Solution to Problem 1

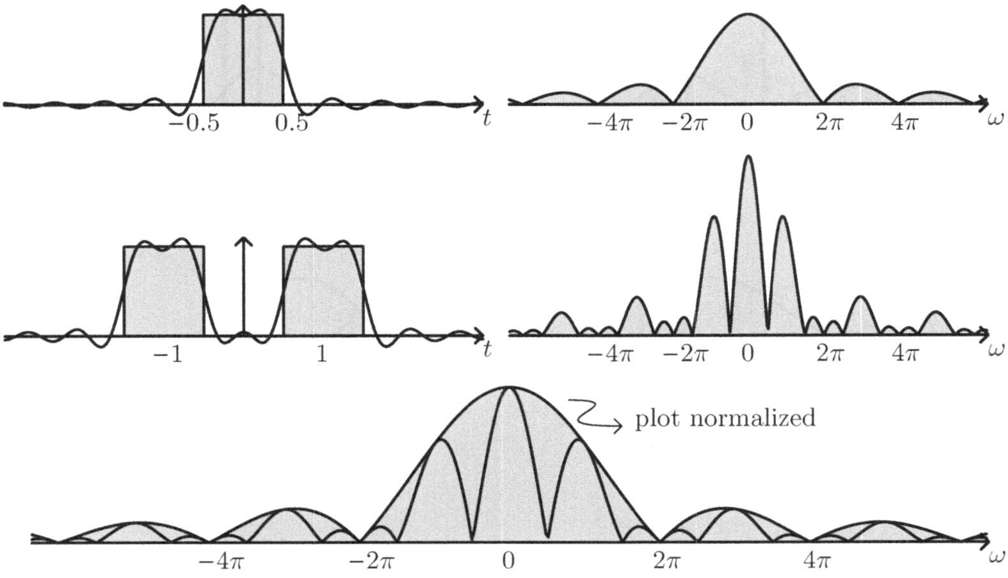

Fig. 13.17 Solution to Problem 4

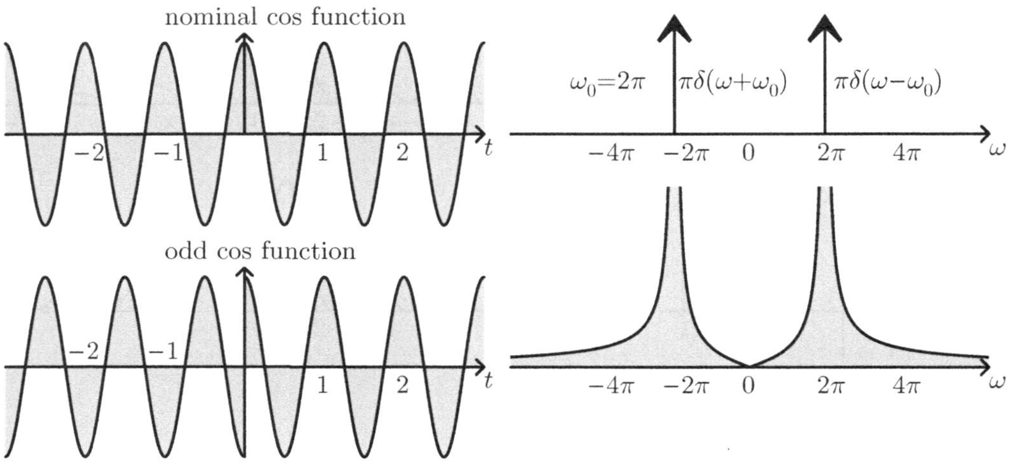

Fig. 13.18 Solution to Problem 5

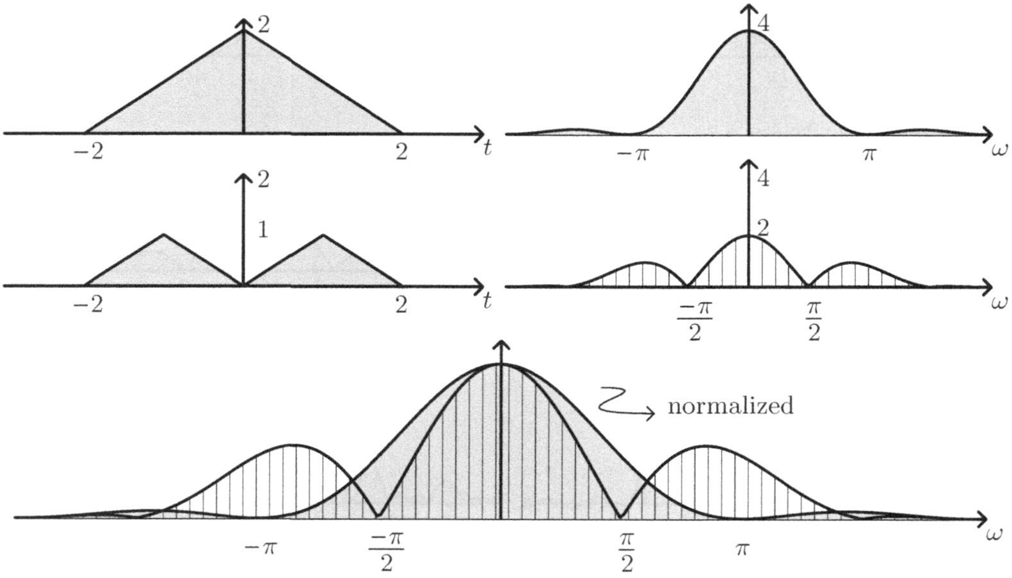

Fig. 13.19 Solution to Problem 6

14 Laplace Transform

14.1 Introduction

With eight chapters on Fourier series and transform, we are both exhausted and yearning for a change! Not to say the prior material was not worth the time; on the contrary, as we will see throughout the rest of the text, it really is the foundation for what is coming up next. The most logical continuation after covering Fourier analysis is the Laplace transform. While the name is different, the underlying machinery is really similar. The only difference is that the Laplace transform tweaks the signal *before* going back and applying the Fourier transform to it. But why all the trouble? As we saw in the Fourier transform chapter(s), we can find frequency representation of rather a large class of functions. However, we run into problems when those functions are not integrable; for example the function $tu(t)$. To remedy this, the Laplace transform is used.

14.2 Idea Behind Laplace Transform

If a function $f(t)$ is not integrable Laplace transform, and in, how about we multiply it by a decaying function, so that the product is integrable? For example, take the function $f(t) = tu(t)$ as shown in Fig. 14.1. Clearly this function is not integrable. Meaning

$$\int_0^\infty tdt = \infty \tag{14.1}$$

Now multiply it by $e^{-\sigma t}$ and suddenly the function $e^{-\sigma t}tu(t)$ is integrable, as shown in the right side of the figure. Now we can find the frequency representation of the product function, using the FT. We can also find the inverse FT. To get the original function $f(t)$ we simply multiply the inverse transform by $e^{\sigma t}$. That is all there is to it! Notice that the sign of σ is positive now.

To recap, starting with any arbitrary signal, take that signal and multiply by $e^{-\sigma t}$ to make the product integrable. Now take the Fourier transform of the product and arrive at the spectrum (which typically would depend on the choice of σ). So now we have a frequency representation of a signal that presumably had no Fourier transform. Now to go back from the frequency domain to the time domain we simply perform inverse transform on the spectrum just achieved. The resulting time signal is not quite what we want, since it still has $e^{-\sigma t}$ in it. To remedy this and regain our original time signal we simply multiply the inverse Fourier transform (which by now is a time signal) by $e^{\sigma t}$ (notice $+\sigma$). With that we have gone full circle from the time domain to the frequency domain.

F. Badrieh, *Spectral, Convolution and Numerical Techniques in Circuit Theory*,
https://doi.org/10.1007/978-3-319-71437-0_14

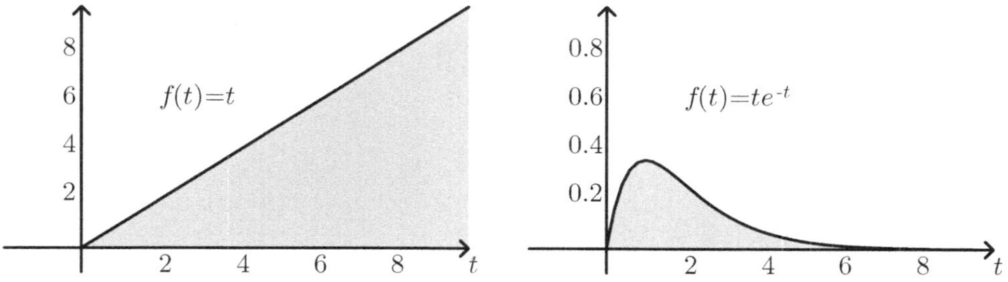

Fig. 14.1 Function t and $te^{-\sigma t}$

14.3 Laplace Transform Defined

Assume for now that the signal of interest is limited to positive time. For a signal $f(t)$ in the time domain, the Laplace transform in the frequency domain is defined as

$$\boxed{\mathcal{L}[f(t)] = F(s) = \int_0^{\infty} f(t)e^{-st}dt} \tag{14.2}$$

where the "complex" frequency s is defined as

$$\boxed{s = \sigma + j\omega} \tag{14.3}$$

The inverse LT in terms is defined as

$$\boxed{f(t) = \frac{1}{2\pi j}\int_{\sigma-j\infty}^{\sigma+j\infty} F(s)e^{st}ds} \tag{14.4}$$

Notice that in both cases we have not defined σ. This parameter is chosen depending on the function at hand, and ensuring that the product $e^{-\sigma t}f(t)$ is integrable over time. Once σ is set in the time domain, the same σ has to be used in the frequency domain.

14.4 The Laplace Transform Flow

The overall flow of the Laplace transform, and inverse, is illustrated in Fig. 14.2. Assume our starting function is

$$f(t) = u(t)t \tag{14.5}$$

This function grows beyond bound, and it is not integrable; hence it does not have a Fourier transform. To make it integrable, let's multiply by

$$g(t) = f(t)e^{-\sigma t}, \qquad \sigma > 0 \tag{14.6}$$

Now let us find the Fourier transform of the new function $g(t)$:

$$G(\omega) = \int_0^{\infty}\left[f(t)e^{-\sigma t}\right]e^{-j\omega t}dt \tag{14.7}$$

To reiterate, the function $te^{-\sigma t}$ is integrable, and it does have a Fourier transform; we call this transform the Laplace transform! As will be shown later, the Laplace transform of the function $f(t) = t$ comes out

$$F(s) = \mathcal{L}[t] = \frac{1}{s^2} \tag{14.8}$$

Let's expand this in terms of $s = \sigma + j\omega$

$$F(\sigma + j\omega) = \frac{1}{(\sigma + j\omega)^2} \tag{14.9}$$

From the Fourier transform chapters (Eq. (9.112)), we know that this in fact is the Fourier transform of $te^{-\sigma t}$; that is

$$te^{-\sigma t} \to \frac{1}{(\sigma + j\omega)^2} \tag{14.10}$$

Now, knowing the Laplace transform of t, or equivalently the Fourier transform of $te^{-\sigma t}$, how can we do inverse transform and recover the original signal? For a starter, we do the inverse Fourier transform of the Laplace transform as

$$\frac{1}{2\pi}\int_{-\infty}^{\infty}\frac{1}{(\sigma + j\omega)^2}e^{j\omega t}d\omega = te^{-\sigma t} \tag{14.11}$$

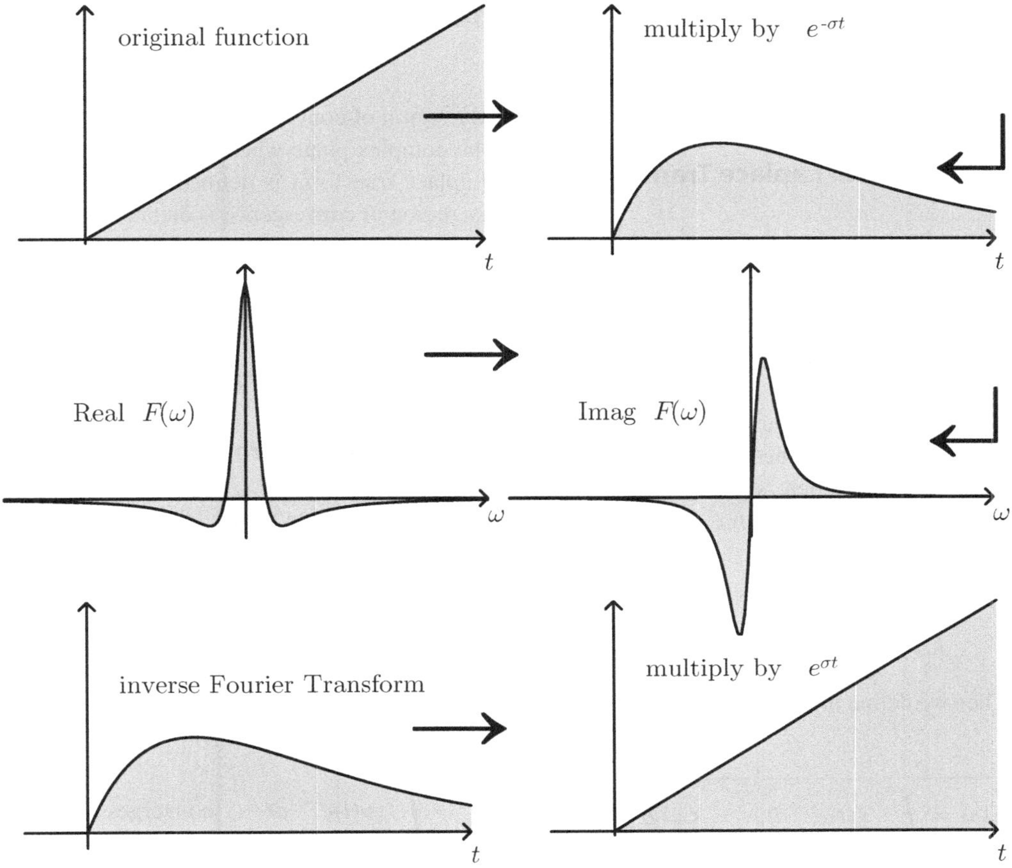

Fig. 14.2 Laplace transform flow starting from original function in time domain and ending in it after finding inverse transform

Now multiply this result by $e^{\sigma t}$ (notice positive σ) and get

$$te^{-\sigma t} \cdot e^{\sigma t} = t \tag{14.12}$$

So we have at last recovered the original function, which in this case is t; that is, do inverse Fourier transform, then multiply by $e^{\sigma t}$. This is what we call, then, the inverse Laplace transform

$$f(t) = e^{\sigma t} \frac{1}{2\pi} \int_{-\infty}^{\infty} F(\omega) e^{j\omega t} d\omega \tag{14.13}$$

Now let

$$s = \sigma + j\omega, \qquad \text{such that} \qquad ds = jd\omega \tag{14.14}$$

Notice that σ is constant, and only ω is variable. Then the integral becomes

$$f(t) = e^{\sigma t} \frac{1}{2\pi j} \int_{\sigma - j\infty}^{\sigma + j\infty} F(\omega) e^{j\omega t} ds \tag{14.15}$$

Pull the outer $e^{\sigma t}$ inside the integral (which is integrated over frequency) and get

$$f(t) = \frac{1}{2\pi j} \int_{\sigma - j\infty}^{\sigma + j\infty} F(\omega) e^{j\omega t} e^{\sigma t} ds \tag{14.16}$$

Collect variables, and use $s = \sigma + j\omega$ and finally get

$$f(t) = \frac{1}{2\pi j} \int_{\sigma - j\infty}^{\sigma + j\infty} F(s) e^{st} ds \tag{14.17}$$

So we have derived the inverse Laplace transform; it really is basically the inverse Fourier transform, times $e^{\sigma t}$!!!

14.5 Bilateral Laplace Transform

So far we have assumed that the signal is zero for negative time; hence our integration was

$$F(s) = \int_0^{\infty} f(t)e^{-st}dt, \qquad s = \sigma + j\omega \quad (14.18)$$

Assume this is not the case, and that instead we have the signal defined from $-\infty$ to ∞. We can use the same idea, such that we multiply the function by $e^{-\sigma t}$ to ensure that the resulting function is integrable

$$\int_{-\infty}^{\infty} f(t)e^{-\sigma t}dt \qquad \text{finite} \qquad (14.19)$$

Then we define the bilateral Laplace transform as

$$\boxed{F(s) = \int_{-\infty}^{\infty} f(t)e^{-st}dt, \qquad s = \sigma + j\omega;} \qquad (14.20)$$

however, the selection of the damping factor σ becomes more difficult.

14.6 Region of Convergence (ROC)

By region of convergence we mean the region in the complex plane where σ resides such that the Laplace transform is defined (convergent). That is, region of convergence is dictated by the selection of the damping factor σ. The main idea again is to multiply the (non-convergent) function by $e^{-\sigma t}$ to make the result integrable over time (be it negative, positive or all time). Here are a few examples demonstrating the selection of σ.

14.6.1 Example Where σ Is Positive

The unit step function $u(t)$ and as shown in Fig. 14.3 is none integrable is the sense

$$\int_0^{\infty} u(t)dt \qquad \text{blows up} \qquad (14.21)$$

However, multiplying by $e^{-\sigma t}$ for $\sigma > 0$ makes the integral convergent.

$$\int_0^{\infty} u(t)e^{-\sigma t}dt \qquad \text{converges} \qquad (14.22)$$

Hence our region of convergence is all $\sigma > 0$.

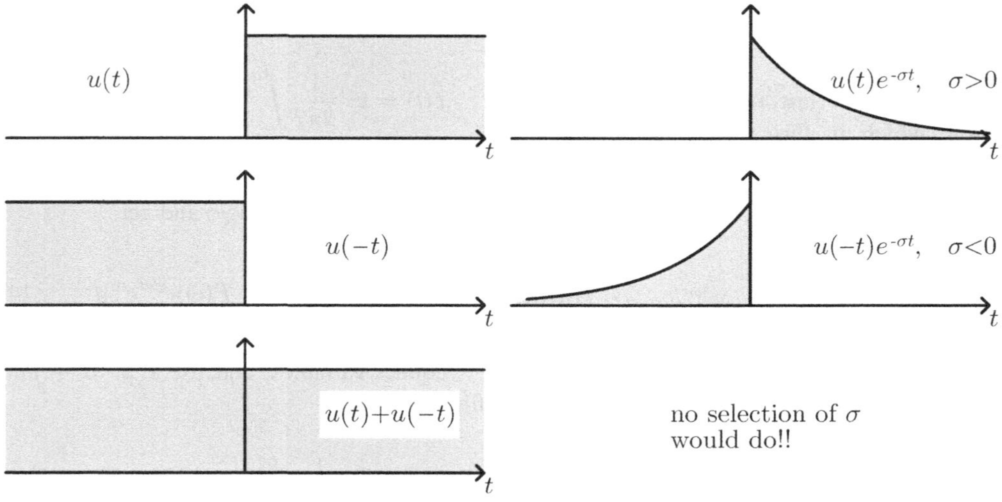

Fig. 14.3 Demonstration of damping factor σ selection

14.6.2 Example Where σ Is Negative

The flipped unit step function (Fig. 14.3) $u(-t)$ is also none integrable

$$\int_{-\infty}^{0} u(t)dt \qquad \text{blows up} \tag{14.23}$$

However, multiplying by $e^{-\sigma t}$ for $\sigma < 0$ makes the integral convergent.

$$\int_{-\infty}^{0} u(t)e^{-\sigma t}dt \qquad \text{converges for} \qquad \sigma < 0 \tag{14.24}$$

Hence our region of convergence is all $\sigma < 0$.

14.6.3 Example with No Applicable σ

The DC function (Fig. 14.3) is none integrable

$$\int_{-\infty}^{\infty} 1dt \qquad \text{blows up} \tag{14.25}$$

In this exceptional case, there is no selection of σ that would make this function integrable over all time. If we choose σ positive, the positive time integration converges, but the negative one does not; and vice versa. Hence, this function has no Laplace transform! Or the ROC is nonexistent! With this out of the way, let us focus back on signals defined between zero and positive time and test the Laplace transform on some important signals.

14.7 Laplace Transform of Delta Function

To find the Laplace transform of the delta function, let

$$f(t) = \delta(t) \tag{14.26}$$

and use the Laplace transform formula

$$F(s) = \int_{0}^{\infty} f(t)e^{-st}dt \tag{14.27}$$

Then we get

$$F(s) = \int_{0}^{\infty} \delta(t)e^{-st}dt = e^{-st}\Big|_{t=0} = \boxed{1} \tag{14.28}$$

That is the LT of the delta function is simply 1! This is similar to the Fourier transform of the delta function. Since the delta function is so abrupt in time, it samples all frequencies (equally).

14.8 Laplace Transform of Unit Step Function

The unit step function is defined as

$$f(t) = u(t) = \begin{cases} 0 & t < 0 \\ 1 & t \geq 0 \end{cases} \tag{14.29}$$

The LT is then

$$\begin{aligned} F(s) &= \int_{0}^{\infty} u(t)e^{-st}dt = \int_{0}^{\infty} e^{-st}dt \\ &= -\frac{1}{s}e^{-st}\Big|_{0}^{\infty} = \boxed{\frac{1}{s}} \end{aligned} \tag{14.30}$$

provided that the real part of s is larger than zero; that is

$$\Re(s) = \sigma > 0 \tag{14.31}$$

This condition is needed to force the following convergence

$$\lim_{t\to\infty} e^{-st} = \lim_{t\to\infty} e^{(-\sigma-j\omega)t} = \lim_{t\to\infty} e^{-\sigma t} \lim_{t\to\infty} e^{-j\omega t} = 0 \times \lim_{t\to\infty} e^{-j\omega t} = 0 \tag{14.32}$$

That is we rely on the $e^{-\sigma\infty}$ to go to zero, and that would cause the integration of the function to become bound.

14.9 Plotting the Laplace Transform

To put things in perspective, the Fourier transform had both a real part and an imaginary part as the dependent variables, and ω as the independent variable. So typically we ended up with two graphs, each with ω on the x-axis, and $\Re$ and $\Im$ on the y-axis. In the case of the Laplace transform, the two dependent variables (real and imaginary) remain the same; but the single independent variable (used to be ω) now becomes *two* independent variables—σ and $j\omega$. So, when plotting the real and imaginary parts of the Laplace transform, we would end up with 3D plots; on the x axis we have σ, on the y-axis we have $j\omega$, and finally on the z-axis we would have the Laplace transform (either real part or imaginary one). A sample 3D plot of the Laplace transform (for the unit step function) is shown in Fig. 14.4. The plot was constructed as follows:

First start with

$$F(s) = \frac{1}{s} \tag{14.33}$$

Next plug in for s

$$F(s) = \frac{1}{\sigma + j\omega} \tag{14.34}$$

Now derive the real and imaginary parts

$$F(s) = \frac{\sigma}{\sigma^2 + \omega^2} - j\frac{\omega}{\sigma^2 + \omega^2} \tag{14.35}$$

The first term on the right is the real part, while the second term is the imaginary part. Each depends on σ and ω, which would form the two planar axis. It is extremely important to familiarize oneself with these sorts of plot. It is paramount to discern between σ, ω, $j\omega$, s, the real part, and the imaginary part of the Laplace transform. Between these five variables lies the key to reconstruct the time signal. These are not some fictional quantities! If used in tandem they will recreate the time signal which is all that matters.

14.10 Laplace Transform of Ramp Function $u(t)t$

The ramp function defined zero for negative time and t for positive time can be written as

$$f(t) = u(t)t \tag{14.36}$$

The Laplace transform is derived as follows:

$$F(s) = \int_0^\infty f(t)e^{-st}dt = \int_0^\infty te^{-st}dt \tag{14.37}$$

$$= -t\frac{e^{-st}}{s}\bigg|_{t=0}^{t=\infty} + \frac{1}{s}\int_0^\infty e^{-st}dt = \frac{1}{s^2}$$

Notice that we have used the fact that

$$\lim_{t\to 0} te^{-st} = 0, \quad \text{and} \tag{14.38}$$

$$\lim_{t\to\infty} te^{-st} = 0 \tag{14.39}$$

In summary we then have

$$\boxed{t \to \frac{1}{s^2}} \tag{14.40}$$

In terms of real and imaginary components we have

$$\begin{aligned} F(s) &= \frac{1}{(\sigma + j\omega)^2} = \frac{1}{[\sigma^2 - \omega^2] + 2j\sigma\omega} \\ &= \frac{[\sigma^2 - \omega^2] - j[2\sigma\omega]}{[\sigma^2 - \omega^2]^2 + 4\sigma^2\omega^2} \end{aligned} \tag{14.41}$$

We can expand the denominator and collect terms to get

$$F(s) = \frac{[\sigma^2 - \omega^2] - j[2\sigma\omega]}{[\sigma^2 + \omega^2]^2} \tag{14.42}$$

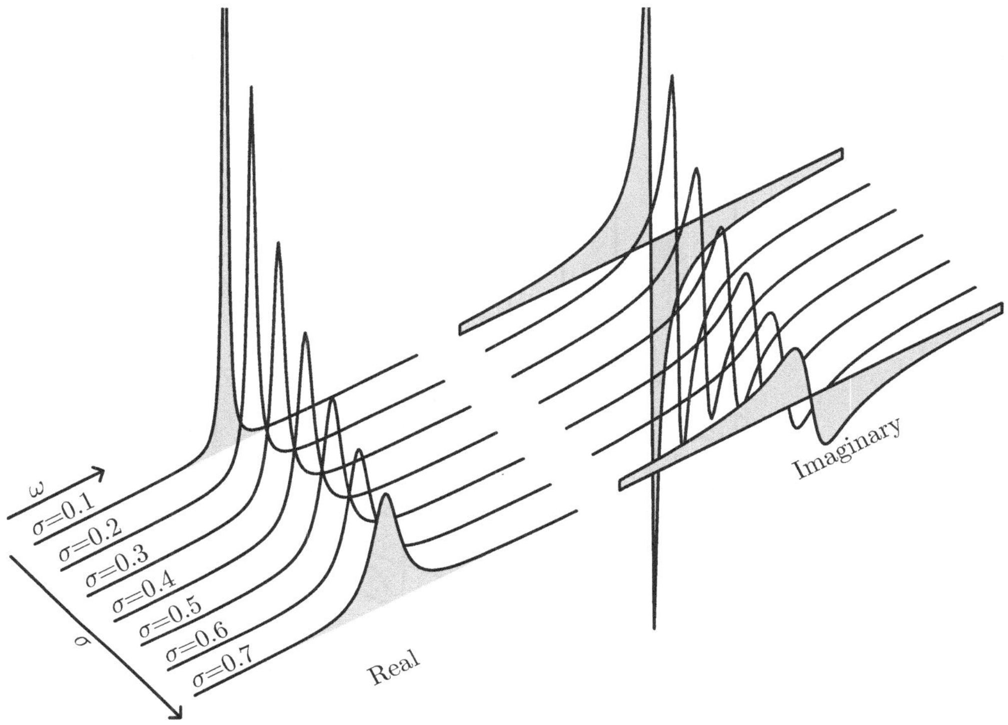

Fig. 14.4 Laplace transform (real and imaginary) versus σ and ω

14.11 Validating the Laplace Transform

Once we have a Laplace transform, how do we know it is right? Let us take the prior example, which is the ramp function, and its Laplace transform

$$t \to \frac{1}{s^2} \tag{14.43}$$

As shown in the prior example, we split this in terms of real and imaginary

$$F(s) = \frac{[\sigma^2 - \omega^2] - j[2\sigma\omega]}{[\sigma^2 + \omega^2]^2} \tag{14.44}$$

Since the real is even (in frequencies) and the imaginary is odd, the inverse Laplace transform comes out

$$f(t) = \frac{e^{\sigma t}}{2\pi}\left[\int_{-\infty}^{\infty} \frac{\sigma^2 - \omega^2}{(\sigma^2 + \omega^2)^2} \cos \omega t d\omega \right.$$

$$\left. + \int_{-\infty}^{\infty} \frac{2\sigma\omega}{(\sigma^2 + \omega^2)^2} \sin \omega t d\omega \right] \tag{14.45}$$

Notice that the integration is carried over only ω and notice that we multiply the frequency integration by $e^{\sigma t}$. Figure 14.5 shows that in fact as we add more harmonics, we recover the ramp function. These particular results have been carried out for $\sigma = 0.1$.

14.12 Choice of σ

In theory, any choice of σ, so long it lies in the ROC (region of convergence) would do. For example, for both the unit step input, and ramp function, any sigma such that

$$\sigma > 0 \tag{14.46}$$

would work. That is, multiplying either of $u(t)$ or $tu(t)$ by $e^{-\sigma t}$ for $\sigma > 0$ would ensure the product dies off for large t, and hence it being integrable.

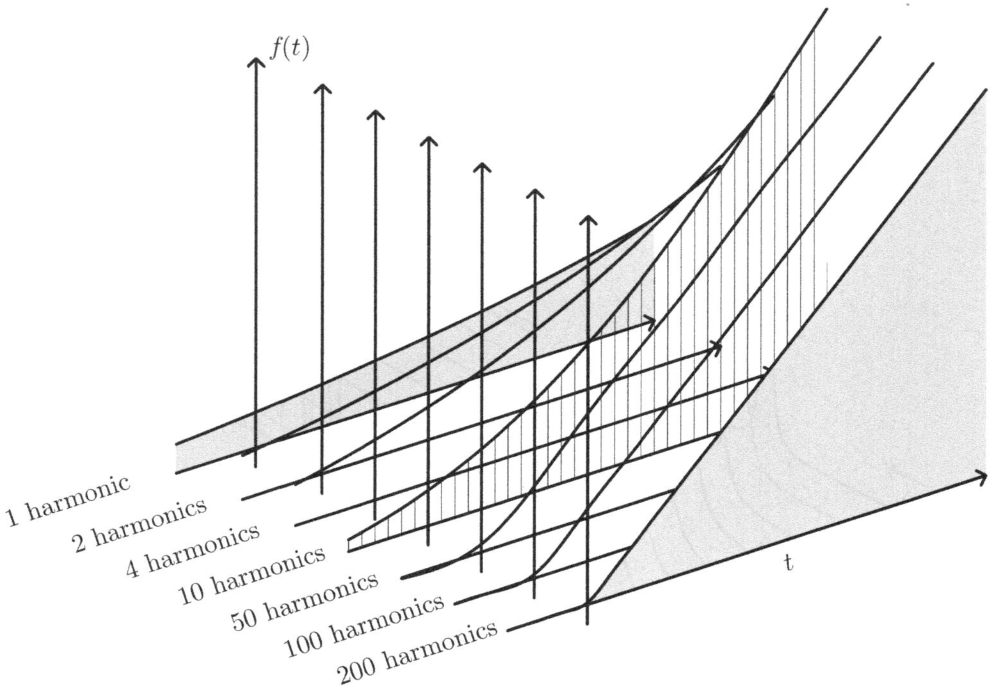

Fig. 14.5 Inverse Laplace transform of $u(t)t$ as a function of number of harmonics (case of $\sigma = 0.1$)

However, in practice, we do observe some bias to some choices. Sticking with the ramp test case, let's do the following sequence of steps:

- Assume some σ (subject to $\sigma > 0$).
- Do frequency integration to figure inverse Laplace transform (build the signal in the time domain).
- Limit the frequency integration to a certain number of harmonic count (for example, include only the first 100 harmonics).
- Repeat for a different σ.

The end result of this is shown in Fig. 14.6. As can be seen from the figure, the best choice (again subject to including a finite set of harmonics) appears to be the one with the smallest σ. This does not mean that the larger σ cases don't work; it only means that for those larger σ cases, we would need to include a larger set of harmonics. This is confirmed in Fig. 14.7. Stated differently, setting σ large tends to amplify any errors at the step when $e^{\sigma t}$ is multiplied by the inverse Fourier (not Laplace) transform. Recall

$$f(t) = e^{\sigma t} \times \text{inverse Fourier transform} \quad (14.47)$$

So, for a given set of harmonics, and having figured the inverse Fourier transform, if that ends up with some minor oscillations, those oscillations would now get amplified by $e^{\sigma t}$ after the multiplication! So, either set the σ smaller, or ensure the error is small to start with, and this typically happens by using a larger set of harmonics.

14.13 Laplace Transform of Negative Exponential

The LT of the negative exponential

$$f(t) = u(t)e^{-at} \quad (14.48)$$

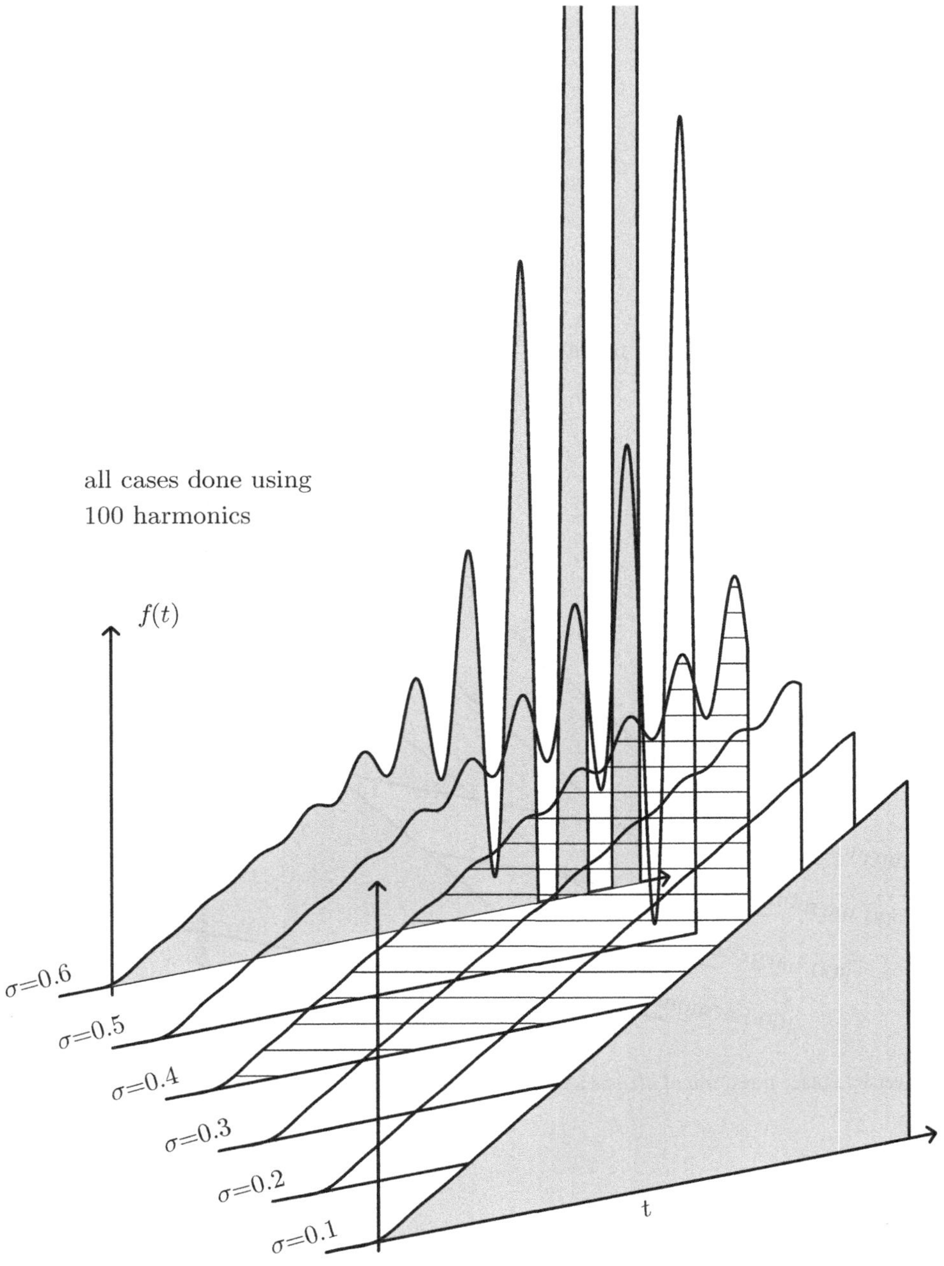

Fig. 14.6 Inverse Laplace transform of $u(t)t$ as a function of σ (case of 100 harmonics)

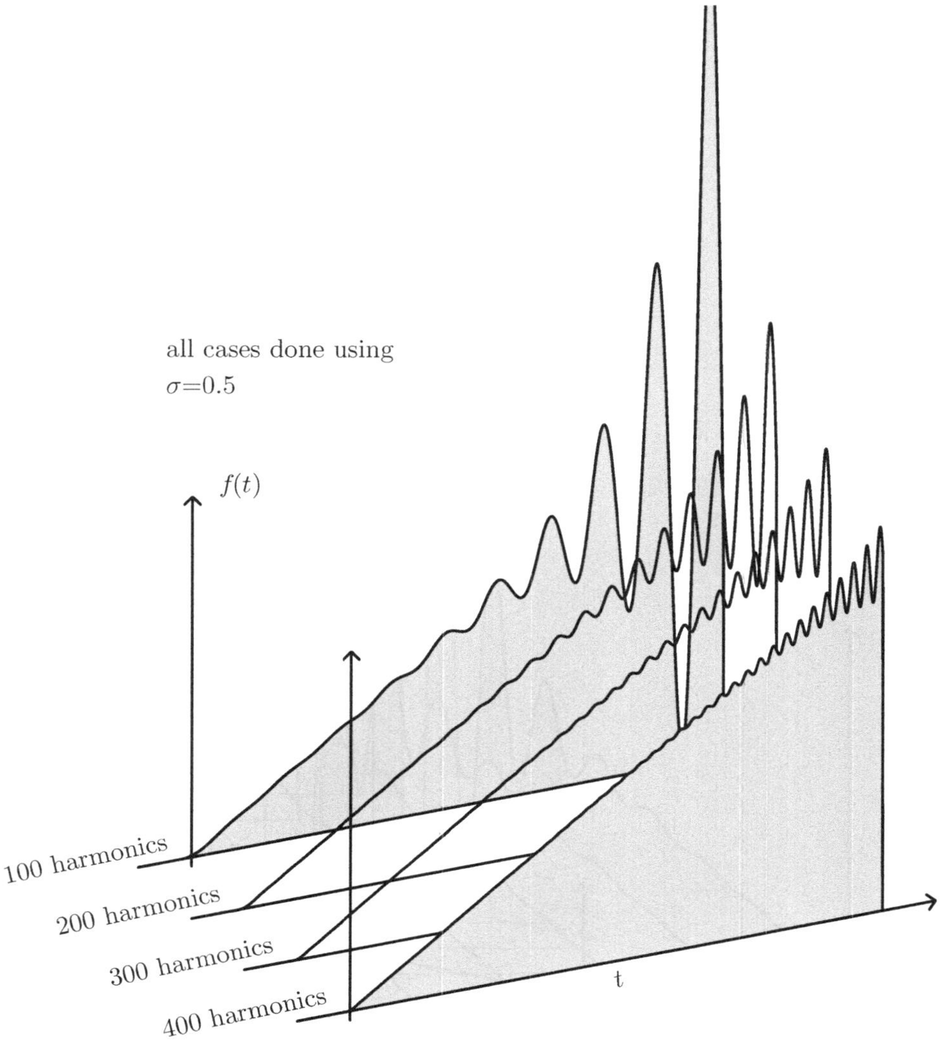

Fig. 14.7 Inverse Laplace transform of $u(t)t$ as a function of number of number of harmonics (case of $\sigma = 0.5$)

is derived as

$$F(s) = \int_0^\infty e^{-at}e^{-st}dt = \int_0^\infty e^{-t(s+a)}dt = -\frac{1}{s+a}e^{-t(a+a)}\bigg|_0^\infty = \boxed{\frac{1}{s+a}} \qquad (14.49)$$

The region of convergence is such that

$$\sigma > -a \tag{14.50}$$

For example, if $a = 1$ such that $f(t) = e^{-t}$ then σ can be anything real such that $\sigma > -1$. For example, if we set $\sigma = -0.25$ (which satisfies the requirement that $\sigma > -a$) then

$$f(t)e^{-\sigma t} = e^{-t}e^{0.25t} = e^{-0.75t} \tag{14.51}$$

which still is integrable (i.e., dies off at large time). But if we choose $\sigma = -1.5$ (which does NOT satisfy the requirement $\sigma < -a$) then

$$f(t)e^{-\sigma t} = e^{-t}e^{1.5t} = e^{0.5t} \tag{14.52}$$

which is *no* longer integrable! Back to the transfer function

$$F(s) = \frac{1}{s+a} \tag{14.53}$$

Notice that for the special case of $a = 0$ we regain the LT of the unit step function

$$\lim_{a\to 0} \frac{1}{s+a} = \frac{1}{s} \tag{14.54}$$

which makes sense noting that

$$\lim_{a\to 0} e^{-at} = u(t) \tag{14.55}$$

In terms of real and imaginary we get

$$\frac{1}{s+a} = \frac{1}{(a+\sigma)+j\omega} = \frac{(a+\sigma)-j\omega}{(a+\sigma)^2+\omega^2} \tag{14.56}$$

The Laplace transform is shown in Fig. 14.8. Notice that at zero the real part converges to

$$\lim_{\omega\to 0} \frac{a+\sigma}{(a+\sigma)^2+\omega^2} = \frac{a+\sigma}{(a+\sigma)^2} = \frac{1}{a+\sigma} \tag{14.57}$$

while the imaginary part goes to zero

$$\lim_{\omega\to 0} \frac{\omega}{(a+\sigma)^2+\omega^2} = \lim_{\omega\to 0} \frac{\omega}{(a+\sigma)^2} = 0 \tag{14.58}$$

Let's see if we do the time series (the inverse LT transform) what we get, for a given σ and as a function of number of harmonics; this is shown in Fig. 14.9.

14.14 Laplace Transform of te^{-at}

The LT of te^{-at} is

$$F(s) = \int_0^\infty te^{-at}e^{-st}dt = \int_0^\infty te^{-t(a+s)}dt \tag{14.59}$$

Using integration by parts we get

$$\boxed{te^{-at} \to \frac{1}{(s+a)^2}} \tag{14.60}$$

This is to be compared to

$$e^{-at} \to \frac{1}{s+a} \tag{14.61}$$

In terms of real and imaginary components we have

$$\begin{aligned}\frac{1}{(s+a)^2} &= \frac{1}{[(\sigma+a)+j\omega]^2} = \frac{1}{[(\sigma+a)^2-\omega^2]+j[2\omega(\sigma+a)]}\\ &= \frac{[(\sigma+a)^2-\omega^2]-j[2\omega(\sigma+a)]}{[(\sigma+a)^2-\omega^2]^2+[2\omega(\sigma+a)]^2}\end{aligned} \tag{14.62}$$

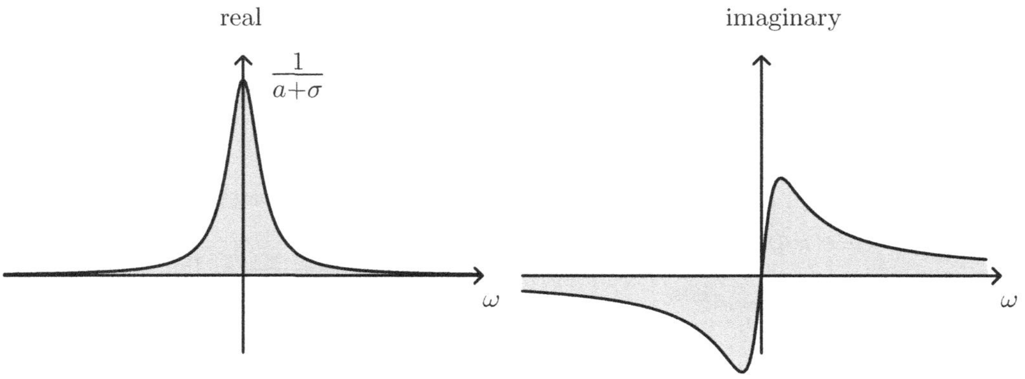

Fig. 14.8 Laplace transform of $u(t)e^{-t}$ (case of $\sigma = -0.2$)

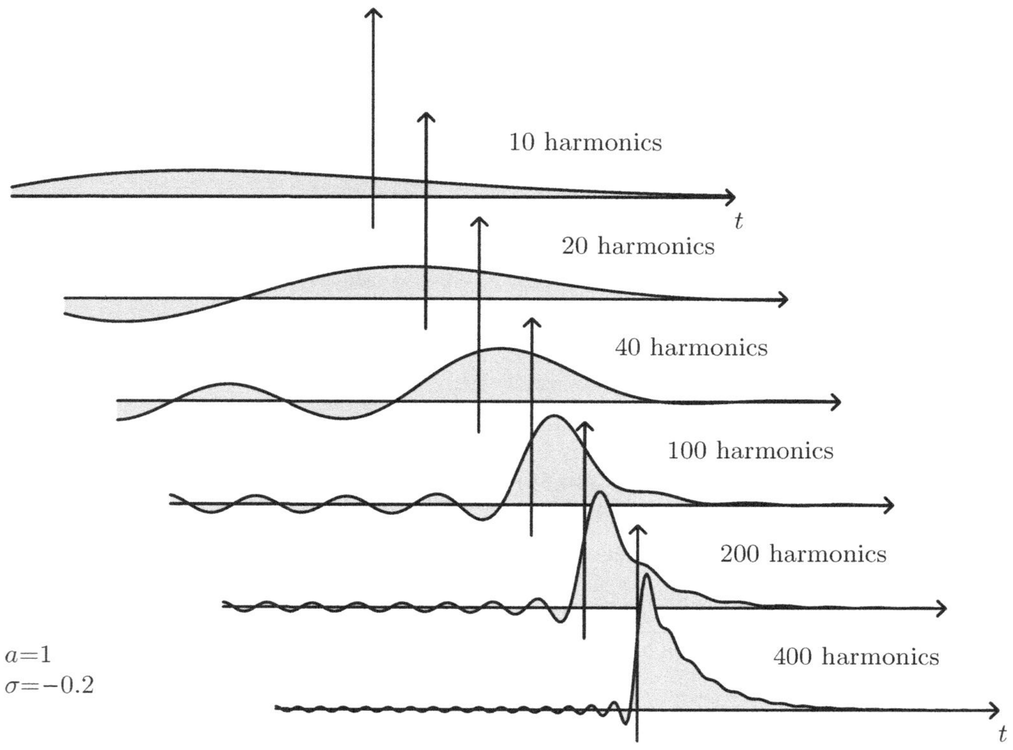

Fig. 14.9 Inverse Laplace transform of $u(t)e^{-t}$ as a function of number of number of harmonics (case of $\sigma = -0.2$)

A plot of this is shown in Fig. 14.10. Notice the definite trend of more smearing with larger σ. Put another way, with smaller sigma, the function in time extends relatively deep in time (since it does not die off as fast); and with time prolongation comes large DC spectrum; hence the spectrum would tend to have a pronounced peak at DC, and die off at high frequency. But with large sigma, the DC spectrum is substantially reduced; hence the spectrum "smears" out! Let's test our analysis by doing an inverse LT; this is shown in Fig. 14.11. Sure enough, the more harmonics we include the better our reconstruction. Notice this process is exactly the same as we have seen so many times when doing inverse Fourier transform. Let's not let the σ blur us from recognizing that the Laplace transform is really the Fourier transform, in disguise!

14.15 Laplace Transform of Single-Sided exp $(-j\omega_0 t)$

Let's find the Laplace transform of the negative complex exponential

$$f(t) = e^{-j\omega_0 t} \tag{14.63}$$

We proceed with the integration

$$\begin{aligned} F(s) &= \int_0^\infty e^{-j\omega_0 t} e^{-st} dt = \int_0^\infty e^{-t(s+j\omega_0)} dt \\ &= -\frac{1}{s+j\omega_0} e^{-t(s+j\omega_0)} \Big|_0^\infty = \boxed{\frac{1}{s+j\omega_0}} \end{aligned} \tag{14.64}$$

provided

$$\Re(s) = \sigma > 0 \tag{14.65}$$

Notice we could have arrived at the same answer by simple setting $a = j\omega_0$ in Eq. (14.49). Similarly

$$\mathcal{L}\left[e^{j\omega_0 t}\right] = \frac{1}{s - j\omega_0} \tag{14.66}$$

Notice again that for the special case $\omega_0 = 0$ we regain the LT of the unit step

$$\lim_{\omega_0 \to 0} \frac{1}{s + j\omega_0} = \frac{1}{s} \tag{14.67}$$

In terms of real and imaginary we have

$$\frac{1}{s+j\omega_0} = \frac{1}{\sigma + j(\omega + \omega_0)} = \frac{\sigma - j(\omega+\omega_0)}{\sigma^2 + (\omega+\omega_0)^2} \tag{14.68}$$

These results are shown in Fig. 14.12. Notice that the real part is neither even nor odd; same thing for the imaginary part. This is the case because the time signal is complex to start with. Notice that the real part has a max at $\omega = -\omega_0$ since the denominator $\sim \frac{1}{\omega+\omega_0}$. The imaginary part also has an event happening at $\omega = \omega_0$. Finally notice that in the limit as $\sigma \to 0$ the real part approaches a delta function $\delta(\omega + \omega_0)$ while the imaginary part approaches $1/(\omega + \omega_0)$.

As to the impact of σ notice from the figure the same pronounced trend of more smearing with larger σ. With zero σ, when a harmonic matches the time signal the resulting integral almost blows up, and that is why we see the large peak around ω_0. But with nonzero σ we are assured that the diluted signal will *not* identically match any harmonic, and hence we avoid the peak, and instead get the smearing. Let's next do the inverse LT and verify that the real part comes out a cosine and that the imaginary one comes out a sine; as shown in Figs. 14.13 and 14.14, this in fact is the case!

It is critical to always remember—at least in the world of the nominal Laplace transform—that while the starting signal is "causal," meaning it is nonzero only for *positive* time, the harmonics used to rebuild the signal (at the inverse Laplace transform step) are nonzero for *all* time! We can see that is the case in both Figs. 14.13 and 14.14. Notice there that for negative time the signal is not exactly zero—in this case because we have not used high enough harmonic count. Point is, those harmonics are defined for all time, and only through the right proportioning that their *sum* for negative times comes out zero!

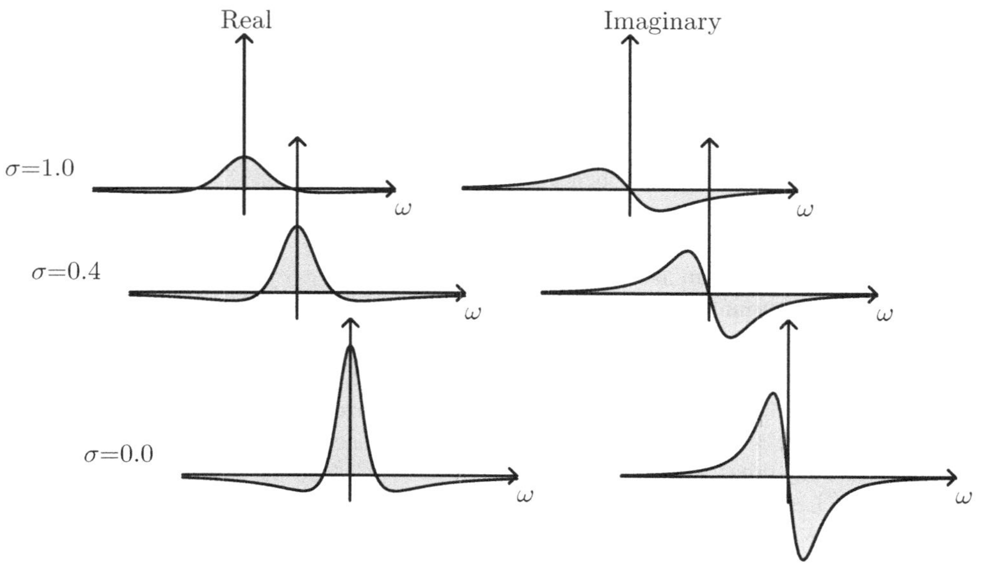

Fig. 14.10 Laplace transform of te^{-at}

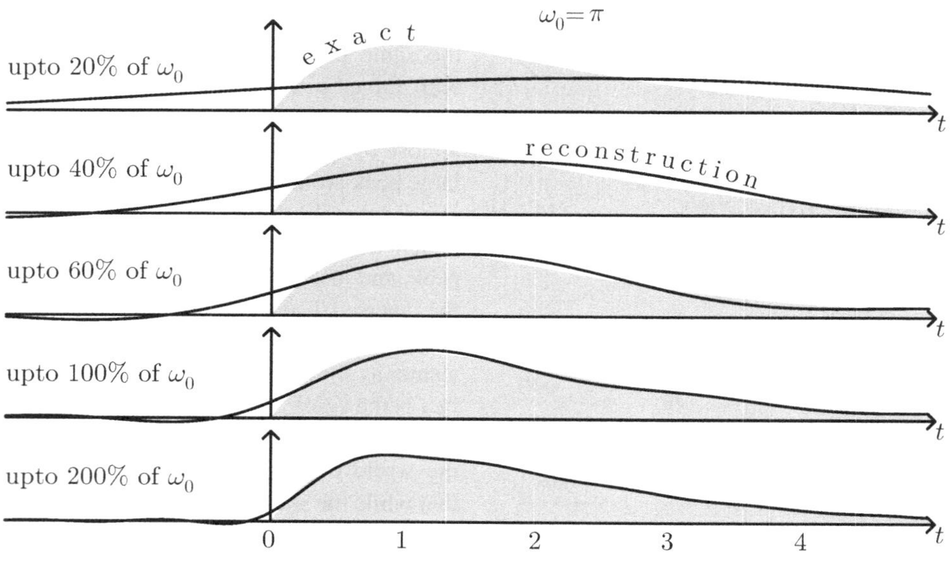

Fig. 14.11 Inverse LT of te^{-at} (case of $\sigma = 0.1$)

14.16 Laplace Transform of the Cosine Function

To find the LT of the cosine function first recall that

$$e^{j\omega_0 t} \to \frac{1}{s - j\omega_0}, \quad \text{and} \quad e^{-j\omega_0 t} \to \frac{1}{s + j\omega_0} \tag{14.69}$$

Next use definition of cosine

$$\cos \omega_0 t = \frac{1}{2}\left[e^{j\omega_0 t} + e^{-j\omega_0 t}\right] \tag{14.70}$$

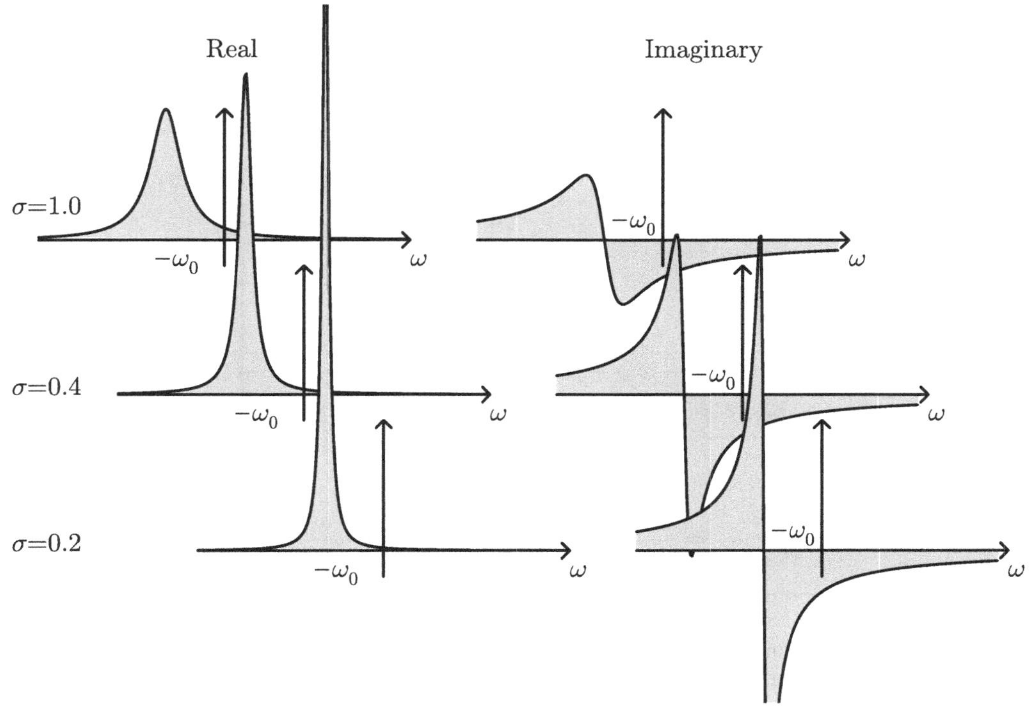

Fig. 14.12 Laplace transform of the negative complex exponential, with different σ values

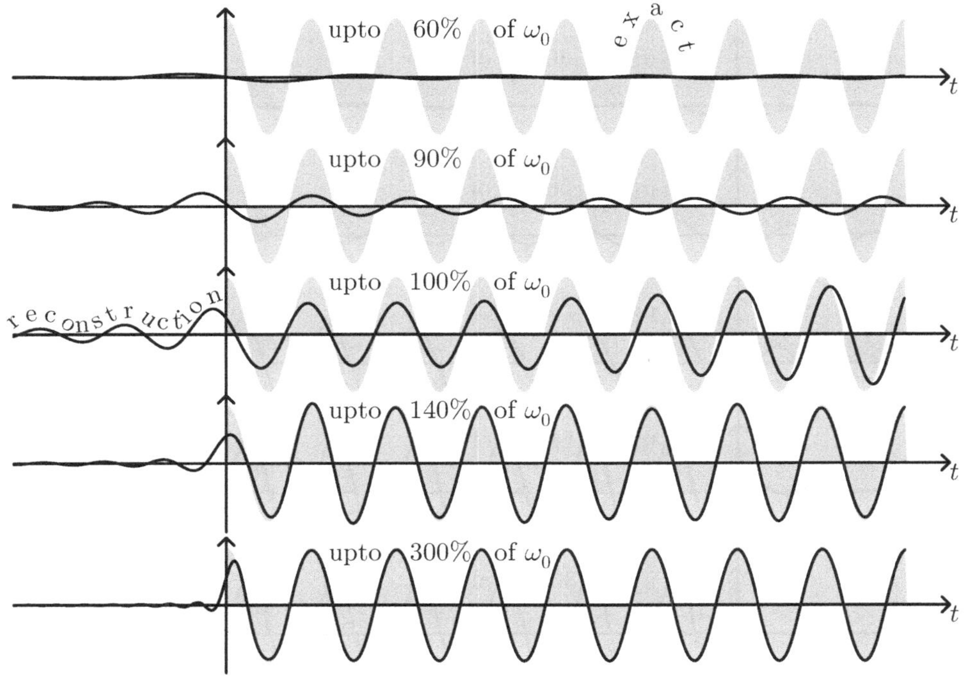

Fig. 14.13 Inverse LT of Eq. (14.68) giving cosine function

Finally using linearity we get

$$\mathcal{L}(\cos\omega_0 t) = \frac{1}{2}\left[\frac{1}{s-j\omega_0} + \frac{1}{s+j\omega_0}\right]$$
$$= \frac{1}{2}\frac{s+j\omega_0+s-j\omega_0}{s^2+\omega_0^2}$$
$$= \boxed{\frac{s}{s^2+\omega_0^2}} \qquad (14.71)$$

In terms of real and imaginary components we get

$$\frac{s}{s^2+\omega_0^2} = \frac{\sigma+j\omega}{(\sigma^2+\omega_0^2-\omega^2)+j2\sigma\omega} = \frac{[\sigma+j\omega][(\sigma^2+\omega_0^2-\omega^2)-j2\sigma\omega]}{(\sigma^2+\omega_0^2-\omega^2)^2+(2\sigma\omega)^2}$$
$$= \frac{[\sigma(\sigma^2+\omega_0^2-\omega^2)+2\sigma\omega^2]+j[\omega(\sigma^2+\omega_0^2-\omega^2)-2\sigma^2\omega]}{(\sigma^2+\omega_0^2-\omega^2)^2+(2\sigma\omega)^2} \qquad (14.72)$$

$$\frac{s}{s^2+\omega_0^2} = \frac{\sigma(\sigma^2+\omega_0^2+\omega^2)+j\omega(-\sigma^2+\omega_0^2-\omega^2)}{(\sigma^2+\omega_0^2-\omega^2)^2+(2\sigma\omega)^2} \qquad (14.73)$$

A plot of this is shown in Fig. 14.15. Notice that in contrast to the complex exponential, this spectrum has two peaks. Of course this makes sense since a cosine combined a negative and positive complex exponential. Let's next test our analysis by doing an inverse LT; this is shown in Fig. 14.16.

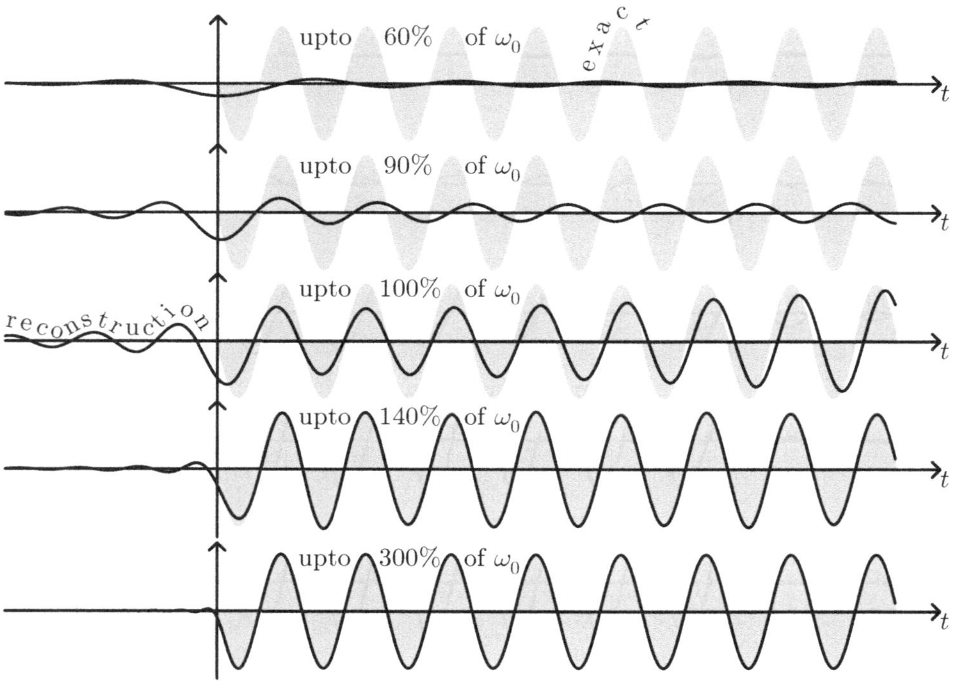

Fig. 14.14 Inverse LT of Eq. (14.68) giving (negative) sine function

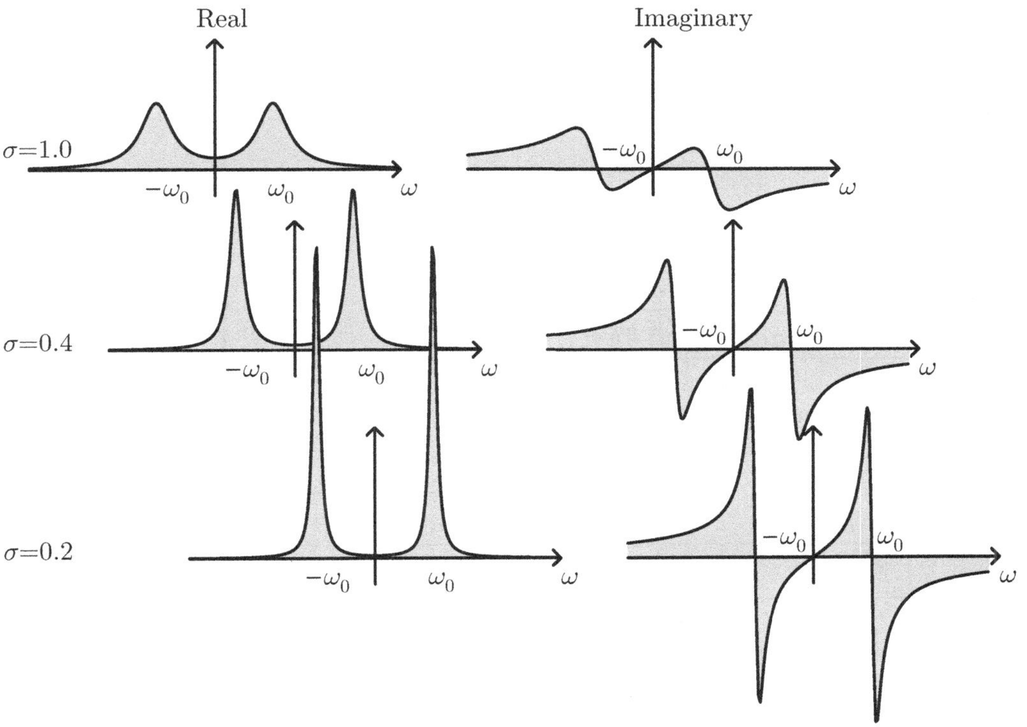

Fig. 14.15 Laplace transform of single-sided cosine function

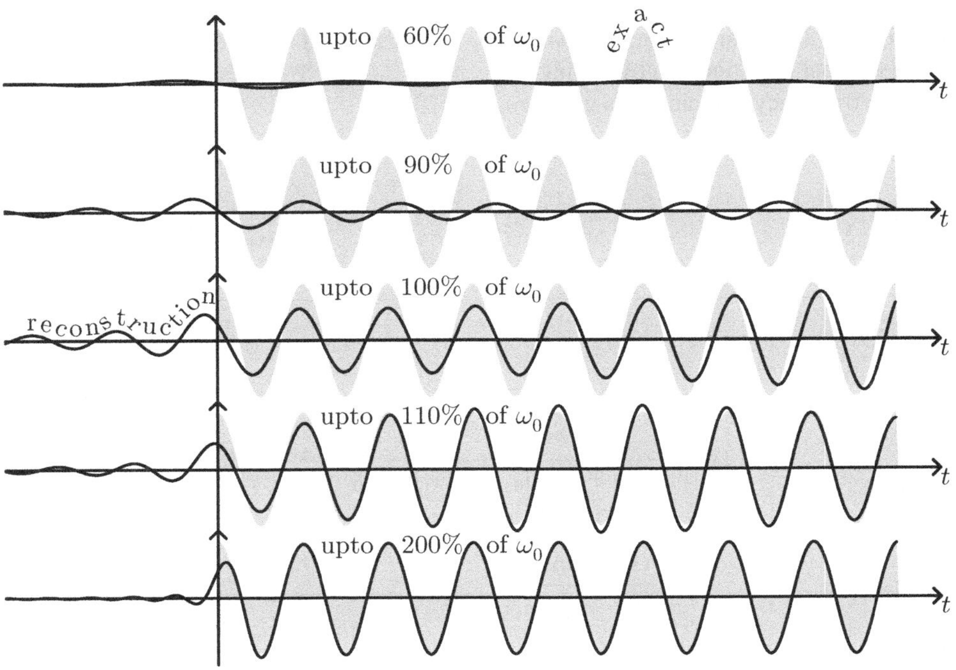

Fig. 14.16 Inverse LT of single-sided cosine function (case of $\sigma = 0.1$)

14.17 Laplace Transform of the Sine Function

Similar to the cosine case, we get for the sine LT

$$\mathcal{L}(\sin\omega_0 t) = \frac{1}{2j}\left[\frac{1}{s-j\omega_0} - \frac{1}{s+j\omega_0}\right] = \frac{1}{2j}\frac{s+j\omega_0-(s-j\omega_0)}{s^2+\omega_0^2}$$
$$= \boxed{\frac{\omega_0}{s^2+\omega_0^2}} \tag{14.74}$$

In terms of real and imaginary components we get

$$\frac{\omega_0}{s^2+\omega_0^2} = \frac{\omega_0}{(\sigma^2+\omega_0^2-\omega^2)+j2\sigma\omega} = \frac{\omega_0(\sigma^2+\omega_0^2-\omega^2)-j2\sigma\omega\omega_0}{(\sigma^2+\omega_0^2-\omega^2)^2+(2\sigma\omega)^2} \tag{14.75}$$

A plot of this is shown in Fig. 14.17. Notice again how the σ smoothes the singularities. Notice also that—as always—the real part of the Laplace transform is even while the imaginary part is odd. Let's next test our derivation by looking at the time series (inverse LT); this is shown in Fig. 14.18. As can be seen, we do in fact recover the original time function by adding harmonics in the frequency domain (for a given σ).

14.18 Laplace Transform of *t* exp $(-j\omega_0 t)$

We proceed with the integration

$$\mathcal{L}\left[te^{-j\omega_0 t}\right] = \int_0^\infty te^{-j\omega_0 t}e^{-st}t = \int_0^\infty te^{-t(s+j\omega_0)}t \tag{14.76}$$

Let

$$u = t, du = 1; \qquad dv = e^{-t(s+j\omega_0)}, v = -\frac{e^{-t(s+j\omega_0)}}{s+j\omega_0} \tag{14.77}$$

Then

$$F(s) = -t\frac{e^{-t(s+j\omega_0)}}{s+j\omega_0}\bigg|_0^\infty + \frac{1}{s+j\omega_0}\int_0^\infty e^{-t(s+j\omega_0)}dt$$
$$= \boxed{\frac{1}{(s+j\omega_0)^2}} \tag{14.78}$$

That is,

$$\text{if} \quad e^{-j\omega_0 t} \to \frac{1}{s+j\omega_0} \quad \text{then} \quad te^{-j\omega_0 t} \to \frac{1}{(s+j\omega_0)^2} \tag{14.79}$$

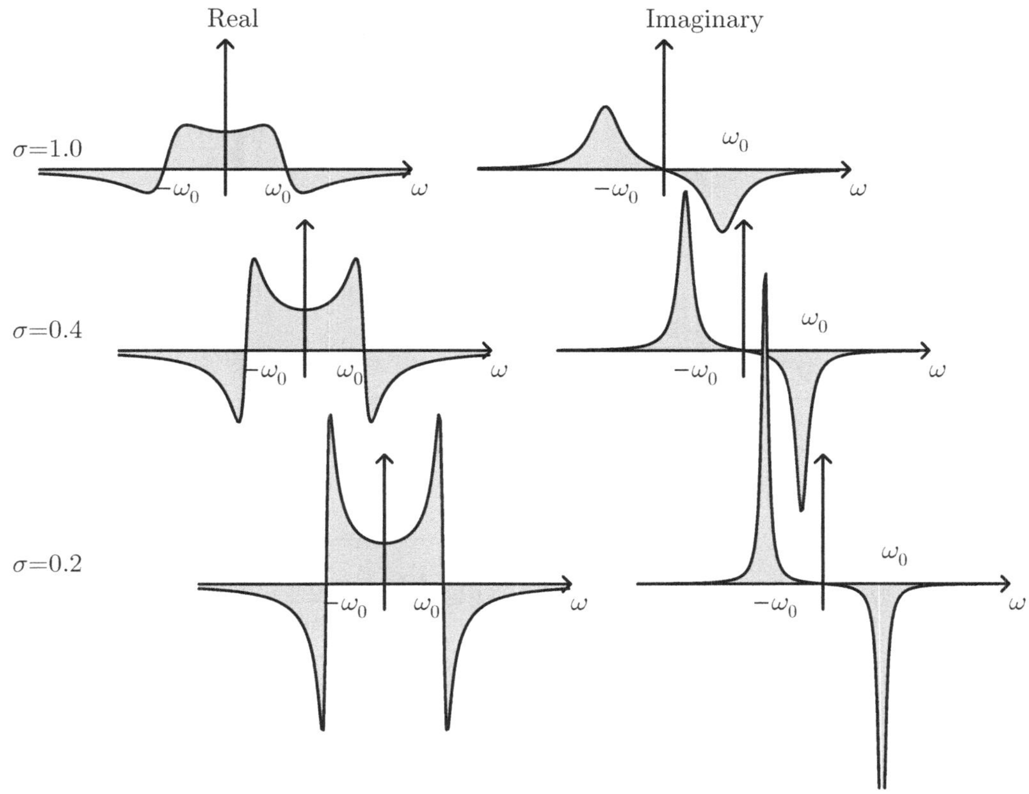

Fig. 14.17 Laplace transform of single-sided sine function

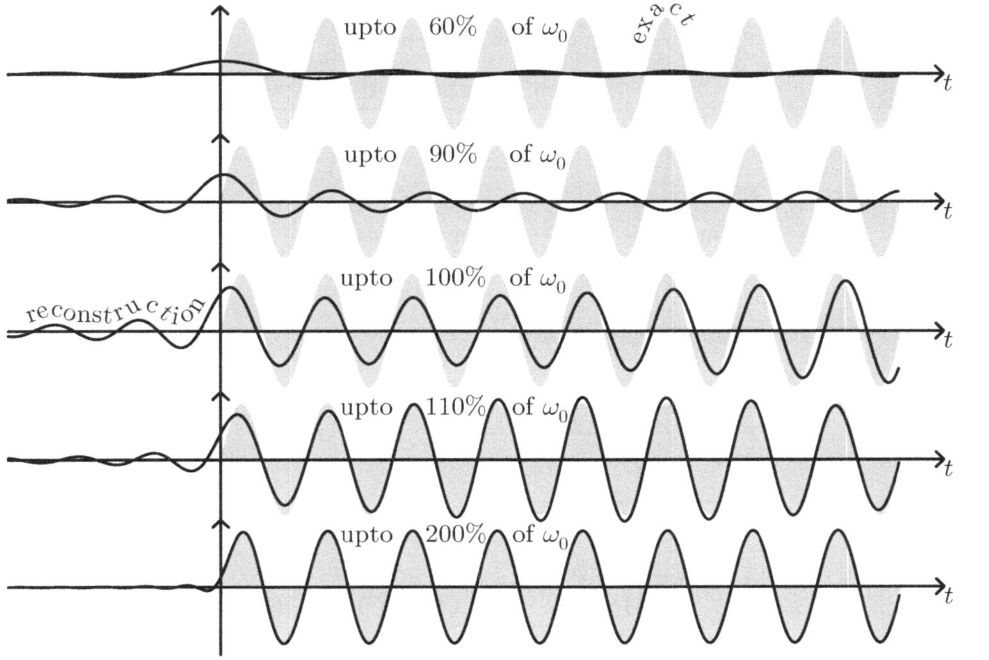

Fig. 14.18 Inverse LT of single-sided sine function (case of $\sigma = 0.1$)

Similarly

$$\mathcal{L}\left[te^{j\omega_0 t}\right] = \frac{1}{(s - j\omega_0)^2} \tag{14.80}$$

14.19 Laplace Transform of *t* sin $\omega_0 t$

Let's find the Laplace transform of the function

$$f(t) = t \sin \omega_0 t \tag{14.81}$$

Rather than doing the time integration we use the following three facts:

$$\sin \omega_0 t = \frac{e^{j\omega_0 t} - e^{-j\omega_0 t}}{2j}, \; te^{j\omega_0 t} \to \frac{1}{(s - j\omega_0)^2}, \quad \text{and} \quad te^{-j\omega_0 t} \to \frac{1}{(s + j\omega_0)^2} \tag{14.82}$$

Then

$$\begin{aligned} F(s) &= \frac{1}{2j}\left[\frac{1}{(s - j\omega_0)^2} - \frac{1}{(s + j\omega_0)^2}\right] = \frac{1}{2j}\frac{(s + j\omega_0)^2 - (s - j\omega_0)^2}{(s - j\omega_0)^2(s + j\omega_0)^2} \\ &= \frac{1}{2j}\frac{\left[s^2 + 2js\omega_0 - \omega_0^2\right] - \left[s^2 - 2js\omega_0 - \omega_0^2\right]}{(s^2 + \omega_0^2)^2} = \boxed{\frac{2s\omega_0}{(s^2 + \omega_0^2)^2}} \end{aligned} \tag{14.83}$$

where we have used

$$\begin{aligned} (s - j\omega_0)^2(s + j\omega_0)^2 &= (s^2 - 2js\omega_0 - \omega_0^2)(s^2 + 2js\omega_0 - \omega_0^2) \\ &= s^4 - 2s^2\omega_0^2 + 4s^2\omega_0^2 + \omega_0^4 \\ &= s^4 + 2s^2\omega_0^2 + \omega_0^4 = (s^2 + \omega_0^2)^2 \end{aligned} \tag{14.84}$$

That is,

$$\text{if} \quad \sin \omega_0 t \to \frac{\omega_0}{s^2 + \omega_0^2} \quad \text{then} \quad t \sin \omega_0 t \to \frac{2s\omega_0}{(s^2 + \omega_0^2)^2} \tag{14.85}$$

In terms of real and imaginary parts we get

$$\frac{2s\omega_0}{[s^2 + \omega_0^2]^2} = 2\omega_0 \frac{\sigma + j\omega}{[(\sigma^2 + \omega_0^2 - \omega^2) + j(2\sigma\omega)]^2} \tag{14.86}$$

Let

$$A = (\sigma^2 + \omega_0^2 - \omega^2), \quad \text{and} \quad B = 2\sigma\omega, \quad \text{then} \tag{14.87}$$

$$\begin{aligned} \frac{2s\omega_0}{[s^2 + \omega_0^2]^2} &= 2\omega_0 \frac{\sigma + j\omega}{[A + jB]^2} = 2\omega_0 \frac{\sigma + j\omega}{(A^2 - B^2) + j2AB} \\ &= 2\omega_0 \frac{[\sigma(A^2 - B^2) + \omega 2AB] + j[-\sigma 2AB + \omega(A^2 - B^2)]}{(A^2 - B^2)^2 + (2AB)^2} \end{aligned} \tag{14.88}$$

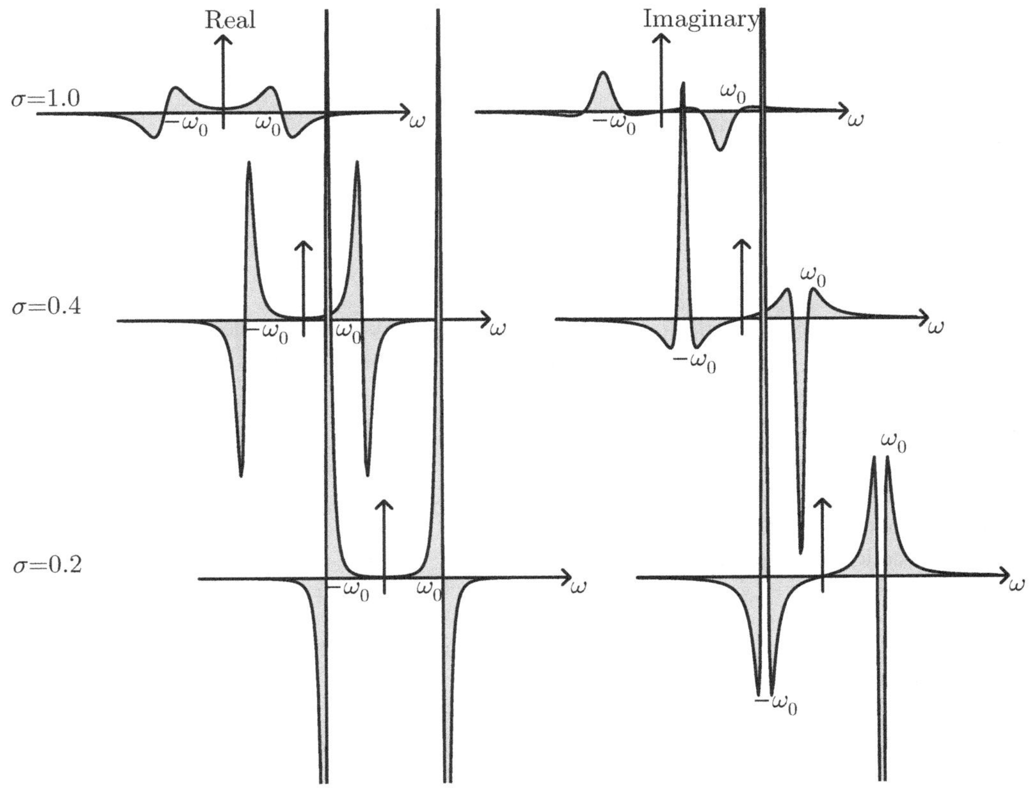

Fig. 14.19 Laplace transform of $t \sin \omega_0 t$

These results are shown in Fig. 14.19. We can verify that the LT is correct by doing inverse LT; this is shown in Fig. 14.20

14.20 Laplace Transform of t^2

We know so far that

$$u(t) \to \frac{1}{s}, \quad \text{and} \tag{14.89}$$

$$tu(t) \to \frac{1}{s^2} \tag{14.90}$$

By direct integration we get

$$F(s) = \int_0^\infty t^2 e^{-st} dt = \frac{t^2 e^{-st}}{-s}\bigg|_0^\infty + \frac{2}{s}\int_0^\infty te^{-st} dt$$

$$= \frac{2}{s}\left[\frac{te^{-st}}{-s}\bigg|_0^\infty + \frac{1}{s}\int_0^\infty e^{-st} dt\right] = \frac{2}{s}\left[\frac{1}{s}\right],$$

so that (14.91)

$$\boxed{t^2 \to \frac{2}{s^3}} \tag{14.92}$$

14.21 Laplace Transform of the Pulse Function

A pulse of width τ, and left edge at time zero, is defined by

$$\text{pulse of width } \tau = \begin{matrix} 1 & t < \tau \\ 0 & t > \tau \end{matrix} \tag{14.93}$$

By direct integration we get

$$F(s) = \int_0^\infty f(t)e^{-st} dt = \int_0^\tau e^{-st} dt = -\frac{1}{s}e^{-st}\bigg|_0^\tau \tag{14.94}$$

$$\boxed{\text{pulse of width } \tau \to \frac{1 - e^{-\tau s}}{s}} \tag{14.95}$$

In terms of real and imaginary we get

$$F(s) = \frac{(1 - e^{-\tau\sigma}\cos\omega\tau) + je^{-\tau\sigma}\sin\omega\tau}{\sigma + j\omega} \tag{14.96}$$

$$F(s) = \frac{\big[(1 - e^{-\tau\sigma}\cos\omega\tau)\sigma + \omega e^{-\tau\sigma}\sin\omega\tau\big] + j\big[-(1 - e^{-\tau\sigma}\cos\omega\tau)\omega + \sigma e^{-\tau\sigma}\sin\omega\tau\big]}{\sigma^2 + \omega^2} \tag{14.97}$$

These results are shown in Fig. 14.21 for different σ values. Notice that in this case σ does not have as pronounced effect as we have seen before (i.e., step, ramp, sin, ...). The reason is that the pulse function is already "behaved" in the sense it does not have a singular behavior; in other words it does not tend to blow up, and hence taming it further has little effect. If we take the inverse transform, we get Fig. 14.22, which shows time-series reconstruction of the pulse function, for the case $\sigma = 0.2$.

14.22 Relation Between Laplace and Fourier Transforms

We know that for non-integrable signals, we cannot do the Fourier transform; and for those we can resort to the Laplace one. In essence we multiply the non-integrable signal by a decaying function $e^{-\sigma t}$, find the Fourier transform of the new signals, then when going back into the time domain (when doing inverse transforms) we do the inverse Fourier transform, and now multiply by $e^{\sigma t}$ (notice σ and not $-\sigma$). In the following two examples we will materially demonstrate the relation between the LT and FT.

14.22.1 The Unit Step Function

We know that the LT of the unit step function is

$$\boxed{\mathcal{L}[u(t)] = \frac{1}{s}} \tag{14.98}$$

But we also know that

$$\mathcal{F}[u(t)] = \boxed{\pi\delta(\omega) + \frac{1}{j\omega}} \tag{14.99}$$

Let's try consolidate the two results. Start with the LT and use the definition of s

$$\mathcal{L}[u(t)] = \frac{1}{\sigma + j\omega} \tag{14.100}$$

Expand in terms of real and imaginary parts

$$F(s) = \frac{\sigma}{\sigma^2 + \omega^2} - j\frac{\omega}{\sigma^2 + \omega^2} \tag{14.101}$$

Now let's take the limit as $\sigma \to 0$. After all when $\sigma = 0$ the Laplace transform *becomes* the Fourier one! For the imaginary part, the limit is straightforward and we get

$$\lim_{\sigma\to 0} -j\frac{\omega}{\sigma^2 + \omega^2} = -j\frac{1}{\omega} = \boxed{\frac{1}{j\omega}} \tag{14.102}$$

which matches exactly the imaginary part of the Fourier transform. Taking the limit on the real part requires some attention. It maybe tempting to simply set

$$\lim_{\sigma\to 0} \frac{\sigma}{\sigma^2 + \omega^2} \overset{???}{=} 0 \tag{14.103}$$

This certainly is the case when $\omega \neq 0$. But when $\omega = 0$ we would get

$$\lim_{\sigma\to 0} \frac{\sigma}{\sigma^2} = \lim_{\sigma\to 0} \frac{1}{\sigma} = \infty \tag{14.104}$$

So, we get 0 when $\omega \neq 0$ and ∞ when $\omega = 0$!! It turns out that this function is nothing other than the delta function!

$$\lim_{\sigma\to 0} \frac{\sigma}{\sigma^2 + \omega^2} = \pi\delta(\omega) \tag{14.105}$$

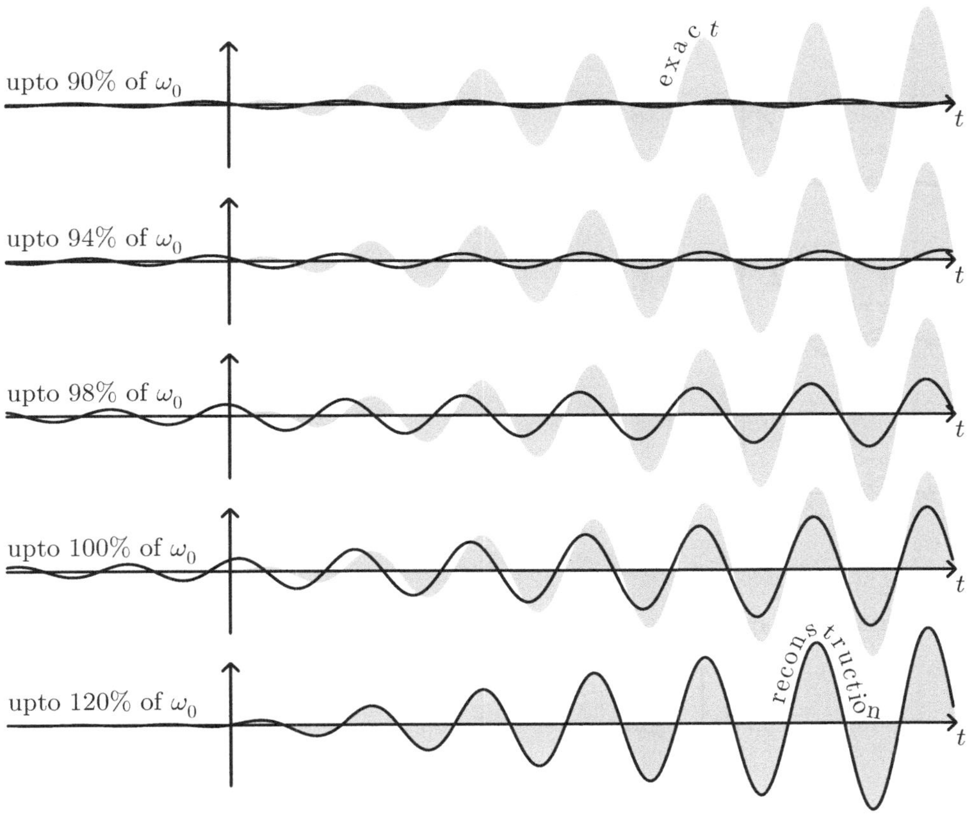

Fig. 14.20 Inverse LT of $t \sin \omega_0 t$ (case of $\sigma = 0.1$)

Combining the real and imaginary parts we certainly arrive back at the Fourier transform of the unit step function

$$\boxed{\lim_{\sigma \to 0} \frac{1}{s} = \pi\delta(\omega) + \frac{1}{j\omega}} \tag{14.106}$$

So we can see that *by using the Laplace transform we get away from using* ***delta*** *functions.* Let's take a look at the transformation between the Laplace and Fourier transforms, as $\sigma \to 0$. Figure 14.23 shows the real and imaginary components of the LT for different σ selection. As is evident, the $\sigma \neq 0$ smears out the transform, but when pushed to zero it does certainly reproduce the delta function (for the real part, and converts to $\frac{1}{j\omega}$ for the imaginary part). This transfusion between the Laplace and Fourier transforms cannot be overemphasized! And the key holder to it all is . . . the σ!

14.22.2 The Single-Side Complex Exponential

From the Fourier transform chapter (Eq. (8.62)) we know that the FT of the single-sided complex exponential (of frequency $-\omega_0$) is

$$\mathcal{F}\left[e^{-j\omega_0 t}\right] = \boxed{\pi\delta(\omega + \omega_0) + \frac{1}{j(\omega + \omega_0)}} \tag{14.107}$$

We also know that the LT of the same function is

$$\mathcal{L}\left[e^{-j\omega_0 t}\right] = \boxed{\frac{1}{s + j\omega_0}} \tag{14.108}$$

Let's try see if we can consolidate the LT to the FS. We start with the LT version and expand s:

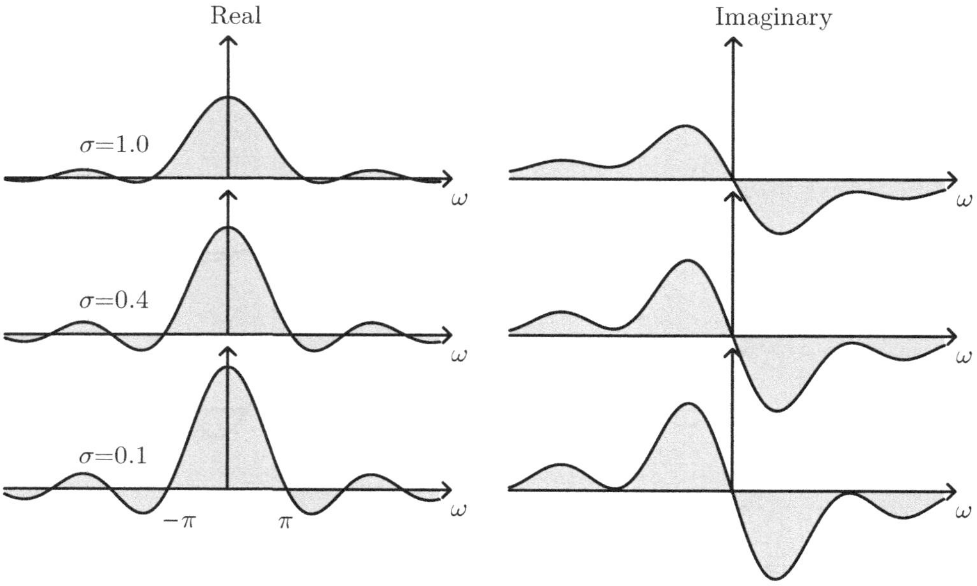

Fig. 14.21 Laplace transform of pulse function

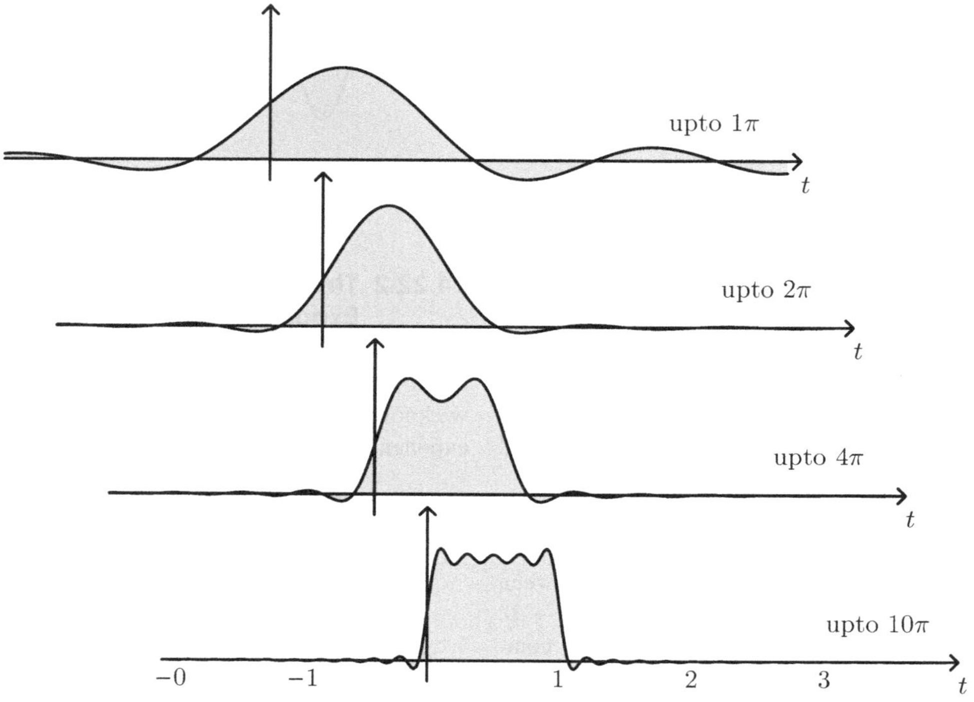

Fig. 14.22 Time series of pulse function ($\sigma = 0.2$)

$$F(s) = \frac{1}{s + j\omega_0} = \frac{1}{\sigma + j\omega + j\omega_0} = \frac{1}{\sigma + j(\omega + \omega_0)} \tag{14.109}$$

Expand in terms of real and imaginary:

$$\frac{1}{\sigma + j(\omega + \omega_0)} = \frac{\sigma}{\sigma^2 + (\omega + \omega_0)^2} - j\frac{(\omega + \omega_0)}{\sigma^2 + (\omega + \omega_0)^2} \tag{14.110}$$

Next we take the limit as $\sigma \to 0$. Again no problem with the imaginary part

$$\lim_{\sigma \to 0} -j\frac{(\omega + \omega_0)}{\sigma^2 + (\omega + \omega_0)^2} = \boxed{\frac{1}{j}\frac{1}{\omega + \omega_0}} \tag{14.111}$$

which agrees with the imaginary part of the FT. For the real part we have to be cautious about how we take the limit. For $\omega \neq -\omega_0$ we get

$$\lim_{\sigma \to 0} \frac{\sigma}{\sigma^2 + (\omega + \omega_0)^2} = \frac{0}{0 + (\omega + \omega_0)^2} = 0, \qquad (\omega \neq -\omega_0) \tag{14.112}$$

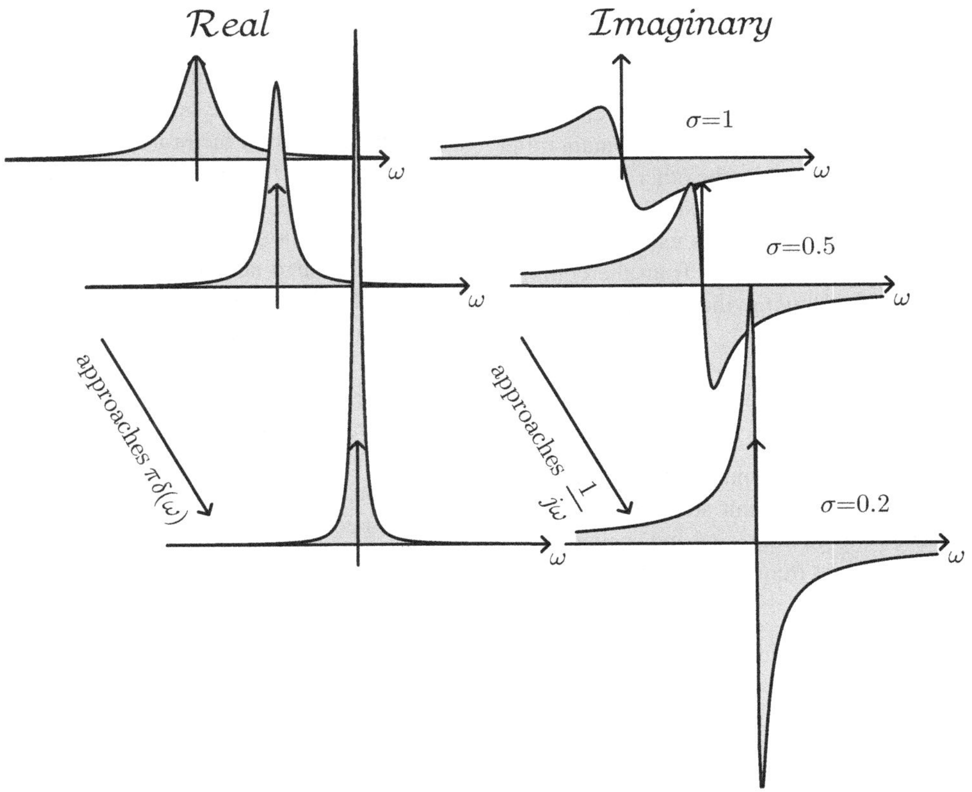

Fig. 14.23 Laplace transform of unit step function and limit as $\sigma \to 0$

For the case $\omega = -\omega_0$ we get

$$\lim_{\sigma\to 0} \frac{\sigma}{\sigma^2 + (\omega + \omega_0)^2} = \lim_{\sigma\to 0} \frac{\sigma}{\sigma^2 + 0}$$

$$= \lim_{\sigma\to 0} \frac{1}{\sigma} = \infty, \qquad (\omega = -\omega_0) \quad (14.113)$$

When combining the last two results we get

$$\lim_{\sigma\to 0} \frac{\sigma}{\sigma^2 + (\omega + \omega_0)^2} = \boxed{\pi\delta(\omega + \omega_0)} \quad (14.114)$$

which matches the real part of the FT. When adding the real and imaginary we finally get

$$\boxed{\lim_{\sigma\to 0} \frac{1}{s + j\omega_0} = \pi\delta(\omega + \omega_0) + \frac{1}{j}\frac{1}{\omega + \omega_0}} \quad (14.115)$$

which matches exactly the Fourier transform results. The transition process, from the LT to the FT, as $\sigma \to 0$ is shown in Fig. 14.24. Notice the remarkable resemblance between this figure and Fig. 14.23 (corresponding to the unit step)! Other than an offset in the frequency domain, both figures look almost identical! Even though they differ substantially in the time domain (after all, one is a unit step while the other is a complex exponential), they look very similar in the frequency domain! The answer is really simple. We can get the unit step function from the complex exponential by allowing the following simple limit:

$$\lim_{\omega_0\to 0} u(t)e^{-j\omega_0 t} = u(t) \qquad (14.116)$$

So from a frequency point of view, the only difference between the unit step and an oscillatory complex exponential (sine and cosine) is that the former has zero frequency!

14.23 Summary

This is the first chapter on the Laplace transform. First we outlined the need for the Laplace transform and that is to deal with non-integrable functions. Functions that don't die off in time (like the ramp, quadratic, ...) cannot be dealt with in the Fourier transform world. Even gentler functions such as the unit step or sines/cosines gave us singular delta functions in the Fourier world. The Laplace transform solves the convergence problems and alleviates the delta functions. Quite simple—tame the function and *then* find its Fourier transform; the result is the Laplace transform! Taming takes place via the function $e^{-\sigma t}$ and σ is figured to ensure convergence in the time domain. We illustrated the LT via many examples, such as the unit step, causal sines/cosines, ramp, and quadratic functions. Implicit in the Laplace transform world is that signals are zero for negative time. For the rare exceptions we can fall back on the bilateral LT. With the introduction of σ we meet for the first time the complex frequency $s = \sigma + j\omega$. While it looks a bit intimidating, it really is a handy device to capture both the real frequency ω and the dissipation factor σ. Remember, when doing frequency integration we only do it over ω; in other words, σ is set to a constant and the material integration happens only over ω. As to the selection of σ and its influence we sought 3D plots to illustrate how the σ takes care of smoothing the frequency spectrum (both real and imaginary parts) and eliminating delta functions. Finally we demonstrated the relation between the Laplace transform and the Fourier one, using a couple of examples, and arrived at the simple conclusion that in the limit $\sigma \to 0$ the Laplace transform collapses to that of Fourier!

14.24 Problems

1. What is the region of convergence (σ selection) to find the Laplace transform of the function $f(t) = e^{4t}$.

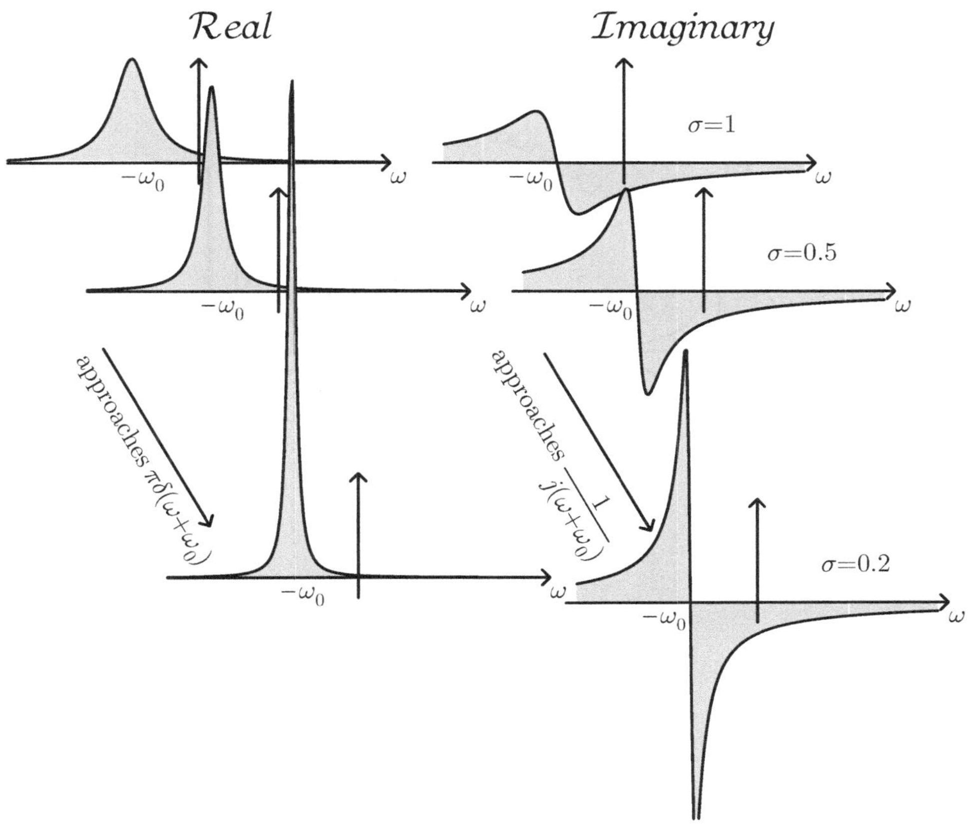

Fig. 14.24 Laplace transform of single-sided complex exponential function and limit as $\sigma \rightarrow 0$

2. What is the region of convergence (σ selection) to find the Laplace transform of the function $f(t) = te^{2t}$.
3. What is the region of convergence to find the Laplace transform of the function $f(t) = t^2$.
4. What is the Laplace transform of the shifted delta function $\delta(t - t_0)$?
5. What is the Laplace transform of the shifted unit step function $u(t - t_0)$?
6. What is the effect of nonzero σ on the real and imaginary parts of the Laplace transform, taking Fig. 14.4 as a reference case? What happens to the singularities in the plots?
7. The Laplace transform of the negative exponential was shown to be

$$u(t)e^{-at} \rightarrow \frac{1}{s+a}$$

How does this compare to the Fourier transform of the same function? For what value of σ does the Laplace transform collapse to the Fourier transform?
8. It was shown that the function

$$te^{-at} \rightarrow \frac{1}{(s+a)^2}$$

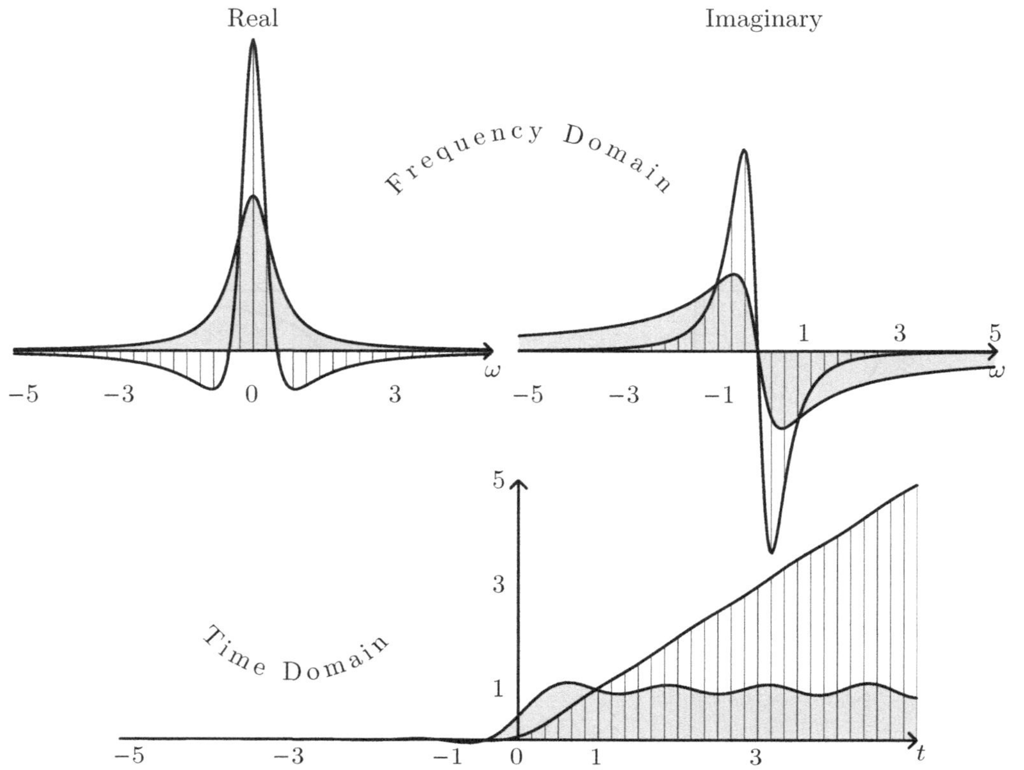

Fig. 14.25 Solution to Problem 12

What happens if $a = 0$? How does this compare to the Laplace transform of the function t?

9. It was shown that the single-sided cosine has the Laplace transform

$$\cos \omega_0 t \rightarrow \frac{s}{s^2 + \omega_0^2}$$

Derive the Laplace transform of unit step function based on the above!

10. How does the spectrum of the cosine function, and as shown in Fig. 14.15 compare to that of the complex exponential which is shown in Fig. 14.12?
11. If we know the Laplace transform of

$$te^{-j\omega_0 t} \rightarrow \frac{1}{(s + j\omega_0)^2}$$

what would be the LT of the function $t \cos \omega_0 t$?

12. Compare the Laplace transform (both real and imaginary) of the unit step and the linear function, for $\sigma = 0.5$ and construct the time series for both. See sample solution in Fig. 14.25.
13. It was shown that the pulse function of width τ has the Laplace transform

$$F(s) = \frac{1 - e^{-\tau s}}{s}$$

Plot the Laplace transform (real and imaginary) for the three cases of $\tau = 1, 4$, and 10. Use sample $\sigma = 0.1$. What is the trend? See sample solution in Fig. 14.26.

14. Compare the Laplace transform of the pulse function of width 10 to that of the step function; choose $\sigma = 0.1$ See sample solution in Fig. 14.27.
15. Repeat Problem 14 but this time use $\sigma = 0.2$. See sample solution in Fig. 14.28.

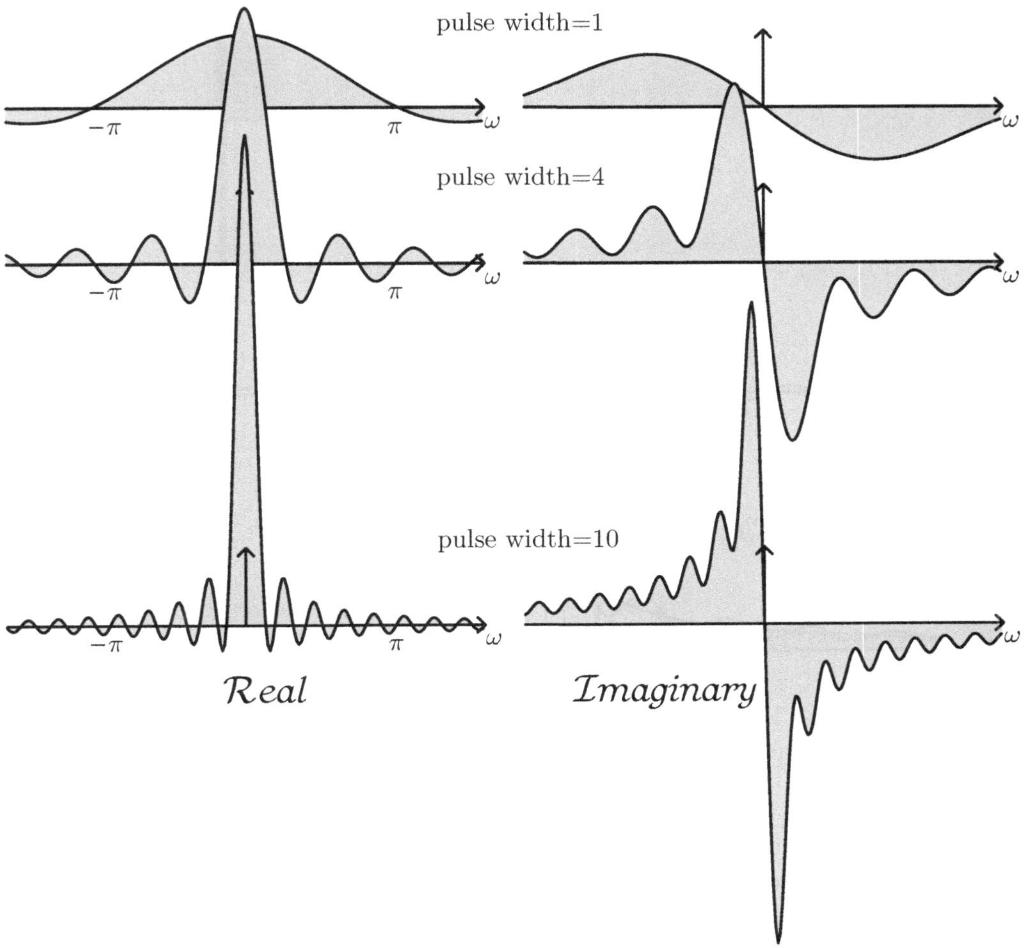

Fig. 14.26 Solution to Problem 13

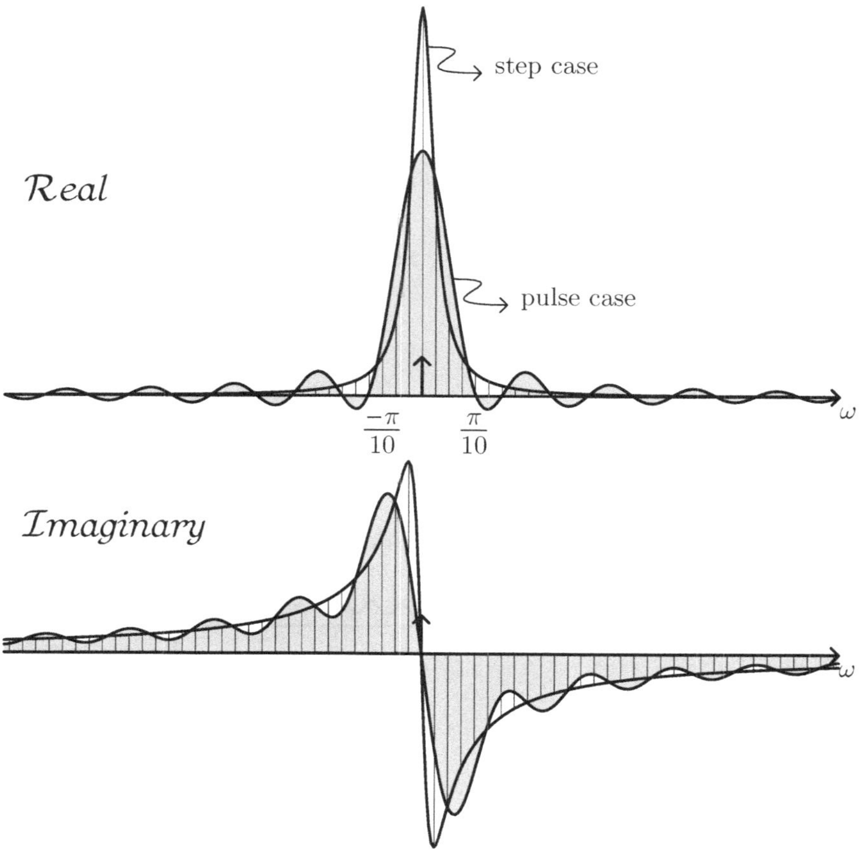

Fig. 14.27 Solution to Problem 14

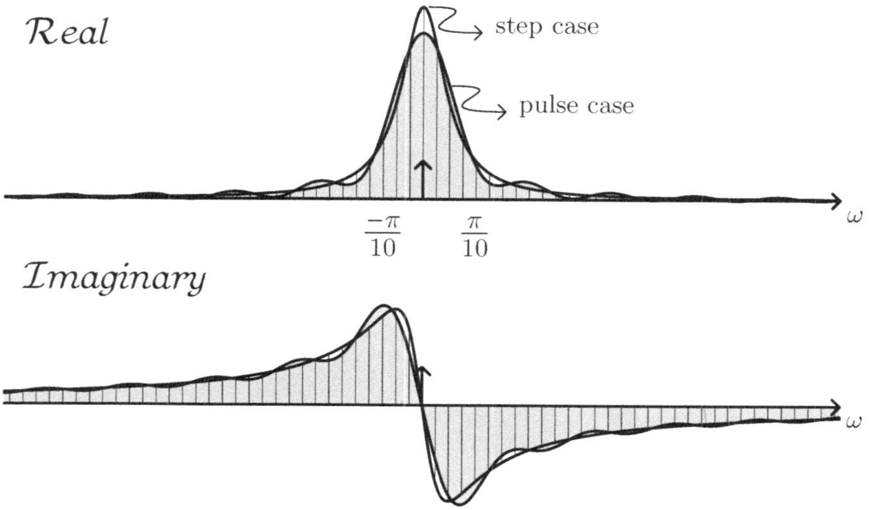

Fig. 14.28 Solution to Problem 15

15 Using Complex Integration to Figure Inverse Laplace Transform

15.1 Introduction

To find the Laplace transform of a function $f(t)$ we do time integration, most often in the form

$$F(s) = \int_0^\infty f(t)e^{-st}dt \tag{15.1}$$

This is a one-dimensional integration, along the t axis, starting from 0 and ending at ∞. To find the inverse Laplace transform we need to evaluate the integral

$$f(t) = \frac{1}{2\pi j}\int_{\sigma-j\infty}^{\sigma+j\infty} F(s)e^{st}ds \tag{15.2}$$

Recalling that

$$s = \sigma + j\omega \tag{15.3}$$

we realize that this amounts to a complex integration. Complex integration is typically taught in a course about complex analysis, and for those needing a refresher Appendix A can be consulted. Having some experience with complex integration and residues, this topic is best learned via practice; so let's jump into it, with as many examples as possible!

15.2 Inverse Laplace Transform of $\frac{1}{s}$

Let's find the inverse LT of the function

$$F(s) = \frac{1}{s} \tag{15.4}$$

Examining this function we find it has a simple pole at $s = 0$; in order for the integral to converge, we need to choose σ such that

$$\sigma > 0 \tag{15.5}$$

Let's for example choose $\sigma = 1$. This sets one side of the contour integration as shown in Fig. 15.1. For time $t < 0$ we need to close the contour by going around the right-hand side of the complex plane, to ensure that e^{st} dies off at large s. Hence we would follow the path labeled "path 1" in the figure. What we want is the integral along the straight edge of the contour, but what we get is the whole contour. We can bank on Jordan's lemma and consider the integration around the arc portion to go to zero. Hence the whole contour integration would equal the straight segment integration. And we know the contour integration equals $2\pi j \sum$ residues. Since there are no residues inside this path we conclude that

$$f(t) = 0, \qquad t < 0 \tag{15.6}$$

F. Badrieh, *Spectral, Convolution and Numerical Techniques in Circuit Theory*,
https://doi.org/10.1007/978-3-319-71437-0_15

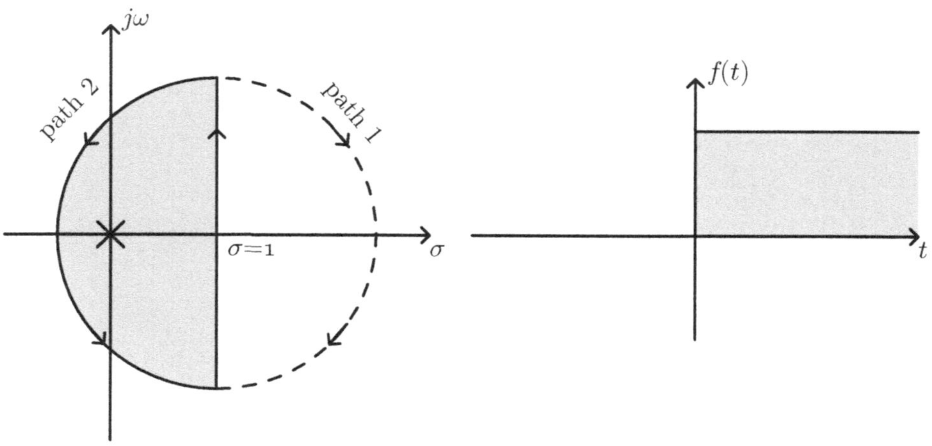

Fig. 15.1 Inverse Laplace transform of $\frac{1}{s}$

For time $t > 0$ we would need to go along the contour covering the left side of the plane, still utilizing the same straight segment, at $\sigma = 1$. Now this path does have a residue, and it is

$$\text{residue} = \frac{e^{st}}{s} \times s\bigg|_{s=0} = 1 \tag{15.7}$$

Our integral then becomes

$$\frac{1}{2\pi j} \times 2\pi j \sum \text{residues} = 1 \tag{15.8}$$

Hence, we have

$$f(t) = 1, \qquad t > 0 \tag{15.9}$$

So in summary we have

$$f(t) = \begin{cases} 0 & t < 0 \\ 1 & t > 0 \end{cases} \tag{15.10}$$

To reiterate, using complex integration we are able to derive the inverse transform of the function $\frac{1}{s}$. Typically standard signals and systems texts merely document inverse transforms via lookup tables. While this is fine for most cases, being versed in performing the actual integration may come in handy one of those times, especially if the desired function is not in the table. Also, familiarizing ourselves with the complex plane, poles, pole order, and impact of σ will only get us deeper understanding of the Laplace domain and beyond.

15.3 Inverse Laplace Transform of $\frac{1}{s^2}$

We already know that the ramp function has the LT

$$u(t)t \rightarrow \frac{1}{s^2} \tag{15.11}$$

Let's use complex analysis to prove this. First notice that the LT has a pole at $s = 0$, and specifically a *double* pole. When we do the frequency integration we would have to follow a path that does not pass by this point, and specifically a path such that the real part of s lies to the *right* of the pole (the origin here). Let's for example choose the point $\sigma = 1$; this would give us the upright branch shown in Fig. 15.2. Now would have to decide how to complete the contour path. We have two cases:

1. Negative time: For $t < 0$ would go around the right half plane, to ensure that e^{st} dies of at infinity. We label this as "path 1." Since this path encloses no residues, we conclude that

$$f(t) = 0, \qquad t < 0 \tag{15.12}$$

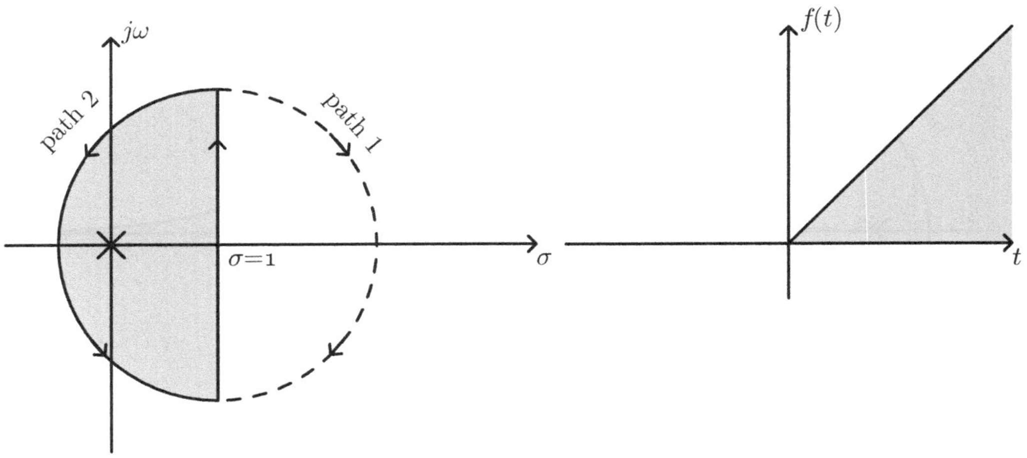

Fig. 15.2 Inverse Laplace transform of $\frac{1}{s^2}$

2. Positive time: For $t > 0$ would go around the left half plane, again to ensure that e^{st} dies of at negative infinity. We label this as "path 2." Now we do pick a residue, at $s = 0$. It is given by

$$\text{residue at } s = 0 = \frac{d}{ds}\left[\frac{e^{st}}{s^2} \times s^2\right]_{s=0} = \frac{d}{ds}\left[e^{st}\right]_{s=0} = te^{st}|_{s=0} = t \tag{15.13}$$

So the complex integration comes out to

$$\int \frac{s^{st}}{s^2} ds = 2\pi jt \tag{15.14}$$

Finally, the inverse Laplace transform comes out

$$f(t) = \frac{1}{2\pi j} \times 2\pi j \times t = t \tag{15.15}$$

So in summary we have

$$f(t) = \begin{cases} 0 & t < 0 \\ t & t > 0 \end{cases} \tag{15.16}$$

It is becoming apparent that in order to do complex integration we need to find the poles, determine their order, pick up the path of integration, and figure the residues. For more details and practice about these items, see the appendix or pick up your favorite text on complex analysis!

15.4 Inverse Laplace Transform of $\frac{1}{s+a}$

We know that the negative exponential e^{-at} has the LT

$$u(t)e^{-at} \rightarrow \frac{1}{s+a} \tag{15.17}$$

Let's use complex analysis to verify this. This LT has a simple pole at $s = -a$, where a is real and positive. The upright branch of the integration contour needs to lie to the *right* of this point ($s = -a$); for example we can choose the origin point; this is shown in Fig. 15.3. For negative time we choose the right half plane, and that encloses no residues. For positive time we pick a single residue given by

$$\text{residue at } s = -a = \frac{e^{st}}{s+a} \times (s+a)\Big|_{s=-a} = e^{st}|_{s=-a} = e^{-at} \tag{15.18}$$

Hence we get

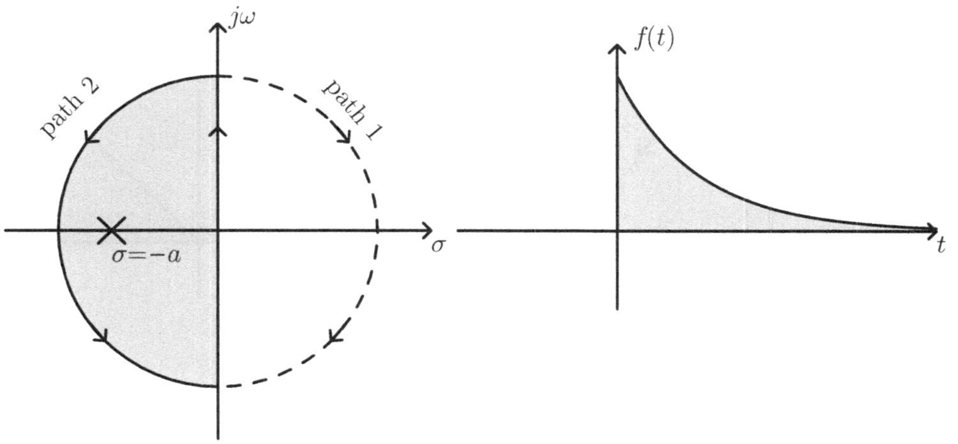

Fig. 15.3 Inverse Laplace transform of $\frac{1}{(s+a)}$

$$f(t) = \begin{cases} 0 & t < 0 \\ e^{-at} & t > 0 \end{cases} \tag{15.19}$$

Pretty systematic process: find the poles, their order, construct the integration path, find the residues, and figure the inverse transform by adding the residues.

15.5 Inverse Laplace Transform of $\frac{1}{(s+a)^2}$

We know that the function te^{-at} has the LT

$$te^{-at} \rightarrow \frac{1}{(s+a)^2} \tag{15.20}$$

Let's use complex analysis to verify. The LT has a *double* pole at $s = -a$; so we would need the upright segment of the contour to lie to the right of $s = -a$; again how about the origin! For negative time, we evaluate the contour integration on the right half plane, and we pick *no* residues, so the integration gives zero. For positive time we pick the left path, and pick a residue at $s = -a$; it is given by

$$\text{residue at } s = -a = \frac{d}{ds}\left[\frac{s^{st}}{(s+a)^2} \times (s+a)^2)\right]_{s=-a}$$
$$= te^{st}|_{s=-a} = te^{-at} \tag{15.21}$$

So we get

$$f(t) = \begin{cases} 0 & t < 0 \\ te^{-at} & t > 0 \end{cases} \tag{15.22}$$

This is shown in Fig. 15.4.

15.6 Inverse Laplace Transform of $\frac{1}{s(s+a)^2}$

Let us try to find the inverse Laplace transform of the function

$$F(s) = \frac{1}{s(s+a)^2} \tag{15.23}$$

This function has two poles—one (simple) at $s = 0$, and the other (double) at $s = -a$. We construct our contour path such that it lies to the right of both poles; how about $s = 1$; this is shown in Fig. 15.5. For negative time we choose path 1, and that yields no residues. For positive time we have two residues:

$$\text{first residue at } s = 0 = \left[\frac{e^{st}}{s(s+a)^2} \times s\right]_{s=0}$$

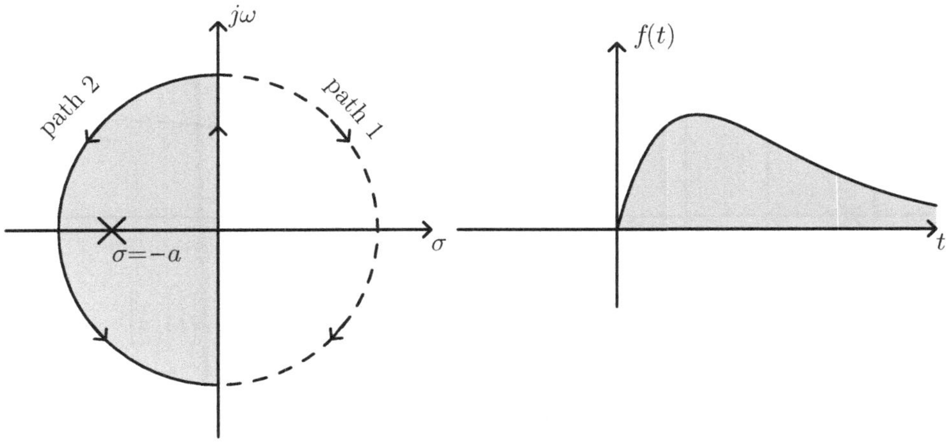

Fig. 15.4 Inverse Laplace transform of $\frac{1}{(s+a)^2}$

$$= \frac{e^{st}}{(s+a)^2}\bigg|_{s=0} = \frac{1}{a^2} \tag{15.24}$$

The second residue is

$$\begin{aligned}
&\text{second residue at } s=-a \\
&= \frac{d}{ds}\left[\frac{e^{st}}{s(s+a)^2} \times (s+a)^2\right]_{s=-a} \\
&= \left[\frac{te^{st}}{s} - \frac{e^{st}}{s^2}\right]_{s=-a} \\
&= -\frac{e^{-at}}{a}\left[t + \frac{1}{a}\right]
\end{aligned} \tag{15.25}$$

So our inverse transform becomes

$$f(t) = \begin{cases} 0 & t<0 \\ \frac{1}{a^2} - \frac{e^{-at}}{a}\left[t+\frac{1}{a}\right] & t>0 \end{cases} \tag{15.26}$$

We can double check this as follows:

$$\frac{1}{a^2} \rightarrow \frac{1}{a^2}\frac{1}{s} \tag{15.27}$$

$$-\frac{1}{a^2}e^{-at} \rightarrow -\frac{1}{a^2}\frac{1}{s+a}, \quad \text{and} \tag{15.28}$$

$$-\frac{1}{a}te^{-at} \rightarrow -\frac{1}{a}\frac{1}{(s+a)^2} \tag{15.29}$$

If we add these three results we get

$$\begin{aligned}
&\frac{1}{a^2}\left[\frac{1}{s} - \frac{1}{s+a} - \frac{a}{(s+a)^2}\right] \\
&= \frac{1}{a^2}\frac{(s+a)^2 - s(s+a) - as}{s(s+a)^2} \\
&= \frac{1}{a^2}\frac{s^2 + 2as + a^2 - s^2 - as - as}{s(s+a)^2} \\
&= \frac{1}{s(s+a)^2}
\end{aligned} \tag{15.30}$$

which is the transfer function we started with in Eq. (15.23). Results are shown in Fig. 15.5.

15.7 Inverse Laplace Transform of $\frac{s}{(s+a)^2}$

The transfer function

$$F(s) = \frac{s}{(s+a)^2} \tag{15.31}$$

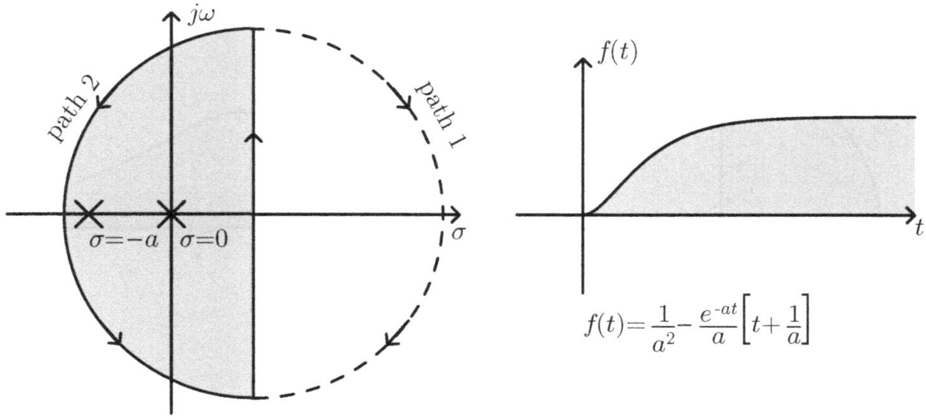

Fig. 15.5 Inverse Laplace transform of $\frac{1}{s(s+a)^2}$

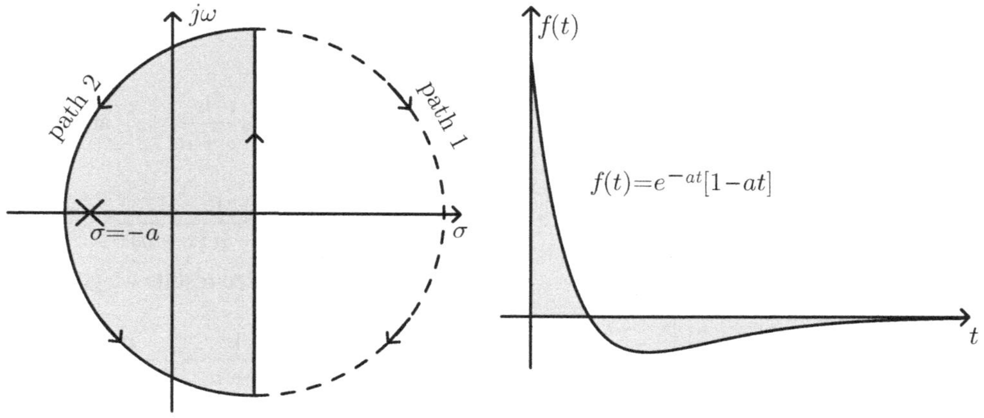

Fig. 15.6 Inverse Laplace transform of $\frac{s}{(s+a)^2}$

has a double pole at $s = -a$. We will choose a σ such that $\sigma > -a$ as shown in Fig. 15.6. When time is $t < 0$ we take the right path, and pick no residues. For positive time we take the left contour, and pick a residue at $s = -a$; it is given by

$$\begin{aligned}
&\text{residue at } s = -a \\
&= \frac{d}{ds}\left[\frac{se^{st}}{(s+a)^2} \times (s+a)^2\right]_{s=-a} \\
&= \frac{d}{ds}\left[se^{st}\right]_{s=-a} \\
&= \left[e^{st} + tse^{st}\right]_{s=-a} \\
&= e^{-at}\left[1 - at\right]
\end{aligned} \tag{15.32}$$

This gets multiplied by $2\pi j$, to get the frequency integration, then divided by the same factor (part of LT inverse transform), so that in the end we get

$$f(t) = \begin{cases} 0 & t < 0 \\ e^{-at}\left[1 - at\right] & t > 0 \end{cases} \tag{15.33}$$

Results are shown in Fig. 15.6.

15.8 Inverse Laplace Transform of $\frac{1}{(s+a)(s+b)}$

We'd like to find the inverse Laplace transform of the function

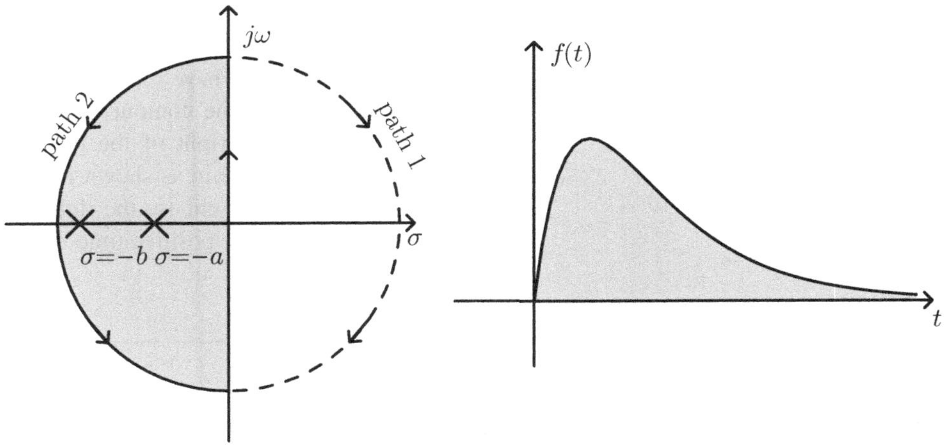

Fig. 15.7 Inverse Laplace transform of $\frac{1}{(s+a)(s+b)}$

$$F(s) = \frac{1}{(s+a)(s+b)} \tag{15.34}$$

Assume that both a and b are positive, and that $b > a$. The transfer function has two poles—one at $s = -a$ and the other at $s = -b$. In doing the frequency integration in the inverse LT

$$f(t) = \frac{1}{2\pi} \int_{\sigma-j\omega}^{\sigma-j\omega} F(s)e^{st}ds \tag{15.35}$$

we set σ such that it is to the right of the *right-most* pole; in this case we set $\sigma > -a$. Let's for example set $\sigma = 0$ and that defines the upright segment of the contour integration, as shown in Fig. 15.7. For negative time, we choose the right path to ensure that $e^{-\sigma t}$ goes to zero; we also pick no residues, and hence the integration (and the inverse LT) goes to zero. For positive time, we pick the left path such that again the real exponential goes to zero at large enough a radius. Here we pick two residues:

residue at $s = -a$

$$= \left[\frac{e^{st}}{(s+a)(s+b)} \times (s+a)\right]_{s=-a}$$

$$= \left[\frac{e^{st}}{(s+b)}\right]_{s=-a} = \frac{e^{-at}}{b-a} \tag{15.36}$$

Similarly

residue at $s = -b$

$$= \left[\frac{e^{st}}{(s+a)(s+b)} \times (s+b)\right]_{s=-b}$$

$$= \left[\frac{e^{st}}{(s+a)}\right]_{s=-b} = \frac{e^{-bt}}{a-b} \tag{15.37}$$

We multiply the sum of residues by $2\pi j$, then divide by the same constant, and finally get

$$f(t) = \begin{cases} 0 & t < 0 \\ \frac{1}{a-b}\left[e^{-bt} - e^{-at}\right] & t > 0 \end{cases} \tag{15.38}$$

The same principle but different case. All lies in identifying the poles. Whether a simple pole, a double one, multiple poles, or combination thereof—the poles hold the key for finding the inverse Laplace transform. In simple cases the poles are evident; but in other cases they may have to be solved for numerically. Either way, they need to be known. So far all the poles we have dealt with happened to be real; but as shown in the next section that does not have to be the case.

15.9 Inverse Laplace Transform of $\frac{1}{s^2+\omega_0^2}$

We'd like to find the inverse Laplace transform of the function

$$F(s) = \frac{1}{s^2 + \omega_0^2} \tag{15.39}$$

We can rewrite this as

$$F(s) = \frac{1}{(s + j\omega_0)(s - j\omega_0)} \tag{15.40}$$

So we have two poles—one at $s = -j\omega_0$ and the other at $s = j\omega_0$. Those are shown in Fig. 15.8. Both poles have real part of 0, so the upright segment of the contour integration can lie anywhere to the right of the $\sigma = 0$ axis; how about $\sigma = 1$, again as shown in the figure. For negative time we choose the right path, and incur no residues. For positive time we pick two residues

$$\text{residue (at } s = -j\omega_0) = \left[\frac{e^{st}}{(s + j\omega_0)(s - j\omega_0)} \times (s + j\omega_0)\right]_{s=-j\omega_0} = \left.\frac{e^{st}}{(s - j\omega_0)}\right|_{s=-j\omega_0} = -\frac{e^{-j\omega_0}}{2j\omega_0} \tag{15.41}$$

$$\text{residue (at } s = j\omega_0) = \left[\frac{e^{st}}{(s + j\omega_0)(s - j\omega_0)} \times (s - j\omega_0)\right]_{s=j\omega_0} = \left.\frac{e^{st}}{(s + j\omega_0)}\right|_{s=j\omega_0} = \frac{e^{j\omega_0}}{2j\omega_0} \tag{15.42}$$

When adding together we get

$$\sum \text{residues} = \frac{1}{j2\omega_0}\left[e^{j\omega_0 t} - e^{-j\omega_0 t}\right] = \frac{1}{\omega_0}\sin\omega_0 t \tag{15.43}$$

so that we finally get

$$f(t) = \frac{1}{\omega_0}\sin\omega_0 t \tag{15.44}$$

which we know to be true remembering that (Eq. (14.74))

$$\sin\omega_0 t \rightarrow \frac{\omega_0}{s^2 + \omega_0^2} \tag{15.45}$$

Results are shown in Fig. 15.8.

15.10 Inverse Laplace Transform of $\frac{1}{s(s^2+\omega_0^2)}$

This function has a pole at $s = 0$ and two at $s = j\omega_0$ and $s = -j\omega_0$. For negative time, we take the right path in Fig. 15.9 and sustain no residues. Then

$$f(t) = 0, \qquad t < 0 \tag{15.46}$$

For positive time we take the left path and pick three residues. The pole at $s = 0$ gives a residue of

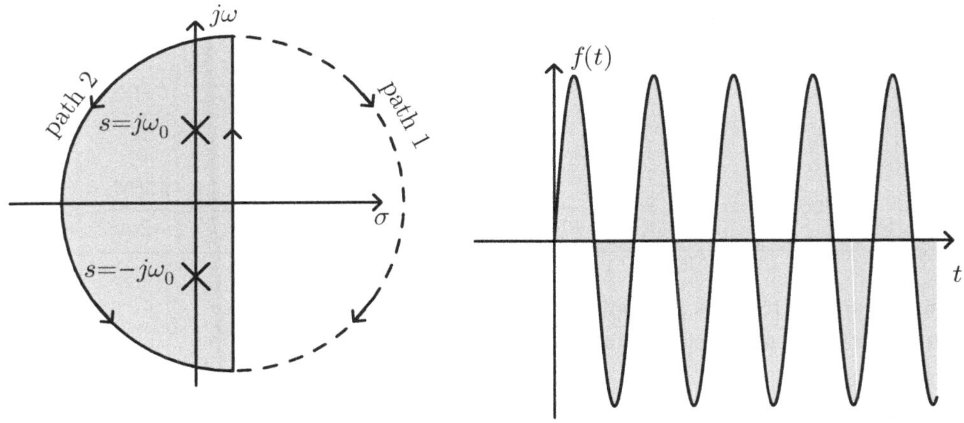

Fig. 15.8 Inverse Laplace transform of $\frac{1}{s^2+\omega_0^2}$

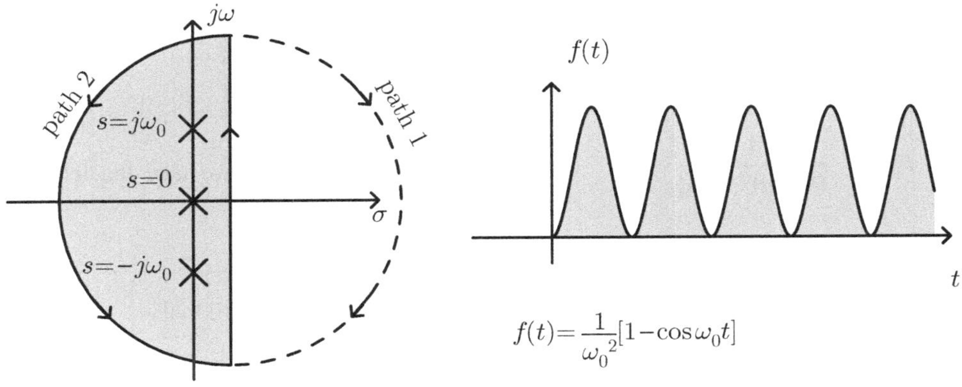

Fig. 15.9 Inverse Laplace transform of $\frac{1}{s(s^2+\omega_0^2)}$

$$\text{residue at } s = 0 = \frac{e^{st}}{s(s^2+\omega_0^2)} \times s \bigg|_{s=0} = \frac{1}{\omega_0^2} \tag{15.47}$$

To find the other two residues we need to factor

$$s^2 + \omega_0^2 = (s + j\omega_0)(s - j\omega_0) \tag{15.48}$$

such that

$$F(s) = \frac{1}{s(s + j\omega_0)(s - j\omega_0)} \tag{15.49}$$

$$\text{residue at } s = j\omega_0 = \left.\frac{e^{st}}{s(s+j\omega_0)(s-j\omega_0)} \times (s-j\omega_0)\right|_{s=j\omega_0} = \frac{e^{j\omega_0 t}}{-2\omega_0^2} \tag{15.50}$$

$$\text{residue at } s = -j\omega_0 = \left.\frac{e^{st}}{s(s+j\omega_0)(s-j\omega_0)} \times (s+j\omega_0)\right|_{s=-j\omega_0} = \frac{e^{-j\omega_0 t}}{-2\omega_0^2} \tag{15.51}$$

When we sum the residues, multiply by $2\pi j$, and then divide by the same ratio, we arrive at

$$f(t) = \frac{1}{\omega_0^2}\left[1 - \cos\omega_0 t\right], \qquad t > 0 \tag{15.52}$$

15.11 Inverse Laplace Transform of $\frac{1}{(s+a)^2+\omega_0^2}$

The transfer function

$$F(s) = \frac{1}{(s+a)^2 + \omega_0^2} \tag{15.53}$$

can be rewritten as

$$F(s) = \frac{1}{[(s+a)+j\omega_0][(s+a)-j\omega_0]} \tag{15.54}$$

It has two poles—one at $s = -a - j\omega_0$, and one at $s = -a + j\omega_0$. Both are located to the left of the complex plane. For negative time we take the right contour path in Fig. 15.10 and incur no residues; hence

$$f(t) = 0, \qquad t < 0 \tag{15.55}$$

For positive time we take the left path and incur the two residues.

$$\begin{aligned}\text{residue at } (s = -a - j\omega_0) &= \left.\frac{e^{st}}{(s+a+j\omega_0)(s+a-j\omega_0)} \times (s+a+j\omega_0)\right|_{s=-a-j\omega_0} \\ &= \frac{e^{-at}e^{-j\omega_0 t}}{-2j\omega_0}\end{aligned} \tag{15.56}$$

$$\begin{aligned}\text{residue at } (s = -a + j\omega_0) &= \left.\frac{e^{st}}{(s+a+j\omega_0)(s+a-j\omega_0)} \times (s+a-j\omega_0)\right|_{s=-a+j\omega_0} \\ &= \frac{e^{-at}e^{j\omega_0 t}}{2j\omega_0}\end{aligned} \tag{15.57}$$

The sum of the residues is then

$$\sum \text{residues} = \frac{1}{\omega_0}e^{-at}\sin\omega_0 t \tag{15.58}$$

Hence

$$f(t) = \frac{1}{\omega_0}e^{-at}\sin\omega_0 t, \qquad t > 0 \tag{15.59}$$

as shown in Fig. 15.10.

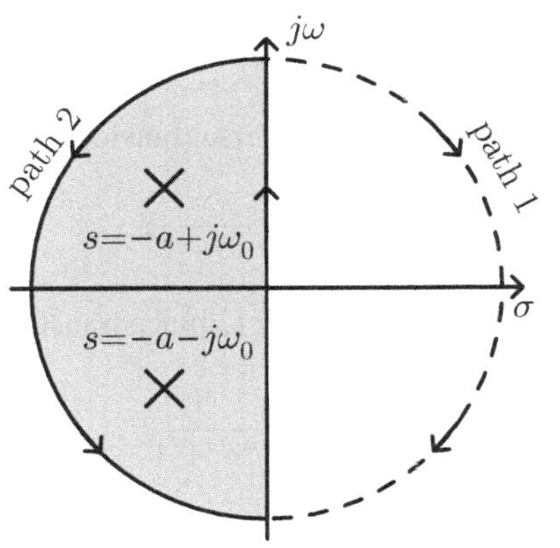

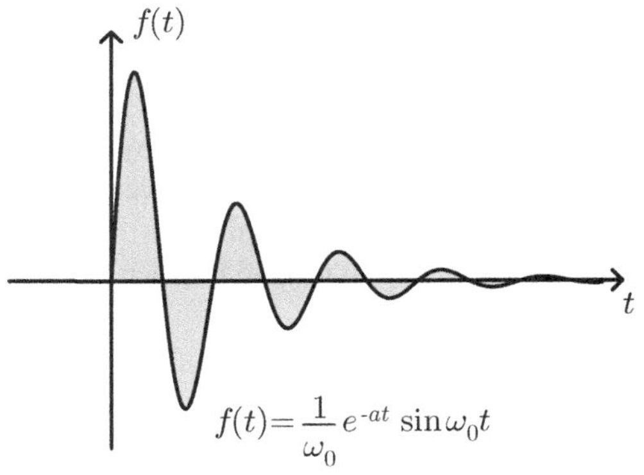

Fig. 15.10 Inverse Laplace transform of $\frac{1}{(s+a)^2+\omega_0^2}$

15.12 Summary

Having introduced the Laplace transform in the last chapter, this chapter deals with finding the inverse transform using complex integration. The inverse transform calls for an integration from $s = \sigma - j\infty$ to $s = \sigma + j\infty$, which is straight line for a given σ. The Laplace transform, it being a complex function, means its inverse integral is complex too. As a way how to pick the integration from negative to positive infinite frequency we can replace the straight integral path with a semicircular one, assuming the integration around the arc goes to zero (valid for most functions). Having recast the integration in the form of a contour one we are now in a position to harness the calculus of residues and evaluate the integral simply by reading the residues. The residues are evaluated systematically first by finding the location of the poles, and then by applying a straightforward formula, depending on the pole order. Appendix A has more details about complex integration. In this chapter we worked through many examples and showed how to locate the poles on the complex plane, how to pick the integration path (depending on σ), how to evaluate the residues, sum them, and then arrive at the inverse transform. While using Laplace transform lookup tables may suffice most of the time, an understanding of the foundations of evaluating the complex inverse Laplace transform will be of great use, especially when we encounter the method of partial fractions in evaluating inverse Laplace transforms.

15.13 Problems

1. What is the inverse transform of the function

$$F(s) = \frac{1+s}{s}$$

2. What is the inverse transform of the function

$$F(s) = \frac{1+s}{s^2}$$

   Use complex integration to find the answer.
   Answer:

$$f(t) = u(t)\,[t+1]$$

3. Locate on the complex plane the poles of the function

$$F(s) = \frac{1+s}{2+s}$$

   What σ should we choose to ensure that the inverse Laplace transform is convergent?

4. How many poles does the following function have?

$$F(s) = \frac{1}{s^2(s+1)}$$

What is the order of each pole? What σ selection is necessary to ensure convergence of the inverse Laplace transform?

5. We know the following inverse Laplace transform relation

$$\frac{1}{s^2+1} \rightarrow \sin t$$

What would be the inverse transform of

$$F(s) = \frac{1}{s^2-1}?$$

6. We know the following inverse Laplace transform relation

$$\frac{s}{s^2+1} \rightarrow \cos t$$

What would be the inverse transform of

$$F(s) = \frac{s}{s^2-1}?$$

7. Consider the following Laplace transform

$$F(s) = \frac{1}{s^3+1}$$

Assume we already know the three poles, which are

$$s_1 = -1$$
$$s_2 = \frac{1}{2}\left[1 + j\sqrt{3}\right]$$
$$s_3 = \frac{1}{2}\left[1 - j\sqrt{3}\right]$$

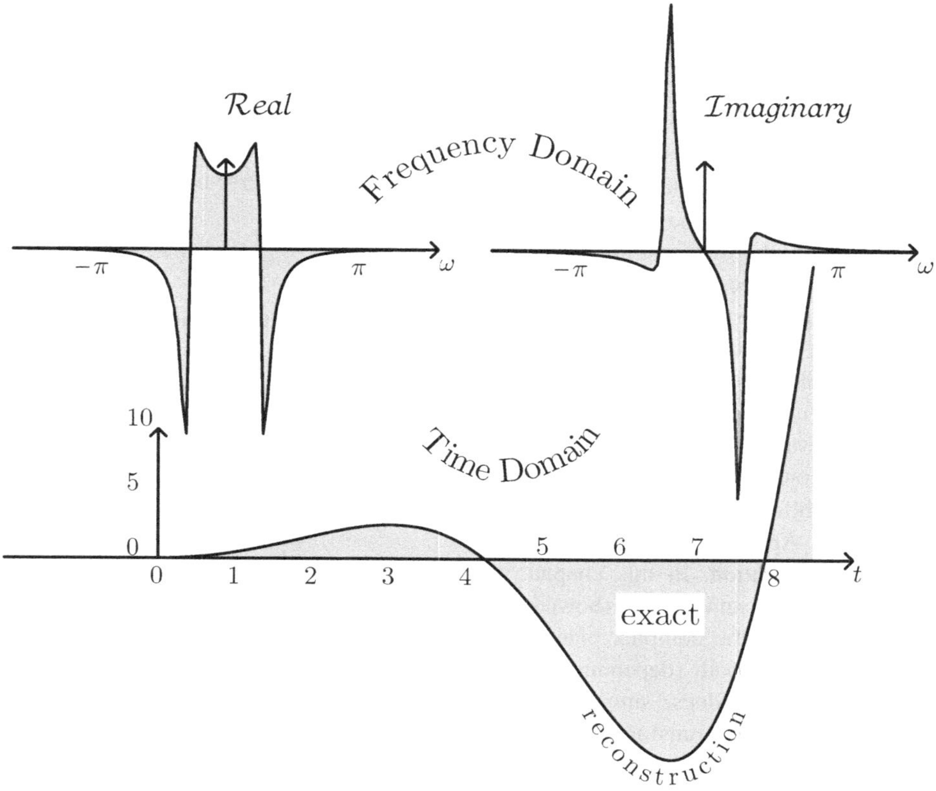

Fig. 15.11 Solution to Problem 7 ($\sigma = 0.6$ case)

First, verify indeed that these are the actual poles (by direct substitution). Next show that the transfer function can be written as

$$F(s) = \frac{1}{[s+1]\left[s-\frac{1}{2}\left(1+j\sqrt{3}\right)\right]\left[s-\frac{1}{2}\left(1-j\sqrt{3}\right)\right]}$$

Next find the three residues of the function

$$F(s) = \frac{e^{st}}{[s+1]\left[s-\frac{1}{2}\left(1+j\sqrt{3}\right)\right]\left[s-\frac{1}{2}\left(1-j\sqrt{3}\right)\right]}$$

See sample solution in Fig. 15.11.

Answer:

$$f(t) = \frac{1}{3}\left[e^{-t}+e^{t/2}\left(\sqrt{3}\sin\frac{\sqrt{3}}{2}t-\cos\frac{\sqrt{3}}{2}t\right)\right]$$

16 Properties of Laplace Transform

16.1 Introduction

Similar to the Fourier transform, the Laplace transform has many important properties that can facilitate frequency analysis and manipulations considerably. Those properties enable us to find the Laplace transform of many functions based on other, simpler ones. While many of the properties share their origin with those of the Fourier transform, some do have subtle differences. Also, while presenting the properties we will take advantage of the opportunity and present some useful application examples.

16.2 Linearity

This states that if

$$\mathcal{L}[f_1(t)] = F_1(s) \qquad \text{and} \qquad \mathcal{L}[f_2(t)] = F_2(s) \tag{16.1}$$

then

$$\mathcal{L}[f_1(t) + f_2(t)] = F_1(s) + F_2(s) \tag{16.2}$$

16.3 Scaling

This states that if

$$\mathcal{L}[f(t)] = F(s) \tag{16.3}$$

then

$$\mathcal{L}[f(at)] = \frac{1}{|a|} F\left(\frac{s}{a}\right) \tag{16.4}$$

We can prove this as follows:

$$\mathcal{L}[f(at)] = \int_0^\infty f(at)e^{-st}dt \tag{16.5}$$

Define $u = at$ such that $t = u/a$ and $dt = du/a$.

$$\mathcal{L}[f(at)] = \frac{1}{a}\int_0^\infty f(u)e^{-\frac{s}{a}u}du \tag{16.6}$$

But the right side is nothing more than $\frac{1}{a}F\left(\frac{s}{a}\right)$; hence the proof is complete. The $||$ sign is needed in case a was negative.

16.4 Time Shifting

This states that if $\mathcal{L}[f(t)] = F(s)$ then

$$\boxed{\mathcal{L}[f(t - t_0)] = F(s)e^{-st_0}} \tag{16.7}$$

The proof is as follows:

$$\mathcal{L}[f(t - t_0)] = \int_0^\infty f(t - t_0)e^{-st}dt \tag{16.8}$$

Let $u = t - t_0$ such that $t = u + t_0$ and $du = dt$; then

$$\mathcal{L}[f(t - t_0)] = \int_0^\infty f(u)e^{-s(u+t_0)}du \tag{16.9}$$

F. Badrieh, *Spectral, Convolution and Numerical Techniques in Circuit Theory*,
https://doi.org/10.1007/978-3-319-71437-0_16

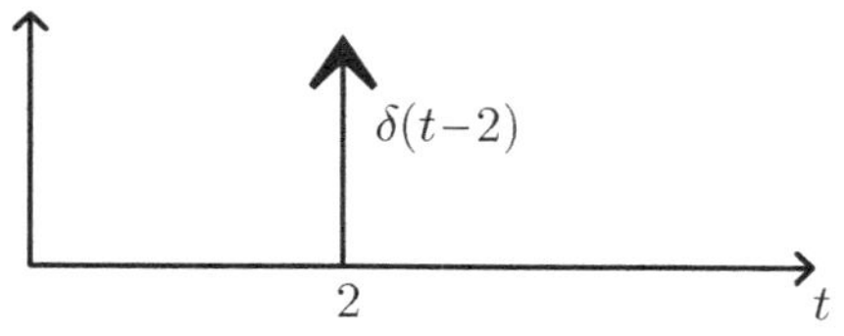

Fig. 16.1 Shifted delta function

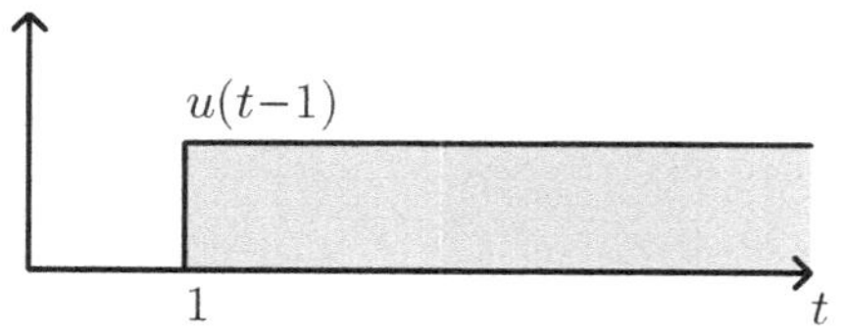

Fig. 16.2 Shifted unit step function

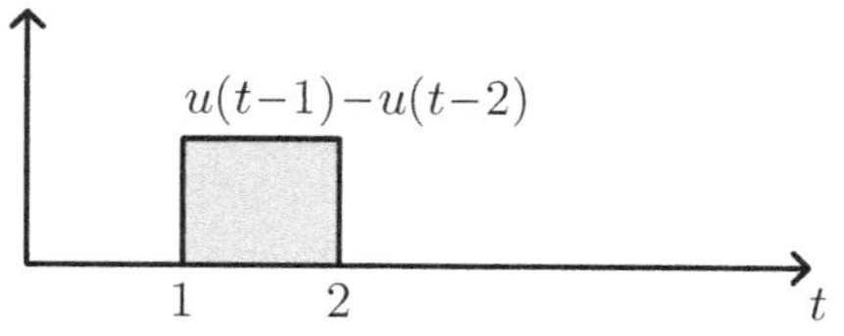

Fig. 16.3 Shifted pulse function

Pull out the e^{-st_0} since it has no u dependence and have

$$\mathcal{L}[f(t-t_0)] = e^{-st_0}\int_0^{\infty} f(u)e^{-su}du = e^{-st_0}F(s) \tag{16.10}$$

and the proof is complete.

Example Find LT of shifted impulse function $f(t) = \delta(t-2)$ as shown in Fig. 16.1: We know that $\delta(t) \to 1$; then

$$\boxed{\delta(t-2) \to e^{-2s}} \tag{16.11}$$

Example Find LT of shifted unit step function $f(t) = u(t-1)$ as shown in Fig. 16.2. We know $u(t) \to \frac{1}{s}$ then

$$\boxed{u(t-1) \to \frac{e^{-s}}{s}} \tag{16.12}$$

Example Find LT of the single pulse as shown in Fig. 16.3. We can construct this pulse out of two unit steps as follows:

$$\text{pulse}(1, 2, t) = u(t-1) - u(t-2) \tag{16.13}$$

We know the shifted LT of each unit step function; hence

$$\boxed{\text{pulse}(1, 2, t) \to \frac{e^{-s} - e^{-2s}}{s}} \tag{16.14}$$

16.5 Frequency Shifting

This states that if $\mathcal{L}[f(t)] = F(s)$ then

$$\boxed{\mathcal{L}\left[f(t)e^{-s_0t}\right] = F(s+s_0)} \tag{16.15}$$

We can prove this as follows:

$$\mathcal{L}\left[f(t)e^{-s_0t}\right] = \int_0^{\infty} f(t)e^{-s_0t}e^{-st}dt = \int_0^{\infty} f(t)e^{-t(s+s_0)}dt = F(s+s_0) \tag{16.16}$$

Example Find the LT of te^{-at} as shown in Fig. 16.4. We know $t \to \frac{1}{s^2}$; then using the frequency shifting property we get

$$\boxed{te^{-at} \to \frac{1}{(s+a)^2}} \tag{16.17}$$

Example Find the LT of $e^{-at}\sin\omega_0 t$ as shown in Fig. 16.5. We know $\sin\omega_0 t \to \frac{\omega_0}{s^2+\omega_0^2}$; then by the frequency shifting property we get

$$\boxed{e^{-at}\sin\omega_0 t \to \frac{\omega_0}{(s+a)^2+\omega_0^2}} \tag{16.18}$$

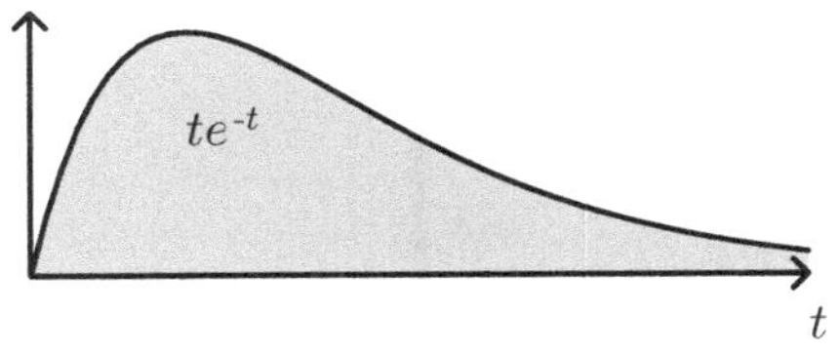

Fig. 16.4 Function te^{-at}

Example Find the LT of $te^{-j\omega_0 t}$. We know $t \to \frac{1}{s^2}$; then using the frequency shifting property we get

$$\boxed{te^{-j\omega_0 t} \to \frac{1}{(s+j\omega_0)^2}} \tag{16.19}$$

Similarly we get

$$\boxed{te^{j\omega_0 t} \to \frac{1}{(s-j\omega_0)^2}} \tag{16.20}$$

Example Find the LT of $t\sin\omega_0 t$ (Fig. 16.6). We can rewrite our function as

$$t\sin\omega_0 t = t\frac{e^{j\omega_0 t} - e^{-j\omega_0 t}}{2j} \tag{16.21}$$

We already know the LT of each of the terms on the RHS of the above equation. By linearity we proceed

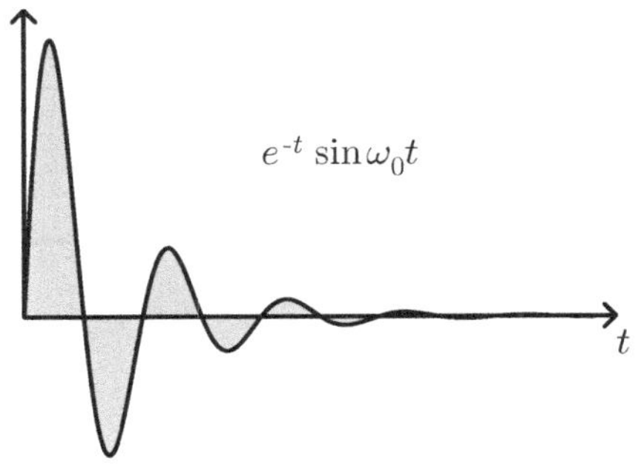

Fig. 16.5 Function $e^{-at}\sin\omega_0 t$

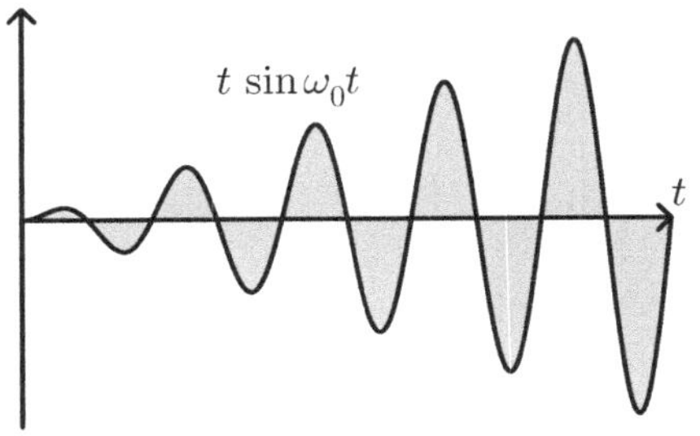

Fig. 16.6 Function $t\sin\omega_0 t$

$$\begin{aligned}\mathcal{L}[t\sin\omega_0 t] &= \frac{1}{2j}\left[\frac{1}{(s-j\omega_0)^2} - \frac{1}{(s+j\omega_0)^2}\right] = \frac{1}{2j}\frac{(s+j\omega_0)^2 - (s-j\omega_0)^2}{(s^2+\omega_0^2)^2}\\ &= \frac{1}{2j}\frac{s^2+2js\omega_0-\omega_0^2-[s^2-2js\omega_0-\omega_0^2]}{(s^2+\omega_0^2)^2} = \frac{1}{2j}\frac{4j\omega_0 s}{(s^2+\omega_0^2)^2}\\ &= \boxed{\frac{2\omega_0 s}{(s^2+\omega_0^2)^2}}\end{aligned} \tag{16.22}$$

16.6 Time Differentiation Property

This states that if $\mathcal{L}[f(t)] = F(s)$ then

$$\boxed{\mathcal{L}\left[\frac{df(t)}{dt}\right] = sF(s) - f(0-)} \tag{16.23}$$

where $f(0-)$ is the value of the function right at time $0-$. To prove this we do the following:

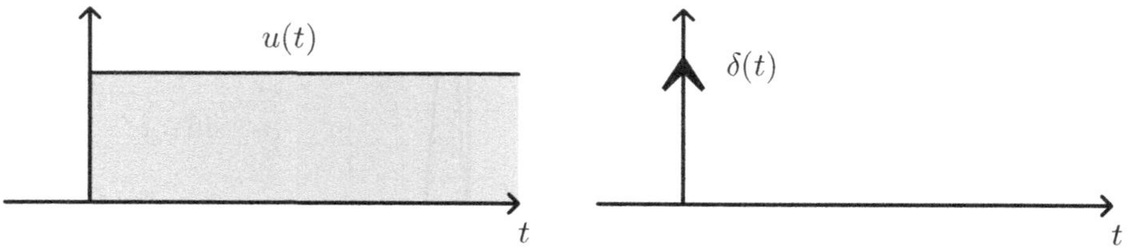

Fig. 16.7 Unit step and derivative

$$\mathcal{L}\left[\frac{df(t)}{dt}\right] = \int_{0-}^{\infty} \frac{df(t)}{dt} e^{-st} dt \qquad (16.24)$$

Do integration by parts

$$\int_{0-}^{\infty} \frac{df(t)}{dt} e^{-st} dt = f(t)e^{-st}\Big|_{0-}^{\infty} + s\int_{0-}^{\infty} f(t)e^{-st} dt \qquad (16.25)$$

Notice that

$$f(t)e^{-st}|_{t=\infty} = 0 \qquad \text{and}$$
$$f(t)e^{-st}|_{0-} = f(0-) \qquad (16.26)$$

Hence

$$\int_{0-}^{\infty} \frac{df(t)}{dt} e^{-st} dt = sF(s) - f(0-) \qquad (16.27)$$

and the proof is complete.

Meaning of $f(0-)$ In almost all cases $f(0-)$ is taken to be zero. The exception is systems with *initial conditions*. Then $f(0-)$ would represent the initial condition of the system. To illustrate this, consider the capacitor. The relation between its current and voltage is

$$i(t) = C\frac{dv(t)}{dt} \qquad (16.28)$$

If we assume voltage of the form $v(t) \sim e^{st}$ then applying the Laplace transform to the above equation gives

$$I(s) = C\left[sV(s) - v(0-)\right] \qquad (16.29)$$

Rearranging we get

$$V(s) = \frac{I(s)}{sC} + \frac{v(0-)}{s} \qquad (16.30)$$

The first term on the right side is nothing other than the current charging the cap. For example, if current is a delta function such that $I(s) = 1$ then that term simply gives a unit step function in time. But there is more! The second term on the right outright gives a unit step function with magnitude $v(0-)$. What this means is that the initial voltage on the cap continues to manifest itself. It is the *sum* of initial voltage plus the active current that finally dictates *total* voltage across the cap!

Example with Zero Initial Conditions Find the LT of the derivative of the unit step function (Fig. 16.7). We know that the LT of the unit step function is $u(t) \to \frac{1}{s}$. The LT of the derivative is

$$\mathcal{L}\left[\frac{d}{dt}u(t)\right] = s\mathcal{L}\left[u(t)\right] - u(0-) = s\frac{1}{s} = \boxed{1} \qquad (16.31)$$

but this is nothing more than the LT of the delta function; i.e.,

$$\delta(t) \to 1 \qquad (16.32)$$

This should come as no surprise since

$$\frac{d}{dt}u(t) = \delta(t) \qquad (16.33)$$

Example Where $f(0-)$ Is Not Zero Consider the step function shown in Fig. 16.8. This function has a nonzero value for negative time; in particular it has $f(0-) = \frac{1}{2}$. If we calculate the LT we get

$$F(s) = \frac{1}{s} \qquad (16.34)$$

which is the same as the typical unit step function. The reason is that the LT only integrates from 0 to ∞ and in that view, both would give the same answer. But when taking the derivative of the shown step function, we would get

$$\frac{df(t)}{dt} = \frac{1}{2}\delta(t) \tag{16.35}$$

rather than a $\delta(t)$. Let's see if using the time differentiation property predicts the correct answer. We get

$$\mathcal{L}[f(t)] = s\frac{1}{s} - \frac{1}{2} = 1 - \frac{1}{2} = \frac{1}{2} \tag{16.36}$$

which is exactly the LT of the half delta function

$$\frac{1}{2}\delta(t) \rightarrow \frac{1}{2} \tag{16.37}$$

Example with Zero Initial Conditions Find the LT of the derivative of the ramp function (Fig. 16.9). We know that the LT of the ramp function is $t \rightarrow \frac{1}{s^2}$. The LT of the derivative is

$$\mathcal{L}\left[\frac{d}{dt}t\right] = s\mathcal{L}(t) - 0 = s\frac{1}{s^2} = \boxed{\frac{1}{s}} \tag{16.38}$$

but this is nothing more than the LT of the unit step function; this is no surprise since

$$\frac{d}{dt}t = u(t) \tag{16.39}$$

Example with Zero Initial Conditions Find the LT of the derivative of the sine function $\sin\omega_0 t$ (Fig. 16.10). Recall that

$$\mathcal{L}[\sin\omega_0 t] = \frac{\omega_0}{s^2+\omega_0^2} \tag{16.40}$$

The LT of the derivative of this function is then

$$\mathcal{L}\left[\frac{d\sin\omega_0 t}{dt}\right] = s\mathcal{L}(\sin\omega_0 t) - \sin(0) = \boxed{\omega_0\frac{s}{s^2+\omega_0^2}} \tag{16.41}$$

But this is nothing more than the LT of the cosine function:

$$\omega_0\cos\omega_0 t \rightarrow \omega_0\frac{s}{s^2+\omega_0^2} \tag{16.42}$$

This should be no surprise since

$$\frac{d\sin\omega_0 t}{dt} = \omega_0\cos\omega_0 t \tag{16.43}$$

Example with Zero Initial Conditions Find the LT of the derivative of the (causal) cosine function $\cos\omega_0 t$ (Fig. 16.11). Notice we called it "causal" assuming that right at time zero (−), it has a value of 0. Then it jumps abruptly to $\cos(0) = 1$. Nonetheless, the value at time zero (−) is 0! We start with

$$\mathcal{L}[\cos\omega_0 t] = \frac{s}{s^2+\omega_0^2} \tag{16.44}$$

The LT of the derivative of the cosine is

$$\mathcal{L}\left[\frac{d\cos\omega_0 t}{dt}\right] = s\mathcal{L}(\cos\omega_0 t) - 0 = \boxed{\frac{s^2}{s^2+\omega_0^2}} \tag{16.45}$$

But hold on—this does not look like the LT of the sine function, which we would have expected. Let's simplify further. Using long division we get

$$\frac{s^2}{s^2+\omega_0^2} = 1 - \frac{\omega_0^2}{s^2+\omega_0^2} \tag{16.46}$$

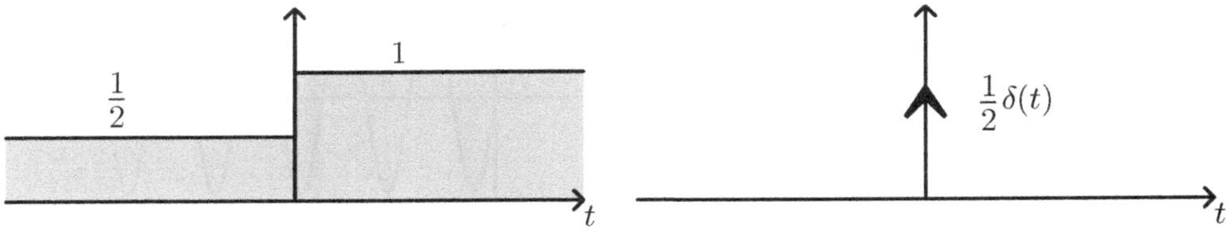

Fig. 16.8 Modified unit step and derivative

Now the inverse LT of this is

$$1 - \frac{\omega_0^2}{s^2 + \omega_0^2} \rightarrow \delta(t) - \omega_0 \sin \omega_0 t \qquad (16.47)$$

Now we pick the sine function, but additionally we are picking up a delta function! The reason we pick the delta function is because of the abrupt discontinuity right at time zero. Since we assumed the cosine starts 0 at time zero, then the derivative will pick the delta function right there. What this means is that the derivative of the cosine function is not simply a sine function—it is a sine plus a delta function. And this in fact is what the application of the time differentiation rule has given us in Eq. (16.45).

Example with Nonzero Initial Conditions Let's tweak the example just presented but this time assume the cosine is continuous for all time. Again we know

$$\cos \omega_0 t \rightarrow \frac{s}{s^2 + \omega_0^2} \qquad (16.48)$$

Noting that now $f(0-)$ is *not* zero, and in fact

$$f(0-) = 1 \qquad (16.49)$$

Application of the time differentiation property yields

$$\mathcal{L}\left[\frac{d \cos \omega_0 t}{dt}\right] = s\mathcal{L}(\cos \omega_0 t) - 1 = \frac{s^2}{s^2 + \omega_0^2} - 1 = \boxed{\frac{-\omega_0^2}{s^2 + \omega_0^2}} \qquad (16.50)$$

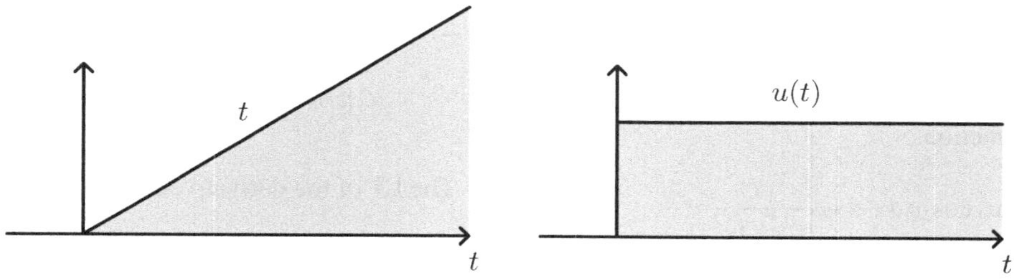

Fig. 16.9 Ramp function and derivative

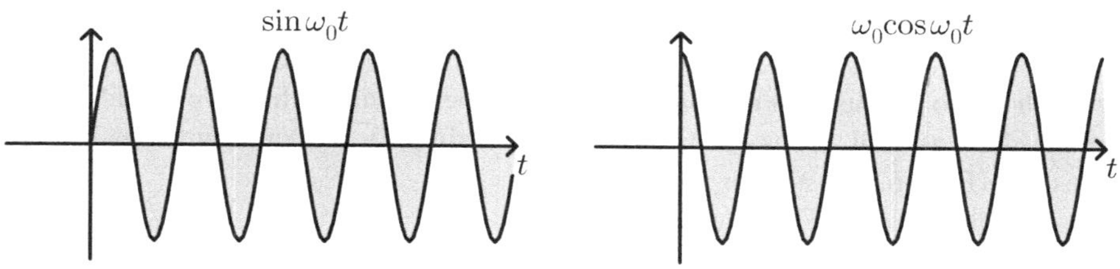

Fig. 16.10 Sine function and derivative

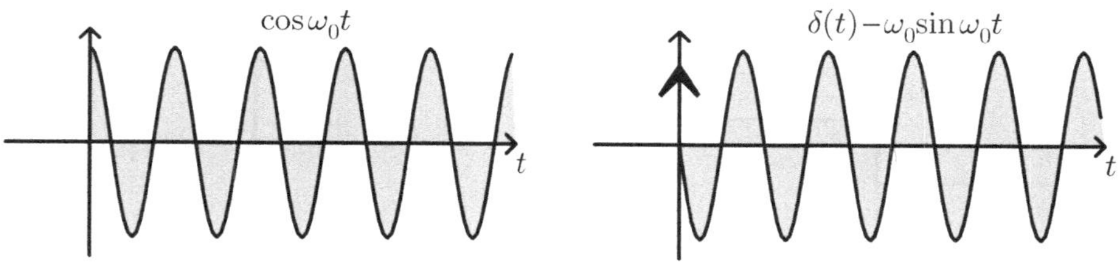

Fig. 16.11 Causal cosine function and derivative

which is nothing other than the LT of the sine function

$$\frac{\omega_0^2}{s^2+\omega_0^2} \rightarrow \omega_0 \sin \omega_0 t \qquad (16.51)$$

So in *this* case we get a straightforward mapping between the cosine and the sine, and without picking any delta functions. The reason is because the assumed cosine function here is continuous and does *not* have a discontinuity at time zero; this is shown in Fig. 16.12.

The time differentiation property will give us exactly what we throw at it. If there is a discontinuity at time zero we will get a delta function. Notice that a discontinuity may exist because the system was not initialized *or* partially initialized! What this means if we plot the transient signal and observe *any* discontinuity around time zero, then we should expect some sort of a delta function. On the other hand, if there is no discontinuity whatsoever around time zero, then we will not get a delta function.

Example with Zero Initial Conditions Find the LT of the derivative of the negative exponential function e^{-at} (Fig. 16.13). Recall that

$$e^{-at} \rightarrow \frac{1}{s+a} \qquad (16.52)$$

Then

$$\mathcal{L}\left[\frac{de^{-at}}{dt}\right] = s\mathcal{L}(e^{-at}) - 0 = \frac{s}{s+a} \qquad (16.53)$$

We can expand this via partial fractions to get

$$\frac{s}{s+a} = \boxed{1 - \frac{a}{s+a}} \qquad (16.54)$$

We can identify the right two terms above as the LT of

$$1 - \frac{a}{s+a} \rightarrow \delta(t) - ae^{-at} \qquad (16.55)$$

That is, the LT of the derivative gave a negative exponential (scaled by $-a$) *and* a delta function. We should expect the latter due to the discontinuity at time zero.

16.7 Time Integration Property

This states that the LT of the integral of $f(t)$ is the LT of $f(t)$ divided by s; that is

$$\boxed{\mathcal{L}\left[\int_0^t f(\tau)d\tau\right] = \frac{F(s)}{s}} \qquad (16.56)$$

(Luckily, we don't pick up the $f(0-)$ issue and corresponding time delta functions we struggled with in the time *differentiation* property. The formula is mere division by frequency—no offset factors!) We can prove this as follows:

$$\mathcal{L}\left[\int_0^t f(\tau)d\tau\right] = \int_0^\infty \left[\int_0^t f(\tau)d\tau\right] e^{-st}dt \qquad (16.57)$$

Let

$$\begin{aligned} u &= \textstyle\int_0^t f(\tau)d\tau, \text{ and } dv = e^{-st}, \text{ such that} \\ du &= f(t), \qquad \text{and } v = -\tfrac{1}{s}e^{-st} \end{aligned} \qquad (16.58)$$

Then using integration by parts $\int u dv = uv - \int v du$, we get

$$\mathcal{L}\left[\int_0^t f(\tau)d\tau\right] = \left[-\frac{1}{s}e^{-st}\int_0^t f(\tau)d\tau\right]_0^\infty - \int_0^\infty f(t)\frac{-1}{s}e^{-st}dt \tag{16.59}$$

We notice that the first (bracketed) terms on the right side go to zero both at $t = 0$ and $t = \infty$; so we are left with the second term. But this is nothing other than

$$-\int_0^\infty f(t)\frac{-1}{s}e^{-st}dt = \frac{1}{s}\int_0^\infty f(t)e^{-st}dt = \frac{1}{s}F(s) \tag{16.60}$$

Hence the proof is complete!

Example Find the LT of the integral of $\delta(t)$ (Fig. 16.14). We proceed as follows:

$$\mathcal{L}\left[\int_0^t \delta(\tau)d\tau\right] = \frac{1}{s}\mathcal{L}\left[\delta(t)\right] = \boxed{\frac{1}{s}} \tag{16.61}$$

which makes sense since

$$\int_0^t \delta(\tau)d\tau = u(t) \tag{16.62}$$

and

$$\mathcal{L}[u(t)] = \frac{1}{s} \tag{16.63}$$

Example Find the LT of the integral of the unit step function $u(t)$ (Fig. 16.15). We proceed as follows:

$$\mathcal{L}\left[\int_0^t u(\tau)d\tau\right] = \frac{1}{s}\mathcal{L}\left[u(t)\right] = \boxed{\frac{1}{s^2}} \tag{16.64}$$

which makes sense since

$$\int_0^t u(\tau)d\tau = t \tag{16.65}$$

and

$$\mathcal{L}[t] = \frac{1}{s2} \tag{16.66}$$

Example Find the LT of the integral of the negative exponential function e^{-at} (Fig. 16.16). We proceed as follows:

$$\mathcal{L}\left[\int_0^t e^{-a\tau}d\tau\right] = \frac{1}{s}\mathcal{L}\left[e^{-at}\right] = \boxed{\frac{1}{s}\frac{1}{s+a}} \tag{16.67}$$

which makes sense since

$$\int_0^t e^{-a\tau}d\tau = \frac{1}{a} - \frac{1}{a}e^{-at}, \quad \text{and} \tag{16.68}$$

$$\mathcal{L}\left[\frac{1}{a} - \frac{1}{a}e^{-at}\right] = \frac{1}{as} - \frac{1}{a}\frac{1}{s+a} = \frac{1}{a}\frac{s+a-s}{s(s+a)} = \frac{1}{s(s+a)} \tag{16.69}$$

Example Find the LT of the integral of the cosine function $\cos\omega_0 t$. We proceed as follows

$$\mathcal{L}\left[\int_0^t \cos\omega_0\tau d\tau\right] = \frac{1}{s}\mathcal{L}\left[\cos\omega_0 t\right] = \frac{1}{s}\frac{s}{s^2+\omega_0^2} = \boxed{\frac{1}{s^2+\omega_0^2}} \tag{16.70}$$

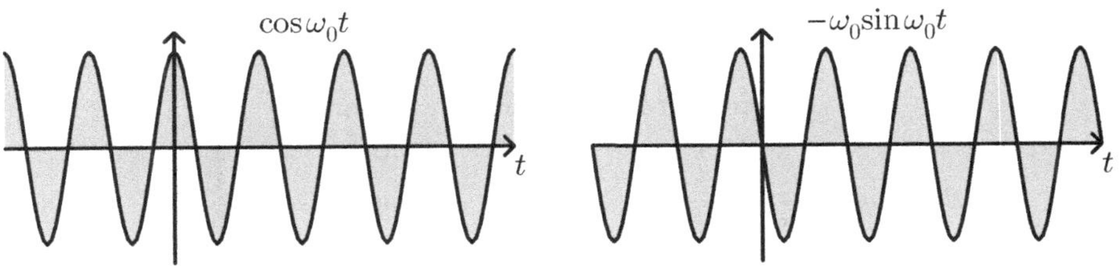

Fig. 16.12 Continuous cosine function and derivative

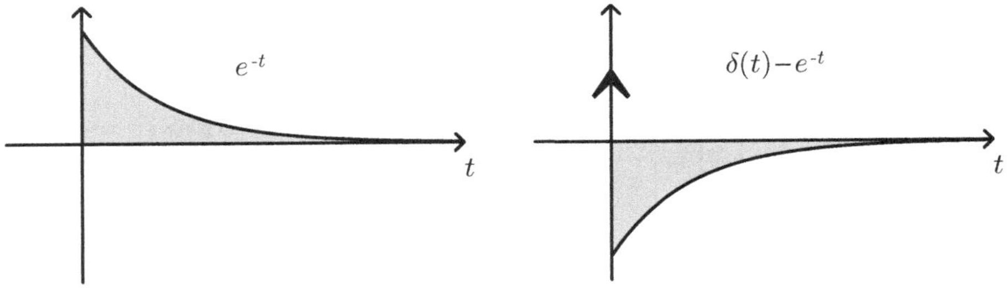

Fig. 16.13 Negative exponential and derivative

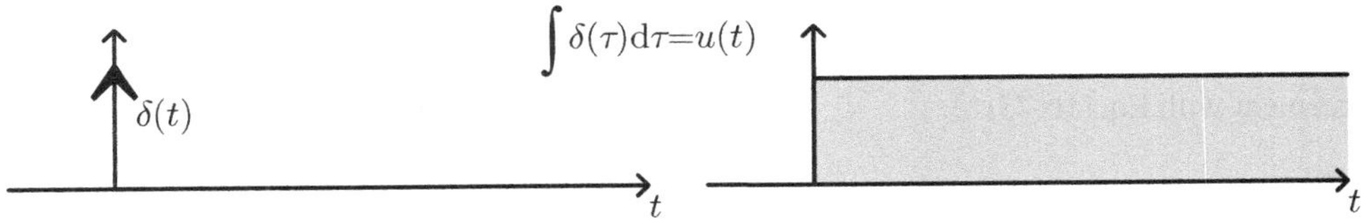

Fig. 16.14 Delta function and time integration

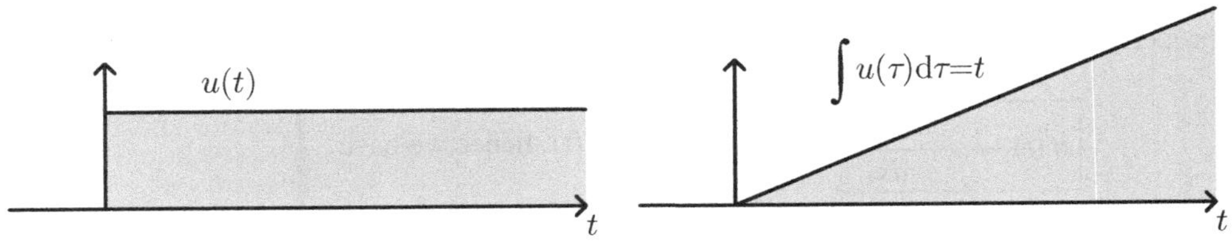

Fig. 16.15 Unit step function and time integration

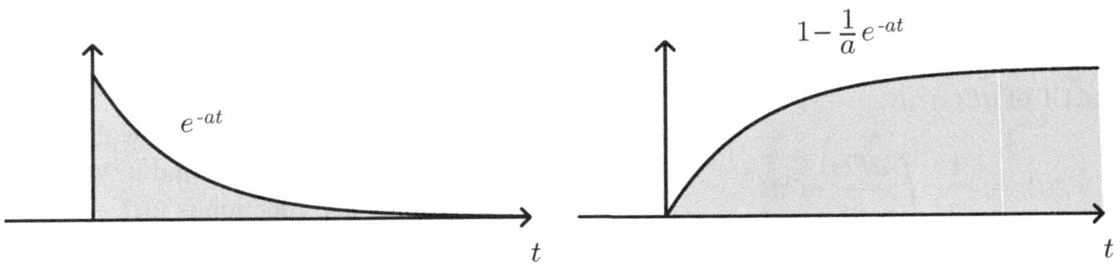

Fig. 16.16 Negative exponential and time integration

This makes sense since

$$\int_0^t \cos\omega_0\tau d\tau = \frac{1}{\omega_0}\sin\omega_0 t, \quad \text{and} \tag{16.71}$$

$$\mathcal{L}[\sin\omega_0] = \frac{\omega_0}{s^2+\omega_0^2}, \quad \text{so that} \quad \frac{1}{\omega_0}\sin\omega_0 t \rightarrow \frac{1}{s^2+\omega_0^2} \tag{16.72}$$

Example Find the LT of the integral of the sine function $\sin\omega_0 t$ (Fig. 16.17). We proceed as follows:

$$\mathcal{L}\left[\int_0^t \sin\omega_0\tau d\tau\right] = \frac{1}{s}\mathcal{L}[\sin\omega_0 t] = \boxed{\frac{1}{s}\frac{\omega_0}{s^2+\omega_0^2}} \tag{16.73}$$

This is not as easy to see but we can prove as follows. Perform the time integration to get

$$\int_0^t \sin\omega_0\tau d\tau = -\frac{1}{\omega_0}\cos\omega_0\tau\Big|_0^t = \frac{1}{\omega_0}(1-\cos\omega_0 t) \tag{16.74}$$

Now take the LT of this last result and get

$$\mathcal{L}\left[\frac{1}{\omega_0}(1-\cos\omega_0 t)\right] = \frac{1}{\omega_0}\left[\frac{1}{s} - \frac{s}{s^2+\omega_0^2}\right] = \frac{1}{\omega_0}\frac{s^2+\omega_0^2-s^2}{s(s^2+\omega_0^2)} = \frac{\omega_0}{s(s^2+\omega_0^2)} \tag{16.75}$$

in agreement with Eq. (16.73).

16.8 Frequency Differentiation Property

This property states that if $f(t) \rightarrow F(s)$ then

$$\boxed{tf(t) \rightarrow -\frac{dF(s)}{ds}} \tag{16.76}$$

provided $F(s)$ goes to zero as $|s| \rightarrow \infty$. This last condition is almost guaranteed—most signals have a finite frequency bandwidth, and as such die off at high enough frequency. The property can be proven as follows. We start with the inverse LT of $dF(s)/ds$:

$$g(t) = \frac{1}{2\pi j}\int \frac{dF(s)}{ds}e^{st}ds \tag{16.77}$$

Notice we called this $g(t)$ and not $f(t)$. Doing integration by parts we get

$$g(t) = F(s)e^{st}\Big|_{s=-\infty}^{\infty} - t\int F(s)e^{st}ds \tag{16.78}$$

Since we assumed that $F(s)$ goes to zero for both $-\infty$ and ∞, the first term on the right side goes to zero and we end up with

$$g(t) = -t\int F(s)e^{st}ds \tag{16.79}$$

But this integral term is nothing more than $f(t)$; hence we have

$$g(t) = -tf(t) \tag{16.80}$$

In other words,

$$-\frac{dF(s)}{ds} \rightarrow tf(t) \tag{16.81}$$

Example Assume we don't know the LT of the function t (Fig. 16.18); let's find it. We know the LT of the unit step function is $u(t) \rightarrow \frac{1}{s}$. By the frequency differentiation property we get

$$tu(t) \rightarrow -\frac{d}{ds}\frac{1}{s} = \frac{1}{s^2} \tag{16.82}$$

as expected.

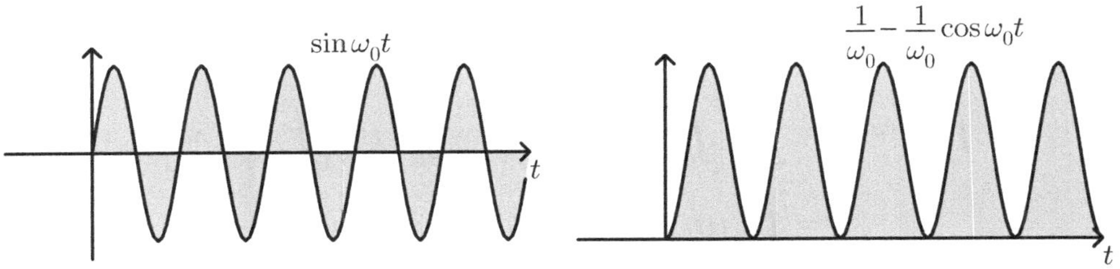

Fig. 16.17 Sine function and time integration

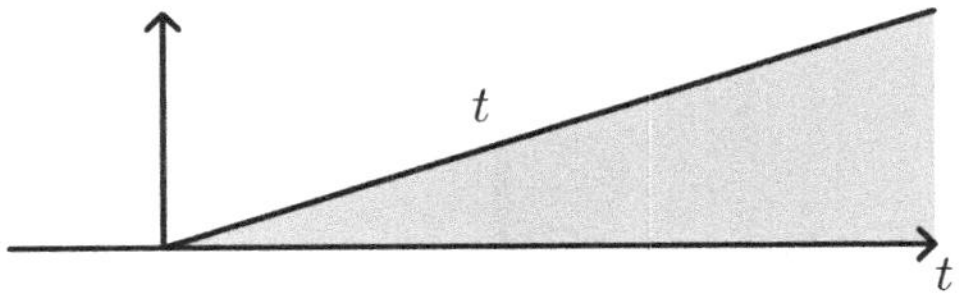

Fig. 16.18 Function t as $u(t) \times t$

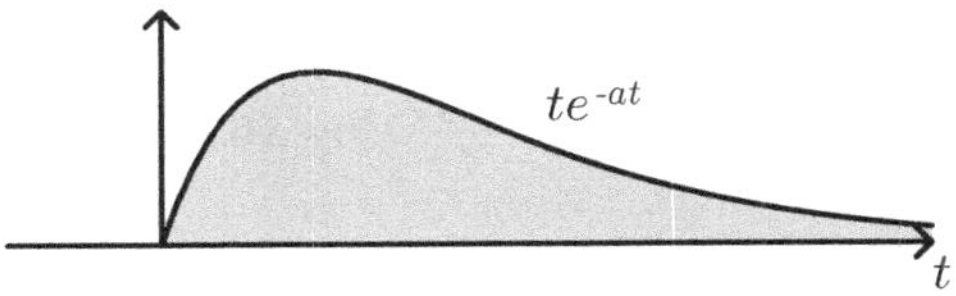

Fig. 16.20 Function te^{-at} as $t \times e^{-at}$

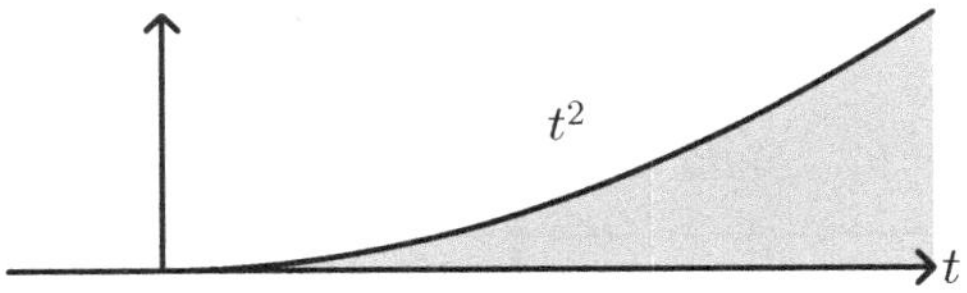

Fig. 16.19 Function t^2 as $t \times t$

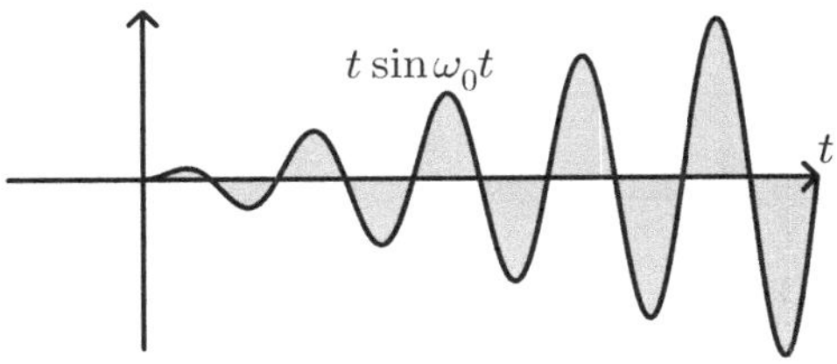

Fig. 16.21 Function $t \sin \omega_0 t$ as $t \times \sin \omega_0 t$

Example Find LT of t^2 (Fig. 16.19). We know the LT of the t function is $t \rightarrow \frac{1}{s^2}$. By the frequency differentiation property we get

$$t^2 \rightarrow -\frac{d}{ds}\frac{1}{s^2} = \frac{2}{s^3} \tag{16.83}$$

Example Find LT of te^{-at} (Fig. 16.20). We know LT of e^{-at} is $\frac{1}{s+a}$; then by the frequency differentiation property we get

$$te^{-at} \rightarrow -\frac{d}{ds}\frac{1}{s+a} = \boxed{\frac{1}{(s+a)^2}} \tag{16.84}$$

Example Find LT of $t \sin \omega_0 t$ (Fig. 16.21). Recall

$$\sin \omega_0 t \rightarrow \frac{\omega_0}{s^2 + \omega_0^2} \tag{16.85}$$

In this case we get

$$t \sin \omega_0 t \rightarrow -\frac{d}{ds}\frac{\omega_0}{s^2 + \omega_0^2} = \boxed{\frac{2s\omega_0}{(s^2 + \omega_0^2)^2}} \tag{16.86}$$

16.9 Frequency Integration Property

This property states that if $f(t) \rightarrow F(s)$ then

$$\boxed{\frac{f(t)}{t} \rightarrow \int_s^\infty F(u)du} \tag{16.87}$$

This can be proved as follows:

$$\int_s^\infty F(u)du = \int_s^\infty \left[\int_0^\infty f(t)e^{-ut}dt\right] du = \int_0^\infty f(t)\left[\int_s^\infty e^{-ut}du\right] dt$$
$$= \int_0^\infty f(t)\left[-\frac{e^{-ut}}{t}\bigg|_s^\infty\right] dt = \int_0^\infty f(t)\left[0+\frac{e^{-st}}{t}\right] dt$$
$$= \int_0^\infty \frac{f(t)}{t}e^{-st}dt = \mathcal{L}\left[\frac{f(t)}{t}\right] \tag{16.88}$$

Example Find LT of $\frac{\sin\omega_0 t}{t}$ (Fig. 16.22). Recall

$$\sin\omega_0 t \rightarrow \frac{\omega_0}{s^2+\omega_0^2} \tag{16.89}$$

Then by the frequency integration property we should expect

$$\frac{\sin\omega_0 t}{t} \rightarrow \int_s^\infty \frac{\omega_0}{u^2+\omega_0^2}du \tag{16.90}$$

To do the integration we get

$$u = \omega_0\tan\theta, \qquad u^2+\omega_0^2 = \omega_0^2\sec^2\theta, \qquad du = \omega_0\sec^2\theta d\theta \tag{16.91}$$

Then our integral becomes

$$\int_s^\infty \frac{\omega_0}{u^2+\omega_0^2}du = \int \frac{\omega_0^2\sec^2\theta}{\omega_0^2\sec^2\theta}d\theta = \int d\theta = \theta \tag{16.92}$$

Now we plug back for θ and get

$$\int_s^\infty \frac{\omega_0}{u^2+\omega_0^2}du = \text{atan}\frac{u}{\omega_0}\bigg|_s^\infty = \text{atan}(\infty) - \text{atan}\frac{s}{\omega_0} = \boxed{\frac{\pi}{2} - \text{atan}\frac{s}{\omega_0}} \tag{16.93}$$

Results are shown in Fig. 16.22.

Example We know that the LT of the ramp function is

$$u(t)t \rightarrow \frac{1}{s^2} \tag{16.94}$$

If we do the frequency integration of this function we get

$$\int_s^\infty \frac{1}{u^2}du = -\frac{1}{u}\bigg|_s^\infty = \frac{1}{s} \tag{16.95}$$

But this is nothing more than the LT of

$$\frac{u(t)t}{t} = u(t)t \rightarrow \frac{1}{s} \tag{16.96}$$

Hence we have verified the frequency integration property. Results are shown in Fig. 16.23.

Example We know that

$$u(t)e^{-at} \rightarrow \frac{1}{s+a} \tag{16.97}$$

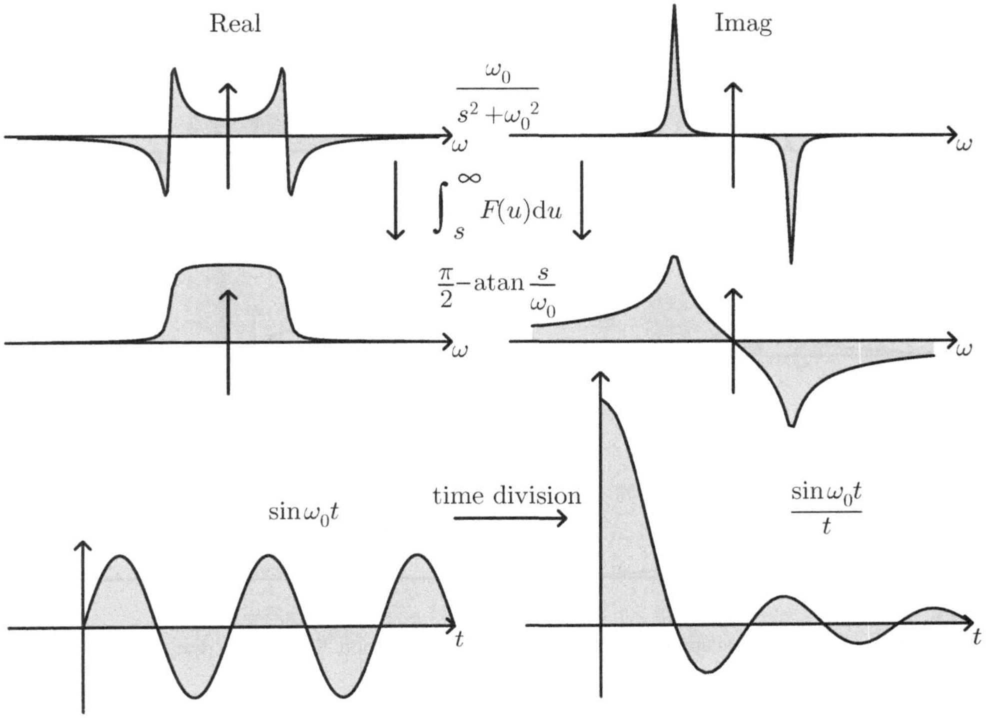

Fig. 16.22 Frequency integration applied to the $\sin\omega_0 t$ function

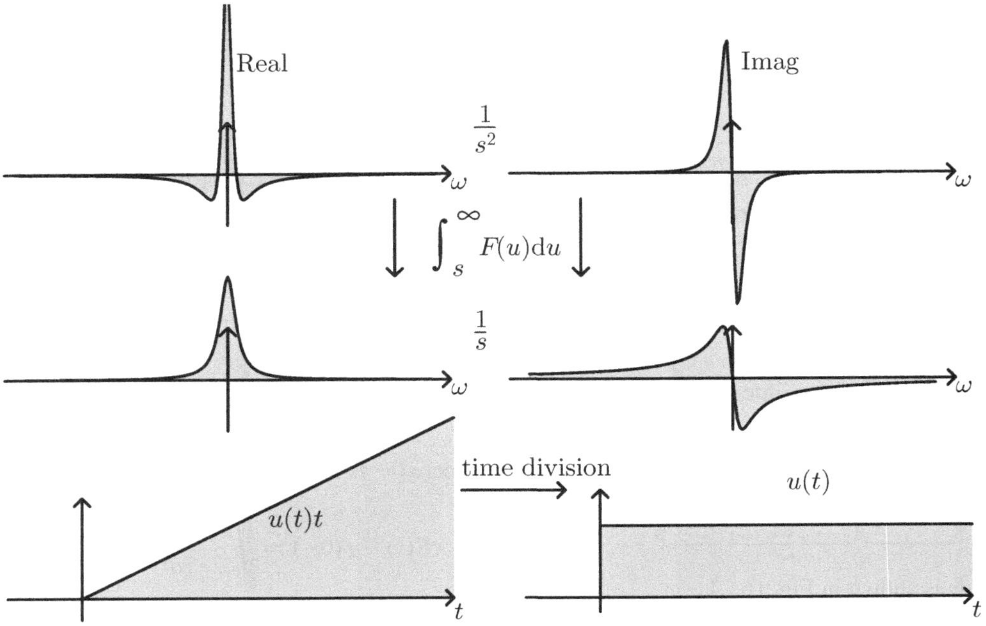

Fig. 16.23 Frequency integration applied to the $u(t)t$ function

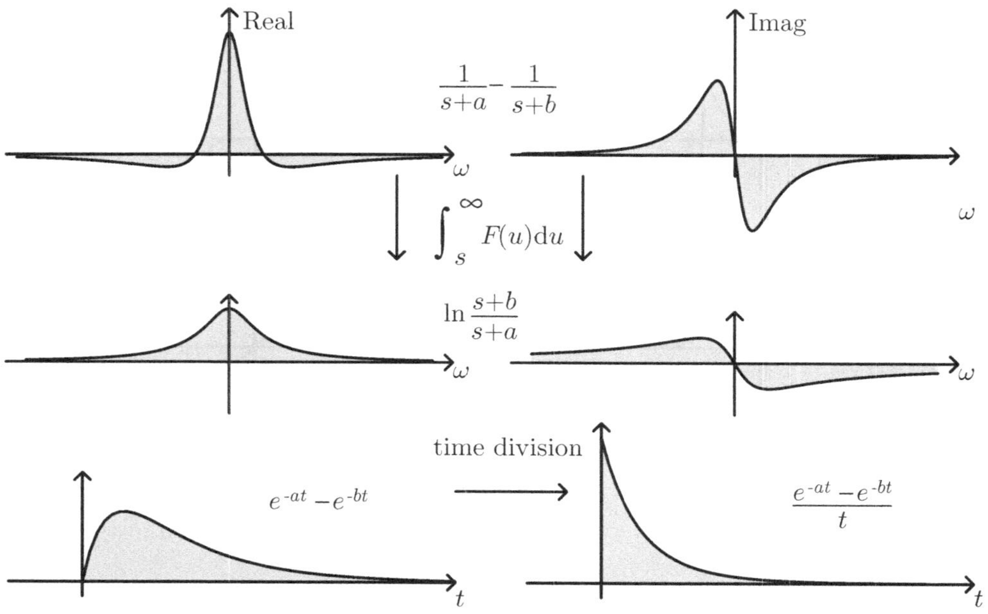

Fig. 16.24 Frequency integration applied to the $u(t)\left[e^{-at} - e^{-bt}\right]$ function. Case of $a = 1$ and $b = 3$

We also know that

$$u(t)e^{-bt} \rightarrow \frac{1}{s+b} \tag{16.98}$$

If we take the difference we arrive at

$$u(t)\left[e^{-at} - e^{-bt}\right] \rightarrow \frac{1}{s+a} - \frac{1}{s+b} \tag{16.99}$$

If we next integrate the resulting transfer function we arrive at

$$\int_s^\infty \frac{1}{u+a} - \frac{1}{u+b} du = \ln \frac{s+b}{s+a} \tag{16.100}$$

From the frequency integrate property we then conclude that

$$\boxed{u(t)\frac{e^{-at} - e^{bt}}{t} \rightarrow \ln \frac{s+b}{s+a}} \tag{16.101}$$

This is shown in Fig. 16.24.

16.10 Initial Value Theorem

This theorem relates the value of the function $f(t)$ at time $0+$ to the value of $sF(s)$ in the limit at s goes to ∞; in particular it states that if $f(t) \rightarrow F(s)$ then

$$\boxed{\lim_{s\to\infty} sF(s) = f(0+)} \tag{16.102}$$

where $f(0+)$ is the value of the function right at time $0+$. This can be proved as follows. From the time differentiation property we know that

$$\frac{df(t)}{dt} \rightarrow sF(s) - f(0-) \tag{16.103}$$

where $0-$ is time just before zero. Hence

$$sF(s) - f(0-) = \int_{0-}^{\infty} \frac{df(t)}{dt} e^{-st} dt \tag{16.104}$$

Split the integral as follows:

$$sF(s) - f(0-) = \int_{0-}^{0+} \frac{df(t)}{dt} e^{-st} dt + \int_{0+}^{\infty} \frac{df(t)}{dt} e^{-st} dt \quad (16.105)$$

The first integral is evaluated right around time zero; hence $e^{-st} \rightarrow 1$. Then we would integrate a derivative and that gives

$$sF(s) - f(0-) = f(0+) - f(0-) + \int_{0+}^{\infty} \frac{df(t)}{dt} e^{-st} dt \quad (16.106)$$

We see that $f(0-)$ subtracts out. Next take the $s \rightarrow \infty$ limit. Under this limit

$$\lim_{s\rightarrow\infty} e^{-st} \rightarrow 0 \quad (16.107)$$

Then we end up with

$$sF(s) - \cancel{f(0-)} = f(0+) - \cancel{f(0-)} + \cancelto{0}{\int_{0+}^{\infty} \frac{df(t)}{dt} e^{-st} dt} \quad (16.108)$$

Example We know that the LT of the unit step function is $u(t) \rightarrow \frac{1}{s}$. Then we have

$$\lim_{s\rightarrow\infty} sF(s) = \lim_{s\rightarrow\infty} s\frac{1}{s} = 1 \quad (16.109)$$

which is in fact u(0+).

Example We know that the LT of the ramp function is $t \rightarrow \frac{1}{s^2}$. Then we have

$$\lim_{s\rightarrow\infty} sF(s) = \lim_{s\rightarrow\infty} s\frac{1}{s^2} = \lim_{s\rightarrow\infty} \frac{1}{s} = 0 \quad (16.110)$$

which is in fact the function t at time zero (+).

Example We know that the LT of the negative exponential function is $e^{-at} \rightarrow \frac{1}{s+a}$. Then we have

$$\lim_{s\rightarrow\infty} sF(s) = \lim_{s\rightarrow\infty} s\frac{1}{s+a} = 1 \quad (16.111)$$

which is in fact e^0

Example We know that the LT of the cosine function is $\cos\omega_0 t \rightarrow \frac{s}{s^2+\omega_0^2}$. Then we have

$$\lim_{s\rightarrow\infty} sF(s) = \lim_{s\rightarrow\infty} s\frac{s}{s^2+\omega_0^2} = 1 \quad (16.112)$$

which is in fact cos(0).

16.11 Final Value Theorem

This states that if $f(t) \rightarrow F(s)$ then

$$\boxed{\lim_{s\rightarrow 0} sF(s) = f(\infty)} \quad (16.113)$$

This can be proved as follows. Start again with the time differentiation property of the LT:

$$\frac{df(t)}{dt} \rightarrow sF(s) - f(0-) \quad (16.114)$$

which means

$$\int_{0-}^{\infty} \frac{df(t)}{dt} e^{-st} dt = sF(s) - f(0-) \quad (16.115)$$

If we now take the limit $s \rightarrow 0$ we notice that $e^{-st} \rightarrow 1$. Then we have

$$\int_{0-}^{\infty} \frac{df(t)}{dt} dt = sF(s) - f(0-) \quad (s \rightarrow 0 \text{ limit}) \quad (16.116)$$

The left side can be directly integrated and gives

$$f(\infty) - \cancel{f(0-)} = sF(s) - \cancel{f(0-)} \quad (16.117)$$

The $f(0-)$ cancels out and we end up with

$$\lim_{s\rightarrow 0} sF(s) = f(\infty) \quad (16.118)$$

Example We know LT of the unit step function is $u(t) \to \frac{1}{s}$. Then

$$\lim_{s\to 0} s\frac{1}{s} = 1 = u(\infty) \tag{16.119}$$

Example We know LT of the negative exponential function is $e^{-at} \to \frac{1}{s+a}$. Then

$$\lim_{s\to 0} s\frac{1}{s+a} = 0 = e^{-\infty t} \tag{16.120}$$

Example We know LT of the function is $1 - e^{-at} \to \frac{a}{s(s+a)}$. Then

$$\lim_{s\to 0} s\frac{a}{s(s+a)} = \lim_{s\to 0}\frac{a}{(s+a)} = 1 = 1 - e^{-\infty t} \tag{16.121}$$

Example We know LT of the ramp function is $t \to \frac{1}{s^2}$. Then

$$\lim_{s\to 0} s\frac{1}{s^2} = \lim_{s\to 0}\frac{1}{s} = \infty = t|_\infty \tag{16.122}$$

16.12 Time Convolution Property

This states that if $f(t) \to F(s)$ and $g(t) \to G(s)$ then

$$\boxed{f(t) * g(t) \to F(s)G(s)} \tag{16.123}$$

Proof: Write the convolution integral and the LT one

$$\mathcal{L}\left[f(t) * g(t)\right] = \int_0^\infty \left[\int_0^\infty f(\tau)g(t-\tau)d\tau\right] e^{-st}dt \tag{16.124}$$

Exchange the order of integration and collect terms

$$\mathcal{L}\left[f(t) * g(t)\right] = \int_0^\infty f(\tau)\left[\int_0^\infty g(t-\tau)e^{-st}dt\right] d\tau \tag{16.125}$$

Let's look a bit closer at the integral inside the bracket. Let

$$u = t - \tau; \qquad du = dt \tag{16.126}$$

Then the integral becomes

$$\int_0^\infty g(t-\tau)e^{-st}dt = \int_0^\infty g(u)e^{-s(u+\tau)}du = e^{-s\tau}\int_0^\infty g(u)e^{-su}du \tag{16.127}$$

So the integral inside the bracket is nothing more than $G(s)e^{-s\tau}$. Hence we have

$$\mathcal{L}\left[f(t) * g(t)\right] = G(s)\int_0^\infty f(\tau)e^{-s\tau}d\tau \tag{16.128}$$

Now the right side is nothing more than $F(s)$; hence

$$\mathcal{L}\left[f(t) * g(t)\right] = G(s)F(s) \tag{16.129}$$

Example Let's find the LT of the convolution of the delta function with the unit step function (Fig. 16.25):

$$\mathcal{L}\left[\delta(t) * u(t)\right] = \mathcal{L}[\delta(t)] \times \mathcal{L}[u(t)] = 1 \times \frac{1}{s} = \boxed{\frac{1}{s}} \tag{16.130}$$

That is, the LT of the convolution of the delta function and the unit step function gives the LT of the unit step! In fact, *the Laplace transformof the convolution of the delta function and* ***any*** *other function gives the LT of the other function.*

Example Let's find the LT of the convolution of the unit step function with itself (Fig. 16.26):

$$\mathcal{L}\left[u(t) * u(t)\right] = \mathcal{L}[u(t)] \times \mathcal{L}[u(t)] = \frac{1}{s} \cdot \frac{1}{s} = \boxed{\frac{1}{s^2}} \tag{16.131}$$

But this is nothing more than the LT of the ramp function, which is the integral of the unit step function. It turns out that *the Laplace transform of the unit step function convolved with any other function gives the LT of the integral of the other function.*

Example Let's find the LT of the convolution of the unit step function with the ramp function (Fig. 16.27):

$$\mathcal{L}\left[u(t) * t\right] = \mathcal{L}[u(t)] \times \mathcal{L}[t] = \frac{1}{s} \cdot \frac{1}{s^2} = \boxed{\frac{1}{s^3}} \tag{16.132}$$

But the result is nothing more than the LT of the function

$$\frac{t^2}{2} \to \boxed{\frac{1}{s^3}} \tag{16.133}$$

Hence we conclude that

$$u(t) * t = \frac{t^2}{2} \tag{16.134}$$

which agrees with our conclusion that convolving a function with the unit step function amounts to integrating the function; that is, $\int t dt = t^2/2$.

Example Find the LT of the convolution of the unit step function with the negative exponential (Fig. 16.28). We know that

$$u(t) \to \frac{1}{s}, \qquad \text{and} \qquad e^{-at} \to \frac{1}{s+a} \tag{16.135}$$

Hence

$$\mathcal{L}\left[u(t) * e^{-at}\right] = \frac{1}{s}\frac{1}{s+a} \tag{16.136}$$

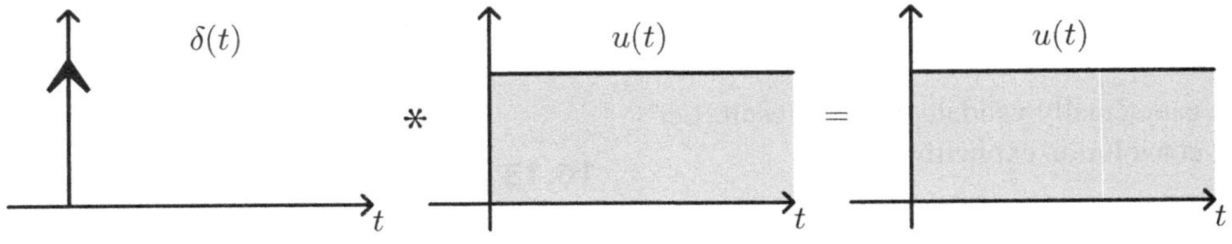

Fig. 16.25 Convolution between delta and unit step functions

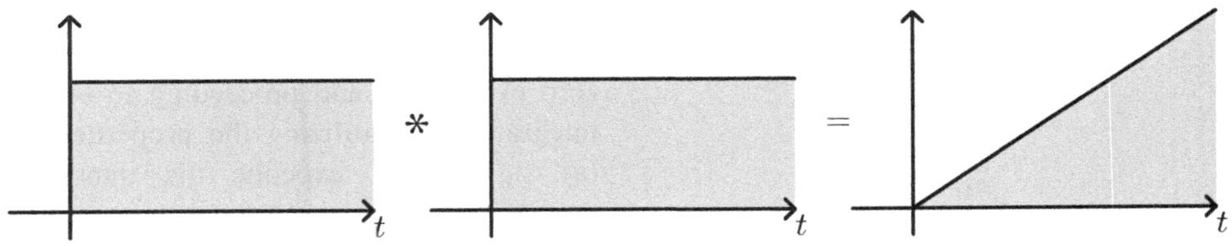

Fig. 16.26 Convolution between unit step function and itself

We can decompose this (using partial fractions) and get

$$\mathcal{L}\left[u(t) * e^{-at}\right] = \boxed{\frac{1}{a}\left[\frac{1}{s} - \frac{1}{s+a}\right]} \quad (16.137)$$

which would imply that

$$u(t) * e^{-at} = \frac{1}{a}\left[u(t) - e^{-at}\right] \quad (16.138)$$

which is nothing more than the integral of e^{-at}!

Example Find the LT of the convolution of two negative exponentials (Fig. 16.29). We know that

$$e^{-at} \rightarrow \frac{1}{s+a}, \qquad \text{and} \qquad e^{-bt} \rightarrow \frac{1}{s+b} \quad (16.139)$$

$$\mathcal{L}\left[e^{-at} * e^{-bt}\right] = \frac{1}{s+a}\frac{1}{s+b} \quad (16.140)$$

We can decompose this (using partial fractions) and get

$$\mathcal{L}\left[e^{-at} * e^{-bt}\right] = \boxed{\frac{1}{b-a}\left[\frac{1}{s+a} - \frac{1}{s+b}\right]} \quad (16.141)$$

This would imply that

$$e^{-at} * e^{-bt} = \frac{1}{b-a}\left[e^{-at} - e^{-bt}\right] \quad (16.142)$$

We can actually validate this last result. Let's do the convolution explicitly

$$\begin{aligned} e^{-at} * e^{-bt} &= \int_0^t e^{-a\tau} e^{-b(t-\tau)} d\tau \\ &= e^{-bt} \int_0^t e^{-a\tau} e^{b\tau} d\tau \\ &= e^{-bt} \int_0^t e^{\tau(b-a)} d\tau \\ &= \frac{e^{-bt}}{b-a} e^{\tau(b-a)} \Big|_0^t \\ &= \frac{1}{b-a}\left[e^{-at} - e^{-bt}\right] \end{aligned} \quad (16.143)$$

which agrees with Eq. (16.142).

Example Find the LT of the convolution of the ramp function with the negative exponential (Fig. 16.30). We know that

$$t \rightarrow \frac{1}{s^2}, \qquad \text{and} \qquad e^{-at} \rightarrow \frac{1}{s+a} \quad (16.144)$$

Then

$$\mathcal{L}\left[t * e^{-at}\right] = \frac{1}{s^2}\frac{1}{s+a} \quad (16.145)$$

We can rewrite this (using partial fractions) as

$$\mathcal{L}\left[t * e^{-at}\right] \boxed{\frac{1}{a^2}\left[-\frac{1}{s} + \frac{a}{s^2} + \frac{1}{s+a}\right]} \quad (16.146)$$

This would imply that

$$t * e^{-at} = \frac{1}{a^2}\left[-u(t) + at + e^{-at}\right] \quad (16.147)$$

Results are shown in Fig. 16.30. How do we know that the convolution results are good? Let's write a quick script to numerically find the convolution between t and e^{-at} for the case $a = 0.5$. On the one hand we plot Eq. (16.147) and on the other hand we plot the numerical integration output. Results are shown in Fig. 16.31. As can be seen in the figure we get excellent match, which means our convolution statement is most likely sound!

16.13 Summary

Equally as important in the case of the Fourier transform, mastering the properties of the Laplace transform is almost a must! Again the premise is that instead of starting from grounds zero every time and proceeding to brute force integration, by utilizing the properties we can (a) significantly expedite the transformation process, (b) better understand the machinery of the Laplace transform, and (c) grasp the various relations between the time domain and frequency one. Though the properties greatly resemble

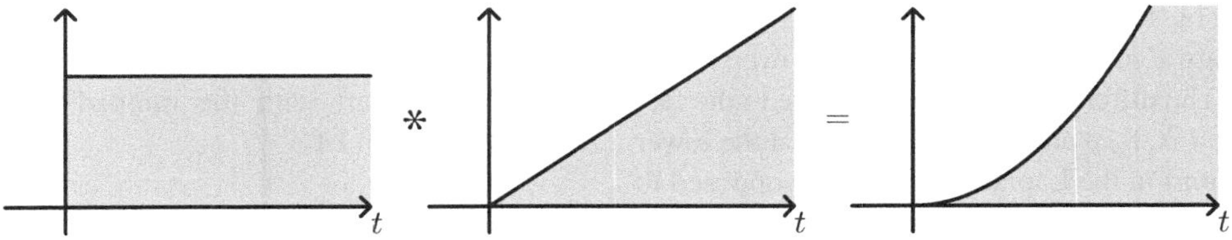

Fig. 16.27 Convolution between unit step and ramp function

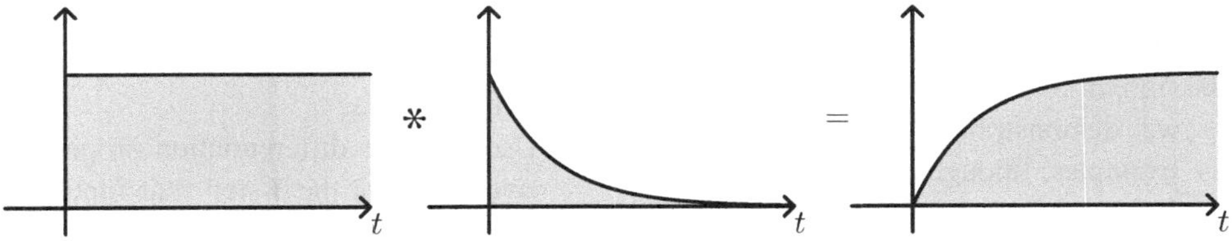

Fig. 16.28 Convolution between unit step and negative exponential functions

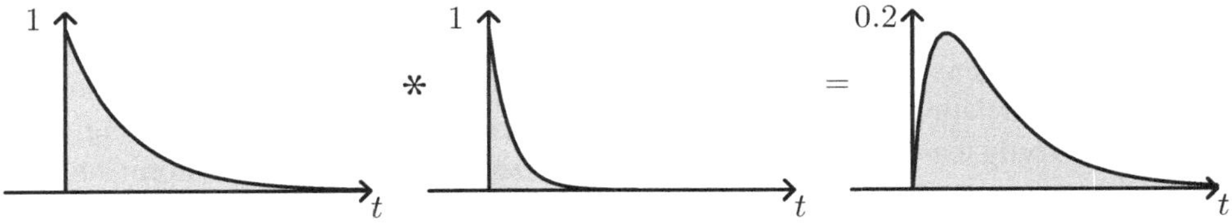

Fig. 16.29 Convolution between two negative exponentials

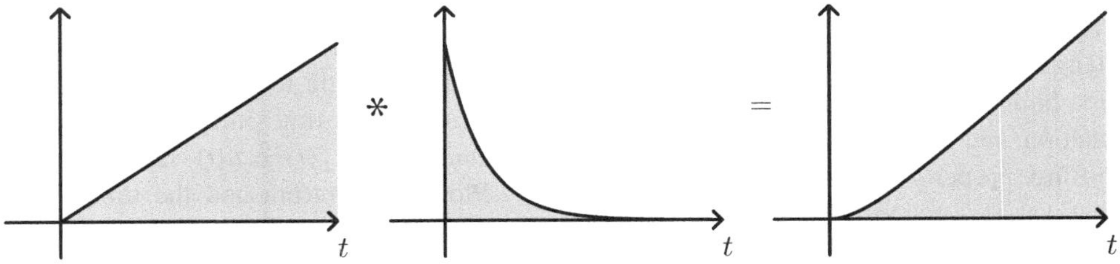

Fig. 16.30 Convolution between ramp and negative exponential

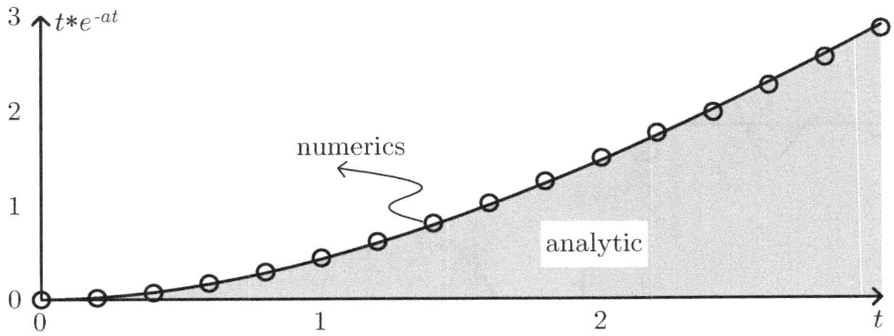

Fig. 16.31 Convolution between ramp and negative exponential—comparison between Eq. (16.147) and numerical results

those of their Fourier transform counterparts there are some subtle differences and even newer ones. The differences arise mainly due to the fact that $s = \sigma + j\omega$ and due to the fact that the lower time limit in the Laplace case is 0 (as opposed to $-\infty$). In addition to the conventional properties of linearity, scaling, time shifting, frequency shifting, time differentiation, time integration, time convolution, and frequency differentiation, we covered the frequency integration property and the initial and final value theorems. As always we demonstrated the properties with various examples, though truth is we barely scratched the surface. The more effort is put in, the richer the relations that arise tying signals to signals, to their time derivative or time integration, and so forth. Finally beware of negative signs and the complex j when going back and forth between Fourier and Laplace properties due to the relation $s = \sigma + j\omega$; that is, s does not map directly to ω!

16.14 Problems

1. Derive the Laplace transform of the upright triangle using linearity and time shifting. First build the function in terms of the ramp function and unit step one, then use time shifting property to get the LT. Plot the time series version. See sample solution in Fig. 16.32.
2. Use the time differentiation property to derive the LT of the pulse function of pulse width 1. Start with the ramped unit step function with LT

$$\text{ramped unit step function} \rightarrow \frac{1 - e^{-s}}{s^2}$$

Take the time derivative and then the LT thereof. Plot the Laplace transform and time series of both. See sample solution in Fig. 16.33.
3. Use the time differentiation property to derive the LT of the 3-step stair function. First take the time derivative of the stair function to end up with 4 delta functions. Next find the LT of those delta functions, utilizing the time shifting property. Lastly divide by s to arrive at the LT of the stair function. Plot the spectrum (for $\sigma = 0.05$) and plot the time series. See sample solution in Fig. 16.34.
4. Derive the Laplace transform of the pulse function using the time integration property and starting with $f(t) = \delta(t) - \delta(t-1)$. Plot the spectrum and the time series; see sample solution in Fig. 16.35.
5. Derive the Laplace transform of the symmetric triangle function, with center at $t = 1$ using the time integration property and starting with $f(t) = u(t) - 2u(t-1) + u(t-2)$. Plot the spectrum and the time series; see sample solution in Fig. 16.36.

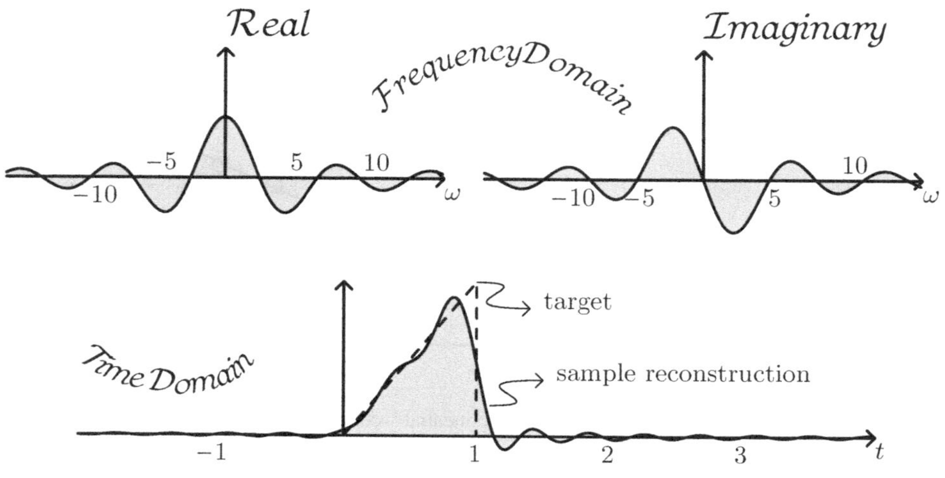

Fig. 16.32 Solution to Problem 1 (case of $\sigma = 0.1$)

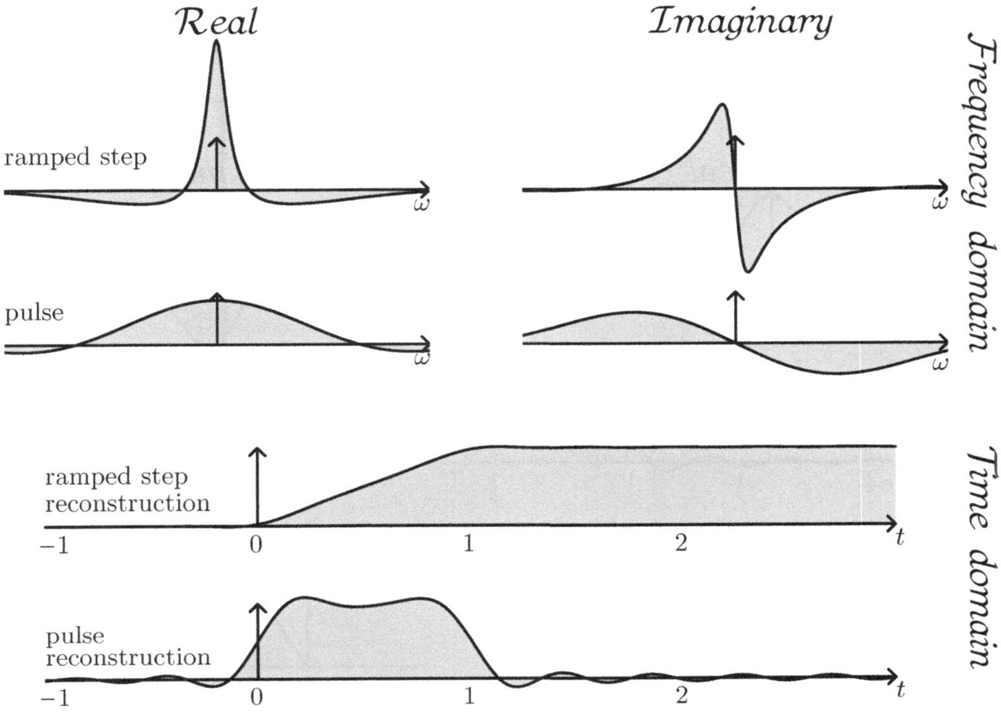

Fig. 16.33 Solution to Problem 2 (case of $\sigma = 0.3$)

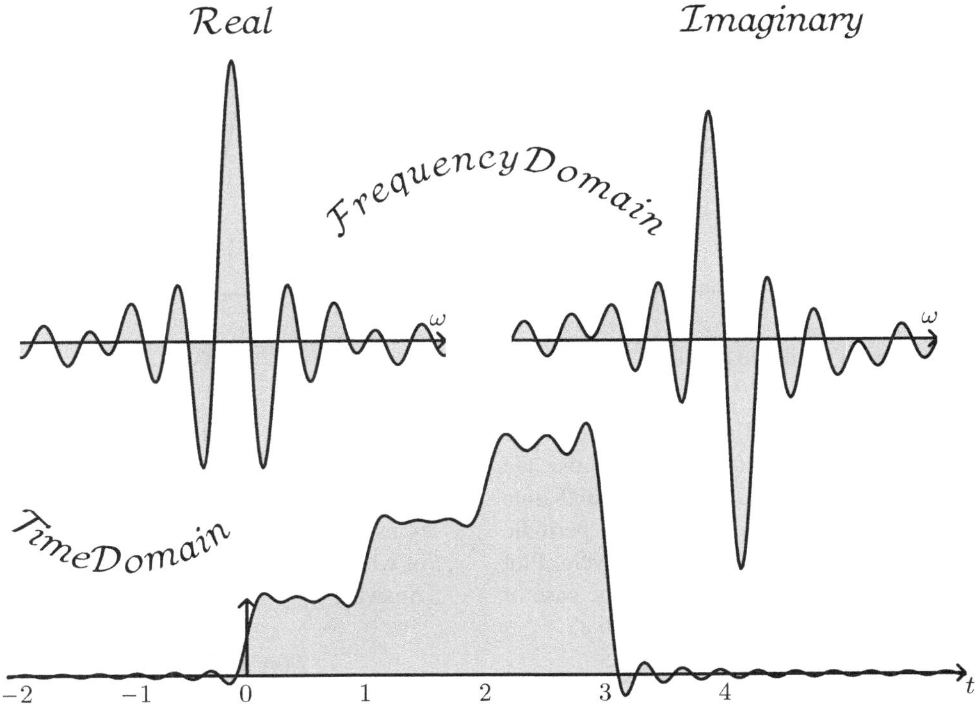

Fig. 16.34 Solution to Problem 3

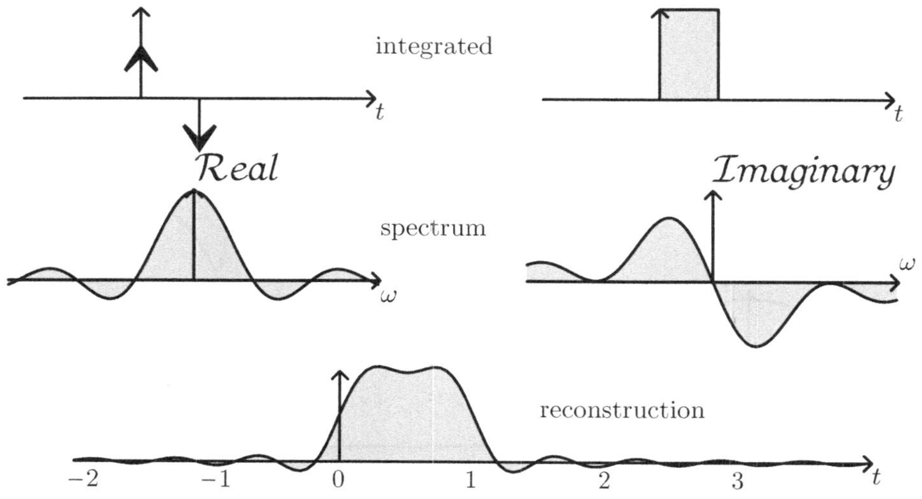

Fig. 16.35 Sample solution to Problem 4

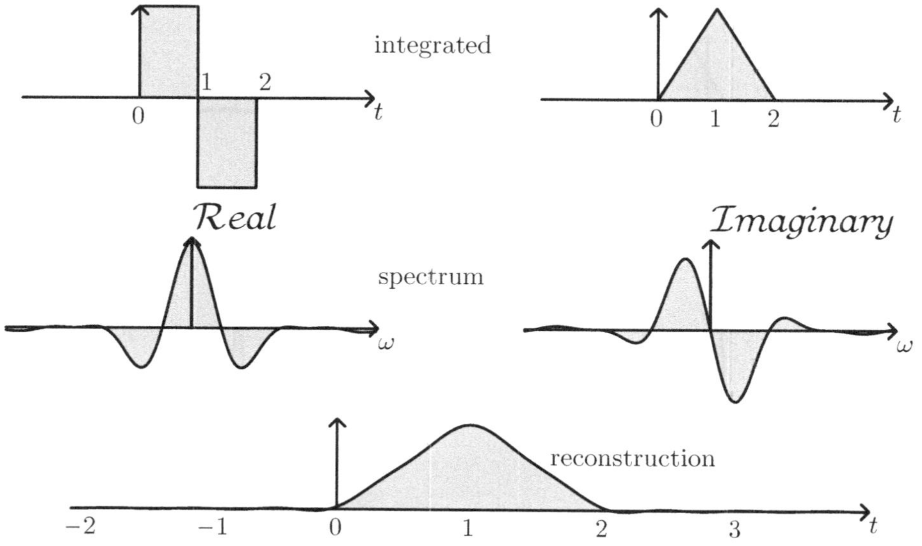

Fig. 16.36 Sample solution to Problem 5

6. Starting with the Laplace transform of the pulse function $F(s) = \frac{1-e^{-s}}{s}$, use the frequency differentiation property and time shifting to derive the LT of the periodic triangular pulse with 100% duty cycle. Plot the spectrum and time series for the case of three periods. See sample in Fig. 16.37.

7. Start with the negative pulse, differentiate it in the frequency domain, and then add it to nominal pulse to arrive at the Laplace transform of the upright triangular function of width 1; see sample results in Fig. 16.38. Answer:

$$F(s) = \frac{1}{s} - \frac{1}{s^2} + \frac{e^{-s}}{s^2}$$

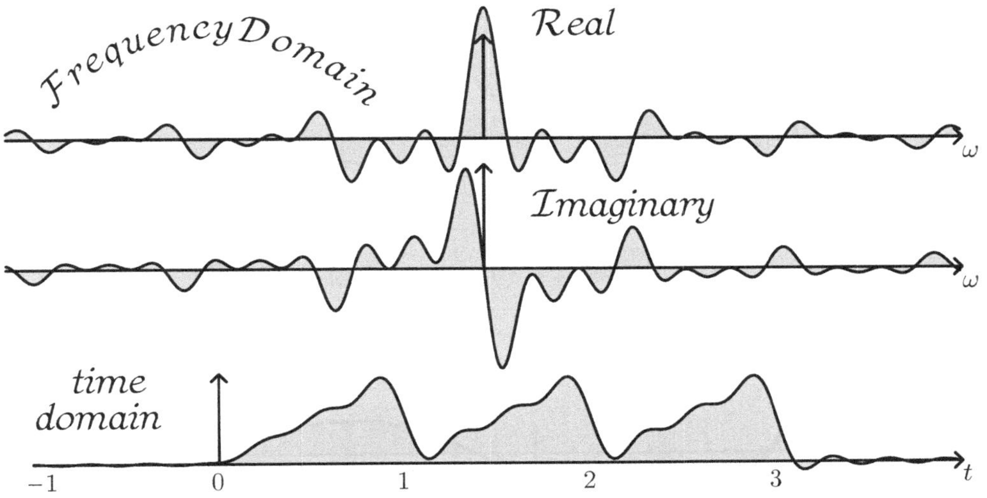

Fig. 16.37 Sample solution to Problem 6; case of $\sigma = 0.1$

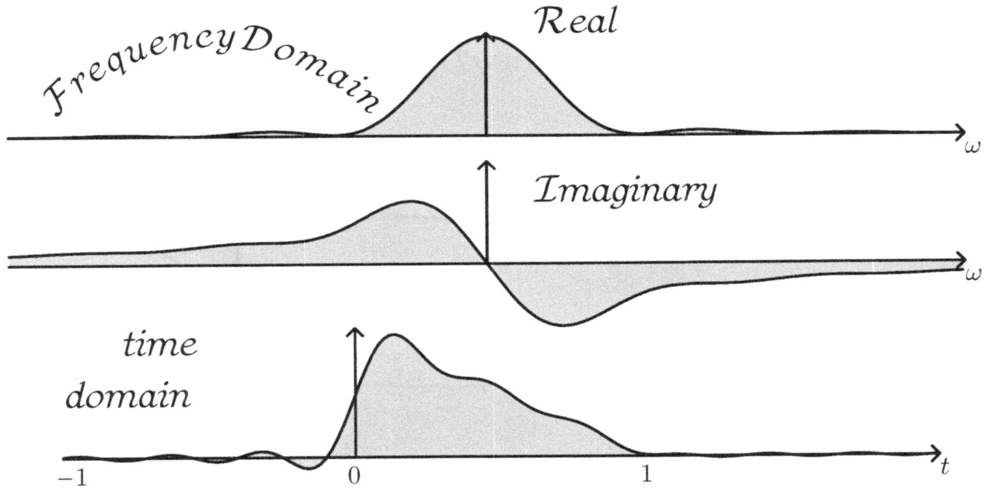

Fig. 16.38 Sample solution to Problem 7

8. Find the LT of the function

$$f(t) = \frac{\cos t - e^{-t}}{t}$$

utilizing the frequency integration property. Plot spectrum and time series; see sample solution in Fig. 16.39.

Answer:

$$F(s) = -\frac{1}{2}\ln(s^2 + 1) + \ln(s + 1)$$

9. Use the initial value theorem to predict $f(0)$ of the function with Laplace transform $F(s) = \frac{1}{s^2+1}$; what is $f(t)$?

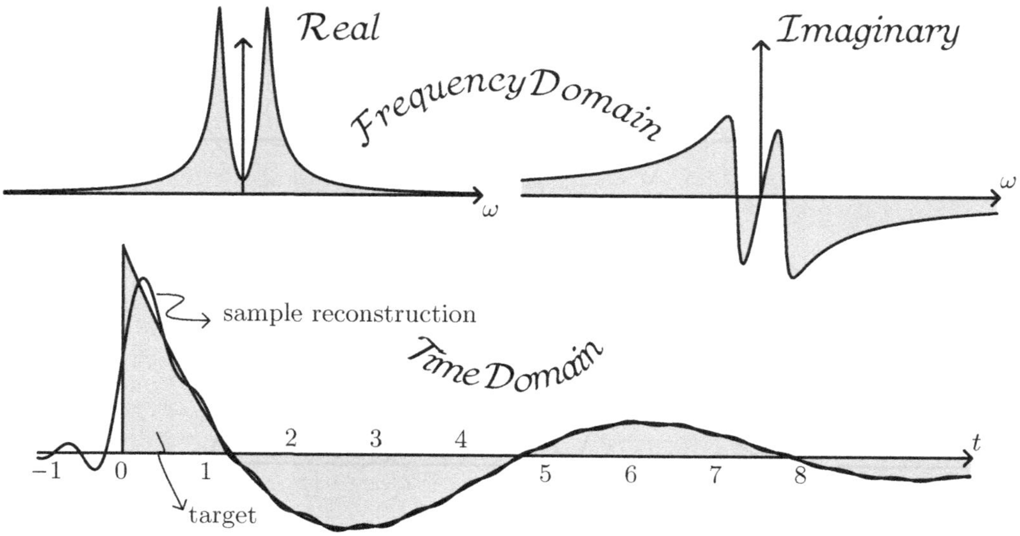

Fig. 16.39 Sample solution to Problem 8

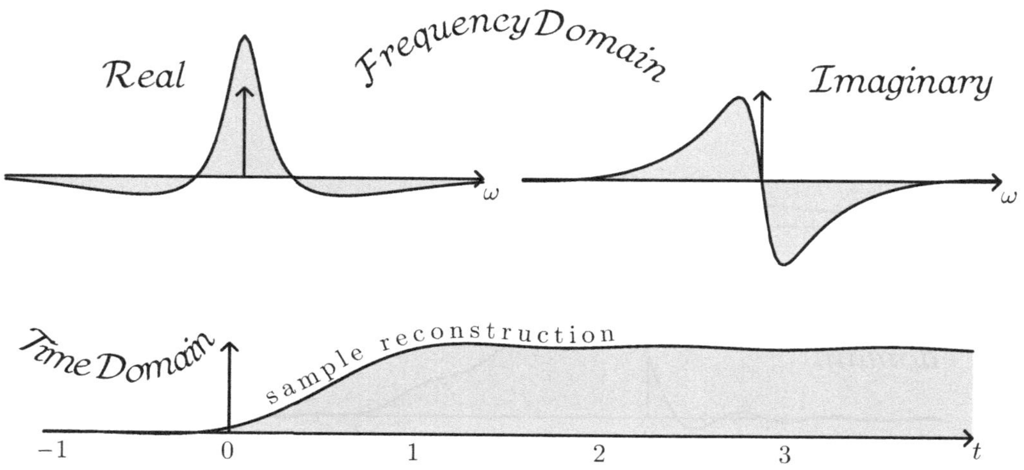

Fig. 16.40 Sample solution to Problem 15

10. Use the initial value theorem to predict $f(0)$ of the function with Laplace transform $F(s) = \frac{1}{s+1} + \frac{1}{s+2}$; what is $f(t)$?
11. Use the initial value theorem to predict $f(0)$ of the function with Laplace transform $F(s) = \frac{1}{s+1} - \frac{1}{s+2}$; what is $f(t)$?
12. The unit step function $u(t)$ has the Laplace transform $F(s) = \frac{1}{s}$; applying the initial value theorem we conclude that $f(0) = 1$. How about the shifted unit step function $u(t-1)$—what would the initial value theorem product at time zero?
13. The function $f(t) = te^{-t}$ has the LT $F(s) = \frac{1}{(s+1)^2}$; use the final value theorem to predict $f(\infty)$.
14. The function $f(t) = \frac{e^{-at}-e^{-bt}}{t}$ has the LT $F(s) = \ln\frac{s+b}{s+a}$; use the final value theorem to predict $f(\infty)$.
15. Derive the Laplace transform of the ramped unit step function, with ramp time 1 by convolving the unit step function with the pulse one (with width 1), and then by using the convolution property. Plot the spectrum and time series; see sample solution in Fig. 16.40.

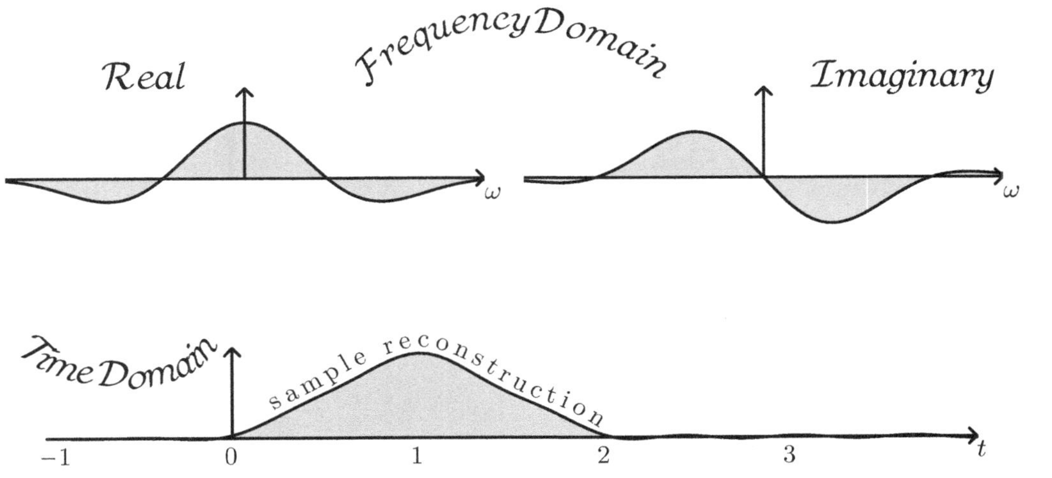

Fig. 16.41 Sample solution to Problem 16

16. Derive the Laplace transform of the symmetric triangular function with center at $t = 1$ convolving the pulse function (of width 1) with itself and then by using the convolution property. Plot the spectrum and time series; see sample solution in Fig. 16.41.

17 Laplace Transform of Periodic Functions

17.1 Introduction

The Laplace transform is most often used for "single-timer" functions, such as the unit step function, the pulse one, or the ramp function. As will be shown below, we can extend the coverage for periodic functions. By periodic we mean periodic after time zero; that is, the function is still zero for negative time, but becomes periodic for positive time. In other words, it is a *causal* periodic function!

17.2 Derivation

Assume that a function has the following LT:

$$f(t) \rightarrow F(s) \tag{17.1}$$

Furthermore assume that the function is repeated every T seconds as shown in Fig. 17.1. We know the LT of the first burst. We can find the LT of the second burst by using the time shifting property of the LT.

$$\mathcal{L}\left[f(t-t_0)\right] = e^{-t_0 s}F(s) \tag{17.2}$$

Specifically

$$\mathcal{L}\left[\text{Second burst}\right] = e^{-Ts}F(s) \tag{17.3}$$

Similarly the third and fourth bursts have Laplace transforms of

$$\begin{aligned}\mathcal{L}\left[\text{3rd burst}\right] &= e^{-2Ts}F(s)\\ \mathcal{L}\left[\text{4th burst}\right] &= e^{-3Ts}F(s)\end{aligned} \tag{17.4}$$

and so on. Since our periodic input is the sum of all these bursts, then the LT of the periodic input would be the sum of the individual burst LT.

$$\begin{aligned}&\mathcal{L}\left[\text{periodic burst}\right]\\ &= F(s)+e^{-Ts}F(s)+e^{-2Ts}F(s)+e^{-3Ts}F(s)+\cdots\\ &= F(s)\left[1+e^{-Ts}+e^{-2Ts}+e^{-3Ts}+\cdots\right]\\ &= \boxed{F(s)\sum_{n=0}^{\infty} e^{-nTs}}\end{aligned} \tag{17.5}$$

If we recall the following series expansion

$$\frac{1}{1-x} = 1+x+x^2+x^3+\cdots \tag{17.6}$$

we finally arrive at the LT of the period signal in terms of the aperiodic one.

$$\boxed{\mathcal{L}\left[\text{period burst}\right] = \frac{F(s)}{1-e^{-sT}}} \tag{17.7}$$

F. Badrieh, *Spectral, Convolution and Numerical Techniques in Circuit Theory*,
https://doi.org/10.1007/978-3-319-71437-0_17

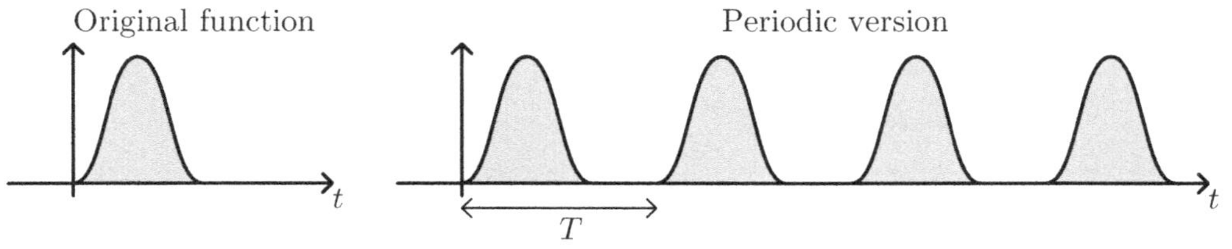

Fig. 17.1 Original function and periodic version thereof

This is such a simple, elegant, yet very powerful relation. Notice that for the first time the denominator is no longer of the form s, s^2, and so forth; instead it is a transcendental function! With a simple symbolic manipulation we opened the door for pretty much an infinite selection of periodic signals. By simply varying the original function $F(s)$ and the period T we are guaranteed a wide selection of interesting signals. Notice the special case of infinite period which yields the aperiodic version of the function, as expected.

$$\lim_{T\to\infty} \frac{F(s)}{1-e^{-sT}} = F(s) \qquad (17.8)$$

Let us then get some practice in generating periodic signals out of their aperiodic counterparts.

17.3 Periodic Pulse

Let us find the LT of the periodic (causal) pulse function with pulse width τ and period T. We know that the single pulse of width τ has the LT of

$$\text{single pulse} \to \frac{1-e^{-s\tau}}{s} \qquad (17.9)$$

The periodic pulse will then have the LT

$$\text{periodic pulse} \to \frac{\mathcal{L}[\text{single pulse}]}{1-e^{-sT}} \qquad (17.10)$$

$$\boxed{\mathcal{L}[\text{periodic pulse}] = \frac{1}{s}\frac{1-e^{-s\tau}}{1-e^{-sT}}} \qquad (17.11)$$

Notice that for the special case where period equals pulse width we get

$$F(s)_{T=\tau} = \frac{1}{s}\frac{1-e^{-s\tau}}{1-e^{-s\tau}} = \frac{1}{s} \qquad (17.12)$$

which is nothing but the LT of the unit step function; that is, if we start with the pulse function, repeat it every pulse width then we end up with the continuous unit step function!

Figure 17.2 shows the spectrum (both real and imaginary) of the single-timer pulse, and the periodic one. Figure 17.3 shows the spectrum for smaller period. We can see that the more periodic the pulse is, the more discretized its spectrum becomes. In other words, the spikes get taller and more spaced out. *It is extremely important to grasp the transformation in the transform function as the signals transition from being non-periodic to periodic.* The overall envelope remains the same; but the inner details differ in the sense the transform becomes spiked and almost discretized. The more periodic the function becomes, the more *selective* its spectrum becomes; in other words, for a strongly periodic signal we rely on select frequencies to represent the signal. Figure 17.4 shows inverse Laplace transform for the larger period case, while Fig. 17.5 shows inverse Laplace transform results for the smaller period case.

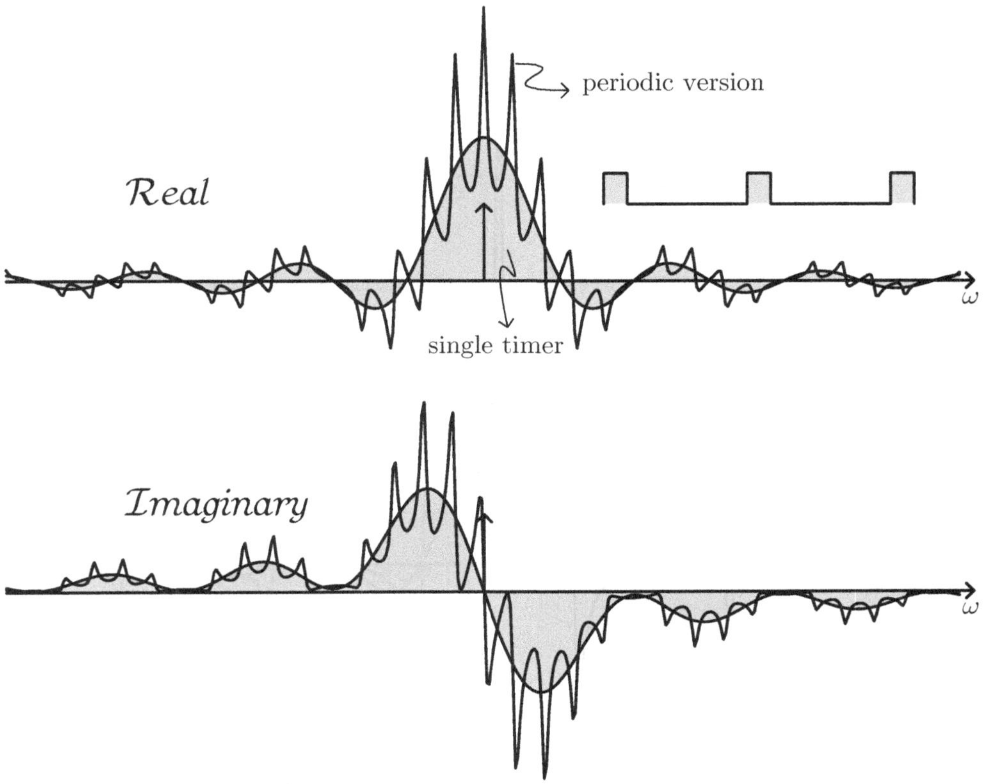

Fig. 17.2 Laplace transform of single-timer and periodic pulse ($\sigma = 0.15$ and $T = 5$)

17.4 Periodic Upright Triangle

The upright triangle of width τ and height 1 has the LT

$$\mathcal{L}\,[\text{upright triangle}\,] = \frac{1}{\tau}\frac{1-e^{-\tau s}}{s^2} - \frac{e^{-\tau s}}{s}$$
$$= \boxed{\frac{1}{\tau}\frac{1-e^{-\tau s}\,[1+\tau s]}{s^2}} \quad (17.13)$$

The LT of the periodic triangle, based on our theorem, is then

$$\boxed{\mathcal{L}\,[\text{periodic upright triangle}\,] = \frac{1}{\tau}\frac{1-e^{-\tau s}\,[1+\tau s]}{s^2(1-e^{-Ts})}} \quad (17.14)$$

Figure 17.6 shows the spectrum for the single-timer and periodic one; notice that the periodic one follows the other, but is a "discretized" version thereof. Figure 17.7 shows the time series for both cases.

17.5 Intermixing Two Periodic Signals

Let's find the LT of the periodic signal, of period T, comprised of a periodic pulse (width τ_1) and periodic upright triangle (width τ_2). The pulse on its own has a LT

$$\mathcal{L}[\text{pulse of width}\tau_1] = \frac{1-e^{-\tau_1 s}}{s} \quad (17.15)$$

The LT of the upright triangle is

$$\mathcal{L}[\text{upright triangle of width}\tau_2] = \frac{1}{\tau_2}\frac{1-e^{-\tau_2 s}\,[1-\tau_2 s]}{s^2} \quad (17.16)$$

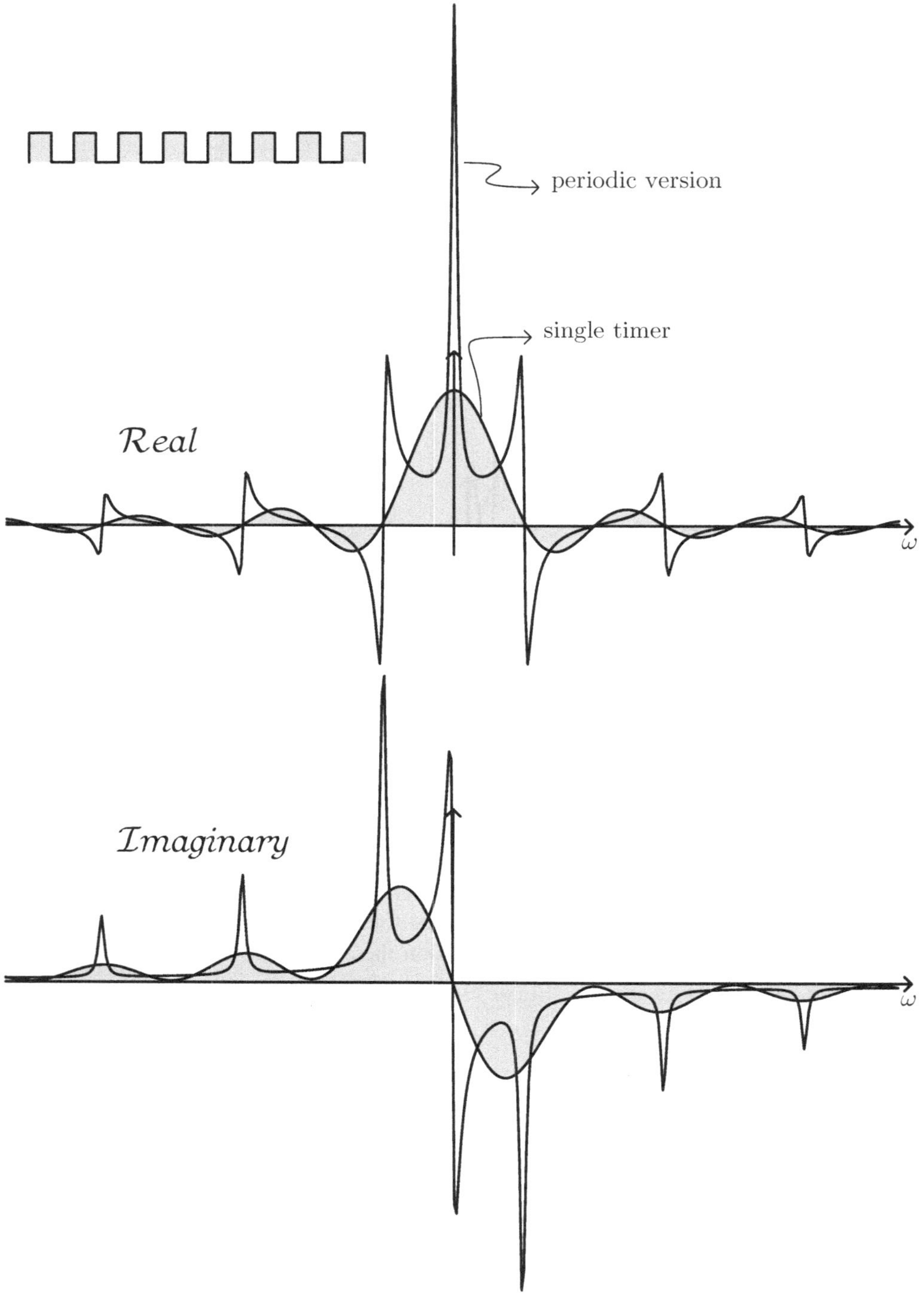

Fig. 17.3 Laplace transform of single-timer and periodic pulse ($\sigma = 0.15$ and $T = 2$)

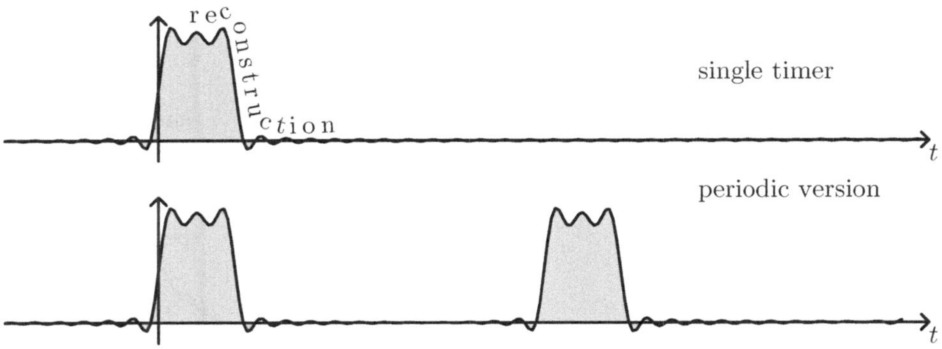

Fig. 17.4 Time series (inverse LT) of single-timer and periodic pulse ($\sigma = 0.15$ and $T = 5$)

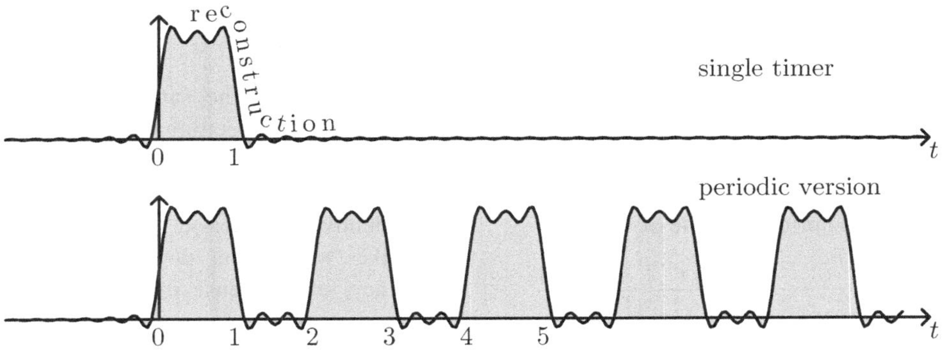

Fig. 17.5 Time series (inverse LT) of single-timer and periodic pulse ($\sigma = 0.15$ and $T = 2$)

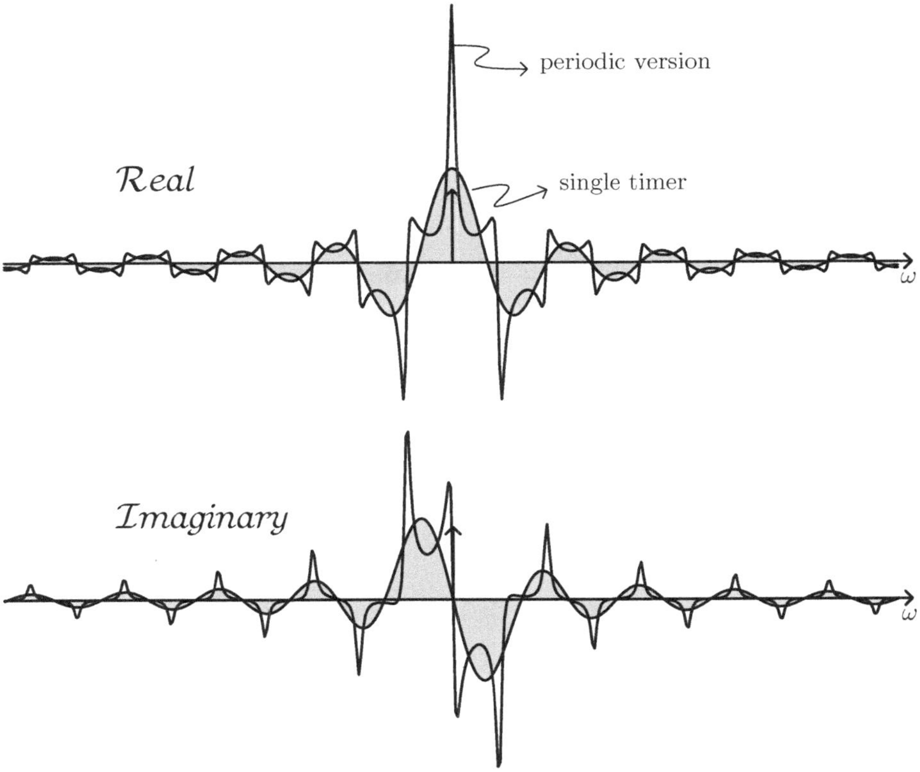

Fig. 17.6 Laplace transform of single-timer and periodic upright triangle ($\sigma = 0.15$)

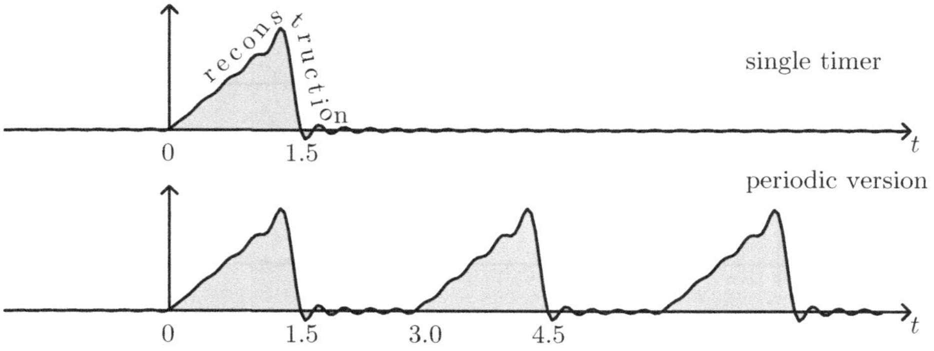

Fig. 17.7 Time series (inverse LT) of single-timer and periodic upright triangle ($\tau = 1.5$ and $T = 3$)

But we would need to shift this in time to align half way between the pulses. The time shift comes out $(T + \tau_1 - \tau_2)/2$; hence

$$\mathcal{L}[\text{shifted upright triangle of width}\tau_2] = \frac{e^{-(T+\tau_1-\tau_2)s/2}}{\tau_2} \frac{1 - e^{-\tau_2 s}[1 - \tau_2 s]}{s^2} \tag{17.17}$$

Now add both functions, and then divide by $1 - e^{-Ts}$ to get the LT of the mixed periodic signals

$$\boxed{\mathcal{L}[\text{mixed periodic signal}] = \frac{1 - e^{-\tau_1 s}}{s(1 - e^{-Ts})} + \frac{e^{-(T+\tau_1-\tau_2)s/2}}{\tau_2} \frac{1 - e^{-\tau_2 s}[1 - \tau_2 s]}{s^2(1 - e^{-Ts})}} \tag{17.18}$$

Figure 17.8 shows the spectrum for both single-timer and periodic signals. Notice the undeniable recurring signature in transitioning from the aperiodic to periodic worlds—a spiky spectrum that has the same envelope as that of the aperiodic spectrum. It is expected that as we make the period T smaller, the spikes grow in amplitude, but space out in frequency. Equivalently, if the period T is made larger, then the spikes would tame down and get closer in frequency, in the sense of closely resembling the continuous spectrum corresponding to the aperiodic signal. Figure 17.9 shows the inverse LT in the form of time series.

So we get a feel of how we can construct different flavors of a periodic signal by (a) decomposing it into the bare bone aperiodic traits; (b) offsetting the individual signatures; and (c) using the periodicity property to construct the conglomerate final product.

17.6 Single-Cycle Sine and Cosine Functions

Assume we want to find the LT of the single-cycle sine function of frequency ω_0. That is, the function $\sin \omega_0 t$ defined nonzero only between 0 and $T = \frac{2\pi}{\omega_0}$. The hard way about this would be

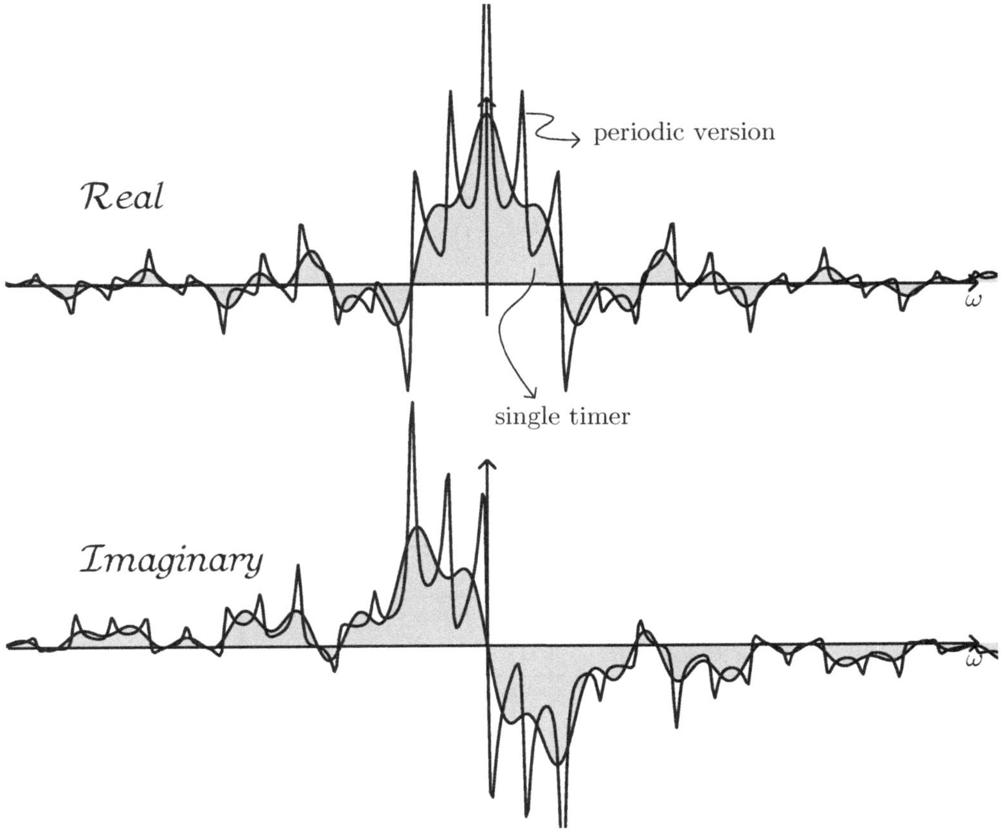

Fig. 17.8 Laplace transform of mixed periodic signal

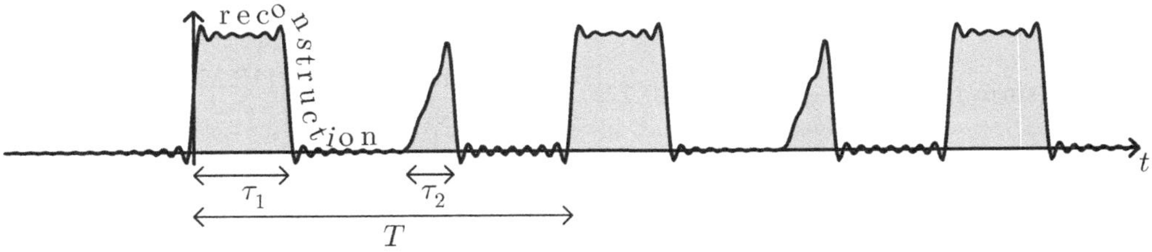

Fig. 17.9 Time series (inverse LT) of mixed periodic signal comprised of pulse and upright triangle

$$F(s) = \int_0^T \sin\omega_0 t e^{-st} dt = \frac{1}{2j}\int_0^T \left[e^{j\omega_0 t} - e^{-j\omega_0 t}\right] e^{-st} dt$$
$$= \frac{1}{2j}\left[-\frac{1}{s-j\omega_0}e^{-t(s-j\omega_0)} + \frac{1}{s+j\omega_0}e^{-t(s+j\omega_0)}\right]_0^T$$
$$= \frac{1}{2j}\left[-\frac{1}{s-j\omega_0}\left(e^{-T(s-j\omega_0)}-1\right) + \frac{1}{s+j\omega_0}\left(e^{-T(s+j\omega_0)}-1\right)\right] \quad (17.19)$$

Notice that

$$e^{-Tj\omega_0} = e^{-Tj\frac{2\pi}{T}} = e^{-j2\pi} = 1 \quad (17.20)$$

Hence we have

$$F(s) = \frac{1}{2j}\left[-\frac{1}{s-j\omega_0}\left(e^{-Ts}-1\right) + \frac{1}{s+j\omega_0}\left(e^{-Ts}-1\right)\right]$$
$$= \frac{1-e^{-Ts}}{2j}\left[\frac{1}{s-j\omega_0} - \frac{1}{s+j\omega_0}\right] = \frac{1-e^{-Ts}}{2j}\frac{s+j\omega_0-(s-j\omega_0)}{s^2+\omega_0^2}$$
$$(17.21)$$

$$F(s) = \boxed{\frac{\omega_0}{s^2+\omega_0^2}\left[1-e^{-Ts}\right]} \quad (17.22)$$

The easy way about this is to start with the LT of the periodic sine function

$$\mathcal{L}\left[\sin\omega_0 t\right] = \frac{\omega_0}{s^2+\omega_0^2} \quad (17.23)$$

We know that this function is tied to the single period function via

$$\mathcal{L}\left[\sin\omega_0 t\right] = \frac{\mathcal{L}[\text{single period sine}]}{1-e^{-sT}} \quad (17.24)$$

Solving for the numerator we get

$$\mathcal{L}[\text{single period sine}] = \mathcal{L}[\sin\omega_0 t]\left[1-e^{-sT}\right], \text{ or} \quad (17.25)$$

$$\boxed{\mathcal{L}[\text{single period sine}] = \frac{\omega_0}{s^2+\omega_0^2}\left[1-e^{-sT}\right]} \quad (17.26)$$

in agreement with Eq. (17.22). By similar argument we conclude that

$$\boxed{\mathcal{L}[\text{single period cosine}] = \frac{s}{s^2+\omega_0^2}\left[1-e^{-sT}\right]} \quad (17.27)$$

This is an amazing development. Rather than taking a single-timer function and making it periodic, we are doing the opposite: take a periodic function and make it aperiodic! We can see that we can go both directions—from the aperiodic to periodic or from the periodic to the aperiodic—riding on the same theory. How can we tell if our analysis is correct? One simple way is to simply take the resulting Laplace transform and build its time series. Figure 17.10 shows such time series (inverse LT of Eq. (17.22)). The figure shows in gray the periodic version and in solid the aperiodic one. As can be seen including enough frequency content duplicates the intended signal.

17.7 The Periodic Impulse Function

Let's find the LT of the periodic delta function, of period T. We know that the single delta function has the LT

$$\delta(t) \rightarrow 1 \tag{17.28}$$

By the periodicity theorem we then get

$$\mathcal{L}[\text{periodic impulse function}] = \mathcal{L}[\delta(t)] \times \frac{1}{1-e^{-sT}} = \boxed{\frac{1}{1-e^{-sT}}} \tag{17.29}$$

How simple can this get? Let's take a look how this spectrum looks like—see Fig. 17.11. Since our starting LT function is 1, then the "sampled" one in the frequency domain comes out also a train of impulses! This is a sort of similar to the impulse train we saw back in Sect. 11.9. Let's test this result by finding the inverse Laplace transform. Figure 17.12 shows such inverse transform, and shows that we get localized pulses around integers of T. Notice that away from nT the function does not completely go to zero, as expected; instead it does so on the average. For all purposes, we can treat the results in Fig. 17.12 as a train of delta functions, with energy localized around nT and zero outside.

17.8 Stair Function and Limit to Ramp Function

Start with the unit step function with LT

$$u(t) \rightarrow \frac{1}{s} \tag{17.30}$$

Now scale it by T, where T is some number; then

$$Tu(t) \rightarrow \frac{T}{s} \tag{17.31}$$

Now make this function periodic in T, such that

$$f(t) = T\left[u(t) + u(t-T) + u(t-2T) + u(t-3T) + \cdots\right] \tag{17.32}$$

The results would look like Fig. 17.13. Notice that this function resembles the (discretized version of the) ramp function! That is, when time has progressed $10T$ units, the function would have elevated to also $10T$ units, such that the slope is always 1! But this is the same behavior as the ramp function. As we decrease T we would better approximate the real ramp as shown in Fig. 17.13.

From the periodicity theory we can derive the LT of the stair function as

$$\boxed{\mathcal{L}\left[\text{stair function}\right] = \frac{T}{s}\frac{1}{1-e^{-Ts}}} \tag{17.33}$$

Let us see what happens as we let the step size (T) go to zero; expanding and keeping the first two terms of the exponential gives

$$\lim_{T\rightarrow 0} F(s) = \frac{T}{s}\frac{1}{1-[1-Ts]} = \frac{T}{s}\frac{1}{Ts} = \frac{1}{s^2} \tag{17.34}$$

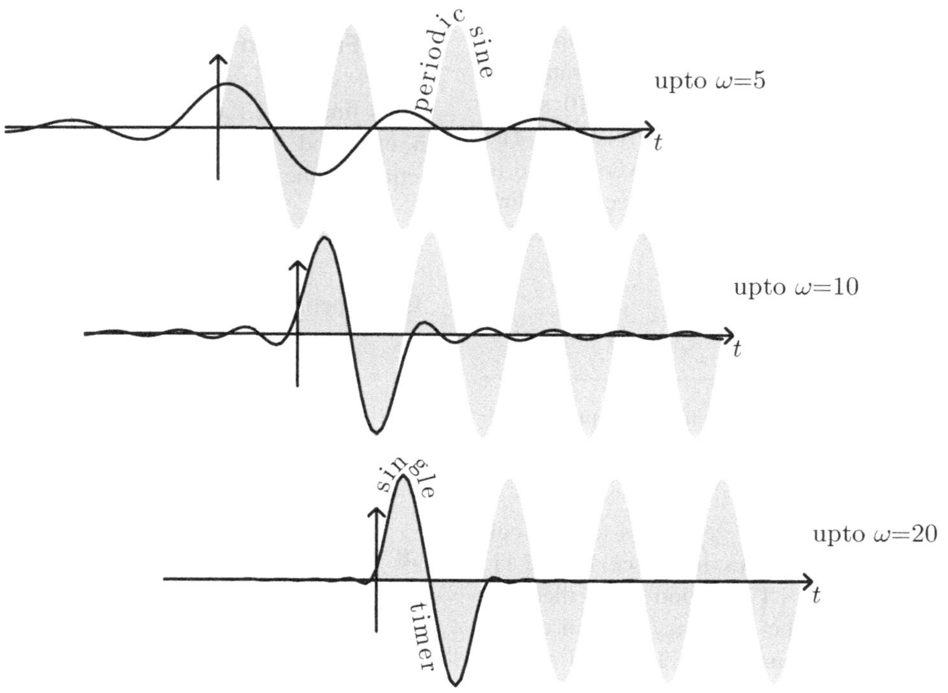

Fig. 17.10 Time series (inverse LT) of single-timer sine function and comparison to periodic one

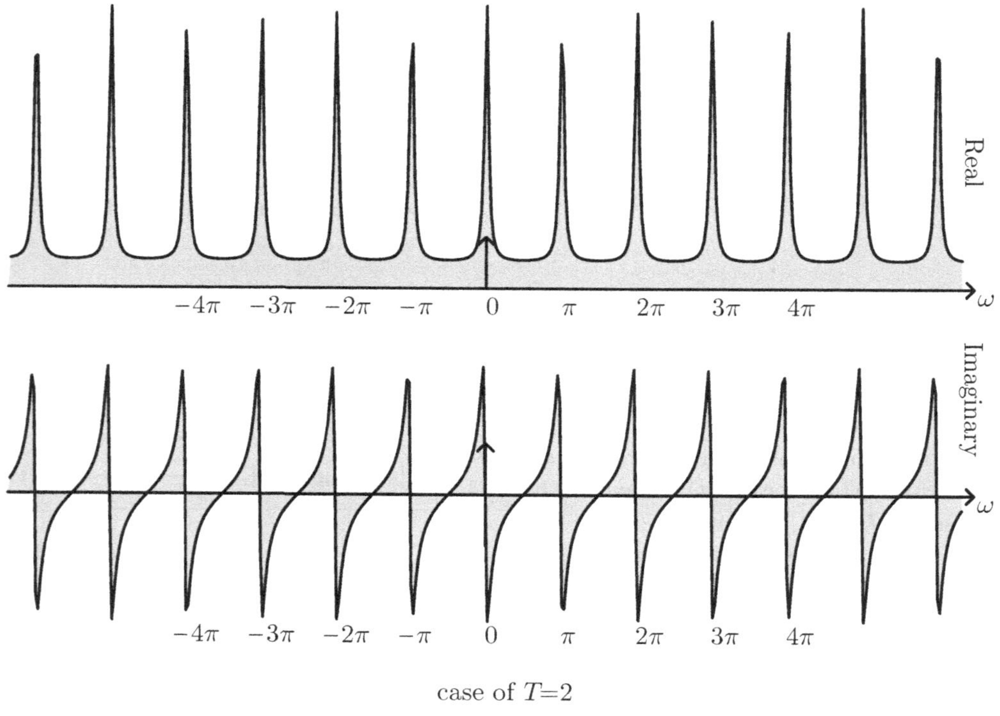

Fig. 17.11 Laplace transform of (causal) periodic delta function, with period $T = 2$ (case of $\sigma = 0.1$)

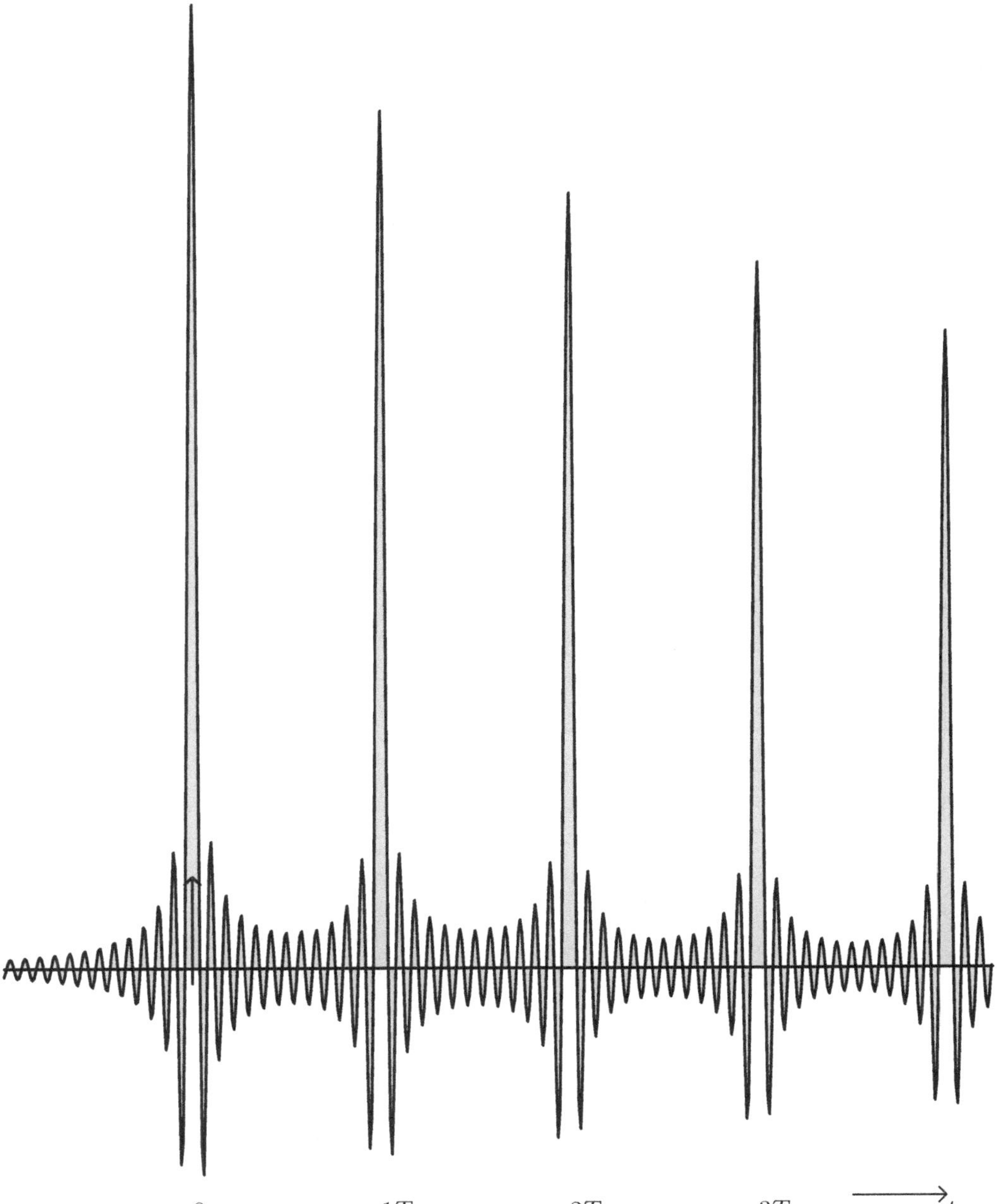

Fig. 17.12 Time series (inverse LT) of periodic delta function (period $T = 2$)

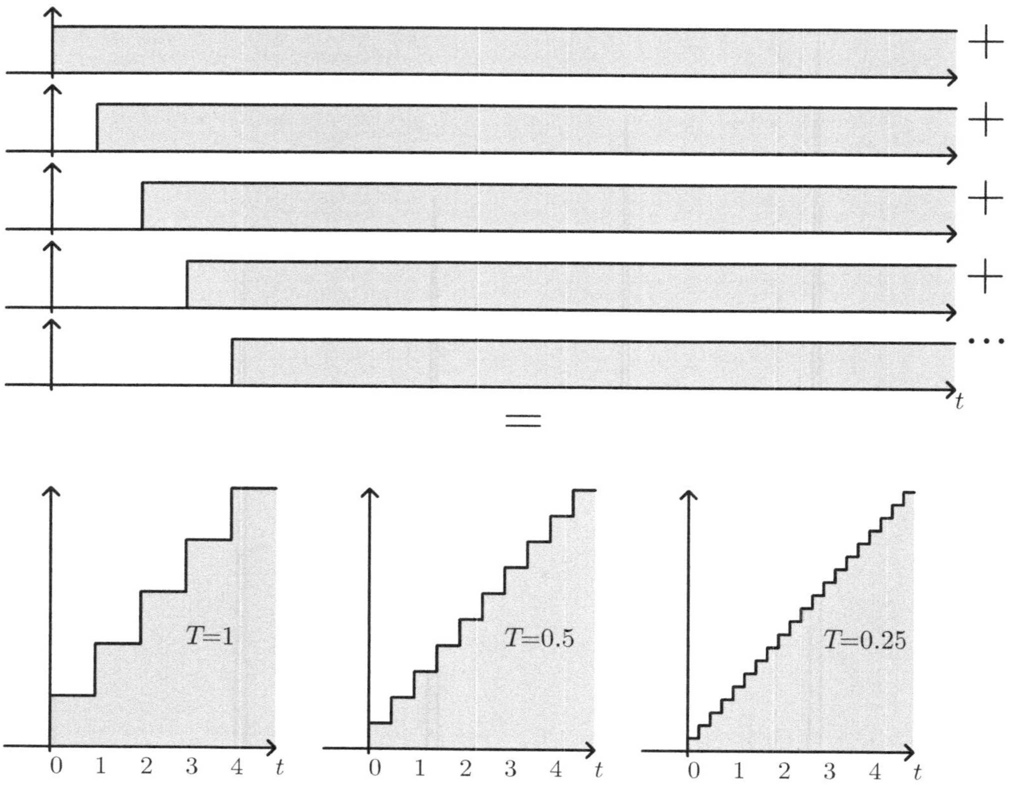

Fig. 17.13 Periodic unit step function, with offset T, and usage to construct the discretized ramp function

But this is nothing more than the LT of the ramp function!

$$u(t)t \rightarrow \frac{1}{s^2} \tag{17.35}$$

So we have succeeded in arriving at the LT of the smooth ramp function, as a result of adding infinitely smaller, offset unit step functions. This adds confidence in the periodicity theorem, and sheds light into superposition and the underlying principles behind signal representation, both in frequency and time domain.

17.9 Summary

The most common usage of the Laplace transform is to find the frequency representation of time finite-extent signal, such as a pulse or the negative exponential. But even then we have passed by signals that are not time-limited, such as the unit step, ramp, or even the sinusoids. The new development in this chapter is a class of functions which are periodic in time. In fact we can take any non-periodic single-timer signal (such as the pulse) and make it periodic by simply repeating the signal every T. And that exact step has a mirror in the frequency domain, embodied by the most important equation in this chapter—Eq. (17.7). What this means is that after having analyzed the spectrum of the single-timer signal we can readily find the modified spectrum for the periodic version of the same signal. Mathematically the steps are those in the above equation, but physically and qualitatively the end outcome is a spiky spectrum with the same outline (envelope) as that of the single-timer signal. Furthermore, the spacing between the spikes, as well as their heights, becomes larger with smaller time period. We illustrated the spectrum on a few examples, and did the time series for verification; and in all

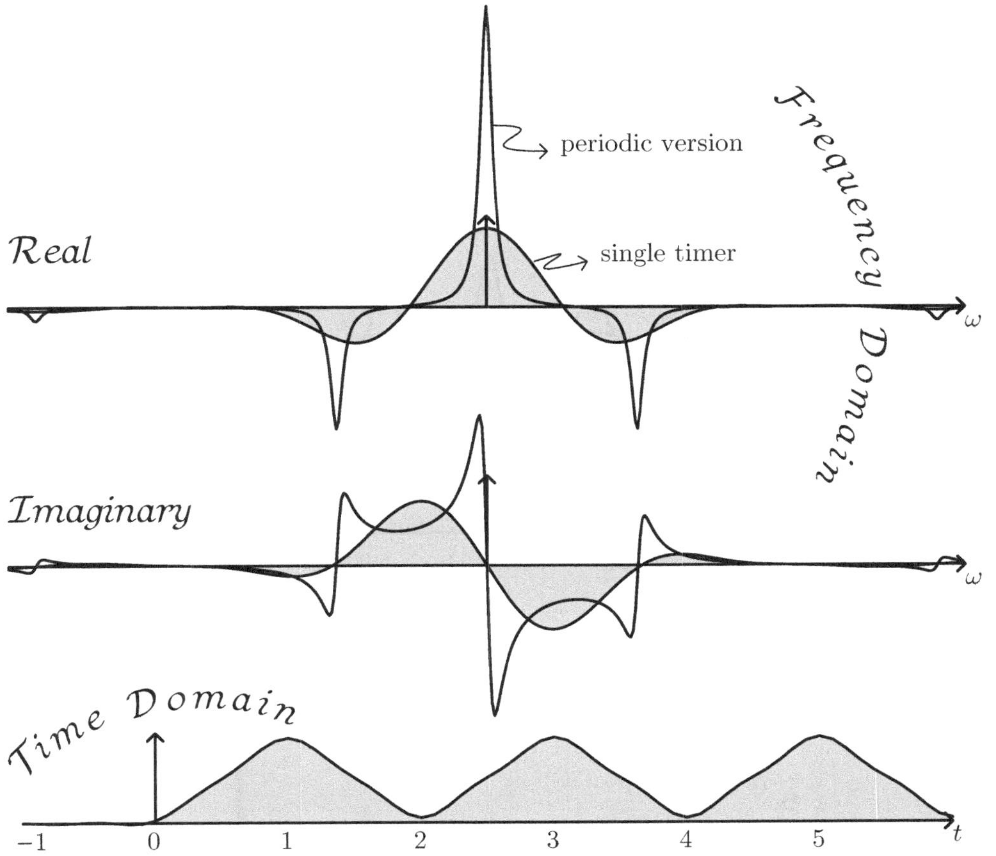

Fig. 17.14 Sample solution to Problem 1

cases we observed the excellent applicability of the theorem. And even more we were able to take a periodic signal and strip it of its periodicity to generate its single-timer version of the signal. So, and equipped with the periodicity theorem, we are able to go both ways in the frequency domain, from periodic to non-periodic and vice versa.

17.10 Problems

1. The symmetric triangle function with width 2, and center at $t = 1$ has the Laplace transform

$$F(s) = \left[\frac{1 - e^{-s}}{s}\right]^2$$

Find the Laplace transform of the periodic version of this function, with period $T = 2$. Plot the spectrum for both cases (using for example $\sigma = 0.15$), and the time series; see sample solution in Fig. 17.14.
Answer:

$$F(s) = \left[\frac{1 - e^{-s}}{s}\right]^2 \frac{1}{1 - e^{-Ts}}, \qquad T = 2$$

2. Repeat Problem 1 for the cases $T = 4$ and $T = 8$. Compare the spectrum (magnitude) for all three cases (again using for example $\sigma = 0.15$). What is the trend as the period T grows large? What is the limit as $T \to \infty$? See sample solution in Fig. 17.15.

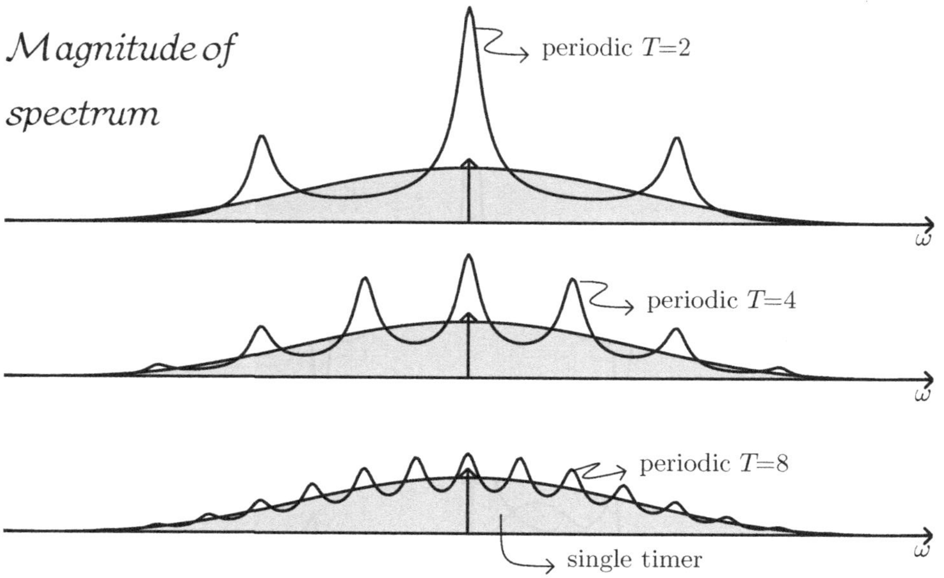

Fig. 17.15 Sample solution to Problem 2

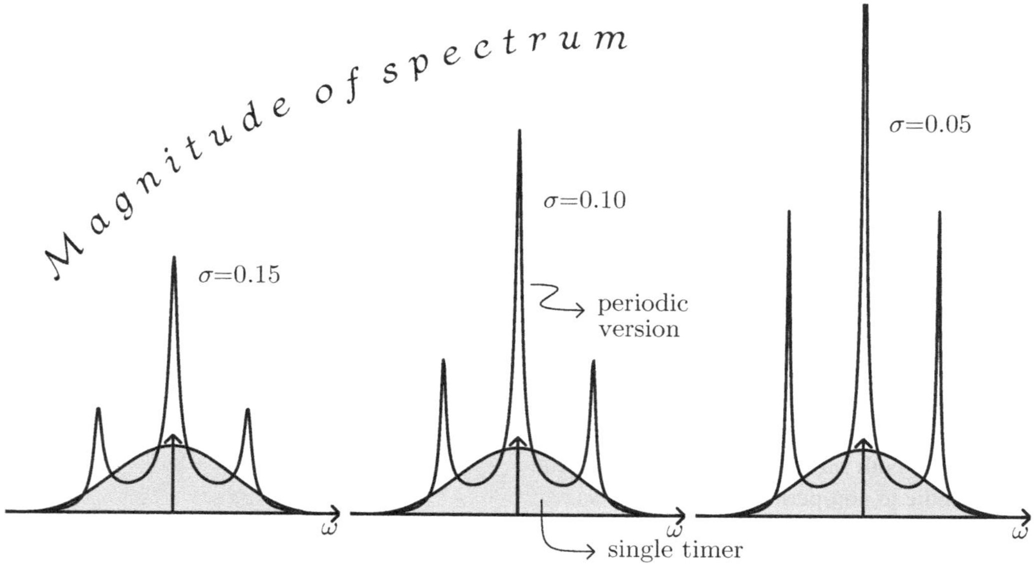

Fig. 17.16 Sample solution to Problem 3

3. Repeat Problem 1 which was done using $\sigma = 0.15$ and instead use $\sigma = 0.10$ for one case and $\sigma = 0.05$ for the other. Plot the spectrum (magnitude) and compare all three cases. What is the trend for smaller σ? Does the number of peaks change? Does the peak spacing change? See sample solution in Fig. 17.16.
4. The Laplace transform of the parabolic pulse, centered at $t = 1$, is

$$F(s) = 2\left[(1 + e^{-2s})/s^2 + (-1 + e^{-2s})/s^3\right]$$

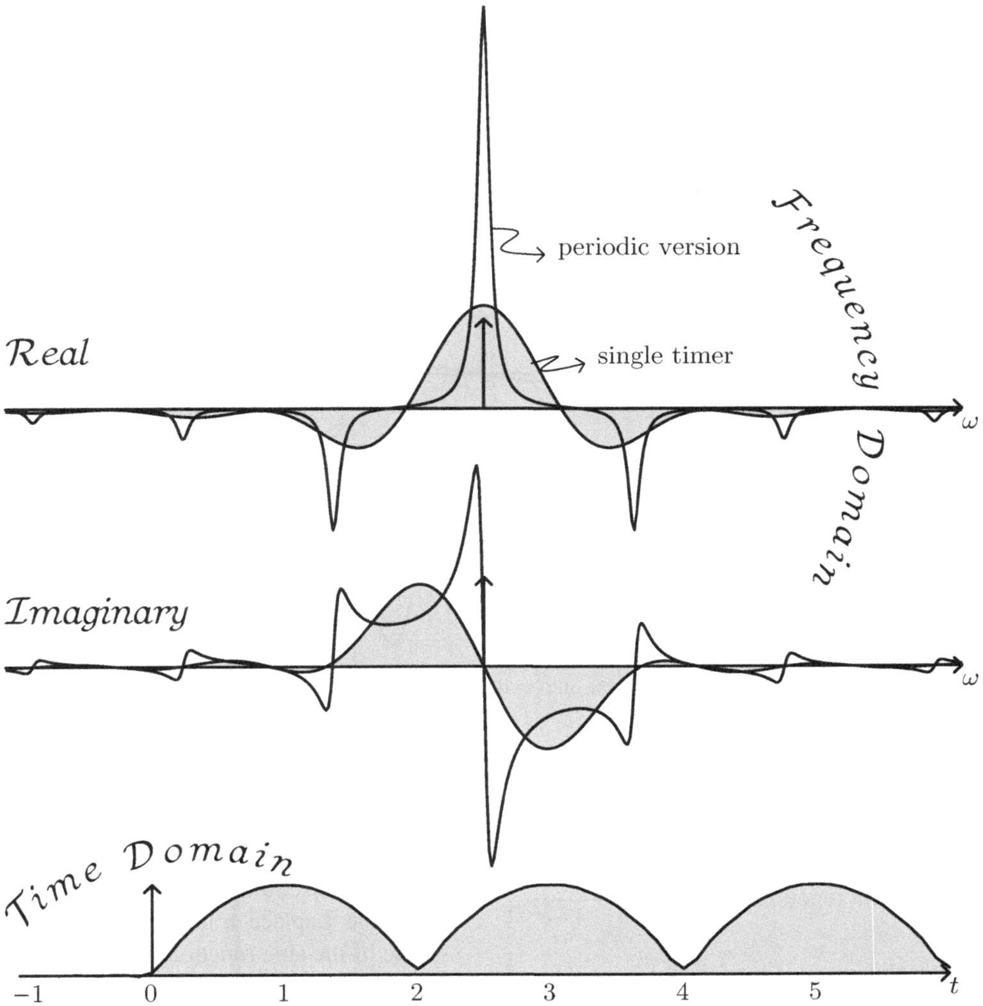

Fig. 17.17 Sample solution to Problem 4 (case of $\sigma = 0.15$)

First prove this to be true: Hint—the single timer function defined between 0 and 2 is given by

$$f(t) = 2t - t^2$$

Use the time differentiation property to arrive at the transform. Next, find the Laplace transform of the periodic version thereof, with period $T = 2$ and plot the spectrum and time series; see sample solution in Fig. 17.17.

5. When deriving the compact form of the Laplace transform of periodic function we ended up utilizing the following series expansion:

$$\frac{1}{1-x} = 1 + x + x^2 + x^3 + \cdots$$

Derive this formula using Taylor series; how about the related expression $\frac{1}{1+x}$?

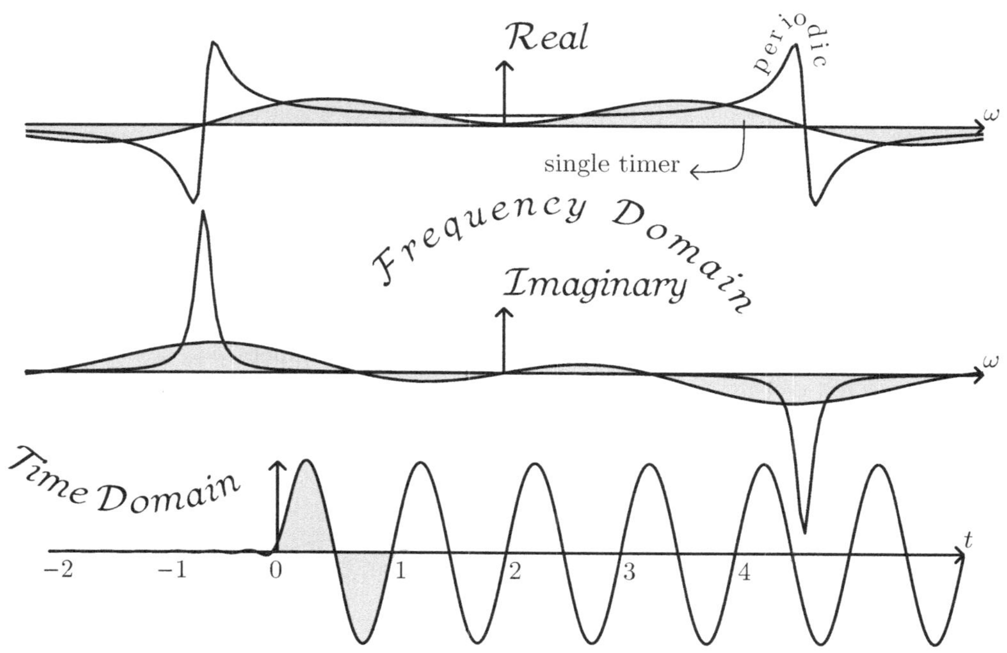

Fig. 17.18 Sample solution to Problem 7 (case of $\sigma = 0.2$)

6. It was shown in the text that the Laplace transform of the periodic delta function of period T is

$$\text{train of delta functions at period } T \rightarrow \frac{1}{1-e^{-Ts}}$$

   What is the inverse transform of $1 - e^{-Ts}$ (call it $g(t)$)? What is the time convolution between the periodic delta function (still of period T) and $g(t)$?
7. Fill in some of the details in Sect. 17.6. Plot the spectrum (both real and imaginary) of both the periodic sine and the single-timer (both with $\omega_0 = 2\pi$). Then plot the time series; see sample solution in Fig. 17.18.
8. What function in time needs to be convolved with the pulse function (of width 1) to make a periodic pulse function, of period 2?
9. Find the Laplace transform of the absolute value of the sine function, with $\omega_0 = 1$ using the periodicity property. Plot the spectrum and time series; see sample solution in Fig. 17.19.

   Answer:

$$F(s) = \frac{1}{s^2+1} \frac{1+e^{-\pi s}}{1-e^{-\pi s}}$$

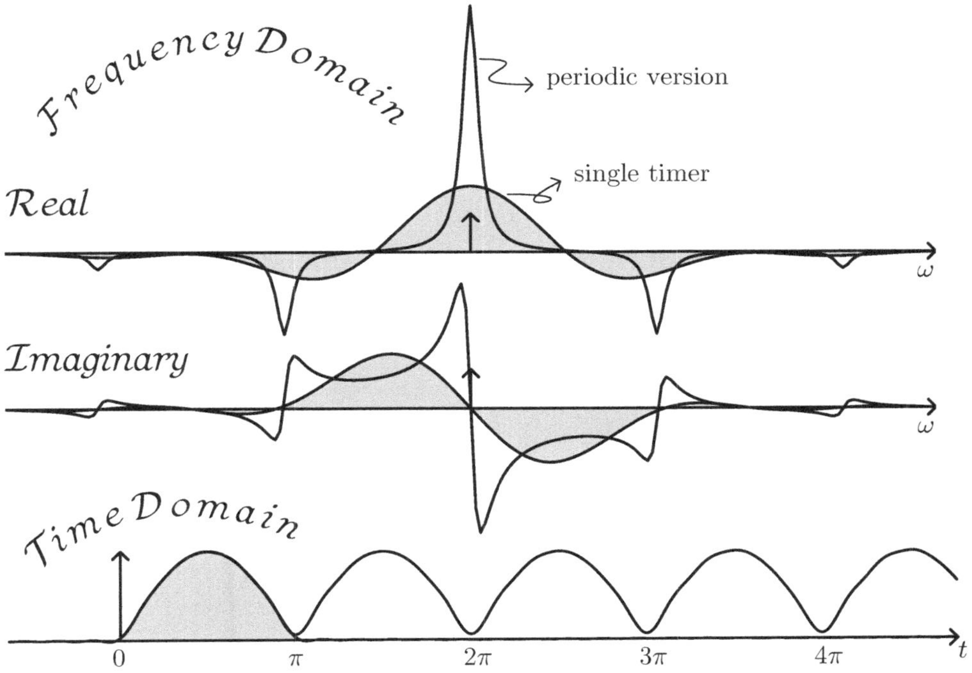

Fig. 17.19 Sample solution to Problem 9 (case of $\sigma = 0.1$)

Finding Inverse Laplace Transform via Partial Fractions

18

18.1 Introduction

So far we have assumed the signal is known in time and we were interested in finding its Laplace (or Fourier) transform in frequency. While this is very important, it is not the only scenario. Quite often the transform is known and we are interested in finding the inverse transform. For example, the transfer function could be measured, or simulated, or fit and the task is to find its inverse transform. In theory the transform (or transfer function) can assume any form. And in theory we can always figure the inverse transform by brute force numerical integration. But doing that could become quite expensive. To get some work underway we need to simplify, at least for a starter. If the transfer function is simple enough, we can guess the inverse transform using a lookup table (or from memory). If not, we learned before how to do complex integration, but even that sometimes could get quite lengthy. As a way about simplifying things, the complex integration flow was reformulated to what is now referred to as partial fractions. In this flow it is assumed that the transfer function has the following form

$$F(s) = \frac{a_0 + a_1 s + a_2 s^2 + \cdots}{b_0 + b_1 s + b_2 s^2 + \cdots} \tag{18.1}$$

Rather than do the full blown complex integration, while keeping track of the carrier signal e^{st}, the partial fraction flow focuses only on the frequency domain; i.e., there is no mention of t. The rational function polynomial is factored into a simple form, in particular a series summation of simple or higher order poles. The premise is that if we can factor out the transfer function in terms of individual poles, we can figure the inverse transform keeping in mind the simple relation between a pole in the frequency domain and its inverse transform in the time domain. For example, in the simplest case a pole of the form

$$F(s) = \frac{1}{s+a} \tag{18.2}$$

has the simple inverse transform of

$$f(t) = e^{-at} \tag{18.3}$$

while a double pole of the form

$$F(s) = \frac{1}{(s+a)^2} \tag{18.4}$$

has the inverse transform

$$f(t) = te^{-at} \tag{18.5}$$

and so forth. So the task is recast in the form of finding the pole expansion of the transfer function as

$$F(s) = \sum_p \sum_n \frac{A_n}{(s+B_n)^p} \tag{18.6}$$

F. Badrieh, *Spectral, Convolution and Numerical Techniques in Circuit Theory*,
https://doi.org/10.1007/978-3-319-71437-0_18

Here B_n is the location of the pole; A_n is the magnitude of the pole; and p is the *order* of the pole. That is, sometimes we could have a pole of higher order than 1, such as 2, 3, or even higher. And sometimes, we could have poles of different orders at the same frequency. That's why we have the index p. So the task is to find the poles, identify their order and their magnitude, and then rewrite the transfer function as a summation of these poles. While it may be more efficient and elegant to cover all scenarios analytically and symbolically, let's just get some practice by doing. And in the course of practice we will pick up the core of the idea.

18.2 Plotting the Transfer Function

As an aid to learning and as a means of confirming the series expansion of the transfer function we will be plotting the transfer function—both magnitude and phase—as a function of frequency. Typically the magnitude plot is on a log-log scale while the phase plot is on a log(frequency)-normal(phase) scale. We will be observing certain trends in both magnitude and phase. The magnitude typically would stay constant until a pole (or zero) is encountered. At each pole the magnitude deflects by an amount determined by the pole order. For simple poles, the magnitude tends to go down one decade per decade of frequency. For a double pole, the magnitude goes down 2 decades per decade of frequency, and so forth. Notice that once two poles are active, the decay rate becomes 2 decades per decade of frequency, since each pole contributes 1 decade of decay per decade of frequency, and so forth. Zeros enact exactly the opposite—rather than decaying the transfer function grows, at the corresponding grow rate. The phase, on the other hand, sustains a shift of *negative* 90° for a simple pole and negative 180° for a double pole, and so forth. Notice that two simple poles would sustain a phase shift of −180° since each simple pole contributes −90°, and so forth. The phase would change *positive* for zeroes, and the change would be exactly the same as that of the pole (albeit negative its sign). Notice that for both poles and zeroes we are assuming that they are located in the negative complex plane. If not, then the phase gets flipped. Let's take a couple of examples. Assume the pole resides in the negative complex plane and is of the form

$$F(s) = \frac{1}{s+1} \tag{18.7}$$

Substitute in for s (assuming $\sigma = 0$ for now)

$$F(\omega) = \frac{1}{1+j\omega} \tag{18.8}$$

Expand

$$F(\omega) = \frac{1-j\omega}{\omega^2+1} \tag{18.9}$$

The phase is simply

$$\text{Phase} = \text{atan}\left[\frac{-\omega}{1}\right] \tag{18.10}$$

which at high frequency becomes

$$\text{Phase} = \text{atan}(-\infty) = -90^\circ \tag{18.11}$$

But had the pole been in the positive complex plane, such as

$$F(s) = \frac{1}{s-1} \tag{18.12}$$

then the phase would have been

$$\text{Phase} = \text{atan}(\infty) = 90^\circ \tag{18.13}$$

Similarly assume we have a transfer function with a zero in the negative complex plane

$$F(s) = 1 + s \tag{18.14}$$

After simplifying we get

$$F(\omega) = 1 + j\omega \tag{18.15}$$

The phase then is

$$\text{Phase} = \text{atan}\left[\frac{\omega}{1}\right] \tag{18.16}$$

which would be 90° at high frequency, and −90° had the zero resided in the positive complex plane, such as the case for $F(s) = s - 1$. For almost all of the cases we will be concerned with poles and zeroes that lie in the negative complex plane.

18.3 Decibels

Rather than continue using phrases such as "1 decade of decay per decade of frequency" we can switch to decibels which are defined as

$$\text{dB} = 20 \log \text{something} \tag{18.17}$$

In our context if a function goes down by 1 decade per 1 decade of frequency, such as

$$F(s) \sim \frac{1}{s} \tag{18.18}$$

then when a frequency decade had passed the magnitude would have dropped to 0.1 of the reference value; taking the dB of this we get

$$20 \log(0.1) = 20 \times (-1) = -20\,\text{dB} \tag{18.19}$$

So we say the function goes down by 20 dB/dec. Had the pole been of order 2, then we would have 40 dB since

$$20 \log(0.01) = 20 \times (-2) = -40\,\text{dB} \tag{18.20}$$

and so forth.

18.4 Transfer Function of the Form $\frac{1}{(s+a)(s+b)}$

In this case the LT has the form

$$F(s) = \frac{1}{(s+a)(s+b)} \tag{18.21}$$

We assume we can write the fraction as

$$\frac{1}{(s+a)(s+b)} = \frac{A}{s+a} + \frac{B}{s+b} \tag{18.22}$$

To find A multiply both sides by $(s+a)$ and evaluate at $s = -a$

$$A = \frac{1}{s+b}\Big|_{s=-a} = \frac{1}{b-a} \tag{18.23}$$

To find B multiply both sides by $(s+b)$ and evaluate at $s = -b$

$$B = \frac{1}{s+a}\Big|_{s=-b} = \frac{1}{a-b} \tag{18.24}$$

Then

$$\boxed{F(s)\frac{1}{b-a}\left[\frac{A}{s+a} - \frac{B}{s+b}\right]} \tag{18.25}$$

For the case $a = 1$ and $b = 10$ we get

$$F(s) = \frac{1}{9}\left[\frac{1}{s+1} - \frac{1}{s+10}\right] \tag{18.26}$$

Figure 18.1 shows the resulting transfer function. At zero frequency we get $F(0) = 0.1$ and

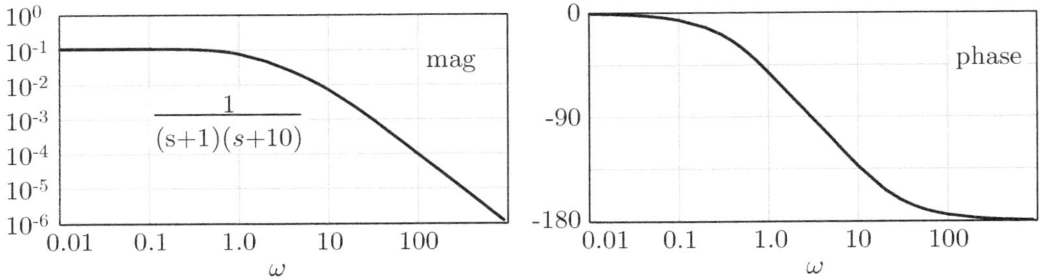

Fig. 18.1 Transfer function with two real poles

the phase is zero. In this case at high frequency we can see that the transfer function is going down at the rate of 2 decades per decade of frequency (−40 dB/dec), since we have two simple poles. Also at high frequency the phase shift has saturated to −180° again since we have two simple poles.

18.5 Transfer Function of the Form $\frac{1}{s(s+a)}$

In this case the LT has the form

$$F(s) = \frac{1}{s(s+a)} \tag{18.27}$$

We have two poles—one at 0, and the other at $s = -a$. We expand the function as

$$\frac{1}{s(s+a)} = \frac{A}{s} + \frac{B}{s+a} \tag{18.28}$$

To find A we do

$$\begin{aligned} A = sF(s)\Big|_{s=0} &= s\frac{1}{s(s+a)}\Big|_{s=0} \\ = \frac{1}{s+a}\Big|_{s=0} &= \frac{1}{a} \end{aligned} \tag{18.29}$$

To find B we do

$$\begin{aligned} B = (s+a)F(s)\Big|_{s=-a} &= (s+a)\frac{1}{s(s+a)}\Big|_{s=-a} \\ = \frac{1}{s}\Big|_{s=-a} &= -\frac{1}{a} \end{aligned} \tag{18.30}$$

Hence

$$\boxed{\frac{1}{s(s+a)} = \frac{1}{a}\left[\frac{1}{s} - \frac{1}{s+a}\right]} \tag{18.31}$$

A plot of this transfer function is shown in Fig. 18.2. Notice that the first zero has already forced the decay rate at 1 decade per decade of frequency (−20 dB/dec), and the phase already started at −90°. Notice also once the second pole is active the decay rate is 2 decades per 1 decade of frequency (−40 dB/dec), and the final phase is −180° since we have two poles.

18.6 Transfer Function of the Form $\frac{s}{s+1}$

In this case the LT has the form

$$F(s) = \frac{s}{s+1} \tag{18.32}$$

Since the numerator has same order as denominator we have to use long division. To that end we have

$$\begin{array}{r} 1 \\ s+1\,\big)\ \ s \\ -s-1 \\ \hline -1 \end{array} \tag{18.33}$$

Hence

$$\boxed{F(s) = 1 - \frac{1}{s+1}} \tag{18.34}$$

A plot of this transfer function is shown in Fig. 18.3. Notice that initially we have a zero, and hence the growth rate of 1 decade per decade of frequency (20 dB/dec) and initial phase is 90°. When the pole is active, it nulls the effect of the zero, and we fall back at a constant magnitude and zero phase.

18.7 Transfer Function of the Form $\frac{1+s+s^2/10}{s}$

In this case the LT has the form

$$F(s) = \frac{1+s+s^2/10}{s} \tag{18.35}$$

This we can factor right away

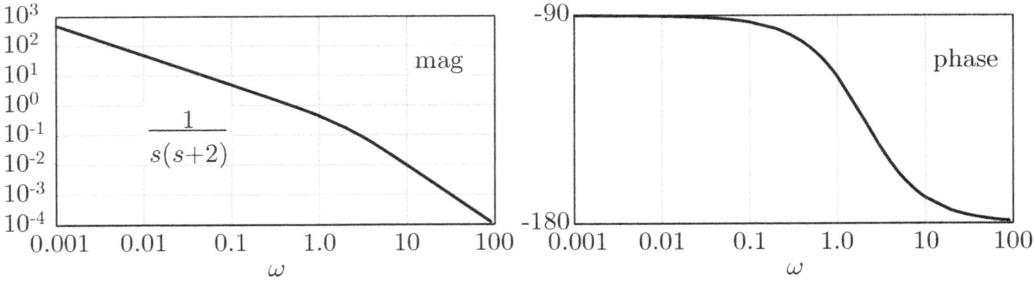

Fig. 18.2 Transfer function with two real poles, one at zero

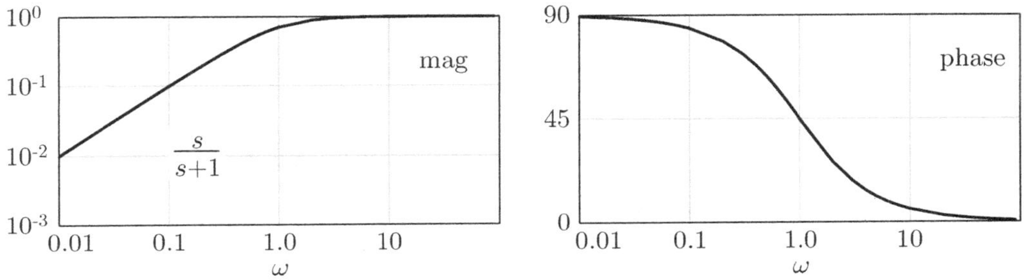

Fig. 18.3 Transfer function with zero, followed by a pole

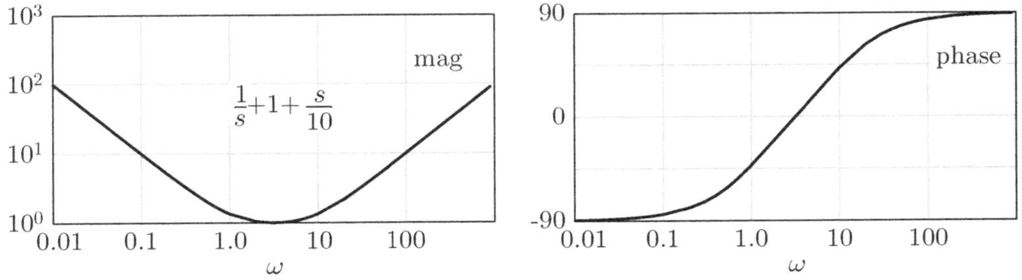

Fig. 18.4 Transfer function with pole, followed by double zeroes

$$\boxed{F(s) = \frac{1}{s} + 1 + \frac{s}{10}} \tag{18.36}$$

A plot of this transfer function is shown in Fig. 18.4. This transfer function starts with a pole (−20 dB/dec and −90°), followed by a zero (0 dB/dec and 0°) followed by another zero (20 dB/dec and 90°).

18.8 Transfer Function of the Form $\frac{1}{(s+1)(s+2)(s+3)}$

Assume our LT has the form

$$F(s) = \frac{1}{(s+1)(s+2)(s+3)} \tag{18.37}$$

We expand in the following form:

$$F(s) = \frac{A}{s+1} + \frac{B}{s+2} + \frac{C}{s+3} \tag{18.38}$$

The various coefficients are figured as follows:

$$\begin{aligned} A &= \left.\frac{(s+1)}{(s+1)(s+2)(s+3)}\right|_{s=-1} = \frac{1}{2} \\ B &= \left.\frac{(s+2)}{(s+1)(s+2)(s+3)}\right|_{s=-2} = -1 \\ C &= \left.\frac{(s+3)}{(s+1)(s+2)(s+3)}\right|_{s=-3} = \frac{1}{2} \end{aligned} \tag{18.39}$$

Then our function is

$$\boxed{F(s) = \frac{1}{2}\left[\frac{1}{s+1} - \frac{2}{s+2} + \frac{1}{s+3}\right]} \tag{18.40}$$

Results are shown in Fig. 18.5. Notice we start with 0 dB/dec and zero phase. Then when all poles are active (i.e., frequency is much larger than each of the poles) the decay rate is −60 dB/dec and the phase is −270° since we have three simple poles.

18.9 Transfer Function of the Form $\frac{1}{(s-j)(s+j)}$

In this case the LT has the form

$$F(s) = \frac{1}{(s-j)(s+j)} \tag{18.41}$$

Assume

$$\frac{1}{(s-j)(s+j)} = \frac{A}{s-j} + \frac{B}{s+j} \tag{18.42}$$

To find A multiply both sides by $(s-j)$ and evaluate at $s = j$

$$A = \frac{1}{2j} \tag{18.43}$$

To find B multiply both sides by $(s+j)$ and evaluate at $s = -j$

$$B = -\frac{1}{2j} \tag{18.44}$$

Our LT becomes

$$\boxed{F(s) = \frac{1}{2j}\left[\frac{1}{s-j} - \frac{1}{s+j}\right]} \tag{18.45}$$

Results are shown in Fig. 18.6. Notice we start at 0 dB/dec and zero phase. When both poles are active we have a decay rate of −40 dB/dec and a final phase of −180° since we have two simple poles. Just a remark in passing we observe that the inverse transform of this is

$$f(t) = \frac{1}{2j}\left[e^{jt} - e^{-jt}\right] = \sin(t) \tag{18.46}$$

18.10 Transfer Function of the Form $\frac{a+s}{b+s}$

Assume transfer function has the form

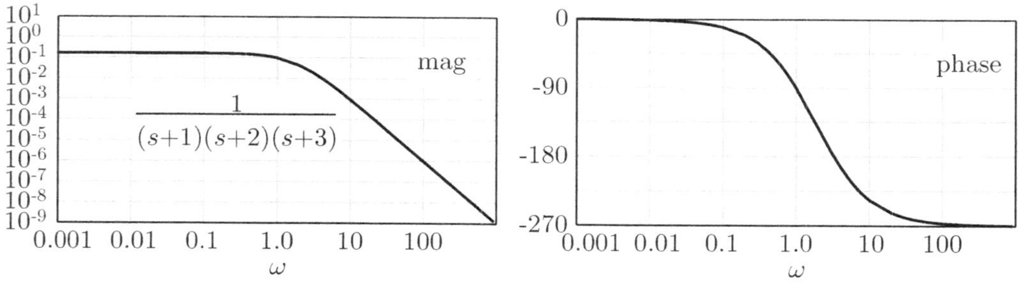

Fig. 18.5 Transfer function with three poles

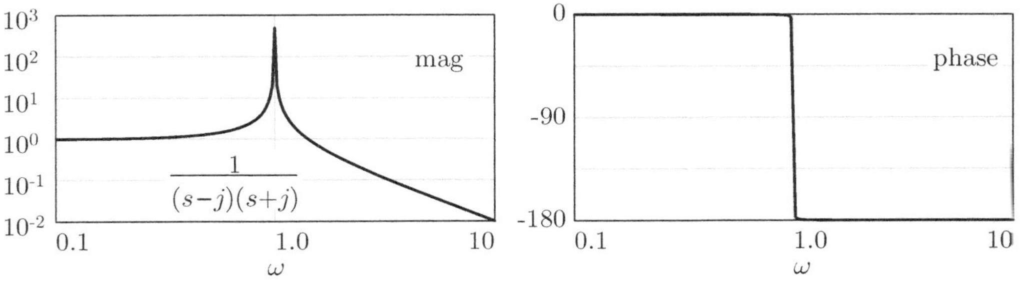

Fig. 18.6 Transfer function with two complex conjugate poles

$$F(s) = \frac{s+10}{s+1} \tag{18.47}$$

Since the order of the numerator is not less than the denominator we have to use long division. To that end we get

$$\begin{array}{r|rr} & 1 & \\ \hline s+1) & s+10 & \\ & -s\ -1 & \\ \hline & 9 & \end{array} \tag{18.48}$$

so that our function can be written as

$$\boxed{F(s) = 1 + \frac{9}{s+1}} \tag{18.49}$$

Results are shown in Fig. 18.7. We start with 0 dB/dec and zero phase. Then the pole kicks in (is activated) and we pick a −20 dB/dec and −90°. Finally the zero kicks in and it nulls the pole for a final decay of 0 dB/dec and final phase of 0°. Notice that due to the proximity of the zero to the pole, we don't really reach the full −90°.

18.11 Transfer Function of the Form $\frac{s+1}{s(s+10)}$

Assume that our transfer function has the form

$$F(s) = \frac{s+1}{s(s+10)} \tag{18.50}$$

We have two poles—one at 0 and one at −10 and we have a zero at $s = -1$; hence we can write

$$F(s) = \frac{A}{s} + \frac{B}{s+10} \tag{18.51}$$

The coefficients are evaluated as

$$A = F(s)s|_{s=0} = \frac{s+1}{s+10}|_{s=0} = \frac{1}{10} \tag{18.52}$$

$$B = F(s)(s+10)|_{s=-10} = \frac{s+1}{s}|_{s=-10} = \frac{9}{10} \tag{18.53}$$

Hence

$$\boxed{F(s) = \frac{1}{10}\left[\frac{1}{s} + \frac{9}{s+10}\right]} \tag{18.54}$$

Results are shown in Fig. 18.8. Sine we have a pole at zero, we already start with −20 dB/dec and a phase of −90°. The zero kicks and it nulls the first pole to get 0 dB/dec and zero phase. Finally the second pole kicks in and we go back to −20 dB/dec and −90°.

18.12 Transfer Function of the Form $\frac{s+1}{(s+2)(s+3)}$

The LT here has the form

$$F(s) = \frac{s+1}{(s+2)(s+3)} \tag{18.55}$$

Assume that our LT can be written as

$$\frac{s+1}{(s+2)(s+3)} = \frac{A}{s+2} + \frac{B}{s+3} \tag{18.56}$$

To find A multiply both sides by $(s+2)$ and evaluate at $s = -2$

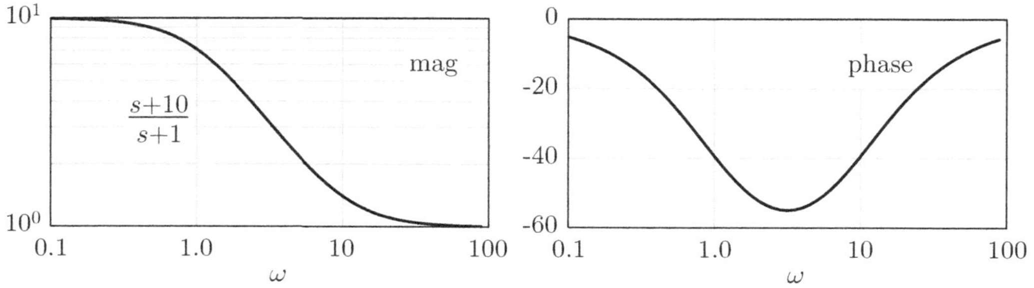

Fig. 18.7 Transfer function with early pole and later zero

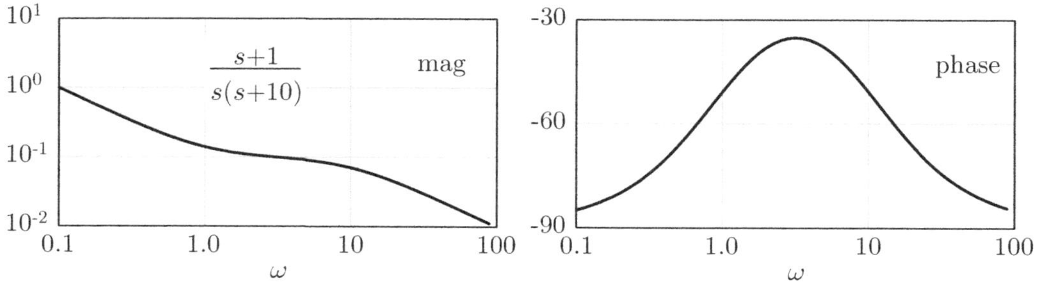

Fig. 18.8 Transfer function with two poles and one zero

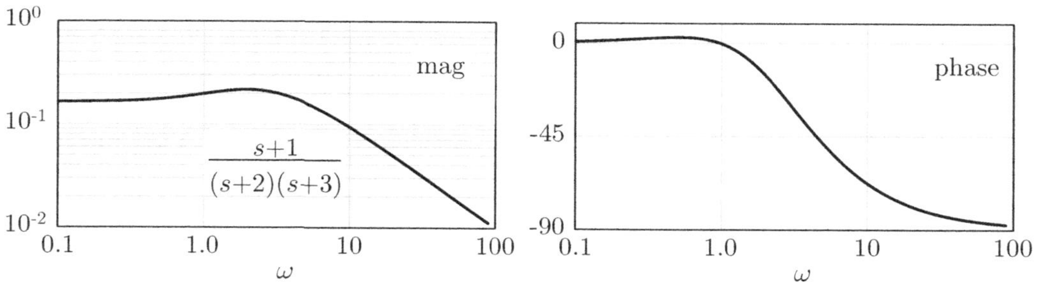

Fig. 18.9 Transfer function with two poles and one zero

$$A = \left.\frac{s+1}{(s+3)}\right|_{s=-2} = \frac{-2+1}{-2+3} = -1 \quad (18.57)$$

To find B multiply both sides by $(s+3)$ and evaluate at $s = -3$

$$B = \left.\frac{s+1}{(s+2)}\right|_{s=-3} = \frac{-3+1}{(-3+2)} = 2 \quad (18.58)$$

So that our LT function can now be written as

$$\boxed{F(s) = \frac{-1}{s+2} + \frac{2}{s+3}} \quad (18.59)$$

Results are shown in Fig. 18.9. In addition to the two poles, this system has a zero at $s = -1$. So we start with 0 dB/dec and zero phase. Then the zero kicks in and gives us 20 dB/dec and 90°. Next comes the first pole and that nulls the zero for a 0 dB/dec and zero phase. Finally the second pole kicks in and we go back to −20 dB/dec and −90°. Notice that due to the proximity of the poles and the zero the decay/growth rate as well as the intermediate phases did not assume their full values! A note in passing—the inverse transform is simply

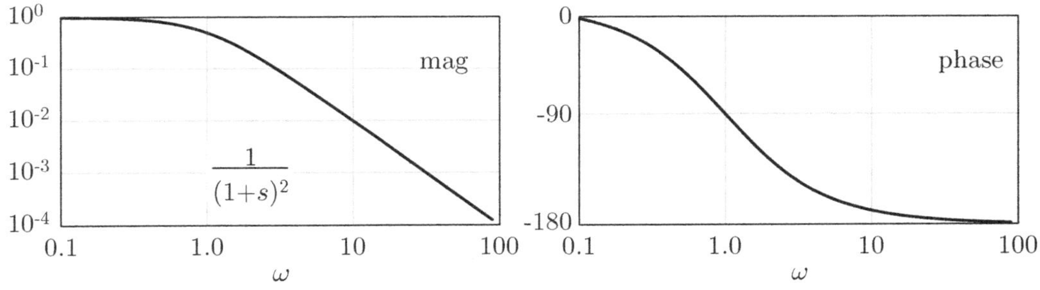

Fig. 18.10 Transfer function with pole of order two

$$f(t) = -e^{-2t} + 2e^{-3t} \tag{18.60}$$

18.13 Transfer Function of the Form $\frac{1}{(s+a)^2}$

In this case the LT has the form

$$F(s) = \frac{1}{(s+a)^2} \tag{18.61}$$

This cannot be simplified any further, and instead has to be dealt with using the following three facts:

$$\frac{1}{s+a} \rightarrow u(t)^{-at}, \tag{18.62}$$

$$\frac{1}{(s+a)^2} = -\frac{d}{ds}\frac{1}{s+a} \tag{18.63}$$

and the frequency differentiation property of the LT:

$$\text{If} \quad f(t) \rightarrow F(s)$$

$$\text{then} \quad -tf(t) \rightarrow \frac{dF(s)}{ds} \tag{18.64}$$

As such our desired function comes out

$$\boxed{f(t) = tu(t)e^{-at}} \tag{18.65}$$

Another method to get the same answer is to recall that

$$t \rightarrow \frac{1}{s^2} \tag{18.66}$$

and the frequency shifting property:

$$\text{If} \quad f(t) \rightarrow F(s),$$

$$\text{then} \quad f(t)e^{-at} \rightarrow F(s+a) \tag{18.67}$$

Results are shown in Fig. 18.10. We start with 0 dB/dec and zero phase. When the pole—of order 2—is activated we switch to −40 dB/dec and −180°.

18.14 Transfer Function of the Form $\frac{s}{(s+1)^2}$

Assume our LT has the form

$$F(s) = \frac{s}{(s+1)^2} \tag{18.68}$$

We have a double pole at $s = -1$. Assume we can expand as

$$F(x) = \frac{A}{s+1} + \frac{B}{(s+1)^2} \tag{18.69}$$

To find B we use

$$B = \left[\frac{s}{(s+1)^2}(s+1)^2\right]_{s=-1} = -1 \tag{18.70}$$

To find A we apply the formula

$$A = \frac{d}{ds}\left[\frac{s}{(s+1)^2}(s+1)^2\right]_{s=-1}$$

$$= \frac{d}{ds}s\Big|_{s=-1} = 1 \tag{18.71}$$

Then our function becomes

$$\boxed{F(x) = \frac{1}{s+1} - \frac{1}{(s+1)^2}} \tag{18.72}$$

Results are shown in Fig. 18.11. Notice that we have a zero at the origin; hence we start at 20 dB/dec and 90°. When the double pole kicks in, not only does it null the zero; it even reverses its direction. The end result is −20 dB/dec and −90°.

18.15 Transfer Function of the Form $\frac{s}{(s+1)^3}$

Assume our LT is given by

$$F(s) = \frac{s}{(s+1)^3} \tag{18.73}$$

We assume that the function can be expanded as

$$F(s) = \frac{A}{(s+1)} + \frac{B}{(s+1)^2} + \frac{C}{(s+1)^3} \tag{18.74}$$

To find C we apply

$$C = \left[\frac{s}{(s+1)^3}(s+1)^3\right]_{s=-1} = -1 \tag{18.75}$$

To find B we apply

$$B = \frac{d}{ds}\left[\frac{s}{(s+1)^3}(s+1)^3\right]_{s=-1} = 1 \tag{18.76}$$

And finally to get A we apply

$$A\frac{1}{2} = \frac{d^2}{ds^2}\left[\frac{s}{(s+1)^3}(s+1)^3\right]_{s=-1} = 0 \tag{18.77}$$

so that

$$\boxed{F(s) = \frac{1}{(s+1)^2} - \frac{1}{(s+1)^3}} \tag{18.78}$$

Results are shown in Fig. 18.12. Again we start at 20 dB/dec and 90° due to the zero at the

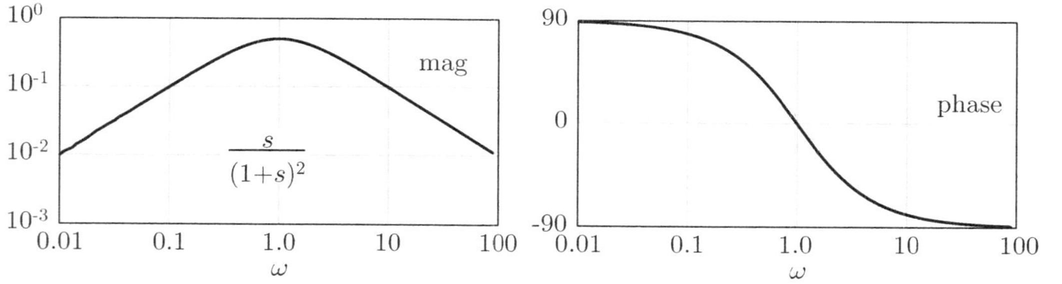

Fig. 18.11 Transfer function with zero and pole of order two

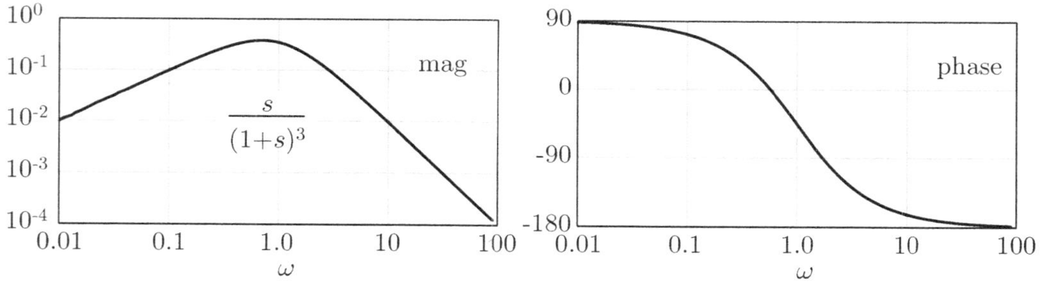

Fig. 18.12 Transfer function with zero and pole of order three

origin. Next, when the pole (of order 3) kicks in, it nulls the zero, but then acts as a double pole to bring the decay back at −40 dB/dec and the final phase at −180°.

18.16 Transfer Function of the Form $\frac{s^2}{(s+1)^3}$

Assume our LT is given by

$$F(s) = \frac{s^2}{(s+1)^3} \tag{18.79}$$

We assume we can write this in the form

$$F(s) = \frac{A}{s+1} + \frac{B}{(s+1)^2} + \frac{C}{(s+1)^3} \tag{18.80}$$

To evaluate C we do

$$C = (s+1)^3 F(s)|_{s=-1} = 1 \tag{18.81}$$

To evaluate B we do

$$B = \frac{d}{ds}(s+1)^3 F(s)|_{s=-1} = \frac{d}{ds}s^2|_{s=-1} = 2s|_{s=-1} = -2 \tag{18.82}$$

To evaluate A we do

$$A = \frac{1}{2}\frac{d^2}{ds^2}(s+1)^3 F(s)|_{s=-1} = \frac{1}{2}\frac{d^2}{ds^2}s^2|_{s=-1} = 1 \tag{18.83}$$

Then our function becomes

$$\boxed{F(s) = \frac{1}{s+1} - \frac{2}{(s+1)^2} + \frac{1}{(s+1)^3}} \tag{18.84}$$

Results are shown in Fig. 18.13. Since we have s^2 in the numerator, which qualifies as a double zero, we start at 40 dB/dec and a phase of 180°. Once the pole (or order 3) kicks in, it nulls the double zero, and forces an effective single pole; hence we end up with −20 dB/dec and final phase of −90°.

18.17 Transfer Function of the Form $\frac{1}{(s+a)(s^2+\omega_0^2)}$

Assume the input transfer function has the form

$$F(s) = \frac{1}{(s+a)(s^2+\omega_0^2)} \tag{13.85}$$

We have a single root at $s = -a$ and complex roots at $j\omega_0$ and $-j\omega_0$. Hence we can write the function as

$$F(s) = \frac{A}{s+a} + \frac{B}{s+j\omega_0} + \frac{C}{s-j\omega_0} \tag{18.86}$$

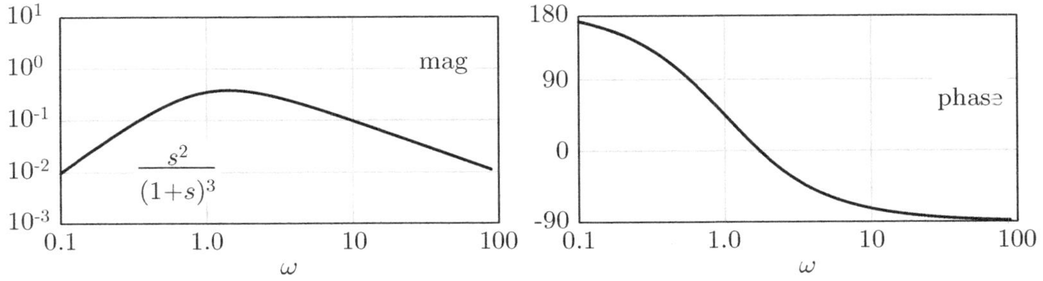

Fig. 18.13 Transfer function with zero of order 2 and pole of order three

We find the various coefficients as follows:

$$A = (s+a)F(s)\Big|_{s=-a} = \frac{1}{s^2+\omega_0^2}\Big|_{s=-a} = \frac{1}{a^2+\omega_0^2}$$
$$B = (s+j\omega_0)F(s)\Big|_{s=-j\omega_0} = \frac{1}{(s+a)(s-j\omega_0)}\Big|_{s=-j\omega_0} = \frac{1}{(a-j\omega_0)(-2j\omega_0)}$$
$$C = (s-j\omega_0)F(s)\Big|_{s=j\omega_0} = \frac{1}{(s+a)(s+j\omega_0)}\Big|_{s=j\omega_0} = \frac{1}{(a+j\omega_0)(2j\omega_0)}$$

Hence

$$F(s) = \frac{1}{a^2+\omega_0^2}\frac{1}{s+a} - \frac{1}{2j\omega_0}\frac{1}{a-j\omega_0}\frac{1}{s+j\omega_0} + \frac{1}{2j\omega_0}\frac{1}{a+j\omega_0}\frac{1}{s-j\omega_0} \tag{18.87}$$

We simplify as follows.

$$-\frac{1}{a-j\omega_0}\frac{1}{s+j\omega_0} + \frac{1}{a+j\omega_0}\frac{1}{s-j\omega_0} = \frac{-(a+j\omega_0)(s-j\omega_0)+(a-j\omega_0)(s+j\omega_0)}{(a-j\omega_0)(a+j\omega_0)(s+j\omega_0)(s-j\omega_0)}$$
$$= \frac{-as-\omega_0^2+j\omega_0(a-s)+as+\omega_0^2+j\omega_0(a-s)}{(a^2+\omega_0^2)(s^2+\omega_0^2)} = \frac{2j\omega_0(a-s)}{(a^2+\omega_0^2)(s^2+\omega_0^2)} \tag{18.88}$$

Then our transfer function becomes

$$\boxed{\frac{1}{(s+a)(s^2+\omega_0^2)} = \frac{1}{a^2+\omega_0^2}\left[\frac{1}{s+a} + \frac{a-s}{s^2+\omega_0^2}\right]} \tag{18.89}$$

Results are shown in Fig. 18.14 for the case $a = 1$ and $\omega_0 = 100$; notice scaling by 1000. We start flat at 0 dB/dec and zero phase. When the first pole kicks in around 1, the decay rate changes to −20 dB/dec and the phase becomes −90°. When the complex pole kicks in around 100 the decay rate switches to −60 dB/dec because now we have effectively a pole of order 3. Also, the phase should pick another −180° for a total of −270°, but what we are seeing is +90°; the explanation is that −270° is really 90° (try visualize that).

18.18 Transfer Function of the Form $\frac{s^2+1}{s^2+4}$

Here the transfer function has the form

$$F(s) = \frac{s^2+1}{s^2+4} \tag{18.90}$$

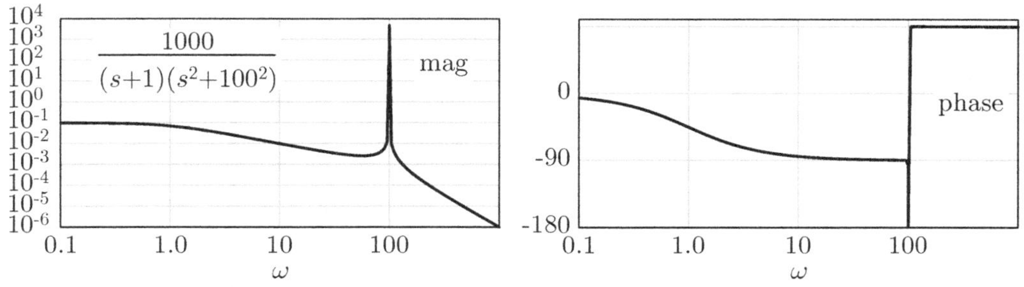

Fig. 18.14 Transfer function with real pole, and two complex conjugate poles

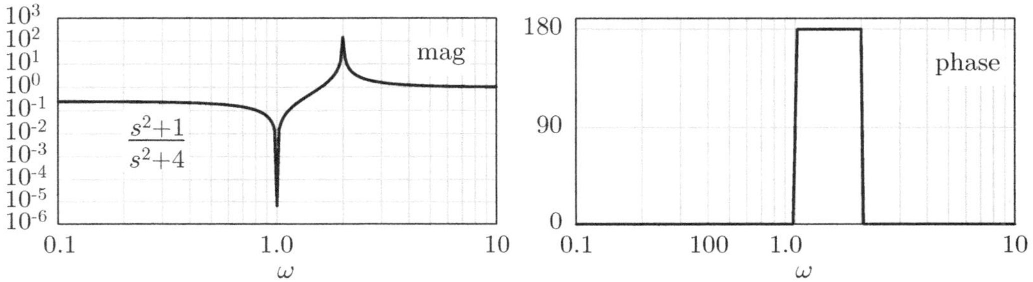

Fig. 18.15 Transfer function with complex conjugate zero pair and complex conjugate pole pair

Notice that the degree of the numerator (2) is the same as the denominator. We will have to use long division as follows:

$$\begin{array}{r|l} & \overline{\quad\quad\quad 1} \\ s^2+4) & s^2+1 \\ & \underline{-s^2-4} \\ & \quad\quad -3 \end{array} \tag{18.91}$$

Hence our function can be written as

$$\boxed{F(s) = 1 - \frac{3}{s^2+4}} \tag{18.92}$$

Notice that we don't have to simplify any more, since we can figure out the inverse transform readily (it will be a delta function and a sine one). Results are shown in Fig. 18.15. Notice initially we start flat at 0 dB/dec and zero phase. When the complex conjugate zero pair kicks in around 1, the growth rate picks up at the rate 40 dB/dec with a phase of 180°. Notice this happens after the "anti-resonance" takes place right at $\omega = 1$. Once the complex conjugate pole pair kicks in around 2 the rate of change goes back to flat at 0 dB/dec and the phase goes back by 180° for a settling value of 0°. Again notice this happens after the "resonance" right at $\omega = 2$.

18.19 Transfer Function of the Form $\frac{s^3+s^2+1}{s^2+4}$

Here the transfer function has the form

$$F(s) = \frac{s^3+s^2+1}{s^2+4} \tag{18.93}$$

Notice that here the degree of the numerator (3) is larger than that of the denominator (2). We will have to use long division:

$$\begin{array}{r|l} & s+1 \\ \hline s^2+4 & s^3+s^2+1 \\ & -s^3 \quad -4s \\ \hline & s^2-4s+1 \\ & -s^2 \quad -4 \\ \hline & -4s-3 \end{array} \tag{18.94}$$

Hence our function can be written as

$$\boxed{F(s) = s + 1 - \frac{4s+3}{s^2+4}} \tag{18.95}$$

Results are shown in Fig. 18.16. At DC we have a constant $F(0) \sim$ const and hence the change rate is 0 dB/dec and the phase is zero. At high frequency the function follows the asymptote $F(\infty) \sim s$ which behaves like a zero, with rate of change 20 dB/dec and phase of 90°. Both of these observations are evident in the plot. While we can simplify the transfer function further, we already can tell the inverse transform:

$$f(t) = \frac{d\delta(t)}{dt} + \delta(t) - \frac{3}{2}\sin 2t - 4\cos 2t \tag{18.96}$$

18.20 Transfer Function of the Form $\frac{1}{a+bs+cs^2+ds^3}$

Assume we have a transfer function of the form

$$\boxed{F(s) = \frac{1}{6+11s+6s^2+s^3}} \tag{18.97}$$

We can factor the polynomial as

$$6+11s+6s^2+s^3 = (s+1)(s+2)(s+3) \tag{18.98}$$

Then

$$F(s) = \frac{1}{(s+1)(s+2)(s+3)} \tag{18.99}$$

We can express this as

$$F(s) = \frac{A}{s+1} + \frac{B}{s+2} + \frac{C}{s+3} \tag{18.100}$$

The various coefficients are

$$A = \left.\frac{1}{(s+2)(s+3)}\right|_{s=-1} = \frac{1}{2}$$

$$B = \left.\frac{1}{(s+1)(s+3)}\right|_{s=-2} = -1$$

$$C = \left.\frac{1}{(s+1)(s+2)}\right|_{s=-3} = \frac{1}{2} \tag{18.101}$$

Then our transfer function is

$$\boxed{F(s) = \frac{1/2}{s+1} - \frac{1}{s+2} + \frac{1/2}{s+3}} \tag{18.102}$$

Results are shown in Fig. 18.17. Notice we start at 0 dB/dec and zero phase. When all three poles have been activated the decay rate becomes 60 dB/dec (20 per pole) and the phase −270° (−90 per pole).

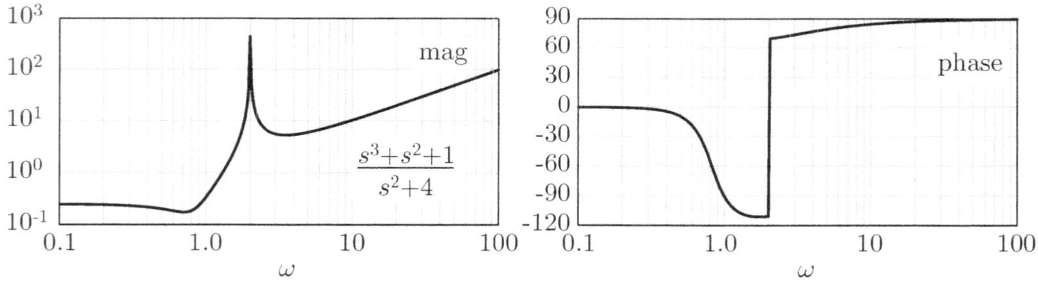

Fig. 18.16 Transfer function with three zeroes and complex conjugate pole pair

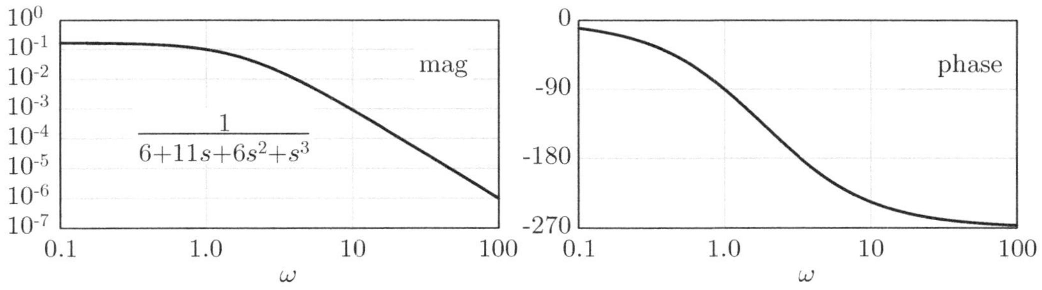

Fig. 18.17 Transfer function with three distinct real poles

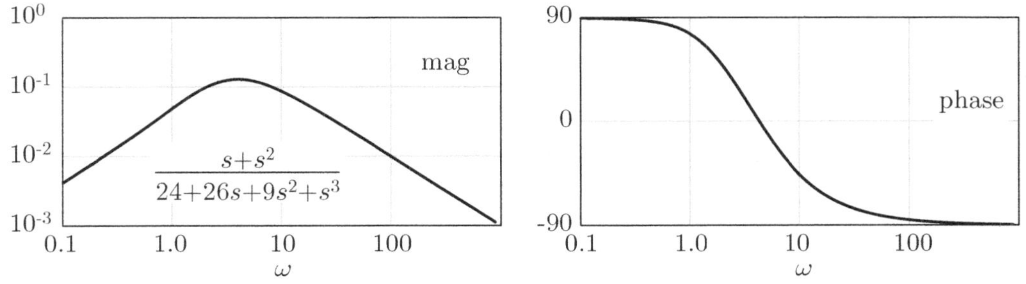

Fig. 18.18 Transfer function with two zeroes and three distinct real poles

18.21 Transfer Function of the Form $\frac{s+s^2}{a+bs+cs^2+ds^3}$

Assume we have a transfer function of the form

$$\boxed{F(s) = \frac{s+s^2}{24+26s+9s^2+s^3}} \tag{18.103}$$

We can factor the polynomial as

$$24+26s+9s^2+s^3 = (s+2)(s+3)(s+4) \tag{18.104}$$

Then

$$F(s) = \frac{s+s^2}{(s+2)(s+3)(s+4)} \tag{18.105}$$

We can express this as

$$F(s) = \frac{A}{s+2} + \frac{B}{s+3} + \frac{C}{s+4} \tag{18.106}$$

The various coefficients are

$$A = \left.\frac{s+s^2}{(s+3)(s+4)}\right|_{s=-2} = 1$$
$$B = \left.\frac{s+s^2}{(s+2)(s+4)}\right|_{s=-3} = -6$$
$$C = \left.\frac{s+s^2}{(s+2)(s+3)}\right|_{s=-4} = 6 \tag{18.107}$$

So finally our transfer function can be written as

$$\boxed{F(s) = \frac{1}{s+2} - \frac{6}{s+3} + \frac{6}{s+4}} \tag{18.108}$$

Results are shown in Fig. 18.18. Since we have a zero at the origin we start at 20 dB/dec and a phase of 90°. When all three poles have been activated, we have the competition between three poles and two zeroes which is won by the former for a net rate of change of −20 dB/dec and a settling phase of −90°.

18.22 Transfer Function of the Form $\frac{1}{(s+a)(s+b)^2}$: Repeated Roots

Assume we have a transfer function of the form

$$\boxed{F(s) = \frac{1}{(s+1)(s+2)^2}} \tag{18.109}$$

Since we have repeated roots we can assume

$$F(s) = \frac{A}{s+1} + \frac{B}{s+2} + \frac{C}{(s+2)^2} \tag{18.110}$$

The various coefficients can be figured as follows:

$$A = \left.\frac{1}{(s+2)^2}\right|_{s=-1} = 1$$

$$B = \left.\frac{d}{ds}\frac{1}{s+1}\right|_{s=-2} = -1$$

$$C = \left.\frac{1}{s+1}\right|_{s=-2} = -1 \tag{18.111}$$

Then finally our transfer function assumes the form

$$\boxed{F(s) = \frac{1}{s+1} - \frac{1}{s+2} - \frac{1}{(s+2)^2}} \tag{18.112}$$

Results are shown in Fig. 18.19. We start at 0 dB/dec and zero phase. When the first pole (at 1) kicks in the rate of change goes to −20 dB/dec and the phase goes to −90°. Then the pole of order 2 kicks in at 2 and pushes the decay rate to −60 dB/dec and the final phase to −270°.

18.23 Transfer Function of the Form $\frac{1}{(s^2+1)^2}$: Repeated Roots

Assume our function is of the form

$$F(s) = \frac{1}{(s^2+1)^2} \tag{18.113}$$

We have roots at j and $-j$, each of order 2; that is

$$F(s) = \frac{1}{[(s-j)(s+j)]^2} = \frac{1}{(s-j)^2(s+j)^2} \tag{18.114}$$

We assume that we can expand this in the form

$$F(s) = \frac{A}{s-j} + \frac{B}{(s-j)^2} + \frac{C}{s+j} + \frac{D}{(s+j)^2} \tag{18.115}$$

The various constants are evaluated as follows:

$$B = F(s)(s-j)^2|_{s=j} = -\frac{1}{4} \tag{18.116}$$

$$D = F(s)(s+j)^2|_{s=-j} = -\frac{1}{4} \tag{18.117}$$

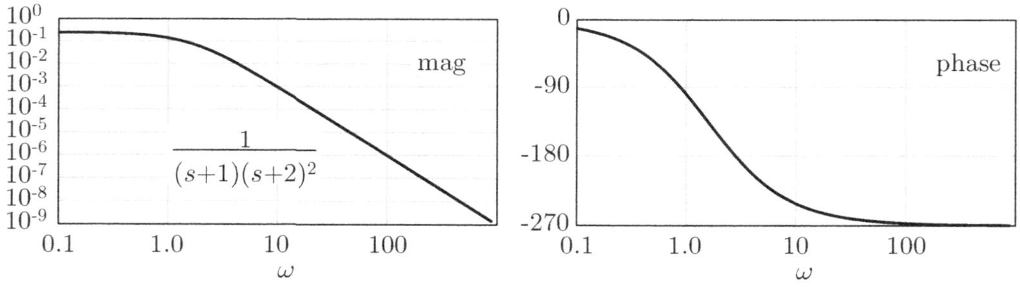

Fig. 18.19 Transfer function with two poles: one simple and the other of order 2

$$A = \frac{d}{ds} F(s)(s-j)^2|_{s=j} = \frac{d}{ds} \frac{1}{(s+j)^2}|_{s=j}$$
$$= -2\frac{1}{(s+j)^3}|_{s=j} = -2\frac{1}{-8j} = \frac{1}{4j} \tag{18.118}$$

$$C = \frac{d}{ds} F(s)(s+j)^2|_{s=-j} = \frac{d}{ds} \frac{1}{(s-j)^2}|_{s=-j}$$
$$= -2\frac{1}{(s-j)^3}|_{s=-j} = -2\frac{1}{(-2j)^3} = \frac{-1}{4j} \tag{18.119}$$

Then we get

$$\boxed{\begin{aligned} F(s) &= \frac{1}{4j}\frac{1}{s-j} - \frac{1}{4}\frac{1}{(s-j)^2} \\ &- \frac{1}{4j}\frac{1}{s+j} - \frac{1}{4}\frac{1}{(s+j)^2} \end{aligned}} \tag{18.120}$$

Results are shown in Fig. 18.20. We start at 0 dB/dec and zero phase. When all poles (four of them) have been activated we transition to −80 dB/dec (20 per pole) and final phase of −360° (−90 per pole). In passing notice that the inverse LT would be

$$\boxed{f(t) = \frac{1}{2}\left[\sin t - t\cos t\right]} \tag{18.121}$$

18.24 Transfer Function of the Form $\frac{s}{(s^2+1)^2}$: Repeated Roots

Assume our function is of the form

$$F(s) = \frac{s}{(s^2+1)^2} \tag{18.122}$$

Notice that this differs from the prior section in that the numerator has an s rather 1. Again we have roots at j and $-j$, each of order 2; that is

$$F(s) = \frac{s}{[(s-j)(s+j)]^2}$$
$$= \frac{s}{(s-j)^2(s+j)^2} \tag{18.123}$$

We assume that we can expand this in the form

$$F(s) = \frac{A}{s-j} + \frac{B}{(s-j)^2} + \frac{C}{s+j} + \frac{D}{(s+j)^2} \tag{18.124}$$

The various constants are evaluated as follows:

$$B = F(s)(s-j)^2|_{s=j} = \frac{s}{(s+j)^2}|_{s=j} = -\frac{j}{4} \tag{18.125}$$

$$D = F(s)(s+j)^2|_{s=-j} = \frac{s}{(s-j)^2}|_{s=-j} = +\frac{j}{4} \tag{18.126}$$

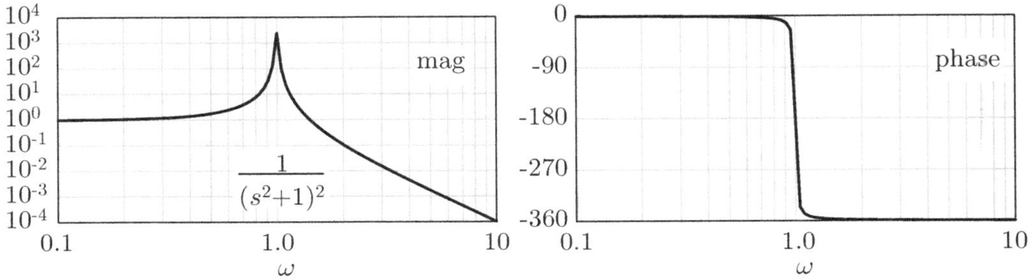

Fig. 18.20 Transfer function a complex conjugate pole pair, each of order 2

$$A = \frac{d}{ds}F(s)(s-j)^2|_{s=j} = \frac{d}{ds}\frac{s}{(s+j)^2}|_{s=j}$$

$$= -2\frac{s}{(s+j)^3} + \frac{1}{(s+j)^2}|_{s=j}$$

$$= -\frac{2j}{-8j} - \frac{1}{4} = \frac{1}{4} - \frac{1}{4} = 0 \qquad (18.127)$$

$$C = \frac{d}{ds}F(s)(s+j)^2|_{s=-j} = \frac{d}{ds}\frac{s}{(s-j)^2}|_{s=-j}$$

$$= -2\frac{s}{(s-j)^3} + \frac{1}{(s-j)^2}|_{s=-j}$$

$$= -\frac{-2j}{8j} - \frac{1}{4} = \frac{1}{4} - \frac{1}{4} = 0 \qquad (18.128)$$

So that

$$\boxed{F(s) = \frac{1}{4j}\left[\frac{1}{(s-j)^2} - \frac{1}{(s+j)^2}\right]} \qquad (18.129)$$

Results are shown in Fig. 18.21. Since we have a zero at the origin we start already at 20 dB/dec and phase of 90°. When the two poles (each of order 2) kick in, we pick −80 dB/dec for a total of −60 dB/dec and −360° for a settling phase value of −270°. In passing it is worth noting that the inverse LT would be

$$\boxed{f(t) = \frac{1}{2}t\sin t} \qquad (18.130)$$

18.25 Transfer Function of the Form $\frac{1}{s^2(s+a)}$

In this case the LT has the form

$$F(s) = \frac{1}{s^2(s+a)} \qquad (18.131)$$

We have a single pole at $s = -a$ and double one at $s = 0$. We can assume an expansion form

$$F(s) = \frac{A}{s} + \frac{B}{s^2} + \frac{C}{s+a} \qquad (18.132)$$

The various coefficients are evaluated as follows

$$A = \frac{d}{ds}s^2F(s)\Big|_{s=0} = \frac{d}{ds}\frac{1}{s+a}\Big|_{s=0}$$
$$= -\frac{1}{(s+a)^2}\Big|_{s=0} = -\frac{1}{a^2}$$
$$B = s^2F(s)\Big|_{s=0} = \frac{1}{s+a}\Big|_{s=0} = \frac{1}{a}$$
$$C = (s+a)F(s)\Big|_{s=-a} = \frac{1}{s^2}\Big|_{s=-a} = \frac{1}{a^2}$$
$$(18.133)$$

Hence our function can be written as

$$\boxed{\frac{1}{s^2(s+a)} = \frac{1}{a^2}\left[-\frac{1}{s} + \frac{a}{s^2} + \frac{1}{s+a}\right]} \qquad (18.134)$$

Results show in Fig. 18.22.

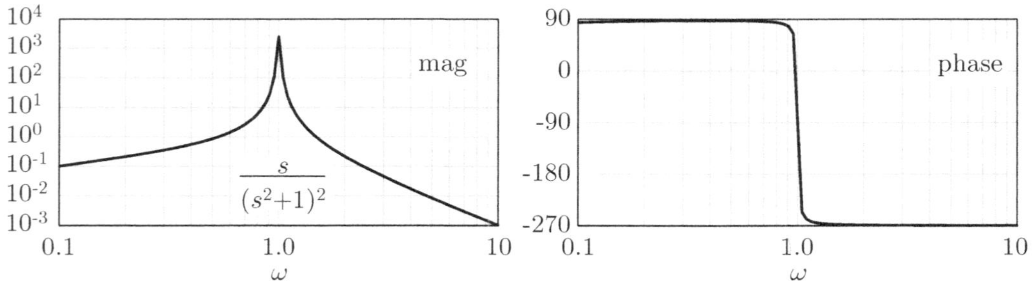

Fig. 18.21 Transfer function with single zero and complex conjugate pole pair—each of order 2

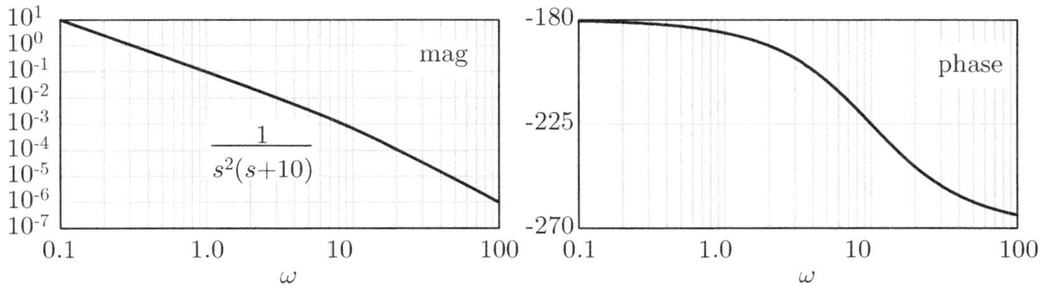

Fig. 18.22 Transfer function with simple pole and another of order 2

18.26 Transfer Function of the Form $\frac{\omega_0}{s(s^2+\omega_0^2)}$

In this case the LT has the form

$$F(s) = \frac{\omega_0}{s(s^2+\omega_0^2)} \tag{18.135}$$

We have a single pole at $s = 0$ and conjugate poles at $s = \pm j\omega_0$. Then we can expand as

$$F(s) = \frac{A}{s} + \frac{B}{s+j\omega_0} + \frac{C}{s-j\omega_0} \tag{18.136}$$

The coefficients are figured as follows:

$$A = sF(s)\Big|_{s=0} = \frac{\omega_0}{s^2+\omega_0^2}\Big|_{s=0} = \frac{1}{\omega_0}$$

$$B = (s+j\omega_0)F(s)\Big|_{s=-j\omega_0} = \frac{\omega_0}{s(s-j\omega_0)}\Big|_{s=-j\omega_0} = -\frac{1}{2\omega_0}$$

$$C = (s-j\omega_0)F(s)\Big|_{s=+j\omega_0} = \frac{\omega_0}{s(s+j\omega_0)}\Big|_{s=j\omega_0} = -\frac{1}{2\omega_0} \tag{18.137}$$

Hence our transfer function can be written as

$$\frac{\omega_0}{s(s^2+\omega_0^2)} = \frac{1}{\omega_0}\left[\frac{1}{s} - \frac{1}{2}\frac{1}{s+j\omega_0} - \frac{1}{2}\frac{1}{s-j\omega_0}\right] \tag{18.138}$$

This can be further simplified to

$$\boxed{\frac{\omega_0}{s(s^2+\omega_0^2)} = \frac{1}{\omega_0}\left[\frac{1}{s} - \frac{s}{s^2+\omega_0^2}\right]} \tag{18.139}$$

Results are shown in Fig. 18.23. We start with a pole at the origin and that gives −20 dB/dec and a starting phase of −90°. When the double conjugate poles kick in we pick and additional −40 dB/dec for a total of −60 dB/dec and another −180° for a final settling phase of −270°.

18.27 Transfer Function of the Form $\frac{\omega_0}{s^2(s^2+\omega_0^2)}$

In this case the LT has the form

$$F(s) = \frac{\omega_0}{s^2(s^2+\omega_0^2)} \tag{18.140}$$

Rather than doing the full work we can leverage results from prior section. By inspection we then have

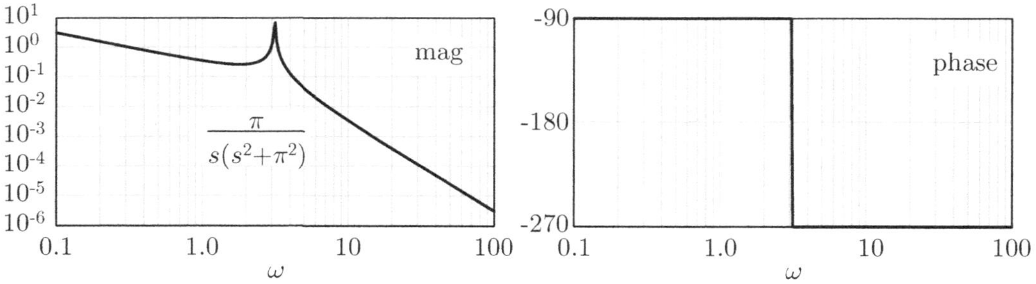

Fig. 18.23 Transfer function with three poles—one at zero and two imaginary ones

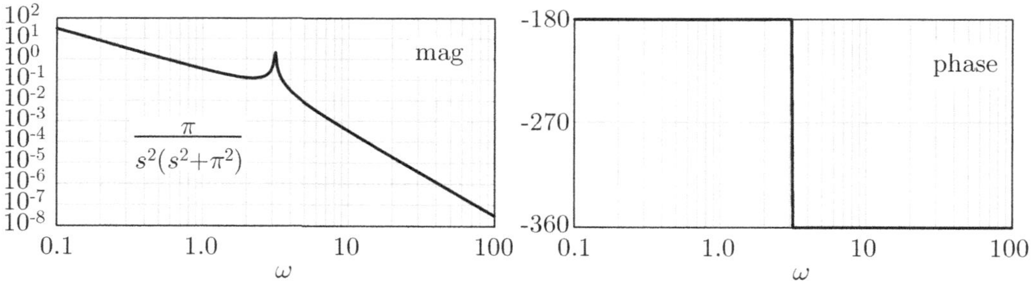

Fig. 18.24 Transfer function with three poles—one at zero (of order 2) and two imaginary ones

$$\boxed{\frac{\omega_0}{s^2(s^2+\omega_0^2)} = \frac{1}{\omega_0}\left[\frac{1}{s^2} - \frac{1}{s^2+\omega_0^2}\right]} \tag{18.141}$$

Results are shown in Fig. 18.24. Since we have a pole at the origin, of order 2, we start with −40 dB/dec and a phase of −180°. Once the complex conjugate pole pair kicks in we pick an additional −40 dB/dec for a total of −80 dB/dec and another −180° for a final phase of −360°.

18.28 Transfer Function of the Form $\frac{1}{s^2+a^2}\frac{1}{s^2+b^2}$

Assume the input transfer function has the form

$$\frac{1}{s^2+a^2}\frac{1}{s^2+b^2} \tag{18.142}$$

Here we have four distinct roots: $s = \pm ja$ and $s = \pm jb$. So in theory we can expand the function as

$$F(s) = \frac{A}{s+ja} + \frac{B}{s-ja} + \frac{C}{s+jb} + \frac{D}{s-jb} \tag{18.143}$$

and we would have to evaluate the four constants how we have been doing all along. But upon further reflection we notice that in fact we can expand as

$$F(s) = \frac{A}{s^2+a^2} + \frac{B}{s^2+b^2} \tag{18.144}$$

Carrying on we get

$$\begin{aligned}\frac{1}{s^2+a^2}\frac{1}{s^2+b^2} &= \frac{A(s^2+b^2)+B(s^2+a^2)}{(s^2+a^2)(s^2+b^2)}\\ &= \frac{s^2(A+B)+Ab^2+Ba^2}{(s^2+a^2)(s^2+b^2)}\end{aligned} \tag{18.145}$$

Equating coefficients we conclude that

$$A + B = 0; \qquad Ab^2 + Ba^2 = 1 \tag{18.146}$$

This solves for

$$B = \frac{1}{a^2 - b^2}, \quad \text{and} \quad A = -\frac{1}{a^2 - b^2} \tag{18.147}$$

So finally we have

$$\boxed{\frac{1}{s^2 + a^2}\frac{1}{s^2 + b^2} = \frac{1}{b^2 - a^2}\left[\frac{1}{s^2 + a^2} - \frac{1}{s^2 + b^2}\right]} \tag{18.148}$$

Results are shown in Fig. 18.25. We start flat with 0 dB/dec and zero phase. When the first complex conjugate pole pair kicks in we pick −40 dB/dec and −180°. When the second complex conjugate pole pair kicks in we pick the same amounts for a final decay rate of −80 dB/dec and final phase of −360°.

18.29 Long Division

Sometime we end up with a polynomial whose numerator is higher (or equal) degree to that of its denominator; then we can use long division to simplify. Here are a few examples:

Example

$$f(s) = \frac{s + 2}{s + 1} \tag{18.149}$$

We setup the skeleton shown below. We put the numerator $(s + 2)$ on the right and the denominator $(s + 1)$ on the left.

$$\begin{array}{r|l} & \;\;1 \\ \hline s+1) & \;\;s+2 \\ & -s-1 \\ \hline & \;\;\;\;\;1 \end{array} \tag{18.150}$$

We divide the s in the numerator by the s in the denominator and get 1. We put the 1 above. Then multiply the 1 by the denominator to get $s + 1$. Change the sign to $-s - 1$ and put under the numerator. Add last two expressions to get 1, which is shown towards the bottom. No further steps. Hence our answer is

$$\frac{s + 2}{s + 1} = 1 + \frac{1}{s + 1} \tag{18.151}$$

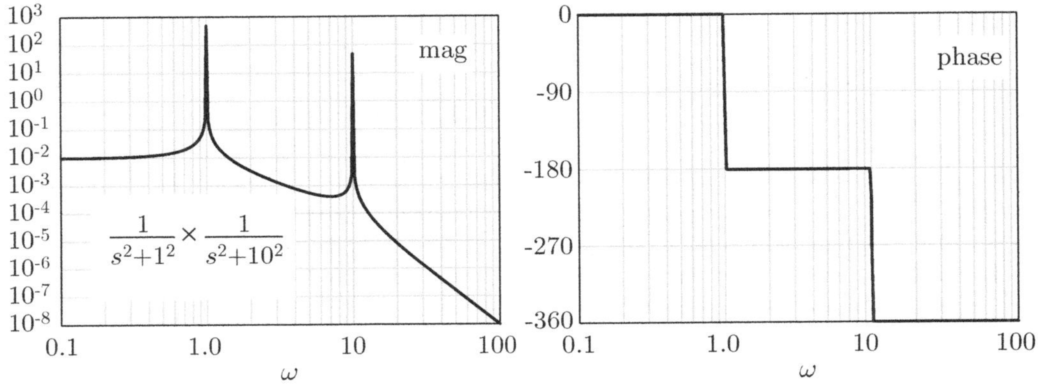

Fig. 18.25 Transfer function with four poles—two complex conjugate pole pairs

Example Let's factor the function

$$f(s) = \frac{s^2 + 1}{s^2 + 2} \tag{18.152}$$

which we saw in Sect. 18.18.

$$\begin{array}{r|l} & \quad\quad\;\; 1 \\ \hline s^2 + 2) & s^2 + 1 \\ & -s^2 - 2 \\ \hline & \quad\quad -1 \end{array} \tag{18.153}$$

Again we put the numerator $s^2 + 1$ on the right and denominator $s^2 + 2$ on the left. We start with the s^2 and divide that by the s^2 in the denominator; this gives 1 which we put on top. Now multiply the 1 by the denominator to get $s^2 + 2$; change sign and then add to numerator to get -1 which we see towards the bottom. We're done; hence our answer is

$$\frac{s^2 + 1}{s^2 + 2} = 1 - \frac{1}{s^2 + 2} \tag{18.154}$$

Example Let's factor the function

$$f(s) = \frac{s^3 + s^2 + 1}{s^2 + 4} \tag{18.155}$$

The needed steps follow:

$$\begin{array}{r|l} & \quad\quad\quad\quad\; s + 1 \\ \hline s^2 + 4) & s^3 + s^2 \quad\quad + 1 \\ & -s^3 \quad\quad -4s \\ \hline & \quad\;\; s^2 - 4s + 1 \\ & \quad -s^2 \quad\quad -4 \\ \hline & \quad\quad -4s - 3 \end{array} \tag{18.156}$$

First we divide s^3 by s^2 to get s and put on upper left; multiply this by the denominator to get $s^3 + 4s$. Invert sign and add to numerator to get $s^2 - 4s + 1$. Now divide the first term of this (s^2) by s^2 to get 1 and put this on top right. Multiply the 1 by the denominator and invert sign to get $-s^2 - 4$. Add to the intermediate result $s^2 - 4s + 1$ to get $-4s$ -3. Since the highest power in this result is lower than that of denominator we stop. Our answer is then

$$\frac{s^3 + s^2 + 1}{s^2 + 4} = s + 1 - \frac{4s + 3}{s^2 + 4} \tag{18.157}$$

18.30 Finding Poles Numerically

Quite often we have a transfer function in the form

$$H(s) = \frac{a_0 + a_1 s + a_2 s^2 + \cdots}{b_0 + b_1 s + b_2 s^2 + \cdots} \tag{18.158}$$

and we are interested in finding the poles, which are the zeroes of the denominator. The pole location is extremely important, for many reasons, including being able to factor the transfer function in simpler terms (such as partial fractions) and being able to find whether the pole lies in the left- or right-hand side of the complex plane. For the case of second order denominator, we can use closed-form solutions, and perhaps third order degree. But beyond that it becomes exceptionally difficult to find the zeroes of the denominator, and to that we resort to numerical techniques. These latter techniques have a field of their own, quite often using elaborate software tools; but in this chapter we will outline a simple technique that can be followed using even the simplest of math tools.

18.30.1 Main Idea

The main idea for finding the zeroes of the denominator is to plot the magnitude of it, locate the zeroes, and then read them out! Quite simple, and the best way to explain this method is via a few examples.

18.30.2 Example

Consider the transfer function given by

$$H(s) = \frac{1}{s^2 + 1} \tag{18.159}$$

Clearly we know how to factor the denominator, which has the solution

$$s^2 + 1 = (s + j)(s - j) \tag{18.160}$$

But assume that we did not know this a priori! How can we find the zeroes? We will rely on constructing a 3D surface plot with the x axis being the real part of s, y-axis imaginary part, and z-axis the magnitude of the denominator. In this case the magnitude has the form

$$\begin{aligned} f(x, y) &= |s^2 + 1| \\ &= \left|(x + jy)^2 + 1\right| \\ &= \left|x^2 - y^2 + 2jxy + 1\right| \\ &= \sqrt{(x^2 - y^2 + 1)^2 + (2xy)^2} \end{aligned} \tag{18.161}$$

If we plot this function we get Fig. 18.26. Notice that the function is always positive, and tends towards zero around the $x = 0$ axis; meaning the real part is zero. If we now take slices and narrow on the zero regions as shown in Fig. 18.27, we get

$$z_1 = 0 - j, \qquad z_2 = 0 + j \tag{18.162}$$

So that our transfer function can be written as

$$H(s) = \frac{1}{(s + j)(s - j)} \tag{18.163}$$

as we already know it should be!

18.30.3 Second Example

Assume here that our transfer function has the form

$$H(s) = \frac{1}{s^2 + s + 1} \tag{18.164}$$

We are interested in finding the poles, which are the zeroes of

$$s^2 + s + 1 \tag{18.165}$$

Again we define the magnitude function as follows:

$$\begin{aligned} f(x, y) &= |s^2 + s + 1| \\ &= \left|(x + jy)^2 + (x + jy) + 1\right| \\ &= \left|(x^2 - y^2 + x + 1) + j(2xy + y)\right| \\ &= \sqrt{(x^2 - y^2 + x + 1)^2 + (2xy + y)^2} \end{aligned} \tag{18.166}$$

If we plot this function we get Fig. 18.28. If we decompose into slices, as shown in Fig. 18.29, we get the zeroes

Fig. 18.26 Magnitude of function in Eq. (18.161)

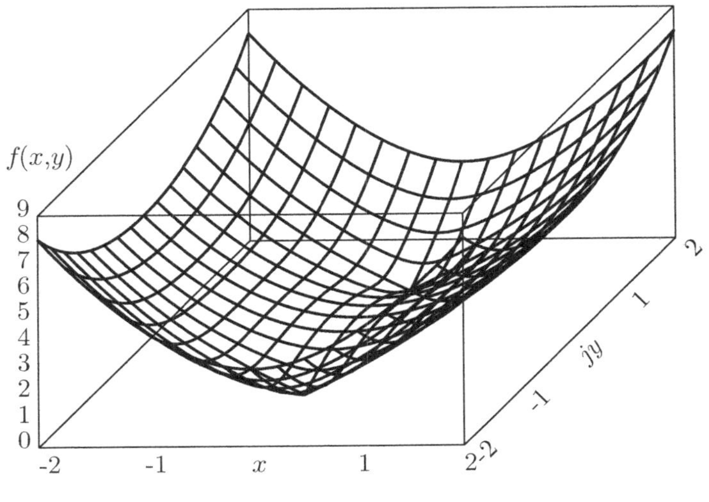

$$z_1 = -0.5 - j\sqrt{0.75}$$
$$z_2 = -0.5 + j\sqrt{0.75} \qquad (18.167)$$

Let's test if this makes sense. We form the product

$$\begin{aligned}(s - z_1)(s - z_2) &= (s + 0.5 + j\sqrt{0.75})(s + 0.5 - j\sqrt{0.75}) \\ &= s^2 + 0.5s - js\sqrt{0.75} + 0.5s + 0.5^2 - j0.5\sqrt{0.75} \\ &+ js\sqrt{0.75} + j0.5\sqrt{0.75} + 0.75 \\ &= s^2 + s + 0.25 + 0.75 \\ &= s^2 + s + 1 \end{aligned} \qquad (18.168)$$

in agreement with Eq. (18.165).

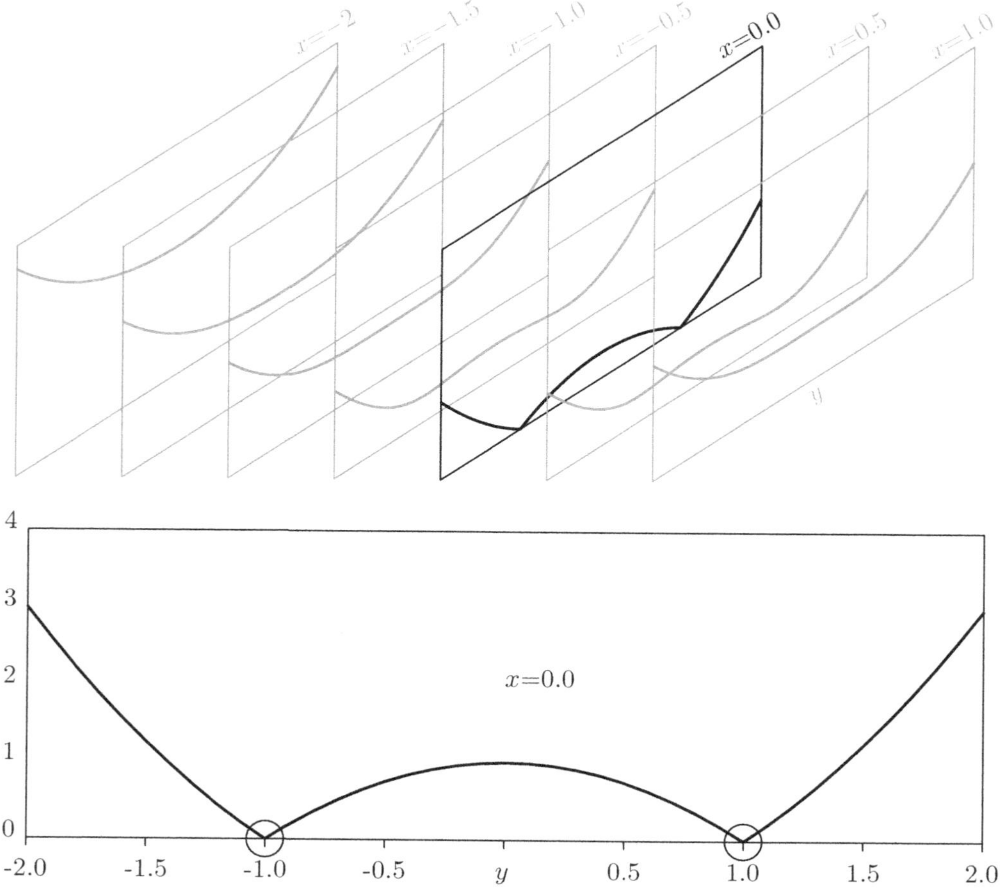

Fig. 18.27 Zoom in version of Fig. 18.26

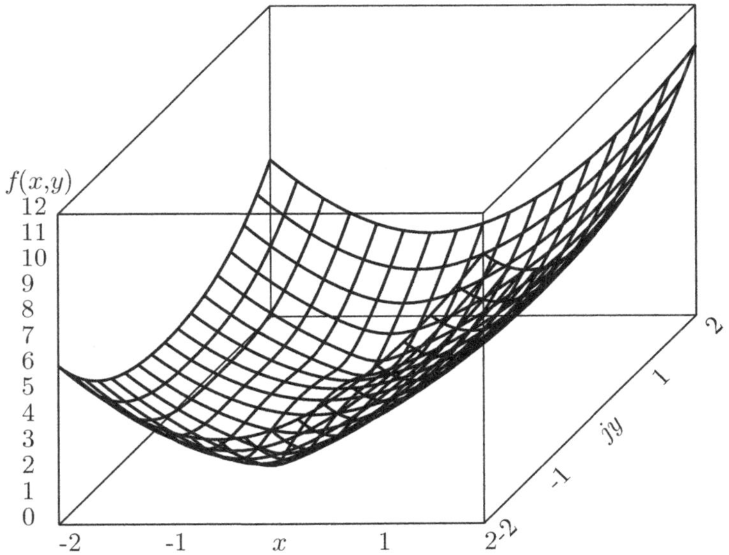

Fig. 18.28 Magnitude of function in Eq. (18.166)

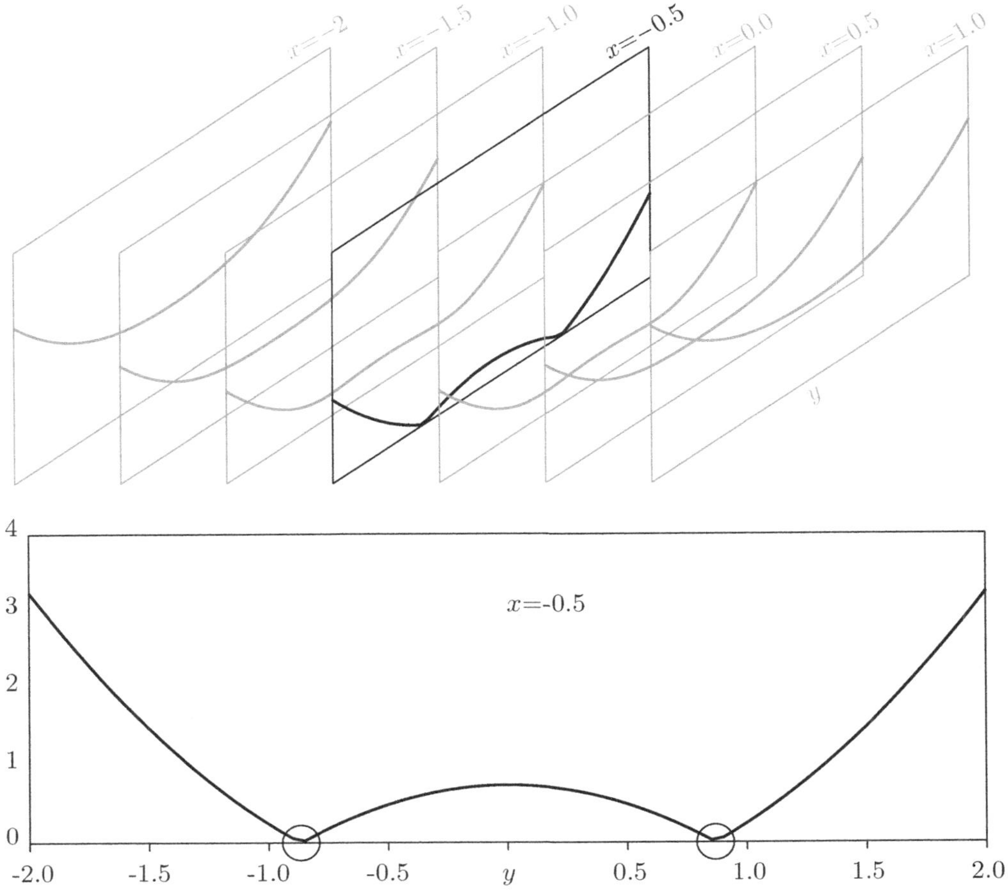

Fig. 18.29 Zoom in version of Fig. 18.28

18.30.4 Third Example

The prior two examples were simplistic in the sense of us knowing their solution a priori. This example is a little bit more complicated, and has a degree of three:

$$H(s) = \frac{1}{s^3 + s^2 + s + 1} \tag{18.169}$$

Again we are interested in finding the poles, which are the zeroes of

$$s^3 + s^2 + s + 1 \tag{18.170}$$

We define the magnitude function as follows:

$$\begin{aligned} f(x, y) &= |s^3 + s^2 + s + 1| \\ &= \left|(x + jy)^3 + (x + jy)^2 + (x + jy) + 1\right| \\ &= \left|(x + jy)(x^2 - y^2 + 2jxy) + x^2 - y^2 + x + 1 + j(2xy + y)\right| \\ &= \left|x^3 - xy^2 + 2jx^2y + jx^2y - jy^3 - 2xy^2 \right. \\ &\quad \left. + \, x^2 - y^2 + x + 1 + j(2xy + y)\right| \\ &= \sqrt{(x^3 - 3xy^2 + x^2 - y^2 + x + 1)^2 + (-y^3 + 3x^2y + 2xy + y)^2} \end{aligned} \tag{18.171}$$

If we plot this function we get Fig. 18.30. If we decompose into slices, we get Fig. 18.31 and from there decipher the three zeroes as:

$$\begin{aligned} z_1 &= 0.0 - j \\ z_2 &= 0.0 + j \\ z_3 &= -1.0 \end{aligned} \tag{18.172}$$

Then our transfer function can be written as

$$H(s) = \frac{1}{(s + 1)(s - j)(s + j)} \tag{18.173}$$

Again this is a very simple method just to give an idea where the poles reside and to get some practice in manipulating and plotting complex numbers. Most likely it is the case that the reader has access to some form of mathematical package (or calculator) that is able to figure complex roots. In that case, finding the roots is the first step, and the subsequent steps would follow in accordance with the text.

18.31 Summary

In this chapter we got introduced to the method of partial fractions which aims at taking a frequency dependent transfer function and reducing it into a series summation of distinct pole functions. The poles can be simple (of order 1) or of higher order. They can be real, imaginary, or complex. The method of partial fraction is really a derivative of the field of complex integration (via residues) but it simplifies things by not having to carry on the e^{st} function at every step of the analysis. Once a rather arbitrary complex transfer function—of the form of a rational polynomial divided by another rational polynomial—has been simplified in terms of pole fractions, the inverse Laplace transform follows easily since we know exact mapping between a single pole fraction and its equivalent in the time domain (which would be variants of e^{-at}, etc.). We illustrated the method via plenty of examples and in each case plotted the magnitude and phase of the transfer function.

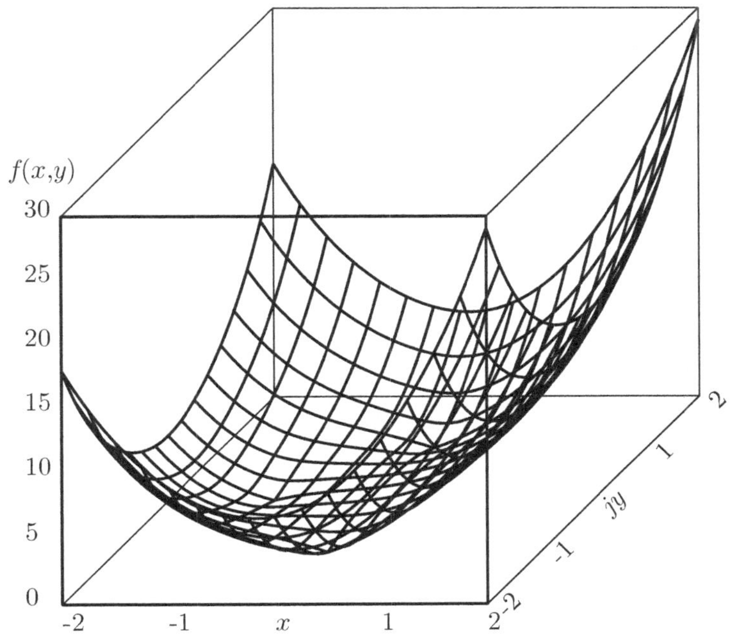

Fig. 18.30 Magnitude of function in Eq. (18.171)

We also spent a good amount of time analyzing the magnitude and phase behavior and explained the notion of 20 dB/dec and 90° per decade of frequency. Typically poles and zeroes have opposite effects in terms of dB/dec and phase shift. We wrapped the chapter with a simple visual method of identifying and locating poles which are a must for the subsequent steps in the partial faction method, and in fact for other areas such as stability analysis. This material will be of use especially in the upcoming chapter on transfer functions.

18.32 Problems

1. What is the partial fraction expansion of the function $F(s) = \frac{1}{(s+1)(s+2)}$?
2. Prove that the phase of the function $F(s) = \frac{1}{(1+s)^2}$ goes to $-180°$ at high frequency.
3. Prove that the phase of the function $F(s) = \frac{1}{(s+1)(s+2)}$ goes to $-180°$ at high frequency.
4. Find the partial fraction expansion of the function $F(s) = \frac{100}{(s+10)(s+100)}$.
   Answer: $F(s) = \frac{10}{9}\left[\frac{1}{s+10} - \frac{1}{s+100}\right]$
5. Find the partial fraction expansion of the function $F(s) = \frac{s}{s+2}$.
   Answer: $F(s) = 1 - \frac{2}{s+2}$
6. Find the partial fraction expansion of the function $F(s) = \frac{s}{(s+j)(s-j)}$.
   Answer: $F(s) = \frac{1}{2}\left[\frac{1}{s+j} + \frac{1}{s-j}\right]$
7. Plot magnitude and phase of the following two functions:

$$F_1(s) = \frac{s+10}{s+1}, \qquad F_2(s) = \frac{s+1}{s+10}$$

   See sample solution in Fig. 18.32.
8. What is the inverse Laplace transform of the function $F(s) = \frac{1}{(s+1)^3}$; show the steps. Plot the spectrum and time series; see sample solution in Fig. 18.33.
   Answer: $f(t) = \frac{1}{2}t^2 e^{-t}$
9. Do we need to use partial fractions to figure the inverse Laplace transform of the function $F(s) = \frac{s^{-s}}{s+1}$? What is the inverse transform? Plot the spectrum and time series; see sample solution in Fig. 18.34.
10. We derived in this chapter (Sect. 18.14) the following partial fraction expansion:

$$\frac{s}{(s+1)^2} = \frac{1}{s+1} - \frac{1}{(s+1)^2}$$

   implying that the inverse transform is

$$f(t) = e^{-t} - te^{-t}$$

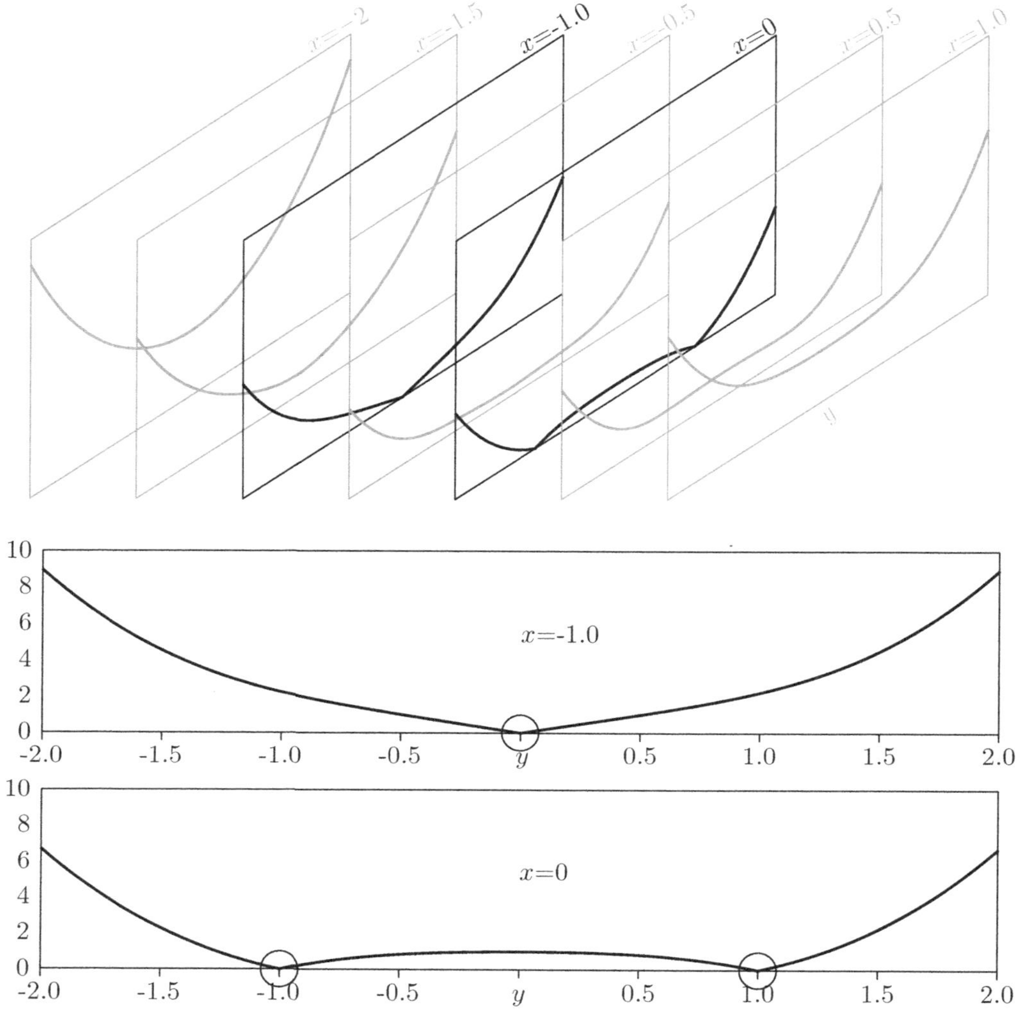

Fig. 18.31 Zoom in version of Fig. 18.30

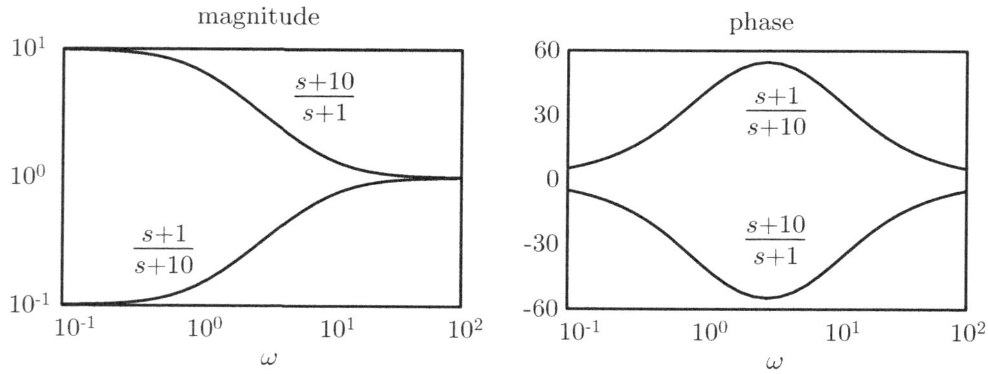

Fig. 18.32 Sample solution to Problem 7

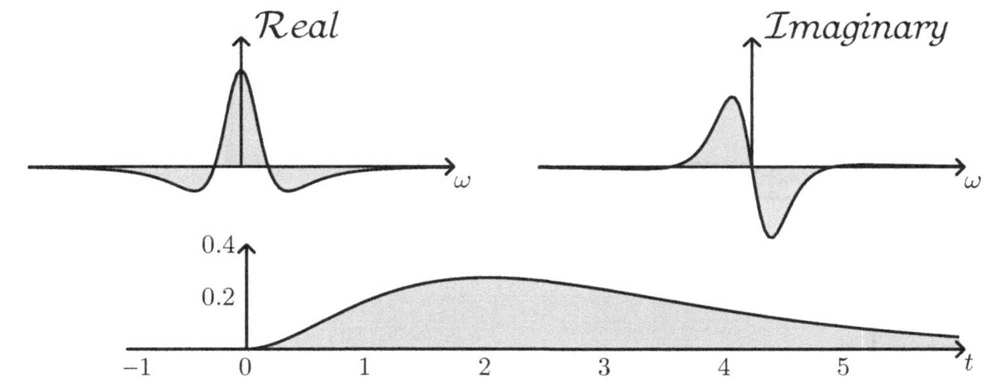

Fig. 18.33 Sample solution to Problem 8

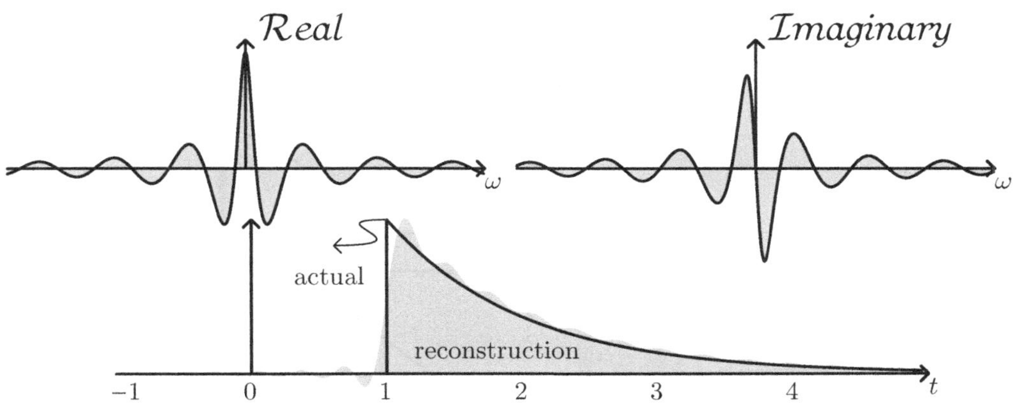

Fig. 18.34 Sample solution to Problem 9

Arrive at the same conclusion using the fact that $\frac{1}{(s+1)^2} \to te^{-t}$ and using the time differentiation property!

11. Verify, by direct substitution and expansion, results in Sect. 18.16

$$\frac{s^2}{(s+1)^3} = \frac{1}{s+1} - \frac{2}{(s+1)^2} + \frac{1}{(s+1)^3}$$

12. Consider the transfer function

$$F(s) = 10000\frac{s+10}{(s+100)(s+10000)}$$

Without plotting it, how many zeroes does this function have? How many poles? What is the initial phase at small frequency? Final phase at large frequency? Plot the magnitude and spectrum and confirm your predictions. See sample solution in Fig. 18.35.

13. A function looks like that in Fig. 18.36. Try predict the function? Use location of zeroes and poles to figure the overall behavior. Use phase as guide. Remember, when we hit a pole, the function inflects down by 20 dB/dec; when we hit a zero, it inflects upwards, again with 20 dB/dec. If we hit a double zero, the ramp rate becomes 40 dB/dec, and so forth! When the function is dominated by a pole, the phase is around $-90°$; when the function is dominated by a zero, the phase is around $90°$, and so forth.
14. A function looks like that in Fig. 18.37. Try predict the function? Use hints in Problem 13.
15. All of the following three transfer functions have something special happening around 10:

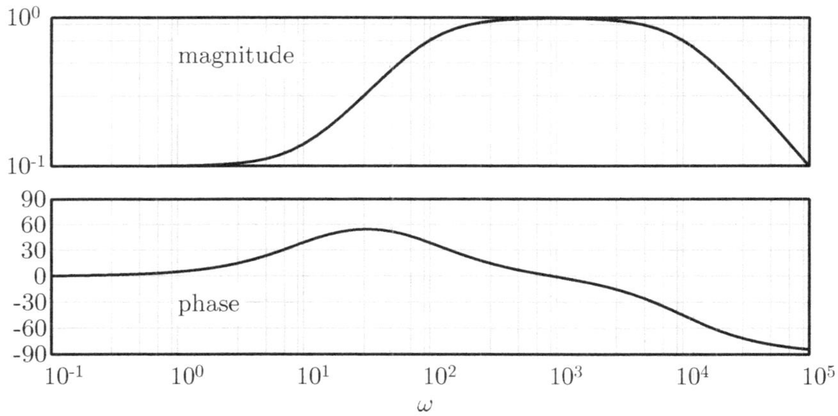

Fig. 18.35 Sample solution to Problem 12

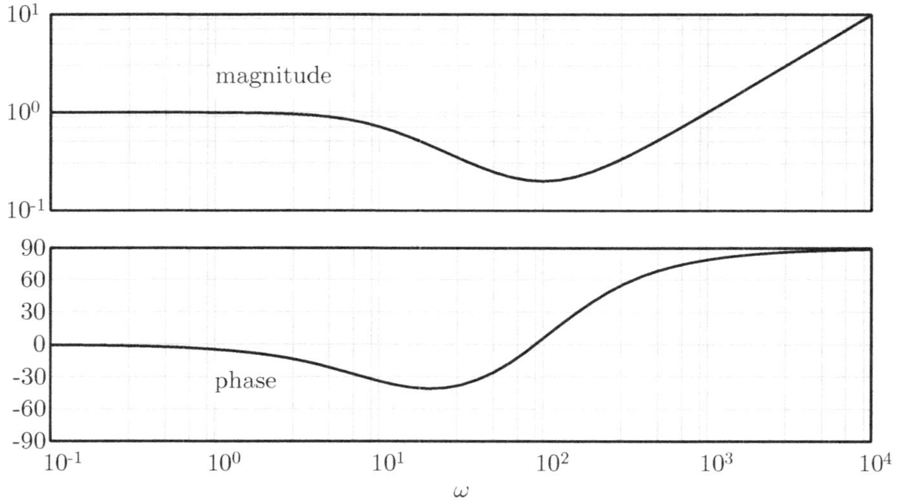

Fig. 18.36 Problem 13

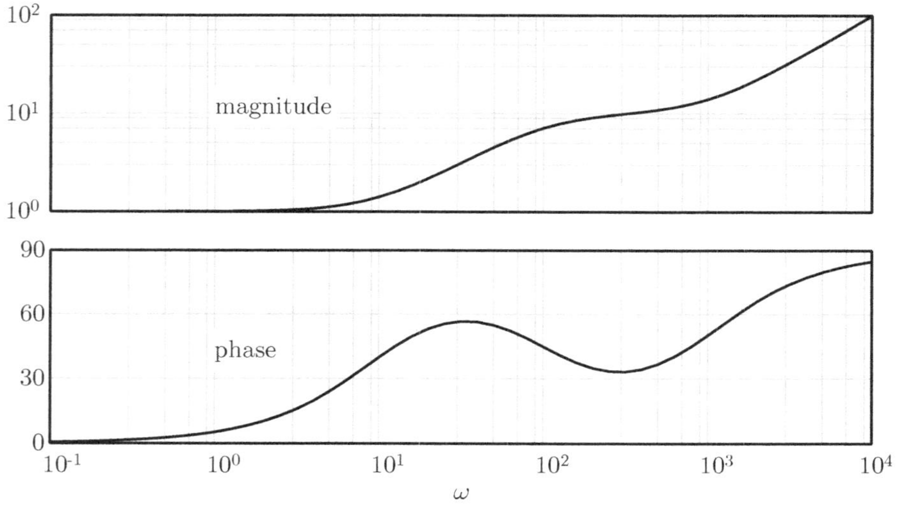

Fig. 18.37 Problem 14

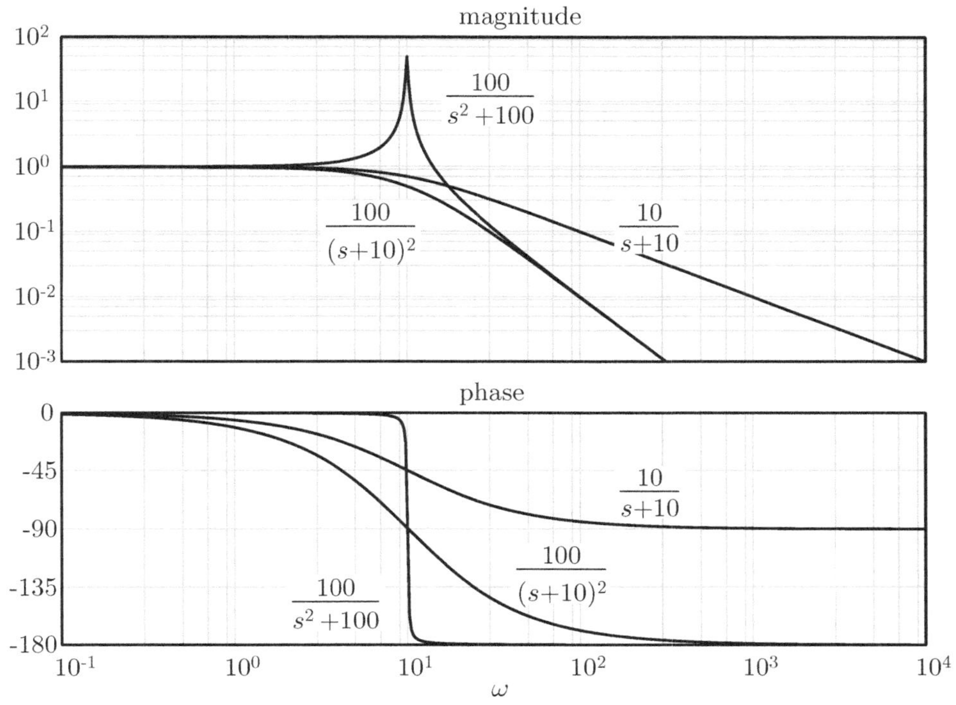

Fig. 18.38 Sample solution to Problem 15

$$F_1(s) = \frac{10}{s+10}, \quad F_2(s) = \frac{100}{(s+10)^2},$$

$$F_3(s) = \frac{100}{s^2+10}$$

Plot the three functions, both magnitude and phase, and explain the different behaviors. See sample solution in Fig. 18.38.

16. Consider the transfer function given by

$$F(s) = \frac{s+s^2}{24+26s+9s^2+s^3}$$

and whose magnitude and phase are shown in solid, in Fig. 18.39. As a very crude approximation, approximate this function with one that has a *single zero*, and *single pole of order 2*. Compare the approximation to the actual case, as shown in sample solution in the same Fig. 18.39.

17. Expand the transfer function $F(s)$ as follows:

$$F(s) = \frac{1}{s(s^2+1)} = \frac{A}{s} + \frac{B}{s^2+1} + \frac{Cs}{s^2+1}$$

What would be A, B, and C? Show the steps! Answer: $F(s) = \frac{1}{s} - \frac{s}{s^2+1}$

18. Using the following facts, and using both the frequency differentiation property and the time integration one, verify Eq. (18.121).

$$\sin t \rightarrow \frac{1}{s^2+1}, \qquad \cos t \rightarrow \frac{s}{s^2+1}$$

19. Consider the transfer function shown in Fig. 18.40. Try to predict it! Remember, when we hit a pole of order 2, phase flips by $-180°$; similarly when we hit a zero, phase flips the other way.
20. Prove that Eq. (18.130) is correct by starting with the following fact and then using the frequency differentiation property:

$$\sin t \rightarrow \frac{1}{s^2+1}$$

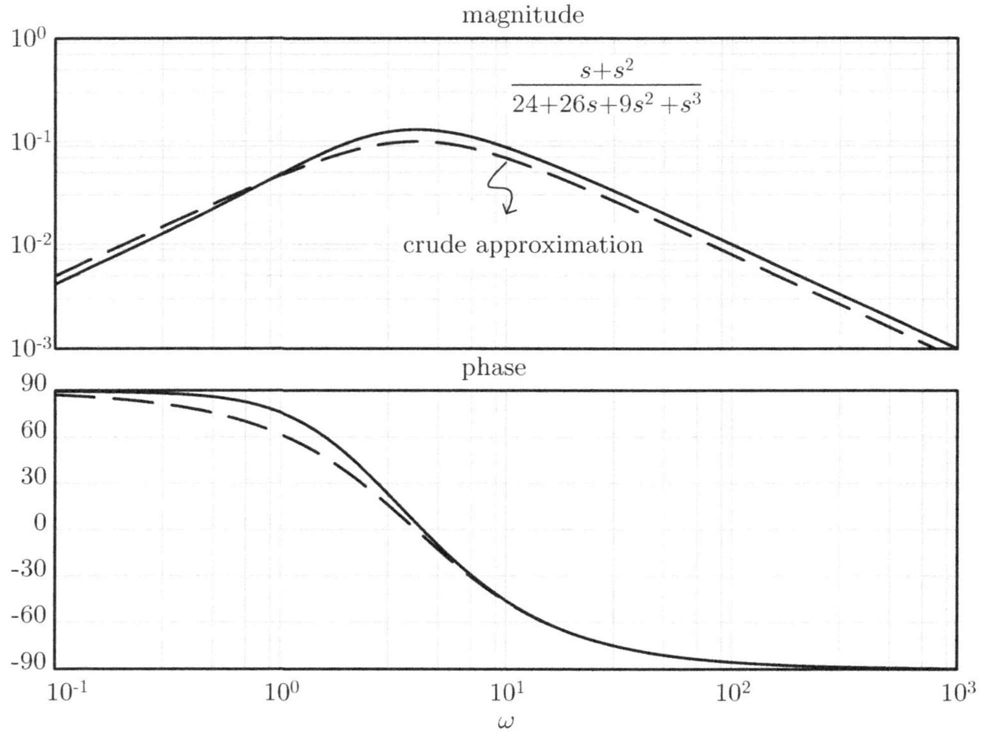

Fig. 18.39 Sample solution to Problem 16

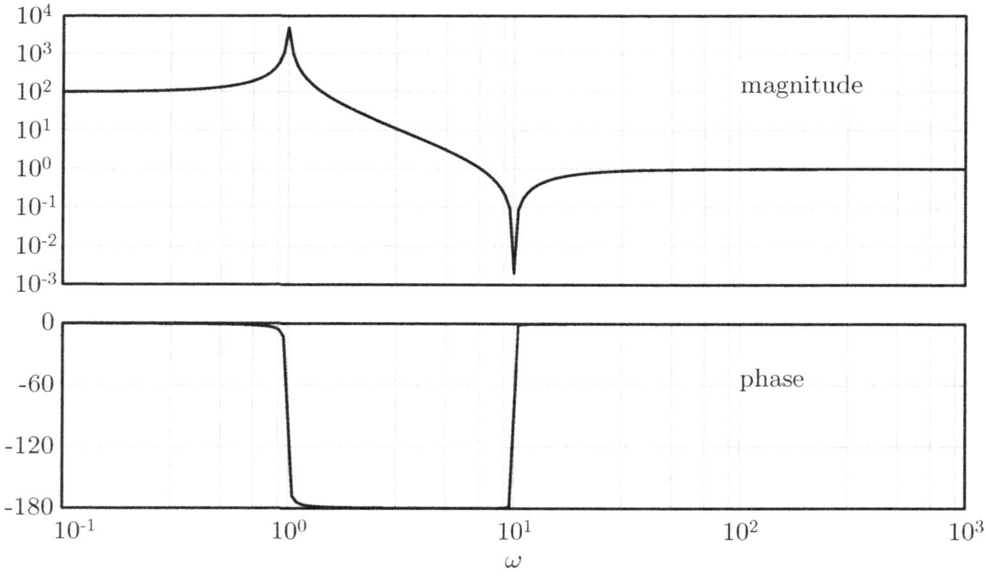

Fig. 18.40 Problem 19

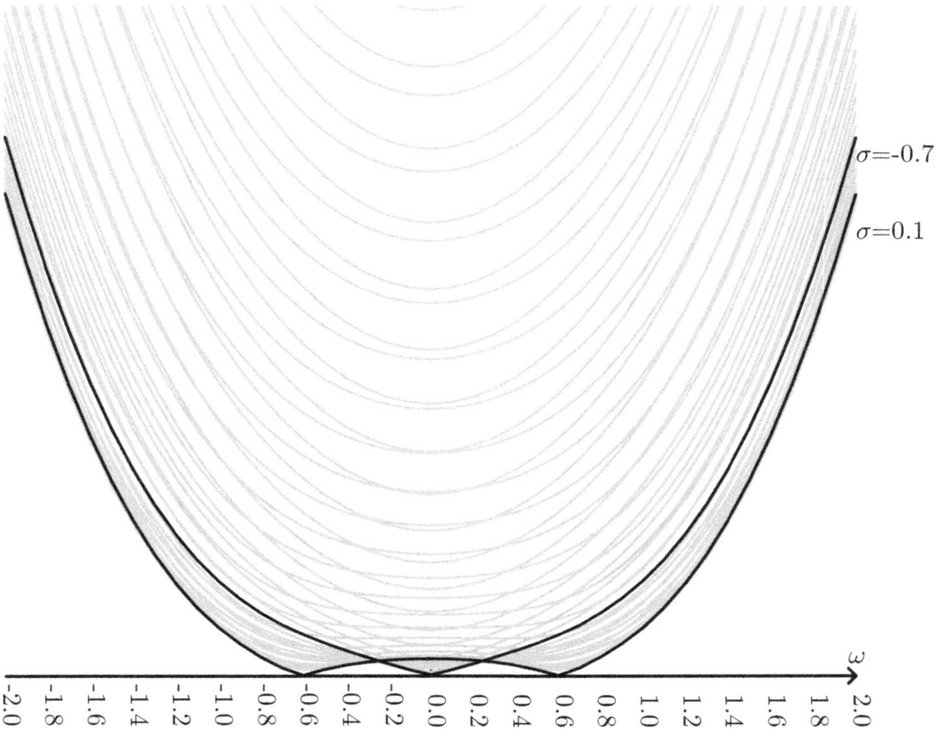

Fig. 18.41 Sample solution to Problem 22

21. Examine Fig. 18.22, back in Sect. 18.25, which showed the transfer function of $F(s) = \frac{1}{s^2(s+10)}$. For small s, how many decades does F go down per decade of s? How about for large s?
22. Find graphically the zeroes of the function

$$F(s) = 1 + s + 2s^2 + 4s^3$$

Vary σ from -5 to 5, and ω from -2 to 2. See sample solution in Fig. 18.41.
23. Find graphically the zeroes of the function

$$F(s) = 5 + 2s + s^2 + 2s^3 + 0.5s^4$$

Vary σ from -5 to 5, and ω from -2 to 2. See sample solution in Fig. 18.42.

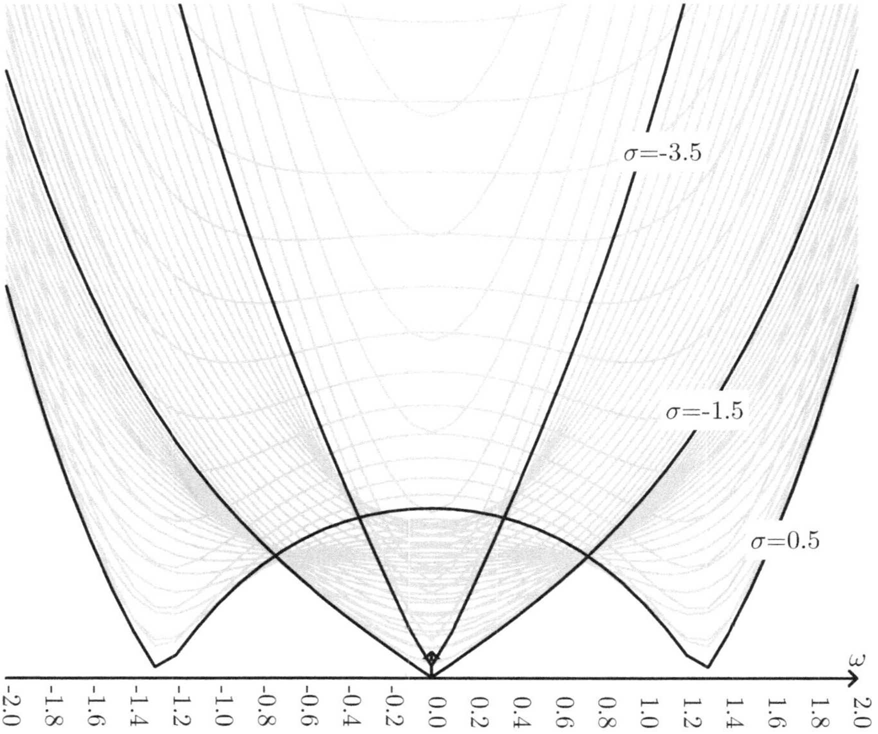

Fig. 18.42 Sample solution to Problem 23

19 Convolution

19.1 Introduction

Put this way—if spectral analysis was one side of the coin, the convolution would be the other side! Really other than basic calculus where derivatives and integrals are used, spectral and convolution techniques would be the next level up. No longer do we deal just with the function at a particular time—instead we decompose it in terms of harmonics, or in terms of convolution integrals with other functions. In both spectral and convolution worlds we seek to express the function in terms of something else. Why so much trouble? In the spectral world if we know the response of a system to a harmonic, and if we can decompose the signal in terms of harmonics we already know the system response to that signal. Similarly and as will be shown over and over if we know the system response to a "select" signal (be it an impulse, unit step, ramp, or others) *and* if we know how to construct an arbitrary signal in terms of these "select" signals—which is the purpose of convolution—then again we already know the system response to that signal. So the same end goal (which is system response) but two different means. Though similar in goal the intermediate steps are quite different. As a simple analogy, if spectral techniques can be coined the parallel method, then convolution techniques would be coined the series method, or something to that effect.

19.2 Definition of Convolution

Two signals $f(t)$ and $g(t)$ are said to be convolved if

$$f(t) * g(t) = \int_{-\infty}^{\infty} f(\tau)g(t-\tau)d\tau \qquad (19.1)$$

Notice the following items:

1. the result of convolution is a *function* (namely of the variable t) and *not* a number.
2. the integration variable τ cancels out and the remaining independent variable is t.

19.3 Convolution of Causal Signals

For the special case of causal signals, which are zero for negative time, the convolution integral simplifies to

$$f(t) * g(t) = \int_{0}^{t} f(\tau)g(t-\tau)d\tau \qquad (19.2)$$

That is, the τ integration starts from zero and ends at t. Unless stated otherwise, in this chapter we will assume all signals to be causal, and as such use the above definition for convolution. The convolution process is next illustrated with many examples.

F. Badrieh, *Spectral, Convolution and Numerical Techniques in Circuit Theory*,
https://doi.org/10.1007/978-3-319-71437-0_19

19.4 Convolution Between Rectangle and Triangle

Start with two functions $f(t)$ and $g(t)$ as shown in Fig. 19.1. Notice that both functions are a function of time t. Change t to τ, leave f as is, but flip g and shift it by t. Next sweep t and for each t value multiply $f(\tau)$ times $g(t-\tau)$ and find the *area* under the resulting curve; that would be the convolution result for *that* t. Read the area and plot versus t. Repeat and sweep over t. This is illustrated graphically in Fig. 19.2. The reader is advised to trace each time step, examine the overlapping (scaled) area, and ensure that area is what is being reported in the convolution plot (versus t). Also, take a look for negative and large time and ensure the overlapping area there is zero.

19.5 Convolution Between Two Rectangles of Unequal Widths

Consider the convolution $f(t) * g(t)$ where

$$f(t) = \begin{cases} 2 & |t| < 2 \\ 0 & |t| > 2 \end{cases} \tag{19.3}$$

$$g(t) = \begin{cases} 1 & |t| < 1 \\ 0 & |t| > 1 \end{cases} \tag{19.4}$$

as shown in Fig. 19.3.

The convolution is carried as follows:

1. Time $-3 < t < -1$: Here the integral gives

$$f(t) * g(t) = 2[(t+1)+2] \tag{19.5}$$

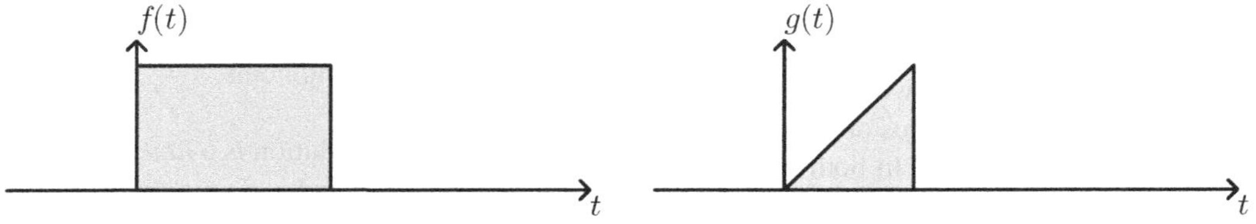

Fig. 19.1 Two functions $f(t)$ and $g(t)$ ready to be convolved

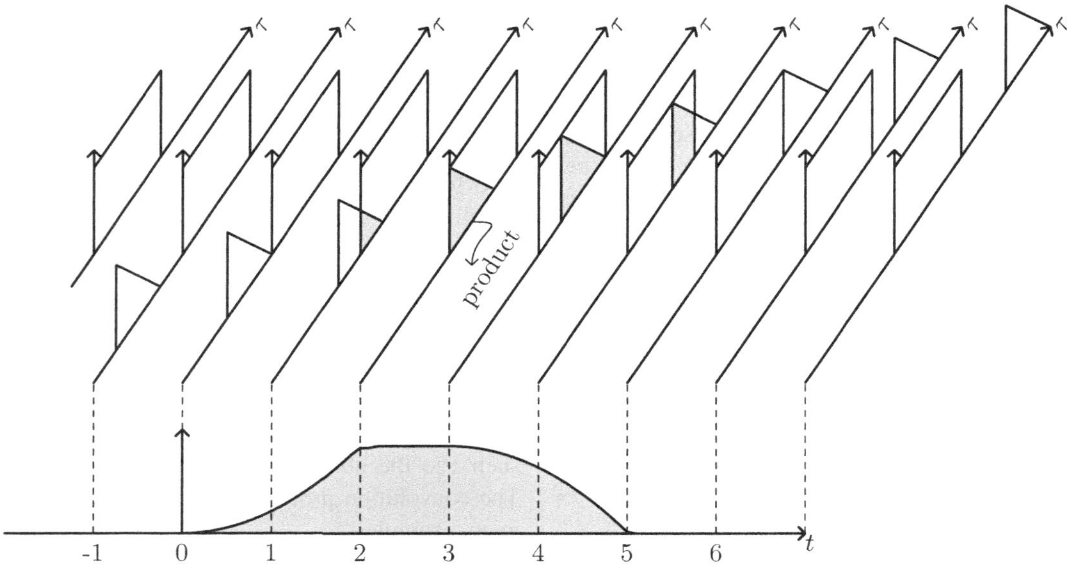

Fig. 19.2 Convolution between $f(t)$ and $g(t)$

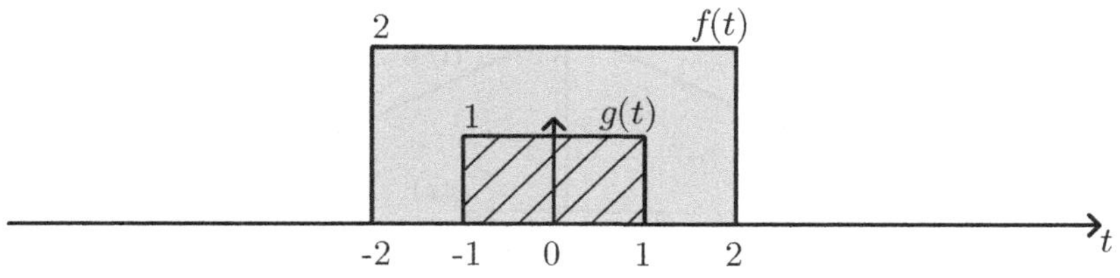

Fig. 19.3 Two rectangles of unequal widths (and heights) ready for convolution

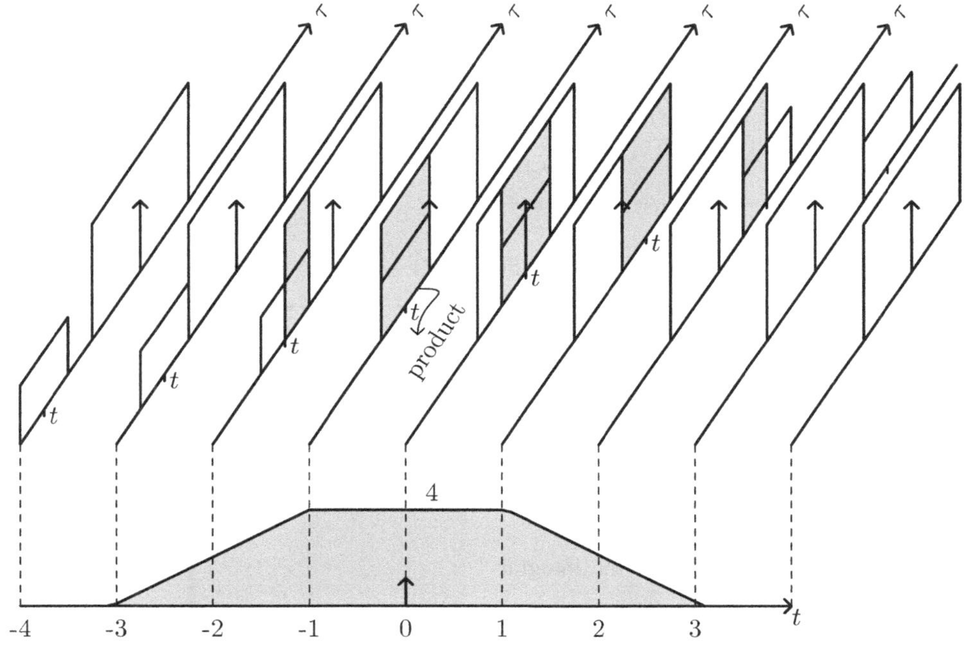

Fig. 19.4 Convolution between two rectangles

2. Time $-1 < t < 1$: Here the integral gives

$$f(t) * g(t) = 4 \tag{19.6}$$

3. Time $1 < t < 3$: Here the integral gives

$$f(t) * g(t) = 2[2 - (t - 1)] \tag{19.7}$$

The results are shown in Fig. 19.4. Again the main theme is sweeping t and for each t identifying the product area. Of course to get any area the two functions have to overlap first. So first we identify the overlap area, and then calculate the actual area. The τ domain is a temporary one; when integration is competed the new axis becomes t—time.

19.6 Convolution Between Symmetric Triangle and Rectangle

Consider the convolution $f(t) * g(t)$ where

$$f(t) = \begin{cases} 5 - |t| & |t| < 5 \\ 0 & |t| > 5 \end{cases}, \quad \text{and} \tag{19.8}$$

$$g(t) = \begin{cases} 1 & |t| < 1 \\ 0 & |t| > 1 \end{cases} \tag{19.9}$$

as shown in Fig. 19.5. So we have a large symmetric triangle to be convolved with a small rectangle. In order to do the integration analytically we will need the following facts:

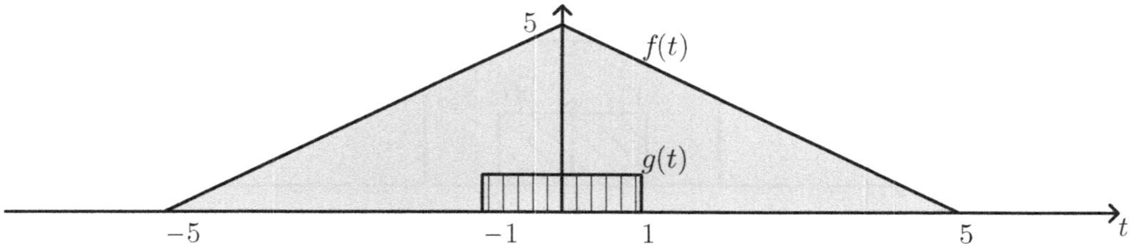

Fig. 19.5 Triangle and rectangle functions ready to be convolved

$$\int (5+t)dt = 5t + \frac{t^2}{2}, \quad \text{and} \tag{19.10}$$

$$\int (5-t)dt = 5t - \frac{t^2}{2} \tag{19.11}$$

The integration is carried as follows:

1. Time range $-6 < t < -4$: Here the integral gives

$$f(t) * g(t) = 5\tau + \left.\frac{\tau^2}{2}\right|_{-5}^{t+1} = 5(t+1) + \frac{(t+1)^2}{2} - 5(-5) - \frac{5^2}{2} \tag{19.12}$$

2. Time range $-4 < t < -1$: Here the integral gives

$$f(t) * g(t) = 5\tau + \left.\frac{\tau^2}{2}\right|_{t-1}^{t+1} = 5(t+1) + \frac{(t+1)^2}{2} - 5(t-1) - \frac{(t-1)^2}{2} \tag{19.13}$$

3. Time range $-1 < t < 1$: Here the integral gives

$$f(t) * g(t) = 5\tau + \left.\frac{\tau^2}{2}\right|_{t-1}^{0} + 5\tau - \left.\frac{\tau^2}{2}\right|_{0}^{t+1} = -5(t-1) - \frac{(t-1)^2}{2} + 5(t+1) - \frac{(t+1)^2}{2} \tag{19.14}$$

4. Time range $1 < t < 4$: Here the integral gives

$$f(t) * g(t) = 5\tau - \left.\frac{\tau^2}{2}\right|_{t-1}^{t+1} = +5(t+1) - \frac{(t+1)^2}{2} - 5(t-1) + \frac{(t-1)^2}{2} \tag{19.15}$$

5. Time range $4 < t < 6$: Here the integral gives

$$f(t) * g(t) = 5\tau - \left.\frac{\tau^2}{2}\right|_{t-1}^{5} = +5(5) - \frac{(5)^2}{2} - 5(t-1) + \frac{(t-1)^2}{2} \tag{19.16}$$

Results are shown in Fig. 19.6. Notice that the result is symmetric (since both inputs are) and that the maximum happens at time 0. Notice also that the convolution is zero for both $t < -6$ and $t > 6$ since there is no overlap there. Remember it is not the overlap area we seek—instead, it is the *product* area!

19.7 Convolution Between Unit Step Function and Negative Exponential

Let's find the convolution between the unit step function and the negative exponential:

$$u(t) * e^{-at} = \int_0^{\infty} u(\tau) e^{-a(t-\tau)} d\tau \qquad (19.17)$$

as shown in Fig. 19.7. The steps are as follows:

$$\int_0^{\infty} u(\tau) e^{-a(t-\tau)} d\tau = \int_0^t e^{-a(t-\tau)} d\tau = e^{-at} \int_0^t e^{a\tau} d\tau = e^{-at} \left[\frac{1}{a} e^{a\tau} \right]_0^t$$

$$= e^{-at} \frac{1}{a} \left[e^{at} - 1 \right] = \boxed{\frac{1}{a} \left[1 - e^{-at} \right]} \qquad (19.18)$$

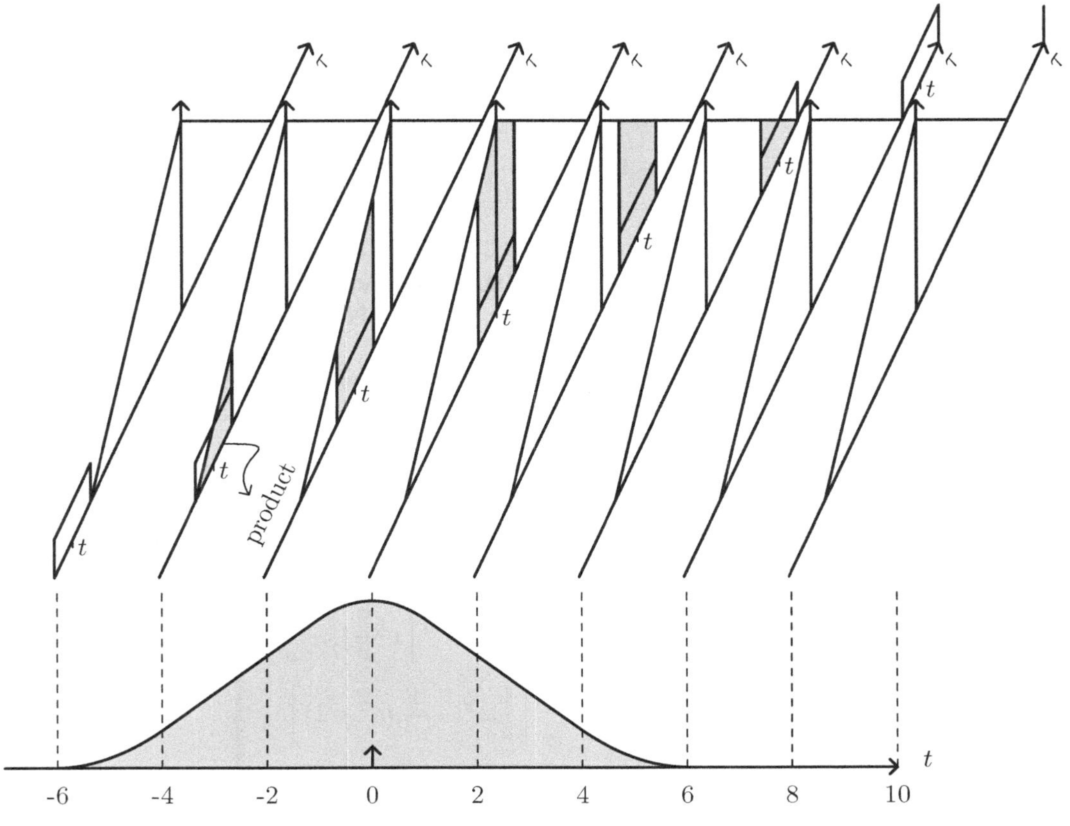

Fig. 19.6 Convolution between triangle and rectangle

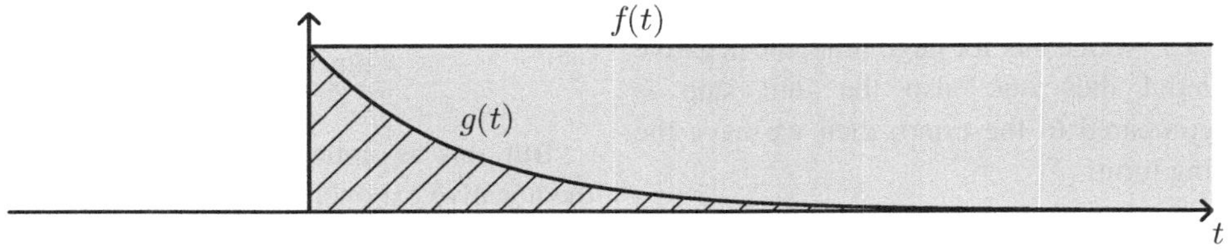

Fig. 19.7 Unit step function and negative exponential one ready to be convolved

The convolution process is shown in Fig. 19.8. For each t value, we arrange the two functions as shown, multiply them, and find the area under the product (the shaded region). Notice that convolving a function with the unit step function is tantamount to integrating the function. Notice also that in the limit of large time, the convolution result goes to $1/a$

$$\lim_{t\to\infty} u(t) * e^{-at} = \frac{1}{a} \tag{19.19}$$

which is nothing more than the total area under the negative exponential. That is, the convolution area reaches a constant and that is evident in the unchanging shaded area as shown in the figure.

19.8 Convolution Between Two Negative Exponentials

Let's find the convolution between two negative exponentials

$$e^{-at} * e^{-bt} = \int_0^\infty e^{-a\tau} e^{-b(t-\tau)} d\tau \tag{19.20}$$

as shown in Fig. 19.9. The steps are as follows:

$$\int_0^\infty e^{-a\tau} e^{-b(t-\tau)} d\tau = e^{-bt} \int_0^t e^{-\tau(a-b)} d\tau = e^{-bt} \left[-\frac{1}{a-b} e^{-\tau(a-b)} \right]_0^t$$

$$= -e^{-bt} \frac{1}{a-b} \left[e^{-t(a-b)} - 1 \right] = \boxed{\frac{1}{b-a} \left[e^{-at} - e^{-bt} \right]} \tag{19.21}$$

The convolution process is shown in Fig. 19.10. Notice that in the limit of large time, the convolution result goes to zero

$$\lim_{t\to\infty} e^{-at} * e^{-bt} = 0 \tag{19.22}$$

That is, the area formed by the product of the (offsetted) graphs goes to zero, as shown in the figure.

19.9 Convolution Between Ramp and Negative Exponential

Let's find the convolution between the ramp function and the negative exponential one

$$t * e^{-at} = \int_0^\infty \tau e^{-a(t-\tau)} d\tau \tag{19.23}$$

as shown in Fig. 19.11. The steps are as follows:

$$\int_0^\infty \tau e^{-a(t-\tau)} d\tau = e^{-at} \int_0^t \tau e^{a\tau} d\tau = e^{-at} \left[\tau \frac{e^{a\tau}}{a} \Big|_0^t - \frac{1}{a} \int_0^t e^{a\tau} d\tau \right]$$

$$= e^{-at} \left[\frac{1}{a} t e^{at} - \frac{1}{a^2} \left(e^{a\tau} \right)_0^t \right] = e^{-at} \left[\frac{1}{a} t e^{at} - \frac{1}{a^2} \left(e^{at} - 1 \right) \right]$$

$$= e^{-at} \left[\frac{1}{a} t e^{at} - \frac{1}{a^2} e^{at} + \frac{1}{a^2} \right] = \boxed{\frac{1}{a^2} \left[-u(t) + at + e^{-at} \right]} \tag{19.24}$$

The convolution process is shown in Fig. 19.12. Notice that for large time, the negative exponential dies out; also the unit step is small compared to the ramp; then we have the following limit:

$$\lim_{t\to\infty} t * e^{-at} = \frac{1}{a} t \tag{19.25}$$

But this is nothing more than the area of the negative exponential $(1/a)$, scaled linearly in time!

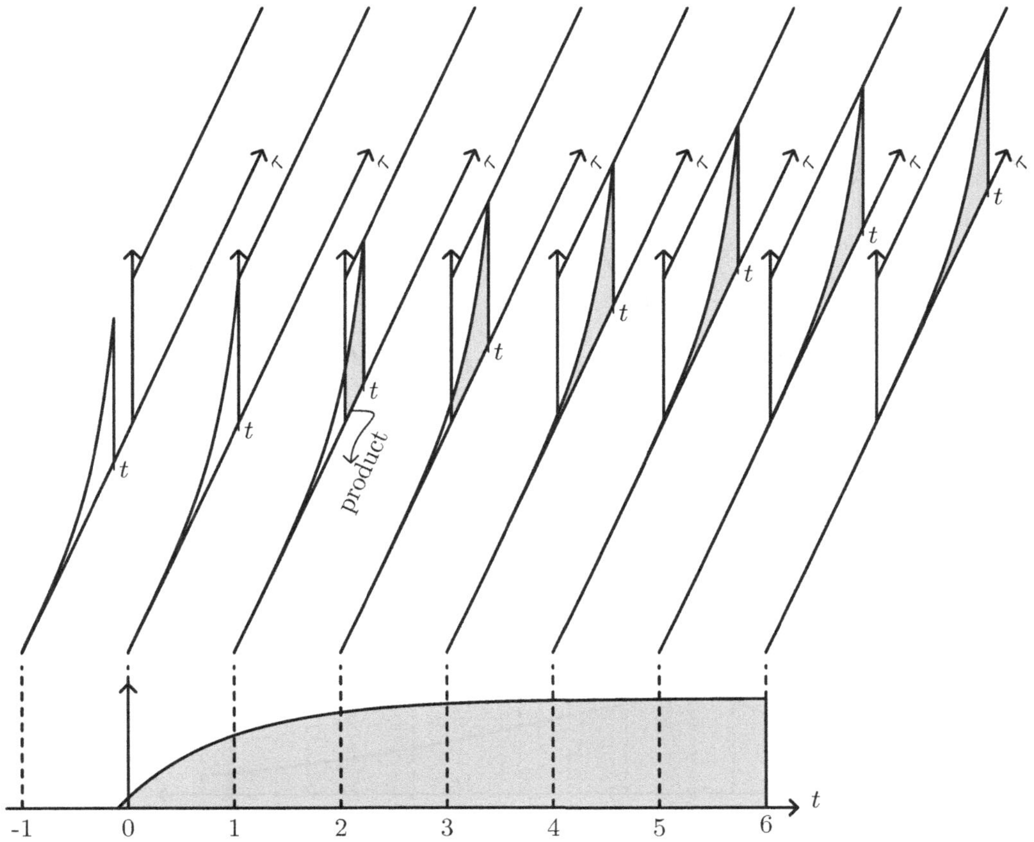

Fig. 19.8 Convolution of unit step function and negative exponential

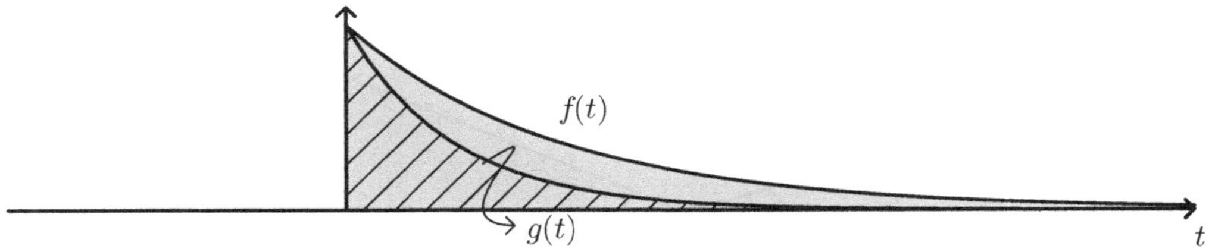

Fig. 19.9 Two negative exponentials ready to be convolved

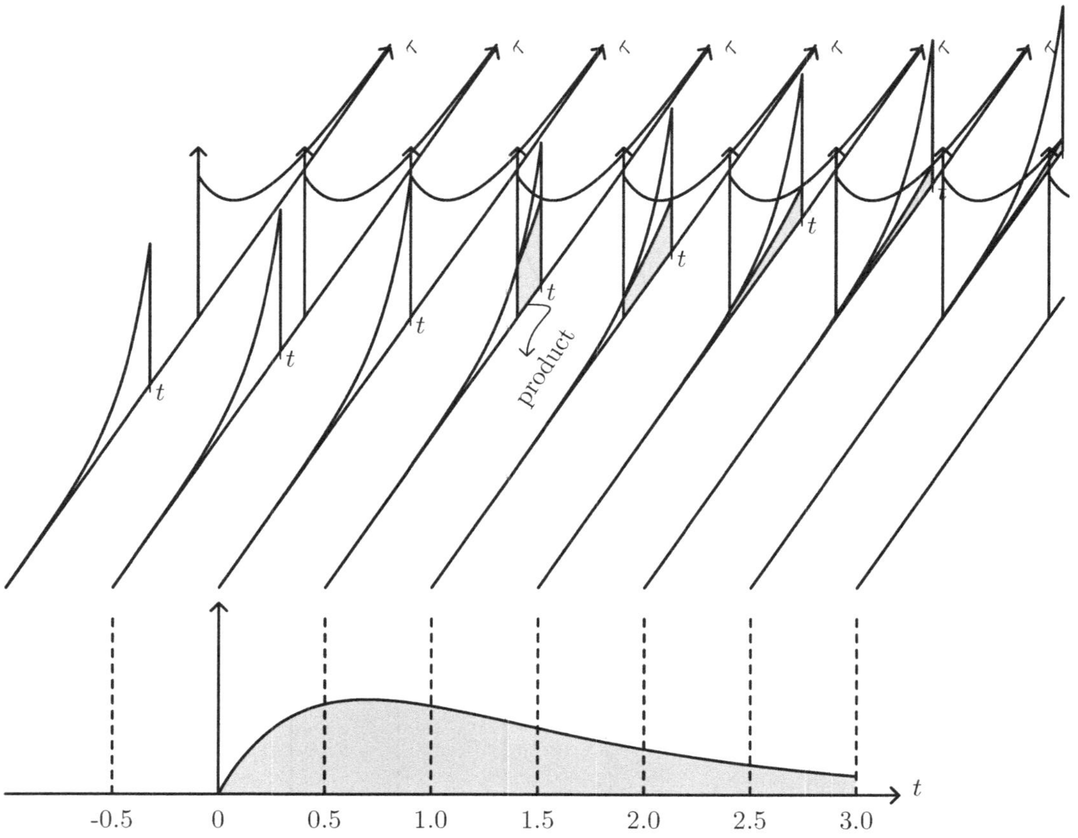

Fig. 19.10 Convolution between two negative exponentials

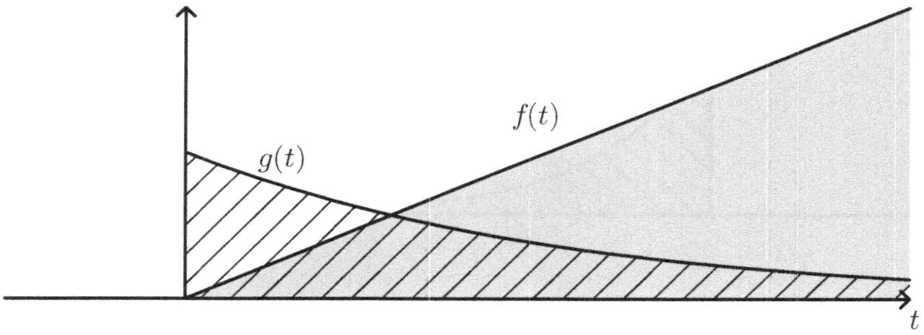

Fig. 19.11 Ramp function and negative exponential one ready to be convolved

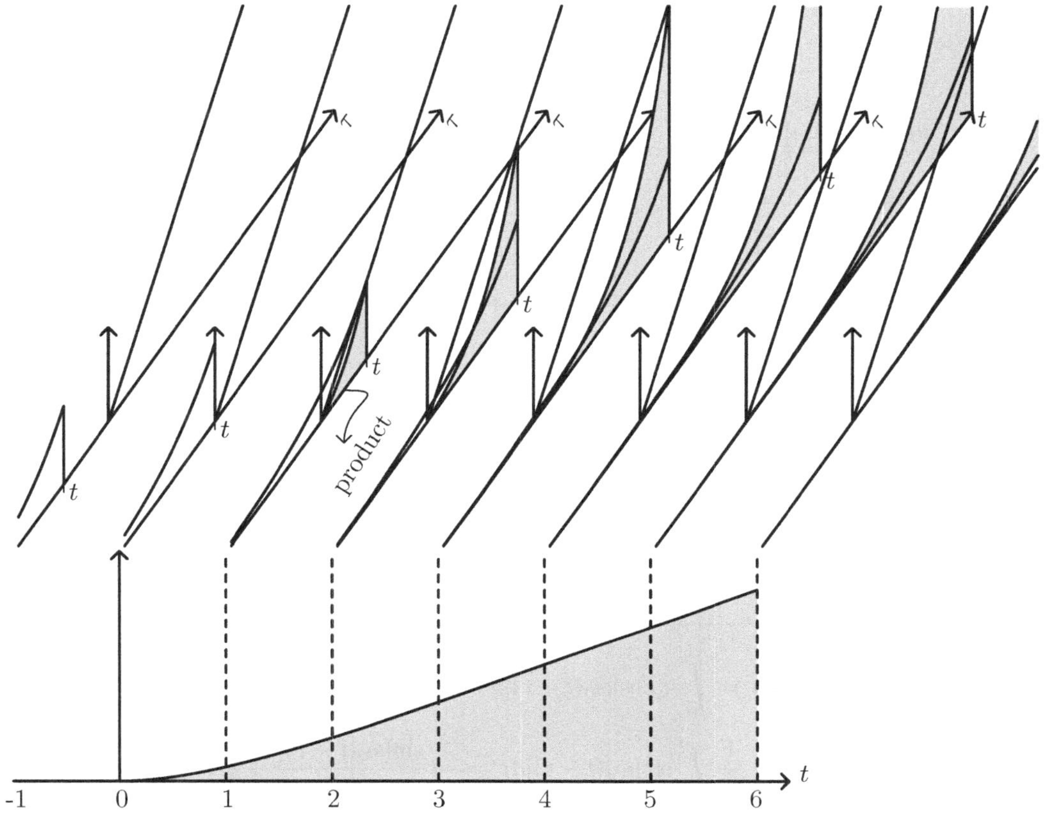

Fig. 19.12 Convolution between ramp and negative exponential

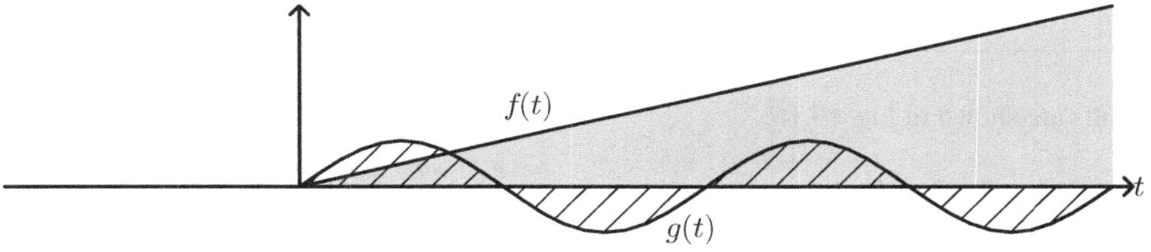

Fig. 19.13 Ramp function and sine one ready to be convolved

19.10 Convolution Between Sine and Ramp Function

Let's find the convolution between the ramp function and the sine function

$$t * \sin \omega_0 t = \int_0^{\infty} \tau \sin[\omega_0(t-\tau)] d\tau \quad (19.26)$$

as shown in Fig. 19.13. The steps are as follows:

$$\int_0^\infty \tau \sin[\omega_0(t-\tau)]d\tau = \int_0^t \tau \sin[\omega_0(t-\tau)]d\tau$$
$$= \tau\frac{\cos[\omega_0(t-\tau)]}{\omega_0} - \frac{1}{\omega_0}\int_0^t \cos[\omega_0(t-\tau)]d\tau = \tau\frac{\cos[\omega_0(t-\tau)]}{\omega_0} + \frac{1}{\omega_0^2}\sin[\omega_0(t-\tau)]\Big|_0^t$$
$$= \boxed{\frac{t}{\omega_0} - \frac{1}{\omega_0^2}\sin\omega_0 t} \tag{19.27}$$

Results are shown in Fig. 19.14.

19.11 Convolution Between Cosine and Ramp Function

Let's find the convolution between the ramp function and the cosine function

$$t * \cos\omega_0 t = \int_0^\infty \tau \cos[\omega_0(t-\tau)]d\tau \tag{19.28}$$

as shown in Fig. 19.15. The steps are as follows

$$\int_0^\infty \tau \cos[\omega_0(t-\tau)]d\tau = \int_0^t \tau \cos[\omega_0(t-\tau)]d\tau$$
$$= -\tau\frac{\sin[\omega_0(t-\tau)]}{\omega_0} + \frac{1}{\omega_0}\int_0^t \sin[\omega_0(t-\tau)]d\tau = -\tau\frac{\sin[\omega_0(t-\tau)]}{\omega_0} + \frac{1}{\omega_0^2}\cos[\omega_0(t-\tau)]\Big|_0^t$$
$$= \boxed{\frac{1}{\omega_0^2}[1-\cos\omega_0 t]} \tag{19.29}$$

Results are shown in Fig. 19.16.

19.12 Convolution Between Negative Exponential and Complex One

Let's find the convolution between the negative exponential and the complex one

$$e^{-at} * e^{j\omega_0 t} = \int_0^\infty e^{-a\tau}e^{j\omega_0(t-\tau)}d\tau \tag{19.30}$$

We proceed as follows:

$$\int_0^\infty e^{-a\tau}e^{j\omega_0(t-\tau)}d\tau = e^{j\omega_0 t}\int_0^t e^{-a\tau}e^{-j\omega_0\tau}d\tau = e^{j\omega_0 t}\int_0^t e^{-\tau(a+j\omega_0)}d\tau$$
$$= -e^{j\omega_0 t}\frac{1}{a+j\omega_0}e^{-\tau(a+j\omega_0)}\Big|_0^t = -\frac{e^{j\omega_0 t}}{a+j\omega_0}\left[e^{-t(a+j\omega_0)}-1\right]$$
$$= \boxed{\frac{e^{j\omega_0 t}-e^{-at}}{a+j\omega_0}} \tag{19.31}$$

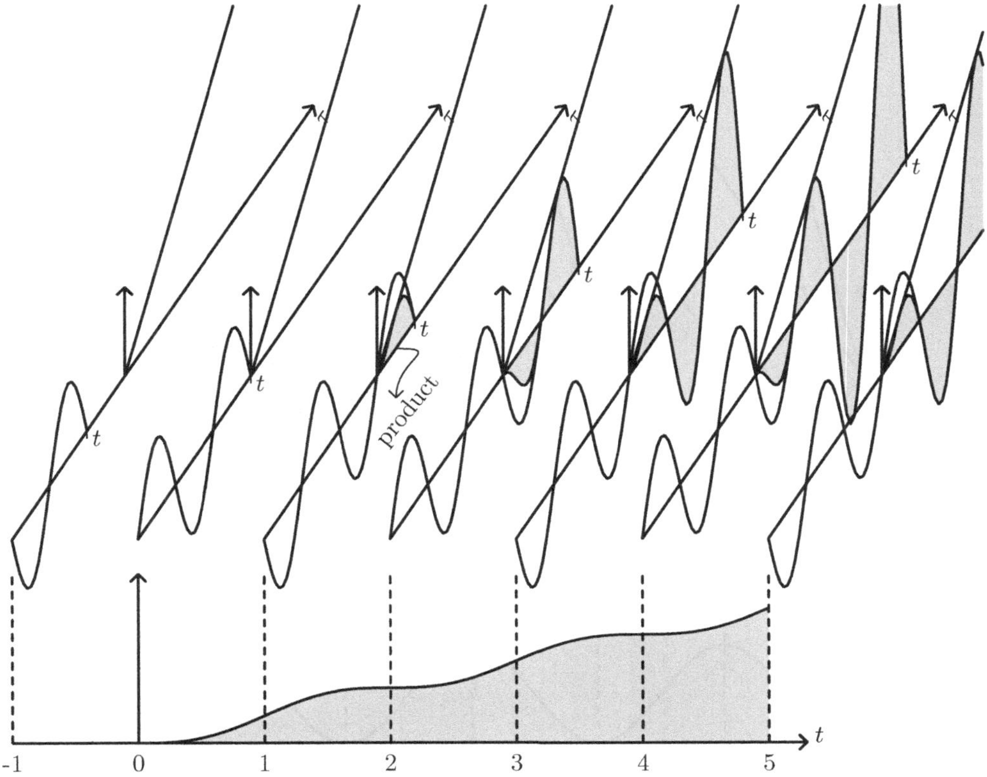

Fig. 19.14 Convolution between ramp and sine function

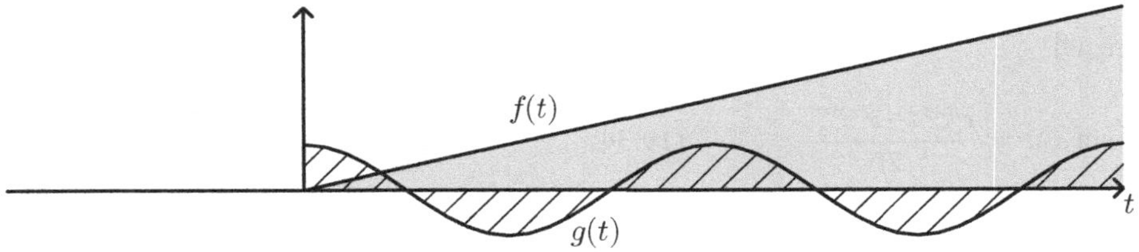

Fig. 19.15 Ramp function and cosine one ready to be convolved

19.13 Convolution Between Sine and Negative Exponential

Let's find the convolution between the sine function and the negative exponential one.

$$e^{-at} * \sin \omega_0 t = \int_0^{\infty} e^{-\tau} \sin[\omega_0(t-\tau)]d\tau \tag{19.32}$$

as shown in Fig. 19.17. Rather than doing the integration, we know from prior section the convolution between the negative exponential and the complex one. Since we can tie a sine to a complex exponential, we hope we can get the sine results easily. First recall Eq. (19.31)

$$e^{-at} * e^{j\omega_0 t} = \frac{e^{j\omega_0 t} - e^{-at}}{a + j\omega_0} \tag{19.33}$$

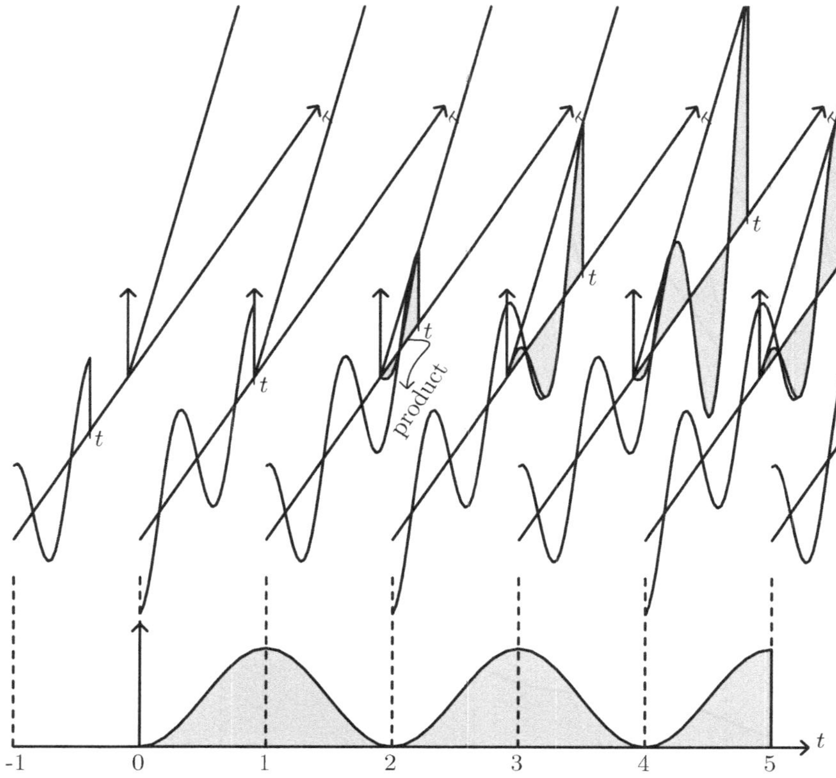

Fig. 19.16 Convolution between ramp and cosine function

Next recall

$$\sin \omega_0 t = \frac{e^{j\omega_0 t} - e^{-j\omega_0 t}}{2j} \tag{19.34}$$

Then we should expect

$$e^{-at} * \sin \omega_0 t = \frac{1}{2j}\left[\frac{e^{j\omega_0 t} - e^{-at}}{a + j\omega_0} - \frac{e^{-j\omega_0 t} - e^{-at}}{a - j\omega_0}\right]$$

$$= \frac{1}{2j}\frac{(a - j\omega_0)(e^{j\omega_0 t} - e^{-at}) - (a + j\omega_0)(e^{-j\omega_0 t} - e^{-at})}{a^2 + \omega_0^2}$$

$$= \frac{1}{2j}\frac{a(e^{j\omega_0 t} - e^{-j\omega_0 t}) - j\omega_0(e^{j\omega_0 t} + e^{-j\omega_0 t} - 2e^{-at})}{a^2 + \omega_0^2}$$

$$= \boxed{\frac{1}{a^2 + \omega_0^2}\left[\omega_0 e^{-at} + a \sin \omega_0 t - \omega_0 \cos \omega_0 t\right]} \tag{19.35}$$

Results are shown in Fig. 19.18.

19.14 Convolution Between Two Sines with Different Frequencies

Find the convolution between two sine functions, with different frequencies—ω_0 and ω_1.

$$I = \sin \omega_0 t * \sin \omega_1 t$$

$$= \int_0^{\infty} \sin \omega_0 \tau \sin \omega_1 (t - \tau) d\tau$$

$$= \int_0^{t} \sin \omega_0 \tau \sin \omega_1 (t - \tau) d\tau \tag{19.36}$$

as shown in Fig. 19.19. We need to use the following identity:

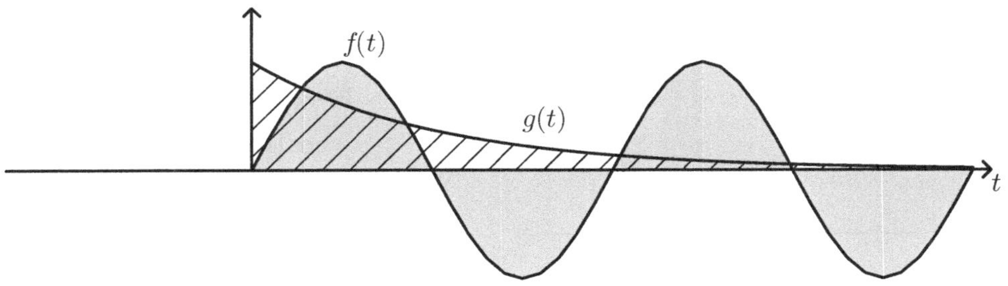

Fig. 19.17 Negative exponential and sine function ready to be convolved

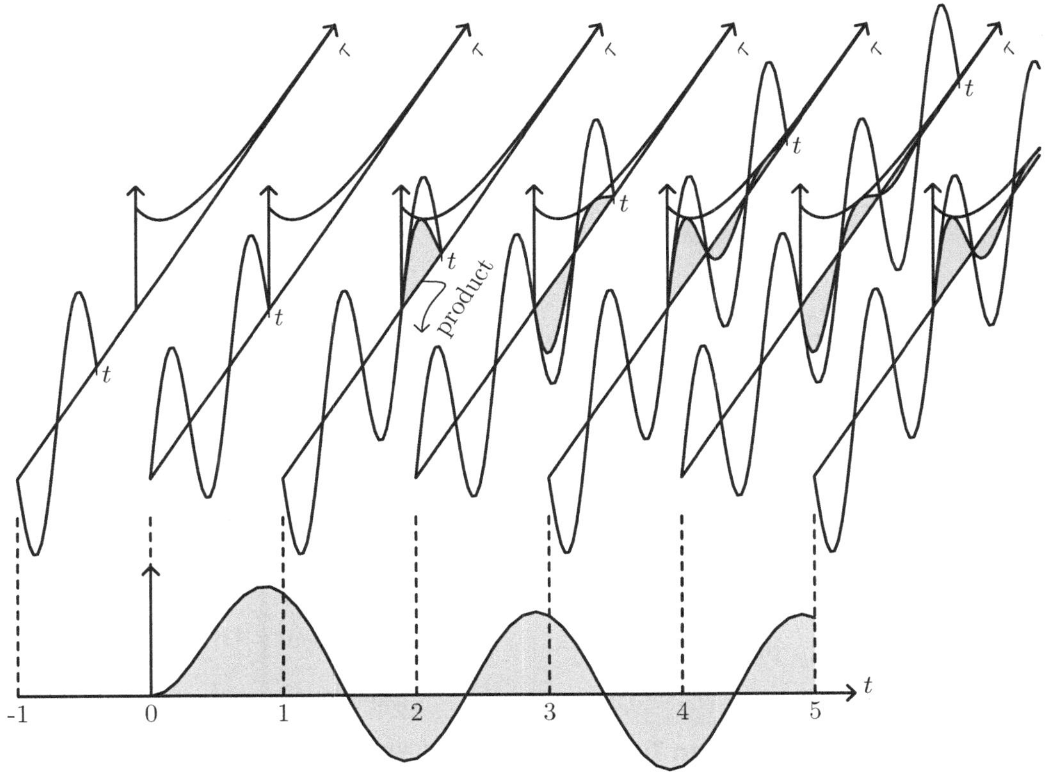

Fig. 19.18 Convolution between negative exponential and sine function

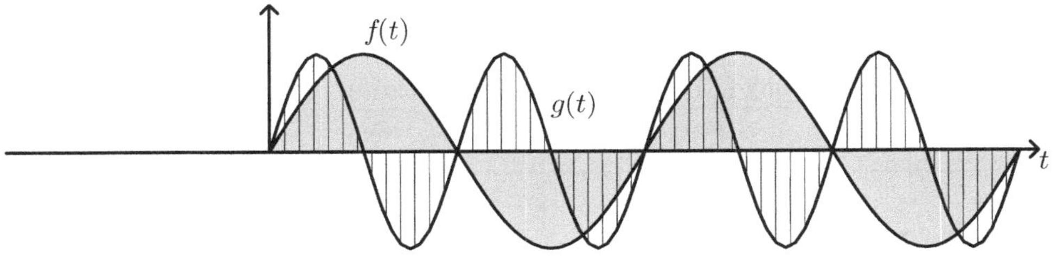

Fig. 19.19 Two sine functions with different frequencies ready to be convolved

$$\sin A \sin B = \frac{1}{2}\{\cos(A-B) - \cos(A+B)\} \tag{19.37}$$

which gives

$$\sin\omega_0\tau \sin\omega_1(t-\tau) = \frac{1}{2}\{\cos[\tau(\omega_0+\omega_1)-\omega_1 t] - \cos[\tau(\omega_0-\omega_1)+\omega_1 t]\} \tag{19.38}$$

When this gets integrated we end up with

$$\begin{aligned}
2I &= \frac{1}{\omega_0+\omega_1}\sin[\tau(\omega_0+\omega_1)-\omega_1 t]_0^t - \frac{1}{\omega_0-\omega_1}\sin[\tau(\omega_0-\omega_1)+\omega_1 t]_0^t \\
&= \frac{1}{\omega_0+\omega_1}[\sin\omega_0 t + \sin\omega_1 t] - \frac{1}{\omega_0-\omega_1}[\sin\omega_0 t - \sin\omega_1 t] \\
&= \sin\omega_0 t\left[\frac{1}{\omega_0+\omega_1} - \frac{1}{\omega_0-\omega_1}\right] + \sin\omega_1 t\left[\frac{1}{\omega_0+\omega_1} + \frac{1}{\omega_0-\omega_1}\right] \\
&= -\frac{2\omega_1}{\omega_0^2-\omega_1^2}\sin\omega_0 t + \frac{2\omega_0}{\omega_0^2-\omega_1^2}\sin\omega_1 t \\
&= \frac{2\omega_1}{\omega_1^2-\omega_0^2}\sin\omega_0 t - \frac{2\omega_0}{\omega_1^2-\omega_0^2}\sin\omega_1 t
\end{aligned} \tag{19.39}$$

This finally gives

$$\boxed{\sin\omega_0 t * \sin\omega_1 t = \frac{1}{\omega_1^2-\omega_0^2}[\omega_1\sin\omega_0 t - \omega_0\sin\omega_1 t]} \tag{19.40}$$

Results are shown in Fig. 19.20.

19.15 Convolution Between Two Sines with Same Frequency

Let's find the convolution of two sines with the same frequency

$$I = \int_0^t \sin\omega_0\tau \sin\omega_0(t-\tau)d\tau \tag{19.41}$$

From last section (rework of Eq. (19.38)) we get

$$\sin\omega_0\tau \sin\omega_0(t-\tau) = \frac{1}{2}\{\cos[\omega_0(2\tau-t)] - \cos\omega_0 t\} \tag{19.42}$$

Integrating from 0 to t gives

$$2I = \frac{\sin[\omega_0(2\tau-t)]}{2\omega_0} - \tau\cos\omega_0 t\Big|_0^t = \frac{\sin\omega_0 t}{\omega_0} - t\cos\omega_0 t \tag{19.43}$$

This finally gives

$$\boxed{\sin\omega_0 * \sin\omega_0 t = \frac{1}{2}\left[\frac{\sin\omega_0 t}{\omega_0} - t\cos\omega_0 t\right]} \tag{19.44}$$

Results are shown in Fig. 19.21. Notice how the results are some form of growing oscillation. We can tell from the above equation that the settling value would be $\sim t\cos\omega_0 t$.

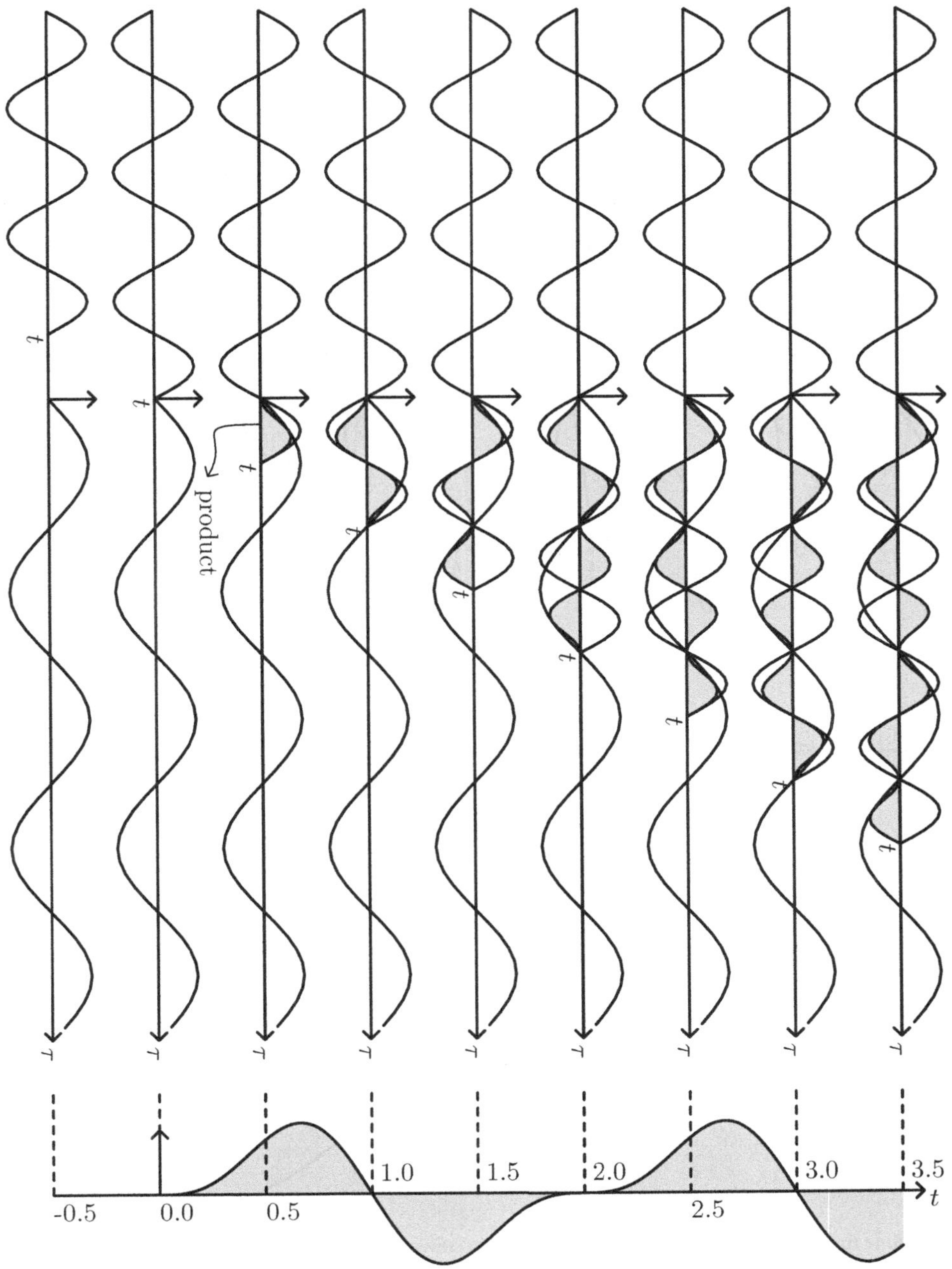

Fig. 19.20 Convolution between two sines with different frequencies

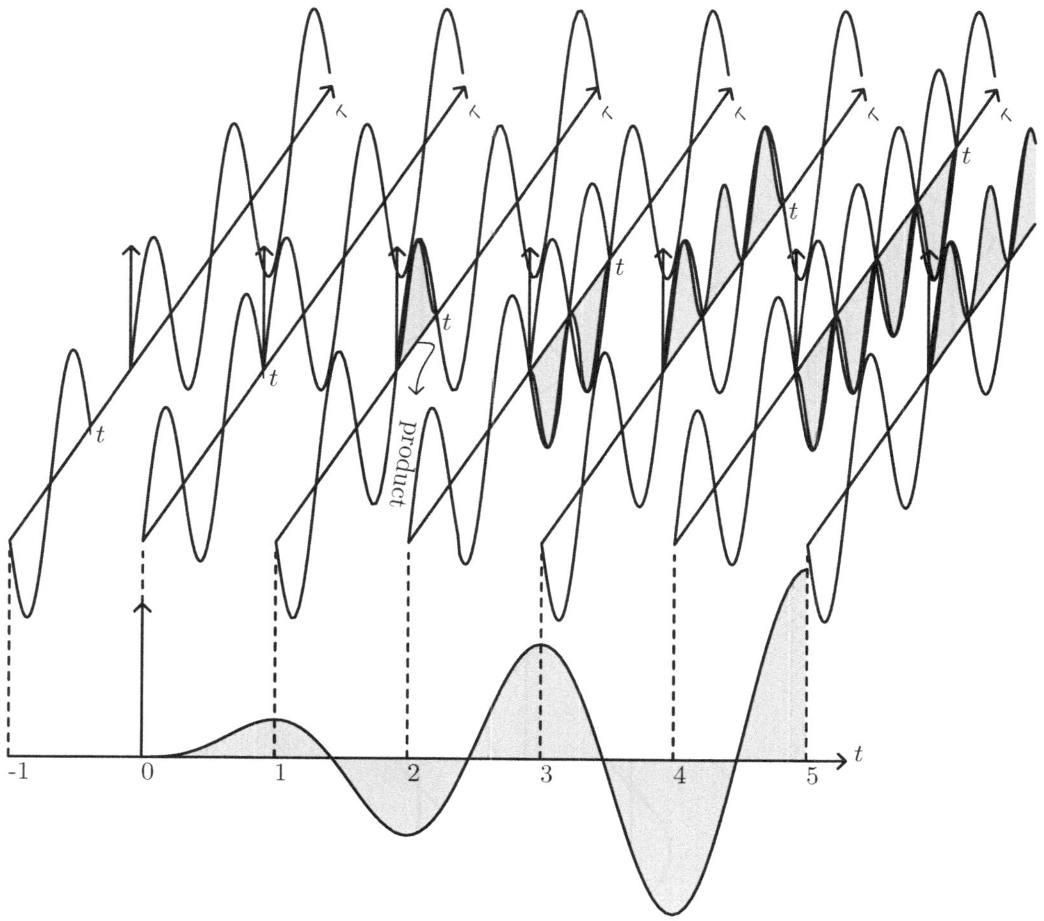

Fig. 19.21 Convolution between two sines with same frequency

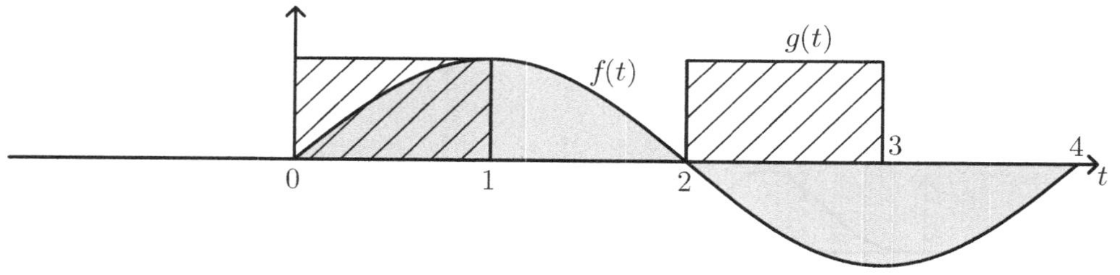

Fig. 19.22 Sine function and periodic pulse one ready to be convolved

19.16 Convolution Between Periodic Pulse and Sine Function

Let's find the convolution between a sine function of frequency $\omega_0 = \frac{\pi}{2}$ and a periodic pulse with fundamental frequency $\omega_1 = \pi$ as shown in Fig. 19.22. First we can decompose the pulse as a Fourier series;

$$\text{pulse} = \frac{1}{2} + \sum_{n=1,3,5,\cdots}^{\infty} \frac{2}{n\omega_1} \sin n\omega_1 t \qquad (19.45)$$

Next we know from Eq. (19.40) that

$$\sin\omega_0 * \sin\omega_1 t = \frac{1}{\omega_1^2 - \omega_0^2}[\omega_1 \sin\omega_0 t - \omega_0 \sin\omega_1 t] \tag{19.46}$$

We also know that

$$\frac{1}{2} * \sin\omega_0 t = \frac{1-\cos\omega_0 t}{2\omega_0} \tag{19.47}$$

By superposition we then get

$$\text{pulse} * \sin\omega_0 t = \frac{1-\cos\omega_0 t}{2\omega_0} + \sum_{n=1,3,5,\cdots} \frac{2}{n\omega_1}\frac{1}{(n\omega_1)^2 - \omega_0^2}[(n\omega_1)\sin\omega_0 t - \omega_0 \sin(n\omega_1 t)] \tag{19.48}$$

Results are shown in Fig. 19.23.

19.17 Convolution Between Periodic Pulse and Unit Step Function

Let's find the convolution between the unit step function and the periodic pulse as shown in Fig. 19.24.

Our strategy is to figure the convolution due to a single pulse, then use superposition to fold in the rest of the pulse train. We can write the single pulse in terms of the unit function as

$$\text{single pulse} = u(t) - u(t-1) \tag{19.49}$$

The convolution of this with the unit step function, which we coin $G(t)$, is

$$G(t) = \text{single pulse} * u(t) = tu(t) - (t-1)u(t-1) \tag{19.50}$$

All that has to be done now is to add the results of the rest of the pulse train;

$$\text{pulse train} * u(t) = \sum_{n=0,2,4,\cdots} G(t-n) \tag{19.51}$$

Results are shown in Fig. 19.25.

19.18 Convolution Between Periodic Pulse (with 0 Average) and Unit Step Function

Let's repeat the last exercise, but this time use a periodic pulse with zero average, as shown in Fig. 19.26. Instead of finding the results due to a single pulse, we recast the answer in terms of unit steps. We can express our input as

Fig. 19.23 Convolution between periodic pulse and sine function

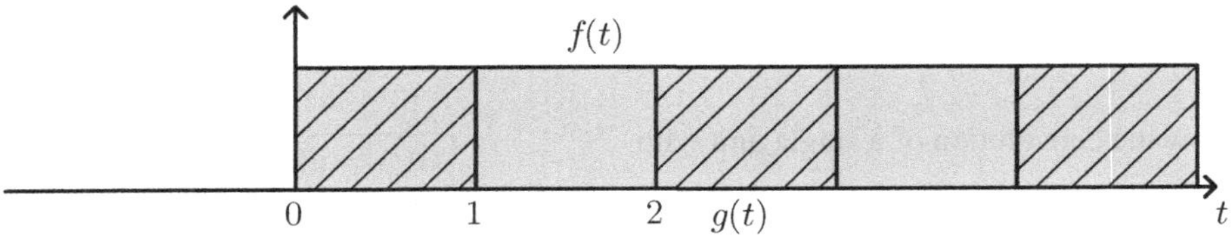

Fig. 19.24 Unit step function and periodic pulse one ready to be convolved

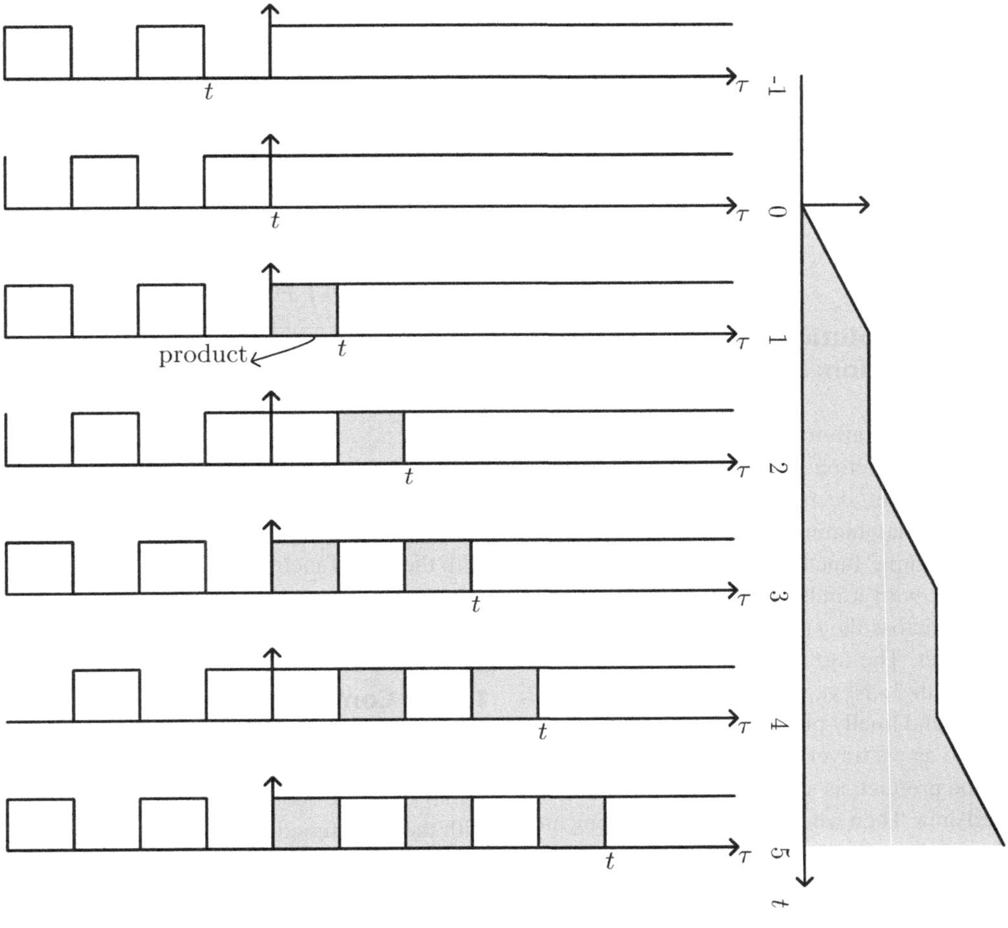

Fig. 19.25 Convolution between periodic pulse and unit step function

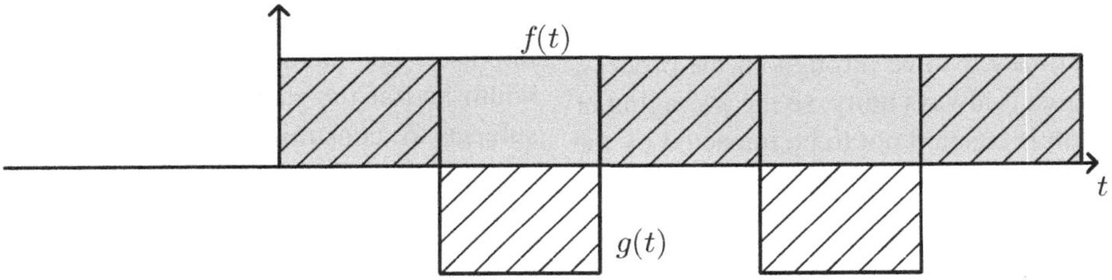

Fig. 19.26 Unit step function and periodic pulse one (with zero average) ready to be convolved

$$\text{periodic pulse with 0 avg} = u(t) - 2u(t-1) + 2(t-2) - 2u(t-3) + \cdots \tag{19.52}$$

We know the convolution of a single unit with another as

$$\boxed{G(t) = u(t) * u(t) = tu(t)} \tag{19.53}$$

It follows that the convolution of the train pulse is

$$\boxed{\text{pulse train with 0 avg} * u(t) = G(t) - 2G(t-1) + 2G(t-2) - 2G(t-3) + \cdots} \tag{19.54}$$

Results are shown in Fig. 19.27.

19.19 Convolution with the Delta Function: Basic Idea

We start with an arbitrary function and multiply it with a pulse function. We incrementally increase the pulse height, decrease its width, while at the same time maintaining unity area. Figure 19.28 shows a sample function in the process of being convolved with a pulse. By convolving we step the pulse across the $f(\tau)$, and find the area under the product. The top figure shows the case with pulse width 1. Next, pulse width 0.5; next 0.2, then 0.1, and finally pulse width of 0.05. For each case, and as we traverse the pulse along $f(\tau)$, we form the product, as shown in the corresponding shaded area. Then we find the corresponding area of the product. So, convolving with the pulse in essence finds the average of the signal around the pulse. As the pulse width is made incrementally smaller and smaller, the averaging happens closer and closer around the point t. In the limit, and as pulse width goes to zero, we sample the average exactly at time t. Notice throughout the process, the pulse area is always unity; so the averaging of the function is ensured not to be impacted by the averaging of the pulse.

So, integrating the product of a function times and impulse gives us the function exactly at the location of the impulse; mathematically we have

$$\boxed{f(t) = \int f(\tau)\delta(t-\tau)d\tau} \tag{19.55}$$

But this is nothing other than the definition of convolution:

$$\boxed{f(t) = f(t) * \delta(t)} \tag{19.56}$$

That is, the function equals itself convolved with the delta function!

19.20 Convolution with Delta Function: Example

As an example let's convolve the sine function with the delta function

$$\int_0^t \sin \omega_0 \tau \delta(t-\tau)d\tau \tag{19.57}$$

To illustrate the process we will numerically convolve the sine with a pulse (rather than an impulse), but incrementally decrease the pulse width and at the same time increase its height, subject to constraint that pulse area remains unity. In essence, the pulse will always have a "strength" of one, and it would incrementally approach a delta function. Figure 19.29 shows

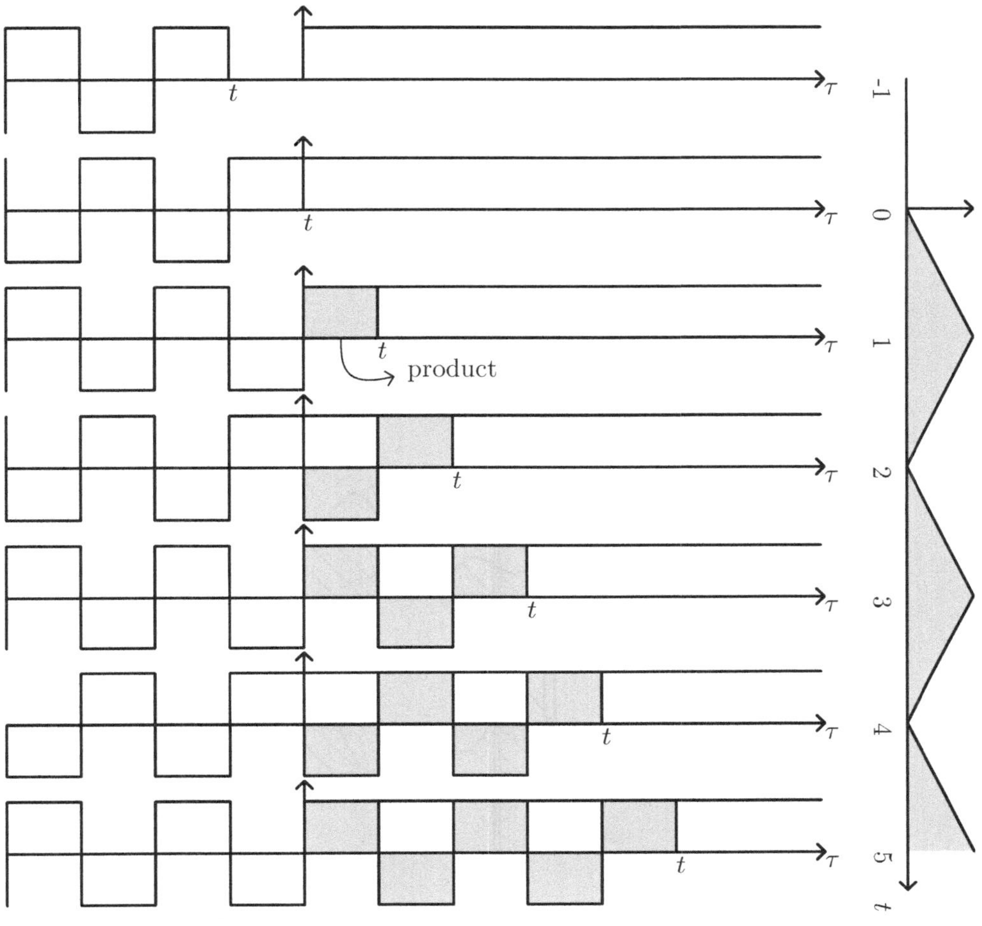

Fig. 19.27 Convolution between periodic pulse (with 0 avg) and unit step function

the starting function and that obtained via convolution (again between the function and the pulse). Results are shown for different pulse widths, with top-most having widest pulse (and shortest height) and bottom-most having narrowest pulse width (but tallest height). As can be seen, as pulse approaches an impulse, the convolution outcomes does reproduce the original function! Hence we have shown how convolving a function with a delta function results in the original function.

19.21 Summary

This is the first chapter about convolution. In it we presented the convolution equation and explained the convolution process. Convolution happens between two functions and the result is a function on its own. Convolution is a basic integration process, but the calculated area is not a number—rather it is a series of numbers, with each area number corresponding to a particular snapshot of a scaled area of overlap between two

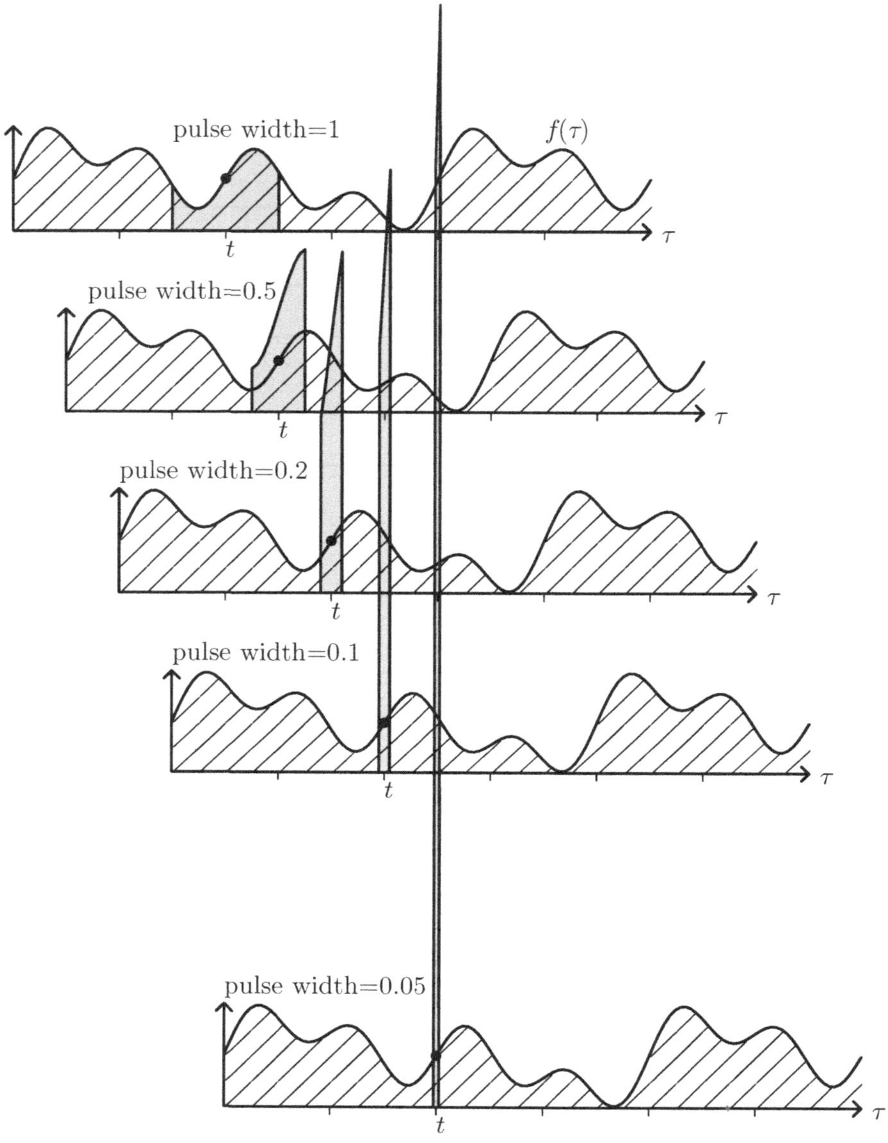

Fig. 19.28 Function and sampling by delta function; preliminary to delta convolution

functions. As we step time t we offset (the flipped version of) a function with respect to the other, form the product, and calculate the corresponding area. For causal signals this area typically extends from time zero to time t. Keep in mind that the integration variable is not time; it is τ. Think of convolution as the successor of integration! It is a fancier version of integration. While the convolution equation is succinct it is best understood by example and plenty visualization. To that end we presented many examples and for each example showed the convolution process in action, as well as of course the end result. We also folded in Fourier series for a sample case of convolution between a periodic pulse and a sine function. We were able to do that by first figuring an analytic

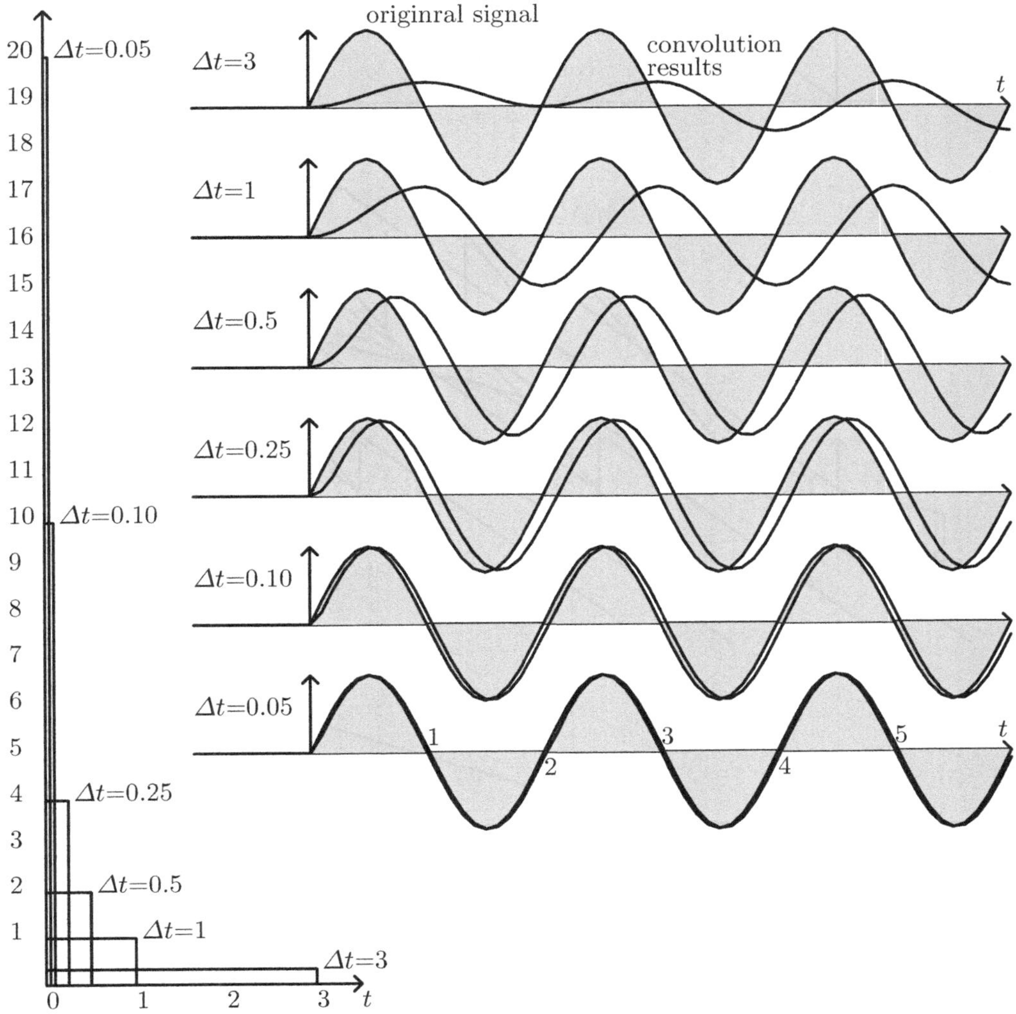

Fig. 19.29 Convolution with delta function as limit of convolution with pulse function

result governing convolution between two sines, and then representing the periodic pulse as a sine series. Clearly we can extend this principle to more complicated signals. Finally we wrapped the chapter with the important topic of convolution with the delta function and showed a corresponding numerical example. It is important—in this chapter and others—to always remember that fancy mathematical equations must always boil down to numbers. So when the equation does not make sense try to discretize it and approximate it, then simply manipulate the numbers. As the discretization/approximation is made better, the exact outcome of the intended equation will show up!

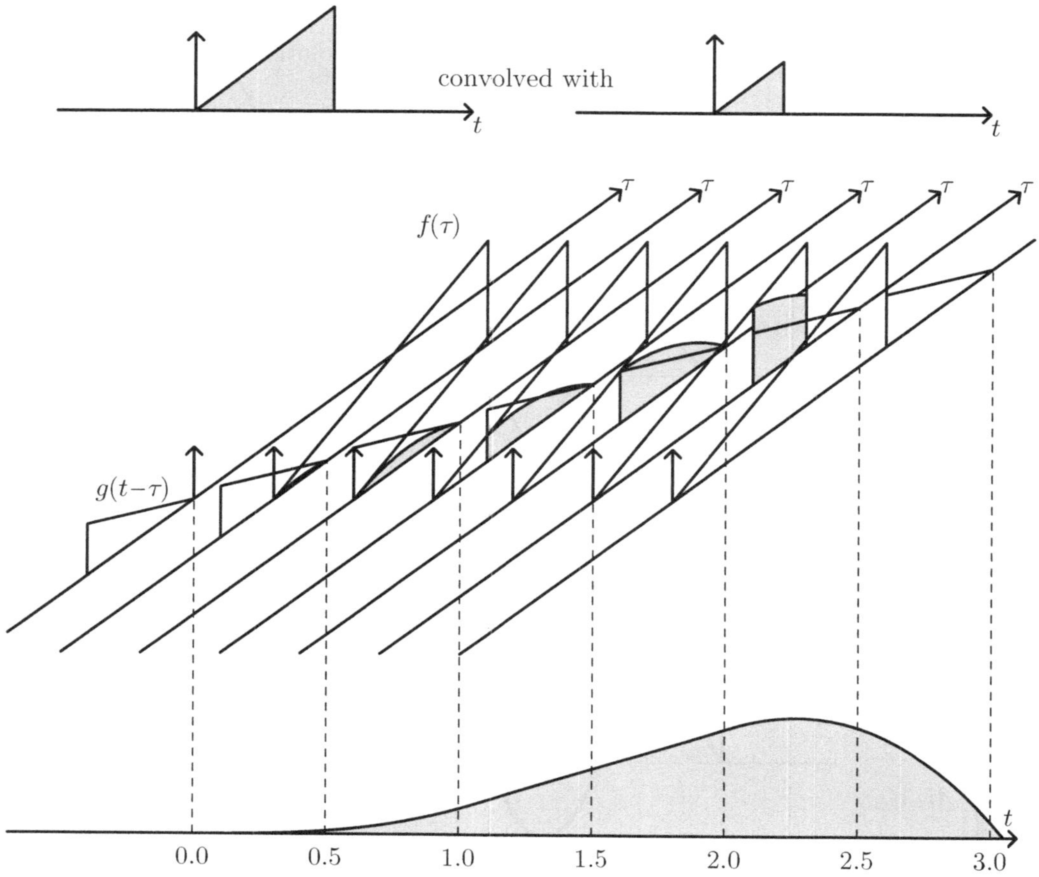

Fig. 19.30 Statement and sample solution to Problem 1

19.22 Problems

1. Fill in the steps to find and plot the convolution between two triangular functions, one with width 2 and the other 1, as shown in Fig. 19.30.
2. Find the convolution between a pulse of width 1 and the unit step function; see sample solution in Fig. 19.31 and fill in the details.
3. Consider convolving the upright triangle with the unit step function; show results for both variants of the triangle (both of width 1), as shown in Fig. 19.32; show intermediate steps.
4. Convince yourself that the convolution between the 0-averaged triangular shape shown at the top of Fig. 19.33 and the unit step function results in the single inverted parabola, of width 2 and height 0.5, also shown in the same figure.
5. What is the convolution between the periodic pulse, with zero average, but asymmetric period, as shown at the top of Fig. 19.34 with the unit step function? Show the steps and use solution in same figure as guide.
6. Convolve the sine function with the pulse one; vary pulse width T from 1 all the way to ∞; plot results and compare to sample solution in Fig. 19.35. What is the meaning of the special case $T \to \infty$?
7. Show that the convolution between the periodic zigzag pattern and a pulse results in the inverted parabola, as shown in Fig. 19.36.

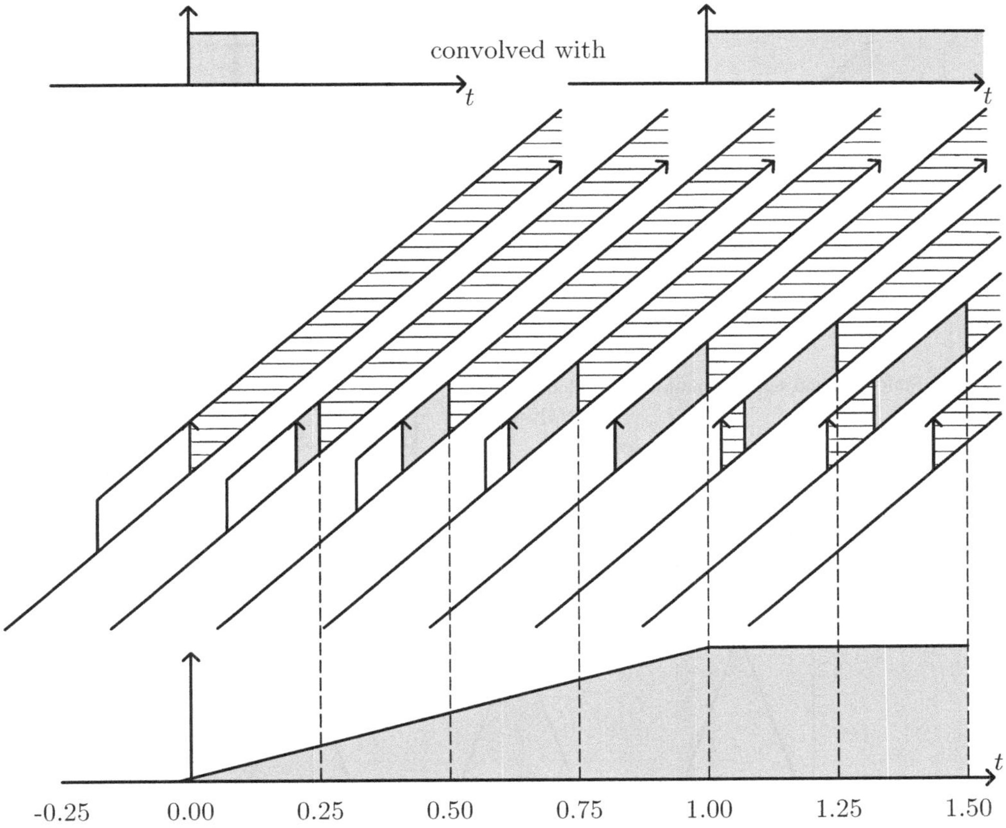

Fig. 19.31 Statement and sample solution to Problem 2

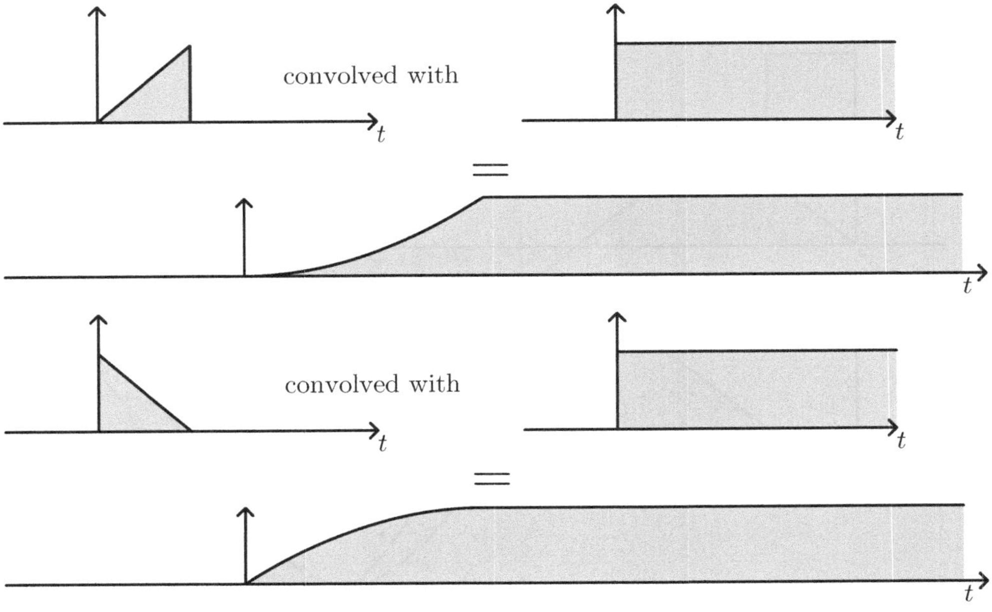

Fig. 19.32 Statement and sample solution to Problem 3

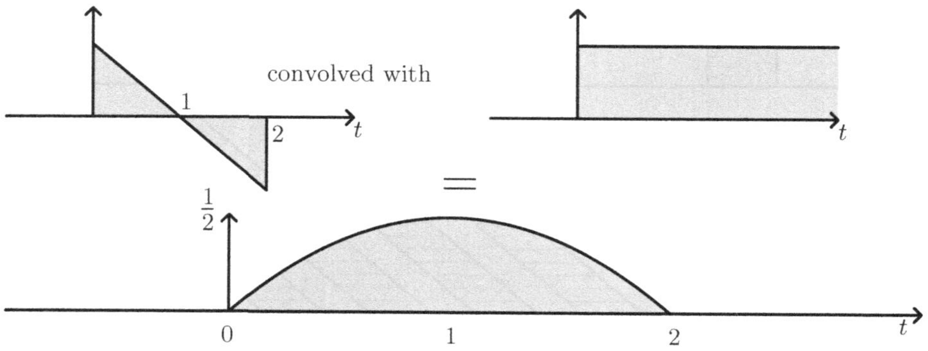

Fig. 19.33 Statement and sample solution to Problem 4

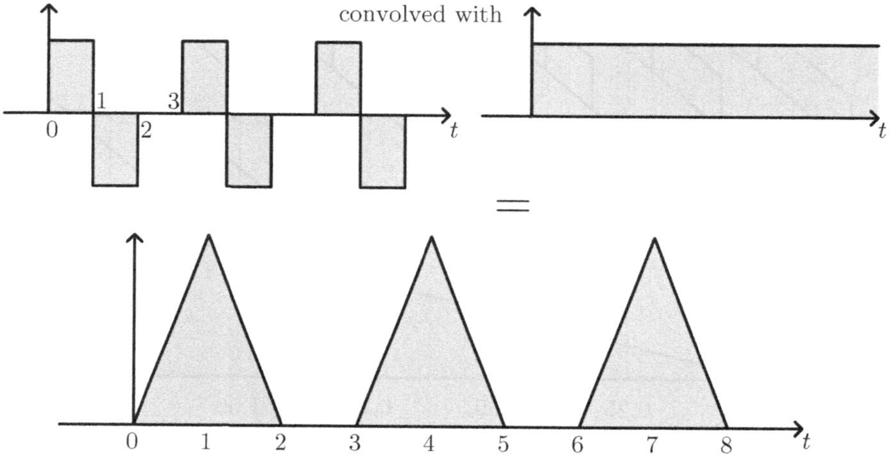

Fig. 19.34 Statement and sample solution to Problem 5

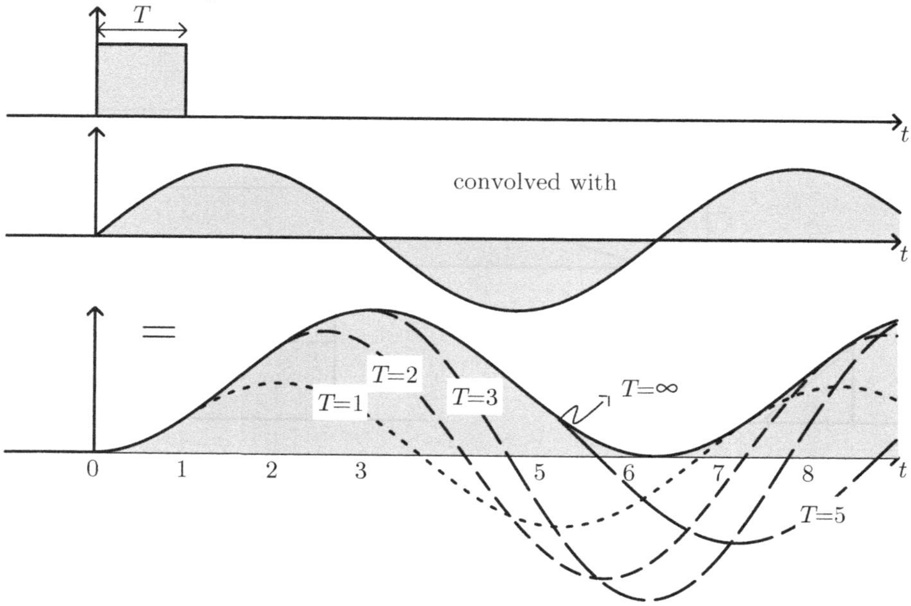

Fig. 19.35 Statement and sample solution to Problem 6

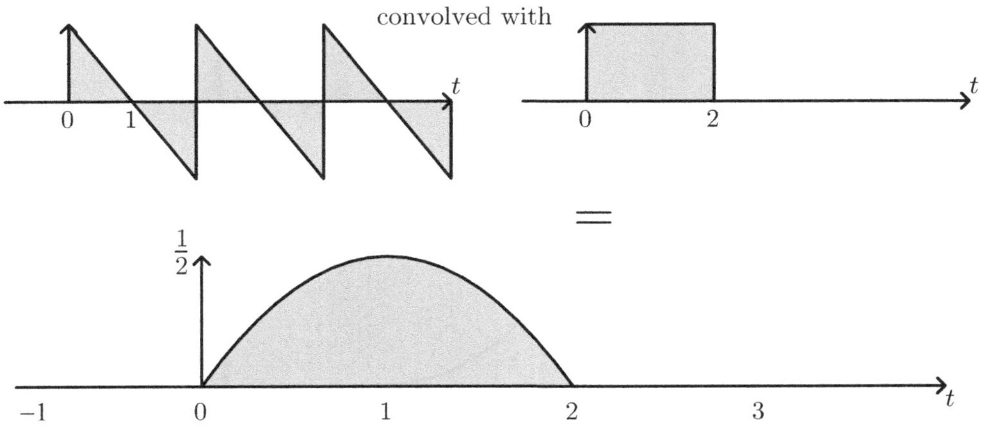

Fig. 19.36 Statement and sample solution to Problem 7

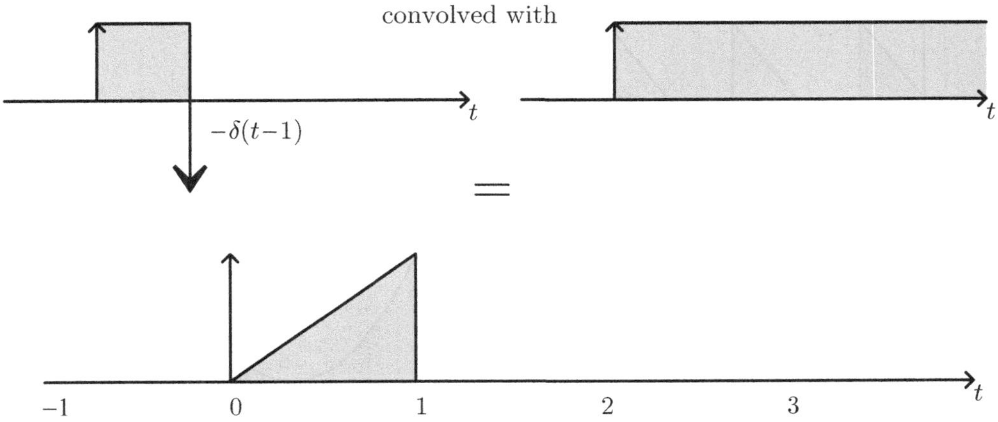

Fig. 19.37 Statement and sample solution to Problem 8

8. Show that the convolution between a pulse followed by a negative delta function, with the unit step function results in the upright triangle as shown in Fig. 19.37.
9. Show that the convolution between a single cycled zigzag function with the unit step function results in the symmetric triangle, with tapered, quadratic edges, as shown in Fig. 19.38.
10. Show that the convolution between the *periodic* zigzag function and the unit step one results in the periodic symmetric triangle, with tapered, quadratic edges, as shown in Fig. 19.39.
11. Show that the convolution between the periodic oscillating pulse, with increasing magnitude, with a single pulse results in a periodic symmetric triangle, with flat head, and as shown in Fig. 19.40.

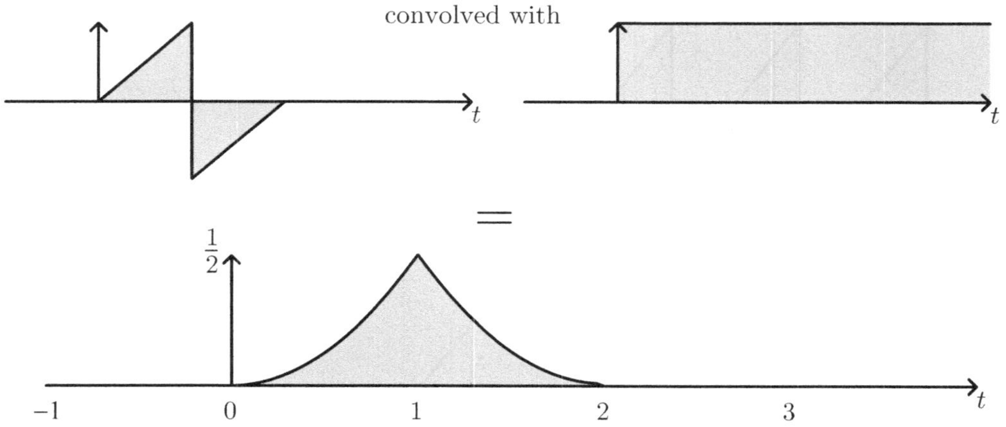

Fig. 19.38 Statement and sample solution to Problem 9

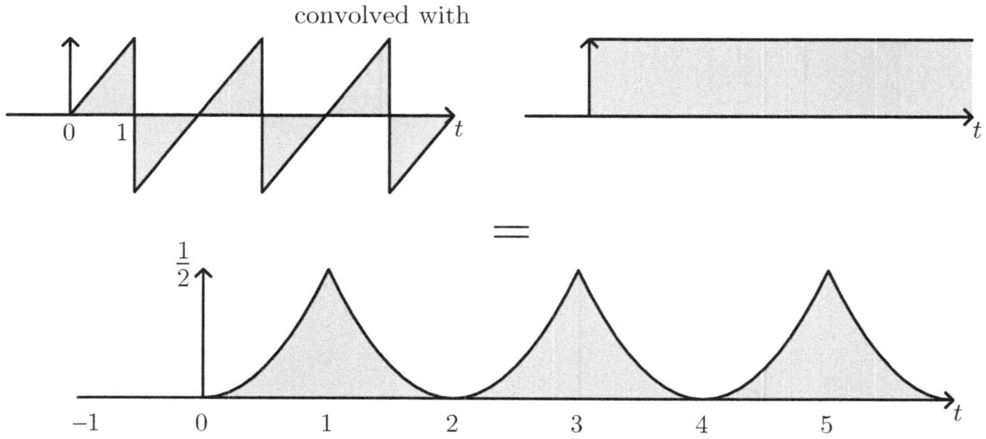

Fig. 19.39 Statement and sample solution to Problem 10

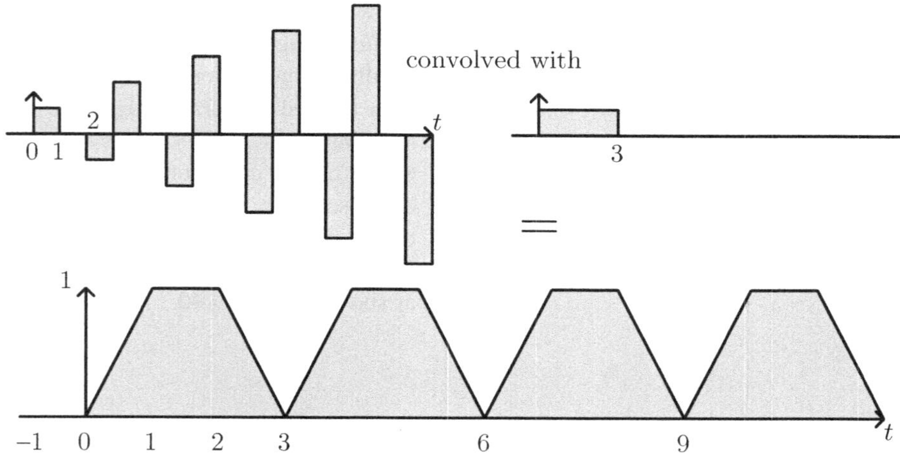

Fig. 19.40 Statement and sample solution to Problem 11

Signal Construction in Terms of Convolution Integrals

20

20.1 Introduction

In the last chapter we introduced the concept of convolution and showed some basic examples. In this chapter we show a sample application of convolution—which is signal construction. We will show how we can use rather arbitrary functions—such as the delta function, unit step one, ramp, quadratic (or even others)—to build an arbitrary function. Why would one want to do something like this from the beginning? Same rationale why we decided to build an arbitrary function in terms of the complex exponentials (the sine/cosine harmonics, and sometimes scaled by the negative exponential). The premise again is that if we know the system response to a simpler signal (such as a delta or unit step one), and if we know how to construct an arbitrary signal in terms of the simpler ones, and using superposition we are guaranteed to find the system response to the arbitrary signal! In the spectral domain, our "simpler" signals were sines/cosines—here they are delta functions, unit step ones, or others. In this chapter we will not concern ourselves with system response—not just yet; instead we will be strictly concerned with the magical equation that is going to take us from the "simpler" signals to the arbitrary one. As a demonstration vehicle let's take the function in Fig. 20.1 as a sample case. It combines some important features, in that it is ahybrid of analytic functions, it is limited, has an abrupt discontinuity, as well as a derivative discontinuity. It is defined as

$$f(t) = \begin{cases} 0 & t < 0 \\ t^2 & 0 < t < 1 \\ 1 & 1 < t < 2 \\ 0 & t > 2 \end{cases} \tag{20.1}$$

20.2 Representation of Signal in Terms of Convolution with Impulse Function

Our goal here is to derive an expression tying an arbitrary signal to a convolution involving the impulse (delta) function. We already saw this before (Sect. 19.19), but the following is a detailed derivation. As to why we would want to do that, it relates to impulse response. In essence, if we know the response of a system to an impulse, we are able to know the response to *any* stimulus! How? Since we know the decomposition of the signal in terms of impulse functions, and we know the response of each impulse, all that has to be done is add up all the impulse responses; this is one flavor/application of what we refer to as convolution. We will get plenty of practice on this, but let's go back to our original goal—how to represent a signal in terms of impulses? First recall definition of the delta function

F. Badrieh, *Spectral, Convolution and Numerical Techniques in Circuit Theory*,
https://doi.org/10.1007/978-3-319-71437-0_20

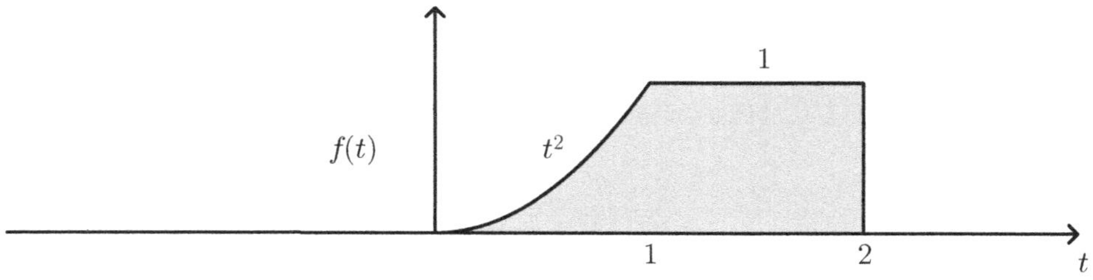

Fig. 20.1 Sample function for demonstrating signal reconstruction using convolution integrals

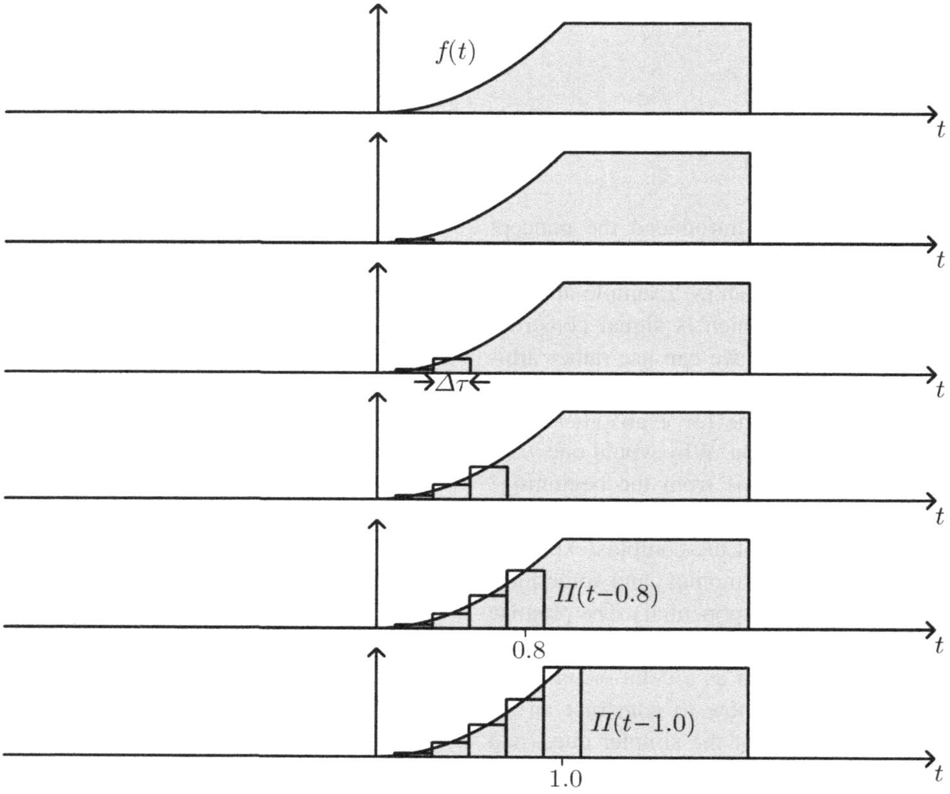

Fig. 20.2 Representation of a signal in terms of impulse functions

$$\delta(t-t_0)=\begin{cases}0 & t\neq t_0\\ \infty & t=t_0\end{cases} \tag{20.2}$$

subject to the limitation

$$\int_{-\infty}^{\infty}\delta(t-t_0)dt=1 \tag{20.3}$$

That is, the impulse is a very tall, but narrow pulse of unity area. Consider an arbitrary signal as shown in Fig. 20.2. We can digitize this signal in terms of train of pulse (not impulse) functions. Each pulse has width $\Delta\tau$ and located at $n\Delta\tau$ where n is some integer.

$$\Pi(t - n\Delta\tau) = \begin{cases} 1 & \left[n\Delta\tau - \frac{\Delta\tau}{2}\right] < t < \left[n\Delta\tau + \frac{\Delta\tau}{2}\right] \\ 0 & \text{else wise} \end{cases} \tag{20.4}$$

The magnitude of each pulse would reflect the value of the function at the time center of the pulse, which is $n\Delta\tau$. Hence we would have

$$f(t) = \sum_n f(n\Delta\tau)\Pi(t - n\Delta\tau) \tag{20.5}$$

That is, the function is represented as a summation of pulse functions, as shown in the figure. The narrower each pulse, the more accurate the approximation. Notice that each pulse is approaching an impulse, in the sense it is pretty narrow, but it does *not yet* qualify as an impulse because the impulse has unity area, but our pulses have

$$\text{area of pulse} = \Delta\tau \tag{20.6}$$

That's not a problem, we can migrate to using impulses, provided we scale the pulse area such that it is unity; multiply and divide by $\Delta\tau$ to get

$$f(t) = \sum_n f(n\Delta\tau) \boxed{\frac{\Pi(t - n\Delta\tau)}{\Delta\tau}} \Delta\tau \tag{20.7}$$

The boxed term now does approach the impulse function, since it is infinitely narrow, but its area is unity. Hence we can now switch and use the impulse function:

$$f(t) = \sum_n f(n\Delta\tau)\delta(t - n\Delta\tau)\Delta\tau \tag{20.8}$$

Now let's take the limit as $\Delta\tau \rightarrow 0$; three things happen:

1. The sequence $n\Delta\tau$ becomes smoother and eventually continuous, and we coin it τ.
2. The pulse width $\Delta\tau$ becomes a differential, which we label $d\tau$; and
3. The summation approaches an integral.

So we get

$$\lim_{\Delta\tau \rightarrow 0} \sum_n f(n\Delta\tau)\delta(t - n\Delta\tau)\Delta\tau = \int f(\tau)\delta(t - \tau)d\tau \tag{20.9}$$

That is,

$$\boxed{f(t) = \int_{-\infty}^{\infty} f(\tau)\delta(t - \tau)d\tau} \tag{20.10}$$

But this is nothing more than the convolution integral! That is, a random function can be represented as a convolution between the function and the delta function! Figure 20.3 shows convolution in practice, and main take is that the narrower the pulse, the more accurate the convolution reconstruction.

So the magical formula taking us from the "simpler" function (in this case the delta function) to the arbitrary function is simply Eq. (20.10). It seems like a simple formula but it cannot be viewed lightly! It has the basics of convolution and all the subsequent convolution integrals—if anything—will be a more complicated version thereof. Hence it is important to feel at ease with this formula and use as a guide for the next sections.

20.3 Representation of Signal in Terms of Convolution with Unit Step Function

Having covered signal construction in terms of delta (impulse) functions the next logical and simple candidate is nothing other than the inte-

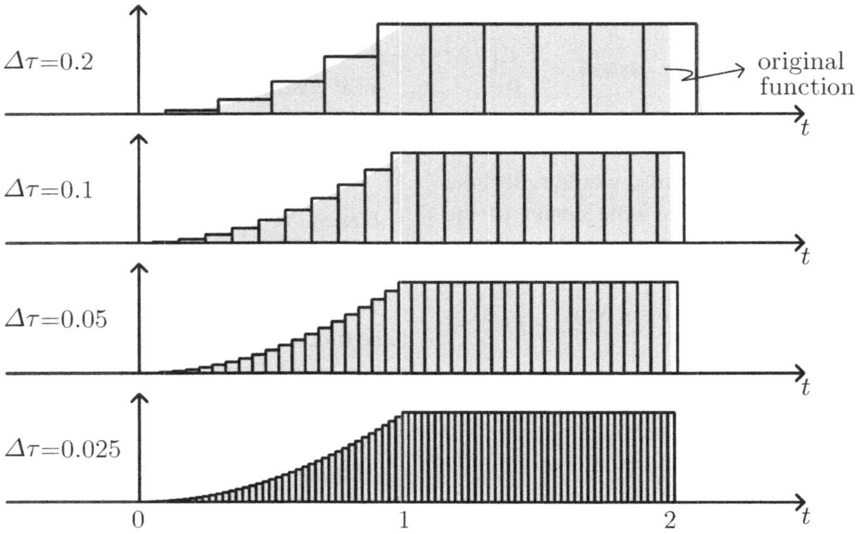

Fig. 20.3 Representation of a signal in terms of impulse functions—practical considerations

gral of the delta function—the unit step one. Let us summarize where we stand and use that as foundation for the unit step convolution derivation. We have shown that we can represent a function in terms of a convolution against the delta function

$$f(t) = \int_{-\infty}^{\infty} f(\tau)\delta(t-\tau)d\tau \tag{20.11}$$

If the function is causal (nonzero only for positive time) then we get

$$f(t) = \int_{0}^{\infty} f(\tau)\delta(t-\tau)d\tau \tag{20.12}$$

If we let $f(t) = u(t)$ we get

$$u(t) = \int_{0}^{\infty} \delta(t-\tau)d\tau \tag{20.13}$$

(see Problem 4). Equivalently we have

$$u(t) = \int_{0}^{t} \delta(\tau)d\tau \tag{20.14}$$

If we take the time derivative of both sides we get

$$\delta(t) = \frac{du(t)}{dt}, \quad \text{or} \tag{20.15}$$

$$\boxed{\delta(t-\tau) = -\frac{d}{d\tau}u(t-\tau)} \tag{20.16}$$

Thence we could enquire, can we use unit step functions as building blocks (via convolution) for signal construction? Let's put back into the delta convolution integral

$$\begin{aligned} f(t) &= \int_{0}^{\infty} f(\tau)\delta(t-\tau)d\tau \\ &= -\int_{0}^{\infty} f(\tau)\frac{d}{d\tau}u(t-\tau)d\tau \end{aligned} \tag{20.17}$$

Using integration by parts we get

$$\begin{aligned} f(t) =& f(0)u(t) - f(\infty)u(t-\infty) \\ &+ \int_{0}^{\infty} f'(\tau)u(t-\tau)d\tau \end{aligned} \tag{20.18}$$

Consider the first two terms

$$E(t) = f(0)u(t) - f(\infty)u(t-\infty) \qquad (20.19)$$

This function represent the end edges, at time zero and infinity. For the more general case where the function f dies off after some time t_1, and so does its derivative the E function becomes

$$E(t) = f(0)u(t) - f(t_1)u(t-t_1), \qquad \text{(assumes } f \text{ dies of after } t_1\text{)} \qquad (20.20)$$

Consider this function: it is really two step functions, one scaled by $f(0)$ and the other by $f(t_1)$. We can pull this back inside the integral in Eq. (20.18) if we use a couple of delta functions. An alternative is to simply forget about this function *but ensure to include the proper delta functions at the edges when differentiating* $f(t)$. That is, if the function has an abrupt start (say at time zero it starts with 1), or if the function ends with an abrupt end (say the function is a pulse, such that at the end of the pulse we have a jump from 1 to zero), then we need to include the delta functions that would result at those two ends. With this in mind we can recast Eq. (20.18) as

$$f(t) = \int_0^\infty f'(\tau)u(t-\tau)d\tau, \qquad \text{(ensure to include delta functions at abrupt end edges)} \qquad (20.21)$$

This will become clear with examples and practice. Since the unit function is zero for $\tau > t$ we can clip the integration limits from ∞ to t and finally arrive at the unit step convolution for signal generation:

$$\boxed{f(t) = \int_0^t f'(\tau)u(t-\tau)d\tau} \qquad (20.22)$$

That is a very important result! In order to construct a signal from a bunch of unit steps we do a summation over these time-shifted unit steps and scale each unit step by the *derivative* of the original function. Notice the keyword—derivative. In the impulse world we scaled each impulse by the value of the function at the location of the impulse; in the unit step world we scale by the derivative. Either way we have arrived at the magical formula tying a function to unit steps. Just like we did tying a function to harmonics or tying it to impulses.

Let's take our sample signal shown on the top of Fig. 20.4. Taking the derivative we end up with signal shown next.

$$\frac{df(t)}{dt} = \begin{cases} 0 & t < 0 \\ 2t & 0 < t < 1 \\ 0 & 1 < t < 2 \\ -\delta(t-2) & t = 2 \\ 0 & t > 2 \end{cases} \qquad (20.23)$$

Notice the delta function above, at time $t = 2$, *which attends to the requirement in Eq.* (20.21). It is critical that any delta function—both at start and end of signal—be captured correctly. In our case the signal started clean 0 at 0, but the ending was abrupt—hence the need for a delta function only at the end. Next we put in the unit step function, flip it and shift it by t as shown in the next 6 subplots. As we shift the unit step function, we multiply by the derivative, and integrate the

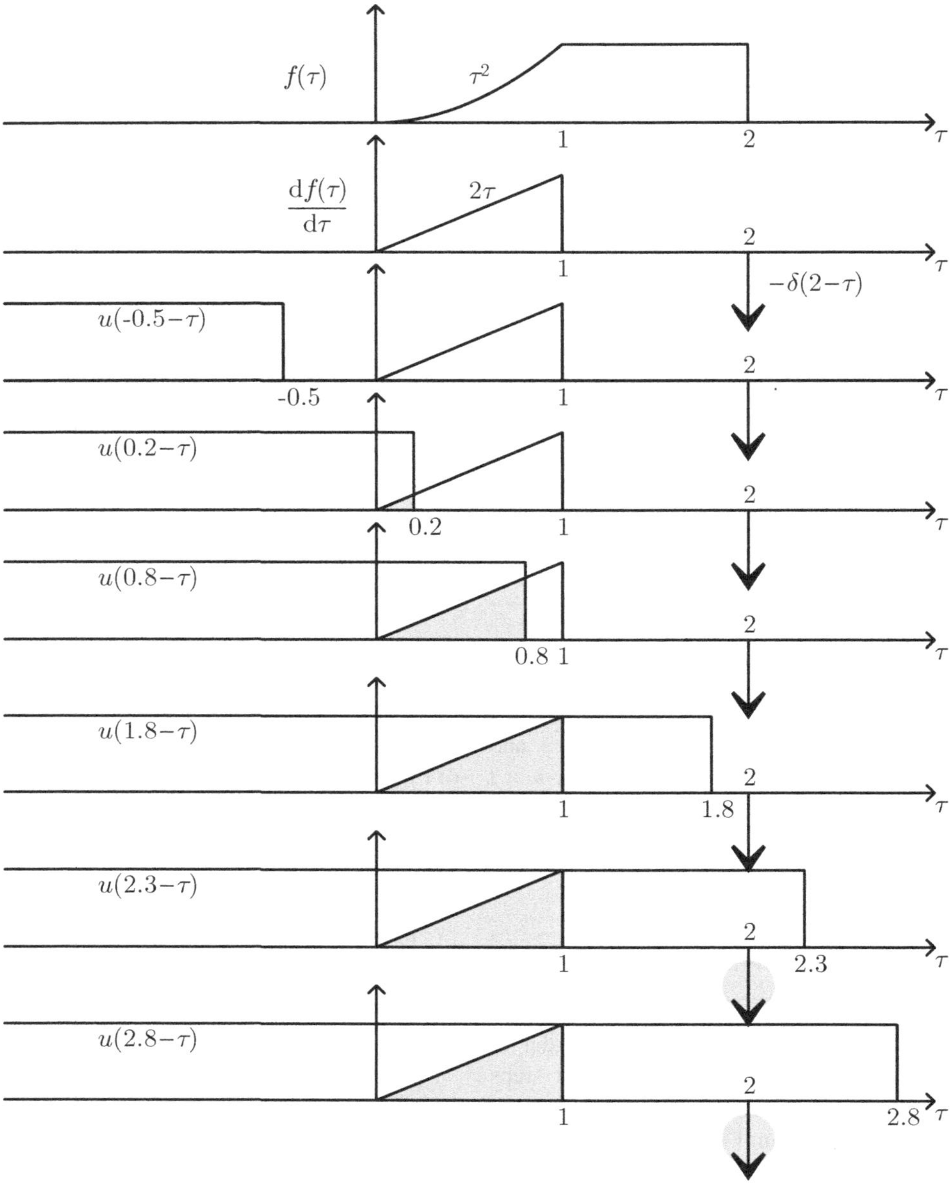

Fig. 20.4 Representation of target function as convolution with unit step function

result as shown in the gray area. The integral is evaluated depending on t. For $t < 0$ we get

$$f(t) = 0, \qquad t < 0 \tag{20.24}$$

For $0 < t < 1$ we get

$$f(t) = \int_0^t 2\tau u(t-\tau)d\tau = \int_0^t 2\tau d\tau = 2\frac{\tau^2}{2} = \tau^2\Big|_0^t = t^2, \qquad 0 < t < 1 \tag{20.25}$$

For $1 < t < 2$ we get

$$f(t) = \int_0^1 2\tau d\tau = 2\frac{\tau^2}{2} = \tau^2\Big|_0^1 = 1, \quad 1 < t < 2 \tag{20.26}$$

As soon as we hit $t = 2$ we pick the delta function and this is denoted by the shaded circle around the impulse. Picking up the integral of the delta function adds a $-u(t-2)$ which subtracts a 1 after time $t = 2$; since we enter with 1 and now take out 1 we end up with zero:

$$f(t) = 0, \qquad t > 2 \tag{20.27}$$

As shown in the last few steps, we do indeed arrive back at our starting function, and this proves that our convolution formula (tying the derivative of the function against the shifted unit step functions) is correct. As is evident in above the convolution process is greatly a visual process; hence it is important to be able to flip signals, shift them around, and find areas of overlapping functions. Figure 20.5 shows practical application of the method; the smaller $\Delta\tau$, the better the resulting approximation.

20.4 Representation of Signal in Terms of Convolution with the Ramp Function

Similar to using delta and unit step functions, we can use the ramp function, via convolution, to reconstruct an arbitrary function. Again the premise here is that sometimes it maybe more convenient to apply a ramp stimulus and measure its response; and if that is available and knowing how the signal is tied to the ramp function we should be able to find the response due to the signal. First recall the ramp function

$$x(t) = \begin{cases} 0 & t < 0 \\ t & t > 0 \end{cases} \tag{20.28}$$

Next recall from last section we got the convolution integral representing a signal in terms of the unit step function

$$f(t) = \int_0^t f'(\tau)u(t-\tau)d\tau \tag{20.29}$$

If we can tie the ramp to the unit function, then we can use the ramp function in the convolution integral. But this is not difficult; all that is needed is

$$u(t-\tau) = \frac{d}{dt}x(t-\tau) \tag{20.30}$$

As such we plop in the ramp function into the prior convolution integral and get

$$f(t) = \int_0^t f'(\tau)\frac{dx(t-\tau)}{d\tau}d\tau \tag{20.31}$$

By integration by parts we finally arrive at

$$\boxed{f(t) = \int_0^t f''(\tau)x(t-\tau)d\tau} \tag{20.32}$$

Again, we need to *make sure that when taking first and second derivatives, at left and right edges, that we pick up the corresponding delta functions (and possibly derivative thereof) if there are any discontinuities*. Equation (20.32) states that an arbitrary function $f(t)$ can be represented as a convolution of ramp functions, with the *key* to each ramp function being equal to the second derivative of the function at the shift point.

To put things in perspective, the "key" in the Fourier time series was the Fourier transform;

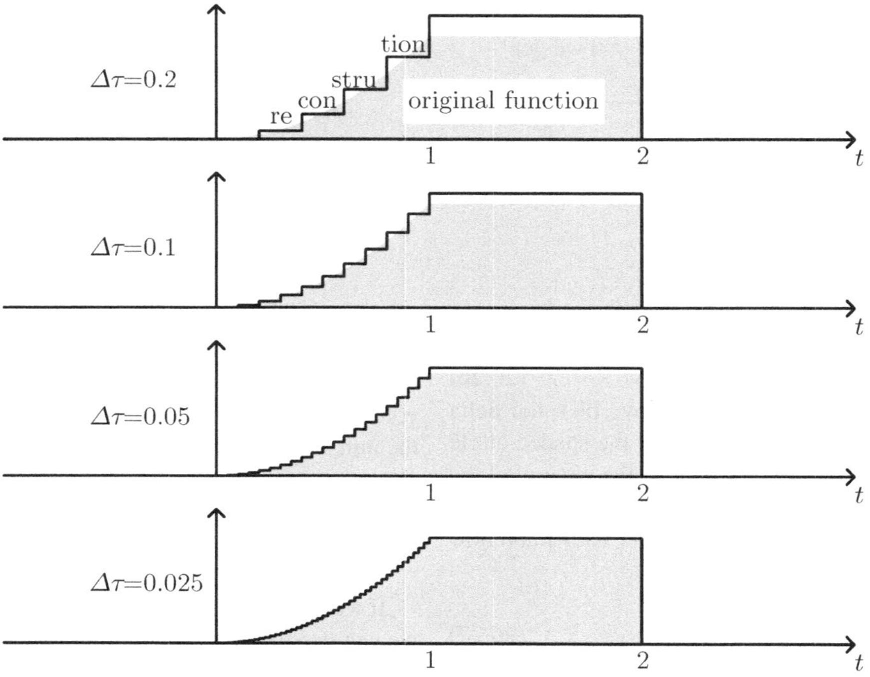

Fig. 20.5 Representation of target function as convolution with unit step function—practical considerations

that is what we used to scale the sines/cosines was the Fourier transform. On the other hand the "key" for constructing the signal in terms of convolution with the impulse function was the function itself. And the key when doing unit step convolution was the derivative of the function. Finally the key right here, and when representing the function as a convolution with the ramp function is the second derivative of the function. It is important to see the underlying infrastructure (or signature) behind all these attempts to dissect a function in terms of something else; and to do so we need the access to the "keys"!

Back to the ramp convolution, let's apply Eq. (20.32) to an example. Again take our test vehicle function shown on the top of Fig. 20.6. Similar to the unit step function case we start by taking the first derivative, which is shown on the second subplot. Here we pick a delta function at $\tau = 2$. Next we take the second derivative, as shown on the third subplot. Not only do we pick up another delta, but we also pick a *derivative* of a delta function (again at $\tau = 2$); the derivative of the delta function is denoted by the double-headed arrow (as opposed to the normal delta function which is denoted by a single-head arrow). It is very important to keep track of all delta functions (and their derivatives) before we start the convolution process. The convolution process in turn is started on the fourth subplot where the ramp is first flipped and then stepped in time. At each time t we multiply the ramp with the second derivative and find the area under the curve. Then step to a new time t (on the τ scale) and repeat. (Remember t offsets the ramp but τ is the integration variable!) For time $t < 1$ the convolution integral evaluates to

$$\begin{aligned} f(t) &= \int_0^t 2(t-\tau)d\tau = 2\left(t\tau - \frac{\tau^2}{2}\right)_0^t \\ &= 2\left(t^2 - \frac{t^2}{2}\right) = t^2, \quad (0 < t < 1) \end{aligned} \tag{20.33}$$

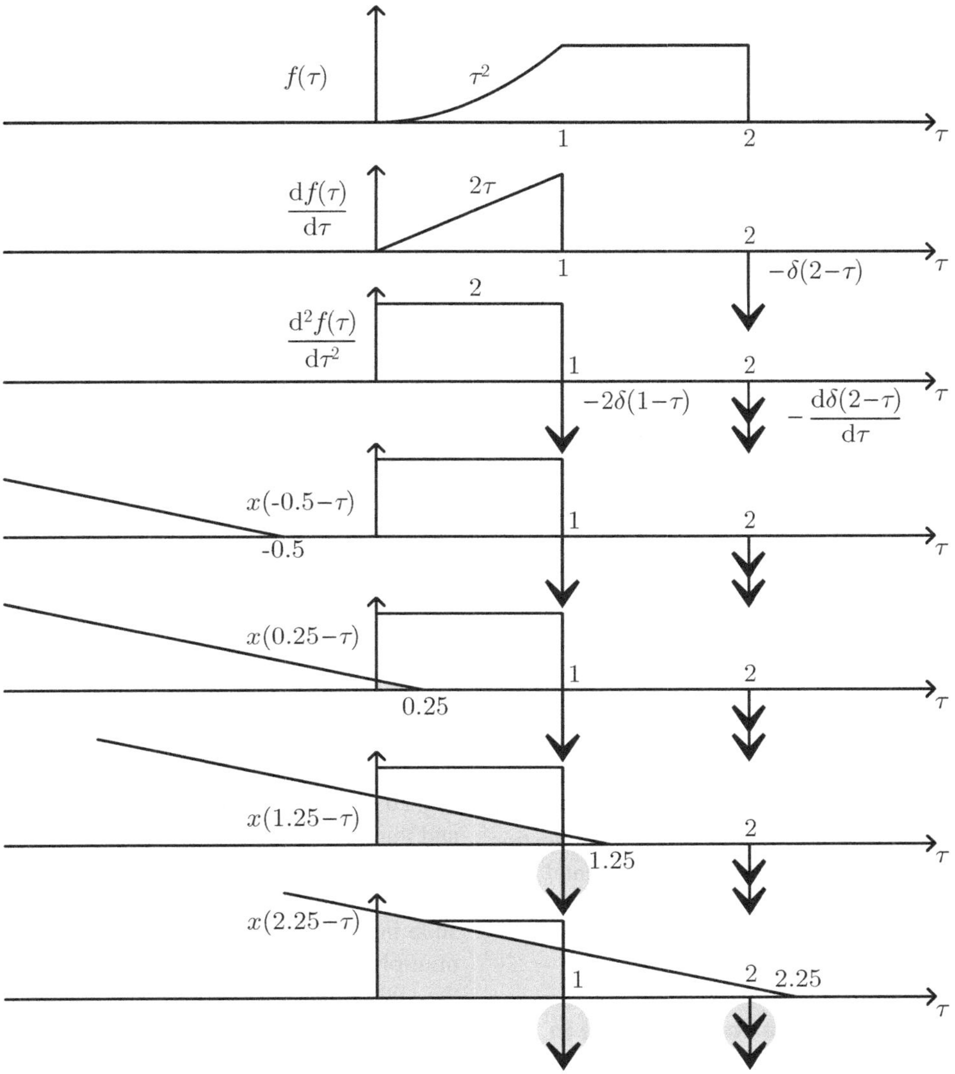

Fig. 20.6 Representation of a signal in terms of convolution with the ramp function

For time $1 < t < 2$ two things happen. First we pick a delta function and that gives

$$\int -2\delta(\tau-1)(t-\tau) = -2(t-1) = -2t+2 \quad (20.34)$$

Second the upper integration limit becomes 1 so that

$$\int_0^1 2(t-\tau)d\tau = 2\left(t\tau - \frac{\tau^2}{2}\right)_0^1 = 2\left(t - \frac{1^2}{2}\right) = 2t-1 \quad (20.35)$$

When added together we finally get

$$f(t) = 2t - 1 - 2t + 2 = 1, \qquad (1 < t < 2) \quad (20.36)$$

Finally when $t > 2$ we pick a derivative of a delta such that

$$-\int \delta'(\tau-2)(t-\tau)d\tau = -1 \quad (20.37)$$

When added to the prior result we get

$$f(t) = -1 + 1 = 0, \qquad (t > 2) \quad (20.38)$$

So we have indeed confirmed that the result of the convolution did reconstruct the starting function. Figure 20.7 shows real life application of the method; the smaller the incremental time between the convolution basis, the better the resulting reconstruction.

20.5 Representation of Signal in Terms of Convolution with the Quadratic Function

The next logical step in convolution basis function selection—having covered the delta function, unit step, and then the ramp one—is the quadratic function define

$$y(t) = \begin{cases} 0 & t < 0 \\ t^2 & t > 0 \end{cases} \quad (20.39)$$

Again using integration by parts we get

$$\boxed{f(t) = \frac{1}{2}\int_0^t f'''(\tau)y(t-\tau)d\tau} \quad (20.40)$$

By now it is implicit that while evaluating the first, second, and now third derivatives we must keep record of all resulting delta functions and derivatives thereof! But why do we pick up a factor of $\frac{1}{2}$ in above? The reason is that in order to get the ramp function out of the quadratic one we end up dividing by 2:

$$x(t) = \frac{1}{2}\frac{d}{dt}y(t) \quad (20.41)$$

Let us test the new set of basis functions on our demo function as shown on the top of Fig. 20.8. We start by taking the first, second, and third derivatives as shown in the figure. Next we pull in the (flipped) quadratic basis function and start sliding it from far left to far right, with slide index t. At each point along the τ axis we multiply the third derivative of the function times the sliding basis function, and integrate (versus τ). The result is our target function, as a function of t. The first derivative comes out

$$f'(\tau) = 2\tau\,[u(\tau) - u(\tau-1)] - \delta(2-\tau) \quad (20.42)$$

The second derivative comes out

$$f''(\tau) = 2\,[u(\tau) - u(\tau-1)] - 2\delta(1-\tau) - \frac{d\delta(2-\tau)}{d\tau} \quad (20.43)$$

Lastly the third derivative comes out

$$f'''(\tau) = 2\delta(\tau) - 2\delta(1-\tau) - 2\frac{d\delta(1-\tau)}{d\tau} - \frac{d^2\delta(2-\tau)}{d\tau^2} \quad (20.44)$$

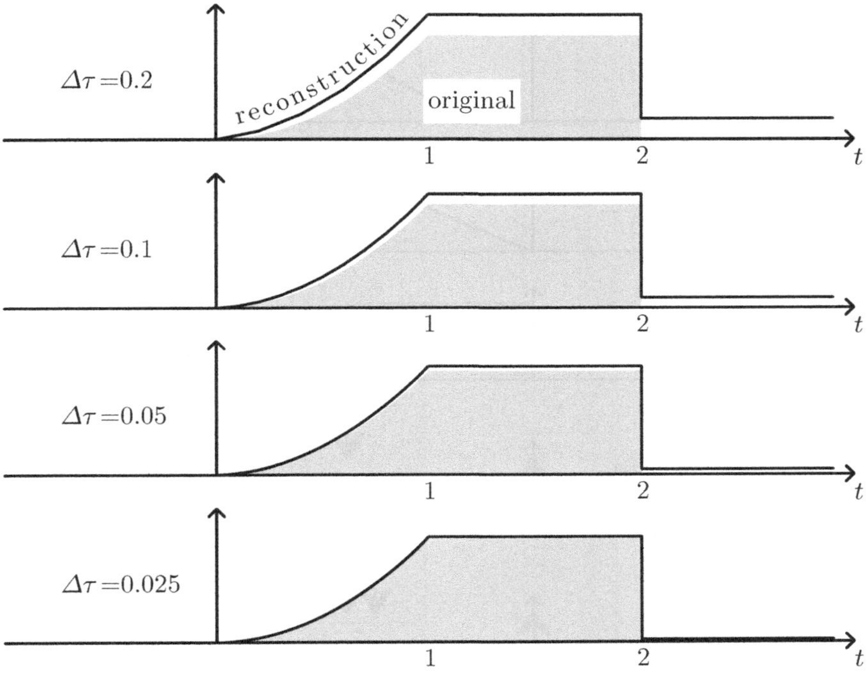

Fig. 20.7 Representation of a signal in terms of convolution with the ramp function—practical application

Putting this back into the convolution integral we arrive at

$$f(t) = y(t) - y(t-1) - 2(t-1)u(t-1) - u(t-2) \tag{20.45}$$

This in fact is our starting function! The above steps are illustrated in Fig. 20.9.

20.6 Recap of Convolution Integrals

As was shown in this chapter, we are able to represent a function as a convolution integral against some basis functions. The basis functions ranged from delta function, unit step one, ramp one, and quadratic function. In fact other higher order basis functions can be used. The *key* to each basis—that is the *weight* of each basis—relates to the original function, in the form of the function value, the first derivative, second, and so forth. Combining the basis functions, which are offset in time, with the key weights, and adding all, we regain the original function!

20.7 Relation Between Spectral Techniques (Fourier Analysis) and Convolution

Frequency spectrum techniques (aka Fourier transform) and time convolution enable us to find the response of a system due to a stimulus. At first glance they appear to be completely independent tracks. But here and there we sensed some relation between them. For example we know the FT of the convolution of two functions results in the product of the individual transforms. As it turns out there is in fact an intimate relation between the two tracks, and in fact it could be argued that one path can be completely derived from the other. Before we establish the relation between the Fourier transform and convolution we need to recall a few facts. The first is the delta function which is defined as

$$\delta(t) = \frac{1}{2\pi}\int_{-\infty}^{\infty} e^{j\omega t} d\omega \tag{20.46}$$

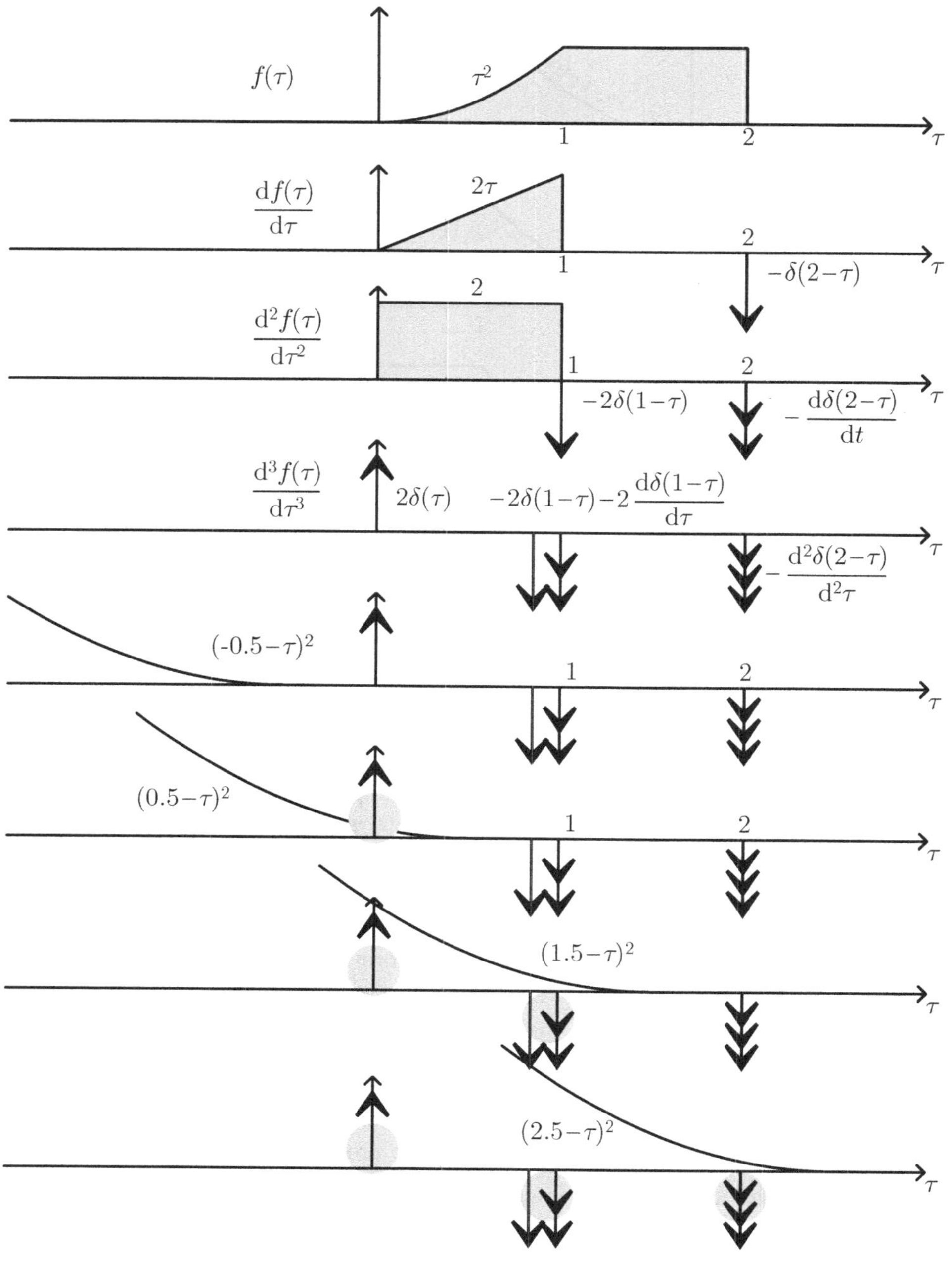

Fig. 20.8 Representation of a signal in terms of convolution with the quadratic function

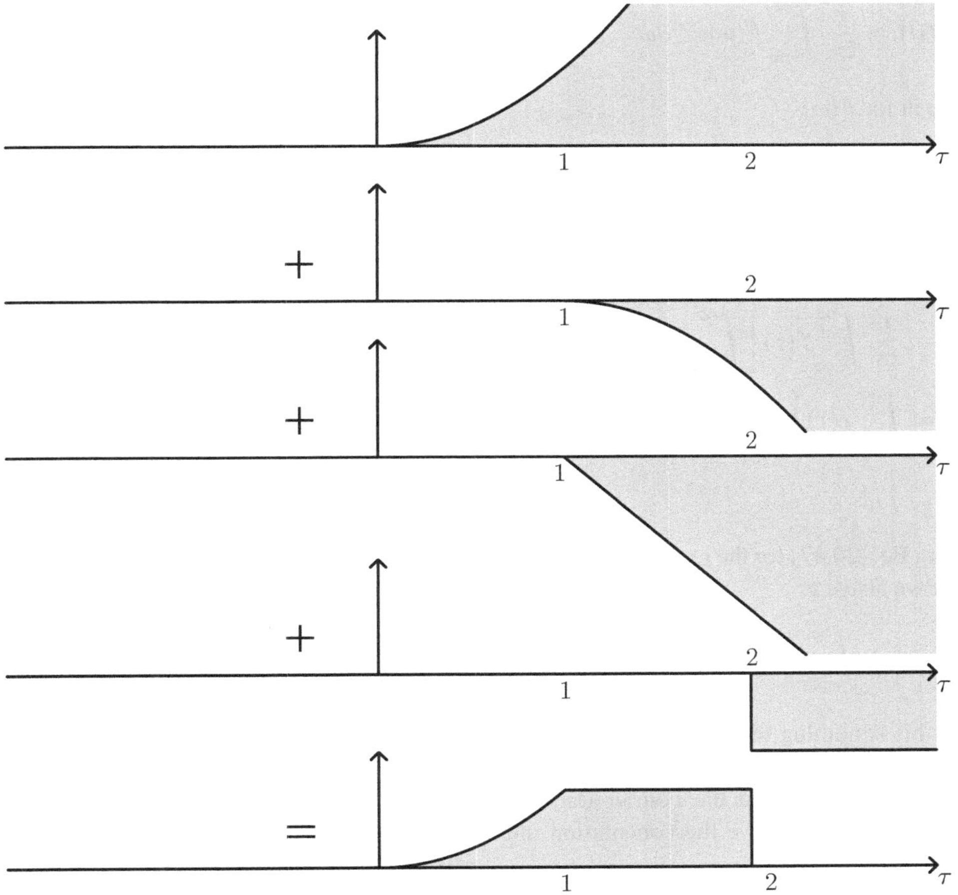

Fig. 20.9 Graphic application of Eq. (20.45)

We can establish this experimentally. The delta function is even in time, which means we only need the cosine terms in the frequency integration. We can verify that the integral of the cosine is zero except when frequency is zero, at which point it blows up. This is the definition of the delta function. It also follows that

$$\delta(t-\tau) = \frac{1}{2\pi}\int_{-\infty}^{\infty} e^{j\omega(t-\tau)}d\omega \qquad (20.47)$$

The next item is just a reminder of the Fourier transform pair

$$F(\omega) = \int_{\infty}^{\infty} f(t)e^{-j\omega t}dt$$

$$F(t) = \frac{1}{2\pi}\int_{-\infty}^{\infty} F(\omega)e^{j\omega t}d\omega \qquad (20.48)$$

The last item is the signal construction in terms of the convolution integral

$$f(t) = \int_{-\infty}^{\infty} f(\tau)\delta(t-\tau)d\tau \qquad (20.49)$$

20.7.1 From Fourier Transform to Convolution Theory

Here we will show that we can derive the convolution integral from the Fourier transform. Start with

$$f(t) = \frac{1}{2\pi}\int_{-\infty}^{\infty} F(\omega)e^{j\omega t}d\omega \quad (20.50)$$

Plug in for $F(\omega)$

$$f(t) = \frac{1}{2\pi}\int_{-\infty}^{\infty}\left[\int_{-\infty}^{\infty} f(\tau)e^{-j\omega\tau}d\tau\right]e^{j\omega t}d\omega \quad (20.51)$$

Change the order of integration

$$\begin{aligned} f(t) &= \frac{1}{2\pi}\int_{-\infty}^{\infty} f(\tau)\left[\int_{-\infty}^{\infty} e^{-j\omega\tau}e^{j\omega t}d\omega\right]d\tau \\ &= \int_{-\infty}^{\infty} f(\tau)\left[\frac{1}{2\pi}\int_{-\infty}^{\infty} e^{j\omega(t-\tau)}d\omega\right]d\tau \end{aligned} \quad (20.52)$$

Using Eq. (20.47) for the expression inside the bracket we arrive at

$$f(t) = \int_{-\infty}^{\infty} f(\tau)\delta(t-\tau)d\tau \quad (20.53)$$

But this is nothing more than the convolution integral as specified in Eq. (20.49). So we have shown that starting from the Fourier transform pair we are able to derive the convolution integral.

20.7.2 From Convolution Integral to Fourier Transform

The starting point here is the convolution integral

$$f(t) = \int_{-\infty}^{\infty} f(\tau)\delta(t-\tau)d\tau \quad (20.54)$$

We replace the delta function with its integral representation

$$f(t) = \int_{-\infty}^{\infty} f(\tau)\left[\frac{1}{2\pi}\int_{-\infty}^{\infty} e^{j\omega(t-\tau)}d\omega\right]d\tau \quad (20.55)$$

Change the order of integration

$$f(t) = \frac{1}{2\pi}\int_{-\infty}^{\infty}\left[\int_{-\infty}^{\infty} f(\tau)e^{-j\omega\tau}d\tau\right]e^{j\omega t}d\omega \quad (20.56)$$

But the term in bracket is nothing more than the FT of the function:

$$F(\omega) = \int_{-\infty}^{\infty} f(\tau)e^{-j\omega\tau}d\tau \quad (20.57)$$

Then Eq. (20.56) becomes

$$f(t) = \frac{1}{2\pi}\int_{-\infty}^{\infty} F(\omega)e^{j\omega t}d\omega \quad (20.58)$$

But this is nothing more than the inverse FT of the signal. So, we have started with the convolution integral and derived the Fourier transform pairs.

20.8 Summary

As a first application of the convolution concept this chapter dealt with signal construction in terms of convolution with basis functions. Just like in the Fourier world there were basis functions (complex exponentials), there are basis functions in the convolution world. In fact one could argue that there could be infinite basis functions for use in convolution; but as a starter and for practical reasons our basis functions ranged from the delta function, unit step one, the ramp and higher orders of t. While the basis functions differ, the general idea is the same. In all cases the starting point was finding the key to the basis functions—that is, what needs to be done to the original function before it gets multiplied by the basis function and integrated. For the delta function basis function the key was simply nothing—just multiply the starting function as is by the delta function and integrate. For the unit step basis function the key was differentiating the starting function; and for the ramp basis function the key was taking the first derivative of the starting function and so forth. As stressed multiple times throughout the chapter, it is critical to keep record of all delta functions (or derivatives thereof) while differentiating the original function; this is especially important at the start and end of the signal. In all cases the basis functions are first flipped, then shifted by t, and for each t value the integral (in τ) against the starting function was calculated. As to why we would want to go through all this trouble is because if we know signal construction in terms of some basis functions, and if we happen to know system response to any of those basis functions, we are guaranteed to find system response to our starting signal. We will learn more about this in the following chapters.

20.9 Problems

1. Consider the hat function defined between -1 and 1; reconstruct it using pulses, with pulse width 0.5, 0.3, and 0.1, consecutively. Make sure the interface between adjacent pulses is taken care off correctly. See sample solution in Fig. 20.10.
2. Consider the inverted parabola function defined between -1 and 1; reconstruct it using pulses, with pulse width 0.5, 0.3, and 0.1, consecutively. See sample solution in Fig. 20.11.
3. Consider the single-cycled sine function, defined between 0 and 2π; reconstruct it using pulses, with pulse width 0.5, 0.3, and 0.1, response. See sample solution in Fig. 20.12.
4. Verify Eq. (20.13). Pick a few values of t, starting from negative and then going positive. For each value, plot the delta function location on the τ scale, then integrate it from 0 to ∞. For each of the assumed t values, plot the corresponding integration results. Convince yourself that the resulting function is in fact a unit step one!
5. Consider the hat function defined between 0 and 2. First find the derivative, and then use that to express the function in terms of convolution with the unit step function. Plot results for a couple of cases of different spacings between the unit steps used; see sample solution in Fig. 20.13.
6. Consider the single-cycled cosine function defined between 0 and 2π. First find the derivative (paying attention to delta functions at edges), and then reconstruct the function in terms of convolution with the unit step function. Plot results for a couple of samples of unit step spacings; see sample solution in Fig. 20.14.
7. Consider the tapered triangle of top width 1 and bottom width 3, centered at 1.5. Reconstruct it using convolution with the unit step function. Find the derivative, then draw the convolution process, and convince yourself that the area under the convolution results in the original function. See sample solution in Fig. 20.15.

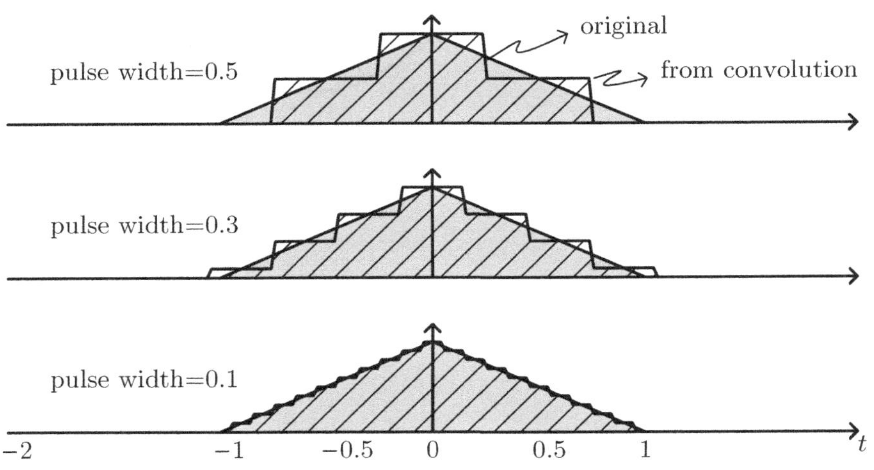

Fig. 20.10 Sample solution to Problem 1

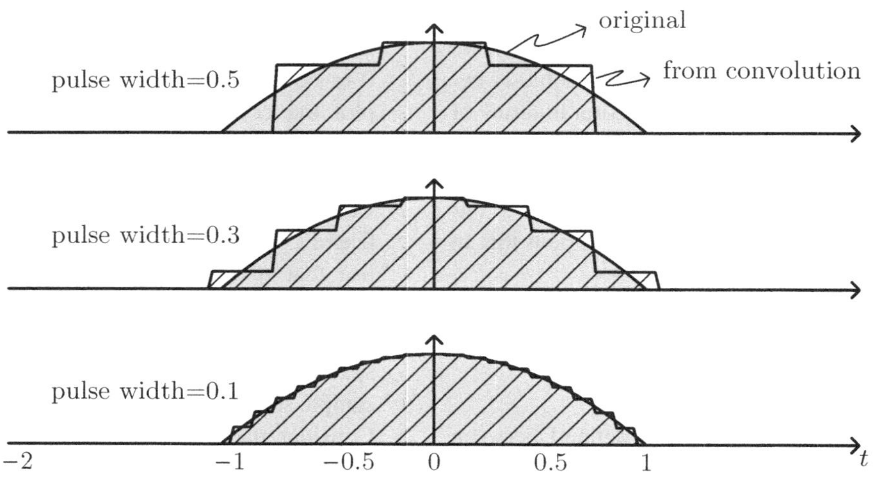

Fig. 20.11 Sample solution to Problem 2

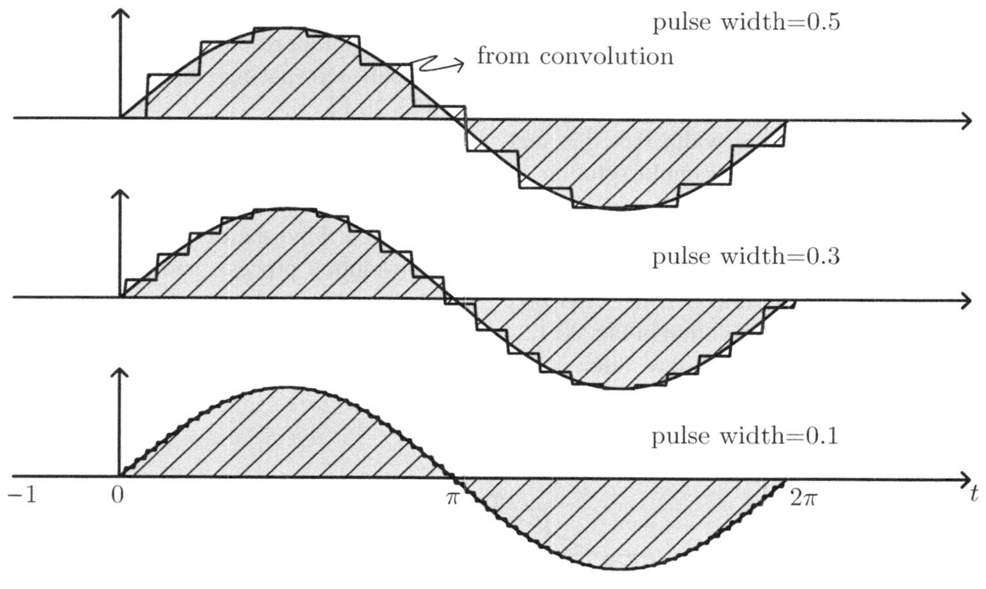

Fig. 20.12 Sample solution to Problem 3

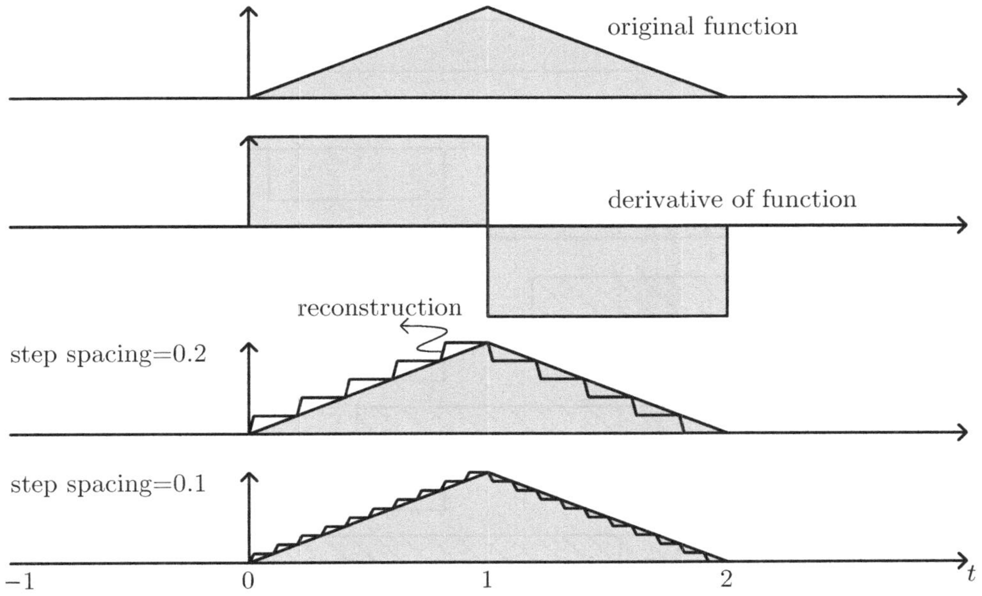

Fig. 20.13 Sample solution to Problem 5

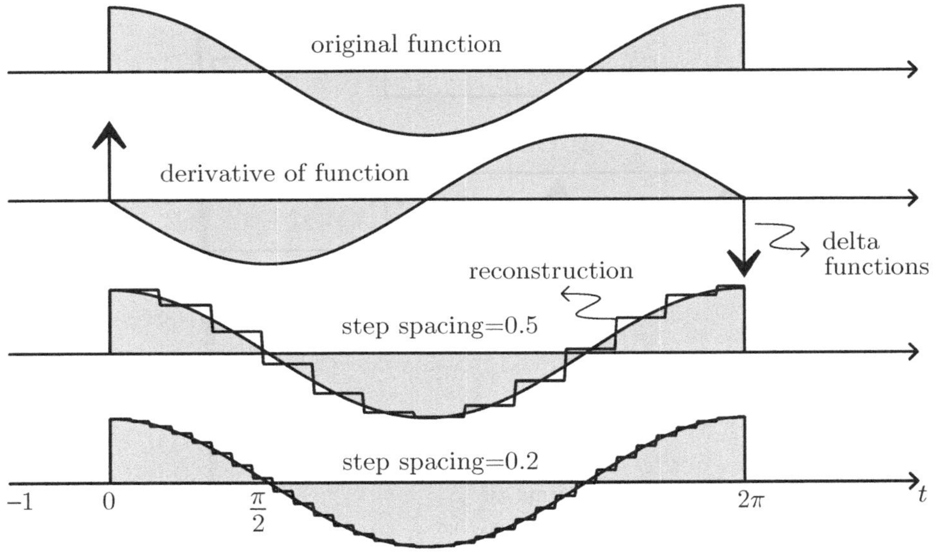

Fig. 20.14 Sample solution to Problem 6

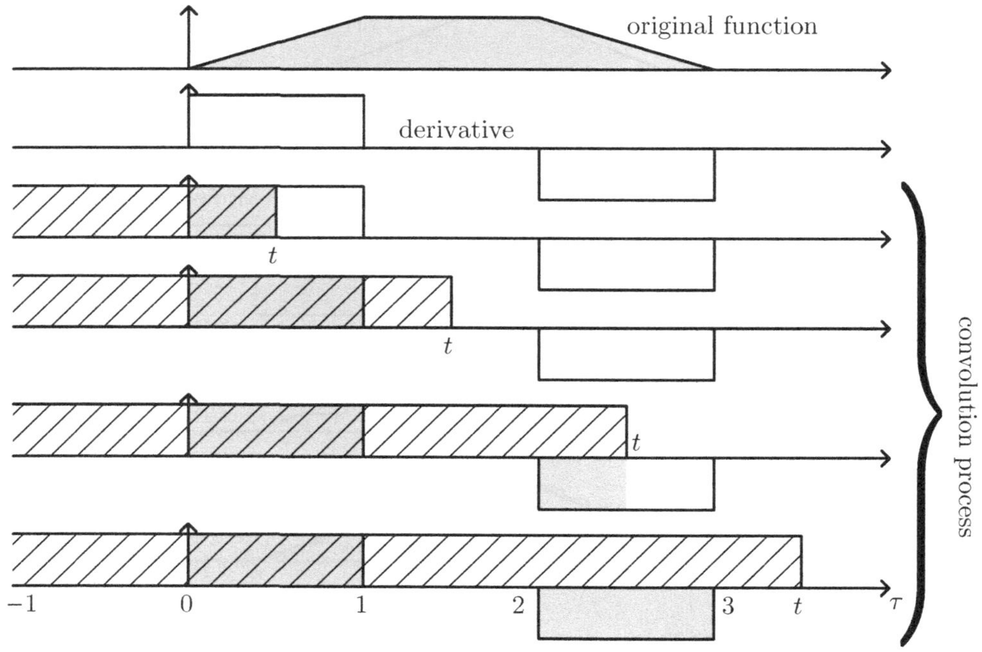

Fig. 20.15 Sample solution to Problem 7

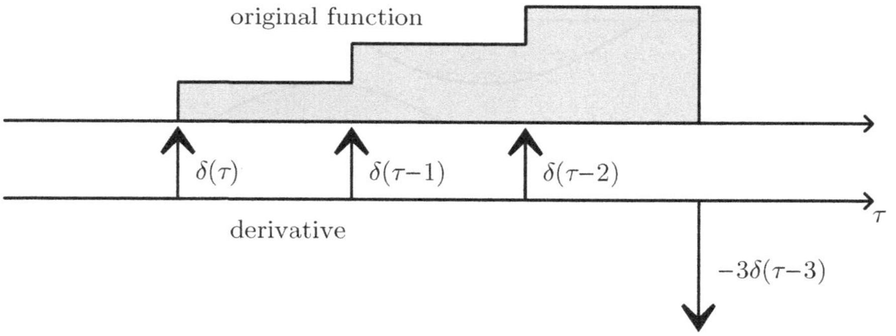

Fig. 20.16 Sample solution to Problem 8

8. Express the three-step ladder in terms of convolution with the unit step function; see sample solution in Fig. 20.16.
9. Consider the inverted parabola, defined between 0 and 2, and zero elsewhere. Reconstruct it using convolution with the ramp function and prove that

$$f(t) = \begin{cases} 2t - t^2 & 0 < t < 2 \\ 0 & t > 2 \end{cases}$$

See sample solution in Fig. 20.17.

10. Consider the function e^t defined between 0 and 1, and zero otherwise. Reconstruct it using convolution with the ramp function;

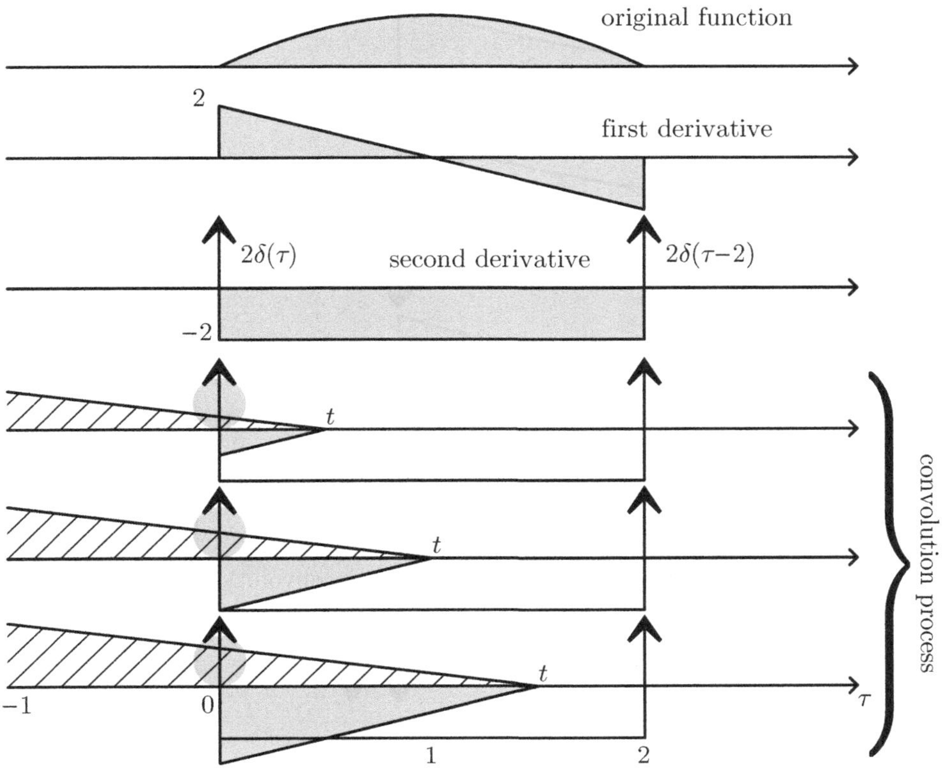

Fig. 20.17 Sample solution to Problem 9

see partial solution in Fig. 20.18. Hint: the convolution integral, for $0 < t < 1$ is

$$
\begin{aligned}
f(t) &= (t+1) + \int_0^t e^{\tau}(t-\tau)d\tau \\
&= (t+1) + te^{\tau}\Big|_0^t - \int_0^t \tau e^{\tau} d\tau \\
&= (t+1) + (te^t - t) - \tau e^{\tau}\Big|_0^t + \int_0^t e^{\tau} d\tau \\
&= (te^t + 1) - te^t + e^{\tau}\Big|_0^t = 1 + e^t - 1 = \boxed{e^t}
\end{aligned}
$$

For $t > 1$

$$
\begin{aligned}
f(t) &= (t+1) + \int_0^1 e^{\tau}(t-\tau)d\tau - e^1\left[(t-1)+1\right] \\
&= (t+1) + te^{\tau}\Big|_0^1 - \int_0^1 \tau e^{\tau} d\tau - te^1 \\
&= (t+1) + (te^1 - t) - \tau e^{\tau}\Big|_0^1 + \int_0^1 e^{\tau} d\tau - te^1 \\
&= 1 - e^1 + e^{\tau}\Big|_0^1 = 1 - e^1 + e^1 - 1 = \boxed{0}
\end{aligned}
$$

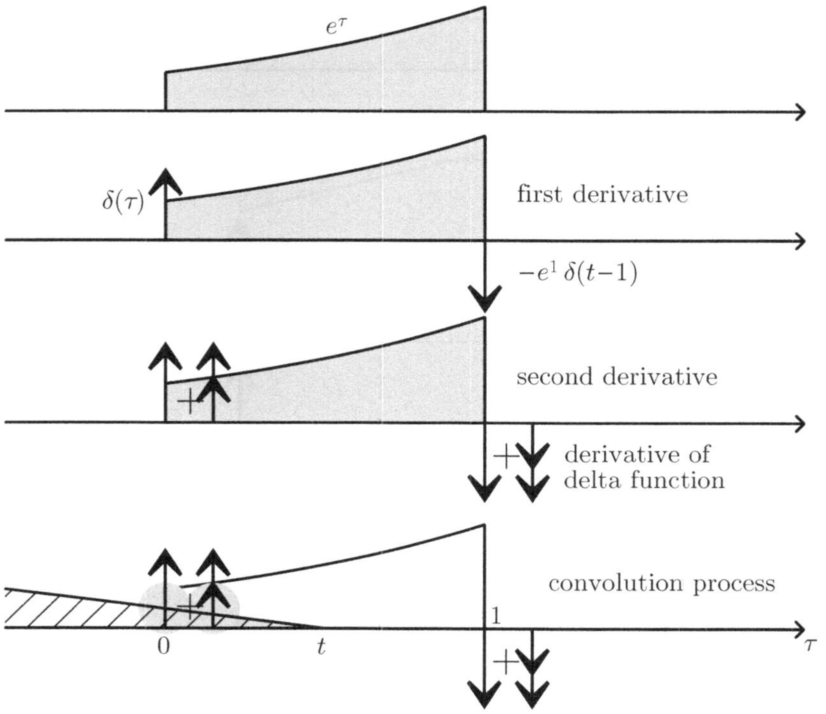

Fig. 20.18 Partial solution to Problem 10

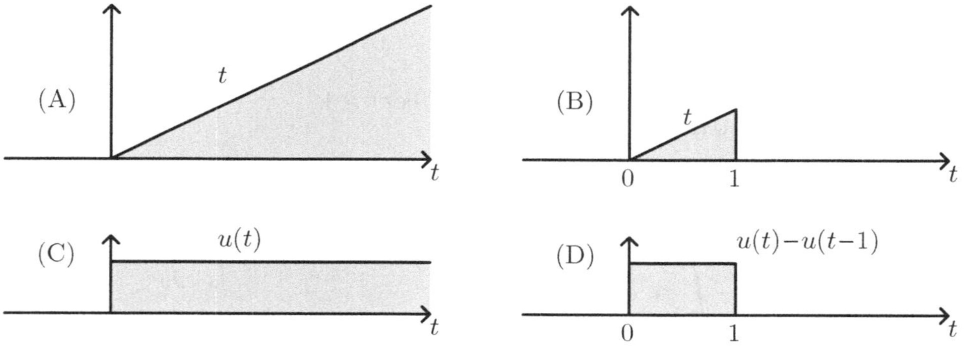

Fig. 20.19 Statement to Problem 11

11. Consider the four functions shown in Fig. 20.19; express them in terms of convolution integrals with the ramp function.

Answer (A):

$$f(t) = \int_0^t \delta(\tau)[t-\tau]d\tau = t$$

Answer (B):

$$f(t) = \int_0^t \left[\delta(\tau) - \frac{d\delta(\tau-1)}{d\tau} - \delta(\tau-1)\right][t-\tau]$$

$$= tu(t) - (t-1)u(t-1) - u(t-1)$$

Answer (C):

$$f(t) = \int_0^t \frac{d\delta(\tau)}{d\tau}[t-\tau] = u(t)$$

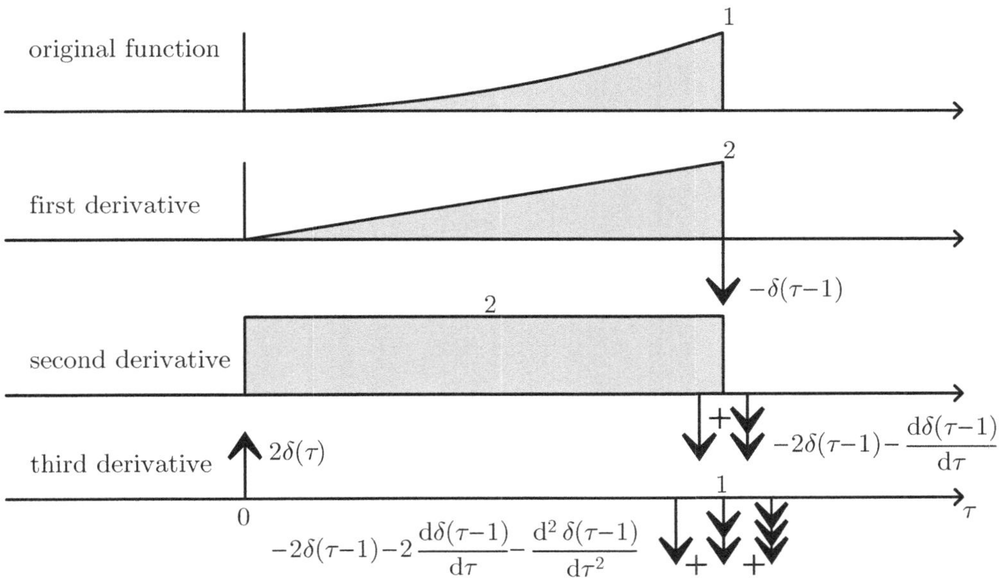

Fig. 20.20 Partial solution to Problem 12

Answer (D):

$$f(t) = \int_0^t \left[\frac{d\delta(\tau)}{d\tau} - \frac{d\delta(\tau-1)}{d\tau} \right] [t-\tau]$$
$$= u(t) - u(t-1)$$

12. Consider the quadratic function defined between 0 and 1, and is zero otherwise; reconstruct it using convolution with the quadratic function. See sample solution in Fig. 20.20. Answer (A): $f(t) = y(t) - y(t-1) - 2x(t-1) - u(t-1)$

21 The Delta Function

21.1 Introduction

The delta function is an extremely important function in many areas of math, physics, and engineering. It has many applications, too many to summarize here; but for a starter it is used as a sampling function, in Fourier and Laplace transforms, transfer functions, impulse response, time convolution, point charge, point mass, point load, and more. In this chapter we study this function in little bit more detail, so that the reader has a better feel of what this function looks like and how it acts in different scenarios. Think of the delta function as an extreme function, and although kind of out of the norm we must deal with it. Or we can think of it as a limiting function in the sense while it on its own is difficult to deal with, making it a limit of some other function (such as a narrow, tall pulse) takes away some of the associated difficulty. Like many other things the more practice we invest in a topic, the easier the underlying concepts become.

21.2 Basic Properties of the Delta Function

The basic *three* properties of the delta function are as follows:

1. It is zero everywhere, except at a single point (application point), conveniently taken at zero; that is

$$\delta(t) = 0 \qquad \text{for all} \qquad t \neq 0 \tag{21.1}$$

2. It is infinite at the application point; that is

$$\delta(0) = \infty \tag{21.2}$$

3. And finally, its time integration is unity; that is

$$\int_{-\infty}^{\infty} \delta(t)dt = 1 \tag{21.3}$$

Notice that the first two properties are quite unconventional from the point of view of normal functions; and notice the abrupt discontinuity both before and after the application points; but let's not let those anomalies distract us away from how valuable this function is, and how we can use it.

A byproduct of the above three properties is the *sampling* property, which states that

$$f(t_0) = \int_{-\infty}^{\infty} f(t)\delta(t - t_0)dt \tag{21.4}$$

F. Badrieh, *Spectral, Convolution and Numerical Techniques in Circuit Theory*,
https://doi.org/10.1007/978-3-319-71437-0_21

21.3 How to Get the Delta Function

There are many ways to get functions to behave like a delta function. Below is a sample.

- By taking the derivative of the unit step function.
- By representing the function as a pulse of unity area, contracting the pulse width (along time) and prolonging its height beyond bound, but insuring that the area remains unity.
- By taking the limit of the single-sided negative exponential

$$\delta(t) = \lim_{a\to 0} u(t)\frac{1}{a}e^{-t/a} \tag{21.5}$$

- By taking the limit of the sinc function as follows:

$$\pi\delta(t) = \lim_{\omega\to\infty} \frac{\sin t\omega}{t} \tag{21.6}$$

- By taking the limit

$$\pi\delta(t) = \lim_{\sigma\to 0} \frac{\sigma}{t^2+\sigma^2} \tag{21.7}$$

The last three sequences are shown in Fig. 21.1. Let us then discuss in some detail some of these methods.

21.4 The Delta Function as the Derivative of the Unit Step Function

We can build the delta function by taking the derivative of the unit step function as shown in Fig. 21.2. As the unit step function is made more abrupt, the width of the delta function is decreased, while the height increased, subject to the condition that pulse area remains unity. Alternatively we could think of the delta function as a pulse of width T and height $1/T$ and take the limit as $T \to 0$. Of course we run into trouble *right at* the point $T = 0$; but point is, before that point the resulting function is as close to an ideal delta function as possible! One drawback of either of these methods is that neither of them (unit step or pulse) are defined exactly in terms of analytic functions (like sines, cosine, exponentials, ...). But that need deter us of using them, especially that they have been tested and proven to work. So the main take from here is a kind of geometric, intuitive explanation of the delta function.

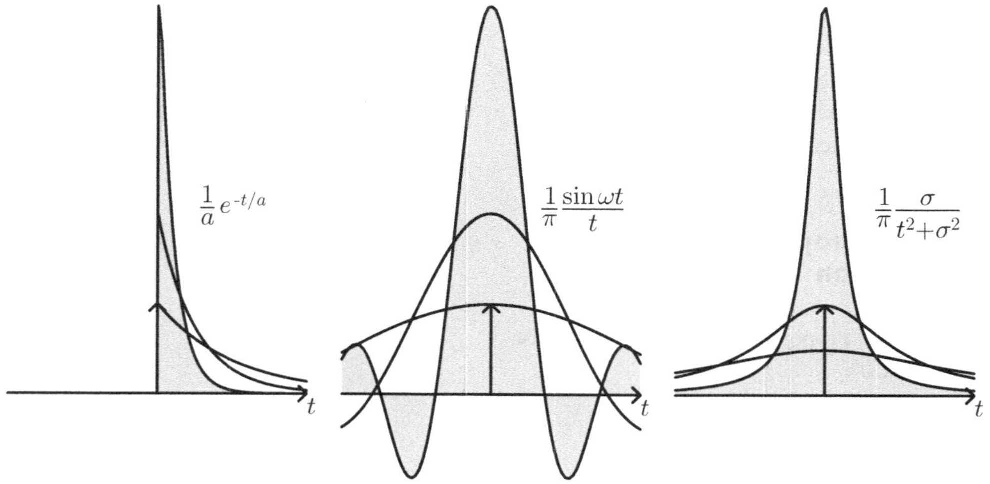

Fig. 21.1 Various sequences to get the delta function

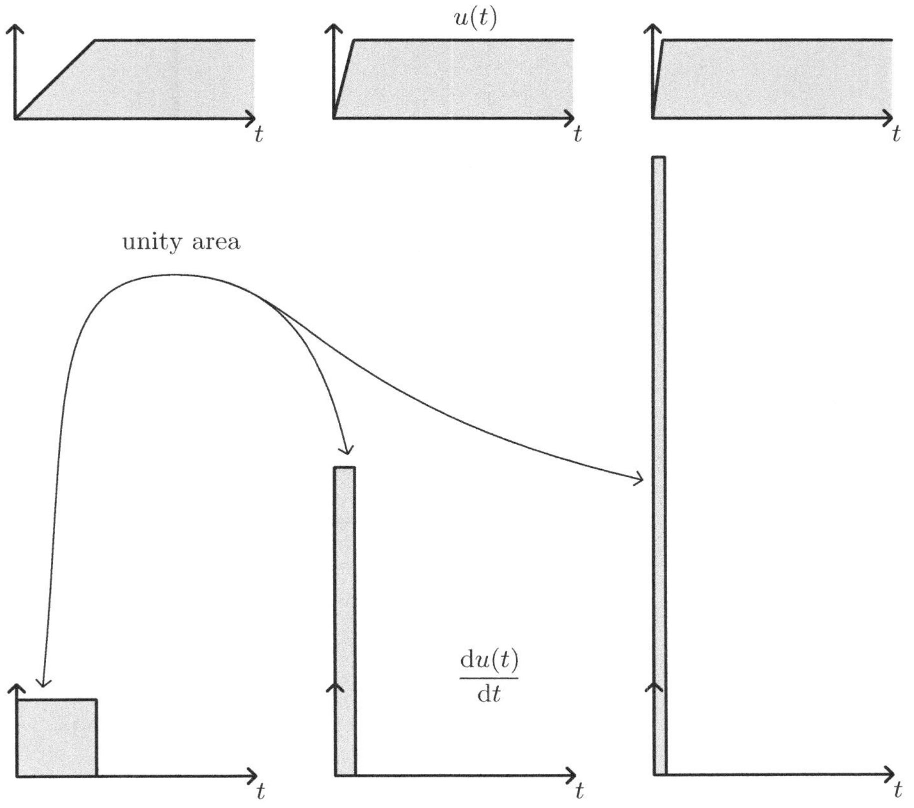

Fig. 21.2 The delta function as the derivative of the unit step function

21.5 The Delta Function as Limit of Negative Exponential

We can also build the delta function from the scaled negative exponential function

$$\delta(t) = \lim_{a \to} \frac{1}{a} e^{-t/a} \tag{21.8}$$

as shown in Fig. 21.3. Notice that the value of this function at time zero is

$$\lim_{a \to 0} \frac{1}{a} e^{-t/a} = \lim_{a \to 0} \frac{1}{a} \to \infty \tag{21.9}$$

giving the required singularity at time zero; and notice that the integral is

$$\int_0^\infty \frac{e^{-t/a}}{a} dt = -e^{-t/a}\Big|_0^\infty = 0 - -1 = 1 \tag{21.10}$$

giving the needed unity integral. Notice though that the function is asymmetric (in time), but the asymmetry appears less as $a \to 0$. Notice also that while the negative exponential is analytic (continuous, smooth, ...) for $t > 0$ it is discontinuous at time zero, just like the unit step function. So the main take here is that as we make $a \to 0$ the negative exponential grows beyond bound around time zero, becomes more localized (in the sense of being zero everywhere else), and still has unity area; all the requirements for a delta function.

21.6 The Delta Function as a Limit of the Sinc Function

Recall that one definition of the delta function is

$$\pi\delta(t) = \lim_{\omega \to \infty} \frac{\sin \omega t}{t} \tag{21.11}$$

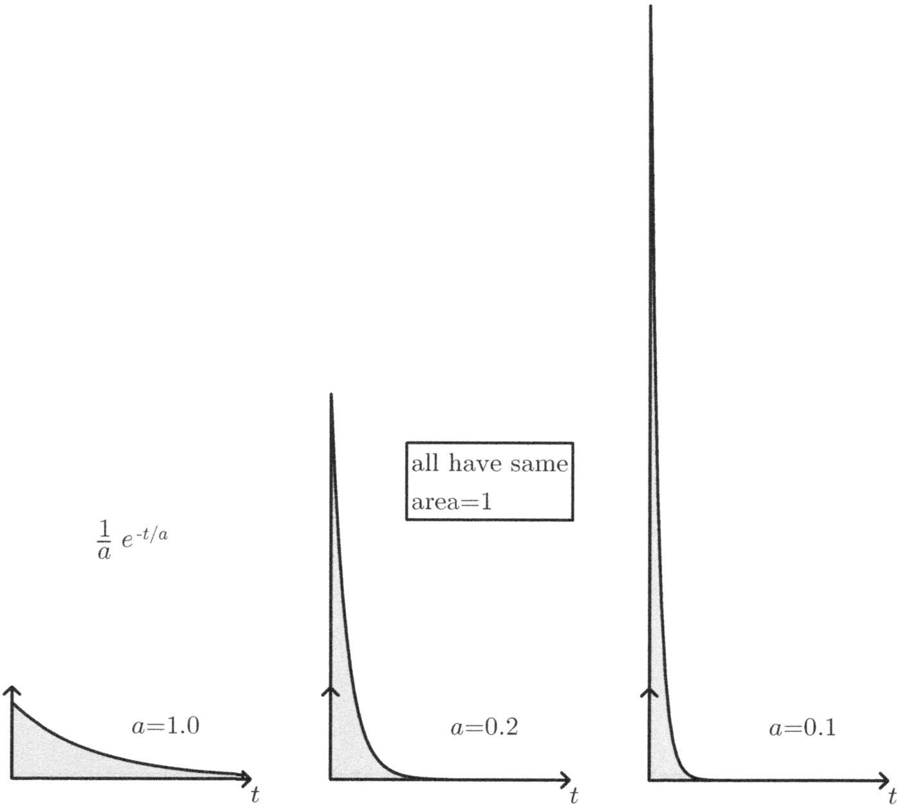

Fig. 21.3 The delta function as a limit of scaled negative exponential function

Now this may require a bit of explanation: how can a sinc function map into a delta function? Consider Fig. 21.4 which shows the sinc function for different strength values. Notice first that as $\omega \to \infty$, the sinc function would oscillate at infinitely high speeds, so that on *average* it appears around zero. However, around zero t things look different. In particular, at $t = 0$ the function evaluates to

$$\lim_{t \to 0} \frac{\sin \omega t}{t} = \omega \tag{21.12}$$

That is, right at zero, the function evaluates to ω. And if ω is pushed to infinity, right at zero the function evaluates to infinity. This is the first signature of the delta function. To recap, away from zero, as ω is pushed to infinity, on average the sinc function goes to zero; but under the same limit, it would go to infinity at zero. This is the second signature of a delta function, which is being 0 everywhere except at 0. The third interesting property of this function is that its integral over time is constant, and independent of ω; that is

$$\int_{-\infty}^{\infty} \frac{\sin \omega t}{t} dt = \pi \tag{21.13}$$

To recap, as we set the limit $\omega \to \infty$ we have the following three behaviors of the sinc function as a function of time:

1. The function on average appears zero away from zero time.
2. The function blows up at time zero.
3. The area under the curve (integral of function) is unity (times π).

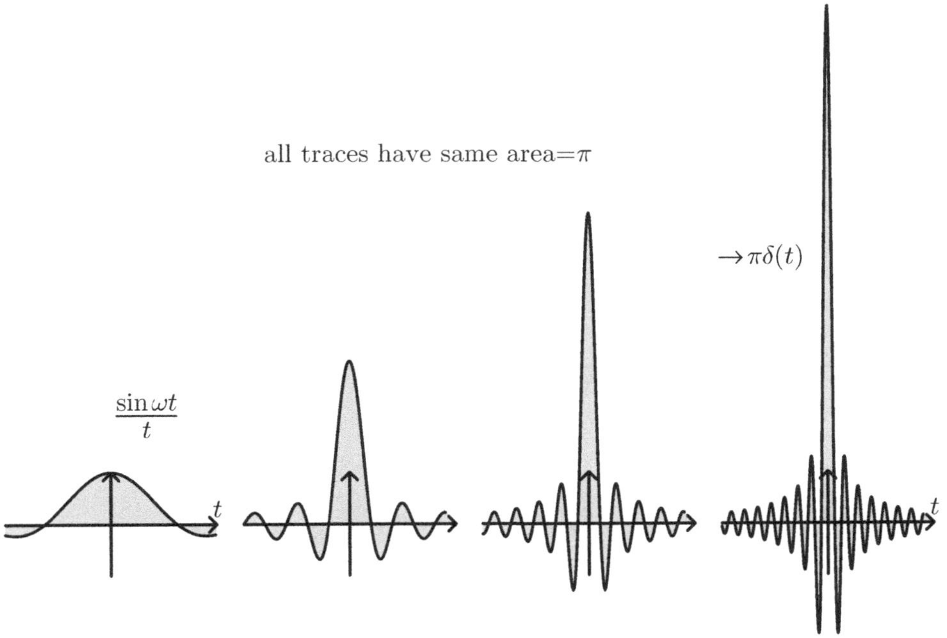

Fig. 21.4 Sinc function and limit to delta function

All these properties belong to a delta function. Notice that of all the above, it is the second item that is not very clear. The possible confusion may arise because we stated that the function "on average" goes to zero for nonzero time. It does NOT go to zero for each time point (except at $t = \infty$). But since the function oscillates so fast, and the value away from the origin is quite smaller than that at the origin, we say that most of the "energy" is centered at the origin. Keep in mind that we rarely deal with a delta function on its own; most of the time we apply the delta function to another, and integrate the result. It is during the integration that the "average" properties of the delta function dominate, and not necessarily the instantaneous ones. Stated another way, if we choose any time t other than zero, and if we take the limit $\omega \rightarrow \infty$, then integrate the sinc function around t we will get about zero. The only exception is when $t = 0$; there, and no matter how large ω is the integral evaluates to unity (time π). See also Problem 4.

21.7 The Delta Function as a Limit of the Pulse Function and the Sampling Property

We want to show here a simple and intuitive proof of the sampling theory, in the framework of the pulse approximation to the delta function. Let's take an arbitrary function as shown in Fig. 21.5. First we approximate the delta function with a pulse having width 1 and height 1 as shown in the upper left figure. After multiplying both functions, we get the figure on the upper right corner. Notice that most of the function $f(t)$ dropped already due to the sampling property

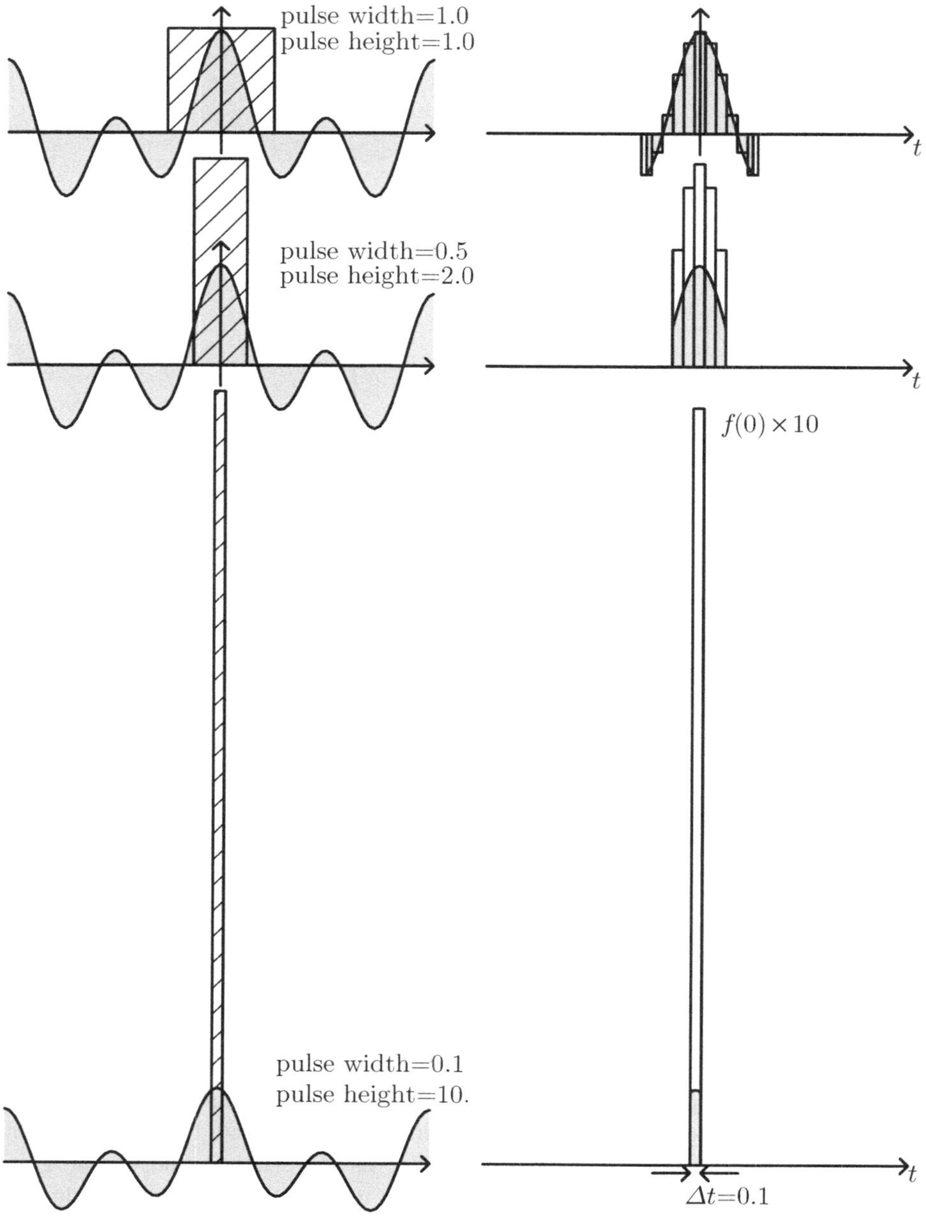

Fig. 21.5 The delta function and the sampling property

of the pulse function. Now let's find the area. For demonstration purposes, let's divide the time axis in terms of $\Delta t = 0.1$. Then the first integral can be approximated by

$$I = \frac{1}{1} \times 0.1 \Big[f(-0.5) + f(-0.4) + f(-0.3) + f(-0.2) + f(-0.1) + f(0.0) + f(0.1) + f(0.2) + f(0.3) + f(0.4) + f(0.5) \Big] \tag{21.14}$$

Now if we make the pulse narrower (0.5) and higher (2.0), still maintaining itsarea at unity, and as shown in the middle of Fig. 21.5, the integral becomes

$$I = \frac{1}{0.5} \times 0.1 \Big[f(-0.2) + f(-0.1) + f(0.0) + f(0.1) + f(0.2) \Big] \tag{21.15}$$

Finally if we make the pulse even narrower (0.1) and higher (10.0), still maintaining its area at unity, and as shown in the bottom of the figure, the integral becomes

$$\begin{aligned} I &= \frac{1}{0.1} \times 0.1 \Big[f(0.0) \Big] \\ &= f(0) \end{aligned} \tag{21.16}$$

What we have established then is that

$$\int_{-\infty}^{\infty} f(t)\delta(t)dt = f(0) \tag{21.17}$$

which is the sampling property. More generally we have

$$\int \delta(t - t_0) f(t) dt = f(t_0) \tag{21.18}$$

Notice when $t_0 = 0$ we regain the next to prior equation. Verbally, when we integrate the shifted delta function, centered at t_0, against a function, we pick up the value of the function at t_0. This is the sampling property of the delta function. Notice that in this context the delta function is not used on its own; instead the delta function is first multiplied by another function, and then integrated.

21.8 The Derivative of the Delta Function and Its Sampling Property

If we thought the delta function was hard to visualize, imagine its derivative! Nonetheless, the derivative is important and it also has an important sampling property. In particular,

$$\int \delta'(t) f(t) dt = -f'(0). \tag{21.19}$$

and more generally

$$\int \delta'(t - t_0) f(t) dt = -f'(t_0) \tag{21.20}$$

That is, when integrating the time derivative of the delta function, which is centered at t_0 against a function $f(t)$ we pick (negative) the time derivative of the function at t_0. Notice that while this looks very similar to the sampling property of the delta function itself (as opposed to its derivative), there is one important difference, and that is the negative sign! This can be shown as follows. Consider Fig. 21.6 which shows the delta function (top) and its derivative (bottom). Notice that we switch from using a rectangle for the delta function to using a triangle, just for convenience. Notice also that the base of the triangle is $2\Delta t$ (and not Δt) because we always want to ensure that the area under the curve is unity

$$\text{area under triangle} = \frac{1}{2} \times \text{width} \times \text{height} = \frac{1}{2} \times 2\Delta t \times \frac{1}{\Delta t} = 1 \tag{21.21}$$

If we take the derivative of this function we get the second part of Fig. 21.6. This looks like two delta functions, one centered at $-\frac{\Delta t}{2}$ and one centered at $\frac{\Delta t}{2}$. Each of those has a "strength" of $\frac{1}{\Delta t}$; that is, each delta function has an area of $\frac{1}{\Delta t}$. Now when this new set of delta functions is integrated against a function, we pick the function at both $-\frac{\Delta t}{2}$ and $+\frac{\Delta t}{2}$, that latter having a negative sign. That is

$$\int \frac{\delta(t)}{dt} f(t)dt = -\frac{f(\Delta t/2)}{\Delta t} + \frac{f(-\Delta t/2)}{\Delta t} \tag{21.22}$$

But this is nothing more than the (negative of the) derivative of the function at time zero, especially as we take $\Delta t \rightarrow 0$. Hence we demonstrated the meaning of the derivative of the delta function and proved Eq. (21.20). See also Problem 5.

21.9 Summary

Let's just say that the delta function is a cornerstone of signal construction and system response. It is such a simple concept but it is also a subtle one. When we exchange terms such as zero, infinity, and unity all within the same construct we need to be careful about not crossing the line between the real and the illusion! The delta function is a different kind of function and at first hard to visualize (let alone deal with). But as we think of it as the limit of another (easier) function we get closer to understanding how the delta function is generated and how it acts (on other functions). In a nutshell the delta function is a localized function with very short duration (ideally zero), with very large magnitude (ideally infinite) but with finite area (ideally one). It can

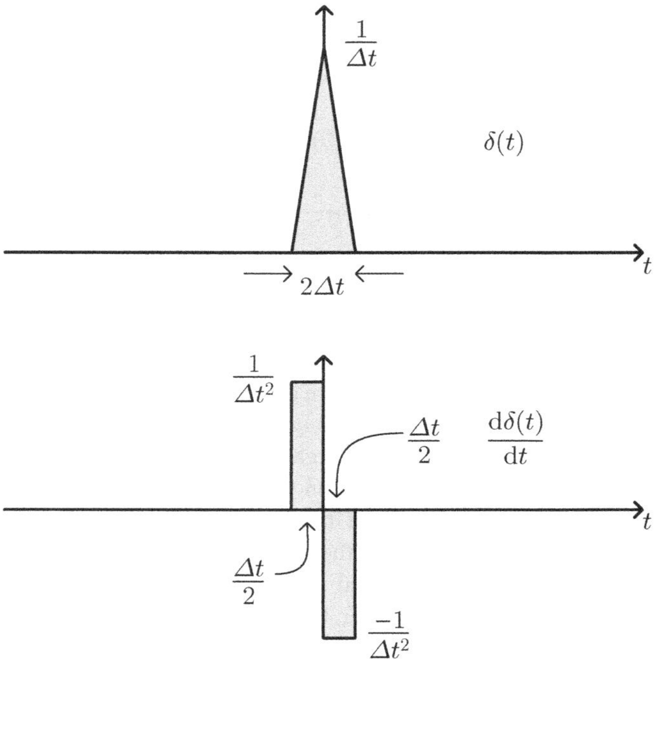

Fig. 21.6 The derivative of the delta function and its sampling property

be generated from other functions (such as the negative exponential or the sinc one) and by different means (simple limit or derivative). But in all cases the end outcome is a function with immense localized concentration that when applied to another signal, and integrated over yields the value of the other signal at the application point; and hence the sampling property of the delta function. Additionally we learned about the *derivative* of the delta function, and *its* sampling properties. In the next chapters we will see how the delta function disturbs a system and how knowing the system response to a delta input (aka impulse response) can be used to generate the system response to *any* other signal.

21.10 Problems

1. Express the delta function in terms of the ramp function $x(t) = t$.

   Answer: $\delta(t) = \dfrac{d^2x(t)}{dt^2}$

2. Express the delta function in terms of the single-sided negative exponential $f(t) = e^{-t}$.

   Answer: $\delta(t) = f(t) + \dfrac{df(t)}{dt}$

3. Express the delta function in terms of the single-sided sine function $f(t) = \sin t$. See sample solution in Fig. 21.7.

   Answer: $\delta(t) = f(t) + \dfrac{d^2f(t)}{dt^2}$

4. One of the delta function definitions was

$$\pi\delta(t) = \lim_{\omega\to\infty} \frac{\sin \omega t}{t}$$

   In this problem we want to show that for $t \neq 0$, this function is *on the average* zero. Start by plotting the function for $\omega = 1$. Then plot it for increasing values of ω. Let's pick the region between $t = \pi$ and $t = 2\pi$ as a sand box. Find the integral of the function, numerically, in this region and prove to yourself that even though the function itself is not zero for all time (in the aforementioned time interval), on the average, and in particular as $\omega \to \infty$ the average is! See sample solution in Figs. 21.8 and 21.9.

5. Prove the sampling property of the derivative of the delta function using integration by parts

$$\int f(t)\frac{d\delta(t)}{dt}dt = f(t)\delta(t) - \int \delta(t)\frac{df(t)}{dt}dt$$

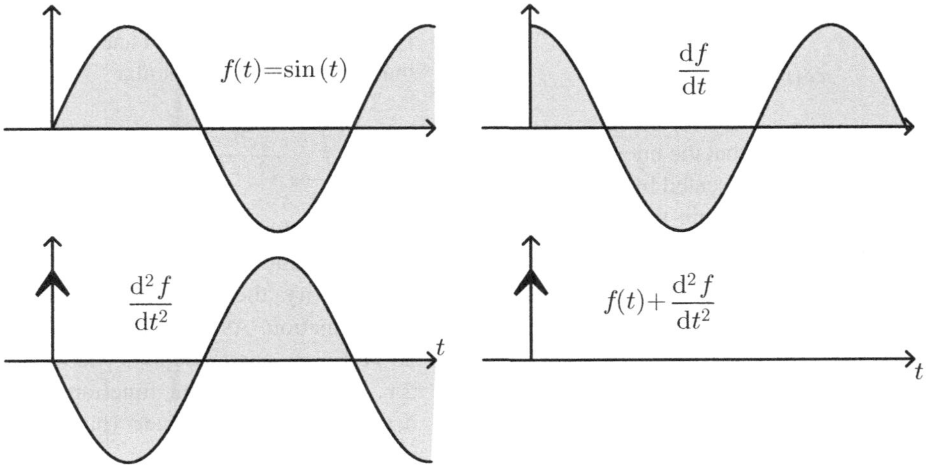

Fig. 21.7 Sample solution to Problem 3

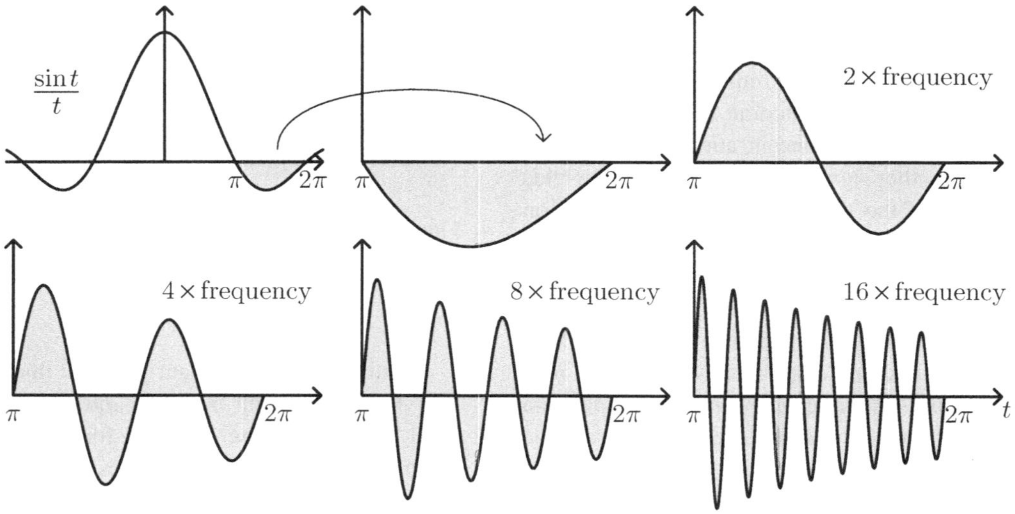

Fig. 21.8 Sample solution (1/2) to Problem 4

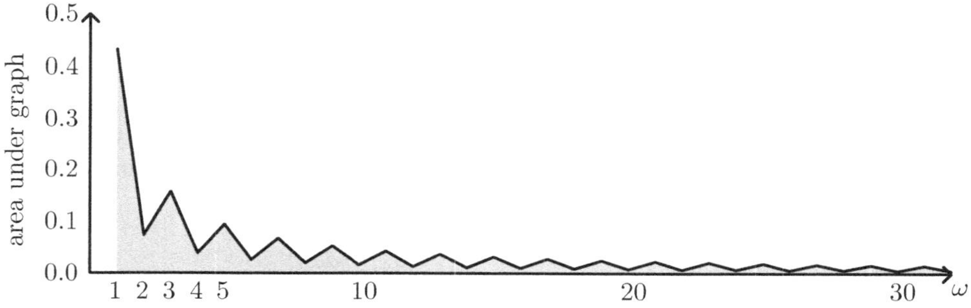

Fig. 21.9 Sample solution (2/2) to Problem 4

6. If the delta function is defined as

$$\pi\delta(t) = \lim_{\omega\to\infty} \frac{\sin \omega t}{t}$$

then it must be a fact that the integral of $\frac{\sin \omega t}{t}$ is the unit step function (scaled by π). Prove this by numerically integrating the sinc function; try a couple of ω values, such that the delta function is localized. See sample solution in Fig. 21.10.

7. If the delta function is defined as

$$\pi\delta(t) = \lim_{\sigma\to 0} \frac{\sigma}{t^2+\sigma^2}$$

then it must be a fact that the integral of $\frac{\sigma}{t^2+\sigma^2}$ is the unit step function (scaled by π). Show that this is true; in particular

$$\int_{-\infty}^{t} \frac{\sigma}{\tau^2+\sigma^2} d\tau = \frac{\pi}{2} + \text{atan}\frac{t}{\sigma}$$

Plot this function for a few values of σ. See sample solution in Fig. 21.11.

8. Let's verify the sampling property of the delta function. Assume the sampled function is $f(t) = \sin t$, defined between 0 and 2π. Assume the delta function is given by $\delta(t) = \frac{1}{\pi}\frac{\sin 40t}{t}$. Evaluate (numerically) the integral

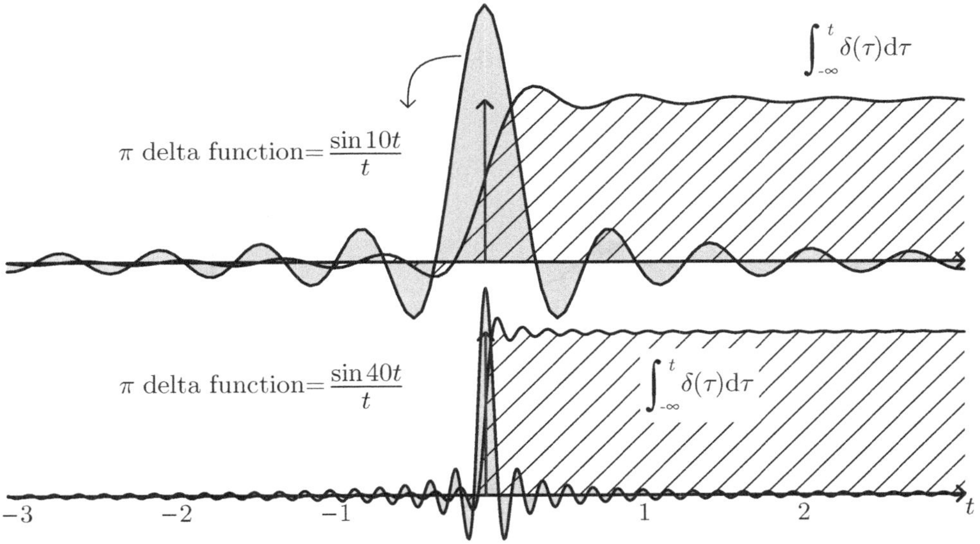

Fig. 21.10 Sample solution to Problem 6. Delta functions not to scale

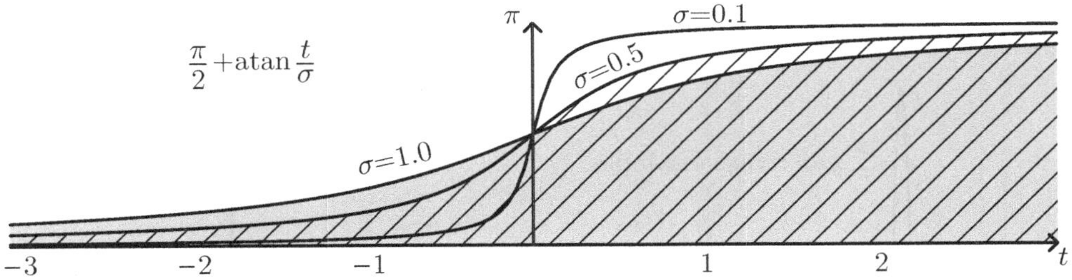

Fig. 21.11 Sample solution to Problem 7

$$I(t) = \int_0^{2\pi} f(\tau)\delta(t-\tau)d\tau$$

for a 10 t-values. Plot the results atop of the sampled function. See sample solution in Fig. 21.12.

9. Let's verify the sampling property of the *derivative* of the delta function. Assume the sampled function is $f(t) = \sin t$, defined between 0 and 2π. Assume the delta function is given by $\delta(t) = \frac{1}{\pi}\frac{\sin 40t}{t}$. First find the derivative of the delta function, analytically. Then evaluate (numerically) the integral

$$I(t) = \int_0^{2\pi} f(\tau)\frac{d}{d\tau}\delta(t-\tau)d\tau$$

for a 10 t-values. Plot the results atop of the sampled function. See sample solution in Fig. 21.13.

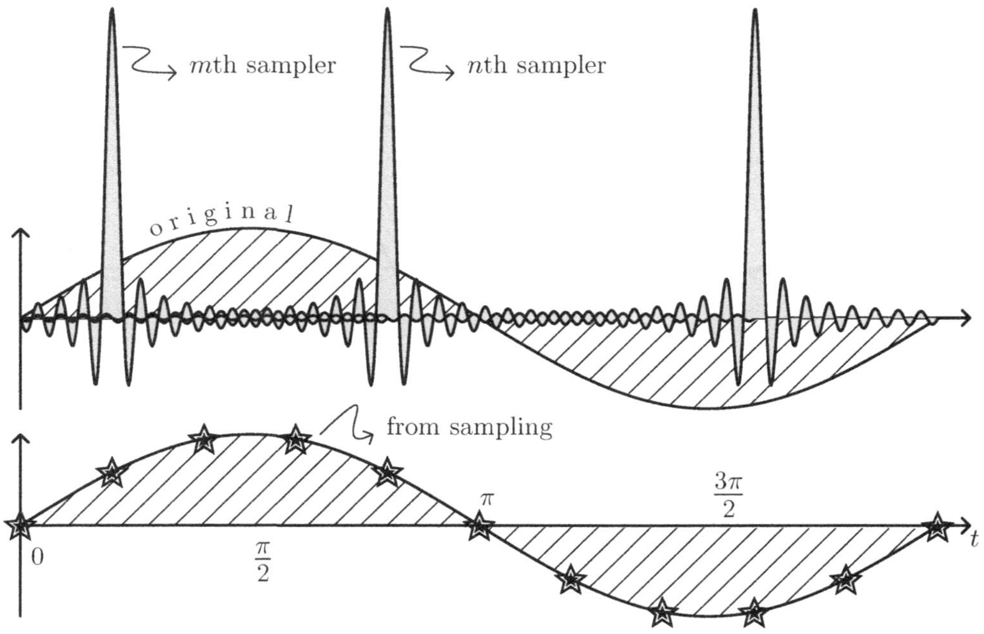

Fig. 21.12 Sample solution to Problem 8

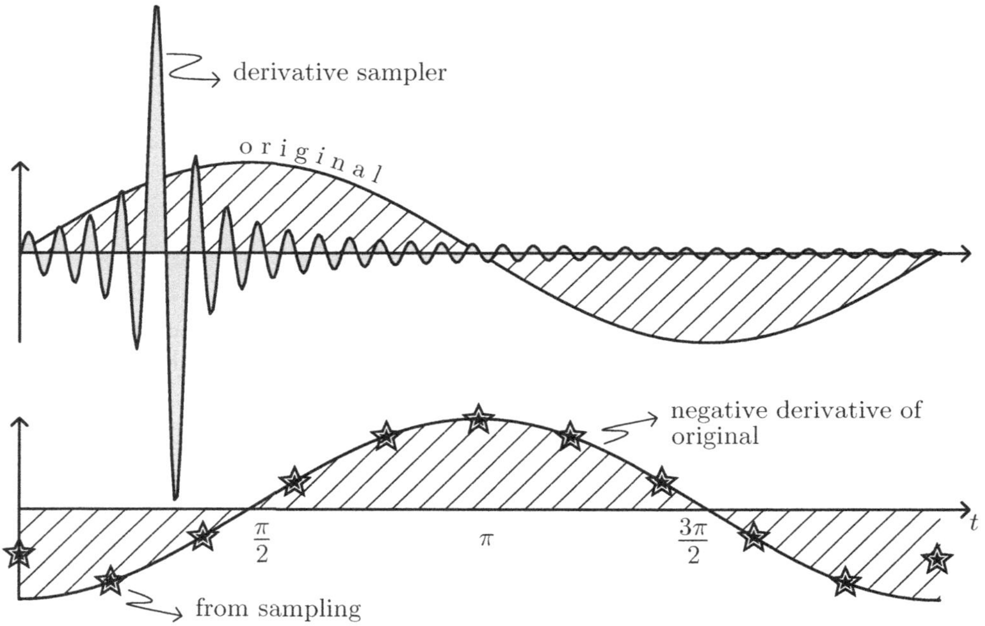

Fig. 21.13 Sample solution to Problem 9

22 Impulse Response

22.1 Introduction

Having covered the delta (impulse) function in some detail in the last chapter we move a step forward to the impulse *response*. The impulse response is extremely important, in circuit theory, signal processing, mechanics, dynamics, and many other fields. Its concept is very important and its applications are far reaching. Yet many a times it is not well understood, and viewed vaguely, and its power is not fully harnessed! The impulse response is also intimately tied to frequency analysis; after all, it is the inverse transform of the transfer function! In this chapter we take a look at the impulse response completely independent of frequency analysis; later we tie them together.

22.2 The Impulse Response as Limiting Case of Pulse Response

The impulse response—as the name implies—is the response of the network (or system) to an abrupt, infinitely high stimulus. Right there one is tempted to discard this idea, as we don't like to deal with infinite concepts! The trick is even though the impulse is infinite in height, it is infinitely narrow in duration; the end result is that the *area* under the impulse is finite—and namely unity. Again, if we were to take the impulse, multiply its height by its width we get a finite number, that of one. As was shown in the last chapter there are many methods to generate an impulse function, ranging from negative exponentials to sinc functions. For ease of applications we choose the pulse method. This then suggests that we can tackle the impulse response by first deriving the response to a finite pulse—of equal area—and then incrementally decreasing the pulse width, while at the same time increasing the pulse height, so in the end the area under the curve remains the same. The premise is that what really matters is the area under the input curve; so long that is equal the response will most likely come out equal. In essence the area under the curve is the "energy" applied, and for the same amount of applied energy we should expect the same output. Let us then take a few representative circuits, figure their pulse response, then take the limit as pulse width goes to zero, height to infinity while product remains unity.

22.3 First Example: Parallel *RC* Network Voltage Response Due to Impulse Current

As a demonstration of deriving the impulse response, let's take the parallel *RC* network shown in Fig. 22.1 with input as current and output as voltage. We are set to derive the impulse

F. Badrieh, *Spectral, Convolution and Numerical Techniques in Circuit Theory*,
https://doi.org/10.1007/978-3-319-71437-0_22

Fig. 22.1 Parallel RC network driven by pulse

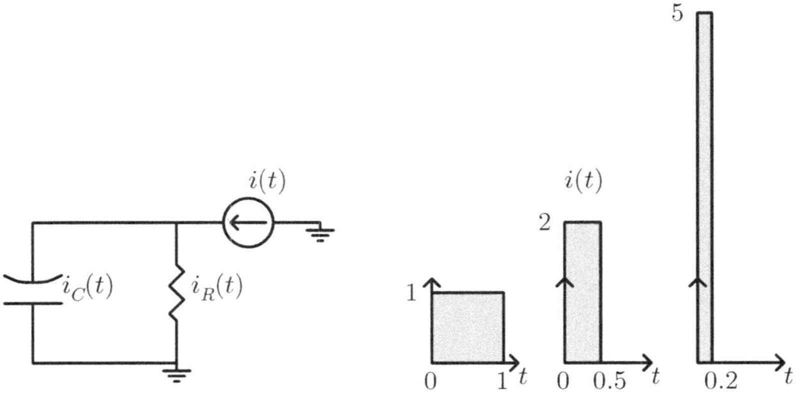

response. First let's find the response to a pulse, of unity height and unity width. That is, input is

$$i(t) = \begin{cases} 1 & 0 < t < 1 \\ 0 & t > 1 \end{cases} \tag{22.1}$$

We know the voltage across the resistor is

$$v_R = i_R R \tag{22.2}$$

and that across the cap is

$$v_C = \frac{1}{C} \int_t i_C dt \tag{22.3}$$

Since both nodes are in parallel we should expect both voltages to be equal:

$$v = v_R = v_C \tag{22.4}$$

so that

$$i_R R = \frac{1}{C} \int_t i_C dt \tag{22.5}$$

Furthermore, we know that the sum of both currents must equal the input current; that is

$$i_R + i_C = i(t) \tag{22.6}$$

Putting this back into the voltage equation we get

$$i_R R = \frac{1}{C} \int (i - i_R) dt \tag{22.7}$$

If we take the time derivative of both sides we get

$$\frac{di_R}{dt} = \frac{1}{RC} [i - i_R] \tag{22.8}$$

For the pulse current at hand, which is 1 between 0 and 1, we then have

$$\frac{di_R}{dt} = \frac{1}{RC} [1 - i_R], \qquad 0 < t < 1 \tag{22.9}$$

Rearranging we get

$$\frac{di_R}{dt} + \frac{1}{RC} i_R = \frac{1}{RC} \tag{22.10}$$

A particular solution for this is

$$i_{R.\text{particular}}(t) = 1 \tag{22.11}$$

The homogeneous solution satisying Eq. (22.10) with zero driving function is

$$i_{R.\text{homogeneous}}(t) = Ae^{-t/RC} \tag{22.12}$$

The total solution is the same for both solutions

$$i_R(t) = 1 + A^{-t/RC} \tag{22.13}$$

This solution does not yet satisfy the initial conditions. To satisfy the initial condition of $i(0) = 0$, A has to be -1 and we finally arrive at the total solution which also satisfies initial conditions.

$$\boxed{i_R(t) = 1 - e^{-t/RC}, \qquad 0 < t < 1} \tag{22.14}$$

Hence the voltage across the resistor (and the cap) is simply

$$\boxed{v(t) = R\left[1 - e^{-t/RC}\right], \qquad 0 < t < 1} \tag{22.15}$$

For time $t > 1$ input current drops to zero and our differential equations reduce to

$$\frac{di_R}{dt} = -\frac{i_R}{RC} \tag{22.16}$$

A solution of this gives us

$$i_R(t) = Ae^{-t/RC} \tag{22.17}$$

To find the initial conditions we simply evaluate the current in Eq. (22.14) at time $t = 1$

$$i_R(1) = 1 - e^{-1/RC} \tag{22.18}$$

Applying initial conditions we get

$$1 - e^{-1/RC} = Ae^{-1/RC}, \qquad \text{which implies} \tag{22.19}$$

$$A = e^{1/RC} - 1, \qquad \text{and the current coming out at} \tag{22.20}$$

$$i_R(t) = \left(e^{1/RC} - 1\right)e^{-t/RC}, \qquad t > 1 \tag{22.21}$$

Finally the voltage is current scaled by R:

$$\boxed{v(t) = R\left(e^{1/RC} - 1\right)e^{-t/RC}, \qquad t > 1} \tag{22.22}$$

A plot of the above results is shown in Fig. 22.2. Along with the derived voltage we show what we expect as the true impulse response, labeled "impulse response."

Limit of Narrow and High Pulse Next we successively decrease the input current pulse width, while at the same time increase its height, again maintaining the same area. Figures 22.3, 22.4, and 22.5 show the results. We can see a definite trend in output response; in particular

- All pulse responses look the same at large enough time; this is expected, since viewed from large enough time, all cases have same "energy" in them, in terms of equal area under curve.
- As the pulse approaches an impulse, the overall response still looks the same, though the very initial response becomes more aggressive.
- In the limit as pulse width approaches zero, and height approaches infinity, the pulse response becomes the impulse response which is identically

$$v(t) = \frac{1}{C}e^{-t/RC} \tag{22.23}$$

22.4 Formal Derivation of Impulse Response (for Parallel *RC* Network)

We are set to prove that the impulse response of the parallel *RC* network is

$$v(t) = \frac{1}{C}e^{-t/RC} \tag{22.24}$$

We can rigorously derive this by starting with the couple of equations derived above, namely Eqs. (22.15) and (22.22). In general for a pulse of width t_0 those equations can be rewritten as response *during* the pulse

$$\boxed{v_1(t) = \frac{R}{t_0}\left[1 - e^{-t/RC}\right], \qquad 0 < t < t_0} \tag{22.25}$$

and response *after* the pulse

$$\boxed{v_2(t) = \frac{R}{t_0}\left(e^{t_0/RC} - 1\right)e^{-t/RC}, \qquad t > t_0} \tag{22.26}$$

As we make the pulse narrower and narrower, by setting t_0 smaller and smaller, the first part of response becomes narrower and narrower since it is defined only within t_0. To see the limiting

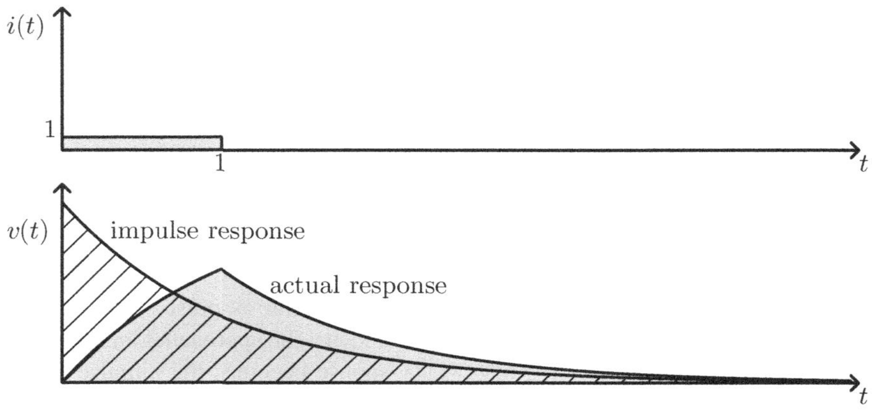

Fig. 22.2 Input current and output voltage of a pulse of 1 s duration

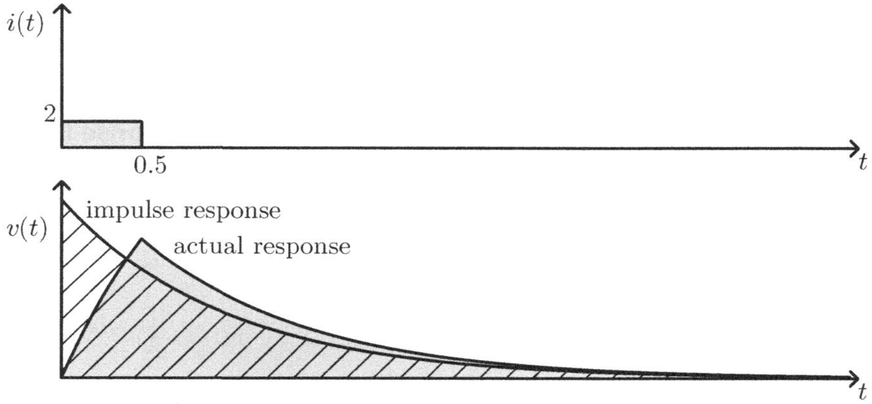

Fig. 22.3 Input current and output voltage of a pulse of 0.5 s duration

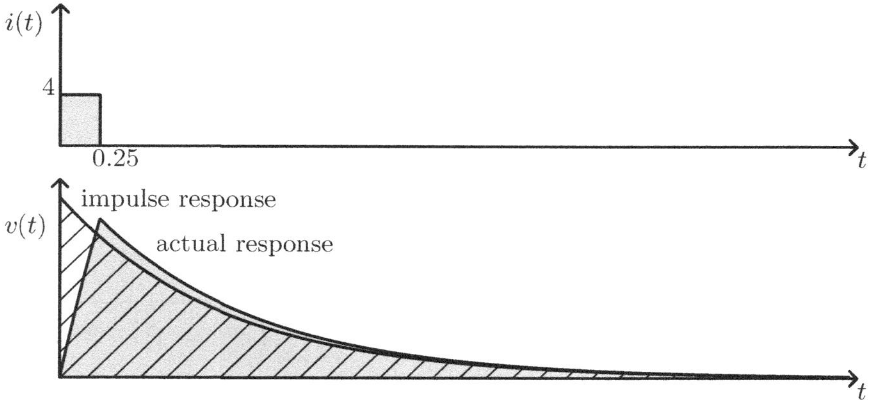

Fig. 22.4 Input current and output voltage of a pulse of 0.25 s duration

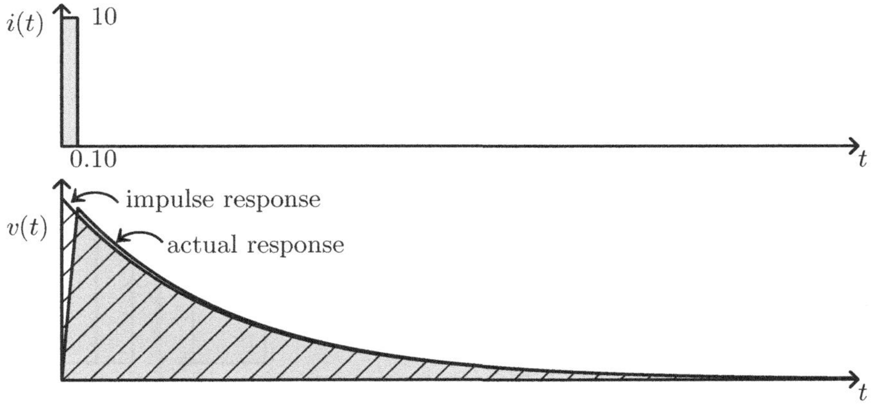

Fig. 22.5 Input current and output voltage of a pulse of 0.1 s duration

trend we expand the exponential by the first two terms and rewrite as

$$v_1(t) = \frac{R}{t_0}\left[1 - 1 + \frac{t}{RC}\right], \qquad 0 < t < t_0$$
$$= \frac{1}{C}\frac{t}{t_0}, \qquad 0 < t < t_0 \qquad (22.27)$$

We can see that by decreasing t_0 the first term approaches

$$v_1(t_0) = \frac{1}{C} \qquad (22.28)$$

Next we take the second voltage term and expand the $e^{t_0/RC}$ since t_0 is small. We get

$$v_2(t) = \frac{R}{t_0}\left(1 + \frac{t_0}{RC} - 1\right)e^{-t/RC}$$
$$= \frac{1}{C}e^{-t/RC} \qquad (22.29)$$

Combining Eqs. (22.28) and (22.29) we do indeed arrive at Eq. (22.24)!

22.5 Second Example: Series *RC* Network Voltage Response Due to Impulse Current

As another demonstration of the impulse response, let's take the series *RC* network shown in Fig. 22.6 and apply an impulse current to it. The resulting voltage is what we call the impulse response. Similar to before, instead of an impulse, we start with a pulse and successively decrease the width and increase the height. Assuming the pulse has duration t_0 and height $1/t_0$, such that the area is 1, the voltage across the resistor is simply

$$v_R(t) = \begin{cases} \frac{R}{t_0} & t < t_0, \\ 0 & t > t_0 \end{cases} \qquad (22.30)$$

The voltage across the cap is related to the integral of the same current, namely

$$v_C = \frac{1}{C}\int_0^t i(\tau)d\tau$$
$$= \begin{cases} \frac{1}{Ct_0}t & t < t_0 \\ \frac{1}{C} & t > t_0 \end{cases} \qquad (22.31)$$

The output voltage is simply the sum of both voltages

$$\boxed{v(t) = \begin{cases} \frac{R}{t_0} + \frac{t}{Ct_0} & t < t_0, \\ \frac{1}{C} & t > t_0 \end{cases}} \qquad (22.32)$$

Notice that while the voltage across the resistor goes to zero after the current shuts off, the voltage across the cap does not! What happens is that the cap has charged to a nonzero value, namely $\frac{1}{C}$ times total charge, which in this case equals unity (since the integral of the impulse current is one). Figures 22.7, 22.8, 22.9, and 22.10 show input current, resistor voltage, cap voltage, and total voltage for successively nar-

Fig. 22.6 Series *RC* network driven by pulse

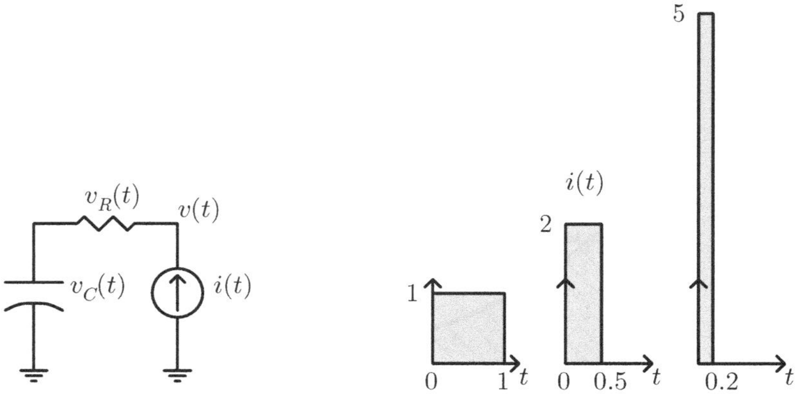

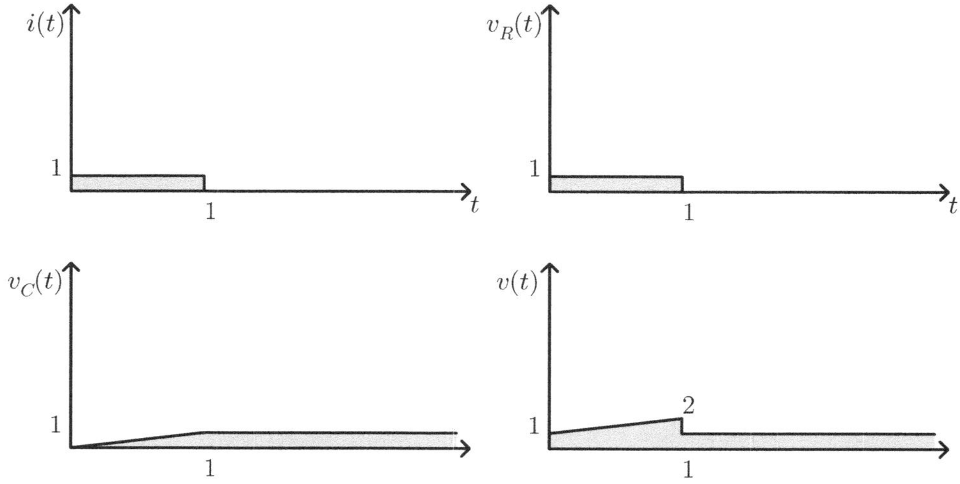

Fig. 22.7 Series *RC* input current and output voltage of a pulse of 1.0 s duration

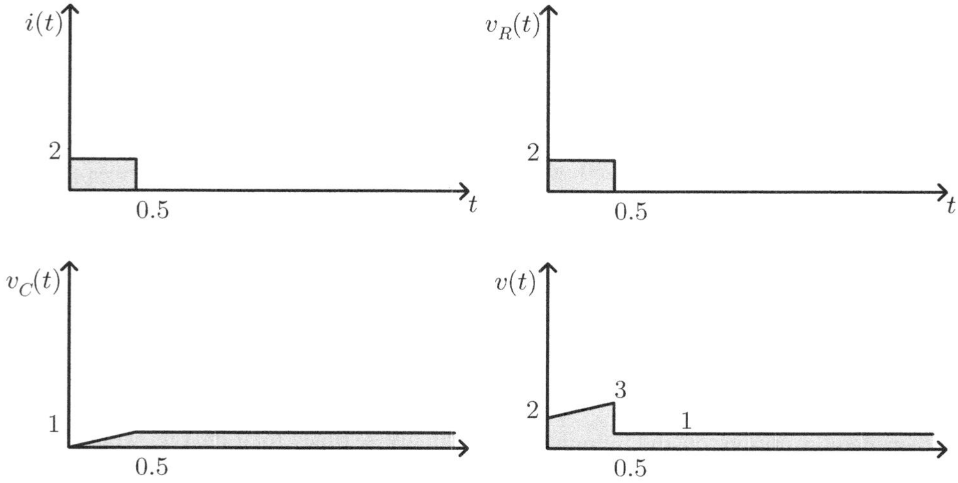

Fig. 22.8 Series *RC* input current and output voltage of a pulse of 0.5 s duration

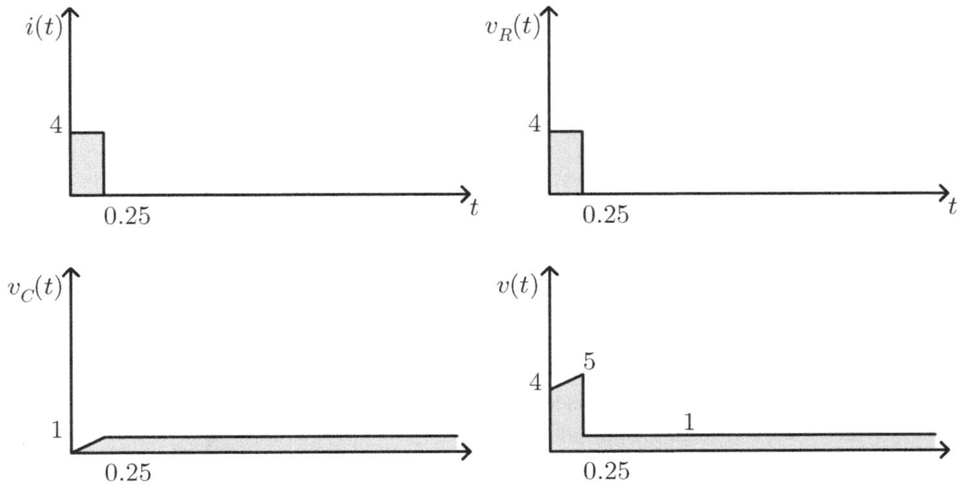

Fig. 22.9 Series *RC* Input current and output voltage of a pulse of 0.25 s duration

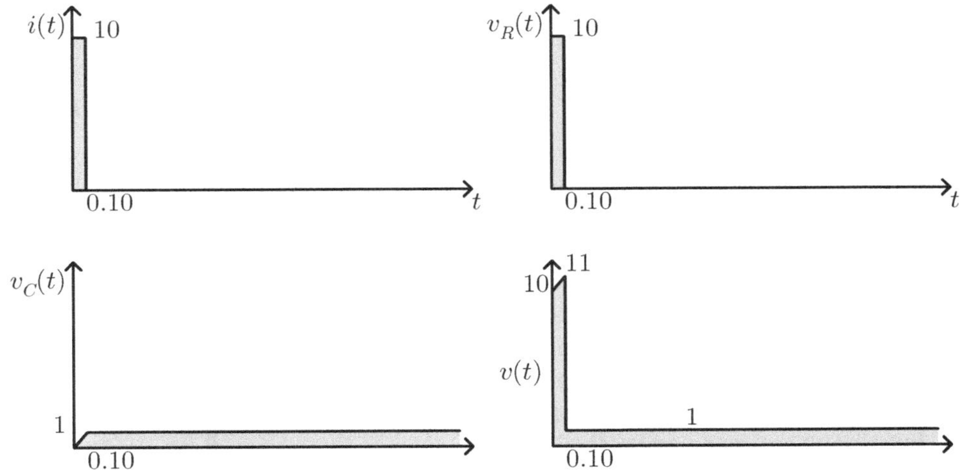

Fig. 22.10 Series *RC* input current and output voltage of a pulse of 0.1 s duration

rower pulse widths, starting from 1 to 0.1 s. Notice that independent of pulse width, the settling voltage of the resistor is zero while that of the cap is unity (divided by cap value). Notice also that voltage across the resistor scales linearly with the pulse height. From the figures, but also from the above equation we can arrive at the conclusion that in the limit as the pulse width t_0 approaches zero, the impulse response will be

$$\boxed{v(t) = R\delta(t) + \frac{1}{C}u(t)} \qquad (22.33)$$

In essence, we have a delta function across the resistor since it absorbs the delta function of the current. The voltage across the resistor dies off to zero immediately after the current ceases. On the other hand, the cap charges immediately to $1/C$ since voltage across cap is charge divided by cap, and since charge is time integral of current (which is unity here). Having charged immediately, the cap voltage stays there for all future time. Thus we have found the impulse response of the series *RC* network. While it looks a bit unconventional, this response is the primary ingredient for finding all sorts of other responses, including step response, ramp response, and in fact any other response!

22.6 Third Example: High-Pass Voltage Filter

We can also derive the impulse response from a series of manipulations to the unit step response. As a sample we consider the voltage high-pass filter shown in Fig. 22.11 which is comprised of a series RC network where voltage is applied at the input and response measured as voltage across the resistor.

Step Response We know that the unit step response of the high-pass filter is given by

$$v_{o,u}(t) = e^{-t/RC}u(t) \tag{22.34}$$

That is, voltage is zero for negative time and decaying exponential for positive time. In other words, immediately after the application of the step input, the cap shorts and all the voltage (unity) lands at the resistor (which is the output terminal). From there onwards, the cap impedance increases (i.e., low frequency) and starts to absorb the voltage; in the end, all the voltage is across the cap and none across the resistor. Initially we get max current ($1/R$) and gradually that current charges the cap and eventually the current goes to zero, the iR across the resistor is zero, and voltage across cap is max (1 V).

Pulse Response We can form the pulse response from the step response by adding a step to a negative, shifted one. Assume we want the pulse to have a width of T and height $1/T$, so that the area is one; then

$$p(t) = \frac{1}{T}\left[u(t) - u(t-T)\right] \tag{22.35}$$

Via superposition and linearity we can conclude that the pulse response would be

$$\begin{aligned} v_{o,p}(t) &= \frac{1}{T}\left[v_{o,u}(t) - v_u(t-T)\right] \\ &= \frac{1}{T}\left[e^{-t/RC} - u(t-T)e^{-(t-T)/RC}\right] \end{aligned} \tag{22.36}$$

We can split this in terms of two time regimes

$$\begin{aligned} v_{o,p}(t) &= \frac{1}{T}e^{-t/RC}, \qquad 0 < t < T \\ &= \frac{1}{T}e^{-t/RC}\left[1 - e^{T/RC}\right], \qquad t > T \end{aligned} \tag{22.37}$$

Limit of High Pulse Now we take the limit as $T \to 0$. We notice that the two time regimes behave as follows:

- $0 < t < T$: This response $1/Te^{-t/RC}$ tends to blow up since we have T in the denominator; but how about its area?

Fig. 22.11 High-pass filter and impulse response

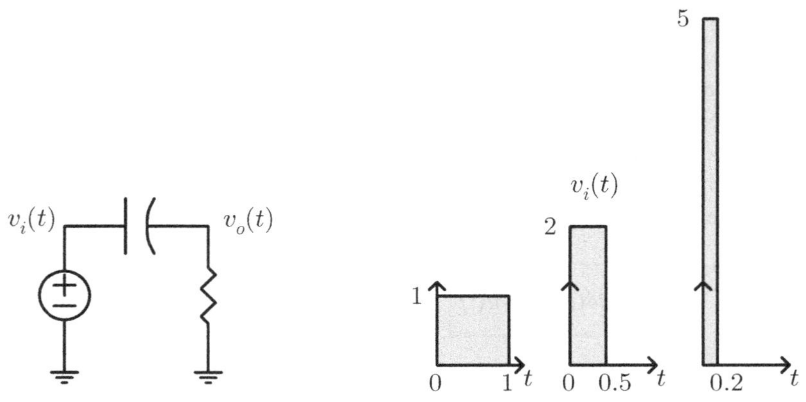

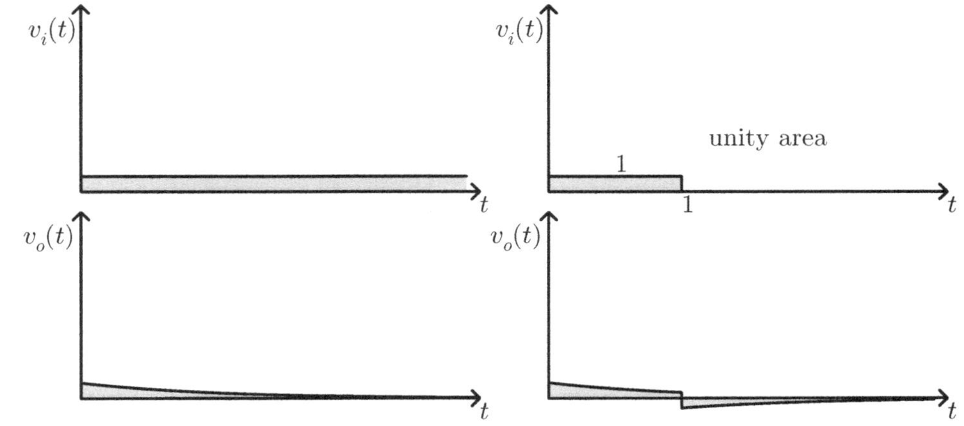

Fig. 22.12 Step and pulse response: top—input; bottom—output

$$\begin{aligned}\frac{1}{T}\int_0^T e^{-t/RC}dt &= -\frac{1}{T}RC\left[e^{-t/RC}\right]_0^T \\ &= -\frac{1}{T}RC\left[e^{-T/RC}-1\right] \\ &= -\frac{1}{T}RC\left[1-\frac{T}{RC}-1\right], \; (T << 1) \\ &= 1 \qquad (22.38)\end{aligned}$$

That is we have a function that blows up at time zero but whose area is unity!! This is nothing but the delta function; hence

$$v_{o,p}(t) = \boxed{\delta(t)}, \qquad \text{(exactly at time zero)} \qquad (22.39)$$

- $t > T$: As shown above the response here is

$$v_{o,p}(t) = \frac{1}{T}e^{-t/RC}\left[1-e^{T/RC}\right], \quad 0 < t < T \qquad (22.40)$$

Now we take the limit of $T \to 0$

$$\begin{aligned}\lim_{T\to 0} v_{o,p}(t) &= \frac{1}{T}e^{-t/RC}\left[1-1-\frac{T}{RC}\right] \\ &= \boxed{-\frac{1}{RC}e^{-t/RC}}, \quad \text{(after time zero)}\end{aligned} \qquad (22.41)$$

Hence our total solution (due to *impulse* current) is

$$\boxed{v(t) = \delta(t) - \frac{1}{RC}e^{-t/RC}} \qquad (22.42)$$

This is our impulse response to the high-frequency pass filter. We can rationalize this as follows. Immediately while the impulse is applied, all the voltage gets applied across the (out terminal) resistor since the cap acts like a short. Hence we see the delta function. After the input turns to ground, we now have an impulse current of magnitude $1/R$ that would be applied across the cap, whose other terminal is grounded. We know that an impulse current applied to a *parallel* RC network would cause a negative exponential response in the sense the cap would initially charge to $1/C$ (times total charge ($1/R$ here)) and from there onwards would discharge at the RC rate. But notice that current would be going in opposite direction (away from output) and hence we have the negative sign in the total solution. That is, right after the delta passed we have a negative sign, times $1/RC$ times the negative exponential.

Figure 22.12 shows step and pulse responses. Figure 22.13 shows how pulse response converges to impulse one, as pulse width is diminished and height increased.

22.7 Fourth Example: Low-Pass Voltage Filter

The low-pass voltage filter is comprised of a series RC with input applied in the form of voltage and output (again in form of voltage) taken across

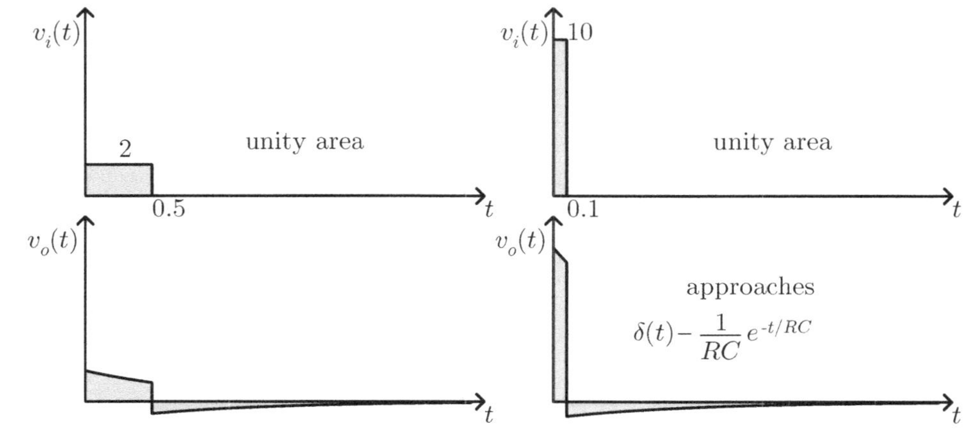

Fig. 22.13 Narrow pulse and impulse response: top—input; bottom—output

Fig. 22.14 Low-pass filter and impulse response

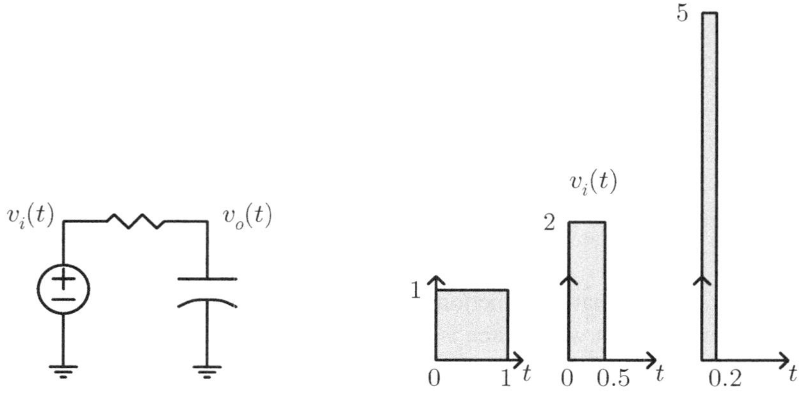

the cap, as shown in Fig. 22.14. We start with the unit step response where a unit step input voltage is applied and output voltage is measured as

$$v_{o,u}(t) = 1 - e^{-t/RC} \tag{22.43}$$

That is, right at time zero the voltage across the cap is zero, since it had no time to charge. Equivalently we say all the voltage is across the resistor, since impedance of cap at high frequency (which is equivalent at instant of turn on) is zero. As time goes by the cap charges and final steady state voltage goes to one. There, current goes to zero. By virtue of current being zero, voltage across the resistor goes to zero too, hence leaving all the input voltage (unity) across the cap.

Next we form the pulse response. We define a pulse of width T and height $1/T$ such that the area is unity. We define the pulse in terms of two unit steps, shifted by T and each scaled by $1/T$. By virtue of linearity and superposition the pulse response would be (for time less than T)

$$v_{o,p}(t) = \frac{1}{T}\left[1 - e^{-t/RC}\right], \quad < 0 < t < T \tag{22.44}$$

and for time greater than T

$$\begin{aligned} v_{o,p}(t) &= \frac{1}{T}\left[1 - e^{-t/RC} - \left(1 - e^{-(t-T)/RC}\right)\right], \quad t > T \\ &= \frac{-1}{T}e^{-t/RC}\left[1 - e^{T/RC}\right], \quad t > T \end{aligned} \tag{22.45}$$

Now we take the limit as $T \to 0$ for the two time regimes

- $0 < t < T$: here and for very small time we can approximate the $e^{-t/RC}$ by $1 - \frac{t}{RC}$; then we have

$$v_{o,p}(t) = \frac{1}{T}\left[1 - 1 + \frac{t}{RC}\right], \qquad 0 < t < T$$
$$= \frac{t}{T}\frac{1}{RC} \qquad (22.46)$$

To first order we can replace t by T, since as we narrow the pulse width, the pulse width becomes comparable to small time. Then we get

$$v_{o,p}(t) = \boxed{\frac{1}{RC}}, \qquad 0 < t < T \quad (22.47)$$

- $t > T$:

Here we expand the $e^{T/RC}$ as $1 + \frac{T}{RC}$ and get

$$v_{o,p}(t) = \frac{-1}{T}e^{-t/RC}\left[1 - \left(1 + \frac{T}{RC}\right)\right], \qquad t > T$$
$$= \boxed{\frac{1}{RC}e^{-t/RC}} \qquad (22.48)$$

Then our total solution as we set the limit $T \to 0$ becomes

$$\boxed{v(t) = \frac{1}{RC}e^{-t/RC}} \qquad (22.49)$$

That is, immediately as the impulse is applied all the impulse lands across the resistor (infinite amount) but a tiny portion ($1/RC$) lands across the cap. That is, since the cap is very low impedance at high frequency it assumes only a tiny portion ($1/RC$) of the impulse input. Notice that while this amount is tiny, it is not zero! After the impulse passes the cap would simply discharge through the R via the RC time constant.

Notice as a byproduct that immediately after time zero the voltage across the cap $1/RC$ is proportional to the inverse of R and C. That is, the larger the R, the less that cap voltage which makes sense by sheer impedance voltage division. Similarly the larger the C, and hence lower the cap impedance, the lower the initial cap voltage. Finally notice that both solutions $0 < t < T$ and $t > T$ give the same value $1/RC$ at time zero; hence we have continuity.

22.8 Summary

In this chapter we learned how to derive the impulse response from the unit step response, or from the pulse one. The process is as simple as forming a pulse, finding its response, then continuously shrinking the pulse width, and increasing its height, while at the same time keeping pulse area unity. We could also derive the impulse response by first figuring the step response and then taking its time derivative; see Problems section for examples on this. The impulse response is a cornerstone in the convolution world in the sense that finding the impulse response is the first step in finding the response to any stimulus, via the mechanics of convolution. The next chapter will show how to do just that. The impulse itself is quiet attractive because its frequency spectrum is constant—that is, the Fourier transform of the impulse is unity. Hence, it simplifies the frequency analysis, which is a plus.

22.9 Problems

1. Show that taking the response of a pulse, of width T, and height $1/T$ really is equivalent to taking the derivative of the step function response, especially as $T \to 0$.
2. Consider the RC network shown in Fig. 22.15. The input current is of the form of a unit step function. Find the corresponding step current response across the output cap, C_1. Next find step output voltage response. Lastly find the *impulse* output voltage response by taking the time derivative of this latter one. Plot results and compare to SPICE. Assume $R = 1$, $C_1 = 2$, and $C_2 = 5$. What is the impulse response for the two limits $C_2 = \infty$ and $C_2 = 0$? See sample solution in Fig. 22.16.

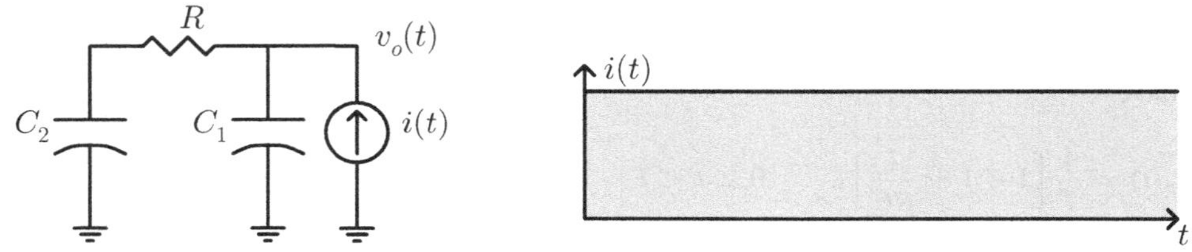

Fig. 22.15 Schematics for Problem 2

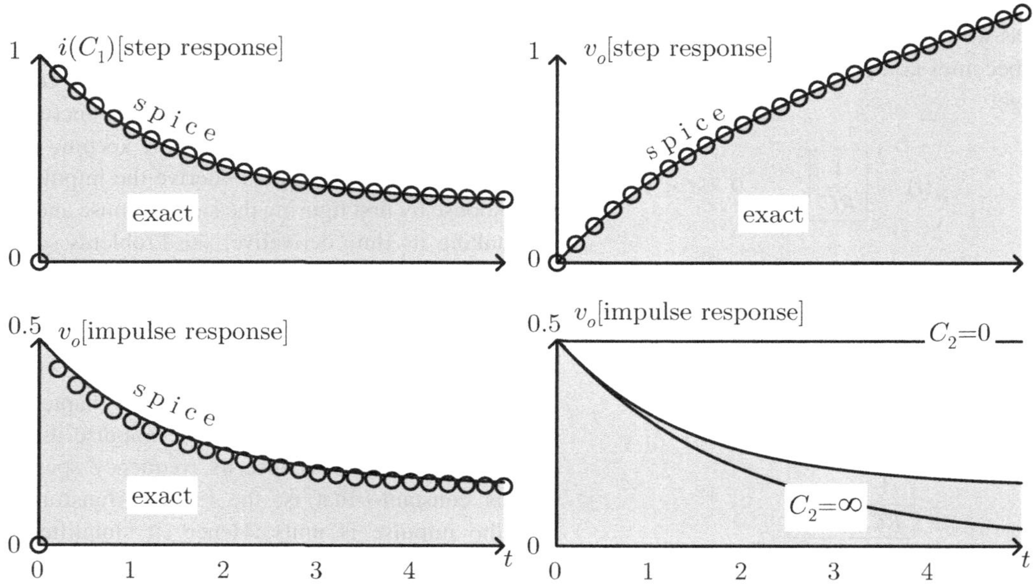

Fig. 22.16 Sample solution to Problem 2

$$\text{Answer: } i(t) = \left[1 - \frac{C_1}{C_1 + C_2}\right] e^{-t/(RC||)} + \frac{C_1}{C_1 + C_2}, \qquad C|| = \frac{C_1 C_2}{C_1 + C_2}$$

$$v_o(t)[\text{step}] = \frac{1}{C_1}\left\{RC|| \times \left[1 - \frac{C_1}{C_1 + C_2}\right]\left[1 - e^{-t/(RC||)}\right] + \frac{C_1}{C_1 + C_2} t\right\}$$

$$v_o(t)[\text{impulse response}] = \frac{1}{C_1}\left\{\left[1 - \frac{C_1}{C_1 + C_2}\right] e^{-t/(RC||)} + \frac{C_1}{C_1 + C_2}\right\}$$

---

3. Consider the *LC* network in Fig. 22.17 subject to pulse input current (width 2, height 0.5). Find the cap current, and then output voltage. Next, change pulse width to 0.5 and height 2, followed by pulse width 0.1 and pulse height 10. What is the limit as pulse width approaches 0 and height ∞? Use $C = 1$ and $L = 1$ as a sample. See sample solution in Fig. 22.18.

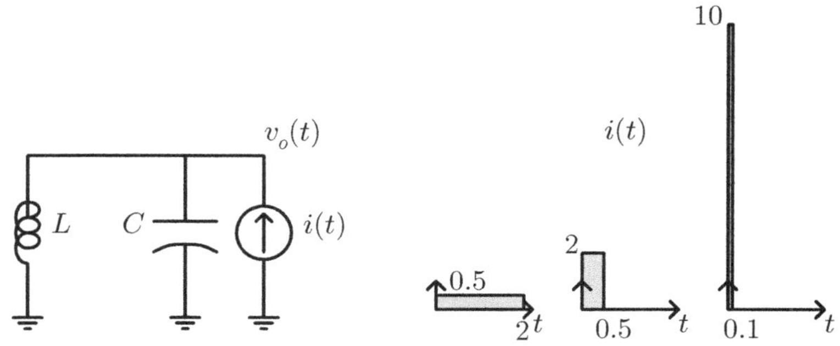

Fig. 22.17 Schematics for Problem 3

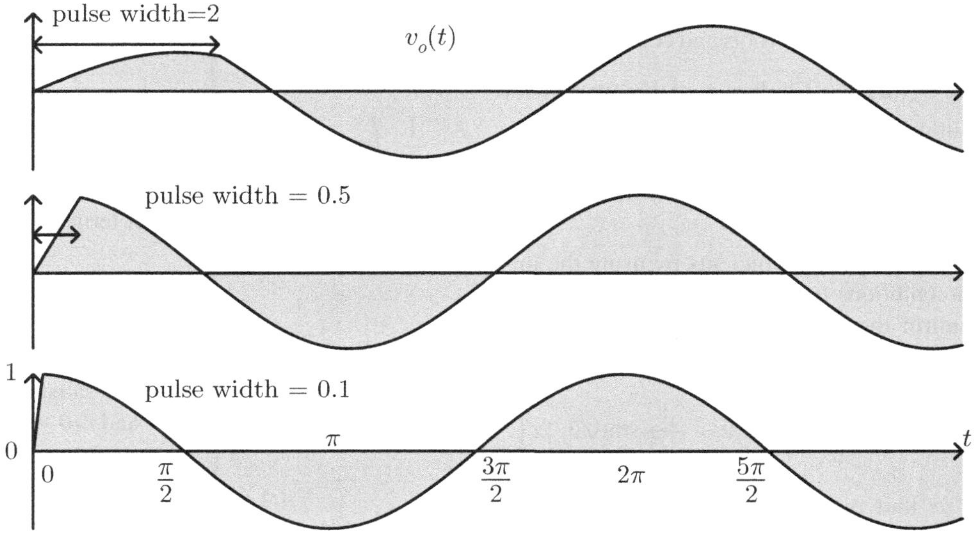

Fig. 22.18 Sample solution for Problem 3

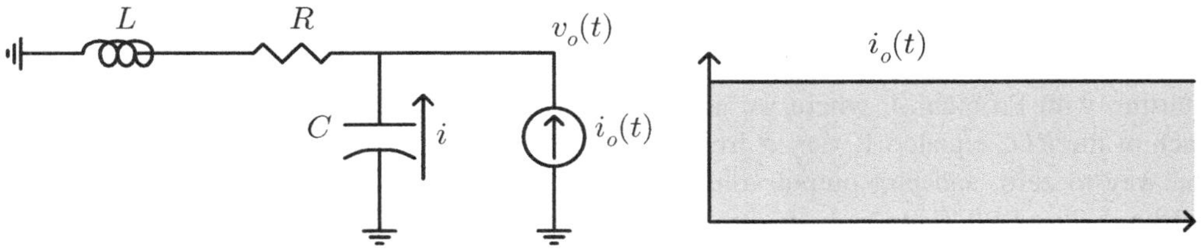

Fig. 22.19 Schematics for Problem 4

4. Consider the *RLC* network shown in Fig. 22.19. A unit step current is applied and we want to find the impulse (voltage) response. Assume for simplicity that each of the *RLC* is 1. First, setup KVL around the *RLC* and arrive at the differential equation for the cap current:

$$i + \frac{di}{dt} + \frac{d^2i}{dt^2} = 0$$

Next assume a solution of the form e^{st} and arrive at the requirement

$$1 + s + s^2 = 0$$

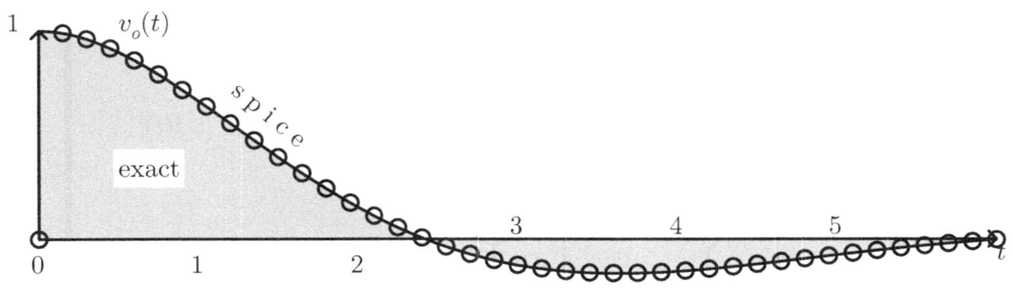

Fig. 22.20 Sample solution for Problem 4

Solve for the two roots and confirm they are $s = -\frac{1}{2} \pm j\frac{\sqrt{3}}{2}$. Confirm next that the general solution assumes the form

$$i(t) = e^{-t/2}\left[A\cos\sqrt{3}/2t + B\sin\sqrt{3}/2t\right]$$

Next find the coefficients by using the initial conditions $i(0) = 1$ and $di/di|_{t=0+} = 0$. Confirm then that the final solution for cap current is

$$i(t) = e^{-t/2}\left[\cos\sqrt{3}/2t + \frac{1}{\sqrt{3}}\sin\sqrt{3}/2t\right]$$

To find output voltage we integrate this current and divide by cap (assumed 1 here); then to find impulse response we differentiate this; so really in the end $v_o(t) = i(t)$! Plot the solution and compare to SPICE. See sample solution in Fig. 22.20.

5. Starting with Problem 4, where we assumed each of the *RLC* equaled 1, vary *R* from 1 all the way to zero, and plot output voltage accordingly; use SPICE for analysis. See sample solution in Fig. 22.21.
6. Consider the *LC* network in Fig. 22.22, where $C_1 = C_2 = 2$ and $L = 1$. A unit step current is applied; find current through C_1, then output voltage. Then take derivative of the latter to find impulse response. Start by deriving KVL as

$$\frac{1}{C_1}\int idt - \frac{1}{C_2}\int[1-i]dt - L\frac{d[1-i]}{dt} = 0$$

Plug in for the *LC*, collect terms, and then take the derivative:

$$i + i'' = \frac{1}{2}$$

Show that the solution to this satisfying the initial conditions $i(0+) = 1$ and $i'(0+) = 0$ is

$$i(t) = \frac{1}{2}\cos t + \frac{1}{2}$$

Integrate and then divide by $C_2 = 2$ to get voltage step response

$$g(t) = \frac{1}{2}\left[\frac{1}{2}\sin t + \frac{t}{2}\right]$$

The impulse response is simply the derivative of this

$$h(t) = \frac{1}{2}\left[\frac{1}{2}\cos t + \frac{1}{2}\right]$$

Plot results and compare to SPICE; see sample solution in Fig. 22.23.

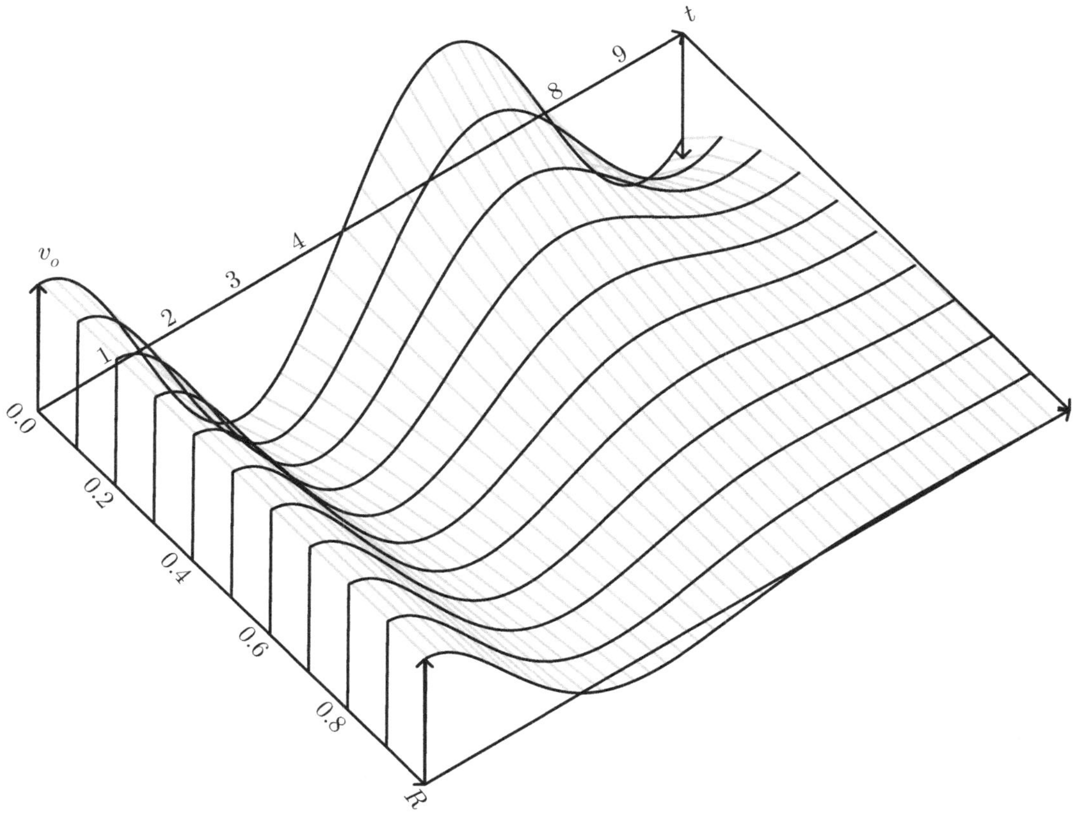

Fig. 22.21 Sample solution for Problem 5

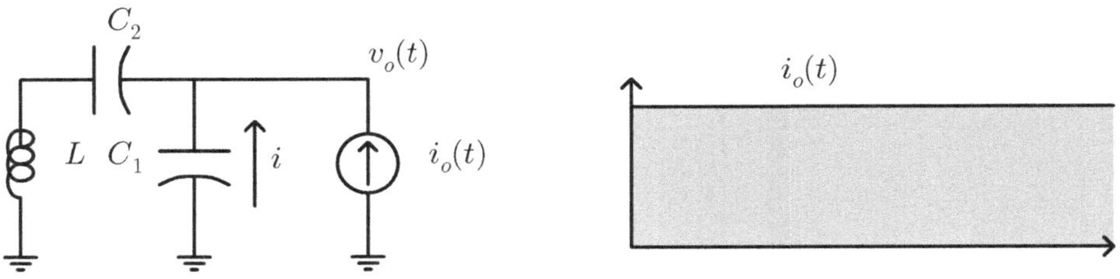

Fig. 22.22 Setup for Problem 6

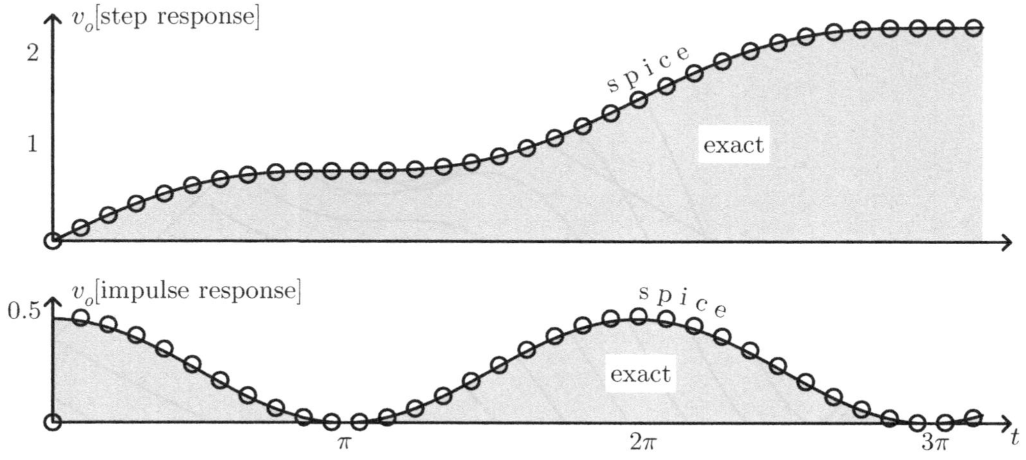

Fig. 22.23 Sample solution for Problem 6

Time Convolution with Impulse Response

23

23.1 Introduction

Having covered convolution and signal construction in terms of the convolution integral, having studied in some detail the delta function and finally having studied impulse response we are now at a point where we can put convolution into real use. In this flow first the impulse response is figured. Then this impulse response is convolved with any input stimuli to yield the system response to that stimulus. In many cases, such as the last chapter, we can derive the impulse response directly; in other cases, however, we may not be able to do so. But assume *somehow* we have access to the impulse response by means of measurement, simulations, or some other means. Once we have the impulse response, how exactly can we use it to predict the response to any other input?

23.2 Main Idea

The main idea of convolution with the impulse response is as follows. We know each signal can be decomposed in terms of impulses. That is, by staging a train of impulses, shifted in time, each with magnitude scaled depending on the original function, we can regain the function. This was covered in Chap. 20. Now if we know the impulse response, can't we simply stage that response (and scale it) to get total response? The answer is yes. To best explain this we will run a few examples.

23.3 Parallel *RC* Network and Unit Step Response

The parallel RC network driven by a current source is shown in Fig. 23.1. We already know the impulse response to be

$$\boxed{h(t) = \frac{1}{C}u(t)e^{-t/RC}} \tag{23.1}$$

What this says is that with the application of an impulse current, with unity area, such that total applied charge is 1, the output voltage initially charges to $1/C$, and later discharge with a time constant RC.

Knowing the impulse response in theory we should be able to predict the response due to any input. How about the unit step input? We can find the answer analytically, or empirically.

F. Badrieh, *Spectral, Convolution and Numerical Techniques in Circuit Theory*,
https://doi.org/10.1007/978-3-319-71437-0_23

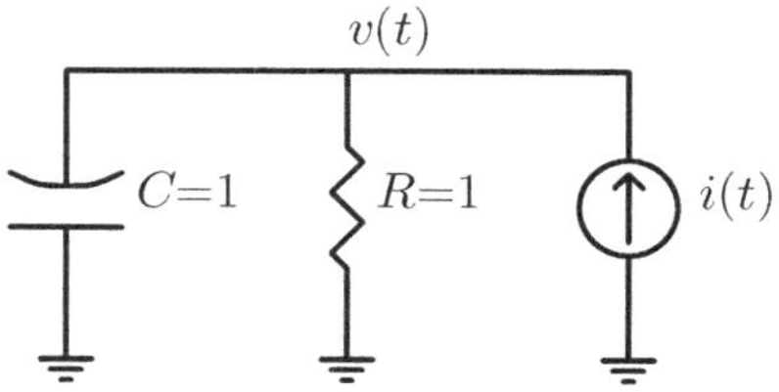

Fig. 23.1 Parallel RC network driven by current source

Analytically the step response is simply the time integral of the impulse response; that is

$$g(t) = \int_0^t h(\tau)d\tau = \int \frac{1}{C}e^{-\tau/RC}d\tau = \boxed{R\left[1 - e^{-t/RC}\right]} \tag{23.2}$$

That is, initially output voltage is zero and gradually it builds up to unity (assuming $R = 1$) with a time constant RC. But *assume for now we are unable to derive the unit step response* for reasons such as not being able to integrate analytically, or because we don't actually have an expression for the impulse response, but we know what it looks like. How would we then proceed? The idea is that a unit step input is really comprised of a sequence of miniature impulses each of strength $1 \times \Delta t$ where Δt is the spacing between the sequence, as shown in Fig. 23.2. Assume for example that we set the spacing to 0.25; then a step input can be built using

$$u(t) = \Delta t \sum_n \delta(t - n\Delta t), \qquad \Delta t = 0.25 \tag{23.3}$$

Notice the Δt in front of the summation. The reason is that each of the pulses in Fig. 23.2 is *not really* a full delta function—instead it is a *scaled* delta function, namely $\Delta t\delta(t)$. That is, the area of each of the pulses is Δt (and not 1). Since we know the response to each delta function (Eq. (23.1)) we would expect the step response to be simply the sum of the impulse functions, spaced out by Δt

$$g(t) = \Delta t \sum_n \frac{1}{C}e^{-(t-n\Delta t)/RC} \tag{23.4}$$

That is, the step response is simply the summation of the impulse response sequence. Figure 23.3 shows how we incrementally build the total response by adding the shifted sequence of impulse responses. Figure 23.4 shows convolution results as compared to exact response, using finer spacing between impulses. As we shrink the impulse width, we better mimic exact results.

Though simply put, the process followed in this example is extremely important and will be repeated many a times for all sorts of examples and problems. Embedded in this simple example are four critical steps:

1. Impulse response is known, by whatever means—analytic, simulated, measured, or fitted!
2. Impulse response is scaled by actual response. That is, impulse response at time t should be scaled by value of stimulus at that exact time.
3. Impulse response is shifted in time. That is, impulse response at time t *starts* at that time and is zero before.
4. The scaled and shifted impulse responses are added together. That is, we need to bookkeep the responses due to all the impulses scaled and shifted as per above.

These steps can be cast succinctly as

$$v(t) = \int_0^t i(\tau)h(t - \tau)d\tau \tag{23.5}$$

where $h(t)$ is the impulse response, $i(t)$ is the input stimulus, and $v(t)$ is the system response.

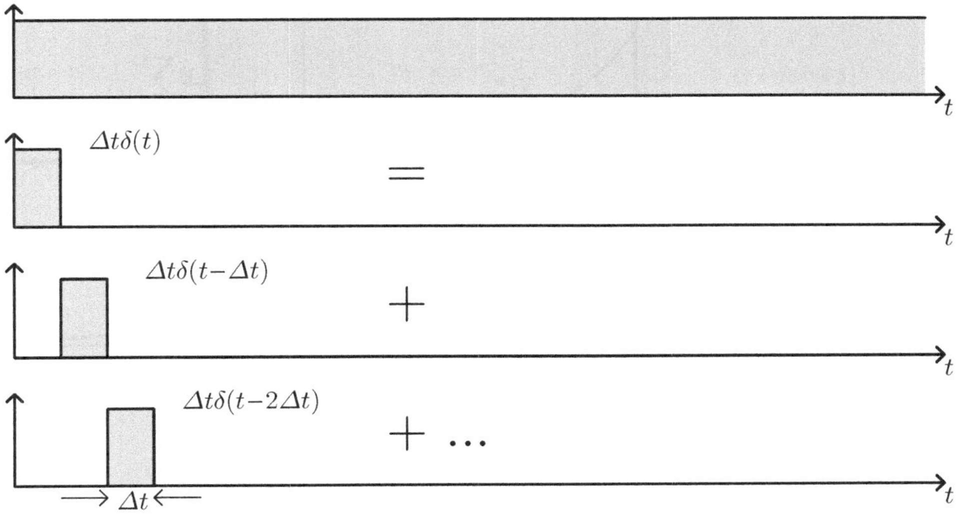

Fig. 23.2 Unit step in terms of sequence of impulse functions

Notice how each impulse response $h(t-\tau)$ is first scaled by the stimulus at that time $i(\tau)$; is shifted by t; and in the end all impulse responses are added (via the integration operation) to yield system response.

23.4 Parallel *RC* Network and Periodic Pulse

As a second example of convolution with the impulse response assume the same parallel *RC* network as above and instead of a unit step apply now a periodic current pulse. Find the corresponding voltage. While the input changed (from unit to periodic pulse), the impulse response did not. *No* new impulse response calculations are needed. All that needs to be done is convolve the impulse response with the periodic pulse.

$$v(t) = \int_0^t i(\tau)h(t-\tau)d\tau \tag{23.6}$$

where again $v(t)$ is output voltage, $i(t)$ is input current, and $h(t)$ is impulse response. The steps are shown in Fig. 23.5. As we make the convolution time step smaller, we get better resolution as shown in Fig. 23.6.

23.5 Parallel *RC* Network and Causal Sinusoid Input

As a third example of convolution with impulse response let's apply a causal sinusoid input current to the same parallel *RC* network. Recall again that the impulse current is

$$h(t) = \frac{1}{C}e^{-t/RC} \tag{23.7}$$

The new input current is

$$i(t) = \sin(\omega_0 t) \tag{23.8}$$

Then by the convolution theory the response would be

$$v(t) = \frac{1}{C}\int_0^t \sin\omega_0\tau e^{-(t-\tau)/RC}d\tau \tag{23.9}$$

Figure 23.7 shows calculated response and comparison to exact solution for the case $\omega_0 = \frac{\pi}{2}$. As seen, the smaller the delta step in the τ integration, the closer the computed result to the exact one.

So whether it be a unit step, a periodic pulse, or a sinusoidal input, we saw above that we

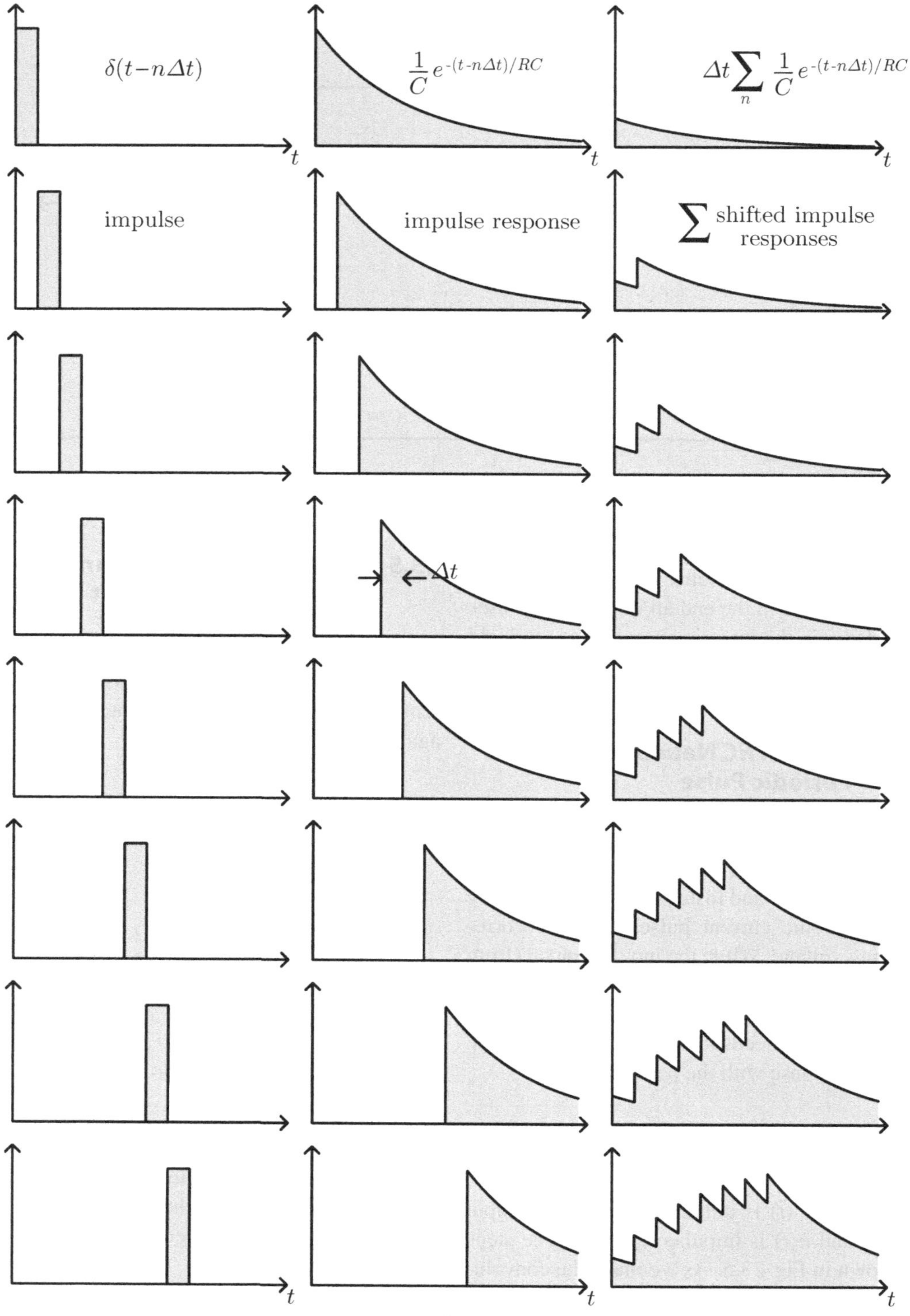

Fig. 23.3 Linear response via convolution with impulse response: left—impulse current; center—impulse response; right—total response

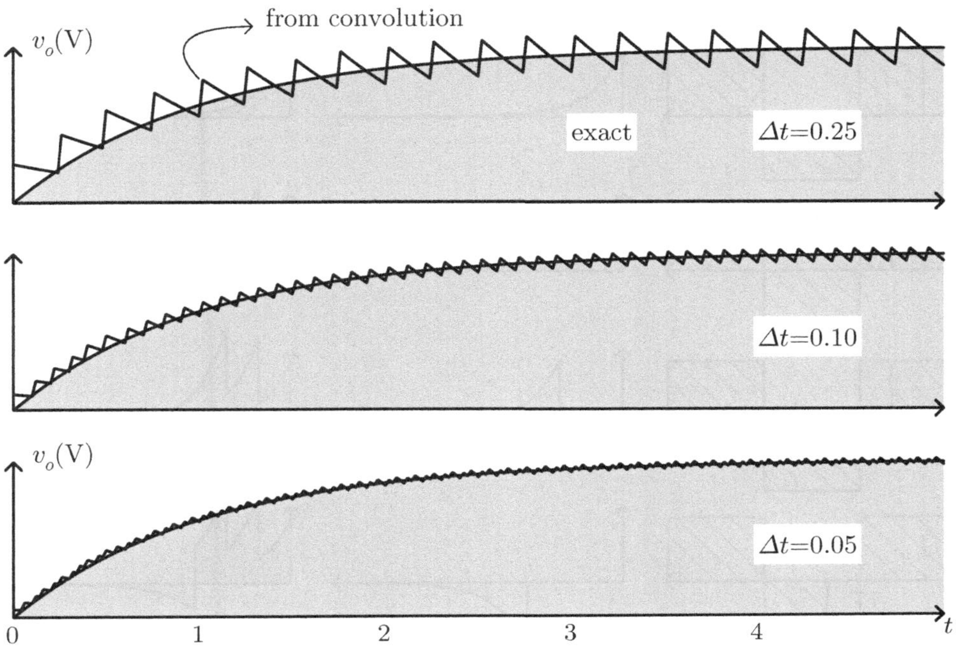

Fig. 23.4 Step response of parallel *RC* network using finer resolution convolution results and comparison to exact answer

can obtain the solution by simply convolving the impulse response (which was calculated only once) with the input stimulus at hand. Really what looks like the bottleneck is being able to do the convolution correctly. The impulse response is a one-timer effort; but the convolution has to be redone for every new stimulus. Luckily the exact steps—be them numerical or analytical—should apply for all convolution cases. So if we can get one convolution example working correctly we can simply swap in the new stimulus and repeat!

23.6 Impulse Response Due to Parallel *RLC* Network

Consider the parallel *RLC* network in Fig. 23.8 which is driven by a current source. Before we find the output voltage due to an arbitrary current source we are set to first figure the impulse response which is the output voltage due to an impulse current. The impulse response is given by

$$h(t) = \frac{1}{C}e^{-at}\left[\cos\omega_0 t - \frac{a}{\omega_0}\sin\omega_0 t\right], \quad a = \frac{1}{2RC}, \quad \omega_{LC}^2 = \frac{1}{LC}, \quad \omega_0^2 = \omega_{LC}^2 - a^2 \tag{23.10}$$

and is shown in Fig. 23.9. [As to how this was derived, there is more than one way to it, but the one we used is based on the transfer function for this circuit, as shown in Eq. (26.47). Once the transfer function is known, and using partial fractions, we can simply find the inverse transform; the reader will master the intermediate steps once we tackle transfer functions in Chap. 26.] This impulse response will now be used as basis to generate the response due to all other stimuli; a sample application is shown in the next section.

As to the meaning of the impulse response here, we see that the cap is first fully charged (to $\frac{1}{C}$ here). After that it discharges onto the inductor and back and forth with an angular frequency ω_0. As the current goes back and forth dissipation happens across the resistor and that eventually kills the current! The associated time constant is simply $t_{RC} = 2RC$.

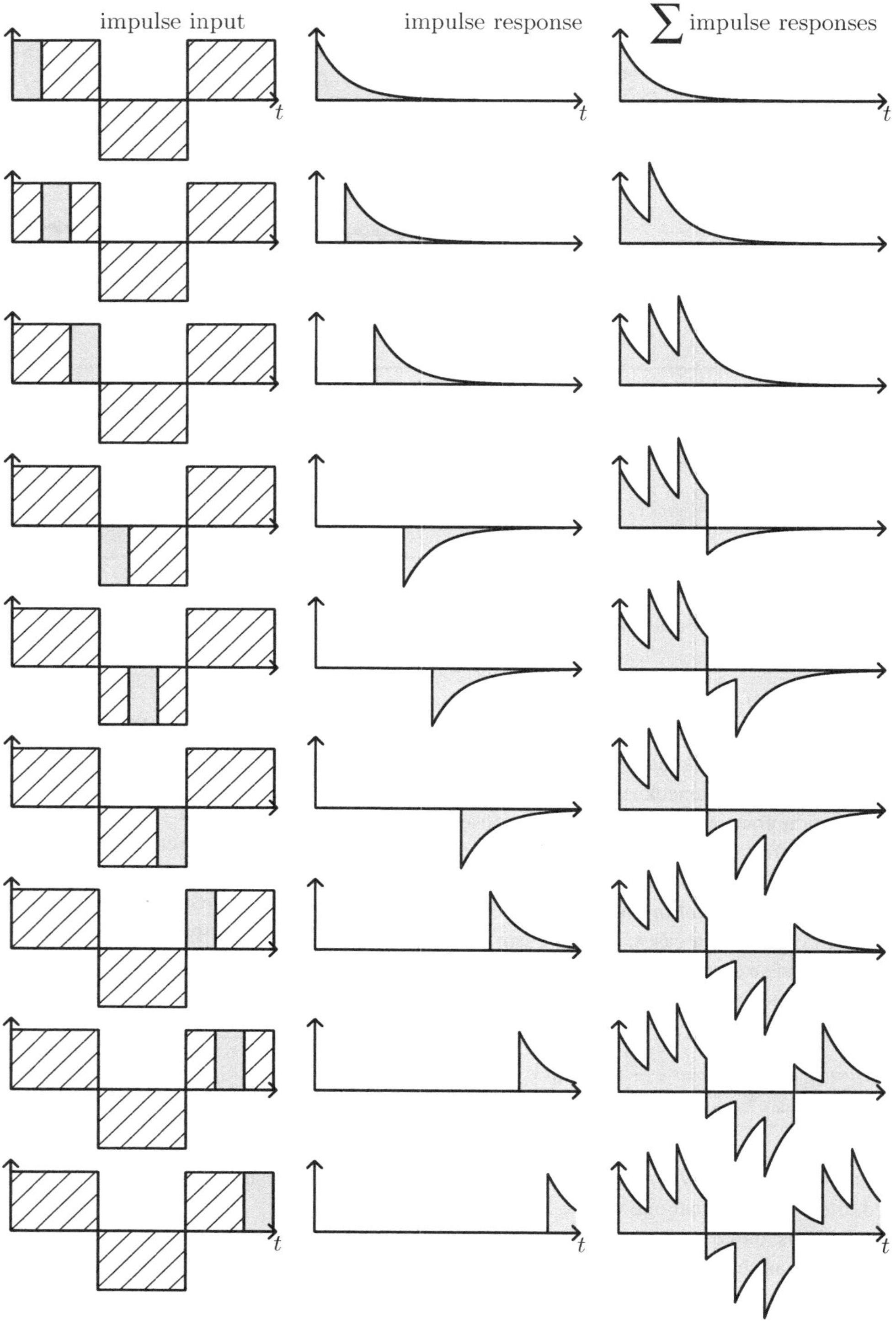

Fig. 23.5 Linear response via convolution with impulse response: left—impulse current; center—impulse response; right—total response

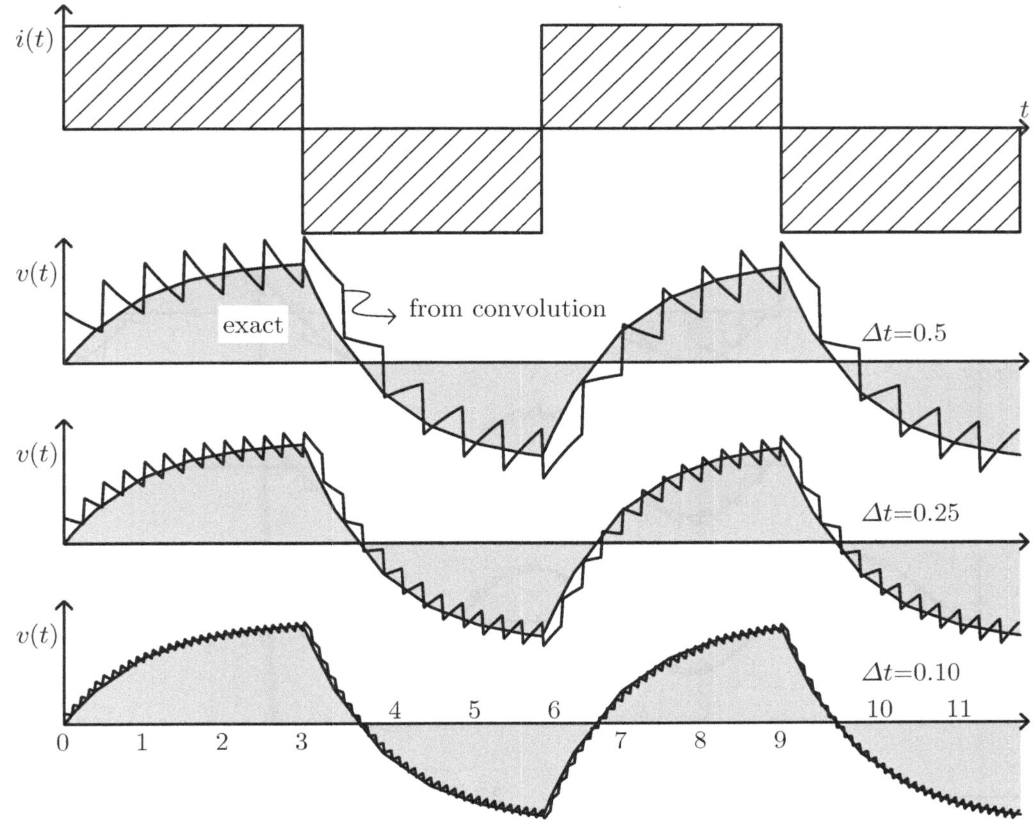

Fig. 23.6 Periodic pulse response to parallel *RC* network as a function of convolution time step, and comparison to exact solution

23.7 Periodic Pulse Response Due to Parallel *RLC* Network

Knowing the impulse response of the parallel *RLC* network we are able to figure the response due to any other input; for example let's find the response due to a periodic pulse input. The pulse has a period of 6, and toggles between 1 and −1. Results are shown in Fig. 23.10. As shown, with smaller convolution step we get better accuracy.

Another success story! The periodic pulse is really a very generic stimulus and baked in it are many harmonics; and the fact that convolution gets us the correct answer is very encouraging. Notice that even with a relatively coarse Δt selection, which is the assumed impulse width, we still get relatively good accuracy. Of course using smaller Δt gets us better results, but even worst case and with coarse Δt we will get *some* answer as opposed to nothing as could be the case when a simulator does not converge or worst simulations blow up!

23.8 Summary

Knowing the impulse response enables us to find system response due to any input (provided the system is linear). The impulse response is simulated/measured by applying a delta stimulus and measuring the corresponding system response. To find the response due to any other input, all that needs to take place is convolving the impulse response with the particular stimulus at

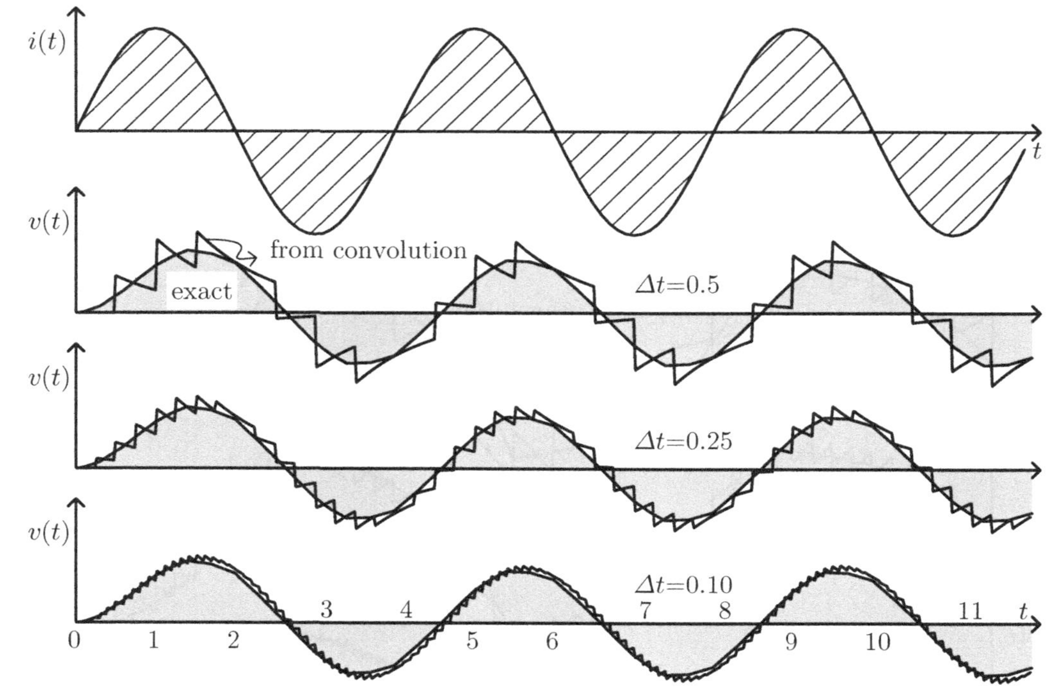

Fig. 23.7 Causal sinusoid response (with different convolution time step intervals) and comparison to exact solution

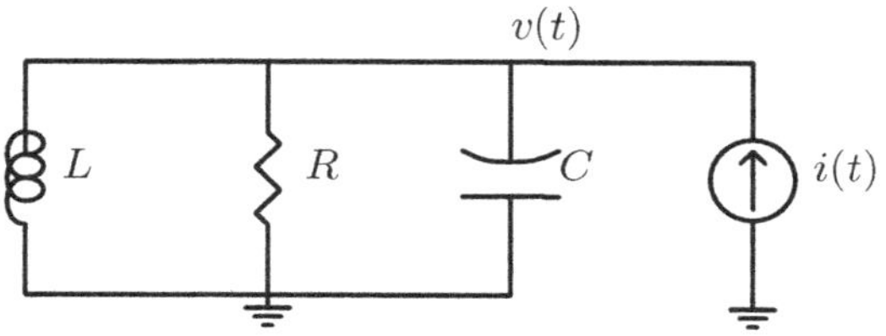

Fig. 23.8 Parallel *RLC* network

hand. This was demonstrated in this chapter for a couple of circuit topologies (parallel *RC* and parallel *RLC* circuits) and for multiple stimuli, including unit step, periodic pulse, and causal sinusoid. With this we have come full circle in analyzing and demonstrating the idea and use of convolution. Starting with the basics of convolution, leading into signal construction via convolution, the impulse function and finally impulse response, this chapter completes the circle by predicting system response via convolution with the impulse response. In the next chapter we redo what was done in this chapter but rather than using the impulse response we use the *step* response.

23.9 Problems

1. The impulse response to the *LC* network shown in Fig. 23.11 is

$$h(t) = \frac{1}{C}\cos\omega_0 t, \qquad \omega_0^2 = \frac{1}{LC}$$

   Assume $C = 1$, and $L = \frac{1}{4\pi^2}$; (A) find the unit step response, and (B) find the response due to the input current

$$i(t) = u(t) + u(t - 0.5)$$

   Explain this last result; in particular how come there is a continuous input yet output ceased! See sample results in Fig. 23.12.
2. Repeat Problem 1 and find voltage response due to (A) pulse of width 1, and (B) pulse of width 0.5. See sample results in Fig. 23.13.
3. Consider the parallel *RC* network with voltage response due to impulse current input as

$$h(t) = \frac{1}{C}e^{-t/RC}$$

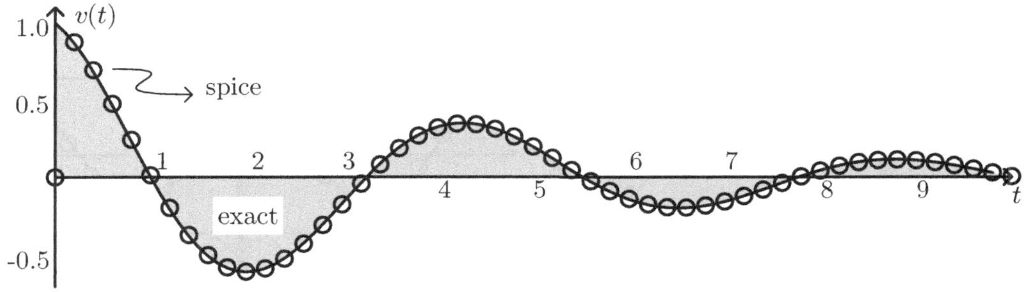

Fig. 23.9 Impulse response of Parallel *RLC* network. Case of $R = 2\,\Omega$, $C = 1\,\text{F}$ and $L = 0.5\,\text{H}$

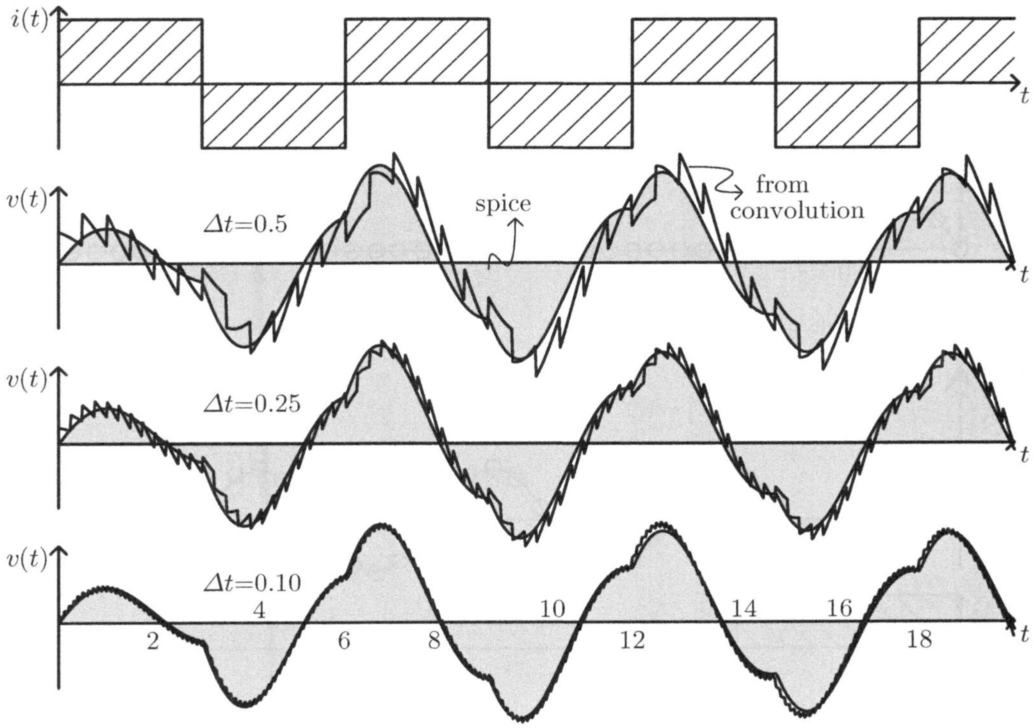

Fig. 23.10 Periodic pulse response of Parallel *RLC* network

Fig. 23.11 Setup for Problem 1

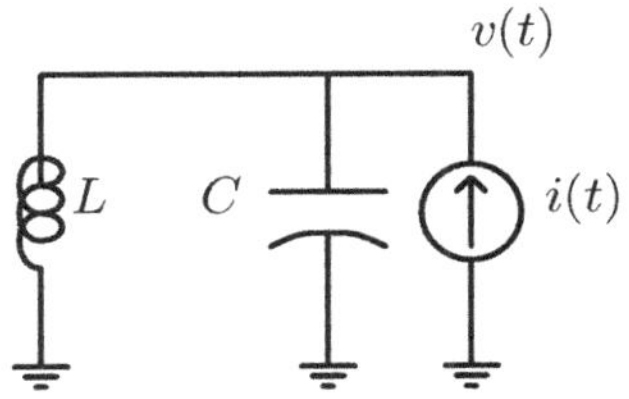

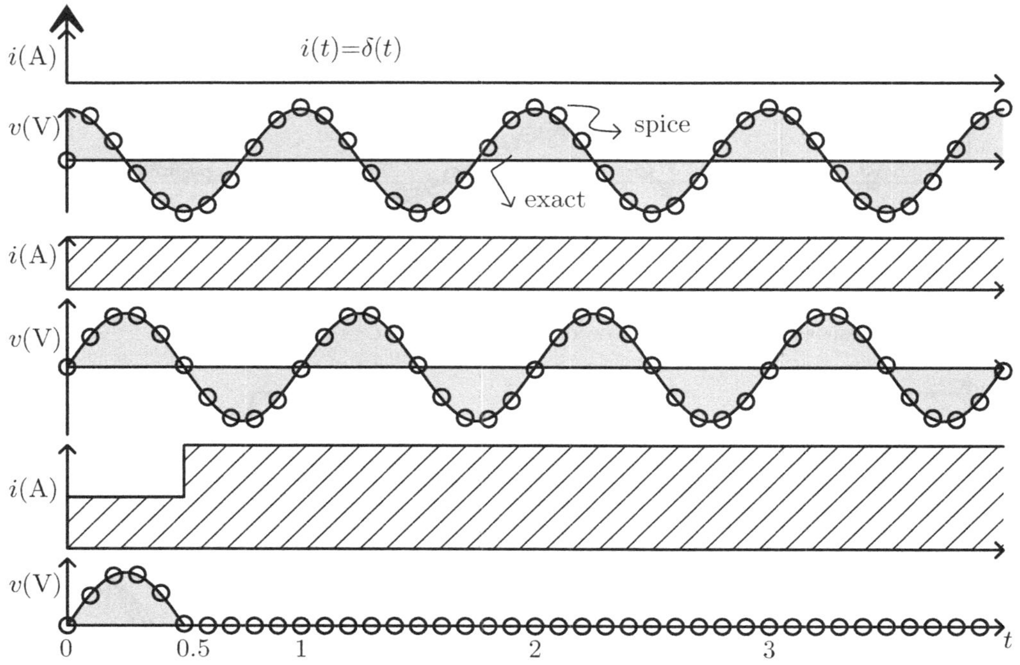

Fig. 23.12 Sample results for Problem 1

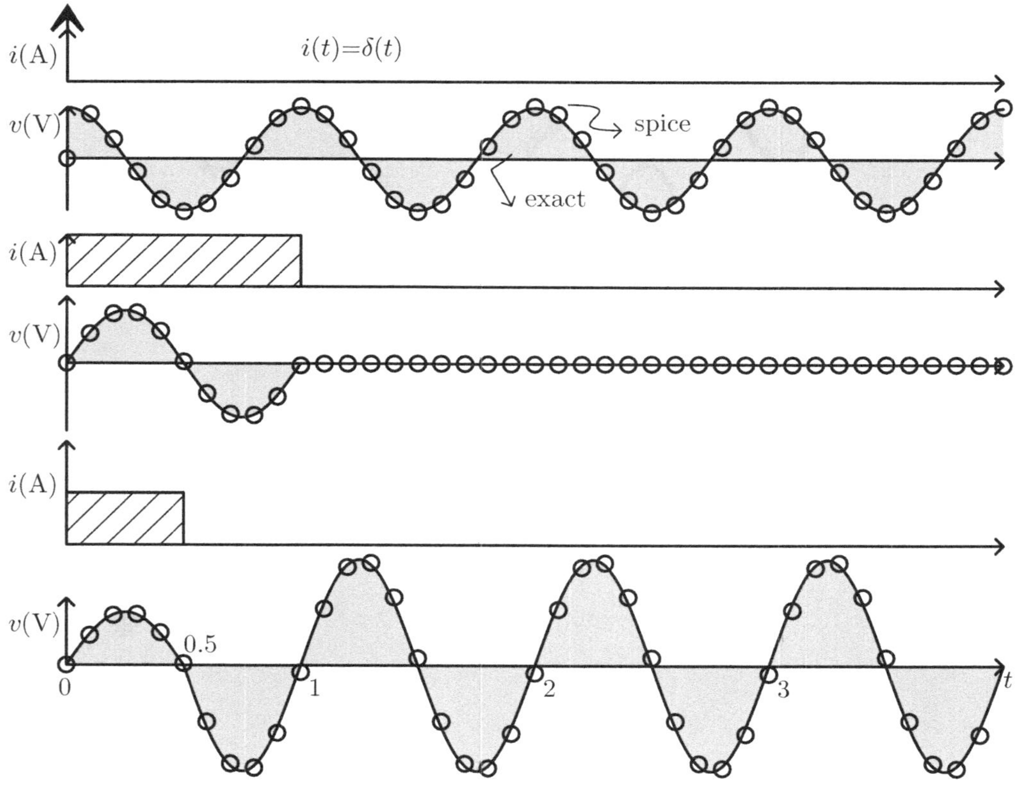

Fig. 23.13 Sample results for Problem 2

Find the response due to input current $i(t) = \sin(t^2)$ by doing the convolution numerically. Then compare to SPICE; see sample results in Fig. 23.14. Note: may have to input current to SPICE as a PWL (piecewise linear).

4. Many a times the impulse response is not known analytically, and rather is either measured or simulated. In either of the last two formats, the impulse response available is already discretized. If the discretization time step is Δt, then we can discretize time, impulse response, and output into

$$t = \Delta t\,[0, 1, 2, \ldots N] \quad \text{, time}$$
$$h = [h_0, h_1, h_2, \ldots h_N] \quad \text{, impulse response}$$
$$I = [I_0, I_1, I_2, \ldots I_N] \quad \text{, new current}$$
$$v = [v_0, v_1, v_2, \ldots v_N] \quad \text{, new voltage}$$

Write an algorithm to take the impulse response h, and an arbitrary current I to generate the corresponding voltage v. Code the algorithm into any language, execute the program, plot output voltage, and compare to SPICE for case of input current periodic of pulse width 3.8, period 10, and rise/fall times 0.1. The impulse current is shown below, where entry i corresponds to time $t = \Delta t \times i$ and where sampling happened for $\Delta t = 0.1$. In other words, at time zero impulse response was 1 while at time 10 it was -0.01. See also sample solution in Fig. 23.15

1.00	1.01	0.97	0.91	0.82	0.72	0.60	0.48	0.36	0.23
0.11	-0.00	-0.11	-0.20	-0.28	-0.34	-0.39	-0.43	-0.45	-0.46
-0.45	-0.43	-0.40	-0.36	-0.32	-0.27	-0.22	-0.17	-0.11	-0.06
-0.01	0.03	0.07	0.10	0.13	0.15	0.17	0.18	0.18	0.18
0.17	0.16	0.14	0.13	0.11	0.08	0.06	0.04	0.02	-0.00
-0.02	-0.04	-0.05	-0.06	-0.07	-0.08	-0.08	-0.08	-0.08	-0.08
-0.07	-0.07	-0.06	-0.05	-0.04	-0.03	-0.02	-0.01	-0.00	0.00
0.01	0.02	0.02	0.03	0.03	0.03	0.03	0.03	0.03	0.03
0.02	0.02	0.02	0.01	0.01	0.01	0.00	-0.00	-0.00	-0.01
-0.01	-0.01	-0.01	-0.01	-0.01	-0.01	-0.01	-0.01	-0.01	-0.01
-0.01									

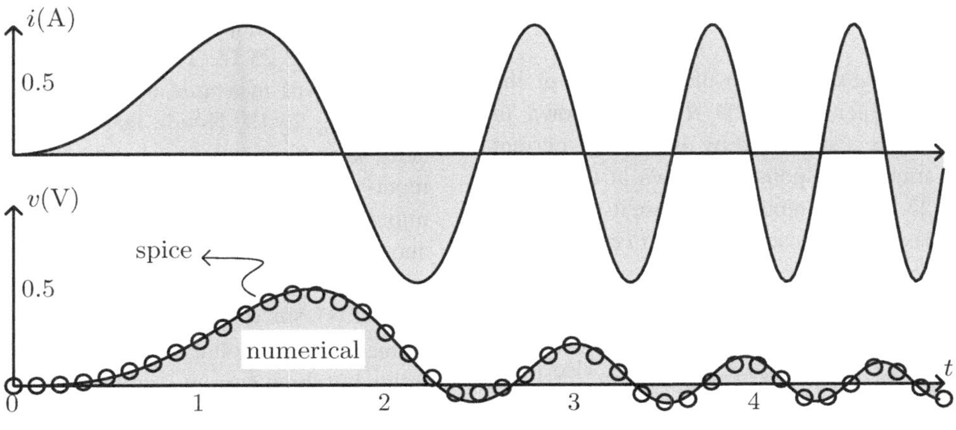

Fig. 23.14 Sample results for Problem 3

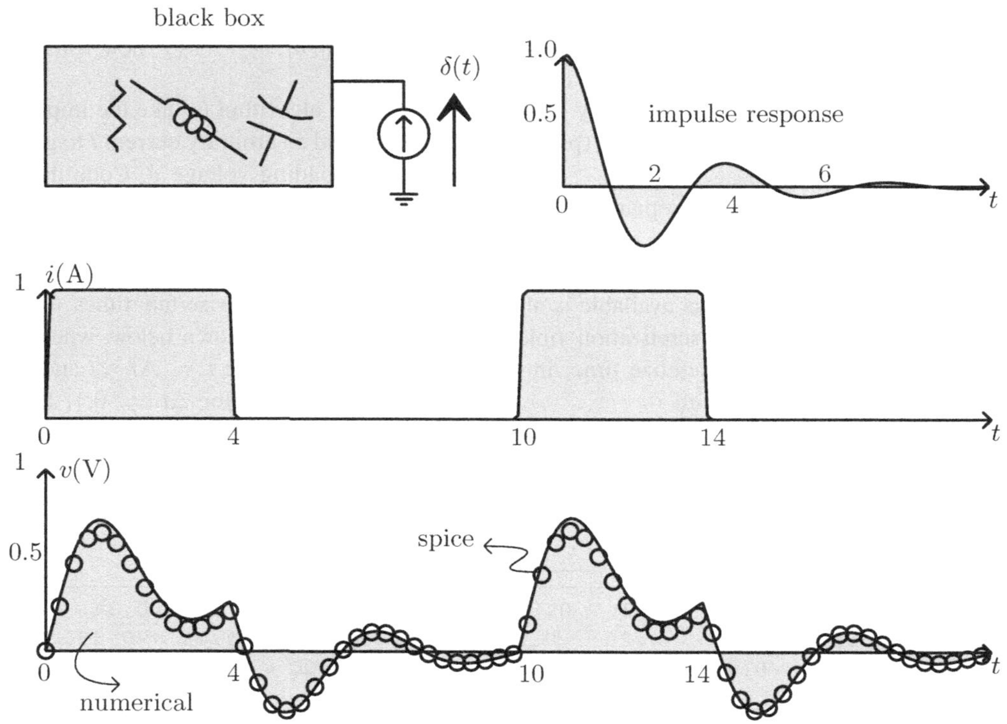

Fig. 23.15 Sample results for Problem 4

Answer:

For $i = 0; i <= 100; i = i + 1$
$\Sigma = 0$
For $j = 0; j <= i; j = j + 1$
$\Sigma = \Sigma + I[j] \times h[i - j]$
End j loop
$v[i] = \Delta t \times \Sigma$
Plot $v[i]$ versus $T[i]$
End i loop

5. This problem relies on the algorithm of the prior problem. The 2D *RC* grid shown in Fig. 23.16 is simulated by an impulse current and impulse response is shown at the top of Fig. 23.17. The impulse response itself is tabulated below, and each data point i corresponds to $i\Delta t$, where $\Delta t = 0.1$. Find response to an input current comprised of a triangle, followed by negative square as shown in the middle of Fig. 23.17 and compare to SPICE. Use $R = 5.0\,\Omega$ and $C = 0.1$ F. The same figure has sample solution.

6. A transmission line has cap/unit length and inductance/unit length of 1. It has a length 10 and is open terminated; this means any signal arriving at the far end will be fully reflected. At the input side, it is terminated by its characteristic impedance (1 here), which means any signal coming back from the far end and impinging on the close end would be fully absorbed. In parallel with the input termination an impulse current is applied, as shown in Fig. 23.18. The voltage response at the input was measured and is shown at the top of Fig. 23.19. Notice that we get half a delta since current splits equally between the input termination and the T-line itself. The impulse response shows an initial delta, and another after 20 s. This time is what it takes to travel across the T-line (10 s) and come back (total 20 s), since velocity $\frac{1}{\sqrt{LC}}$ is 1, and length 10 m. Find the voltage response due to the following three current stimuli: (A) unit step, (B) upright triangle of width 2, and (C) sine of angular frequency $\frac{1}{2}\pi$. See sample solution in Fig. 23.19; explain the results.

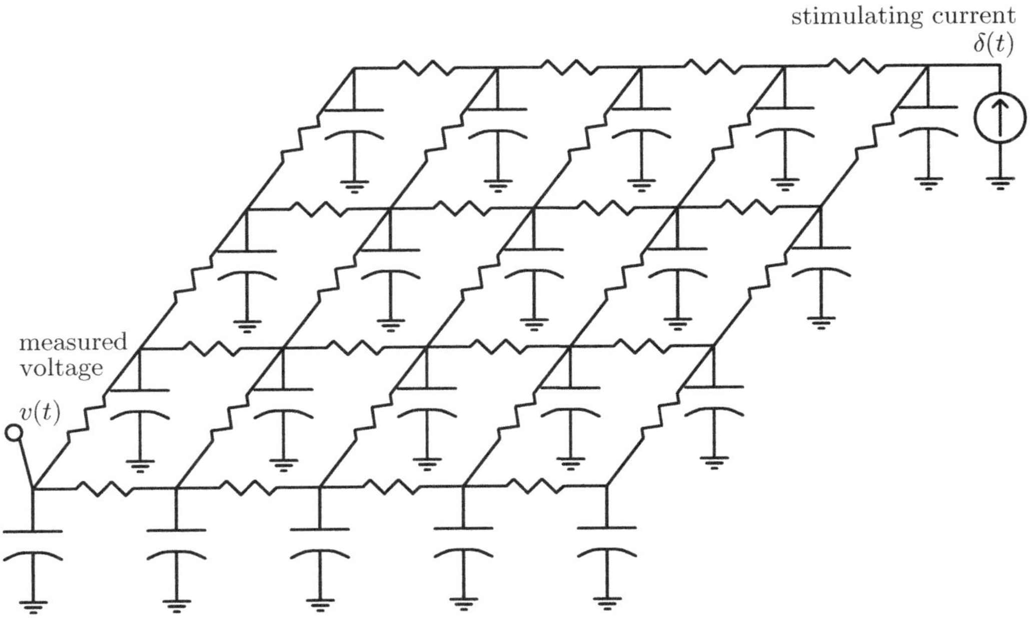

Fig. 23.16 Statement to Problem 5

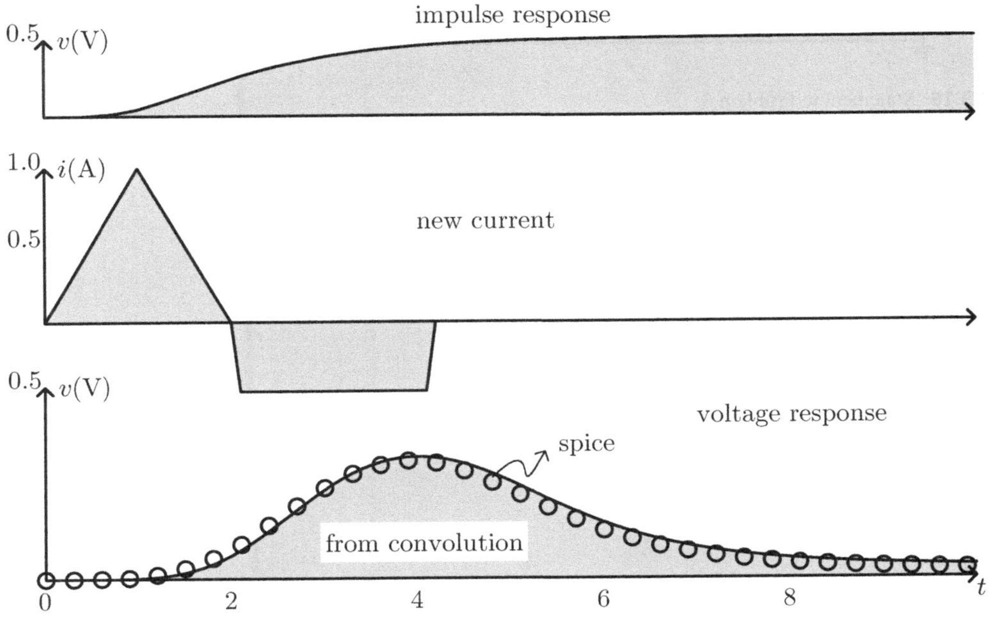

Fig. 23.17 Sample solution to Problem 5

```
0.00 0.00 0.00 0.00 0.00 0.00 0.01 0.01 0.02 0.03
0.04 0.06 0.08 0.10 0.12 0.14 0.16 0.18 0.20 0.22
0.24 0.26 0.28 0.29 0.31 0.32 0.34 0.35 0.36 0.37
0.38 0.39 0.40 0.41 0.42 0.43 0.43 0.44 0.44 0.45
0.45 0.46 0.46 0.47 0.47 0.47 0.47 0.48 0.48 0.48
0.48 0.49 0.49 0.49 0.49 0.49 0.49 0.49 0.50 0.50
0.50 0.50 0.50 0.50 0.50 0.50 0.50 0.50 0.50 0.50
0.50 0.50 0.50 0.50 0.50 0.51 0.51 0.51 0.51 0.51
0.51 0.51 0.51 0.51 0.51 0.51 0.51 0.51 0.51 0.51
0.51 0.51 0.51 0.51 0.51 0.51 0.51 0.51 0.51 0.51
0.51
```

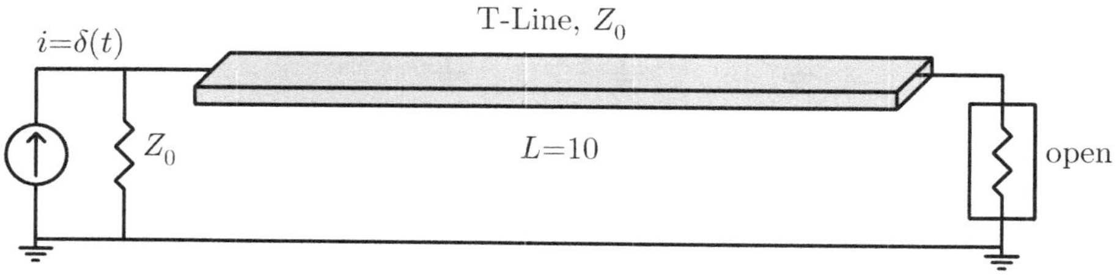

Fig. 23.18 Statement to Problem 6

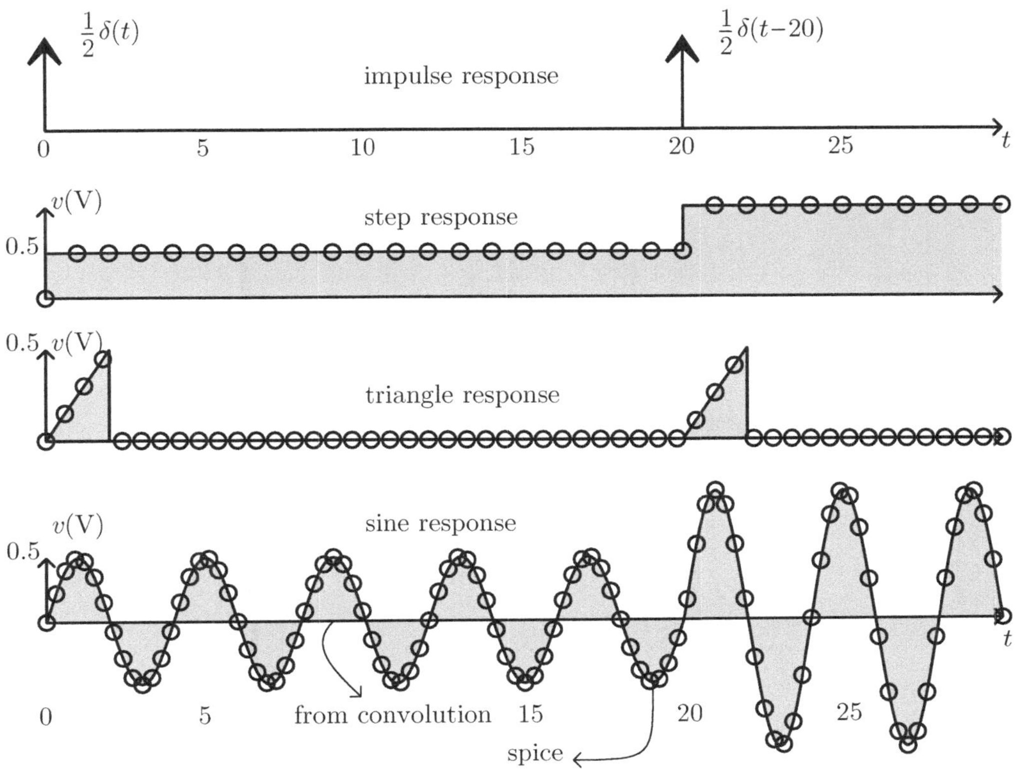

Fig. 23.19 Sample solution to Problem 6

24 Time Convolution with the Unit Step Response

24.1 Introduction

This chapter is an exact mirror of the prior chapter with the exception of using the unit step response as opposed to the impulse one. As a reminder the unit step function is one of the most important functions in circuit theory (as well as many other areas). The unit step truly exercises the circuit in the sense of full load. Not only does it capture the DC and low-frequency response, but it also samples the high frequency spectrum as well (though not at strongly as a delta function). The unit step can also be used to build many other signals, such as the pulse, periodic pulse, ramp, and so on. So assuming we know the unit step response, how do we go about extracting the response to a different stimulus?

24.2 The Unit Step as Building Block for Other Signals

One of the most important properties of the unit step function is that it can be used to build other signals. Why this is important? Because if we know the response to the unit step, we then know the response of an arbitrary signal if we know the decomposition of that signal in terms of the unit step. So let's plunge into some samples showing how this is done.

The Pulse in Terms of Step Function In this first example we demonstrate how to use the unit step function to build a pulse function. Figure 24.1 shows that we can get a pulse function, of unity width as

$$\text{pulse} = u(t) - u(t-1) \qquad (24.1)$$

F. Badrieh, *Spectral, Convolution and Numerical Techniques in Circuit Theory*,
https://doi.org/10.1007/978-3-319-71437-0_24

The Ramped Unit Step Function We can construct a ramped unit step function from an ideal one as shown in Fig. 24.2. Assuming $N = 5$ we get a first order approximation for a ramped unit step function as follows:

$$\text{ramped step} = \frac{1}{5}\left[u(t) + u(t-0.2) + u(t-0.4) + u(t-0.6) + u(t-0.8)\right] \quad (24.2)$$

If we refine N to 10 and 20 we get bottom right approximations in the figure.

The Periodic Pulse Unlike the single pulse described before, the periodic pulse needs an infinite sequence of unit step functions

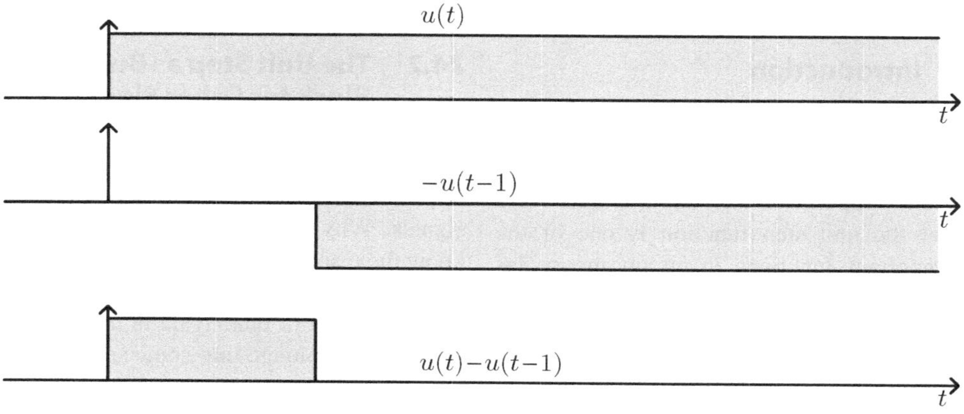

Fig. 24.1 Pulse function in terms of unit step ones

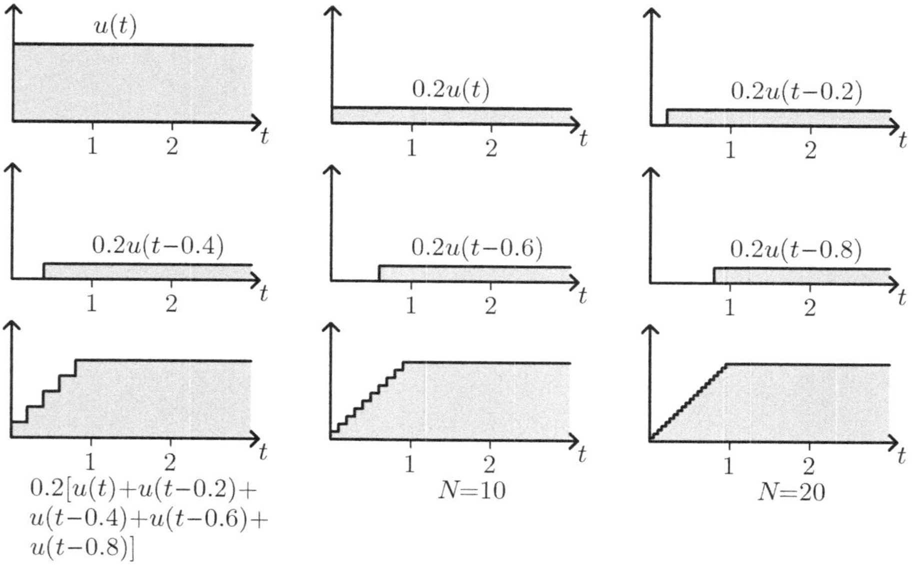

Fig. 24.2 Ramped unit step function in terms of ideal ones

$$\text{periodic pulse} = \sum_{n=0,2,\ldots}^{\infty} u(t-nT)-u(t-(n+1)T) \tag{24.3}$$

where T is the period (assumed 1 here). This is shown in Fig. 24.3. As can be seen above we can represent other signals easily in terms of the unit step function. Most of the times this amounts to scaling and offsetting. In the next section we get a formal definition of this process.

24.3 Generic Equation for Signal Construction in Terms of Convolution with Unit Step Function

As was shown in Sect. 20.3, the formal relation between a signal and its reconstruction in terms of unit step functions is

$$\boxed{f(t) = \int_0^t \frac{df(\tau)}{d\tau} u(t-\tau)d\tau} \tag{24.4}$$

That is, we can construct a signal in terms of shifted and scaled unit step functions provided the unit step functions are scaled by the *derivative* of the original function (at the shift point). The signals assumed here are causal in the sense they are zero for negative time. Also remember to keep track of *any and all* resulting delta functions (or derivatives thereof) while taking the time derivative of the function, especially at the start and end of the function. That is any discontinuity in the function would result in delta functions and those need to be carried throughout the convolution process. With the technical details out of the way let's demonstrate the convolution process. But first we need to devise a unit step response in the first place! Let's create such a unit step response as shown next.

24.4 Low-Pass Filter Step Response

Consider the low-pass filter shown in Fig. 24.4. Input voltage is applied across the RC circuit and output voltage is measured across the cap. For a unit step input voltage, output voltage is given by

$$\boxed{g(t) = 1 - e^{-t/RC}} \tag{24.5}$$

That is, at time zero and shortly thereafter, voltage across the cap remains zero, as it started uncharged. Slowly and in time the cap charges, and eventually (after a few time constants RC) the cap fully charges. After that, current ceases and the whole voltage is across the cap; that is, no IR voltage drop across the resistor. With Eq. (24.5) at hand we are now ready to find the response due to *any* input voltage. Let's try a few such cases.

24.5 Low-Pass Filter Periodic Pulse Response

Now assume instead of unit step we apply a periodic pulse of width 3 s and period 6 s. Without rerunning the simulations, how can we predict the solution? Since we know how to build the periodic pulse function out of the unit step function

$$[\text{periodic pulse of width 3 and period 6}] = u(t) - u(t-3) + u(t-6) - u(t-9)|\cdots \tag{24.6}$$

the corresponding solution would then be superposition of the scaled unit step responses

$$v(t) = g(t) - g(t-3) + g(t-6) - g(t-9)|\cdots \tag{24.7}$$

This is shown in Fig. 24.5. Notice that in this case we did not have to formally use Eq. (24.4); instead we just used intuition to build the periodic pulse out of the unit step one. In reality our intuition did exactly what this equation proclaims

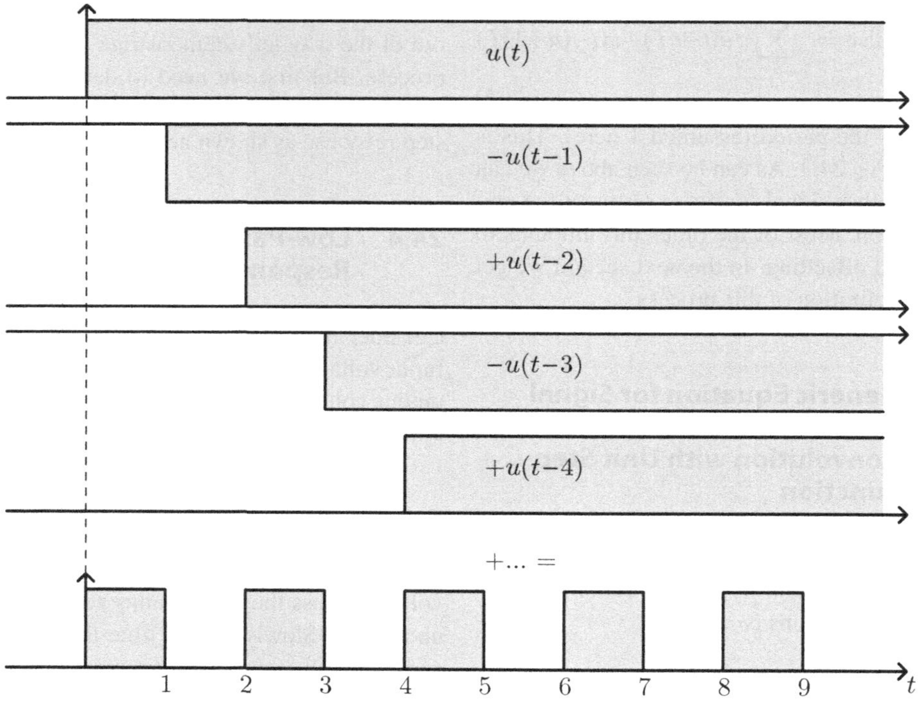

Fig. 24.3 Periodic pulse in terms sequence of unit step ones

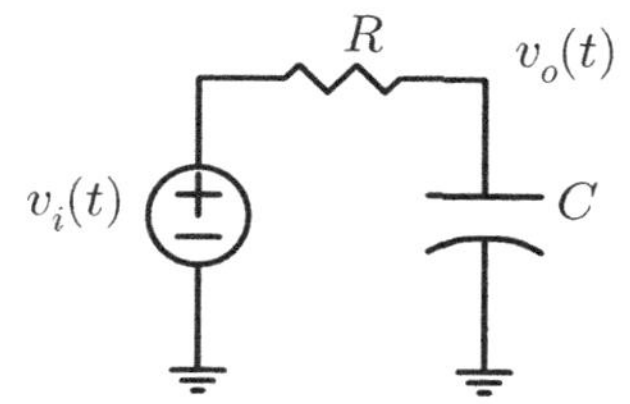

Fig. 24.4 Simple low pass *RC* network

to do. This, however, will not always be the case; most of the time full use of Eq. (24.4) will be required.

24.5.1 Impact of Frequency

We can try different pulse frequencies as shown in Figs. 24.6 and 24.7. Notice that in all cases the cap charges on average to a half, since duty cycle is 50%.

Notice the distinct difference between Figs. 24.5 and 24.7. In the former the charging (and discharging) process was slow enough that we end up with *one* unique signature which is an event happening every 6 s. But in the latter, the charging (and discharging) was happening too fast that we end up with *two* unique signatures. The first of course is the charge/discharge happening every 1 s. The second is the low-frequency ramp happening between 0 and—say—3 s. Notice that during that range the response is not yet fully periodic. But after 3 s the response becomes fully periodic. In other words, now we have a *transient* response (0–3 s) and a steady state one with oscillatory period 1 (after 3 s). Pretty much every system response has this trait.

24.5.2 Impact of Duty Cycle

Let's take the case of unity period and vary the duty cycle from 0 to 100%. Again we will use the step response to build the total response. By adjusting the shift in step responses, we can build any duty cycle. Figure 24.8 shows sample results and comparison to SPICE simulations. Notice the following observations:

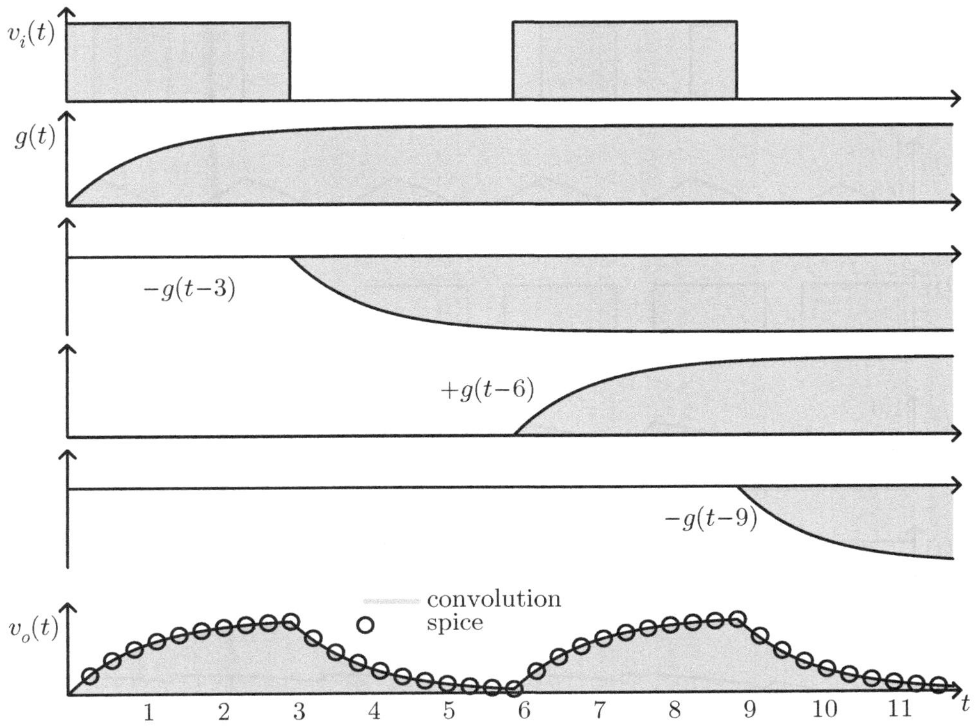

Fig. 24.5 Low-pass filter response due to periodic pulse

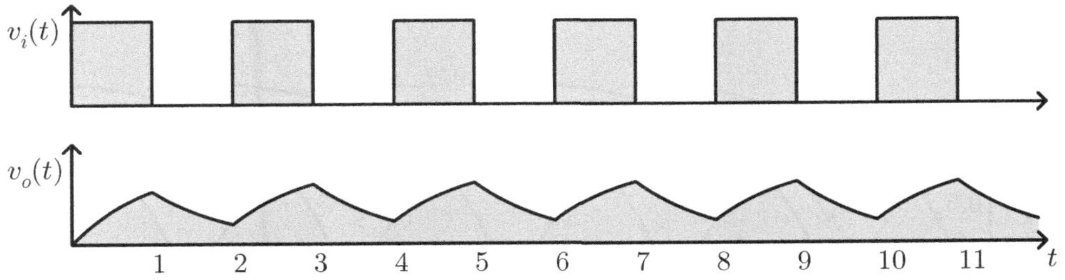

Fig. 24.6 Low-pass filter response due to periodic pulse (3× frequency in Fig. 24.5)

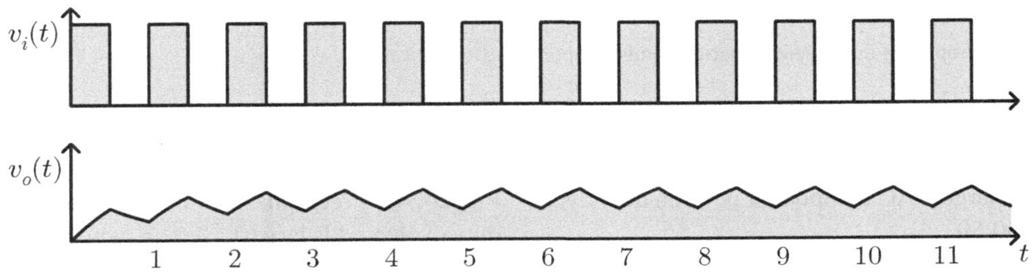

Fig. 24.7 Low-pass filter response due to periodic pulse (6× frequency in Fig. 24.5)

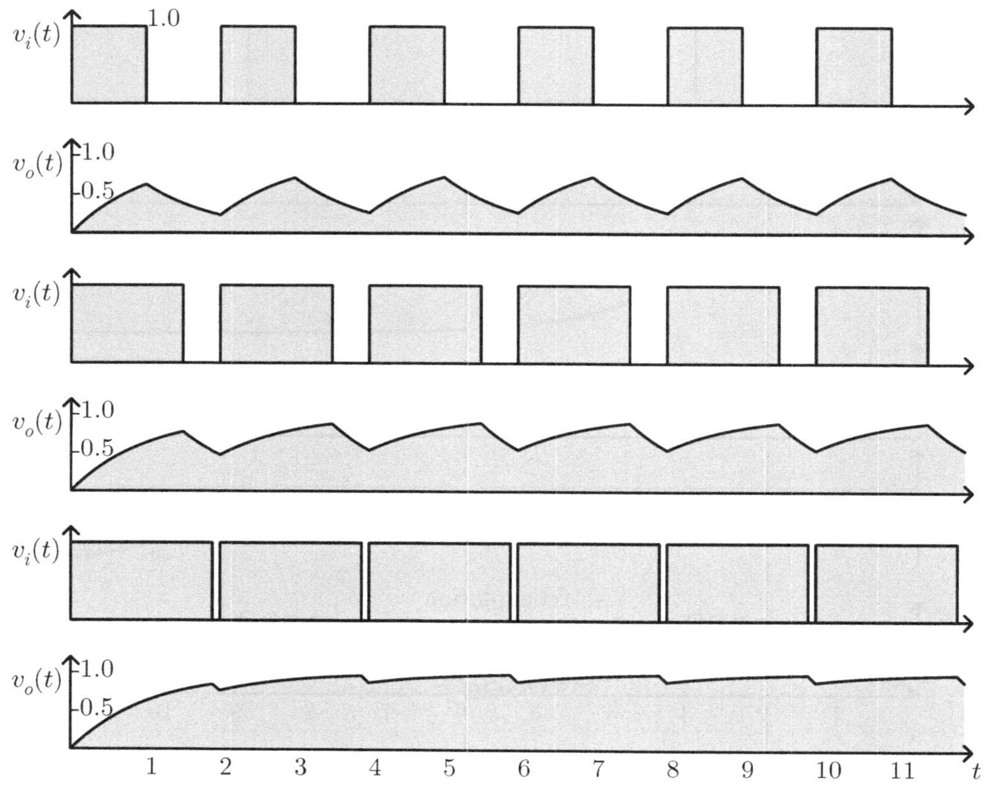

Fig. 24.8 Impact of duty cycle on periodic pulse response

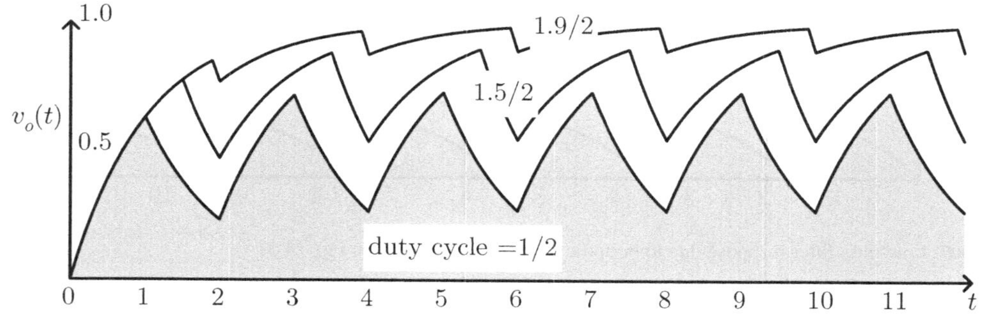

Fig. 24.9 Impact of duty cycle on periodic pulse response—overlaid version

1. The larger the duty cycle, the higher the settling voltage (on average).
2. The largest AC component is when duty cycle is 50/50.

Notice for the 100% duty cycle case we should recover the full unit step response as outlined in Eq. (24.5) It would be interesting to overlay the *envelope* of all three cases; this is shown in Fig. 24.9. As reemphasized in this figure, larger duty cycle yields largest average, while the 50/50 one yields the largest variations. It is also interesting to notice that all three responses look exactly the same at the start (say from time 0 to 1s);

that is, the system does *not* know a prior what will happen in the future! It cannot tell that this signal is going to end up a pulse while the other one as a unit step; it only sees the same input voltage during that time and hence puts out the same output voltage!

24.6 Low-Pass Filter Response to Periodic Triangular Pulse

Let us instead of a perfect square pulse assume as input voltage a periodic symmetric triangular pulse as shown in Fig. 24.10. To use the convolution formula we need to take the first derivative of the input voltage, also shown in the same figure. The resulting derivative is next convolved with the unit step response

$$v(t) = \int_0^t \frac{dv_i(\tau)}{d\tau} g(t-\tau)d\tau \tag{24.8}$$

We will illustrate this numerically. Assume the period of the signal is 1 (50% duty cycle). Assume further that our convolution time step is $\Delta\tau = 0.25$. Then we would have

$$\begin{aligned} v(t) = \Delta\tau\,[&g(t) + g(t-0.25) - g(t-0.50) - g(t-0.75) \\ &+ g(t-1.0) + g(t-1.25) - g(t-1.50) - g(t-1.75) \\ &+ g(t-2.0) + g(t-2.25) - g(t-2.50) - g(t-2.75) + \cdots] \end{aligned} \tag{24.9}$$

As we make $\Delta\tau$ smaller, the summation approaches an integration and results become more accurate. Figure 24.11 shows convolution results and comparison to SPICE for different $\Delta\tau$. As observed we get very good match. True smaller Δt gives better accuracy; but even the coarser selection still gave some *reasonable* solution. What this means is that—just like including more harmonics in a Fourier series—including finer resolution in convolution only *refines* the solution—it does not significantly alter it.

Notice that unlike last section where we used intuition, in this section we had to use formula Eq. (24.4). All that needed to be done was to first take the derivative of the input (periodic triangle here) and *then* convolve that with the unit step response. To put things in perspective, had we had the *impulse* response then we would have convolved the stimulus *directly* with the impulse response. In either case the convolution itself is the same; it is just a matter what are we convolving! Finally notice that due to the nature of the input signal the derivative did *not* produce any delta functions; had we had any abrupt edges in the input then we would have gotten delta functions. Not that it is a bad thing; it's just that delta function convolution (which is actually pretty easy) needs to be accounted for.

24.7 Low-Pass Filter Response to Ramp Input

Assume next that input voltage is a ramp one;

$$v_i(t) = t \tag{24.10}$$

We already know the exact solution for this case, and it is simply the integral of the unit step response; namely

$$v_o(t) = \int_0^{d\tau} g(\tau)d\tau = \int_0^{d\tau} \left[1 - e^{-\tau/RC}\right] d\tau \tag{24.11}$$

$$\boxed{v_o(t) = t + RC\left[e^{-t/RC} - 1\right]} \tag{24.12}$$

But let's get some practice applying the convolution equation. The derivative of input voltage is simply

$$\frac{dv_i(t)}{dt} = 1 \tag{24.13}$$

so that our convolution formula becomes

$$
\begin{aligned}
v_o(t) &= \int_0^t g(t-\tau)d\tau = \int_0^t \left[1 - e^{-(t-\tau)/RC}\right]d\tau \\
&= t - e^{-t/RC}\int_0^t e^{\tau/RC}d\tau = t - e^{-t/RC}RC\left[e^{t/RC} - 1\right] \\
&= t - RC + RCe^{-t/RC} = t + RC\left[e^{-t/RC} - 1\right] \qquad (24.14)
\end{aligned}
$$

which matches exactly Eq. (24.12). In this particular case we were lucky; but this integral could have come out non-integrable and that would have necessitated numerical integration. Towards getting practice with this latter let's do the convolution numerically. Results in Fig. 24.12. As can be seen we get reasonable results with coarse Δt and excellent ones with fine Δt. Again since the input is smooth and since it started at zero we did not get any delta functions while evaluating the derivative.

24.8 Low-Pass Filter Response to Causal Sinusoid

As a final example assume now that input is a sine function. Remember, this is a causal sinusoid, as opposed to a continuous one; that is, it starts

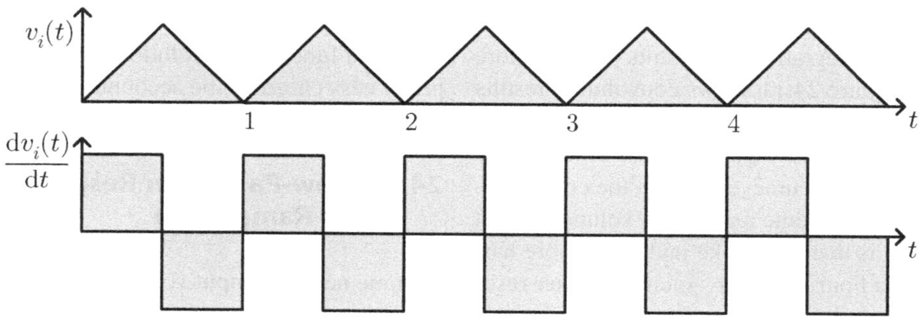

Fig. 24.10 Periodic triangular pulse as input for low-pass filter

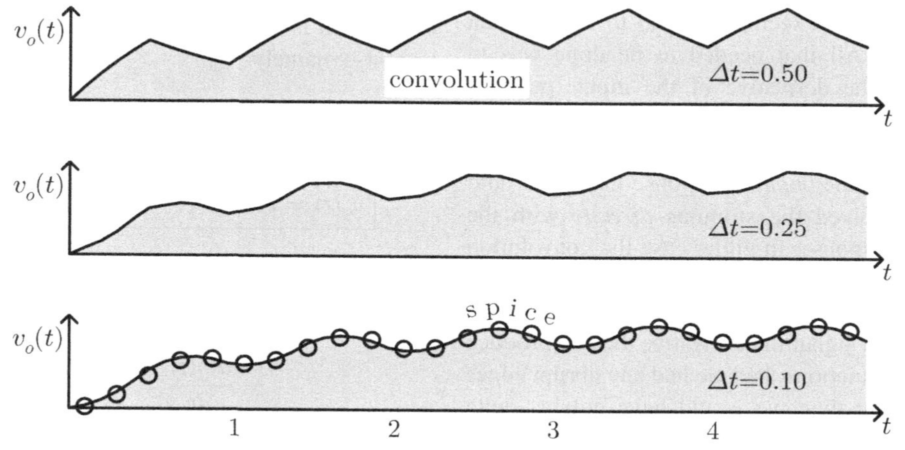

Fig. 24.11 Low-pass filter convolution response to periodic triangular pulse and comparison to SPICE

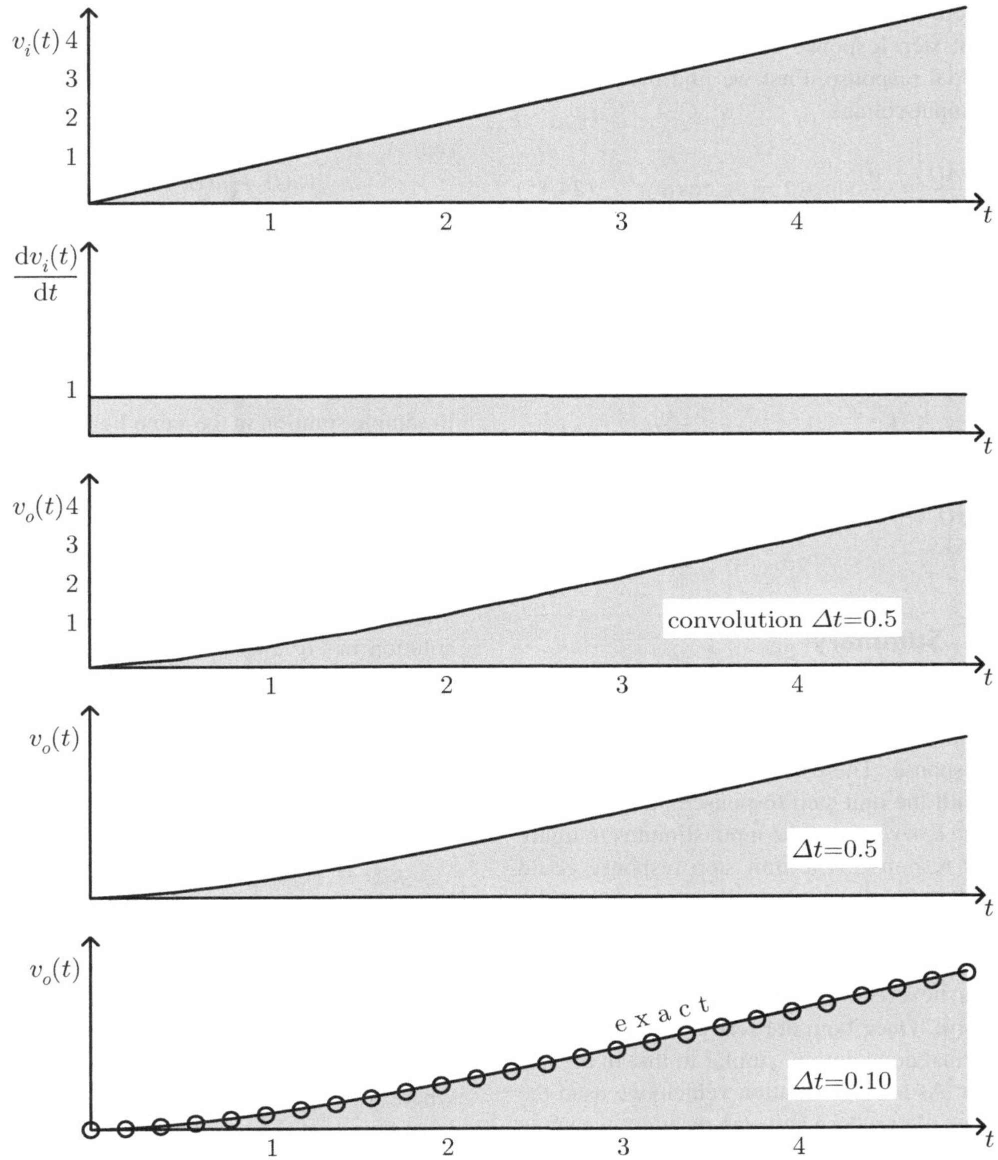

Fig. 24.12 Ramp response in terms of convolution with the unit step response

from zero (rather than from $-\infty$). Again we use the unit step response as building block to the sinusoidal response. First we find the derivative of the input voltage

$$\frac{dv_i(t)}{dt} = \frac{d}{dt}\sin\omega_0 t = \omega_0\cos\omega_0 t \qquad (24.15)$$

Again no delta functions during differentiation because the function is smooth and starts at zero. Then we just use the convolution integral; in particular

$$v(t) = \int_0^t \omega_0\cos\omega_0\tau g(t-\tau)d\tau \qquad (24.16)$$

with $g(t) = 1 - e^{-t/RC}$. Results are shown in Fig. 24.13.

24.9 Summary

Just like how the impulse response can be used to generate response to any input, so does the unit step response. The essence of this chapter is to start with the unit step response and convolve it with the *derivative* of the input stimulus to figure system response. The unit step response could have been figured analytically, through simulations or from measurements—it does not matter! Once it is ready it is not to be regenerated! What matters next is the new stimulus—namely its derivative. Once both are ready what remains is the normal convolution, similar to that in the last chapter. As a demonstration vehicle we used the low-pass filter whose unit step response we know a priori. Then we tried various cases, ranging from periodic pulse to a causal sinusoid. In each case we applied convolution and observed excellent match with finer convolution time step. We also shed some light on impact of duty cycle on the low-pass filter. Up next is our last stop in the time-domain world and that relates to sampling. After that we dive into transfer functions and how to get transient responses out of them.

24.10 Problems

1. The series RC network in Fig. 24.14 is excited via a step current and the step response comes out

$$v_o(t) = u(t) + u(t)t$$

as shown in the same figure. Find the voltage response to (a) an input current comprised of three pulses, of width 1 and period 2, and as shown in Fig. 24.15; and (b) an input current comprised of 5 alternating pulses, as shown in the same figure. Compare to SPICE, as shown in sample solution in the same figure. Assume $R = 1\,\Omega$ and $C = 1$ F.

2. Consider again the RC network in Fig. 24.14. Find the response due to (a) sinusoid current of angular frequency 1, that is applied for 1.5 periods; and (b) same current applied for 1.25 periods. Compare to SPICE. See sample solution in Fig. 24.16. Hint: when taking the derivative of input current, pay attention to any arising delta functions!
3. Consider the RL network shown in Fig. 24.17. The voltage step response (due to a unit step current) is given by

$$g(t) = 0.724e^{-0.382t} + 0.276e^{-2.618t}$$

Find the response due to (a) pulse of width 1, and (b) repeated pulse of width 1, and period 2. Compare to SPICE; see sample solution in Fig. 24.18. Use $R = 1\,\Omega$ and $L = 1$H.

4. Consider the feedback problem in Fig. 24.19 where $R = C = 1$. Output voltage is regulated based on reference voltage. The gain of feedback A is set to 5. The reference voltage is stepped from 0 to 1; show that the output voltage is

$$v_o(t) = \frac{A}{A+1}\left[1 - e^{-t(A+1)}\right]$$

Next, and using convolution, find response to reference voltage being (a) periodic pulse of

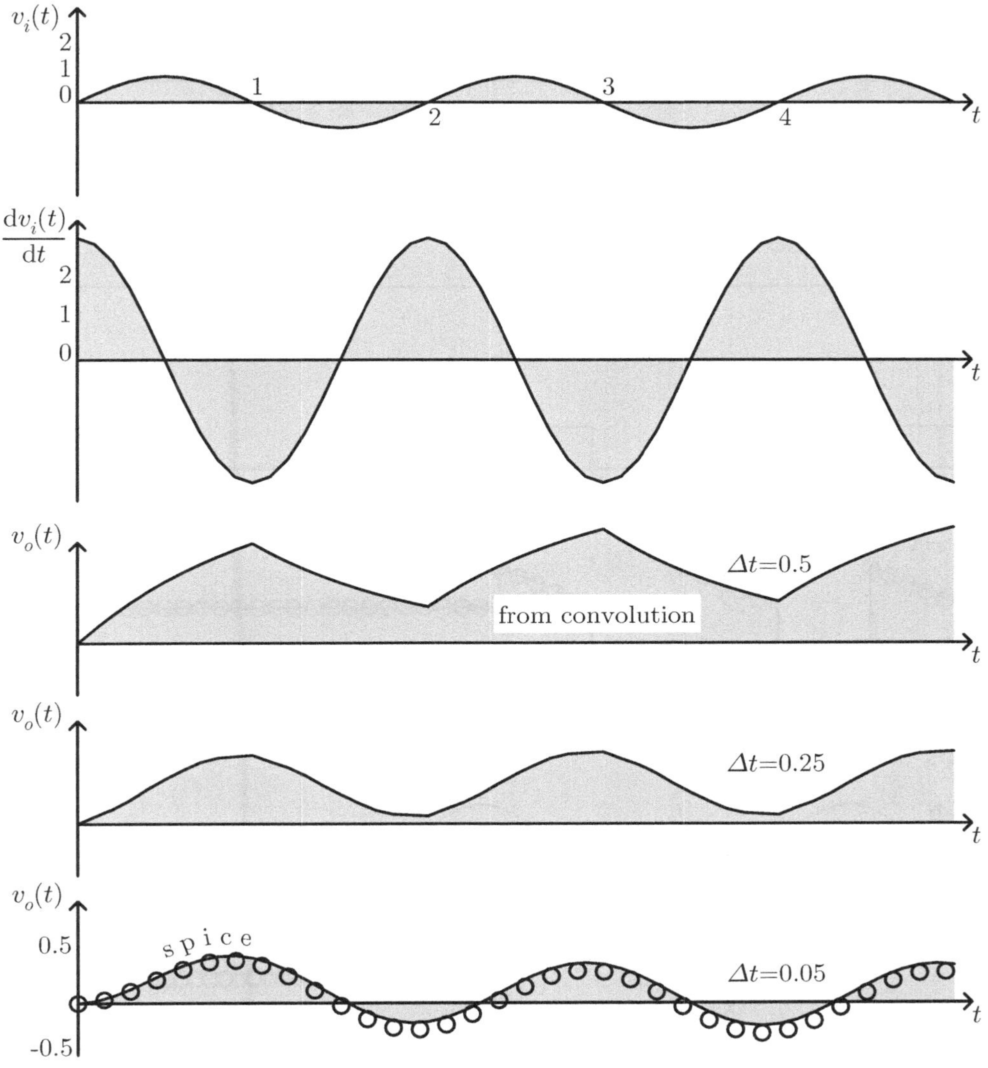

Fig. 24.13 Sinusoid response in terms of unit step response

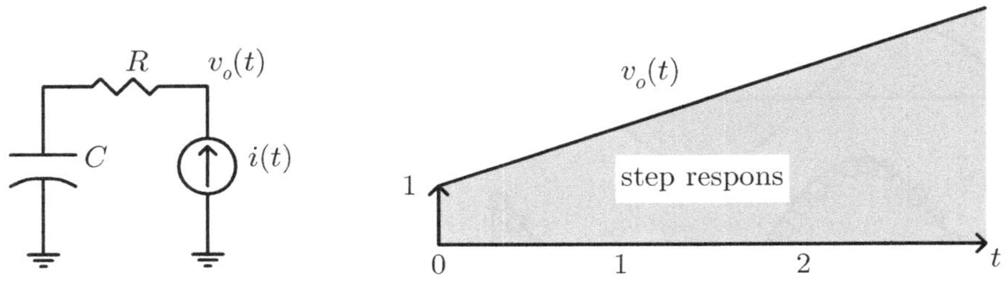

Fig. 24.14 Statement to Problem 1

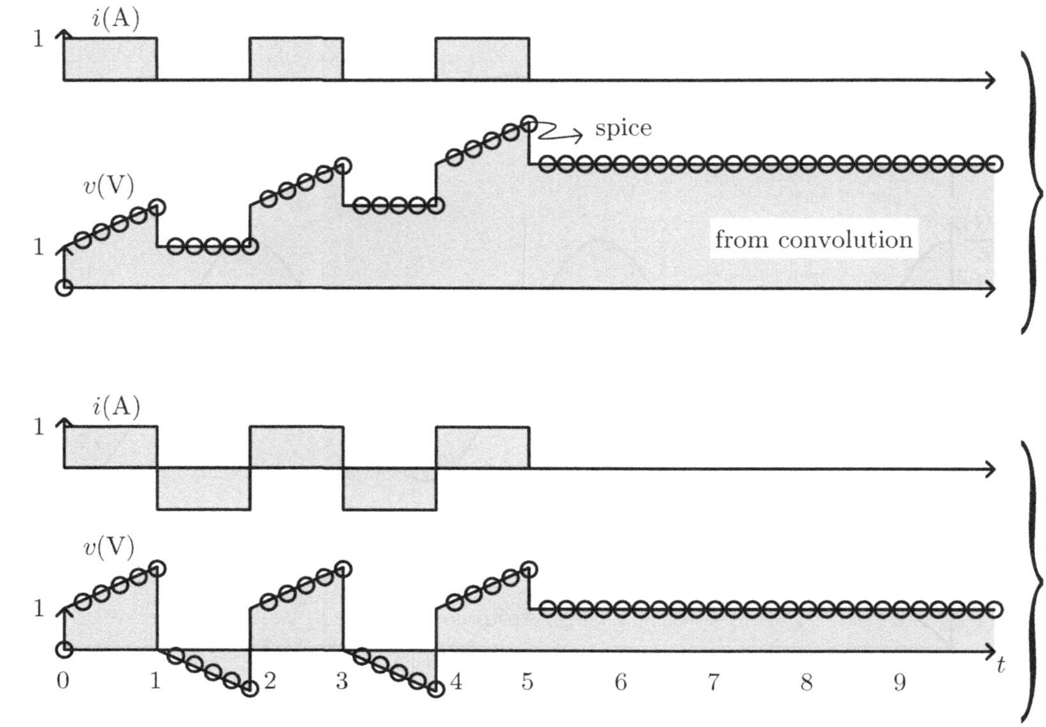

Fig. 24.15 Sample solution to Problem 1

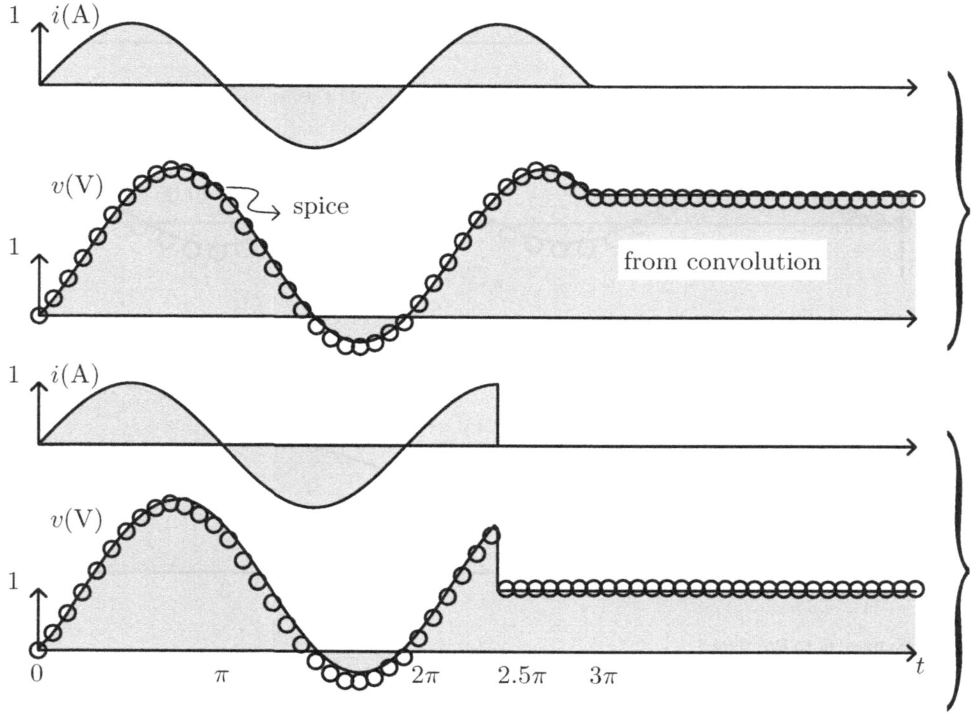

Fig. 24.16 Sample solution to Problem 2

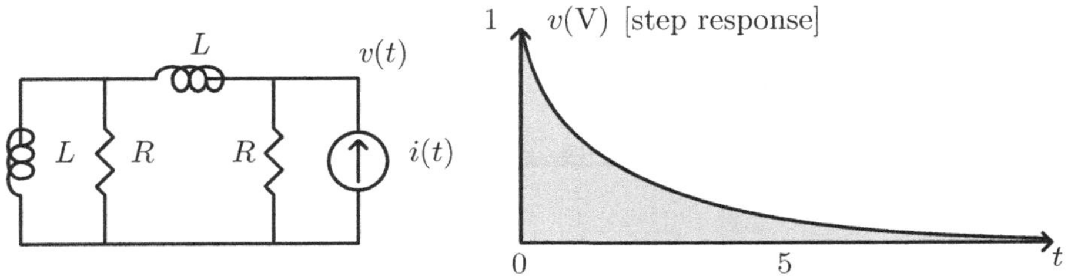

Fig. 24.17 Statement to Problem 3

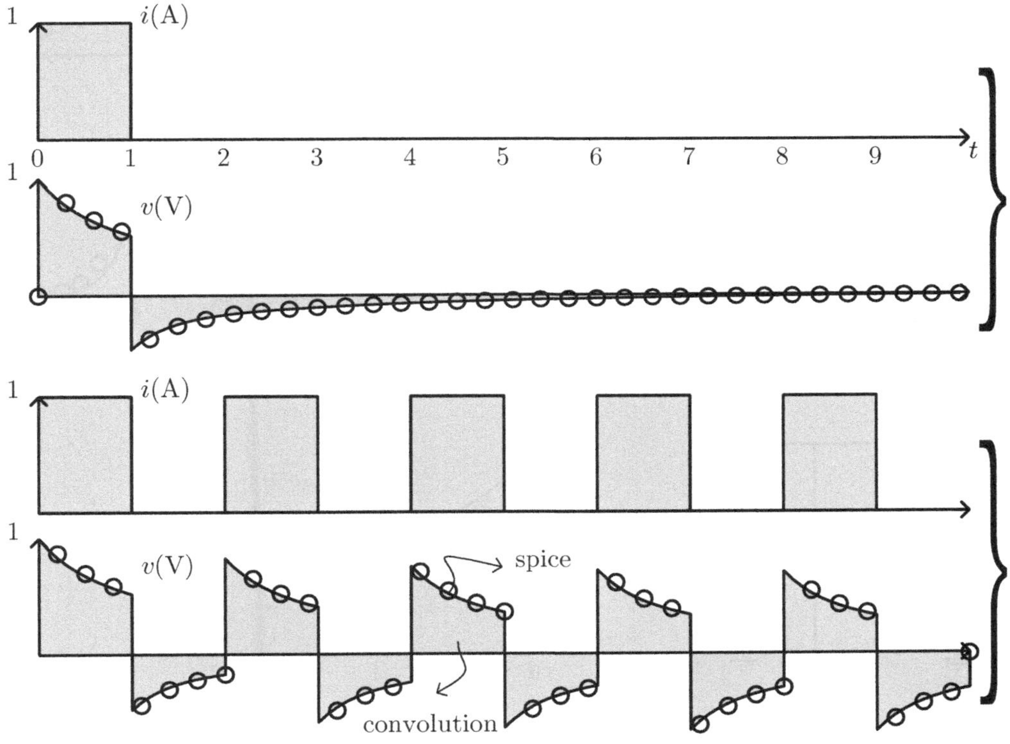

Fig. 24.18 Sample solution to Problem 3

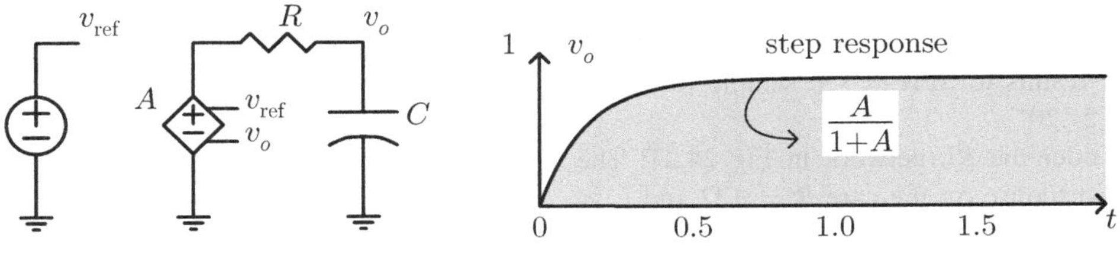

Fig. 24.19 Statement to Problem 4

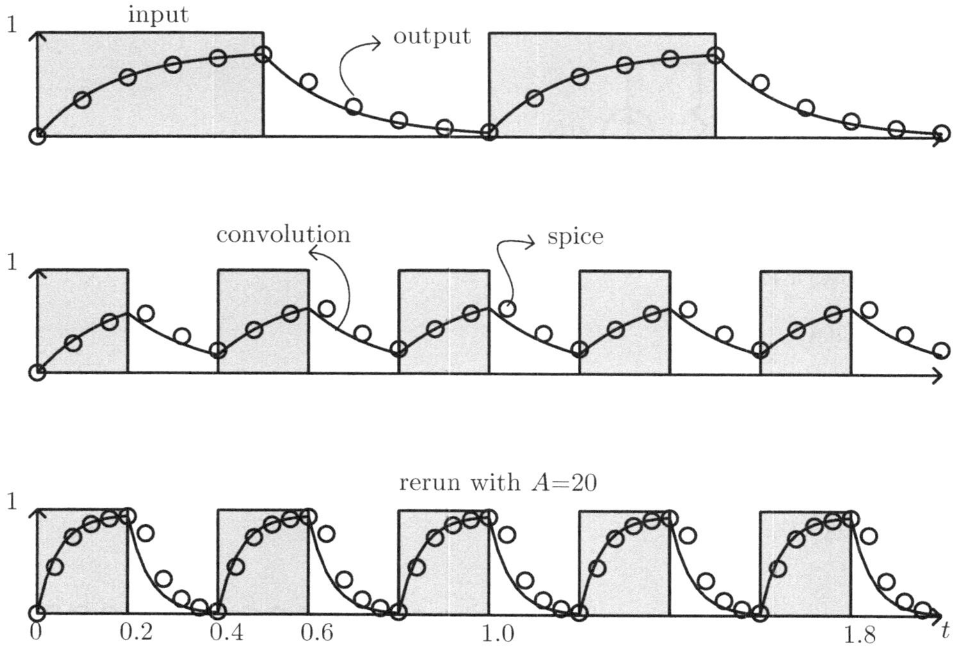

Fig. 24.20 Sample solution to Problem 4

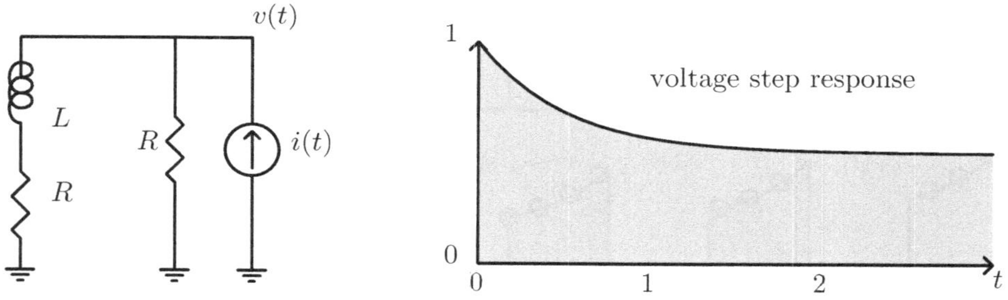

Fig. 24.21 Statement to Problem 5

width 0.5 and period 1; and then (b) periodic pulse of width 0.2 and period 0.4. Repeat the last step using larger gain $A = 20$. Compare results to SPICE; see sample results in Fig. 24.20.

5. Consider the *RL* network in Fig. 24.21. The step response for the case $R = 1\,\Omega$ and $L = 1\text{H}$ is given by

$$v(t) = \frac{1}{2}\left[1 + e^{-2t}\right]$$

Find the voltage response due to (a) single-cycled sine function with angular frequency 2π; (b) single-cycled sine function with angular frequency 2π, followed by a single-cycled sine function with angular frequency 4π; and (c) single-cycled sine function with angular frequency 2π, followed by a single-cycled sine function with angular frequency 4π, followed by a single-cycled sine function with angular frequency 6π! Compare to SPICE; see sample solution in Fig. 24.22

6. Consider the *LC* network in Fig. 24.23. The step response is measured to be

$$v(t) = 1.22\sin(2\pi \times 0.07t) + 0.28\sin(2\pi \times 0.19t)$$

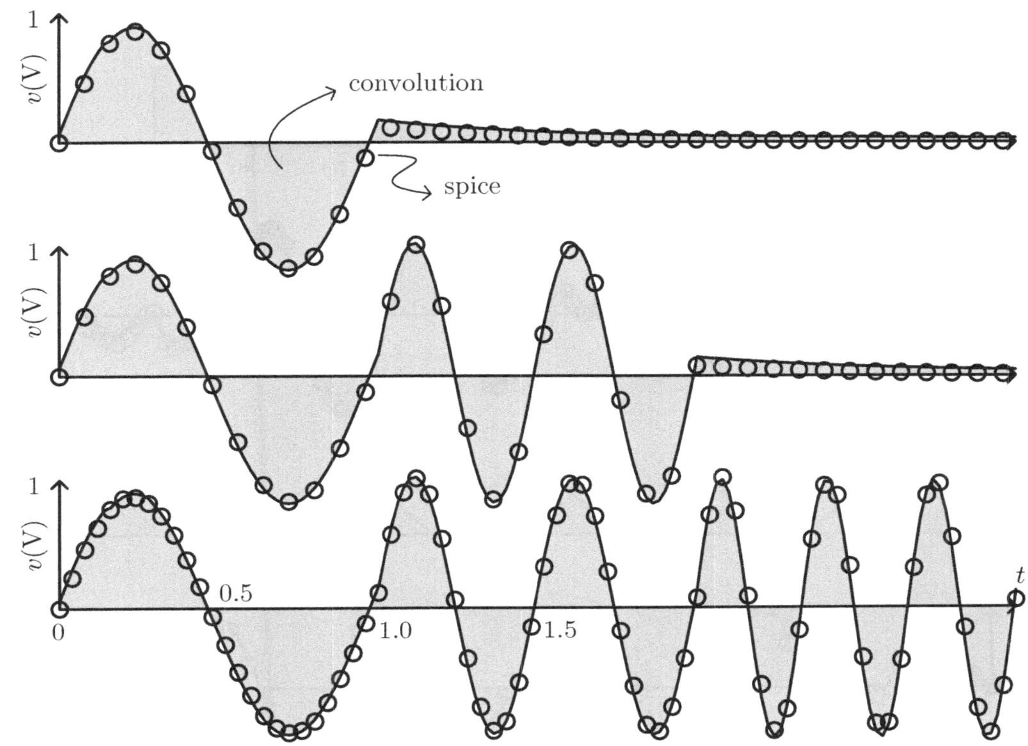

Fig. 24.22 Sample solution to Problem 5

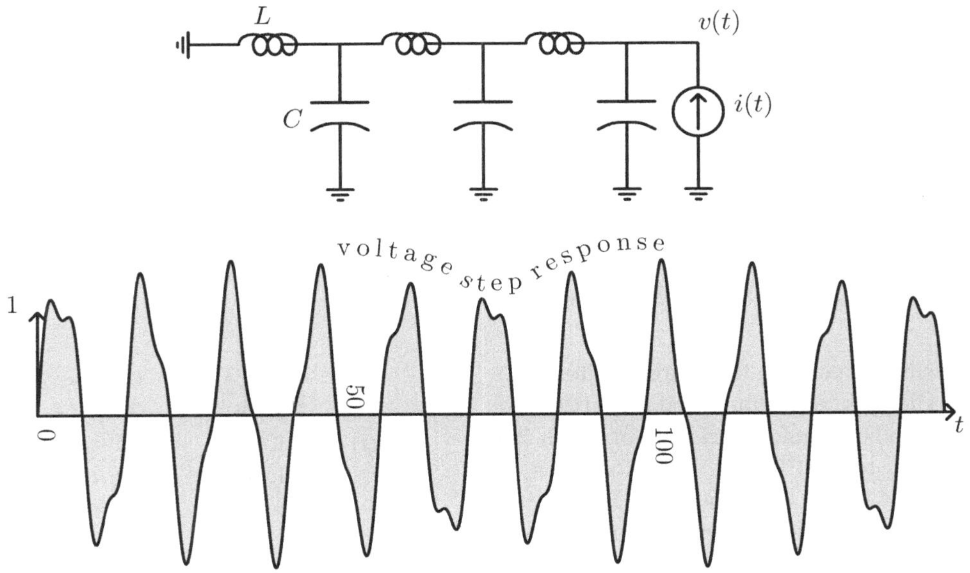

Fig. 24.23 Statement to Problem 6

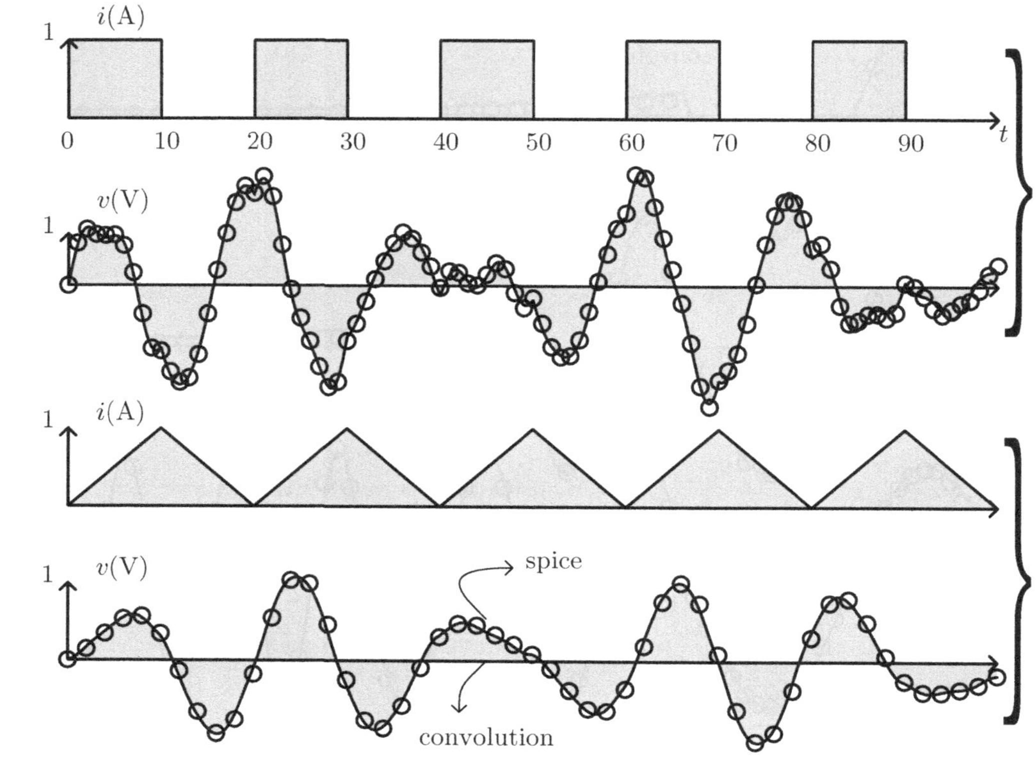

Fig. 24.24 Sample solution to Problem 6

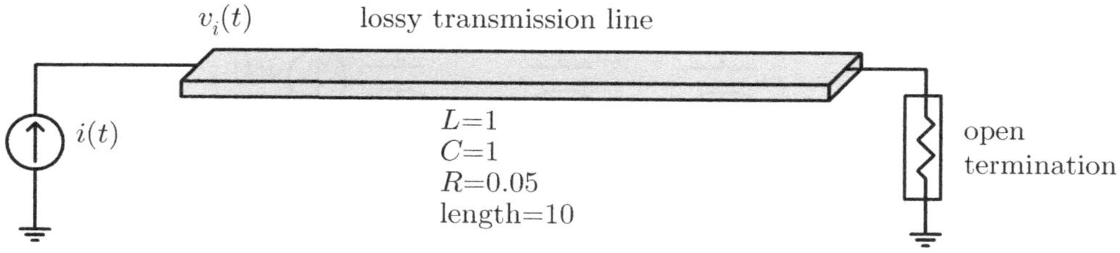

Fig. 24.25 Statement to Problem 7. *RLC* numbers are per meter

Using convolution find response due to (a) periodic pulse of width 10 and period 20; and (b) periodic symmetric triangle of width 20. Compare to SPICE; see sample solution in Fig. 24.24 Use $L = 1$H and $C = 1$F.

7. Consider the lossy transmission line shown in Fig. 24.25 which is open terminated on both sides. The voltage step response was measured and can be fit as

$$v(t) = \begin{cases} 1 + \frac{t}{40} & 0 < t < 20 \\ 1.7 + \frac{t}{20} & 20 < t < 40 \\ 1.6 + \frac{t}{14} & 40 < t < 60 \\ 1.3 + \frac{t}{12} & 60 < t < 80 \\ 0.9 + \frac{t}{11} & 80 < t < 100 \end{cases}$$

The step response is shown at the top of Fig. 24.26. Using convolution, find the response due to a sinusoid input of angular

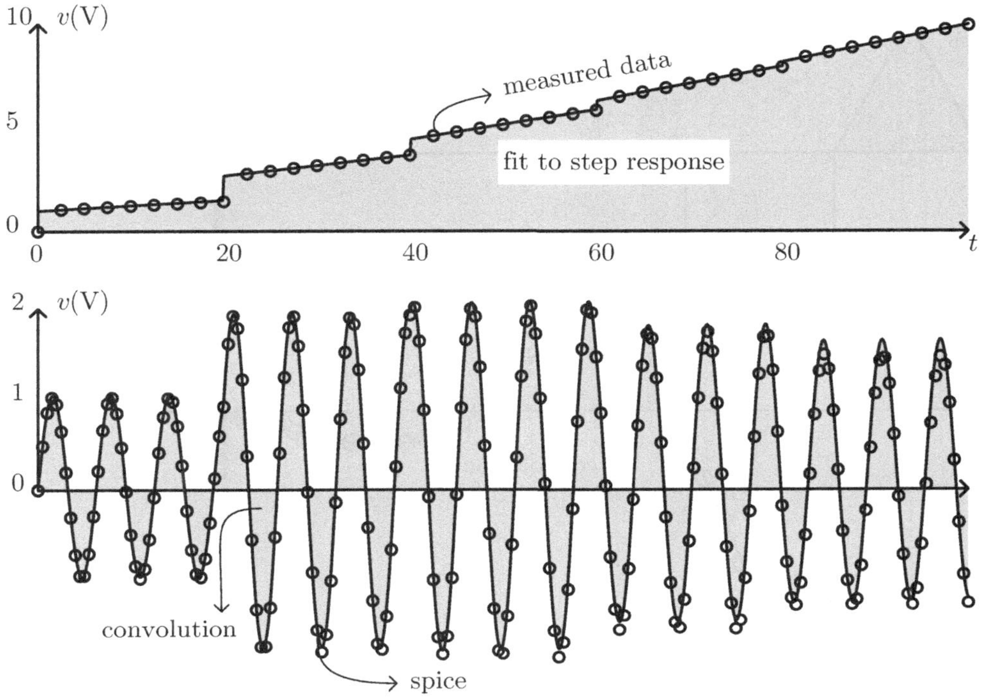

Fig. 24.26 Sample solution to Problem 7

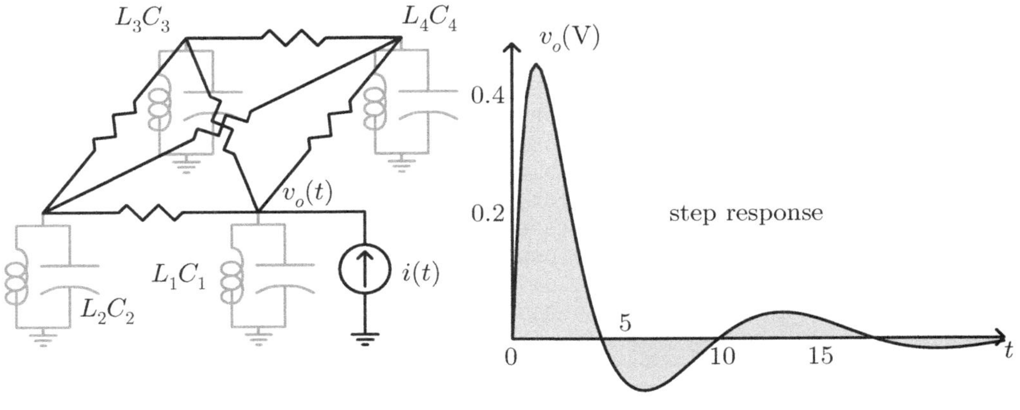

Fig. 24.27 Statement to Problem 8

frequency 1. Compare to SPICE; see sample solution in Fig. 24.26.

8. The *RLC* network in Fig. 24.27 has the unit step response shown on the right side. Assume the following

$$C_i = i, \qquad L_i = i, \qquad R = 2$$

The step response was measured and tabulated below. Each data point i corresponds to a time interval $t_i = i \times 0.25$.

Find the response due to a symmetric triangle, followed by a negative pulse as shown at the top of Fig. 24.28, which also shows sample solution. Compare to SPICE.

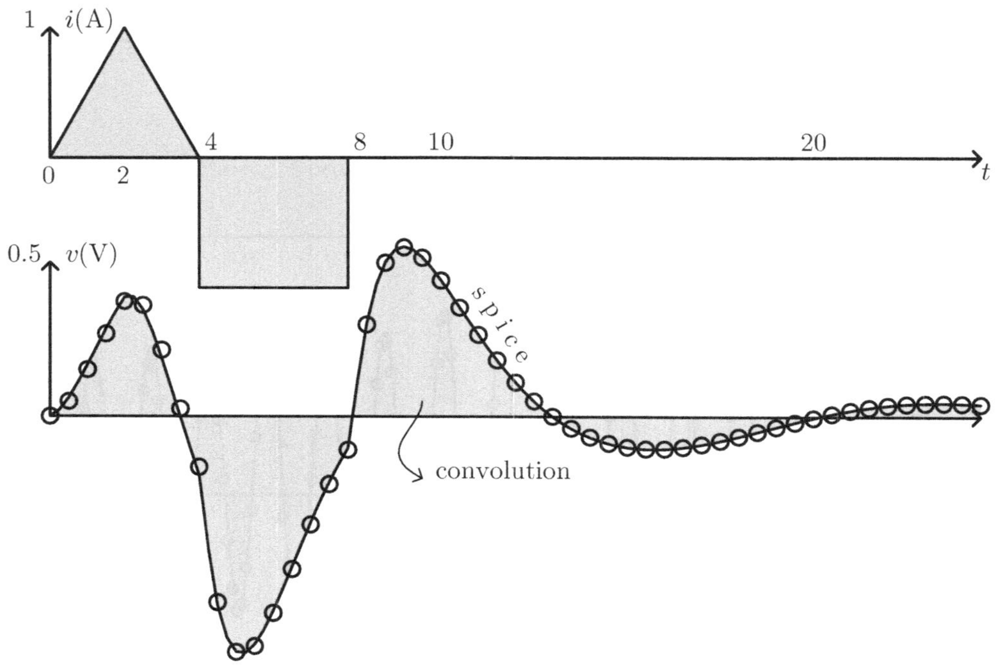

Fig. 24.28 Sample solution to Problem 8

```
0.000    0.172  0.319  0.408  0.453  0.466  0.455  0.429  0.392  0.349
0.304    0.257  0.212  0.168  0.128  0.091  0.058  0.029  0.003 -0.019
-0.037 -0.052 -0.064 -0.073 -0.080 -0.085 -0.087 -0.088 -0.087 -0.084
-0.081 -0.076 -0.070 -0.064 -0.058 -0.051 -0.043 -0.036 -0.028 -0.021
-0.014 -0.007  0.000  0.006  0.012  0.018  0.023  0.027  0.031  0.035
0.038    0.040  0.042  0.043  0.044  0.044  0.044  0.043  0.042  0.040
0.038    0.036  0.034  0.031  0.028  0.026  0.022  0.019  0.016  0.013
0.010    0.007  0.004  0.001 -0.002 -0.004 -0.006 -0.008 -0.010 -0.012
-0.014 -0.015 -0.016 -0.017 -0.017 -0.018 -0.018 -0.018 -0.018 -0.017
-0.017 -0.016 -0.015 -0.014 -0.013 -0.012 -0.011 -0.010 -0.009 -0.007
-0.006
```

25 Sampling and the Sampling Theorem

25.1 Introduction

No text about spectral, convolution, and numerical techniques would be complete without some coverage of the sampling theorem. In the process of going back and forth between continuous and discrete signals sampling happens, even if not totally evident. For example, when we do Fourier/Laplace transforms we sample the signal in the time domain at specific time points:

$$F(\omega) = \int f(t)e^{-j\omega t}dt \tag{25.1}$$

That is the Fourier transform at a specific frequency ω is the summation of the product of $f(t)$ times $e^{-j\omega t}$ at *specific* time points. Of course in theory we would have infinite time points, but in practice a finite number would suffice. Similarly when finding the inverse transform

$$f(t) = \frac{1}{2\pi}\int F(\omega)e^{j\omega t}d\omega \tag{25.2}$$

we are sampling the Fourier transform at specific ω points. It goes unsaid that if we pick enough t (or ω) data points we ought to expect an accurate spectrum (time or frequency wise). But how big of a sample rate is enough? To derive the sampling theorem we fall back on the delta train function. Really when we transform (say from time to frequency) we are replacing the continuous function with a delta train with each delta having the magnitude $f(t)\Delta t$. Having replaced the function with a sampled delta train one, we proceed with the transform, but this time on the scaled delta train. So first we refresh our memory about the delta train function and its transform.

25.2 The Delta Pulse Train

We can think of sampling in the time domain as multiplying the signal with the delta pulse train function. This function which is periodic in T is defined as

F. Badrieh, *Spectral, Convolution and Numerical Techniques in Circuit Theory*,
https://doi.org/10.1007/978-3-319-71437-0_25

$$\begin{aligned}\text{pulse train} &= \cdots + \delta(t+3T) + \delta(t+2T) + \delta(t+T) + \delta(t) \\ &\quad + \delta(t-T) + \delta(t-2T) + \delta(t-3T) + \cdots \\ &= \sum_{-\infty}^{\infty} \delta(t-nT) \end{aligned} \tag{25.3}$$

The sampling process is shown in Fig. 25.1. Of course the finer the sampling, the closer the delta samples to each other. Remember a delta function has infinite magnitude but unity area. These delta function still have infinite magnitude but their area is $f(t)$. Notice that the spacing between the delta functions Δt or T does not yet come into the picture, but will come back later; see Eq. (25.8) below.

25.3 Fourier Transform of Train Pulse Function

We know from Sect. 11.9 that a periodic delta function in the time domain (with period T) transforms also to a periodic delta function in the frequency domain, in accordance with

$$\text{train pulse in } T \rightarrow \omega_s \sum_{-\infty}^{\infty} \delta(\omega - n\omega_s), \quad \omega_s = \frac{2\pi}{T} \tag{25.4}$$

Notice that the transform has the proportionality factor ω_s and is periodic in ω_s. That is, the longer the sampling period T in time, the closer the pulses in frequency domain; and vice versa.

25.4 Multiplication in the Time Domain Equates to Convolution in Frequency Domain

Having covered the delta train function and its Fourier transform we move to the tying link between sampling in the time domain and impact on the frequency domain. Assume we know the Fourier transform of the sampled signal $f(t)$ as $F(\omega)$; then multiplying this signal in the time domain with the train pulse equates to convolving its Fourier transform with the Fourier of the train pulse. That is

$$\mathcal{F}[\text{train pulse} \times f(t)] = \frac{1}{2\pi} F(\omega) * \omega_s \sum_{-\infty}^{\infty} \delta(\omega - n\omega_s) \tag{25.5}$$

Again sampling the signal in the time domain is tantamount to multiplying it by the delta train function. Now that the signal has been digitized we move to the frequency domain. The spectrum of the sampled signal behaves in accordance with the above equation. No claim is made that this spectrum is the same as that of the original signal $f(t)$; in fact it won't! The catch is can we still get the spectrum of the original signal from this new spectrum or has it be lost (distorted) forever?

25.5 Assumption That Sampled Signal Is Bandwidth Limited

For the sake of simplicity, assume for now that the continuous time signal is bandwidth limited; that is, it has frequency content up to some frequency ω_b shown in Fig. 25.2. This assumption becomes important as will be shown next.

25.6 Frequency Spectrum of Sampled Signal

Now that we know the frequency spectrum of the continuous signal and that of the train pulse, we do *convolution* in the frequency domain. Recall the important fact that *convolving with a delta function amounts to duplicating the function*

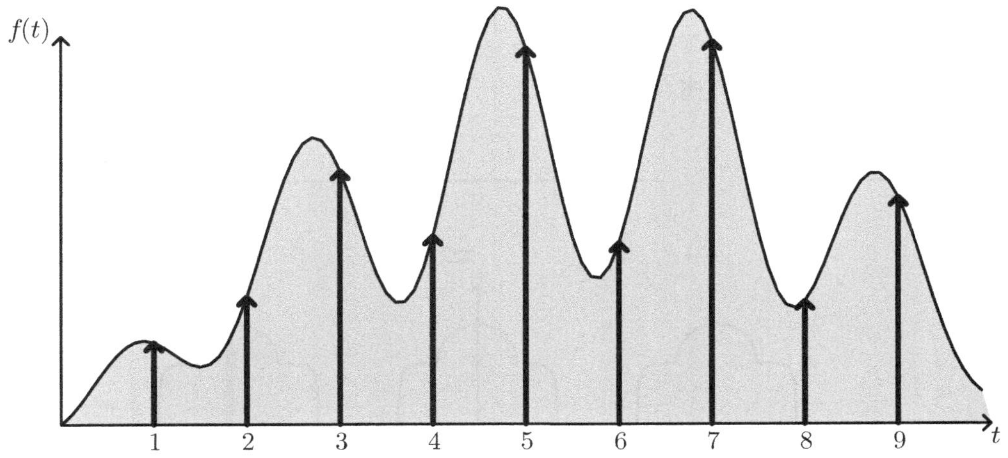

Fig. 25.1 Sampling continuous signal every T seconds

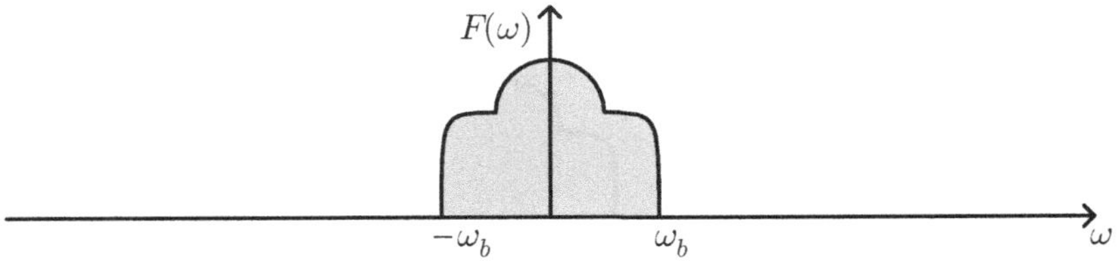

Fig. 25.2 Signal to be sampled is bandwidth limited

around the delta function; that is, the convolution result is the same signal but now centered around the delta function. With this in mind, we predict that the sampled signal would have the Fourier transform as shown in Fig. 25.3.

Notice and as was proclaimed above the frequency spectrum of the sampled signals does *not* equate to the spectrum of the original signal. It has great resemblance but it does not match identically. More specifically the spectrum of the sampled signal is a *periodic* version of the spectrum of the original signal. It is inevitable—the moment we discretized the original signal the moment its spectrum becomes periodic. The trick though is to either make the period infinite—in which case the sampling rate would be infinite (impractical); or figure a way to extract the spectrum of the original signal from this periodic one.

25.7 Sampling Theorem

Looking closer at Fig. 25.3 we may ask the question: what happens if we make ω_s smaller and smaller, which is equivalent to making the sampling period T larger and larger? Clearly if ω_s is smaller than twice the bandwidth of the continuous signal, then we will get *mixing* in the frequency domain as shown in Fig. 25.4. We are now *unable to filter the spectrum of the sampled signal*, and hence will lose some information about the original signal. However, if the sample frequency ω_s is large enough then we can simply use a low-pass filter to *filter* the spectrum of the sampled signal, and upon doing inverse transform we are guaranteed to regain the original signal as shown in Fig. 25.5.

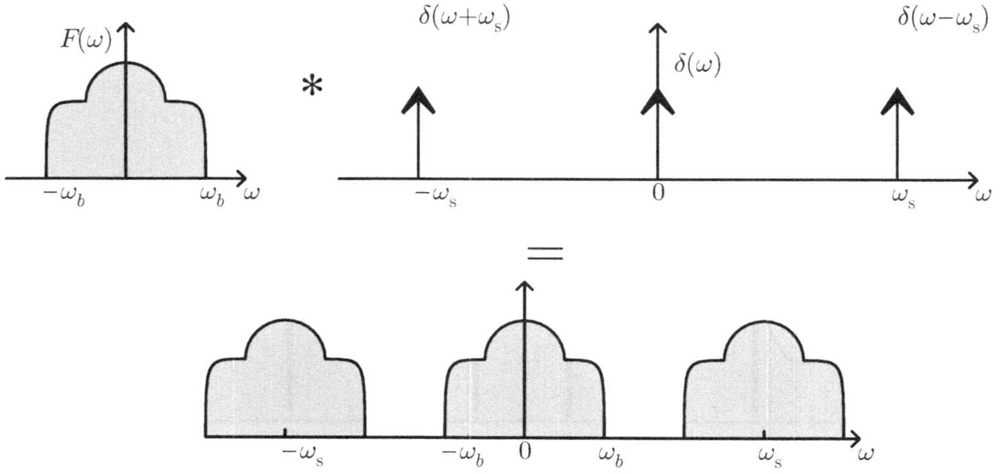

Fig. 25.3 Frequency spectrum of sampled signal

Fig. 25.4 Continuous signal not sampled enough

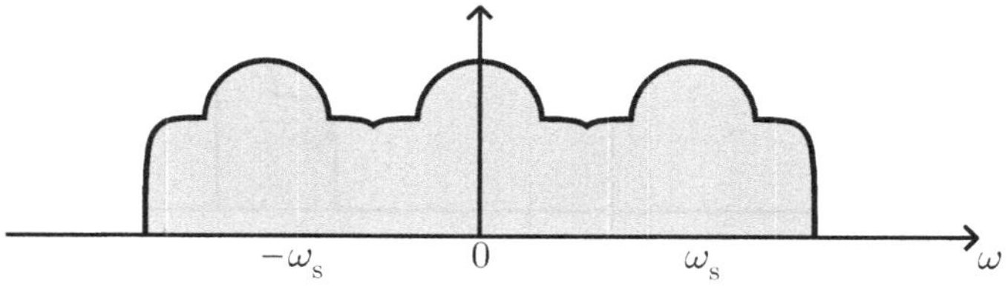

Fig. 25.5 Continuous signal sampled more than enough, and high-frequency filtering

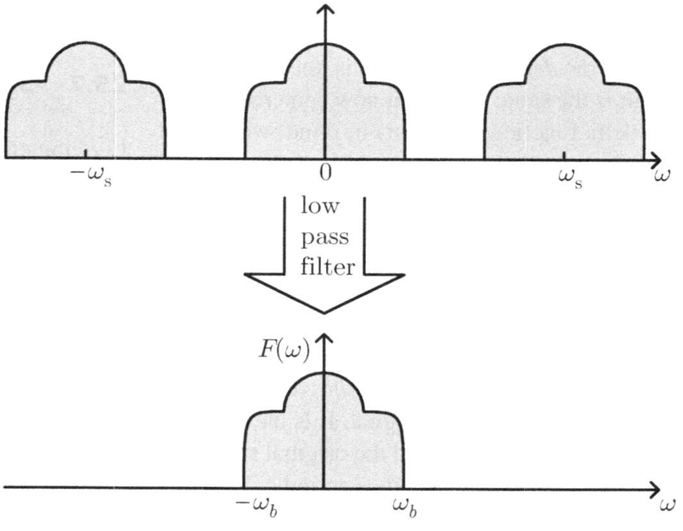

We then finally arrive at the sampling theorem: *In order to be able to regenerate a sampled signal, it has to be sampled at a rate that is at least twice its bandwidth*! That is

$$\boxed{\omega_s \geq 2\omega_b} \tag{25.6}$$

For example assume that a signal has bandwidth $\omega_b = \pi$. This would imply that $\omega_s \geq 2\pi$. Recalling $\omega_s = \frac{2\pi}{T}$ this would imply that the spacing between the impulse samples, in the time domain, T has to be less than 1!

25.8 Low-Pass Filtering

We've established that multiplying in the time domain amounts to convolution in the frequency domain. Since we are convolving with a delta train we end up with duplicates of the spectrum of the signal at the location of the deltas in the frequency domain. It would seem then that in order to retrieve the spectrum of the signal all that needs to be done is low-pass filtering. Namely filter the convolved signal from $-\omega_s/2$ to $\omega_s/2$. We are almost there, but we need to worry about a proportionality constant. In particular, and in accordance with Eq. (25.5) we have to divide the filtered signal by ω_s and multiply by 2π. So we have

$$\boxed{\text{spectrum of original signal} = \frac{2\pi}{\omega_s} \times \text{low-pass filter of spectrum of convolved signal}} \tag{25.7}$$

Or equivalently in the time domain

$$\boxed{\begin{array}{l}\text{original signal} = T\times \\ \text{inverse transform of [low-pass filter of spectrum of convolved signal]}\end{array}} \tag{25.8}$$

where T is the time sampling duration, and where we have used $\omega_s = \frac{2\pi}{T}$. So we have come full circle from sampling a signal, finding the sampled signal spectrum, low-pass filtering, and finally finding inverse transform. That is, assume we have a continuous signal. We don't wish to use a trillion (!) data points, so we sample the signal every T. We next find the Fourier transform of the sampled signal. What the sampling theorem says is that we should be now able to regain the original signal if we made T small enough, or equivalently $\omega_s = \frac{2\pi}{T}$ large, at least $\omega_s \geq 2\omega_b$. In that case we can simply low-pass filter the new spectrum and regain the original signal (after doing inverse transform).

It is crucial to remember that if we don't do the low-pass filtering—in other words if we use the full spectrum of the sampled signal, and if we do inverse transform we will *not* get the original signal; rather, we will get a spiky signal (comprised of sharp delta functions) which has the same envelope as the original signal (but not exactly the same). That is, we will get back the *sampled signal*—and *not the signal itself*! So doing filtering in the frequency domain amounts to doing *smearing* in the time domain, such that the spiky delta functions smear out and disperse back into the continuous original signal.

25.9 Example Application of the Sampling Theorem: Sine Function

Consider a transient signal simply given by

$$f(t) = \sin 2\pi t \tag{25.9}$$

It is a sine with period 1! In this example the bandwidth is simply $\omega_b = 2\pi$. Now let's sample it at ω_s angular sampling frequency. Next low-pass filter the resulting spectrum and do inverse transform. How does the reconstructed signal look like as a function of sampling frequency? Figure 25.6 shows original signal (gray), samples points (dark circles), and reconstructed signal (after filtering, hollow circles). As evident from the graphs, the reconstructed signal does in fact resemble the sampled points—but does it resemble the actual signal? The answer is no, unless the

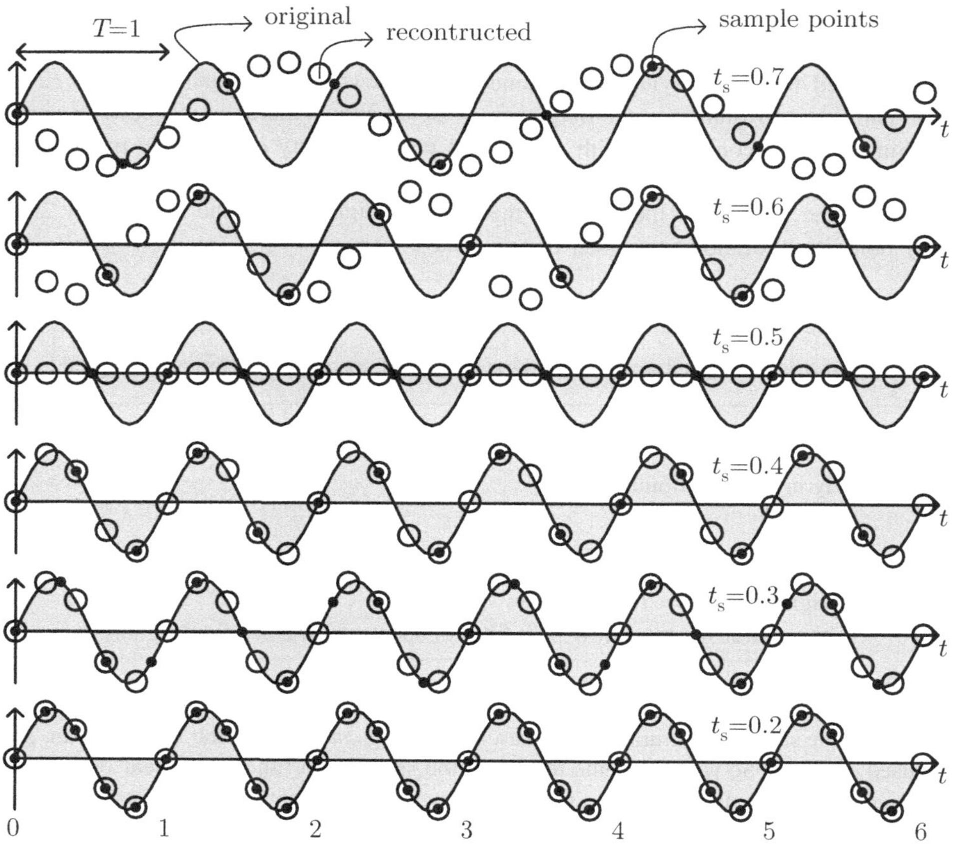

Fig. 25.6 Application of sampling theorem on sine function

sampling frequency is at least 2× the bandwidth of the original signal. In this case we need

$$\omega_s > 4\pi, \quad \text{or} \quad \frac{2\pi}{t_s} > 4\pi, \quad \text{or} \quad t_s < 0.5 \tag{25.10}$$

That is, if the sampling time t_s is less than 0.5 we are guaranteed we can reconstruct the signal. That is why for the top three graphs ($t_s > 0.5$) the reconstructed signal does *not* look like the original one, but for the bottom three graphs ($t_s < 0.5$) the reconstructed signal *matches identically* the original one! This is a first-hand and applied proof how the sampling theorem works. It's not like if we don't sample enough we would get a completely bogus signal! As shown above, with under-sampling we still get a signal that *crosses* the original signal at the *original* sampling points. But that alone does not guarantee that the reconstructed signal matches the original one for *all* time! To get match for all time we need to sample fine enough.

25.10 Second Example Application of the Sampling Theorem: Pulse Function

The pulse function (defined between −0.5 and 0.5) is not really a periodic function, so in theory it has an infinite bandwidth. Nonetheless we can apply the sampling theorem to it taking advantage that while its bandwidth is infinite, the spectrum dies off rapidly in frequency. So in theory, if we sample at high enough rate, we could bank on the fact that the spacing between the two sampling frequencies at some point would dwarf the bandwidth of the pulse; i.e., the case where $\omega_s \gg \omega_b$. To demonstrate this, take the pulse (of width 1) and as shown in Fig. 25.7 and crudely sample it using only three samples. The spectrum of the samples function would be

$$F_3(\omega) = 1 + 2\cos\omega t_s, \qquad t_s = 0.5 \quad (25.11)$$

where we have used

$$\delta(t) \rightarrow 1, \quad \text{and} \quad \delta(t - t_0) \rightarrow e^{-j\omega t_0} \quad (25.12)$$

If we were now to filter this function [between $-\omega_s/2$ and $\omega_s/2$ (and multiply by $2\pi/\omega_s$)], then take the inverse transform we would get the reconstructed signal as shown on the right of top row. If we were to reduce the sample size to 0.25, as shown in the second row, such that we have 5 total samples, the resulting spectrum would be

$$F_5(\omega) = 1 + 2\cos\omega t_s + 2\cos\omega 2t_s, \quad t_s = 0.25 \quad (25.13)$$

Taking the inverse transform we get results on the right of second row. Continuing on, if we were to reduce the sample size to 0.10, as shown in the third row, such that we have a total of 11 samples, the resulting spectrum would be

$$F_{11}(\omega) = 1 + 2\cos\omega t_s + 2\cos\omega 2t_s + 2\cos\omega 3t_s + 2\cos\omega 4t_s + 2\cos\omega 5t_s \\ t_s = 0.10 \quad (25.14)$$

Taking the inverse transform we get results on the right of third row. It becomes evident that the more samples we apply, the more accurate the reconstructed signal. In fact it is apparent that not too many samples are needed to get a reasonable duplicate of the starting signal. Again, this signal had "infinite bandwidth," but nonetheless the sampling theorem proved very useful as applied to this case.

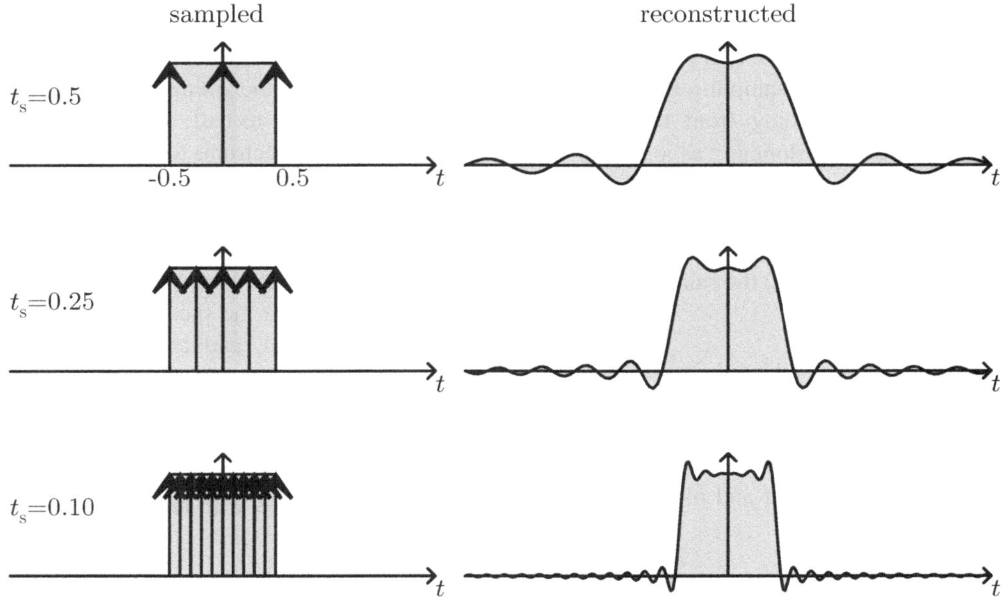

Fig. 25.7 Application of sampling theorem on pulse function

25.11 Summary

The sampling theorem is extremely important and it states that starting with a continuous time signal if we sample it via the train delta function, find the Fourier transform, then low-pass filter it we should be able to regain the original signal provided the sampling frequency ω_s is at least $2\times$ the bandwidth of the starting signal ω_b. In essence it states that if we are to faithfully preserve a *time* signal in the *frequency* domain, then we need to time-sample it fine enough. If we do so then we should be able to take the spectrum of the sampled signal and filter out of it the high frequency content, and the resulting spectrum would be identical to the spectrum of the original signal (as if were not sampled). If we don't then the spectrum of the sampled signal will get mixed and we will not be able to regain the spectrum of the original signal, nor be able to reconstruct the original signal via inverse transform.

We demonstrated the sampling theorem on a couple of examples where we showed very clearly the impact of under-sampling and over-sampling. Under-sampling does provide a signal; but that signal—while it matches the original signal *only* at the sampled time points—does not match the original signal for all time. Over-sampling, on the other hand, is guaranteed to reproduce the starting signal. It is worth noting that the whole logic of the sampling theorem would apply equally well going from the frequency domain to the time domain, as opposed to—and in conjunction with the current treatment—going from the time domain to the frequency one. Time or frequency—they are all (at least mathematically speaking) the same; just a matter of variable change!

25.12 Problems

1. Consider the pulse function shown at the top of Fig. 25.8. Sample it and plot the spectrum having used 3, 5, 7, 9, and 11 samples. Compare it to the analytic spectrum of a pulse. What is the trend of the spectrum of the sampled function as the number of samples increases? See sample solution in the same figure.
2. Consider the spectrum of the sampled signal again in Fig. 25.8. As evident the spectrum is periodic in angular frequency. By examining the period of the spectrum, deduce what the sampling time t_s must have been. What is the trend in t_s as the angular frequency period is increased?
3. Consider the odd pulse function shown at the top of Fig. 25.9. Sample it and plot the spectrum having used 3, 5, 7, 9, and 11 samples. Compare it to the analytic spectrum of the odd pulse. What is the trend of the spectrum of the samples function as the number of samples increases? See sample solution in the same figure.
4. Consider the sinc function shown at the top of Fig. 25.10. Sample it and plot the spectrum having used 11, 21, and 31 samples. What should the spectrum approach as we sample more and more—what is the limit? See sample solution in the same figure.
5. Consider the pulse function of width 2 as shown at the top of Fig. 25.11. Sample it using five samples and plot spectrum. Repeat but this time reduce the strength of the two samples at the edges to half, as shown in figure. Compare both spectrums to actual spectrum of pulse—which is better? Why?
6. Consider the absolute sine function defined between -3 and 3 with angular frequency π and as shown at the top of Fig. 25.12. Sample it every 0.5 s, plot the spectrum, calculate the inverse transform, and compare to original signal. Repeat but this time use time step of 0.25 s. See sample solution in the same figure.
7. Consider the absolute sine function defined between -3 and 3 with angular frequency π

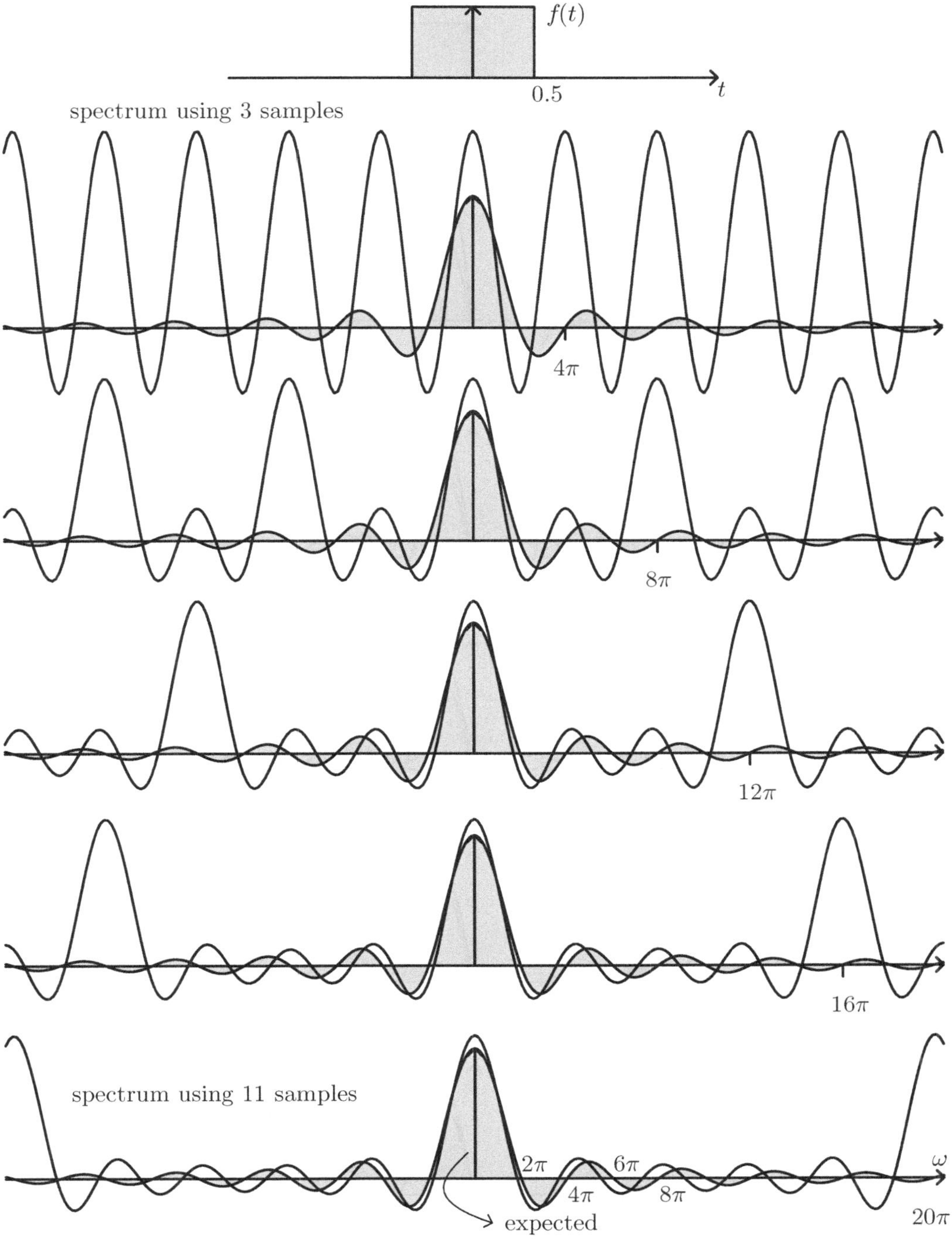

Fig. 25.8 Sample solution to Problem 1

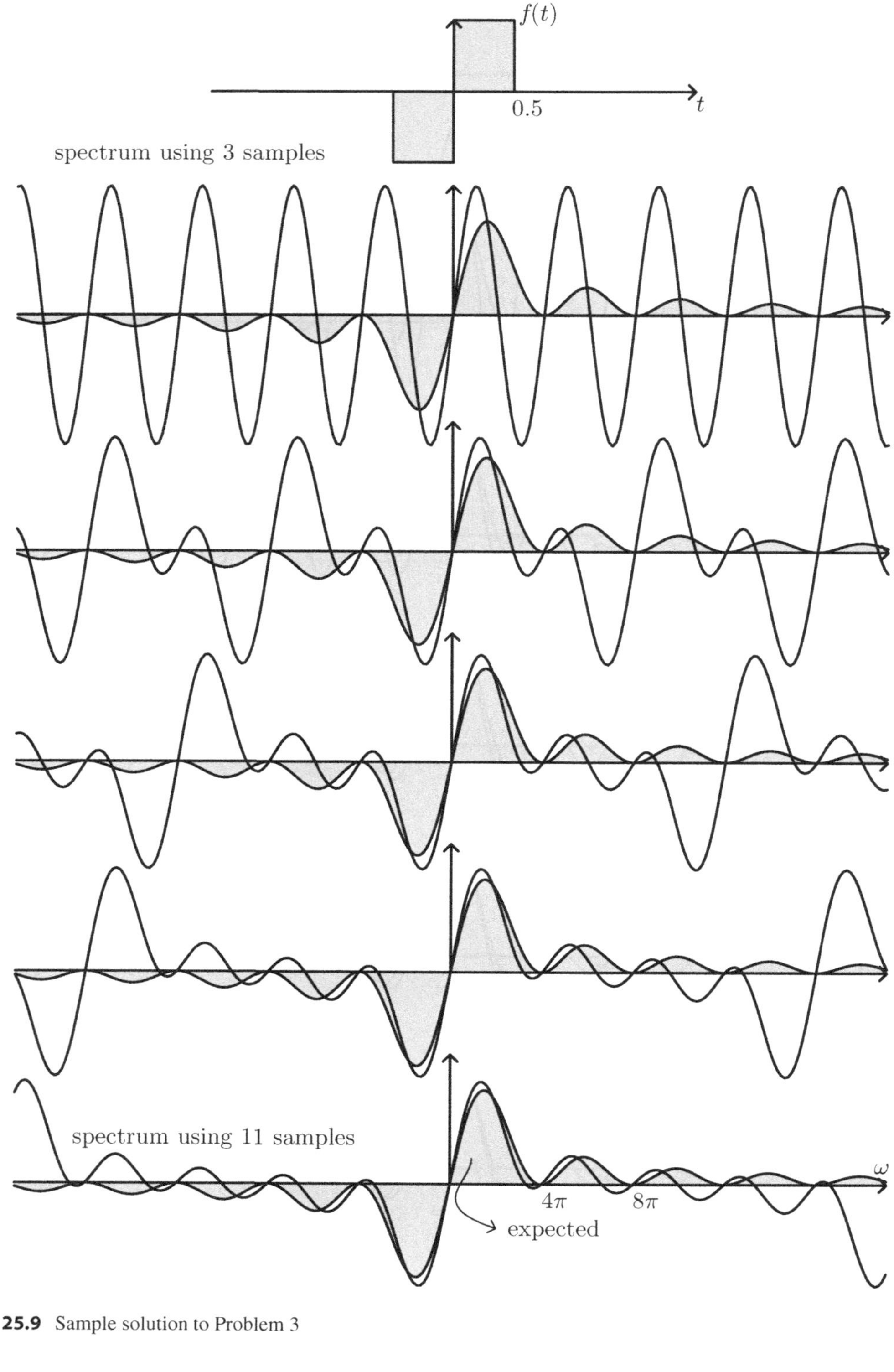

Fig. 25.9 Sample solution to Problem 3

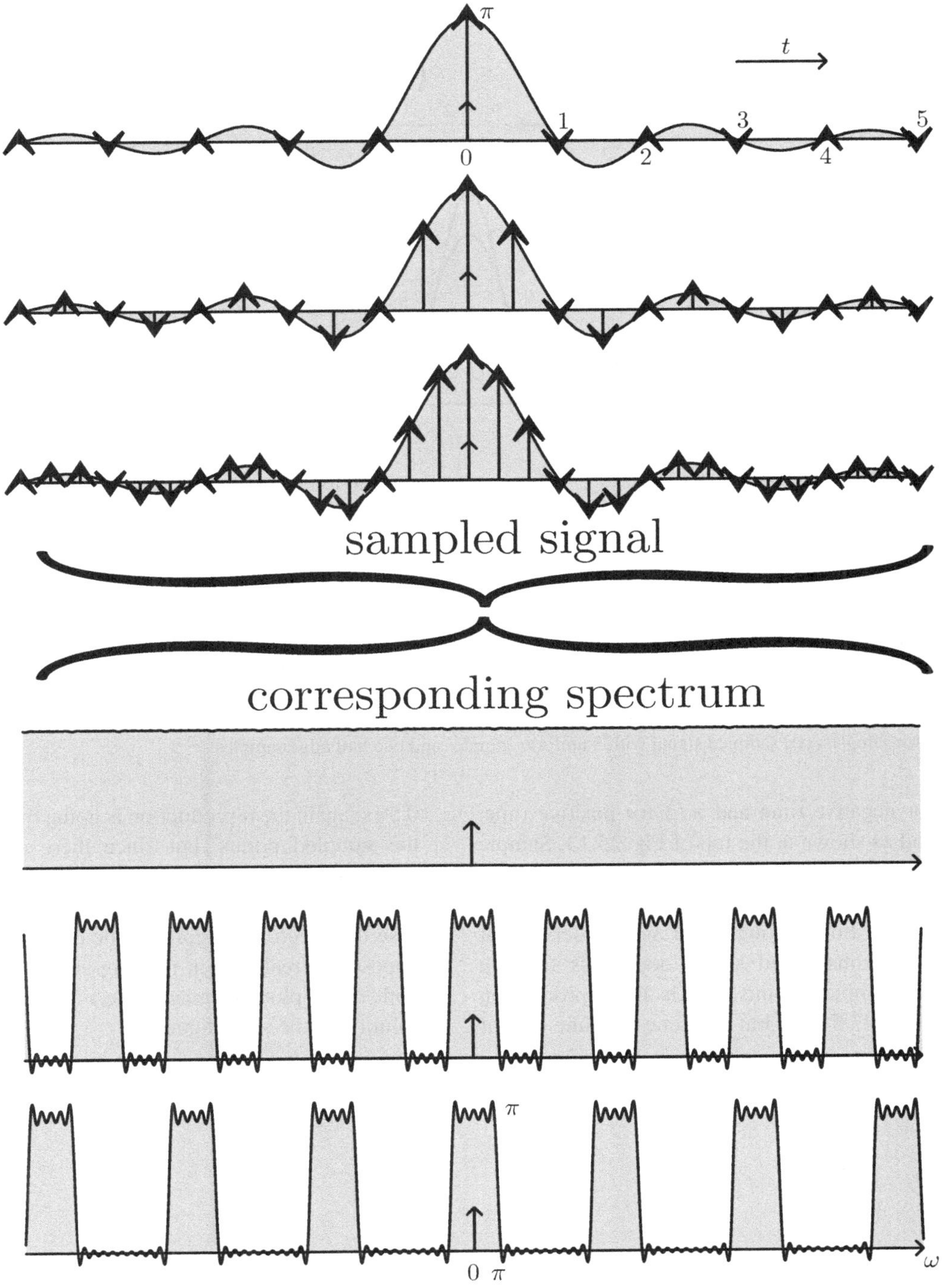

Fig. 25.10 Sample solution to Problem 4

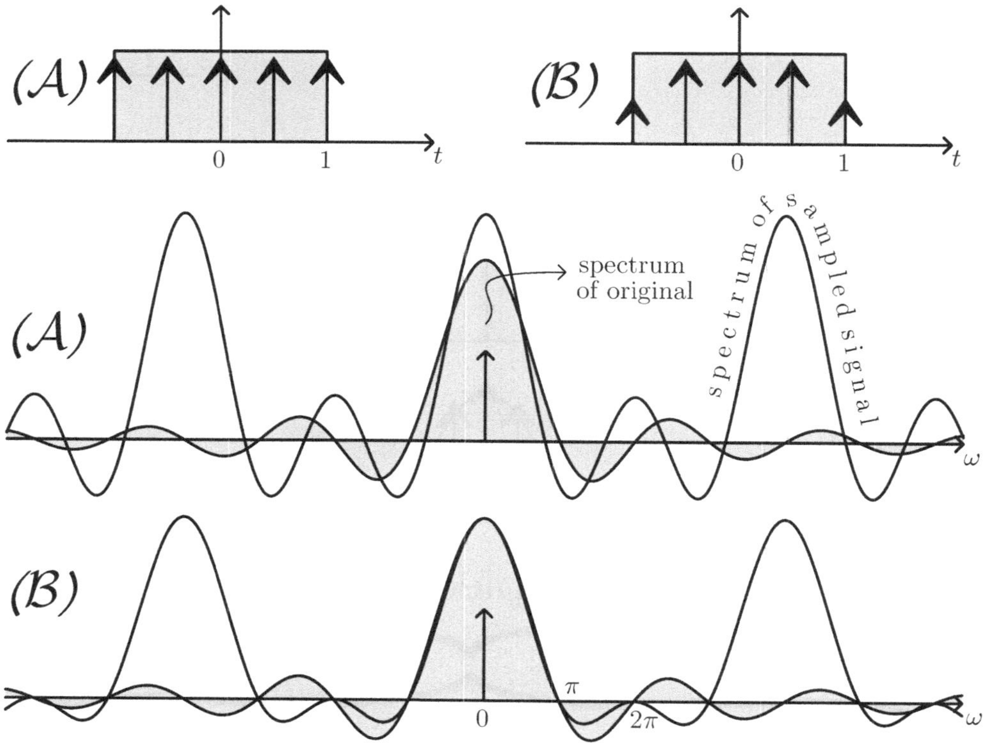

Fig. 25.11 Sample solution to Problem 5. Top – time domain; bottom – frequency domain. (A) sampled signal with 5 uniform samples; (B) sampled signal with 3 uniform samples and two half edge samples

for negative time and $\pi/3$ for positive time, and as shown at the top of Fig. 25.13. Sample it every 1.5 s, plot the spectrum, calculate the inverse transform, and compare to original signal. Sure enough you will observe that the reconstructed signal does pass through the sampled points, but is the reproduction faithful? Repeat but this time use time step of 0.50 s; again the reproduction is going through the sampled points, but since there are so many of them, the overall signal is becoming closer to the original signal. (Since in this last case the spectrum is going to be complex (as opposed to real), when plotting spectrum for both cases plot the magnitude.) See sample solution in the same figure.

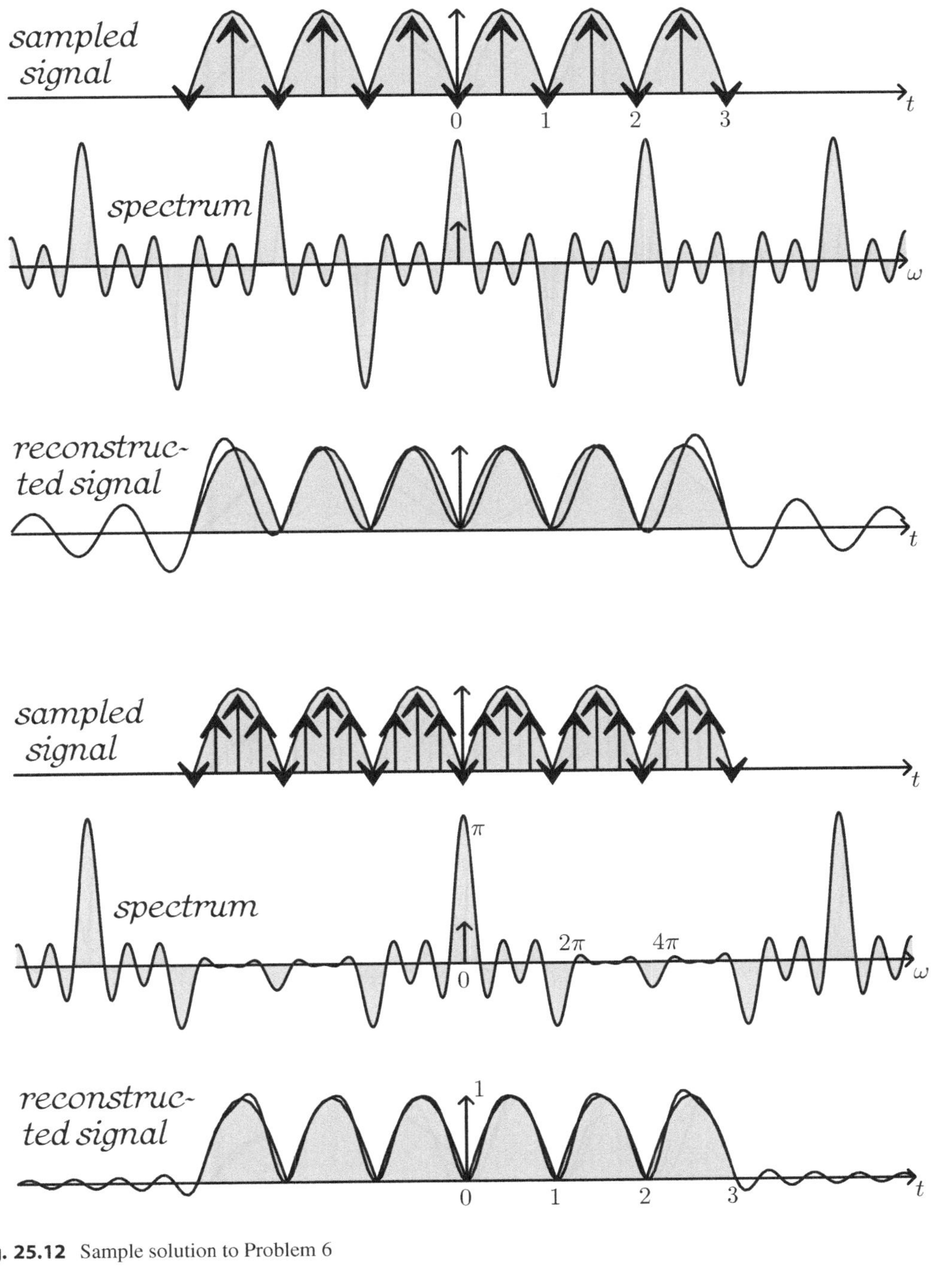

Fig. 25.12 Sample solution to Problem 6

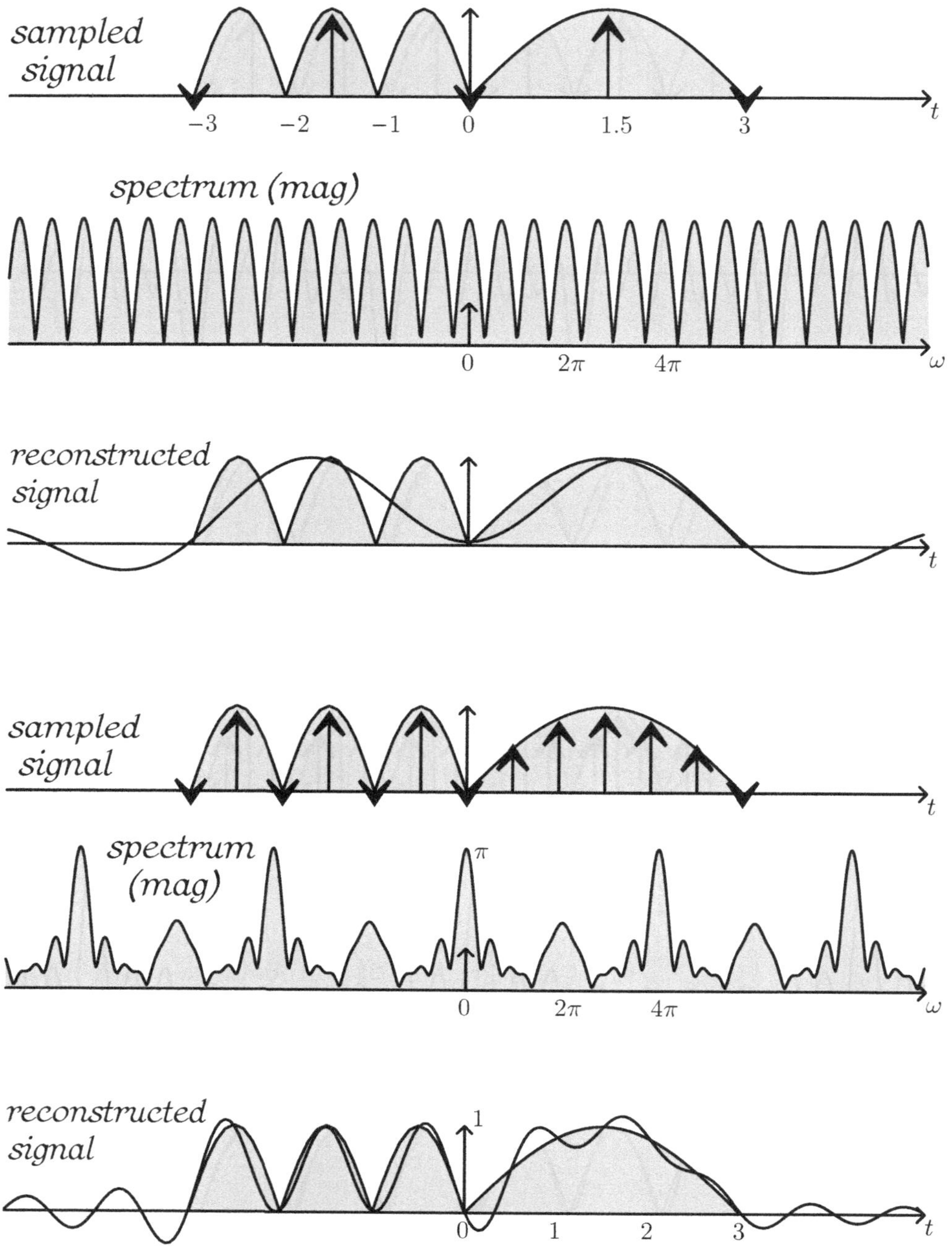

Fig. 25.13 Sample solution to Problem 7

26 Transfer Functions

26.1 Introduction

With the exception of the introductory few chapters, most of the material covered so far dealt abstractly either in the frequency domain (Fourier/Laplace transforms) or in the time domain (convolution). We dealt with signals more or less as abstract quantities. Starting from this chapter and onwards we will start putting those concepts/tools into practice. And what better start than transfer functions. A transfer function ties something to something! For example a transfer function tying input current to output voltage is called an *impedance* transfer function. On the other hand a transfer function tying input voltage to output current is called an *admittance* transfer function and so forth. A transfer function is not a number—it is a function! Most commonly it is a function in the frequency domain. Also, and as will be shown in a later chapter a transfer function does not necessarily tie a single input to a single output. For example in multi-port networks it is the practice to tie a single port input to all other ports output and so forth. But how exactly do we go from the time domain with operating conditions governed by differential equations to a transfer function in the frequency domain? We need a carrier signal!

26.2 Carrier Signal

In order to measure something, we need to apply something. We could for example apply an impulse current, or a step one and measure output voltage. However, soon enough we will see that transient characteristics of output don't much resemble those of input, at least for storage devices, such as capacitors and inductors. For example, applying an impulse current to a cap results in a step voltage; trying to relate a step to impulse is not straight forward.

One particular stimulus, the (complex) exponential, proves to be very useful as a carrier vehicle to derive transfer functions. That is

F. Badrieh, *Spectral, Convolution and Numerical Techniques in Circuit Theory*,
https://doi.org/10.1007/978-3-319-71437-0_26

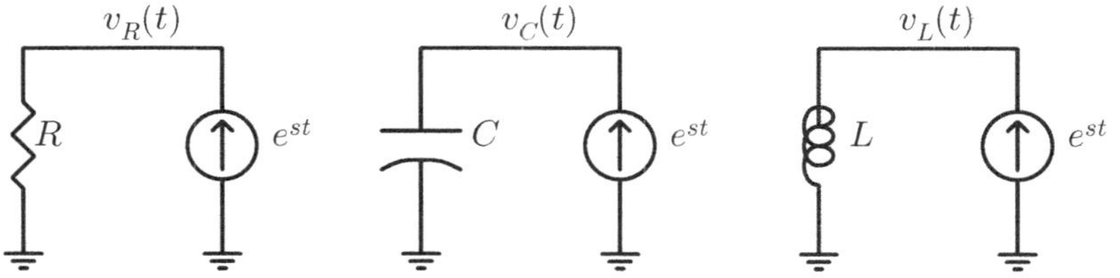

Fig. 26.1 Complex exponential current applied to *RLC* elements

$$\text{input stimulus in the form } \boxed{e^{st}}, \quad s = \sigma + j\omega_0 t \tag{26.1}$$

This is shown in Fig. 26.1. For example, for the case of a simple resistor of magnitude R

$$\text{if input current is} \quad e^{st} \quad \text{then output voltage is} \quad Re^{st} \tag{26.2}$$

Another example, for the case of an inductor of magnitude L

$$\text{if input current is} \quad e^{st} \quad \text{then output voltage is} \quad sLe^{st} \tag{26.3}$$

And finally, for the case of a capacitor of magnitude C

$$\text{if input current is} \quad e^{st} \quad \text{then output voltage is} \quad \frac{1}{sC}e^{st} \tag{26.4}$$

This last one warrants some elaboration. If input current is $i(t) = e^{st}$ then output voltage is

$$v(t) = \frac{1}{C}\int_{-\infty}^{t} i(\tau)d\tau = \frac{1}{C}\int_{-\infty}^{t} e^{s\tau}d\tau = \frac{1}{sC}e^{s\tau}\Big|_{-\infty}^{t} = \frac{1}{sC}e^{st} \tag{26.5}$$

where we have made use of the fact that

$$\lim_{t\to-\infty} e^{st} = 0, \qquad \sigma > 0 \tag{26.6}$$

26.3 Impedance Transfer Function

Notice that in all cases, the *output voltage looks exactly like the input current*, albeit scaled (by R for the resistor, sL for the inductor case, or divided by sC for the cap one). Since the output has the same form as the input, then might we not drop the e^{st} from the picture, and simply focus on the scaling factor? This scaling factor does not depend on time, and in this case depends only on frequency. That is,

$$\begin{aligned} &\text{transfer function of resistor} && Z_R(s) = R, \\ &\text{transfer function of inductor} && Z_L(s) = sL, \\ &\text{transfer function of capacitor} && Z_C(s) = \frac{1}{sC} \end{aligned} \tag{26.7}$$

Again, when we refer to the cap as having $Z_C(s) = \frac{1}{sC}$, this means

$$\text{if input current to cap is} \quad e^{st} \quad \text{then output voltage of cap is} \quad \frac{1}{sC}e^{st} \tag{26.8}$$

26.4 Relation Between Transfer Function and Impulse Response

Without loss of generality, let's set σ to zero such that

$$s = j\omega_0 \tag{26.9}$$

Let's take a sample case that of the inductor; we know that

$$Z_L(\omega_0) = j\omega_0 L, \qquad \text{meaning} \tag{26.10}$$

$$\text{if input current to inductor is} \quad e^{j\omega_0 t} \quad \text{then output voltage of inductor is} \quad j\omega_0 L e^{j\omega_0 t} \tag{26.11}$$

How about if input current was $e^{j\omega_1 t}$? Then we would have

$$i(t) = e^{j\omega_1 t} \Rightarrow v(t) = j\omega_1 L e^{j\omega_1 t} \tag{26.12}$$

How about if input current was $e^{j\omega_2 t}$? Then we would have

$$i(t) = e^{j\omega_2 t} \Rightarrow v(t) = j\omega_2 L e^{j\omega_2 t} \tag{26.13}$$

What about input current having two exponentials $e^{j\omega_0 t}+e^{j\omega_1 t}$? By superposition we would have

$$i(t) = e^{j\omega_0 t}+e^{j\omega_1 t} \Rightarrow v(t) = j\omega_0 L e^{j\omega_0 t}+j\omega_1 L e^{j\omega_1 t} \tag{26.14}$$

How about input current having many exponentials?

$$i(t) = \sum_n e^{j\omega_n t} \Rightarrow v(t) = \sum_n j\omega_n L e^{j\omega_n t} \tag{26.15}$$

Finally, how about input current having infinitely many exponentials? Then we would have

$$i(t) = \int e^{j\omega t} d\omega \Rightarrow v(t) = \int j\omega L e^{j\omega t} d\omega \tag{26.16}$$

Scale both sides by $1/(2\pi)$

$$i(t) = \frac{1}{2\pi}\int e^{j\omega t} d\omega \Rightarrow v(t) = \frac{1}{2\pi}\int j\omega L e^{j\omega t} d\omega \tag{26.17}$$

What does the current term mean?

$$\frac{1}{2\pi}\int e^{j\omega t} d\omega = ? \tag{26.18}$$

From prior experience we immediately identify this as the definition of the delta function $\delta(t)$!! This means that the response of an impulse current is

$$i(t) = \delta(t) \Rightarrow v(t) = \frac{1}{2\pi}\int j\omega L e^{j\omega t} d\omega \tag{26.19}$$

How about the integral term on the right side; what does that represent?

$$\frac{1}{2\pi}\int j\omega L e^{j\omega t} d\omega = ? \tag{26.20}$$

Again from prior experiment we immediately identify this as the inverse Fourier transform of some function (the transfer function which in this case is $j\omega L$) which we call $h(t)$

$$h(t) = \frac{1}{2\pi}\int [j\omega L] e^{j\omega t} d\omega \tag{26.21}$$

But this $h(t)$ is nothing more than the impulse response!! That is

$$\text{if input current is } \delta(t) \text{ then output voltage is } h(t) \tag{26.22}$$

Hence

the Fourier/Laplace transform of the impulse response is the transfer function!! (26.23)

Put another way

$$\mathcal{F}[h(t)] = H(\omega), \qquad \text{where} \quad H(\omega) \quad \text{is the transfer function} \tag{26.24}$$

Hence we have identified the relation between the impulse response and the transfer function. If we know one, we know the other (by forward or inverse transforms). Let's use some examples to solidify the meaning of the transfer function, applied here in the form of impedance which ties input current to output voltage.

26.5 Series *RC* Impedance Transfer Function

Consider the simple series *RC* shown in Fig. 26.2; the total impedance is

$$Z(s) = R + \frac{1}{sC} = \frac{1 + RCs}{sC} \tag{26.25}$$

The impedance transfer function has the following limits:

$$Z(0) \sim \infty, \qquad \text{and} \tag{26.26}$$

$$Z(\infty) = R \tag{26.27}$$

In other words at DC the impedance is infinite because the cap is open; at high frequency the cap shorts and we fall back on R. The impedance transfer function is shown in Fig. 26.3. Notice that the phase starts at -90 (capacitor dominated) and ends up at zero (resistor dominated). The inflection point happens when

$$RCs = 1, \quad \text{or when} \quad s = \frac{1}{RC} \tag{26.28}$$

For the case at hand ($R = 1\,\Omega$ and $C = 1\,\text{mF}$) this happens at $s = 1000\,\text{rad/s}$, and this is confirmed in the figure.

Before plunging into plethora of examples let's reflect on what was accomplished in this simple example since the underlying operation here will carry on for all subsequent examples. First we assigned the individual impedances to each circuit element. Since the elements were in series we simply added them, in the frequency domain. We got the transfer function in Eq. (26.25) in terms of s. This transfer function *most often* will be *complex*! To plot the function we must unwind s in terms of actual frequency; for simplicity here we set $\sigma = 0$ such that $s = j\omega$. Next we plotted the *magnitude* and *phase* of the transfer function, though sometimes we may plot real and imaginary. Most often the magnitude is plotted on a log/log scale while

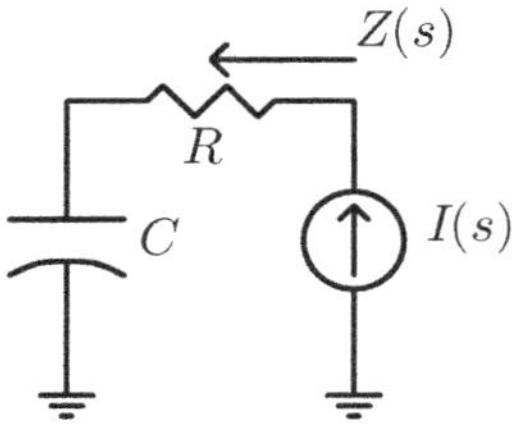

Fig. 26.2 Series *RC* and input impedance

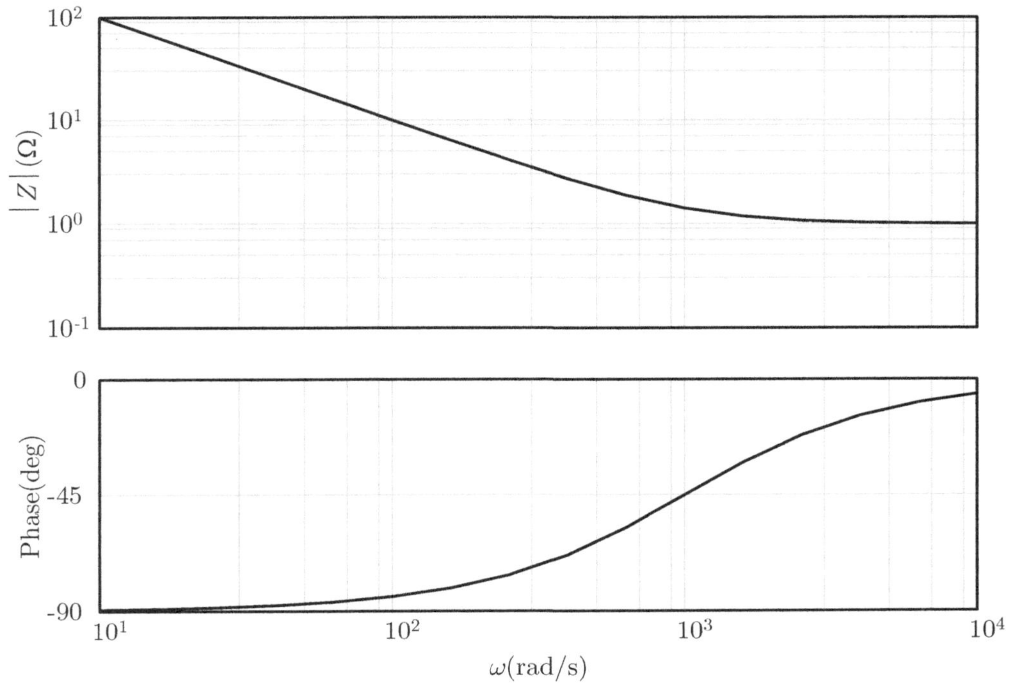

Fig. 26.3 Transfer function of series *RC* network (case of $R = 1$ and $C = 1$ mF)

phase on log(x)/linear(y). Both magnitude and phase will vary in frequency. As we will see over and over, typically caps result in linear decrease in magnitude (on a log/log scale) while sustaining a $-90°$ phase. Inductors act the opposite—linear ramp and positive 90°. Resistors act flat on magnitude and result in no phase change. As we will see the overall shape of magnitude and phase can assume pretty much anything. But the crucial thing to take from the transfer function is that it *tells us everything we need to know about the operation of the circuit*! Equipped with it, we are assured that we can figure output not only for a complex exponential input, but for *any input*!

26.6 Parallel *RC* Impedance Transfer Function

Consider the simple parallel *RC* shown in Fig. 26.4; We know the relevant impedances of the two elements:

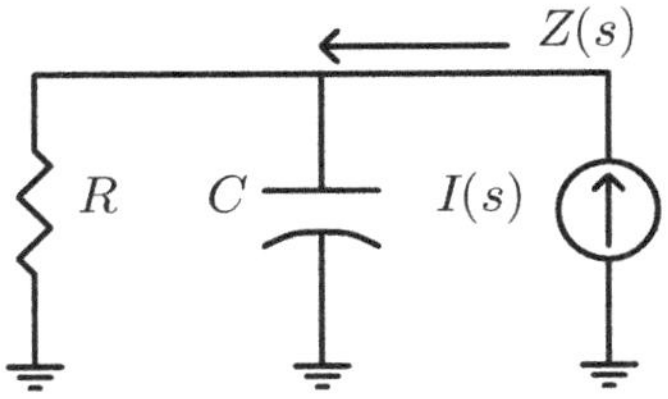

Fig. 26.4 Parallel *RC* and input impedance

$$Z_R(s) = R; \qquad Z_C(s) = \frac{1}{sC} \tag{26.29}$$

The net impedance of the two blocks is the parallel one;

$$Z = \frac{R\frac{1}{sC}}{R + \frac{1}{sC}} = \frac{R}{1 + RCs} \tag{26.30}$$

Factor an *RC* from the denominator and get

$$\boxed{Z = \frac{1}{C}\frac{1}{s + \frac{1}{RC}}} \tag{26.31}$$

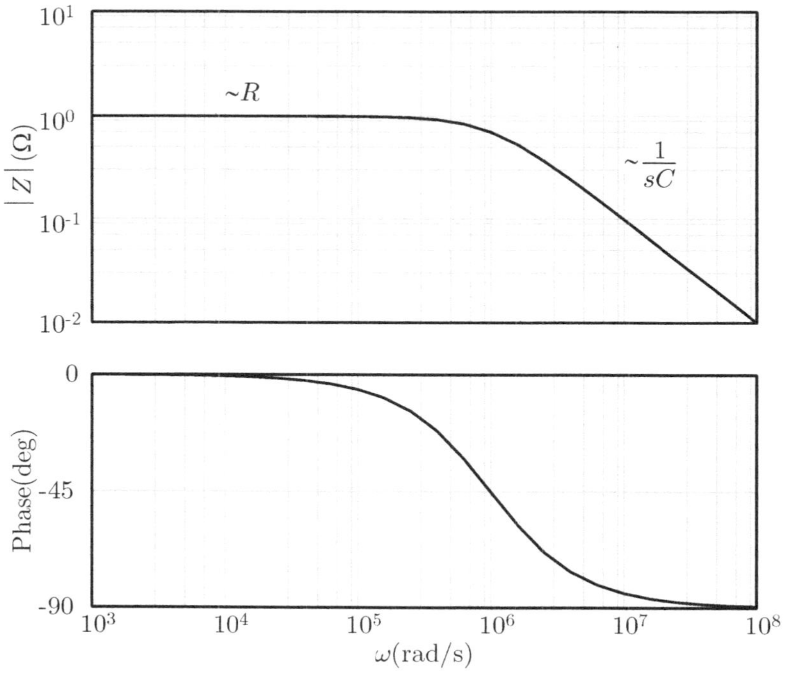

Fig. 26.5 Transfer function of parallel *RC* network (case $R = 1\,\Omega$ and $C = 1\,\mu\text{F}$)

This is a complex function with magnitude and phase. Notice that at low frequency we get

$$Z(0) = \frac{1}{C}\frac{1}{0 + \frac{1}{RC}} = R \tag{26.32}$$

That is at DC the cap is open and we are simply left with R. On the other hand, at high frequency we get

$$Z(\text{high frequency}) = \frac{1}{sC} \tag{26.33}$$

which is nothing other than the impedance of the cap. And finally at really high frequency we get

$$Z(\infty) \sim 0 \tag{26.34}$$

meaning at high frequency the cap shorts and we end up with zero impedance. That is if current is changing so fast, the cap never really charges! And hence output voltage would be zero. Recalling Ohm's law

$$V = IZ \tag{26.35}$$

we conclude that at high frequency impedance is in fact zero (since V is zero, while I not). The transfer function (for the case $R = 1$ and $C = 1\,\mu\text{F}$) is shown in Fig. 26.5. Notice the DC limit matches R and the slope of the high frequency is $1/C$. Since the DC limit is R we start at 0 phase. At high frequency the cap takes control and we revert to $-90°$. Notice at the inflection frequency

$$s = \frac{1}{RC} = \frac{1}{1 \times 10^{-6}} = 1 \times 10^6\,\text{rad/s} \tag{26.36}$$

the phase is exactly $-45°$. The impact of R is shown in Fig. 26.6. Notice that with larger R we get larger values at DC, but the frequency inflection point moves to the left. In the limit of very large R we recover the $1/s$ behavior, which is that due to the cap (since the R is open by now).

The impact of C is shown in Fig. 26.7. Notice that the DC value is independent of C (as it depends only on R). Notice also that with larger cap the roll off frequency shifts to the left; that is, larger cap results in lower bandwidth. Finally notice that larger cap results in lower impedance at high frequency (which makes sense).

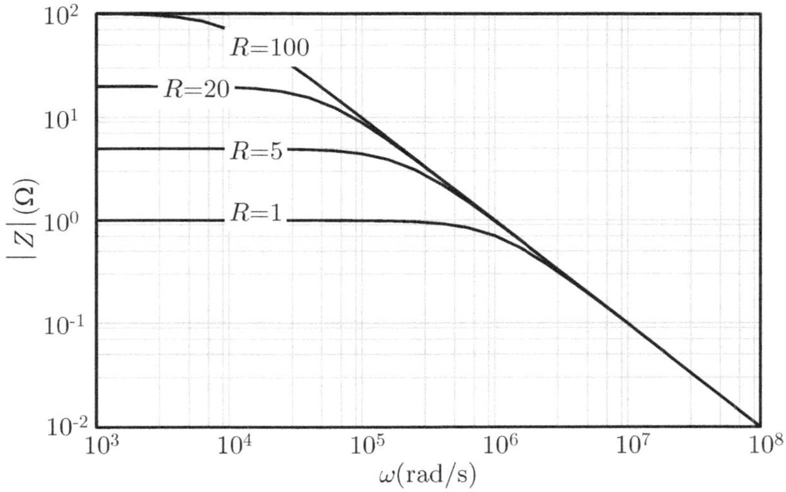

Fig. 26.6 Impact of R on transfer function of parallel RC network ($C = 1\,\mu$F)

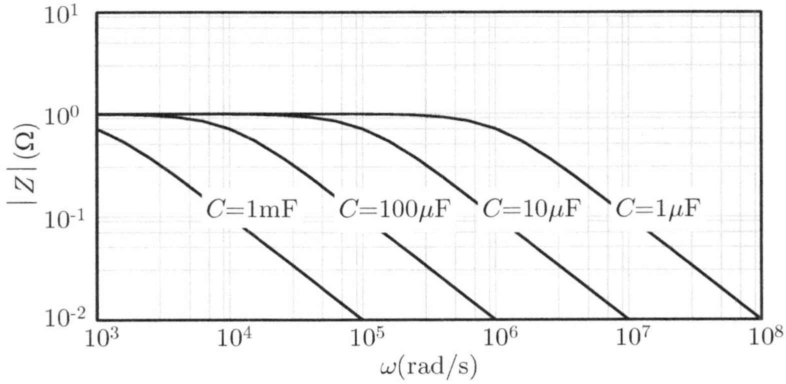

Fig. 26.7 Impact of C on transfer function of parallel RC network ($R = 1\,\Omega$)

26.7 Parallel *LC* Impedance Transfer Function

Consider next the simple parallel LC shown in Fig. 26.8; we wish to find its impedance transfer function. The individual impedances are

$$Z_L = sL; \qquad Z_C = \frac{1}{sC} \tag{26.37}$$

The effective, parallel impedance is

$$Z(s) = \frac{sL\frac{1}{sC}}{sL + \frac{1}{sC}} = \frac{sL}{1 + LCs^2} \tag{26.38}$$

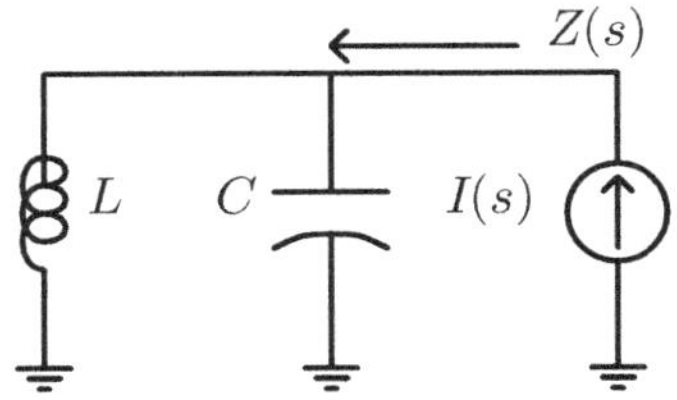

Fig. 26.8 Parallel LC and input impedance

Factor an LC from the denominator and get

$$Z(s) = \frac{1}{C}\frac{s}{s^2 + \frac{1}{LC}} = \boxed{\frac{1}{C}\frac{s}{s^2 + \omega_0^2}, \quad \omega_0^2 = \frac{1}{LC}} \tag{26.39}$$

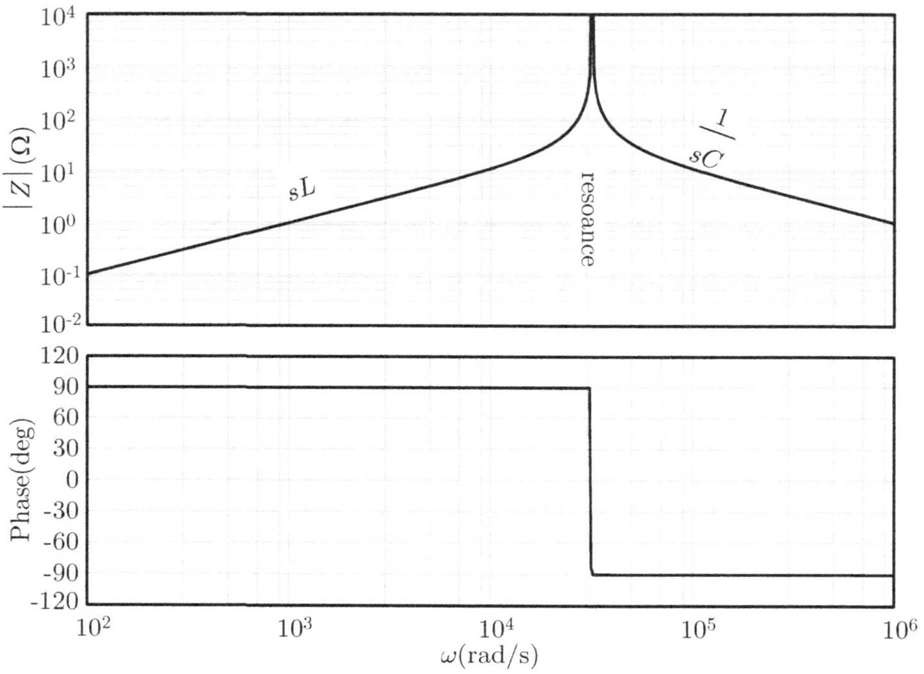

Fig. 26.9 Transfer function of parallel LC network; $C = 1\,\mu$F and $L = 1$ mH

Notice that at low frequency we get the limit

$$\lim_{s\to 0} Z(s) = \frac{1}{C}\frac{s}{\omega_0^2} = \frac{1}{C}sLC = sL \qquad (26.40)$$

which is nothing more than the inductive impedance; that is, at low frequencies the network acts like an inductor. The phase there is $+90°$. At high frequency we have the limit

$$\lim_{s\to\infty} Z(s) = \frac{1}{C}\frac{1}{s} = \frac{1}{sC} \qquad (26.41)$$

which is nothing more than the capacitive impedance; that is, at high frequency the impedance is dominated by the capacitor. The phase there would be $-90°$. The impedance transfer function is shown in Fig. 26.9. Notice the formation of a *resonance* right at the point

$$s^2 = \omega_0^2 \Rightarrow \omega = \frac{1}{\sqrt{LC}} \qquad (26.42)$$

For the case $L = 1$ mH and $C = 1\,\mu$F we get

$$\omega = \frac{1}{\sqrt{10^{-9}}} = 31622\,\text{rad/s} \qquad (26.43)$$

The impedance literally explodes there because we have a zero in the denominator:

$$Z(j\omega_0) \sim \frac{1}{(j\omega_0)^2 + \omega_0^2} = \frac{1}{-\omega_0^2 + \omega_0^2} = \frac{1}{0} = \infty \qquad (26.44)$$

It will be only dissipated by a nonzero σ selection. We will speak more of this but this really means that the voltage will incrementally grow beyond bound. As the input current continues oscillating (definition of complex exponential) the output voltage continues to grow (that is peak-to-peak) till it blows up! Equivalently we say the network has a high Q. The phase transition (from 90° to $-90°$) happens right at the resonance point.

26.8 Parallel *RLC* Impedance Transfer Function

Knowing the parallel *LC* impedance we move to the parallel *RLC* one as shown in Fig. 26.10. We know that

$$Z_{LC}(s) = \frac{1}{C}\frac{s}{s^2+\omega_0^2}, \qquad \omega_0^2 = \frac{1}{LC} \tag{26.45}$$

To find total impedance we put Z_{LC} in parallel with R:

$$Z(s) = \frac{R\frac{1}{C}\frac{s}{s^2+\omega_0^2}}{R+\frac{1}{C}\frac{s}{s^2+\omega_0^2}} = \frac{\frac{1}{C}\frac{s}{s^2+\omega_0^2}}{1+\frac{1}{RC}\frac{s}{s^2+\omega_0^2}} \tag{26.46}$$

$$\boxed{Z(s) = \frac{1}{C}\frac{s}{s^2+\frac{1}{RC}s+\omega_0^2}} \tag{26.47}$$

Notice that at low frequency impedance approaches that of the inductor

$$\lim_{s\to 0} Z(s) = \frac{1}{C}\frac{s}{\omega_0^2} = \frac{1}{C}\frac{s}{\frac{1}{LC}} = sL \tag{26.48}$$

On the other hand at that high frequency impedance approaches that of the cap

$$\lim_{s\to\infty} Z(s) = \frac{1}{C}\frac{s}{s^2} = \frac{1}{sC} \tag{26.49}$$

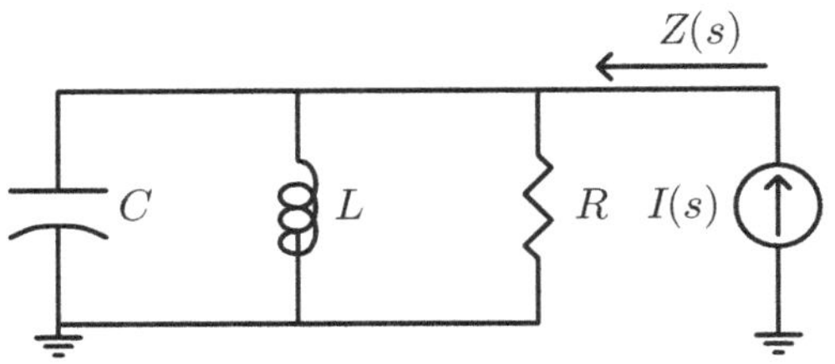

Fig. 26.10 Parallel *RLC* and input impedance

So far as impact of R, notice that for very large R we fall back on the ideal parallel *LC* impedance

$$\lim_{R\to\infty} Z(s) = \frac{1}{C}\frac{s}{s^2+\omega_0^2} \tag{26.50}$$

Taking the other limit of R we notice that as it approaches zero we get the impedance of R

$$\lim_{R\to 0} Z(s) = \frac{1}{C}\frac{s}{\frac{s}{RC}} = R \tag{26.51}$$

(This would hold so long as R is smaller than sL; if that is not the case, the sL would dominate.) Figure 26.11 shows impedance function for various R values. The phase always starts at 90° which is that of an inductor and terminates at −90° which is that of a capacitor. The change happens around ω_0. Notice from the figure that the case of small R tends to negate the impact of both the inductor and the cap; we can see that by observing the flat region in the magnitude plot and the zero phase region in the phase plot.

26.9 Series *LC* Impedance Transfer Function

The series *LC* is shown in Fig. 26.12; we wish to find its impedance transfer function. By inspection we have

$$Z(s) = sL + \frac{1}{sC} = \frac{1+s^2LC}{sC} \tag{26.52}$$

At low frequency we get the cap impedance since that dominates total impedance:

$$\lim_{s\to 0} Z(s) = \frac{1}{sC} \tag{26.53}$$

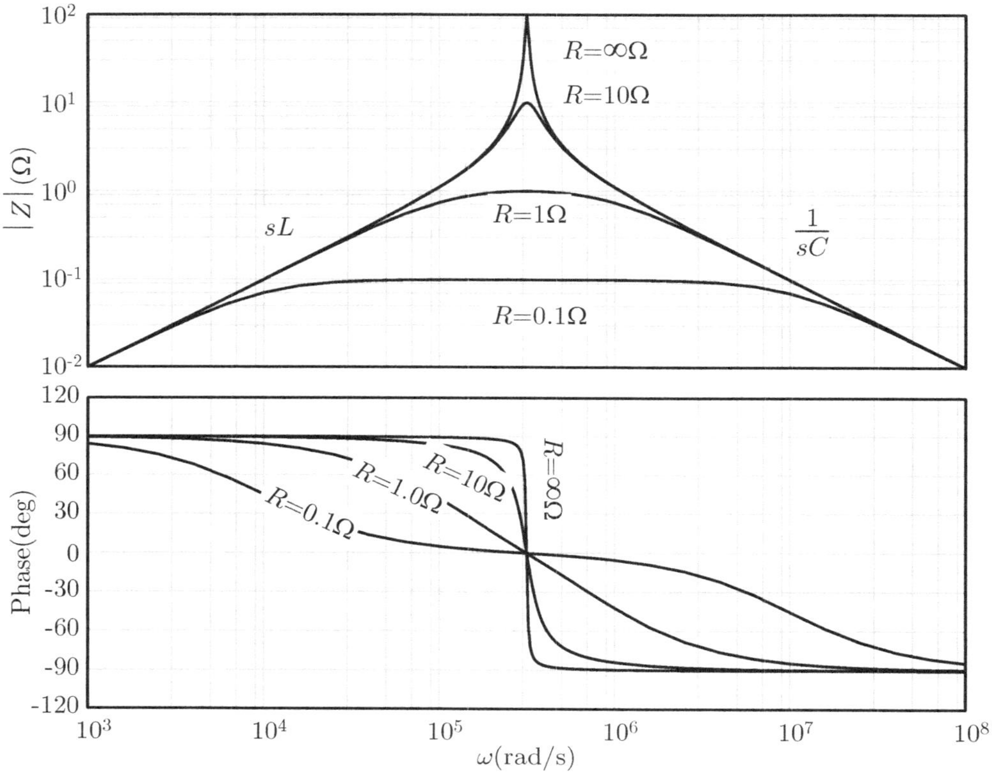

Fig. 26.11 Parallel RLC impedance function: $C = 1\,\mu\text{F}$, $L = 10\,\mu\text{H}$, and various R values

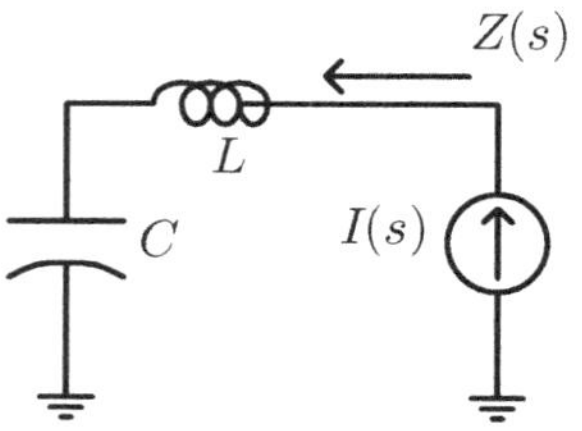

Fig. 26.12 Series LC and input impedance

Also the phase there is $-90°$ due to the cap. At high frequency, the cap shorts and we fall back on the inductive impedance:

$$\lim_{s\to\infty} Z(s) = sL \qquad (26.54)$$

The phase there flips to $+90°$ due to the inductor. At the resonance impedance, we get zero impedance:

$$\omega_0^2 = \frac{1}{LC}; \qquad Z(\omega_0) = 0 \qquad (26.55)$$

The impedance transfer function is shown in Fig. 26.13. Notice that this impedance transfer function looks like the (upside down) mirrored version of the parallel LC one as shown earlier in Fig. 26.9.

26.10 Series *RLC* Impedance Transfer Function

Consider next the series RLC shown in Fig. 26.14; we wish to find its impedance transfer function. By inspection total impedance is

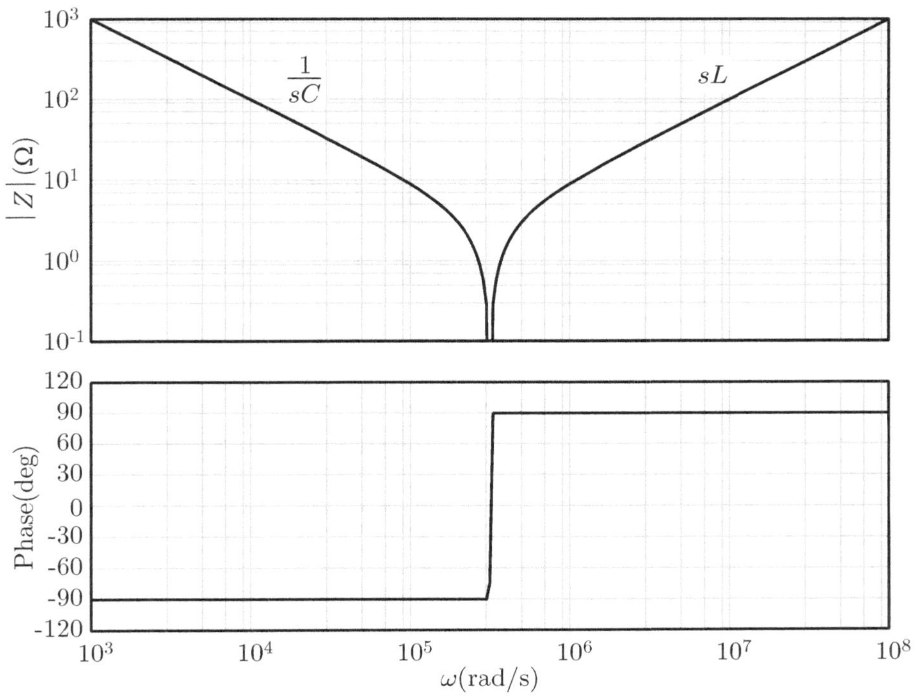

Fig. 26.13 Transfer function of series *LC* network; $L = 10\,\mu$H and $C = 1\,\mu$F

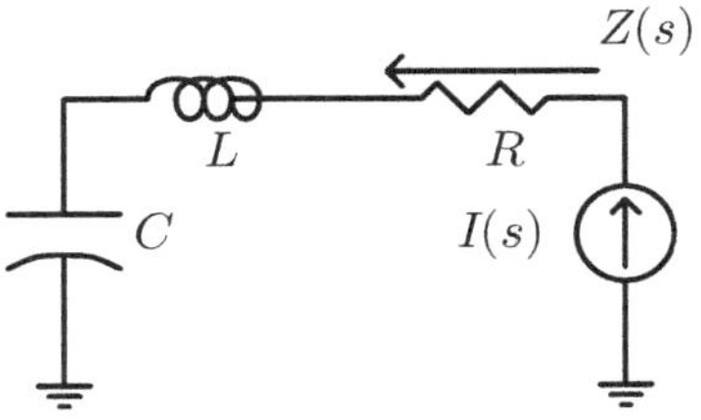

Fig. 26.14 Series *RLC* and input impedance

$$Z(s) = R + sL + \frac{1}{sC} = \frac{1 + s^2LC + sRC}{sC} \tag{26.56}$$

Notice that at low frequency the cap impedance dominates:

$$\lim_{s\to 0} Z(s) = \frac{1}{sC} \tag{26.57}$$

That is at low frequency the impedance of R and L is small compared to that of the cap, and as such the larger one dominates. Also at low frequency phase is $-90°$, due to the cap. On the other hand, at high frequency the inductance dominates

$$\lim_{s\to\infty} Z(s) = sL \tag{26.58}$$

and the phase flips to $+90°$. At resonance the impedance collapses to

$$Z(\omega_0) = R, \qquad \omega_0^2 = \frac{1}{LC} \tag{26.59}$$

That is, at resonance the impedance of the cap cancels that of the inductor and the left over is simply R. The phase at resonance is zero. Figure 26.15 shows impedance transfer function as a function of frequency.

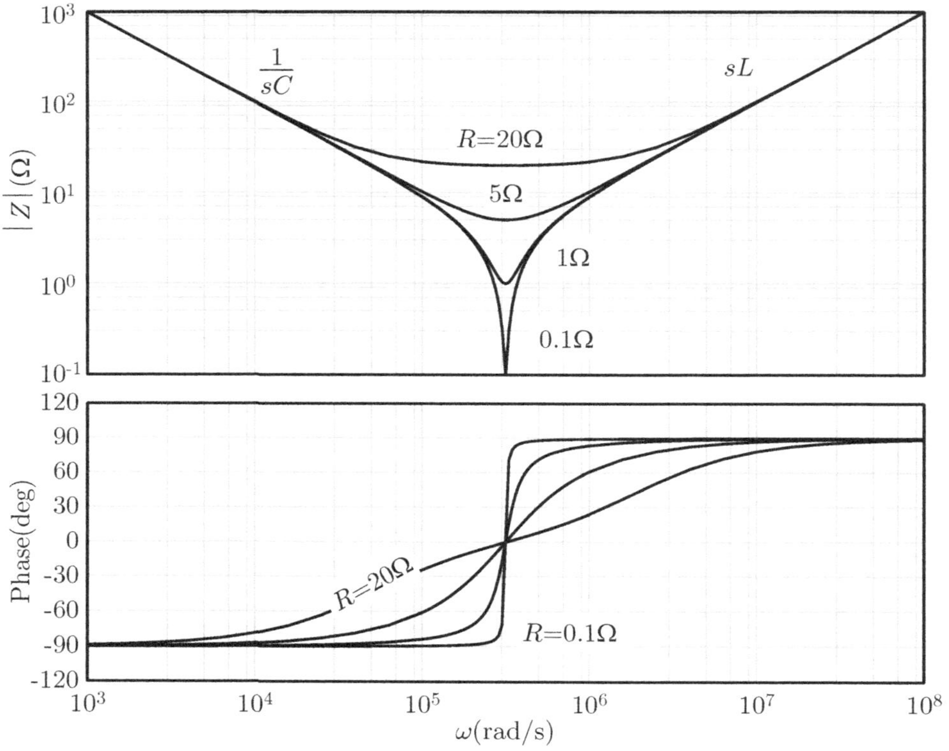

Fig. 26.15 Series *RLC* impedance function: $C = 1\,\mu$F, $L = 10\,\mu$H, and various R values

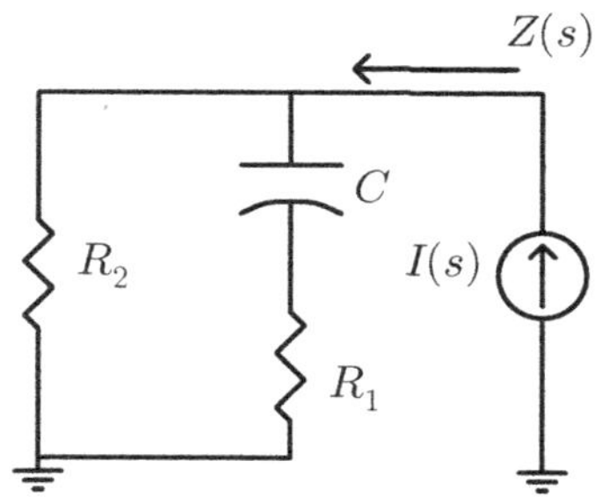

Fig. 26.16 Series *RC* in parallel with *R* and input impedance

26.11 Series *RC* in Parallel with *R* Impedance Transfer Function

Consider the series *RC* in parallel with *R* circuit shown in Fig. 26.16; we wish to find the input impedance transfer function. The series *RC* branch has an impedance of

$$Z_{R_1C}(s) = \frac{1 + sR_1C}{sC} \tag{26.60}$$

Total impedance is a parallel combination of this and R_2:

$$Z(s) = \frac{R_2 \frac{1+sR_1C}{sC}}{R_2 + \frac{1+sR_1C}{sC}} = \boxed{\frac{R_2 + sR_1R_2C}{1 + sC(R_1 + R_2)}} \tag{26.61}$$

Notice that at low frequency this converges to R_2

$$\lim_{s \to 0} Z(s) = R_2 \tag{26.62}$$

since the cap there would be open. Conversely, at high frequency we get the following limit:

$$\lim_{s\to\infty} Z(s) = \frac{R_2R_2}{R_1 + R_2} \tag{26.63}$$

which is nothing more than the parallel resistance of R_1 and R_2; in other words, since the cap shorts at high frequency, we end up with $R_1||R_2$. Next we examine impact of R_2; to do this we rewrite the transfer function as

$$Z(s) = R_2\frac{1 + sR_1C}{1 + sC(R_1 + R_2)} \tag{26.64}$$

If we set R_2 very large, we get the following limit:

$$Z(s)_{R_2\to\infty} = R_2\frac{1 + sR_1C}{sCR_2} = \frac{1 + sR_1C}{sC} \tag{26.65}$$

which is nothing more than the series impedance of R_1 and C. That is, when R_2 is very large the left branch simply opens and we fall back on the right branch. Conversely, if we take the limit as R_2 goes to zero we get zero impedance:

$$Z(s)_{R_2\to 0} = 0 \tag{26.66}$$

That is, if R_2 is zero, and it being in parallel with the right branch renders the whole network at zero impedance! Finally we examine impact of R_1. In the limit this goes to ∞ and we get

$$Z(s)_{R_1\to\infty} = R_2\frac{sR_1C}{sR_1C} = R_2 \tag{26.67}$$

which makes sense. That is, if R_1 is open the right branch drops out and we fall back on the left branch which is merely R_2. And in the limit as R_1 goes to zero we get

$$Z(s)_{R_1\to 0} = \frac{R_2}{1 + sCR_2} \tag{26.68}$$

which is nothing more than the impedance of $R_2||C$. That is, if R_1 is short, it cancels out and we end up with C in parallel with R_2. Figure 26.17

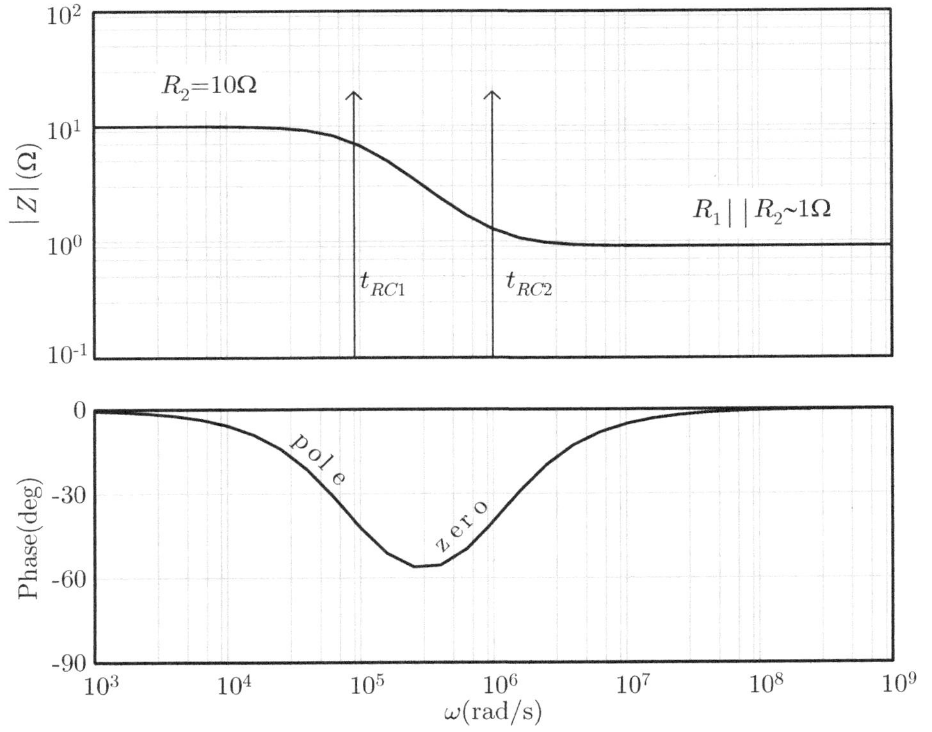

Fig. 26.17 Series RC in parallel with R impedance function: $C = 1\,\mu\text{F}, R_1, R_2 = 1, 10\,\Omega$

shows the impedance function versus frequency. Notice that we start at R_2, roll down (due to cap), and settle at $R_1||R_2$. Notice also we have two characteristics frequencies:

$$t_{RC1} = C(R_1 + R_2), \qquad t_{RC2} = CR_1 \quad (26.69)$$

Since the former is larger, its characteristic frequency happens earlier, as shown in the figure. The phase starts at zero due to R_2 since the cap in the right branch takes that branch out of the picture. As the impedance of the cap drops, the phase starts going negative with the aim of getting to $-90°$. However, due to the presence of R_1 in series with the cap, the phase reverses direction and settles to 0 at very high frequency. That is, at very high frequency and since we end up with pure resistive impedance (that of $R_1||R_2$) the phase naturally goes to zero.

26.12 Series *RC* in Parallel with Series *RC*

The series RC, in parallel with another series RC, is shown in Fig. 26.18. The total impedance is the parallel combination of both impedances:

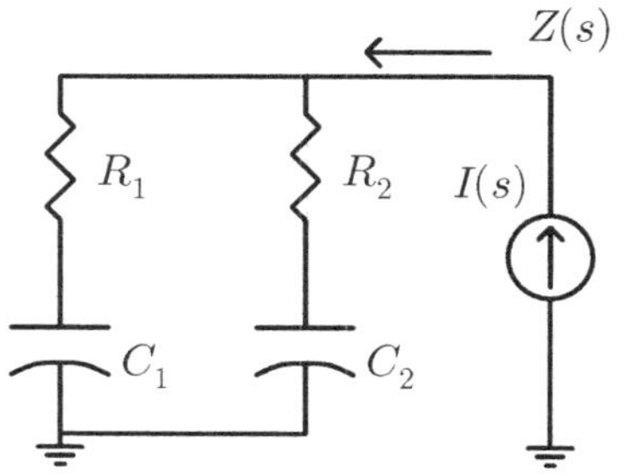

Fig. 26.18 Series RC in parallel with another series RC

$$Z(s) = \left(R_1 + \frac{1}{sC_1}\right) || \left(R_2 + \frac{1}{sC_2}\right)$$

$$= \frac{\left(R_1 + \frac{1}{sC_1}\right)\left(R_2 + \frac{1}{sC_2}\right)}{R_1 + R_2 + \frac{1}{s}\frac{C_1+C_2}{C_1C_2}} \quad (26.70)$$

Multiply denominator and numerator by $s^2C_1C_2$

$$Z(s) = \frac{(1 + sR_1C_1)(1 + sR_2C_2)}{s(C_1 + C_2) + s^2(R_1 + R_2)C_1C_2} \quad (26.71)$$

Factor an $s(R_1 + R_2)C_1C_2$ from the denominator

$$Z(s) = \frac{1}{s(R_1 + R_2)C_1C_2} \frac{(1 + sR_1C_1)(1 + sR_2C_2)}{\frac{C_1+C_2}{(R_1+R_2)C_1C_2} + s} \quad (26.72)$$

$$\boxed{Z(s) = \frac{R_1R_2}{s(R_1 + R_2)} \frac{\left(s + \frac{1}{R_1C_1}\right)\left(s + \frac{1}{R_2C_2}\right)}{\frac{1}{(R_1 + R_2)C_s} + s}, \qquad C_s = \frac{C_1C_2}{C_1 + C_2}} \quad (26.73)$$

Let's estimate some limits. At DC we have

$$\lim_{s\to 0} \sim \frac{R_1R_2}{s(R_1 + R_2)} \frac{\frac{1}{R_1C_1}\frac{1}{R_2C_2}}{\frac{C_1+C_2}{(R_1+R_2)C_1C_1}} = \frac{1}{s(C_1 + C_2)} \quad (26.74)$$

which states that the low-frequency limit behaves as a cap of total cap $C_1 + C_2$. That is, at low frequency, we see the parallel combination of both caps. The other limit, that of high frequency is

$$\lim_{s\to\infty} Z(s) \sim \frac{R_1R_2}{s(R_1 + R_2)} \frac{s^2}{s} = \frac{R_1R_2}{R_1 + R_2} \quad (26.75)$$

which states that the high-frequency limit behaves like a resistor whose magnitude equals the parallel combination of both resistors. A sample impedance profile for this network is shown in Fig. 26.19. As shown in the figure when C dominates phase tends to move towards $-90°$. On the other hand when R dominates—as evident with flat regions in the magnitude plot—phase tends to move towards 0°. Finally looking closer at Eq. (26.73) tells us we have the following pole/zero combination:

1. Pole at $\omega = 0$
2. Pole at $\omega = \frac{1}{(R_1+R_2)C_s}$
3. Zero at $\omega = \frac{1}{R_1C_1}$ and
4. Zero at $\omega = \frac{1}{R_2C_2}$

Later when calculating impulse response we would need to simplify the impedance expression. Let

$$A = \frac{R_1R_2}{R_1+R_2}, \qquad B = \frac{1}{(R_1+R_2)C_s}, \qquad C = \frac{1}{R_1C_1}, \qquad D = \frac{1}{R_2C_2} \tag{26.76}$$

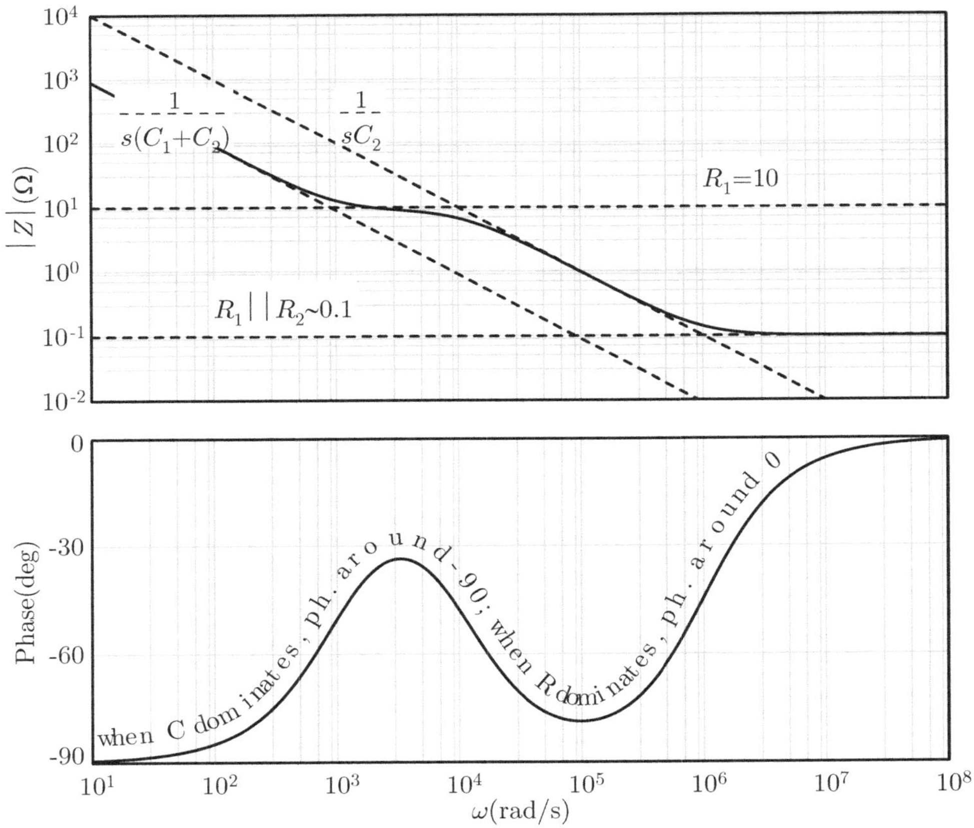

Fig. 26.19 Impedance vs. frequency for series RC in parallel with another series RC; case of $R_1 = 10\,\Omega$, $R_2 = 0.1\,\Omega$, $C_1 = 100\,\mu F$, and $C_2 = 10\,\mu F$

Then the impedance assumes the form

$$Z(s) = \frac{A}{s}\frac{(s+C)(s+D)}{s+B} \quad (26.77)$$

Expand

$$Z(s) = A\frac{s^2 + s(C+D) + CD}{Bs + s^2} \quad (26.78)$$

Since the order of the numerator is the same as that of the denominator, we will need to use polynomial division

$$Z(s) = A\left[1 + \frac{(C+D-B)s + CD}{Bs + s^2}\right] \quad (26.79)$$

$$Z(s) = A + \frac{A}{s}\frac{(C+D-B)s + CD}{B+s} \quad (26.80)$$

We see that we have a pole at zero and one at $-B$. The one at zero gives the coefficient

$$A\frac{CD}{B} = \frac{R_1R_2}{R_1+R_2}\frac{1}{R_1R_2C_1C_2}(R_1+R_2)\frac{C_1C_2}{C_1+C_2} = \frac{1}{C_1+C_2} \quad (26.81)$$

The one at $-B$ gives the coefficient

$$\frac{A}{-B}\left[(C+D-B)(-B) + CD\right] = A\left[C+D-B-CD/B\right] \quad (26.82)$$

$$Z(s) = \frac{R_1R_2}{R_1+R_2} + \frac{1}{C_1+C_2}\frac{1}{s} + \frac{A(C+D-B-CD/B)}{s + \frac{1}{(R_1+R_2)C_s}} \quad (26.83)$$

While not in its most compact form, this last expression will evaluate numerically.

26.13 Ladder of Series *RC* Branches

The circuit under considerations is comprised of three parallel branches of series *RC* blocks as shown in Fig. 26.20. The *RC* values of each branch are different. We have here three parallel branches, so cumulative impedance is the parallel combination of the three impedances. The algebra is quite extensive, so along the way we need to define new variables to minimize the typing! First start with

$$G_1 = \frac{1}{C_1}, G_2 = \frac{1}{C_2}, G_3 = \frac{1}{C_3} \quad (26.84)$$

Total impedance is then

$$Z(s) = \frac{(R_1 + G_1/s)\frac{(R_2+G_2/s)(R_3+G_3/s)}{R_2+R_3+(G_2+G_3)/s}}{R_1 + G_1/s + \frac{(R_2+G_2/s)(R_3+G_3/s)}{R_2+R_3+(G_2+G_3)/s}}$$

Expand the fractions

$$Z(s) = \frac{\frac{(R_1+G_1/s)(R_2R_3 + 1/s(R_2G_3+R_3G_2) + G_2G_3/s^2)}{R_2+R_3+(G_2+G_3)/s}}{R_1 + G_1/s + \frac{R_2R_3 + 1/s(R_2G_3+R_3G_2) + G_2G_3/s^2}{R_2+R_3+(G_2+G_3)/s}} \quad (26.85)$$

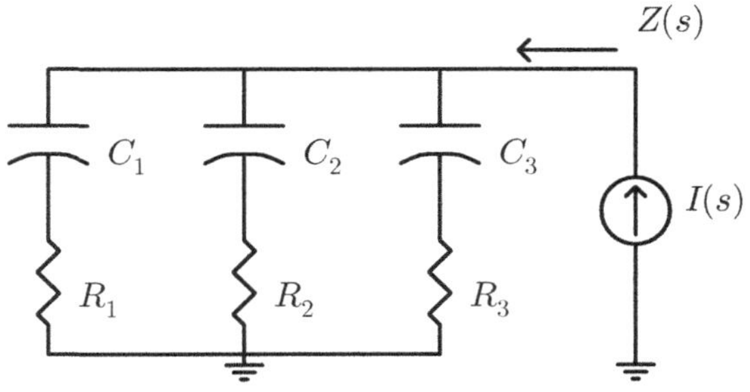

Fig. 26.20 Ladder of series *RC* branches

Multiply by s/s

$$Z(s) = \frac{\dfrac{(R_1 s + G_1)(R_2R_3 + 1/s(R_2G_3 + R_3G_2) + G_2G_3/s^2)}{R_2 + R_3 + (G_2 + G_3)/s}}{R_1 s + G_1 + \dfrac{(R_2R_3)s + (R_2G_3 + R_3G_2) + G_2G_3/s}{R_2 + R_3 + (G_2 + G_3)/s}} \tag{26.86}$$

Multiply the top and bottom fractions of the numerator by s/s

$$Z(s) = \frac{\dfrac{(R_1 s^2 + G_1 s)(R_2R_3 + 1/s(R_2G_3 + R_3G_2) + G_2G_3/s^2)}{(R_2 + R_3)s + (G_2 + G_3)}}{R_1 s + G_1 + \dfrac{(R_2R_3)s^2 + (R_2G_3 + R_3G_2)s + G_2G_3}{(R_2 + R_3)s + (G_2 + G_3)}} \tag{26.87}$$

Multiply bottom and top by $(R_2 + R_3)s$ $+(G_2 + G_3)$

$$Z(s) = \frac{(R_1 s + G_1)(R_2R_3 s + (R_2G_3 + R_3G_2) + G_2G_3/s)}{\{(R_1 s + G_1)[(R_2 + R_3)s + G_2 + G_3] + R_2R_3 s^2 + (R_2G_3 + R_3G_2)s + G_2G_3\}} \tag{26.88}$$

Multiply top and bottom by s

$$Z(s) = \frac{(R_1 s + G_1)(R_2R_3 s^2 + (R_2G_3 + R_3G_2)s + G_2G_3)}{\{(R_1 s^2 + G_1 s)[(R_2 + R_3)s + G_2 + G_3] + R_2R_3 s^3 + (R_2G_3 + R_3G_2)s^2 + G_2G_3 s\}} \tag{26.89}$$

Collect terms in denominator

$$Z(s) = \frac{(R_1 s + G_1)(R_2 R_3 s^2 + (R_2 G_3 + R_3 G_2)s + G_2 G_3)}{\begin{array}{c}\{s(G_2G_3 + G_1(G_2 + G_3)) + s^2(R_1(G_2 + G_3) \\ + G_1(R_2 + R_3) + (R_2G_3 + R_3G_2)) + s^3(R_1(R_2 + R_3) + R_2R_3)\}\end{array}} \tag{26.90}$$

Expand numerator

$$Z(s) = \frac{\begin{array}{c}R_1R_2R_3s^3 + R_1(R_2G_3 + R_3G_2)s^2 + R_1G_2G_3s + \\ G_1R_2R_3s^2 + G_1(R_2G_3 + R_3G_2)s + G_1G_2G_3\end{array}}{\begin{array}{c}s(G_2G_3 + G_1(G_2 + G_3)) + s^2(R_1(G_2 + G_3) \\ +G_1(R_2 + R_3) + (R_2G_3 + R_3G_2)) + s^3(R_1(R_2 + R_3) + R_2R_3)\end{array}} \tag{26.91}$$

Collect terms in numerator

$$Z(s) = \frac{\begin{array}{c}G_1G_2G_3 + s(R_1G_2G_3 + G_1(R_2G_3 + R_3G_2)) \\ +s^2(R_1(R_2G_3 + R_3G_2) + G_1R_2R_3) + s^3(R_1R_2R_3)\end{array}}{\begin{array}{c}s(G_2G_3 + G_1(G_2 + G_3)) + s^2(R_1(G_2 + G_3) \\ +G_1(R_2 + R_3) + (R_2G_3 + R_3G_2)) + s^3(R_1(R_2 + R_3) + R_2R_3)\end{array}} \tag{26.92}$$

To simplify make the following substitutions:

$$A = G_1G_2G_3$$
$$B = R_1G_2G_3$$
$$C = R_2G_3 + R_3G_2$$
$$D = G_1R_2R_3$$
$$E = R_1R_2R_3$$
$$F = G_2G_3 + G_1(G_2 + G_3)$$
$$G = R_1(G_2 + G_3) + G_1(R_2 + R_3) + C$$
$$H = R_1(R_2 + R_3) + R_2R_3$$
$$I = B + G_1C$$
$$J = R_1C + D \tag{26.93}$$

Then our transfer function becomes

$$Z(s) = \frac{A + Is + Js^2 + Es^3}{Fs + Gs^2 + Hs^3} \tag{26.94}$$

By long division we get

$$Z(s) = \frac{E}{H} + \frac{A + (I - \frac{EF}{H})s + (J - \frac{EG}{H})s^2}{Fs + Gs^2 + Hs^3} \tag{26.95}$$

Let us assume some value; say

$$R_1 = 10\,\Omega, R_2 = 5\,\Omega, R_3 = 1\,\Omega, C_1 = 100\,\mu\text{F}, C_2 = 1\,\mu\text{F}, C_3 = 0.1\,\mu\text{F} \tag{26.96}$$

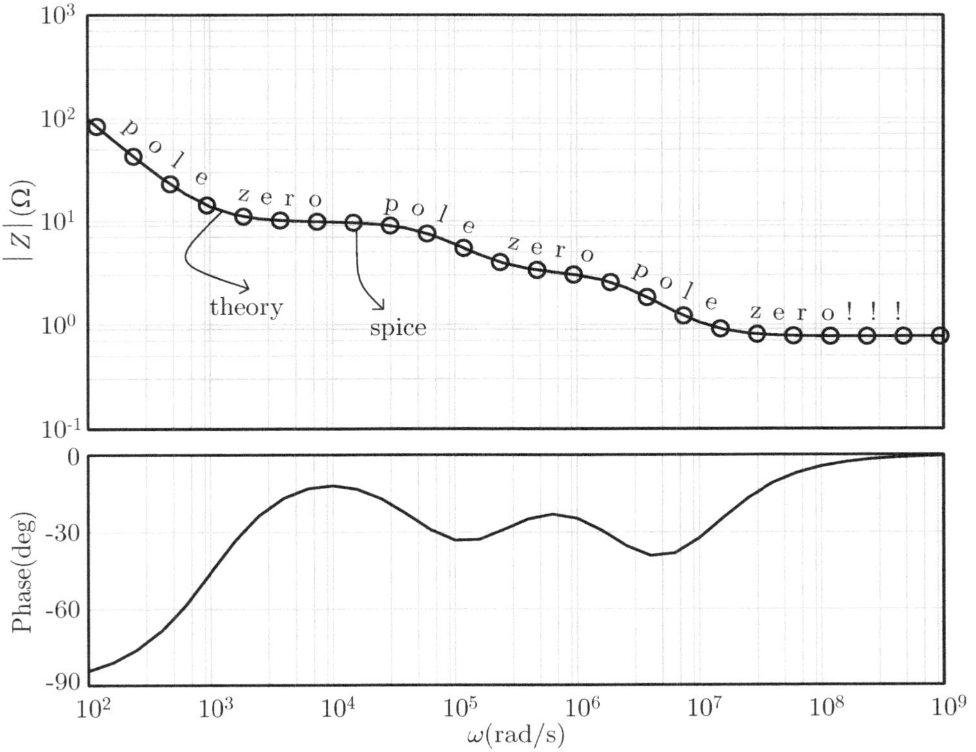

Fig. 26.21 Impedance transfer function of ladder *RC* network and comparison to SPICE

The denominator has three zeroes given by

$$z_0 = 0$$

$$z_1 = -64448$$
$$z_2 = -2413398 \tag{26.97}$$

and using partial fraction we get

$$\boxed{Z(s) = \frac{E}{H} + \frac{9891}{s} + \frac{434200}{s + 64448} + \frac{5496795}{s + 2413398}} \tag{26.98}$$

We can test this solution against SPICE as shown in Fig. 26.21. As shown in the figure we get identical match. As seen from the figure we have three poles; each pole wants to change the phase by $-90°$. Had it not been for the zeroes we would have accumulated $-270°$. But we do have zeroes, and in particular three of them! So the end phase is zero; that is we start at $-90°$ since we have a pole at the origin and end up at zero, once the three zeroes neutralized the three poles! Each pole wants to add -20 dB to the decay rate, and each zero 20 dB. So again we start as -20 dB and end at 0 dB decay since at really high frequency the transfer function settles to a number (not frequency-dependent).

Let us recap what we have done to ensure we were not lost just doing algebra. We started with a multi-branch network. Using circuit rules we extracted the transfer function symbolically having introduced some simplifying variables along the way. Once we had the transfer function Eq. (26.94) we are able to tell how many poles and how many zeroes we have. At this point we can directly plot the transfer function (magnitude and phase). But if we wanted to further extract the time-domain response we need to factor the transfer function. To that end we started with long division and then partial fractions. In order to do the latter we had to figure the roots of the denominator. This is a step best suited for a calculator of mathematical package. Once we have factored the transfer function Eq. (26.98) we are ready for the last step which is finding the inverse transform (left for the reader). It is very beneficial to traverse back and forth between the transfer function and between its graphic representation in order to ensure pole/zero count makes sense and decay rate and phase values make sense too.

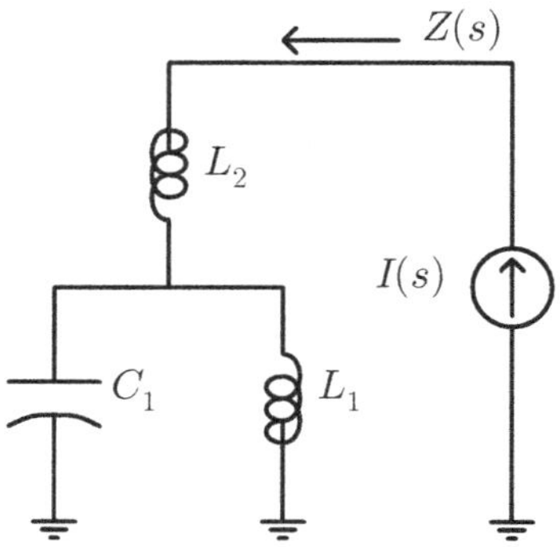

Fig. 26.22 Parallel LC in series with L network

26.14 Parallel *LC* in Series with *L* Circuit

The parallel LC, series L circuit is shown in Fig. 26.22. The impedance of the parallel part is given by

$$Z'(s) = \frac{1}{C}\frac{s}{s^2+\omega_1^2}, \qquad \omega_1^2 = \frac{1}{CL_1} \tag{26.99}$$

When added in series with L_2 we get

$$\begin{aligned} Z(s) &= \frac{1}{C}\frac{s}{s^2+\omega_1^2} + sL_2 = \frac{\frac{s}{C} + sL_2\omega_1^2 + s^3L_2}{s^2+\omega_1^2} = \frac{s\left(1/C + L_2\omega_1^2\right) + s^3L_2}{s^2+\omega_1^2} \\ &= \frac{s\frac{1+L_2C\omega_1^2}{C} + s^3L_2}{s^2+\omega_1^2} = \frac{s\left[\frac{1+L_2C\omega_1^2}{C} + s^2L_2\right]}{s^2+\omega_1^2} \end{aligned} \tag{26.100}$$

We see that we have a complex conjugate pole pair at ω_1, a zero at 0, and another complex conjugate zero pair when

$$s^2L_2 = -\frac{1+L_2C\omega_1^2}{C} \tag{26.101}$$

This evaluates to

$$\begin{aligned} \omega^2 &= \frac{1+L_2C\omega_1^2}{LC_2} = \frac{1+\frac{\omega_1^2}{\omega_2^2}}{1/\omega_2^2} = \omega_2^2 + \omega_1^2, \\ \omega_2^2 &= \frac{1}{L_2C} \end{aligned} \tag{26.102}$$

Impedance profile shown in Fig. 26.23. As seen from the figure due to the first zero we start at 20 dB/dec and 90°. When the complex conjugate pole pair kicks in we fall back to −20 dB/dec and −90°. Finally when the complex conjugate *zero* pair kicks in we go back to 20 dB/dec and 90°. Remember a simple zero adds 20 dB/dec and 90°; a double zero or a complex conjugate pair one adds 40 dB/dec and 180°. Varying L_2 on the low side is shown in Fig. 26.24. Varying L_2 on the high side is shown in Fig. 26.25 Notice that at very high frequency C_1 shorts and takes out L_1; this leaves us only with L_2. So at high

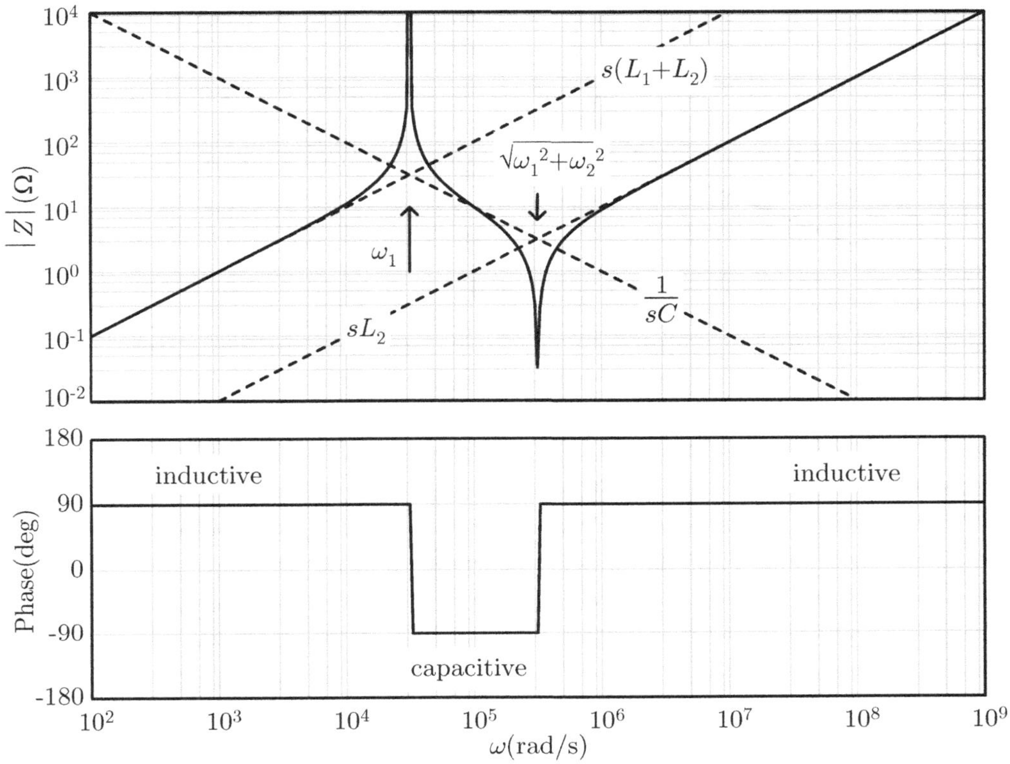

Fig. 26.23 Transfer function of parallel LC in series with L network. $L_1 = 1\,\text{mH}$, $C = 1\,\mu\text{F}$, $L_2 = 10\,\mu\text{H}$

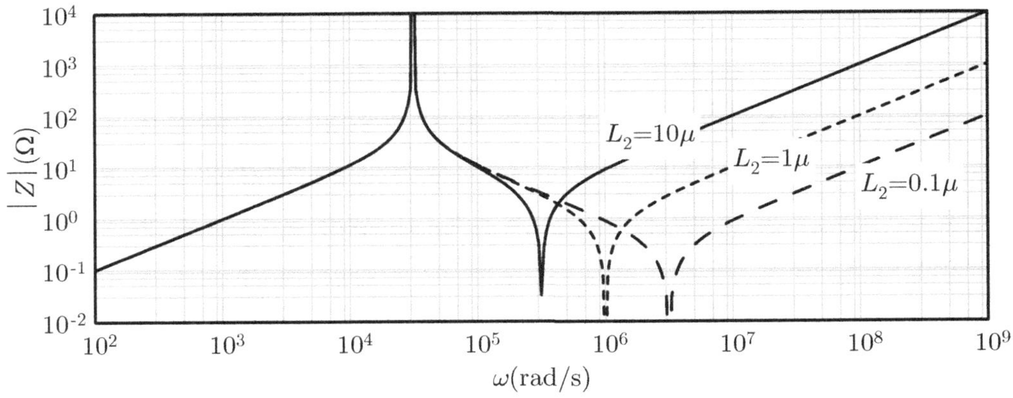

Fig. 26.24 Transfer function of parallel LC in series with L network (impact of L_2 on the low side)

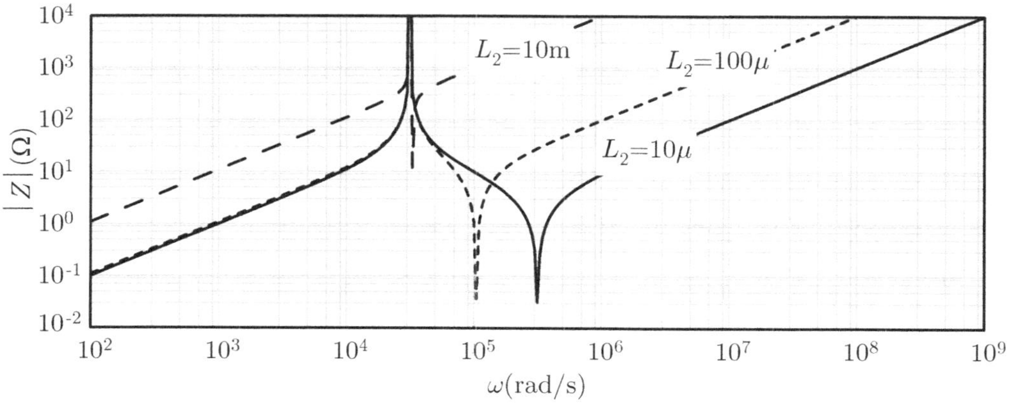

Fig. 26.25 Transfer function of parallel LC in series with L network (impact of L_2 on the high side)

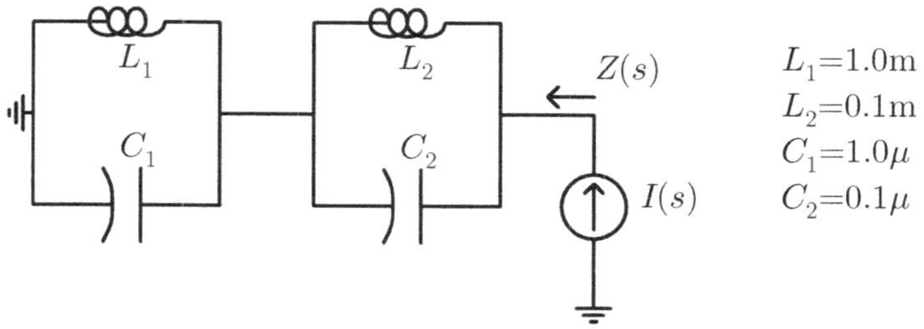

Fig. 26.26 Parallel LC in series with another parallel LC network

frequency the impedance is dictated by L_2 and that is why increasing L_2 tenfold results in an equivalent increase in total impedance.

26.15 Parallel *LC* in Series with Another Parallel *LC* Circuit

The parallel LC, in series with another LC circuit, is shown in Fig. 26.26. The impedance of the first parallel part is given by

$$Z_1(s) = \frac{1}{C_1}\frac{s}{s^2+\omega_1^2}, \qquad \omega_1^2 = \frac{1}{L_1C_1} \tag{26.103}$$

The impedance of the second parallel part is given by

$$Z_2(s) = \frac{1}{C_2}\frac{s}{s^2+\omega_2^2}, \qquad \omega_2^2 = \frac{1}{L_2C_2} \tag{26.104}$$

Total impedance is simply the sum of both parts

$$Z(s) = \frac{s/C_1}{s^2+\omega_1^2} + \frac{s/C_2}{s^2+\omega_2^2} = s\frac{(s^2+\omega_2^2)/C_1 + (s^2+\omega_1^2)/C_2}{(s^2+\omega_1^2)(s^2+\omega_2^2)} \tag{26.105}$$

$$\boxed{Z(s) = \frac{s}{C_1C_2}\frac{s^2(C_1+C_2)+C_2\omega_2^2+C_1\omega_1^2}{(s^2+\omega_1^2)(s^2+\omega_2^2)}} \tag{26.106}$$

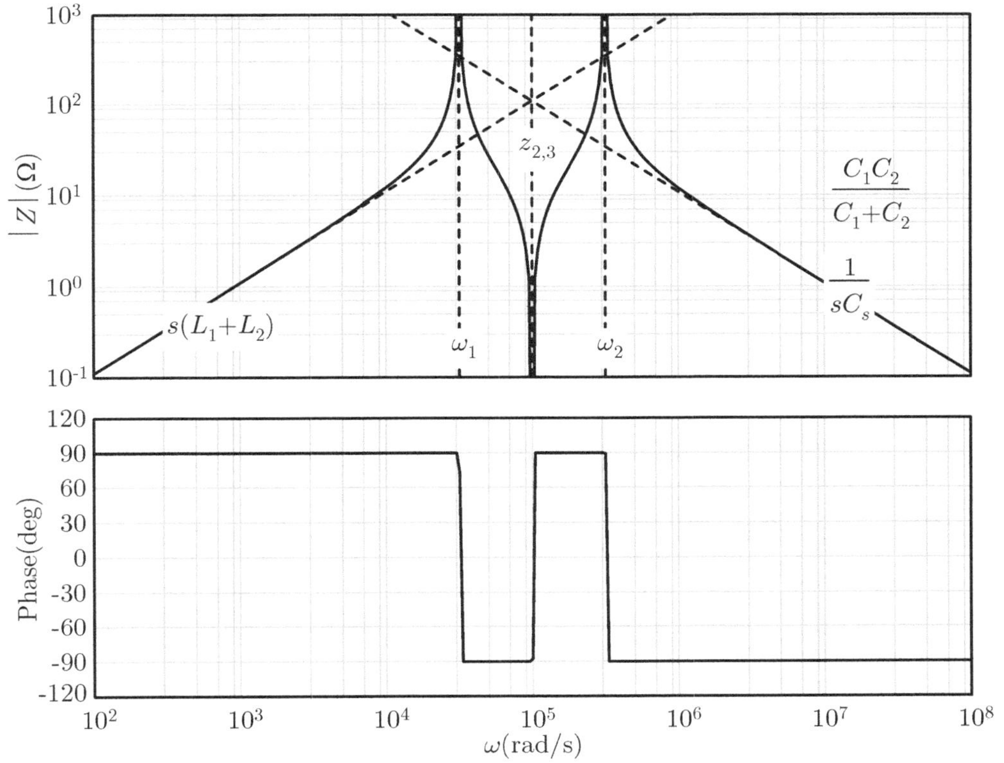

Fig. 26.27 Impedance transfer function for parallel *LC* in series with another parallel *LC* network. $L_1 = 1\,\text{mH}$, $L_2 = 0.1\,\text{mH}$, $C_1 = 1\,\mu\text{F}$, and $C_2 = 0.1\,\mu\text{F}$

We see that we have three zeroes: one at zero and a complex conjugate pole pair

$$\omega^2 = \frac{C_2\omega_2^2 + C_1\omega_1^2}{C_1 + C_2} \tag{26.107}$$

We also have four poles—one complex conjugate pair at ω_1 and another complex conjugate pair at ω_2. Impedance results are shown in Fig. 26.27. Notice that due to the zero at the origin we start with 20 dB/dec and 90°. At very high frequency the transfer function assumes the form $\sim\frac{1}{s}$ and that results in −20 dB/dec and −90°; or equivalently we say that at high frequency each of the caps shorts the parallel inductor and we end up effectively with a capacitive network. All of this can be verified in the figure.

26.16 Series *RL*/Parallel *C* Circuit

The series *RL*, parallel *C* circuit is shown in Fig. 26.28. The output impedance is the series combination of *R* and *L*, in parallel with *C*:

$$Z(s) = (R+sL)||\frac{1}{sC} = \frac{(R+sL)\frac{1}{sC}}{R+sL+\frac{1}{sC}} \tag{26.108}$$

Multiply both numerator and denominator by sC

$$Z(s) = \frac{R+sL}{1+sRC+s^2LC} \tag{26.109}$$

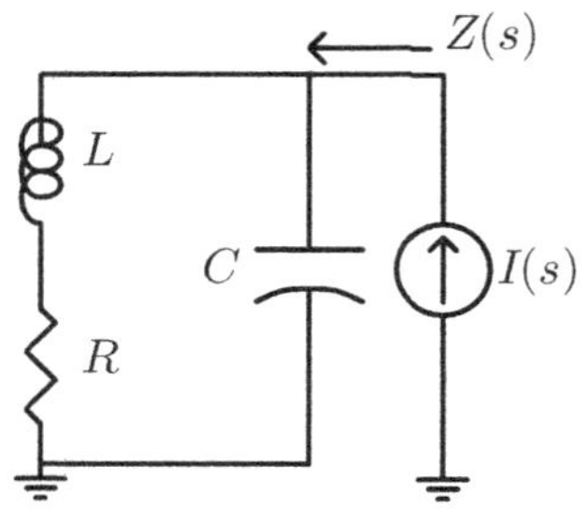

Fig. 26.28 Series *RL*/parallel *C* circuit

Factor an L from the numerator and LC from the denominator

$$Z(s) = \frac{L}{LC}\frac{\frac{R}{L}+s}{s^2+s\frac{R}{L}+\frac{1}{LC}} = \frac{1}{C}\frac{\frac{R}{L}+s}{s^2+s\frac{R}{L}+\omega_{LC}^2} \tag{26.110}$$

where we have defined

$$\omega_{LC}^2 = \frac{1}{LC} \tag{26.111}$$

Define the constant

$$a = \frac{R}{2L} \tag{26.112}$$

and the impedance becomes

$$Z(s) = \frac{1}{C}\frac{2a+s}{s^2+s2a+\omega_{LC}^2} \tag{26.113}$$

Complete the square in the denominator

$$Z(s) = \frac{1}{C}\frac{2a+s}{(s+a)^2+\left(\omega_{LC}^2-a^2\right)} \tag{26.114}$$

Define the new frequency

$$\omega_0^2 = \omega_{LC}^2 - a^2 \tag{26.115}$$

and the new expression for impedance (along with the various abbreviations) becomes

$$\boxed{\begin{aligned} Z(s) &= \frac{1}{C}\frac{2a+s}{(a+s)^2+\omega_0^2} \\ \omega_{LC}^2 &= \frac{1}{LC} \\ a &= \frac{R}{2L} \\ \omega_0^2 &= \omega_{LC}^2 - a^2 \end{aligned}} \tag{26.116}$$

$$\boxed{Z(s) = \frac{1}{C}\frac{\frac{R}{L}+s}{\left(\frac{R}{2L}+s\right)^2+\frac{1}{LC}-\left(\frac{R}{2L}\right)^2}} \tag{26.117}$$

Figure 26.29 shows sample results and comparison to SPICE. Notice that at DC we get the following limit:

$$Z(0) = \frac{1}{C}\frac{R/L}{(R/2L)^2+1/LC-(R/2L)^2} = L\frac{R}{L} = R \tag{26.118}$$

as expected. That is, at DC the cap is open, the inductor is short, and we fall back on the resistor only. At large frequency we have

$$Z(\infty) \sim \frac{1}{C}\frac{s}{s^2} = \frac{1}{sC} \tag{26.119}$$

which is nothing other than the cap impedance. That is, at high frequency the inductor opens and we only see the cap. Phase-wise we start at 0 due to the resistor and then transition to 90° due to the inductor. At high frequency the cap kicks in and moves the phase to −90°, or equivalently we say that the complex conjugate pole pair shifts the phase by −180°. It is important to become familiar and at home probing into transfer functions and deciphering the various traits, be it number of poles/zeroes, rates of decay/growth and phase behavior. The essence of the circuit is buried in the transfer function. In other words, looking at the actual schematics of a circuit is tantamount to looking at a transfer function—they both (should) convey the same amount of information!

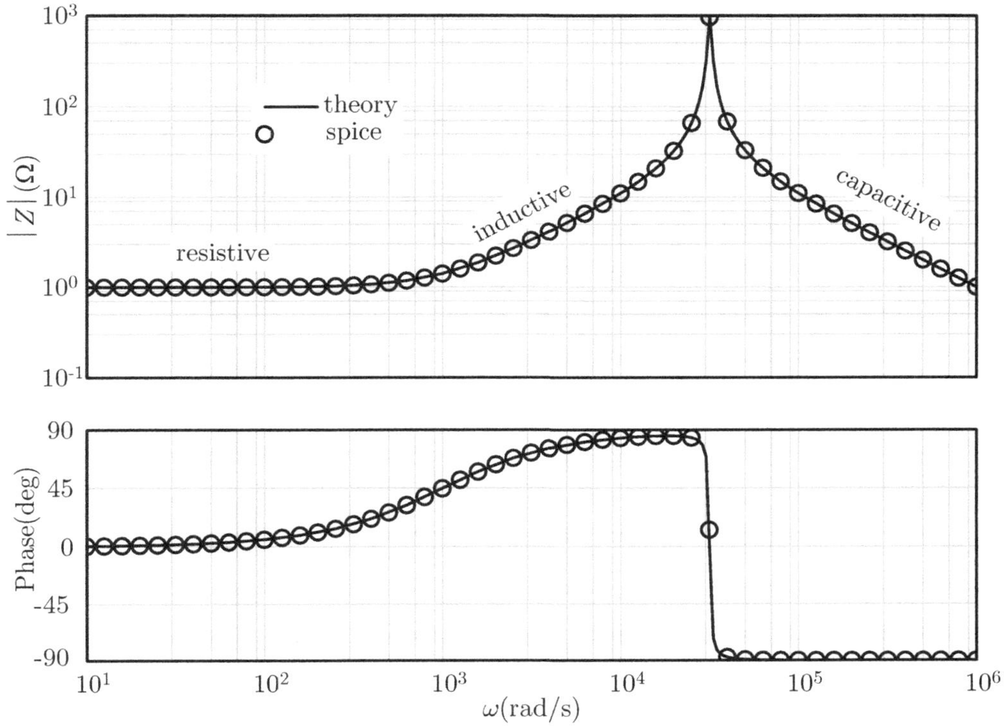

Fig. 26.29 Series *RL*/parallel *C* impedance (magnitude) versus frequency, and comparison to SPICE ($R = 1\,\Omega$, $L = 1\,\text{mH}$, and $C = 1\,\mu\text{F}$)

For the special case of small R we can estimate impedance at resonance

$$Z(\omega_{LC}) \sim \frac{1}{C}\frac{s}{R/Ls + s^2 + 1/LC} = \frac{1}{C}\frac{s}{R/Ls} = \frac{1}{R}\frac{L}{C} = \frac{Z_0^2}{R} \quad (26.120)$$

where we define the characteristic impedance as

$$Z_0 = \sqrt{\frac{L}{C}} \quad (26.121)$$

Just how small does R need to be? We need it such that

$$\frac{R}{L} \ll \frac{1}{\sqrt{LC}}, \quad \frac{R^2}{L^2} \ll \frac{1}{LC}, \quad R^2 \ll \frac{L}{C}, \quad \text{or} \quad (26.122)$$

$$R \ll Z_0 \quad (26.123)$$

To summarize,

$$\boxed{\begin{array}{lcc} \text{if} & R \ll Z_0 & \text{then} \\ Z(\omega_{LC}) & \sim & \dfrac{Z_0^2}{R} \end{array}} \quad (26.124)$$

Figure 26.30 shows sample results for different R values. Notice that the above approximation (Eq (26.124)) is valid even for non-negligible R values, as shown in circles in the figure. For this particular case the characteristic impedance comes out to $Z_0 = \sqrt{1000} = 31.6\,\Omega$.

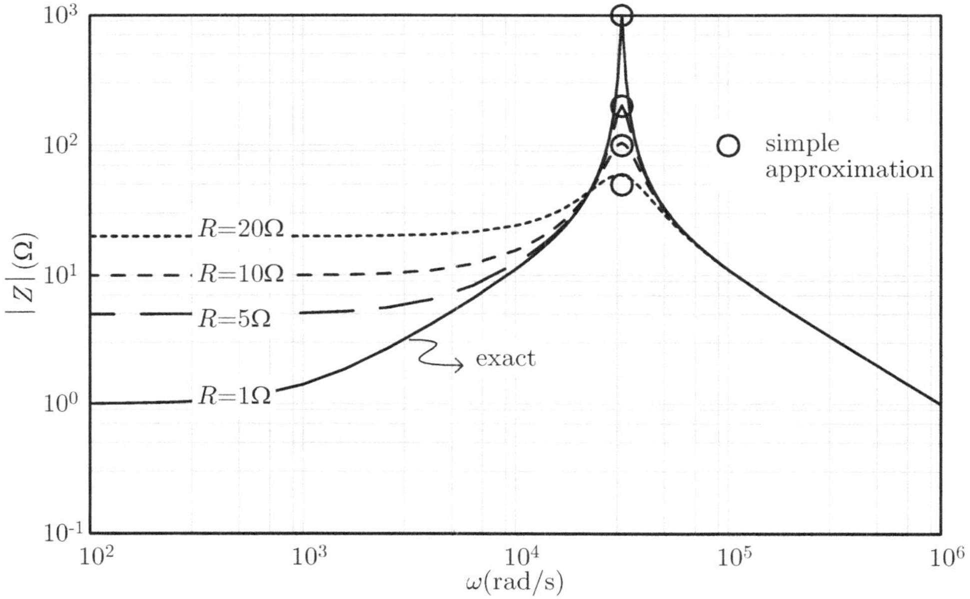

Fig. 26.30 Series RL/parallel C impedance (magnitude) versus frequency, and impact of R ($L = 1$ mH and $C = 1\,\mu$F)

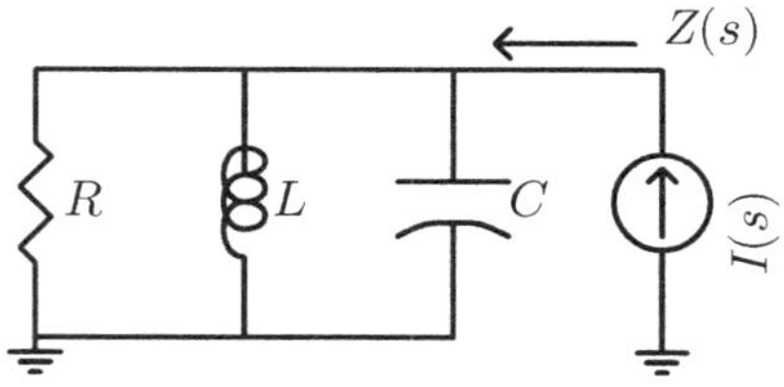

Fig. 26.31 Parallel RLC network

26.17 Parallel *RLC* Circuit

The parallel RLC circuit is shown in Fig. 26.31. To find the cumulative impedance we first find the parallel impedance of the R and L and then find the parallel impedance of that and the C. First the parallel impedance of R and L is

$$Z_{RL} = \frac{RsL}{R + sL} \tag{26.125}$$

Next we find the parallel impedance of this and the cap

$$\begin{aligned} Z(s) &= \frac{\frac{RsL}{R+sL}\frac{1}{sC}}{\frac{RsL}{R+sL} + \frac{1}{sC}} = \frac{RL}{C}\frac{\frac{1}{R+sL}}{\frac{RsL}{R+sL} + \frac{1}{sC}} = \frac{RL}{C}\frac{1}{RsL + \frac{R+sL}{sC}} \\ &= RL\frac{s}{Rs^2LC + sL + R} = \frac{1}{C}\frac{s}{s^2 + \frac{s}{RC} + \frac{1}{LC}} \end{aligned} \tag{26.126}$$

Define

$$\omega_{LC}^2 = \frac{1}{LC}, \quad \text{and} \quad a = \frac{1}{2RC} \tag{26.127}$$

then

$$Z(s) = \frac{1}{C}\frac{s}{s^2 + 2as + \omega_{LC}^2} \tag{26.128}$$

Complete the square

$$Z(s) = \frac{1}{C}\frac{s}{(s+a)^2 + \omega_{LC}^2 - a^2} \tag{26.129}$$

Define

$$\omega_0^2 = \omega_{LC}^2 - a^2 \tag{26.130}$$

and finally get

$$\boxed{\begin{aligned} Z(s) &= \frac{1}{C}\frac{s}{(a+s)^2 + \omega_0^2} \\ \omega_{LC}^2 &= \frac{1}{LC} \\ a &= \frac{1}{2RC} \\ \omega_0^2 &= \omega_{LC}^2 - a^2 \end{aligned}} \tag{26.131}$$

Figure 26.32 shows the impedance transfer function. Notice that impedance starts low, hits a max, and then decays. The initial increase in impedance is due to the inductor, while the final reduction is due to the cap. The inductive phase comes with a 90° phase while the capacitive one comes with a −90° phase. Physically we say that at low frequency the inductor has the lower impedance and hence it dominates; at high frequency the cap assumes that role. Figure 26.33 shows impact of varying R. In essence, the impedance at resonance equals R; so, bigger R results in larger resonance. So far as phase is concerned, varying R only changes the transition rate—it does not change the starting or ending values.

26.18 Series *RC*, Parallel *L* Network

The series RC, parallel L network is shown in Fig. 26.34. The impedance as seen from the current source side is the parallel combination of the inductor and the sum of resistor and capacitor impedance:

$$Z(s) = sL || \left[R + \frac{1}{sC}\right] = \frac{sL(R + \frac{1}{sC})}{R + sL + \frac{1}{sC}} = \frac{sL + s^2RLC}{1 + sRC + s^2LC} \tag{26.132}$$

$$Z(s) = \frac{1}{C}\frac{s + s^2RC}{s^2 + s\frac{R}{L} + \frac{1}{LC}} \tag{26.133}$$

Notice that at DC we get

$$Z(0) = 0 \tag{26.134}$$

due to the inductor shorting. On the other hand, at high frequency we get

$$Z(\infty) \sim \frac{1}{C}\frac{s^2RC}{s^2} = R \tag{26.135}$$

due to the inductor opening, and the cap shorting. Figure 26.35 shows results and confirms our predictions. Notice the formation of a resonance at the frequency

$$\omega = \frac{1}{\sqrt{LC}} = \frac{1}{\sqrt{1 \times 10^{-3} \times 1 \times 10^{-6}}} = \sqrt{1.E9} = 31.6 \text{ krad/s} \tag{26.136}$$

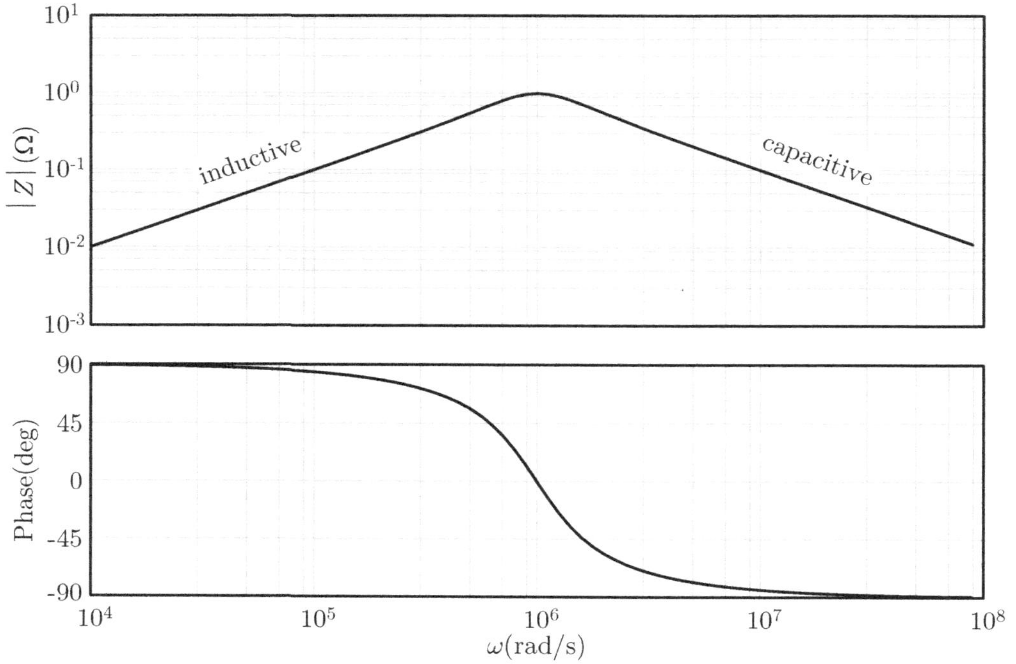

Fig. 26.32 Parallel *RLC* impedance versus frequency ($R = 1\,\Omega$, $L = 1\,\mu\text{H}$, and $C = 1\,\mu\text{F}$)

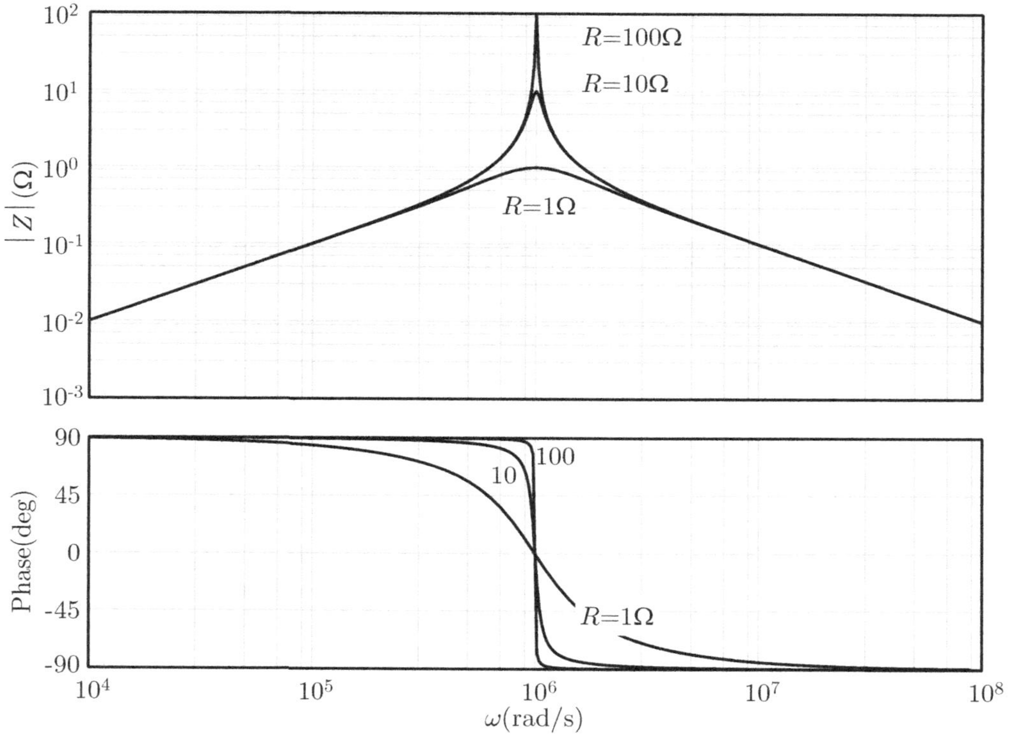

Fig. 26.33 Parallel *RLC* impedance versus frequency as a function of *R* ($L = 1\,\mu\text{H}$ and $C = 1\,\mu\text{F}$)

Fig. 26.34 Series *RC*, parallel *L* network

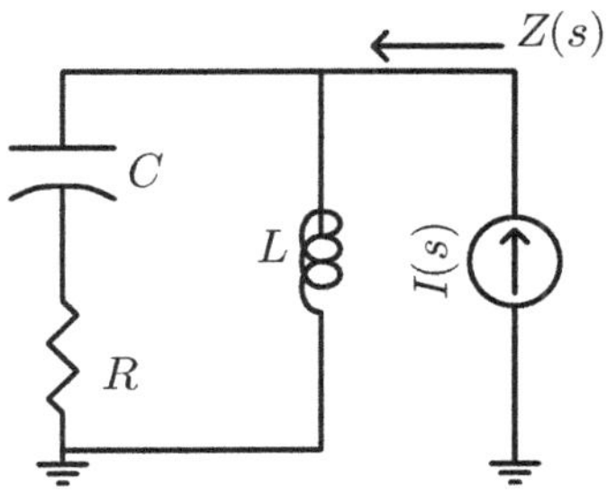

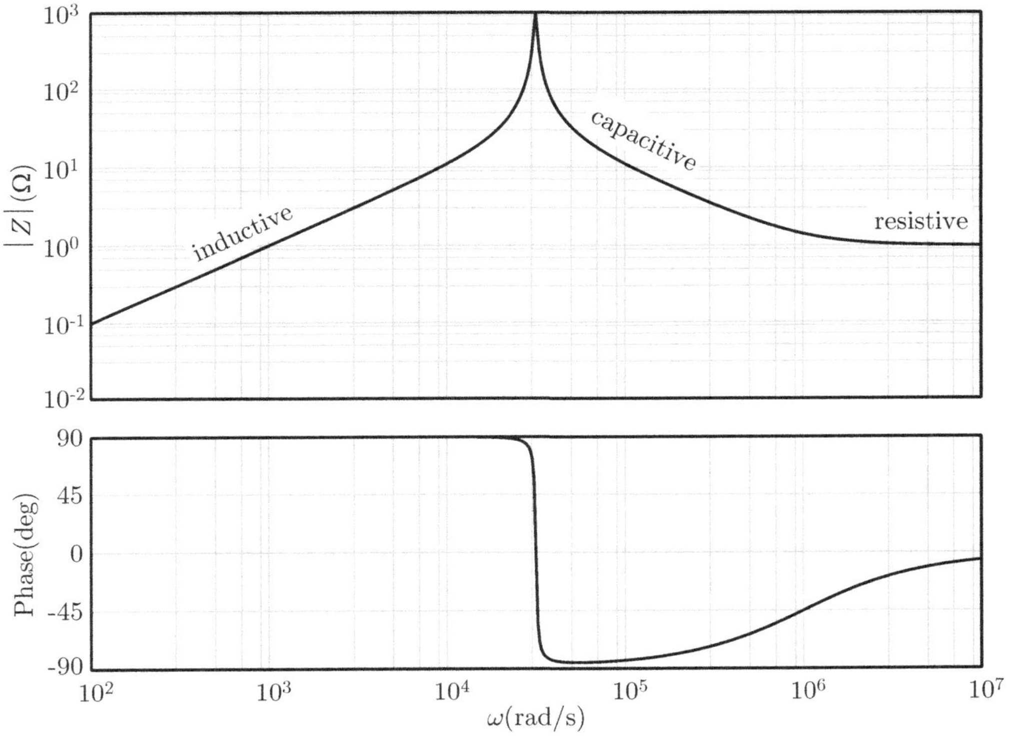

Fig. 26.35 Series *RC*, parallel *L* ($R = 1\,\Omega$, $L = 1$ mH, and $C = 1\,\mu$F)

The phase starts at 90° due to the inductor, flips to −90° due to the cap, and settles at 0° due to the resistor. Figure 26.36 shows impact of *R* on transfer function. It is clear that the settling value follows *R*. Also, for large enough *R* we bypass the resonance from the start and we fall back on a parallel *RL* topology where the inductor assumes the leading role at low frequency followed by the resistor at high frequency.

26.19 Series *RL* in Parallel with Series *RC* Circuit

The series *RL* in parallel with series *RC* is shown in Fig. 26.37. We derive the impedance transfer function as follows:

$$Z(s) = (R_1 + sL)||(R_2 + \frac{1}{sC})$$
$$= \frac{(R_1 + sL)(R_2 + \frac{1}{sC})}{R_1 + R_2 + sL + \frac{1}{sC}} \quad (26.137)$$

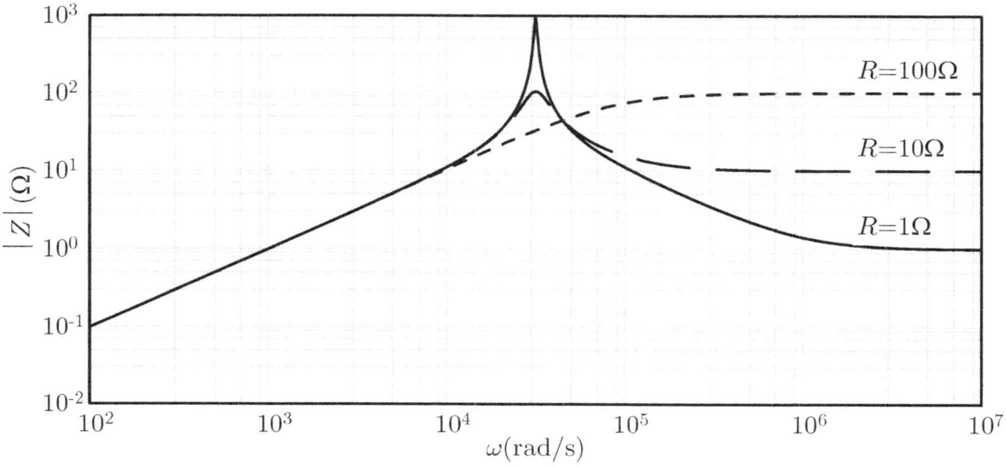

Fig. 26.36 Impact of R on series RC, parallel L ($L = 1$ mH and $C = 1\,\mu$F)

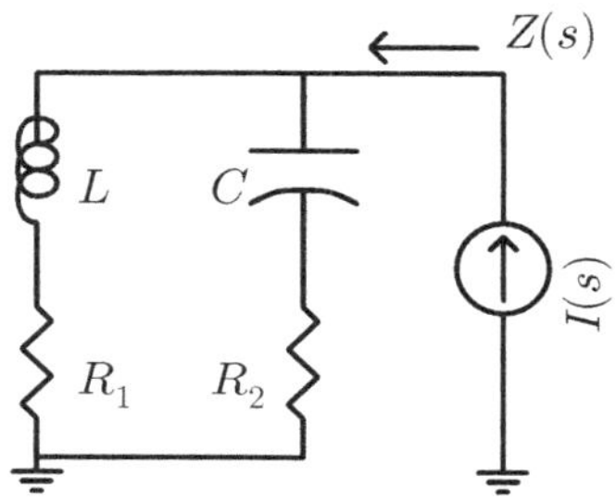

Fig. 26.37 Series RL in parallel with series RC

Multiply top and bottom by sC to get

$$Z(s) = \frac{(R_1 + sL)(1 + R_2 sC)}{(R_1 + R_2)sC + s^2 LC + 1} \quad (26.138)$$

Collect terms

$$Z(s) = \frac{R_1 + (L + R_1 R_2 C)s + R_2 LCs^2}{(R_1 + R_2)sC + s^2 LC + 1} \quad (26.139)$$

Factor out LC from denominator

$$Z(s) = \frac{1}{LC} \frac{R_1 + (L + R_1 R_2 C)s + R_2 LCs^2}{\frac{R_1+R_2}{L}s + s^2 + \frac{1}{LC}} \quad (26.140)$$

Define $\omega_{LC}^2 = \frac{1}{LC}$ and get

$$Z(s) = \omega_{LC}^2 \frac{R_1 + (L + R_1 R_2 C)s + R_2 LCs^2}{\frac{R_1+R_2}{L}s + s^2 + \omega_{LC}^2} \quad (26.141)$$

To ease the algebra, define the following constants:

$$\begin{aligned} a &= \frac{R1 + R2}{2L} \\ D &= \omega_{LC}^2 R_1 \\ E &= \omega_{LC}^2 (L + R_1 R_2 C) \\ \omega_0^2 &= \omega_{LC}^2 - a^2 \end{aligned} \quad (26.142)$$

and get

$$Z(s) = \frac{D + Es + R_2 s^2}{2as + s^2 + \omega_{LC}^2} \quad (26.143)$$

Doing long division gives

$$Z(s) = R_2 + \frac{D - R_2\omega_{LC}^2 + (E - 2R_2 a)s}{2as + s^2 + \omega_{LC}^2} \quad (26.144)$$

Complete square in denominator and get

$$Z(s) = R_2 + \frac{D - R_2\omega_{LC}^2 + (E - 2R_2 a)s}{(a + s)^2 + \omega_0^2} \quad (26.145)$$

where again $\omega_0^2 = \omega_{LC}^2 - a^2$. Figure 26.38 shows impedance across frequency. Notice that the DC impedance is that of R_1, and that the high-frequency limit is R_2.

Note to get the inverse Laplace transform of Eq. (26.145) we need to add and subtract terms to the numerator such that

$$Z(s) = R_2 + \frac{D - R_2\omega_{LC}^2 - (E - 2R_2a)a + (E - 2R_2a)(s + a)}{(a + s)^2 + \omega_0^2} \tag{26.146}$$

Now we can identify the following inverse transforms:

$$\begin{aligned} R_2 &\to \sim \delta(t) \\ \frac{D - R_2\omega_{LC}^2 - (E - 2R_2a)a}{(a + s)^2 + \omega_0^2} &\to \sim e^{-at}\sin\omega_0 t \\ \frac{(E - 2R_2a)(s + a)}{(a + s)^2 + \omega_0^2} &\to \sim e^{-at}\cos\omega_0 t \end{aligned} \tag{26.147}$$

26.20 Summary

This chapter is a critical one in transitioning from the abstract Fourier/Laplace (spectral) domain into the applied circuit domain. In circuits we deal with *RLC* elements that are stimulated by currents or voltages and whose output also has the form of currents/voltages. The first critical step in molding the spectral world in the framework of circuits is to identify the "carrier signal". This is the "mother" signal that will be always assumed to be running in the background. Every signal, be it input, output, or an internal node/branch, will be assumed to run versus time with the dependence e^{st}. The catch is what is the scaling factor? The scaling factor, which will be frequency dependent, is what we typically call the *transfer function*! Transfer functions tie something to something. In this introductory chapter we focus on the *impedance* transfer function which ties input current to output voltage. For example, for a resistor of value R it is simply R; for an inductor of value L it is sL; and for a cap of value C it is $\frac{1}{sC}$. For more complicated cases we fall back on KVL/KCL, use the above elemental transfer functions, and derive the corresponding one for internal or external node voltages. If we stimulate the network by a current source (again of the form e^{st}) and figure the output voltage, which will be of the form $Z(s)e^{st}$, then $Z(s)$ will be our impedance transfer function. When doing KVL/KCL manipulations we don't bother carrying on the e^{st} at all steps; instead we replace it with unity. As such we derive $Z(s)$ and *then* tack in the e^{st} to yield output voltage in the time domain. As with most concepts they are best demonstrated via examples, and towards that end the chapter had multitude of real life *RLC* examples. For each case we started with KVL/KCL relations, derived $Z(s)$, analyzed the transfer function so far as number of poles/zeroes, plotted both magnitude and phase, and finally rationalized both the decay rate (in terms of dB/dec) and the phase values (in terms of 0 for resistors, 90° for inductors, and −90° for capacitors). This process of circuit manipulations in the frequency domain is the heart of spectral techniques—it is the ability to work in the frequency domain and deal with algebraic equations (as opposed to working in the time domain and dealing with integro/differential equations) to arrive at the response in the frequency domain, ready to be inverse-transformed to provide the answer in the time domain.

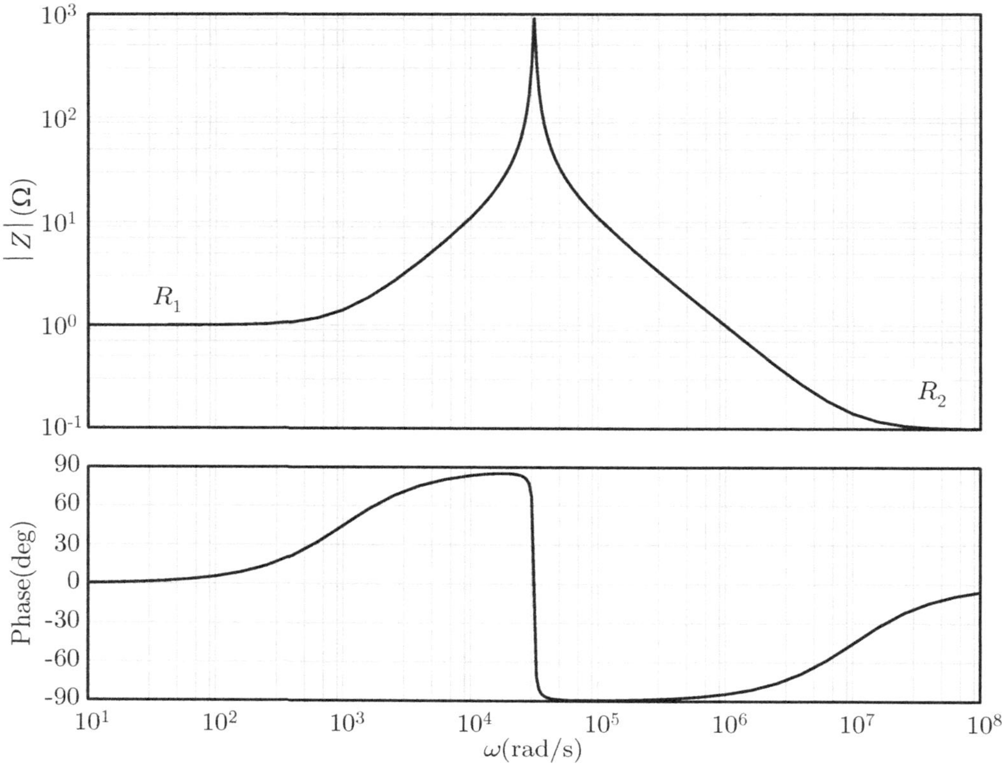

Fig. 26.38 Series RL in parallel with series RC: $R_1 = 1\,\Omega$, $R_2 = 0.1\,\Omega$, $L = 1$ mH and $C = 1\,\mu$F

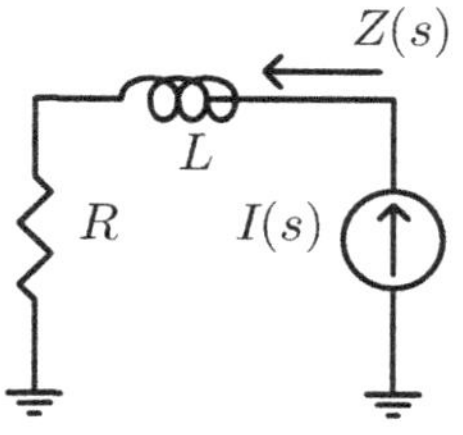

Fig. 26.39 Statement to Problem 1

26.21 Problems

1. The series RL is shown in Fig. 26.39; Derive and plot the transfer function (both magnitude and phase). What is the low frequency limit? High one? Set $R = 1\,\Omega$ and $L = 1$ mH. See sample solution in Fig. 26.40.
2. Consider the simple parallel RL circuit shown in Fig. 26.41; find its impedance transfer function. Plot it using $R = 1\,\Omega$ and $L = 1$ mH. What is the low-frequency limits? High one? Explain. See sample solution in Fig. 26.42.

   Answer:

$$Z(s) = R\frac{s}{s + \frac{R}{L}}$$

3. Consider the series RC/parallel C circuit shown in Fig. 26.43. Find input impedance. What is the limit at DC? At infinite frequency? For the case of $R = 1$, $C_1 = 1\,\mu$, and $C_2 = 10$n plot impedance and compare to SPICE; see sample solution in Fig. 26.44. How many poles/zeroes does this network have? Identify them on the plot.

   Answer:

$$Z(s) = \frac{1}{sC_2}\frac{s + \dfrac{1}{RC_1}}{s + \dfrac{1}{RC_s}}, \qquad C_s = \frac{C_1C_2}{C_1 + C_2}$$

4. The parallel RC/ series R network looks like that in Fig. 26.45. Find the impedance as seen from the current source. What is the

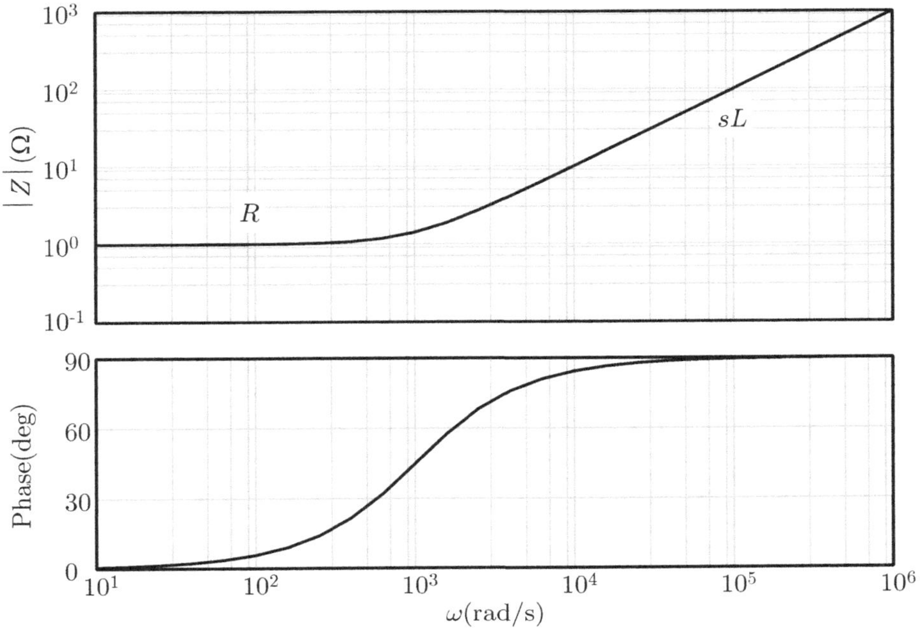

Fig. 26.40 Sample solution to Problem 1

Fig. 26.41 Statement to Problem 2

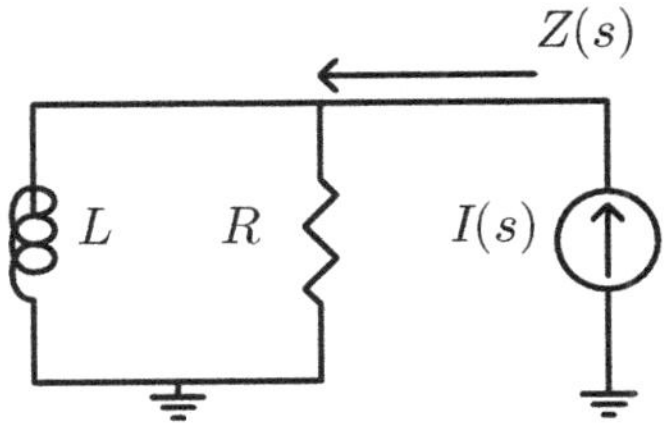

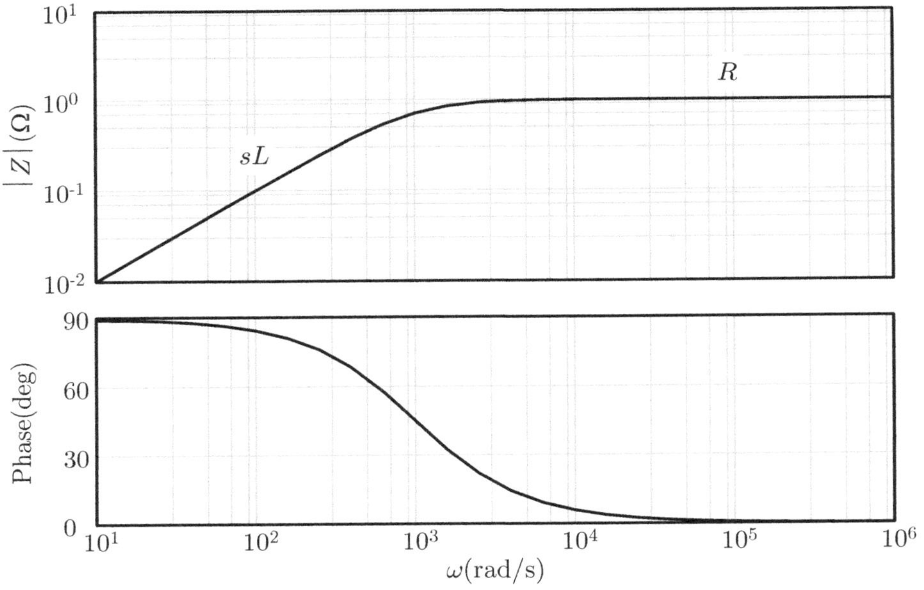

Fig. 26.42 Solution to Problem 2

Fig. 26.43 Series RC/parallel C circuit

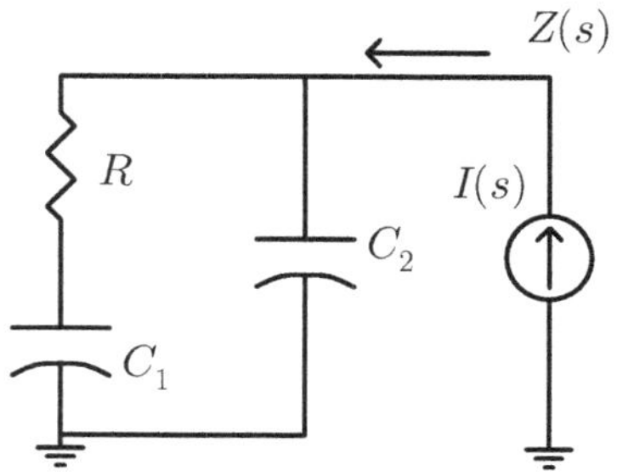

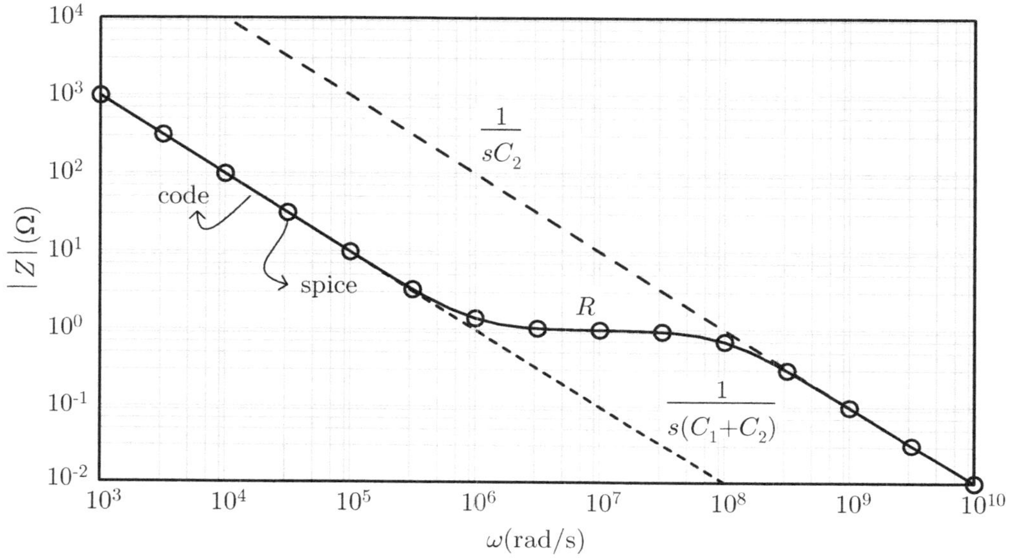

Fig. 26.44 Sample solution to Problem 3

Fig. 26.45 Statement to Problem 4

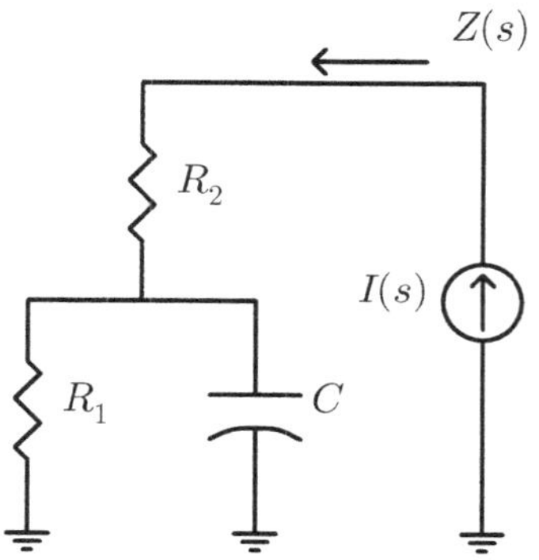

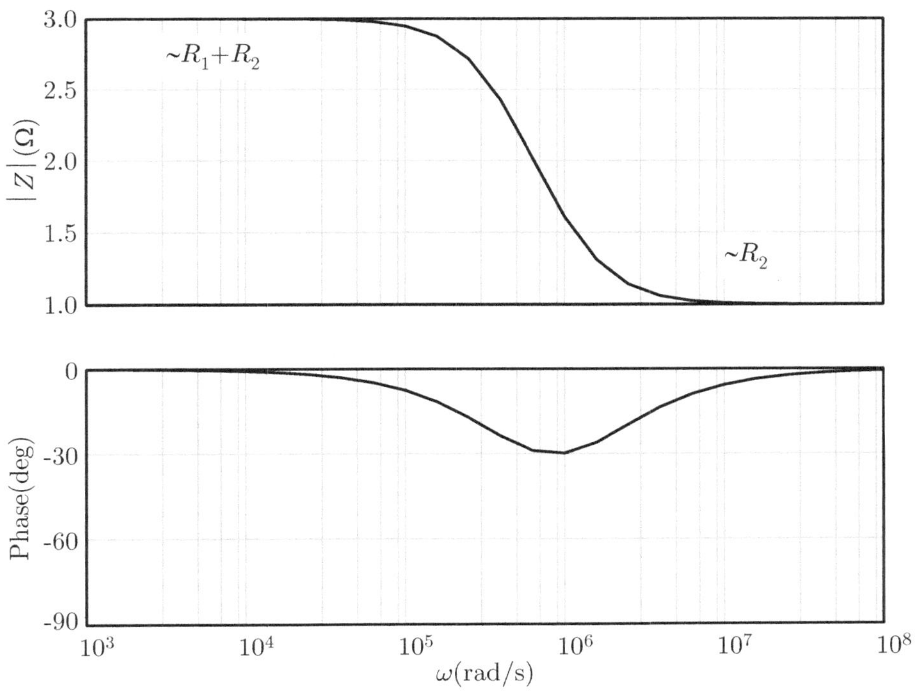

Fig. 26.46 Solution to Problem 4

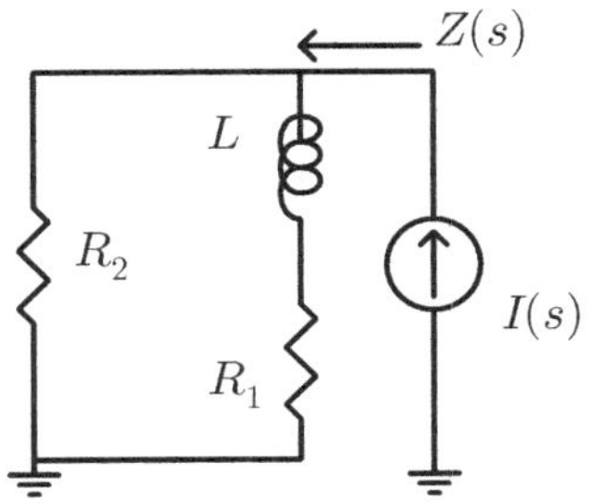

Fig. 26.47 Statement to Problem 5

DC limit? What is the hi-f one? The infinite one? Plot and compare to SPICE for the case $R_1 = 2, R_2 = 1\ \Omega$ and $C = 1\ \mu$F; see sample solution in Fig. 26.46. Where is the location of the pole? The zero? Explain the initial and final phases.

Answer:

$$Z(s) = R_2 + \frac{R_1}{1 + sR_1C}$$

$$Z(0) = R_2 + R_1$$

$$Z(s)_{\text{large frequency}} = R_2 + \frac{1}{sC}$$

$$Z(\infty) = R_2$$

5. The series RL, parallel R is shown in Fig. 26.47. Derive the impedance transfer function. What is the DC limit? High frequency one? Plot and compare to SPICE for the case $R_1 = 1,\ R_2 = 2\,\Omega$, and $L = 1$ mH. See sample solution in Fig. 26.48.
Answer:

$$Z(s) = \frac{R_2}{L}\frac{R_1 + sL}{s + a}, \quad a\frac{R_1 + R_2}{L}$$

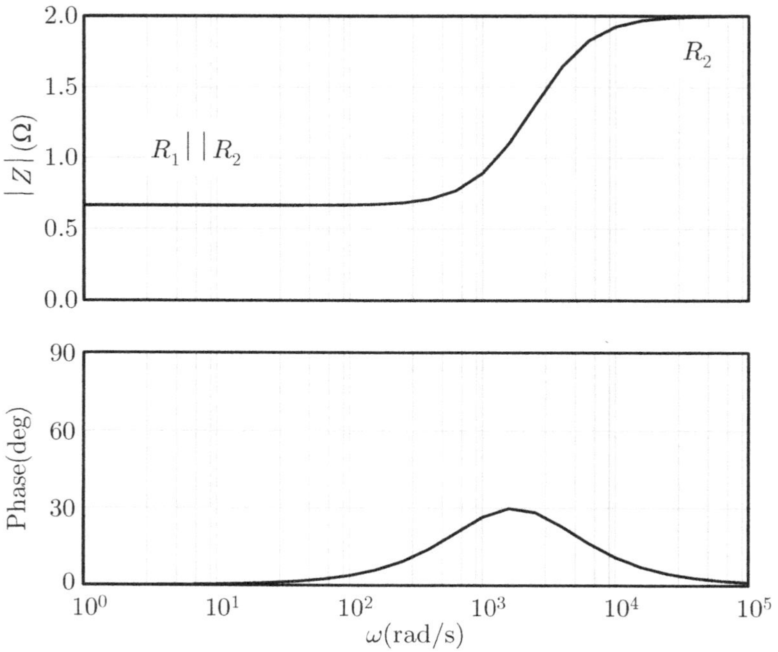

Fig. 26.48 Sample solution to Problem 5

6. Consider the distributed *RC* problem in Fig. 26.49. Find the input impedance as seen from the right side. While this problem can be done purely algebraically let's try solving it algorithmically, with a little help from the computer. Define the various impedances as seen in the figure; in particular

$$z_1 = R_1 + \frac{1}{sC_1}, \qquad z_{12} = z_1 || \frac{1}{sC_2}, \qquad z_2 = R_2 + z_{12}$$

$$z_{23} = z_2 || \frac{1}{sC_3}, \qquad z_3 = R_3 + z_{23}, \qquad Z = z_3$$

For each frequency calculate the above quantities and finally plot magnitude and phase of input impedance $Z(s)$. Compare to SPICE for case

$$C_1 = 1 \times 10^{-3}; C_2 = 10 \times 10^{-6}; C_3 = 1 \times 10^{-6}; R_1 = 100; R_2 = 10; R_3 = 1$$

How many poles does this network have? How many zeroes? Explain the starting and ending phase values. See sample solution in Fig. 26.50.

7. Use the same method as described in Problem 6 to find input impedance of *LC* network shown in Fig. 26.51. Compare to SPICE; see sample solution in Fig. 26.52. How many poles/zeroes are there?
8. Consider the *RC* network in Fig. 26.53. Unity AC current is injected and we want to find input impedance, which would be voltage at current terminal. Do KVL/KCL around the network and show that the resulting matrix is

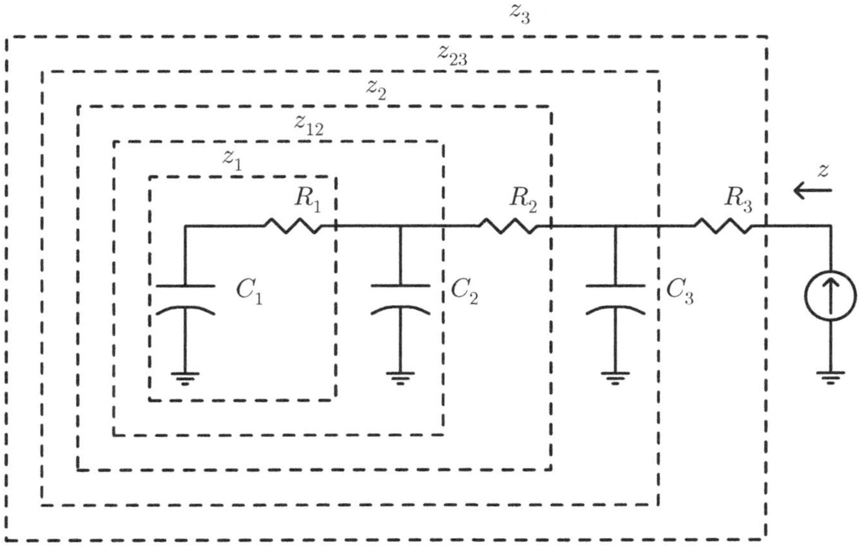

Fig. 26.49 Statement to Problem 6

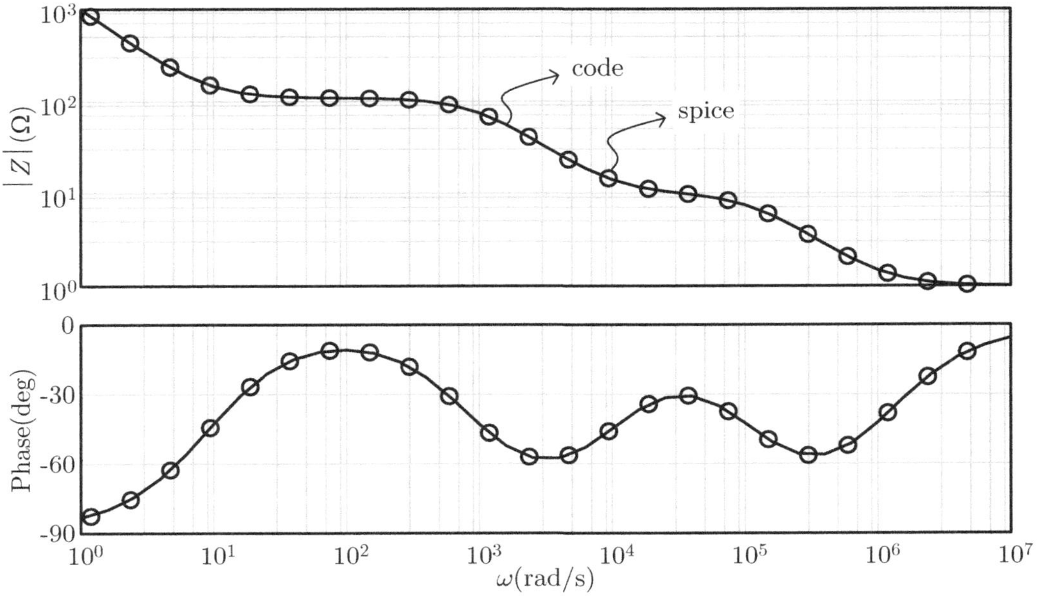

Fig. 26.50 Sample solution to Problem 6

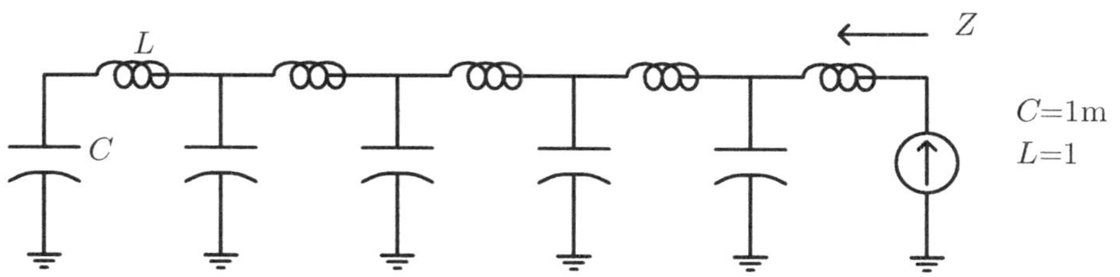

Fig. 26.51 Statement to Problem 7

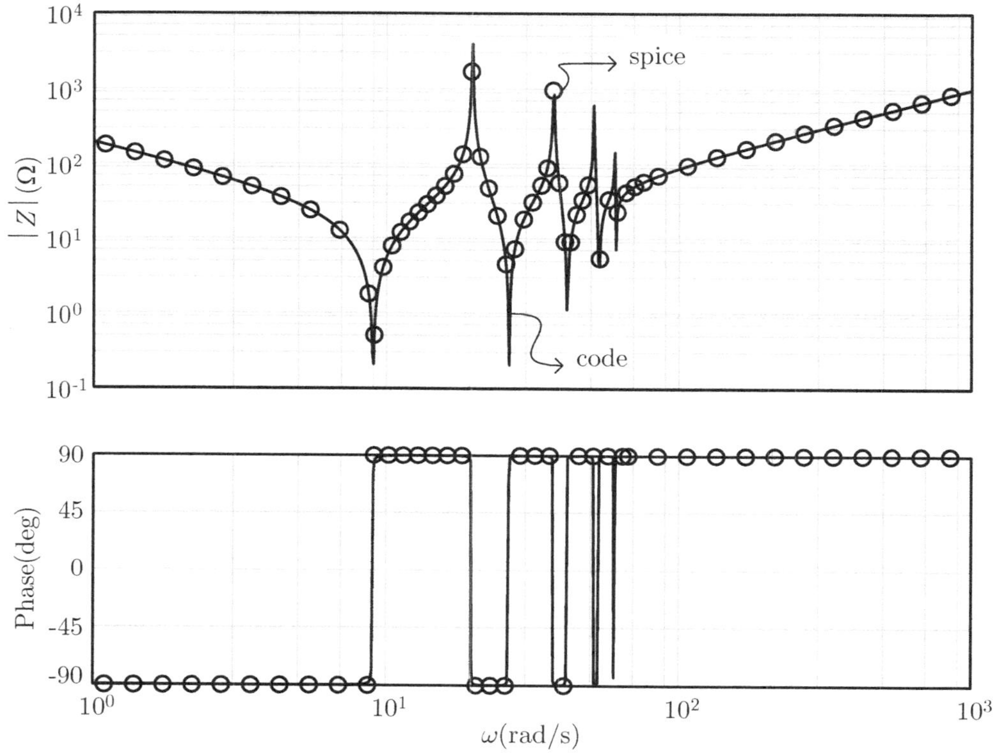

Fig. 26.52 Sample solution to Problem 7

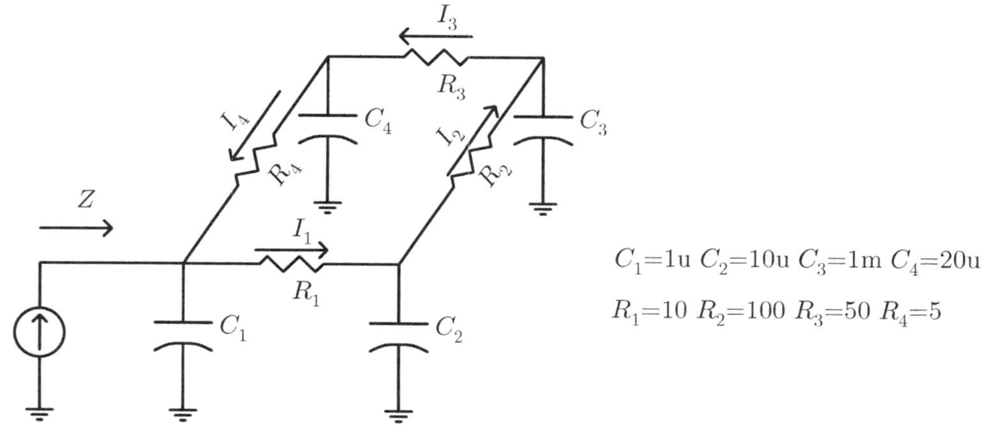

Fig. 26.53 Statement to Problem 8

$$\begin{bmatrix} -\frac{1}{s}\left(\frac{1}{C_1}+\frac{1}{C_2}\right)-R_1 & \frac{1}{sC_2} & 0 & \frac{1}{sC_1} \\ \frac{1}{sC_2} & -\frac{1}{s}\left(\frac{1}{C_2}+\frac{1}{C_3}\right)-R_2 & \frac{1}{sC_3} & 0 \\ 0 & \frac{1}{sC_3} & -\frac{1}{s}\left(\frac{1}{C_3}+\frac{1}{C_4}\right)-R_3 & \frac{1}{sC_4} \\ \frac{1}{sC_1} & 0 & \frac{1}{sC_4} & -\frac{1}{s}\left(\frac{1}{C_4}+\frac{1}{C_1}\right)-R_4 \end{bmatrix}$$

$$\times \begin{bmatrix} I_1 \\ I_2 \\ I_3 \\ I_4 \end{bmatrix} = \begin{bmatrix} -\frac{1}{sC_1} \\ 0 \\ 0 \\ \frac{1}{sC_1} \end{bmatrix}$$

Next, for each frequency value, solve this system either using a software or by Gaussian elimination and figure the various currents. Knowing those we ought to figure output voltage, which equals impedance here

$$Z(s) = \frac{1}{sC_1}\left[1 + I_4 - I_1\right]$$

Plot impedance and compare to SPICE; see sample solution in Fig. 26.54. What would you expect initial and final phase to be—why?

9. Consider the *RLC* network in Fig. 26.55. An AC current is injected into node v_1; measure input impedance Z_1 which is voltage v_1 and mutual impedance Z_2 which is voltage v_2.

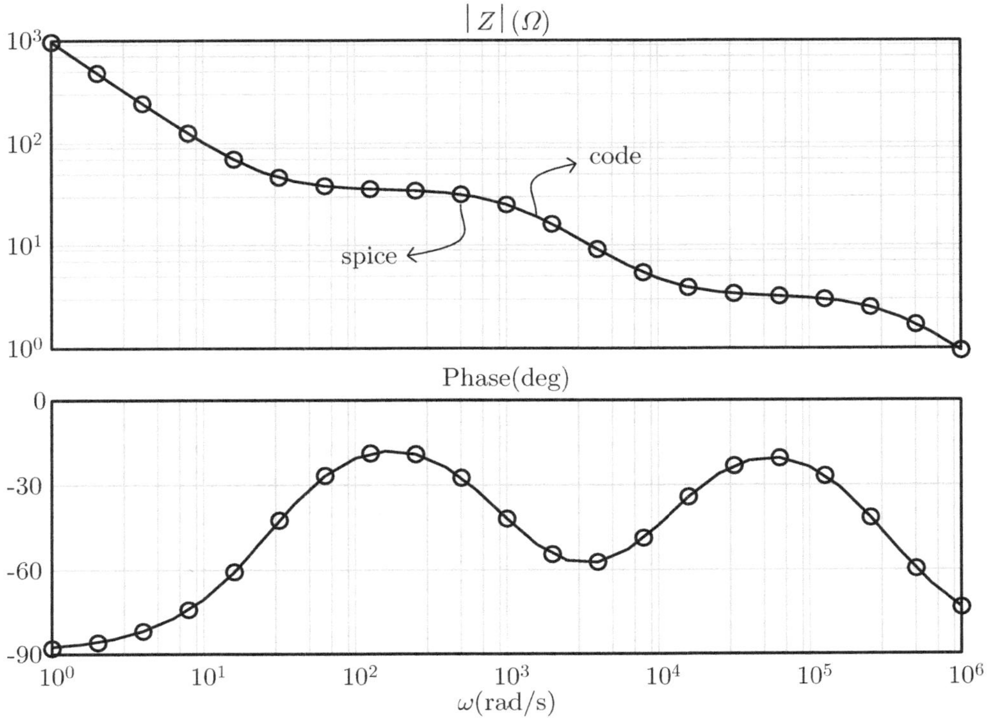

Fig. 26.54 Sample solution to Problem 8

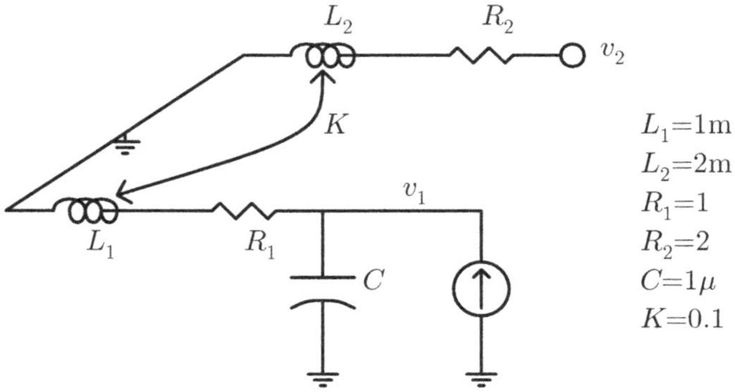

Fig. 26.55 Statement to Problem 9

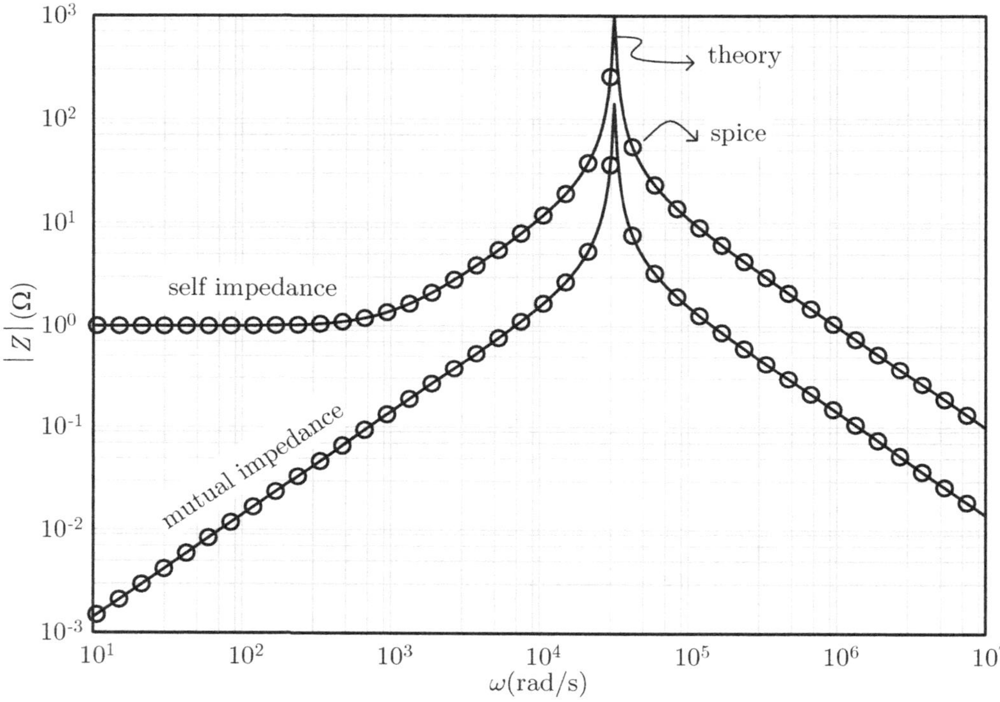

Fig. 26.56 Sample solution to Problem 9

Plot both and compare to SPICE; see sample solution in Fig. 26.56. Why is the mutual impedance low at low frequency? How about high frequency?
Answer:

$$Z_1(s) = \frac{1}{L_1 C} \frac{R_1 + sL_1}{\frac{1}{L_1 C} + s\frac{R_1}{L_1} + s^2}$$

$$Z_2(s) = \frac{Z_1}{R_1 + sL_1} \times s \times M, \quad M = K\sqrt{L_1 L_2}$$

10. Consider the *RLC* network in Fig. 26.57. Figure input impedance and mutual one. This can be achieved by injecting an AC current of unity magnitude onto node v_1 and measuring v_1 for self-impedance and v_2 for mutual impedance. Plot results and compare to SPICE; see sample solution in Fig. 26.58.

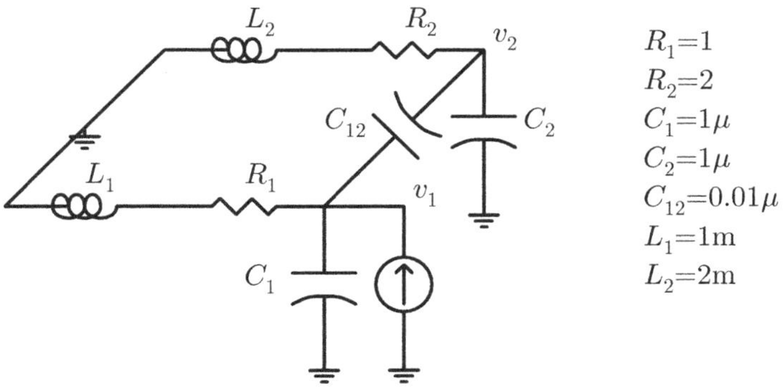

Fig. 26.57 Statement to Problem 10

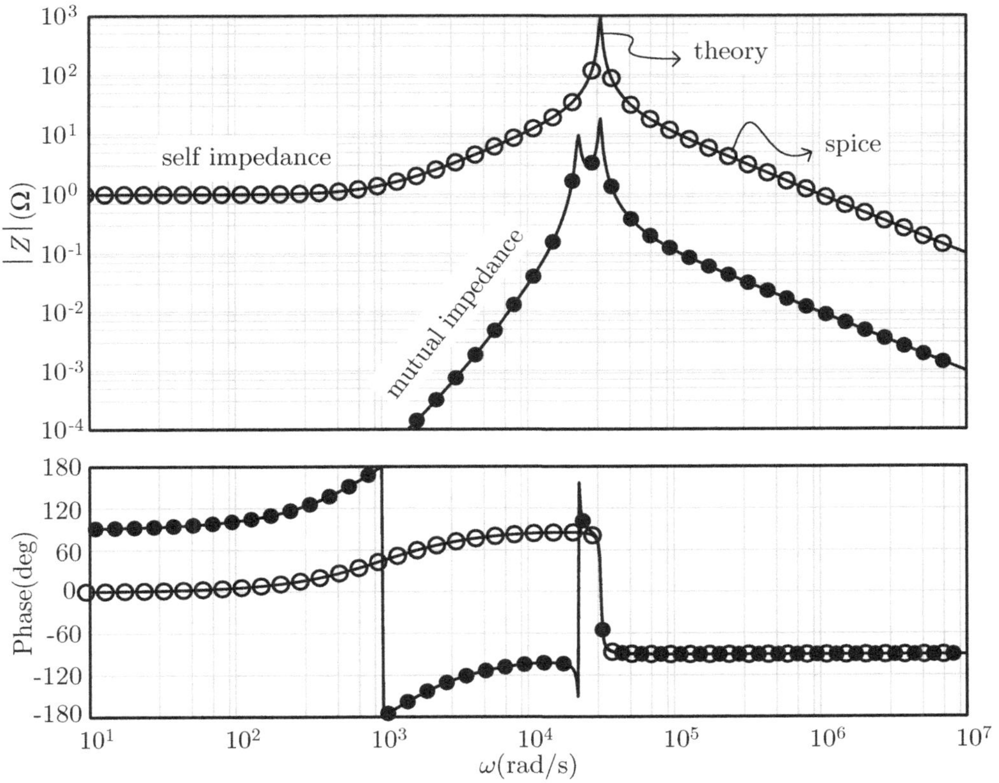

Fig. 26.58 Sample solution to Problem 10

Answer:

$$z_1 = \frac{1}{sC_1}, \qquad z_2 = R_1 + sL_1, \qquad z_3 = \frac{1}{sC_2}, \qquad z_4 = R_2 + sL_2$$

$$z_5 = \frac{1}{sC_{12}}, \qquad z_{12} = z_1||z_2, \qquad z_{34} = z_3||z_4, \qquad z_6 = z_{34} + z_5$$

$$z_{\text{self}} = z_{12}||z_6, \qquad z_{\text{mut}} = \frac{z_{\text{self}}}{z_5 + z_{34}} \times z_{34}$$

27 The Phase

27.1 Introduction

No one would argue that the imaginary part of the transfer function is as important as the real part! We could easily say that if the real part is the one side of a coin, then the imaginary part would be the other (Fig. 27.1)!

Really, without knowing the two sides, much information would be missing. Yet when it comes to the same comparison between the magnitude and phase (which are the other form of representing complex numbers), the magnitude is almost guaranteed to have preferential treatment. Consider for example the transfer function

$$H_1(s) = \frac{1}{s+10} \tag{27.1}$$

whose magnitude and phase are shown in Fig. 27.2. Looking at the form of the function we may conclude that at DC we get perfect transmission (with a gain of 1/10) and that at high (angular) frequency (compared to 10) the gain goes down, eventually to zero. Somehow just by looking at the form, and by corroborating with the plot, we almost have made our mind about the behavior of this function. So why worry about the phase?

27.2 Why the Phase Is Important

Sticking with the example from the prior section, consider next a very similar one

$$H_2(s) = \frac{1}{s-10} \tag{27.2}$$

It is almost identical to Eq (27.1) with the exception that the pole is a right-hand one! The magnitude and phase of this are shown in Fig. 27.3. Judging only by the magnitude we arrive at the conclusion that *both have the same magnitude*! So what else matters? There is much more at stake. The phase is quite different, as seen in the same figure. So what does that all mean?

Fact is, while the first function points to a *stable* system

$$h_1(t) = \frac{1}{s+10} \rightarrow e^{-10t} \tag{27.3}$$

the second one points to an *unstable* system!

$$h_2(t) = \frac{1}{s-10} \rightarrow e^{10t} \tag{27.4}$$

That makes all the difference. What we've learned then is that looking at the magnitude

F. Badrieh, *Spectral, Convolution and Numerical Techniques in Circuit Theory*,
https://doi.org/10.1007/978-3-319-71437-0_27

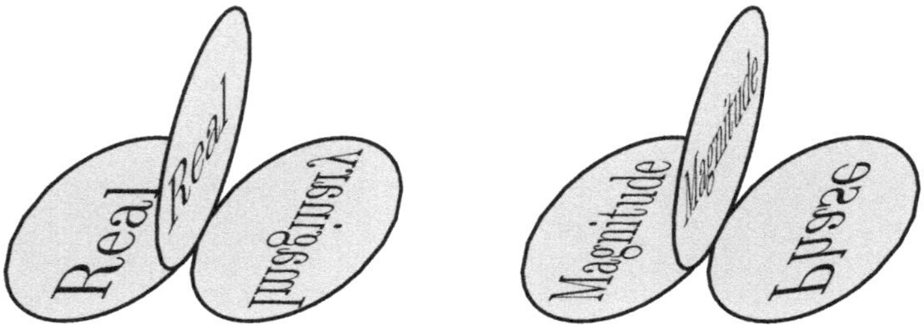

Fig. 27.1 Real/imaginary and magnitude/phase aspects

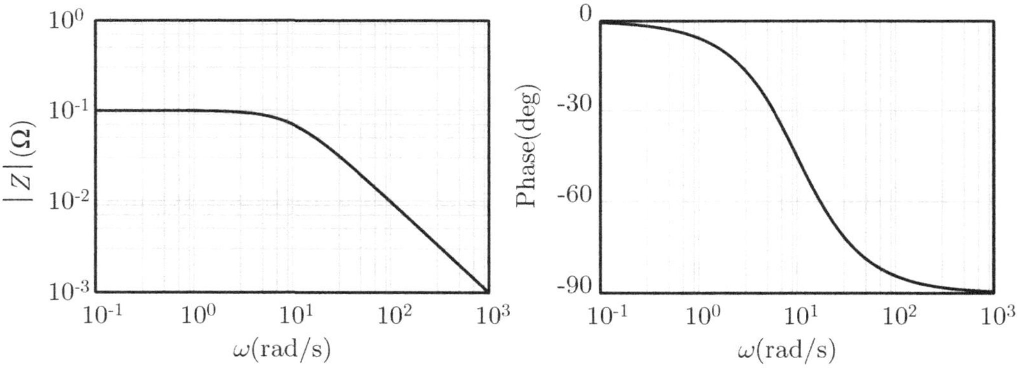

Fig. 27.2 Magnitude and phase of $\frac{1}{s+10}$

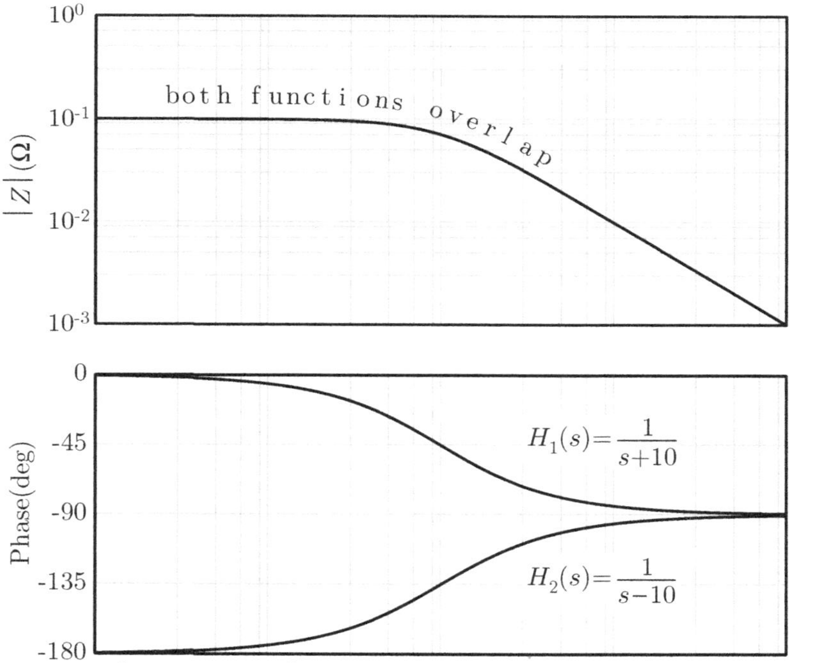

Fig. 27.3 Magnitude and phase of $\frac{1}{s+10}$ and $\frac{1}{s-10}$

alone—while very useful—is not sufficient to fully understand the system under consideration. While predicting the behavior of the magnitude is relatively simple, predicting the phase is a bit more complicated—and perhaps that is the reason why phase analysis is typically postponed (or totally ignored). To overcome this limitation we next examine how to predict phase behavior.

27.3 Transfer Function with Sable Pole

As a first example, let's deal with a stable system of single pole. Again the transfer function is

$$H(s) = \frac{1}{s+10} \tag{27.5}$$

We want to predict the phase at DC and that at ∞. Let's assume σ is zero here, and use $s = j\omega$; we then have

$$H(\omega) = \frac{1}{10+j\omega} \tag{27.6}$$

Expand to get

$$H(\omega) = \frac{10-j\omega}{\alpha} \sim 10-j\omega \tag{27.7}$$

where α is a real number which we don't care about; it will scale both the real and imaginary parts equally, and hence won't impact the phase. Notice next what happens. When $\omega \rightarrow 0$ we arrive at the approximation

$$H(0) \sim 10 \tag{27.8}$$

which is purely real; that is the DC limit is a real one, and is indicated by the arrow in Fig. 27.4a. On the other hand, at high frequency we arrive at the approximation

$$H(\infty) \sim -j\omega \tag{27.9}$$

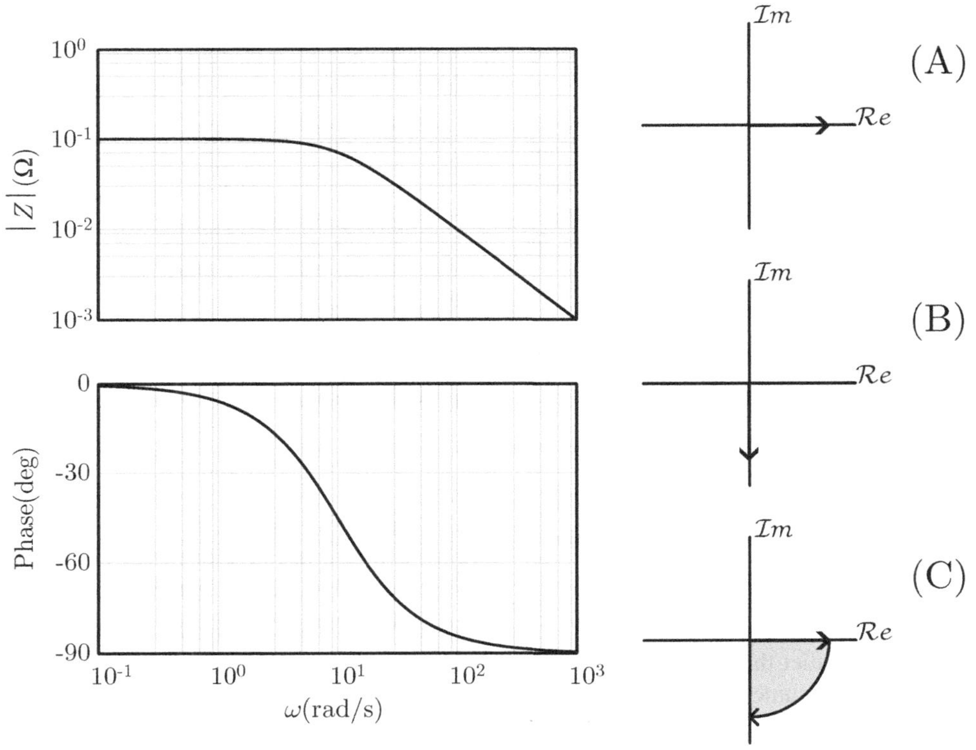

Fig. 27.4 Analysis of phase of $\frac{1}{s+10}$

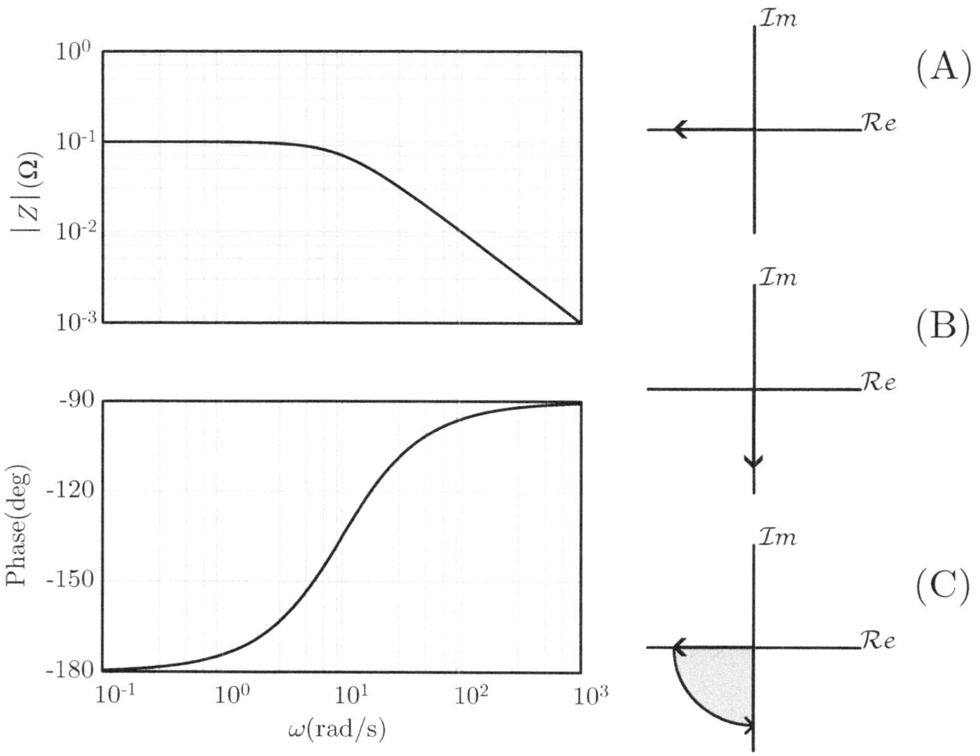

Fig. 27.5 Analysis of phase of $\frac{1}{s-10}$

which is not only purely imaginary, but a negative one; hence we arrive at the visual representation in Fig. 27.4b. Forming a path between the limits we arrive then at Fig. 27.4c, which states that the phase starts at zero, and then goes to $-90°$. This in fact is the case as shown on the left side of the figure!

27.4 Transfer Function with Unstable Pole

As a second example, let's deal with the unstable system of single pole. Again the transfer function is

$$H(s) = \frac{1}{s-10} \tag{27.10}$$

We want to predict the phase at DC and that at ∞. Again let's assume σ is zero here, and use $s = j\omega$; we then have

$$H(\omega) = \frac{1}{-10+j\omega} \tag{27.11}$$

Expand to get

$$H(\omega) = \frac{-10-j\omega}{\alpha} \sim -10 - j\omega \tag{27.12}$$

where again α is a real number which we don't care about and which will scale both the real and imaginary parts equally, and hence won't impact the phase. When $\omega \to 0$ we arrive at the approximation

$$H(0) \sim -10 \tag{27.13}$$

which is purely real and negative; that is the DC limit is a real negative one, and is indicated by the arrow in Fig. 27.5a. On the other hand, at high frequency we arrive at the approximation

$$H(\infty) \sim -j\omega \tag{27.14}$$

which is a purely negative imaginary one; hence we arrive at the visual representation in Fig. 27.5b. Forming a path between the limits we arrive then at Fig. 27.5c, which states that the phase starts at $-180°$, and then goes to $-90°$. This in fact is the case as shown on the left side of the figure! Notice that the *ending* phase of both this section and the prior one is the same, namely $-90°$. Actually the reason behind this is that in both cases the limit of the function as $s \to \infty$ is the same, namely $\frac{1}{s}$; as such we would expect the phase (and the magnitude) to act the same. The starting phase, on the other hand, differs between the two transfer functions because both transfer functions behave differently at DC (one ending up with inverting sign).

27.5 Transfer Function with Pole and Zero

Consider next the transfer function

$$H(s) = \frac{s}{s+10} \tag{27.15}$$

which is stable, and has a single pole *and* a single zero. Plugging in for s we get

$$H(\omega) = \frac{j\omega}{10 + j\omega} \tag{27.16}$$

Expand to get

$$H(\omega) = \frac{j\omega(10 - j\omega)}{\alpha} \sim \omega^2 + j10\omega \tag{27.17}$$

At low frequency the ω^2 dies out faster and we end up with

$$H(0) \sim j10\omega \tag{27.18}$$

which is purely imaginary and positive; hence we end up with Fig. 27.6a. On the other hand, at high frequency the ω^2 term dominates and we end up with the approximation

$$H(\infty) \sim \omega^2 \tag{27.19}$$

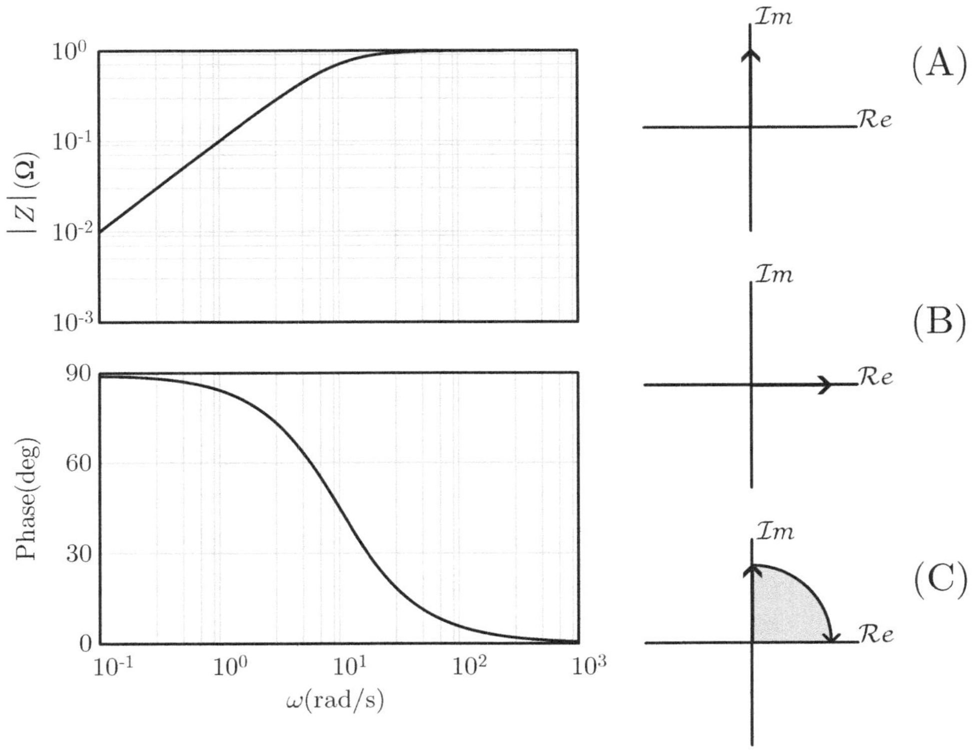

Fig. 27.6 Analysis of phase of $\frac{s}{s+10}$

which is purely real and positive, as shown in Fig. 27.6b. When we trace the path we end up with Fig. 27.6c. That is, phase starts at 90° and then goes to zero, as confirmed in the phase plot on the same figure. Incidentally this transfer function mimics that of a parallel *RL* network impedance transfer function. At low frequency the inductor dominates and furnishes the 90°, and at high frequency the resistor takes control and brings the phase back to zero.

27.6 Transfer Function with Two Poles

Consider next the transfer function

$$H(s) = \frac{1}{(s+1)(s+10)} \tag{27.20}$$

which has two poles, one at $s = -1$ and the other at $s = -10$. Expand the denominator to get

$$H(s) = \frac{1}{s^2 + 11s + 10} \tag{27.21}$$

Plugging in for s we get

$$H(\omega) = \frac{1}{(10-\omega^2) + j11\omega} \tag{27.22}$$

Expand to get

$$H(\omega) = \frac{(10-\omega^2) - j11\omega}{\alpha} \sim (10-\omega^2) - j11\omega \tag{27.23}$$

At low frequency we get the approximation

$$H(0) \sim 10 \tag{27.24}$$

which is purely real and positive; hence we end up with Fig. 27.7a. On the other hand, at high frequency the $-\omega^2$ term dominates and we end up with the approximation

$$H(\infty) \sim -\omega^2 \tag{27.25}$$

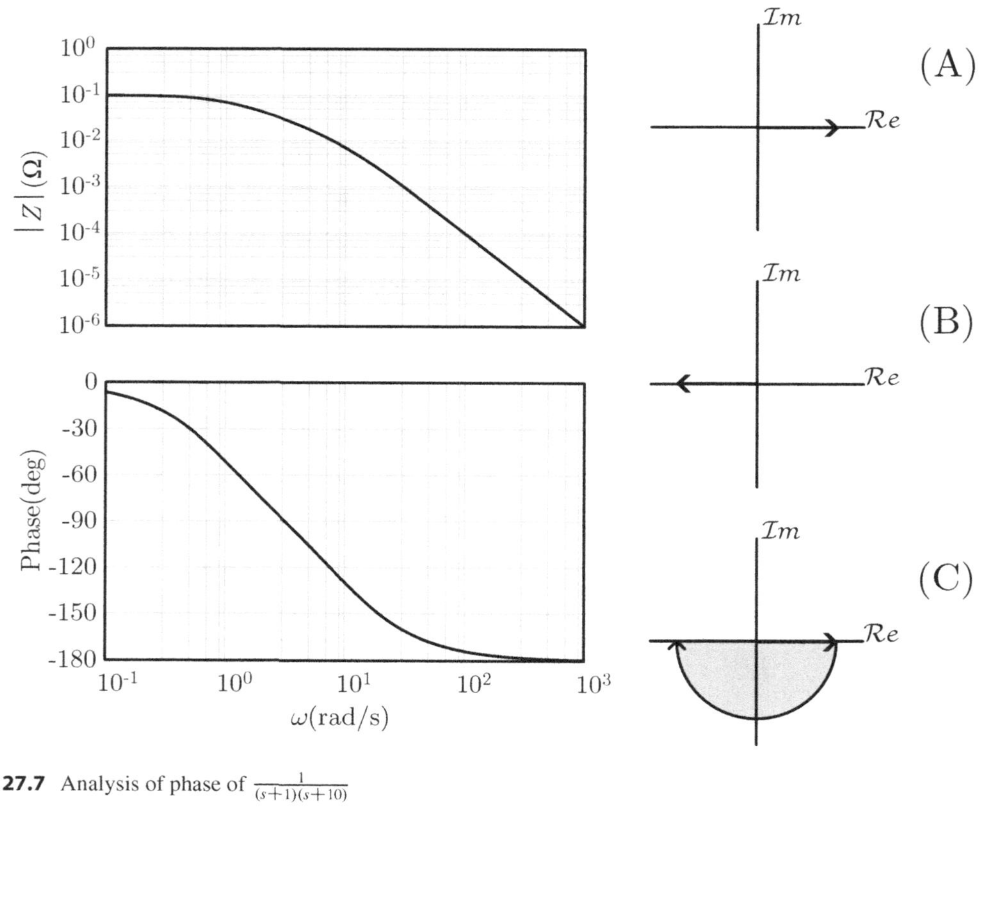

Fig. 27.7 Analysis of phase of $\frac{1}{(s+1)(s+10)}$

which is again purely real but negative, as shown in Fig. 27.7b. Question now, going from 0 to 180°—do we traverse the upper complex plane, or the lower one? To answer this we need an intermediate frequency point. Let's reexamine the transfer function

$$H(\omega) \sim (10 - \omega^2) - j11\omega \qquad (27.26)$$

We can see that the imaginary part is *always* negative; hence we live in the lower complex plane. So finally the answer would be start at zero, traverse the lower complex plane, and arrive at −180° as shown in Fig. 27.7c. This is confirmed in the phase plot on the left of the figure.

27.7 Summary

In this brief and introductory chapter on the phase we stress the importance and relevance of the transfer function phase. While it is almost universal that when the real part of the transfer function is plotted, it is accompanied by the imaginary one, the case is different when it comes to plotting the magnitude and the phase of the same transfer function. While true the magnitude trace gives a lot of information about the system under consideration—such as the gain, bandwidth, filtering ...; the phase cannot be compromised. As shown towards the very start of the chapter two systems with identical magnitude plots have drastically different behavior in the time domain due to a difference in their phase (one being stable while the other unstable). Most of the time the phase relation to the magnitude is pretty straightforward, especially for stable systems. But that does not warrant omitting the step of actually looking at the phase plot and rationalizing it in terms of poles/zeroes and potential inversion issues (sign flip). We showed a simple method of predicting the starting and ending phase, simply by unfolding the s in terms of $j\omega$, taking the asymptotic limit of the transfer function, identifying the corresponding real and imaginary parts, and mapping the limits in terms of starting and ending arrows on the complex plane. Having identified those points we can simply trace a line showing the phase behavior as a function of frequency. If ever in doubt we can plug in additional intermediate points to guide us whether we are going in one direction as opposed to another.

27.8 Problems

1. What is the phase of the following transfer functions?
   (a) $H(s) = 1$.
   (b) $H(s) = 2$.
   (c) $H(s) = -1$.
   (d) $H(s) = -0.5$.
2. What is the phase of the following transfer functions?
   (a) $H(s) = j$.
   (b) $H(s) = 3j$.
   (c) $H(s) = -j$.
   (d) $H(s) = -0.25j$.
3. What is the DC and high frequency limit of the phase in the following transfer functions?
   (a) $H(\omega) = 1 + \omega$.
   (b) $H(\omega) = 1 - \omega$.
   (c) $H(s) = -1 + 0.5\omega$.
   (d) $H(s) = -3 - 2\omega$.
4. What is the DC and high frequency limit of the phase in the following transfer functions?
   (a) $H(\omega) = j + \omega$.
   (b) $H(\omega) = j - \omega$.
   (c) $H(s) = -j + 0.5\omega$.
   (d) $H(s) = -j - 2\omega$.
5. What is the DC and high frequency limit of the phase in the following transfer functions?
   (a) $H(\omega) = 1 + j\omega$.
   (b) $H(\omega) = 1 - j\omega$.
   (c) $H(s) = j + j\omega$.
   (d) $H(s) = -j + \omega$.
6. Plot the magnitude and phase of the transfer function $H_1(s) = 10 + s$ and $H_2(s) = 10 - s$; assume $\sigma = 0.0$ in both cases and explain results. See sample solution in Fig. 27.8.
7. Plot the magnitude and phase for each of the following transfer functions:

$$H_1(s) = \frac{1+s}{10+s}$$

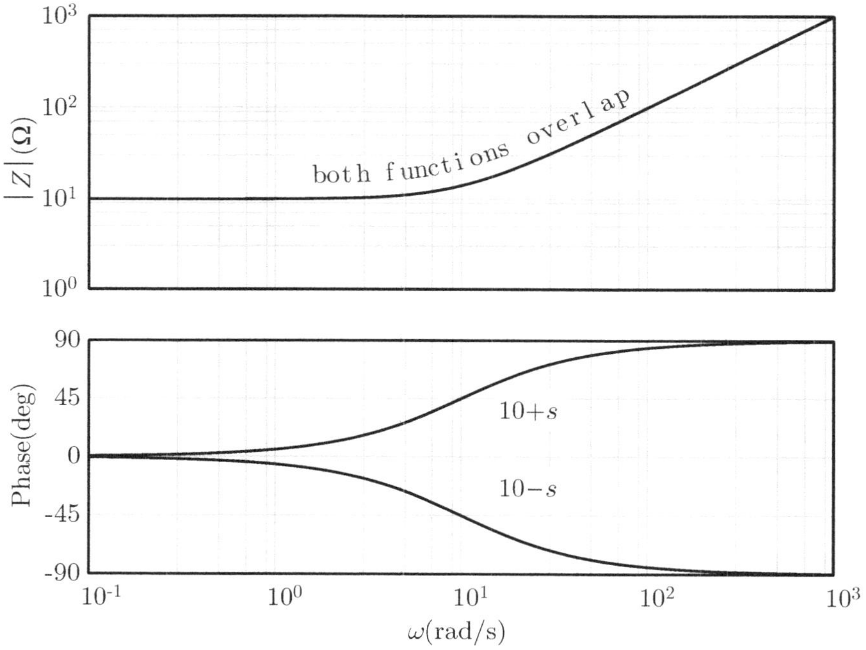

Fig. 27.8 Sample solution to Problem 6

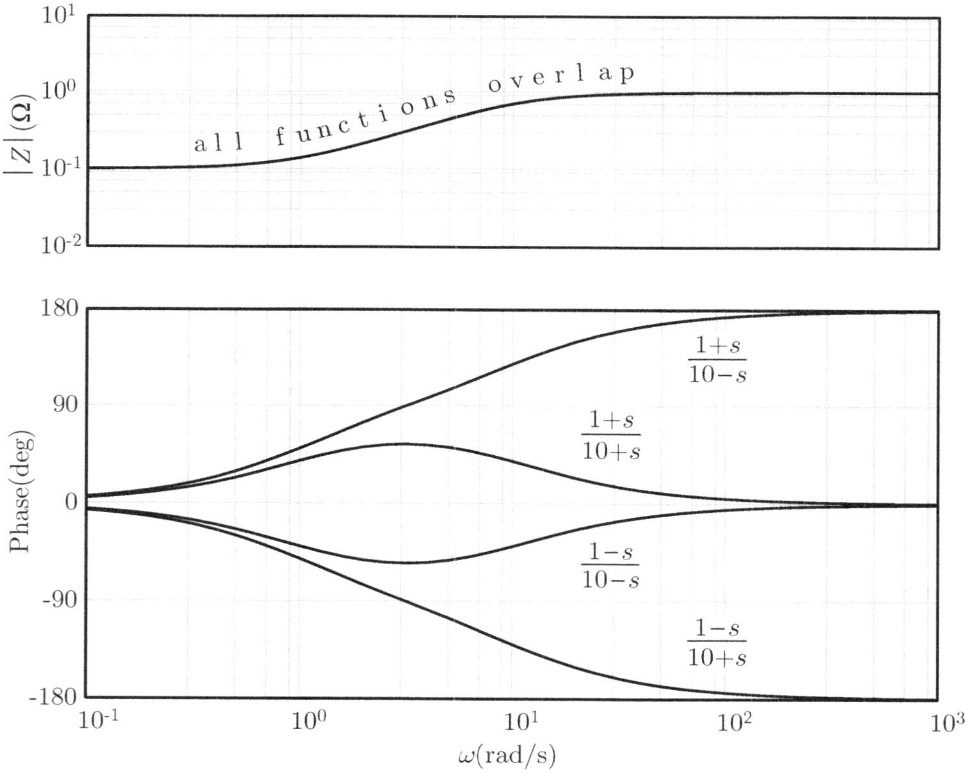

Fig. 27.9 Sample solution to Problem 7

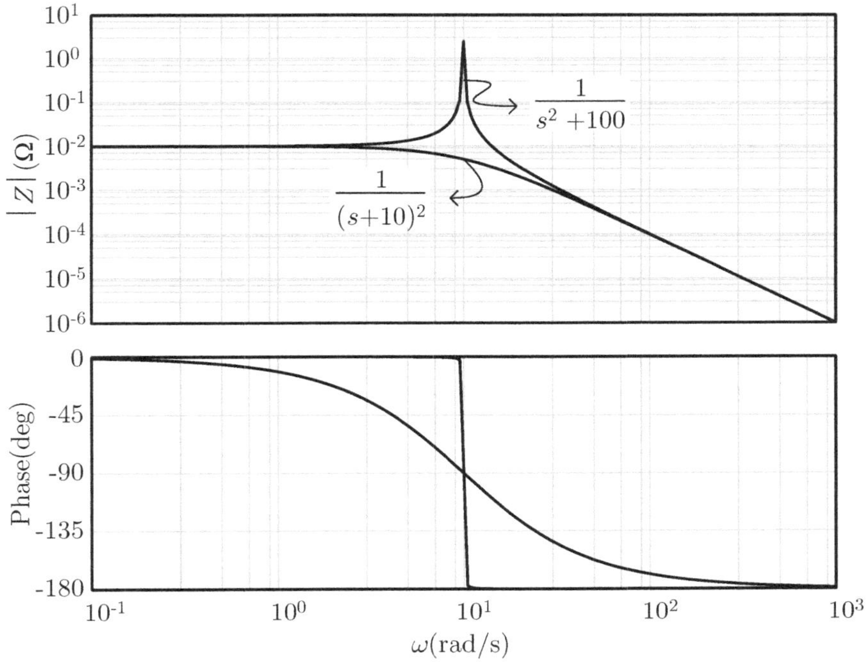

Fig. 27.10 Sample solution to Problem 8

$$H_2(s) = \frac{1+s}{10-s}$$

$$H_3(s) = \frac{1-s}{10+s}$$

$$H_4(s) = \frac{1-s}{10-s}$$

See sample solution in Fig. 27.9; explain results.

8. The two functions below have the same DC and high frequency limits—both in magnitude and phase. Plot them and identify which is which; see sample solution in Fig. 27.10.

$$H_1(s) = \frac{1}{(s+10)^2}$$

$$H_2(s) = \frac{1}{s^2+100}$$

28 Stability and Relation to Poles Placements

28.1 Introduction

Often is the case that our transfer function (tying output Y to input X) is given in the form of

$$H(s) = \frac{a_0 + a_1 s + a_2 s^2 + \cdots}{b_0 + b_1 s + b_2 s^2 + \cdots} \tag{28.1}$$

We are interested in finding a criterion on stability based on the characteristics of the transfer function. As we have sensed all along and as will be shown next, stability depends on the location of poles. If the poles reside on the left-hand side of the complex plane then the system is stable; else it is not. For an unstable system the response simply blows up. Rather than digging too much into the theory let us sample some stable and unstable transfer functions and examine in more detail the particulars of their frequency and time response.

28.2 Example of Stable System

To demonstrate the stability criterion let's take a sample system which has a transfer function

$$H(s) = \frac{4+s}{(s+1)(s+2)(s+3)} \tag{28.2}$$

The transfer function has three poles at

$$s_1 = -1, \qquad s_2 = -2, \qquad s_3 = -3 \tag{28.3}$$

As shown in Fig. 28.1 all three poles reside on the left-hand side of the complex plane.

We can expand the transfer function as

$$H(s) = \frac{A}{s+1} + \frac{B}{s+2} + \frac{C}{s+3} \tag{28.4}$$

The coefficients are evaluated as

$$A = \frac{4+s}{(s+1)(s+2)(s+3)} \times (s+1)\bigg|_{s=-1} = \frac{3}{2} \tag{28.5}$$

$$B = \frac{4+s}{(s+1)(s+2)(s+3)} \times (s+2)\bigg|_{s=-2} = \frac{2}{(-1)(1)} = -2 \tag{28.6}$$

$$C = \frac{4+s}{(s+1)(s+2)(s+3)} \times (s+3)\bigg|_{s=-3} = \frac{1}{(-2)(-1)} = \frac{1}{2} \tag{28.7}$$

So that

$$\boxed{H(s) = \frac{3/2}{s+1} - \frac{2}{s+2} + \frac{1/2}{s+3}} \tag{28.8}$$

F. Badrieh, *Spectral, Convolution and Numerical Techniques in Circuit Theory*,
https://doi.org/10.1007/978-3-319-71437-0_28

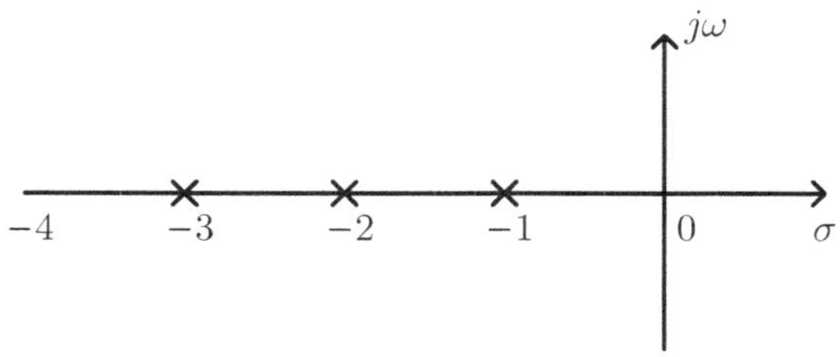

Fig. 28.1 Pole placement of transfer function $\frac{4+s}{(s+1)(s+2)(s+3)}$

A plot of this transfer function is shown in Fig. 28.2. Notice that at DC we get the value $H(0) = \frac{4}{6}$ and at high frequency we get 0. Also notice that the starting phase is zero and ending one is $-180°$ since we have three poles and a single zero. Finally notice at high frequency the decay rate is -40 dB/dec due to the $\sim \frac{1}{s^2}$ asymptotic limit. Applying inverse transform we arrive at the impulse response:

$$\boxed{h(t) = \frac{3}{2}e^{-t} - 2e^{-2t} + \frac{1}{2}e^{-3t}} \qquad (28.9)$$

Since each transient terms is a dying exponential, we conclude that the impulse response is a dying one too! Hence the system is stable. Figure 28.3 shows transient results and confirms stability. To recap, since the system had all poles residing on the left-hand side of the complex plane the impulse response in time does not blow up; hence the system is stable.

28.3 Example of Unstable System

Consider a similar transfer function, but with last pole shifted in sign:

$$H(s) = \frac{4+s}{(s+1)(s+2)(s-3)} \qquad (28.10)$$

The transfer function has three poles at

$$s_1 = -1, \qquad s_2 = -2, \qquad s_3 = 3 \qquad (28.11)$$

As shown in Fig. 28.4, not all poles now reside on the left-hand side of the complex plane; instead there is a single pole residing on the right-hand side of the complex plane—namely at $s = 3$.

We can expand the transfer function as

$$H(s) = \frac{A}{s+1} + \frac{B}{s+2} + \frac{C}{s-3} \qquad (28.12)$$

The coefficients are evaluated as

$$A = \frac{4+s}{(s+1)(s+2)(s-3)} \times (s+1)\bigg|_{s=-1} = \frac{3}{(1)(-4)} = -\frac{3}{4} \qquad (28.13)$$

$$B = \frac{4+s}{(s+1)(s+2)(s-3)} \times (s+2)\bigg|_{s=-2} = \frac{2}{(-1)(-5)} = \frac{2}{5} \qquad (28.14)$$

$$C = \frac{4+s}{(s+1)(s+2)(s-3)} \times (s-3)\bigg|_{s=3} = \frac{7}{(4)(5)} = \frac{7}{20} \qquad (28.15)$$

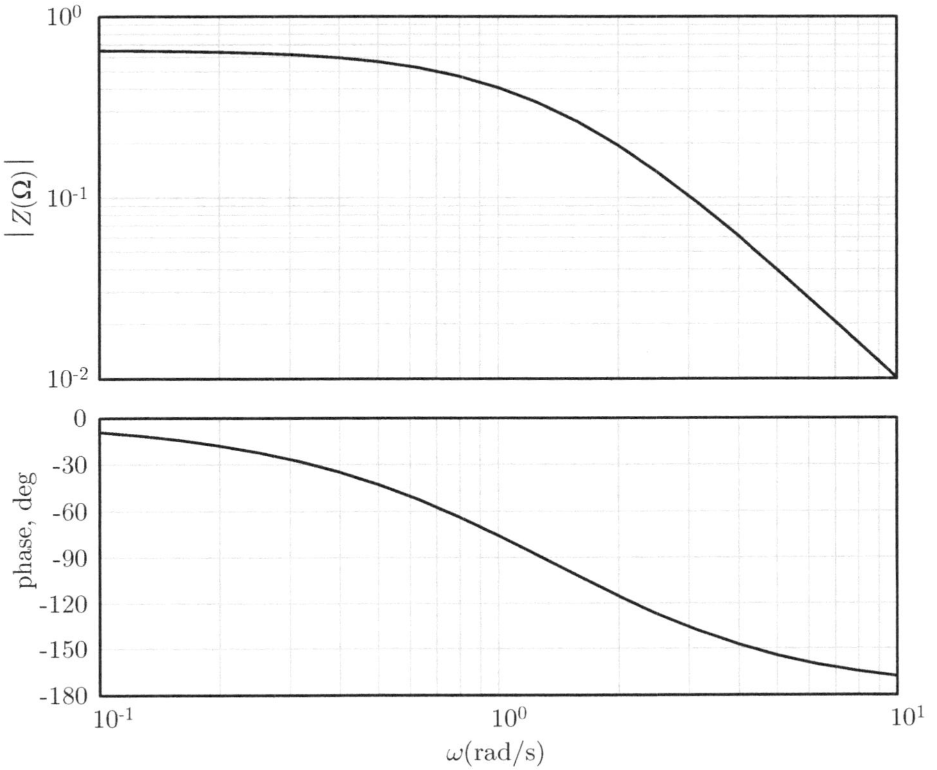

Fig. 28.2 Stable transfer function (frequency domain)

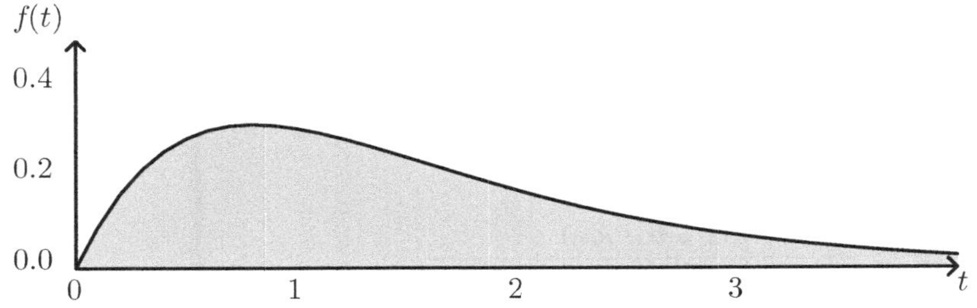

Fig. 28.3 Stable transfer function (time domain)

Fig. 28.4 Pole placement of transfer function $\frac{4+s}{(s+1)(s+2)(s-3)}$

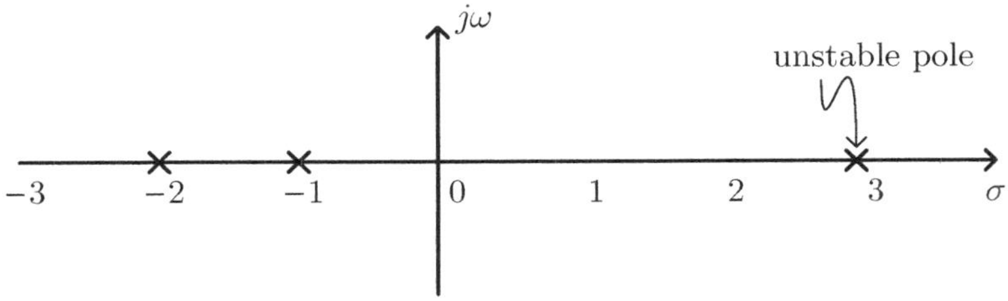

so that

$$\boxed{H(s) = -\frac{3/4}{s+1} + \frac{2/5}{s+2} + \frac{7/20}{s-3}} \qquad (28.16)$$

This transfer function is shown in Fig. 28.5. Notice that the magnitude is the same as that of Fig. 28.2 but the phase is quite different. The impulse response is then

$$\boxed{h(t) = -\frac{3}{4}e^{-t} + \frac{2}{5}e^{-2t} + \frac{7}{20}\boxed{e^{3t}}} \qquad (28.17)$$

Notice that while two of the three exponentials die off, the third one does not! Since one of the exponentials blows up, the impulse function also will blow up; hence we deem this system as being *unstable*. Figure 28.6 shows how the impulse response diverges in time. To recap, this system had all poles other than one in the left-half of the complex plane; since one of the poles resided in the right-side of the complex plane that was enough to cause the impulse response to blow up, in time.

28.4 Example of an Oscillatory, but Stable System

The prior two sections assumed the zeroes of the denominator (or the poles of the transfer function) to be real. Here we relax that requirement and allow some to be complex. Suppose that our system has the following transfer function:

$$H(s) = \frac{1}{(s+a)(s+b-j\omega_0)(s+b+j\omega_0)} \qquad (28.18)$$

We assume we can expand as

$$H(s) = \frac{A}{s+a} + \frac{B}{s+b-j\omega_0} + \frac{C}{s+b+j\omega_0} \qquad (28.19)$$

We solve for the constant as follows:

$$\begin{aligned} A &= \left.\frac{(s+a)}{(s+a)(s+b-j\omega_0)(s+b+j\omega_0)}\right|_{s=-a} \\ &= \frac{1}{(-a+b-j\omega_0)(-a+b+j\omega_0)} = \frac{1}{(a-b)^2+\omega_0^2} \end{aligned} \qquad (28.20)$$

$$\begin{aligned} B &= \left.\frac{(s+b-j\omega_0)}{(s+a)(s+b-j\omega_0)(s+b+j\omega_0)}\right|_{s=-b+j\omega_0} \\ &= \frac{1}{(-b+j\omega_0+a)(-b+j\omega_0+b+j\omega_0)} = \frac{1}{((a-b)+j\omega_0)(2j\omega_0)} \\ &= \frac{1}{-2\omega_0^2+j2(a-b)\omega_0} = \frac{-2\omega_0^2-j2(a-b)\omega_0}{4\omega_0^4+4(a-b)^2\omega_0^2} \end{aligned} \qquad (28.21)$$

$$\begin{aligned} C &= \left.\frac{(s+b+j\omega_0)}{(s+a)(s+b-j\omega_0)(s+b+j\omega_0)}\right|_{s=-b-j\omega_0} \\ &= \frac{1}{(-b-j\omega_0+a)(-b-j\omega_0+b-j\omega_0)} = \frac{1}{((a-b)-j\omega_0)(-2j\omega_0)} \\ &= \frac{1}{-2\omega_0^2-j2(a-b)\omega_0} = \frac{-2\omega_0^2+j2(a-b)\omega_0}{4\omega_0^4+4(a-b)^2\omega_0^2} \end{aligned} \qquad (28.22)$$

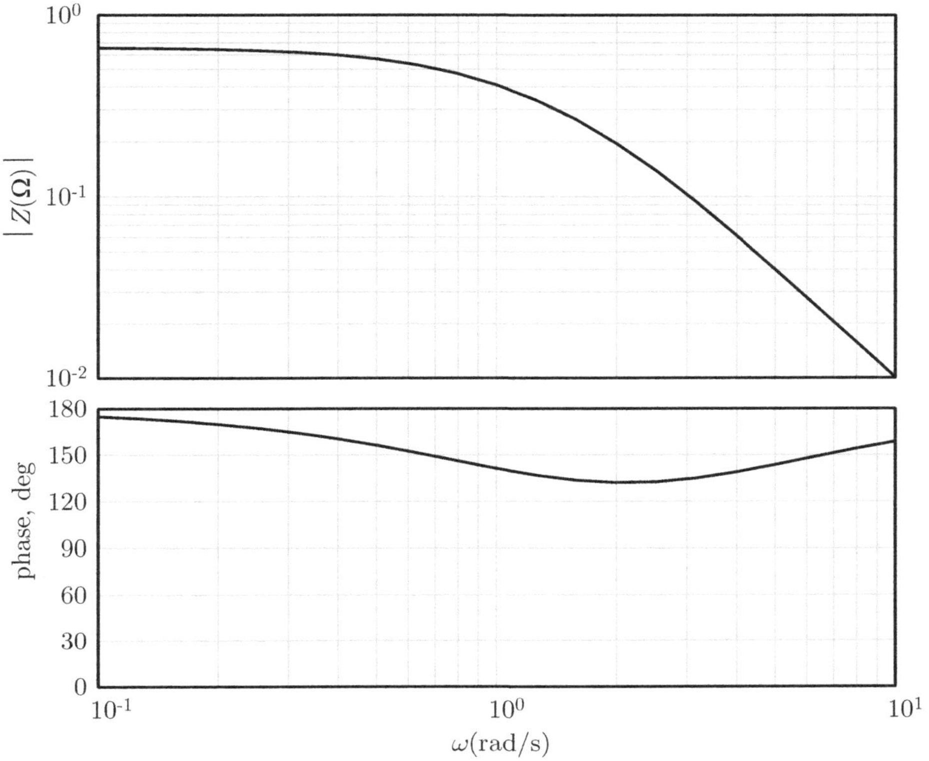

Fig. 28.5 Unstable transfer function (frequency domain)

Fig. 28.6 Unstable transfer function (time domain)

The transfer function is shown in Fig. 28.7 (for sample case of $a = 2$, $b = 1$, and $\omega_0 = 50$).

Notice the DC value comes out at

$$H(0)=\frac{1}{a(b^2+\omega_0^2)}=\frac{1}{2(1+50^2)}=\frac{1}{5002}\sim 2\times 10^{-4} \tag{28.23}$$

Notice also that at high frequency the decay rate is -60 dB/dec due to the $\frac{1}{s^3}$ asymptotic limit, and that the phase there is $-270°$. Finally notice the resonance exactly at $\omega = \omega_0 = 50$. (It is always a good practice to test the magnitude/phase plots against some sanity check points, just to ensure one is looking at the right plot!)

Our solution is then

$$h(t) = \frac{1}{(a-b)^2+\omega_0^2}e^{-at} + \frac{-2\omega_0^2 - j2(a-b)\omega_0}{4\omega_0^4 + 4(a-b)^2\omega_0^2}e^{-bt+j\omega_0 t} + \frac{-2\omega_0^2 + j2(a-b)\omega_0}{4\omega_0^4 + 4(a-b)^2\omega_0^2}e^{-bt-j\omega_0 t} \tag{28.24}$$

Simplify

$$\begin{aligned} h(t) &= \frac{1}{(a-b)^2+\omega_0^2}e^{-at} \\ &+e^{-bt}\left[\frac{-2\omega_0^2 - j2(a-b)\omega_0}{4\omega_0^4 + 4(a-b)^2\omega_0^2}(\cos\omega_0 t + j\sin\omega_0 t)\right. \\ &+\left.\frac{-2\omega_0^2 + j2(a-b)\omega_0}{4\omega_0^4 + 4(a-b)^2\omega_0^2}(\cos\omega_0 t - j\sin\omega_0 t)\right] \end{aligned} \tag{28.25}$$

$$\begin{aligned} h(t) &= \frac{1}{(a-b)^2+\omega_0^2}e^{-at} \\ &+e^{-bt}\left[\frac{-\omega_0^2\cos\omega_0 t + (a-b)\omega_0\sin\omega_0 t}{\omega_0^4 + (a-b)^2\omega_0^2}\right] \end{aligned} \tag{28.26}$$

Notice that since $a > 0$ and $b > 0$ this response will also die out in time. That is, the system is stable. So even if some of the roots come out complex, what matters is the *real part of the pole being negative*. The impulse response is shown in Fig. 28.8.

Notice that at time zero we get

$$\begin{aligned} h(0) &= \frac{1}{(a-b)^2+\omega_0^2} - \frac{\omega_0^2}{\omega_0^4 + (a-b)^2\omega_0^2} \\ &= \frac{1}{(a-b)^2+\omega_0^2} - \frac{1}{\omega_0^2 + (a-b)^2} = 0 \end{aligned} \tag{28.27}$$

We could have also predicted this by using the initial value theorem

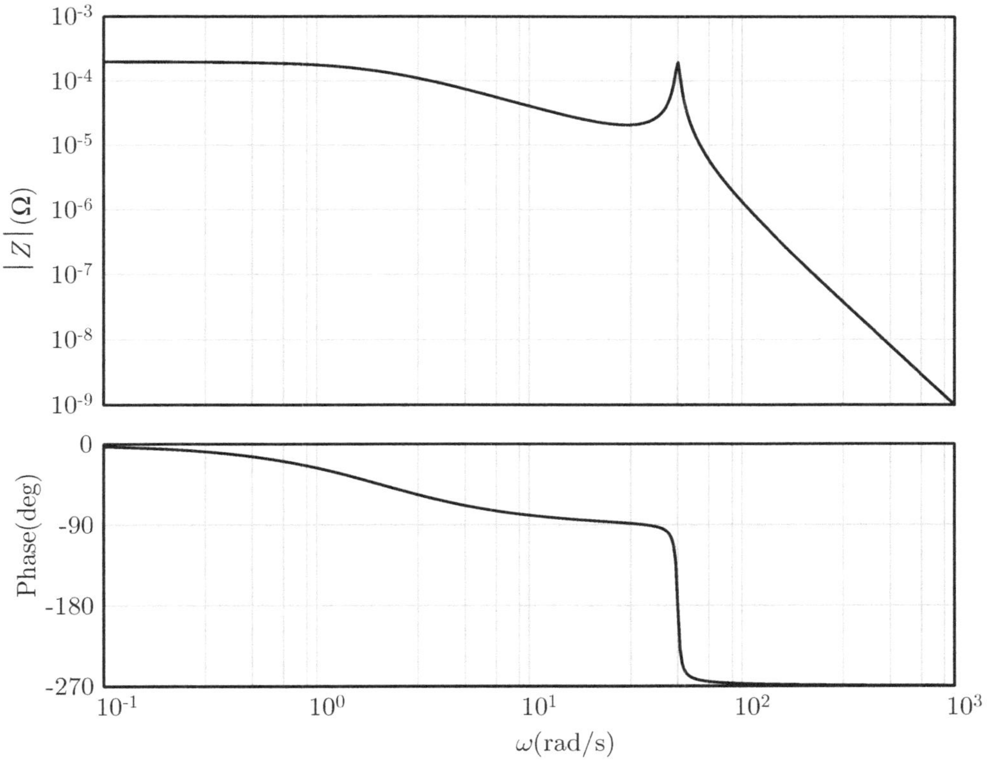

Fig. 28.7 Oscillatory, but stable system (frequency domain)

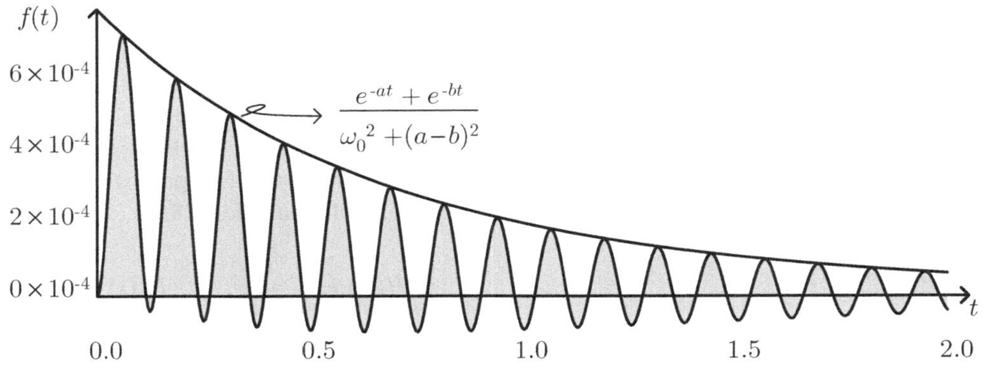

Fig. 28.8 Oscillatory, but stable system (time domain)

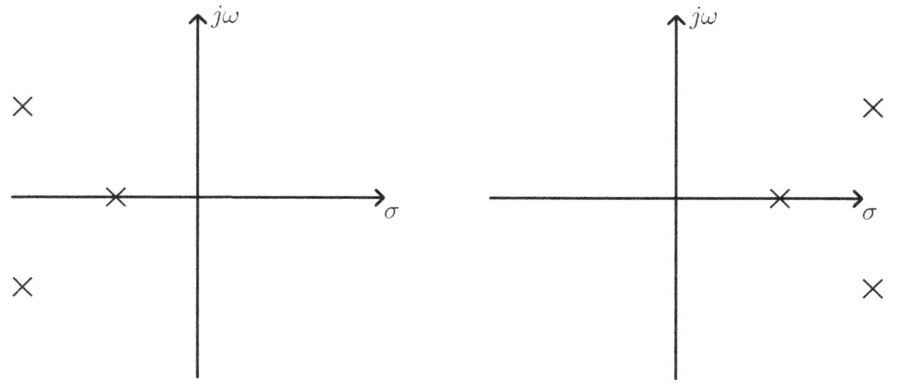

Fig. 28.9 Stability and relation to pole placement

$$h(0) = \lim_{s\to\infty} sH(s) = 0 \tag{28.28}$$

Also notice that the envelop follows the shape of

$$\text{envelope} \sim \frac{e^{-at} + e^{-bt}}{\omega_0^2 + (a-b)^2} \tag{28.29}$$

Finally notice that the oscillation period is around 0.125 which is nothing other than

$$T = \frac{2\pi}{\omega_0} = \frac{2\pi}{50} = 0.1257 \tag{28.30}$$

Again similar to sanity checking the transfer function it is beneficial to sanity check the impulse response. After all, they both carry in them the exact amount of information!

28.5 Summary

It seems that everything revolves around finding the poles of the transfer function! Not only do we need to find the poles in order to do any sort of partial fraction expansion, and subsequently be able to find the inverse transform straightforwardly; but knowing the location of the poles also enables us to know a priori whether the system under consideration is stable or not, even without knowing the exact form of the impulse response. Poles could be real, imaginary, or combination thereof. Stated verbally we can say a system is stable if the real part of all the poles resides on the left-hand side of the complex plane; and equivalently a system is unstable if *any* pole has its real part residing on the right-hand side of the plane. Visually we simply refer to Fig. 28.9.

28.6 Problems

1. Both functions below have the same magnitude versus frequency but different phase. The stable one sustains a $-90°$ when the pole occurs; the unstable one, the opposite—a $+90°$. Plot both functions and study this problem in preparation for next problems. See sample solution in Fig. 28.10.

$$H_1(s) = \frac{1}{s+10}, \qquad H_2(s) = \frac{1}{s-10}$$

2. Plot the magnitude and phase of each of the following two functions; which one of them is stable and which one is unstable? Why is the DC phase of $H_2(s)$ 180° and not $-180°$? Show by expanding the function around DC that this is the case. See sample solution in Fig. 28.11. Verify for an intermediate frequency point that $H_2(s)$ always lies in the upper half of the complex plane; that is the phase is always between 0 and 180°.

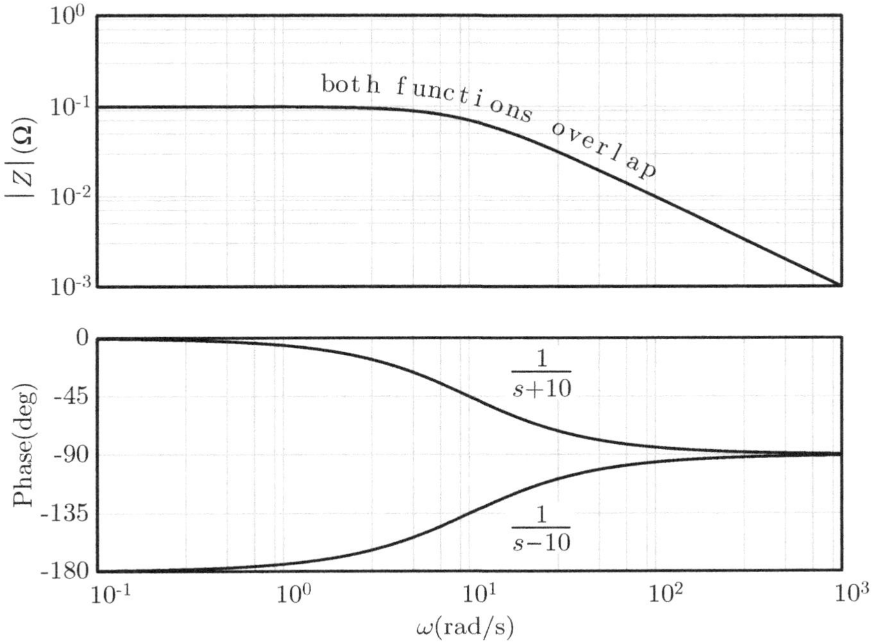

Fig. 28.10 Sample solution to Problem 1

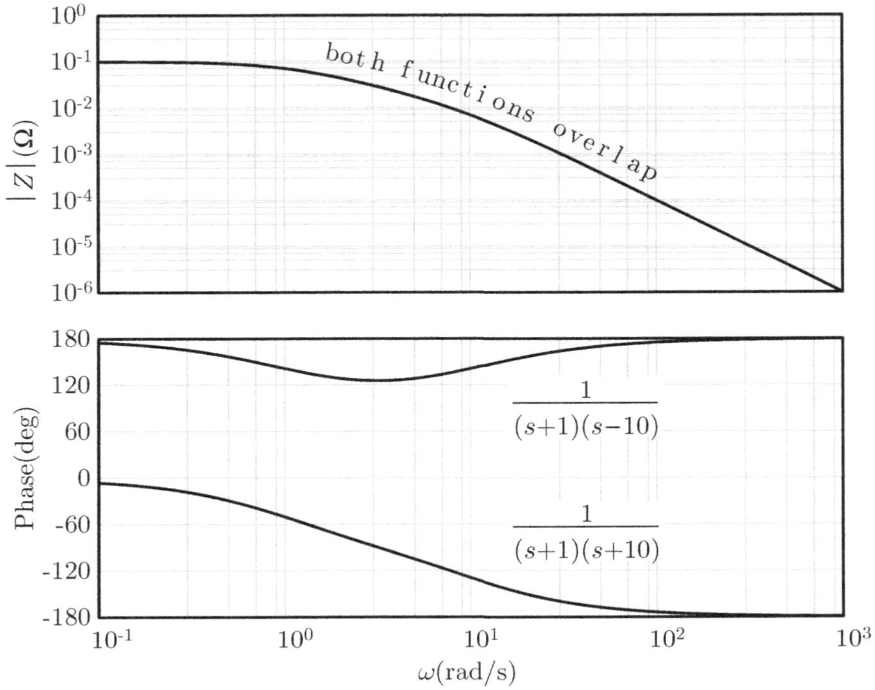

Fig. 28.11 Sample solution to Fig. 2

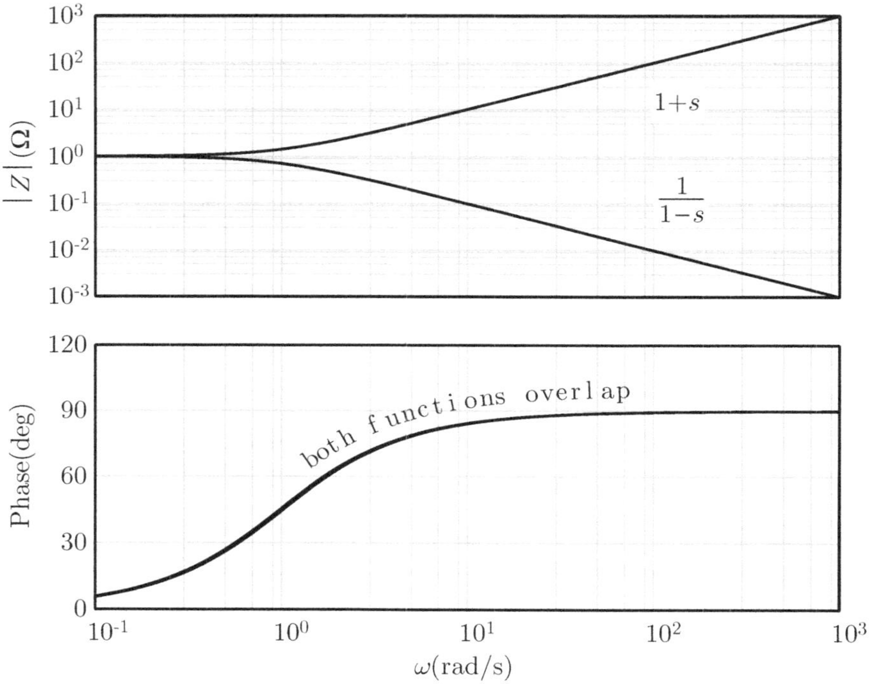

Fig. 28.12 Sample solution to Fig. 3

$$H_1(s) = \frac{1}{(s+1)(s+10)},$$

$$H_2(s) = \frac{1}{(s+1)(s-10)}$$

3. The two functions below have different magnitude, but the *same phase*; plot them and confirm this. While the former is stable, the latter is not—why? What is the inverse Laplace transform of each? See sample results in Fig. 28.12.

$$H_1(s) = 1 + s, \qquad H_2(s) = \frac{1}{1-s}$$

Answer:

$$f_1(t) = \delta(t) + \frac{d\delta(t)}{dt}, \qquad f_2(t) = -e^t$$

4. A transfer function is given by

$$H(s) = \frac{s+1}{1+4s+3s^2+s^3}$$

Confirm that the following are the three roots of the denominator:

$$p_1 = -0.3177 + j0.000,$$
$$p_2 = -1.3412 + j1.162,$$
$$p_3 = -1.3412 - j1.162$$

Plot the roots on the complex plain; is this a stable function? See sample solution in Fig. 28.13.

5. A transfer function is given by

$$H(s) = \frac{s+1}{1 + 1s - 3s^2 + s^3}$$

Confirm that the following are the three roots of the denominator:

$$p_1 = -0.4142 - j0, \quad p_2 = 1.0000 + j0,$$
$$p_3 = 2.4142 - j0$$

Plot the roots on the complex plain; is this a stable function? See sample solution in Fig. 28.13.

6. Is the transfer function in Fig. 28.14 stable? Why?
7. Is the transfer function in Fig. 28.15 stable? Why?
8. Is the transfer function in Fig. 28.16 stable? Why?

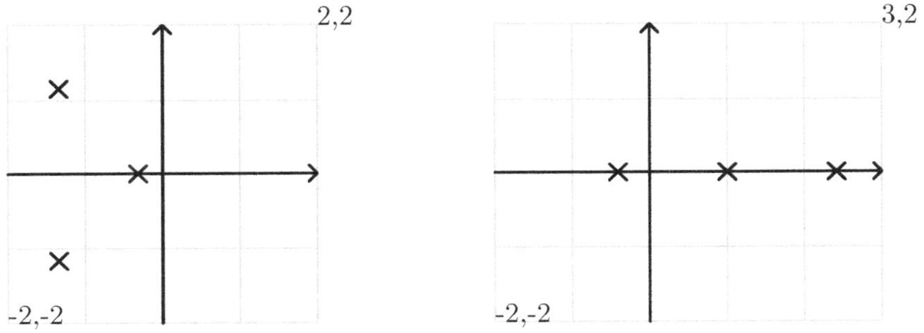

Fig. 28.13 Sample solution to Problems 4 and 5

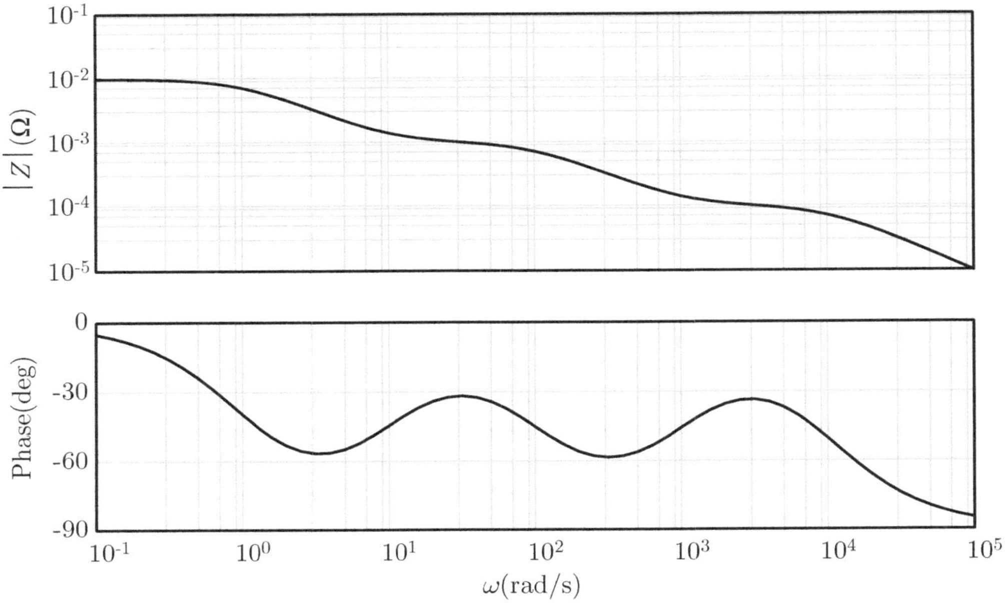

Fig. 28.14 Statement to Problem 6

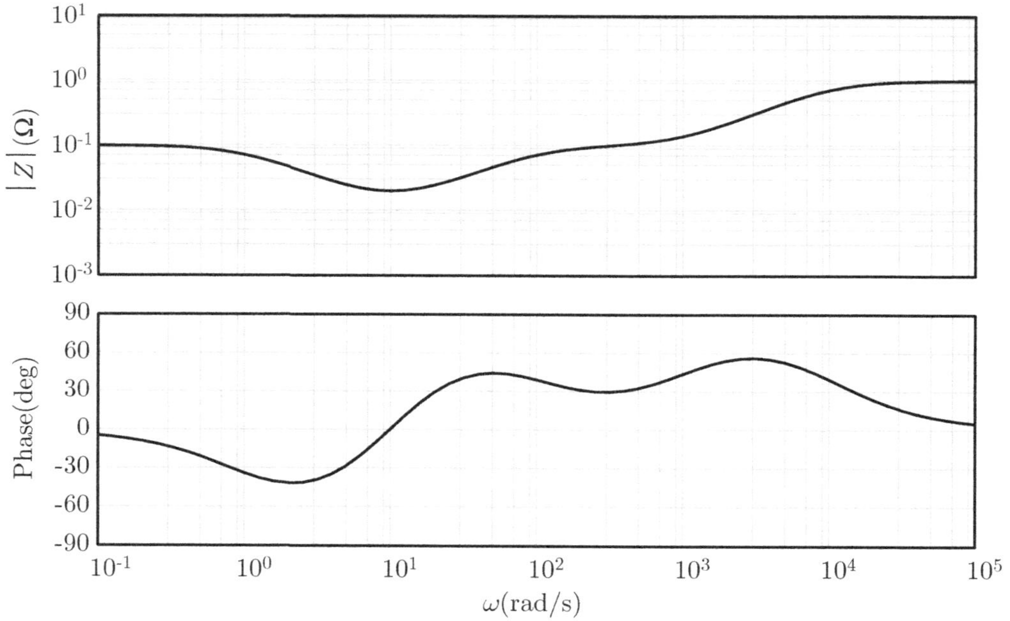

Fig. 28.15 Statement to Problem 7

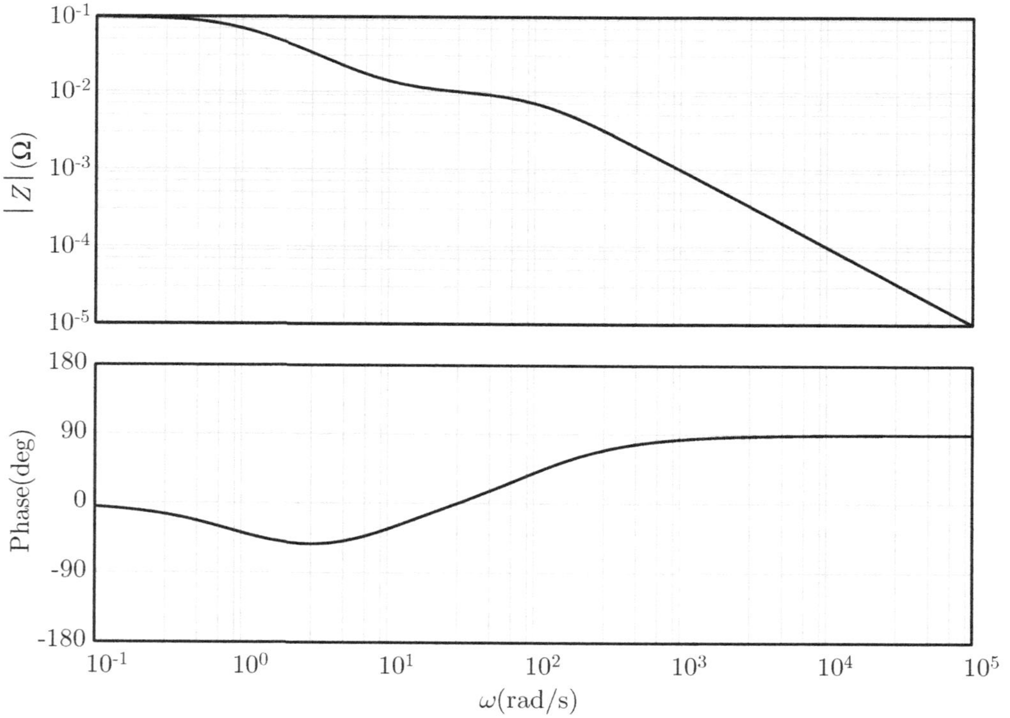

Fig. 28.16 Statement to Problem 8

29 Impulse Response as Figured from Inverse Transform

29.1 Introduction

Having dealt with impulse response directly in the time domain, and as outlined in Chap. 22, we now move to figuring the impulse response directly from the transfer function. The prior treatment in the time domain was just for illustration; the most common method of generating the impulse response comes via inverse transforming the transfer function. Whether the transfer function is measured, simulated, or figured by doing KVL/KCL around the circuit, it is now time to unfold it from the frequency domain and generate its counterpart (the impulse response) in the time domain. We may argue that this very process is the very reason why we went to great length in doing spectral analysis and in defining the concept of transfer functions. In the end we need to look at things in the time domain! Why is the impulse response so important? For one thing if we know it we already know the response due to all other stimuli, simply by convolving the impulse response by the new stimulus! So the three-step process in finding a typical stimulus is (a) first find the transfer function, (b) find the impulse response, and (c) convolve this with the stimulus—done! In this chapter we illustrate the first two steps; as for the convolution part we already had practice in it in Chap. 23. So let's jump to some sample cases where the transfer function is known and simply figure its inverse transform.

29.2 Series *RC* Network

The series *RC* network is shown in Fig. 29.1. The impedance of the network is

$$Z(s) = R + \frac{1}{sC} \tag{29.1}$$

The impulse response of the transfer function is

$$h(t) = R\delta(t) + \frac{1}{C}u(t) \tag{29.2}$$

That is, an impulse current through the resistor generates an impulse voltage (scaled by R) and an impulse current through the cap generates a step output voltage (scaled by $\frac{1}{C}$). This is simple as it gets; there isn't really much more to say about it other than to add the impulse response of the elements in series.

F. Badrieh, *Spectral, Convolution and Numerical Techniques in Circuit Theory*,
https://doi.org/10.1007/978-3-319-71437-0_29

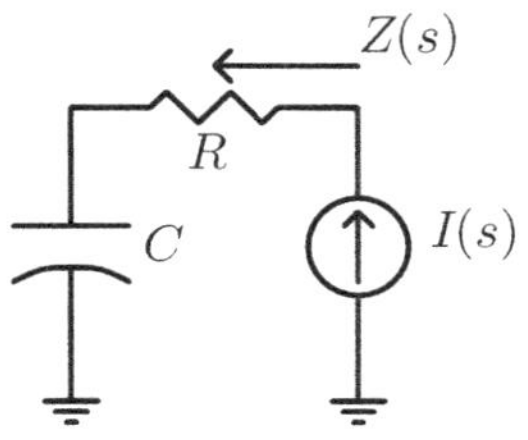

Fig. 29.1 Series RC network

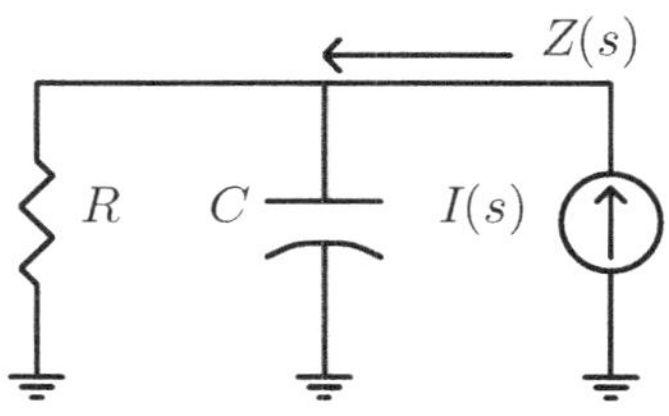

Fig. 29.2 Parallel RC network

29.3 Parallel *RC* Network

The parallel RC network is shown in Fig. 29.2. The impedance transfer function is

$$\boxed{Z(s) = \frac{1}{C}\frac{1}{s + \frac{1}{RC}}} \quad (29.3)$$

The impulse response, in the form of voltage across the network, is simply the inverse transform of the above equation:

$$\boxed{h(t) = \frac{1}{C}e^{-\frac{t}{RC}}} \quad (29.4)$$

Notice that right after the application of the impulse current the voltage is

$$v(0+) = \frac{1}{C} \quad (29.5)$$

That is, all the current is absorbed by the cap, such that all the charge is dumped on the cap. Since the charge, which is the integral of current is unity, the voltage is simply

$$v(0+) = \frac{Q}{C} = \frac{1}{C} \quad (29.6)$$

As time progresses, the charge now is discharged through the resistor, at a *time constant*

$$t_{RC} = RC \quad (29.7)$$

Figure 29.3 shows the impulse response for the sample case $R = 1\,\Omega$ and $C = 1\,\text{mF}$. Figure 29.4 shows impact of R. As can be observed, varying R has no impact on the starting voltage, which we established to be $1/C$. Larger R results in a *longer* time constant, and hence less discharge. In fact as we push R large enough, the cap would never discharge; i.e., it would become an *ideal cap*. Figure 29.5 shows impact of varying C. Larger C results in two things. First, the starting voltage goes down with larger C since as just mentioned it is tied to $1/C$. In other words, given the same amount of charge (which is 1 here), a bigger cap would charge less than a smaller cap. The second outcome of larger cap is longer time constant, and hence longer time to discharge. That is, a larger cap takes longer to discharge.

Not to get side tracked with what we originally set out to achieve, let's summarize what was done. We started with the impedance transfer function which was derived in the earlier chapters, or can easily be derived by simply putting the R in parallel with the $\frac{1}{sC}$. Having the transfer function we now flex our muscles having had extensive practice in finding inverse transforms as was taught in the prior chapters. Once the inverse transform is done we have the impulse response and that we can now analyze and rationalize. We can also use it to figure other responses, but for the sake of space we stop here. So now we see the abstract concepts and elaborate math come to life.

29.4 Series *RC*/Parallel *R* Circuit

The series RC, parallel R is shown in Fig. 29.6. The starting point for finding the impulse response is the impedance transfer function

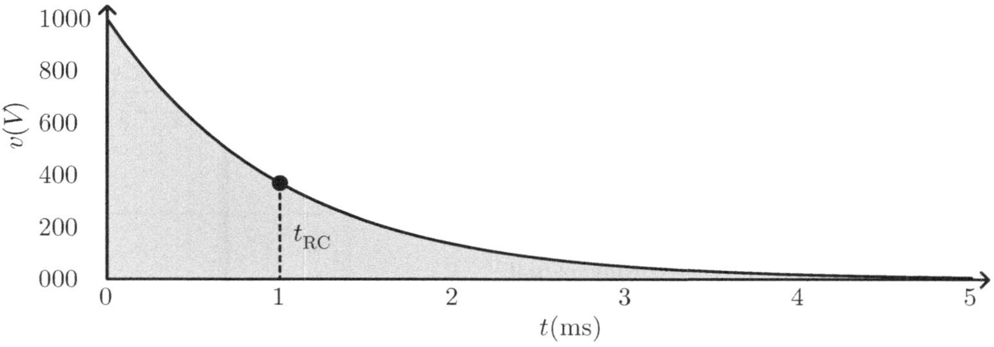

Fig. 29.3 Impulse response of parallel *RC* network ($R = 1\ \Omega$ and $C = 1$ mF)

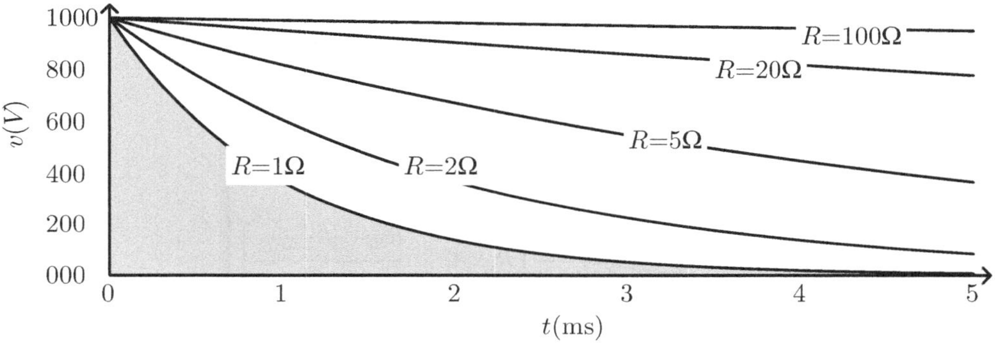

Fig. 29.4 Impact of *R* on impulse response of parallel *RC* network ($C = 1$ mF)

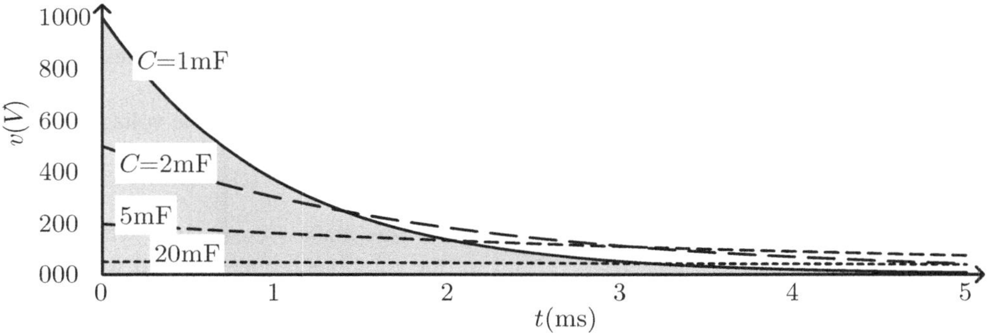

Fig. 29.5 Impact of *C* on impulse response of parallel *RC* network ($R = 1\ \Omega$)

Fig. 29.6 Series *RC* in parallel with *R* and input impedance

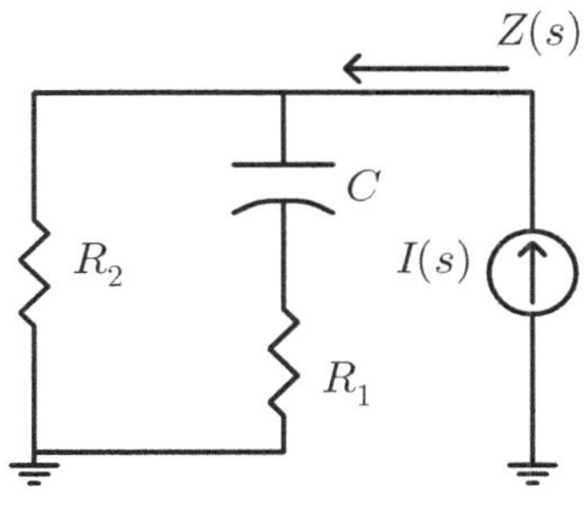

$$Z(s) = \frac{R_2\frac{1+sR_1C}{sC}}{R_2 + \frac{1+sR_1C}{sC}} = \frac{R_2 + R_1R_2Cs}{1+(R_1+R_2)Cs} = \frac{1}{(R_1+R_2)C}\frac{R_2 + R_1R_2Cs}{\frac{1}{(R_1+R_2)C}+s}$$
$$= a\frac{R_2 + R_1R_2Cs}{s+a} \tag{29.8}$$

We will need to simplify this using partial fractions. To that end we recall the following equality:

$$\frac{s}{s+a} = 1 - \frac{a}{s+a} \tag{29.9}$$

Then our impedance becomes

$$Z(s) = a\left[R_2\frac{1}{s+a} + R_1R_2C - R_1R_2C\frac{a}{s+a}\right]$$
$$= a\left[R_1R_2C + \frac{R_2 - R_1R_2Ca}{s+a}\right]$$
$$= \frac{1}{(R_1+R_2)C}\left[R_1R_2C + \frac{R_2 - \frac{R_1R_2C}{(R_1+R_2)C}}{s+\frac{1}{(R_1+R_2)C}}\right] \tag{29.10}$$

The inverse transform is obtained straightforwardly:

$$v(t) = \frac{R_1R_2}{R_1+R_2}\delta(t) + \frac{R_2 - \frac{R_1R_2}{R_1+R_2}}{(R_1+R_2)C}e^{-\frac{t}{(R_1+R_2)C}}$$

which can be simplified to

$$\boxed{v(t) = \frac{R_1R_2}{R_1+R_2}\delta(t) + \frac{R_2^2}{(R_1+R_2)^2C}e^{-\frac{t}{(R_1+R_2)C}}} \tag{29.11}$$

Let us test the above results against some known limits

- $R_2 = 0$: in this limit the above expression reduces to

$$v(t) \sim 0 \tag{29.12}$$

  This makes sense since setting R_2 to zero shorts the whole network, and no voltage happens across a short impedance!
- $R_1 = 0$: in this limit the above expression reduces to

$$v(t) = \frac{1}{C}e^{-\frac{t}{R_2C}} \tag{29.13}$$

  But this makes sense, since in the limit as $R_1 \to 0$ we end up with a simple parallel RC network, whose impulse response we just derived in Eq. (29.4).
- $R_1 \to \infty$: In this limit the voltage reduces to

$$v(t) = R_2\delta(t) \tag{29.14}$$

  This makes sense, since when $R_1 \to \infty$ the cap branch opens up and we end up with the simple network of R_2, whose impulse response is simply the delta function (scaled by resistance).
- $R_2 \to \infty$: In this limit the voltage expression reduces to

$$v(t) = R_1\delta(t) + \frac{1}{C}e^{-\frac{t}{\infty}} = R_1\delta(t) + \frac{1}{C}u(t) \tag{29.15}$$

  This says that the left branch (R_2) opens and factors out, and we end up with R_1 in series with C; and the impulse response there would be an impulse (times R) and a step function (scaled by $1/C$, charging the cap), and as was derived in Eq. (29.2).

Figure 29.7 shows results for the case $R = 1\,\Omega$, $R_2 = 2\,\Omega$, and $C = 1\,\mathrm{F}$, and comparison to SPICE.

29.5 Series *RC*/Parallel *C* Circuit

The series *RC*/parallel C is shown in Fig. 29.8. Our transfer function is (see Chap. 26, Problem 3)

$$Z(s) = \frac{1}{sC_2}\frac{s+\frac{1}{RC_1}}{s+\frac{1}{RC_s}}, \quad C_s = \frac{C_1C_2}{C_1+C_2} \tag{29.16}$$

We have a pole at 0 and one at $-1/(RC_s)$. The residue at 0 is

$$\left.\frac{1}{C_2}\frac{s+\frac{1}{RC_1}}{s+\frac{1}{RC_s}}\right|_{s=0} = \frac{1}{C_2}\frac{C_s}{C_1} = \frac{1}{C_1C_2}\frac{C_1C_2}{C_1+C_2}$$
$$= \frac{1}{C_1+C_2} \tag{29.17}$$

The residue at the second pole $-1/(RC_s)$ is

$$\frac{1}{sC_2}\left[s+\frac{1}{RC_1}\right]_{s=-1/(RC_s)}$$
$$= \frac{-1}{C_2}RC_s\left[-\frac{1}{RC_s}+\frac{1}{RC_1}\right]$$
$$= -\frac{R}{C_2}\frac{C_1C_2}{C_1+C_2}\left[\frac{1}{RC_1}-\frac{1}{R}\frac{C_1+C_2}{C_1C_2}\right]$$
$$= -\frac{RC_1}{C_1+C_2}\left[\frac{1}{RC_1}-\frac{1}{RC_2}-\frac{1}{RC_1}\right]$$
$$= \frac{C_1}{C_2(C_1+C_2)} \tag{29.18}$$

So our transfer function can be written as

$$\boxed{Z(s) = \frac{1}{s}\frac{1}{C_1+C_2} + \frac{C_1}{C_2(C_1+C_2)}\frac{1}{s+\frac{1}{RC_s}}} \tag{29.19}$$

Thus we can find its inverse LT to get the impulse response

$$\boxed{h(t) = \frac{1}{C_1+C_2}u(t) + \frac{C_1}{C_2(C_1+C_2)}e^{-t/(RC_e)}, \qquad C_s = \frac{C_1C_2}{C_1+C_2}} \tag{29.20}$$

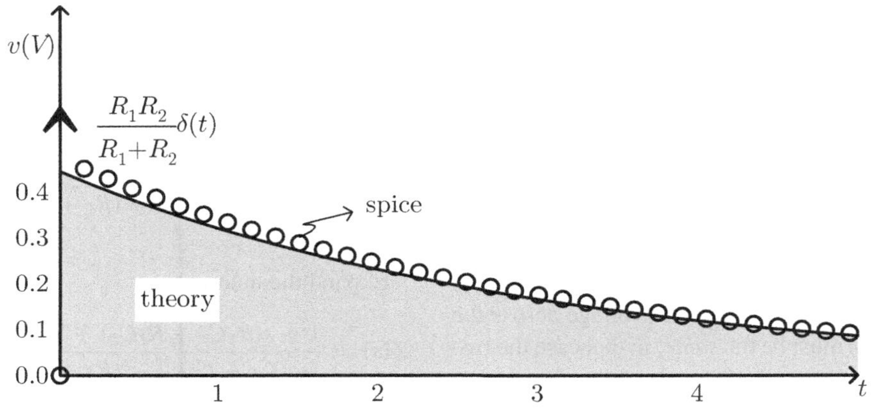

Fig. 29.7 Impulse response of series *RC*, parallel *R* ($R_1 = 1$, $R_2 = 2\,\Omega$, and $C = 1\,\mathrm{F}$)

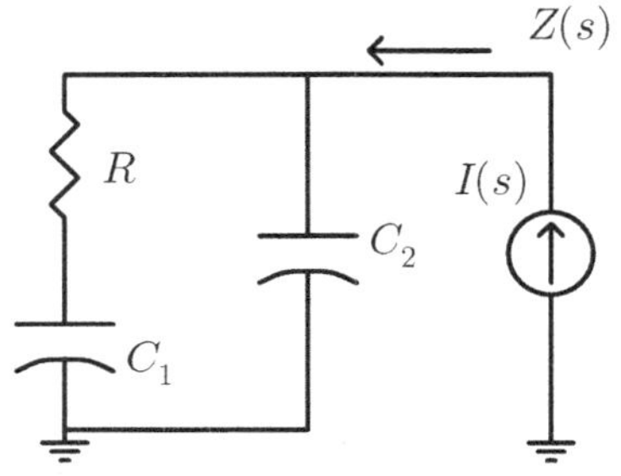

Fig. 29.8 Series RC/parallel C circuit

Notice the following observations:

- At time zero we get

$$v(0+) = \frac{1}{C_1 + C_2} + \frac{C_1}{C_2(C_1 + C_2)} = \frac{C_2 + C_1}{C_2(C_1 + C_2)} = \frac{1}{C_2} \tag{29.21}$$

  That is, at time 0+ only C_2 manifests itself, and all charge is dumped upon it. The R in the second branch hides the other cap.
- The solution is comprised of two components: a steady state one (the unit step) and a transient (dying) one (the decaying exponential).
- The time constant is R times an equivalent C which is the series combination of C_1 and C_2. It is tempting to think of the equivalent capacitance that being of the parallel combination of the two caps! But fact is, once C_2 charges and the impulse source is gone (out of the picture) any current movement around the CRC loop would be a serial one; hence the two caps would be in series, and so would the R.
- When things settle down (i.e., when the exponential dies out) the settling value of voltage is simply $1/(C_1 + C_2)$ which is a unit charge dumped on a cap whose value is the sum of C_1 and C_2; that is C_1 in parallel with C_2. That is, once C_2 is charged, there is no way for the charge to exist the circuit. The charge must somehow split between the two caps. Under equilibrium, there is no current across R and the two caps would have settled. Since there is no current across the R the voltage across the two caps must be the *same*; in that case the two caps act in *parallel*! Lastly, being in parallel does *not* mean each cap has the same amount of charge! In fact, they won't—the smaller cap will assume smaller charge, and vice versa; in the end $\frac{Q_1}{C_1}$ equals $\frac{Q_2}{C_2}$.

Figure 29.9 shows results for a couple of cap values (both with $R = 1\,\Omega$). Since in both cases the sum of cap is the same, the settling value is the same too (1/3 here). With smaller C_2, however, we get larger initial value, and in accordance with Eq. (29.21).

29.6 Series *RC* in Parallel with Series *RC*

The series RC in parallel with another series RC is shown in Fig. 29.10. The transfer function was derived back in Sect. 26.12 but was not fully simplified; here we re-derive it in a slightly different way, and arrive at the full expansion. Our starting point is Eq. (26.71), retyped here for convenience

$$Z(s) = \frac{(1 + sR_1C_1)(1 + sR_2C_2)}{s(C_1 + C_2) + s^2(R_1 + R_2)C_1C_2} \tag{29.22}$$

Expand the numerator

$$Z(s) = \frac{1 + s(R_1C_1 + R_2C_2) + s^2R_1R_2C_1C_2}{s(C_1 + C_2) + s^2(R_1 + R_2)C_1C_2} \tag{29.23}$$

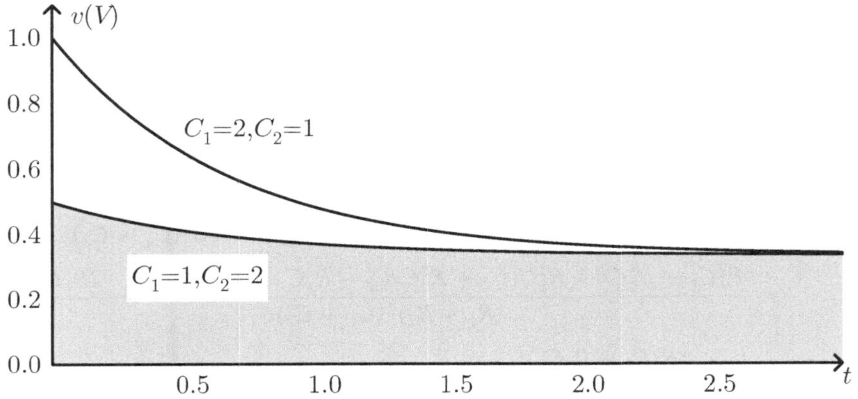

Fig. 29.9 Impulse response of series *RC*/parallel *C* with $R = 1\,\Omega$

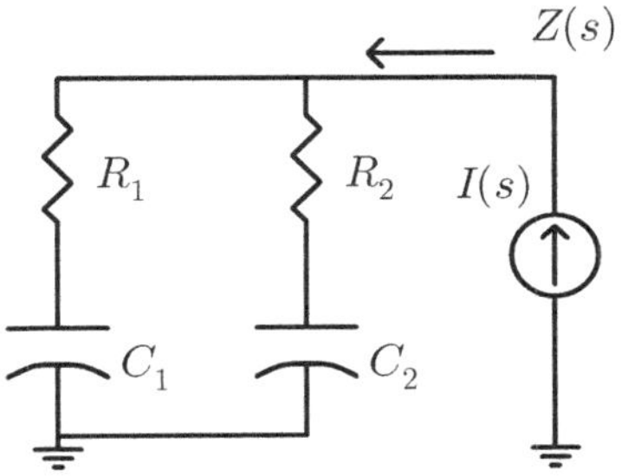

Fig. 29.10 Series *RC* in parallel with another series *RC*

Carry on long division

$$Z(s) = \frac{R_1R_2}{R_1 + R_2} + \frac{1 + s\left[R_1C_1 + R_2C_2 - \frac{R_1R_2}{R_1+R_2}(C_1 + C_2)\right]}{s(C_1 + C_2) + s^2(R_1 + R_2)C_1C_2} \tag{29.24}$$

$$Z(s) = \frac{R_1R_2}{R_1 + R_2} + \frac{1 + s\frac{(R_1C_1+R_2C_2)(R_1+R_2)-R_1R_2(C_1+C_2)}{R_1+R_2}}{s(C_1 + C_2) + s^2(R_1 + R_2)C_1C_2} \tag{29.25}$$

$$Z(s) = \frac{R_1R_2}{R_1 + R_2} + \frac{1 + s\frac{R_1^2C_1+R_2^2C_2}{R_1+R_2}}{s(C_1 + C_2) + s^2(R_1 + R_2)C_1C_2} \tag{29.26}$$

$$Z(s) = \frac{R_1R_2}{R_1 + R_2} + \frac{1}{s}\frac{\frac{1}{(R_1+R_2)C_1C_2} + s\frac{R_1^2C_1+R_2^2C_2}{(R_1+R_2)^2C_1C_2}}{\frac{C_1+C_2}{(R_1+R_2)C_1C_2} + s} \tag{29.27}$$

We have a pole at zero and one at $-\frac{C_1+C_2}{(R_1+R_2)C_1C_2}$. The first residue comes out

$$\text{first residue} = \frac{1}{C_1 + C_2} \tag{29.28}$$

and the second one

$$\begin{aligned}
\text{second residue} &= -\frac{(R_1+R_2)C_1C_2}{C_1+C_2}\left[\frac{1}{(R_1+R_2)C_1C_2} - \frac{C_1+C_2}{(R_1+R_2)C_1C_2}\frac{R_1^2C_1+R_2^2C_2}{(R_1+R_2)^2C_1C_2}\right] \\
&= \frac{R_1^2C_1+R_2^2C_2}{(R_1+R_2)^2C_1C_2} - \frac{1}{C_1+C_2} = \frac{(R_1^2C_1+R_2^2C_2)(C_1+C_2)-(R_1+R_2)^2C_1C_2}{(R_1+R_2)^2(C_1+C_2)C_1C_2} \\
&= \frac{R_1^2C_1^2+R_2^2C_2^2+R_1^2C_1C_2+R_2^2C_1C_2-R_1^2C_1C_2-R_2^2C_1C_2-2R_1R_2C_1C_2}{(R_1+R_2)^2(C_1+C_2)C_1C_2} \\
&= \frac{(R_1C_1-R_2C_2)^2}{(R_1+R_2)^2(C_1+C_2)C_1C_2}
\end{aligned} \tag{29.29}$$

So doing partial fractions gives us

$$\boxed{Z(s) = \frac{R_2R_2}{R_1+R_2} + \frac{1}{s}\frac{1}{C_1+C_2} + \frac{\frac{(R_1C_1-R_2C_2)^2}{(R_1+R_2)^2(C_1+C_2)C_1C_2}}{s+\frac{C_1+C_2}{(R_1+R_2)C_1C_2}}} \tag{29.30}$$

We can right away find the inverse transform which comes out to

$$\boxed{\begin{aligned}
h(t) &= \frac{R_1R_2}{R_1+R_2}\delta(t) + \frac{1}{C_1+C_2}u(t) + \frac{(R_1C_1-R_2C_2)^2}{(R_1+R_2)^2(C_1+C_2)C_1C_2}e^{-t/\tau} \\
\tau &= \frac{C_1C_2(R_1+R_2)}{C_1+C_2}
\end{aligned}} \tag{29.31}$$

Sample results are shown in Fig. 29.11 for two sets of RC numbers. Notice the following observations:

- The very initial response is given by a delta function of magnitude

$$v(0) = \frac{R_1R_2}{R_1+R_2} \tag{29.32}$$

This is nothing but the parallel combination of the two resistors, independent of the caps. That is, right at application point, the caps are short, and we have two Rs in parallel; so voltage would be current (delta function) times impedance (parallel Rs).

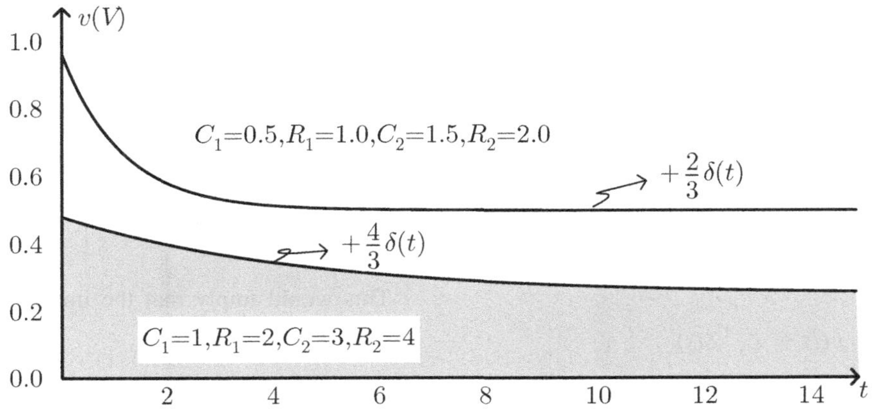

Fig. 29.11 Impulse response of series *RC* in parallel with another series *RC*

- After the delta function has passed by we get at time 0+

$$
\begin{aligned}
v(0+) &= \frac{1}{C_1 + C_2} + \frac{(R_1C_1 - R_2C_2)^2}{(R_1 + R_2)^2(C_1 + C_2)C_1C_2} \\
&= \frac{(R_1 + R_2)^2C_1C_2 + (R_1C_1 - R_2C_2)^2}{(R_1 + R_2)^2(C_1 + C_2)C_1C_2} \\
&= \frac{R_1^2C_1C_2 + R_2^2C_1C_2 + 2R_1R_2C_1C_2 + R_1^2C_1^2 + R_2^2C_2^2 - 2R_1C_1R_2C_2}{(R_1 + R_2)^2(C_1 + C_2)C_1C_2} \\
&= \frac{R_1^2C_1(C_1 + C_2) + R_2^2C_2(C_1 + C_2)}{(R_1 + R_2)^2(C_1 + C_2)C_1C_2} = \frac{R_1^2C_1 + R_2^2C_2}{(R_1 + R_2)^2C_1C_2}
\end{aligned}
\tag{29.33}
$$

- The final settling voltage is given by

$$v(\infty) = \frac{1}{C_1 + C_2} \tag{29.34}$$

That is, when things settle down, we have no currents through the resistors, and we end up with two caps in parallel; the combined cap is simply the sum of both, which is $C_1 + C_2$. If we use the relation $V = Q/C$ we arrive at the above equation.

- The settling time constant is

$$\tau = \frac{C_1C_2(R_1 + R_2)}{C_1 + C_2} \tag{29.35}$$

Notice unit-wise, it looks like an *RC* product; the *R* is $R_1 + R_2$ and the *C* is the series cap of both:

$$C_{\text{series}} = \frac{C_1C_2}{C_1 + C_2} \tag{29.36}$$

So the time constant is really the series resistance times the series capacitance.

29.7 Series *RL* Circuit

The series *RL* is shown in Fig. 29.12. The impedance transfer function is simply

$$Z(s) = R + sL \tag{29.37}$$

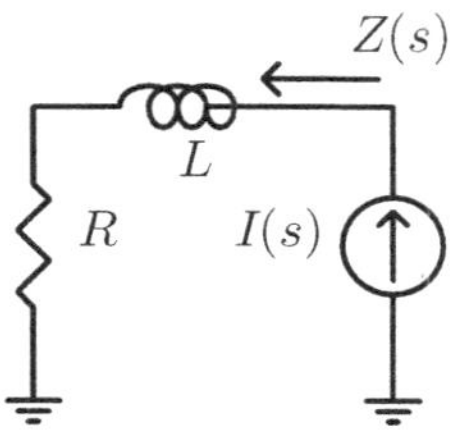

Fig. 29.12 Series RL and input impedance

The impulse response is then

$$\begin{aligned} h(t) &= \mathcal{L}^{-1}Z(s) \\ &= \mathcal{L}^{-1}[R + sL] \\ &= \boxed{R\delta(t) + L\frac{d\delta(t)}{dt}} \end{aligned} \quad (29.38)$$

This states that applying an impulse current to a series RL network results is an impulse voltage across the resistors (scaled by R) and a *derivative* of an impulse voltage across the inductor (again scaled by L). Notice that a derivative impulse is even more abrupt than an impulse function!

29.8 Parallel *RL* Circuit

The parallel RL network is shown in Fig. 29.13. The impedance transfer function is

$$\begin{aligned} Z(s) &= \frac{RsL}{R + sL} = R\frac{s}{s + \frac{R}{L}} \\ &= R\frac{s}{s + a} \end{aligned} \quad (29.39)$$

We can find the inverse of this function using two methods:

First Method We can use long division

$$\begin{array}{r} 1 \\ s + a\,\overline{)\; s \quad} \\ \underline{-s - a} \\ -a \end{array} \quad (29.40)$$

such that

$$\frac{s}{s + a} = 1 - \frac{a}{s + a} \quad (29.41)$$

Then

$$Z(s) = R\left[1 - \frac{a}{a + s}\right] \quad (29.42)$$

This would imply that the impulse response voltage is

$$v(t) = R\left[\delta(t) - ae^{-at}\right], \quad \text{or}$$

$$\boxed{v(t) = R\delta(t) - R\frac{R}{L}e^{-tR/L}} \quad (29.43)$$

Second Method The second method is to start with the fact that

$$\frac{1}{a + s} \to e^{-at} \quad (29.44)$$

and recall the time differentiation property:

$$\frac{df(t)}{dt} \to sL(s) - f(0) \quad (29.45)$$

or

$$\frac{df(t)}{dt} + f(0)\delta(t) \to sL(s) \quad (29.46)$$

This would imply that

$$\begin{aligned} R\frac{s}{a + s} &\to R\left[\frac{de^{-at}}{dt} + e^{0}\delta(t)\right] \\ &\to R\left[\delta(t) - ae^{-at}\right] \end{aligned} \quad (29.47)$$

in agreement with Eq. (29.43). Figure 29.14 shows results. Notice that right after the delta function had passed, the 0+ time value is simply

$$v(0+) = -\frac{R^2}{L} \quad (29.48)$$

and the time constant is $\tau = \frac{L}{R}$.

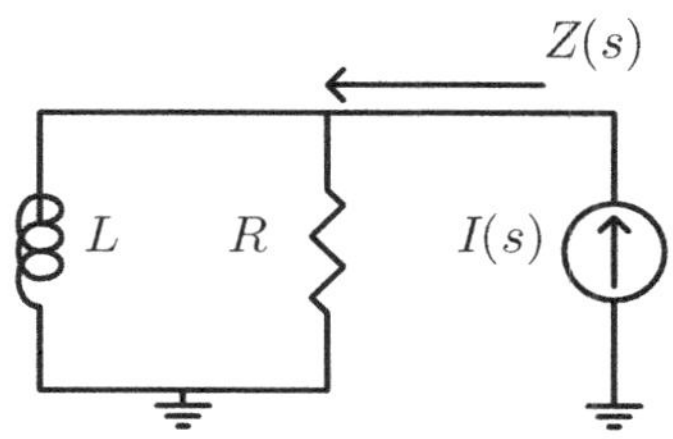

Fig. 29.13 Parallel *RL* and input impedance

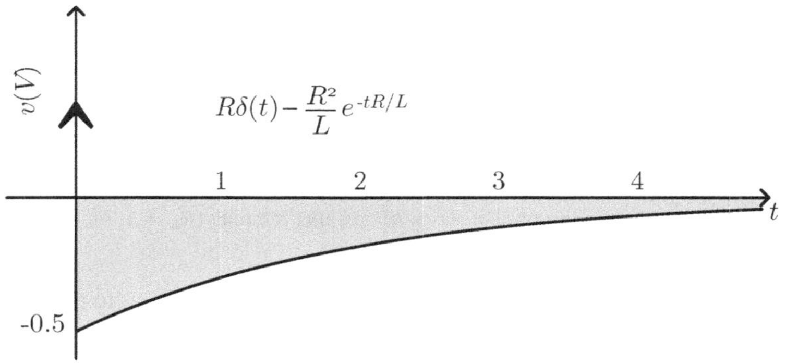

Fig. 29.14 Impulse response of parallel *RL* circuit ($R = 1\,\Omega$ and $L = 2\,\text{H}$)

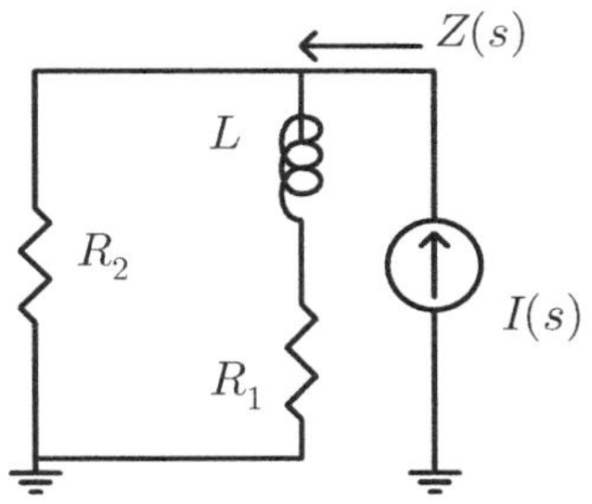

Fig. 29.15 Series *RL*, parallel *R* circuit

29.9 Series *RL*, Parallel *R* Circuit

The series *RL*, parallel *L* is shown in Fig. 29.15. Its impedance transfer function was derived in Problem 5 of Chap. 26:

$$Z(s) = \frac{R_2}{L}\frac{R_1 + sL}{s+a}, \quad a = \frac{R_1 + R_2}{L} \tag{29.49}$$

We can rewrite the impedance as

$$Z(s) = \frac{R_2 R_1}{L}\frac{1}{s+a} + R_2\frac{s}{s+a} \tag{29.50}$$

Using long division we get

$$\frac{s}{s+a} = 1 - \frac{a}{s+a} \tag{29.51}$$

Hence the impedance becomes

$$Z(s) = R_2\left[\frac{R_1}{L}\frac{1}{s+a} + 1 - \frac{a}{s+a}\right] = R_2\left[1 + \frac{R_1/L - a}{s+a}\right] \tag{29.52}$$

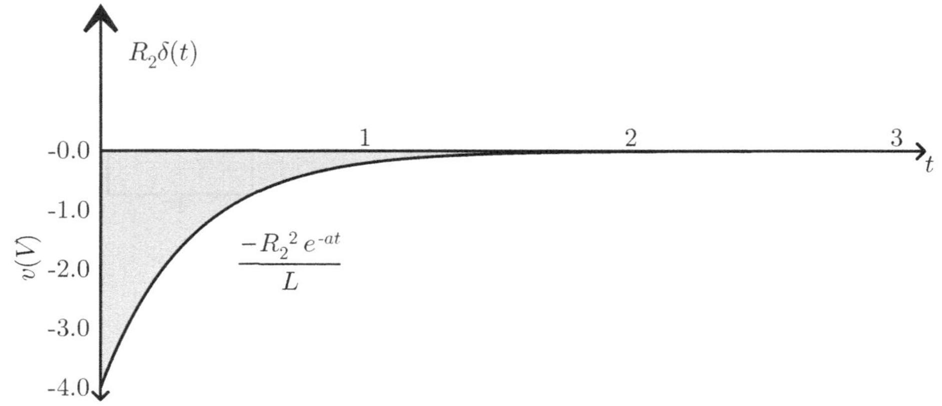

Fig. 29.16 Impulse response of series RL, parallel R circuit ($R_1 = 1, R_2 = 2\,\Omega$ and $L = 1$ H)

Expand

$$\frac{R_1}{L} - a = \frac{R_1}{L} - \frac{R_1 + R_2}{L} = -\frac{R_2}{L} \tag{29.53}$$

so finally

$$\boxed{Z(s) = R_2\left[1 - \frac{R_2}{L}\frac{1}{s+a}\right]} \tag{29.54}$$

We can read the impulse response directly from this as

$$\boxed{v(t) = R_2\delta(t) - \frac{R_2^2}{L}e^{-at}, \qquad a = \frac{R_1 + R_2}{L}} \tag{29.55}$$

The delta function across R_2 is due to the delta current diverted wholly to R_2 since the inductor completely rejected it! The time constant is simply the inductor divided by the sum of the resistors:

$$\tau = \frac{L}{R_1 + R_2} \tag{29.56}$$

Finally the initial voltage value (once the delta had passed) is simply

$$v(0+) = -\frac{R_2^2}{L} \tag{29.57}$$

Results are shown in Fig. 29.16 for the sample case $R_1 = 1\,\Omega$, $R_2 = 2\,\Omega$, and $L = 1$ H.

29.10 Series *RL*/Parallel *C* Circuit

The series RL, parallel C circuit is shown in Fig. 29.17. The impedance transfer function was derived back in Eq. (26.116), repeated here for convenience:

$$\boxed{\begin{aligned} Z(s) &= \frac{1}{C}\frac{2a+s}{(a+s)^2+\omega_0^2} \\ \omega_{LC}^2 &= \frac{1}{LC} \\ a &= \frac{R}{2L} \\ \omega_0^2 &= \omega_{LC}^2 - a^2 \end{aligned}} \tag{29.58}$$

The inverse transform is straightforward and gives

$$\boxed{v(t) = \frac{1}{C\omega_0}\left[ae^{-at}\sin(\omega_0 t)u(t) + \omega_0 e^{-at}\cos(\omega_0 t)u(t)\right]} \tag{29.59}$$

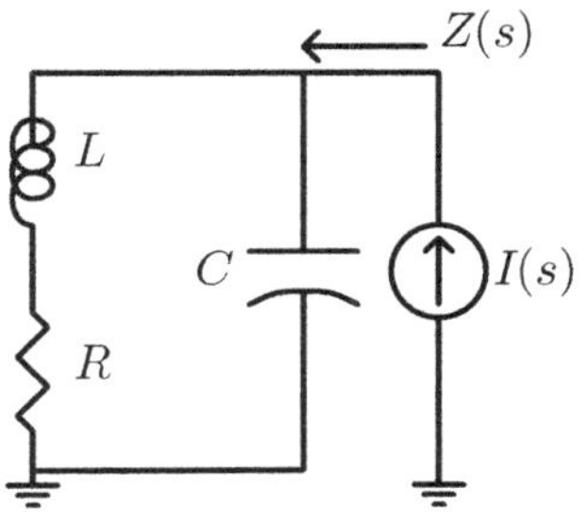

Fig. 29.17 Series *RC*/parallel *L* circuit

Notice that voltage starts at

$$v(0) = \frac{1}{C\omega_0}[0 + \omega_0] = \frac{1}{C} \quad (29.60)$$

which is nothing other than the unit charge dumped on the cap, since the *RL* branch would open at high frequency. Notice also that the time constant (for real ω_0) is

$$\text{time constant} = \frac{2L}{R} \quad (29.61)$$

Figure 29.18 shows results of sample run with $R = 0.5\,\Omega$, $C = 0.5\,\text{F}$, and $L = 0.5\,\text{H}$; and comparison to SPICE. Notice at time zero we get $v(0) = \frac{1}{C} = 2$. Notice too that the period is around π and that can be derived as follows. First, for the given *LC* values we get $\omega_{LC} = 2$; then for the given *R* we get $a = \frac{R}{2L} = \frac{0.5}{1} = 0.5$; finally we get $\omega_0 = \sqrt{\omega_{LC}^2 - a^2} = \sqrt{4 - 0.25} = \sqrt{3.75} = 1.94$. From ω_0 we get period $T = \frac{2\pi}{\omega_0} = \frac{2\pi}{1.94} = 3.2$. Lastly notice the envelope follows the relation $\sim e^{-at}$.

Notice that so far we have assumed that ω_0 is real; that is

$$\omega_0^2 > 0 \quad (29.62)$$

For the special case of

$$\omega_0 = 0 \quad (29.63)$$

the transfer function becomes

$$Z(s) = \frac{1}{C}\frac{2a + s}{(a + s)^2} \quad (29.64)$$

Notice that this happens when

$$\frac{1}{LC} = \frac{R^2}{4L^2} \Rightarrow R^2 = \frac{4L}{C} \Rightarrow R = 2\sqrt{\frac{L}{C}} \quad (29.65)$$

For this case we can expand the impedance transfer function as

$$Z(s) = \frac{1}{C}\left[\frac{1}{s + a} + \frac{a}{(s + a)^2}\right], \quad \text{(special case } \omega_0 = 0) \quad (29.66)$$

The inverse transform is then

$$\boxed{v(t) = \frac{1}{C}u(t)e^{-at}[1 + at]}, \quad (29.67)$$

$$\text{case of } \omega_0 = 0, \qquad a = \frac{R}{2L}$$

Results for this case are shown in Fig. 29.19. Notice that this case exhibits no oscillations. The last possible case for ω_0 being imaginary can be done following similar steps to above.

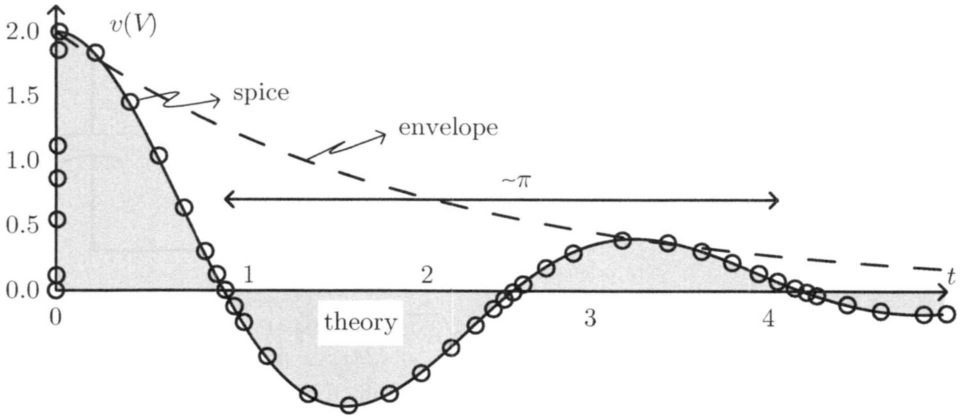

Fig. 29.18 Series *RL*/parallel *C* network impulse response ($R = 0.5\,\Omega$, $L = 0.5\,\text{H}$, and $C = 0.5\,\text{F}$)

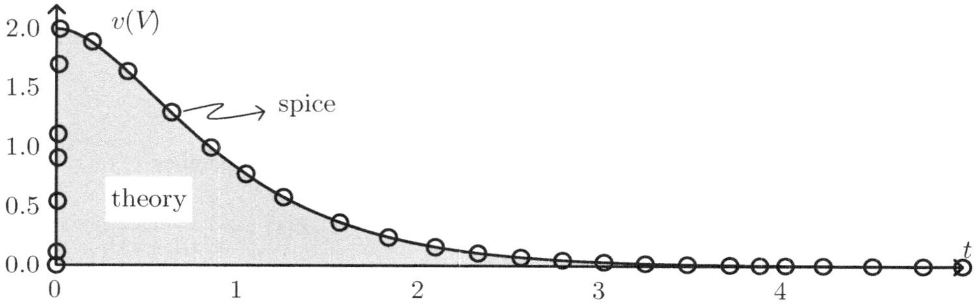

Fig. 29.19 Series *RL*/parallel *C* network impulse response ($R = 2.0\,\Omega$, $L = 0.5\,\text{H}$, and $C = 0.5\,\text{F}$)

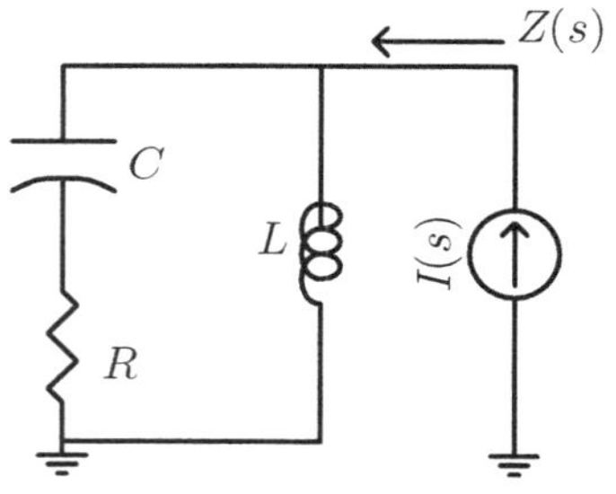

Fig. 29.20 Series *RC*, parallel *L* network

29.11 Series *RC*, Parallel *L* Network

The series *RC*, parallel *L* network is shown in Fig. 29.20. The impedance transfer function was derived in Eq. (26.133), repeated here for convenience:

$$Z(s) = \frac{1}{C}\frac{s + s^2RC}{s^2 + s\frac{R}{L} + \frac{1}{LC}} \tag{29.68}$$

Using long division we get

$$Z(s) = R + \frac{1}{C}\frac{s\left(1 - \frac{R^2C}{L}\right) - \frac{R}{L}}{s^2 + s\frac{R}{L} + \frac{1}{LC}} = R + \frac{1}{LC}\frac{s(L - R^2C) - R}{s^2 + s\frac{R}{L} + \frac{1}{LC}} \tag{29.69}$$

Complete the square and define $\omega_{LC}^2 = \frac{1}{LC}$ and $\omega_0^2 = \omega_{LC}^2 - \left(\frac{R}{2L}\right)^2$

$$Z(s) = R + \omega_{LC}^2 \frac{s(L - R^2C) - R}{(s + \frac{R}{2L})^2 + \omega_0^2} \quad (29.70)$$

Manipulate the numerator

$$Z(s) = R + \omega_{LC}^2 \frac{(L - R^2C)\left(s + \frac{R}{2L}\right) - R - (L - R^2C)\frac{R}{2L}}{\left(s + \frac{R}{2L}\right)^2 + \omega_0^2} \quad (29.71)$$

and finally get

$$\boxed{Z(s) = R + \omega_{LC}^2 \frac{(L - R^2C)\left(s + \frac{R}{2L}\right) - \frac{3LR - R^3C}{2L}}{\left(s + \frac{R}{2L}\right)^2 + \omega_0^2}} \quad (29.72)$$

By inspection we get the impulse response

$$\boxed{\begin{aligned} h(t) &= R\delta(t) \\ &+ \omega_{LC}^2(L - R^2C)e^{-Rt/2L}\cos(\omega_0 t) \\ &- \frac{\omega_{LC}^2(3LR - R^3C)}{2L\omega_0} e^{-Rt/2L}\sin(\omega_0 t) \end{aligned}} \quad (29.73)$$

Figure 29.21 shows results and comparison to SPICE. Notice that for small R the starting value (other than the delta function) is

$$h(0) \sim \omega_{LC}^2 L = \frac{L}{LC} = \frac{1}{C}, \qquad \text{case small } R \quad (29.74)$$

as shown in the first figure. Notice that for the special case of $L - R^2C = 0$ the starting value (again ignoring the delta function) would be zero

$$h(0) = 0, \qquad \text{case } L - R^2C = 0 \quad (29.75)$$

as shown in the second figure.

29.12 Summary

This is the first *applied* chapter in using inverse transform to go from the frequency domain (transfer function) to the time domain (impulse response). We saw how to derive the transfer function for various circuits in Chap. 26. We also saw the impulse response back in Chap. 22, but that was without the aid of transfer functions. Lastly we have practiced—it would seem—meticulously in finding inverse transforms. Now is the time to merge all three flows by first figuring the transfer function, next getting the inverse transform, and last—and as a byproduct—getting the impulse response. Needless to say once the impulse response of a circuit is known, then all there is to know about the circuit is already known! By the simple act of *convolving* the impulse response with *any* other input (and as was shown in Chap. 23) we can figure the circuit response to that other input. And that is all there is to it! To demonstrate this flow we practiced on many *RLC* circuits whose transfer function was the *impedance* transfer function and figured output voltage as a result of impulse current. Of course the same flow can be applied to figure output *current* as a function of input *voltage*, but there we would use the *admittance* transfer function, and so forth. Thus we have come full circle in patching abstract mathematical concepts (Fourier/Laplace transforms), with system concepts (transfer functions) and finally into real time response in the form of impulse response ready to be convolved with any other input to predict circuit response for all inputs. The next chapter repeats this flow, but applied on the *unit step response*.

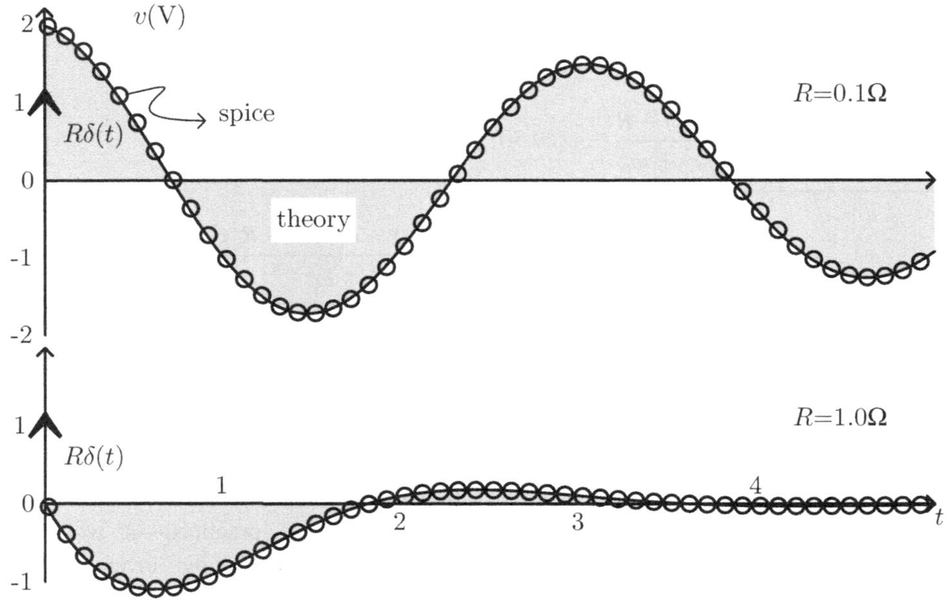

Fig. 29.21 Impulse response of series RC, parallel L ($L = 0.5\,\mathrm{H}$, $C = 0.5\,\mathrm{F}$, and R of 0.1 and 1.0 Ω)

29.13 Problems

1. The parallel RC network has the impedance transfer function

$$Z(s) = \frac{1}{C}\frac{1}{s + \frac{1}{RC}}$$

For the case $R = 1$ and $C = 1$ we get

$$Z(s) = \frac{1}{s + 1}$$

Plot this transfer function and then integrate it to find inverse transform. Compare to exact solution; see sample solution in Figs. 29.22 and 29.23.

2. The parallel RC, series R is shown in Fig. 29.24. Derive the impedance transfer function and impulse response.
Answer:

$$Z(s) = R_2 + \frac{1}{C}\frac{1}{\frac{1}{R_1C} + s}$$

$$h(t) = R_2\delta(t) + u(t)\left[\frac{1}{C}e^{-\frac{t}{R_1C}}\right]$$

3. Plot the transfer function of the series RC in parallel with R (Sect. 29.4) for the values $C = 1\mathrm{m}$, $R_1 = 1$, and $R_2 = 10$. What is the DC limit? What is the high frequency one? What is the time constant? See sample solution in Fig. 29.25.
4. The impulse response of the series RC, parallel C was derived in Eq. (29.20). Use that response to figure the response corresponding to two consecutive delta functions, of opposite sign, as shown in Fig. 29.26. Use $R_1 = 1\,\Omega$, $C_1 = 2\,\mathrm{F}$, and $C_2 = 1\,\mathrm{F}$. Compare to SPICE as shown in the same figure.
5. The impulse response of the series RC in parallel with series RC was derived in Eq. (29.31). Use that response to figure the response due to a pulse input current of width 4 and compare to SPICE; see sample solution in Fig. 29.27. Use sample case with $R_1 = 1.0\,\Omega$, $R_2 = 2.0\,\Omega$, $C_1 = 0.5\,\mathrm{F}$ and $C_2 = 1.5\,\mathrm{F}$.
6. The impulse response of the parallel RL circuit was derived in Eq. (29.43). Knowing the output voltage figure the inductor current for the case $R = 4$ and $L = 2$; explain; see sample results in Fig. 29.28.

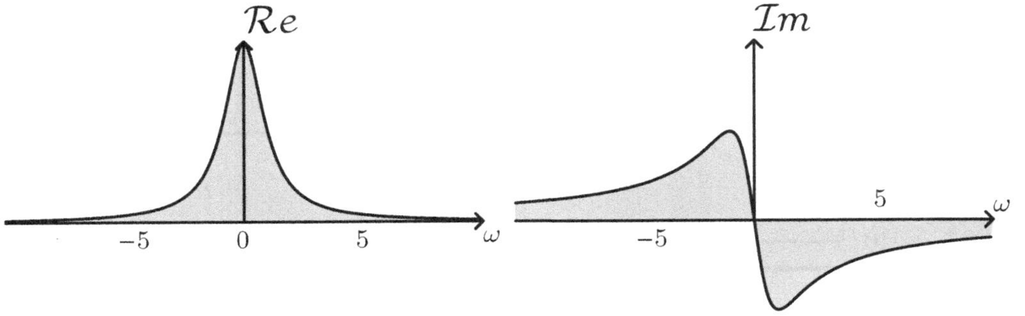

Fig. 29.22 Sample solution to Problem 1 (part 1/2)

Fig. 29.23 Sample solution to Problem 1 (part 2/2)

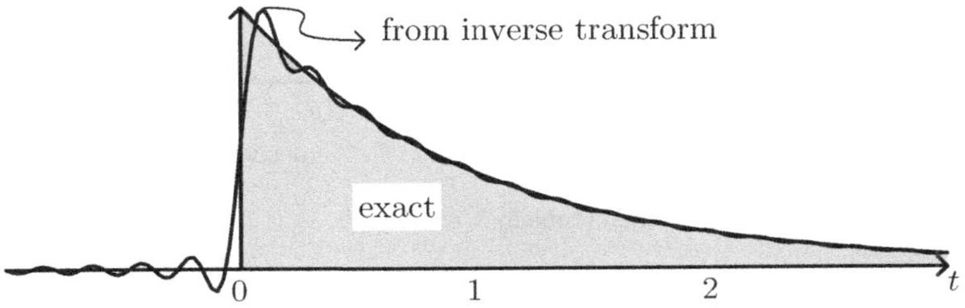

Fig. 29.24 Parallel *RC* in series with *R* network (Problem 2)

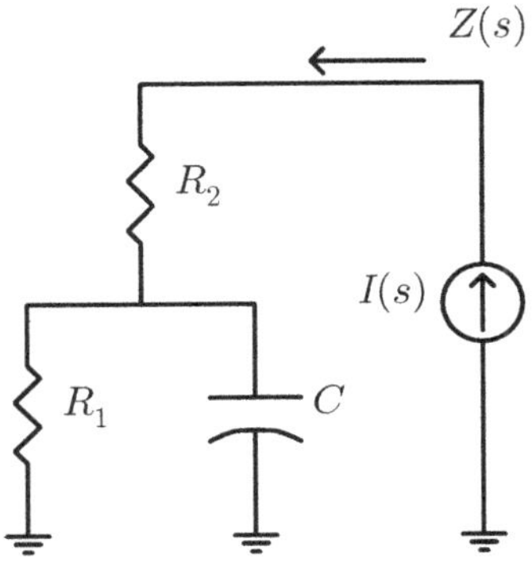

7. The impulse response of the parallel *RL* was derived in Eq. (29.43). Use convolution to figure response to a sine input of angular frequency 2π, and compare to SPICE for the case $R = 4$ and $L = 2$; see sample results in Fig. 29.29.
8. The impulse response of the series *RL*, in parallel with *R*, was derived in Eq. (29.55). Find the step response for the case $R_1 = 1\,\Omega$, $R_2 = 2\,\Omega$ and $L = 1\,\text{H}$; and compare to SPICE; see sample solution in Fig. 29.30.
9. Repeat Problem 8 but this time assume input is periodic pulse of width 1 and period 2; see sample solution in Fig. 29.31.
10. Prove that Eq. (29.59) is in fact the inverse transform of the transfer function in Eq. (29.58).
11. For the series *RL*, parallel *C*, start with the transfer function in Eq. (29.58) and assume $R = 0.5\,\Omega$, $C = 0.5\,\text{F}$, and $L = 0.5\,\text{H}$. Plot it and determine relevant frequency range; see sample solution in Fig. 29.32. Next integrate in the frequency domain to find the inverse transform and compare to exact solution; see sample solution in Fig. 29.33.
12. The impulse response of the series *RC*, parallel *L* was derived in Eq. (29.73). Use that to derive the unit step response and compare

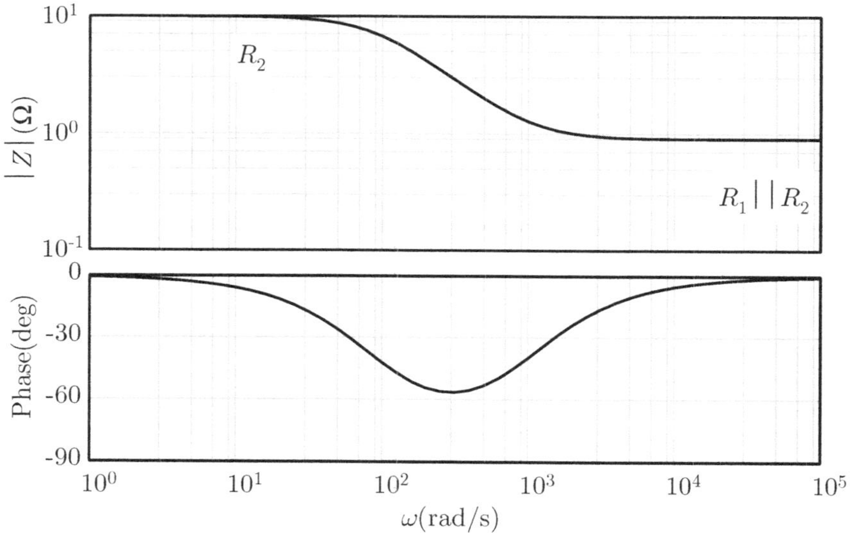

Fig. 29.25 Sample solution to Problem 3

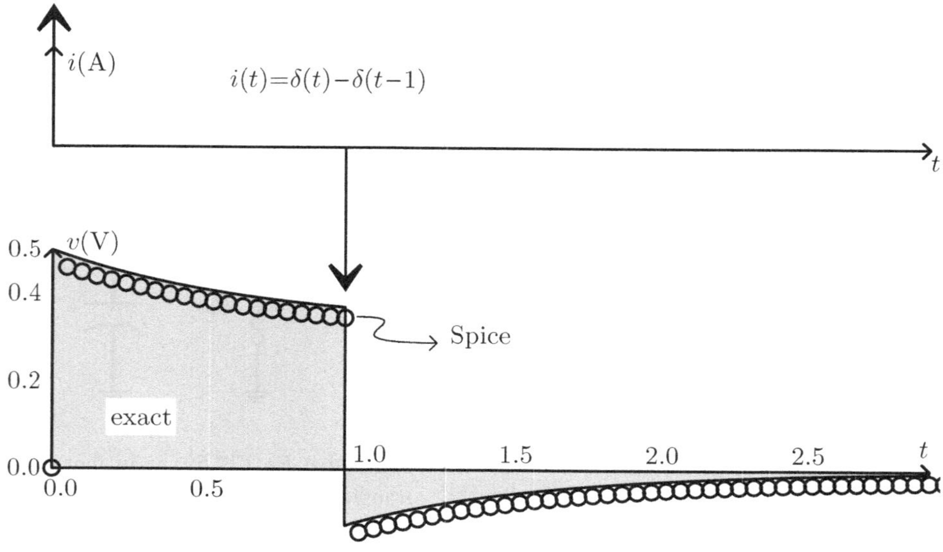

Fig. 29.26 Sample solution to Problem 4

to SPICE for the case $R = 0.5\,\Omega$, $C = 0.5\,\text{F}$, and $L = 0.5\,\text{H}$; see sample solution in Fig. 29.34.

13. Again the impulse response of the series RC, parallel L was derived in Eq. (29.73). Use that to derive the response due to the 3-step input current shown in Fig. 29.35. See sample solution in the same figure and compare to SPICE for the case $R = 0.5\,\Omega$, $C = 0.5\,\text{F}$ and $L = 0.5\,\text{H}$. (Hint: may need to utilize steps in prior Problem 12.)

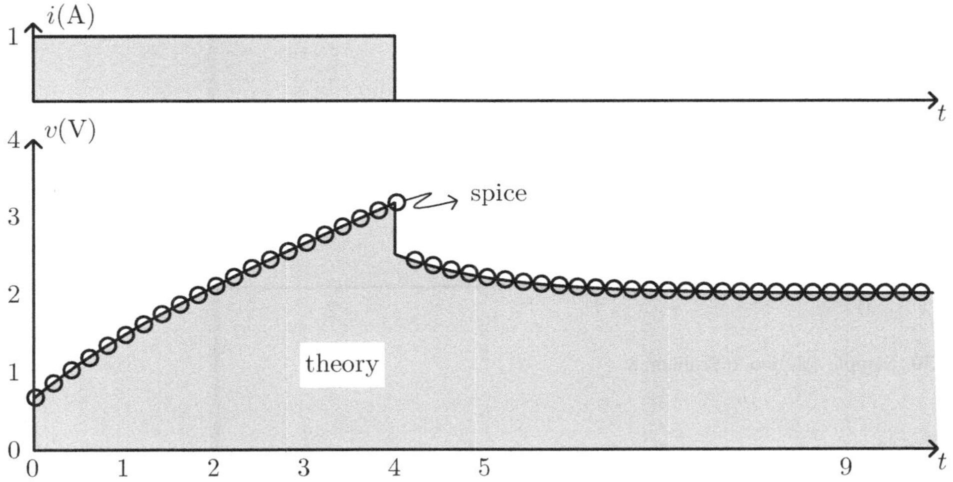

Fig. 29.27 Sample solution to Problem 5

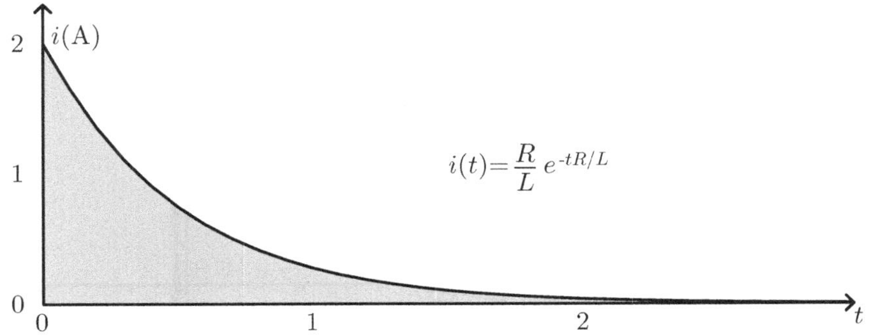

Fig. 29.28 Sample solution to Problem 6

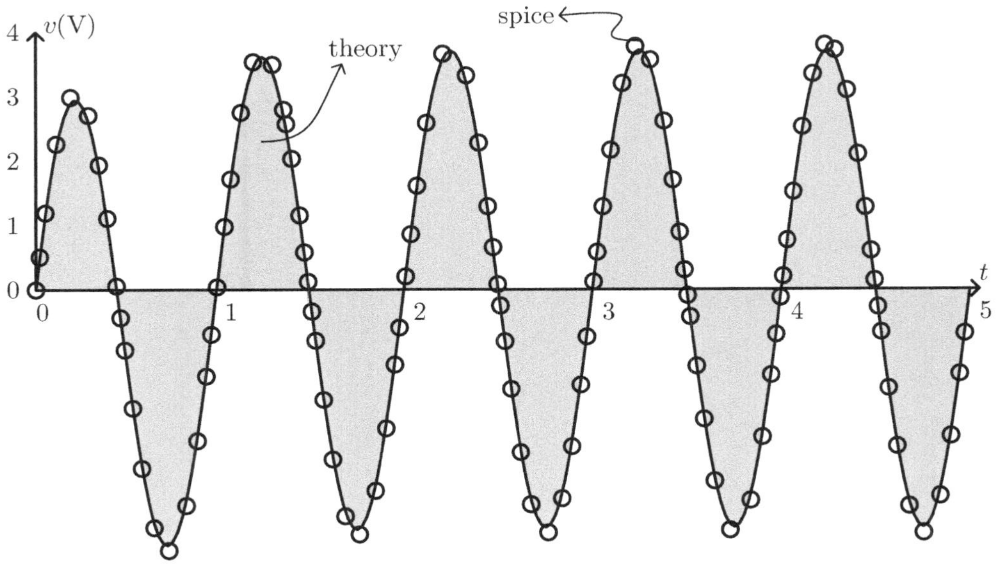

Fig. 29.29 Sample solution to Problem 7

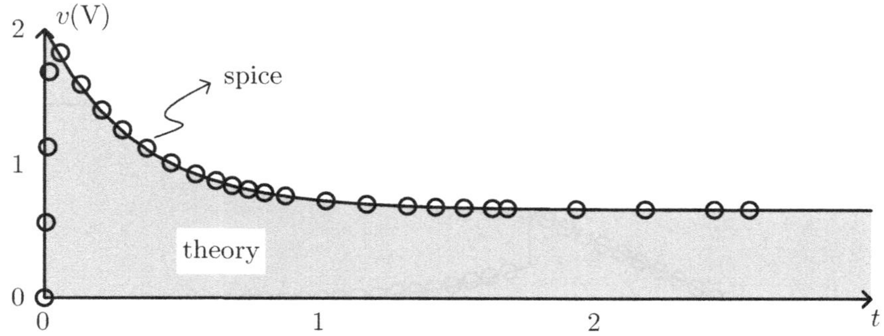

Fig. 29.30 Sample solution to Problem 8

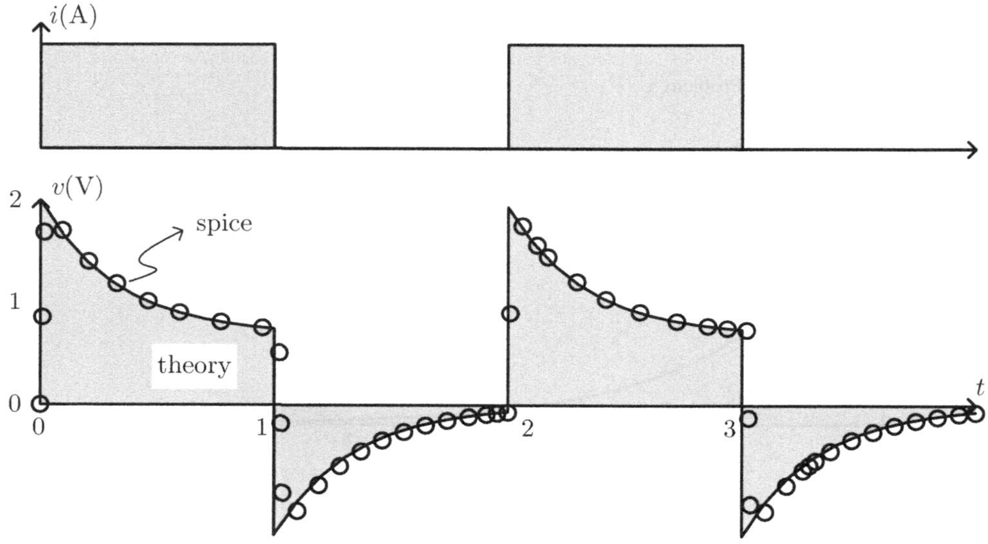

Fig. 29.31 Sample solution to Problem 9

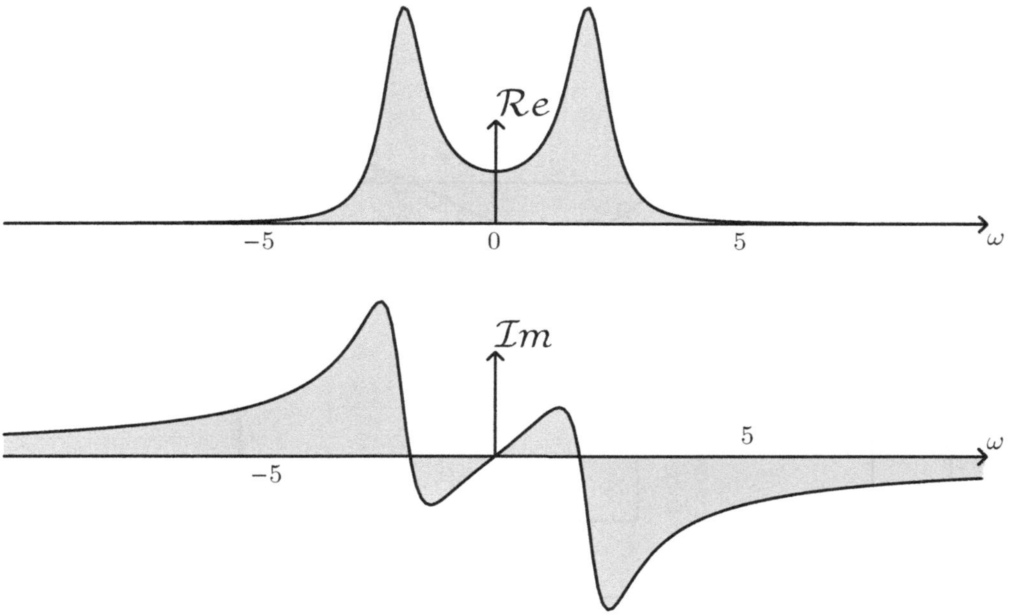

Fig. 29.32 Sample solution to Problem 11, part (1)

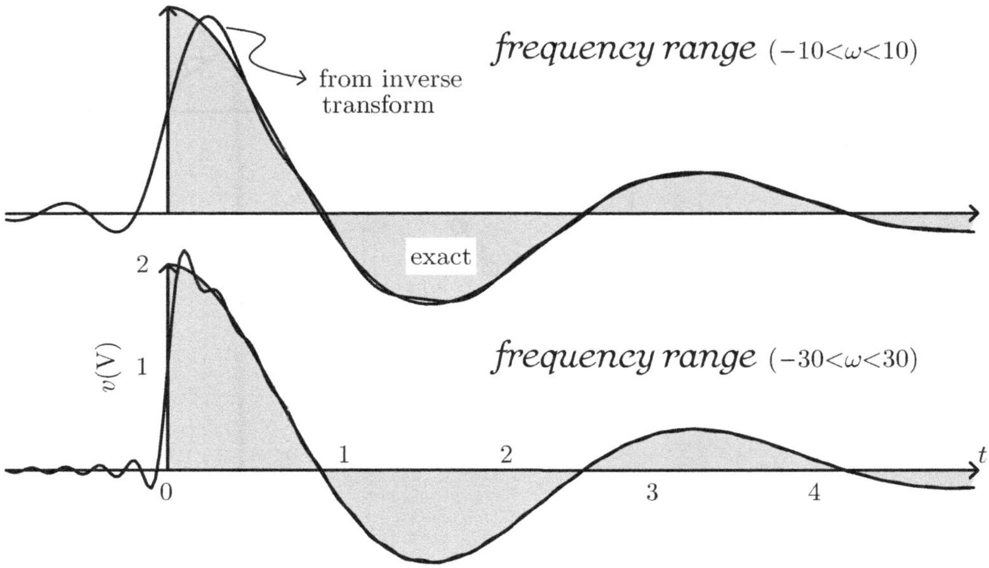

Fig. 29.33 Sample solution to Problem 11, part (2)

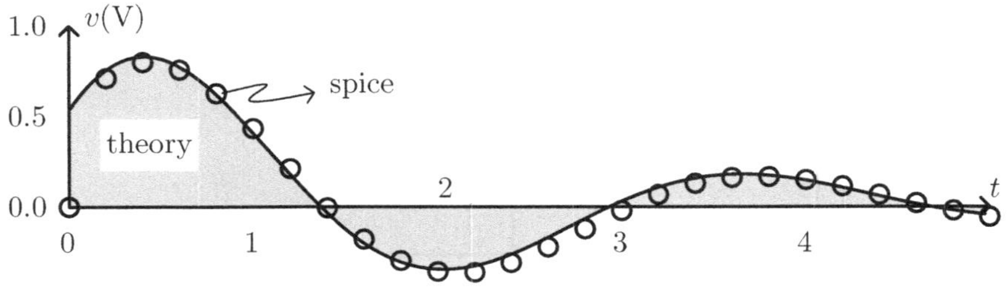

Fig. 29.34 Sample solution to Problem 12

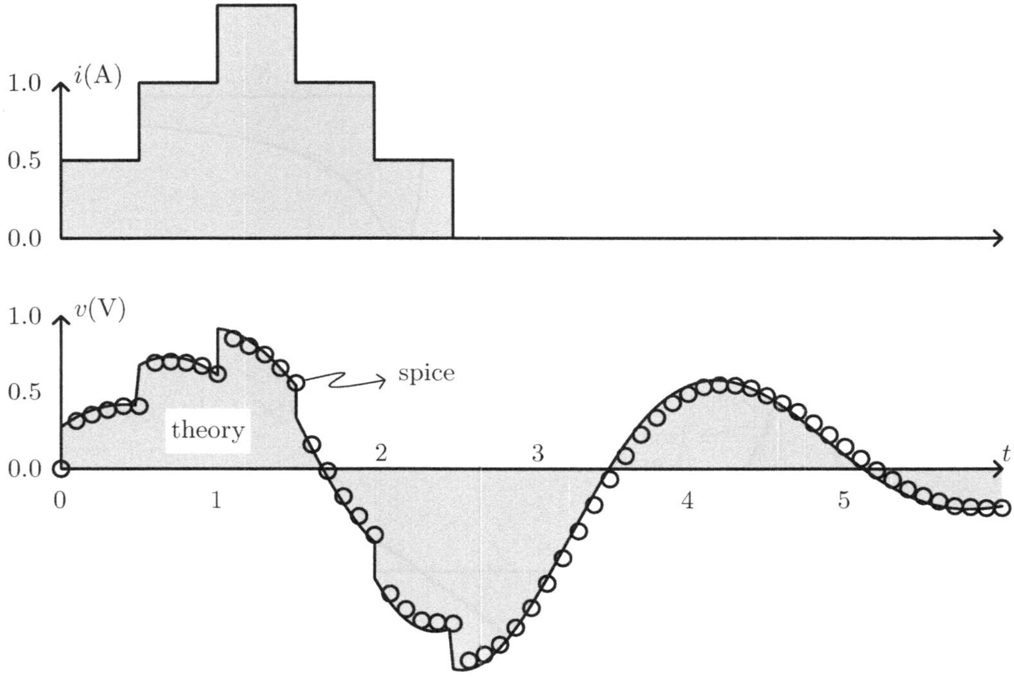

Fig. 29.35 Sample solution to Problem 13

30 Unit Step Response as Figured from Inverse Transform

30.1 Introduction

We already dealt with the unit step response and convolution therewith in Chap. 24 but that was purely from a time-stand point of view. In this chapter we derive the unit step response from the transfer function, similar to what was done in the preceding chapter (on impulse response). That is our starting point will be the transfer function in the *frequency* domain and our task is to use it to extract the unit step response in the *time* domain. Once we have that, and similar to the impulse response, we can use it as is, or use it as a building block to figure the response due to any other input, via the machinery of *convolution*. As with most of the text we illustrate this concept with some examples, and the premise is to apply this learning for other samples/problems as needed.

30.2 Series *RC* Network

As a first example consider the series *RC* network in Fig. 30.1. The impedance transfer function of this network is simply

$$Z(s) = R + \frac{1}{sC} \tag{30.1}$$

To find the unit step response we apply a unit step current and find the voltage transfer function

$$V(s) = Z(s)\frac{1}{s} = \frac{R}{s} + \frac{1}{C}\frac{1}{s^2} \tag{30.2}$$

Finding the inverse transfer we arrive at

$$g(t) = Ru(t) + \frac{1}{C}tu(t) \tag{30.3}$$

That is, the voltage drop across the resistor due to a step input is simply the step input scaled by R. On the other hand the voltage drop across the capacitor due to the step input current is a ramp function, scaled by $\frac{1}{C}$. Figure 30.1 shows the results.

30.3 Parallel *RC* Network

The parallel *RC* is shown in Fig. 30.2. The impedance transfer function for the parallel *RC* network is

$$Z(s) = \frac{1}{C}\frac{1}{s + 1/RC} \tag{30.4}$$

The input current in the frequency domain is

$$I(s) = \frac{1}{S} \tag{30.5}$$

The output voltage is then

$$V_o(s) = I(s)Z(s) = \frac{1}{C}\frac{1}{s}\frac{1}{s+a}, \qquad a = 1/RC \tag{30.6}$$

F. Badrieh, *Spectral, Convolution and Numerical Techniques in Circuit Theory*,
https://doi.org/10.1007/978-3-319-71437-0_30

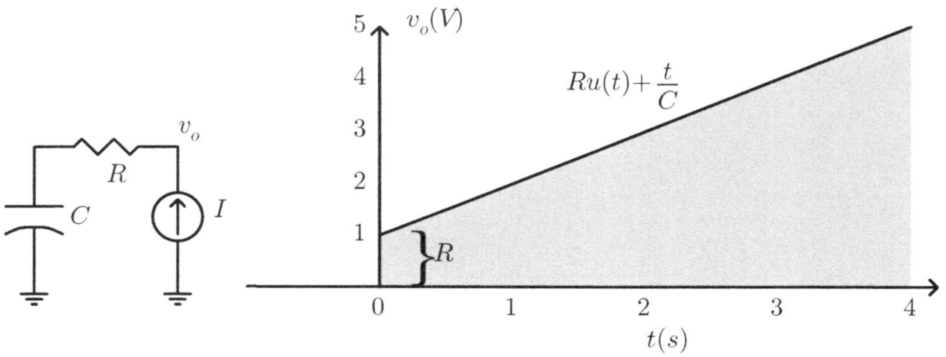

Fig. 30.1 Unit step response of series RC ($R = 1\,\Omega$ and $C = 1$ F)

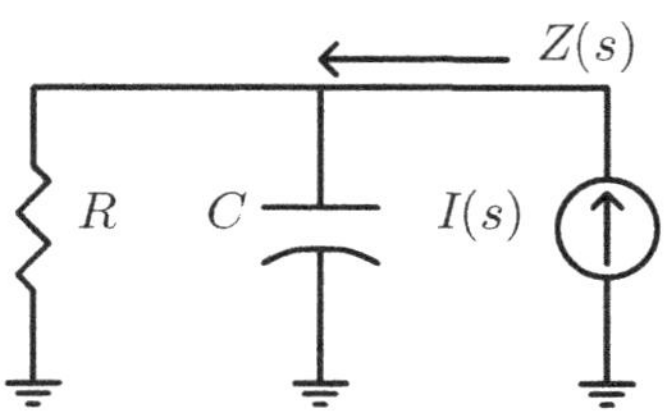

Fig. 30.2 Parallel RC network

We have a pole at 0 and another at $-a$. Using partial fractions we can expand as

$$V_o(s) = \frac{1}{aC}\left[\frac{1}{s} - \frac{1}{s+a}\right] = R\left[\frac{1}{s} - \frac{1}{s+a}\right] \tag{30.7}$$

The inverse transform simply gives

$$\boxed{v_o(t) = Ru(t)\left[1 - e^{-t/RC}\right]} \tag{30.8}$$

We see that right after time zero, the output voltage is still zero, because we cannot charge the cap instantaneously—that is not without using impulse current sources, which is not the case here! After a while, the cap fully charges to a value equal to input current times R. Since the R is in parallel with the cap, the final voltage across R equals that across the cap. When things settle down, no current flows across the cap and all flows across R. The relevant time constant is RC. Figure 30.3 shows the step response for various C values; as expected, larger C delays the charging time, but the settling value is not impacted. Figure 30.4 shows the step response for various R values; larger R delays charging time *and* increases the settling value.

30.4 Parallel *RC* in Series with *R*

The parallel RC in series with R network is shown in Fig. 30.5. We can find the step response in at least two ways. The first is by starting with the transfer function and multiplying by the LT of the step input

$$\begin{aligned} V(s) &= \left[R_2 + \frac{1}{C_1}\frac{1}{\frac{1}{R_1C_1} + s}\right]\frac{1}{s} \\ &= \frac{R_2}{s} + \frac{1}{C_1}\frac{1}{s}\frac{1}{1/\tau + s}, \qquad \tau = R_1C_1 \\ &= \frac{R_2}{s} + \frac{\tau}{C_1}\left[\frac{1}{s} - \frac{1}{1/\tau + s}\right] \\ &= \frac{1}{s}\left[R_2 + \frac{\tau}{C_1}\right] - \frac{\tau}{C_1}\frac{1}{1/\tau + s} \\ &= \frac{1}{s}\left[R_2 + R_1\right] - R_1\frac{1}{s + 1/R_1C_1} \end{aligned} \tag{30.9}$$

The inverse LT is then

$$\boxed{v(t) = u(t)\left[R_2 + R_1 - R_1e^{-\frac{t}{R_1C_1}}\right]} \tag{30.10}$$

The second method is by direct integration of the impulse response

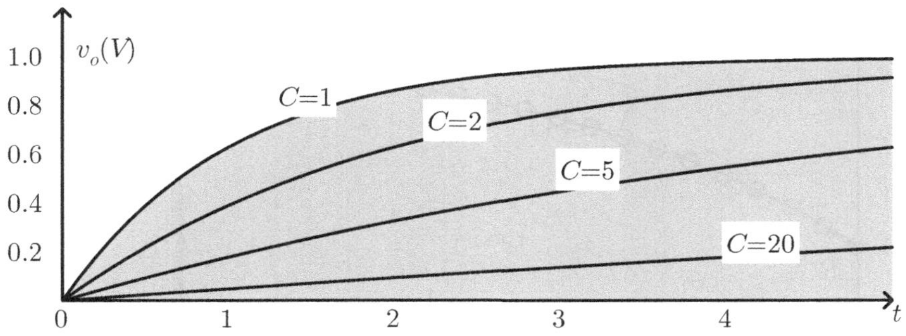

Fig. 30.3 Unit step response of parallel RC ($R = 1\ \Omega$)

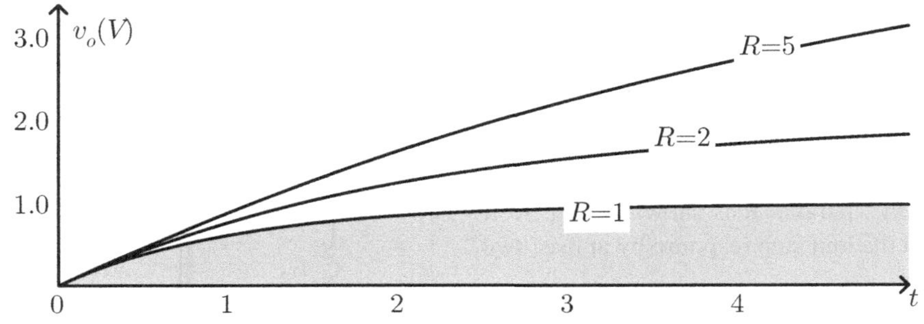

Fig. 30.4 Unit step response of parallel RC ($C = 1$ F)

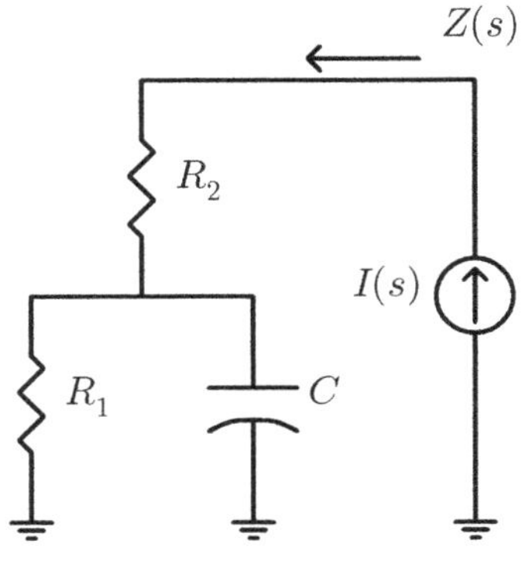

Fig. 30.5 Parallel RC in series with R network

$$
\begin{aligned}
v(t) &= \int_0^t \left[R_2\delta(\tau) + \frac{1}{C_1} e^{-\frac{\tau}{R_1 C_1}} \right] d\tau \\
&= u(t)R_2 - R_1 e^{\frac{-\tau}{R_1 C_1}} \Big|_0^t \\
&= u(t)R_2 + u(t)R_1 \left[1 - e^{-\frac{t}{R_1 C_1}} \right]
\end{aligned}
\tag{30.11}
$$

which is in agreement with Eq. (30.10). Notice the following two limits

- Right after time zero: here we get voltage drop of only R_2 (times current, which is unity here). Right when current is turned on, which is a high frequency event the cap acts like a short, and all the current flows through it (as opposed to R_1). Since the cap is pre-charged to zero, the lower node of R_2 is at zero voltage. Hence it follows that initial voltage is simply the IR product across R_2.
- DC limit: after a long time, we saturate to a voltage drop of $R_2 + R_1$; that is, at DC the cap is open and drops out of the picture; then we simply have a DC current (unity) flowing through the net impedance of $R_2 + R_1$.

So we should expect initial voltage of R_2 followed by an exponential climb to $R_2 + R_1$ with a time constant $R_1 C_1$. Figure 30.6 shows sample results.

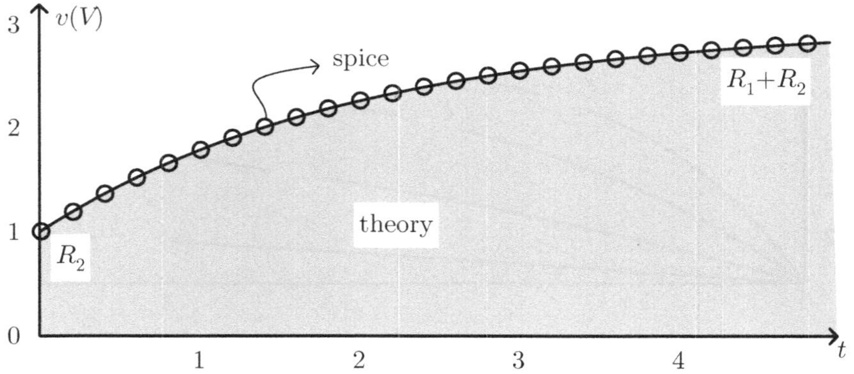

Fig. 30.6 Step response of parallel RC in series with R network and comparison to SPICE; case of $R_1 = 2, R_2 = 1\,\Omega$, and $C_1 = 1\,\mathrm{F}$

30.5 Series *RC*/Parallel *R* Circuit

The series RC, parallel R is shown in Fig. 30.7. We can find the unit step response by at least two methods:

Time Integration Method The unit step response is simply the integral of the impulse response, which was derived in Eq (29.11).

$$v(t) = \int_0^t \frac{R_1R_2}{R_1+R_2}\delta(\tau) + \frac{R_2^2}{(R_1+R_2)^2C}e^{-\frac{\tau}{(R_1+R_2)C}}d\tau = \frac{R_1R_2}{R_1+R_2}u(t) + u(t)\frac{R_2^2}{R_1+R_2}\left[1-e^{-\frac{t}{(R_1+R_2)C}}\right] \tag{30.12}$$

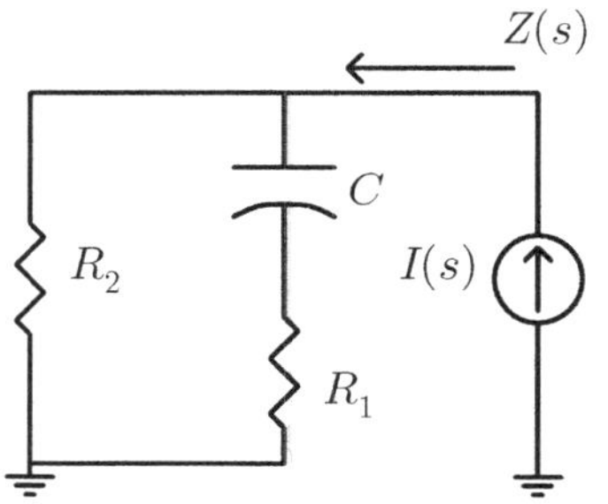

Fig. 30.7 Series RC in parallel with R and input impedance

$$\boxed{v(t) = R_2u(t) - \frac{R_2^2}{R_1+R_2}e^{-\frac{t}{(R_1+R_2)C}}} \tag{30.13}$$

Inverse Transform Method Here we start with the transfer function derived in Eq. (26.61) and multiply by the transform of the unit step function:

$$I(s)Z(s) = \boxed{\frac{R_2 + sR_1R_2C}{1 + sC(R_1 + R_2)}}\frac{1}{s} = \frac{1}{C(R_1 + R_2)}\frac{1}{s}\frac{R_2 + sR_1R_2C}{s + \frac{1}{C(R_1+R_2)}} \tag{30.14}$$

We have a pole at 0 and one at $-\frac{1}{C(R_1+R_2)}$. The pole at zero gives

$$\text{first residue} = R_2 \tag{30.15}$$

and the second pole has the residue

$$\text{second residue} = -\frac{1}{C(R_1 + R_2)}C(R_1 + R_2)\left[R_2 - \frac{R_1R_2C}{C(R_1 + R_2)}\right]$$
$$= -\left[R_2 - \frac{R_1R_2}{R_1 + R_2}\right] = -\frac{R_1R_2 + R_2^2 - R_1R_2}{R_1 + R_2} = -\frac{R_2^2}{R_1 + R_2} \tag{30.16}$$

Hence we can write our transfer function as

$$V(s) = \frac{R_2}{s} - \frac{R_2^2}{R_1 + R_2}\frac{1}{s + \frac{1}{(R_1+R_2)C}} \tag{30.17}$$

Taking the inverse LT of this leads straightaway to Eq. (30.13)

$$v(t) = R_2u(t) - \frac{R_2^2}{R_1 + R_2}e^{-\frac{t}{(R_1+R_2)C}} \tag{30.18}$$

The starting voltage level at time 0+ is

$$v(0+) = \frac{R_1R_2}{R_1 + R_2} \tag{30.19}$$

which is the parallel combination of both resistors, with the cap out of the picture since right after current application the cap pretty much shorts out. Furthermore, notice that the saturation value is simply

$$v(\infty) = R_2 \tag{30.20}$$

which simply reflects the fact that the DC current must all pour into R_1 since at DC (long time) the cap opens. Sample results are shown in Fig. 30.8.

30.6 Series *RL* Circuit

The series *RL* network is shown in Fig. 30.9. The impedance transfer function is simply

$$Z(s) = R + sL \tag{30.21}$$

Applying a unit step current results in voltage

$$V(s) = \frac{R}{s} + L \tag{30.22}$$

The inverse transform of this is simply

$$\boxed{v(t) = Ru(t) + L\delta(t)} \tag{30.23}$$

The response to a step input current gives a first component due to the resistor which also has the form of a unit step (scaled by R); and a second component due to the inductor which is an impulse function (scaled by L) at the point where the input transitions from zero to one. This makes sense; we'd expect the resistor voltage to follow input current, and inductive voltage only when input current changes, which is right at the edge of the step input. Sample results are shown in Fig. 30.10.

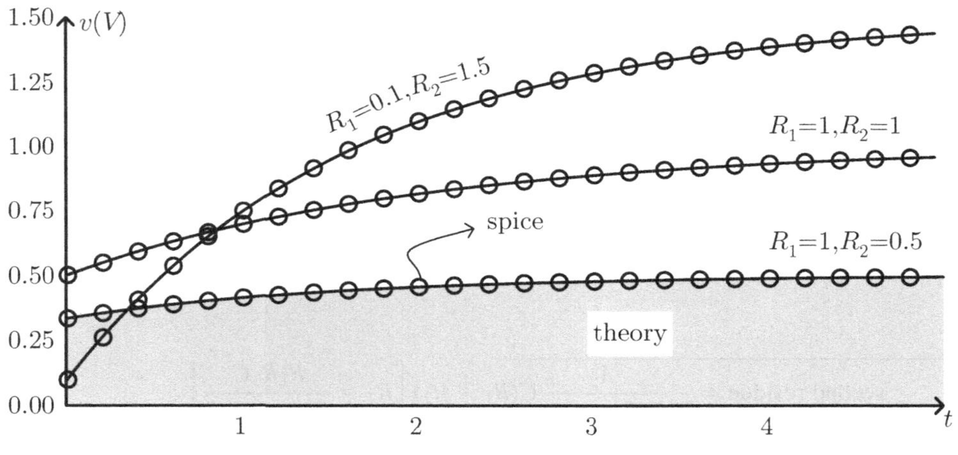

Fig. 30.8 Step response to series *RC*/parallel *R* network and comparison to SPICE ($C = 1$)

Fig. 30.9 Series *RL* network

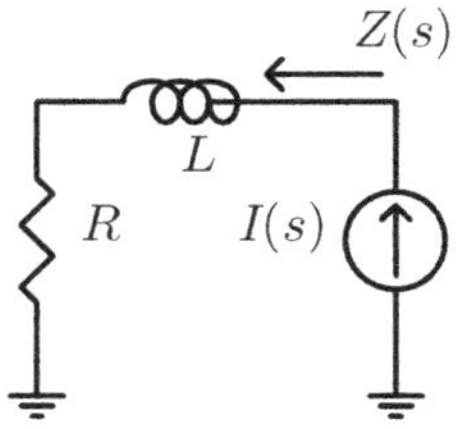

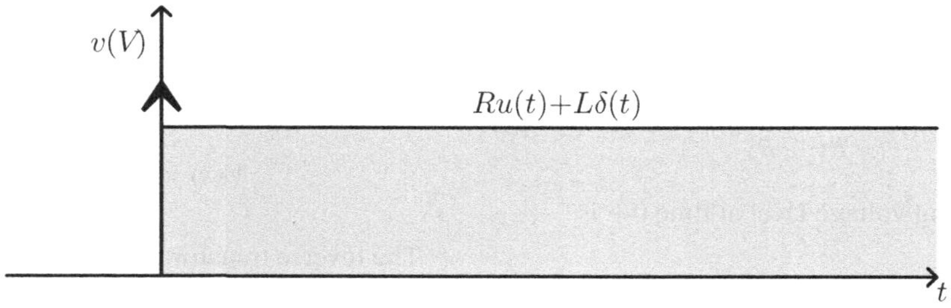

Fig. 30.10 Step response to series *RL*; sample case of $R = 1\,\Omega$ and $L = 1$ H

30.7 Parallel *RL* Circuit

The parallel *RL* network is shown in Fig. 30.11. The impedance transfer function is simply

$$Z(s) = R\frac{s}{s + \frac{R}{L}} \tag{30.24}$$

The output voltage is the input current times impedance

$$V(s) = R\frac{s}{s + R/L}\frac{1}{s} = \boxed{\frac{R}{s + R/L}} \tag{30.25}$$

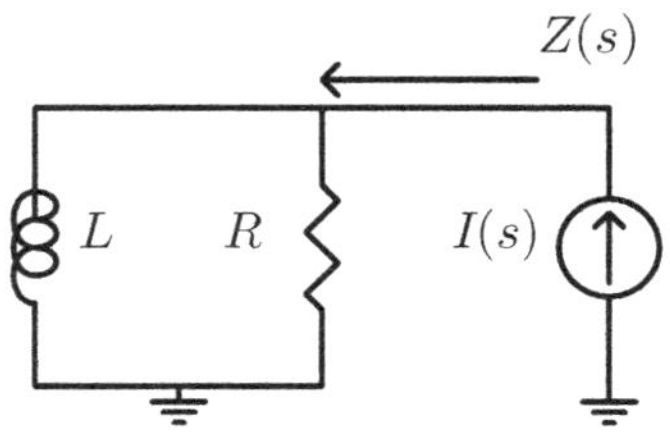

Fig. 30.11 Parallel *RL* and input impedance

The inverse transform is simply

$$\boxed{v(t) = Re^{-tR/L}} \tag{30.26}$$

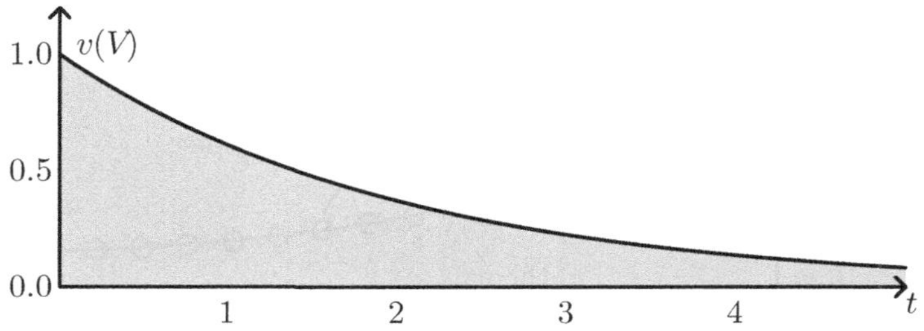

Fig. 30.12 Step response to parallel RL; case of $R = 1\,\Omega$ and $L = 2\,\text{H}$

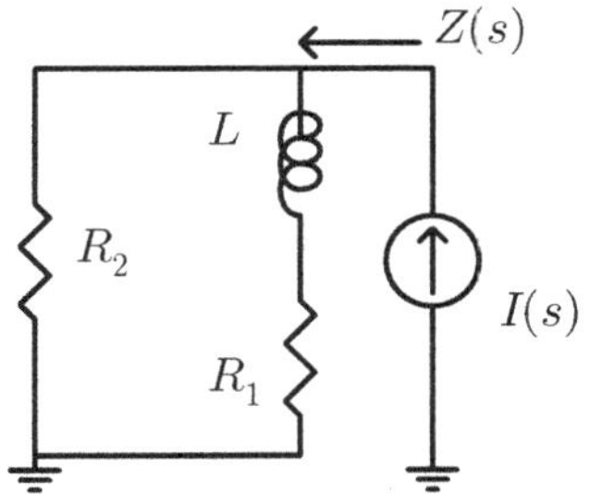

Fig. 30.13 Series RL, parallel R circuit

Notice that at time $0+$, we get $v(0+) = R$ which is input current times R; that is, right after the step is applied, the inductor acts like open, and all current goes through R. Notice also that when things settle down $v(\infty) = 0$; that is, at large time, the inductor shorts, and all the current goes through it; hence zero current goes through R, and the iR product gives zero voltage. This is confirmed in Fig. 30.12.

30.8 Series *RL*, Parallel *R* Circuit

The series RL, parallel L is shown in Fig. 30.13. The impedance transfer function tying output voltage to input current, first shown in Eq. (5), is

$$Z(s) = \frac{R_2}{L}\frac{R_1 + sL}{s+a}, \quad a = \frac{R_1 + R_2}{L} \tag{30.27}$$

The input current input is simply

$$I(s) = \frac{1}{s} \tag{30.28}$$

Output voltage is then

$$V(s) = \frac{R_2}{L}\frac{1}{s}\frac{R_1 + sL}{s+a} \tag{30.29}$$

Using partial fraction we get

$$\begin{aligned} V(s) &= \frac{R_2}{L}\left[\frac{R_1}{a}\frac{1}{s} - \frac{R_1 - aL}{a}\frac{1}{s+a}\right] \\ &= \frac{R_2}{R_1 + R_2}\left[R_1\frac{1}{s} + R_2\frac{1}{s+a}\right] \end{aligned} \tag{30.30}$$

By inspection we read out the inverse transform as

$$\boxed{g(t) = \frac{R_2}{R_1 + R_2}\left[R_1 u(t) + R_2 e^{-\frac{R_1+R_2}{L}t}\right]} \tag{30.31}$$

Notice that if $R_1 = 0$ then we get

$$g(t)\big|_{R_1=0} = R_2 e^{-tR_2/L} \tag{30.32}$$

which is the response of the parallel R_2L circuit. Notice also that if R_1 is open then we simply end up with

$$g(t)\big|_{R_1=\infty} = R_2 u(t) \tag{30.33}$$

which is the response of R_2. Finally notice that if R_2 is zero then we get zero response since impedance is zero

$$g(t)\big|_{R_2=0} = 0 \tag{30.34}$$

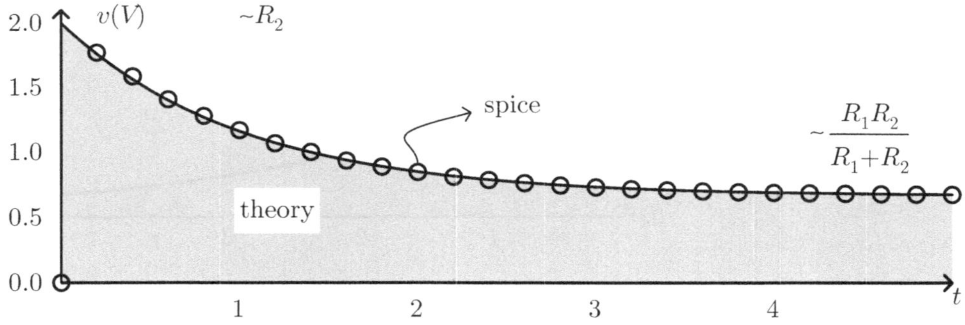

Fig. 30.14 Step response to series *RL*, parallel *R* ($R_1 = 1$, $R_2 = 2\,\Omega$, and $L = 3\,\text{H}$)

The unit step response given in Eq. (30.31) starts at

$$g(0) = \frac{R_2}{R_1 + R_2}\left[R_1 + R_2\right] = R_2 \qquad (30.35)$$

which makes sense; there all the high frequency current flows through R_2 and bypasses the inductor. The other limit of the unit step response is after a long time; then from Eq. (30.31) we get

$$g(\infty) = \frac{R_1 R_2}{R_1 + R_2} \qquad (30.36)$$

which is nothing but the parallel combination of R_1 and R_2; that is, after a long time the inductor shorts and we end up with two resistors in parallel. These results as well as SPICE simulations are shown in Fig. 30.14.

30.9 Series *RL*/Parallel *C* Circuit

The series *RL*, parallel *C* circuit is shown in Fig. 30.15. The impedance transfer function as derived before (Eq. (26.116)) is

$$\boxed{\begin{aligned} Z(s) &= \frac{1}{C}\frac{2a+s}{(a+s)^2+\omega_0^2} \\ \omega_{LC}^2 &= \frac{1}{LC} \\ a &= \frac{R}{2L} \\ \omega_0^2 &= \omega_{LC}^2 - a^2 \end{aligned}} \qquad (30.37)$$

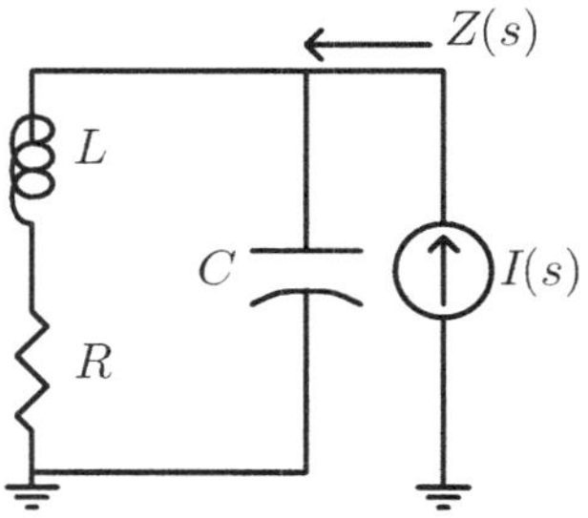

Fig. 30.15 Series *RL*/parallel *C* circuit

The voltage in frequency domain is then

$$\begin{aligned} V(s) &= \frac{1}{C}\frac{2a+s}{(a+s)^2+\omega_0^2}\frac{1}{s} \\ &= \frac{1}{sC}\frac{2a}{(a+s)^2+\omega_0^2} + \frac{1}{C}\frac{1}{(a+s)^2+\omega_0^2} \\ &= V_1(s) + V_2(s) \end{aligned} \qquad (30.38)$$

We know that $V_2(s)$ is going to give us

$$v_2(t) = \frac{1}{\omega_0 C}e^{-at}\sin\omega_0 t u(t) \qquad (30.39)$$

To find $v_1(t)$ we would have to do partial fraction expansion:

$$\frac{A}{s} + \frac{B + Cs}{(a+s)^2+\omega_0^2} = \frac{1}{s}\frac{2a}{(a+s)^2+\omega_0^2} \qquad (30.40)$$

Expanding we get

$$\frac{1}{s}\frac{A(a^2+2as+s^2)+A\omega_0^2+Bs+Cs^2}{(a+s)^2+\omega_0^2}=\frac{1}{s}\frac{2a}{(a+s)^2+\omega_0^2} \tag{30.41}$$

Collecting terms we get

$$(Aa^2+A\omega_0^2)+(2Aa+B)s+(A+C)s^2=2a \tag{30.42}$$

Equating zero order s we get

$$A=\frac{2a}{a^2+\omega_0^2} \tag{30.43}$$

Equating first order s we get

$$B=-2Aa=-\frac{4a^2}{a^2+\omega_0^2} \tag{30.44}$$

And finally equating second order s we get

$$C=-A=-\frac{2a}{a^2+\omega_0^2} \tag{30.45}$$

Then $V_1(s)$ becomes

$$V_1(s)=\frac{1}{C}\frac{2a}{a^2+\omega_0^2}\left[\frac{1}{s}+\frac{-2a-s}{(a+s)^2+\omega_0^2}\right] \tag{30.46}$$

Converting to time domain we get

$$v_1(t)=\frac{2a}{C(a^2+\omega_0^2)}u(t)\left[1-\frac{a}{\omega_0}e^{-at}\sin\omega_0 t-e^{-at}\cos\omega_0 t\right] \tag{30.47}$$

Our final solution is then

$$\begin{aligned}v(t) &= \frac{2a}{C(a^2+\omega_0^2)}u(t)\left[1-\frac{a}{\omega_0}e^{-at}\sin\omega_0 t-e^{-at}\cos\omega_0 t\right]\\ &+ \frac{1}{\omega_0 C}e^{-at}u(t)\sin\omega_0 t\end{aligned} \tag{30.48}$$

$$v(t)=\frac{2a}{C(a^2+\omega_0^2)}u(t)\left[1-e^{-at}\cos\omega_0 t\right]+\frac{1}{\omega_0 C}\frac{\omega_0^2-a^2}{a^2+\omega_0^2}e^{-at}\sin\omega_0 t \tag{30.49}$$

$$\boxed{v(t)=\frac{2a}{C(a^2+\omega_0^2)}u(t)\left[1-e^{-at}\cos\omega_0 t+\frac{\omega_0^2-a^2}{2a\omega_0}e^{-at}\sin\omega_0 t\right]} \tag{30.50}$$

Notice that the initial voltage is

$$v(0+)=0 \tag{30.51}$$

since right then the inductor is open and the cap is pre-charged to zero. On the other hand the final voltage is

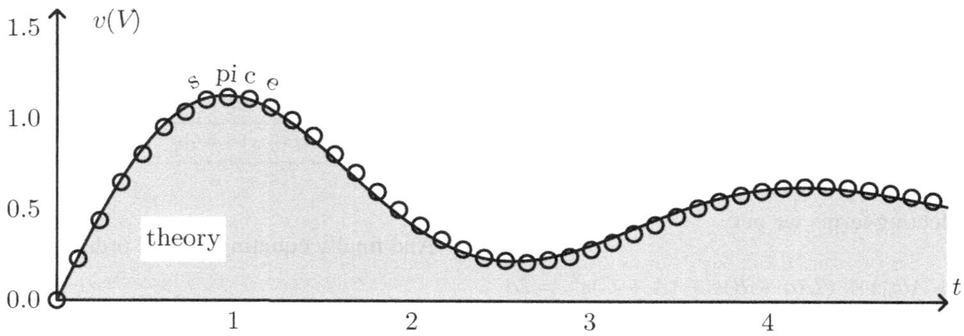

Fig. 30.16 Series *RL*/parallel *C* step response ($R = 0.5\,\Omega$, $L = 0.5\,\text{H}$, and $C = 0.5\,\text{F}$)

$$v(\infty) \sim \frac{2a}{C(a^2+\omega_0^2)} = \frac{2a}{C\omega_{LC}^2} = \frac{2a}{C\frac{1}{LC}} = 2aL = 2L\frac{R}{2L} = R \tag{30.52}$$

That is, when things settle down the inductor is short and the cap is open; all the current goes through the resistor and the corresponding voltage is simple R! Figure 30.16 shows our results and comparison to SPICE.

30.10 Summary

The unit step response is very important for system and circuit characterization. It shares some features with the *impulse* response, in the sense it *samples all frequencies*, but it differs in the sense once applied it *continues* to stress the network as opposed to the impulse which is then *bygone*. The unit step has the simple Laplace transform of $\frac{1}{s}$; so all that has to be done is multiply the system/circuit transfer function $H(s)$ by $\frac{1}{s}$. For the case of *impedance* transfer function $H(s) = Z(s)$, output voltage due to a unit step input current, and in the frequency domain is simply $V(s) = \frac{Z(s)}{s}$. All that remains after that is simply figuring inverse Laplace transform of $V(s)$ to get $v(t)$. The same can be done with other kinds of transfer functions, such as *admittance* ones. A different route which bypasses the frequency domain altogether is simply take the impulse response and time *integrate* it; that directly gives voltage in the time domain. We illustrated the first flow with multitude of examples, where we started with $Z(s)$, figured $V(s)$, figured $v(t)$, plotted results, and rationalized the response. This last step is important in the sense of sanitizing the results and ensuring they make sense; for example an inductor *shorts* in the long time span, while a capacitor *opens* and so forth. And as a second reminder, once we know the unit step response of a system we know *all there is to know* about this system! To find the response to any other input we simply *convolve* the unit step response with the *derivative* of the stimulus, as outlined before in Chap. 24.

30.11 Problems

1. A system has the transfer function

$$H_1(s) = \frac{1}{1+s}$$

What is the unit step response, both in frequency and in time? Plot and compare to transfer function and impulse response; see sample solution in Fig. 30.17.

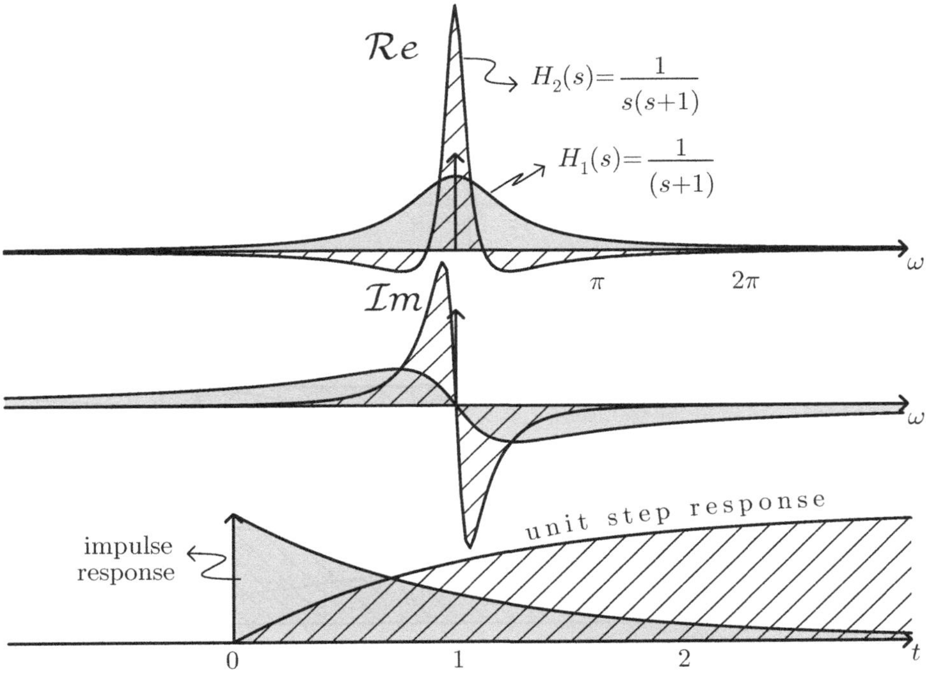

Fig. 30.17 Sample solution to Problem 1, with $\sigma = 0.3$

2. A system has the transfer function

$$H_1(s) = \frac{1}{(s+1)(s+2)}$$

What is the unit step response, both in frequency and in time? Plot and compare to transfer function and impulse response; see sample solution in Fig. 30.18.

3. A system has the transfer function

$$H_1(s) = \frac{1 - e^{-1.5s}}{s+3}$$

What is the unit step response, both in frequency and in time? Plot and compare to transfer function and impulse response; see sample solution in Fig. 30.19.

4. A system has the transfer function

$$H_1(s) = \frac{1 - e^{-s}}{s}$$

What is the unit step response, both in frequency and in time? Plot and compare to transfer function and impulse response; see sample solution in Fig. 30.20.

5. A system has the transfer response

$$H_1(s) = \frac{2\pi}{s^2 + (2\pi)^2}\left[1 - e^{-s}\right]$$

What is the unit step response, both in frequency and in time? Plot and compare to transfer function and impulse response; see sample solution in Fig. 30.21.

6. A system has the transfer response

$$H_1(s) = \frac{\omega_0}{(s+a)^2 + \omega_0^2}$$

What is the unit step response, both in frequency and in time? Plot and compare to transfer function and impulse response for the case $a = 1$ and $\omega_0 = 2\pi$; see sample solution in Fig. 30.22. Hint: the step response in the frequency domain is

$$H_2(s) = \frac{1}{s}\frac{\omega_0}{(s+a)^2 + \omega_0^2}$$

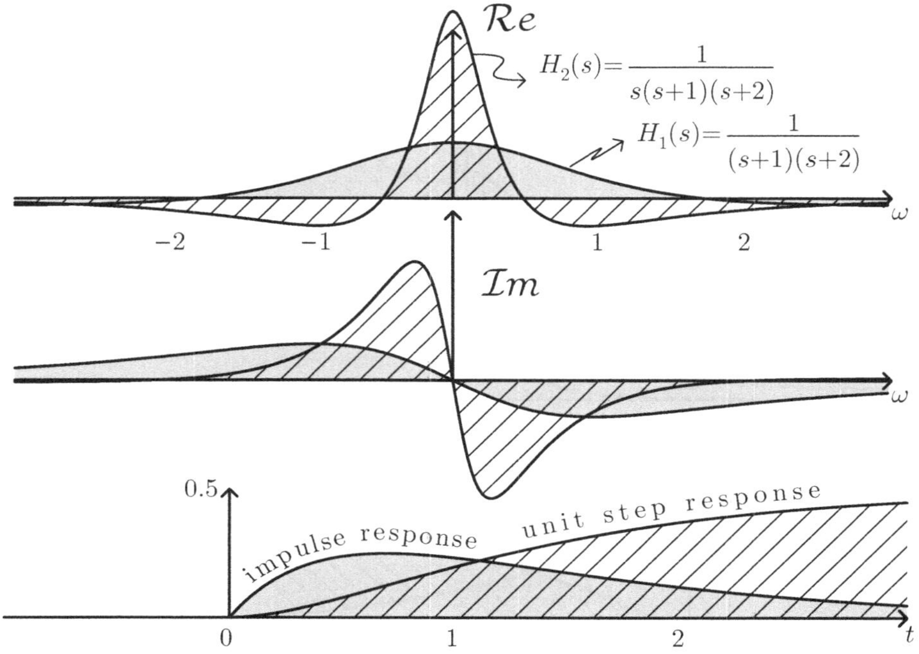

Fig. 30.18 Sample solution to Problem 2, with $\sigma = 0.3$

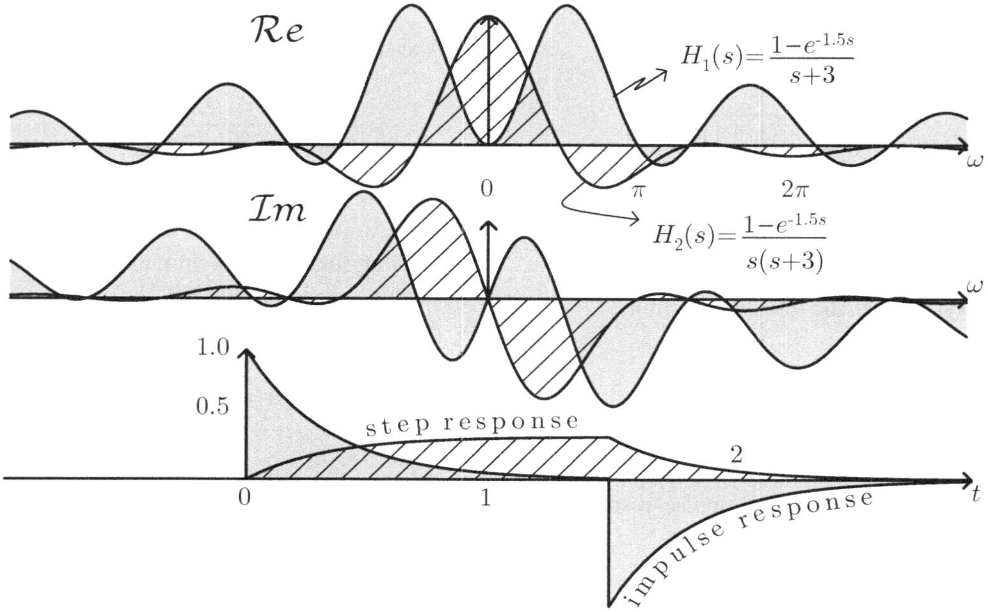

Fig. 30.19 Sample solution to Problem 3, with $\sigma = 0.0$

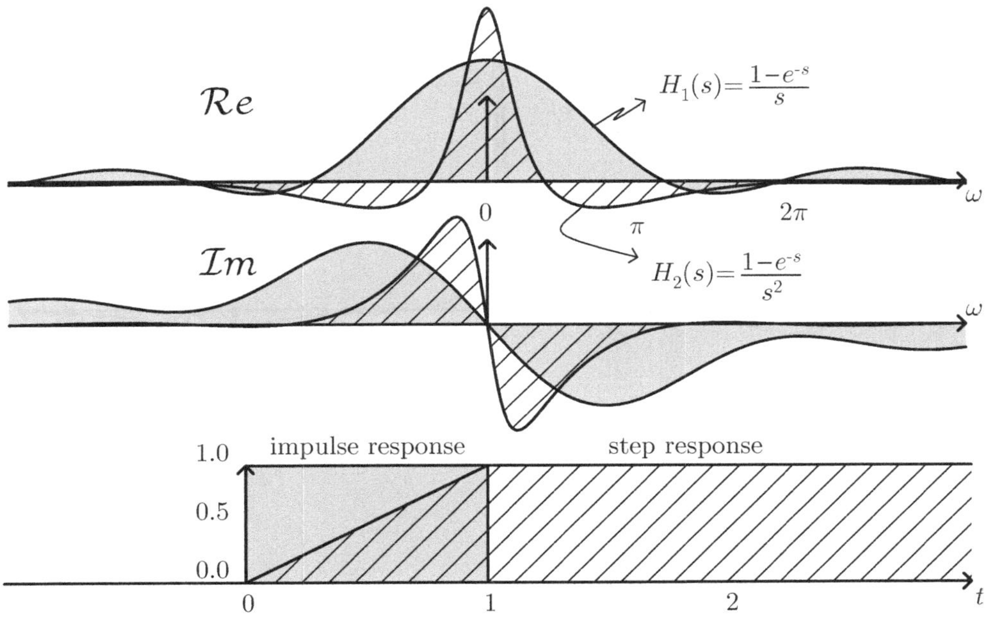

Fig. 30.20 Sample solution to Problem 4, with $\sigma = 0.7$

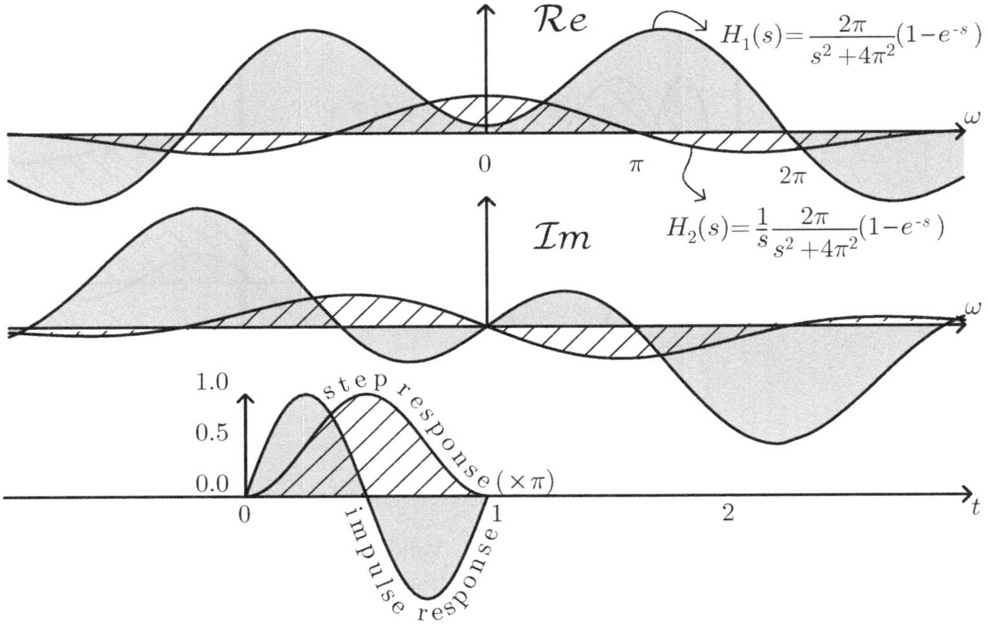

Fig. 30.21 Sample solution to Problem 5, with $\sigma = 0.2$

We can rewrite as

$$H_2(s) = \frac{1}{s}\frac{\omega_0}{(s+a-j\omega_0)(s+a+j\omega_0)}$$

Clearly we have three poles: one at 0, the other at $-a + j\omega_0$, and the last at $-a - j\omega_0$; hence we can write the transfer function as

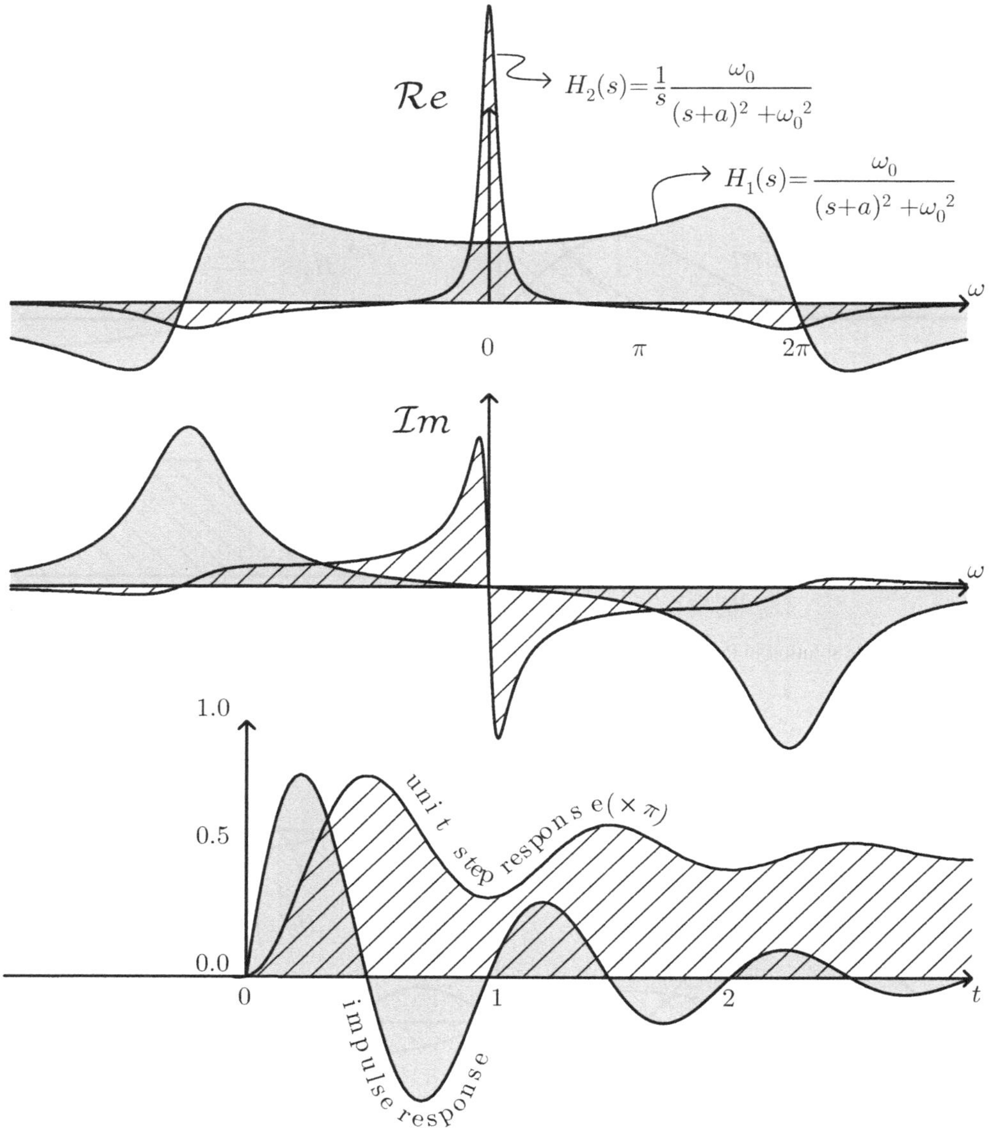

Fig. 30.22 Sample solution to Problem 6, with $\sigma = 0.2$

$$H_2(s) = \frac{A}{s} + \frac{B}{s + a - j\omega_0} + \frac{C}{s + a + j\omega_0}$$

Now use partial fractions to figure A, B, and C. Once those are known, finding the inverse transform should be straightforward.

7. Validate Eq. (30.51) by first starting with the transfer function in Eq. (30.37), multiplying by $\frac{1}{s}$ then and using the *initial* value theorem.
8. Validate Eq. (30.52) by first starting with the transfer function in Eq. (30.37), multiplying by $\frac{1}{s}$ then and using the *final* value theorem.

31 Pulse Response

31.1 Introduction

So far we have covered the impulse and unit step responses as configured from inverse transforms. Another important stimulus is the one-timer pulse, which is to be differentiated from the periodic pulse. In the most general sense, and by arranging the pulse width and height accordingly, we can retrieve from the pulse response the response to a unit step and that to an impulse. But for simplicity we will focus here on the simple pulse, of finite width and height. The one-timer pulse stresses the system with an initial strength just like that of the unit step; but unlike the latter, the input ceases after a predetermined time interval, which is the pulse width. So we would expect the initial response to the pulse to match that of the unit step.

31.2 Transform of Pulse

Let's recall the Laplace transform of the pulse function of width T as derived earlier in Eq. (14.95):

$$I(s) = \frac{1 - e^{-Ts}}{s} \tag{31.1}$$

To put things in perspective recall the Laplace transform of the impulse which was simply 1, and that of the unit step function which was simply $\frac{1}{s}$. This current will next be applied against the impedance transfer function to figure voltage response. As such let us practice the pulse response by applying it to a few representative circuits.

31.3 Parallel *RL* Circuit

The parallel *RL* network is shown in Fig. 31.1. The impedance transfer function is

$$Z(s) = R\frac{s}{s + \frac{R}{L}} \tag{31.2}$$

The output voltage is simple input current times impedance

$$V(s) = I(s)Z(s) = \frac{1 - e^{-Ts}}{s} R \frac{s}{s + R/L} \tag{31.3}$$

$$\boxed{V(s) = R\frac{1 - e^{-Ts}}{s + R/L}} \tag{31.4}$$

To find the inverse LT of this first recall that

$$\frac{1}{s + a} \to e^{-at} \tag{31.5}$$

and recall the time shifting property:

$$f(t - t_0) \to e^{-st_0} F(s) \tag{31.6}$$

This would then imply that

$$\boxed{v(t) = R\left[u(t)e^{-tR/L} - u(t - T)e^{-(t-T)R/L}\right]} \tag{31.7}$$

F. Badrieh, *Spectral, Convolution and Numerical Techniques in Circuit Theory*,
https://doi.org/10.1007/978-3-319-71437-0_31

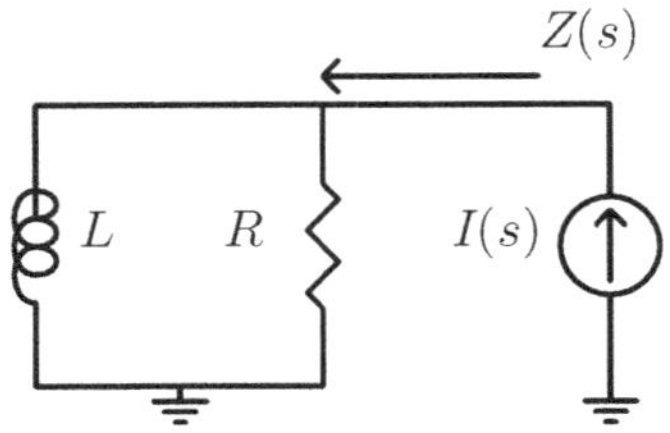

Fig. 31.1 Parallel *RL* and input impedance

Figure 31.2 shows the results and comparison to SPICE.

31.4 Parallel *RC* Circuit

The parallel *RC* is shown in Fig. 31.3. The impedance transfer function for the parallel *RC* network is

$$Z(s) = \frac{1}{C}\frac{1}{s + 1/RC} \tag{31.8}$$

The input current in the frequency domain is

$$I(s) = \frac{1 - e^{-sT}}{s} \tag{31.9}$$

The output voltage is then

$$V(s) = \frac{1}{C}\frac{1}{s+a}\frac{1 - e^{-sT}}{s}, \qquad a = \frac{1}{RC} \tag{31.10}$$

Using partial fractions we know

$$\frac{1}{s}\frac{1}{s+a} = \frac{1}{a}\left[\frac{1}{s} - \frac{1}{s+a}\right] \tag{31.11}$$

Then

$$\boxed{V(s) = R\left[\frac{1}{s} - \frac{1}{s + 1/RC}\right]\left[1 - e^{-sT}\right]} \tag{31.12}$$

The inverse transform is then

$$\boxed{v(t) = R\left[u(t) - e^{-t/RC} - u(t-T) + u(t-T)e^{-(t-T)/RC}\right]} \tag{31.13}$$

This can be rewritten as

$$v(t) = \begin{cases} R\left[1 - e^{-t/RC}\right] & 0 < t < T \\ Re^{-t/RC}\left[-1 + e^{T/RC}\right] & t > T \end{cases} \tag{31.14}$$

Results and comparison to SPICE are shown in Fig. 31.4.

31.5 Parallel *LC* Circuit

The parallel *LC* circuit is shown in Fig. 31.5. The impedance transfer function is

$$Z(s) = \frac{1}{C}\frac{s}{s^2 + \omega_0^2}, \qquad \omega_0^2 = \frac{1}{LC} \tag{31.15}$$

Input current is

$$I(s) = \frac{1 - e^{-Ts}}{s} \tag{31.16}$$

Output voltage is

$$V(s) = \frac{1}{C}\frac{s}{s^2 + \omega_0^2}\frac{1 - e^{-Ts}}{s} = \boxed{\frac{1}{C}\frac{1 - e^{-Ts}}{s^2 + \omega_0^2}} \tag{31.17}$$

By inspection we get

$$\boxed{v(t) = \frac{1}{C\omega_0}\left[u(t)\sin\omega_0 t - u(t-T)\sin\omega_0(t-T)\right]} \tag{31.18}$$

Results, for different pulse width T, are shown in Fig. 31.6.

This example had $C = 0.05\,\mathrm{F}$ and $L = 0.10\,\mathrm{H}$ such that

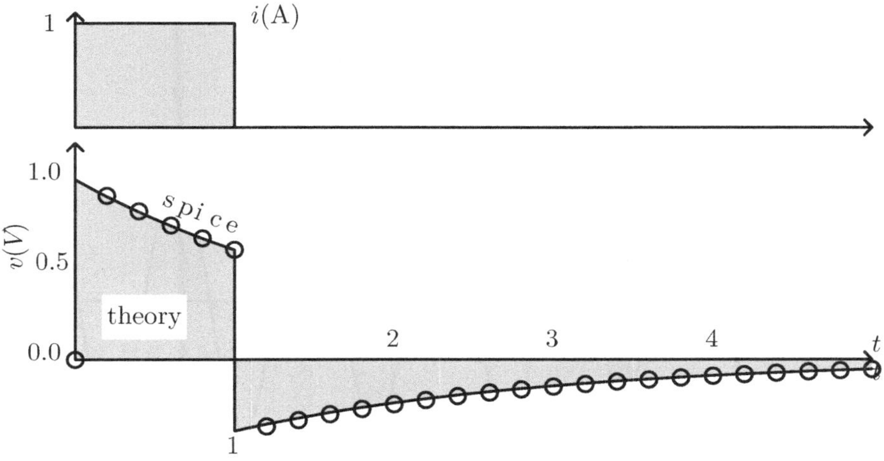

Fig. 31.2 Parallel *RL* pulse response ($R = 1\,\Omega$ and $L = 2\,\text{H}$)

Fig. 31.3 Parallel *RC* network

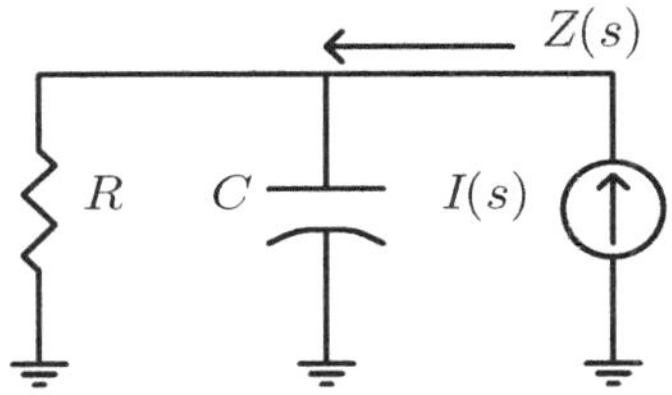

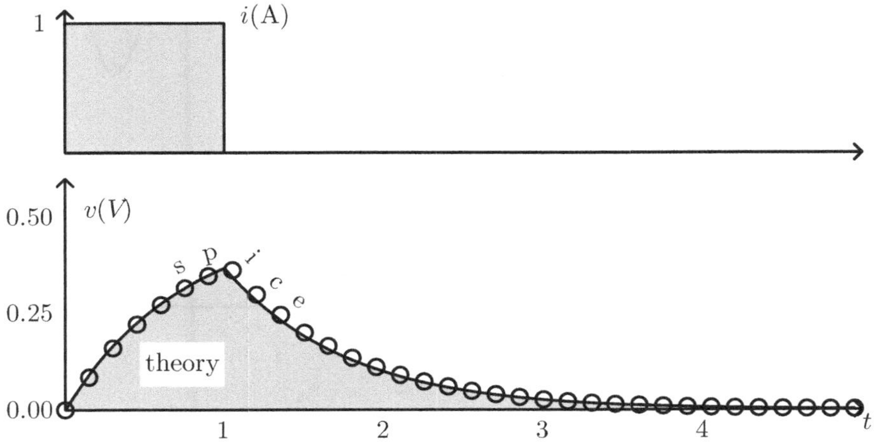

Fig. 31.4 Parallel *RC* pulse response ($R = 0.5\,\Omega$ and $C = 1.5\,\text{F}$)

Fig. 31.5 Parallel *LC* circuit

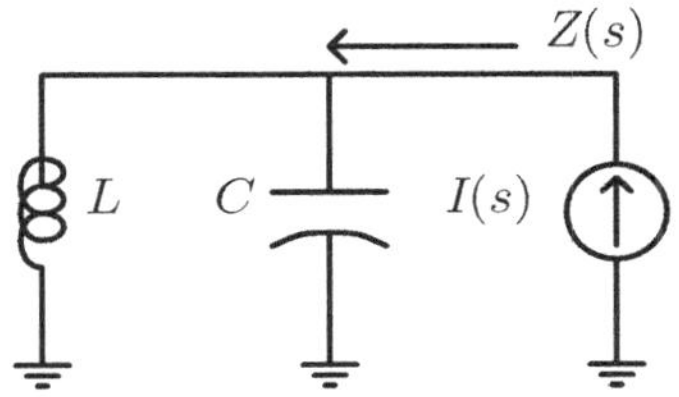

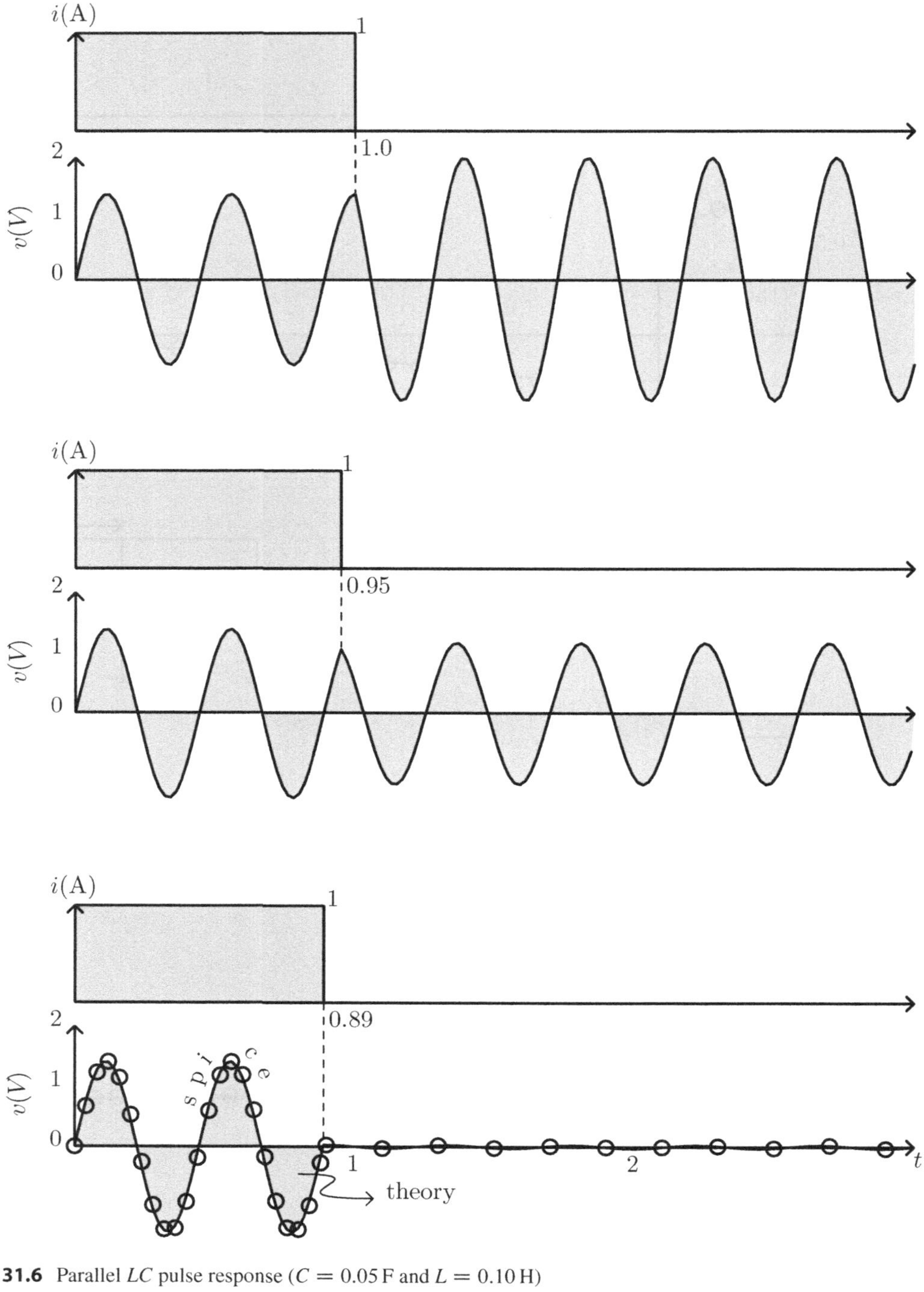

Fig. 31.6 Parallel LC pulse response ($C = 0.05\,\text{F}$ and $L = 0.10\,\text{H}$)

$$\omega_0 = \frac{1}{\sqrt{LC}} = \frac{1}{\sqrt{0.05 \times 0.1}} = \frac{1}{\sqrt{0.005}} = 14.142, \qquad \text{period} = \frac{2\pi}{\omega_0} = \frac{2\pi}{14.142} = 0.444 \tag{31.19}$$

It is interesting to notice that once the pulse width is an integer times the period just derived, the response dies out completely after the input current ceases. This has to do with lining up the falling edge just right with the rising part of the pulse response. What we have here is a resonant network; by lining the input to enforce the oscillations continuously we can cause the response to blow up (not shown here); and conversely if we line up the input to quench the response we can cause the oscillations to die (as is the case here).

31.6 Parallel *RLC* Circuit

The parallel *RLC* circuit is shown in Fig. 31.7. Our starting point is the transfer function derived in Eq. (26.47) and repeated below:

$$Z(s) = \frac{1}{C}\frac{s}{s^2 + \frac{1}{RC}s + \omega_{LC}^2}, \qquad \omega_{LC}^2 = \frac{1}{LC} \tag{31.20}$$

Define

$$a = \frac{1}{2RC}, \qquad \text{and} \qquad \omega_0^2 = \omega_{LC}^2 - a^2 \tag{31.21}$$

and complete the square to get

$$Z(s) = \frac{1}{C}\frac{s}{(s+a)^2 + \omega_0^2} \tag{31.22}$$

The *unit step* response is then

$$W(s) = \frac{1}{C}\frac{1}{(s+a)^2 + \omega_0^2} \tag{31.23}$$

And the *pulse* response follows:

$$\boxed{V(s) = \frac{1}{C}\frac{1}{(s+a)^2 + \omega_0^2}\left[1 - e^{-Ts}\right]} \tag{31.24}$$

Recalling the following transform pair

$$\frac{\omega_0}{(s+a)^2 + \omega_0^2} \to e^{-at}\sin\omega_0 t \tag{31.25}$$

and using the time shifting property we finally get the inverse transform of $V(s)$:

$$\boxed{v(t) = \frac{1}{C\omega_0}\left[u(t)e^{-at}\sin\omega_0 t - u(t-T)e^{-a(t-T)}\sin\omega_0(t-T)\right]} \tag{31.26}$$

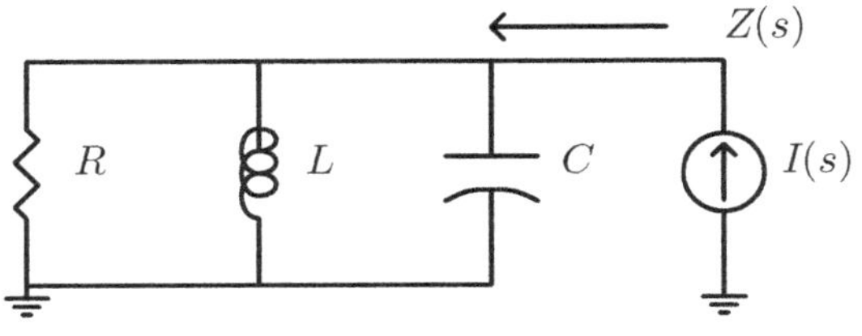

Fig. 31.7 Parallel *RLC* circuit

Our results and comparison to SPICE are shown in Fig. 31.8 for various pulse widths. Notice that in this case we have

$$a = \frac{1}{2RC} = \frac{1}{2 \times 2 \times 0.5} = \frac{1}{2}, \qquad \omega_{LC} = \frac{1}{\sqrt{LC}} = \frac{1}{\sqrt{0.5 \times 0.5}} = 2, \qquad \text{and}$$

$$\omega_0 = \sqrt{\omega_{LC}^2 - 1^2} = \sqrt{4 - 0.25} = \sqrt{3.75} \sim 2 \tag{31.27}$$

This then gives the expected oscillation period of around

$$\text{period of oscillation} = \frac{2\pi}{\omega_0} \sim \frac{2\pi}{2} = \pi \tag{31.28}$$

which is the observed value.

31.7 Network with Feedback

Consider the RC network with feedback shown in Fig. 31.9. The details to derive the transfer function will have to wait till we cover feedback networks and multi-source ones, in the later chapters. But suffices for now to say output voltage depends on two sources: v_{ref} and $I(s)$. For now we will take the former to be a constant, and its contribution on output is a constant too. Specifically

$$v(t) = v_1(t) + v_2(t) \tag{31.29}$$

where

$$v_1(t) = \frac{A}{A+1} \tag{31.30}$$

where A is the gain of the voltage-controlled voltage source.

Focusing now on $v_2(t)$ which is output voltage due to input current we derive the transfer function as

$$V_2(s) = \frac{1}{C}\frac{1}{s+a}, \qquad a = \frac{A+1}{RC} \tag{31.31}$$

The *unit step* response in the frequency domain is then

$$\begin{aligned} G_2(s) &= \frac{1}{sC}\frac{1}{s+a} = \frac{1}{C}\frac{1}{a}\left[\frac{1}{s} - \frac{1}{s+a}\right] \\ &= \frac{R}{A+1}\left[\frac{1}{s} - \frac{1}{s+a}\right] \end{aligned} \tag{31.32}$$

The *pulse* response is then

$$V_2(s) = \frac{R}{A+1}\left[\frac{1}{s} - \frac{1}{s+a}\right]\left[1 - e^{-Ts}\right] \tag{31.33}$$

or in the time domain

$$v_2(t) = \frac{R}{A+1}\left\{\left[u(t) - e^{-at}\right] - \left[u(t-T) - e^{-a(t-T)}\right]\right\} \tag{31.34}$$

So that finally after adding contribution from $v_1(t)$ we get total response

$$\boxed{v(t) = \frac{A}{A+1} + \frac{R}{A+1}\left\{\left[u(t) - e^{-at}\right] - \left[u(t-T) - e^{-a(t-T)}\right]\right\}} \tag{31.35}$$

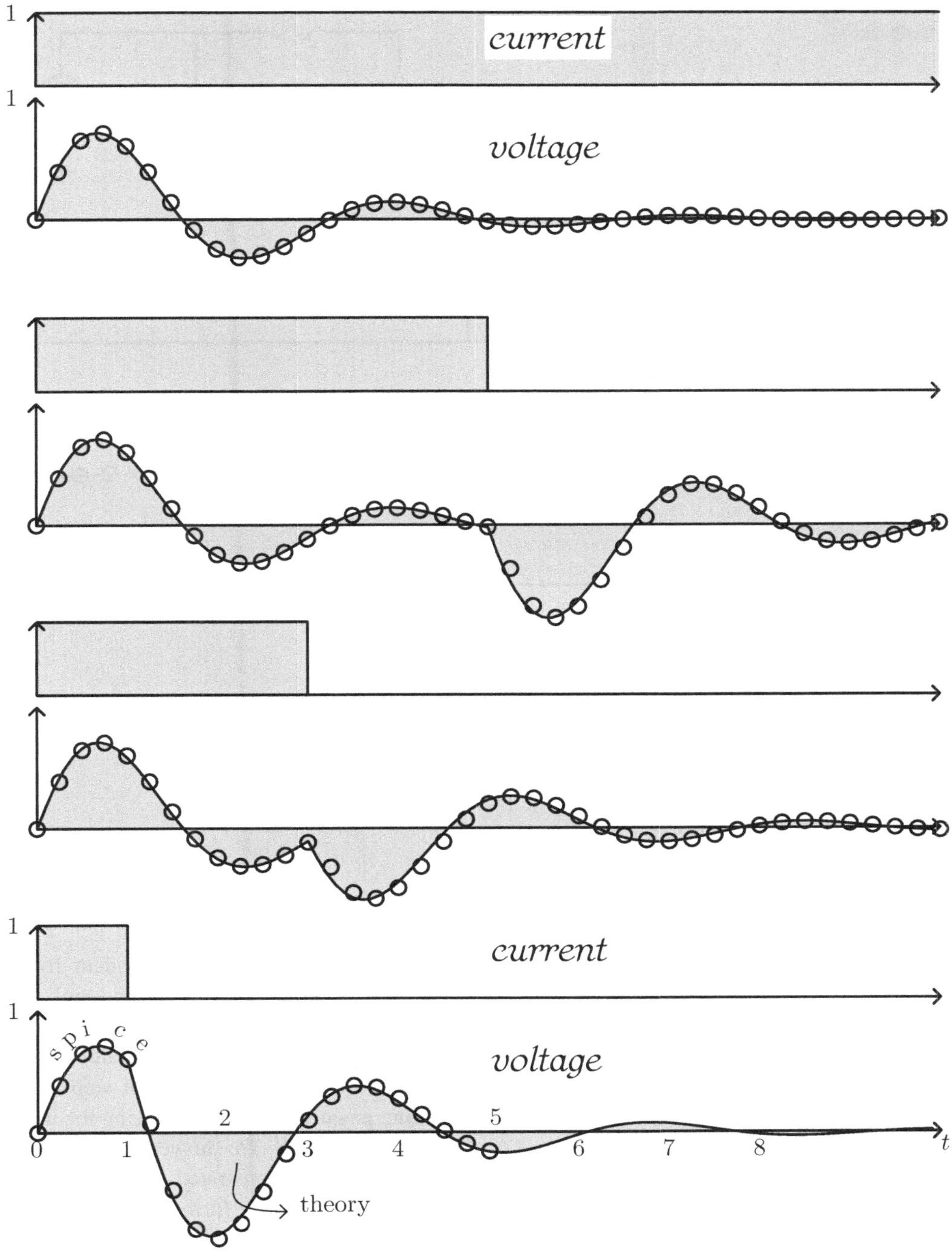

Fig. 31.8 Pulse response of parallel RLC circuit ($R = 2$, $C = 0.5$, and $L = 0.5$)

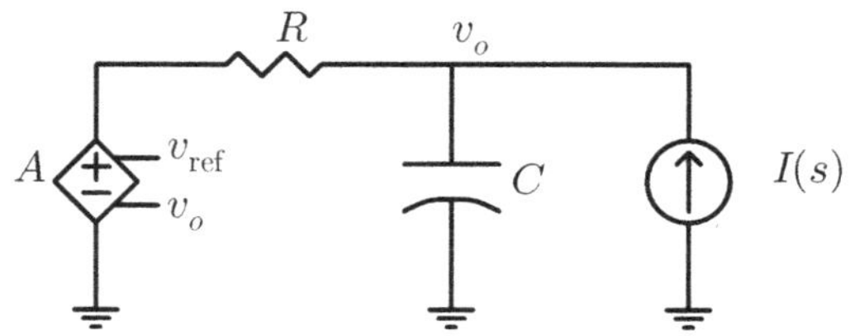

Fig. 31.9 *RC* network with feedback

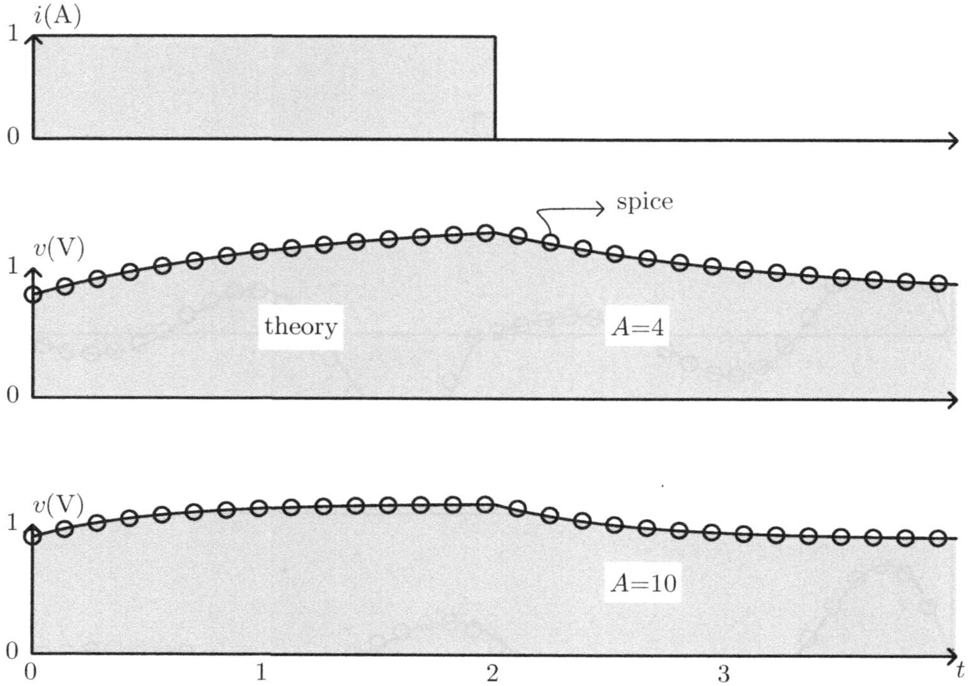

Fig. 31.10 Pulse response of network with feedback; case of $R = 3$ and $C = 2$

Notice that the starting and ending values are

$$v(0) = v(\infty) = \frac{A}{A+1} \tag{31.36}$$

Figure 31.10 shows results and comparison to SPICE for two different gain values.

31.8 Summary

Building on the impulse response and the unit step one we turned next in this chapter to the single-timer pulse response. We mention "single-timer" to discern this pulse from the conventional periodic pulse one, which will be treated on its own in a later chapter. As is evident from the treatment in this chapter we can think of the response to the one-timer pulse simply in terms of two responses: one due to the unit step, and the other due to the *shifted* unit step response (bearing negative sign). So knowing the unit step response says it all. We showed a few illustrative examples and demonstrated excellent match to SPICE results. Some further interesting topics such as visual analysis of the pulse spectrum and that of its response, deriving the impulse response from the pulse one, deriving the step response from the pulse one as well as numerically figuring the pulse response from a measured transfer function will be dealt with in the Problems section.

31.9 Problems

1. Plot the Laplace transform of both the unit step function and the pulse one (with width 1) for the two σ values: 0.3 and 1.0; what is the trend as σ becomes larger? Explain? See sample solution in Fig. 31.11
2. Plot the Laplace transform of both the unit step function and the pulse one for three pulse widths: 1, 2, and 5; what is the trend as the width becomes large? Explain? See sample solution in Fig. 31.12
3. A system has the transfer function below: figure the pulse response (with pulse width 1), both in time and in frequency; compare to impulse one. See sample solution in Fig. 31.13.

$$H(s) = \frac{1}{(s+1)(s+2)}$$

Answer:

$$f(t) = u(t)\left[0.5 - e^{-t} + \frac{1}{2}e^{-2t}\right] - u(t-1)\left[0.5 - e^{-(t-1)} + \frac{1}{2}e^{-2(t-1)}\right]$$

4. A system has the transfer function below: figure the pulse response (with pulse width 1), both in time and in frequency; compare to impulse one. See sample solution in Fig. 31.14.

$$H(s) = \frac{1}{s^2(s+1)}$$

Answer: let

$$g(t) = \frac{t^2}{2} - t + 1 - e^{-t}$$

Then

$$f(t) = u(t)g(t) - u(t-1)g(t-1)$$

5. Given system transfer function $H(s)$; what is the response (in frequency) due to two pulses, one of width 1 and the other 2, with second starting at $t = 3$, as shown in Fig. 31.15?
6. Assume that we are given the Laplace transform of a *pulse* response (not impulse one), $K(s)$. What steps in the frequency domain are

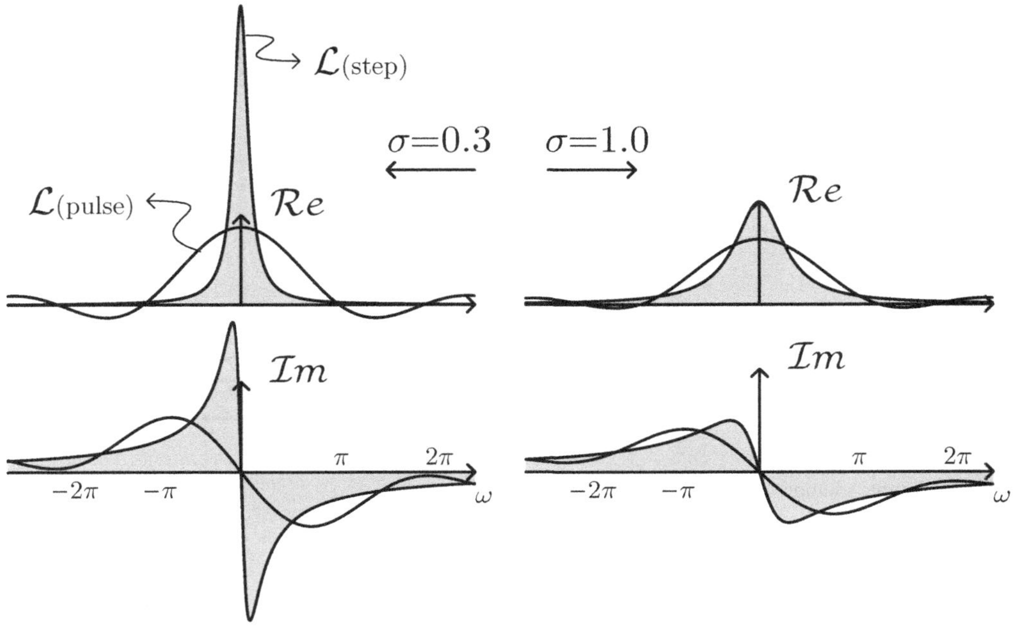

Fig. 31.11 Sample solution to Problem 1

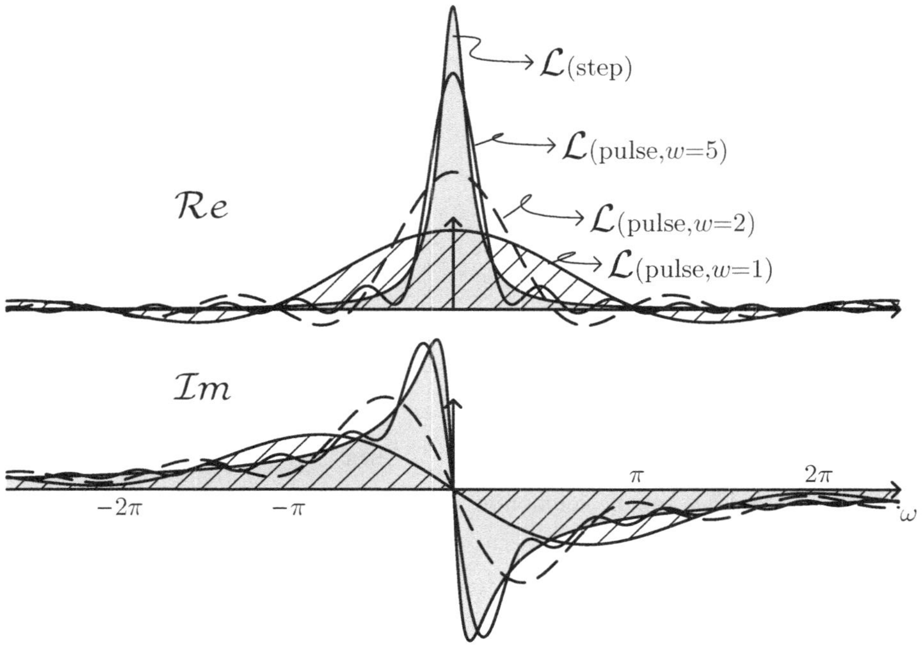

Fig. 31.12 Sample solution to Problem 2; case of $\sigma = 0.3$

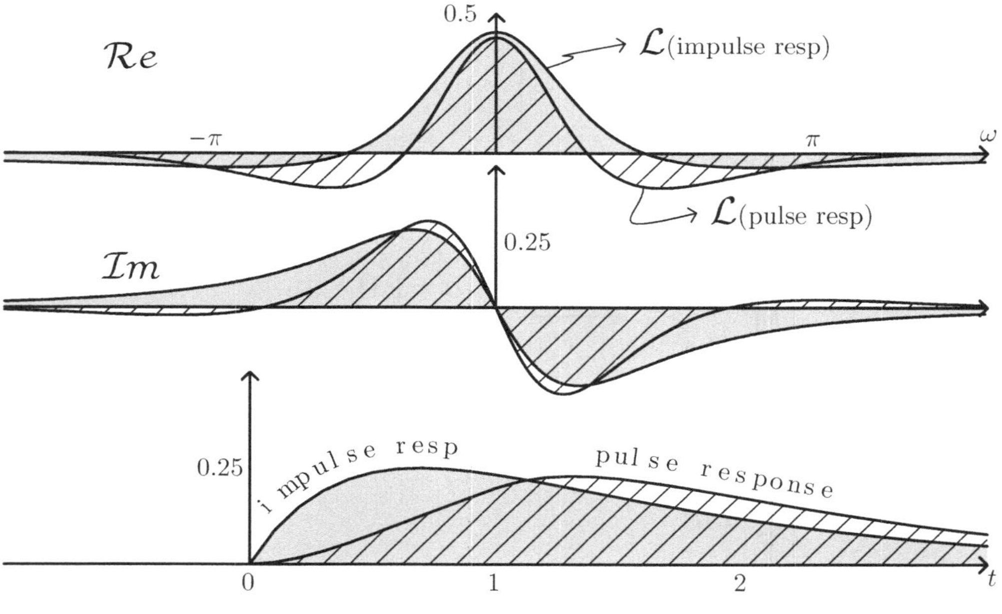

Fig. 31.13 Sample solution to Problem 3 ($\sigma = 0.1$)

needed to get the transfer function (Laplace transform of impulse response) $H(s)$ from $K(s)$. These frequency steps have counterparts in the time domain—what are they? Starting with a pulse of width 1, apply these time steps to the pulse and plot intermediate as well as final outcome—what is the final time signal? Does it make sense? See sample solution in Fig. 31.16.

7. What is the inverse Laplace transform of the following function?

$$H(s) = \frac{1}{1 - e^{-Ts}}$$

8. Assume we are given pulse response (of width 1) as tabulated below. Construct—in the time domain—the unit step response; see sample solution in Fig. 31.17.

t	f(t)	t	f(t)	t	f(t)
0.00	0.00	3.40	-0.39	6.80	0.18
0.10	0.05	3.50	-0.40	6.90	0.19
0.20	0.15	3.60	-0.42	7.00	0.19
0.30	0.25	3.70	-0.42	7.10	0.19
0.40	0.34	3.80	-0.43	7.20	0.19
0.50	0.44	3.90	-0.42	7.30	0.18
0.60	0.52	4.00	-0.42	7.40	0.18
0.70	0.61	4.10	-0.41	7.50	0.17
0.80	0.69	4.20	-0.40	7.60	0.16
0.90	0.76	4.30	-0.38	7.70	0.16
1.00	0.83	4.40	-0.36	7.80	0.15
1.10	0.84	4.50	-0.34	7.90	0.13
1.20	0.79	4.60	-0.32	8.00	0.12
1.30	0.74	4.70	-0.29	8.10	0.11
1.40	0.69	4.80	-0.26	8.20	0.10
1.50	0.63	4.90	-0.23	8.30	0.08
1.60	0.56	5.00	-0.20	8.40	0.07
1.70	0.50	5.10	-0.17	8.50	0.06
1.80	0.43	5.20	-0.14	8.60	0.04
1.90	0.36	5.30	-0.11	8.70	0.03
2.00	0.30	5.40	-0.08	8.80	0.02
2.10	0.23	5.50	-0.06	8.90	0.01
2.20	0.16	5.60	-0.03	9.00	-0.01
2.30	0.10	5.70	0.00	9.10	-0.02
2.40	0.03	5.80	0.03	9.20	-0.03
2.50	-0.03	5.90	0.05	9.30	-0.04
2.60	-0.08	6.00	0.07	9.40	-0.05
2.70	-0.14	6.10	0.09	9.50	-0.05
2.80	-0.19	6.20	0.11	9.60	-0.06
2.90	-0.23	6.30	0.13	9.70	-0.07
3.00	-0.27	6.40	0.14	9.80	-0.07
3.10	-0.31	6.50	0.16	9.90	-0.08
3.20	-0.34	6.60	0.17	10.00	-0.08
3.30	-0.37	6.70	0.18		

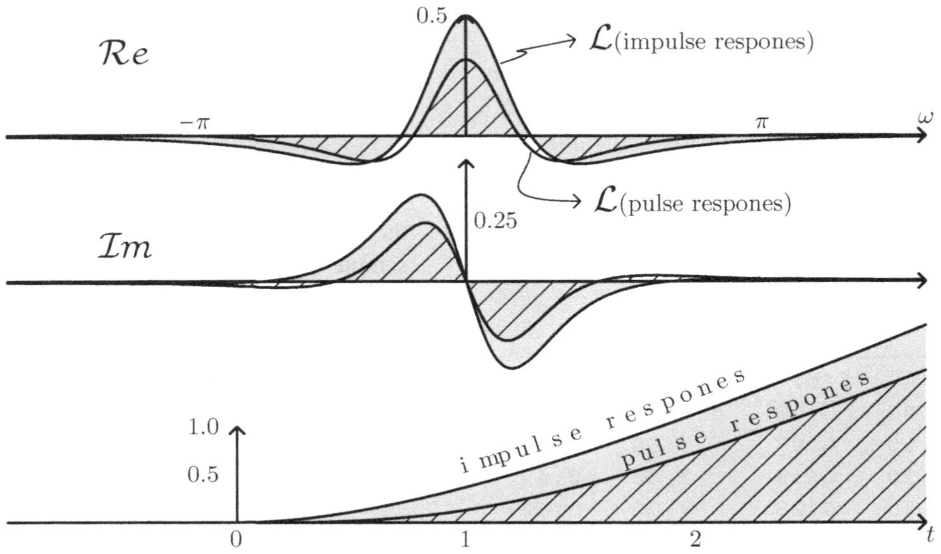

Fig. 31.14 Sample solution to Problem 4 ($\sigma = 1.0$)

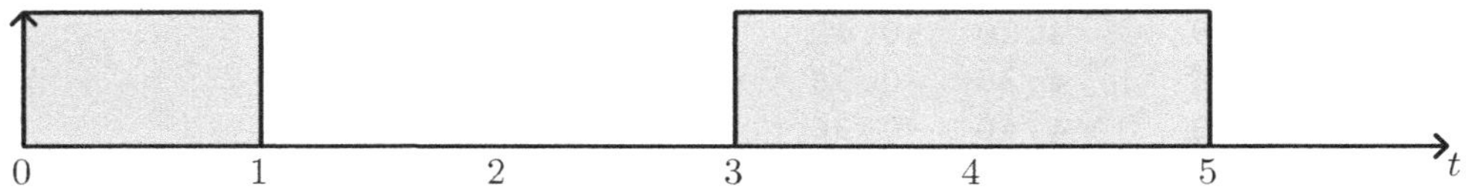

Fig. 31.15 Statement to Problem 5

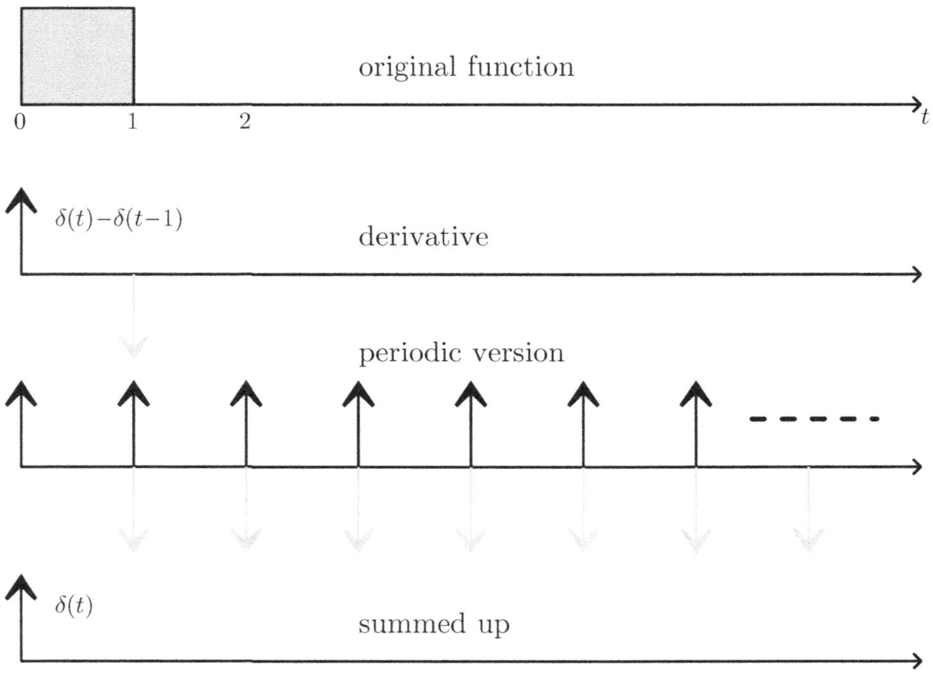

Fig. 31.16 Sample solution to Problem 6

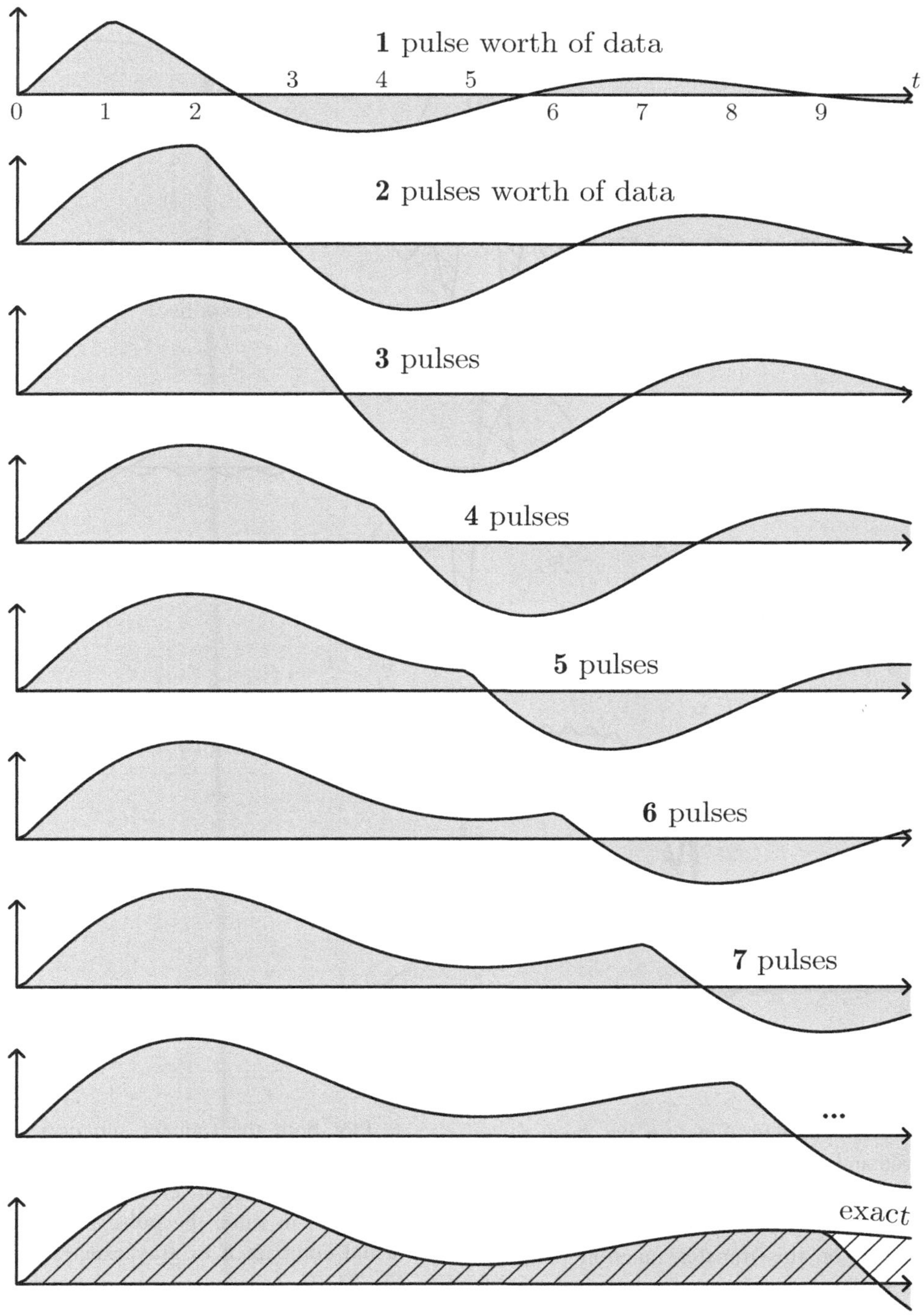

Fig. 31.17 Sample solution to Problem 8

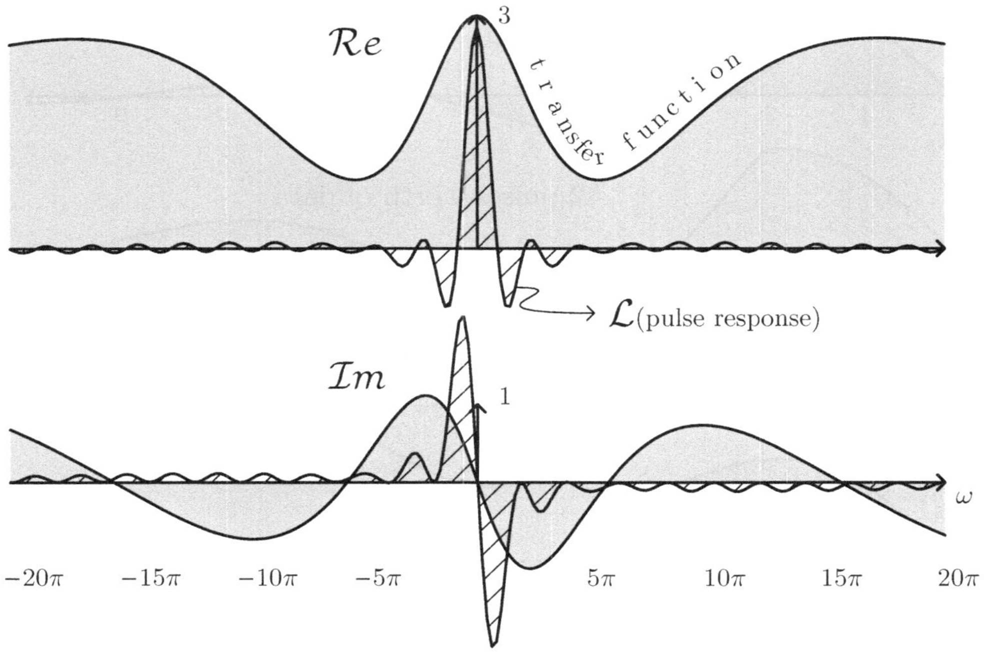

Fig. 31.18 Sample solution to Problem 9, part 1/2

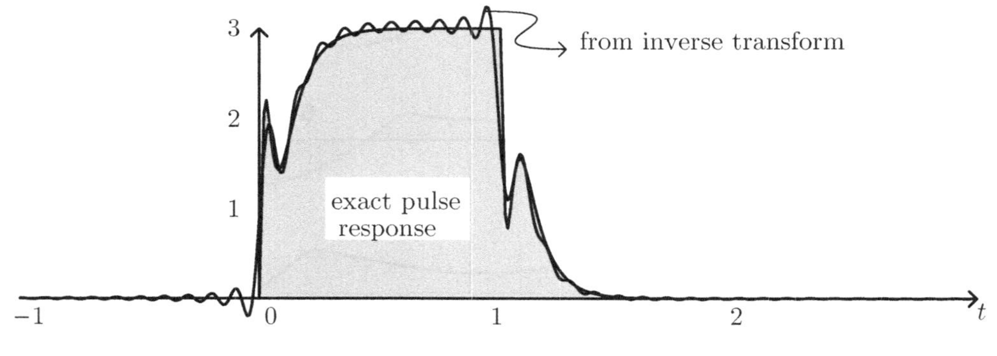

Fig. 31.19 Sample solution to Problem 9, part 2/2

9. A system transfer function has been measured and tabulated as shown below (0-σ assumption). The table shows frequency, real, and imaginary parts of the transfer function. Operate on this transfer function to figure the Laplace transform of a pulse of width 1. Plot both the transfer function and the Laplace transform, as shown in Fig. 31.18. Next, perform the inverse transform of $L(s)$ to figure, in time, the pulse response. Plot it and compare it to gray exact solution in Fig. 31.19. Use $\sigma = 0.1$ throughout the reconstruction.

f	R	I	f	R	I	f	R	I
0.00	3.00	0.00	3.40	1.07	0.39	6.80	2.52	0.33
0.10	2.99	-0.17	3.50	1.11	0.44	6.90	2.55	0.30
0.20	2.95	-0.33	3.60	1.15	0.48	7.00	2.57	0.26
0.30	2.89	-0.49	3.70	1.20	0.52	7.10	2.58	0.23
0.40	2.81	-0.63	3.80	1.25	0.56	7.20	2.60	0.20
0.50	2.71	-0.76	3.90	1.29	0.60	7.30	2.62	0.16
0.60	2.59	-0.87	4.00	1.34	0.63	7.40	2.63	0.13
0.70	2.46	-0.96	4.10	1.40	0.65	7.50	2.65	0.09
0.80	2.33	-1.02	4.20	1.45	0.67	7.60	2.66	0.06
0.90	2.19	-1.07	4.30	1.50	0.69	7.70	2.67	0.03
1.00	2.05	-1.10	4.40	1.55	0.71	7.80	2.68	-0.01
1.10	1.92	-1.11	4.50	1.61	0.72	7.90	2.69	-0.04
1.20	1.78	-1.11	4.60	1.66	0.73	8.00	2.69	-0.08
1.30	1.66	-1.08	4.70	1.71	0.73	8.10	2.70	-0.11
1.40	1.54	-1.05	4.80	1.76	0.73	8.20	2.70	-0.14
1.50	1.43	-1.00	4.90	1.81	0.73	8.30	2.71	-0.18
1.60	1.33	-0.94	5.00	1.86	0.73	8.40	2.71	-0.21
1.70	1.24	-0.88	5.10	1.91	0.72	8.50	2.71	-0.24
1.80	1.16	-0.81	5.20	1.96	0.71	8.60	2.71	-0.28
1.90	1.10	-0.73	5.30	2.00	0.70	8.70	2.71	-0.31
2.00	1.04	-0.65	5.40	2.05	0.69	8.80	2.71	-0.34
2.10	0.99	-0.57	5.50	2.09	0.67	8.90	2.71	-0.37
2.20	0.95	-0.48	5.60	2.14	0.65	9.00	2.70	-0.40
2.30	0.93	-0.40	5.70	2.18	0.63	9.10	2.70	-0.43
2.40	0.91	-0.31	5.80	2.22	0.61	9.20	2.70	-0.46
2.50	0.90	-0.23	5.90	2.25	0.59	9.30	2.69	-0.49
2.60	0.89	-0.15	6.00	2.29	0.56	9.40	2.69	-0.52
2.70	0.89	-0.07	6.10	2.33	0.54	9.50	2.68	-0.55
2.80	0.90	0.00	6.20	2.36	0.51	9.60	2.67	-0.57
2.90	0.92	0.08	6.30	2.39	0.48	9.70	2.66	-0.60
3.00	0.94	0.15	6.40	2.42	0.45	9.80	2.65	-0.63
3.10	0.97	0.21	6.50	2.45	0.42	9.90	2.65	-0.65
3.20	1.00	0.27	6.60	2.48	0.39	10.00	2.64	-0.68
3.30	1.03	0.33	6.70	2.50	0.36			

10. As of now we all know the Laplace transform of the pulse function, with pulse width T is

$$H(s) = \frac{1 - e^{-Ts}}{s}$$

Derive from this the Laplace transform of the unit step function, by taking the limit $T \to \infty$. Derive the LT of a pulse of zero width, by taking the limit $T \to 0$. Finally derive the LT of the *impulse* as pulse width $T \to 0$, *but pulse height* $\frac{1}{T}$ goes to ∞. Hint: for the last step need to keep at least two terms in the Taylor series expansion of e^x.

32 Causal Cosine and Sine Response

32.1 Introduction

So far we have dealt with the impulse, unit step, and pulse responses. The next important class of stimulus functions are the causal sine and cosine functions. It is important to distinguish between a *causal* sine function and a continuous one; the latter is defined for both negative and positive times, but the former is defined zero for negative times and sine for positive times. The causal sine and cosine function responses are of great importance for at least two reasons. The first is that they represent a periodic input stimulus, and the second is that if we know the response to a sine and a cosine input, and using the concepts from the Fourier series, then in theory we should be able to figure the response to any input stimulus (subject to the same convergent conditions of the Fourier series). While the causal sine and cosine share a lot in common, they differ in one main feature, which is the starting point. The causal sine is zero for negative time and zero right after time zero; that is at time zero it is still continuous. On the other hand, while the causal cosine is also zero for negative time it is one right after time zero; hence at time zero it is *non*-continuous and hence causes a discontinuity effect on the network. In other words the very initial response of a sine input will intrinsically differ from that of a cosine, albeit after thingssettle down the response of the two will have a lot of resemblance. Staying in line with the prior treatments we next analyze the causal sine/cosine response via a few illustrative examples.

32.2 Series *RC* Network

The series *RC* network is shown in Fig. 32.1

Our stimulus here is the causal cosine function

$$i(t) = u(t)\cos\omega_0 t \tag{32.1}$$

The Laplace transform of the input is

$$I(s) = \frac{s}{s^2 + \omega_0^2} \tag{32.2}$$

The voltage in frequency domain is then current times impedance:

$$\begin{aligned} V(s) = I(s)Z(s) &= \frac{s}{s^2+\omega_0^2}\left[R + \frac{1}{sC}\right] \\ &= V_1(s) + V_2(s) \end{aligned} \tag{32.3}$$

It is clear that the inverse transform of the resistive voltage is simply the causal cosine, scaled by *R*:

$$v_1(t) = Ru(t)\cos\omega_0 t \tag{32.4}$$

F. Badrieh, *Spectral, Convolution and Numerical Techniques in Circuit Theory*,
https://doi.org/10.1007/978-3-319-71437-0_32

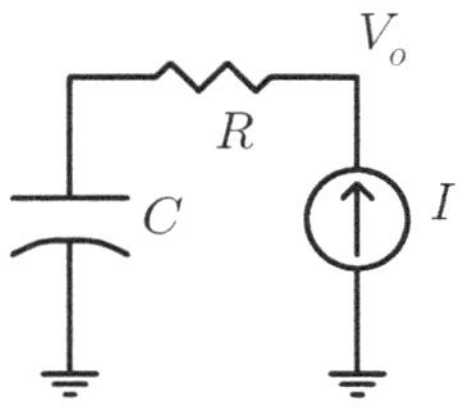

Fig. 32.1 Series *RC* network

The inverse transform of the capacitive voltage, on the other hand, is figured as follows:

$$v_2(t) = \mathcal{L}^{-1}\left[\frac{s}{s^2+\omega_0^2}\frac{1}{sC}\right] = \frac{1}{C}\frac{1}{s^2+\omega_0^2}$$

$$= \frac{1}{\omega_0 C}\frac{\omega_0}{s^2+\omega_0^2} = \frac{u(t)\sin\omega_0 t}{\omega_0 C} \quad (32.5)$$

Hence total voltage is

$$\boxed{v_o(t) = u(t)\left[R\cos\omega_0 t + \frac{1}{\omega_0 C}\sin\omega_0 t\right], \text{ cosine input current}} \quad (32.6)$$

The voltage across the resistor is simply a scaled version of the input current, while that across the cap is (a) phase shifted (from a cosine to a sine) and (b) scaled by $1/(\omega_0 C)$. The larger the cap, the smaller the voltage across it; same for applied frequency ω_0. Notice that in this case there is no *relaxation* behavior. Since the current is forced through each impedance component (R and C) the resulting voltage is simply the sum of both voltages, which do not individually get to behave in a relaxed matter! Figure 32.2 shows voltage across R, C, and output voltage. Notice that since cap is small (or equivalently high impedance) output voltage is biased towards cap voltage. As we increase the cap (or equivalently decrease impedance), and as shown in Fig. 32.3, cap voltage goes down and output voltage collapses to that across R.

32.3 Parallel *RC* Network

The parallel RC network is shown in Fig. 32.4 The impedance transfer function is

$$Z(s) = \frac{1}{C}\frac{1}{s+1/RC} \quad (32.7)$$

Our stimulus here is the causal cosine function

$$i(t) = u(t)\cos\omega_0 t \quad (32.8)$$

The Laplace transform of the input is

$$I(s) = \frac{s}{s^2+\omega_0^2} \quad (32.9)$$

Output voltage is then

$$V(s) = \frac{1}{C}\frac{1}{s+a}\frac{s}{s^2+\omega_0^2}, \qquad a = \frac{1}{RC} \quad (32.10)$$

Assume we can write the voltage transfer function as

$$V(s) = \frac{1}{C}\left[\frac{A}{s+a} + \frac{Bs+D}{s^2+\omega_0^2}\right] \quad (32.11)$$

Expand to get

$$V(s) = \frac{1}{C}\left[\frac{A(s^2+\omega_0^2)+(s+a)(D+sB)}{(s+a)(s^2+\omega_0^2)}\right]$$

$$= \frac{1}{C}\left[\frac{s^2(A+B)+s(D+aB)+(A\omega_0^2+aD)}{(s+a)(s^2+\omega_0^2)}\right] \quad (32.12)$$

Equating constant terms we get

$$aD + A\omega_0^2 = 0 \qquad \Rightarrow D = -\frac{A\omega_0^2}{a} \quad (32.13)$$

Equating s^2 terms we get

$$A + B = 0 \qquad \Rightarrow B = -A \quad (32.14)$$

Finally equating s terms we get

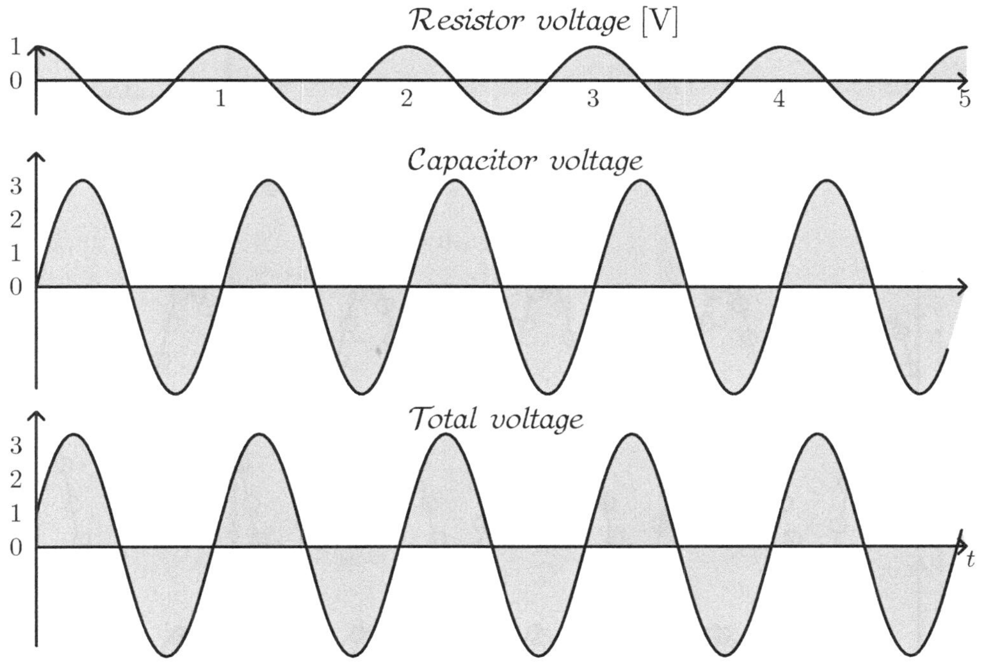

Fig. 32.2 Causal cosine response of series *RC* network ($R = 1\,\Omega$ and $C = 0.05\,\text{F}$)

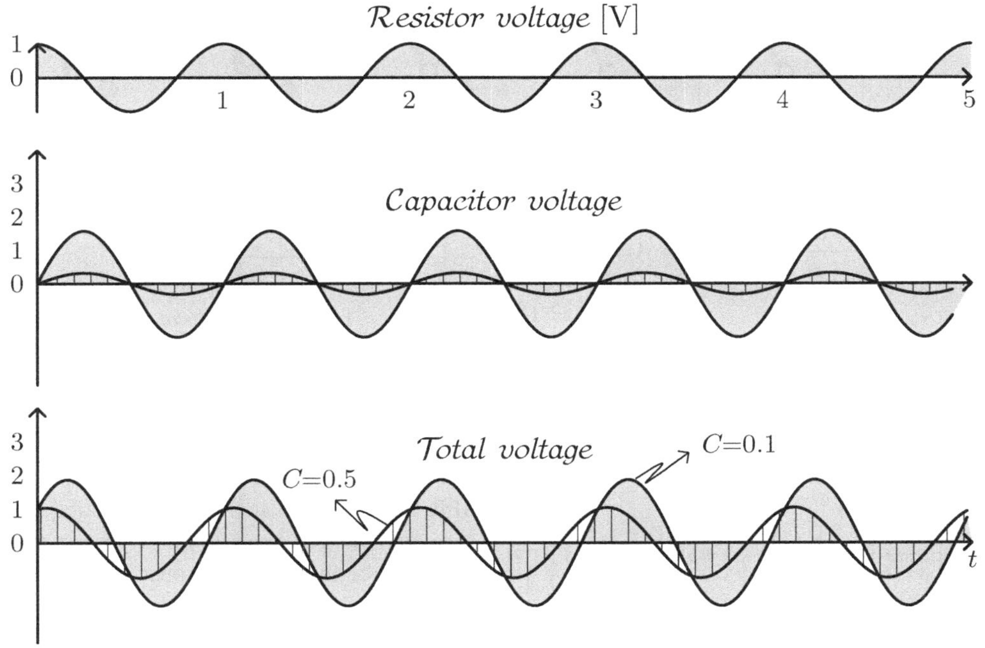

Fig. 32.3 Causal cosine response of series *RC* network ($R = 1\,\Omega$ and $C = 0.1$ and $0.5\,\text{F}$)

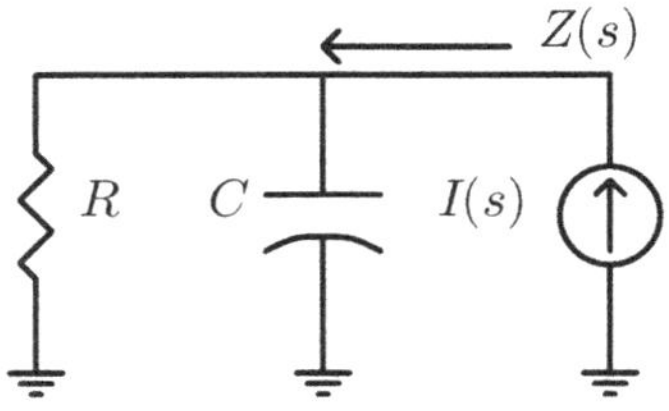

Fig. 32.4 Parallel *RC* network

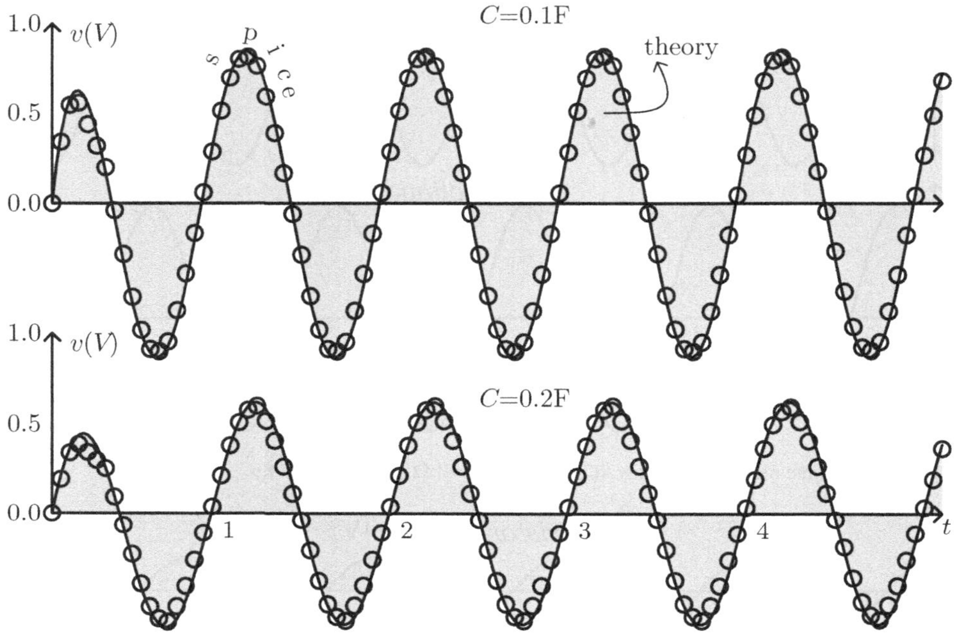

Fig. 32.5 Parallel *RC* network response to causal cosine ($\omega_0 = 2\pi$, $R = 1\,\Omega$, and $C = 0.1$ and 0.2 F)

$$D+aB=1 \Rightarrow -\frac{A\omega_0^2}{a}-aA=1 \Rightarrow A=\frac{-a}{a^2+\omega_0^2} \tag{32.15}$$

Then output voltage can then now be written as

$$\boxed{V(s)=\frac{1}{C}\frac{1}{a^2+\omega_0^2}\left\{-a\left[\frac{1}{s+a}\right]+a\left[\frac{s}{s^2+\omega_0^2}\right]+\omega_0^2\left[\frac{1}{s^2+\omega_0^2}\right]\right\}} \tag{32.16}$$

Taking the inverse transform we finally get

$$\boxed{v(t)=\frac{1}{C}\frac{1}{a^2+\omega_0^2}\left[-ae^{-at}+a\cos\omega_0 t+\omega_0\sin\omega_0 t\right]},$$

$$a=\frac{1}{RC} \tag{32.17}$$

A couple sample runs and comparison to SPICE are shown in Fig. 32.5. Notice that the case with larger cap (a) takes longer to settle down, since the *RC* time constant is larger; and (b) the settling peak-to-peak value is smaller since the impedance of the cap is now smaller and so is total parallel impedance.

32.4 Parallel *RC* in Series with *R*

The parallel *RC* in series with *R* is shown in Fig. 32.6. The impedance transfer function is

$$Z(s) = R_2 + \frac{1}{C}\frac{1}{s+1/\tau}, \tau = R_1 C \quad (32.18)$$

In this case we apply a causal sine input and look at both the transient and steady state response

$$i(t) = \sin \omega_0 t \quad (32.19)$$

Our stimulus has the LT of

$$I(s) = \frac{\omega_0}{s^2+\omega_0^2} \quad (32.20)$$

Our *IV* product (in frequency space) is then

$$V(s) = \left[R_2 + \frac{1}{C}\frac{1}{s+1/\tau}\right]\frac{\omega_0}{s^2+\omega_0^2} \quad (32.21)$$

We will use the following formula for partial fraction expansion

$$\frac{1}{(a+s)(s^2+\omega_0^2)} = \frac{1}{a^2+\omega_0^2}\left[\frac{1}{a+s} + \frac{-s+a}{s^2+\omega_0^2}\right] \quad (32.22)$$

Then our voltage becomes

$$\boxed{\begin{aligned} V(s) &= R_2\frac{\omega_0}{s^2+\omega_0^2} \\ &+ \frac{\omega_0}{C}\frac{1}{1/\tau^2+\omega_0^2}\left[\frac{1}{1/\tau+s} + \frac{-s+1/\tau}{s^2+\omega_0^2}\right] \end{aligned}} \quad (32.23)$$

Our voltage in time domain becomes

$$\boxed{\begin{aligned} v(t) &= R_2 u(t)\sin\omega_0 t + \frac{\omega_0}{C}\frac{1}{1/\tau^2+\omega_0^2}u(t) \\ &\left[e^{-t/\tau} - \cos\omega_0 t + \frac{1/\tau}{\omega_0}\sin\omega_0 t\right], \\ \tau &= R_1 C \end{aligned}} \quad (32.24)$$

To test this we run SPICE simulations for a couple of cases as shown in Fig. 32.7. In both cases the cap is set at 0.5 F. As can be seen from both cases we are able to predict both the *transient* response and *steady state* one. In other words, we are able to predict the *total solution* and remedy the gap assumed in Chap. 2 which is able to only predict the steady state response using phasors. We have the general solution!

32.5 Series *RC*/Parallel *R* Circuit

The series *RC*, parallel *R* is shown in Fig. 32.8. The impedance transfer function previously derived in Eq. (26.61) is

$$Z(s) = \boxed{\frac{R_2 + sR_1R_2C}{1+sC(R_1+R_2)}} \quad (32.25)$$

Notice that at DC impedance is $Z(0) \sim R_2$ and that at high frequency $Z(\infty) \sim \frac{R_1R_2}{R_1+R_2}$ which is the parallel combination of both resistors.

Our sinusoidal input has the transfer function

$$I(s) = \frac{\omega_0}{s^2+\omega_0^2} \quad (32.26)$$

The voltage (both transient and steady state) is derived from the *IZ* product in the frequency domain

$$V(s) = \frac{R_2 + ACs}{1+BCs}\frac{\omega_0}{s^2+\omega_0^2} \quad (32.27)$$

where we have defined $A = R_1R_2$ and $B = (R_1 + R_2)$ to simplify following algebra. Let's pull out the *BC* term from the denominator and rewrite as

$$\begin{aligned} V(s) &= \frac{\omega_0}{BC}\frac{R_2+ACs}{1/BC+s}\frac{1}{s^2+\omega_0^2} \\ &= D\omega_0\frac{R_2+ACs}{D+s}\frac{1}{s^2+\omega_0^2}, \\ D &= \frac{1}{BC} \end{aligned} \quad (32.28)$$

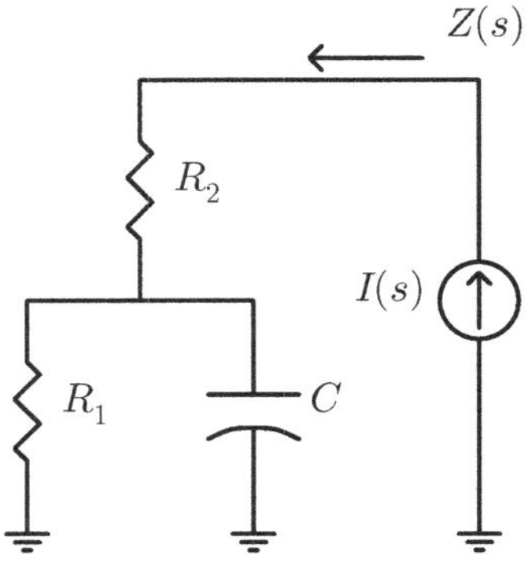

Fig. 32.6 Parallel RC in series with R network

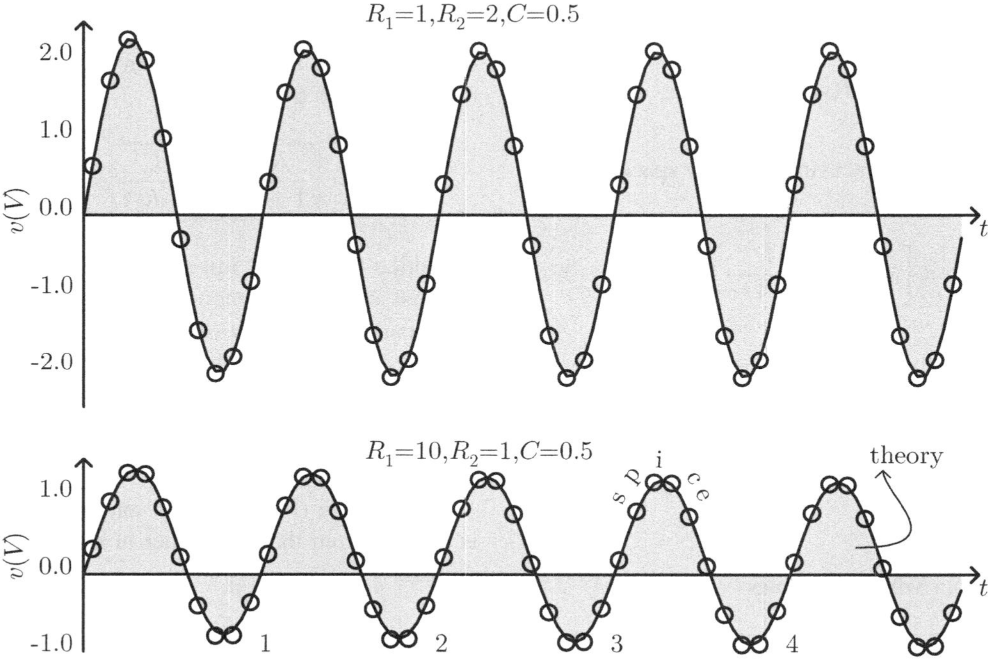

Fig. 32.7 Causal sine response of parallel RC, series R for different R_1 and R_2 values; case of $\omega_0 = 2\pi$

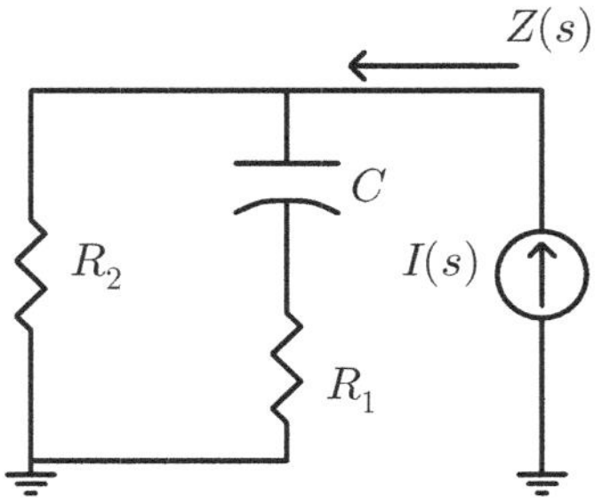

Fig. 32.8 Series RC in parallel with R and input impedance

We will use partial fractions to reduce this to something more manageable. Assume we can write the expression as

$$V(s) = D\omega_0 \left[\frac{\alpha}{D+s} + \frac{\beta + \delta s}{s^2 + \omega_0^2} \right] \quad (32.29)$$

Expanding we get

$$V(s) = D\omega_0 \left[\frac{\alpha(s^2+\omega_0^2)+(D+s)(\beta+\delta s)}{(D+s)(s^2+\omega_0^2)} \right]$$

$$= D\omega_0 \left[\frac{(\alpha\omega_0^2+D\beta)+(D\delta+\beta)s+(\alpha+\delta)s^2}{(D+s)(s^2+\omega_0^2)} \right] \quad (32.30)$$

Equating second order coefficients we get

$$\alpha + \delta = 0, \qquad \Rightarrow \alpha = -\delta \quad (32.31)$$

Equating zero order coefficients we get

$$\alpha\omega_0^2 + D\beta = R_2, \Rightarrow \qquad \beta = \frac{R_2 + \delta\omega_0^2}{D} \quad (32.32)$$

Equating first order coefficients we get

$$D\delta + \beta = AC \quad (32.33)$$

If we put in the value of β we get

$$D\delta + \frac{R_2 + \delta\omega_0^2}{D} = AC \quad (32.34)$$

Collect terms

$$\delta\left(D + \frac{\omega_0^2}{D}\right) = AC - \frac{R_2}{D}$$

$$\delta\frac{D^2 + \omega_0^2}{D} = \frac{ACD - R_2}{D} \quad (32.35)$$

which gives

$$\delta = \frac{ACD - R_2}{D^2 + \omega_0^2} \quad (32.36)$$

We back substitute this to find β:

$$\beta = \frac{R_2}{D} + \frac{\omega_0^2}{D}\frac{ACD - R_2}{D^2 + \omega_0^2} \quad (32.37)$$

and to find α:

$$\alpha = \frac{-ACD + R_2}{D^2 + \omega_0^2} \quad (32.38)$$

Back to our original expansion assumption, repeated here for convenience:

$$V(s) = D\omega_0 \left[\frac{\alpha}{D+s} + \frac{\beta + \delta s}{s^2 + \omega_0^2} \right] \quad (32.39)$$

The inverse transform of this is

$$\boxed{v(t) = D\omega_0 \left[\alpha e^{-Dt} + \frac{\beta}{\omega_0}\sin\omega_0 t + \delta\cos\omega_0 t \right]} \quad (32.40)$$

Results for a couple of *RC* cases alongside comparison to SPICE are shown in Fig. 32.9.

32.6 Series *RL* Network

The series *RL* network is shown in Fig. 32.10.

The impedance transfer function is

$$Z(s) = R + sL \quad (32.41)$$

Input cosine current has transfer function

$$I(s) = \frac{s}{s^2 + \omega_0^2} \quad (32.42)$$

Our voltage in the frequency domain is then simply

$$\boxed{V(s) = \frac{s(R + sL)}{s^2 + \omega_0^2}} \quad (32.43)$$

When we do the inverse transform we can tell straight away that the voltage across the resistor is going to be

$$v_r = \mathcal{L}^{-1}\left[R\frac{s}{s^2 + \omega_0^2} \right] = Ru(t)\cos\omega_0 t \quad (32.44)$$

The voltage across the inductor is going to be the inverse LT as follows:

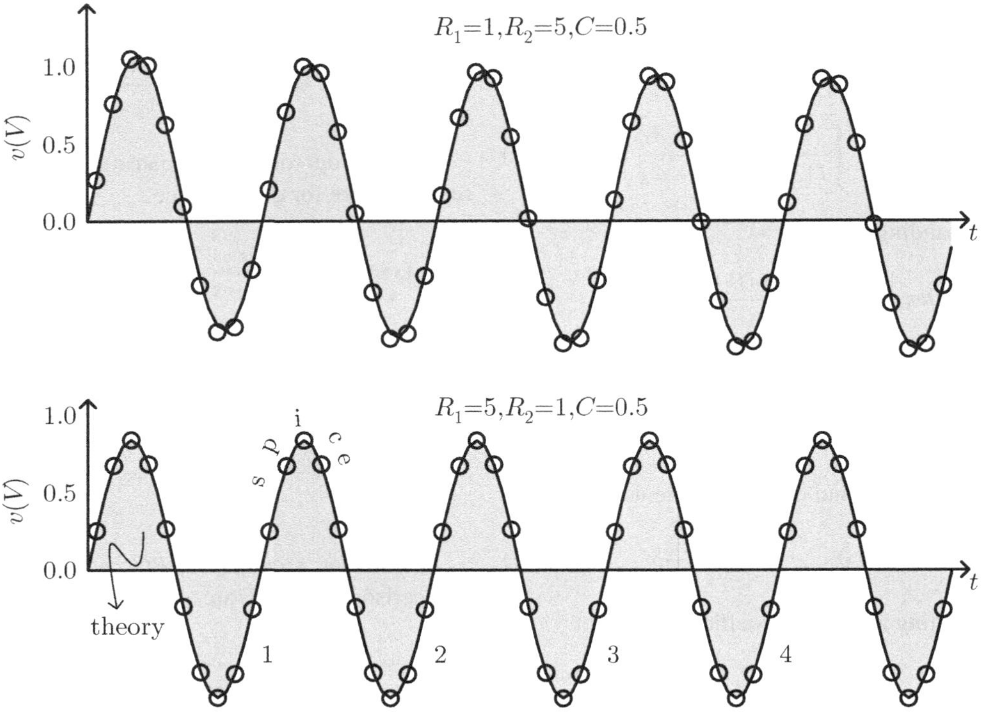

Fig. 32.9 Series RC/parallel R response to sine input; case of $\omega_0 = 2\pi$

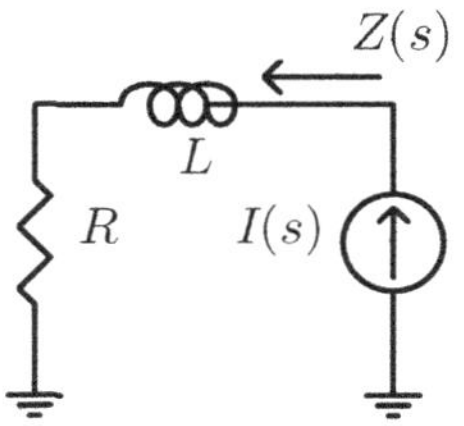

Fig. 32.10 Series RL network

$$v_L(s) = \mathcal{L}^{-1}\frac{Ls^2}{s^2+\omega_0^2} \tag{32.45}$$

We can find the inverse LT of this using at least two methods:

First Inversion Method Using Time Differentiation Property We can rewrite the LT as

$$V_L(s) = Ls\left[\frac{s}{s^2+\omega_0^2}\right] \tag{32.46}$$

Notice that the term in bracket has inverse transform of cosine. The term multiplying the bracket (other than a constant of L) is simply s. We recall that time differentiation property:

$$\mathcal{L}\frac{df(t)}{dt} = sF(s) - f(0+) \tag{32.47}$$

or rewritten in a different form

$$\mathcal{L}\left[\frac{df(t)}{dt} + f(0+)\delta(t)\right] = sF(s) \tag{32.48}$$

In our case $f(0+) = 1$ which would imply

$$\mathcal{L}^{-1}sF(s) = \delta(t) + \frac{d}{dt}\cos\omega_0 t \tag{32.49}$$

That is

$$v_L(t) = L\left[\delta(t) - \omega_0\sin\omega_0 t\right] \tag{32.50}$$

Notice that very important result of having a delta (impulse function) in the response. True the input is a cosine, but it is a causal one. That is, before time zero it was zero, and only after time zero did it jump to a cosine function. It is this jump in input current that results in an impulse in output voltage, due to the di/dt across the inductor. Furthermore, notice that the steady state response of voltage across the inductor (the sine term) is nothing other than the derivative of the current!

Second Inversion Method Using Long Division We rewrite our frequency function as

$$V_L(s) = L\frac{s^2}{s^2+\omega_0^2} \tag{32.51}$$

Using long division

$$\begin{array}{r|l} & 1 \\ \hline s^2+\omega_0^2 & s^2 \\ & -s^2-\omega_0^2 \\ \hline & -\omega_0^2 \end{array} \tag{32.52}$$

we get

$$V_L(s) = L\left[1-\frac{\omega_0^2}{s^2+\omega_0^2}\right] \tag{32.53}$$

The inverse LT of this gives us Eq. (32.50)

Total Response When we add resistor voltage (Eq. (32.44)) to inductor one (Eq. (32.50)) we finally get the total voltage response due to the causal input cosine current:

$$\boxed{v(t) = Ru(t)\cos\omega_0 t + L\left[\delta(t)-\omega_0\sin\omega_0 t\right]} \tag{32.54}$$

Notice that the steady state version of this is

$$\lim_{t\to\infty} v(t) = Ru(t)\cos\omega_0 t - L\omega_0\sin\omega_0 t \tag{32.55}$$

which is what we would expect if steady state analysis is used. Notice that what distinguishes the steady state response from the full response—at least in this case—is that delta (impulse) function. This transient (lapsed) part of the response is a "one-timer," and once it passes, we fall back on the steady state response, which is typically either periodic functions (sines, cosines, ...) or DC values (e.g., constant voltage).

32.7 Parallel *RL* Network

The parallel *RL* network is shown in Fig. 32.11. The impedance transfer function is

$$Z(s) = R\frac{s}{s+\frac{R}{L}} \tag{32.56}$$

Input (causal) cosine has transform

$$\cos(\omega_0 t) \to \frac{s}{s^2+\omega_0^2} \tag{32.57}$$

The voltage in frequency would then be

$$\begin{aligned} V(s) &= \frac{Rs}{s+\frac{R}{L}}\frac{s}{s^2+\omega_0^2} \\ &= \boxed{R\frac{s^2}{(a+s)(s^2+\omega_0^2)}, \qquad a=\frac{R}{L}} \end{aligned} \tag{32.58}$$

To find the inverse LT we will use partial fraction. Assume that the LT can be written as

$$V(s)/R = \frac{A}{a+s} + \frac{Bs+C}{s^2+\omega_0^2} \tag{32.59}$$

Adding the right side we get

$$V(s)/R = \frac{A(s^2+\omega_0^2)+(a+s)(Bs+C)}{(a+s)(s^2+\omega_0^2)}$$

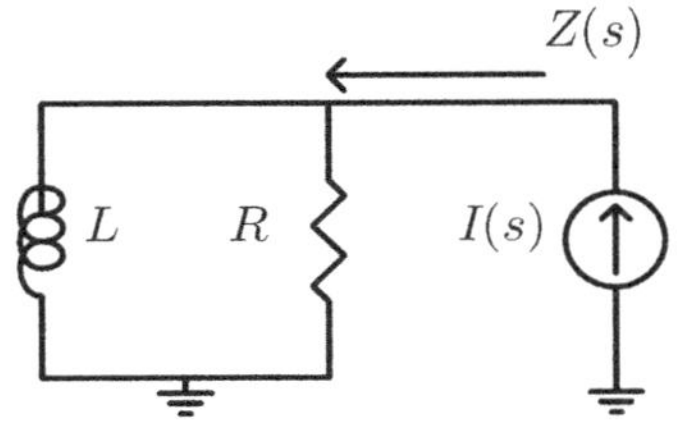

Fig. 32.11 Parallel *RL* circuit

$$= \frac{s^2(A+B) + s(aB+C) + A\omega_0^2 + aC}{(a+s)(s^2+\omega_0^2)} \tag{32.60}$$

Equating the s coefficient we get

$$C = -aB \tag{32.61}$$

Our voltage then becomes

$$V(s) = \frac{s^2(A+B) + A\omega_0^2 - a^2B}{(a+s)(s^2+\omega_0^2)} \tag{32.62}$$

Equating s^0 term we get

$$A\omega_0^2 = a^2B, \quad \text{or} \quad B = \frac{A\omega_0^2}{a^2} \tag{32.63}$$

Finally equating the s^2 term we get

$$A = 1 - B = 1 - A\frac{\omega_0^2}{a^2} \tag{32.64}$$

which gives

$$A = \frac{1}{1+\frac{\omega_0^2}{a^2}} = \frac{a^2}{a^2+\omega_0^2} \tag{32.65}$$

Plugging back for B we get

$$B = \frac{\omega_0^2}{a^2}\frac{a^2}{a^2+\omega_0^2} = \frac{\omega_0^2}{a^2+\omega_0^2} \tag{32.66}$$

and plugging back for C we get

$$C = -a\frac{\omega_0^2}{a^2+\omega_0^2} \tag{32.67}$$

Our voltage finally comes out as

$$\boxed{V(s) = R\frac{a^2}{a^2+\omega_0^2}\left[\frac{1}{a+s}\right] + R\frac{\omega_0^2}{a^2+\omega_0^2}\left[\frac{s}{s^2+\omega_0^2}\right] - R\frac{a\omega_0^2}{a^2+\omega_0^2}\left[\frac{1}{s^2+\omega_0^2}\right]} \tag{32.68}$$

The inverse transform is then

$$\boxed{v(t) = \frac{R}{a^2+\omega_0^2}\left[a^2e^{-at} + \omega_0^2\cos\omega_0 t - a\omega_0\sin\omega_0 t\right]}, \quad a = \frac{R}{L} \tag{32.69}$$

Notice that at time zero the voltage comes out $v(0) = R$ since at high frequency (point of current application) the inductor acts open and all the current (1 here) goes through the resistor. Notice also that if L is very large such that $a \to 0$ then voltage collapses to $R\cos\omega_0 t$ which is nothing other than input current times R (since the inductor is open for all time). Finally notice that if L is very small, such that $a \to \infty$, then voltage collapses to $-R\frac{\omega_0}{a}\sin\omega_0 t$ which comes out $-L\omega_0\sin\omega_0 t$ which is simply L times the derivative of input current, since in this case all current goes through the inductor (it being very small). Results and comparison to SPICE are shown in Fig. 32.12.

32.8 Generic *RLC* Network

The *RLC* network in Fig. 32.13 has an impedance transfer function shown in Fig. 32.14; the impedance transfer function can be approximated by

$$Z(s) = 10000\frac{[1+0.1s][(s+10)^2+70^2]}{[(s+6)^2+20^2][(s+10)^2+330^2]} \tag{32.70}$$

If we apply a sine input, say $\sin 30t$, such that input current is

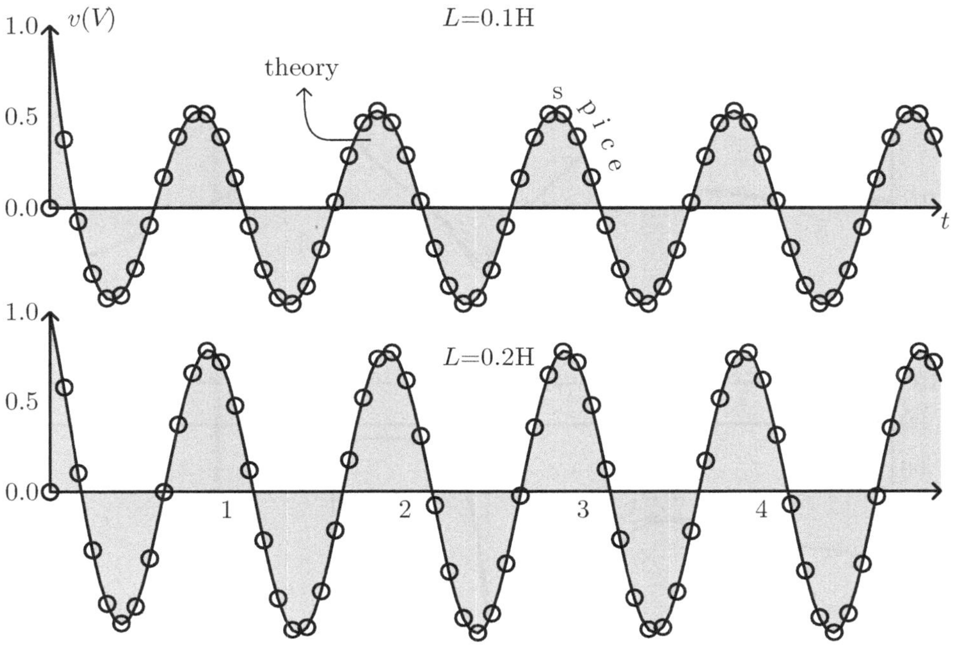

Fig. 32.12 RL circuit response to cosine current demand (case of $\omega_0 = 2\pi$ and $R = 1\ \Omega$)

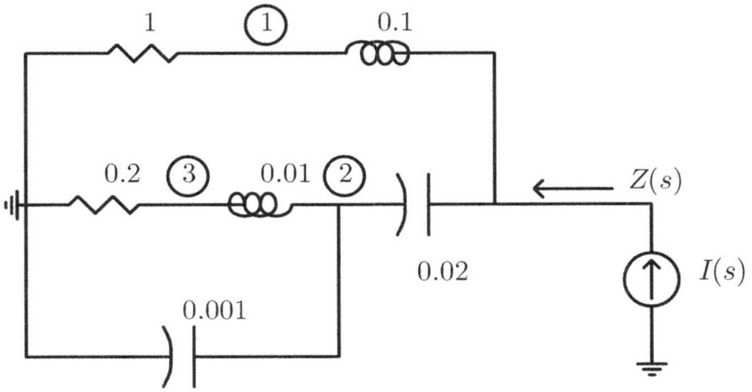

Fig. 32.13 Generic *RLC* network

$$I(s) = \frac{30}{s^2 + 30^2} \tag{32.71}$$

then output voltage would be

$$V(s) = 10000\frac{[1+0.1s][(s+10)^2 + 70^2]}{[(s+6)^2+20^2][(s+10)^2+330^2]} \times \frac{30}{s^2+30^2} \tag{32.72}$$

We can next use partial fractions such that

$$V(s) = \frac{A}{s+6+j20} + \frac{B}{s+6-j20} + \frac{C}{s+10+j330} + \frac{D}{s+10-j330} + \frac{E}{s+j30} + \frac{F}{s-j30} \tag{32.73}$$

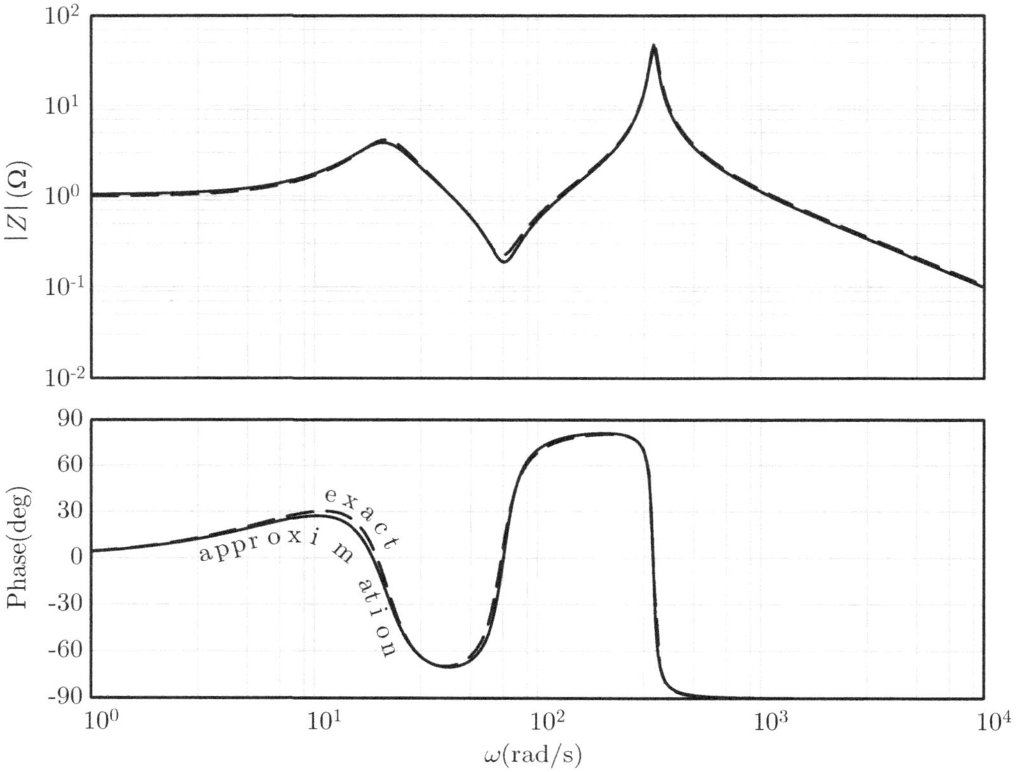

Fig. 32.14 Impedance transfer function of generic *RLC* network in Fig. 32.13

We can find the values of A, B, etc. via partial fractions. The solution in time is then

$$
\begin{aligned}
v(t) = {} & Ae^{-(6+j20)t} + Be^{-(6-j20)t} \\
& + Ce^{-(10+j330)t} + De^{-(10-j330)t} \\
& + Ee^{-j30t} + Fe^{+j30t}
\end{aligned} \tag{32.74}
$$

Plugging in for A, B, etc. and plotting we find that (no surprise) the solution is completely real, as expected. The output voltage, both in frequency and time, is shown in Fig. 32.15.

32.9 Summary

What could be considered more important than the response to causal sines and cosines? These two signals are so flexible and they can represent (via the Fourier series) mountains of other periodic functions! In their response we capture both the transient and the steady state parts of the solution. By varying the application frequency and by varying the split between sine/cosine we can encompass many other functions. So really once

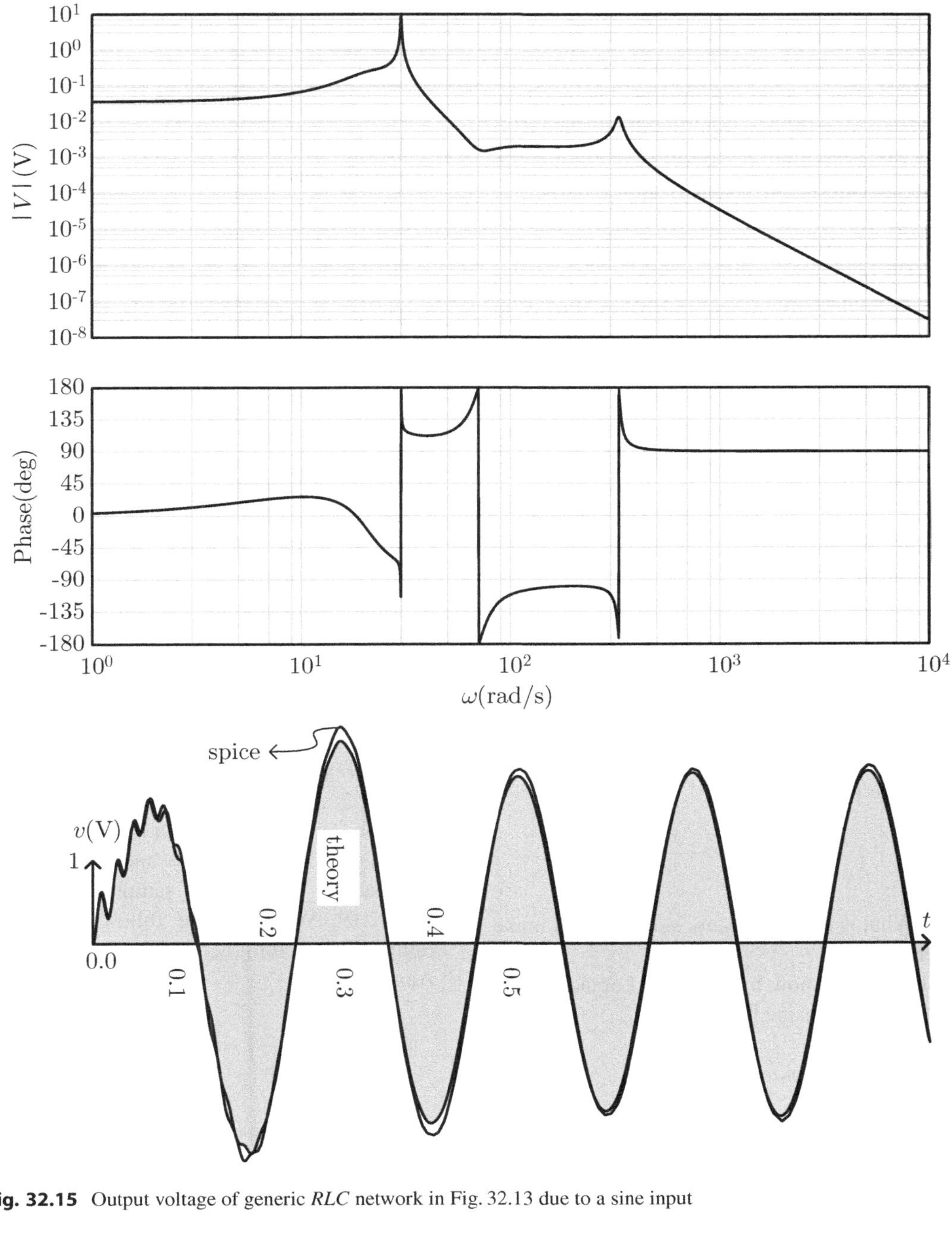

Fig. 32.15 Output voltage of generic *RLC* network in Fig. 32.13 due to a sine input

the causal sine/cosine is at hand it is just a matter of algebra to find the response to other signals, such as the periodic pulse, periodic triangle, and so forth. In this chapter we illustrated how to take the system transfer function (the impedance $Z(s)$ in this case) and multiply by either $\frac{\omega_0}{s^2+\omega_0^2}$ or $\frac{s}{s^2+\omega_0^2}$ to find the causal sine/cosine response in the frequency domain. Once that is known we showed how to use partial fractions to reduce the result in terms of a series of single poles. This series can now be directly inverse transformed to yield the response in the time domain. We also passed by a very interesting scenario where a complex *RLC* network response was approximated by a simple analytic expression. Since the poles for this system are now known, all there is to need to know about this *RLC* network is at hand. As such we then used the standard partial fraction method to figure system response, both in frequency and time domain.

32.10 Problems

1. As we know by now, the Laplace transform of the sine function is

$$\sin \omega_0 t \rightarrow \frac{\omega_0}{s^2 + \omega_0^2}$$

What is the limit as $\omega_0 \rightarrow 0$? Does it make sense?

2. Also as we know by now, the Laplace transform of the cosine function is

$$\cos \omega_0 t \rightarrow \frac{s}{s^2 + \omega_0^2}$$

What is the limit as $\omega_0 \rightarrow 0$? Does it make sense?

3. Plot the Laplace transform of the sine function for the following ω_0 values: 2π, 4π, and 8π. See sample solution in Fig. 32.16.
4. Compare graphically the Laplace transform of the sine to that of cosine, for $\omega_0 = 2\pi$; see sample solution in Fig. 32.17.
5. A system has the transfer function

$$F(s) = \frac{1}{s^2 + 1}$$

A sine input $\sin(2t)$ is applied to the system. Show that the response is

$$H(s) = \frac{2}{(s^2 + 1)(s^2 + 4)}$$

Using partial fraction show that

$$H(s) = \frac{2}{3}\left[\frac{1}{s^2 + 1} - \frac{1}{s^2 + 4}\right]$$

Knowing this figure the time response. Plot the response both in frequency and time; see sample solution in Fig. 32.18.

6. Consider again the series *RC* network in Sect. 32.2. What is the output voltage due to a *sine* input current $i(t) = \sin \omega_0 t$?
Answer:

$$v(t) = Ru(t) \sin \omega_0 t + \frac{1}{C\omega_0} u(t) \left[1 - \cos \omega_0 t\right]$$

7. Consider again the parallel *RL* network in Sect. 32.7. Assume that input current is a *sine* function $i(t) = \sin \omega_0 t$; what is output voltage for $\omega_0 = 2\pi$, $R = 1\,\Omega$, and $L = 0.1\,\text{H}$? Compare to SPICE; see sample solution in Fig. 32.19. What are the following voltage limits: $\lim_{t\rightarrow 0}$; $\lim_{L\rightarrow\infty}$; $\lim_{L\rightarrow 0}$?
Answer:

$$v(t) = \frac{R}{a^2 + \omega_0^2}\left[-a\omega_0 e^{-at} + a\omega_0 \cos \omega_0 t + \omega_0^2 \sin \omega_0 t\right], \quad a = \frac{R}{L}$$

8. The parallel *LC* network (of $L = C = 1$) has an impedance transfer function

$$Z(s) = \frac{s}{s^2 + 1}$$

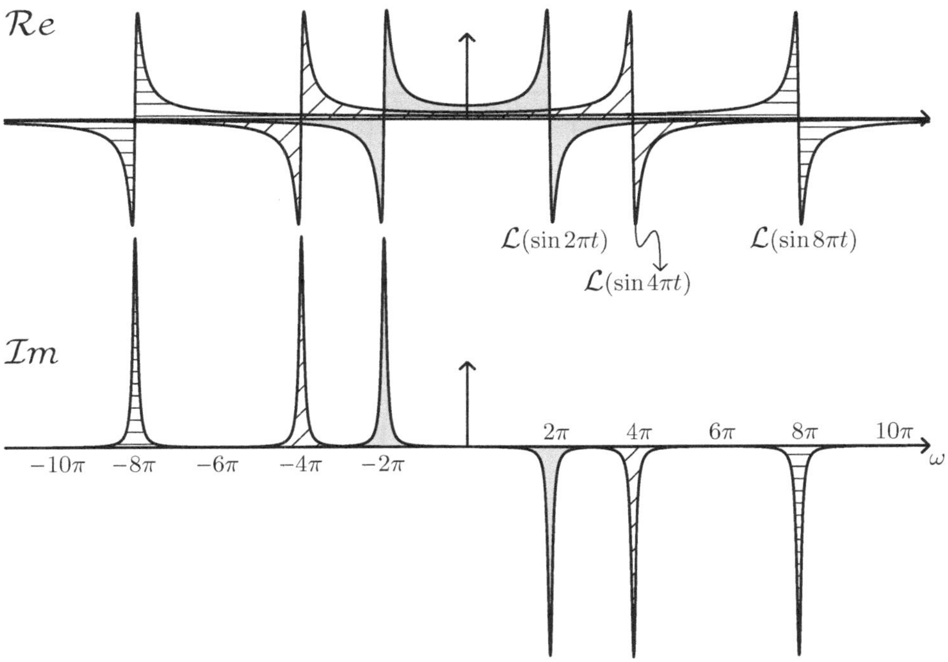

Fig. 32.16 Sample solution to Problem 3; $\sigma = 0.2$

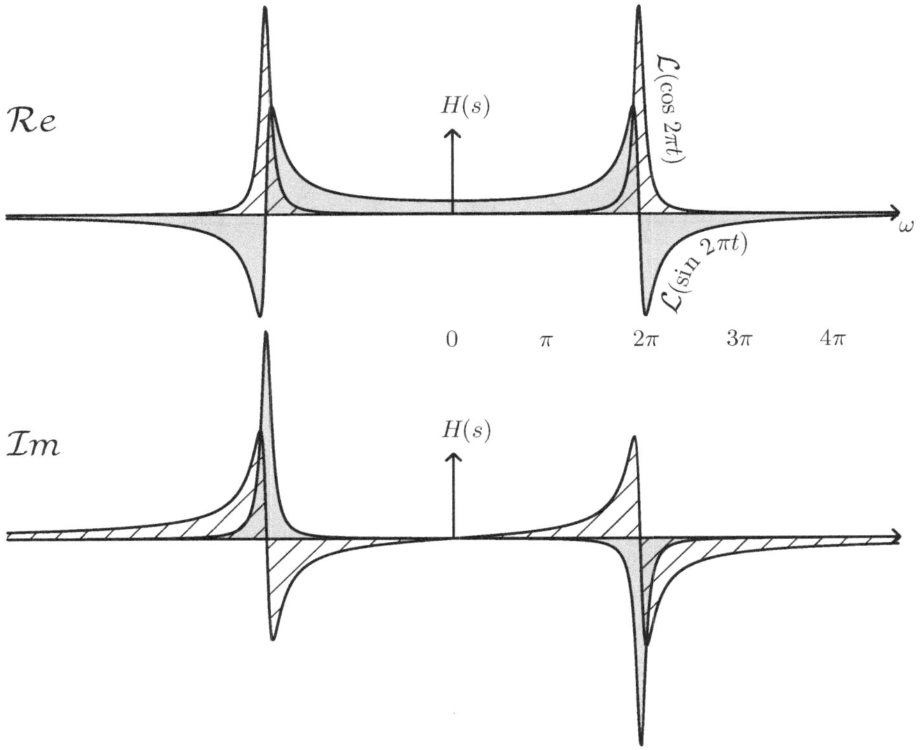

Fig. 32.17 Sample solution to Problem 4; $\sigma = 0.2$

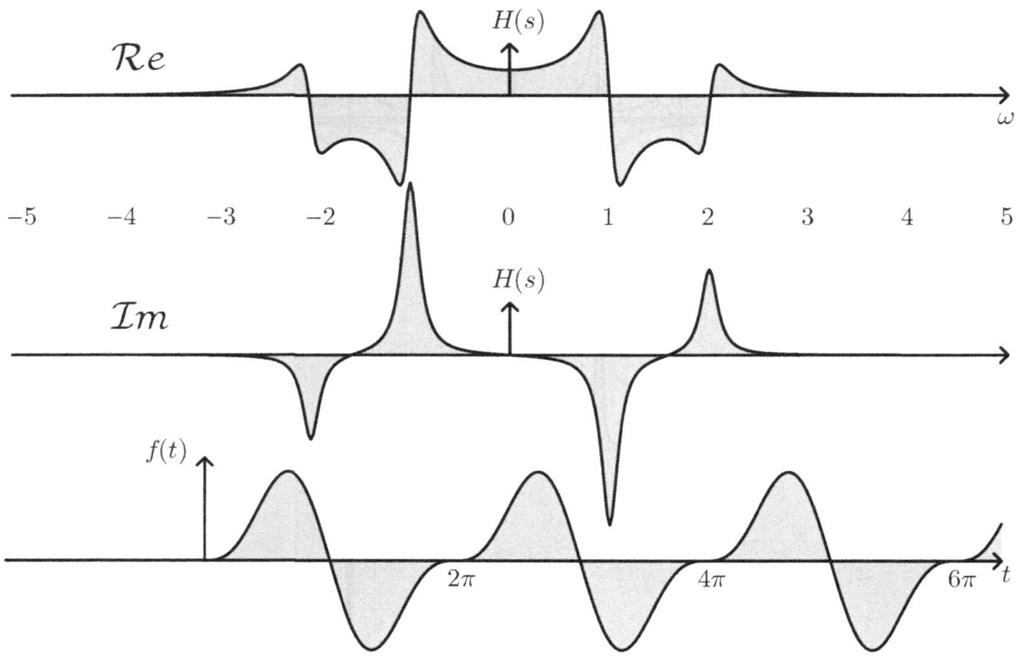

Fig. 32.18 Sample solution to Problem 5 (case of $\sigma = 0.1$)

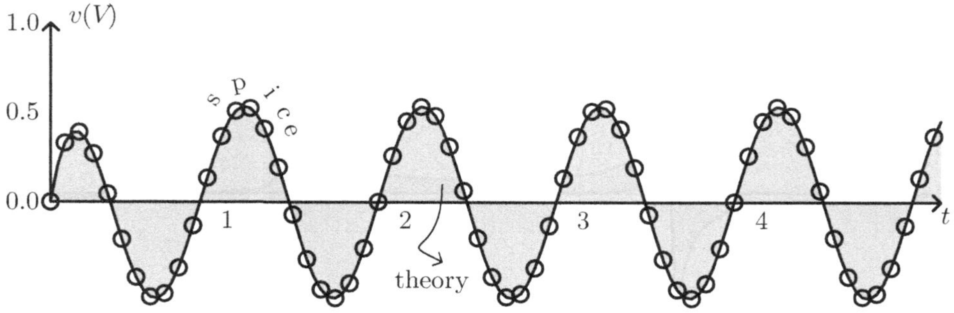

Fig. 32.19 Sample solution to Problem 7

What is the output voltage for an input current of the form $i(t) = \sin 2t$? Plot the solution both in time and frequency domain and compare latter to SPICE; see sample solution in Fig. 32.20. Hint: $V(s) = \frac{2s}{(s^2+1)(s^2+4)}$; then use partial fractions.

9. Repeat Problem 8 for the case of input current is a $i(t) = \sin t$. Hint: use the frequency differentiation property. See sample solution in Fig. 32.21.
10. Repeat the analysis in Sect. 32.8 for the case of input current $i(t) = \sin 50t$; plot output voltage both in frequency and time; compare to SPICE; see sample solution in Fig. 32.22.

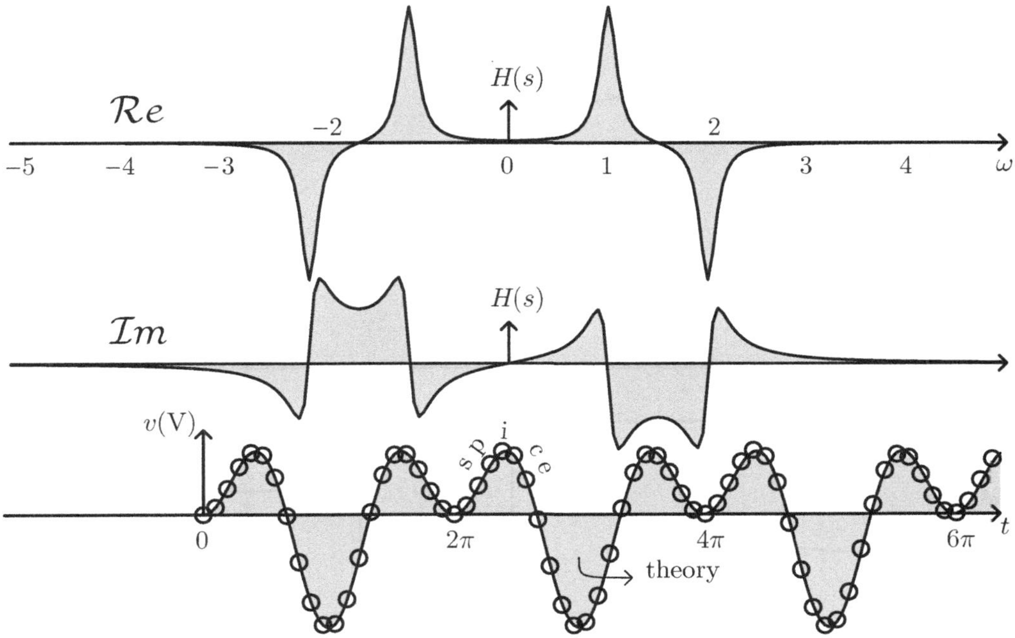

Fig. 32.20 Sample solution to Problem 8

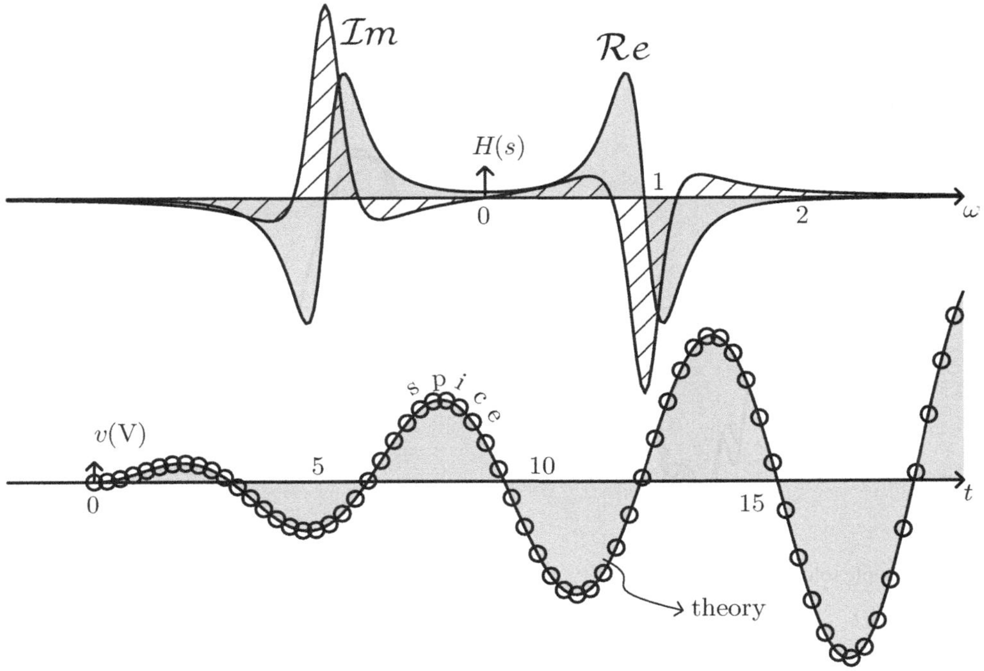

Fig. 32.21 Sample solution to Problem 9

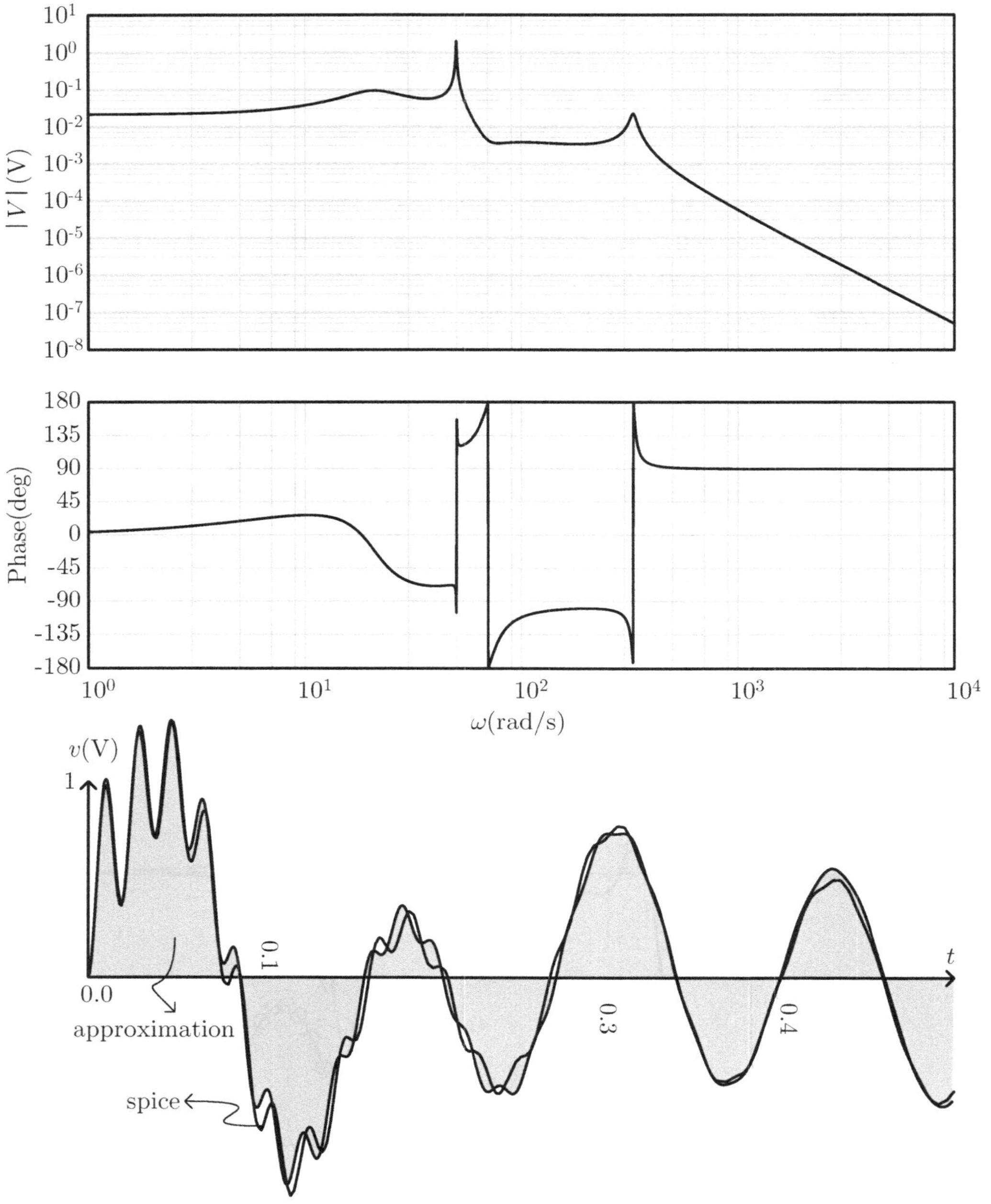

Fig. 32.22 Sample solution to Problem 10

33 Causal Periodic Pulse Response

33.1 Introduction

So far we have covered the impulse, unit step, pulse, and causal sine/cosine responses. The next logical class of stimuli are the periodic pulse ones. This class is extremely important; just think of a free running clock. Pretty much everything in the digital world works with periodic pulses! We have the infinite freedom in setting the pulse width and period; that covers a whole lot of possibilities. Not only so, but we can also mix different widths and periods simply by juxtaposing multiple distinct periodic pulses and so forth.

33.2 Overall Strategy

The strategy we are going to follow is as follows. First decompose the periodic signal as a Fourier series of sines and cosine; then knowing the response to each harmonic (as was done in last chapter), we can simply add the corresponding responses. Notice that all along we would be using causal sines and cosines. For example if we have a symmetric pulse of period T, with peak ± 1, then we can write it in terms of a sine series

$$i(t) = \sum_n b_n \sin \omega_n t, \qquad \omega_0 = \frac{2\pi}{T}, \qquad \omega_n = n\omega_0 \tag{33.1}$$

$$b_n = \frac{2}{T} \int_0^T i(t) \sin \omega_n dt \tag{33.2}$$

For our case we have

$$\begin{aligned} b_n &= \frac{4}{T} \int_0^{T/2} \sin \omega_n dt \\ &= \frac{4}{\omega_n T} [1 - \cos \omega_n T/2] \\ &= \frac{2}{\pi n} [1 - \cos \pi n] \end{aligned} \tag{33.3}$$

Then we have

$$\boxed{i(t) = \sum_{n=1,3,\cdots} \frac{4}{\pi n} \sin \omega_n t, \qquad \omega_n = n\frac{2\pi}{T}} \tag{33.4}$$

The Fourier series reconstruction is shown in Fig. 33.1. All that has to be done now is figure the response to each sine function and add them up in accordance with b_n. Let us then try a couple of examples.

33.3 Series *RC* Network

We know from last chapter (Prob. 6) that the response to an input sine current of the form $i(t) = \sin \omega_0 t$ is

F. Badrieh, *Spectral, Convolution and Numerical Techniques in Circuit Theory*,
https://doi.org/10.1007/978-3-319-71437-0_33

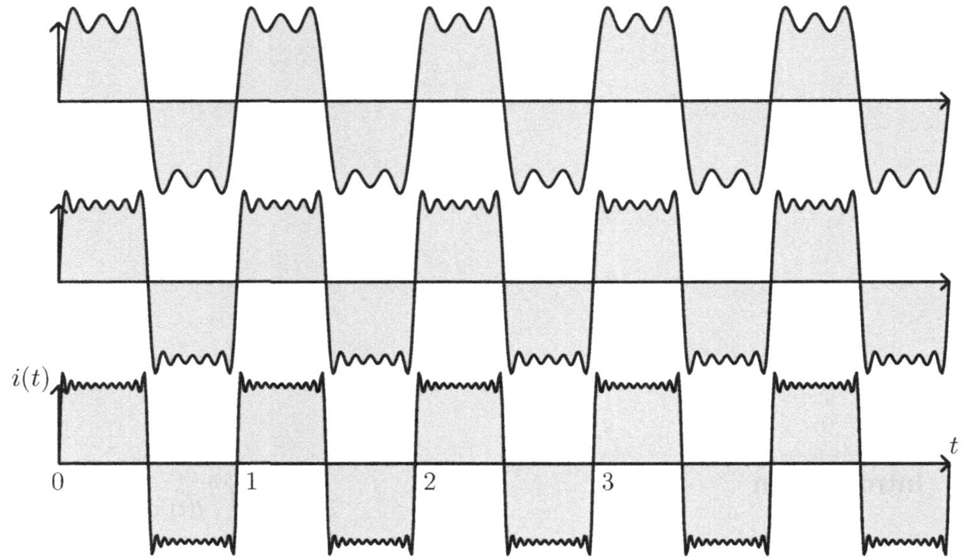

Fig. 33.1 Periodic pulse reconstruction in terms of sinusoids; case of $T = 1$

$$v(t) = Ru(t)\sin\omega_0 t + \frac{1}{C\omega_0}u(t)\left[1 - \cos\omega_0 t\right] \quad (33.5)$$

How about an input of the form $i(t) = \sin\omega_n t$? No problem; let's denote such response by $v_n(t)$:

$$v_n(t) = Ru(t)\sin\omega_n t + \frac{1}{C\omega_n}u(t)\left[1 - \cos\omega_n t\right] \quad (33.6)$$

So we have found the response to a generic sinusoidal current. Now we know from last section that we can represent a periodic pulse current via Eq. (33.4), which is nothing other than a weighted series of sines. But we just found the response to any sine in Eq. (33.6). So the response due to the weighted series is simply the weighted sum of the individual responses:

$$\boxed{v(t) = \sum_{n=1,3,\cdots} \frac{4}{\pi n}\left[Ru(t)\sin\omega_n t + \frac{1}{C\omega_n}u(t)\left[1 - \cos\omega_n t\right]\right]} \quad (33.7)$$

Let us test such solution. Figure 33.2 shows results (for different harmonic count) and comparison to SPICE. Surprisingly even using only a few harmonics results in pretty good approximation.

Let us reflect quickly on what was accomplished—this is truly a feat! By orchestrating our steps carefully, by unfolding the problem into manageable layers, and by simple superposition we are able to deal with unconventional signals (in the sense of not being smooth and continuous) and not only encompass them analytically (via the Fourier series), but also encompass their system response! We are able to figure a circuit response—both transient and steady state—in terms of a series approximation, and we are able to make our approximation better by simply adding more series terms. Notice that in this particular case we don't really have a transient component of the solution, because current is literally forced through the series RC. But, as will be shown in the next section, in the generic case we will have a transient part.

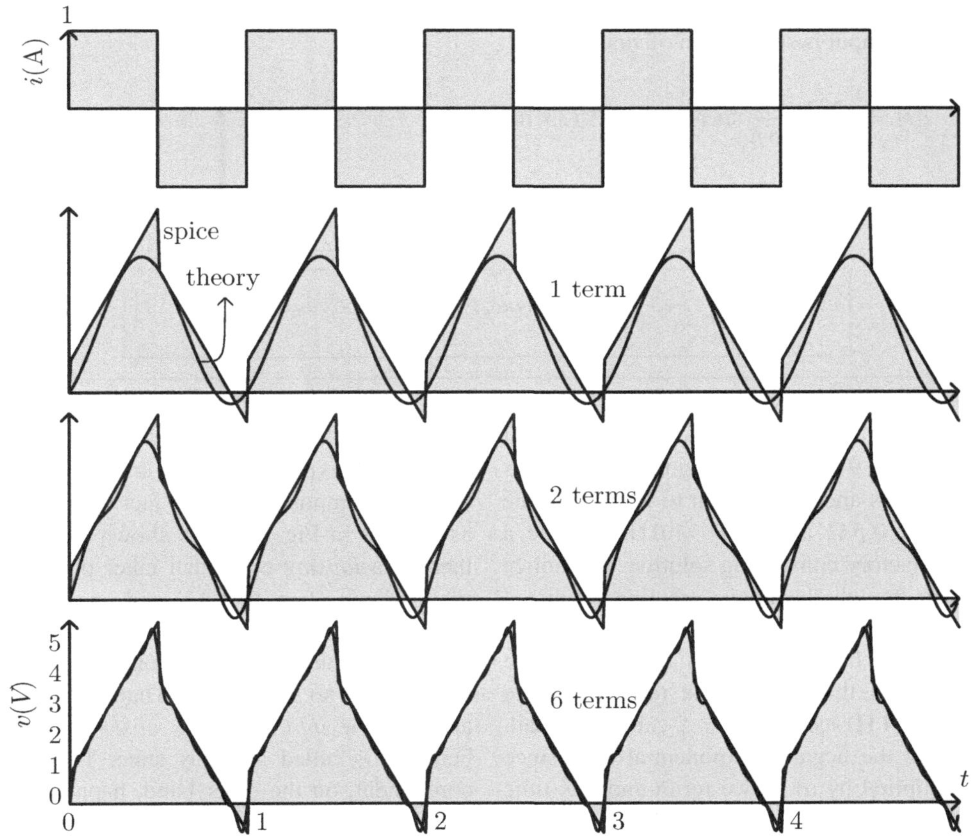

Fig. 33.2 Periodic pulse response to series *RC* network and comparison to SPICE ($T = 1\,\mathrm{s}$, $R = 1\,\Omega$, and $C = 0.1\,\mathrm{F}$)

33.4 Parallel *RL* Circuit

The parallel *RL* network is shown in Fig. 33.3. The impedance transfer function is

$$Z(s) = R\frac{s}{s+a}, \quad a = \frac{R}{L} \tag{33.8}$$

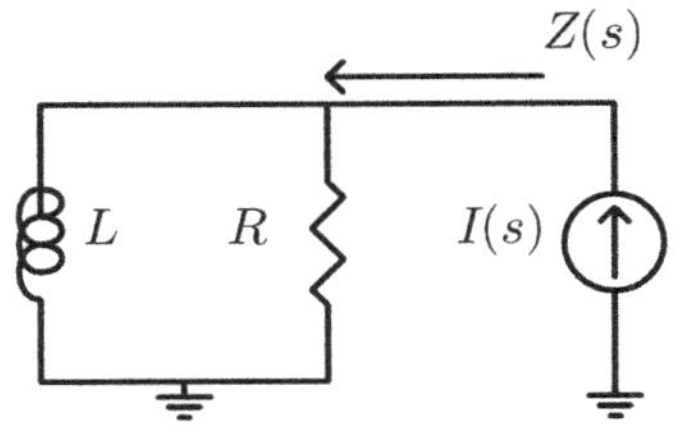

Fig. 33.3 Parallel *RL*

We know from the last chapter (Problem 7) that the response to sine input is

$$v(t) = \frac{R}{a^2+\omega_0^2}\left[-a\omega_0 e^{-at} + a\omega_0 \cos\omega_0 t + \omega_0^2 \sin\omega_0 t\right] \tag{33.9}$$

Since our input is summation of sines

$$i(t) = \sum_{n=1,3,\cdots} \frac{4}{\pi n} \sin \omega_n t \qquad (33.10)$$

then total voltage response would be

$$v(t) = \sum_{n=1,3,\cdots} \frac{R}{a^2 + \omega_n^2} \frac{4}{\pi n} \left[-a\omega_n e^{-at} + a\omega_n \cos \omega_n t + \omega_n^2 \sin \omega_n t \right] \qquad (33.11)$$

Notice that we replaced each occurrence of ω_0 in Eq. (33.9) with ω_n. Figure 33.4 shows sample results and comparison to SPICE for the case $R = 0.5\,\Omega$ and $L = 1.0\,\text{H}$. We see a very satisfactory converging solution. But notice that even though the above solution predicted a *transient* component (the negative exponential one), we can hardly see it in the figure! The reason is that this case had a relatively large inductance (1 H) such that $a = \frac{R}{L}$ came out small, and so did the negative exponential one (since it is multiplied by a). If we rerun and this time use a smaller inductance—say 0.1 H—such that a is bigger, and so is the transient component we should expect to see a more pronounced transient component, and in fact that is the case as evident in Fig. 33.5. As shown in the figure there is a *lapsing* event that takes place *only at the beginning* (say first 1 s) and once it is gone the response falls back on the *steady state* one. Even though the steady state continues to move in time, it does so *predictably*. That is it is periodic and *no new information is observed onwards*; hence it is called a steady state. The transient component, on the other hand, happens only at the beginning and could be thought off as that

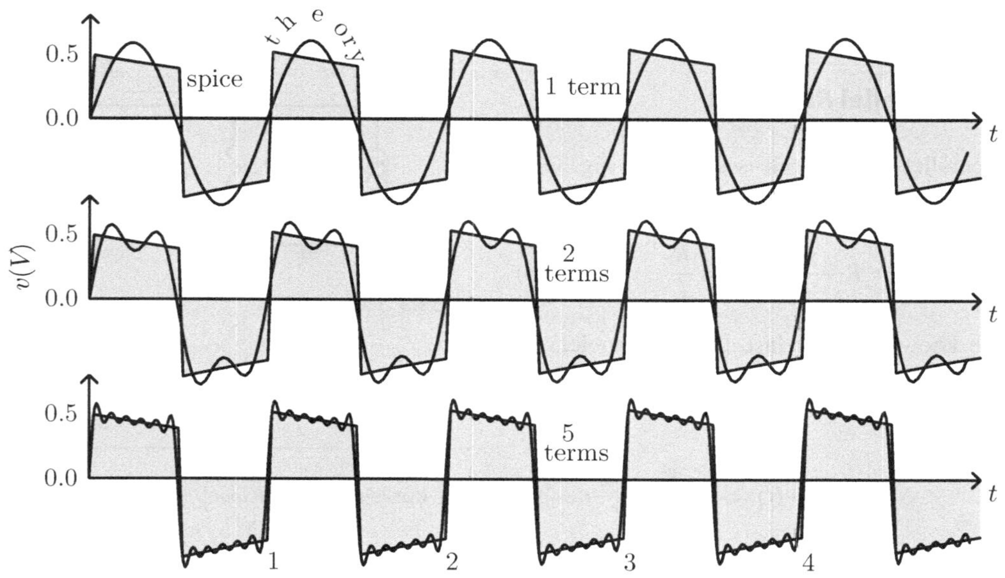

Fig. 33.4 Parallel RL response to causal periodic pulse input ($R_1 = 0.5\,\Omega$ and $L = 1.0\,\text{H}$)

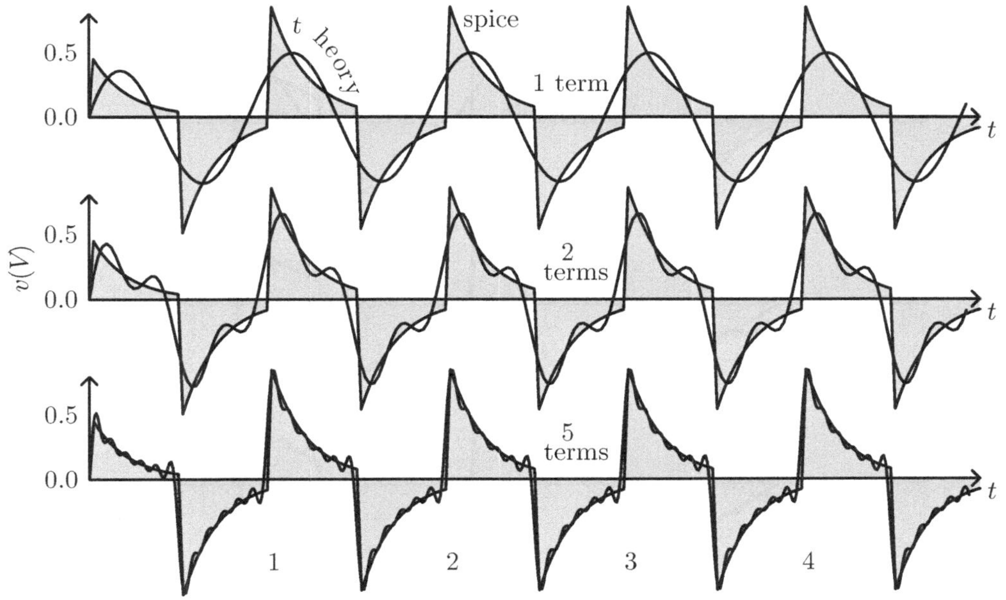

Fig. 33.5 Parallel RL response to causal periodic pulse input ($R_1 = 0.5\,\Omega$ and $L = 0.1$ H)

part of the solution which "assists" the system to migrate from the initial state (zero here) to the final state—the steady state one!

33.5 Series *RC*/Parallel *R* Circuit

The series RC, parallel R is shown in Fig. 33.6. As we saw in the last chapter (Sect. 32.5), if we define

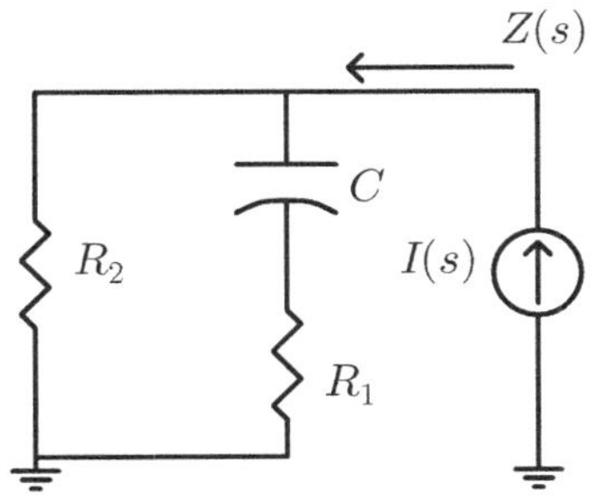

Fig. 33.6 Series RC in parallel with R and input impedance

$$A = R_1R_2, \qquad B = (R_1 + R_2), \qquad D = \frac{1}{BC}, \qquad \delta = \frac{ACD - R_2}{D^2 + \omega_0^2}$$

$$\alpha = \frac{-ACD + R_2}{D^2 + \omega_0^2}, \qquad \beta = \frac{R_2}{D} + \frac{\omega_0^2}{D}\frac{ACD - R_2}{D^2 + \omega_0^2} \tag{33.12}$$

then the response to a sine input current was

$$v(t) = D\omega_0\left[\alpha e^{-Dt} + \frac{\beta}{\omega_0}\sin\omega_0 t + \delta\cos\omega_0 t\right] \tag{33.13}$$

If our periodic pulse input is expanded as a Fourier series

$$i(t) = \sum_{n=1,3,\cdots}\frac{4}{\pi n}\sin\omega_n t, \qquad \omega_n = n\frac{2\pi}{T} \tag{33.14}$$

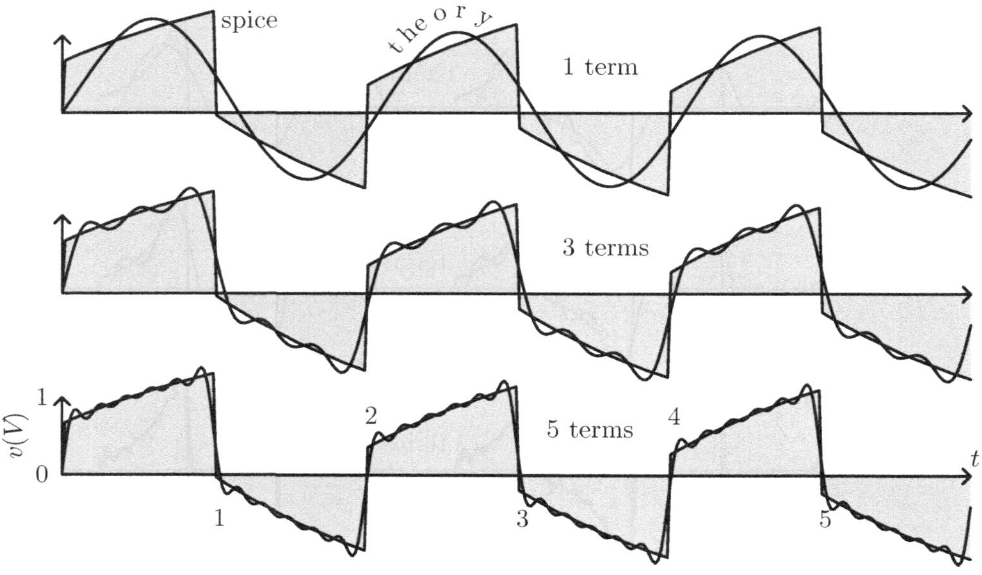

Fig. 33.7 Series *RC*/parallel *R* response to causal periodic pulse input ($R_1 = 1$, $R_2 = 2\,\Omega$, and $C = 0.5\,\text{F}$

then our total solution would be the

$$v(t) = \sum_{n=1,3,\cdots} \frac{4D\omega_n}{\pi n}\left[\alpha_n e^{-Dt} + \frac{\beta_n}{\omega_n}\sin\omega_n t + \delta_n \cos\omega_n t\right] \tag{33.15}$$

Notice that each of α, β, and δ has n index, since the ω_0 in them would need to change to ω_n. Figure 33.7 shows sample results and comparison to SPICE for the case $R_1 = 1$, $R_2 = 2\,\Omega$, and $C = 0.5\,\text{F}$. Figure 33.8 shows another sample results with more pronounced transient component for the case $R_1 = 0.5$, $R_2 = 10\,\Omega$, and $C = 0.5\,\text{F}$. Either way, the more harmonics we include, the better our results become. The whole solution—transient and steady state—is covered!

By now we see the trend. Find the solution due to an input current of the form $i(t) = \sin\omega_0 t$ (or a cosine one, or combination thereof); replace the ω_0 with ω_n; find the periodic pulse in terms of the sines/cosines and make note of the weighing coefficients; and finally sum the n responses weighted in accordance to the weighing coefficients.

33.6 Periodic Pulse of Nonzero Average

So far we have dealt with the causal periodic pulse of *zero average*; let's next try the case of nonzero average. Specifically consider the pulse of value 1, period T and 0.5 average as shown in Fig. 33.9. The method remains the same but the input stimulus now has a slightly different Fourier series.

The Fourier series of this function is similar to that of the zero-average one, with two exceptions: first, all coefficients are scaled down by 2, and second we have a DC term of 0.5; hence

$$i(t) = 0.5u(t) + \sum_{n=1,3,\cdots} \frac{2}{\pi n}\sin\omega_n t, \quad \omega_n = n\frac{2\pi}{T} \tag{33.16}$$

Let's next apply this to the parallel *RL* network in Sect. 33.4. The response due to the sine terms remains the same, albeit divided by 2:

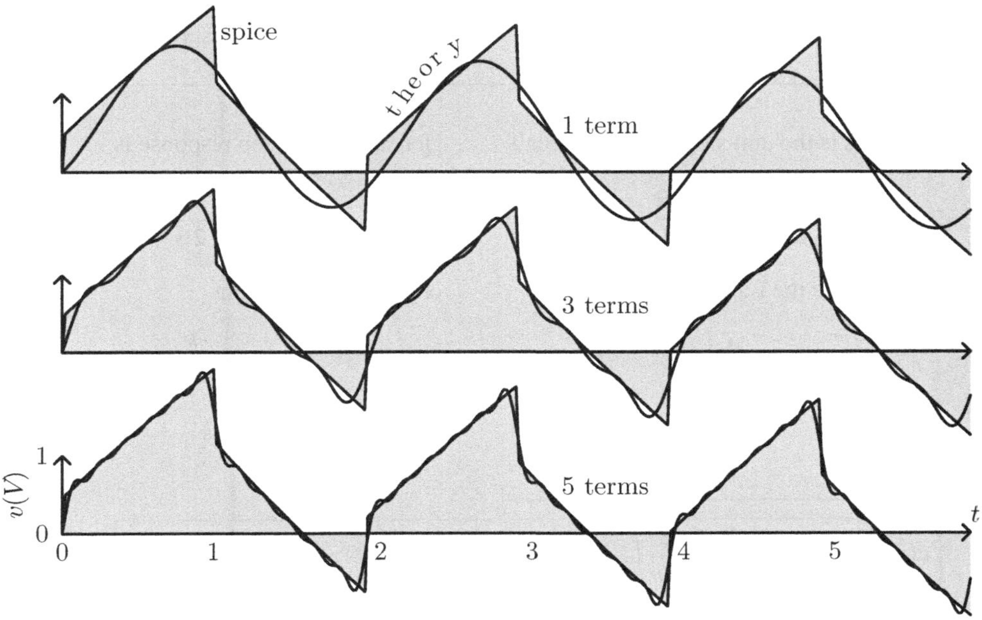

Fig. 33.8 Series *RC*/parallel *R* response to causal periodic pulse input ($R_1 = 0.5$, $R_2 = 10\,\Omega$, and $C = 0.5\,\mathrm{F}$

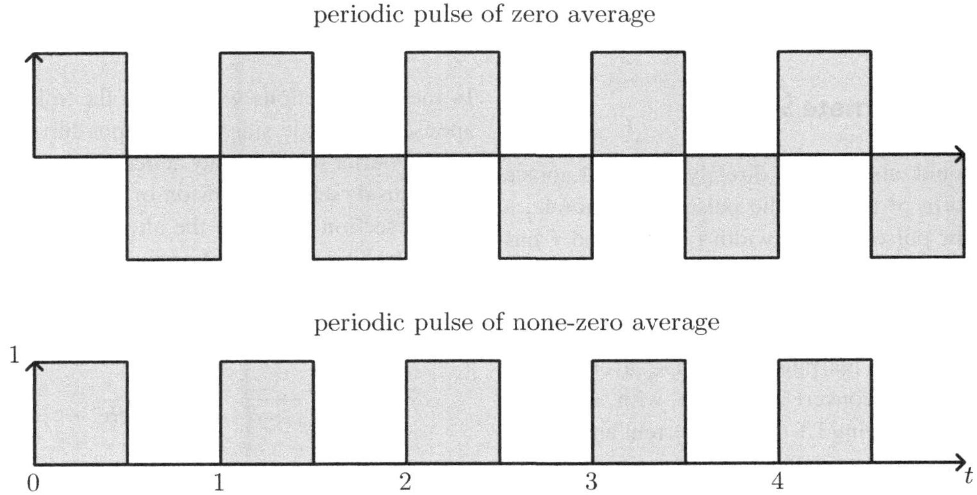

Fig. 33.9 Periodic pulse of nonzero average and comparison to zero-one

$$v_1(t) = \sum_{n=1,3,\cdots} \frac{R}{a^2+\omega_n^2}\frac{2}{\pi n}\left[-a\omega_n e^{-at} + a\omega_n \cos\omega_n t + \omega_n^2 \sin\omega_n t\right] \tag{33.17}$$

What remains is the unit step response; recall

$$Z(s) = R\frac{s}{s+a}, \qquad a = \frac{R}{L} \tag{33.18}$$

Our unit step has the Laplace transform

$$I_2(s) = \frac{1}{2}\frac{1}{s} \tag{33.19}$$

Hence the unit step response is

$$V_2(s) = \frac{R}{2}\frac{1}{s+a} \tag{33.20}$$

or in the time domain

$$v_2(t) = \frac{R}{2}e^{-at} \tag{33.21}$$

So our final solution is

$$\boxed{v(t) = \sum_{n=1,3,\cdots} \frac{R}{a^2+\omega_n^2}\frac{2}{\pi n}\left[-a\omega_n e^{-at} + a\omega_n \cos\omega_n t + \omega_n^2 \sin\omega_n t\right] + \frac{R}{2}e^{-at}} \tag{33.22}$$

Figure 33.10 shows the results and comparison to SPICE.

33.7 Alternate Strategy

We could also resort directly to the Laplace transform of the periodic pulse. For example, a periodic pulse of pulse width τ and period T has the LT

$$I(s) = \frac{1-e^{-s\tau}}{s(1-e^{-Ts})} \tag{33.23}$$

Notice that his pulse has a DC average; we could easily convert it to one with zero DC average. Knowing LT of input current and knowing impedance transfer function we find output voltage simply as

$$V(s) = Z(s)I(s) \tag{33.24}$$

Finding inverse LT of this gives us voltage in time. We will show an example using this method next.

33.8 Alternate Method Applied to Parallel *RC* Network

In the prior sections we utilized the voltage response to a single sine/cosine input current, and using Fourier series simply added all responses to the (causal) series expansion of current at hand. In this section we apply the alternate method of using the LT of the causal periodic pulse directly. The parallel *RC* network is shown in Fig. 33.11.

The impedance transfer function is

$$Z(s) = \frac{1}{C}\frac{1}{s+a}, \qquad a = \frac{1}{RC} \tag{33.25}$$

Input current (for the case pulse width = $T/2$)

$$I(s) = \frac{1-e^{-sT/2}}{s(1-e^{-sT})} \tag{33.26}$$

Output voltage is then

$$\boxed{V(s) = \frac{1}{C}\frac{1}{s(s+a)}\frac{1-e^{-sT/2}}{1-e^{-sT}}}, \qquad a = \frac{1}{RC} \tag{33.27}$$

We know that

$$\frac{1}{s(s+a)} = \frac{1}{a}\left[\frac{1}{s} - \frac{1}{s+a}\right] \rightarrow \frac{1}{a}u(t)\left[1 - e^{-at}\right] \tag{33.28}$$

Using *time shifting* property we conclude that

$$\frac{1 - e^{-sT/2}}{s(s+a)} \rightarrow \frac{1}{a}\Big[u(t)\big(1 - e^{-at}\big) - u(t - T/2)\big(1 - e^{-a(t-T/2)}\big)\Big] \tag{33.29}$$

Using the Laplace transform *theory of periodic functions* we can then tell that

$$\frac{1 - e^{-sT/2}}{s(s+a)}\frac{1}{1 - e^{-sT}} \rightarrow \text{periodic version of}$$
$$\frac{1}{a}\Big[u(t)\big(1 - e^{-at}\big) - u(t - T/2)\big(1 - e^{-a(t-T/2)}\big)\Big] \tag{33.30}$$

Hence

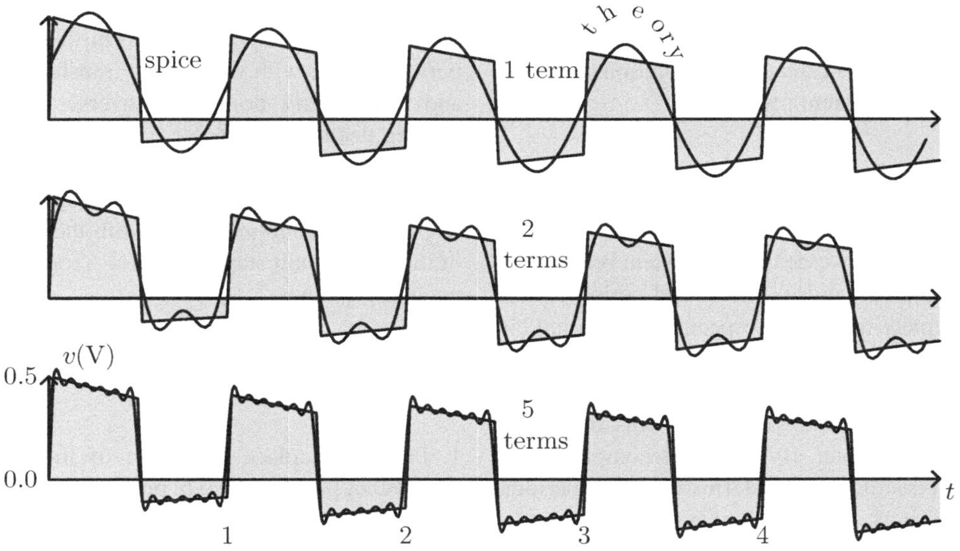

Fig. 33.10 Parallel *RL* response to causal pulse input with nonzero average ($R = 0.5\,\Omega$, and $L = 1.0\,\text{H}$)

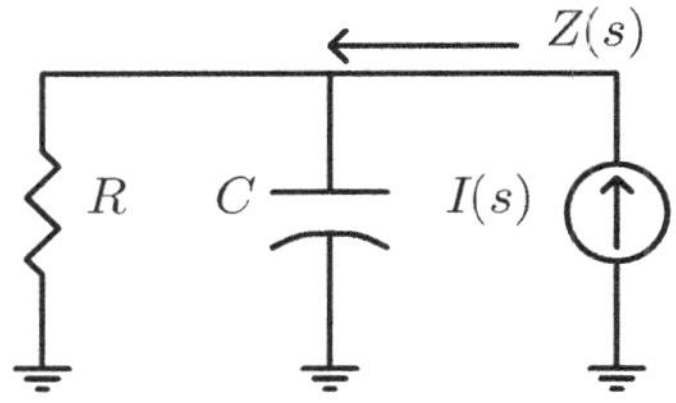

Fig. 33.11 Parallel *RC* circuit

$$v(t) = \text{periodic version of } \frac{1}{aC}\left[u(t)\left(1-e^{-at}\right) - u\left(t-T/2\right)\left(1-e^{-a(t-T/2)}\right)\right], \quad \text{period} = T \tag{33.31}$$

That is, take the solution from Eq. (33.29) and iterate it in time by offsetting it each time by T, and in the end add up all such solutions! That is, if

$$w(t) = \frac{1}{aC}\left[u(t)\left(1-e^{-at}\right) - u\left(t-T/2\right)\left(1-e^{-a(t-T/2)}\right)\right] \tag{33.32}$$

then

$$v(t) = w(t) + w(t-T) + w(t-2T) + \cdots \tag{33.33}$$

Figure 33.12 shows the results and comparison to SPICE. Notice how the single pulse response is used as basis function to build total response, by shifting and adding. The theory works perfectly as indicated by the exact match to simulations. Yet again we are able to predict both transient and steady state solutions; we fully captured the system response!

33.9 Summary

In this chapter we dealt with system response to causal periodic pulse. The causal periodic pulse is extremely important, especially for digital circuits. The assumed strategy was simple and is built assuming that system response to causal sines/cosines is known, and as was dealt with in the last chapter. By simply decomposing the causal periodic pulse in terms of causal periodic sines/cosines, by knowing system response to those, and by simply using superposition we are able to figure system response to any periodic pulse with any period and any duty cycle. The method followed could easily be applied for non-rectangular pulses—for example triangular or parabolic and so forth. We illustrated the flow on a few examples and demonstrated excellent match with SPICE. We finally wrapped the chapter with an alternate strategy—that of using the Laplace transform of the causal periodic pulse directly. In this flow we don't end up using the Fourier series expansion (at least not explicitly). This method basically gives the solution in terms of a periodic elementary one; the elementary solution is iterated infinitely, each time shifted by the period T. Of course we can always fall back on yet a third strategy—that of multiplying the Laplace transform of the causal periodic pulse with the system transfer function and numerically doing the inverse transform. Having dealt with impulse, unit step, one-timer pulse, causal sines/cosines, and finally causal periodic pulse we are almost wrapping up system response to common signals, with the exception of the slanted unit step response—a topic we will uncover right next.

33.10 Problems

1. Plot the Laplace transform of the periodic pulse of pulse width 0.5, period 1, and average 0.5. Compare to the LT of the single pulse (of same width); use $\sigma = 0.4$ in both cases. See sample solution in Fig. 33.13.
2. What is the inverse transform of the function

$$F_1(s) = \frac{1}{s}\frac{1}{1-e^{-s}}$$

Plot it. Next, what is the inverse transform of the function

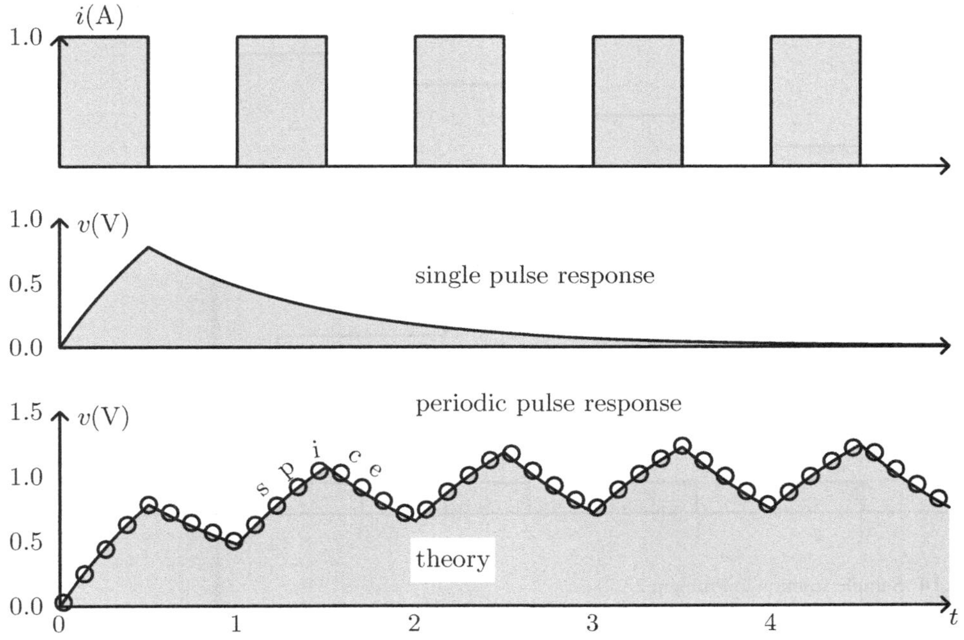

Fig. 33.12 Parallel RC response to periodic pulse ($R = 2\,\Omega$ and $C = 0.5$F)

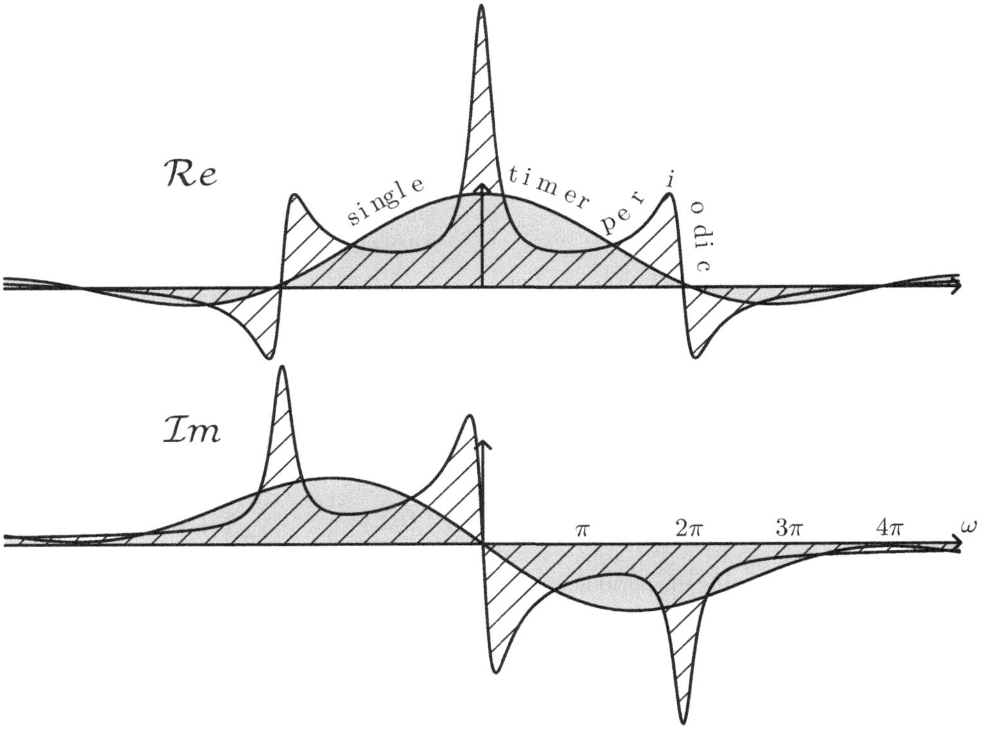

Fig. 33.13 Sample solution to Problem 1

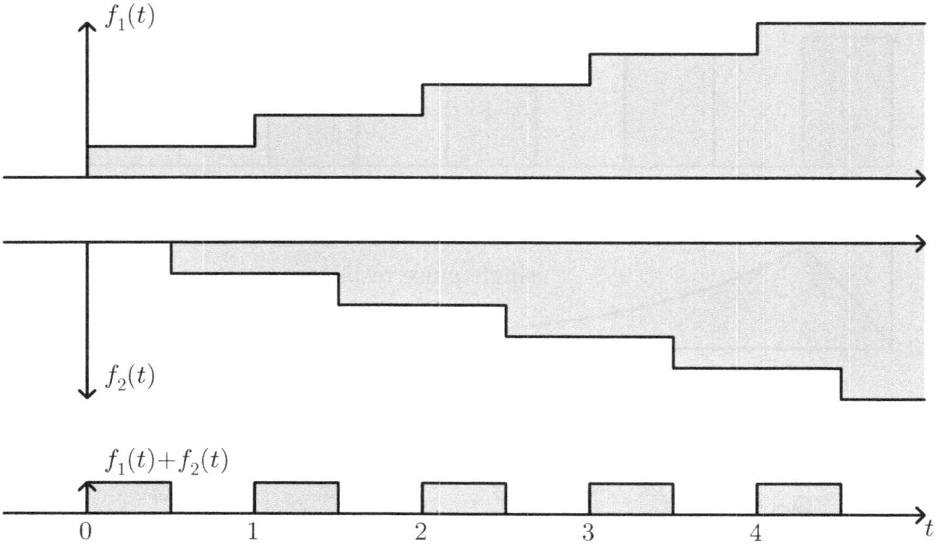

Fig. 33.14 Sample solution to Problem 2

$$F_2(s) = -\frac{e^{-0.5s}}{s}\frac{1}{1-e^{-s}}$$

Again plot it. Finally add both in the frequency domain and time domain, and plot latter. Does it make sense? See sample solution in Fig. 33.14.

3. If we represent the periodic pulse function (of width 0.5, period 1, and average 0.5) in terms of causal sine functions we get

$$f(t) = \frac{1}{2} + \sum_{n=1,3,\cdots} \frac{2}{\pi n} \sin \omega_n t, \qquad \omega_n = 2\pi n$$

If we do the Laplace transform of this we get

$$\boxed{F_1(s) = \frac{1}{2}\frac{1}{s} + \sum_{n=1,3,\cdots} \frac{2}{\pi n}\frac{\omega_n}{s^2+\omega_n^2}, \quad \omega_n = 2\pi n}$$

By the same token, using the periodic property of the Laplace transform we got before

$$\boxed{F_2(s) = \frac{1-e^{-0.5s}}{s(1-e^{-s})}}$$

If things are to add up properly, these last two equations must equate. Plot them in the frequency domain and verify that as $n \to \infty$ they do in fact match; see sample solution in Fig. 33.15.

4. When the pulse width is not exactly half the duty cycle using a sine series is not sufficient to represent the periodic pulse; instead we would need to use the cosine series as well. What is the Fourier series of a periodic pulse of width 0.25 and period 1.0? Plot it; see sample solution in Fig. 33.16.
Answer:

$$\omega_n = 2\pi n$$

$$a_n = 2\frac{\sin \omega_n/4}{\omega_n}$$

$$b_n = 2\frac{1-\cos \omega_n/4}{\omega_n}$$

$$f(t) = \frac{1}{4} + \sum_{n=1,2,\cdots} a_n \cos \omega_n t + b_n \sin \omega_n t$$

5. Let's take the case of parallel *RL* circuit in Sect. 33.4 with input current of the same form as that in the prior problem; find output voltage (for the case $R = 0.5\,\Omega$ and $L = 1\,\text{H}$) and compare to SPICE. See sample solution in Fig. 33.17.
Answer: Using the same a_n and b_n coefficients from last problem and using $a = \frac{R}{L}$ we get

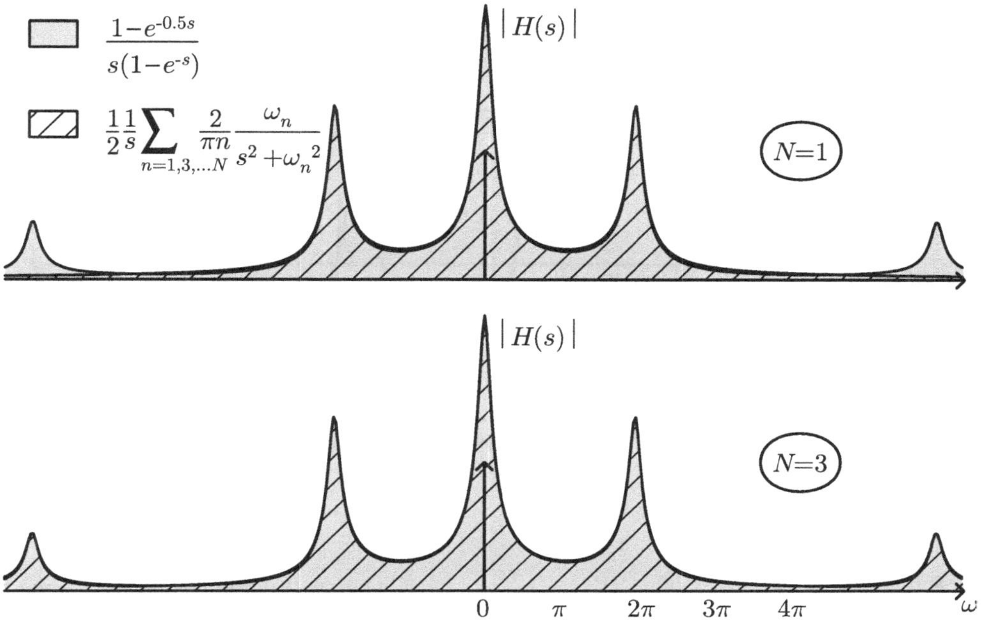

Fig. 33.15 Sample solution to Problem 3; case of $\sigma = 0.25$

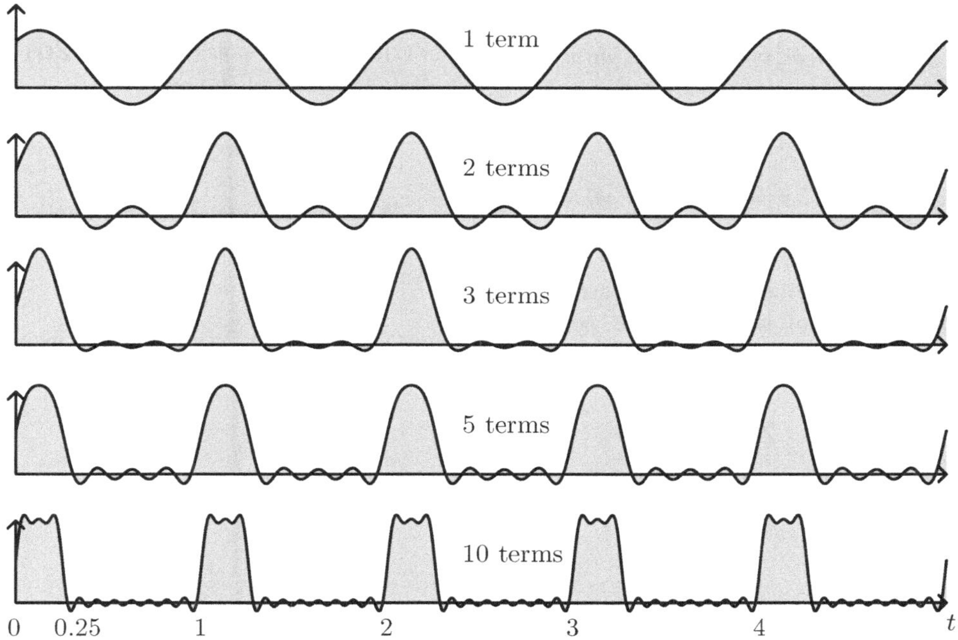

Fig. 33.16 Sample solution to Problem 4

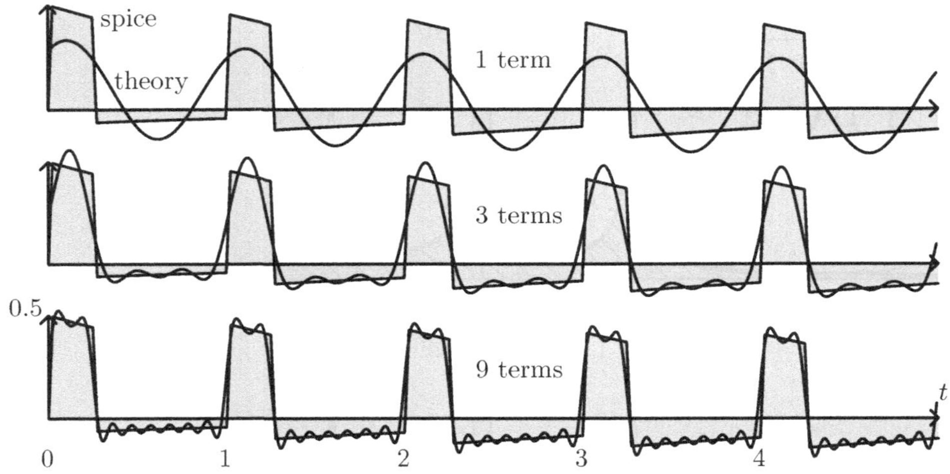

Fig. 33.17 Sample solution to Problem 5

$$v(t) = \frac{R}{4}e^{-Rt/L}$$
$$+ \sum_{n=1,2,\cdots} +a_n \frac{R}{a^2+\omega_n^2}$$
$$\left[a^2e^{-at} + \omega_n^2 \cos\omega_n t - a\omega_n \sin\omega_n t\right]$$
$$+ \sum_{n=1,2,\cdots} +b_n \frac{R}{a^2+\omega_n^2}$$
$$\left[-a\omega_n e^{-at} + a\omega_n \cos\omega_n t + \omega_n^2 \sin\omega_n t\right]$$

6. Assume a system has the transfer function $Z(s)$; assume further that input current is a periodic pulse of width 0.5 and period 1. Hence output voltage will be

$$V(s) = Z(s)\frac{1-e^{-0.5s}}{s(1-e^{-s})}$$

One way to find solution in time is to factor the above expression in terms of partial fractions; knowing those we ought to be able to figure time response directly. But to find the partial fractions we need to find the poles of $V(s)$; and those poles belong to $Z(s)$ and to $I(s)$. The poles of $I(s)$ are

$$\text{zeroes of } s\,(1-e^{-s})$$

What are those zeroes? See sample solution in Fig. 33.18 for $\sigma = 0$ such that $s = j\omega$.
Answer:

zeroes when $\omega = 2\pi n,\ n$ integer, including 0

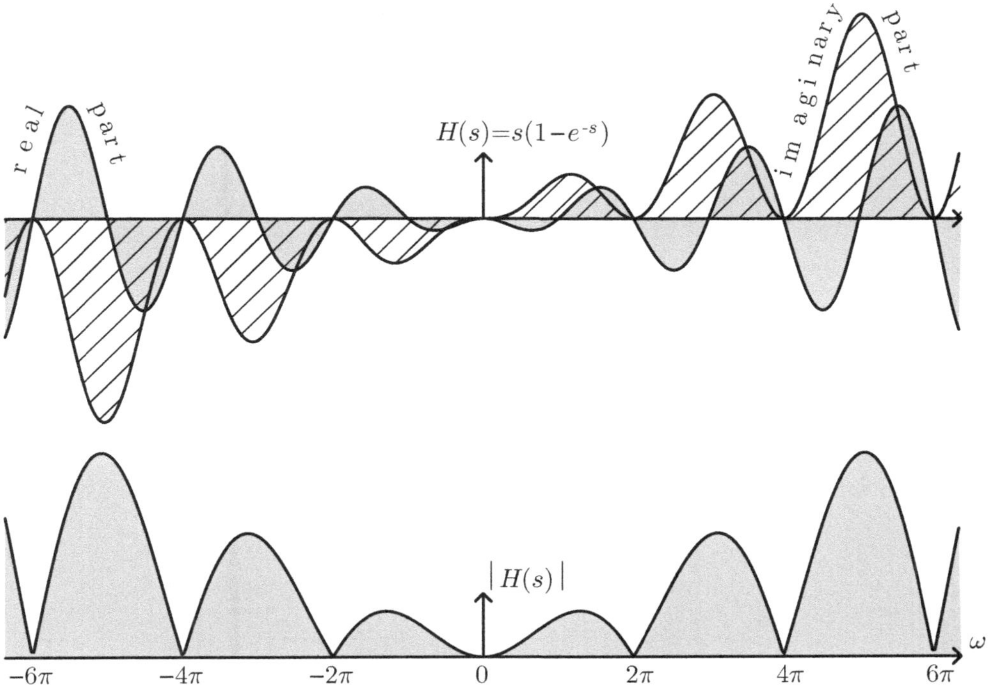

Fig. 33.18 Sample solution to Problem 6

34 Slanted Unit Step Response

34.1 Introduction

The unit step stimulus is one of the most important stimuli, and we've already dedicated a whole chapter to it (Chap. 30). Implicit in the prior treatment is that the edge rate of the step is infinite—that is, the rise time is zero. In reality, this assumption is mostly not valid—there is almost always a finite rise time for the edge. Here we set the rise time to an arbitrary value and derive the response to what we coin as the slanted unit step. Looked at differently if we know the slanted unit step response then we know the ideal unit step response simply by taking the limit as the rise time goes to zero; so from that point of view the slanted unit step response is the more comprehensive one! After we clear up some transform equations we illustrate the slanted unit step response with a few examples.

34.2 Slanted Unit Step Laplace Transform

We know the *ideal* unit step, with zero rise time has the LT

$$u(t) \rightarrow \frac{1}{s} \tag{34.1}$$

We can obtain the LT of the slanted unit step, with rise time t_0 simply by using the convolution theorem. We recognize that we can obtain a slanted unit step simply by convolving an ideal step with a pulse of width t_0:

$$\begin{aligned}&\text{slanted unit step and rise time } t_0\\&= \frac{1}{t_0}\text{ ideal unit step} * \text{pulse of width } t_0\end{aligned} \tag{34.2}$$

and as shown in Fig. 34.1. The pulse of width t_0 and center $t_0/2$ has the LT

$$\text{pulse of width } t_0 \rightarrow \frac{1-e^{-st_0}}{s} \tag{34.3}$$

Hence, and finally, we can obtain the LT of the slanted unit step as

$$\boxed{\text{slanted unit step} \rightarrow \frac{1}{t_0}\frac{1-e^{-st_0}}{s^2}} \tag{34.4}$$

How simple can this get? A very beautiful equation capturing the essence of rise time in the slanted unit. Let's quickly test the validity of this equation. First if we take the limit of rise time t_0 going to zero we should expect to recover the ideal unit step. In fact

$$\lim_{t_0\rightarrow 0}\frac{1}{t_0}\frac{1-e^{-st_0}}{s^2} = \frac{1}{t_0}\frac{1-1+st_0}{s^2} = \frac{1}{t_0}\frac{st_0}{s^2} = \frac{1}{s} \tag{34.5}$$

F. Badrieh, *Spectral, Convolution and Numerical Techniques in Circuit Theory*,
https://doi.org/10.1007/978-3-319-71437-0_34

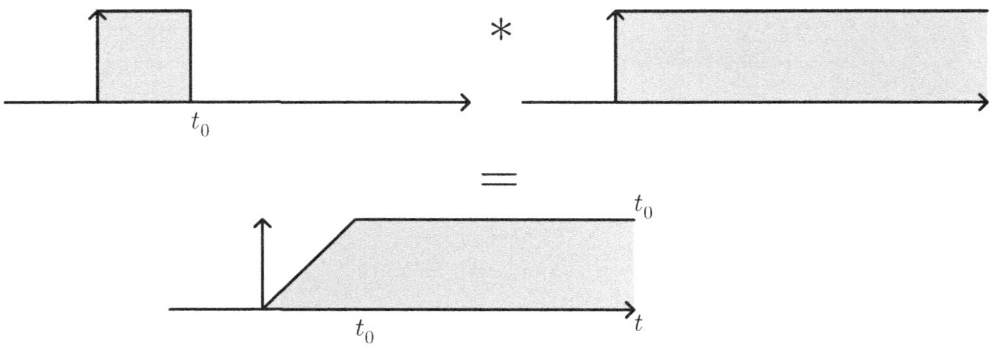

Fig. 34.1 Slanted step obtained by convolving pulse with ideal step

which is what we'd expect. Next imagine that the rise time t_0 goes to infinity, but also multiply the slanted unit step function by t_0 such that the slope is always 1; what function ought we to expect? Of course that would collapse to the ramp function $f(t) = t$ with Laplace transform $\frac{1}{s^2}$. So let's see if we regain this limit:

$$\lim_{t_0\to\infty} t_0 \times \frac{1}{t_0}\frac{1-e^{-st_0}}{s^2} = \lim_{t_0\to\infty}\frac{1-0}{s^2} = \frac{1}{s^2} \tag{34.6}$$

which again is the right limit; notice we made use of the fact that $\lim_{t_0\to 0} e^{-st_0} = 0$. With this out of the way we next jump to some application examples.

34.3 Parallel *RL* Circuit

The parallel RL is shown in Fig. 34.2; The impedance transfer function is

$$Z(s) = R\frac{s}{a+s}, \qquad a = \frac{R}{L} \tag{34.7}$$

The LT of the slanted step function is

$$I(s) = \frac{1}{t_0}\frac{1-e^{-t_0 s}}{s^2} \tag{34.8}$$

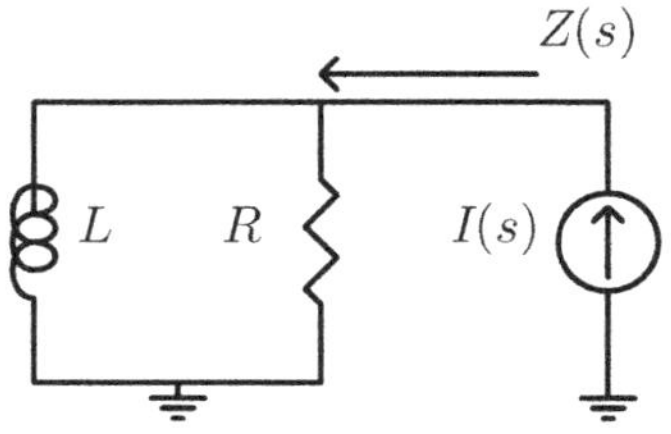

Fig. 34.2 Parallel RL circuit

The IZ product in frequency domain is then

$$\begin{aligned} V(s) = I(s)Z(s) &= \frac{1}{t_0}\frac{1-e^{-t_0 s}}{s^2}R\frac{s}{a+s} \\ &= \frac{R}{t_0}\left(1-e^{-t_0 s}\right)\frac{1}{s(a+s)} \\ &= \frac{R}{t_0 a}\left(1-e^{-t_0 s}\right)\left[\frac{1}{s}-\frac{1}{a+s}\right] \\ &= \frac{L}{t_0}\left(1-e^{-t_0 s}\right)\left[\frac{1}{s}-\frac{1}{a+s}\right], \qquad a = \frac{R}{L} \end{aligned} \tag{34.9}$$

The inverse transform of this is our voltage in time

$$\boxed{v(t) = \frac{L}{t_0}u(t)\left[1-e^{-at}\right] - \frac{L}{t_0}u(t-t_0)\left[1-e^{-a(t-t_0)}\right]} \tag{34.10}$$

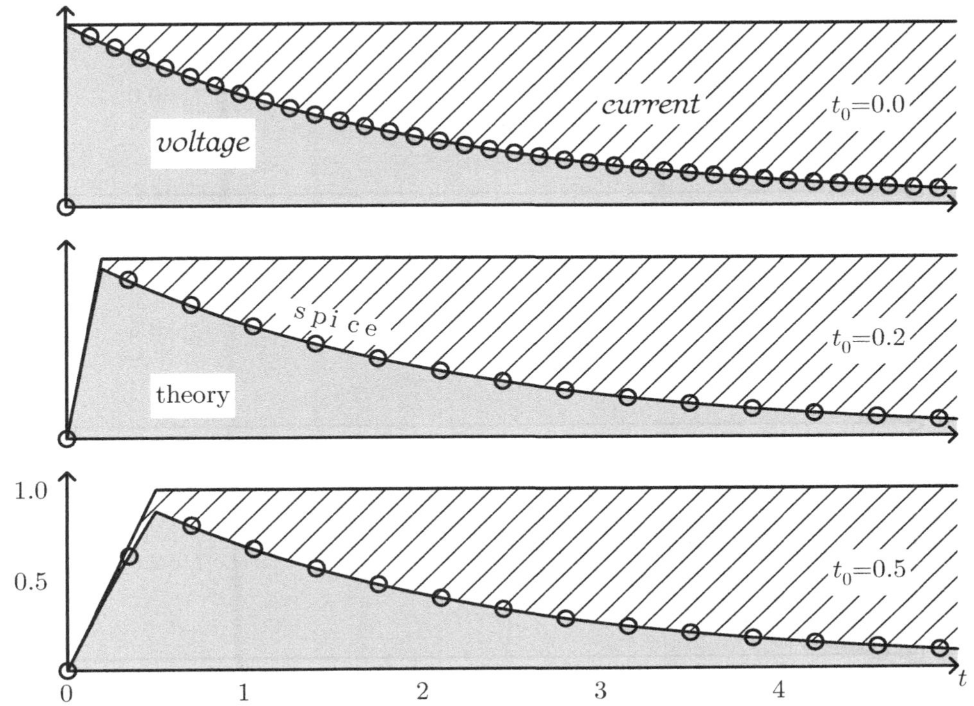

Fig. 34.3 Parallel *RL* slanted step response for different rise time ($R = 1\,\Omega$ and $L = 2\,\text{H}$)

where we have used the time shifting property of the LT. We validate our results against SPICE one for various rise times as shown in Fig. 34.3. Another success story!

34.4 Parallel *RC* Circuit

The parallel *RC* circuit is shown in Fig. 34.4; The impedance transfer function is

$$Z(s) = \frac{1}{C}\frac{1}{s+a}, \quad a = \frac{1}{RC} \tag{34.11}$$

The input current has the LT

$$I(s) = \frac{1}{t_0}\frac{1-e^{-st_0}}{s^2} \tag{34.12}$$

Output voltage is then

$$V(s) = \frac{1-e^{-st_0}}{Ct_0}\frac{1}{s^2(s+a)} \tag{34.13}$$

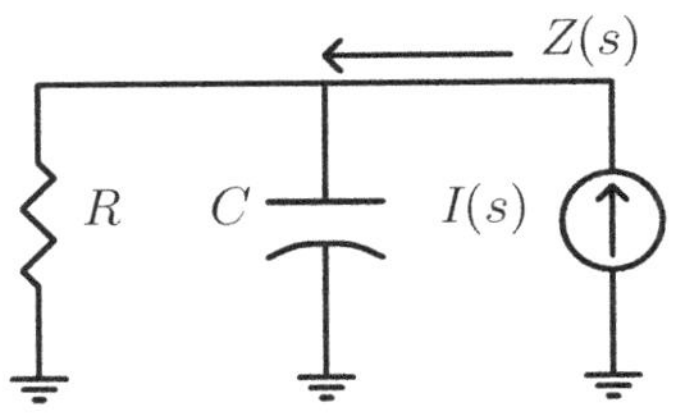

Fig. 34.4 Parallel *RC* and input impedance

We can factor the fraction as

$$\frac{1}{s^2(s+a)} = \frac{1}{a^2}\left[-\frac{1}{s} + \frac{a}{s^2} + \frac{1}{s+a}\right] \tag{34.14}$$

Then our output voltage becomes

$$\boxed{V(s) = \frac{1-e^{-st_0}}{Ct_0a^2}\left[-\frac{1}{s} + \frac{a}{s^2} + \frac{1}{s+a}\right]}, \quad a = \frac{1}{RC} \tag{34.15}$$

Let

$$w(t) = \frac{1}{Ct_0a^2}u(t)\left[-1 + at + e^{-at}\right] \tag{34.16}$$

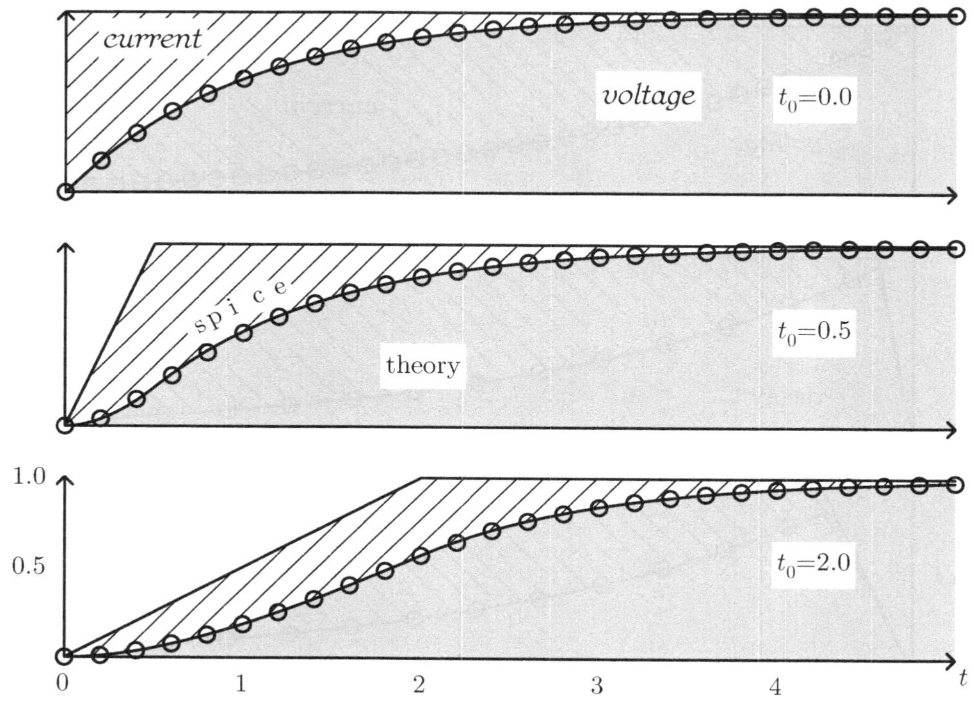

Fig. 34.5 Parallel RC slanted step response for different rise time ($R = 1\,\Omega$ and $C = 1$ F)

Using the time shifting property output voltage can then be written as

$$\boxed{v(t) = w(t) - w(t - t_0)} \tag{34.17}$$

Very nice and succinct ending! Sample results and comparison to SPICE are shown in Fig. 34.5.

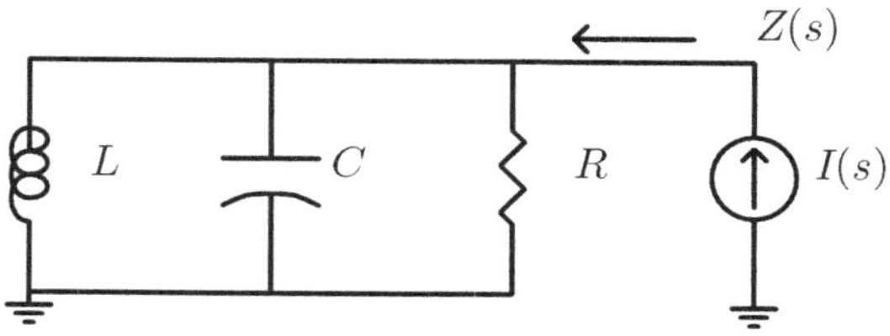

Fig. 34.6 Parallel RLC

34.5 Parallel *RLC* Circuit

Consider next the parallel RLC circuit in Fig. 34.6.

The output impedance has already been derived in Eq. (26.47) and is

$$Z(s) = \frac{1}{C}\frac{s}{s^2 + \frac{1}{RC}s + \omega_{LC}^2}, \qquad \omega_{LC}^2 = \frac{1}{LC} \tag{34.18}$$

which can be rewritten as

$$Z(s) = \frac{1}{C}\frac{s}{s^2 + 2as + \omega_{LC}^2}, \qquad a = \frac{1}{2RC} \tag{34.19}$$

Input current again is

$$I(s) = \frac{1}{t_0}\frac{1 - e^{-t_0 s}}{s^2} \tag{34.20}$$

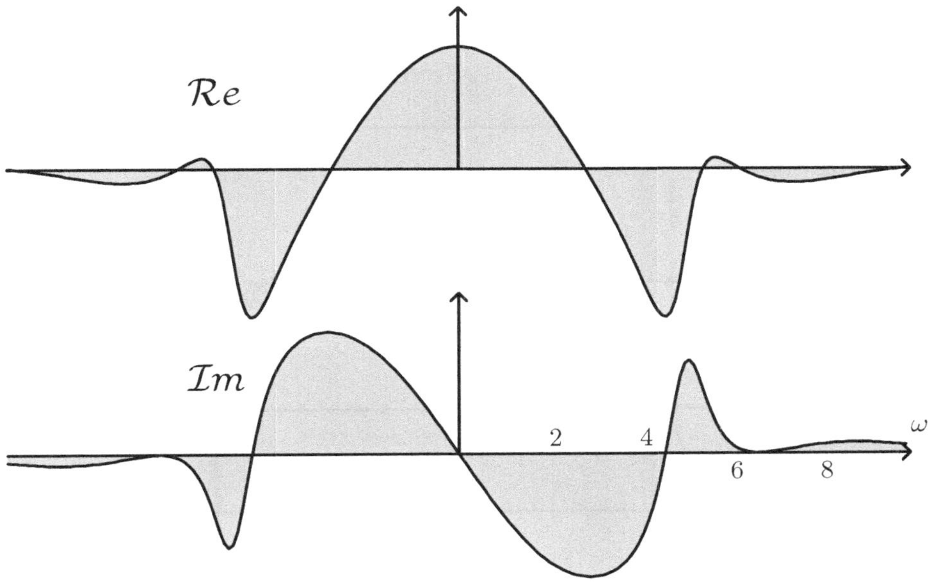

Fig. 34.7 Output voltage of parallel *RLC* network due to a slanted input voltage; case of $R = 5\,\Omega$; $L = 0.2\,\text{H}$ and $C = 0.2\,\text{F}$

Hence output voltage is

$$V(s) = \frac{1}{t_0}\frac{1-e^{-t_0 s}}{s^2}\frac{1}{C}\frac{s}{s^2+2as+\omega_{LC}^2} = \frac{1-e^{-t_0 s}}{t_0 C}\frac{1}{s}\frac{1}{s^2+2as+\omega_{LC}^2} \tag{34.21}$$

Using partial fraction (see Problem 3) we get

$$V(s) = \frac{1-e^{-t_0 s}}{t_0 C}\frac{1}{\omega_{LC}^2}\left[\frac{1}{s} - \frac{2a+s}{s^2+2as+\omega_{LC}^2}\right] \tag{34.22}$$

This can be rewritten as

$$V(s) = \frac{1-e^{-t_0 s}}{t_0 C}\frac{1}{\omega_{LC}^2}\left[\frac{1}{s} - \frac{2a+s}{(s+a)^2+\omega_0^2}\right], \quad \omega_0^2 = \omega_{LC}^2 - a^2 \tag{34.23}$$

This expression is plotted next in Fig. 34.7. It may be mind boggling but in this plot lies the response of the parallel *RLC* network due to a slanted unit step input current. In other words, within the shaded area of this complex (real and imaginary) obscure function lies the answer to this particular system with particular stimulus! All that has to be done to get the equivalent answer in the *time domain* is to do the inverse transform. Worst case we can always do this numerically but of course we are not going to! We know analytically what it would be, and that is (Problem 4)

$$\begin{aligned} w(t) &= \frac{1}{t_0 C\omega_{LC}^2}u(t) \left[1 - e^{-at}\left(\cos\omega_0 t + \frac{a}{\omega_0}\sin\omega_0 t\right)\right] \\ v(t) &= w(t) - u(t-t_0)w(t-t_0) \end{aligned} \tag{34.24}$$

Figure 34.8 shows transient results, for two different ramp rates t_0, and comparison to SPICE; observe exact match!

34.6 Series *RC*/Parallel *C* Circuit

The series *RC*/parallel *C* is shown in Fig. 34.9. The impedance was derived in Problem 3 of Chap. 26, and repeated here for convenience.

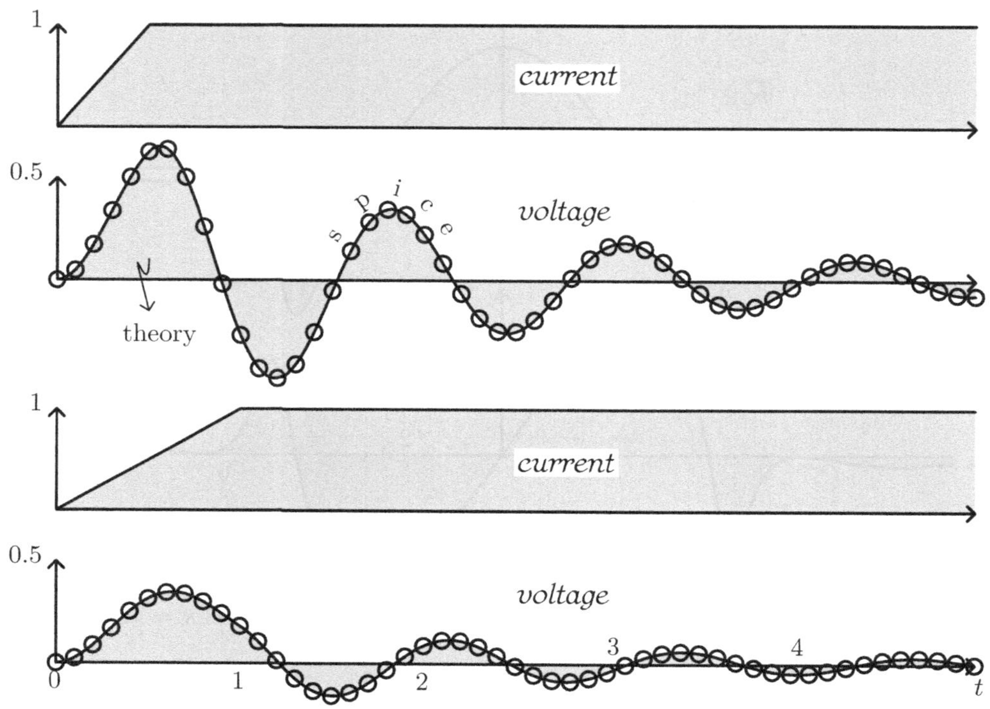

Fig. 34.8 Output voltage of parallel *RLC* network due to a slanted input voltage; case of $R = 5\,\Omega$; $L = 0.2\,\text{H}$ and $C = 0.2\,\text{F}$

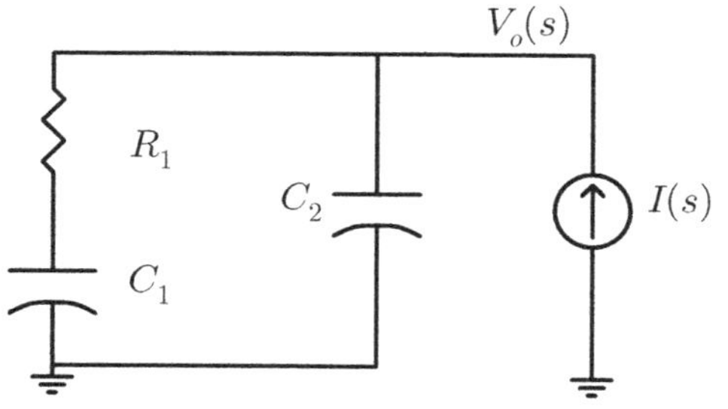

Fig. 34.9 Series *RC*/parallel *C* network

$$Z(s) = \frac{1}{sC_2}\frac{s+\dfrac{1}{R_1C_1}}{s+\dfrac{1}{R_1C_s}}, \qquad C_s = \frac{C_1C_2}{C_1+C_2} \tag{34.25}$$

If we define $a = \frac{1}{R_1C_1}$ and $b = \frac{1}{R_1C_s}$ and if we plug for input current we get

$$V(s) = \frac{1}{sC_2}\frac{s+a}{s+b}\frac{1}{t_0}\frac{1-e^{-t_0s}}{s^2} \tag{34.26}$$

This function is shown in Fig. 34.10. We can rewrite this (Problem 5) as

$$V(s) = \frac{1}{C_2t_0}\left[\frac{A}{s}+\frac{B}{s^2}+\frac{C}{s^3}+\frac{D}{s+b}\right]\left[1-e^{-t_0s}\right] \tag{34.27}$$

where

$$A = -\frac{b-a}{b^3}, \quad B = \frac{b-a}{b^2},$$
$$C = \frac{a}{b}, \quad D = \frac{b-a}{b^3} \tag{34.28}$$

Hence our intermediate solution is

$$\boxed{w(t) = \frac{1}{C_2t_0}u(t)\left[A+Bt+\frac{1}{2}Ct^2+De^{-bt}\right]} \tag{34.29}$$

and our final solution is

$$\boxed{v(t) = w(t) - w(t-t_0)} \tag{34.30}$$

A sample solution and comparison to SPICE are shown in Fig. 34.11.

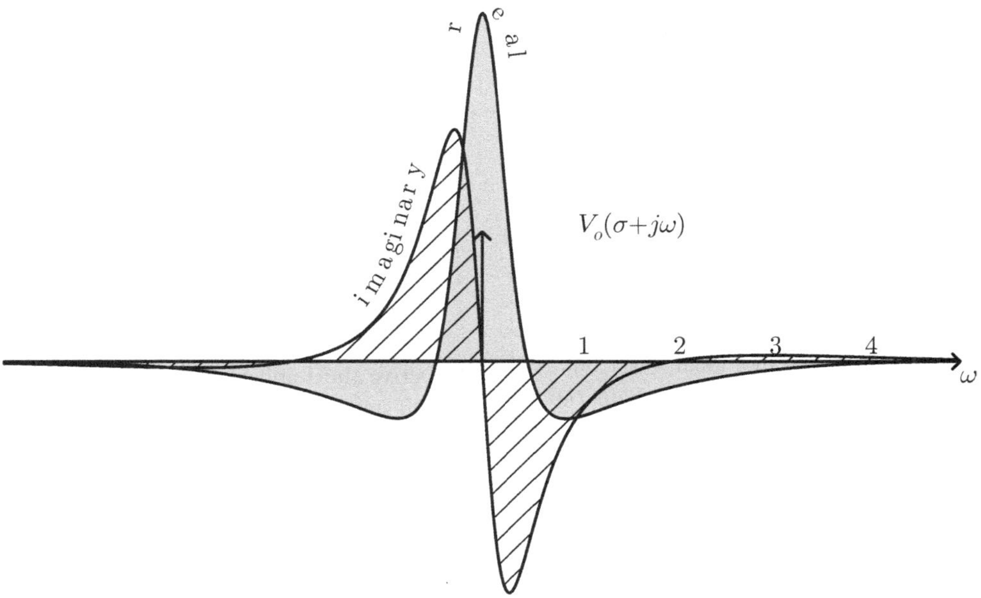

Fig. 34.10 Slanted step frequency response of series *RC*/parallel *C*; case of $R_1 = 1$, $C_1 = 1$, $C_2 = 0.5$, and $t_0 = 1$ (σ was set to 0.5)

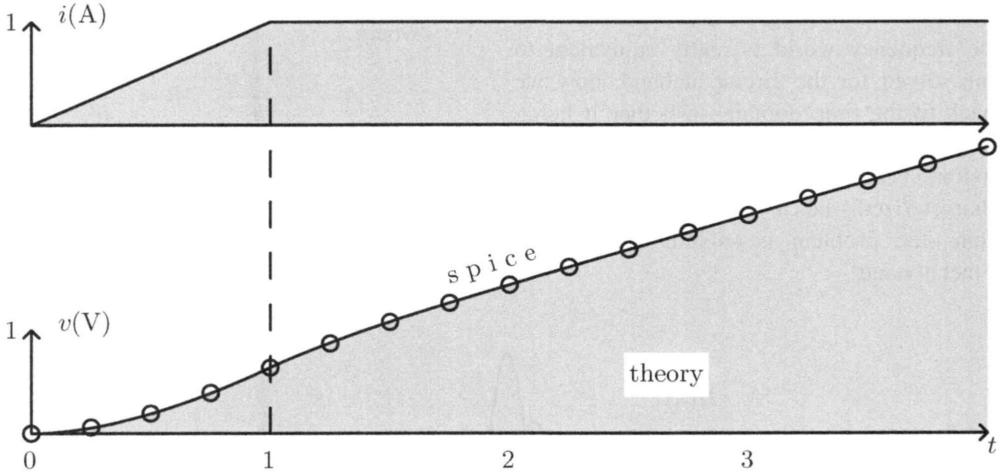

Fig. 34.11 Slanted step response of series *RC*/parallel *C* with $R_1 = 1$, $C_1 = 1$, $C_2 = 0.5$, and $t_0 = 1$

34.7 Summary

In this chapter, we wrapped a series of 6 chapters covering various responses ranging from impulse, unit step, one-timer pulse, causal sine/cosine, causal periodic pulse, and now the slanted unit step response. The slanted unit step response is very important because it also automatically gives us the response due to the *ideal* unit step, simply by taking the limit $t_0 \to 0$. In essence the slanted unit response follows very closely that of the ideal one, but with the exception of somehow *smoothing* out the response. So wherever we saw abrupt response while dealing with the ideal unit step, we should expect to see a more tapered, gradual one with the slanted step. Also whenever we observed oscillations in the response, such as the parallel *RLC* network, we should expect those to die sooner with the slanted step case. Other than that we should expect the low frequency response, or equivalently the long-time response to resemble that of the ideal step one. In this chapter we also emphasized that having obtained the solution in the frequency world is really equivalent to having solved for the circuit at hand; how we go back to the time domain—whether it being done analytically or by brute force numerical inversion—is really secondary. Once the system is characterized—be it in frequency or time domain—the problem is solved; the solution is in fact at hand!

34.8 Problems

1. The Laplace transform of the slanted step was shown to be

$$\text{slanted unit step} \to \frac{1}{t_0}\frac{1-e^{-st_0}}{s^2}$$

What is the inverse transform of $\frac{1}{s^2}$? What is the inverse transform of $\frac{e^{-st_0}}{s^2}$? Plot the inverse transform of each and then subtract latter from former—what do you get?

2. Derive the Laplace transform of the slanted unit step by first starting with a pulse of width t_0 and height $\frac{1}{t_0}$. Find the Laplace transform of the pulse. Then use the time integration property.
3. Fill in the details to derive Eq. (34.22).
4. Fill in the details to derive Eq. (34.24).
5. In Sect. 34.6 we made use of the following partial fraction expansion:

$$\frac{1}{s^3}\frac{s+a}{s+b} = \frac{A}{s} + \frac{B}{s^2} + \frac{C}{s^3} + \frac{D}{s+b}$$

where

$$A = -\frac{b-a}{b^3}, \quad B = \frac{b-a}{b^2},$$

$$C = \frac{a}{b}, \quad D = \frac{b-a}{b^3}$$

Prove this and plot it for the case of $a = 1$ and $b = 5$; see sample solution in Fig. 34.12.

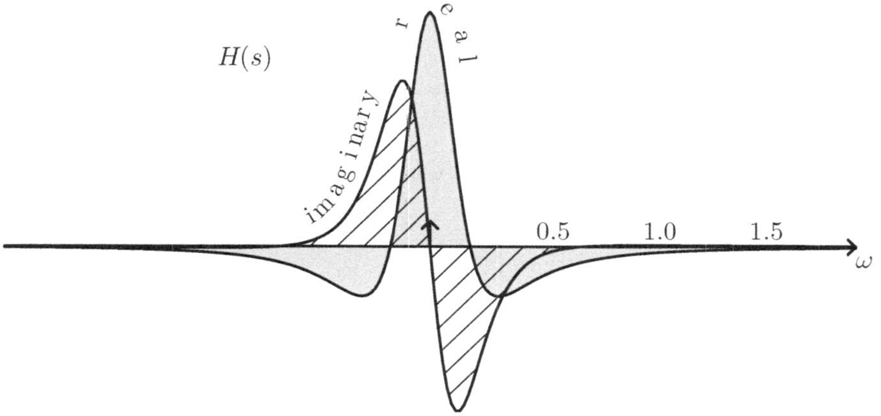

Fig. 34.12 Sample solution to Problem 5

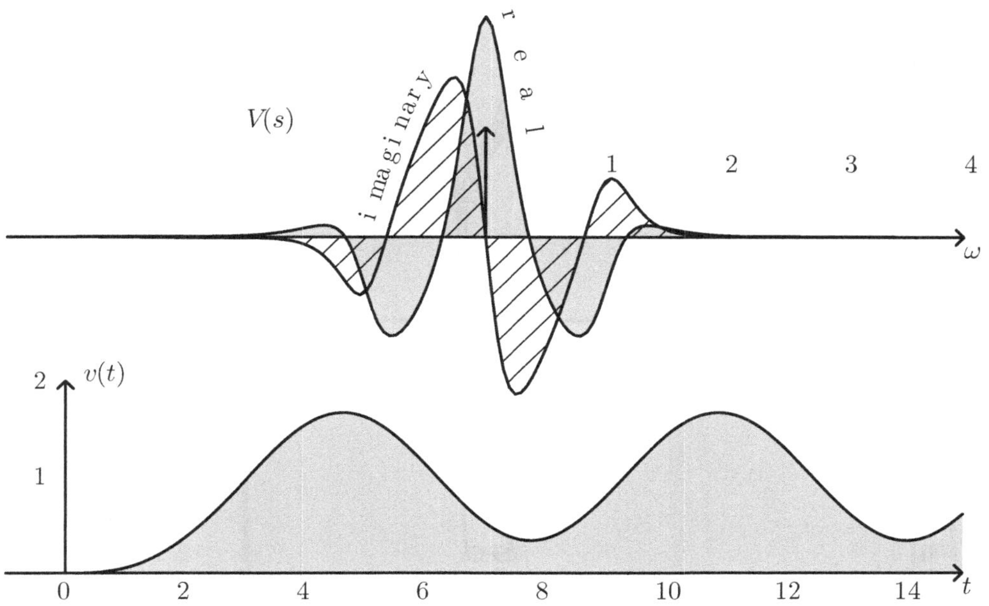

Fig. 34.13 Sample solution to Problem 6; top graph case of $\sigma = 0.3$

6. As system has the transfer function

$$Z(s) = \frac{1}{s^2 + 1}$$

What is the response due to a slanted step, with ramp time 3? Plot answer in frequency and time domain; see sample in Fig. 34.13.
Answer:

$$w(t) = u(t)\left[-\sin(t) + t\right]\frac{1}{3}$$
$$v(t) = w(t) - w(t-3)$$

7. Consider the series *LC*/parallel *L* network in Fig. 34.14.

(A) Find and plot the input impedance function $Z(s)$, and compare to SPICE; (B) find the output voltage due to a slanted unit step input (with $t_0 = 0.1$) and plot mag/phase on log scale; and (C) repeat but plot real/imaginary on linear scale with $\sigma = 1$. See sample solutions in Figs. 34.15, 34.16, and 34.17.
Answer:

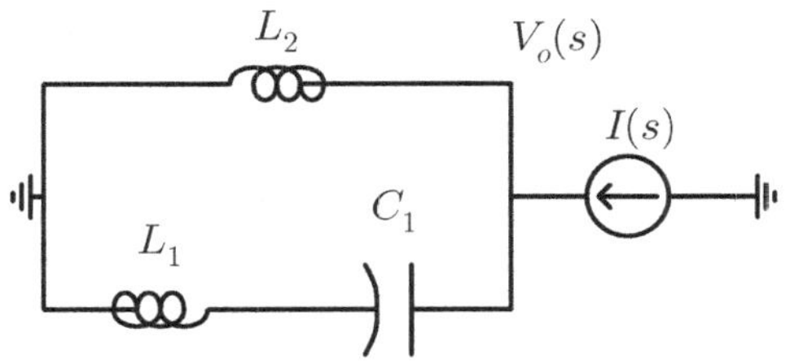

Fig. 34.14 Series *LC*/parallel *L* network for Problem 7; case of $L_1 = 1$ $L_2 = 2$ $C_1 = 0.01$

$$Z(s) = \frac{L_1 L_2}{L_1 + L_2} s \frac{s^2 + a}{s^2 + b},$$
$$a = \frac{1}{C_1 L_1}, \qquad b = \frac{1}{C_1(L_1 + L_2)}$$
$$V(s) = \frac{L_1 L_2}{L_1 + L_2} s \frac{s^2 + a}{s^2 + b} \frac{1}{t_0} \frac{1 - e^{-t_0 s}}{s^2}$$

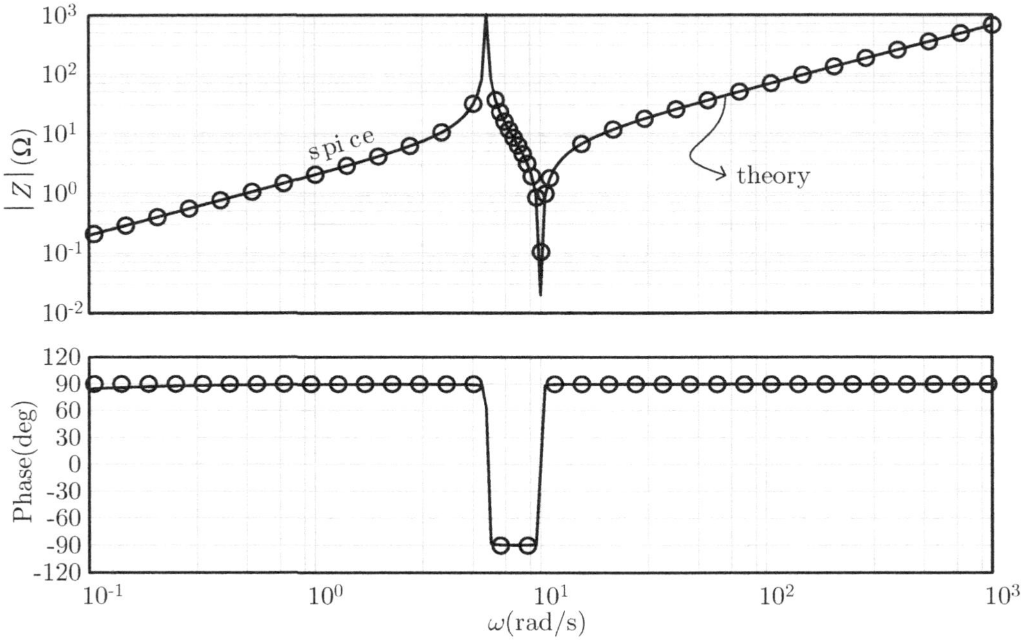

Fig. 34.15 Sample solution to Problem 7 (part (A)); case of $\sigma = 0.01$

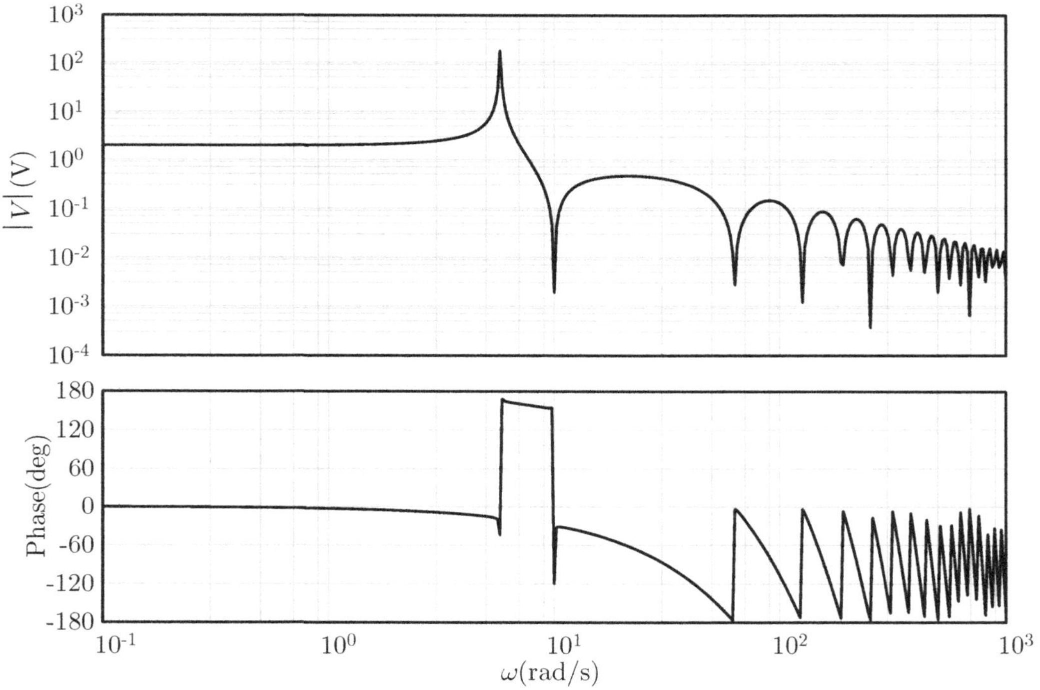

Fig. 34.16 Sample solution to Problem 7 (part (B)). case of $\sigma = 0.01$

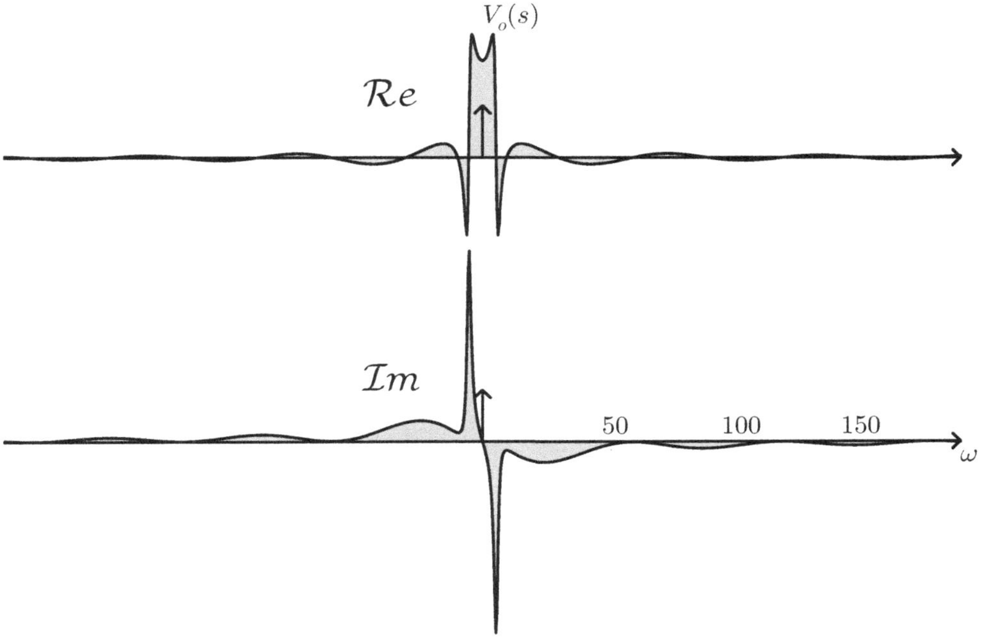

Fig. 34.17 Sample solution to Problem 7 (part (C)). case of $\sigma = 1.00$

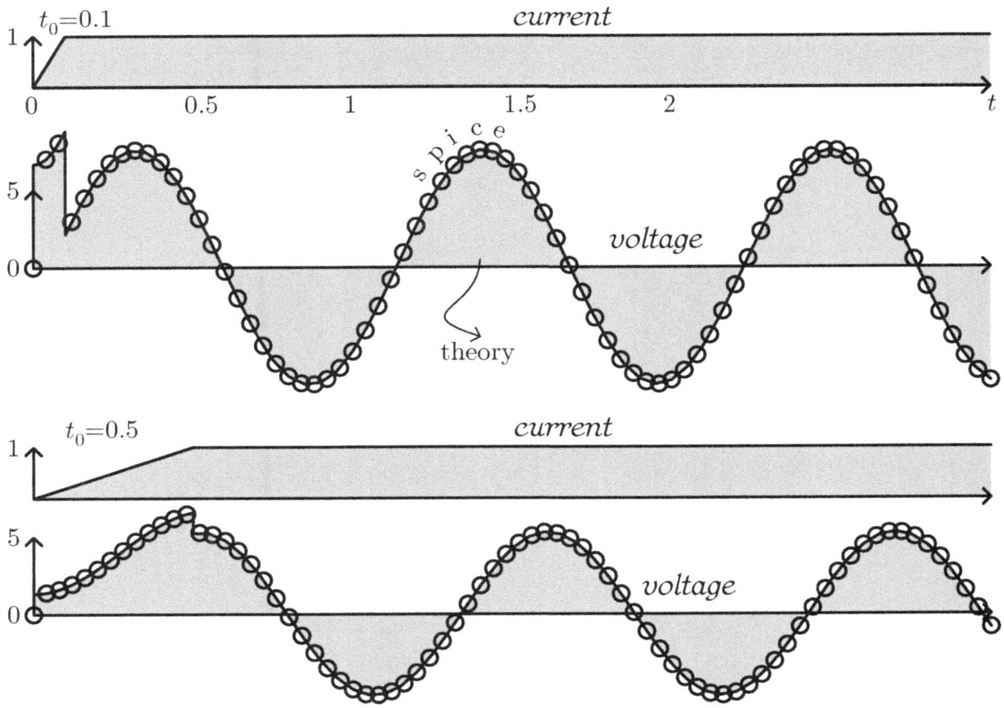

Fig. 34.18 Sample solution to Problem 8; top $t_0 = 0.1$, bottom $t_0 = 1.0$

8. Starting with Problem 7 decompose output function in terms of partial fractions and then find time response for the two cases: $t_0 = 0.1$ and $t_0 = 0.5$; see sample solution in Fig. 34.18
Answer:

$$V(s) = \frac{L_1 L_2}{L_1 + L_2}\left[\frac{As}{s^2 + b} + \frac{B}{s}\right]\frac{1}{t_0}\left[1 - e^{-t_0 s}\right],$$

$$B = \frac{a}{b}, \qquad A = 1 - B$$

$$w(t) = u(t)\frac{L_1 L_2}{L_1 + L_2}\left[A\cos(t\sqrt{b}) + B\right]\frac{1}{t_0}$$

$$v(t) = w(t) - w(t - t_0)$$

35 Voltage/Voltage Filters

35.1 Introduction

The majority of the text so far dealt with problems involving input current and output voltage. Those rendered impedance transfer functions. This, however, does not imply that the techniques developed so far apply only for those cases. In fact the surveyed techniques apply to so much more. This chapter samples another class of problems, where input is voltage and output is voltage too; we call those voltage filters. Really we can assume input is anything and output is anything else! In the end it all boils to symbols and numbers; and so far as spectral and convolution techniques are concerned, numbers remain numbers! In the prior chapters we dealt at least with 6 types of input functions, ranging from impulses, unit step ones, to causal periodic pulses. Let us then sample a few applied voltage filter cases, and test them under one of those 6 classes of input stimuli. Keep in mind the theory works for much more than 6 classes; but we have to start somewhere.

35.2 Series *RC*: Low-Pass Filter

Low-pass filters pass low frequencies and shut off high frequencies; hence the name "low-pass"! The series *RC* low-pass filter has input applied across both *R* and *C* and output measured across the cap. The circuit is shown in Fig. 35.1. The total current is simply input voltage divided by impedance:

$$I(s) = \frac{V_i}{R + \frac{1}{sC}} \tag{35.1}$$

The output voltage is total current times output impedance

$$V_o(s) = \frac{\frac{V_i(s)}{sC}}{R + \frac{1}{sC}} = \frac{V_i}{1 + sRC} = \boxed{\frac{V_i}{RC}\frac{1}{s + \frac{1}{RC}}} \tag{35.2}$$

Clearly when frequency is low enough, output voltage is high, and almost equals to input voltage

$$V_o \sim V_i, \qquad (\omega \quad \text{small}) \tag{35.3}$$

F. Badrieh, *Spectral, Convolution and Numerical Techniques in Circuit Theory*,
https://doi.org/10.1007/978-3-319-71437-0_35

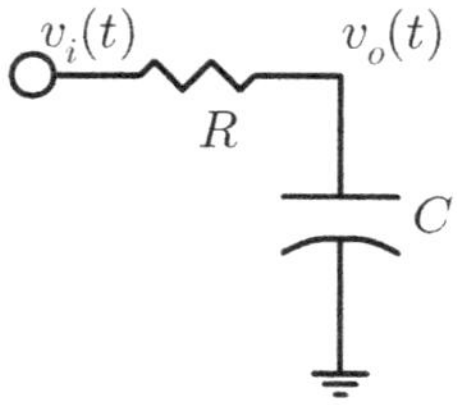

Fig. 35.1 Series RC circuit acting as low-pass filter

On the other hand, for large frequency output voltage dies off

$$V_o \sim 0, \qquad (\omega \quad \text{large}) \tag{35.4}$$

Put another way, what we have here is impedance division. The output is sensed across the cap; so if the cap impedance is large, so will output voltage, and vice versa. But under no circumstance would output voltage exceed that of input one. Clearly at low frequency cap impedance is large; hence output voltage would almost equal that of input one. Conversely at high frequency the cap shorts and so will output voltage. Figure 35.2 shows the results for a couple of cap values. Notice the phase starts at 0, meaning output tracks with input; and it ends at $-90°$, meaning output lags input.

Causal Cosine Input Clearly we can apply a variety of input voltages to the low-pass filter and attempt to figure corresponding output voltage. In this case, however, we are interested in the causal cosine input

$$v_i(t) = u(t) \cos \omega_0 t \tag{35.5}$$

The corresponding voltage in the frequency domain is

$$V_i(s) = \frac{s}{s^2 + \omega_0^2} \tag{35.6}$$

Based on Eq. 35.2, output voltage is then

$$\boxed{V_o(s) = a \frac{s}{s^2 + \omega_0^2} \frac{1}{s + a}, \qquad a = \frac{1}{RC}} \tag{35.7}$$

Using partial fraction (see Problem 1) we get

$$\boxed{V_o(s) = \frac{a^2}{\omega_0^2 + a^2} \left[\frac{-1}{s + a} + \frac{\omega_0^2/a + s}{s^2 + \omega_0^2} \right]} \tag{35.8}$$

The inverse transform of this is

$$\boxed{v_o(t) = \frac{a^2}{a^2 + \omega_0^2} \left[-e^{-at} + \cos \omega_0 t + \frac{\omega_0}{a} \sin \omega_0 t \right]} \tag{35.9}$$

Figure 35.3 shows results for two frequencies: the top half shows low frequency results where output voltage is high (almost equal to input one); conversely the bottom half shows high frequency results where output voltage is lower. And hence the low-pass filter—the low frequencies are passed, but the higher ones are attenuated!

35.3 Cap in Series with Parallel *RC* Voltage Transfer Function

The circuit under study is shown in Fig. 35.4. Voltage is applied at input terminal and voltage is measured at output terminal. The transfer function ties output voltage to input one. We start by finding the total impedance of this network as seen from input side.

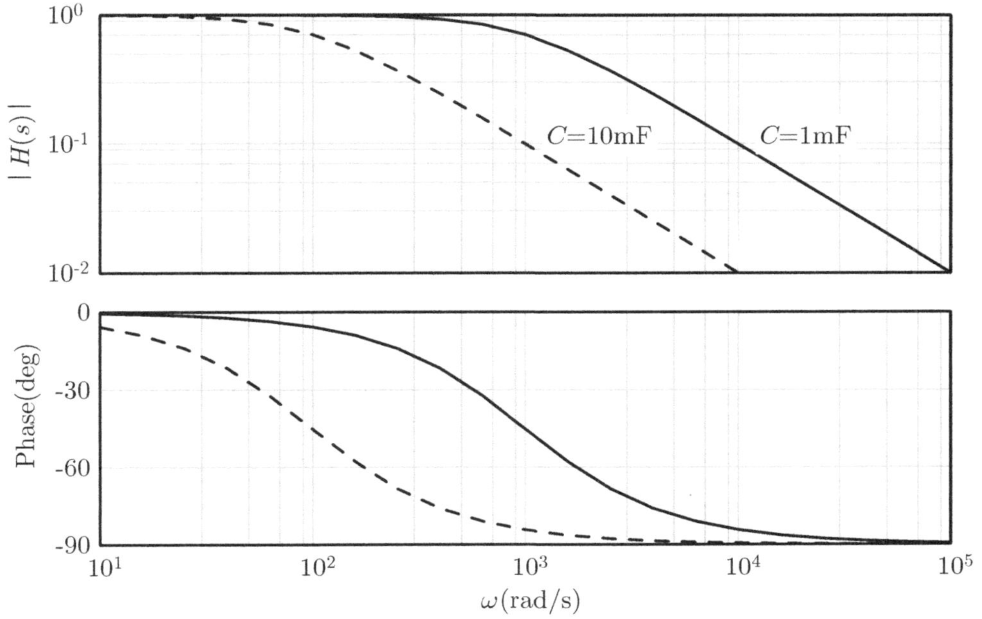

Fig. 35.2 Transfer function of low-pass filter ($R = 1\,\Omega$)

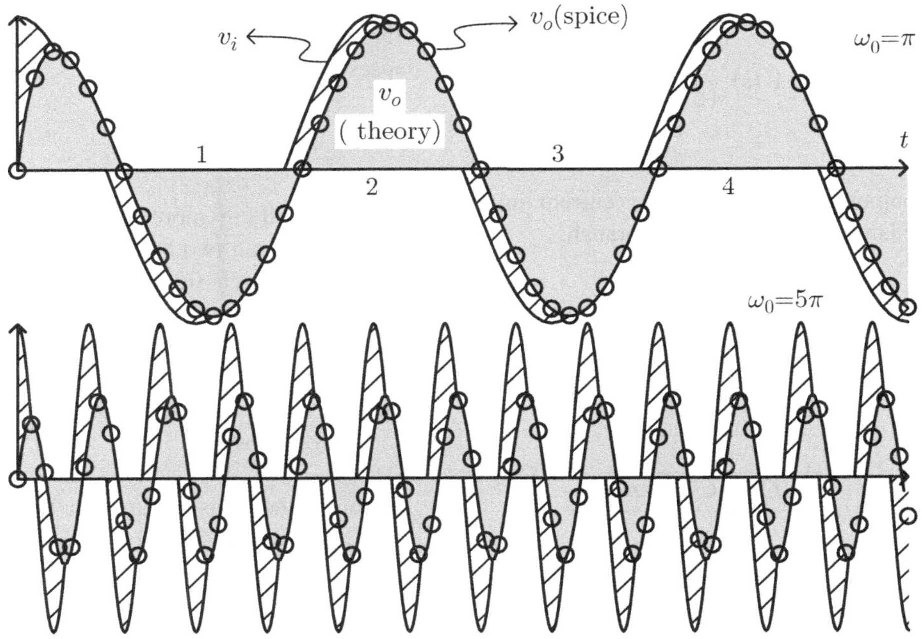

Fig. 35.3 Causal cosine response of low-pass filter ($R = 1\,\Omega$ and $C = 0.1\,$F)

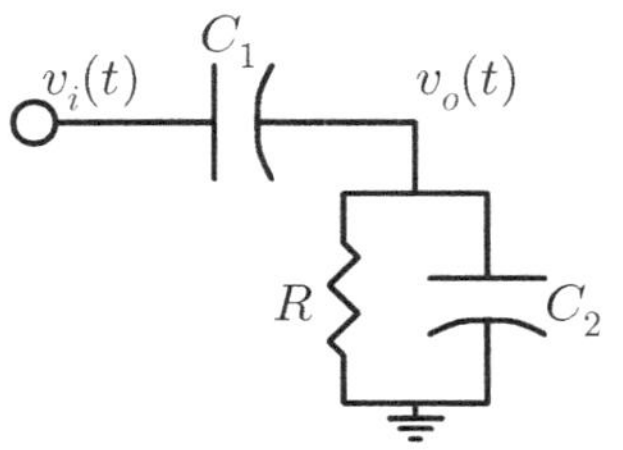

Fig. 35.4 Cap in series with parallel RC network

$$Z(s) = \frac{1}{sC_1} + \frac{1}{C_2}\frac{1}{s+a}, \qquad a = \frac{1}{RC_2}$$
$$= \frac{C_2(s+a) + sC_1}{C_1C_2s(s+a)} \tag{35.10}$$

$$Z(s) = \frac{s(C_1+C_2) + aC_2}{C_1C_2s(s+a)} \tag{35.11}$$

Notice that at zero frequency the impedance is open and that at high frequency it collapses to that of C_1 in series with C_2. The total current (as seen from input side) is simply the inverse of this impedance (times input voltage)

$$I(s) = \frac{V_i(s)}{Z(s)} = V_i(s)\frac{C_1C_2s(s+a)}{s(C_1+C_2)+aC_2} \tag{35.12}$$

The output voltage is simply this current times the impedance of the parallel RC branch:

$$V_o(s) = I(s)\frac{1}{C_2}\frac{1}{s+a}$$
$$= V_i(s)\frac{C_1C_2s(s+a)}{s(C_1+C_2)+aC_2}\frac{1}{C_2}\frac{1}{s+a}$$
$$= \boxed{V_i(s)\frac{C_1s}{s(C_1+C_2)+aC_2}} \tag{35.13}$$

Notice the following two limits:

- Low frequency limit: Here we drop the s term from the denominator and end up with

$$\lim_{s\to 0}\frac{V_o(s)}{V_i(s)} = \frac{C_1s}{aC_2} = \frac{C_1s}{\frac{1}{RC_2}C_2} = RC_1s \tag{35.14}$$

But this is nothing more than the voltage across the resistor when C_2 is dropped out; that is, for the case of no C_2 impedance would be

$$Z(s) = \frac{1}{sC_1} + R = \frac{1+sRC_1}{sC_1} \tag{35.15}$$

and current inverse of that

$$I(s) = V_i(s)\frac{sC_1}{1+sRC_1} \tag{35.16}$$

Then voltage would be current times resistance

$$\frac{V_o(s)}{V_i(s)} = \frac{sRC_1}{1+sRC_1} \tag{35.17}$$

which for low frequency collapses to

$$\lim_{s\to 0}\frac{V_o(s)}{V_i(s)} = sRC_1 \tag{35.18}$$

in agreement with Eq. (35.14).

- High frequency limit: Here Eq. (35.13) collapses to

$$\lim_{s\to\infty}\frac{V_o(s)}{V_i(s)} = \frac{C_1}{C_1+C_2} \tag{35.19}$$

which is nothing more than using voltage division between two caps, in the absence of R; that is at high frequency the impedance of C_2 shunts that of R, and hence R drops out of the picture. Then, voltage across C_2 is obtained from voltage division between two caps: the bigger the other cap (C_1), the bigger the voltage across C_2!

A plot of the transfer function is shown in Fig. 35.5. Notice that initially phase is 90° and then it dies out to 0°.

Impulse Response The impulse response is the inverse Laplace transform of the transfer function, which is rewritten below

$$H(s) = \frac{V_o(s)}{V_i(s)} = \frac{C_1s}{s(C_1+C_2)+\frac{1}{R}} \tag{35.20}$$

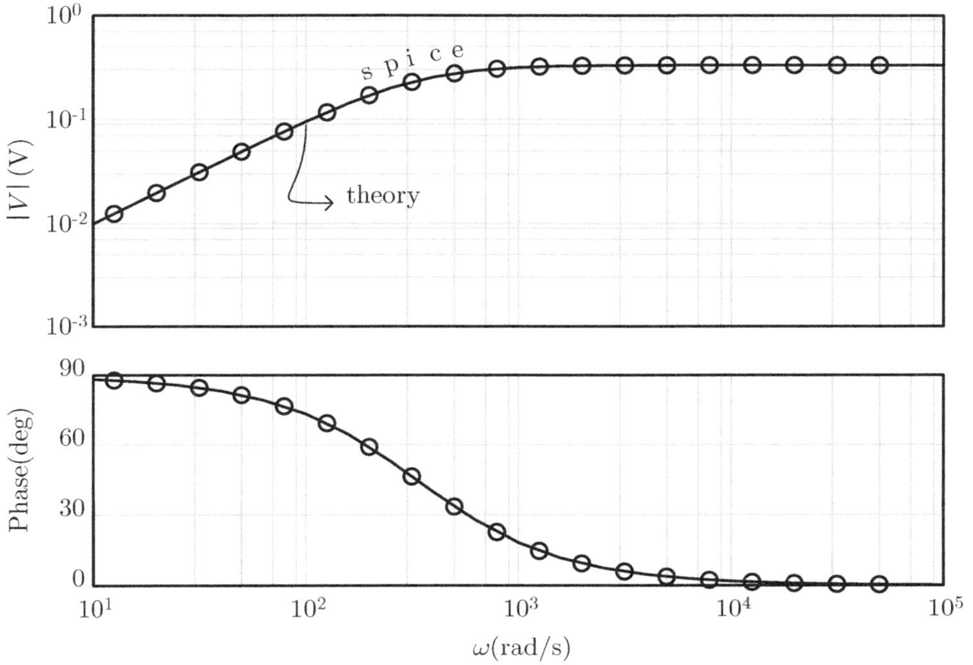

Fig. 35.5 Transfer function of cap in series with parallel *RC* ($C_1 = 1$ mF, $C_2 = 2$ mF, and $R = 1\ \Omega$)

Using long division (see Problem 3) we get

$$H(s) = \frac{C_1}{C_1 + C_2}\left[1 - \frac{1}{R(C_1 + C_2)}\frac{1}{s + \frac{1}{R(C_1+C_2)}}\right] \tag{35.21}$$

Then we can inverse transform to get the impulse response

$$\boxed{h(t) = \frac{C_1}{C_1+C_2}\left[\delta(t) - \frac{1}{R(C_1+C_2)}\exp\left(\frac{-t}{R(C_1+C_2)}\right)\right]} \tag{35.22}$$

Step Response We can get the unit step response by direct integration of the impulse response (Eq. (35.22)); but we can also get it from manipulation of the transfer function. We start again with the transfer function reshown below:

$$H(s) = \frac{V_o(s)}{V_i(s)} = \frac{C_1 s}{s(C_1 + C_2) + \frac{1}{R}} \tag{35.23}$$

The step response is obtained by simply multiplying this by $1/s$

$$\begin{aligned} G(s) &= H(s)\frac{1}{s} \\ &= \frac{C_1}{s(C_1 + C_2) + \frac{1}{R}} \end{aligned} \tag{35.24}$$

The inverse transform of this is easy and given by

$$\boxed{g(t) = \frac{C_1}{C_1 + C_2}\exp\left(\frac{-t}{R(C_1 + C_2)}\right)} \tag{35.25}$$

Notice the following limits:

- Time zero: here the exponential goes to 1 and we get

$$v_o(0) = \frac{C_1}{C_1 + C_2} \tag{35.26}$$

which is nothing more than voltage division between the two caps; that is, right when input

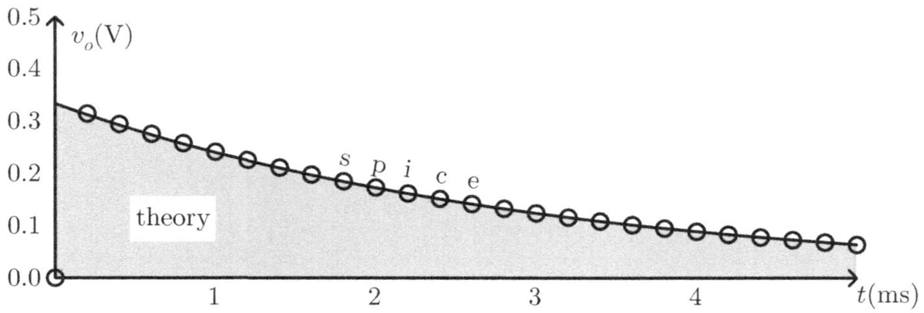

Fig. 35.6 Cap in series with parallel RC unit step response ($C_1 = 1$ mF, $C_2 = 2$ mF, and $R = 1\ \Omega$)

voltage is applied, we have a fast transition, and during this fast time interval, the caps have very low impedance, which renders the R out of the picture; this means that output voltage is simple voltage division between two caps.

- Infinite time: here the exponential goes to zero and we get

$$v_o(\infty) = 0 \tag{35.27}$$

That is, after things have settled down, there is no current across the resistor, which implies its voltage is zero, and hence voltage across C_2. Recall steady state implies caps have constant voltage. For any constant voltage other than zero across C_2 this would imply that there is current across R; but we cannot have current there since any would have to come from either of the two caps, both of which have DC voltage value, which by definition cannot supply current! Hence at steady state, all input voltage is across input cap C_1.

Figure 35.6 shows the results and comparison to SPICE.

35.4 Second Order Low-Pass Filter

Case of study is a second order low-pass filter as shown in Fig. 35.7. Input is applied in the form of a voltage source on the left side, and output is measured (again in form of voltage) at the right side. We will proceed as follows. First find input impedance; second find input current; third find output current; and finally find output voltage!

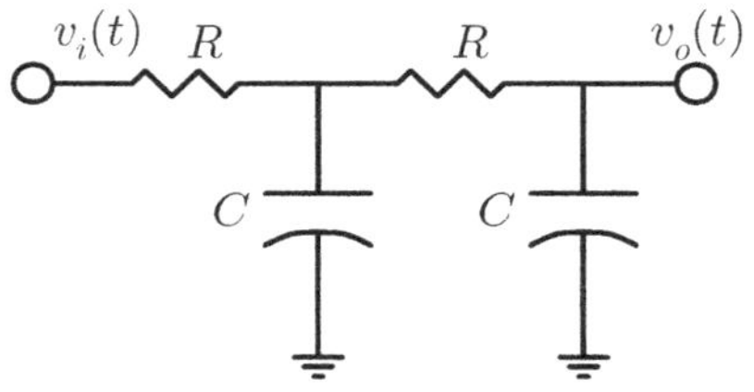

Fig. 35.7 Second order low-pass filter

Input Impedance To find impedance at input side we start at far right side and make our way back. The impedance of the last C and R is

$$Z_1(s) = R + \frac{1}{sC} = \frac{sRC + 1}{sC} \tag{35.28}$$

This impedance goes in parallel with the first cap to get

$$Z_2(s) = \frac{sRC + 1}{sC} || \frac{1}{sC}$$

$$= \frac{\frac{sRC+1}{sC}\frac{1}{sC}}{\frac{sRC+1}{sC} + \frac{1}{sC}} = \frac{1}{C}\frac{sRC+1}{s(sRC+2)} \tag{35.29}$$

This has to be added in series with the left resistor to get

$$Z_i(s) = R + \frac{1}{C}\frac{sRC+1}{s(sRC+2)}$$
$$= \frac{RCs(sRC+2) + sRC + 1}{Cs(sRC+2)} \tag{35.30}$$

$$\boxed{Z(s) = \frac{s^2R^2C^2 + 3sRC + 1}{Cs(sRC+2)}} \tag{35.31}$$

A couple of sanity checks are done. At DC we get the limit

$$\lim_{s\to 0} Z(s) = \frac{1}{2Cs} \tag{35.32}$$

which is nothing more than the impedance of the two caps in parallel. On the other hand at high frequency we get the limit

$$\lim_{s\to\infty} Z(s) = R \tag{35.33}$$

which is simply the left resistor; that is at high frequency the caps short and we see only the first resistor. A plot of input impedance is shown in Fig. 35.8.

Input Current The input current is simply inverse of input impedance; that is

$$I(s) = \frac{s^2RC^2 + 2sC}{s^2R^2C^2 + 3sRC + 1} \tag{35.34}$$

Let's factor an R^2C^2 from the denominator

$$I(s) = \frac{1}{R^2C^2}\frac{s^2RC^2 + 2sC}{s^2 + \frac{3s}{RC} + \frac{1}{R^2C^2}}$$
$$= \frac{\frac{s^2}{R} + 2\frac{s}{R^2C}}{s^2 + \frac{3s}{RC} + \frac{1}{R^2C^2}} \tag{35.35}$$

A couple of sanity checks are done. At DC we get the following limit:

$$\lim_{s\to 0} I(s) = 0 \tag{35.36}$$

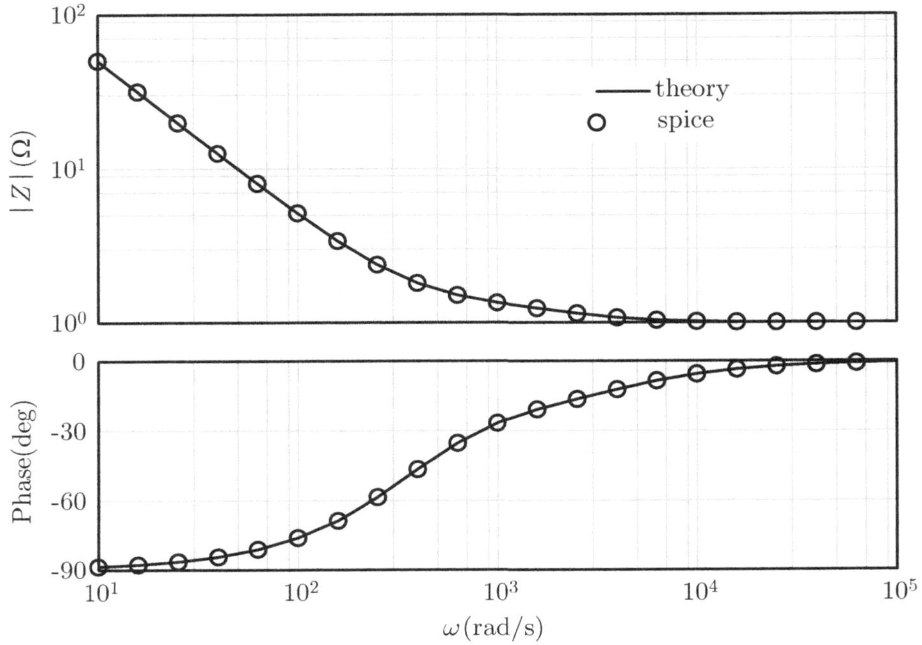

Fig. 35.8 Impedance of second order filter as seen at input terminal ($R = 1\,\Omega$ and $C = 1$ mF)

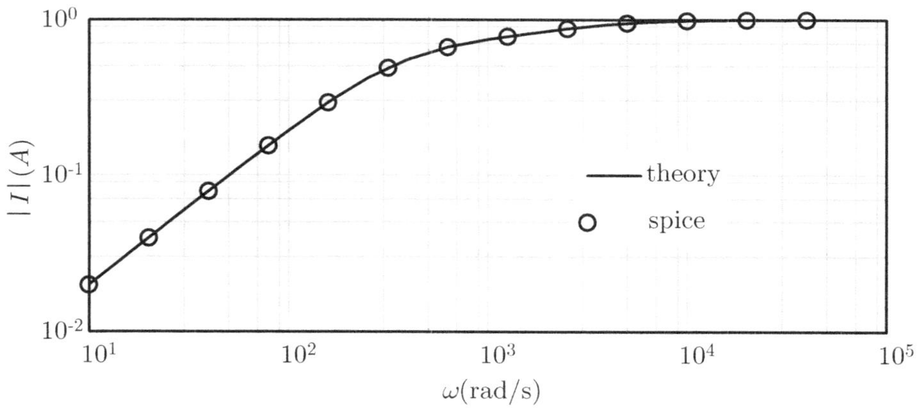

Fig. 35.9 Input current to second order filter ($R = 1\ \Omega$ and $C = 1$ mF)

That is, at DC the caps are open and we get zero current. At high frequency we get the following limit:

$$\lim_{s\to\infty} I(s) = \frac{1}{R} \tag{35.37}$$

which is simply the current across the left resistor assuming that its right terminal is at ground (since the cap shorted). A plot of this current is shown in Fig. 35.9 Next we would want to find the roots of the denominator. To that end we must first factor the denominator.

$$D(s) = s^2 + \frac{3s}{RC} + \frac{1}{R^2C^2} \tag{35.38}$$

Recall that the solutions for the quadratic formula of the form

$$ax^2 + bx + c = 0 \Rightarrow x = \frac{-b \pm \sqrt{b^2 - 4ac}}{2a} \tag{35.39}$$

We can then find the roots of $D(s)$ as

$$s = \frac{-\frac{3}{RC} \pm \sqrt{\frac{9}{R^2C^2} - \frac{4}{R^2C^2}}}{2} = \frac{-\frac{3}{RC} \pm \frac{\sqrt{5}}{RC}}{2} = \frac{-3 \pm \sqrt{5}}{2RC} \tag{35.40}$$

Hence we can write the denominator as

$$D(s) = \left(s + \frac{3+\sqrt{5}}{2RC}\right)\left(s + \frac{3-\sqrt{5}}{2RC}\right) \tag{35.41}$$

So that current becomes

$$I(s) = \frac{\frac{s^2}{R} + 2\frac{s}{R^2C}}{\left(s + \frac{3+\sqrt{5}}{2RC}\right)\left(s + \frac{3-\sqrt{5}}{2RC}\right)} \tag{35.42}$$

Output Current We can now proceed to figure output current using current division. The input current will split at the node to the right of the left resistor as

$$I_o(s) = I(s)\frac{\frac{1}{sC}}{\frac{1}{sC}+R+\frac{1}{sC}}$$
$$= I(s)\frac{1}{2+sRC} = I(s)\frac{1}{RC}\frac{1}{s+\frac{2}{RC}}$$
$$= \frac{1}{RC}\frac{\frac{s^2}{R}+2\frac{s}{R^2C}}{\left(s+\frac{3+\sqrt{5}}{2RC}\right)\left(s+\frac{3-\sqrt{5}}{2RC}\right)}\frac{1}{\left(s+\frac{2}{RC}\right)} \tag{35.43}$$

Let's do a couple of sanity checks. At DC we get

$$\lim_{s\to 0} I_o(s) = 0 \tag{35.44}$$

meaning that the input impedance is infinite and no current passes through. On the other hand, at high frequency we also have the limit

$$\lim_{s\to\infty} I_o(s) = 0 \tag{35.45}$$

which means that since the first (left) cap shorts, no current would pass to the output. This is validated in plot in Fig. 35.10.

Output Voltage Output voltage is output current times impedance of output cap

$$V_o(s) = I_o(s)\frac{1}{sC}$$
$$= \frac{1}{RC}\frac{\frac{s^2}{R}+2\frac{s}{R^2C}}{\left(s+\frac{3+\sqrt{5}}{2RC}\right)\left(s+\frac{3-\sqrt{5}}{2RC}\right)}\frac{1}{\left(s+\frac{2}{RC}\right)}\frac{1}{sC}$$
$$= \boxed{\frac{1}{R^2C^2}\frac{1}{\left(s+\frac{3+\sqrt{5}}{2RC}\right)\left(s+\frac{3-\sqrt{5}}{2RC}\right)}} \tag{35.46}$$

Notice that the denominator has order of 2, and that is why we call this circuit a filter of the order two! A couple of sanity checks are done. At DC we have the following limit:

$$\lim_{s\to 0} V_o(s) = \frac{1}{R^2C^2}\frac{1}{\left(\frac{3+\sqrt{5}}{2RC}\right)\left(\frac{3-\sqrt{5}}{2RC}\right)}$$
$$= \frac{1}{R^2C^2}\frac{1}{\frac{1}{R^2C^2}} = 1 \tag{35.47}$$

which is no surprise! At DC the caps are open and no current flows; hence input voltage equals output voltage. On the other hand, at high frequency we have the following limit:

$$\lim_{s\to\infty} V_o(s) = 0 \tag{35.48}$$

since denominator order is 2. This is expected since we are measuring voltage across a cap

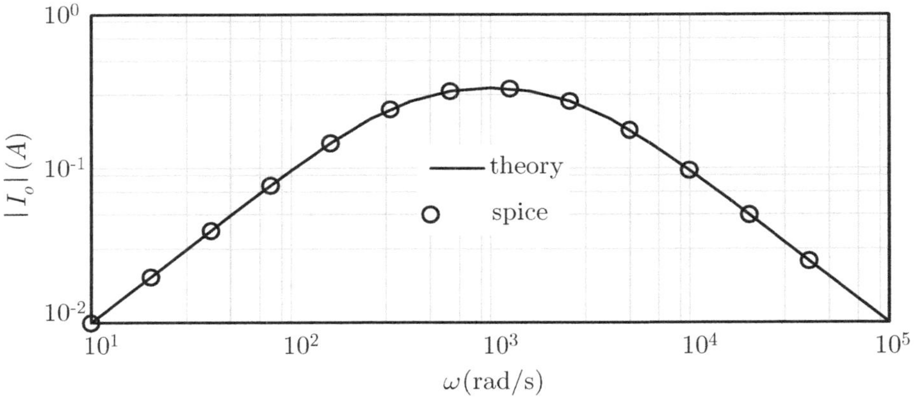

Fig. 35.10 Output current to second order filter ($R = 1\,\Omega$ and $C = 1$ mF)

and cap impedance shorts at high frequency. The phase starts at zero, since output fully tracks input; and it ends at $-180°$ since the high frequency limit is $\sim\frac{1}{s^2}$. This is all validated via simulations as shown in Fig. 35.11. Once the transfer function is known we can now figure various time responses, starting with impulse response, unit step one, and so forth. In all cases we would need to expand the transfer function in terms of a series of single pole transfer functions, using partial fractions, like we've done before. We can also try to fit the complex response in terms of a simpler one (smaller order), then do partial fractions. We can even go bold and evaluate the inverse transform numerically! Any of those methods are left for the motivated reader.

35.5 High-Pass Filter

The series RL high-pass filter is shown in Fig. 35.12. Input impedance is

$$Z(s) = R + sL \tag{35.49}$$

Input current is

$$\begin{aligned} I(s) &= \frac{V_i(s)}{Z(s)} = V_i(s)\frac{1}{R+sL} \\ &= \boxed{V_i(s)\frac{1}{L}\frac{1}{s+R/L}} \end{aligned} \tag{35.50}$$

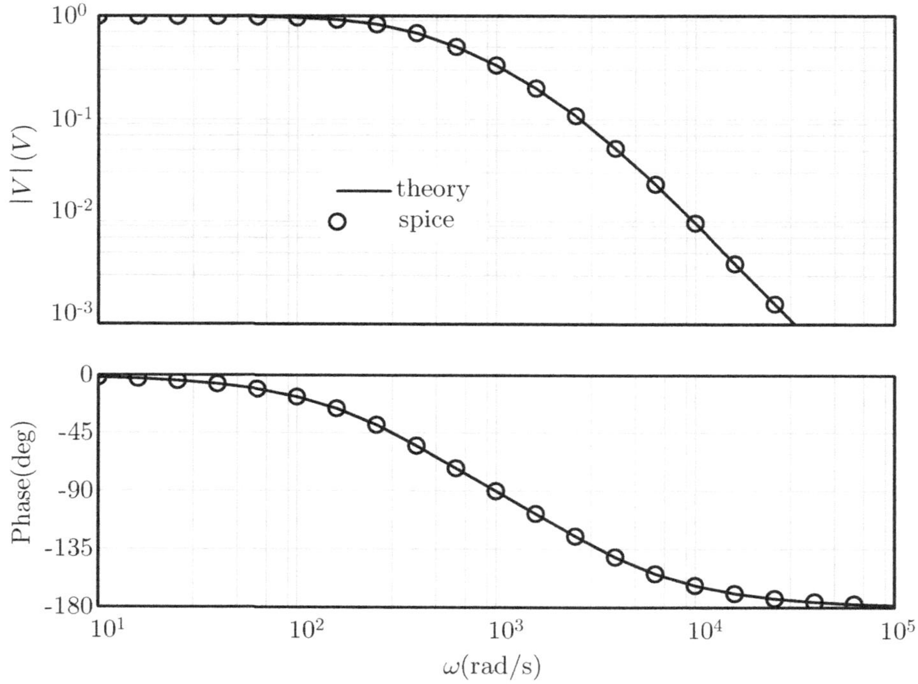

Fig. 35.11 Output voltage of second order filter ($R = 1\ \Omega$ and $C = 1$ mF)

Fig. 35.12 High-pass filter

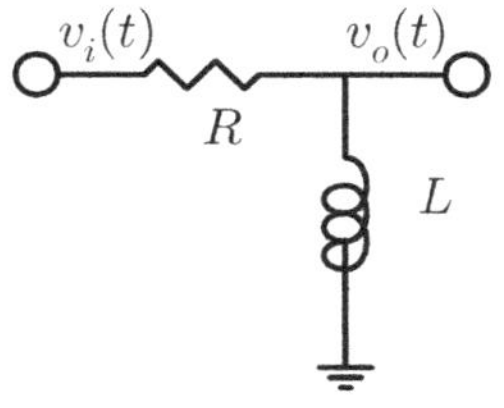

Voltage across the inductor is input current times output impedance

$$V_o(s) = V_i(s)\frac{1}{L}\frac{1}{s+R/L}sL = V_i(s)\frac{s}{s+R/L} \quad (35.51)$$

Using long division we get

$$\boxed{V_o(s) = V_i(s)\left[1 - \frac{R}{L}\frac{1}{s+R/L}\right]} \quad (35.52)$$

Notice that at DC output voltage is zero, since inductor is short; and at high frequency output voltage is unity, since output impedance is infinite. Figure 35.13 shows the transfer function.

Unit Step Response To find the unit step response we set input voltage to

$$V_i(s) = \frac{1}{s} \quad (35.53)$$

Rather than using Eq. (35.52) for the transfer function we use the earlier Eq. (35.51). Then

$$V_o(s) = \frac{1}{s}\frac{s}{s+R/L} = \boxed{\frac{1}{s+R/L}} \quad (35.54)$$

The inverse transform is simply

$$\boxed{v_o(t) = u(t)e^{-tR/L}} \quad (35.55)$$

Notice that right after input voltage application, output is unity; that is, inductor assumes all the voltage. When things settle down ($t \to \infty$) output voltage is zero, since output current is DC. Figure 35.14 shows results and comparison to SPICE.

35.6 Low-Pass *LC* Filter (with Resonance)

A low-pass *LC* filter is shown in Fig. 35.15. Input is applied at the left side of the inductor, and output is measured at the output of the cap. Qualitatively, it is presumed that if the input is

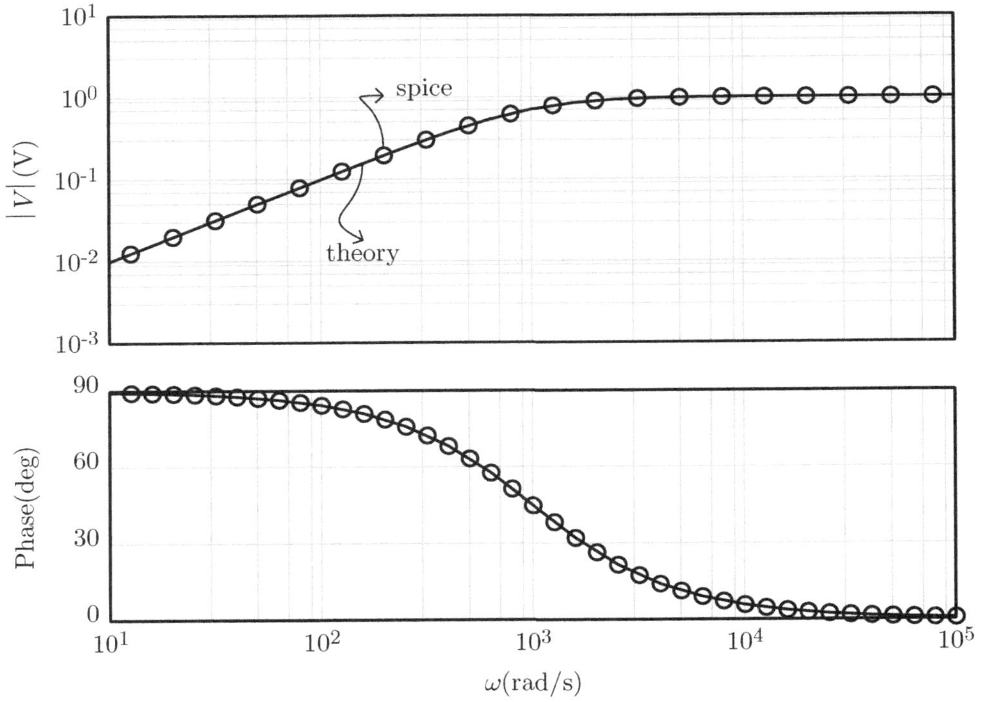

Fig. 35.13 High-pass filter transfer function ($R = 1\,\Omega$ and $L = 1$ mH)

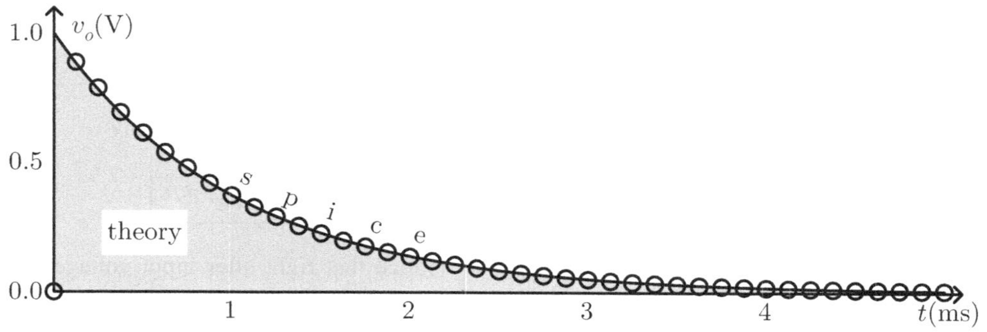

Fig. 35.14 High-pass filter unit step response ($R = 1\ \Omega$ and $L = 1$ mH)

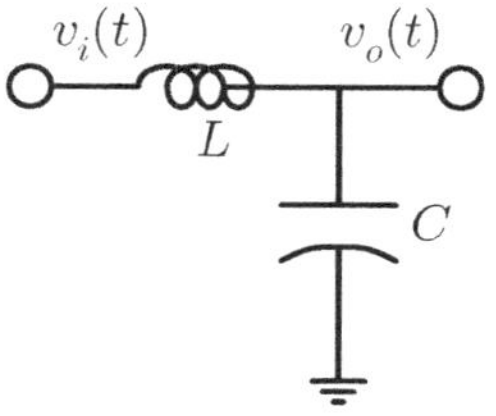

Fig. 35.15 Series LC circuit with output across capacitor

of low-frequency nature, then most of it would pass; and conversely, if input is of high-frequency nature, then most of it would be filtered—i.e., not passed.

35.6.1 Transfer Function

Voltage is applied to the series LC circuit, and voltage is sensed across the cap. The total impedance is

$$Z(s) = sL + \frac{1}{sC} \tag{35.56}$$

The current is given by

$$I(s) = \frac{V_i(s)}{Z(s)} = \frac{V_i(s)}{sL + \frac{1}{sC}} = V_i(s)\frac{sC}{1 + s^2LC} \tag{35.57}$$

$$I(s) = V_i(s)\frac{1}{L}\frac{s}{s^2 + \omega_{LC}^2}, \qquad \omega_{LC}^2 = \frac{1}{LC} \tag{35.58}$$

The output voltage is taken across the cap and would equal

$$V_o(s) = V_i(s)\frac{1}{L}\frac{s}{s + \omega_{LC}^2}\frac{1}{sC} \tag{35.59}$$

$$\boxed{V_o(s) = V_i(s)\frac{\omega_{LC}^2}{s^2 + \omega_{LC}^2}} \tag{35.60}$$

A plot of the magnitude and phase of this (for $V_i(s) = 1$) is shown in Fig. 35.16. Notice that as expected, low-frequency content passes, in the sense output voltage equals input voltage; and high-frequency content is rejected (filtered), in the sense high-frequency content is weakened, or diminished. But additionally, we notice a *resonance* around 1kHz, which is nothing but

$$\omega_{LC} = \sqrt{\frac{1}{LC}} = \sqrt{\frac{1}{1 \times 10^{-3} \times 1 \times 10^{-3}}} = 1000\ \text{rad/s} \tag{35.61}$$

Right at resonance, the transfer function blows up; i.e., it goes to infinity. What this means is that if input voltage is a sinusoid, and it has a frequency of ω_{LC} then output voltage would be a sinusoid at the same frequency, but with AC magnitude of infinity!! Of course this won't happen instantaneously! It would take time for the resonance to "build."

Impulse Response Knowing the transfer function, we are now able to figure impulse response.

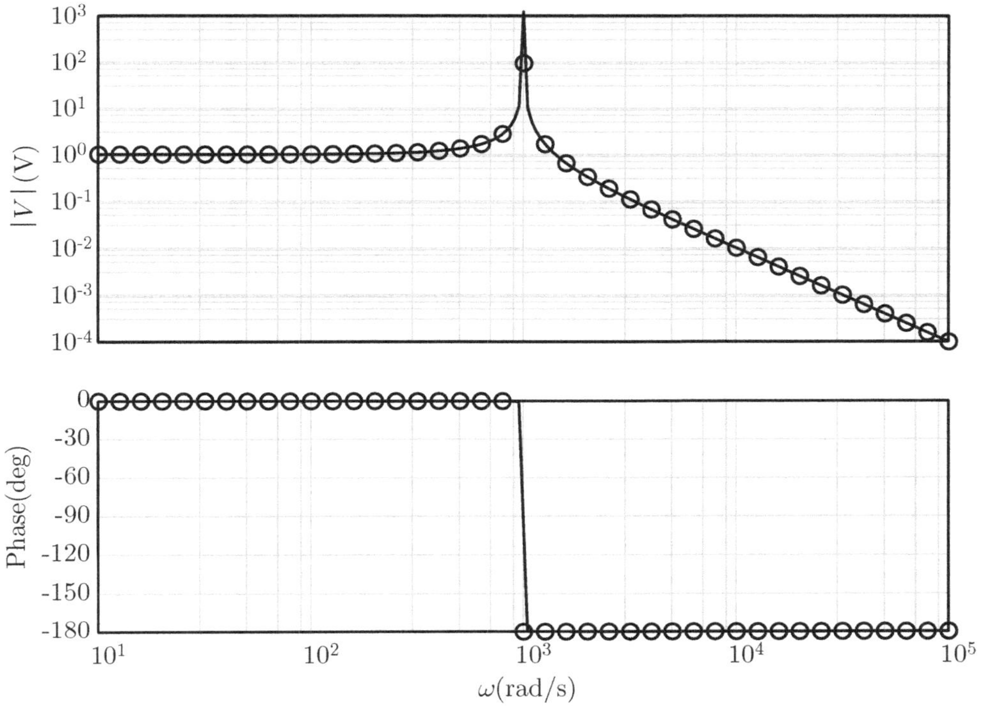

Fig. 35.16 Transfer function of *LC* low-pass filter, with $L = 1$ mH and $C = 1$ mF

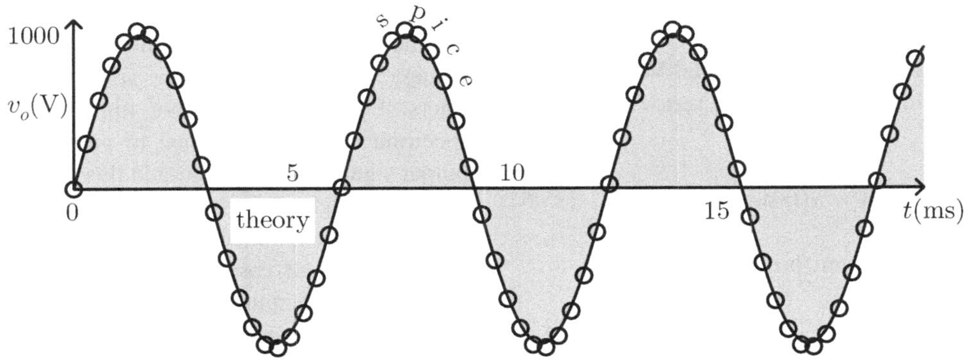

Fig. 35.17 Impulse response of low-pass *LC* circuit ($L = 1$ mH and $C = 1$ mF)

That is, if input is an impulse voltage (of unity strength), what would the output voltage look like? Straight from the transfer function equation we get

$$\boxed{v(t) = u(t)\omega_{LC}\sin(\omega_{LC}t)} \tag{35.62}$$

Notice that output has same frequency of ω_{LC} and that it is scaled by ω_{LC}. Above results as well as those of SPICE are shown in Fig. 35.17.

Step Response Again our starting step is the transfer function

$$V_o(s) = V(s)\frac{\omega_{LC}^2}{s^2 + \omega_{LC}^2} \tag{35.63}$$

With LT of input as $1/s$ we get step response as

$$V_o(s) = \frac{1}{s}\frac{\omega_{LC}^2}{s^2+\omega_{LC}^2} \quad (35.64)$$

Using partial fraction expansion we get

$$\boxed{V_o(s) = \frac{1}{s} - \frac{s}{s^2+\omega_{LC}^2}} \quad (35.65)$$

The inverse LT of this is

$$\boxed{v(t) = u(t)\left[1-\cos(\omega_{LC}t)\right]} \quad (35.66)$$

Notice that we could have equally derived this by simply integrating the impulse response in Eq. (35.62). Our derived results as well as those of SPICE are shown in Fig. 35.18. Notice that again the response is sinusoid, with same ω_{LC} frequency, but now we are picking a net DC component, of one! That is, the average of the step response is not zero (as that of impulse response), but instead it is one. This is nothing but the result of input having a DC value of 1, and that value is simply passing (on average) to the output.

Response to Causal Sinusoidal Voltage In this case the input voltage to the series *LC* circuit is given by

$$v(t) = u(t)\sin(\omega_0)t \quad (35.67)$$

Its Laplace transform is

$$V(s) = \frac{\omega_0}{s^2+\omega_0^2} \quad (35.68)$$

Our output voltage would now be

$$V_o(s) = \frac{\omega_0}{s^2+\omega_0^2}\frac{\omega_{LC}^2}{s^2+\omega_{LC}^2} \quad (35.69)$$

Using partial fractions we rewrite as

$$\boxed{V_o(s) = \frac{\omega_0\omega_{LC}^2}{\omega_{LC}^2-\omega_0^2}\left[\frac{1}{s^2+\omega_0^2} - \frac{1}{s^2+\omega_{LC}^2}\right]} \quad (35.70)$$

The inverse transform of this gives output voltage:

$$\boxed{v_o(t) = \frac{\omega_0\omega_{LC}^2}{\omega_{LC}^2-\omega_0^2}\left[\frac{1}{\omega_0}\sin(\omega_0 t) - \frac{1}{\omega_{LC}}\sin(\omega_{LC}t)\right]} \quad (35.71)$$

Figure 35.19 shows results and comparison to SPICE, for different input frequencies. The first frequency is below ω_{LC}, the second right after, and the third a while after. As seen from the plots, we get excellent agreement.

35.7 Summary

Transfer functions, spectral and convolution methods are not limited to impedance transfer functions, tying output voltage to input current. As shown in the chapter we can tie other outputs to other inputs, such as output voltage to input voltage. In this context we arrive at voltage filters. Filters—as the name imply—filter the spectrum of an input signal to yield a filtered output signal. We demonstrated this on low-pass as well as high-pass filters. We also touched on high order filters, such as the 2nd order one in Sect. 35.4. In all cases we first figured output current, then output voltage. In all cases we end up with a transfer function of the form of a numerator divided by a denominator—both of which are frequency dependent. Once the transfer function is known we can next find various responses, varying from impulse one to unit step one, to causal sinusoidal and more. Pretty much any input whose Laplace transform is known can be used. Once the product of input

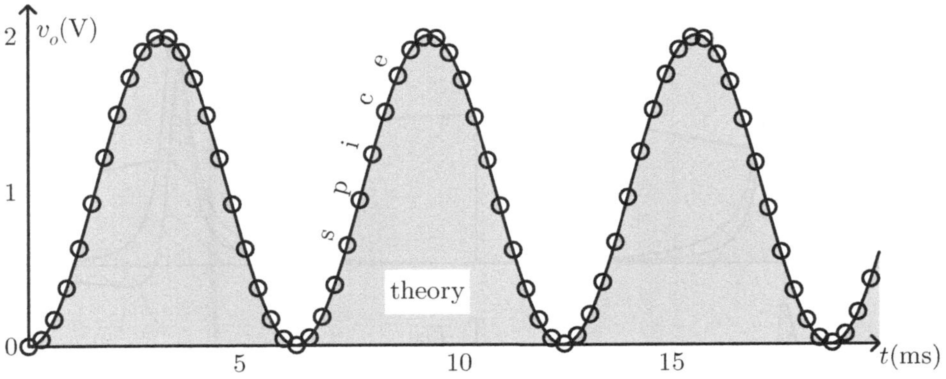

Fig. 35.18 Step response of low-pass LC circuit ($L = 1$ mH and $C = 1$ mF)

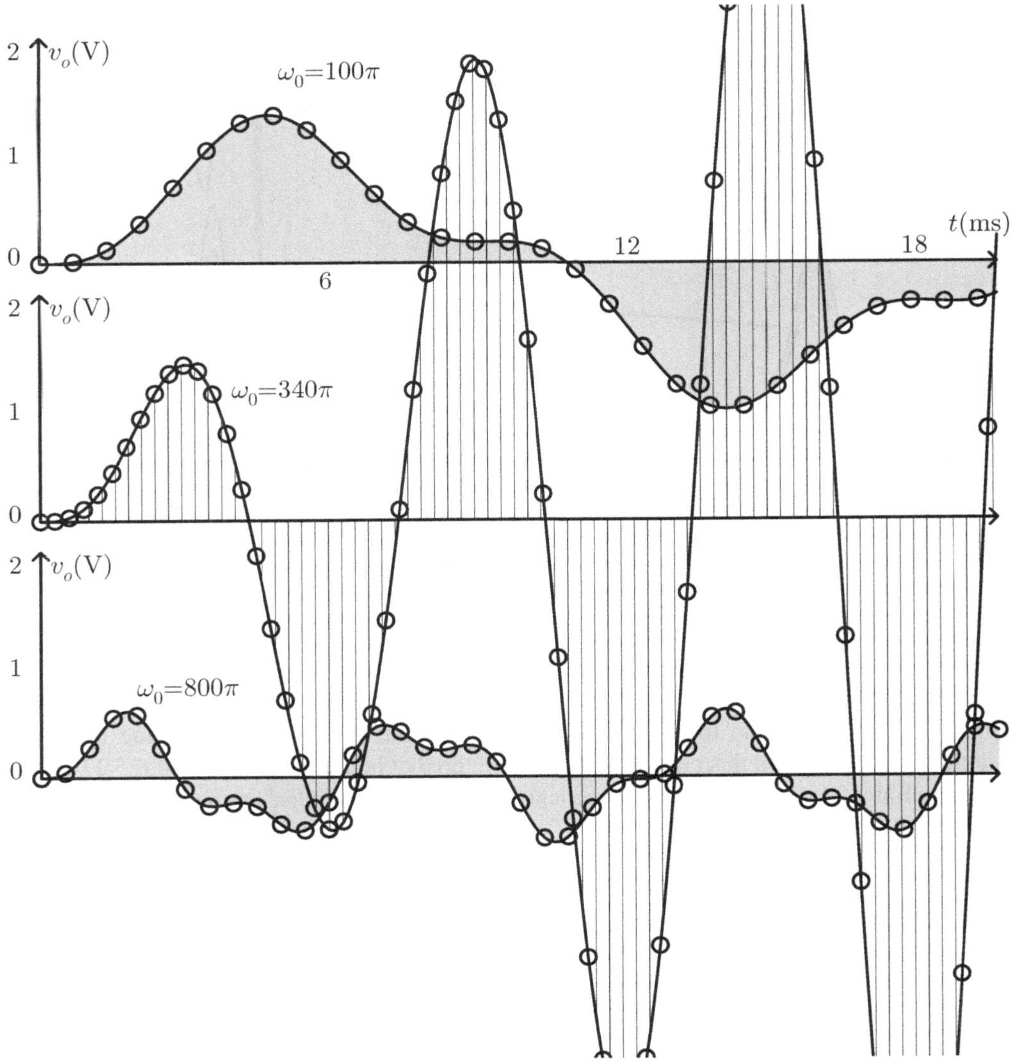

Fig. 35.19 Causal sine response of low-pass LC circuit for various frequencies ($L = 1$ mH and $C = 1$ mF)

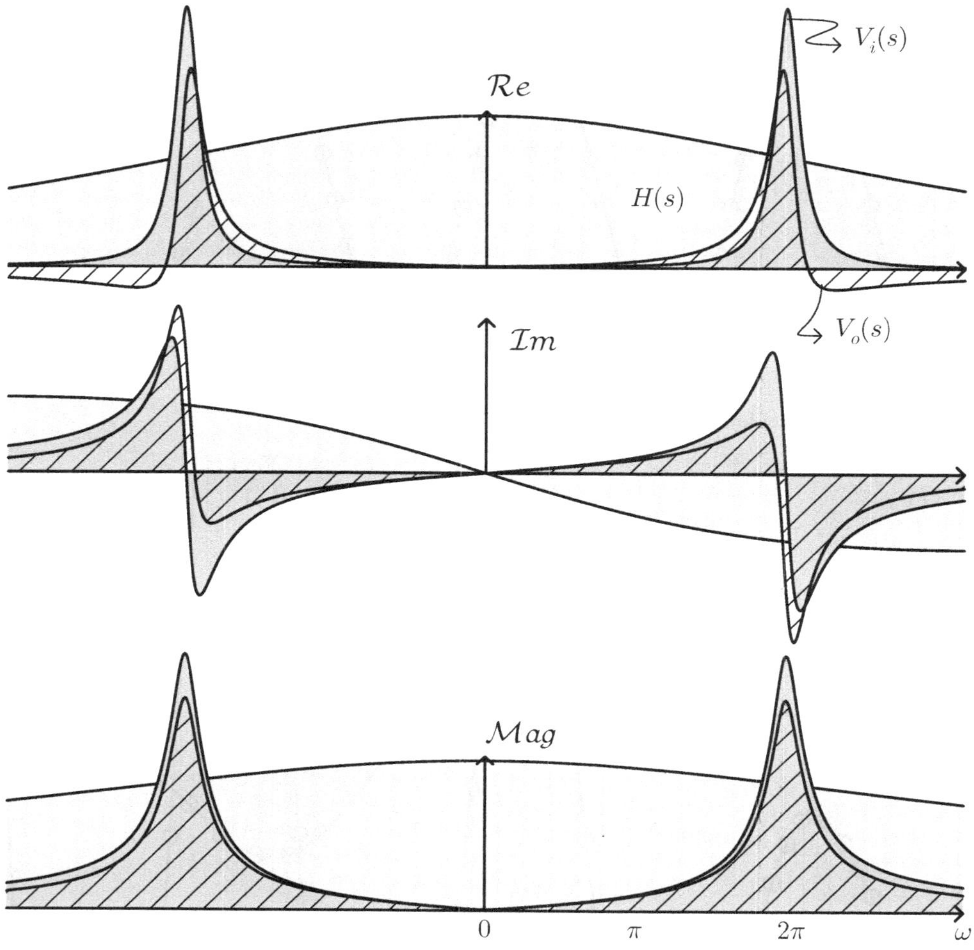

Fig. 35.20 Sample solution to Prob. 2; case of $R = 1\Omega$, $C = 0.1$F and $\omega_0 = 2\pi$

Laplace transform times system transfer function is at hand, we know output response answer in the frequency domain. What remains is simply finding the inverse transform to go back to the time domain. We illustrated this process on a few applied examples and demonstrated excellent match with SPICE.

35.8 Problems

1. Use partial fractions to derive Eq. (35.8).
2. Starting with Sect. 35.2, plot on linear scale the transfer function, input voltage, and output one using $\sigma = 0.3$; see sample solution in Fig. 35.20
3. Use long division to arrive at Eq. (35.21).
4. Use the initial and final value theorems to predict both Eqs. (35.26) and (35.27).
5. The second order low-pass filter in Sect. 35.4 has the transfer function reshown in Fig. 35.21. The transfer function was fit (somehow!) via the following simple expression:

$$H(s) = \frac{1}{\left(1 + \frac{s}{450}\right)\left(1 + \frac{s}{2000}\right)}$$

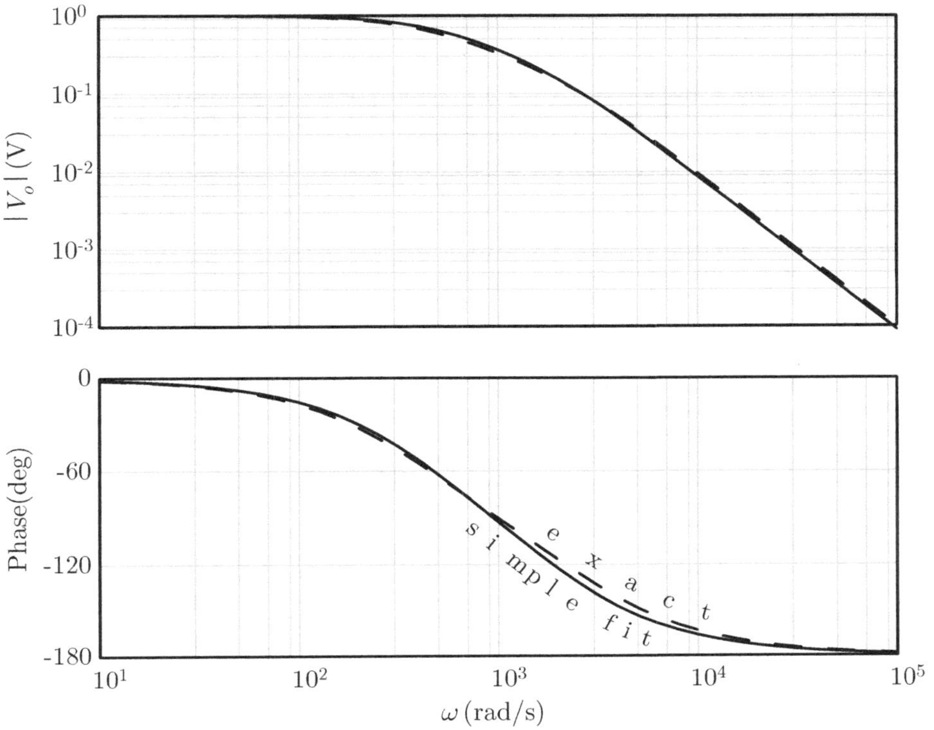

Fig. 35.21 Transfer function of Problem 5

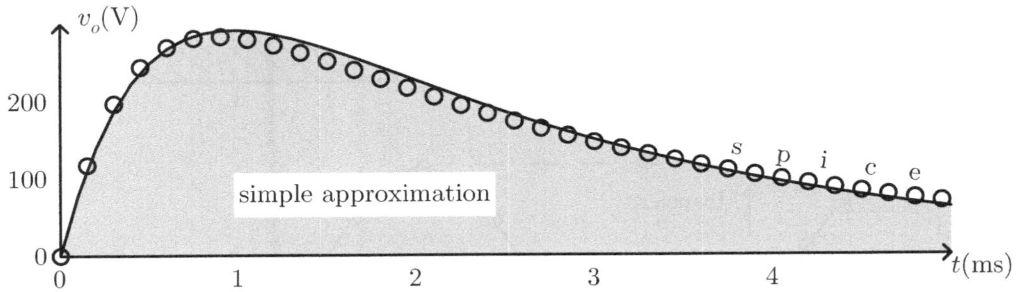

Fig. 35.22 Sample solution to Problem 5

What is the impulse response of this circuit? (That is, if input is an impulse voltage what is output voltage?) Confirm your results by simulating the circuit in SPICE (using $R = 1\,\Omega$ and $C = 1\,\text{mF}$); see sample results in Fig. 35.22.

6. Use the initial and final value theorems to predict the two limits in Fig. 35.14.
7. Starting with the input output transfer function of the high-pass *RL* circuit (Sect. 35.4), derive the output voltage for a sinusoidal input, ofangular frequency π. Plot all of the transfer

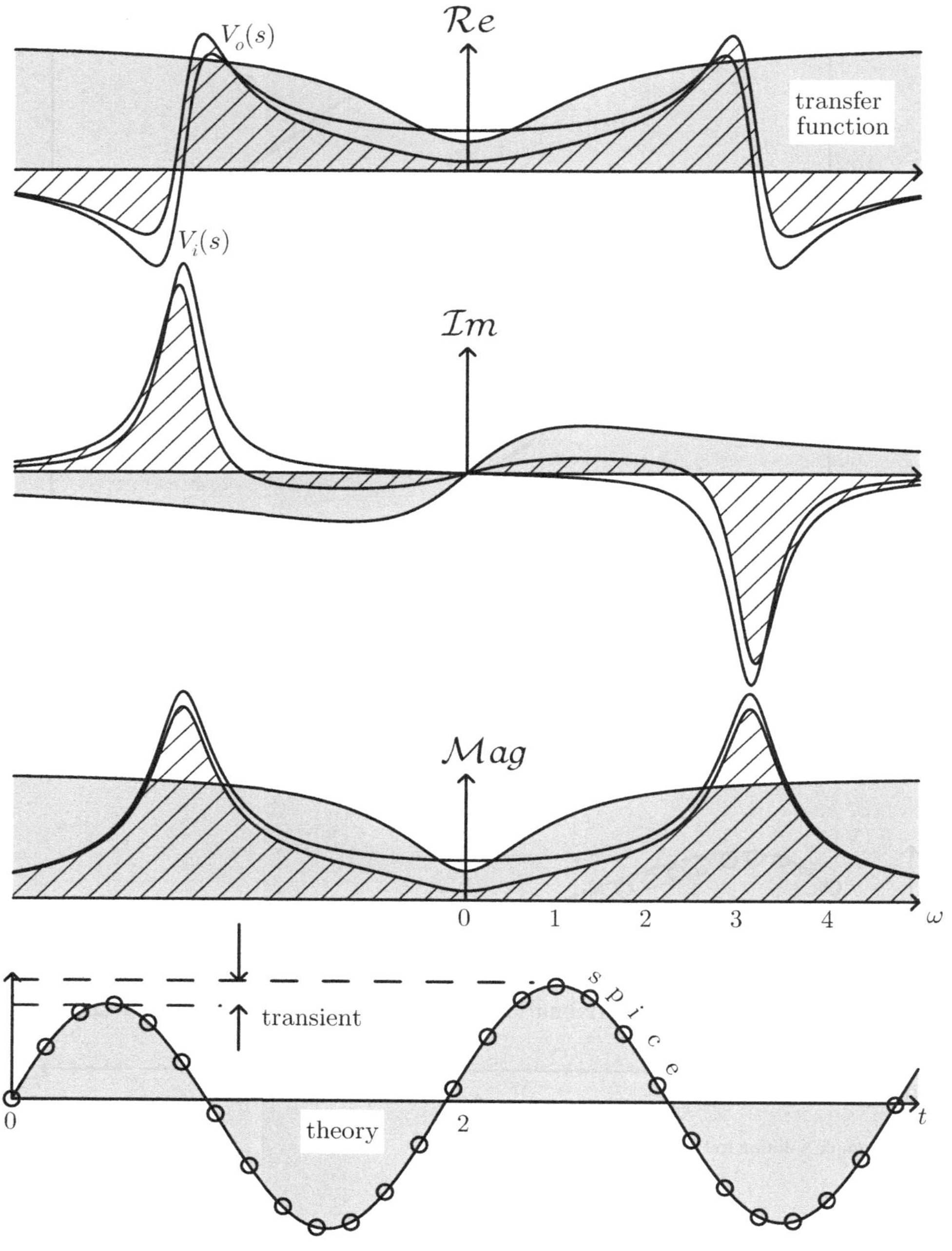

Fig. 35.23 Sample solution to Problem 7. (Case of $\sigma = 0.3$)

function, input voltage, and output voltage in the frequency domain (using $\sigma = 0.3$, $R = 1\,\Omega$, and $L = 1$ H).Then figure response in time domain and compare to SPICE; see sample solution in Fig. 35.23.

RLC Circuits with Feedback

36

36.1 Introduction

Feedback is a very important concept and tool not only in electrical engineering, but in all engineering! The idea of *sensing* something and *making a decision* accordingly has vast opportunities of applications. Of course whole books are dedicated to this subject, and we can only scratch the surface here. In particular we want to show how spectral and convolution techniques apply equally well in this arena of engineering. More specifically we will deal with *RLC* circuits which have controlled voltage sources (with a given gain G) which sense output voltage through a dedicated feedback filter.

36.2 Main Idea Behind Feedback

The main idea behind feedback in circuit theory is to monitor some point and compensate for it. For example, a node voltage can be monitored, and depending on the level, input voltage is altered. Feedback is typically accomplished schematically by controlled sources; for example, a voltage controlled voltage source is a voltage source whose output depends on voltage on some node. Feedback dramatically impacts circuit performance in both time and frequency domains, including input and output impedance, bandwidth, and settling values.

36.3 Simple Feedback Example

Consider the simple resistive circuit in Fig. 36.1. The left side is the no-feedback version while the right side is the feedback version. Let's first treat the no-feedback version. Doing KVL we get

$$v_i - iR = v_o \tag{36.1}$$

Assuming for simplicity $v_i(t) = 1$, $i(t) = 1$, and $R = 1$ we get

$$v_o = 0 \tag{36.2}$$

With feedback things become different; starting now with the right schematics and doing KVL we get

$$v_i + G(v_{\text{ref}} - v_o) - iR = v_o \tag{36.3}$$

Using the same assumptions above, and further assuming $v_{\text{ref}} = 1$ we get

$$1 + G(1 - v_o) - 1 = v_o \tag{36.4}$$

Rearrange

$$v_o(1 + G) = G \tag{36.5}$$

such that

$$v_o = \frac{G}{1 + G} \tag{36.6}$$

F. Badrieh, *Spectral, Convolution and Numerical Techniques in Circuit Theory*,
https://doi.org/10.1007/978-3-319-71437-0_36

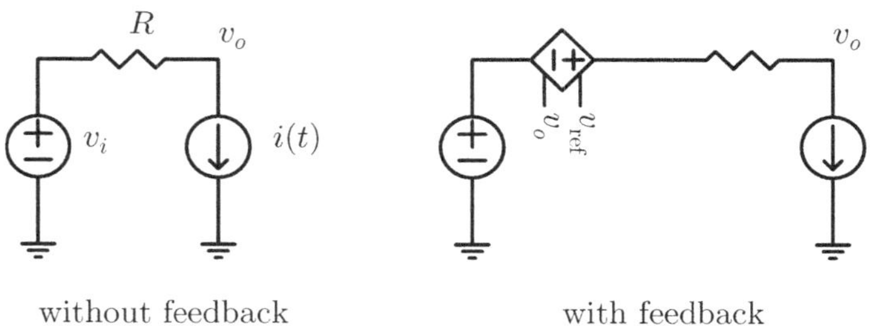

Fig. 36.1 Simple circuit without and with feedback

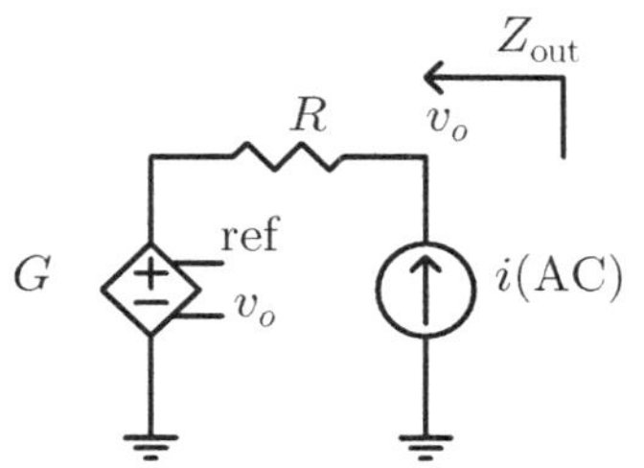

Fig. 36.2 Simple resistor circuit with feedback: setup for measuring impedance as seen from load (current source) side

Let us double check: Redo KVL and this time use $v_o = G/(1+G)$ to get

$$1+G\left(1-\frac{G}{G+1}\right)-1=1+G\frac{1}{G+1}-1=\frac{G}{1+G} \tag{36.7}$$

which in fact is the assumed output voltage. So, in summary, without feedback output voltage was 0 while with feedback it came out $G/(1+G)$. Notice that the larger the gain the closer v_o is to 1:

$$\lim_{G\to\infty} v_o = 1 \tag{36.8}$$

That is with larger gain the typically sustained losses between the input voltage source and output node, as a result of draining current, are cut down!

Output Impedance Let's next find output and input impedance. In the absence of feedback, output impedance would simply be R. But with feedback things change. For impedance calculations we short the input supply, apply an input current as shown in Fig. 36.2, and figure the resulting voltage.

Doing KVL, we arrive at

$$G(v_{\text{ref}} - v_o) + iR = v_0 \tag{36.9}$$

Rearrange

$$Gv_{\text{ref}} + iR = v_0(1+G) \tag{36.10}$$

For AC purposes we set $v_{\text{ref}} = 0$; then we have

$$iR = v_0(1+G) \tag{36.11}$$

such that

$$v_0 = i\frac{R}{1+G} \tag{36.12}$$

So we see that effective impedance is

$$\boxed{Z_{\text{out}}(\text{seen from load}) = \frac{R}{1+G}} \tag{36.13}$$

That is, instead of seeing R, we see a reduced version thereof—namely by $1+G$. For large gain we can approximate output impedance simply by R/G; that is actual impedance divided by the gain.

Input Impedance Next let's figure input impedance as seen from source side. One way to accomplish this is to put an AC voltage source and replace current source with some load impedance Z_L, as shown in Fig. 36.3 Doing KVL we arrive at

$$1 + G(v_{\text{ref}} - v_o) - iR = v_o \tag{36.14}$$

Again for AC set v_{ref} to zero:

$$1 - Gv_o - iR = v_o \tag{36.15}$$

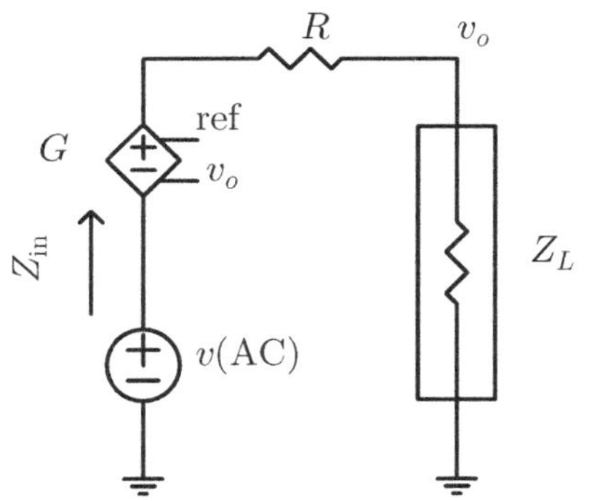

Fig. 36.3 Simple resistor circuit with feedback: setup for measuring impedance as seen from VCVS side

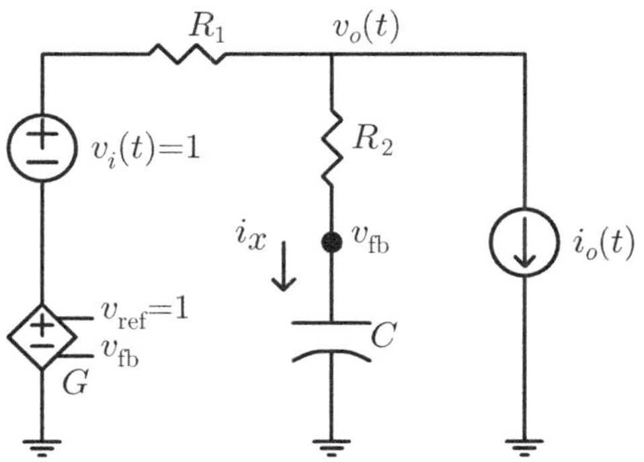

Fig. 36.4 *RC* circuit with feedback

Collect terms

$$v_o(1+G) = 1 - iR \tag{36.16}$$

Replace v_o with iZ_L and get

$$1 = iR + iZ_L(1+G) \tag{36.17}$$

or

$$i = \frac{1}{R + (G+1)Z_L} \tag{36.18}$$

That is, for an AC voltage stimulus of 1 we are getting an AC current of $1/(R+(G+1)Z_L)$. Hence impedance is

$$\boxed{Z_{\text{in}}(\text{as seen from input source}) = R + Z_L(G+1)} \tag{36.19}$$

So, rather than seeing $R+Z_L$, the Z_L gets multiplied by $G+1$. So output load gets magnified as seen from the input side. As seen above, feedback altered what we typically viewed as input and output impedances.

36.4 Network with Low-Pass Feedback Filter: Time Domain Analysis

A more complicated example is shown in Fig. 36.4. This circuit captures a more realistic form of feedback. Initially output voltage equals unity and no current is flowing in the circuit. Current is applied and as a result output voltage would droop; when it does, the voltage across the cap would droop too, albeit after a while. Once that is registered by the controlled source, the input voltage is raised to compensate, and after things settle down, output voltage goes back close to nominal, depending on the gain G. In this section we will deal with the problem in the time domain, for reference; in the next section we will do the problem in the frequency domain, and after that we will confirm that both methods give the same answer.

Assume now we apply a unit step current demand such that

$$i_o(t) = u(t) \tag{36.20}$$

Doing KVL around the RC loop and noting that the voltage at the feedback point is

$$v_{\text{fb}} = \frac{1}{C}\int i_x(t)dt \tag{36.21}$$

we get

$$G\left[1-\frac{1}{C}\int i_x(t)dt\right]+1-R_1(i_x+u(t))-i_xR_2-\frac{1}{C}\int i_x(t)dt=0 \tag{36.22}$$

Notice that right after time zero we can replace the unit step function with 1;

$$G\left[1-\frac{1}{C}\int i_x(t)dt\right]+1-R_1(i_x+1)-i_xR_2-\frac{1}{C}\int i_x(t)dt=0 \tag{36.23}$$

To get rid of the integral we take the time derivative and end up with

$$-\frac{G}{C}i_x(t)-\frac{di_x}{dt}(R_1+R_2)-\frac{1}{C}i_x(t)=0 \tag{36.24}$$

Collect terms

$$\boxed{\frac{di_x(t)}{dt}(R_1+R_2)=-\frac{G+1}{C}i_x(t)} \tag{36.25}$$

This has the solution

$$i_x(t)=Ae^{-t/t_{RC}} \tag{36.26}$$

where the time constant is given by

$$t_{RC}=\frac{(R_1+R_2)C}{G+1} \tag{36.27}$$

To determine A we need to figure initial conditions. Right before current application the cap is charged to 1 and no current flows in the circuit. Right after current application, the input DC source acts like short; the controlled source does not do anything because the cap holds its voltage, so the controlled source is out of the picture. The cap acts as a short and we end up with R_1 in parallel with R_2. Then the current going through R_2 is simply (by current division)

$$A=-\frac{R_1}{R_1+R_2} \tag{36.28}$$

Hence we finally get the total solution

$$\boxed{i_x(t)=-\frac{R_1}{R_1+R_2}e^{-t/t_{RC}}, \qquad t_{RC}=\frac{(R_1+R_2)C}{G+1}} \tag{36.29}$$

Results and comparison to SPICE are shown in Fig. 36.5. In this case we assumed

$$R_1=1, \quad R_2=2, \quad C=2\,\mathrm{m}, \quad G=10 \tag{36.30}$$

Next we solve for output voltage. It is simply the voltage across the cap plus that across R_2:

$$v_o=i_xR_2+\frac{1}{C}\int i_x(t)dt \tag{36.31}$$

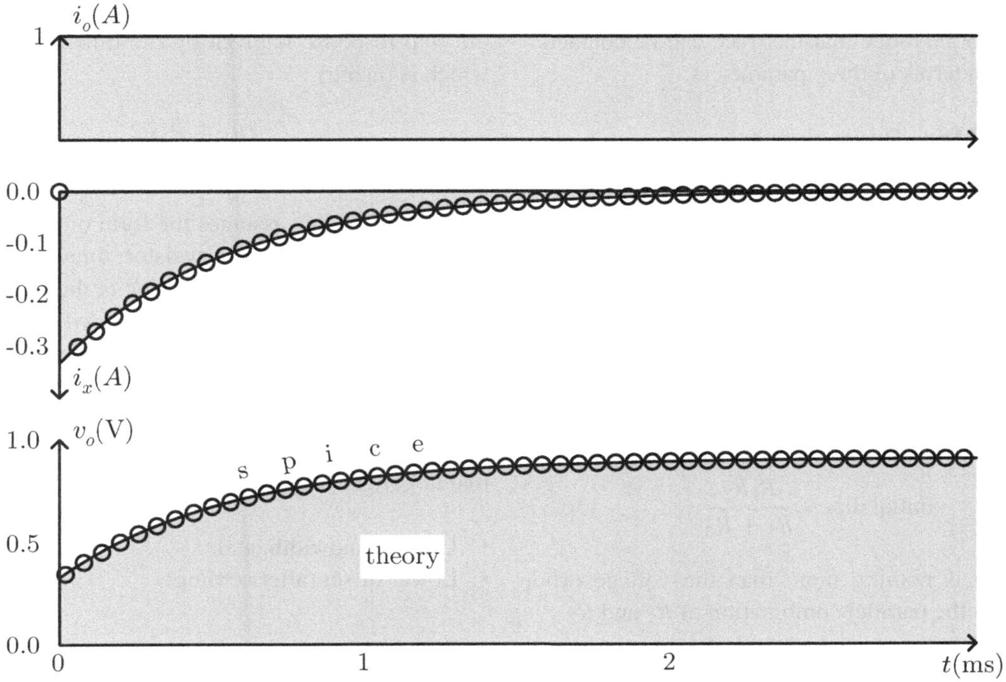

Fig. 36.5 Cap current and output voltage response to unit step output current demand

The voltage across the cap comes out to

$$
\begin{aligned}
v_C(t) &= \frac{1}{C}\frac{-R_1}{R_1+R_2}\int e^{-t/t_{RC}}dt = \frac{1}{C}\frac{R_1}{R_1+R_2}\frac{(R_1+R_2)C}{G+1}e^{-t/t_{RC}} + \text{const} \\
&= \frac{R_1}{G+1}e^{-t/T_{RC}} + \text{const}
\end{aligned}
\tag{36.32}
$$

To evaluate the constant we apply the initial condition such that voltage across the cap started at 1:

$$
\begin{aligned}
v_C(0) &= \frac{R_1}{G+1} + \text{const} = 1 \\
&\Rightarrow \text{const} = 1 - \frac{R_1}{G+1}
\end{aligned}
\tag{36.33}
$$

So finally we arrive at

$$
v_C(t) = 1 - \frac{R_1}{G+1} + \frac{R_1}{G+1}e^{-t/t_{RC}} \tag{36.34}
$$

To this we add the voltage across R_2 and arrive at total output voltage

$$
\boxed{v_o(t) = 1 - \frac{R_1}{G+1} + R_1\left[\frac{1}{G+1} - \frac{R_2}{R_1+R_2}\right]e^{-t/t_{RC}}, \qquad t_{RC} = \frac{(R_1+R_2)C}{G+1}} \tag{36.35}
$$

Results and comparison to SPICE are shown in Fig. 36.5. Notice that the trace can be characterized in terms of three parameters:

Initial Dip This is given by

$$\text{initial dip} = -\frac{R_1}{G+1} + R_1 \left[\frac{1}{G+1} - \frac{R_2}{R_1+R_2}\right] \quad (36.36)$$

For the case of large gain, specifically $G \gg \frac{R_1}{R_2}$, we can drop the $\frac{R_1}{G+1}$ and $\frac{1}{G+1}$ terms and we can approximate the initial dip as

$$\text{initial dip} \sim \frac{R_1 R_2}{R_1+R_2} \quad (36.37)$$

which is nothing more than the voltage droop across the parallel combination of R_1 and R_2.

Final Settling Voltage The offset between initial and final voltage level is

$$\text{voltage offset after settling} = \frac{R_1}{G+1} \quad (36.38)$$

That is, the voltage offset is current times R_1 *scaled down* by $1+G$. Notice that without feedback, the voltage offset would have been much larger—that of R_1 (for unity current).

Time Constant The last characteristic feature of the step response is given by the time constant, which is (again)

$$t_{RC} = \frac{(R_1+R_2)C}{G+1} \quad (36.39)$$

Notice that this assumes the form of a typical time constant, that of a resistor times a cap; the difference is that the resistor is the sum of R_1+R_2, and the RC product gets *scaled down* by $G+1$. That is, the presence of feedback causes the time constant to speed up (recover) much faster—namely by $G+1$. Figure 36.6 shows output voltage versus gain and it demonstrates that larger gain results in

- Larger bandwidth, and
- Lower offset (after settling)

36.5 Analysis in the Frequency Domain

Continuing on with the RC network with low-pass feedback filter from the last section (Fig. 36.4) we now move to the frequency domain. Analysis in the frequency domain ties output voltage to input stimulus. We have to be careful here since our input is really comprised of *three* sources! We have output current $I_o(s)$,

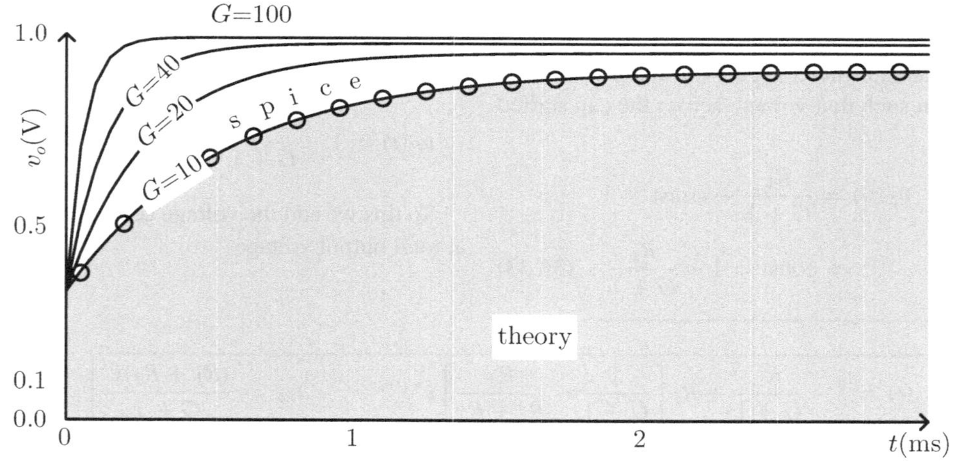

Fig. 36.6 Output voltage versus gain

input voltage $V_i(s)$, and reference voltage V_{ref}. In the most generic case output voltage assumes the form

$$V_o(s) = Z(s)I_o(s) + E_1(s)V_i(s) + E_2(s)V_{\text{ref}}(s) \tag{36.40}$$

For simplicity, however, we will assume that both input and reference voltages are constant. Hence E_1 and E_2 would be frequency independent. Furthermore, assume both are unity. We might as well start by calculating those. To that end, we must first turn off the third source, which is the output current. Doing KVL around the left loop, and keeping in mind that at DC there is no current through the capacitor, and hence no current throughout the network, we get

$$G(1 - V_o) + 1 = V_o \tag{36.41}$$

Notice in above we set $V_{\text{fb}} = V_o$ since there is no current through R_2.

Expanding we get

$$G - GV_o + 1 = V_o \tag{36.42}$$

Collecting terms we finally get

$$\boxed{V_o = 1, \qquad \text{DC and no output current}} \tag{36.43}$$

That is, if no current is drawn, output voltage remains at reference level, which in this case equals one. Next we shut off the two voltage sources, and enable the current one. We apply an impulse in time domain which translates to unity in frequency domain. Notice that in order to derive the output impedance we change the direction of current to pour *into* the network (as opposed to out of the network). Doing KVL and keeping in mind that $v_{\text{fb}} = \frac{I_x}{sC}$ we get

$$G\left[0 - \frac{I_x}{sC}\right] + R_1\,[1 - I_x] - R_2 I_x - \frac{I_x}{sC} = 0 \tag{36.44}$$

Collect terms

$$I_x\left[\frac{G+1}{sC} + R_1 + R_2\right] = R_1, \qquad \text{or} \tag{36.45}$$

$$I_x \frac{G + 1 + sC(R_1 + R_2)}{sC} = R_1 \tag{36.46}$$

Solve for I_x

$$I_x = \frac{sCR_1}{G + 1 + sC(R_1 + R_2)} = \frac{1}{C(R_1 + R_2)} \frac{sCR_1}{s + \frac{G+1}{C(R_1+R_2)}} \tag{36.47}$$

$$\boxed{I_x(s) = \frac{R_1}{R_1 + R_2} \frac{s}{s + 1/t_{RC}}, \qquad t_{RC} = \frac{(R_1 + R_2)C}{G + 1}} \tag{36.48}$$

Once output current is figured many other variables, such as output impedance, now fall easily into place as shown next.

Output Impedance Output voltage due to output current would be the sum of voltage across R_2 (which is $R_2 I_x$) and voltage across the cap (which is $\frac{I_x}{sC}$)

$$V_o(s) = \frac{R_1}{R_1 + R_2} \frac{s}{s + 1/t_{RC}} \left[R_2 + \frac{1}{sC}\right] \tag{36.49}$$

output voltage due to output current, no DC voltage input

Since the input current is unity (in the frequency domain), output voltage translates to output impedance; hence

$$\boxed{Z_o(s) = \frac{R_1}{R_1 + R_2}\frac{s}{s + 1/t_{RC}}\left[R_2 + \frac{1}{sC}\right]} \tag{36.50}$$

Results and comparison to SPICE are shown in Fig. 36.7. From the graph we can identify four critical signatures:

DC Impedance Notice that at DC we get

$$\lim_{s\to 0} Z_o(s) = \frac{R_1}{R_1 + R_2} s t_{RC} \frac{1}{sC} = \frac{R_1}{(R_1 + R_2)C}\frac{(R_1 + R_2)C}{G + 1} = \frac{R_1}{G + 1} \tag{36.51}$$

That is, at DC we see the diluted impedance $R_1/(G + 1)$. At DC the cap is open, and what typically would have appeared as R_1 now appears R_1 reduced by $G + 1$.

High-Frequency Impedance The high-frequency limit of impedance is

$$\lim_{s\to\infty} Z_o(s) = \frac{R_1 R_2}{R_1 + R_2} \tag{36.52}$$

At high frequency feedback cannot cope with output voltage change, so it factors out of the equation. Furthermore the cap shorts out and we end up with R_1 in parallel with R_2, which is nothing other than the equation we just derived.

Feedback Filter Time Constant This is the time constant of the feedback filter, and it is manifested in the frequency domain by the first inflection point. This time constant is related to R_2 and C.

$$\text{low-f time constant} = R_2 C \tag{36.53}$$

Notice that below this frequency point full sensing takes place and output impedance is minimal ($R_1/(G + 1)$). Above this frequency point, sensing is impeded and would be completely lost by the next time constant.

High-Frequency Time Constant The high-frequency time constant is given by

$$t_{RC} = \frac{(R_1 + R_2)C}{G + 1} \tag{36.54}$$

Basically it is the time constant of the sum of all resistors (R_1 and R_2 here), times cap (C), divided by gain ($G + 1$). Above this frequency point, sensing is completely out of the picture, and impedance falls back to the parallel combination of R_1 and R_2 (as if sensing were not there, as well as cap shorting).

Impact of Feedback Filter on Output Impedance The feedback filter is comprised of R_2 and C. The feedback voltage point is determined by voltage division between the R and the C voltages. For better feedback we would want the voltage across the C to be the largest. This is achieved by either lowering R or by lowering the C.

Impact of *R* As mentioned above, the impact of R_2 is to set the low-f inflection point. As we increase R_2, the inflection point moves to the left, as verified in Fig. 36.8. Also, increasing R_2 increases the high-frequency impedance, since that is determined by the parallel combination of R_2 and R_1. Notice that in the limit of very large R_2 output impedance would collapse to R_1, for all frequencies as if the R_2C branch did not exist at all. This makes sense since very large R_2 is equivalent to opening the R_2C branch.

Impact of *C* The cap in the filter sets both low- and high-frequency inflection points. Smaller cap pushes both frequency points to the right, as is confirmed in Fig. 36.9. Recall that inflections happen when resistive impedances (be it R_1 or R_2) become comparable to capacitive impedance; hence a smaller cap with higher impedance re-

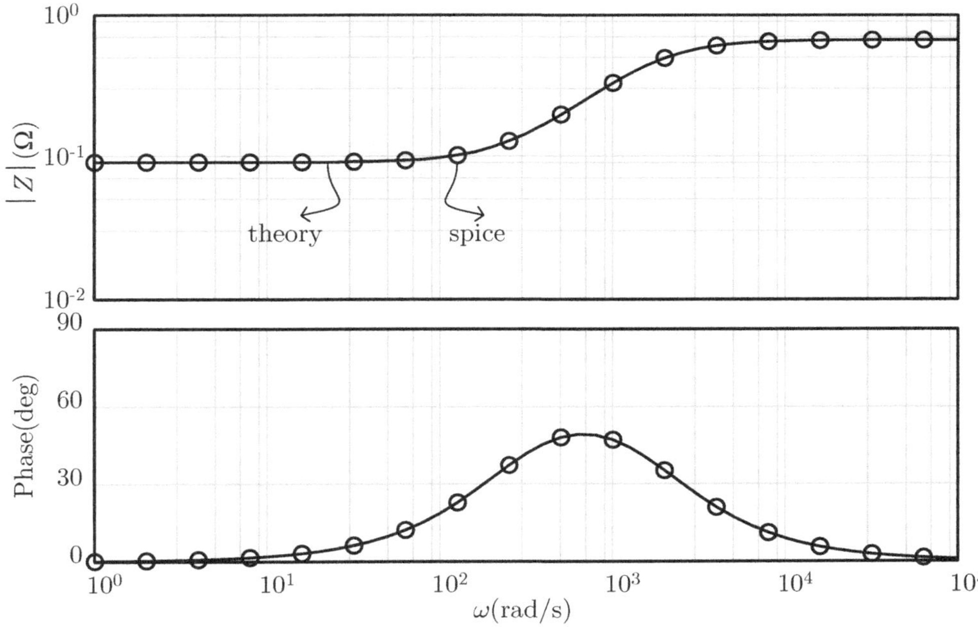

Fig. 36.7 Output impedance as a function of frequency (case of $R_1 = 1$, $R_2 = 2$, $C = 2\,\text{m}$, and $G = 10$

quires higher frequency such that its impedance goes back to being comparable to either of the Rs.

Impact of Gain and DC Resistance on Output Impedance The other two elements in this feedback system (other than the feedback filter) are gain and DC resistance. We will analyze the impact of each next.

Impact of Gain The gain G impacts the impedance at low frequency, and the high-f inflection point. In particular, and as discussed above, the DC impedance is given by

$$Z(0) = \frac{R_1}{G+1} \tag{36.55}$$

So with larger G we should expect to see lower impedance at low frequency and DC. This is confirmed in Fig. 36.10. Furthermore, larger gain decreases the second time constant

$$t_{RC} = \frac{(R_1 + R_2)C}{G+1} \tag{36.56}$$

and hence would increase the band width, determined by the second inflection point on the impedance plot. That is, larger gain enables this system to react faster, or push into higher band width. This too is confirmed in Fig. 36.10. So if we want lower DC droop and higher bandwidth, we would want to choose as large of again as possible (bearing everything else held constant). Finally notice that G has no impact on settling impedance value, at high frequency; as shown below, that depends only on R_1 and R_2.

Impact of DC Resistance (R_1) This resistance impacts low frequency impedance, high frequency impedance, and second inflection point on output impedance plot. In particular, larger R_1 increases the low-f impedance as confirmed in Fig. 36.11. Also, larger R_1 increases the settling impedance at high frequency, again as confirmed in the figure. Finally, larger R_1 increases the time constant

$$t_{RC} = \frac{(R_1 + R_2)C}{G+1} \tag{36.57}$$

which means the second inflection point in the frequency plot shifts to the left (again confirmed in Fig. 36.11).

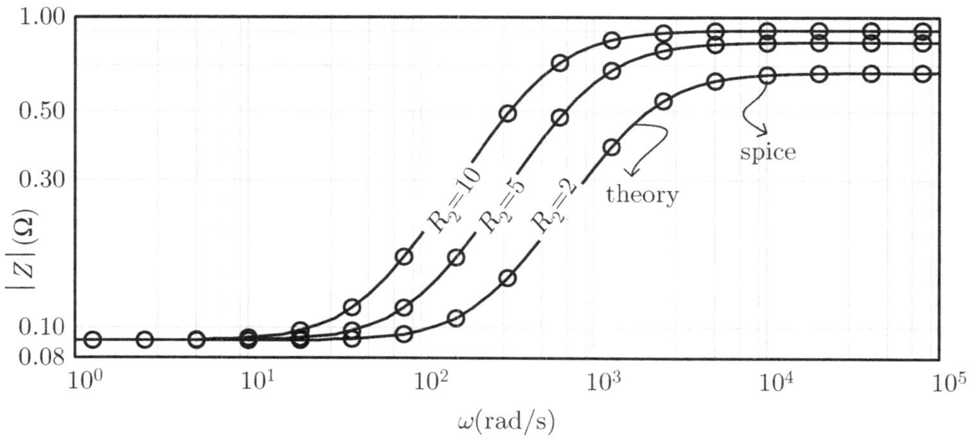

Fig. 36.8 Impact of R_2 on output impedance

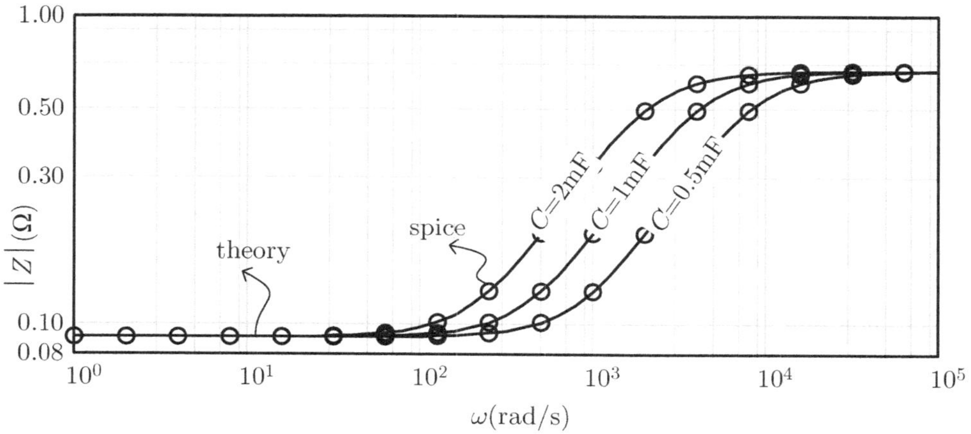

Fig. 36.9 Impact of C on output impedance

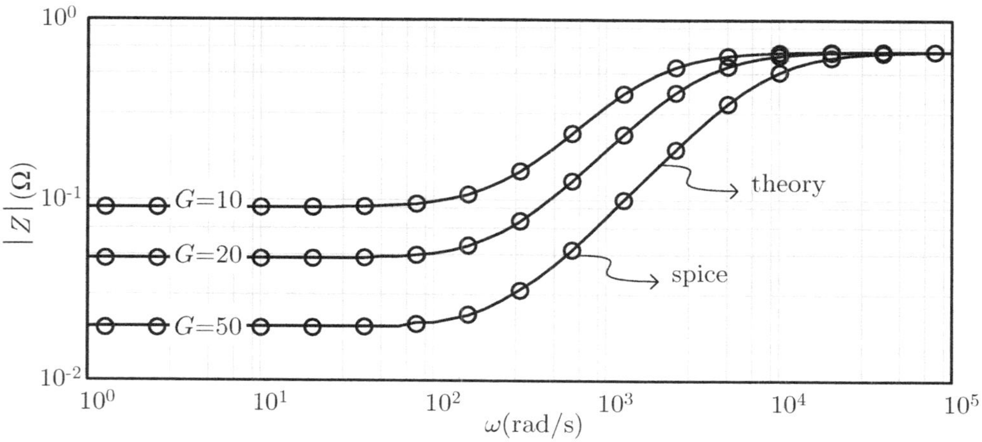

Fig. 36.10 Impact of G on output impedance

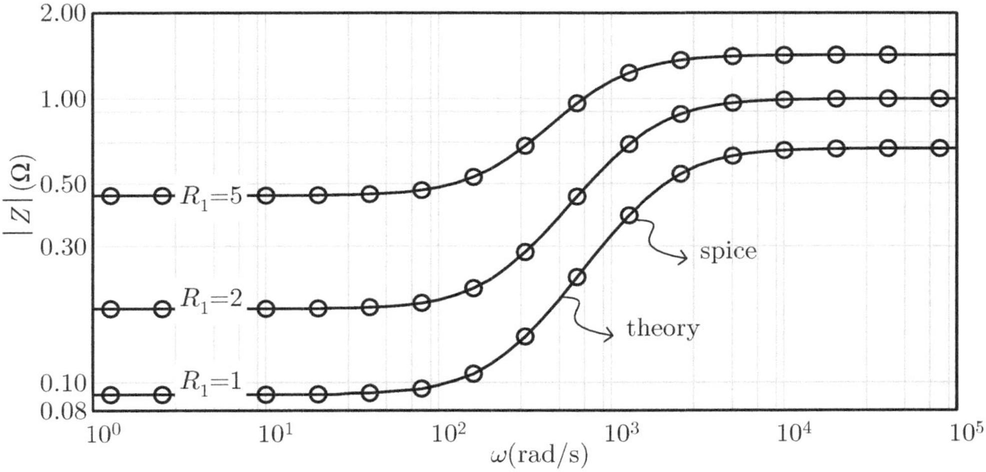

Fig. 36.11 Impact of R_1 on output impedance

36.6 Derivation of Unit Step Response from Frequency Transfer Function

We already derived in Sect. 36.4 the unit step response, but in the time domain and using differential equations. Let's derive it leveraging the frequency response. We already know that

$$v_o(s) = 1 + \text{inverse transform of } Z_o(s)I_o(s) \tag{36.58}$$

The first term reflects dependence of output voltage on input one, while the second ties output voltage to output current via the frequency dependent output impedance. For the case of unity current we have

$$I_o(s) = -\frac{1}{s} \tag{36.59}$$

Notice the negative sign since we are drawing current *out* of the circuit, as opposed to sinking it in. Plugging in for the impedance function (Eq. (36.50)) we have

$$\begin{aligned} I_o(s)Z_o(s) &= -\frac{R_1}{R_1+R_2}\frac{s}{s+1/t_{RC}}\left[R_2+\frac{1}{sC}\right]\frac{1}{s} \\ &= -\frac{R_1}{R_1+R_2}\frac{1}{s+1/t_{RC}}\left[R_2+\frac{1}{sC}\right] \\ &= -\frac{R_1R_2}{R_1+R_2}\frac{1}{s+1/t_{RC}} - \frac{R_1}{R_1+R_2}\frac{1}{s+1/t_{RC}}\frac{1}{sC} \end{aligned} \tag{36.60}$$

If we expand using partial fractions and use

$$\frac{1}{s+a}\frac{1}{s} = \frac{1}{a}\left[\frac{1}{s} - \frac{1}{s+a}\right] \tag{36.61}$$

then the second term in the above equation becomes

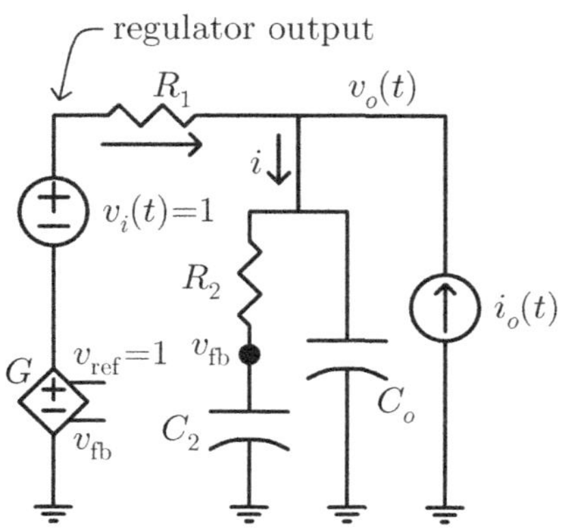

Fig. 36.12 Impact of adding dedicated output cap

$$-\frac{R_1}{(R_1+R_2)C}\frac{1}{s+1/t_{RC}}\frac{1}{s}=-\frac{R_1}{(R_1+R_2)C}\frac{C(R_1+R_2)}{G+1}\left[\frac{1}{s}-\frac{1}{s+1/t_{RC}}\right]$$
$$=\frac{R_1}{G+1}\left[-\frac{1}{s}+\frac{1}{s+1/t_{RC}}\right] \quad (36.62)$$

Plugging back into Eq. (36.60) we get

$$I_o(s)Z_o(s)=\frac{1}{s+1/t_{RC}}\left[-\frac{R_1R_2}{R_1+R_2}+\frac{R_1}{G+1}\right]-\frac{R_1}{G+1}\frac{1}{s} \quad (36.63)$$

If we do the inverse Laplace transform of this, and add 1 (due to input voltage) we finally get

$$\boxed{v(t)=1-\frac{R_1}{G+1}+R_1e^{-t/t_{RC}}\left[\frac{1}{G+1}-\frac{R_2}{R_1+R_2}\right],\quad t_{RC}=\frac{C(R_1+R_2)}{G+1}} \quad (36.64)$$

which matches prior results (Eq. (36.35)). So we have shown how we can work back from the frequency domain into the time domain. This was all accomplished by using the transfer function and multiplying by the Laplace transform of the unit step current.

36.7 Impact of Adding Output Cap

As we saw above, the response to the unit step current amounted to some steep initial droop. We can improve that by adding a dedicated output cap as shown in Fig. 36.12.

To derive output impedance we short both of the input and reference voltages. We set our unknown as i as shown in the figure, which translates to $I(s)$ in the frequency domain. The current going through the RC branch, in terms of $I(s)$, can be figured in terms of current division and comes out

$$I_{C_2}(s)=I(s)\frac{\frac{1}{sC_o}}{R_2+\frac{1}{sC_2}+\frac{1}{sC_o}}$$

$$= \frac{I(s)}{C_o}\frac{1/s}{R_2 + \frac{1}{s}\frac{C_2 + C_o}{C_2 C_o}}$$

$$= \frac{I(s)}{C_o}\frac{1}{sR_2 + \frac{C_2 + C_o}{C_2 C_o}} \quad (36.65)$$

Factor R_2 out and get

$$I_{C_2}(s) = I(s)\frac{1}{R_2 C_o}\frac{1}{s+a}, \qquad a = \frac{C_2 + C_o}{R_2 C_2 C_o} \quad (36.66)$$

Consequently the voltage across C_2

$$V_{C_2}(s) = I(s)\frac{1}{sC_2}\frac{1}{R_2 C_o}\frac{1}{s+a}$$

$$= \frac{I(s)}{sR_2 C_2 C_o}\frac{1}{s+a} \quad (36.67)$$

Doing KVL (and noting that $i_o(t) = \delta(t) \rightarrow 1$) we get

$$-Gv_{\text{fb}} - R_1[-1 + I(s)] - R_2 I_{C_2}(s) - v_{\text{fb}} = 0 \quad (36.68)$$

Collect terms

$$v_{\text{fb}}(1+G) + R_1 I(s) + R_2 I_{C_2}(s) = R_1 \quad (36.69)$$

Plugging in Eqs. (36.66) and (36.67) and keeping in mind $v_{\text{fb}} = V_{C_2}(s)$ we get

$$I(s)\left[\frac{G+1}{sR_2 C_2 C_o}\frac{1}{s+a} + R_1 + \frac{1}{C_o}\frac{1}{s+a}\right] = R_1 \quad (36.70)$$

Carry on the fraction addition

$$I(s)\frac{(G+1)/(R_2 C_2 C_o) + s/C_o + R_1 s(s+a)}{s(s+a)} = R_1 \quad (36.71)$$

$$I(s)\frac{b + (1/C_o + R_1 a)s + R_1 s^2}{s(s+a)} = R_1, \qquad b = \frac{G+1}{R_2 C_2 C_o} \quad (36.72)$$

Finally solve for I

$$\boxed{I(s) = \frac{R_1 s(s+a)}{b + (1/C_o + R_1 a)s + R_1 s^2}, \quad a = \frac{C_2 + C_o}{R_2 C_2 C_o}, \quad b = \frac{G+1}{R_2 C_2 C_o}} \quad (36.73)$$

Figure 36.13 shows current transfer function and comparison to SPICE. Output voltage would be simply this current times the impedance of output cap in parallel with feedback impedance.

$$V_o(s) = I(s)\left[\frac{1}{sC_o}\right] \| \left[R_2 + \frac{1}{sC_2}\right] = I(s)\frac{\frac{1}{sC_o}\left[R_2 + \frac{1}{sC_2}\right]}{R_2 + \frac{1}{s}\frac{C_2+C_o}{C_2 C_o}} \quad (36.74)$$

Multiply all by $s^2 C_2 C_o$

$$V_o(s) = I(s)\frac{1 + sR_2 C_2}{s^2 R_2 C_2 C_o + s(C_2 + C_o)} \quad (36.75)$$

So finally our impedance transfer function is

$$\boxed{Z_o(s) = \frac{R_1 s(s+a)}{b + (1/C_o + R_1 a)s + R_1 s^2}\frac{1 + sR_2 C_2}{s^2 R_2 C_2 C_o + s(C_2 + C_o)}} \quad (36.76)$$

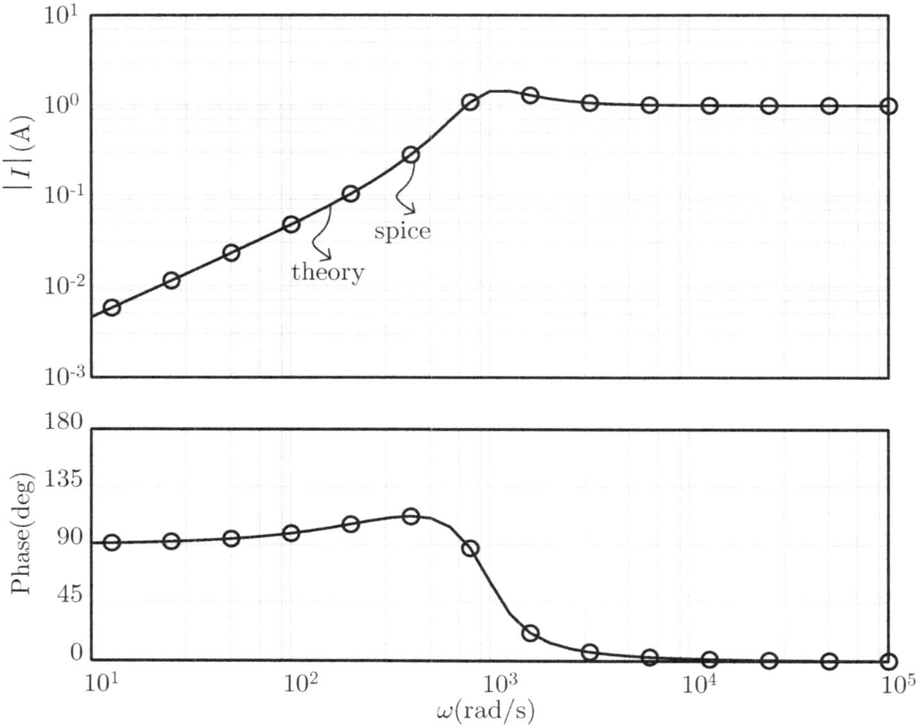

Fig. 36.13 Current transfer function for Fig. 36.12; case of $R_1 = 1\,\Omega$, $R_2 = 2\,\Omega$, $C_2 = 2\,\text{mF}$, $C_o = 3\,\text{mF}$ and $G = 10$

Figure 36.14 shows our results and comparison to SPICE. Notice that at DC we get

$$Z(0) \sim \frac{sR_1 a}{b}\frac{1}{s(C_2 + C_o)} = \frac{R_1 a}{b}\frac{1}{C_2 + C_o}$$
$$= \frac{R_1}{C_2 + C_o}\frac{C_2 + C_o}{R_2 C_2 C_o}\frac{R_2 C_2 C_o}{G+1} = \frac{R_1}{G+1} \tag{36.77}$$

which is the same as we got before (Eq. (36.51)). At high frequency, however, the impedance approaches zero in contrast to Eq. (36.52); the reason, and by design, is the presence of the output cap; this one alleviates the impedance by lowering it. Figure 36.15 shows output impedance for various output cap.

Knowing the impedance we are ready to figure step response. Apply step current demand and calculate output voltage. We already know

$$v_o(t) = 1 + \text{inverse transform of } I_o(s)Z_o(s) \tag{36.78}$$

In this case we have $I_o(s) = -1/s$, so we get

$$v_o(t) = 1 + \text{inverse transform of}$$
$$\left[-\frac{1}{s}\frac{R_1 s(s+a)}{b + (1/C_o + R_1 a)s + R_1 s^2}\frac{1 + sR_2C_2}{s^2 R_2 C_2 C_o + s(C_2 + C_o)}\right] \tag{36.79}$$

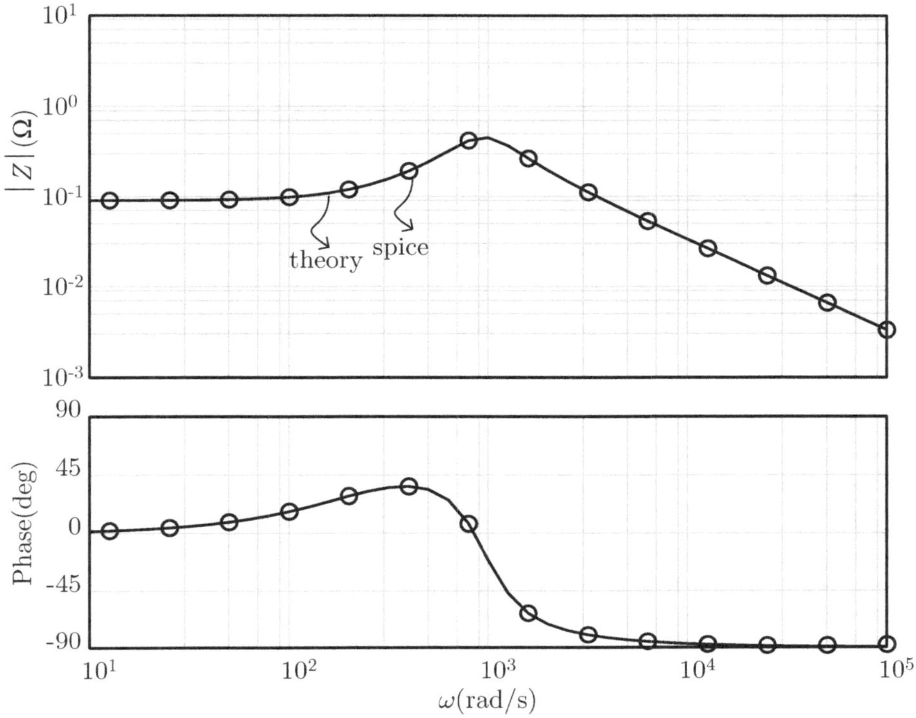

Fig. 36.14 Output impedance of Fig. 36.12; case of $R_1 = 1\,\Omega$, $R_2 = 2\,\Omega$, $C_2 = 2\,\text{mF}$, $C_o = 3\,\text{mF}$, and $G = 10$

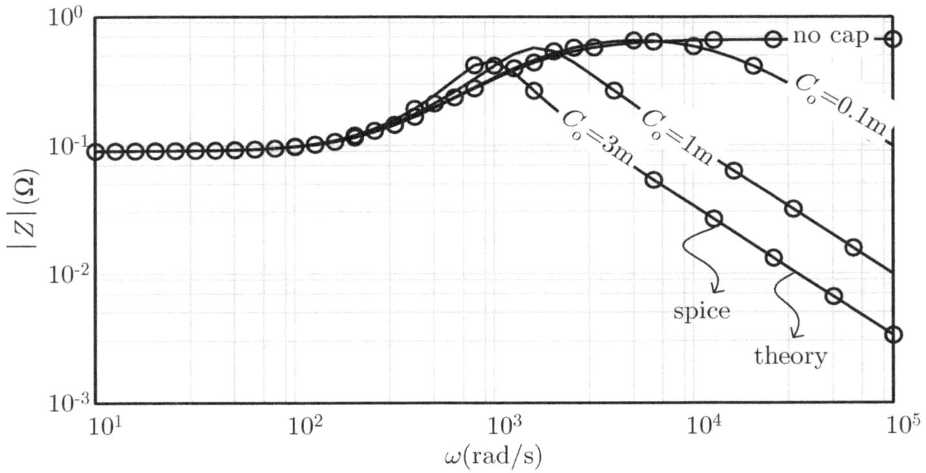

Fig. 36.15 Output impedance of Fig. 36.12 for various output cap values; case of $R_1 = 1\,\Omega$, $R_2 = 2\,\Omega$, $C_2 = 2\,\text{mF}$ and $G = 10$

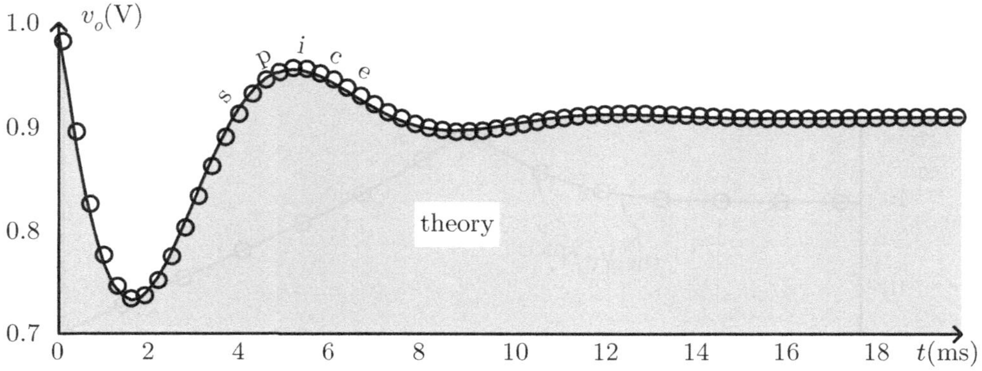

Fig. 36.16 Step response of Fig. 36.12. Assumed parameters are $R_1 = 1$, $R_2 = 2$, $C_2 = 2\,\text{m}$, $C_o = 3\,\text{m}$, and $G = 10$

We can figure the inverse transform via multiple methods, including direct partial fraction expansion, or approximation via polynomial fitting (then partial fraction); but for expediency we opted here for direct numerical evaluation. Results and comparison to SPICE are shown in Fig. 36.16; as seen in the figure we get excellent match. The output of the actual regulator (to the left of R_1) would be this output voltage plus R_1 times the current through R_1 (defined here left to right, and with reference to Fig. 36.12):

$$v_{\text{regulator}} = v_o(t) + \text{inverse transform of } [R_1 \times \text{current through } R_1] \tag{36.80}$$

When output current was an impulse function such that $I_o(s) = 1$ the current through R_1 was simply

$$I_{R_1}(s) = I(s) - 1, \qquad (\text{case of } I_o(s) = 1) \tag{36.81}$$

where $I(s)$ is still given by Eq. (36.73). But here output current is $I_o(s) = -\frac{1}{s}$; hence current through R_1 needs to scale accordingly

$$I_{R_1}(s) = -\frac{1}{s}\left[I(s) - 1\right] = \frac{1}{s} - \frac{I(s)}{s}, \qquad \left(\text{case of } I_o(s) = -\frac{1}{s}\right) \tag{36.82}$$

Hence Eq. (36.80) becomes

$$\boxed{v_{\text{regulator}} = v_o(t) + \text{inverse transform of } R_1\left[\frac{1}{s} - \frac{I(s)}{s}\right]} \tag{36.83}$$

Again doing the inverse transform numerically we get regulator output voltage as shown in Fig. 36.17. Notice that when output dips, output of regulator rises. Notice too that when things settle output of regulator settles to $v_o(t) + R_1$ (for unity current demand).

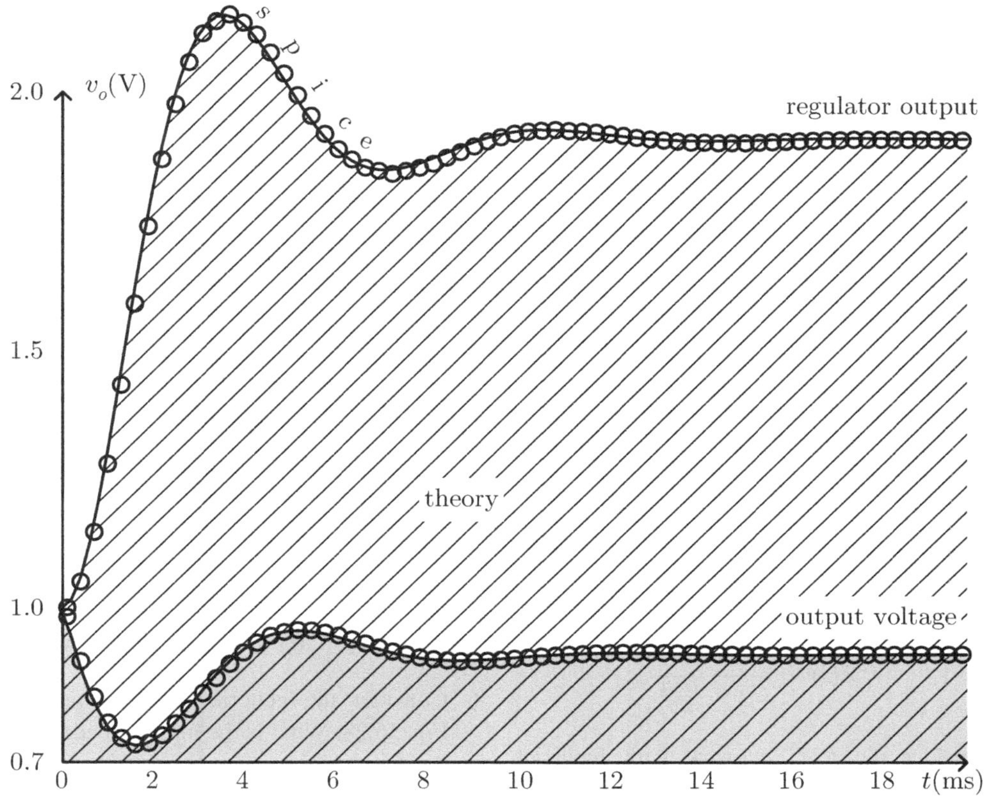

Fig. 36.17 Step response and regulator output voltage of Fig. 36.12

36.8 Summary

There are many configurations for feedback, and this chapter only sampled a very short selection thereof; the end-of-chapter problems will show some more. But in all the examples shown here and in the problems we apply the same underlying techniques of spectral methods, figure the various frequency dependent blocks (be it cap impedance, input impedance, or output one), carry on KVL/KCL and arrive at the final transfer function tying output to input. Once the transfer function is known we can apply our arsenal of techniques to convert to the time domain, be it by straight conversion, numerical inverse transform, polynomial fitting, or even convolution. While we limited ourselves to impulse and step response, similar treatment works for other input stimuli, such as causal periodic functions. We demonstrated feedback on a couple of problems the most important mimicking a regulator function where output is monitored and fed back via a low-pass filter to a controlling function at the input. We examined impact of all of DC resistance, gain, feedback filter, and output cap on output response—both in the frequency domain and time one. We showed perfect match between all of transient analysis, frequency one, and SPICE. Another success story in applying spectral techniques, and more weight added in support of the underlying techniques and their versatility.

36.9 Problems

1. Consider the series *RL* network in Fig. 36.18. What is output impedance for gain $G = 0, 2$, and 5? Plot and compare results to SPICE for $R = 1\,\Omega$ and $L = 1$ mH.; see sample solution in Fig. 36.19.
2. Consider the series *RL* network in Fig. 36.20. What is input impedance as seen from the VCVS for gain $G = 0, 2$, and 5? Plot and compare results to SPICE for $R = 1\,\Omega$ and $L = 1$ mH.; see sample solution in Fig. 36.21.
3. Consider the *RC* network with feedback in Fig. 36.22. What is the input impedance? What are the DC and high-frequency limits? Plot it for the case $R = 1$ and $C = 1$ m, and for the three G values: 0, 2, and 5. Compare to SPICE; see sample solution in Fig. 36.23.
   Answer:

$$Z(s) = \frac{1}{C}\frac{1}{s + \frac{G+1}{RC}}$$

4. Consider the *RLC* network with feedback in Fig. 36.24. What is the input impedance? What are the DC and high-frequency limits? Plot it for the case $R = 1$, $L = 1$ m, and $C = 0.1$ m, and for the three G values: 0, 2, and 5. Compare to SPICE; see sample solution in Fig. 36.25.
   Answer:

$$Z(s) = \frac{R + sL}{(G+1) + sRC + s^2LC}$$

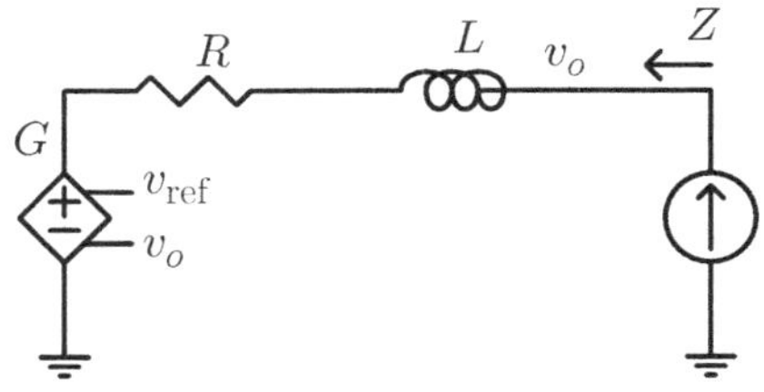

Fig. 36.18 Statement to Problem 1

5. Consider the *RLC* network with feedback in Fig. 36.26. What is the input impedance? What are the DC and high-frequency limits? Plot it for the case $R = 1$, $L = 1$ m, and $C = 0.1$ m, and for the three G values: 0, 2, and 5. Compare to SPICE; see sample solution in Fig. 36.27.
   Answer:

$$Z(s) = \frac{sL(1 + sRC)}{(G+1)(1 + sRC) + s^2LC}$$

6. Consider the generic feedback network in Fig. 36.28. What is the input impedance?
   Answer:

$$Z(s) = \frac{Z_2Z_3 + Z_1Z_2 + Z_1Z_3}{Z_3 + Z_2(1+G)}$$

7. Use results in Problem 6 to figure output impedance for the case $Z_1 = 1\,\Omega$, $Z_3 = 3\,\Omega$ and Z_2 is a cap of magnitude 1 mF. Compare results to SPICE for G values of 0, 2, and 5; see sample results in Fig. 36.29.

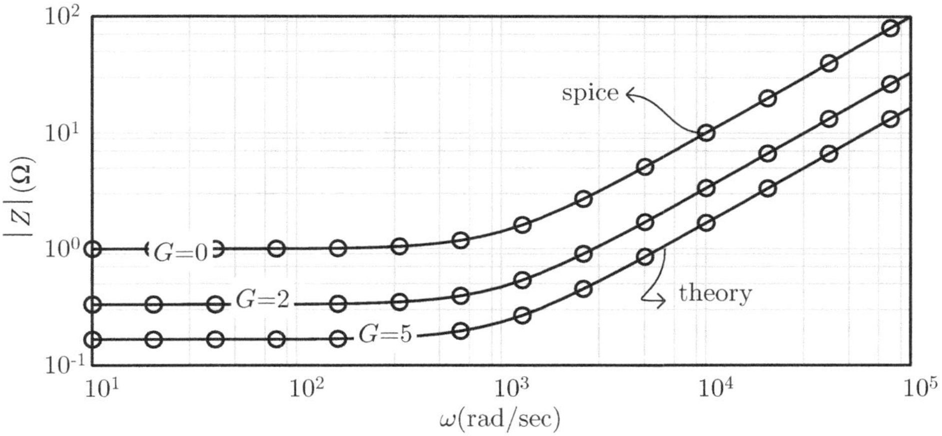

Fig. 36.19 Sample solution to Problem 1

Fig. 36.20 Statement to Problem 2

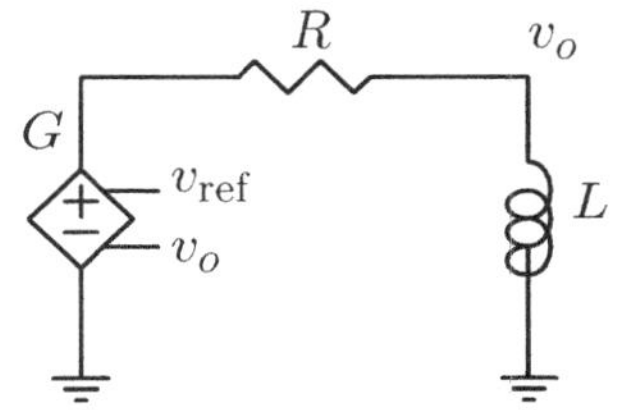

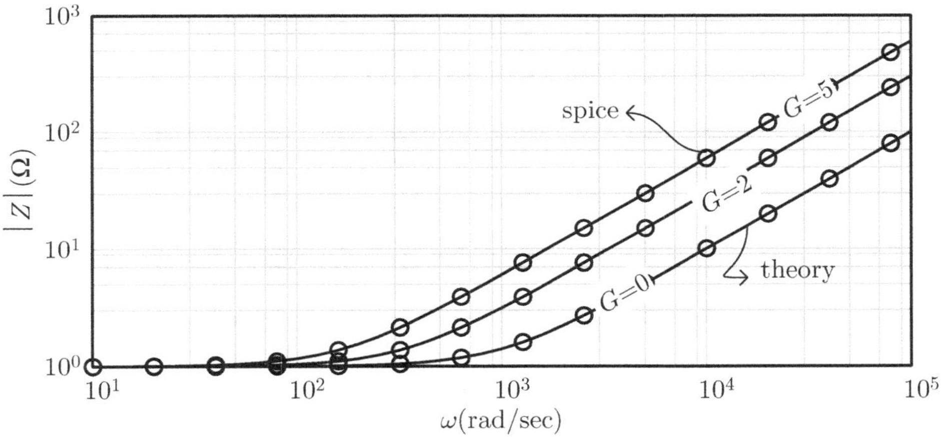

Fig. 36.21 Sample solution to Problem 2

Fig. 36.22 Statement to Problem 3

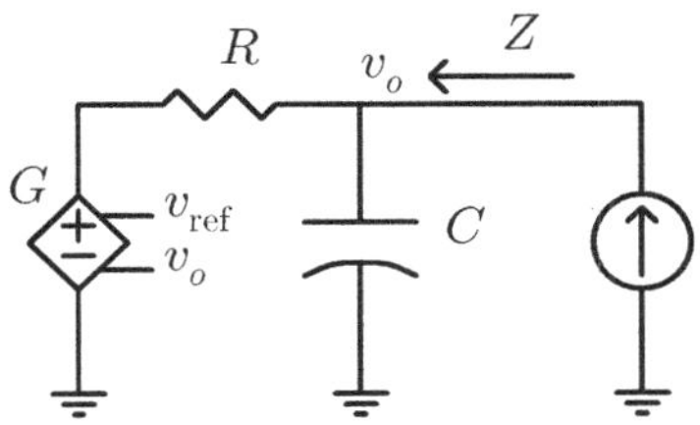

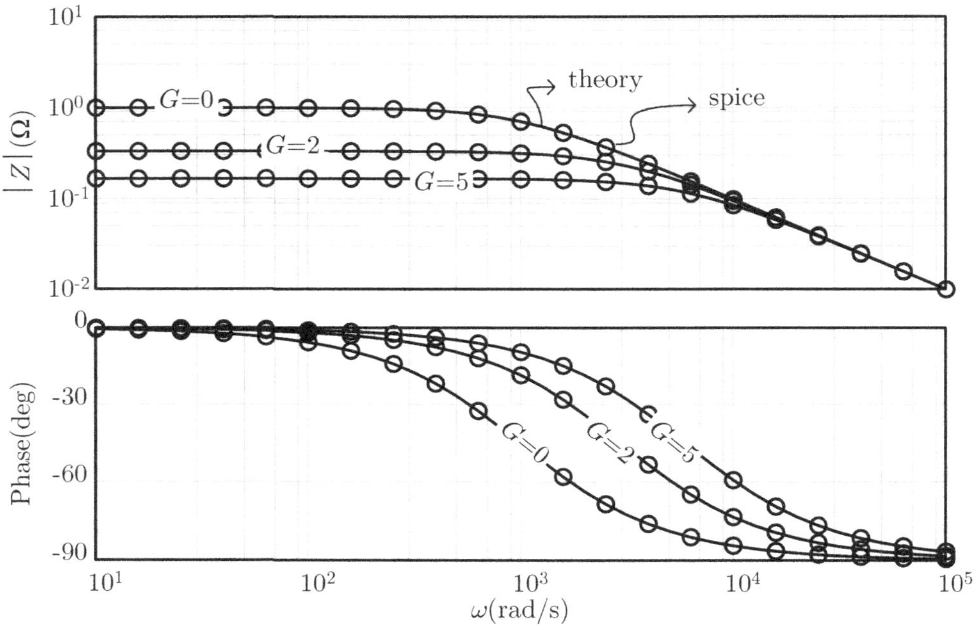

Fig. 36.23 Sample solution to Problem 3

Fig. 36.24 Statement to Problem 4

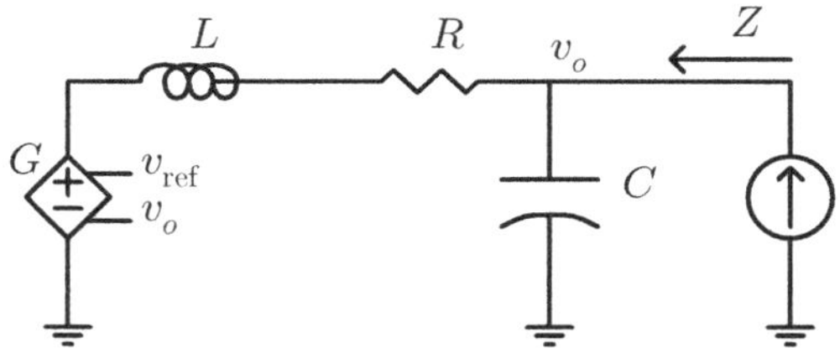

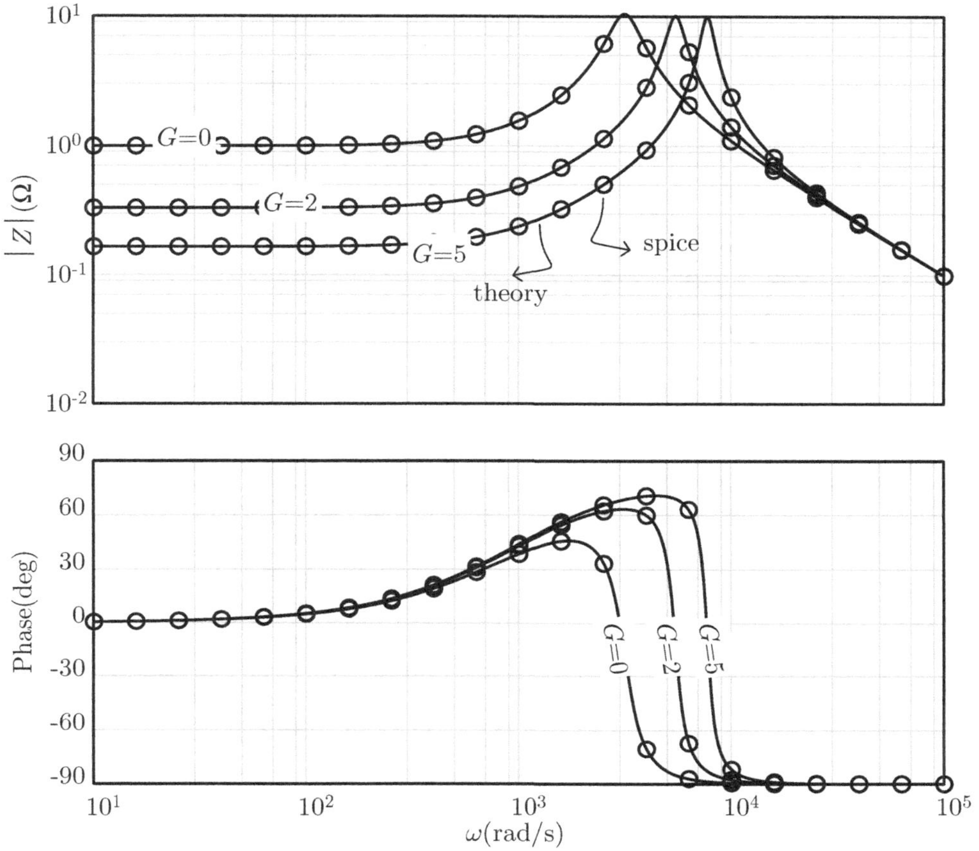

Fig. 36.25 Sample solution to Problem 4

Fig. 36.26 Statement to Problem 5

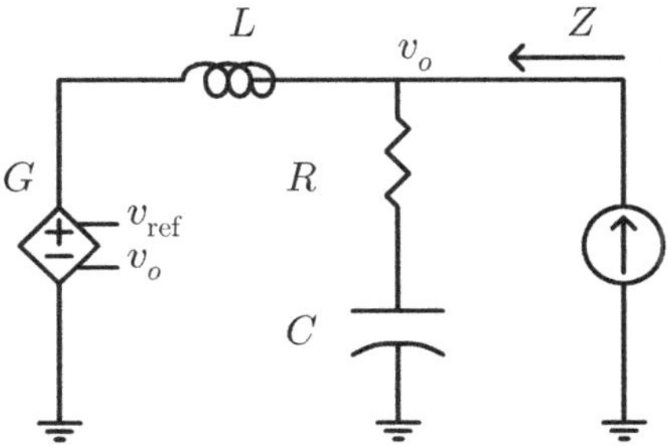

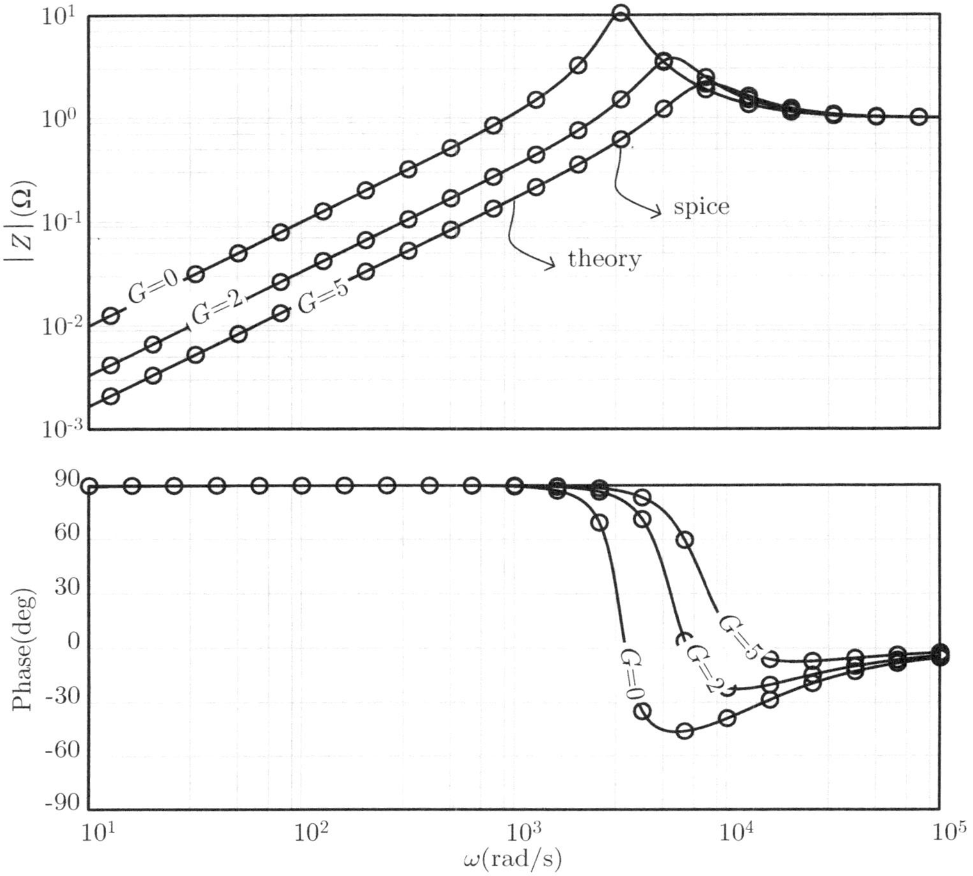

Fig. 36.27 Sample solution to Problem 5

Fig. 36.28 Statement to Problem 6

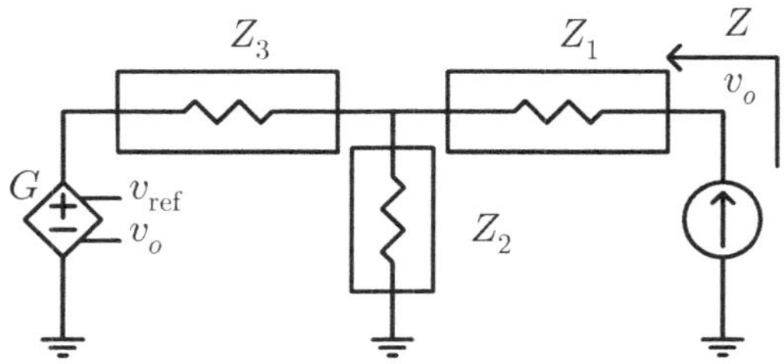

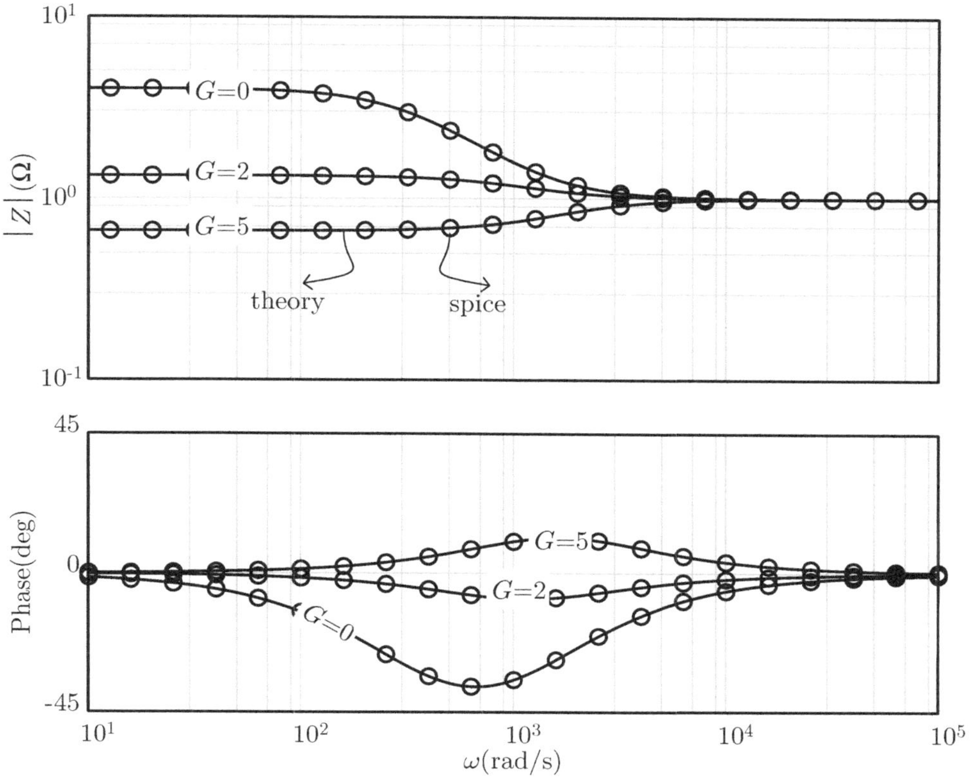

Fig. 36.29 Sample solution to Problem 7

37 Matrix Solution to Multi-Branch Networks

37.1 Introduction

The majority of the text so far has dealt with cases where output was given in terms of input in the form of a transfer function; that is, we had a closed-form solution in the frequency domain, and the remaining work was to find the inverse transform to go back to the time domain. While this flow works efficiently for smaller circuits, it becomes more difficult for larger networks. For these cases, we will have to introduce multiple unknowns, and solve for them, all while still residing in the frequency domain. Upon figuring the frequency solution of the unknown currents/voltages, we can find the individual inverse transforms to go back to time domain. Let's demonstrate the flow with a few examples.

37.2 First Example: Two-Branch *RC* Network

Consider the RC network shown in Fig. 37.1. It is stimulated by an input voltage source on the left and we are interested in figuring output voltage at the right. First we migrate to the frequency domain such that $i(t) \rightarrow I(s)$, $v(t) \rightarrow V(s)$, and the various impedances assume their frequency-domain counterparts. Doing KVL around the right loop we get

$$I_2 \frac{1}{sC_2} + I_2 R_2 - \frac{I_1 - I_2}{sC_1} = 0 \tag{37.1}$$

Collect terms

$$I_2 \left[\frac{1}{s} \left(\frac{1}{C_2} + \frac{1}{C_1} \right) + R_2 \right] - \frac{I_1}{sC_1} = 0 \tag{37.2}$$

$$I_2 \left[\frac{C_1 + C_2}{sC_1 C_2} + R_2 \right] - \frac{I_1}{sC_1} = 0 \tag{37.3}$$

$$\boxed{I_2 \frac{C_1 + C_2 + sR_2 C_1 C_2}{sC_1 C_2} - \frac{I_1}{sC_1} = 0} \tag{37.4}$$

This is the first equation. Next do KVL around the left loop to get

$$\frac{I_1 - I_2}{sC_1} + I_1 R_1 = V_i(s) \tag{37.5}$$

Collect terms

$$I_1 \left[\frac{1}{sC_1} + R_1 \right] - \frac{I_2}{sC_1} = V_i(s) \tag{37.6}$$

$$\boxed{I_1 \frac{1 + sR_1 C_1}{sC_1} - \frac{I_2}{sC_1} = V_i(s)} \tag{37.7}$$

F. Badrieh, *Spectral, Convolution and Numerical Techniques in Circuit Theory*, https://doi.org/10.1007/978-3-319-71437-0_37

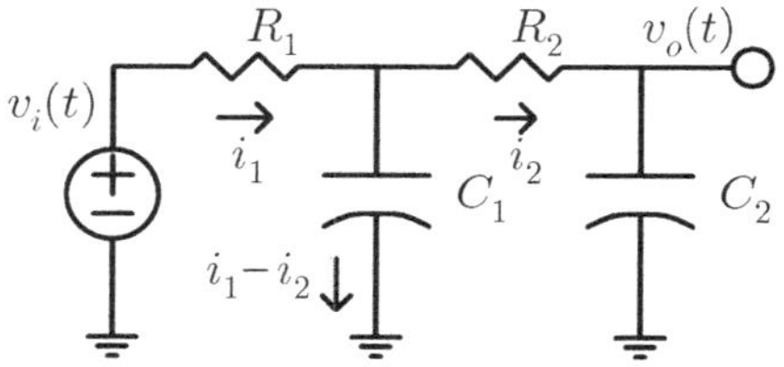

Fig. 37.1 *RC* network comprised of two resistors and two caps

This is the second equation. In matrix form we then have

$$\begin{bmatrix} -\dfrac{1}{sC_1} & \dfrac{C_1 + C_2 + sR_2C_1C_2}{sC_1C_2} \\ \dfrac{1 + sR_1C_1}{sC_1} & -\dfrac{1}{sC_1} \end{bmatrix} \begin{bmatrix} I_1 \\ I_2 \end{bmatrix} = \begin{bmatrix} 0 \\ V_i(s) \end{bmatrix} \tag{37.8}$$

The determinant of this matrix is

$$\begin{aligned} \det &= \frac{1}{s^2C_1^2} - \left[\frac{1 + sR_1C_1}{sC_1}\right]\left[\frac{C_1 + C_2 + sR_2C_1C_2}{sC_1C_2}\right] \\ &= \frac{1}{s^2C_1^2} - \frac{C_1 + C_2 + sR_2C_1C_2 + sR_1C_1(C_1 + C_2) + s^2R_1R_2C_1^2C_2}{s^2C_1^2C_2} \\ &= \frac{1}{s^2C_1^2} - \frac{C_1 + C_2 + s(R_2C_1C_2 + R_1C_1^2 + R_1C_1C_2) + s^2R_1R_2C_1^2C_2}{s^2C_1^2C_2} \end{aligned}$$

$$\boxed{\det(s) = -\frac{C_1 + s(R_2C_1C_2 + R_1C_1^2 + R_1C_1C_2) + s^2R_1R_2C_1^2C_2}{s^2C_1^2C_2}} \tag{37.9}$$

We can find the general solution by inverting the matrix; recall if

$$M = \begin{bmatrix} A & B \\ C & D \end{bmatrix} \tag{37.10}$$

then

$$M^{-1} = \frac{1}{AD - BC}\begin{bmatrix} D & -B \\ -C & A \end{bmatrix} \tag{37.11}$$

We then have

$$\boxed{\begin{bmatrix} I_1 \\ I_2 \end{bmatrix} = \frac{1}{\det(s)} \begin{bmatrix} -\dfrac{1}{sC_1} & -\dfrac{C_1 + C_2 + sR_2C_1C_2}{sC_1C_2} \\ -\dfrac{1 + sR_1C_1}{sC_1} & -\dfrac{1}{sC_1} \end{bmatrix} \begin{bmatrix} 0 \\ V_i(s) \end{bmatrix}} \tag{37.12}$$

In particular

$$I_1(s) = sC_1 \frac{C_1 + C_2 + sR_2C_1C_2}{C_1 + s(R_2C_1C_2 + R_1C_1^2 + R_1C_1C_2) + s^2R_1R_2C_1^2C_2} V_i(s) \tag{37.13}$$

and

$$I_2(s) = sC_1C_2 \frac{1}{C_1 + s(R_2C_1C_2 + R_1C_1^2 + R_1C_1C_2) + s^2R_1R_2C_1^2C_2} V_i(s) \tag{37.14}$$

Thus we have arrived at the most general solution to this network: for any input stimulus ($V_i(s)$) we know I_1 and I_2 both as a function of frequency. We also know output voltage which is simply I_2 times cap impedance:

$$V_o(s) = I_2(s)\frac{1}{sC_2}$$
$$= \frac{C_1}{C_1 + s(R_2C_1C_2 + R_1C_1^2 + R_1C_1C_2) + s^2R_1R_2C_1^2C_2} V_i(s) \tag{37.15}$$

As way of example let's assume some *RC* numbers:

$$R_1 = 1\,\Omega, \quad R_2 = 3\,\Omega, \quad C_1 = 1\,\text{mF}, \quad C_2 = 10\,\text{mF} \tag{37.16}$$

Then current versus frequency results are shown in Fig. 37.2. Notice that since C_2 is much larger than C_1, at low frequency the latter acts like an open and all of I_1 flows into I_2; that is why they appear about the same in the graph. At high frequency, C_1 shorts, $I_1(s)$ assumes the full value [1 here since $V_i(s) = 1\,\text{V}$ and $R_1 = 1\,\Omega$] and $I_2(s)$ is left with nothing! Output voltage is shown in Fig. 37.3. Notice at DC we get full value since the caps are open. What this means is that since the caps are open there is no current; and with no current there is no voltage drop across the resistors; hence output voltage equals input one! At high frequency output voltage decays at the rate −40 dB/dec due to the s^2 in the denominator of Eq. (37.15). The phase also settles at −180° for the same reason. Notice the exact match between theory and SPICE simulations.

Transient Results: Impulse Response Knowing current and voltage in the frequency domain we can now find the inverse transforms to go back to the time domain. For the *RC* values shown in Eq. (37.16), output voltage comes out at

$$V_o(s) = \frac{1.0 \times 10^{-3}}{1.0 \times 10^{-3} + 4.1 \times 10^{-5}s + 3.0 \times 10^{-8}s^2} \tag{37.17}$$

Factor out the 3×10^{-8}

$$V_o(s) = \frac{3.33 \times 10^4}{3.33 \times 10^4 + 1.37 \times 10^3 s + s^2} \tag{37.18}$$

The roots for the denominator are

$$s_1 = -24.8, \qquad s_2 = -1345.2 \tag{37.19}$$

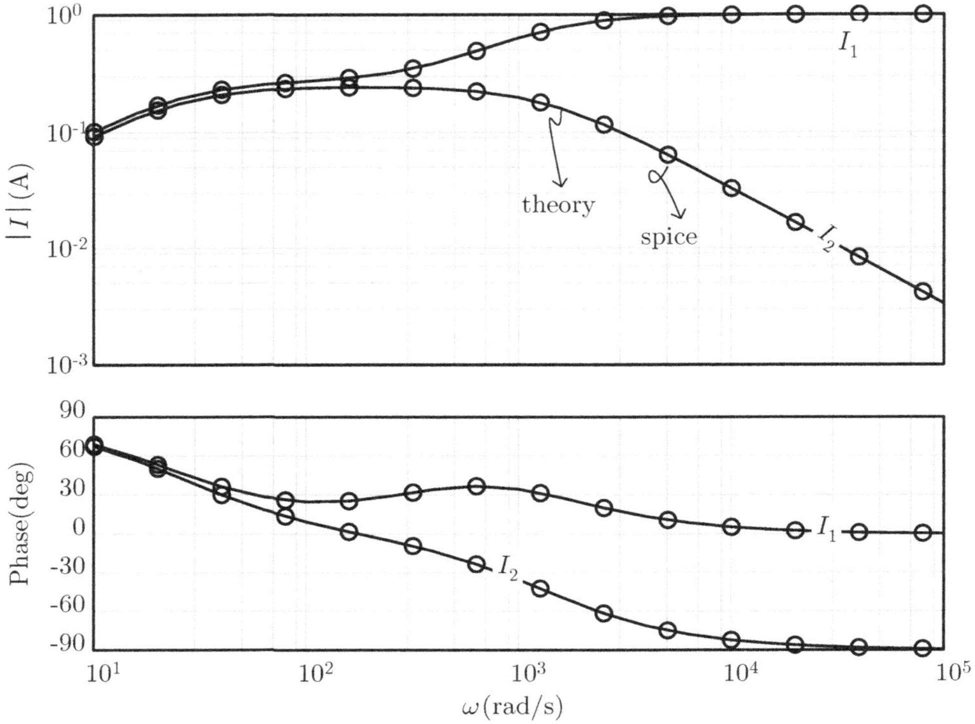

Fig. 37.2 I_1 and I_2 as a function of frequency

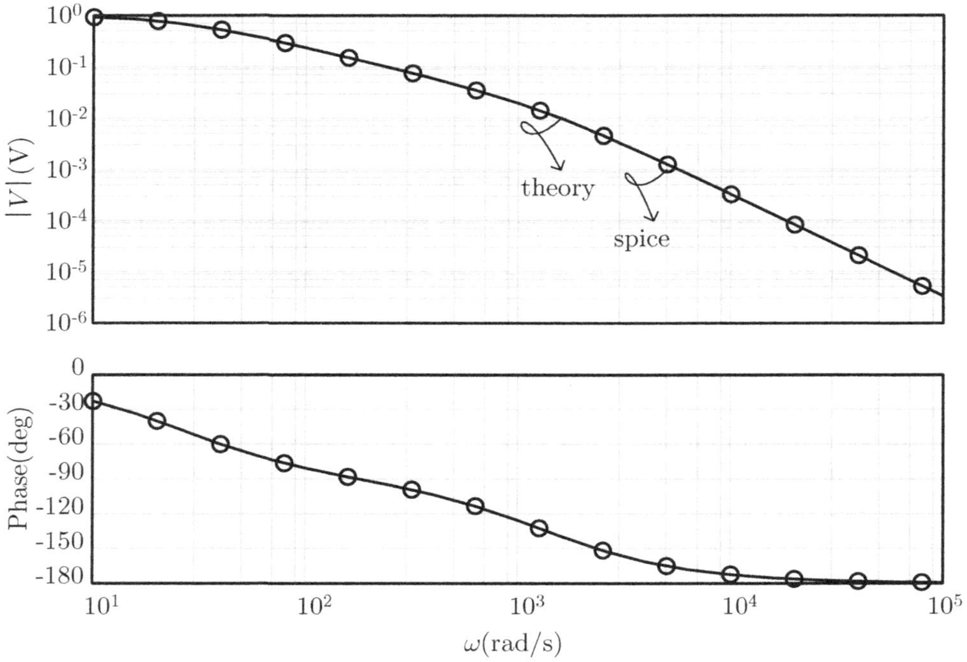

Fig. 37.3 Output voltage transfer function

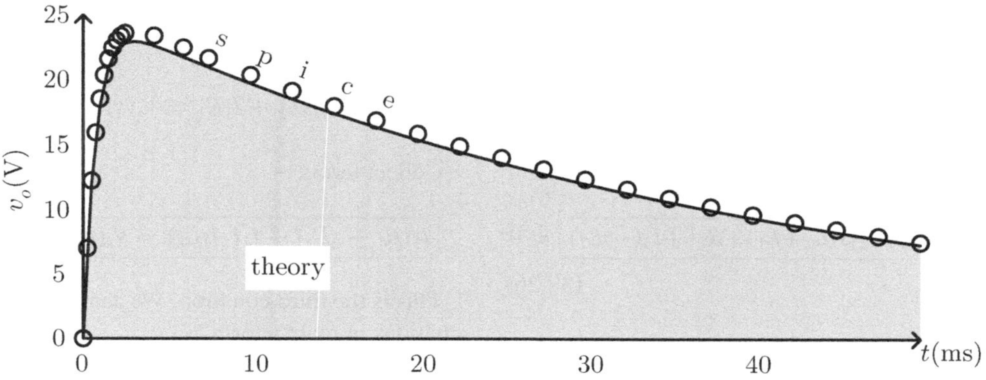

Fig. 37.4 Impulse response and comparison to simulations

Fig. 37.5 Three-loop *RL* network

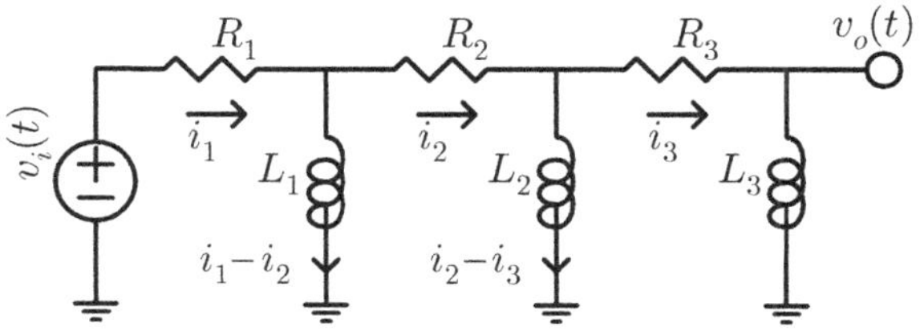

Then our solution becomes

$$V_o(s) = \frac{3.33 \times 10^4}{(s + 24.8)(s + 1345.2)} \tag{37.20}$$

We can factor this out and results are

$$V_o(s) = \frac{25.2}{s + 24.8} - \frac{25.2}{s + 1345.2} \tag{37.21}$$

Taking the inverse transform we arrive at output voltage in time

$$v_o(t) = 25.2\left[e^{-24.8t} - e^{-1345.2t}\right] \tag{37.22}$$

A plot of these results is shown in Fig. 37.4. Notice exact match to simulation. Let's recap what was accomplished. In contrast to the conventional case where upfront we knew the input/output transfer function, in this case we built a linear system comprised of a 2×2 matrix, did the linear algebra (symbolically), figured internal variables then made our way to the desired output voltage, all the way while residing in the frequency domain. Once we figured output voltage in the frequency domain we figured its inverse transform to go back to the time domain. In the next section we expand the analysis to a larger system, but still abiding by the same flow.

37.3 Second Example: Three-Loop *RL* Network

Consider next the 3-loop *RL* network shown in Fig. 37.5. We want to find output voltage across the right-most inductor. Similar to last section we don't have upfront a direct relation between input and output. In other words, it's not straightforward to figure the transfer function by a sequence of series/parallel maneuvering steps. We would have to go one level deeper. Start with KVL on the right branch:

$$I_3 sL_3 + I_3 R_3 - (I_2 - I_3)sL_2 = 0 \tag{37.23}$$

Collect terms

$$I_2(-sL_2) + I_3\left[R_3 + s(L_3 + L_2)\right] = 0 \tag{37.24}$$

This is the first equation. Next do KVL on middle loop

$$(I_2 - I_3)sL_2 + I_2R_2 - (I_1 - I_2)sL_1 = 0 \quad (37.25)$$

Collect terms

$$\boxed{I_1(-sL_1)+I_2\left[s(L_2+L_1)+R_2\right]+I_3(-sL_2)} = 0 \quad (37.26)$$

This is the second equation. Finally do KVL around left loop

$$(I_1 - I_2)sL_1 + I_1R_1 = V_i(s) \quad (37.27)$$

Collect terms

$$\boxed{I_1(R_1 + sL_1) + I_2(-sL_1) = V_i(s)} \quad (37.28)$$

This is the third equation. We can put all three equations in matrix form as

$$\boxed{\begin{bmatrix} 0 & -sL_2 & R_3 + s(L_2 + L_3) \\ -sL_1 & R_2 + s(L_1 + L_2) & -sL_2 \\ R_1 + sL_1 & -sL_1 & 0 \end{bmatrix} \begin{bmatrix} I_1 \\ I_2 \\ I_3 \end{bmatrix} = \begin{bmatrix} 0 \\ 0 \\ V_i(s) \end{bmatrix}} \quad (37.29)$$

At this point we can pass this matrix to a matrix solver (per each frequency), solve it numerically, and call it a day! But we don't necessarily get a closed form solution. As a way of demonstration let's assume we did not have access to such a solver and we wanted to do it the hard way. This method, as of course everyone knows it, is called Gaussian elimination. It is tedious but worth doing at least once (in a lifetime)! The ensuing steps are as follows. Replace first row by itself plus last row

$$\begin{bmatrix} R_1 + sL_1 & -s(L_1 + L_2) & R_3 + s(L_2 + L_3) \\ -sL_1 & R_2 + s(L_1 + L_2) & -sL_2 \\ R_1 + sL_1 & -sL_1 & 0 \end{bmatrix} \begin{bmatrix} I_1 \\ I_2 \\ I_3 \end{bmatrix} = \begin{bmatrix} V_i(s) \\ 0 \\ V_i(s) \end{bmatrix} \quad (37.30)$$

Divide first row by $(R_1 + sL_1)$

$$\begin{bmatrix} 1 & -\dfrac{s(L_1 + L_2)}{R_1 + sL_1} & \dfrac{R_3 + s(L_2 + L_3)}{R_1 + sL_1} \\ -sL_1 & R_2 + s(L_1 + L_2) & -sL_2 \\ R_1 + sL_1 & -sL_1 & 0 \end{bmatrix} \begin{bmatrix} I_1 \\ I_2 \\ I_3 \end{bmatrix} = \begin{bmatrix} \dfrac{V_i(s)}{R_1 + sL_1} \\ 0 \\ V_i(s) \end{bmatrix} \quad (37.31)$$

Divide second row by sL_1

$$\begin{bmatrix} 1 & -\dfrac{s(L_1 + L_2)}{R_1 + sL_1} & \dfrac{R_3 + s(L_2 + L_3)}{R_1 + sL_1} \\ -1 & \dfrac{R_2 + s(L_1 + L_2)}{sL_1} & -\dfrac{sL_2}{sL_1} \\ R_1 + sL_1 & -sL_1 & 0 \end{bmatrix} \begin{bmatrix} I_1 \\ I_2 \\ I_3 \end{bmatrix} = \begin{bmatrix} \dfrac{V_i(s)}{R_1 + sL_1} \\ 0 \\ V_i(s) \end{bmatrix} \quad (37.32)$$

Divide last row by $-(R_1 + sL_1)$

$$\begin{bmatrix} 1 & -\dfrac{s(L_1 + L_2)}{R_1 + sL_1} & \dfrac{R_3 + s(L_2 + L_3)}{R_1 + sL_1} \\ -1 & \dfrac{R_2 + s(L_1 + L_2)}{sL_1} & -\dfrac{sL_2}{sL_1} \\ -1 & \dfrac{sL_1}{R_1 + sL_1} & 0 \end{bmatrix} \begin{bmatrix} I_1 \\ I_2 \\ I_3 \end{bmatrix} = \begin{bmatrix} \dfrac{V_i(s)}{R_1 + sL_1} \\ 0 \\ -\dfrac{V_i(s)}{R_1 + sL_1} \end{bmatrix} \quad (37.33)$$

Replace second row by itself plus first row

$$\begin{bmatrix} 1 & -\frac{s(L_1+L_2)}{R_1+sL_1} & \frac{R_3+s(L_2+L_3)}{R_1+sL_1} \\ 0 & \frac{R_2+s(L_1+L_2)}{sL_1} - \frac{s(L_1+L_2)}{R_1+sL_1} & -\frac{sL_2}{sL_1} + \frac{R_3+s(L_2+L_3)}{R_1+sL_1} \\ -1 & \frac{sL_1}{R_1+sL_1} & 0 \end{bmatrix} \begin{bmatrix} I_1 \\ I_2 \\ I_3 \end{bmatrix} = \begin{bmatrix} \frac{V_i(s)}{R_1+sL_1} \\ \frac{V_i(s)}{R_1+sL_1} \\ -\frac{V_i(s)}{R_1+sL_1} \end{bmatrix} \tag{37.34}$$

Let's simplify the 2 × 2 entry

$$\begin{aligned} \frac{R_2+s(L_1+L_2)}{sL_1} - \frac{s(L_1+L_2)}{R_1+sL_1} &= \frac{(R_1+sL_1)(R_2+s(L_1+L_2)) - sL_1 \cdot s(L_1+L_2)}{sL_1(R_1+sL_1)} \\ &= \frac{R_1R_2+s(R_1L_1+R_1L_2+R_2L_1)}{sL_1(R_1+sL_1)} \end{aligned} \tag{37.35}$$

Next simplify the 2 × 3 entry

$$\begin{aligned} -\frac{L_2}{L_1} + \frac{R_3+s(L_2+L_3)}{R_1+sL_1} &= \frac{-L_2(R_1+sL_1) + L_1(R_3+sL_2+sL_3)}{L_1(R_1+sL_1)} \\ &= \frac{-R_1L_2+R_3L_1+s(L_1L_3)}{L_1(R_1+sL_1)} \end{aligned} \tag{37.36}$$

Our matrix now becomes

$$\begin{bmatrix} 1 & -\frac{s(L_1+L_2)}{R_1+sL_1} & \frac{R_3+s(L_2+L_3)}{R_1+sL_1} \\ 0 & \frac{R_1R_2+s(R_1L_1+R_1L_2+R_2L_1)}{sL_1(R_1+sL_1)} & \frac{-R_1L_2+R_3L_1+s(L_1L_3)}{L_1(R_1+sL_1)} \\ -1 & \frac{sL_1}{R_1+sL_1} & 0 \end{bmatrix} \begin{bmatrix} I_1 \\ I_2 \\ I_3 \end{bmatrix} = \begin{bmatrix} \frac{V_i(s)}{R_1+sL_1} \\ \frac{V_i(s)}{R_1+sL_1} \\ -\frac{V_i(s)}{R_1+sL_1} \end{bmatrix} \tag{37.37}$$

Replace last row by itself plus first row

$$\begin{bmatrix} 1 & -\frac{s(L_1+L_2)}{R_1+sL_1} & \frac{R_3+s(L_2+L_3)}{R_1+sL_1} \\ 0 & \frac{R_1R_2+s(R_1L_1+R_1L_2+R_2L_1)}{sL_1(R_1+sL_1)} & \frac{-R_1L_2+R_3L_1+s(L_1L_3)}{L_1(R_1+sL_1)} \\ 0 & -\frac{sL_2}{R_1+sL_1} & \frac{R_3+s(L_2+L_3)}{R_1+sL_1} \end{bmatrix} \begin{bmatrix} I_1 \\ I_2 \\ I_3 \end{bmatrix} = \begin{bmatrix} \frac{V_i(s)}{R_1+sL_1} \\ \frac{V_i(s)}{R_1+sL_1} \\ 0 \end{bmatrix} \tag{37.38}$$

Multiply the second row by $\dfrac{sL_1(R_1 + sL_1)}{R_1R_2 + s(R_1L_1 + R_1L_2 + R_2L_1)}$

$$\begin{bmatrix} 1 & -\dfrac{s(L_1+L_2)}{R_1+sL_1} & \dfrac{R_3+s(L_2+L_3)}{R_1+sL_1} \\ 0 & 1 & \dfrac{-R_1L_2+R_3L_1+s(L_1L_3)}{L_1(R_1+sL_1)} \times \dfrac{sL_1(R_1+sL_1)}{R_1R_2+s(R_1L_1+R_1L_2+R_2L_1)} \\ 0 & -\dfrac{sL_2}{R_1+sL_1} & \dfrac{R_3+s(L_2+L_3)}{R_1+sL_1} \end{bmatrix} \begin{bmatrix} I_1 \\ I_2 \\ I_3 \end{bmatrix} = \begin{bmatrix} \dfrac{V_i(s)}{R_1+sL_1} \\ \dfrac{V_i(s)}{R_1+sL_1} \times \dfrac{sL_1(R_1+sL_1)}{R_1R_2+s(R_1L_1+R_1L_2+R_2L_1)} \\ 0 \end{bmatrix} \tag{37.39}$$

Simplify the 2×3 entry

$$\begin{aligned} 2 \times 3 \text{ entry} &= \frac{[-R_1L_2 + R_3L_1 + sL_1L_3]s[R_1L_1 + sL_1^2]}{[R_1L_1 + sL_1^2][R_1R_2 + s(R_1L_1 + R_1L_2 + R_2L_1)]} \\ &= \frac{sa + s^2L_1L_3}{R_1R_2 + sb} \end{aligned} \tag{37.40}$$

where

$$a = -R_1L_2 + R_3L_1, \qquad \text{and} \qquad b = R_1L_1 + R_1L_2 + R_2L_1 \tag{37.41}$$

Next simplify the second entry on the right vector

$$\begin{aligned} \text{2nd entry in right vector} &= \frac{[V_i(s)][sL_1(R_1 + sL_1)]}{[R_1 + sL_1][R_1R_2 + s(R_1L_1 + R_1L_2 + R_2L_1)]} \\ &= \frac{V_i(s)sL_1}{R_1R_2 + sb} \end{aligned} \tag{37.42}$$

Our matrix then becomes

$$\begin{bmatrix} 1 & -\dfrac{s(L_1+L_2)}{R_1+sL_1} & \dfrac{R_3+s(L_2+L_3)}{R_1+sL_1} \\ 0 & 1 & \dfrac{sa+s^2L_1L_3}{R_1R_2+sb} \\ 0 & -\dfrac{sL_2}{R_1+sL_1} & \dfrac{R_3+s(L_2+L_3)}{R_1+sL_1} \end{bmatrix} \begin{bmatrix} I_1 \\ I_2 \\ I_3 \end{bmatrix} = \begin{bmatrix} \dfrac{V_i(s)}{R_1+sL_1} \\ \dfrac{V_i(s)sL_1}{R_1R_2+sb} \\ 0 \end{bmatrix} \tag{37.43}$$

Multiply the last row by $\dfrac{R_1 + sL_1}{sL_2}$

$$\begin{bmatrix} 1 - \dfrac{s(L_1+L_2)}{R_1+sL_1} & \dfrac{R_3+s(L_2+L_3)}{R_1+sL_1} \\ 0 & 1 & \dfrac{sa+s^2L_1L_3}{R_1R_2+sb} \\ 0 & -1 & \dfrac{R_3+s(L_2+L_3)}{sL_2} \end{bmatrix} \begin{bmatrix} I_1 \\ I_2 \\ I_3 \end{bmatrix} = \begin{bmatrix} \dfrac{V_i(s)}{R_1+sL_1} \\ \dfrac{V_i(s)sL_1}{R_1R_2+sb} \\ 0 \end{bmatrix} \tag{37.44}$$

Replace last row by itself plus second row

$$\begin{bmatrix} 1 - \dfrac{s(L_1+L_2)}{R_1+sL_1} & \dfrac{R_3+s(L_2+L_3)}{R_1+sL_1} \\ 0 & 1 & \dfrac{sa+s^2L_1L_3}{R_1R_2+sb} \\ 0 & 0 & \dfrac{R_3+s(L_2+L_3)}{sL_2} + \dfrac{sa+s^2L_1L_3}{R_1R_2+sb} \end{bmatrix} \begin{bmatrix} I_1 \\ I_2 \\ I_3 \end{bmatrix} = \begin{bmatrix} \dfrac{V_i(s)}{R_1+sL_1} \\ \dfrac{V_i(s)sL_1}{R_1R_2+sb} \\ \dfrac{V_i(s)sL_1}{R_1R_2+sb} \end{bmatrix} \tag{37.45}$$

Define

$$c = L_2 + L_3 \tag{37.46}$$

The 3×3 entry becomes

$$\begin{aligned} 3 \times 3 \text{ entry} &= \frac{R_3 + sc}{sL_2} + \frac{sa + s^2L_1L_3}{R_1R_2 + sb} \\ &= \frac{[R_3 + sc][R_1R_2 + sb] + [sL_2][sa + s^2L_1L_3]}{sL_2(R_1R_2 + sb)} \\ &= \frac{R_1R_2R_3 + s[cR_1R_2 + bR_3] + s^2[bc + aL_2] + s^3[L_1L_2L_3]}{sL_2(R_1R_2 + sb)} \end{aligned} \tag{37.47}$$

Define

$$d = cR_1R_2 + bR_3, \qquad \text{and} \qquad e = bc + aL_2 \tag{37.48}$$

then our matrix becomes

$$\begin{bmatrix} 1 - \dfrac{s(L_1+L_2)}{R_1+sL_1} & \dfrac{R_3+s(L_2+L_3)}{R_1+sL_1} \\ 0 & 1 & \dfrac{sa+s^2L_1L_3}{R_1R_2+sb} \\ 0 & 0 & \dfrac{R_1R_2R_3 + sd + s^2e + s^3[L_1L_2L_3]}{sL_2(R_1R_2+sb)} \end{bmatrix} \begin{bmatrix} I_1 \\ I_2 \\ I_3 \end{bmatrix} = \begin{bmatrix} \dfrac{V_i(s)}{R_1+sL_1} \\ \dfrac{V_i(s)sL_1}{R_1R_2+sb} \\ \dfrac{V_i(s)sL_1}{R_1R_2+sb} \end{bmatrix} \tag{37.49}$$

From the last row we can figure I_3 as follows:

$$I_3(s) = \frac{V_i(s)sL_1}{R_1R_2 + sb} \times \frac{sL_2(R_1R_2 + sb)}{R_1R_2R_3 + sd + s^2e + s^3[L_1L_2L_3]}$$
$$= \boxed{V_i(s)\frac{s^2L_1L_2}{R_1R_2R_3 + sd + s^2e + s^3L_1L_2L_3}} \quad (37.50)$$

Output voltage is simply output current times L_3 impedance

$$\boxed{V_o(s) = V_i(s)\frac{s^3L_1L_2L_3}{R_1R_2R_3 + sd + s^2e + s^3L_1L_2L_3}} \quad (37.51)$$

Let's set some *RL* values to plot our output current and voltage:

$$R_1 = 1\,\Omega, \quad R_2 = 2\,\Omega, \quad R_3 = 0.1\,\Omega, \quad \text{and}$$
$$L_1 = 1\,\text{mH}, \quad L_2 = 2\,\text{mH}, \quad L_3 = 3\,\text{mH} \quad (37.52)$$

Results are shown in Fig. 37.6. Notice that at low frequency output voltage is zero since no current makes it through to L_3; all current would have sank into L_1. At high frequency all inductors open and all currents go to zero; hence there is no *IR* voltage drop between input and output, and latter becomes equal to former. Notice the exact agreement between theory and SPICE.

Let us try another set of *RL* values

$$R_1 = 1\,\Omega, \quad R_2 = 2\,\Omega, \quad R_3 = 0.1\,\Omega, \quad \text{and}$$
$$L_1{=}5\,\text{mH}, \quad L_2{=}2\,\text{mH}, \quad L_3{=}0.1\,\text{mH} \quad (37.53)$$

Corresponding results are shown in Fig. 37.7. Notice that while results differ materially between this figure and last one, the left- and right-extremes still match; and those are that output voltage is zero at low frequency and unity at high frequency due to the aforementioned reasons. Finally, notice (again) the exact match to SPICE.

Transient Results: Unit Step Response Now that we have at hand output voltage in the frequency domain we move to time domain. Our starting point is Eq. (37.51) which gave output voltage for the 3-loop *RL* circuit, retyped below followed by its constituents

$$V_o(s) = V_i(s)\frac{s^3L_1L_2L_3}{R_1R_2R_3 + sd + s^2e + s^3L_1L_2L_3} \quad (37.54)$$

where

$$\begin{aligned} a &= -R_1L_2 + R_3L_1 \\ b &= R_1L_1 + R_1L_2 + R_2L_1 \\ c &= L_2 + L_3 \\ d &= cR_1R_2 + bR_3 \\ e &= bc + aL_2 \end{aligned} \quad (37.55)$$

Let's assume that our input is a *unit step* function: $V_i(s) = 1/s$; then our output becomes

$$\boxed{V_o(s) = \frac{s^2L_1L_2L_3}{R_1R_2R_3 + sd + s^2e + s^3L_1L_2L_3}} \quad (37.56)$$

Let's assume some *RL* values

$$R_1 = 1\,\Omega, \quad R_2 = 2\,\Omega, \quad R_3 = 3\,\Omega, \quad \text{and}$$
$$L_1 = 1\,\text{mH}, \quad L_2 = 2\,\text{mH}, \quad L_3 = 3\,\text{mH} \quad (37.57)$$

Then output voltage is

$$V_o(s) = \frac{s^2}{1 \times 10^9 + 4.167 \times 10^6 s + 4500 s^2 + s^3} \quad (37.58)$$

Output voltage versus frequency is shown in Fig. 37.8. Again notice exact match to simulations.

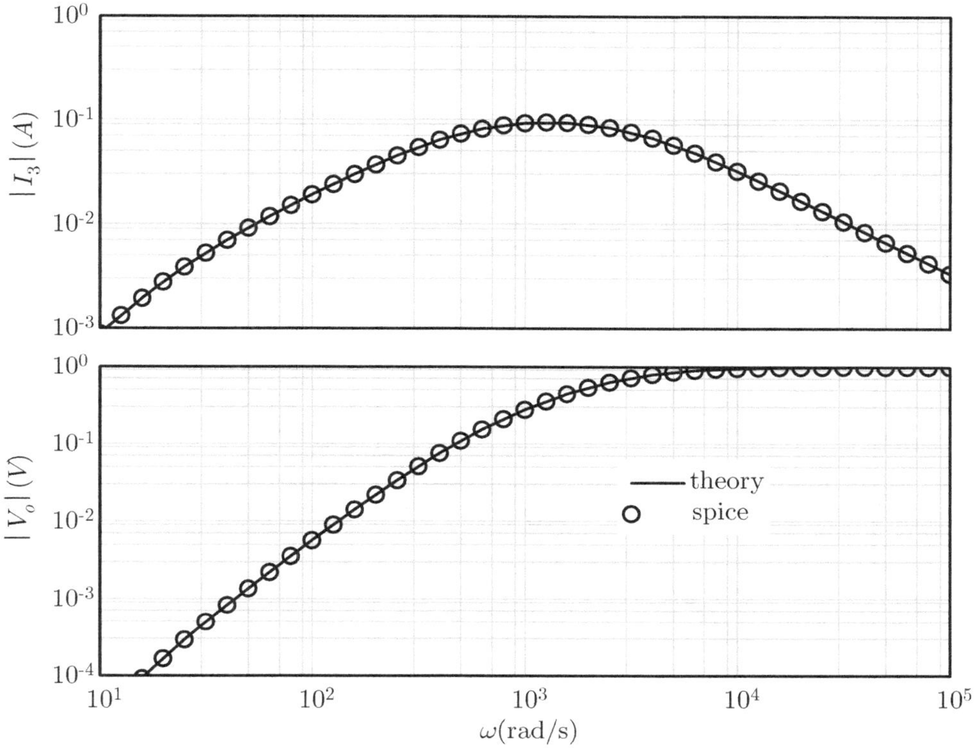

Fig. 37.6 Output current (top) and voltage (bottom) versus frequency for *RL* values as per Eq. (37.52)

The denominator of the transfer function has the following roots:

$$\begin{aligned} s_1 &= -390.00 \\ s_2 &= -767.00 \\ s_3 &= -3343.00 \end{aligned} \tag{37.59}$$

so that

$$\boxed{V_o(s) = \frac{s^2}{(s-s_1)(s-s_2)(s-s_3)}} \tag{37.60}$$

We can write our voltage in the form

$$V_o(s) = \frac{A}{s-s_1} + \frac{B}{s-s_2} + \frac{C}{s-s_3} \tag{37.61}$$

The constants can be evaluated as follows:

$$\begin{aligned} A &= V_o(s)(s-s_1)|_{s=s_1} = 0.137 \\ B &= V_o(s)(s-s_2)|_{s=s_2} = -0.606 \\ C &= V_o(s)(s-s_3)|_{s=s_3} = 1.469 \end{aligned} \tag{37.62}$$

Then we can rewrite

$$V_o(s) = \frac{0.137}{s+390} + \frac{-0.606}{s+767} + \frac{1.469}{s+3343} \tag{37.63}$$

A plot of this is also shown in Fig. 37.8. In time domain we then have

$$\boxed{v_o(t)=0.137e^{-390t}-0.606e^{-767t}+1.469e^{-3343t}} \tag{37.64}$$

Transient results are shown in Fig. 37.9 alongside simulation ones. Notice that right after time zero output voltage equals input one (unity here). The reason is that the sharp event causes the inductors to act as open; as such no current flows through the network and output voltage simply equals input one. On the other hand, notice that

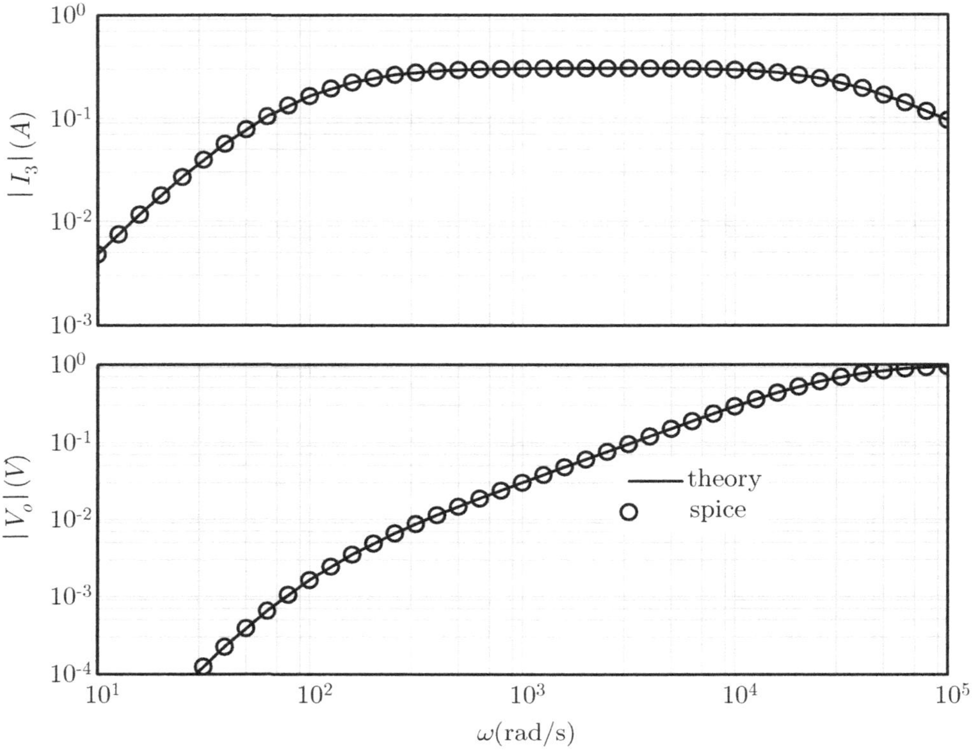

Fig. 37.7 Output current (top) and voltage (bottom) versus frequency for RL values as per Eq. (37.53)

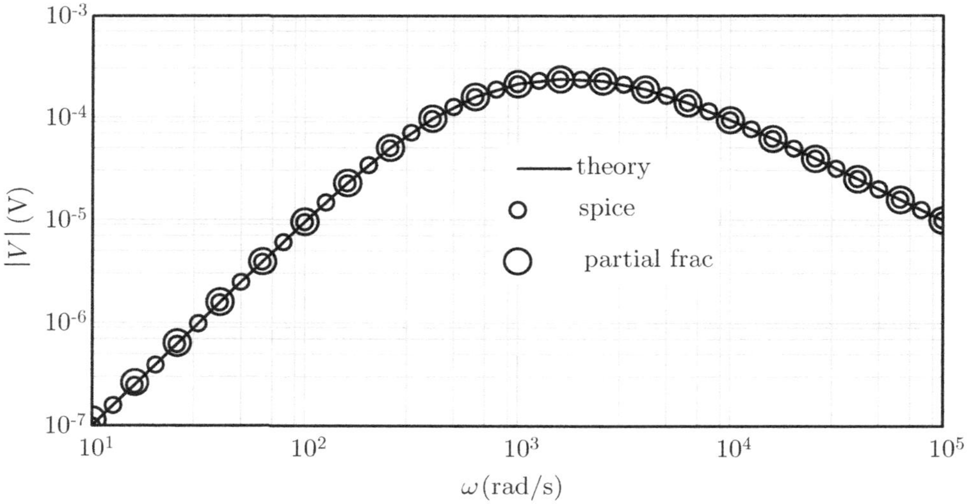

Fig. 37.8 Output voltage due to step input

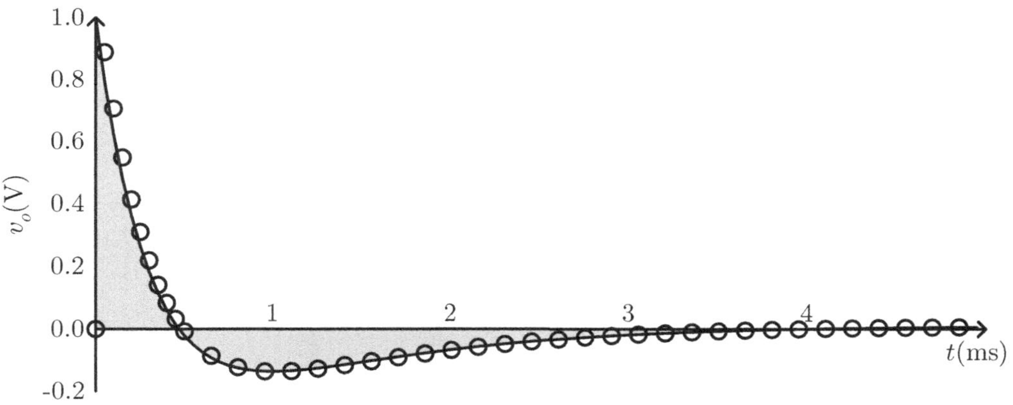

Fig. 37.9 Output voltage due to step input

at large time output voltage goes to zero. This is due to the fact that output inductor acts as a short and all voltage drop occurs across the resistors (the left-most one, to be specific).

So again we succeed in dealing with a complex network by first solving for variables in the frequency domain and then migrating back to the time domain. Similar to the prior section we relied on matrix theory to manage all the internal variables and to eventually furnish output current, which when differentiated yielded output voltage. In this particular instance we assumed that input voltage was an ideal step one. But this is no requirement; we could have easily used a tapered step, a causal sine one, or a causal periodic pulse and so forth. All that changes is using the correct Laplace transform for the input at hand and applying the subsequent alterations to the interim algebra steps.

37.4 Larger Systems

As evident in last section, even a system as small as three unknowns amounted to a good amount of algebra. We certainly don't want to be doing this for larger systems. Hence we must fall back on some mathematical packages that do linear systems. There are plenty such packages or worst case one can write one's own linear solver, pretty much automating the Gaussian eliminations used before. In most cases we end up with a multi-branch network with N unknown currents. We would use KVL/KCL to get a set of N equations which we can cast in the form

$$\begin{bmatrix} A_{11} & A_{12} & A_{1N} \\ A_{21} & A_{22} & A_{2N} \\ \\ A_{N1} & A_{N2} & A_{NN} \end{bmatrix} \begin{bmatrix} I_1 \\ I_2 \\ \\ I_N \end{bmatrix} = \begin{bmatrix} b_1 \\ b_2 \\ \\ b_N \end{bmatrix} \quad (37.65)$$

where generally all of A_{nn} and b_n are frequency dependent. The above system needs to be solved for all of I_n *at each frequency*! Once currents are known, versus frequency, the various input/output terminal voltages can be figured as such, and possibly any variants thereof, such as input/output impedance or admittance. The Problem section will show a few examples for how this is done.

37.5 Summary

In this chapter we dealt with relatively more complex networks whose transfer functions would have been rather difficult to obtain using conventional impedance series/parallel reduction steps. In such cases we resort to unfolding the problem and dealing it at a lower level. What this means is that we had to solve for all internal variables (mostly currents in this case), and then make our way back to the desired outcome (mostly output voltage in this case). But even then we saw that there was nothing to intimidate us from large systems so far as applying spectral techniques (or for that matter convolution) other than correctly setting up the linear system matrices, at each frequency and using a mathematical package to solve for the unknowns (be them currents or voltages). Nothing new so far as the core concepts which are to use the frequency dependent version of the various impedance blocks, in conjunction with KVL and KCL. Of course some familiarity with generating the *complex* matrices, and looping over frequency will become helpful with more practice!

37.6 Problems

1. Consider the network in Fig. 37.10. Show that the governing set of equation is

$$\begin{bmatrix} 0 & -Z_{23} & Z_3 + Z_{23} + Z_{33} \\ -Z_{12} & Z_2 + Z_{12} + Z_{23} & -Z_{23} \\ Z_1 + Z_{12} & -Z_{12} & 0 \end{bmatrix} \begin{bmatrix} I_1 \\ I_2 \\ I_3 \end{bmatrix} = \begin{bmatrix} 0 \\ 0 \\ V_i(s) \end{bmatrix}$$

For the case

$$Z_1 = R_1 = 1\,\Omega, \quad Z_2 = R_2 = 2\,\Omega, \quad Z_3 = R_3 = 0.1\,\Omega, \qquad \text{and}$$
$$Z_{12} = L_1 = 1\,\text{mH}, \quad Z_{23} = L_2 = 2\,\text{mH}, \quad Z_{33} = L_3 = 3\,\text{mH}$$

and assuming input voltage is $V_i(s) = 1$, plot output voltage and compare to SPICE; see sample solution in Fig. 37.11.

2. Consider the *LC* network in Fig. 37.12. Use results/flow from Problem 1 to figure output voltage in the frequency domain. What is the final phase? What is the final decay rate? Explain. Compare to SPICE; see sample results in Fig. 37.13.
3. Consider the network in Fig. 37.14. Open the left current source, and inject 1 AC through the right one. Show that the set of equations governing the 4 denoted currents is as follows:

$$\begin{bmatrix} z_{12} + z_2 + z_{23} & -z_{23} & 0 & 0 \\ -z_{23} & z_{23} + z_3 + z_{34} & -z_{34} & 0 \\ 0 & -z_{34} & z_{34} + z_4 + z_{45} & -z_{45} \\ 0 & 0 & 0 & 1 \end{bmatrix} \begin{bmatrix} I_1 \\ I_2 \\ I_3 \\ I_4 \end{bmatrix} = \begin{bmatrix} z_{12} \\ 0 \\ 0 \\ 0 \end{bmatrix}$$

Once the currents are solved for, figure Z_{11} which translates to output voltage and Z_{12} which translates to input voltage. (Notice the use of upper case for input and output impedances, as opposed to small case for individual elements!) Next open right current and inject unity AC current through the left source and figure Z_{22} which would be input voltage and Z_{21} which would be output voltage.

4. Consider the *RC* network in Fig. 37.15. Use results from Problem 3 to figure Z_{11}, Z_{12}, Z_{22}, and Z_{21}. Plot results and compare to SPICE; see sample solution in Fig. 37.16. Explain both low- and high-frequency results for both of magnitude and phase.
5. Consider the generic network in Fig. 37.17. Apply a unity AC source current and show that the set of governing equations relating the four currents $i_1 \cdots i_4$ is

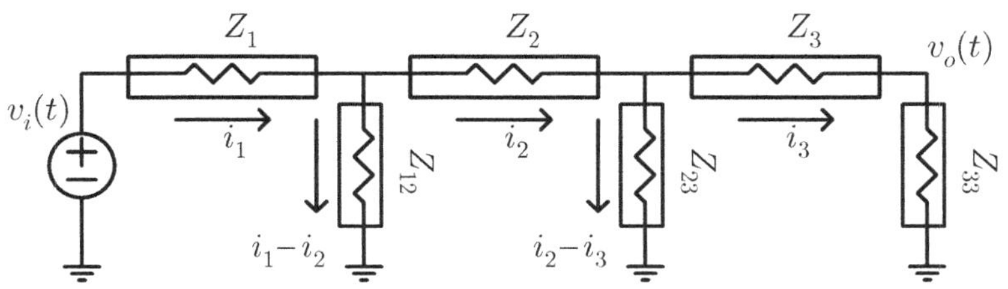

Fig. 37.10 Statement to Problem 1

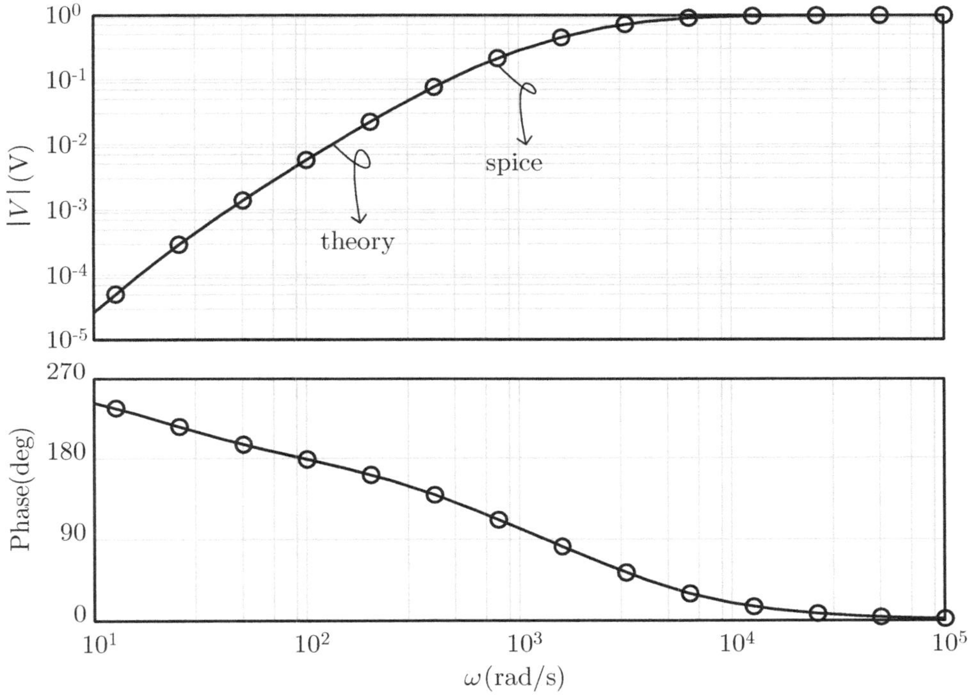

Fig. 37.11 Sample solution to Problem 1

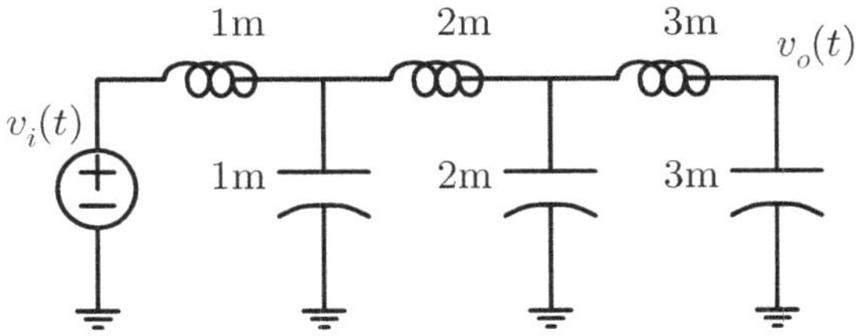

Fig. 37.12 Statement to Problem 2

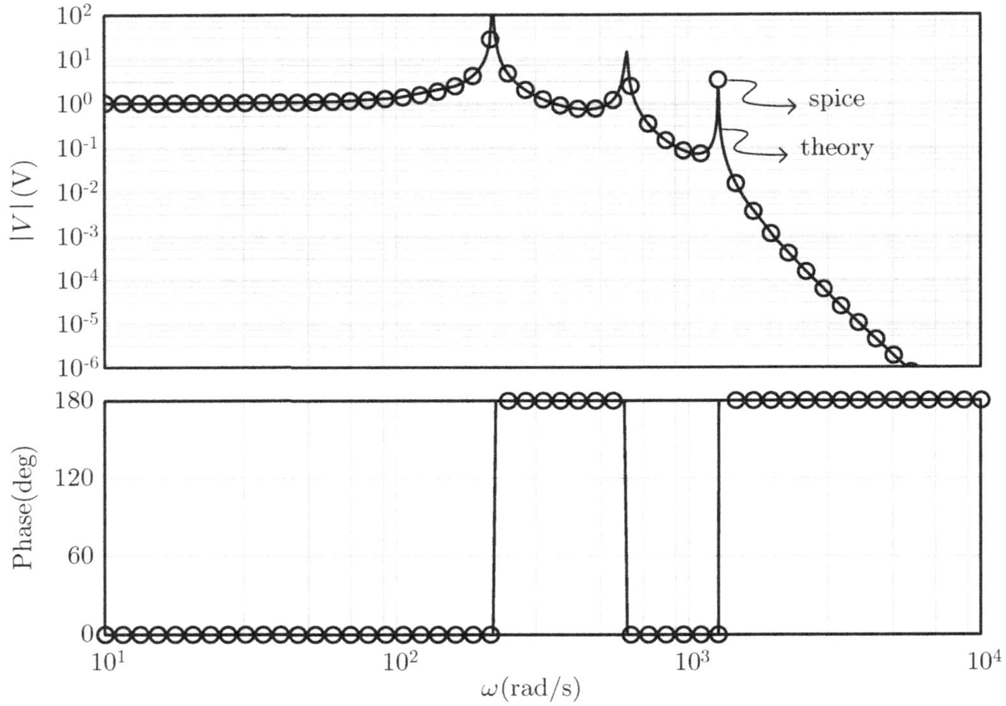

Fig. 37.13 Sample solution to Problem 2

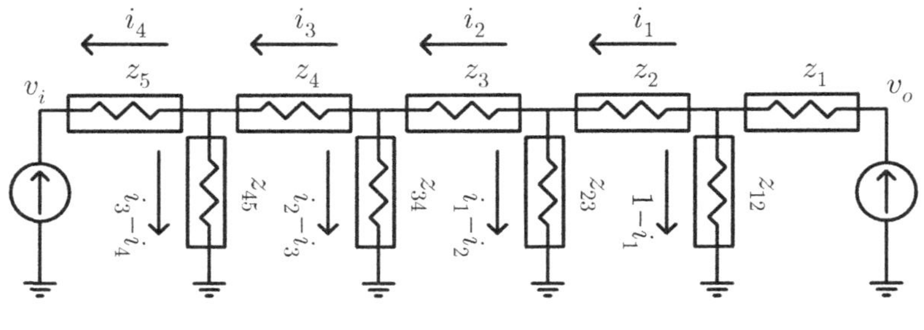

Fig. 37.14 Statement to Problem 3

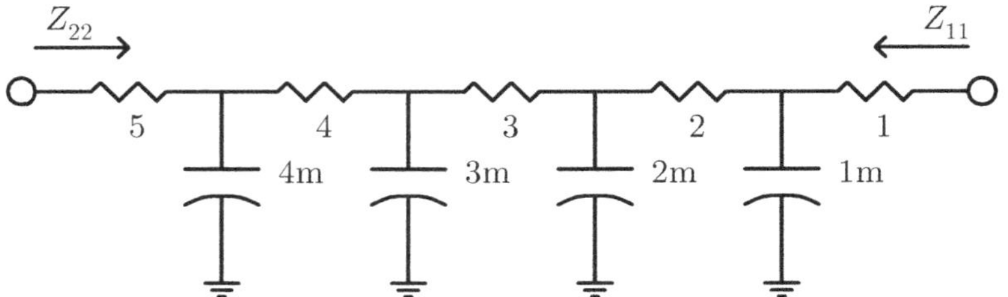

Fig. 37.15 Statement to Problem 4

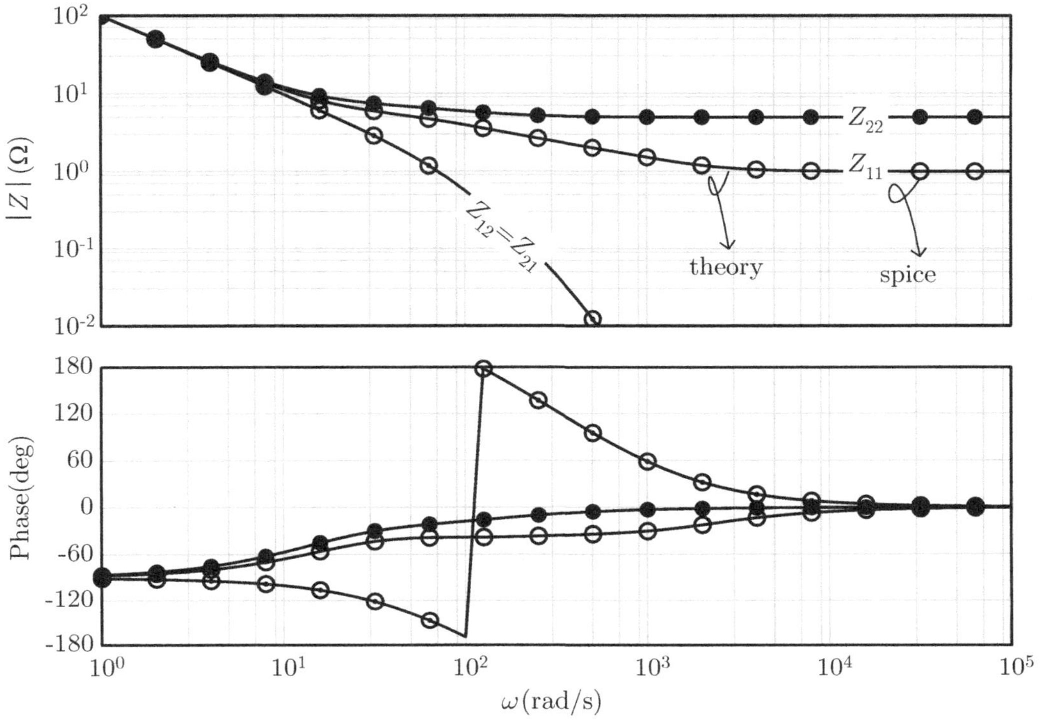

Fig. 37.16 Sample solution to Problem 4

Fig. 37.17 Statement to Problem 5

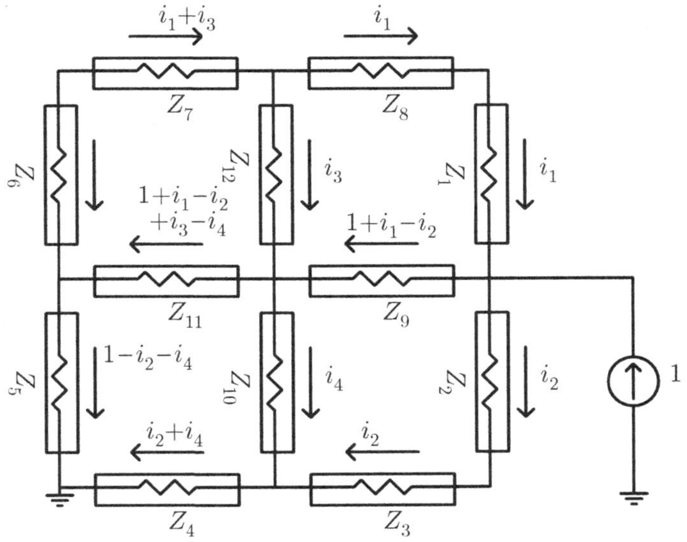

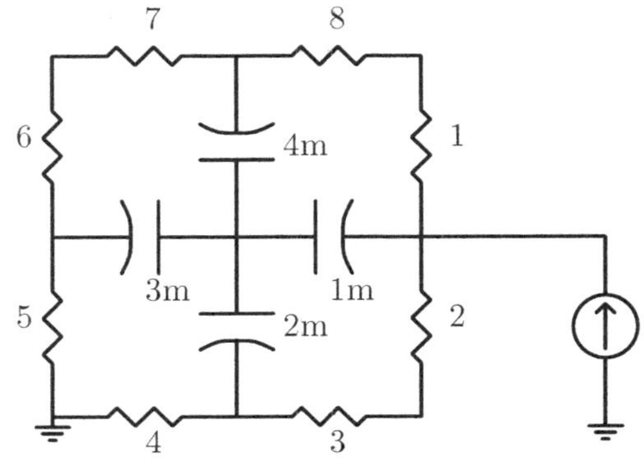

Fig. 37.18 Statement to Problem 6

$$\begin{bmatrix} z_1+z_8+z_9 & -z_9 & -z_{12} & 0 \\ -z_9 & z_2+z_9+z_3 & 0 & -z_{10} \\ z_7+z_6+z_{11} & -z_{11} & z_{12}+z_7+z_6+z_{11} & -z_{11} \\ -z_{11} & z_{11}+z_5+z_4 & -z_{11} & z_{10}+z_{11}+z_5+z_4 \end{bmatrix} \begin{bmatrix} I_1 \\ I_2 \\ I_3 \\ I_4 \end{bmatrix} = \begin{bmatrix} -z_9 \\ z_9 \\ -z_{11} \\ z_{11}+z_5 \end{bmatrix}$$

6. Consider the *RC* network in Fig. 37.18. Use results from Problem 5 to figure input impedance; plot and compare to SPICE; see sample results in Fig. 37.19. Explain the low- and high-frequency limits.
7. Consider the *LC* network in Fig. 37.20. Use results from Problem 5 to figure input impedance; plot and compare to SPICE; see sample results in Fig. 37.21.
8. Consider the *RLC* circuit with feedback in Fig. 37.22. Force unity AC current as shown and figure input impedance. First set up the 3× matrix governing the 3-branch currents as

$$\begin{bmatrix} R_1+\frac{1}{sC_1} & -R_2-sL_2-\frac{1}{sC_2} & 0 \\ 0 & R_2+sL_2+\frac{1}{sC_2} & -R_3-sL_3-\frac{1}{sC_3} \\ R_4+sL_4 & R_4+sL_4 & R_3+R_4+s(L_3+L_4)+\frac{1+G}{sC_3} \end{bmatrix} \begin{bmatrix} I_1 \\ I_2 \\ I_3 \end{bmatrix} = \begin{bmatrix} 0 \\ 0 \\ R_4+sL_4 \end{bmatrix}$$

Once the currents are figured, calculate output voltage which would be input impedance. Plot and compare to SPICE; see sample solution in Fig. 37.23.

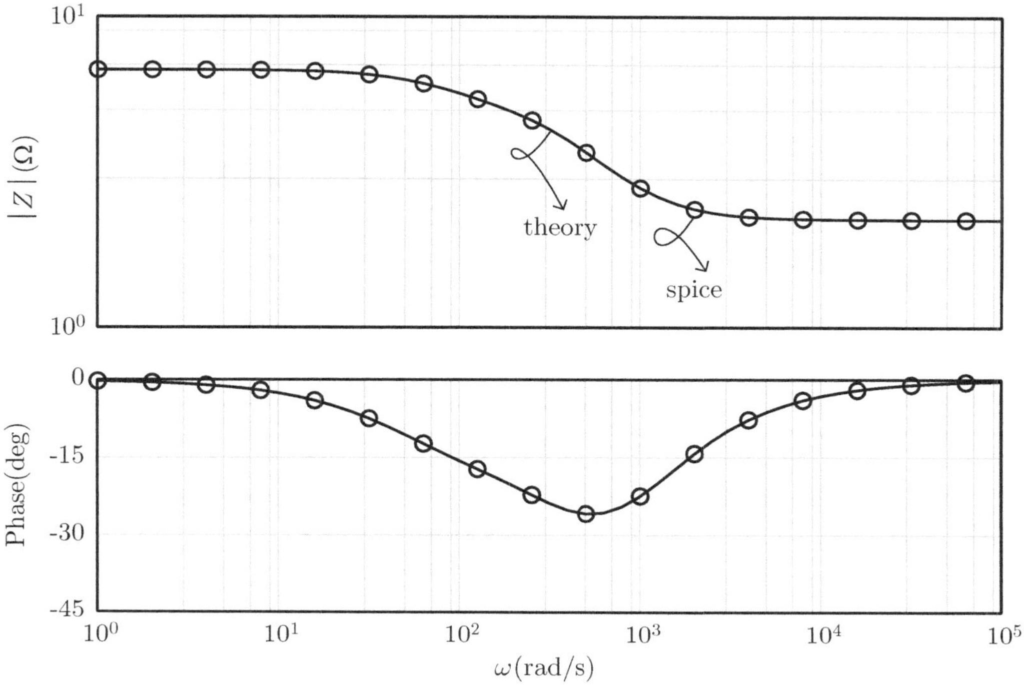

Fig. 37.19 Sample solution to Problem 6

Fig. 37.20 Statement to Problem 7

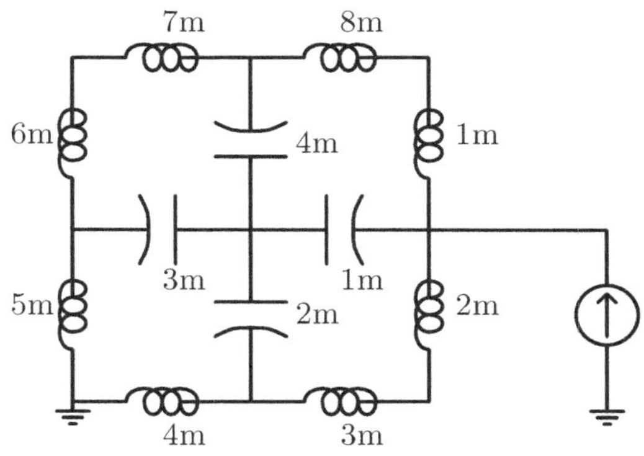

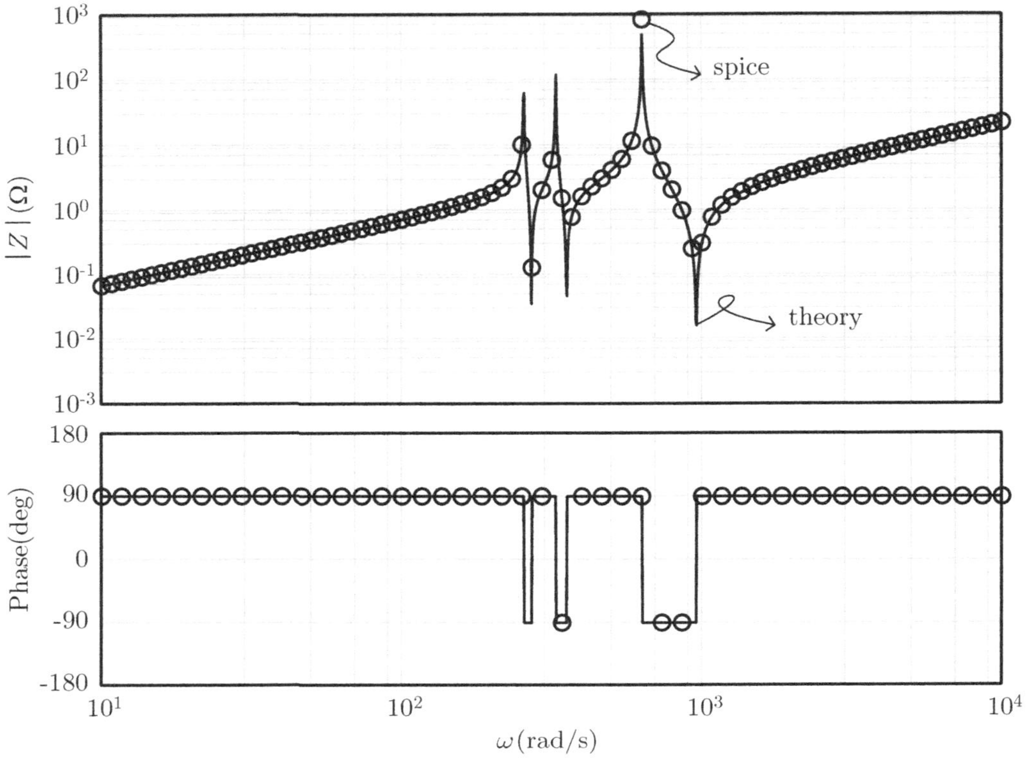

Fig. 37.21 Sample solution to Problem 7

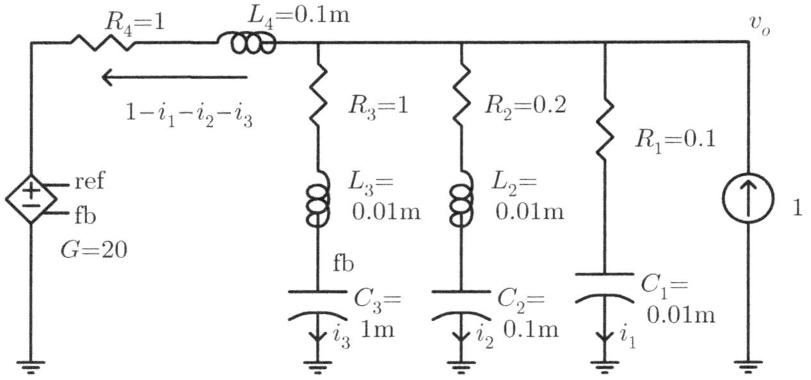

Fig. 37.22 Statement to Problem 8

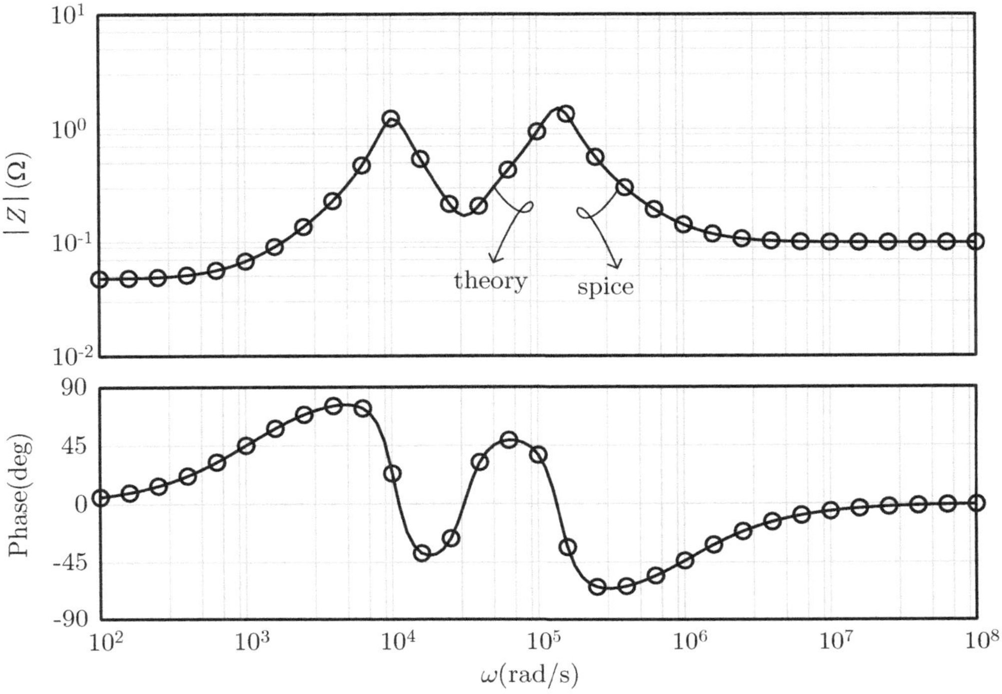

Fig. 37.23 Sample solution to Problem 8

38 Multisource Networks and Superposition

38.1 Introduction

In most of the text we have considered the cases with single stimulus and either single response or multiple responses. For example, input is applied current and output is measured impedance (be itself or mutual); or input is applied voltage and output is measured voltage, and so forth. The multi-output does not really pose any challenge. It is the multisource that requires some attention. We hinted at those when we dealt with feedback systems. But here we give a more systematic treatment thereof. So the goal of this chapter is to define a method for dealing with circuits (or systems) which are excited by more than a single stimulus.

38.2 Reference Case: DC Supplies

As a simple case illustrating the application of superposition to multiple sources consider the DC circuit shown in Fig. 38.1. Given the two source voltages, we want to find output voltage v_o. The conventional approach is to use KCL and KVL to figure the various currents and then various voltages. Let us start with that convention. Doing KVL around left loop we get

$$\begin{aligned} 3(i_1 + i_2) + 2i_2 &= 13 \\ 3i_1 + 5i_2 &= 13 \end{aligned} \tag{38.1}$$

from which we get

$$i_2 = \frac{13 - 3i_1}{5} \tag{38.2}$$

Doing KVL around the vertical loop we get

$$\begin{aligned} 3(i_1 + i_2) + i_1 &= 10 \\ 4i_1 + 3i_2 &= 10 \end{aligned} \tag{38.3}$$

Plugging in for i_2 we get

$$\begin{aligned} 4i_1 + 3\frac{13 - 3i_1}{5} &= 10 \\ 4i_1 + \frac{39}{5} - \frac{9i_1}{5} &= 10 \\ \frac{20i_1 - 9i_1}{5} &= 10 - \frac{39}{5} \\ \frac{11}{5}i_1 &= \frac{11}{5} \end{aligned} \tag{38.4}$$

from which we get for i_1

$$i_1 = 1 \tag{38.5}$$

Now we solve for i_2 and get

$$i_2 = \frac{13 - 3i_1}{5} = \frac{13 - 3}{5} = 2 \tag{38.6}$$

F. Badrieh, *Spectral, Convolution and Numerical Techniques in Circuit Theory*,
https://doi.org/10.1007/978-3-319-71437-0_38

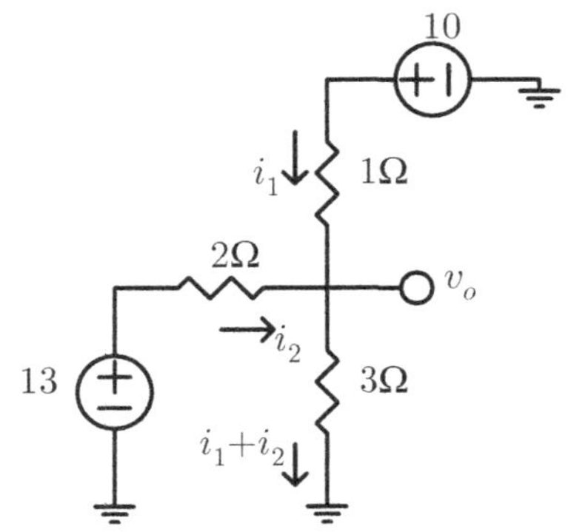

Fig. 38.1 Simple DC circuit with multiple sources

Hence output voltage is

$$v_o = 3(i_1 + i_2) = 3(3) = 9 \tag{38.7}$$

Now let's see if we can re-solve this using the divide and conquer technique!

38.3 Superposition Applied to Reference Case

The idea of superposition is that we *split* the problem into multi parts, where part count equals source count. In this case since we have two sources we split the top problem into two subproblems. For each case, we *leave one source active*, and *disable the other source*. Disabling a source is accomplished depending on the nature of the source: if it is a voltage source it is shorted (i.e., replaced with zero voltage source); on the other hand if it is a current source it is opened (i.e., simply taken out of the network). In our case, and since we have voltage sources, the disabling operation amounts to simply shorting the other source. Once that is done we solve for the various currents and figure corresponding output voltage. Then we repeat for the other case(s). In the end we *add all voltages* (e.g., at the output) and we are assured the correct answer. Let's dive into the details. First we set the left source to zero and get the schematics in Fig. 38.2. From this we figure output current

$$i = \frac{v}{R} = \frac{10}{11/5} = \frac{50}{11}\text{A} \tag{38.8}$$

Output voltage is simply this current times the $\frac{6}{5}$ Ω resistor

$$v_{o1} = \frac{50}{11} \times \frac{6}{5} = \frac{60}{11}\text{V} \tag{38.9}$$

So we have found the response due to the top source. Next we set the top source to zero and reinstate the left source; see Fig. 38.3. Once total impedance is figured (as shown on the right of the figure) we figure corresponding current

$$i = \frac{v}{R} = \frac{13}{11/4} = \frac{52}{11}\text{A} \tag{38.10}$$

Next we use this to solve for output voltage:

$$v_{o2} = \frac{52}{11} \times \frac{3}{4} = \frac{39}{11}\text{V} \tag{38.11}$$

Now we add both voltages to get

$$v_o = v_{o1} + v_{o2} = \frac{60}{11} + \frac{39}{11} = 9 \tag{38.12}$$

in agreement with our earlier answer in Eq. (38.7). So we have shown that by splitting the network into derivative ones, where each derivative has only one power source (be it voltage or current), finding the corresponding solution, then adding all up we do in fact get the correct answer!

38.4 Recap of the Superposition Principle

As we saw in the last section by splitting the problem into individual source ones, solving for the corresponding solution, and then adding all solutions we were able to solve a more complex problem. This is exactly what the superposition principle does. Superposition is extremely important, and not just in circuits but in all fields of engineering. So long the problem is linear we can solve a multi-stimulus case one stimulus at a

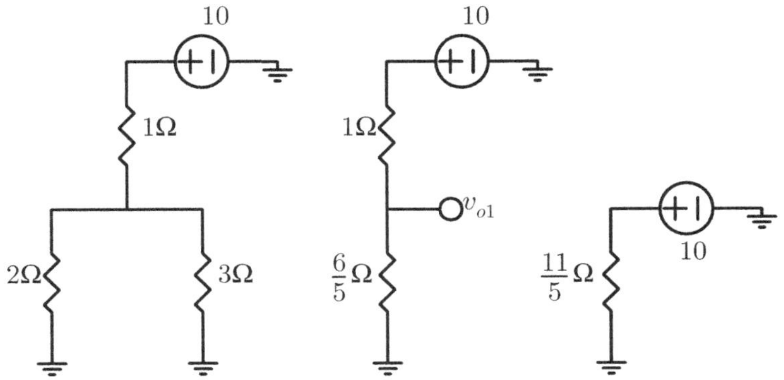

Fig. 38.2 Redraw of Fig. 38.1 with left source set to zero

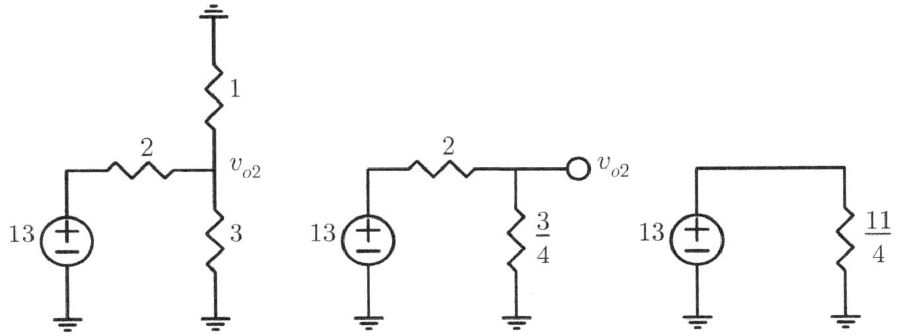

Fig. 38.3 Redraw of Fig. 38.1 with top source set to zero

time and add corresponding solutions. There are so many cases where this simple principle can be used and by doing so it can save a lot of time and energy. And it is not only the resource saving that is at stake here! Quite often there will be multiple sources, and each could be DC, or AC and we are trying to figure various dependencies (such as self and mutual impedances); in these cases it may become blurry just exactly what depends on what! Having a deep understanding of the superposition principle and how it works will help us navigate these sort of complex problems and guide us in the right direction. Let us get more practice with a more complex problem.

38.5 Simple Power Delivery Problem

We have a simple power delivery problem mimicked by a voltage source and current demand as shown in Fig. 38.4. The resistor represents losses to the power source while the decoupling cap

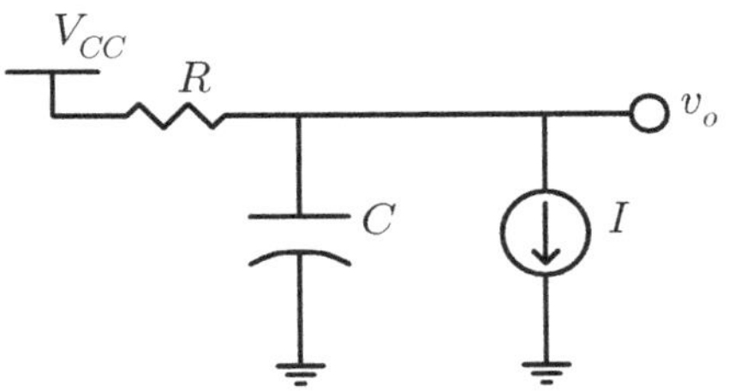

Fig. 38.4 *RC* network with current demand and assumed VCC

represents local power storage. Let us assume that the imposed power source $V_{CC} = 1\,\text{V}$ and that the current demand is delta function for now. This setup furnishes the transfer function of the network, or in the time domain the impulse response. Using superposition we can split the problem into two circuits as shown in Fig. 38.5. The first network is comprised of zero current source and with imposed voltage source V_{CC}, and would give v_{o1}. The second network is comprised of the current source but has shorted V_{CC} and gives v_{o2}.

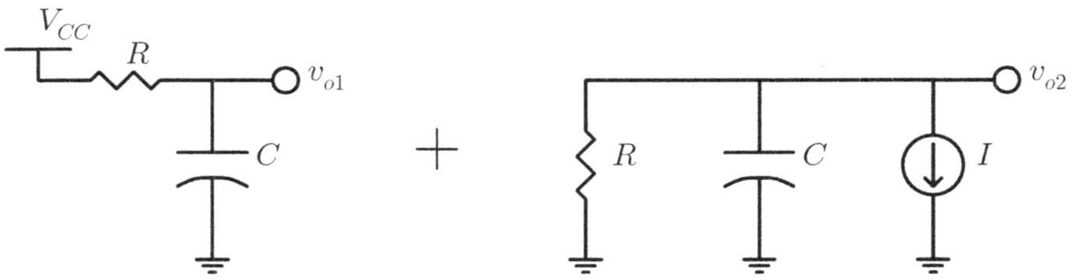

Fig. 38.5 Network in Fig. 38.4 split into two parts

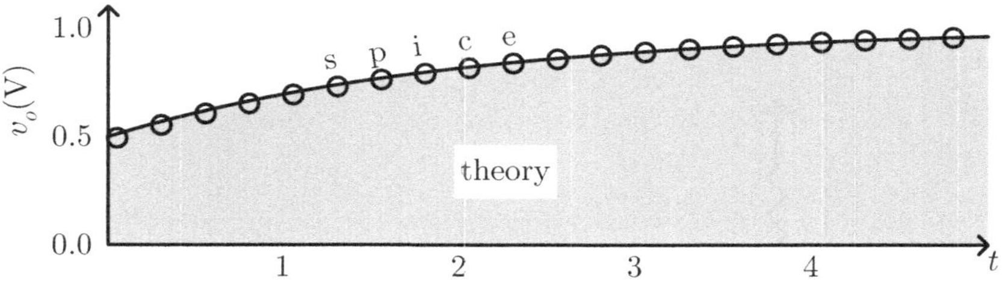

Fig. 38.6 Impulse response of Fig. 38.4 and comparison to SPICE simulations

To solve for v_{o1} we notice that the imposed voltage is DC; this means that the DC solution would be the desired v_{o1}. We know that at DC no currents could flow, and hence the voltage v_{o1} would simply by V_{CC} or simply 1 V:

$$v_{o1} = 1V \tag{38.13}$$

That is, since current is zero, then voltage drop across R_1 would be zero too, and hence v_{o1} would be 0 V offset from V_{CC}. Next we move to v_{o2}. We know that impedance is

$$Z(s) = \frac{\frac{1}{sC}R}{R + \frac{1}{sC}} = \frac{R}{1 + sRC} = \frac{1}{C}\frac{1}{s + 1/RC} \tag{38.14}$$

If we assume current input is a delta function (with LT of 1) we get for v_{o2}

$$v_{o2} = -\frac{1}{C}e^{-t/RC} \tag{38.15}$$

Notice the negative sign since the current is coming *out* of the network. Our total solution is then

$$\begin{aligned} v_o(t) &= v_{o1}(t) + v_{p2}(t) \\ &= \boxed{1 - \frac{1}{C}e^{-t/RC}} \end{aligned} \tag{38.16}$$

A plot of this solution and a comparison to SPICE simulations (case $R = 1\,\Omega$ and $C = 2\,\text{F}$) are shown in Fig. 38.6. Notice that at time zero we get

$$v(0) = 1 - \frac{1}{C} = 1 - \frac{1}{2} = \frac{1}{2} \tag{38.17}$$

The cap before time zero was pre-charged to $V_{CC} = 1$. When the impulse current takes place it discharges the cap immediately by $\frac{1}{C} = \frac{1}{2}$ and leaves output voltage at 0.5 V. On the other hand notice that at large time we get the limit

$$v(\infty) \sim 1 \tag{38.18}$$

That is, once the cap discharged and once the demand current lapsed, the input supply recharges the network and eventually output voltage goes back to $V_{CC} = 1$ V.

Not only did we apply superposition to a case with multiple sources, not only were the sources a mix of DC and transients, and not only were we able to figure response in the frequency domain, we were also able to go back to the time domain (via inverse transform) and figure total solution in real time. We have gone full circle in applying and mixing concepts from spectral methods, superposition, and inverse transforms. All continues to add up!

38.6 Sample Case of Three Transient Current Sources

Let's get more complex by increasing source count to three! Consider the RC network in Fig. 38.7. It is driven by three current sources, here shown in the frequency domain. Our goal is to figure output voltage $V_o(s)$ first in the frequency domain, and then later in the time domain. Our strategy is to incrementally open to all sources leaving only one on, figure the corresponding solution then repeat; in the end add all responses. So start only with $I_1(s)$, opening the other sources, and measure corresponding output voltage first; this comes out

$$V_{o1}(s) = I_1(s)\left[R_1 + \frac{1}{sC}\right] \tag{38.19}$$

Next apply only $I_2(s)$ and measure corresponding output

$$V_{o2}(s) = I_2(s)\frac{1}{sC} \tag{38.20}$$

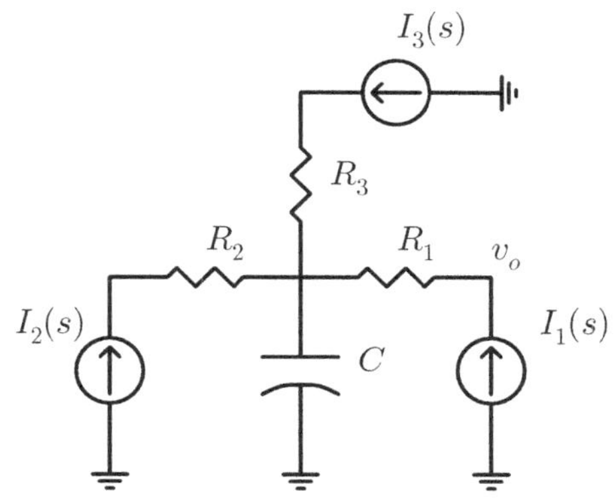

Fig. 38.7 Network with three current sources

Finally apply only $I_3(s)$ and figure corresponding output

$$V_{o3}(s) = I_3(s)\frac{1}{sC} \tag{38.21}$$

Now using superposition we get total output voltage

$$\boxed{V_{o1}(s)=I_1(s)\left[R_1+\frac{1}{sC}\right]+\frac{1}{sC}\left[I_2(s)+I_3(s)\right]} \tag{38.22}$$

Assume for example that the three currents in the time domain are given by

$$i_1(t) = \text{pulse of width } 2, \quad i_2(t) = 2\pi\sin(2\pi t), \quad \text{and} \quad i_3(t) = 2\pi\sin(4\pi t) \tag{38.23}$$

Then in the frequency domain we have

$$I_1(s) = \frac{1-e^{-2s}}{s}, \quad I_2(s) = 2\pi\frac{2\pi}{s^2+(2\pi)^2}, \quad \text{and} \quad I_3(s) = 2\pi\frac{4\pi}{s^2+(4\pi)^2} \tag{38.24}$$

such that the output voltage is

$$\boxed{V_{o1}(s) = \frac{1-e^{-2s}}{s}\left[R_1+\frac{1}{sC}\right]+\frac{2\pi}{sC}\left[\frac{2\pi}{s^2+(2\pi)^2}+\frac{4\pi}{s^2+(4\pi)^2}\right]} \tag{38.25}$$

A plot of output voltage in the frequency domain is shown in Fig. 38.8. Knowing voltage in the frequency domain we can figure it in the time domain as

$$v(t) = u(t)\left[R_1 + \frac{t}{C}\right] - u(t-2)\left[R_1 + \frac{t-2}{C}\right] + \frac{1}{C}\left\{1 - \cos(2\pi t) + \frac{1}{2}\left[1 - \cos(4\pi t)\right]\right\} \quad (38.26)$$

Transient results alongside SPICE ones are shown in Fig. 38.9.

Reflection on Results It maybe hard to believe but the frequency plot in Fig. 38.8 has exactly the same amount of information as that time plot in Fig. 38.9! They both convey exactly the same amount of information. Knowing one gives the other! We continue to marvel about the beauty of back-and-forth transformation between the frequency domain and the time one! Other than that we have shown again the power of superposition and its compatibility with spectral techniques.

38.7 Summary

In this chapter we have shown and demonstrated the principle of superposition both in the frequency and in the time domains via a few simple examples. The main premise is simple, but the applications are immense. Through the simple method of divide-and-conquer we are able to tackle more difficult problems by breaking them into simpler ones. The simpler problems can be attended to serially or in parallel. They can even be done by different people or tools. It does not matter how each subproblem is solved; what matters is that in the end we gather all subsolutions and add them to get total solution. The sub-problems are setup such that each has only one active source; all other sources would have been shut off. To shut off voltage sources we replace them with zero ones (or simply short wire them), and to shut off current sources we take them out (or simply replace them with 1 Meg resistors for example). As demonstrated in the chapter all of the spectral, inverse transform, superposition, and SPICE computer simulations are consistent with each other. Once a solution is obtained, it is a unique one and the only one, no matter how it was derived!

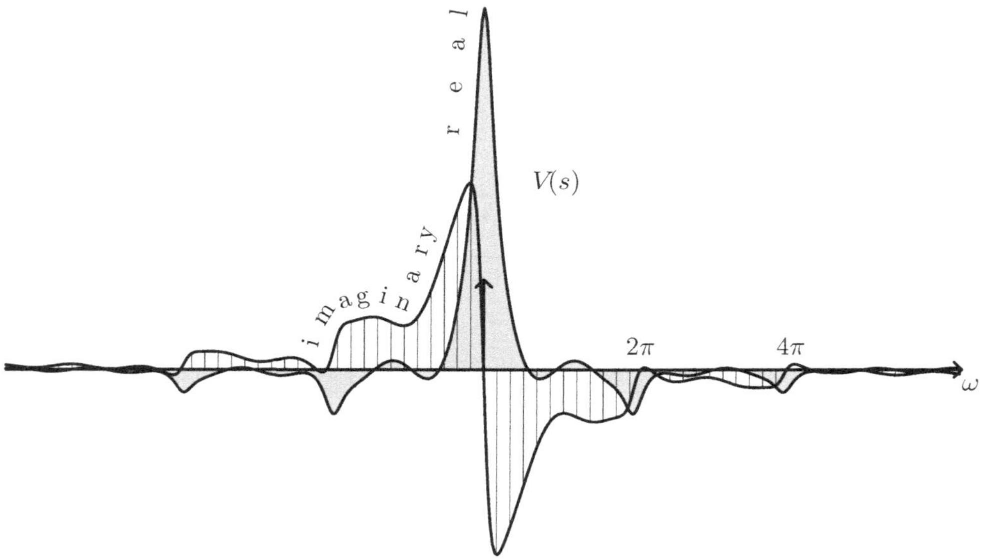

Fig. 38.8 Plot of output voltage of network in Fig. 38.7 versus frequency (case $R = 1\,\Omega$, $C = 2\,\text{F}$, and $\sigma = 0.5$)

38.8 Problems

1. Consider the network on the left side of Fig. 38.10 where $v_i(t) = u(t)$; show by superposition that it can be split into two parts, as shown on the right. Show that the output of the right part is (assuming $V_{CC} = 1$)

$$v_{o2} = \frac{1}{2}$$

Then show that the output voltage of the middle part is

$$V_{o1}(s) = \frac{1}{s + 1/RC}$$

Finally show that total output voltage in time domain is

$$v_o(t) = v_{o1}(t) + v_{o2}(t) = \frac{1}{2} + e^{-t/RC}$$

Plot output voltage for the case $R = 2\,\Omega$ and $C = 1\,\text{F}$ and compare to SPICE; see sample results in Fig. 38.11.

2. Consider the RC network in Fig. 38.12 which is driven by one current source and 2 voltage sources. What is the output voltage in the frequency domain as a function of the three input sources? For the case all of $I_1(s) = V_2(s) = V_3(s) = 1$ (all unity in the frequency domain), plot output voltage (again in the frequency domain) and compare to SPICE; see sample solution in Fig. 38.13.

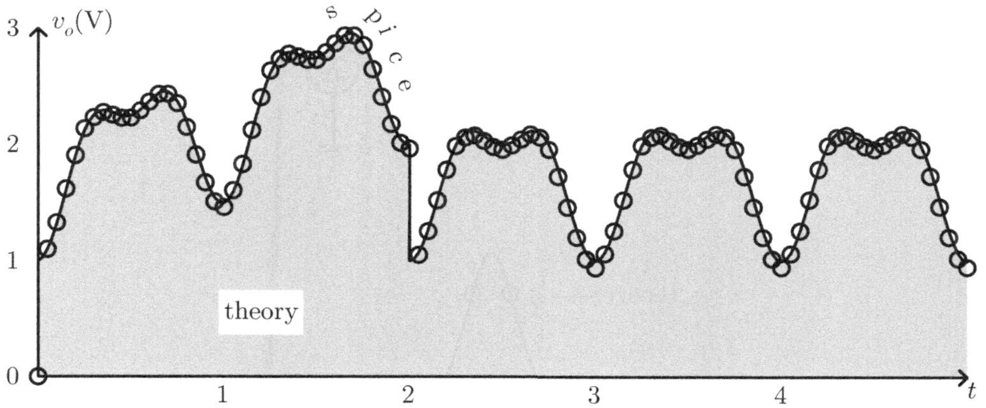

Fig. 38.9 Output voltage for network in Fig. 38.7 and comparison to SPICE; case of $R = 1\,\Omega$ and $C = 2\,\text{F}$

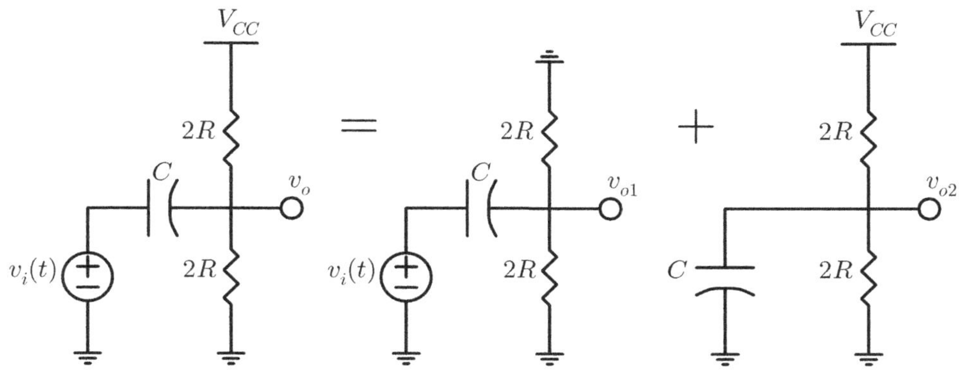

Fig. 38.10 Statement to Problem 1

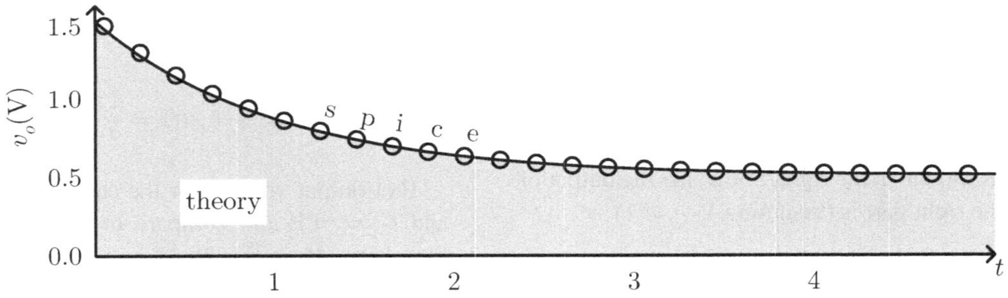

Fig. 38.11 Sample solution to Problem 1

Fig. 38.12 Statement to Problem 2

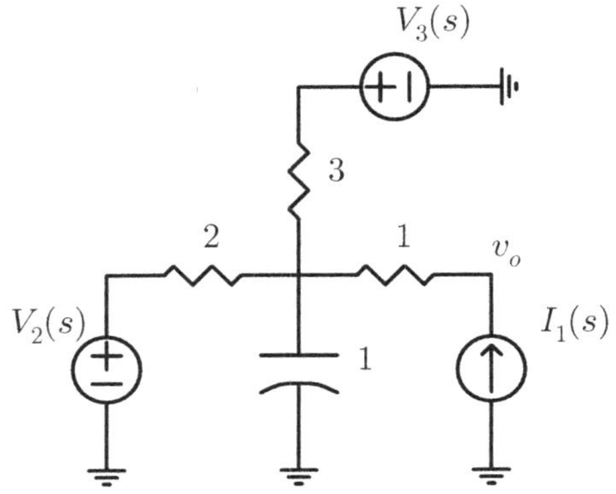

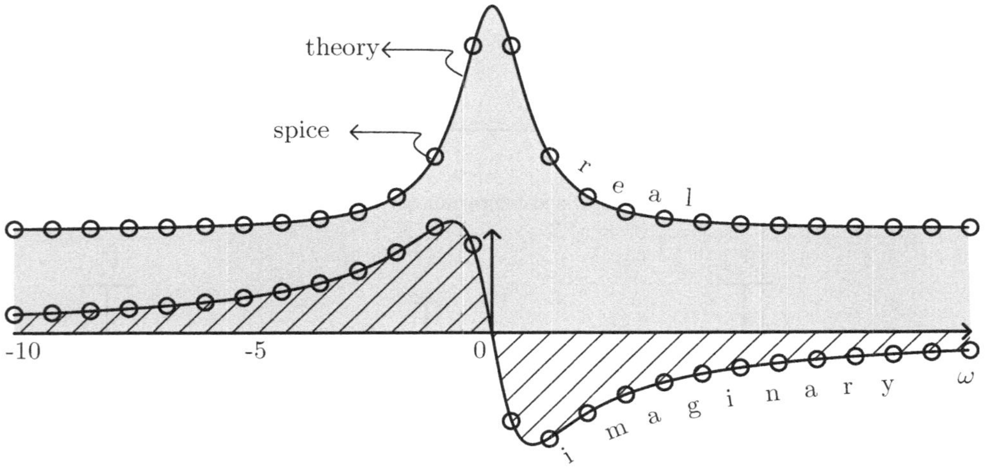

Fig. 38.13 Sample solution of Problem 2; case of $\sigma = 0.01$

Answer:

$$V(s) = \left[1 + \frac{1}{s+5/6}\right] I_1(s) + \frac{1}{2}\frac{1}{s+5/6} V_2(s) + \frac{1}{3}\frac{1}{s+5/6} V_3(s) \quad (38.27)$$

3. Starting with Problem 2 and using the derived transfer function, what is the output voltage, in the frequency domain if

$$i_1(t) = \text{pulse of width 3}, \quad v_2(t) = 2\pi \sin(\pi t), \quad \text{and} \quad v_3(t) = 4\pi \sin(4\pi t)$$

Plot the output voltage in the frequency domain; see sample solution in Fig. 38.14.
Answer:

$$V(s) = \left[1 + \frac{1}{s+5/6}\right]\left[\frac{1-e^{-3s}}{s}\right] + \frac{1}{2}\frac{1}{s+5/6}\left[2\pi \frac{\pi}{s^2+\pi^2}\right] + \frac{1}{3}\frac{1}{s+5/6}\left[4\pi \frac{4\pi}{s^2+(4\pi)^2}\right]$$

4. Take the output voltage obtained in Problem 3, and using partial fraction expand it to sum of fractions. What is the corresponding time domain voltage? Plot and compare to SPICE; see sample solution in Fig. 38.15.
Answer:

$$a = \frac{5}{6}, \qquad \omega_2 = \pi, \qquad \omega_3 = 4\pi$$

$$v_1(t) = u(t)\left[1 + \frac{1}{a}\left(1 - e^{-at}\right)\right] - u(t-3)\left[1 + \frac{1}{a}\left(1 - e^{-a(t-3)}\right)\right]$$

$$v_2(t) = 2\pi\omega_2 \frac{1}{2}\frac{1}{a^2+\omega_2^2}\left[e^{-at} + \frac{a}{\omega_2}\sin\omega_2 t - \cos\omega_2 t\right]$$

$$v_3(t) = 4\pi\omega_3 \frac{1}{3}\frac{1}{a^2+\omega_3^2}\left[e^{-at} + \frac{a}{\omega_3}\sin\omega_3 t - \cos\omega_3 t\right]$$

$$v(t) = v_1(t) + v_2(t) + v_3(t)$$

5. Consider the *RLC* circuit in Fig. 38.16. Write output voltage as

$$V_o(s) = Z(s)I(s) + E(s)V_i(s)$$

What would $Z(s)$ and $E(s)$ be? Plot them, and compare to SPICE; see sample solution in Fig. 38.17.
6. Starting with Problem 5 assume now that the current and voltage sources are given by

$$i(t) = \frac{1}{4} \text{ pulse of duration 5}$$

$$v_i(t) = \frac{1}{2}\sin 2\pi t$$

What is the output voltage in the frequency domain? Plot it and compare to sample solution in Fig. 38.18.

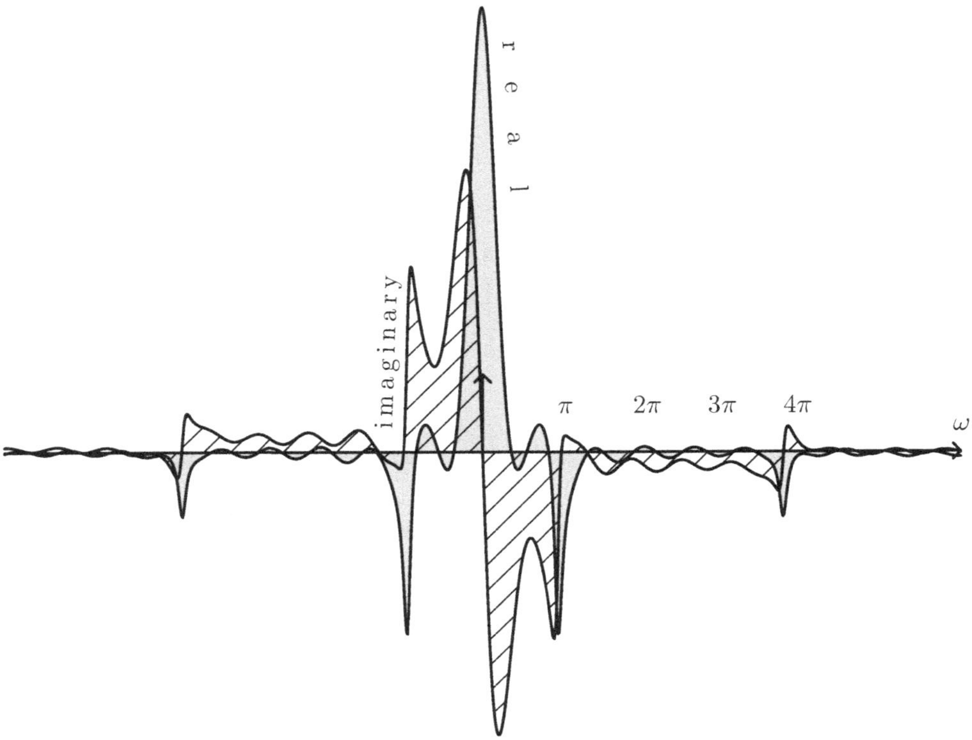

Fig. 38.14 Sample solution of Problem 3; case of $\sigma = 0.2$

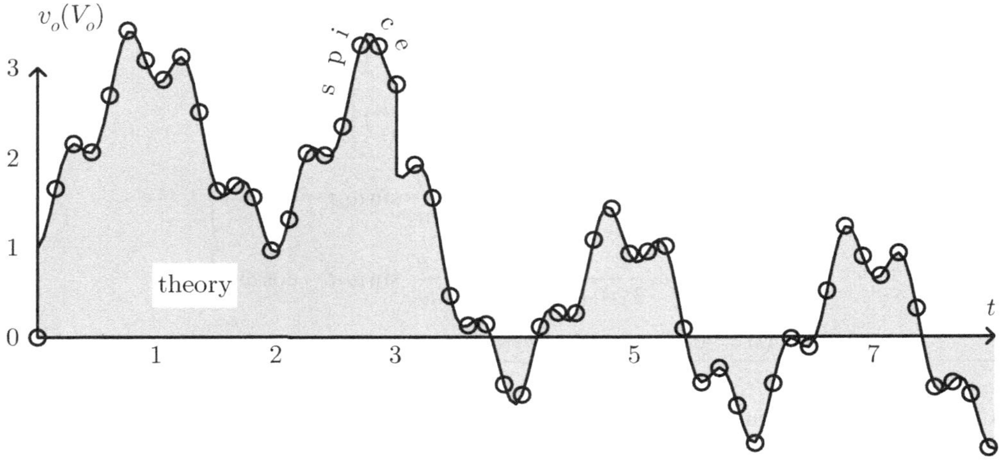

Fig. 38.15 Sample solution of Problem 4

Fig. 38.16 Statement to Problem 5

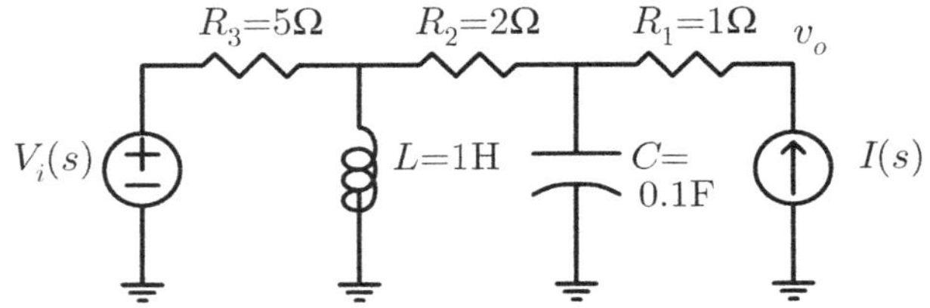

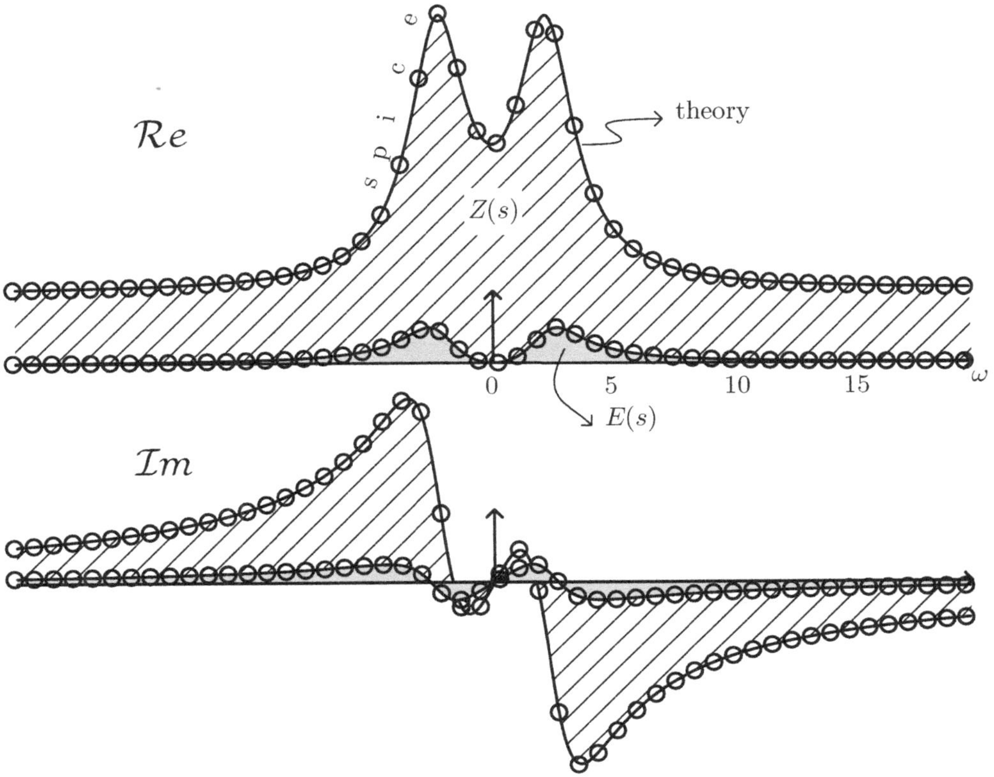

Fig. 38.17 Sample solution to Problem 5; case of $\sigma = 0.01$

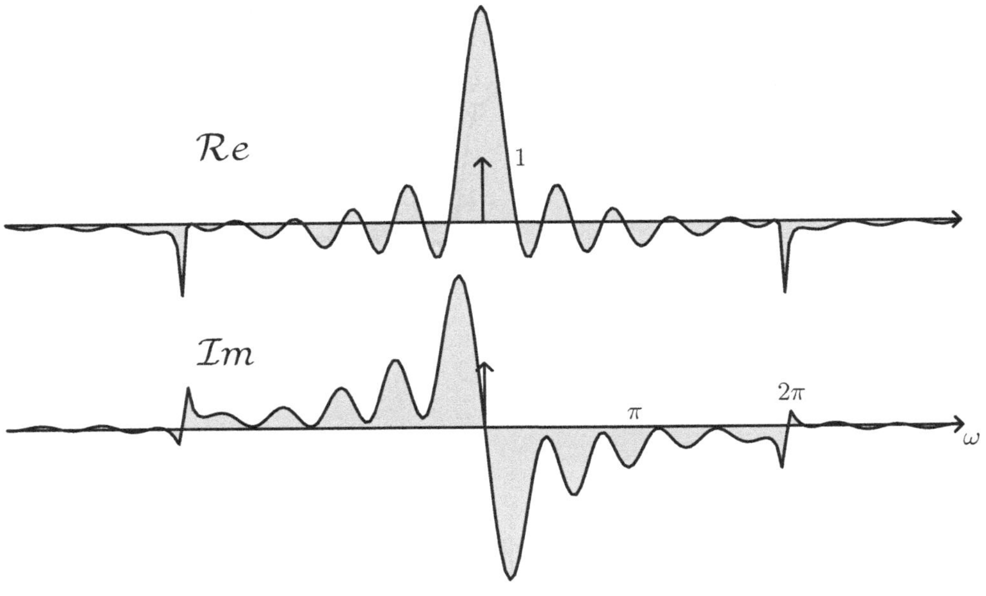

Fig. 38.18 Sample solution to Problem 6; case of $\sigma = 0.05$

7. Starting with Problem 6 and knowing output voltage in the frequency domain—what is it in the time domain? Plot it and compare to SPICE; see sample solution in Fig. 38.19.
8. Consider the RC circuit width feedback in Fig. 38.20. Two input sources are present: a current source and a voltage one (the reference voltage). Derive output voltage, in the frequency domain, as a function of both sources such that

$$V_o(s) = I(s)Z(s) + E(s)V_{\text{ref}}(s)$$

Plot both $Z(s)$ and $E(s)$ and compare to SPICE; see sample solution in Fig. 38.21.

Answer:

$$Z(s) = -\frac{1}{C}\frac{1}{s+\frac{G+1}{RC}}, E(s) = \frac{G}{RC}\frac{1}{s+\frac{G+1}{RC}}$$

9. Starting with Problem 8, what is the voltage in the frequency domain if the two input sources in time domain are

$$i(t) = \text{pulse of width 2 and period 4},$$
$$v_{\text{ref}} = \sin 2\pi t$$

Plot output voltage in the frequency domain; see sample solution in Fig. 38.22.

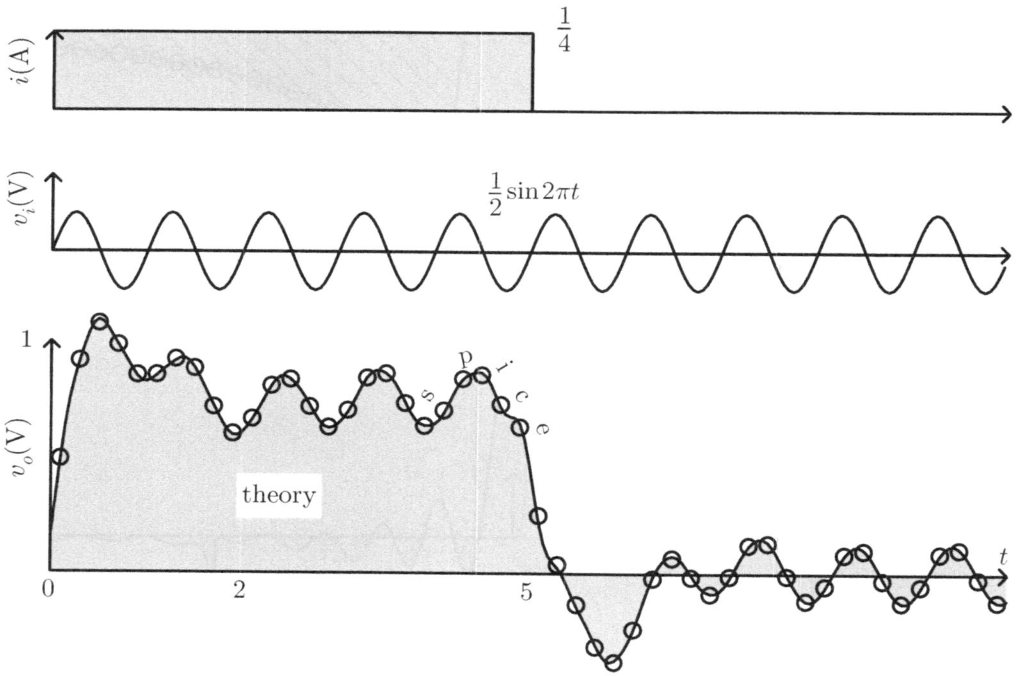

Fig. 38.19 Sample solution to Problem 7

Fig. 38.20 Statement to Problem 8

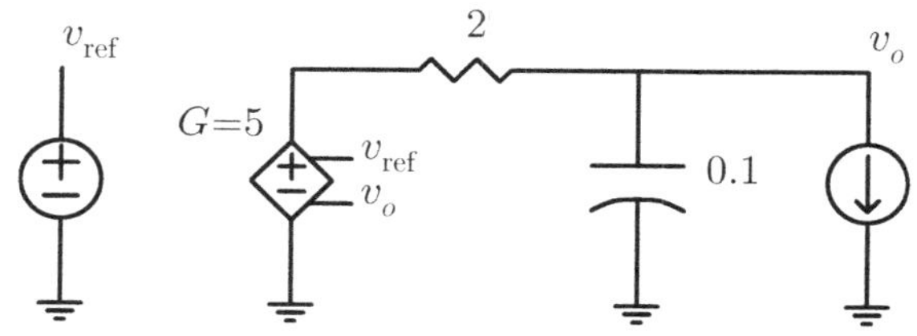

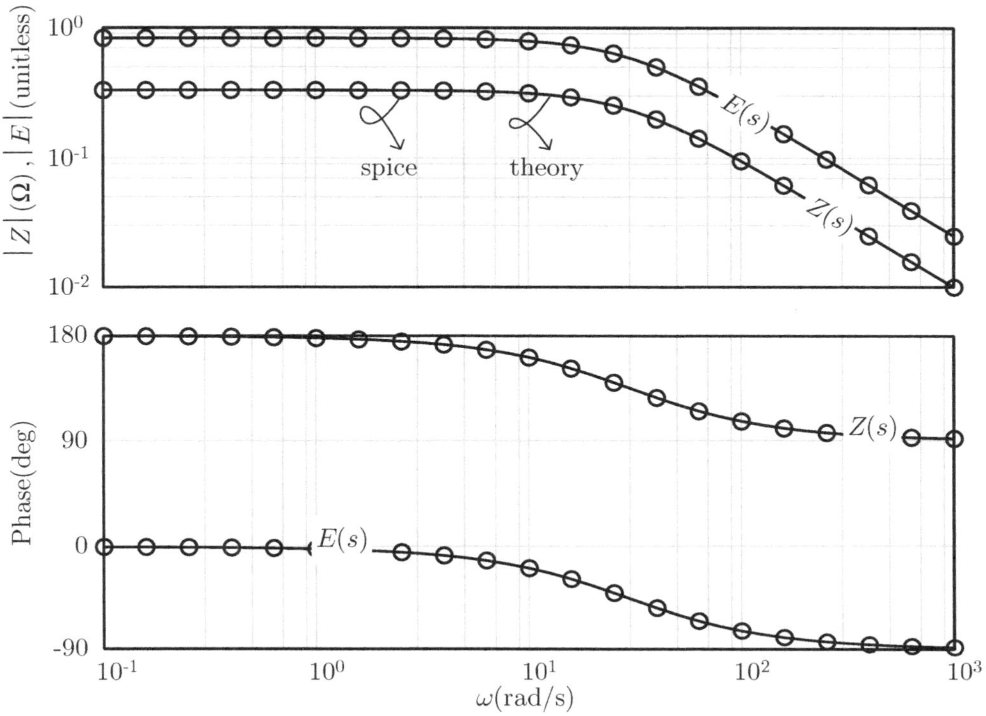

Fig. 38.21 Sample solution to Problem 8; case of $\sigma = 0.01$

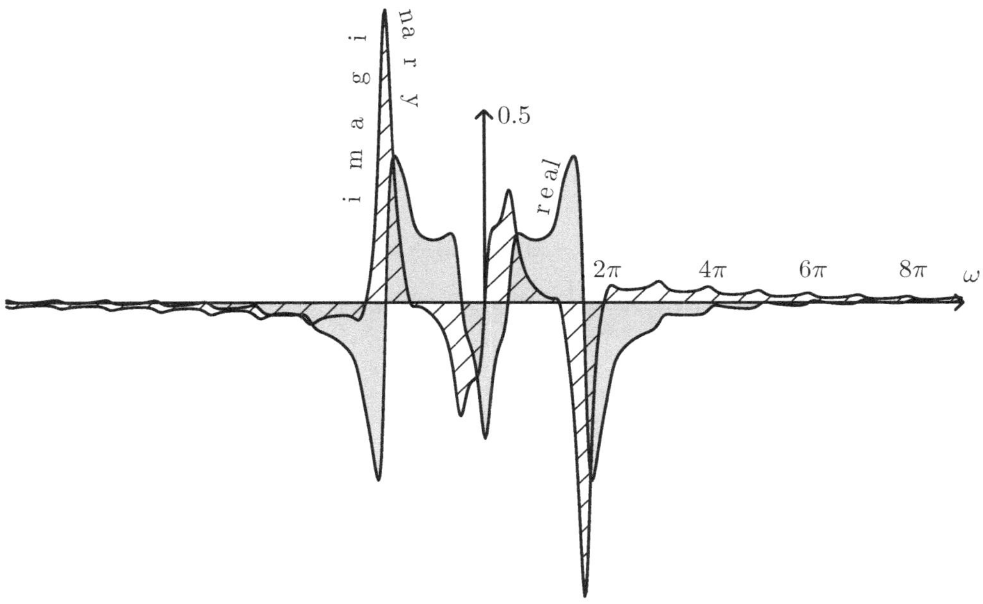

Fig. 38.22 Sample solution to Problem 9; case of $\sigma = 0.5$

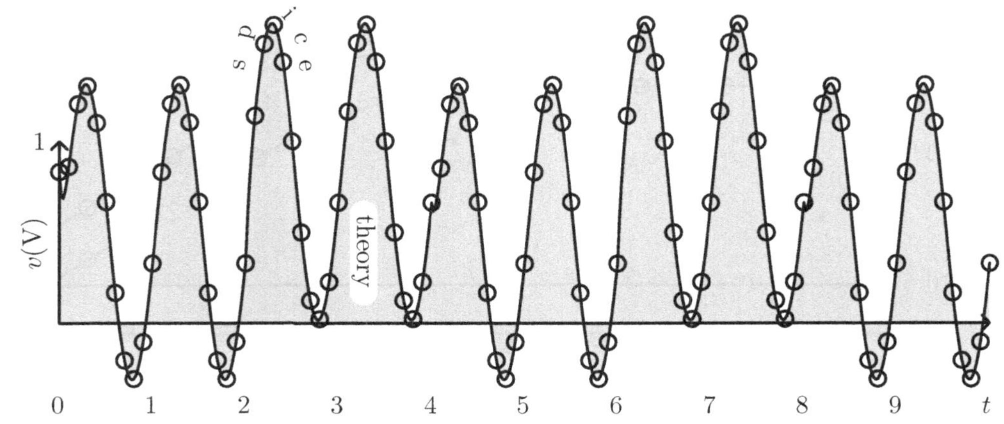

Fig. 38.23 Sample solution to Problem 10

Answer:

$$V_o(s) = \left[-\frac{1}{C} \frac{1}{s + \frac{1+G}{RC}} \frac{1 - e^{-2s}}{s} \frac{1}{1 - e^{-4s}} \right] + \left[\frac{G}{RC} \frac{1}{s + \frac{1+G}{RC}} \frac{2\pi}{s^2 + (2\pi)^2} \right]$$

10. Starting with Problem 9 assume same input stimulus, but now input voltage has a DC average of 1; find output voltage in time and compare to SPICE; see sample solution in Fig. 38.23.
Answer: Let

$$a = \frac{1+G}{RC}, \ \omega_0 = 2\pi, \text{ and } w(t) = -\frac{1}{aC}\left[1 - e^{-at}\right]$$

Then

$$v_1(t) = u(t)w(t) - u(t-2)w(t-2) + u(t-4)w(t-4) - \cdots$$

$$v_2(t) = \frac{G\omega_0}{RC} \frac{1}{a^2 + \omega_0^2} \left[e^{-at} + \frac{a}{\omega_0} \sin \omega_0 t - \cos \omega_0 t \right]$$

$$v(t) = \frac{G}{G+1} + v_1(t) + v_2(t)$$

39 Systems with Initial Conditions

39.1 Introduction

In most of the treatment in this book we dealt with cases with zero initial conditions; that is, voltages across caps are zero and currents across resistors and inductors are zero too. That worked out to be simpler. However, as will be shown below, this is not a condition or requirement. We can in fact use frequency techniques in analyzing systems with nonzero initial conditions.

39.2 Plan

The plan in tackling nonzero initial conditions is illustrated with a pre-charged cap. Consider the cap shown in Fig. 39.1 with some nonzero charge; the cap is not connected to a network, so the charge (and hence voltage) remains on it. We can construct an *equivalent* circuit, but *with zero charged cap* that mimics the original circuit, if we append to the (uncharged) cap a 1 V source as shown on the right side of the same figure. Looking at this circuit from the outside world we ought to see the same thing. Specifically,

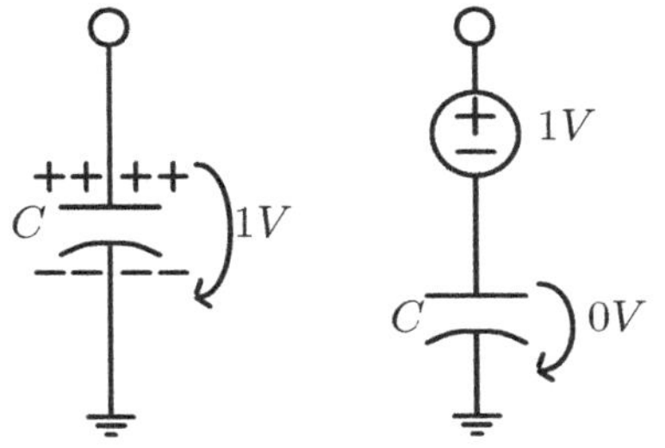

Fig. 39.1 Pre-charged cap and equivalent circuit

1. Both circuits have same terminal voltage, which is 1 V.
2. Both circuits have same terminal current, which is zero.
3. Both circuits have same input impedance, which is $1/sC$.

So, at least for caps, the plan is to *replace each pre-charged cap with a neutral one, in series with a voltage source* (with voltage value equal to the cap initial voltage). This is a very simple strategy, yet very effective one. Let's test it on a sample application.

39.3 Sample Case of Non-driven Pre-charged *RC* Network

Consider the network in Fig. 39.2 where the cap C is pre-charged to 1 V. At some point the open switch is closed and we want to determine

F. Badrieh, *Spectral, Convolution and Numerical Techniques in Circuit Theory*,
https://doi.org/10.1007/978-3-319-71437-0_39

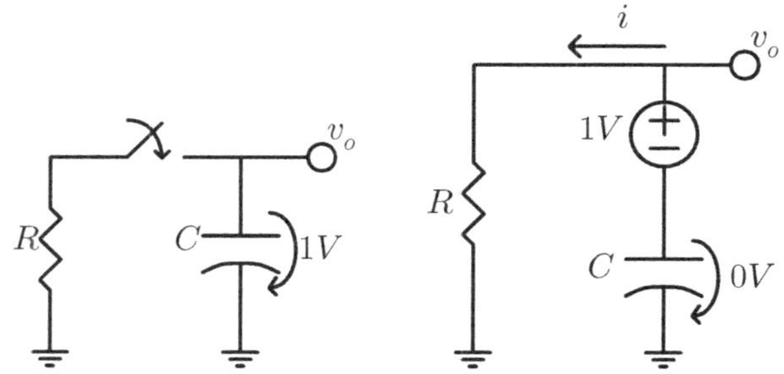

Fig. 39.2 Pre-charged RC network

voltages and currents. Based on the plan built in the prior section we replace the network with that shown on the right side of the same figure.

Doing KVL and starting from the bottom right we have

$$-\frac{I}{sC}+\frac{1}{s}-IR=0 \quad \Rightarrow I\left[R+\frac{1}{sC}\right]=\frac{1}{s} \quad \Rightarrow I\frac{1+sRC}{sC}=\frac{1}{s} \tag{39.1}$$

Notice in passing by the inserted voltage source we put down a $\frac{1}{s}$ term which is the Laplace transform of the unit step function:

$$u(t)\rightarrow\frac{1}{s} \tag{39.2}$$

That is, the inserted voltage source is a unit-step-like kind of source (i.e., causal). It is not one that has been there forever! Solving for $I(s)$ we get

$$I(s)=\frac{C}{1+sRC}=\boxed{\frac{1}{R}\frac{1}{s+1/RC}} \tag{39.3}$$

The inverse LT gives

$$\boxed{i(t)=\frac{1}{R}e^{-t/RC}} \tag{39.4}$$

as we would have expected. That is, immediately after the switch is closed, the current would be voltage (1 V) divided by resistance R. In time (RC time constant), current would decay to zero and the cap would have discharged ending up at zero volt.

To get output voltage we add the voltage across the cap, which is impedance times current, to the 1 V source

$$V(s)=-\frac{1}{sC}I(s)+\frac{1}{s}=-\frac{1}{RC}\frac{1}{s}\frac{1}{s+1/RC}+\frac{1}{s}=-\left[\frac{1}{s}-\frac{1}{s+1/RC}\right]+\frac{1}{s} \tag{39.5}$$

$$V(s)=\boxed{\frac{1}{s+1/RC}} \tag{39.6}$$

The inverse LT gives

$$\boxed{v_o(t)=e^{-t/RC}} \tag{39.7}$$

That is, voltage starts at 1 V (the pre-charged value) and gradually decays (with in an RC time constant) to zero, since the charge is drained from the cap, such that the final cap voltage would be zero. So we have proved that we can arrive at the expected solution, in terms of current and voltage, using the developed method of treating initial conditions. In other words, we can forget about the initial condition complication and leverage our analytic methods simply by inserting transient voltage sources at the right places to take care of those initial conditions.

39.4 Sample Case of Driven Pre-charged *RC* Network

The last example we dealt with was not driven; let's change that. Consider the network comprised of a single input source, R and a pre-charged C (to voltage 1) as shown in Fig. 39.3. At time zero the open switch is closed and we want to know the output voltage across the cap. Similar to before, and in accordance with our recipe plan we replace this network with one which has zero-charged cap, provided we add a source of value $u(t)V_0$ as shown on the right side of figure, where V_0 is the cap initial voltage. Assume for now that V_0 is 1 V and that input is a unit step function of magnitude 2. Doing KVL around right network we get

$$I\frac{1}{sC} + \frac{1}{s} + IR = \frac{2}{s} \tag{39.8}$$

Rearrange to get

$$I\left[R + \frac{1}{sC}\right] = \frac{1}{s} \tag{39.9}$$

Solution to this is

$$\boxed{I(s) = \frac{1}{R}\frac{1}{s + 1/RC}} \tag{39.10}$$

whose time form comes out

$$\boxed{i(t) = \frac{1}{R}e^{-t/RC}} \tag{39.11}$$

This is all not new to us; what is new is that output voltage is not simply I/sC—instead t is

$$V(s) = \frac{I}{sC} + \frac{V_0}{s} \tag{39.12}$$

Integrating current (since we have $1/s$) we finally get for output voltage

$$\begin{aligned} v_o(t) &= V_0 + 1 - e^{-t/RC} \\ &= \boxed{2 - e^{-t/RC}}, \qquad \text{assuming } V_0 = 1\,\text{V} \end{aligned} \tag{39.13}$$

Figure 39.4 shows transient results confirming that output voltage starts at 1 V, ramps up at a time constant RC, and finally settles to 2 V.

39.5 Recap for Circuits Containing Pre-charged Caps

Taking the risk of repeating ourselves, let's reiterate the process for taking care of circuits containing pre-charged caps. First replace the pre-charged cap with a neutral one—simple! Next add in series with this cap a *unit step* voltage source whose value is the pre-charged value. Now run the simulations exactly the same as has been done in all the prior chapters, but keeping in mind that when reporting subsequent voltages across the cap to not only report the new voltage across the cap, but to always include in it the contribution from the artificially inserted voltage source. Remember, the new neutral cap does

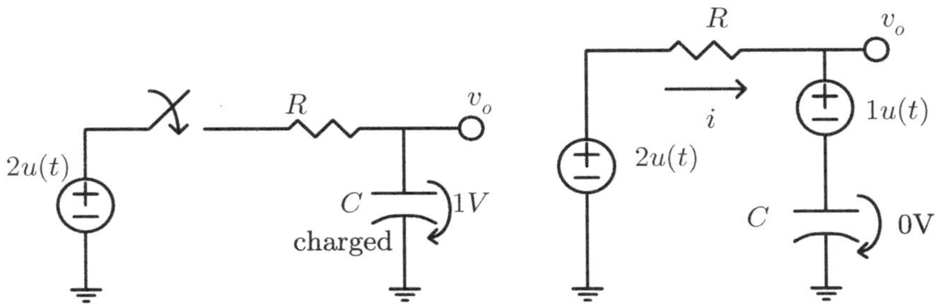

Fig. 39.3 Pre-charged *RC* network with input source

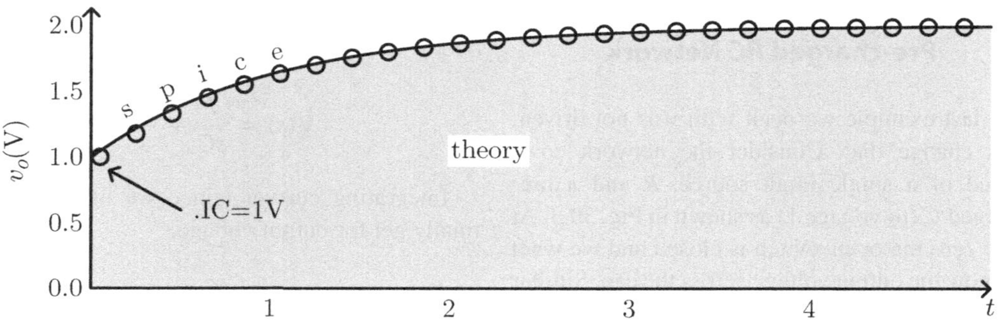

Fig. 39.4 Simulation results of pre-charged RC network with input source; sample case with $R = 1\ \Omega$ and $C = 1$ F

not equal to the old pre-charged cap; what is equal is the new cap *plus* the inserted voltage source! This caveat does not apply for cap current since that flows through both the cap and the voltage source. With this well understood by now we next move to circuits containing pre-charged inductors.

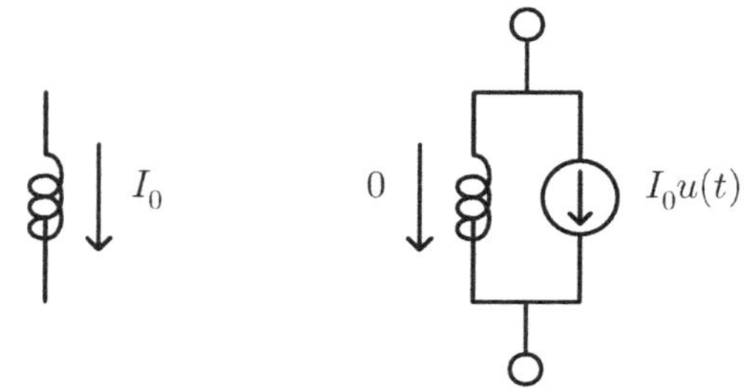

Fig. 39.5 Pre-charged inductor and equivalent circuit

39.6 Pre-charged Inductors

Having covered the pre-charged cap we next move to the pre-charged inductor. Admittedly pre-charged inductors are a bit more difficult to visualize than pre-charged caps. A pre-charged cap is a disconnected device with charge on it. But a pre-charged inductor cannot exist on its own—it must have something else to enable current to flow through it. Consider the pre-charged inductor in Fig. 39.5. We can account for the initial current through the inductor by replacing the inductor with a zero-pre-charged one, and adding a parallel ideal current source which is a unit step function, with current value I_0 which is the initial current through the inductor. This is shown on the right of Fig. 39.5. Notice that the concept is the same as that as the pre-charged cap. The difference is twofold: first, rather than adding a voltage source we add a *current* source; and second rather than adding the source in series with the element (be it C or L) we add it in *parallel*!

The premise is simple and is based on the fact that the two configurations act the same as seen from the outside world. Specifically

1. Both configurations have the same input impedance and that is sL; recall that when doing impedance we disable all sources, which means the added current source is open and has no impact on impedance.
2. Both configurations have the same terminal current, which is I_0 here. Recall that the inductor on the right of Fig. 39.5 has zero current (at least initially).
3. Both configurations have the same input voltage which zero, again at least initially. The left configuration has DC current and hence $Ldi/dt = 0$; the right one has zero inductor current to start with and hence it also has zero initial voltage.

After time zero the current in the left inductor starts to adjust while the current in the right inductor starts to build up. The right inductor sees a step current, and that being a high frequency one is completely rejected initially. Remember we cannot change the current through an inductor instantaneously; we can do that only incrementally. The same concept applies to the cap whose voltage cannot be changed instantaneously, and does so only incrementally and only in due time. (Notice in both cases we are assuming that input stimulus is a typical one, including unit steps; we don't account here for impulsive inputs which can actually change voltages across caps and currents across inductors instantaneously!) Only after time lapses does the current through the right inductor build up. The premise—and as will be tested next—is that so far as the *terminal* current is concerned both configurations sink/source the same current!

39.7 Sample Case of Non-driven Pre-charged *RL* Network

Consider the pre-charged inductor shown on the left of Fig. 39.6. It is pre-charged via a switch (S_1) that initially closed the loop and enabled current flow. Assume after some time we insert a resistor as shown in the middle of the figure and open the initial switch (S_1). We want to find out the evolution of current and voltage. To that end we migrate to the topology shown on the right of the figure.

First we start with KCL which states that

$$I_L + \frac{1}{s} = I, \qquad \text{or} \qquad I_L = I - \frac{1}{s} \tag{39.14}$$

Now KVL around the *RL* loop gives

$$-\left[I - \frac{1}{s}\right] sL - IR = 0, \quad \text{which simplifies to} \tag{39.15}$$

$$I\,[sL + R] = L, \quad \text{or} \tag{39.16}$$

$$I = \frac{L}{sL + R}, \quad \text{which simplifies to} \tag{39.17}$$

$$\boxed{I = \frac{1}{s + R/L}} \tag{39.18}$$

In time domain this has the solution

$$\boxed{i(t) = e^{-t/\tau}, \qquad \tau = \frac{L}{R}} \tag{39.19}$$

That is, initial current across the R is one, and in time it decays to zero, as expected. The time constant is L/R. Knowing resistor current we can figure inductor current *in the equivalent topology* as

$$I_L = I - \frac{1}{s} \tag{39.20}$$

which means in the time domain

$$i_L(t) = i(t) - u(t) \tag{39.21}$$

Plugging in for $i(t)$ we finally get

$$i_L(t) = e^{-t/\tau} - u(t) \tag{39.22}$$

That is, inductor current in the topology on the right of Fig. 39.6 starts at *zero* and ends at -1. When it ends it is absorbing the whole of the current source (in the equivalent topology) which renders the *terminal* current zero, in exact agreement with long time limit of Eq. (39.19). So we have verified that the real inductor current or the *terminal* current of the equivalent circuit act

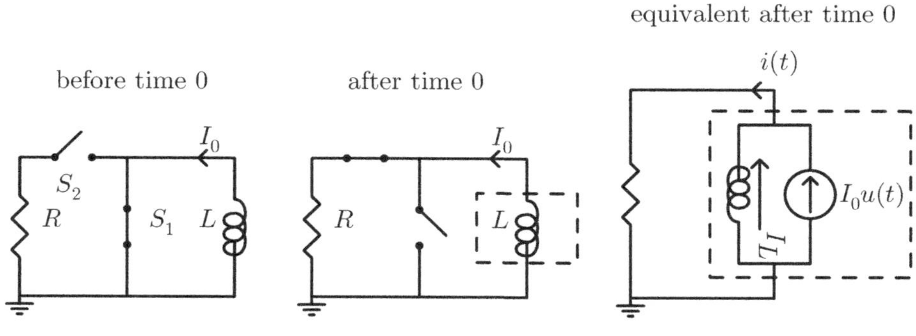

Fig. 39.6 Zero-source pre-charged *RL* network

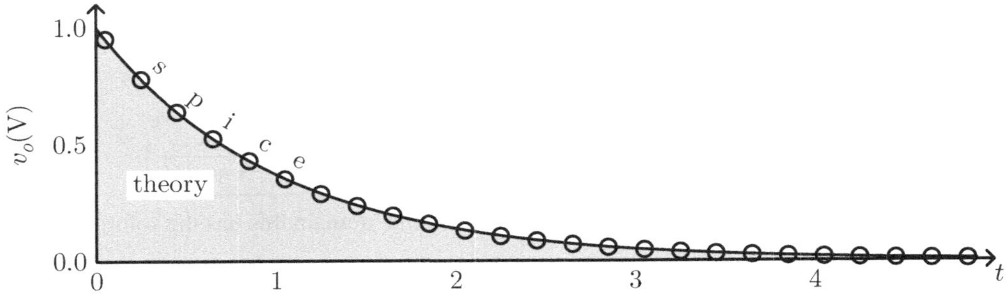

Fig. 39.7 Simulation results of pre-charged *RL* network with zero sources; case of $R = 1\,\Omega$ and $L = 1$ H

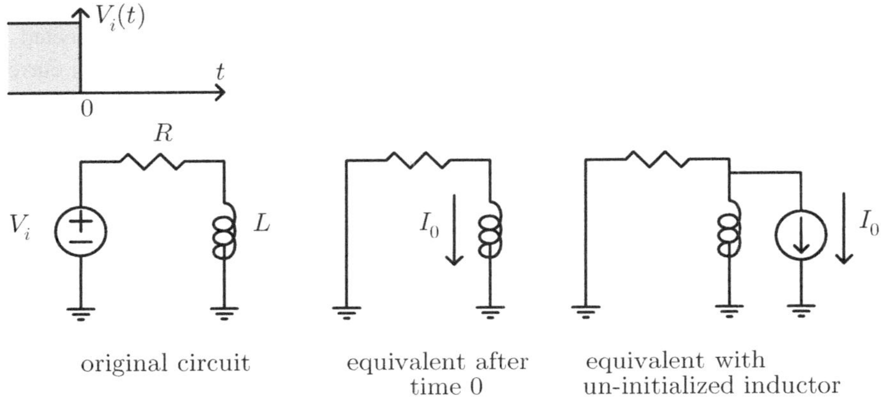

Fig. 39.8 Driven *RL* network

the same; they both start at 1 and end at zero (with the L/R time constant). Output voltage is simply

$$v_o(t) = Ri(t) = Re^{-t/\tau}, \qquad \tau = \frac{L}{R} \tag{39.23}$$

Figure 39.7 shows sample transient simulation results and comparison to SPICE. So we have demonstrated how using artificial current sources in parallel with an un-charged inductor yield the same *effect* as that of the real pre-charged inductor.

39.8 Sample Case of Driven *RL* Network

Consider next the driven *RL* network shown on the left side of Fig. 39.8. The voltage source has been V_0 for all negative time and at time zero it switches to zero. We want to find circuit current (through the resistor).

Notice that for negative time, and while input voltage was V_0, and under steady state there is no voltage across the inductor and all voltage is across the resistor. Hence initial current is simply

$$I_0 = i(0) = \frac{V_0}{R} \tag{39.24}$$

When the input voltage switches to zero, we can replace it with a short and fall back at the circuit in the middle of Fig. 39.8. Now this circuit could be thought of as an un-driven one, but with an initial current across the inductor. As such we can replace it with an equivalent one, with a zero-initialized, but with an additional parallel current source, and as shown on the right of the same figure. If we denote the current across the inductor i_L then doing KVL around the loop gives us

$$sLI_L(s) + R\left(\frac{I_0}{s} + I_L(s)\right) = 0 \tag{39.25}$$

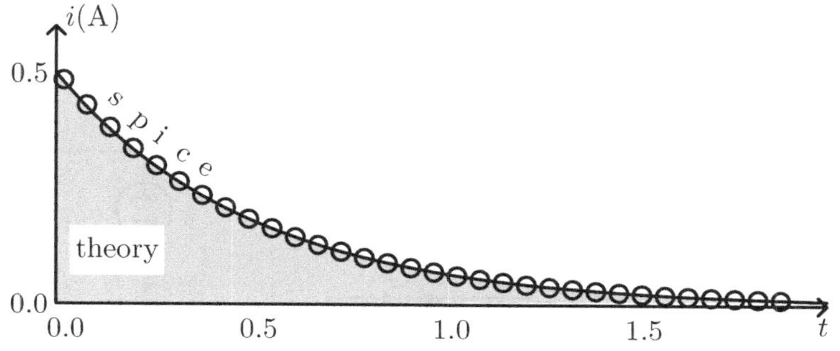

Fig. 39.9 Resistor current of *RL* network; case of $R = 2\omega$ and $L = 1\,\text{H}$

Notice that we used $\frac{I_0}{s}$ since the current source manifests itself as a step function. Collecting terms we get

$$I_L(sL + R) = -I_0\frac{R}{s} \tag{39.26}$$

Assuming $V_0 = 1\,\text{V}$ such that $I_0 = \frac{1}{R}\,\text{A}$ we get

$$I_L(sL + R) = -\frac{1}{s} \tag{39.27}$$

such that

$$I_L(s) = -\frac{1}{s}\frac{1}{sL + R} = -\frac{1}{L}\frac{1}{s}\frac{1}{s + R/L} = -\frac{1}{R}\frac{1}{s} + \frac{1}{R}\frac{1}{s + R/L} \tag{39.28}$$

The current we are after, which is the resistor current, is simply this current plus the added current source:

$$I(s) = I_L(s) + \frac{1}{R}\frac{1}{s} = \frac{1}{R}\frac{1}{s + R/L} \tag{39.29}$$

In the time domain this gives

$$\boxed{i(t) = \frac{1}{R}e^{-tR/L}} \tag{39.30}$$

That is, resistor current starts at $\frac{1}{R}$ (as we established in Eq. (39.24)) and then decays to zero with a time constant $\tau = L/R$. Figure 39.9 shows the results and comparison to SPICE.

39.9 Recap for Circuits Containing Pre-charged Inductors

Let's again reiterate the process for taking care of circuits containing pre-charged inductors. First replace the pre-charged inductor with a neutral one—simple! Next add in parallel with this inductor a *unit step* current source whose value is the pre-charged value. Now run the simulations exactly the same as has been done in all the prior chapters, but keeping in mind that when reporting subsequent currents across the inductor to not only report the new current across the inductor, but to always include in it the contribution from the artificially inserted current source. Remember, the new neutral inductor does not equal to the old pre-charged inductor; what is equal is the new inductor *plus* the inserted current source! This caveat does not apply for inductor voltage since that is the same as that of the current source (since they are in parallel).

39.10 Sample Case with Multiple Pre-charged Capacitors

As a final example consider next the circuit in the upper left side of the Fig. 39.10. The right cap is pre-charged to 1, while the left one to 2 V. At time zero the switch is closed, as shown in the bottom left side of the same figure. Find the settling voltage at time ∞!

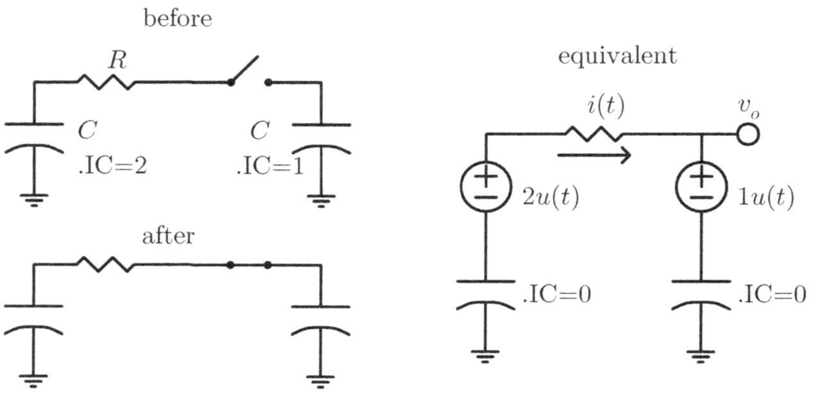

Fig. 39.10 System with pre-charged caps: in this case $C_1 = C_2 = C = 1$ and $R = 2$

Again we can replace each pre-charged cap with an uncharged one, in series with the proper voltage source, as shown on the right side of Fig. 39.10. Doing KVL on the new circuit we arrive at

$$I\left[\frac{2}{sC} + R\right] = \frac{1}{s}, \quad \Rightarrow \quad I\frac{2 + sRC}{sC} = \frac{1}{s} \tag{39.31}$$

Hence system current is

$$I(s) = \frac{C}{2 + sRC} = \frac{1}{R}\frac{1}{s + \frac{2}{RC}} \tag{39.32}$$

Output voltage is simply voltage across the cap, plus $u(t)$:

$$V_o(s) = \frac{1}{sC}I(s) + \frac{1}{s} \tag{39.33}$$

Plugging in we get

$$V_o(s) = \frac{1}{RC}\frac{1}{s}\frac{1}{s + \frac{2}{RC}} + \frac{1}{s} \tag{39.34}$$

Applying partial fractions and using the fact that

$$\frac{1}{s}\frac{1}{s + a} = \frac{1}{a}\left[\frac{1}{s} - \frac{1}{s + a}\right] \tag{39.35}$$

we get

$$V_o(s) = \frac{1}{2}\left[\frac{1}{s} - \frac{1}{s + \frac{2}{RC}}\right] + \frac{1}{s} \tag{39.36}$$

so that finally

$$\boxed{V_o(s) = \frac{3}{2}\frac{1}{s} - \frac{1}{2}\frac{1}{s + \frac{2}{RC}}} \tag{39.37}$$

The solution to this in time domain is simply

$$\boxed{v_o(t) = \frac{3}{2}u(t) - \frac{1}{2}e^{-2t/RC}} \tag{39.38}$$

So at time zero we get $v_o(0) = 1$ which is what we started with, and at time ∞ we get $v_o(\infty) = \frac{3}{2}$ which means that the other cap dumped some charge on the output cap such that in the end both caps are at the same potential. Sample solution and comparison to SPICE are shown in Fig. 39.11

39.11 Summary

In this chapter we dealt with the topic of systems with nonzero initial conditions. In the circuit domain this means caps with initial charge (voltage)

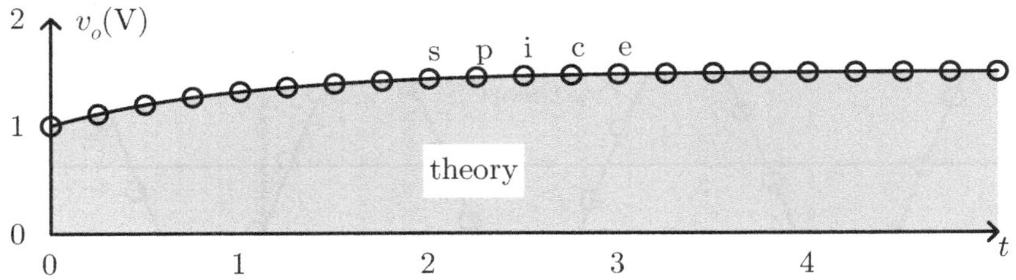

Fig. 39.11 Sample solution to setup in Fig. 39.10

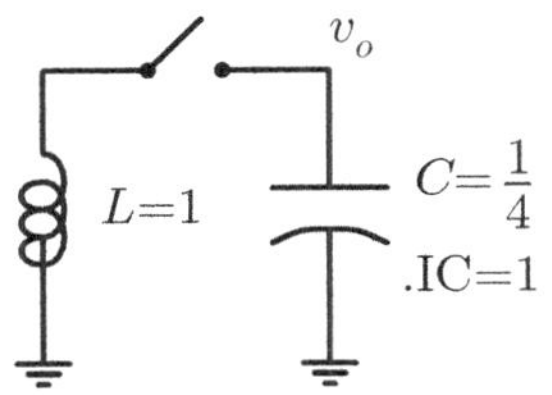

Fig. 39.12 Statement to Problem 1

and inductors with initial magnetic flux (current). Rather than developing an alternate flow we can leverage the mainstream flow used throughout the text by simply replacing the pre-charged elements with uncharged ones and inserting unit step sources at the right locations. For caps we insert unit step voltage sources in series with the caps; for inductors we insert unit step current sources in parallel with the inductors. In both cases the pre-charged element becomes equivalent to the uncharged element in conjunction with the added source *at the terminal* points! It is critical to remember that it is at the terminals of the modified circuit that currents and voltage should be measured and mapped to the original circuit. We demonstrated the flow with a few examples (both source-free and driven) and compared favorably to SPICE simulations. With the exception of this chapter, and unless explicitly stated, though, the majority of this text continues to assume that energy storage elements are zero initialized.

39.12 Problems

1. Consider the *LC* network shown in Fig. 39.12. The cap is pre-charged to 1 V, and at time zero the switch is closed. Determine output voltage and compare to SPICE; see sample solution in Fig. 39.13.
2. A cap of value 1 is pre-charged to 1, and at time zero it is connected to another cap (also of 1 value) as shown in Fig. 39.14. What is the final voltage? What is the final charge (and initial one)? Solve the problem using the methods outlined in the chapter! Confirm results via SPICE.
   Answer: Final voltage of 0.5 V and final charge, per cap of 0.5 C; initial charge on the pre-charged cap was 1 C.
3. Consider the *C* network in Fig. 39.15. The cap on the right is pre-charged to 1, while the other two caps have zero charge. At time zero the switch is closed; what is the output voltage? What is the charge on all three caps? Again use techniques developed in this chapter to arrive at solution.
   Answer: Output voltage (which is voltage across output cap) is 0.5; the other two caps each have 0.25 V across. Output cap has 0.5 Coulomb while each of the other two caps has 0.5 C as well.
4. Consider the *RC* network in Fig. 39.16; the three capacitors are pre-charged to, starting from right to left, 1, 2, and 3 V respectively.

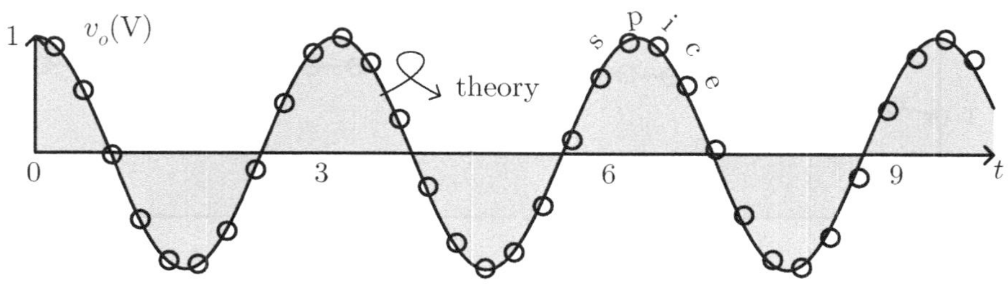

Fig. 39.13 Sample solution to Problem 1

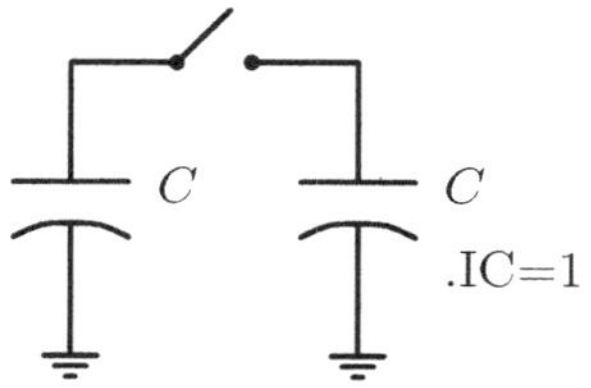

Fig. 39.14 Statement to Problem 2

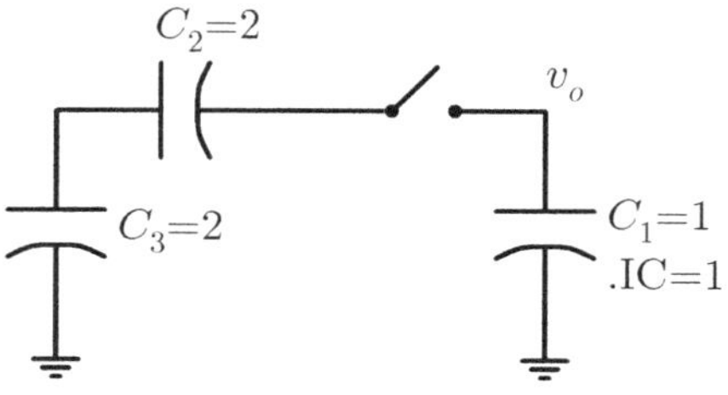

Fig. 39.15 Statement to Problem 3

At time zero the switches are closed. First replace the initial conditions with the appropriate number of voltage sources. Next, write down KVL around the right and left loop and setup the linear system governing $I_1(s)$ and $I_2(s)$. Finally figure the two currents (either analytically or numerically), and once they are known figure output voltage and plot it in the frequency domain; see sample solution in Fig. 39.17.
Answer:

$$\begin{bmatrix} 1+\frac{2}{s} & \frac{1}{s} \\ \frac{1}{s} & 1+\frac{2}{s} \end{bmatrix} \begin{bmatrix} I_1(s) \\ I_2(s) \end{bmatrix} = \begin{bmatrix} \frac{1}{s} \\ -\frac{1}{s} \end{bmatrix},$$

$$V_o(s) = \frac{1}{s} + \frac{I_1(s)}{s}$$

5. Use the initial and final value theorems to predict $v_o(0)$ and $v_o(\infty)$ in Problem 4.
6. Take the frequency dependent output voltage in Problem 4 and use it to figure output voltage in time; accomplish this by whatever means you can, including numerical inversion, approximate pole/zero method, or—if done symbolically—direct inversion. Plot and compare to SPICE; see sample solution in Fig. 39.18. Do the 0 and ∞ limits agree with those of Problem 5?
7. Consider the LC network on the left side of Fig. 39.19 Initially the inductor is precharged with a unity current as shown in the figure. At time zero, the configuration is altered to correspond to that on the right side of the figure. Find output voltage and compare to SPICE; see sample solution in Fig. 39.20.
Answer:

$$V_o(s) = -\frac{1}{s^2+1}, \qquad v_o(t) = -\sin t$$

8. Consider the RLC circuit on the left of Fig. 39.21. The cap is pre-charged to 1. Use SPICE to solve for output voltage using two methods: (a) as is, with cap pre-charged, and ensuring the tool enforces the initial conditions; and (b) with cap uncharged, but with a unit step in series, as shown on the right side of the figure. Compare both results as shown in sample solution in Fig. 39.22

Fig. 39.16 Statement to Problem 4

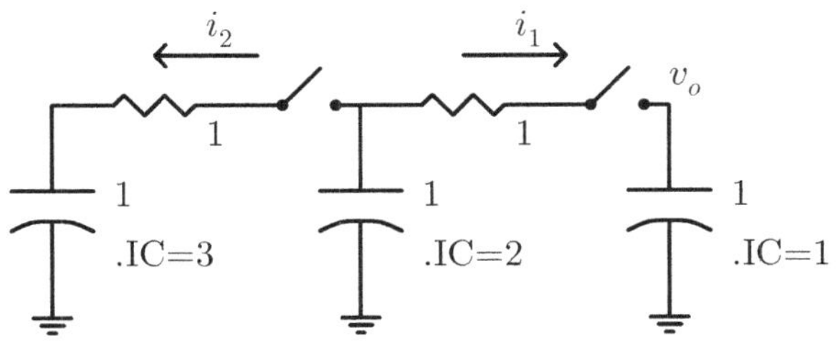

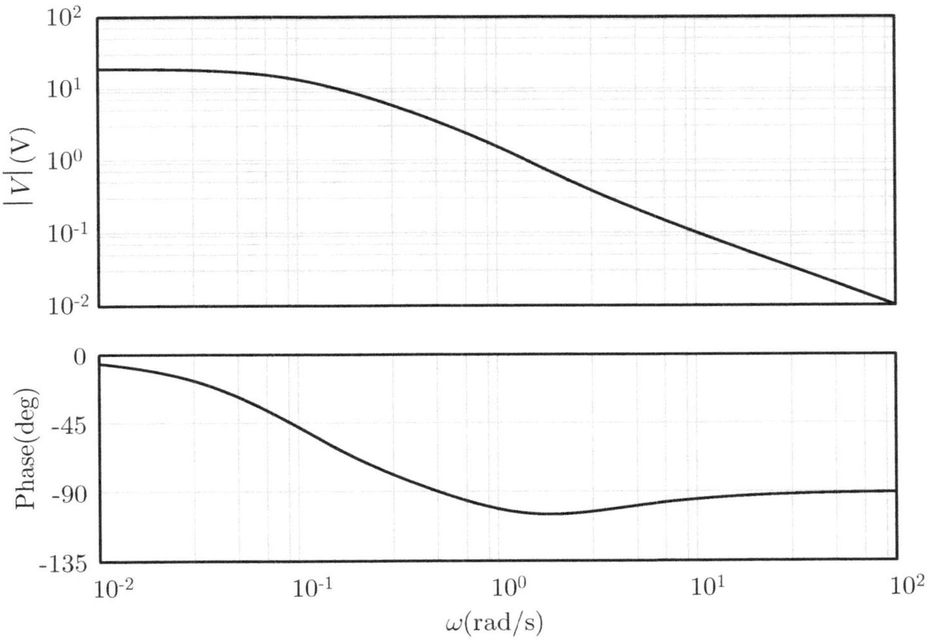

Fig. 39.17 Sample solution to Problem 4; case of $\sigma = 0.1$

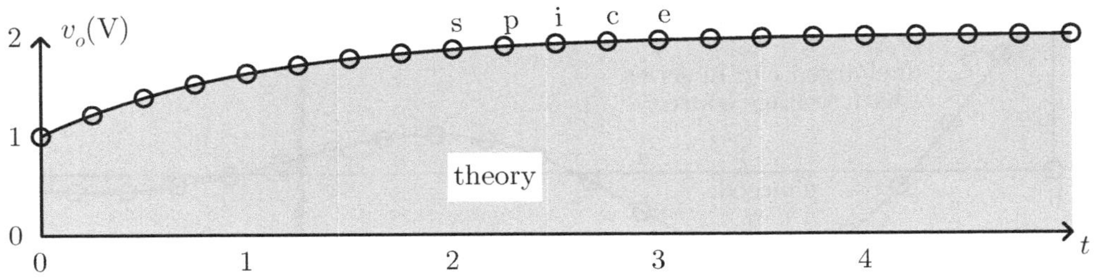

Fig. 39.18 Sample solution to Problem 6

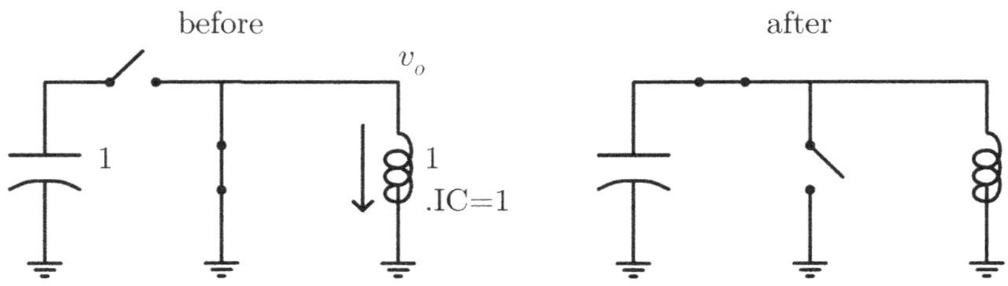

Fig. 39.19 Setup to Problem 7

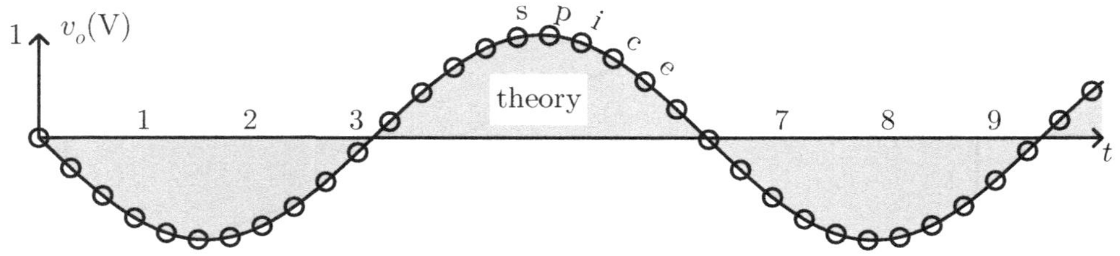

Fig. 39.20 Sample solution to Problem 7

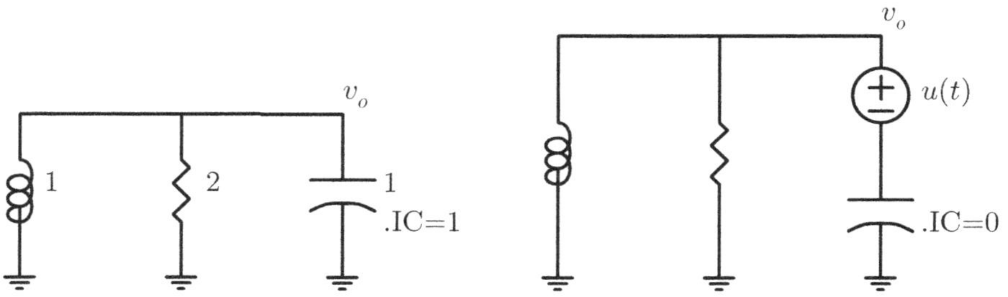

Fig. 39.21 Setup to Problem 8

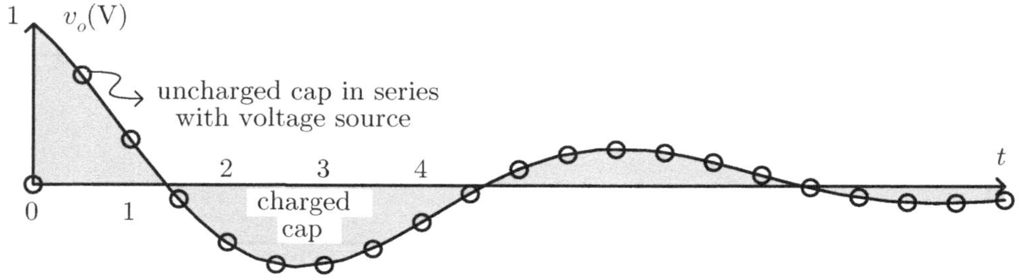

Fig. 39.22 Sample solution to Problem 8

40 Application to Transistor Modeling and Circuits

40.1 Introduction

Next to resistors, capacitors, and inductors, MOSFET devices play a very important role in circuits and electrical engineering. Similar to *RLC* elements, these devices take on voltages and produce currents. The *IV* relation, however, is more complicated. For example, while the *RLC* elements have two terminals, the MOSFET has 4—drain, gate, source, and base. (However, in this brief treatment we will assume the base to be tied to the source.) In other words, while the *RLC* elements have only a single current and voltage, the MOSFET has multiples thereof. Also, while the *RLC* elements have a linear relation between currents and voltages, in the sense that twice the voltage produces twice the current, the MOSFET has a nonlinear relation between current and voltage, sometimes in the form of a quadratic dependence, or cross-terminal products. Nonetheless, around an *operating point* defined by a relatively fixed DC level voltages for the various terminals, we can *linearize* the MOSFET model in the form of an *RC* network with dependent sources, and consequently be able to relate input/output characteristics as a function of frequency. At that point we should be able to map in all the linear tools developed so far, in terms of transfer functions, poles/zeroes, Bode plots, inverse transforms, convolution (and the list goes on) to the MOSFET world! By doing so we are also paving the way for studying more complicated active circuits, such as amplifiers, and afterwards including feedback.

As a final disclaimer note, though, the field of MOSFET modeling and applications is enormous, and what follows is only but a very short and compact introduction to the field, intended mostly to illustrate the extension of spectral techniques to what would appear at first look a non-related field. Once the reader grasped the main points, which are that the small signal model is but a distributed *RC* network (with controlled sources), it is hoped that he/she will apply the text techniques to MOSFET modeling, and circuit design at a deeper, more comprehensive level!

40.2 MOSFET Large Signal Model

The MOSFET large signal model ties input/output currents and voltages. To derive the small signal model we need first the large signal one. As is well known the MOSFET has four terminals: gate, source, drain, and bulk as shown in Fig. 40.1. Assuming bulk at same potential as source (for now), and in the *long-channel approximation*, the *IV* characteristics in the *triode* region is given by

F. Badrieh, *Spectral, Convolution and Numerical Techniques in Circuit Theory*,
https://doi.org/10.1007/978-3-319-71437-0_40

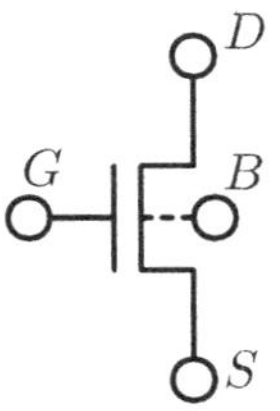

Fig. 40.1 MOSFET symbol and terminals

$$I = KP\frac{W}{L}\left[(V_{GS} - V_{th})V_{DS} - \frac{V_{DS}^2}{2}\right](1 + \lambda V_{DS}), \quad V_{DS} < V_{GS} - V_{th}, \quad \text{(triode region)} \tag{40.1}$$

where W is the device width, L device length, $KP = \frac{\mu\epsilon}{t_{ox}}$, μ is mobility, ϵ is the dielectric constant, t_{ox} is the oxide thickness, and λ is the channel length modulation parameter. V_{GS} is gate-to-source voltage, V_{th} is the threshold voltage, and V_{DS} drain-to-source voltage. When the drain voltage exceeds $V_{GS} - V_{th}$ we get the *saturation* equation

$$I = \frac{KP}{2}\frac{W}{L}(V_{GS} - V_{th})^2\,[1 + \lambda V_{DS}], \quad V_{DS} > V_{GS} - V_{th}, \quad \text{(saturation)} \tag{40.2}$$

A sample plot of current versus voltage is shown in Fig. 40.2. Notice the *nonlinear* dependence of drain current on *drain* voltage. Notice also the nonlinear dependence of drain current on *gate* voltage, in the sense each voltage increment does *not* result in the same current increment. Notice also in the triode region how we have V_{DS} multiplying V_{GS}. So we have nonlinear relations all over the place!

40.3 Transconductance and Output Impedance

Quite often the MOSFET is operated in the saturation region as a current source, especially if output impedance r_o is infinite. That is, it would give the same current independent of voltage (drain to source here), which is the definition of a current source. As such we set out to figure the MOSFET small signal AC parameters, defined by the transconductance and output impedance, in the saturation regime. The dependence of saturation current on V_{GS} is reflected through g_m which is called the *transconductance*. It can be derived as

$$g_m = \frac{dI}{dV_{GS}} = \frac{d}{dV_{GS}}\frac{KP}{2}\frac{W}{L}(V_{GS} - V_{th})^2\,(1 + \lambda V_{DS})$$

$$g_m = KP\frac{W}{L}(V_{GS} - V_{th})\,(1 + \lambda V_{DS}) \tag{40.3}$$

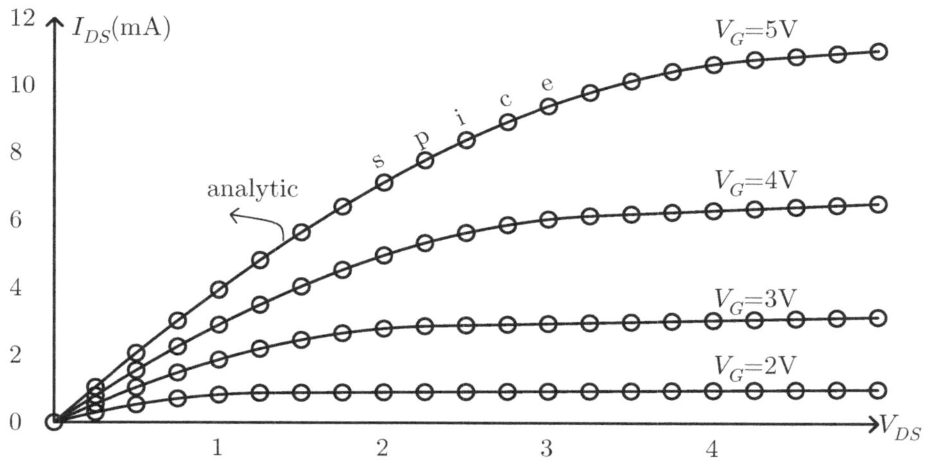

Fig. 40.2 *IV* characteristics of MOSFET ($W = 10$ and $L = 1\,\mu$m)

Fig. 40.3 MOSFET low-frequency small signal model

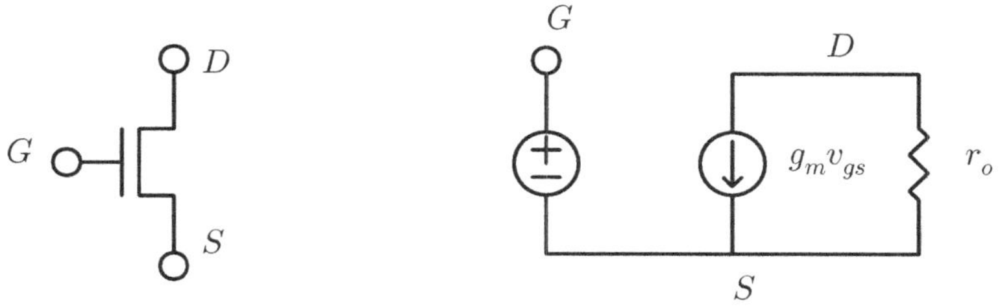

The other DC small signal model parameter is the *output impedance*, which ties drain current to drain voltage, and that is derived as

$$\boxed{r_o^{-1} = \frac{dI}{dV_{DS}} = \frac{KP}{2}\frac{W}{L}(V_{GS} - V_{th})^2\lambda} \qquad (40.4)$$

So (at least for the case of saturation) we have derived the small signal parameters g_m and r_o.

40.4 MOSFET Low-Frequency Small Signal Model

Knowing the bias conditions (e.g., V_{GS} and V_{DS}) we can figure g_m and r_o, and this constitutes what is referred to as the low-frequency small signal model, shown in Fig. 40.3. It is coined low-frequency because we have not yet included the MOSFET parasitic capacitances. To reiterate, if we know the bias conditions, which are the drain/grate/source (and base) voltages, we could approximate the MOSFET small signal operation (i.e., behavior for small voltage/current variations) via the circuit shown in Fig. 40.3. This small signal model does not give us the total voltage (or current) at the various terminals; it only gives us the *deviation* in voltage and current as a function of a stimulus of small magnitude. To find total terminal voltage/current we need to add the operating point voltages/currents (i.e., the DC levels) to the small signal ones derived from the small signal model.

40.5 Sample Application of Low-F Small Signal Model

As a sample demonstration of how to use the low-f small signal model, consider the circuit in Fig. 40.4. The MOSFET has width/length $10\,\mu/1\,\mu$m; $V_{th} = 0.7$, $KP = 100\frac{\mu A}{V^2}$, and $\lambda = 0.1\,V^{-1}$. The first thing we want to do is figure the DC operating point; e.g., what is the

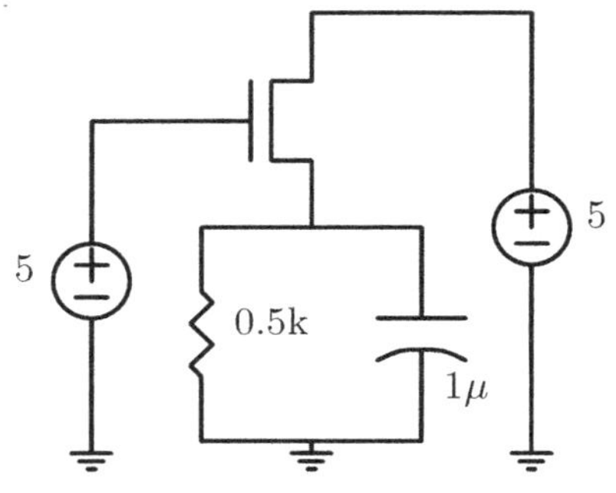

Fig. 40.4 MOSFET with *RC* load

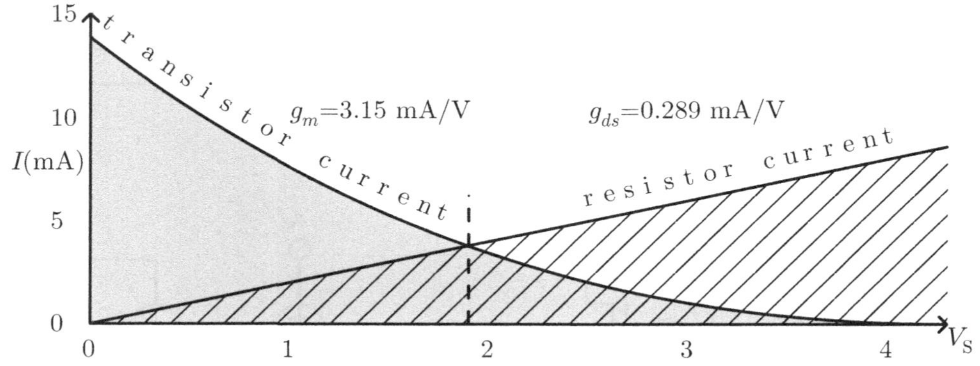

Fig. 40.5 Figuring source voltage in Fig. 40.4

source voltage? To do this we ignore the output cap and insist that the DC current through the load resistor equals that across the MOSFET (from drain to source). We plot both currents as shown in Fig. 40.5 and at the point where both curves meet we read out the source voltage; in this case 1.895 V. Knowing this we next figure g_m and r_o; specifically

$$\begin{aligned} g_m &= KP\frac{W}{L}\left(V_{GS}-V_{th}\right)\left(1+\lambda V_{DS}\right) \\ &= 100\times 10^{-6}\frac{10}{1}\left(3.105-0.7\right)\left(1+0.1\times 3.105\right) \\ &= 3.15\,\text{mA/V} \end{aligned} \tag{40.5}$$

Next we get for output impedance

$$\begin{aligned} r_o &= \left[\frac{KP}{2}\frac{W}{L}\left(V_{GS}-V_{th}\right)^2\lambda\right]^{-1} \\ &= \left[\frac{100\times 10^{-6}}{2}\frac{10}{1}\left(3.105-0.7\right)^2\times 0.1\right]^{-1} \\ &= 3460\,\Omega \end{aligned} \tag{40.6}$$

Next we construct the small signal equivalent circuit as shown in Fig. 40.6. Now we need to solve for output voltage in terms of input one. Let

$$R_p = r_o||0.5\text{k} \tag{40.7}$$

Further let

$$Z(s) = \frac{1}{C}\frac{1}{s+\frac{1}{R_pC}} \tag{40.8}$$

Then doing KCL at the output node gives

$$g_m\left(v_i - v_o\right)Z = v_o \quad \Rightarrow \quad v_o(1+g_mZ) = v_i g_m Z \tag{40.9}$$

$$\boxed{H(s) = \frac{v_o(s)}{v_i(s)} = \frac{g_mZ(s)}{1+g_mZ(s)}} \tag{40.10}$$

A plot of this as well as comparison to SPICE is shown in Fig. 40.7. As shown in the figure we get excellent match. This is fantastic! The very same concepts and tools we used for most of the text, relating to spectral methods, frequency

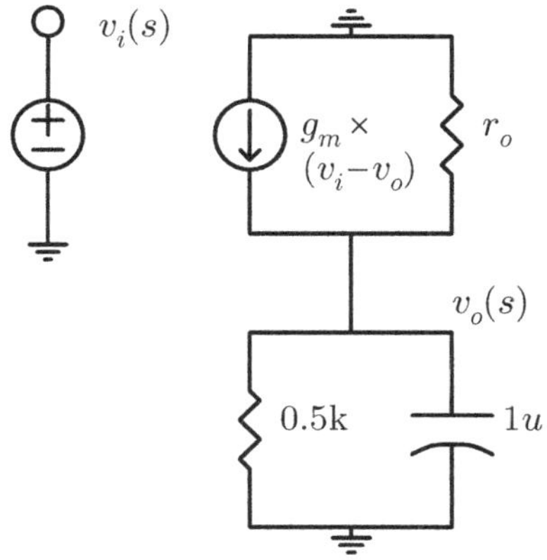

Fig. 40.6 Low-f small signal model of circuit in Fig. 40.4

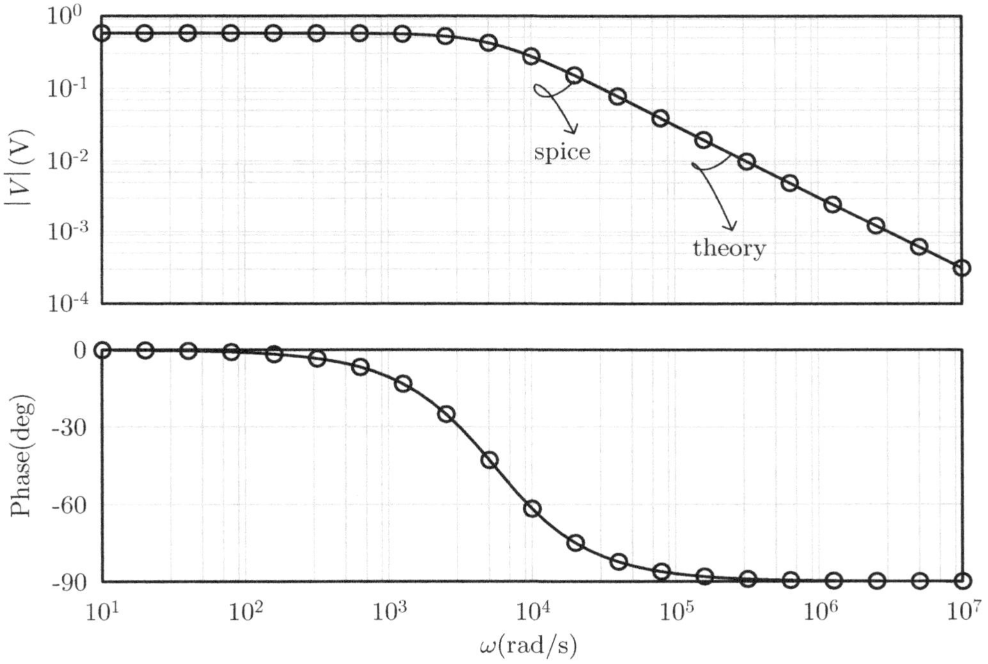

Fig. 40.7 Transfer function of circuit in Fig. 40.4

dependent impedance, transfer functions—all apply equally well in this seemingly unrelated nonlinear problem! Let us take a closer look at the zero frequency limit:

$$H(0) = \frac{g_m Z(0)}{1 + g_m Z(0)} \tag{40.11}$$

From Eq. (40.8) we read $Z(0) = R_p$ such that

$$H(0) = \frac{g_m R_p}{1 + g_m R_p} \tag{40.12}$$

For the derived r_o and g_m values, and for R_p one we get

$$\begin{aligned} H(0) &= \frac{3.15 \times 10^{-3} \times 3460||500}{1 + 3.15 \times 10^{-3} \times 3460||500} \\ &= \frac{3.15 \times 10^{-3} \times 437}{1 + 3.15 \times 10^{-3} \times 437} \\ &= \frac{1.38}{2.38} = 0.58 \end{aligned} \tag{40.13}$$

which is the exact value the plot (and SPICE) provide.

At the other extreme the transfer function predicts that at high frequency output voltage (that is small signal output voltage) goes to zero, in this case due to the cap. In this case we get the limit

$$H(\infty) \sim g_m \frac{1}{sC} \tag{40.14}$$

For example at $\omega = 10^7$ we get

$$\begin{aligned} H(10^7) &= 3.15 \times 10^{-3} \times \frac{1}{10^7 \times 10^{-6}} \\ &= 3.15 \times 10^{-3} \times 0.1 = 3.15 \times 10^{-4} \end{aligned} \tag{40.15}$$

again in agreement with the plot and SPICE. So we are able to at least rationalize the low- and high-frequency limits of this circuit.

40.6 MOSFET High-Frequency Small Signal Model

So far we have dealt with the MOSFET small signal model as a combination of resistors and controlled sources. Next we add in the various node capacitors, the most important being the C_{gs} (gate-to-source cap) and C_{gd} (gate-to-drain cap). This gives the topology shown in Fig. 40.8. Notice that in addition to those caps, there are other caps such as source and drain junction caps, as well as base cap; but in order to make the treatment manageable in this allocated space, and keeping in mind that adding caps does not alter the fundamental treatment, we ignore those caps for now. This model, with the caps, is what is referred to as the high-frequency small signal model of the MOSFET.

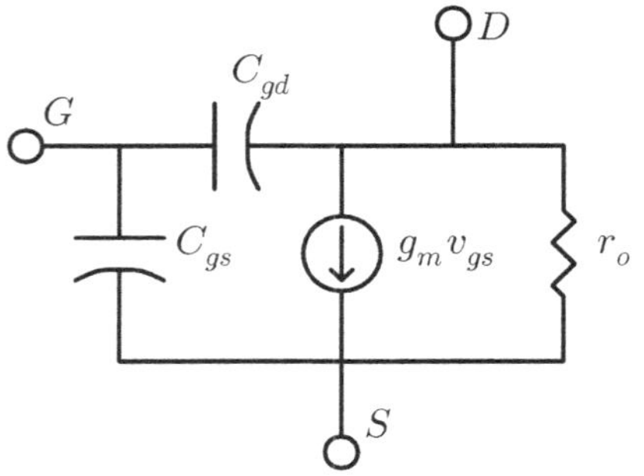

Fig. 40.8 MOSFET small signal model

Notice that the low-frequency limit of the model in Fig. 40.8, where caps open (due to infinite impedance) falls back to that of the low-f model, shown before in Fig. 40.3. So this model is a more generic one. At DC the small signal drain current is given by

$$i_D = g_m v_{gs} + r_o^{-1} v_{ds}, \qquad \text{(DC case)} \tag{40.16}$$

At high frequency we pick *capacitive currents* as well. This will be demonstrated next.

40.7 Sample Application of High-Frequency Small Signal Model

Consider the MOSFET discharging a cap load as shown in Fig. 40.9. The corresponding small signal model is shown on the right. Notice that for this case we dropped C_{gs} since it would be in parallel with the input voltage source; that is, its presence (or for that matter absence) has no bearing on circuit performance! To find the transfer function we do the following steps. First the impedance of the parallel RC branch is

$$Z_{RC} = \frac{r_o \frac{1}{sC_L}}{r_o + \frac{1}{sC_L}} = \frac{r_o}{1 + sr_oC_L} \tag{40.17}$$

The current through the output RC is

$$i_{RC} = v_o \frac{1 + sr_oC_L}{r_o} \tag{40.18}$$

The current through C_{gd} is

$$i_{C_{gd}} = (v_o - v_i)sC_{gd} \tag{40.19}$$

Summing the currents at the output node we get

$$v_o \frac{1 + sr_oC_L}{r_o} + g_m v_i + (v_o - v_i)sC_{gd} = 0 \tag{40.20}$$

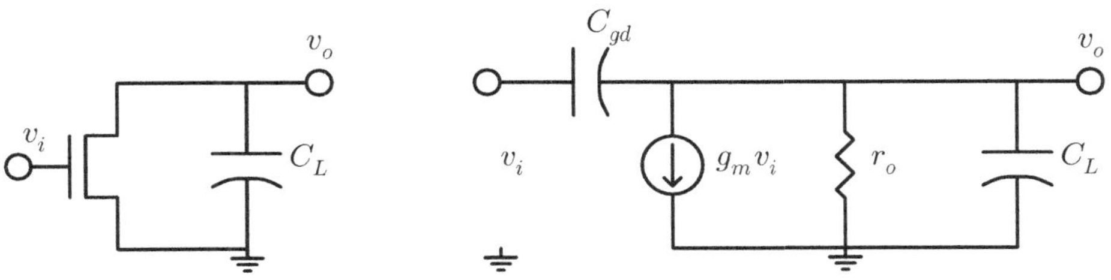

Fig. 40.9 MOSFET driving cap and small signal model

Collecting terms gives

$$v_o\left[\frac{1}{r_o}+s(C_L+C_{gd})\right]+v_i\left[g_m-sC_{gd}\right]=0 \quad (40.21)$$

from which we derive the transfer function

$$\boxed{\frac{v_o}{v_i}=-g_m r_o\frac{1-s\frac{C_{gd}}{g_m}}{1+sr_o(C_L+C_{gd})}} \quad (40.22)$$

Notice that this transfer function has a single pole (left-handed one, as usual) at

$$p_1=-\frac{1}{r_o(C_L+C_{gd})} \quad (40.23)$$

and a single **right hand** zero at

$$z_1=\frac{g_m}{C_{gd}} \quad (40.24)$$

Saturation Model Assume that output voltage is pre-charged to 5 V, while input 1.7 V (one and threshold voltage 0.7 V). Thence the MOSFET would start in the saturation region. Assume the MOSFET has W/L of 10 and 1 μm. Furthermore assume $KP = 100\,\mu\text{A/V}$ and $\lambda = 0.04\,\text{V}^{-1}$. Finally assume oxide thickness of 0.1 nm. The oxide thickness will primarily determine the C_{gs} and C_{gd}. For example, the total oxide cap would be

$$C_{ox}=WL\frac{\epsilon}{t_{ox}}$$

$$=10\times10^{-12}\frac{3.9\times8.854\times10^{-12}}{0.1\times10^{-9}}=3.5\,\text{pF} \quad (40.25)$$

This cap now would split between C_{gs} and C_{gd}, depending on the exact bias. It may also be derated depending whether the device is in depletion or inversion. Assume for our particular case, and with a load cap of $C_L = 1\,\text{pF}$ we get from SPICE

$$C_{gd}=23.\text{E-}15, \qquad g_m=1.2\text{E-}3, \qquad r_o=50\text{k} \quad (40.26)$$

A plot of the transfer function as well as comparison to SPICE simulations is shown in Fig. 40.11. At DC both caps are open and we end up with the setup shown on the left side of Fig. 40.10. The output voltage there is simply

$$\frac{v_o}{v_i}=-g_m r_o=-1.2\times10^{-3}\times50000=60, \quad \text{(at DC)} \quad (40.27)$$

which is confirmed in the transfer function (Fig. 40.11).

At high frequency, on the other hand, the output cap dominates r_o, in the sense of shunting it, and we end up with equivalent circuit shown on the right of Fig. 40.10. Summing currents at output node we get

$$sC_{gd}(v_o-v_i)+g_m v_i+sC_L v_o=0 \quad (40.28)$$

Collect terms

$$v_o s(C_{gd}+C_L)=v_i(sC_{gd}-g_m) \quad (40.29)$$

Form ratio to get transfer function

$$\frac{v_o}{v_i}=\frac{sC_{gd}-g_m}{s(C_{gd}+C_L)} \quad (40.30)$$

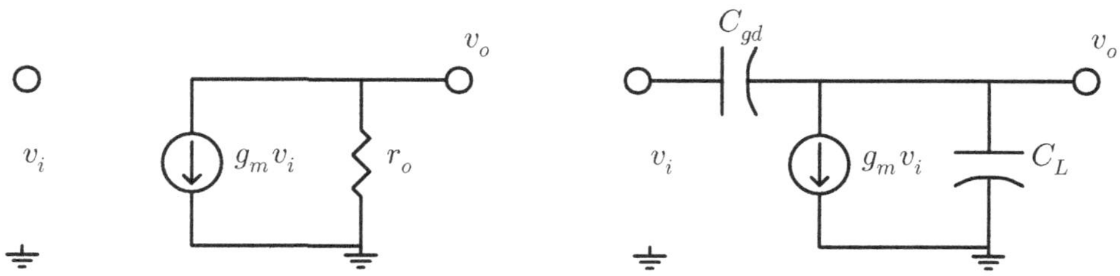

Fig. 40.10 Small signal model of MOSFET driving cap: left—DC; right high frequency

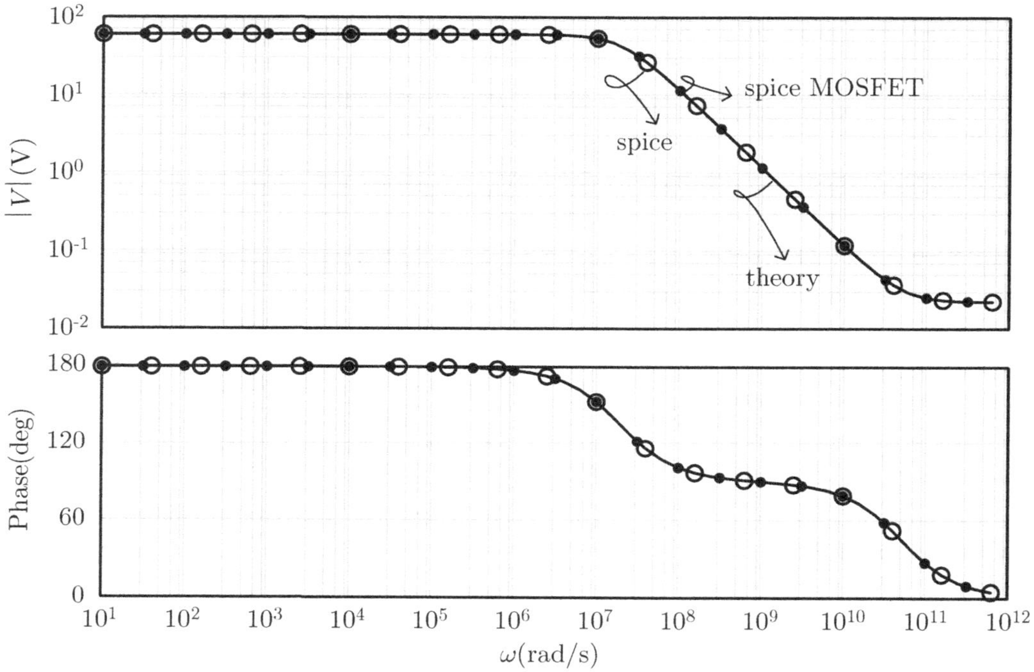

Fig. 40.11 Transfer function of MOSFET driving cap while in saturation. Theory refers to Eq. (40.22); SPICE passive refers to SPICE simulations on circuit in Fig. 40.9; SPICE MOSFET refers to SPICE simulations on the actual MOSFET

At really high frequency we arrive at the approximation

$$\frac{v_o}{v_i} \sim \frac{C_{gd}}{C_{gd} + C_L} = \frac{23}{1000 + 23} = 2.2 \times 10^{-2}, \quad \text{(high frequency)} \tag{40.31}$$

as confirmed in Fig. 40.11. The first pole happens at

$$\omega(\text{pole}) = \frac{1}{r_o(C_L + C_{gd})} \sim \frac{1}{50\,\text{k}\Omega \times 1\,\text{pF}} = 20\,\text{M rad/s} \tag{40.32}$$

which is confirmed in the simulations. The zero happens when

$$\omega(\text{zero}) = \frac{g_m}{C_{gd}} = \frac{1.2\,\text{mA/V}}{23\,\text{fF}} = 52\,\text{G rad/s} \tag{40.33}$$

which is also confirmed in the simulations. So we have confirmed that so long as the MOSFET is in the saturation, we are able to predict the small signal model, and input/output transfer function.

Triode Model As the cap discharges, the drain voltage drops too. At some point the device switches from the saturation region (large V_{DS}) to the triode region (small V_{DS}). For example, if the drain level drops from 5 V to 0.7 V, the device now is in the linear (triode) region; as the device enters the triode region, a few things happen:

1. Transconductance g_m drops down. In this case it drops from 1.2 m to 0.7 mA/V.
2. Output impedance drops down. In this case it drops from 50 k to 3 kΩ.
3. Gate-to-drain cap goes up.

Notice that since the device now is in the linear region, the gate-to-drain cap is expected to be a much larger portion of oxide cap. Assume now that out of total oxide cap of 3.5pF, 0.94pF goes to the drain (while the rest goes to the source). Again the exact split of oxide cap between drain and source is left to SPICE. As the small signal parameters change, so will the small signal model. As such the transfer function will change too. In particular, the DC gain becomes

$$\text{DC gain} = g_m r_o = 0.7\,\text{m} \times 3\,\text{k} = 2.1 \quad (40.34)$$

as confirmed in Fig. 40.12. At high frequency the gain becomes

$$\text{High frequency gain} = \frac{C_{gd}}{C_L + C_{gd}} = \frac{940}{1000 + 940} = 0.5 \quad (40.35)$$

and confirmed in the figure. The first pole happens at

$$\text{Pole location} = \frac{1}{r_o(C_L + C_{gd})} = \frac{1}{3\text{k}(1.0\text{p} + 0.94\text{p})} = 170\,\text{M rad/s} \quad (40.36)$$

As for the zero, the new location is

$$\text{Zero location} = \frac{g_m}{C_{gd}} = \frac{0.7\text{m}}{0.94\text{p}} = 745\,\text{M rad/s} \quad (40.37)$$

Clearly, the small signal transfer function behavior in the triode region is quite different from that in the saturation region. But in all cases, knowing the small signal model with its bias-dependent elements enables us to predict the input/output transfer function which captures the impact of small changes at the input side to the output side.

40.8 Sample Application Utilizing DC, AC, and Transient Simulations

As a wrap up example consider the MOSFET whose source is tied to ground and drain to a load resistor, as shown in Fig. 40.13. Given that R_L is 1 kΩ, our task so to find maximum DC gain.

DC Simulations Before doing gain simulation, which is a form of small signal analysis, we need to do DC simulations to figure the operating

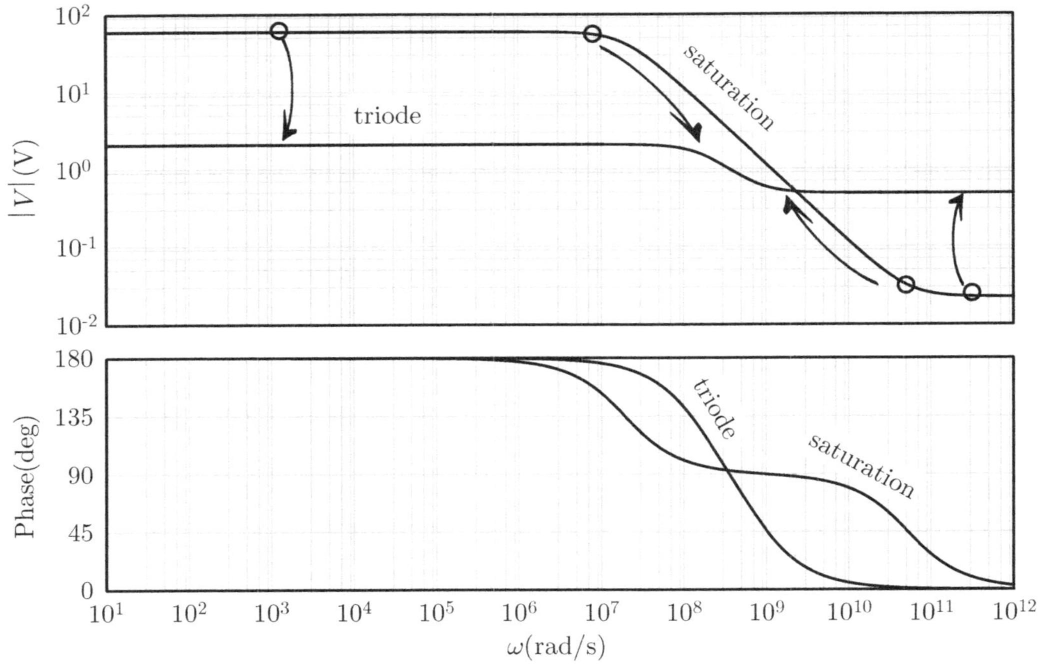

Fig. 40.12 Transfer function of MOSFET driving cap in triode region, and comparison to saturation one

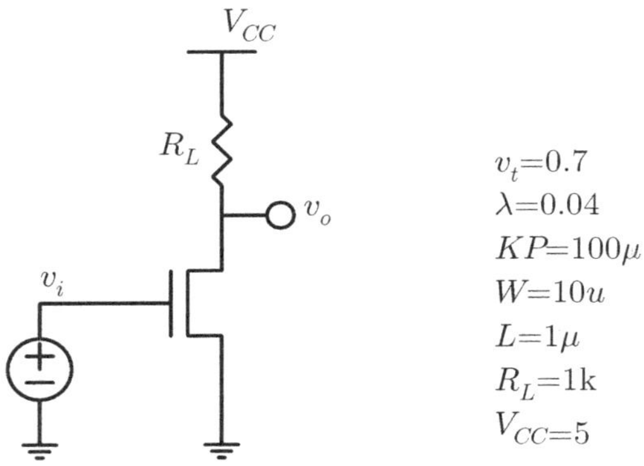

Fig. 40.13 Drain loaded MOSFET

point. The operating point tells us the DC bias from which we can extract the small signal model parameters. The operating point of the circuit which determines v_{out} happens when the drain current of the MOSFET equals that of the load resistor one. Figure 40.14 *IV* simulations show both MOSFET and resistor currents, for different gate voltages. For example, when gate voltage is 1.5 V, the match between both currents happens at $v_{\text{out}} = 4.6$ V. Similarly when gate voltage is 5 V, match happens when $v_{\text{out}} = 1$ V. If we were to, then, plot output voltage versus input voltage we would get results as shown in Fig. 40.15. Notice that these results should match identically with crossing points shown in the prior figure (Fig. 40.14). Again, for example, when input voltage is 1.5 V, output is around 4.6 (confirming results from before).

DC Gain If we set input stimulus as our gate voltage, and output as drain voltage, then we define gain as

$$A = \text{Gain} = \frac{dv_{\text{out}}}{dv_{\text{in}}} \tag{40.38}$$

Notice we are taking the rate of change, and not absolute division. That is, if we take the data in Fig. 40.15, and perform differentiation we get Fig. 40.16. The figure tells us that max gain is around input voltage of 2.9 V. We can also get the gain curve by running AC simulations, and reading out the low-frequency results (for different

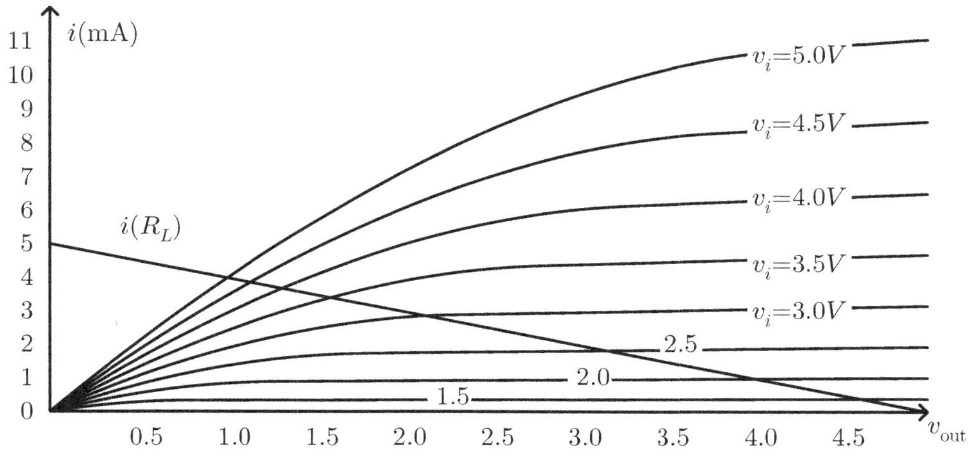

Fig. 40.14 *IV* characteristics of drain load MOSFET, for different gate voltages

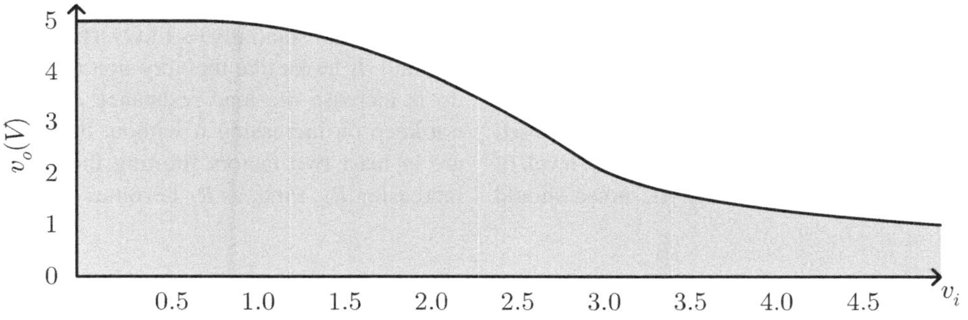

Fig. 40.15 Output voltage versus input voltage for circuit in Fig. 40.14

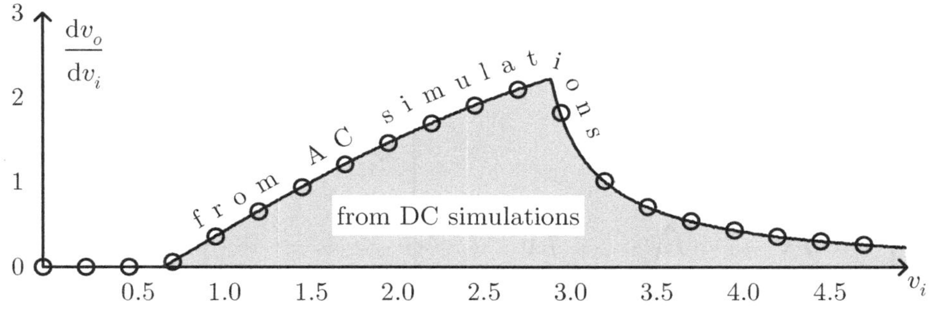

Fig. 40.16 DC gain of transfer function shown in Fig. 40.15

gate voltages). This is also shown in Fig. 40.16 and it matches identically the DC simulations.

Transient Simulations Confirming DC Gain The DC gain results tell us that if we put a low-frequency noise on the input (per DC input voltage) we would get A times that at the output. Notice that input DC level and output DC levels will NOT be equal. For different DC levels of input, we would get different DC levels of output, as shown before in Fig. 40.15; the new discovery relates the AC noise of input to that of output. For example, the DC gain in Fig. 40.16 tells us that if input DC level is 1 V, and we impose on it some small AC signal—say 1 mV—then output would have an AC noise of 0.35 mV. Similarly, if input DC level is 2.9, then output would have an AC noise of 2.2 mV. These results are confirmed in Fig. 40.17 which shows output level in time domain for different DC biases of input node, and with 1 mV of AC noise imposed at the input node. Notice that the DC level of each case is different, but the AC noise should follow the gain prediction, and it does. What we have accomplished here is to confirm the AC (and DC) gain predictions by actually running transient simulations and looking at the ratio of output noise to input one (1 mV here).

Improving DC Gain As seen from the prior subsections, the best we can get in terms of DC gain is about 2.2! Clearly, that is not much of an amplification. We may ask the question, why is that—why not much gain, even across v_{gs}? If we take the specific case of $v_i = 2.9$ V and read out the small signal parameters we get $g_{ds} \sim 1 \times 10^{-4}$ which gives output resistance of $r_o \sim 10\,\mathrm{k}\Omega$. This resistance goes in parallel with the load resistance R_L but since latter is much smaller (1 kΩ), net resistance that multiplies g_m goes down substantially (~1 kΩ). Hence the gain is small. It looks like the very first thing we can do is increase the load resistance R_L. But why not keep on increasing it without bound? There are at least two factors limiting the benefits of increasing R_L. First, as R_L becomes very large, it

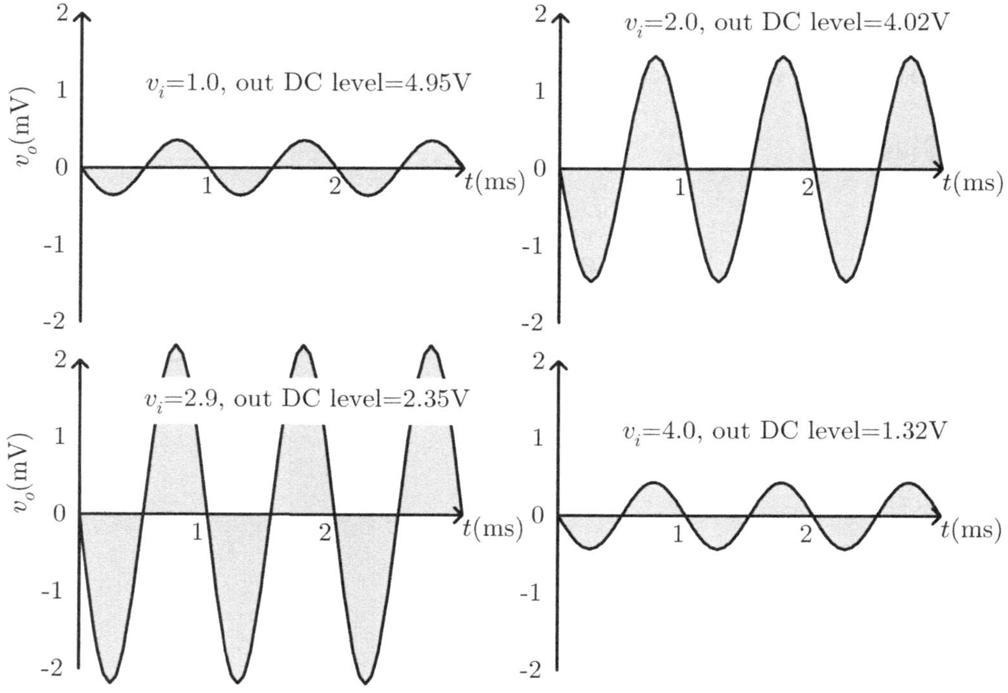

Fig. 40.17 Transient simulations (low-frequency) confirming gain results as shown in Fig. 40.16

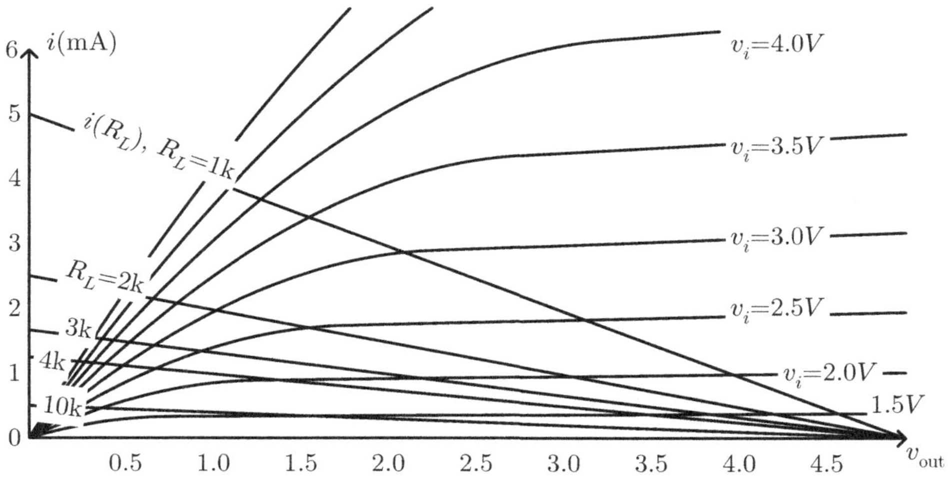

Fig. 40.18 *IV* curves for different gate bias and different load resistance

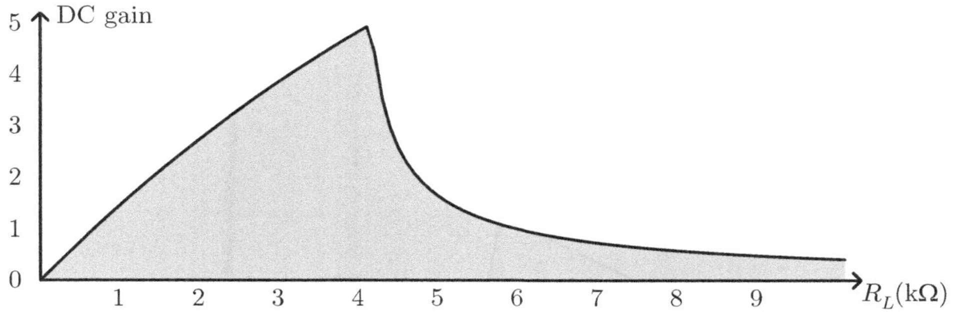

Fig. 40.19 DC gain versus R_L for case $v_i = 2.0\,\text{V}$

being in parallel with transistor output resistance saturates to the latter; that is, we cannot influence the net output resistance anymore. Secondly, as seen in Fig. 40.18, as we make R_L large, the slope of the resistor current decreases, and the intercept of the resistor current curve with that of the transistor happens at earlier voltages, some times too early, resulting in the transistor operating in the linear region—a region where r_o is not as large as it can be. Hence, the $g_m r_o$ product is not maximized.

To demonstrate this, let us take the case with $v_i = 2.0\,\text{V}$ and vary R_L. We get DC gain plot in Fig. 40.19. Notice that initially as we increase R_L the gain improves, and that is due to improvement in $R_L||r_o$. But larger R_L eventually pushes the device into the linear region, and there r_o takes a hit, and the gain degrades. So, gain depends on v_i and R_L.

To get the overall picture we sweep each of these parameters and get plot in Fig. 40.20. We can see from the figure that the best gain is around input voltage of 1, but looks like there the max of R_L effect has not been reached. Let's take that case and sweep R_L even more. We see this in Fig. 40.21. Now we can state that max gain is around 25 and that happens when $v_i = 1.0\,\text{V}$ and $R_L = 100\,\text{k}\Omega$!

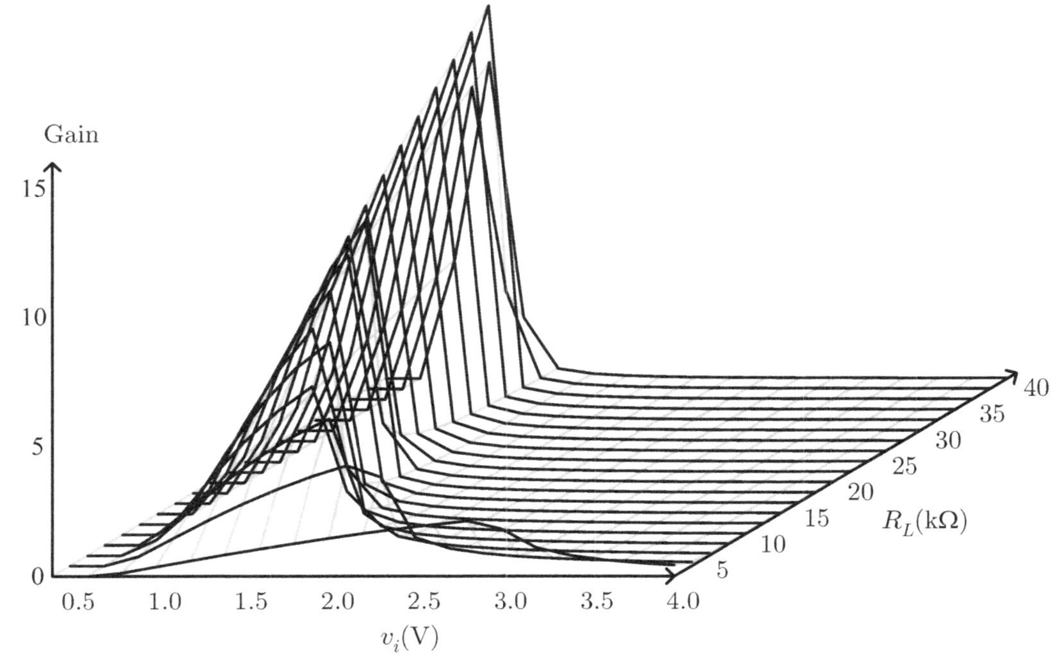

Fig. 40.20 DC gain versus R_L and v_i

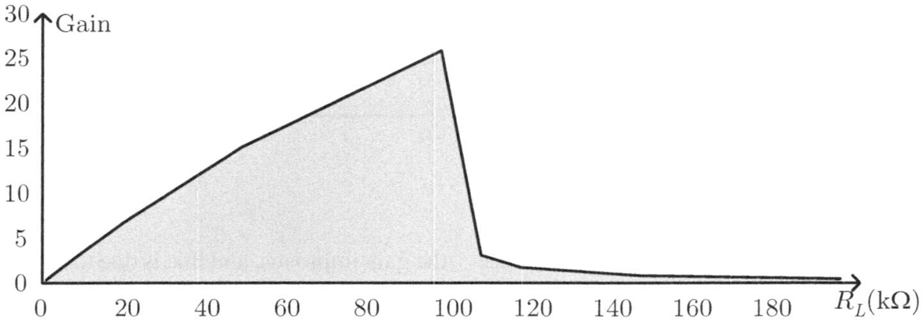

Fig. 40.21 DC gain versus R_L with $v_i = 1.0\,\text{V}$

Transient Simulations Confirming Max Gain From the above analysis we concluded that for max gain we would need

1. Input voltage of $v_i = 1\,\text{V}$.
2. Load impedance of $R_L = 100\,\text{k}\Omega$.

We are told that if we apply those conditions, the AC gain would be somewhere around 25. Figure 40.22 confirms that this in fact is the case.

Notice in the figure there is mention that output DC level is 0.424 V which is pretty low. Let us do a quick sanity check to confirm the DC operating point. We want to verify that the resistor current equals the MOSFET one. The former is simply

$$i_R = \frac{V_{CC} - 0.424}{100000} = 46\,\mu\text{A} \tag{40.39}$$

To evaluate the MOSFET current we need first to know whether it is in the triode region or the saturation one. Since the gate voltage is at 1 V and $V_t = 0.7\,\text{V}$, we have $V_{GS} - V_t = 0.3\,\text{V}$;

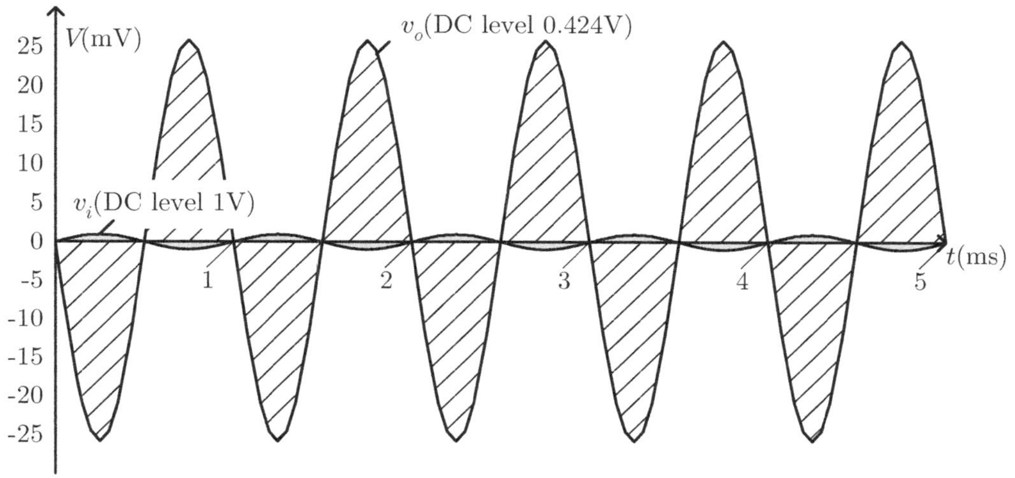

Fig. 40.22 Transient waveforms for $v_i = 1\,\text{V}$ and $R_L = 100\,\text{k}\Omega$

Fig. 40.23 Small signal model of drain loaded MOSFET

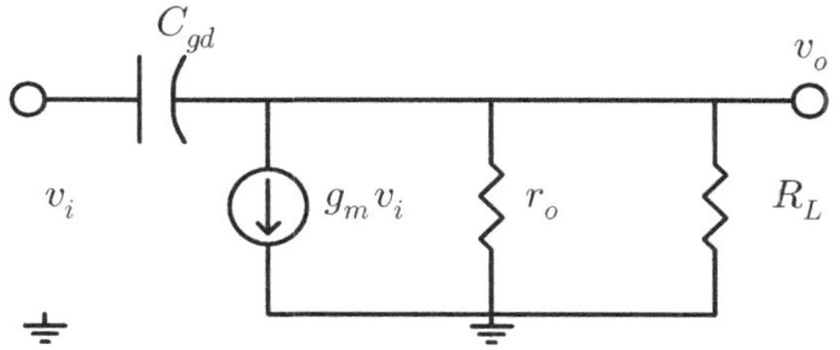

if the drain-to-source voltage V_{DS} is larger than this then we are in saturation; this in fact is the case. Hence we use the saturation current formula (Eq. (40.2))

$$\begin{aligned} I &= \frac{KP}{2}\frac{W}{L}(V_{GS} - V_{th})^2\,[1 + \lambda V_{DS}] \\ &= \frac{100\times 10^{-6}}{2}\times 10\times(0.3)^2\times[1+0.04\times 0.424] \\ &= 46\,\mu\text{A} \end{aligned} \tag{40.40}$$

in agreement with Eq. (40.39). It is critical that we pay attention to the DC operating point. Even though our primary focus is AC, which is the small signal model, we have to remember that the small signal model parameters (gain, transconductance, output impedance, ...) are dependent on the DC operating point, which is mimicked by the device terminal DC voltages and currents.

AC Simulations Showing Bandwidth and Unity Gain We know from DC and transient analysis that max gain is around 25 for the condition $v_i = 1\,\text{V}$ and $R_L = 100\,\text{k}\Omega$. What remains is to assess how this gain behaves versus frequency. The small signal model is shown in Fig. 40.23. We already know the load resistance $R_L = 100\,\text{k}\Omega$. The gate-to-drain cap, as mentioned before, is a ratio of the total gate capacitance, and depends on bias conditions; its value may vary between simulators, but let's assume it comes out 2 fF (just for illustration): $C_{gd} = 2\,\text{fF}$. To get output impedance we use Eq. (40.4) which gives

$$\begin{aligned} r_o^{-1} &= \frac{KP}{2}\frac{W}{L}(V_{GS} - V_{th})^2\lambda \\ &= \frac{100\times 10^{-6}}{2}\times 10\times(0.3)^2\times 0.04 = 1.8\times 10^{-6} \\ r_o &= 0.56\,\text{M}\Omega \end{aligned} \tag{40.41}$$

Finally to get transconductance we use Eq. (40.3) which gives

$$g_m = KP\frac{W}{L}(V_{GS} - V_{th})\,(1 + \lambda V_{DS})$$
$$= 100\times10^{-6}\times10\times(0.3)\times(1+0.04\times0.424)$$
$$= 0.3\,\text{mA/V} \quad (40.42)$$

To locate the first pole we can leverage results from Sect. 40.7 simply by opening C_L and replacing r_o with $r_o||R_L$. The location of the first pole is going to be at

$$\text{location of first pole} = \omega$$
$$= \frac{1}{r_o||R_L\times C_{gd}} = \frac{1}{84\,\text{k}\Omega\times2\,\text{fF}}$$
$$= 6\,\text{G rad/s} \quad (40.43)$$

This is confirmed in Fig. 40.24. From the same section we can also locate the first zero:

$$\text{location of zero} = \frac{g_m}{C_{gd}} = \frac{0.3\,\text{m A/V}}{2\,\text{fF}} = 150\,\text{G rad/s} \quad (40.44)$$

which is again confirmed in Fig. 40.24.

Summary of Drain Loaded MOSFET For the drain loaded MOSFET we were able to adjust gate bias and load resistance to maximize gain. The gain was confirmed by running (low-frequency) transient simulations. Then, and based on earlier derivations, we were able to predict the location of the pole and zero of the transfer function; and those were verified by running SPICE AC simulations.

40.9 Summary

In this quick and short treatment we have shown how the rather complex MOSFET device can be linearized in terms of a small signal model, comprised of capacitors, resistors, and controlled sources. Once in this form, we are able to derive the input/output transfer function in the frequency domain, and we are able to predict the location of the poles and zeroes. Transient simulations were also done to confirm DC and AC analysis. In the end, we are able to apply all the tools developed in the text, from spectral to convolution ones, to this area of electrical engineering. And this is just the beginning; much more elaborate analysis can be done, and to more complex circuits, albeit the very core concepts remain the same. Put another way, once we have converted the circuit into an *RLC* one, with controlled sources, the resulting KVL/KCL yield equations that can be solved via spectral, convolution, and numerical techniques, just like was done in most prior chapters.

40.10 Problems

1. A MOSFET has a width/length $10\,\mu\text{m}/1\,\mu\text{m}$, threshold voltage 0.7 V, and $KP = 100\frac{\mu\text{A}}{\text{V}^2}$. Plot the *IV* characteristics for the two cases: $\lambda = 0$ and $\lambda = 0.2\,\text{V}^{-1}$. Compare to SPICE; see sample solution in Fig. 40.25.
2. Consider the MOSFET circuit shown in Fig. 40.26. The MOSFET has width/length $10\,\mu\text{m}/1\,\mu\text{m}$; $V_{th} = 0.7$, $KP = 100\frac{\mu\text{A}}{\text{V}^2}$, and $\lambda = 0.1\,\text{V}^{-1}$. Plot the output current versus V_{DD} for $V_G = 3\,\text{V}$ and compare to SPICE; see sample solution in Fig. 40.27.
3. Consider the MOSFET circuit shown in Fig. 40.28. The MOSFET has width/length $10\,\mu\text{m}/1\,\mu\text{m}$; $V_{th} = 0.7$, $KP = 100\frac{\mu\text{A}}{\text{V}^2}$, and $\lambda = 0.1\,\text{V}^{-1}$. The cap is initially uncharged; i.e., output voltage is zero. At time zero, the gate is toggled from 0 to 5 V, while the drain is always at 5 V. We want to find output voltage. While there are many ways about accomplishing this consider the following simple algorithm. At time zero, output voltage is zero, and hence V_{DS} is 5 V. Compute the corresponding current I_{D1} and dump that on the cap; hence output voltage would be $I_{D1} \times \Delta t\frac{1}{C}$, where $\Delta t = 0.1\,\text{ms}$ (to be changed later). Now, with the new output voltage, which is V_S, compute the new I_{D2} current, corresponding to this new V_{GS} and

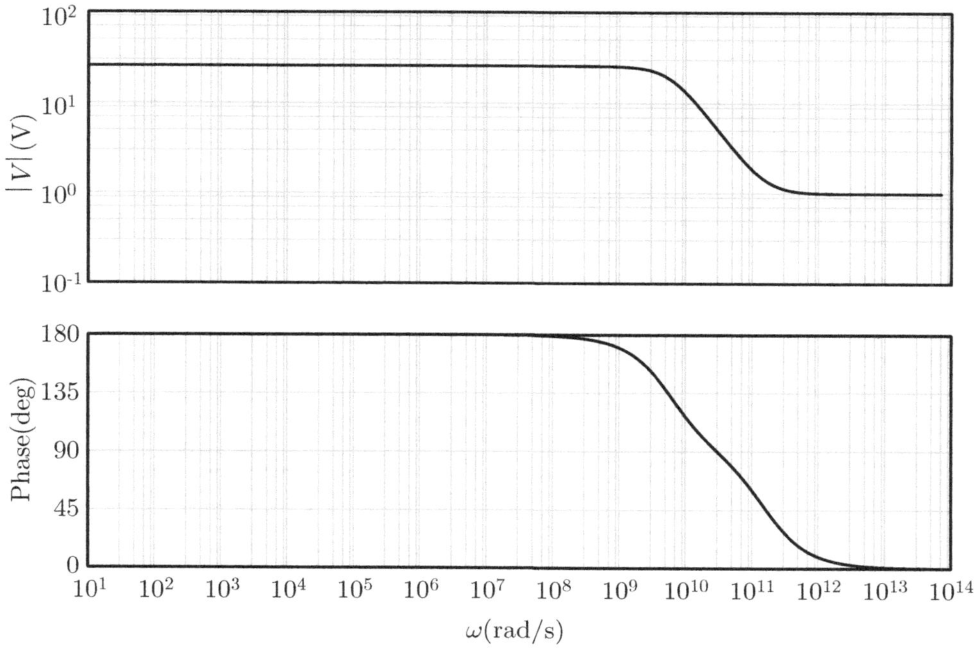

Fig. 40.24 AC gain vs angular frequency for drain load MOSFET

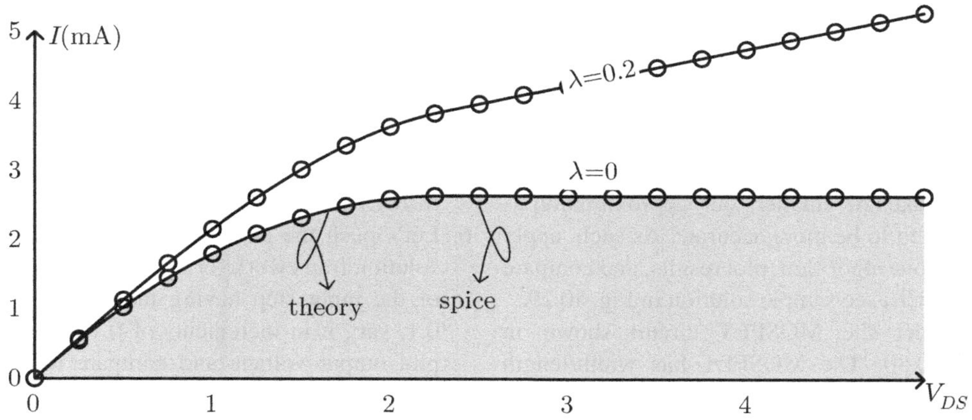

Fig. 40.25 Sample solution to Problem 1

Fig. 40.26 Statement to Problem 2

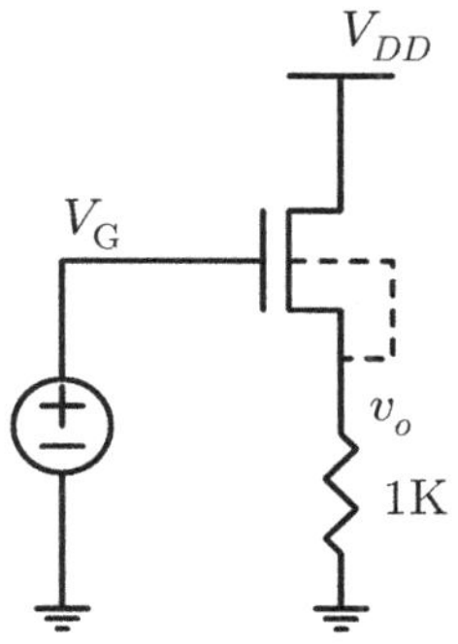

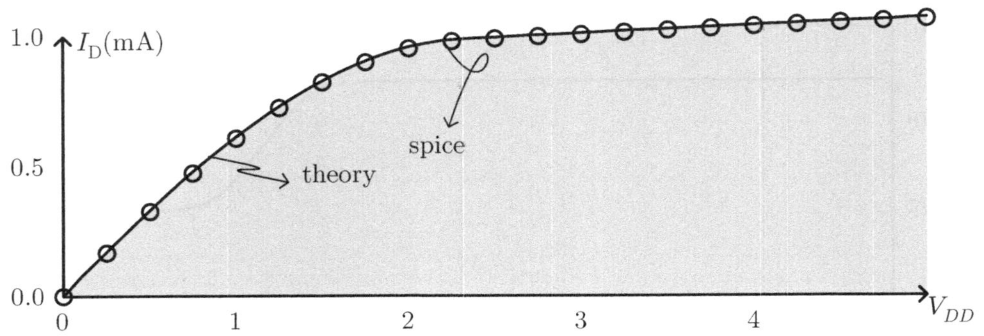

Fig. 40.27 Sample solution to Problem 2

Fig. 40.28 Statement to Problem 3

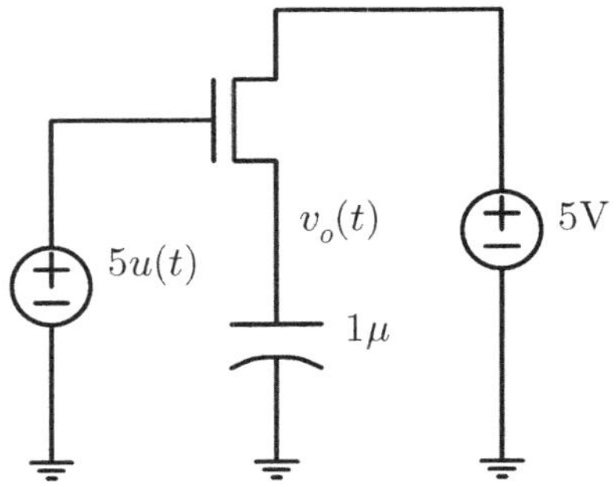

V_{DS}, at time $t = t + \Delta t$. Again dump this on the cap, which would now have the new voltage $(I_{D1} + I_{D2}) \times \Delta t \frac{1}{C}$ and so forth. In the end we would have for each time step the corresponding I_D and output voltage. As we make Δt smaller, our approximation is expected to be more accurate. As such, apply the above algorithm, plot results, and compare to SPICE; see sample solution in Fig. 40.29.

4. Consider the MOSFET circuit shown in Fig. 40.30. The MOSFET has width/length $10\,\mu$m/$1\,\mu$m; $V_{th} = 0.7$, $KP = 100\frac{\mu A}{V^2}$, and $\lambda = 0.1\,V^{-1}$. Vary the source voltage and plot each of g_m and r_o; include in this case the impact of λ on g_m. See sample solution in same figure.
5. Starting with Sect. 40.5, and knowing the input/output transfer function (as was shown in Fig. 40.7), derive the impulse response. Then integrate that to find the step response. That is, what is output voltage if input voltage assumes a step form, *on top* the DC operating gate voltage. Note: since a full step, on top of the gate voltage, may push the operating point way beyond what was used to create the small signal model (e.g., g_m and r_o), instead of applying a full step, apply one of magnitude 0.1 V; that is toggle V_G from 5.0 to 5.1 V Plot the response output voltage and compare to SPICE; see sample solution in Fig. 40.31.
6. Let's push the envelop a bit more. Take the solution framework for Problem 5 and instead of the input step having magnitude of only 0.1, vary it in increments of 0.1 up to 0.5 V; plot output voltage and compare to SPICE, as shown in sample solution Fig. 40.32. (Note that even with a full 0.5 V step, so that gate voltage jumps from 5 to 5.5 V, the approximate solution, built using operating point around gate voltage of 5 V, is still a very good approximation!)
7. Consider the MOSFET circuit shown in Fig. 40.33. The MOSFET has width/length $10\,\mu$m/$1\,\mu$m; $V_{th} = 0.7$, $KP = 100\frac{\mu A}{V^2}$, and $\lambda = 0.1\,V^{-1}$. Find the DC operating point, then g_m and r_o. Once those are ready, construct the small signal equivalent model and figure the transfer function $v_o(s)/v_i(s)$ where $v_i(s)$

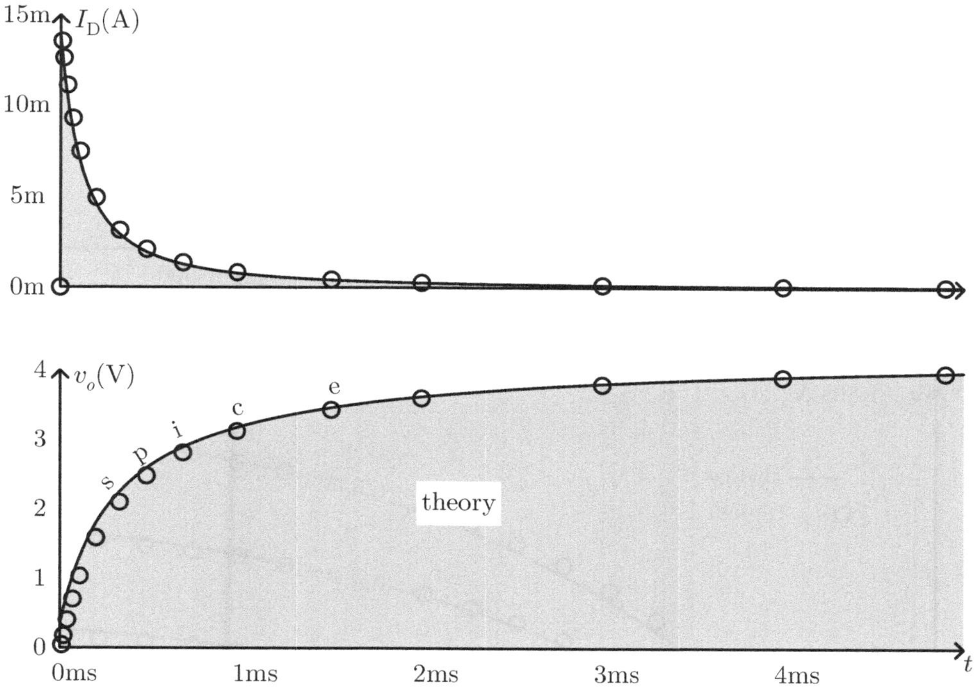

Fig. 40.29 Sample solution to Problem 3

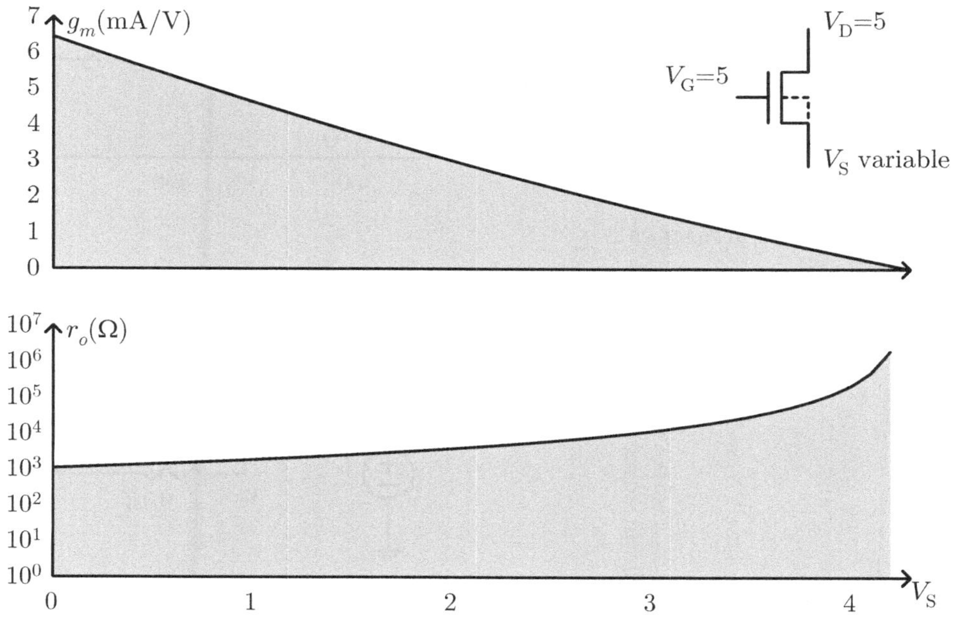

Fig. 40.30 Sample solution for Problem 4

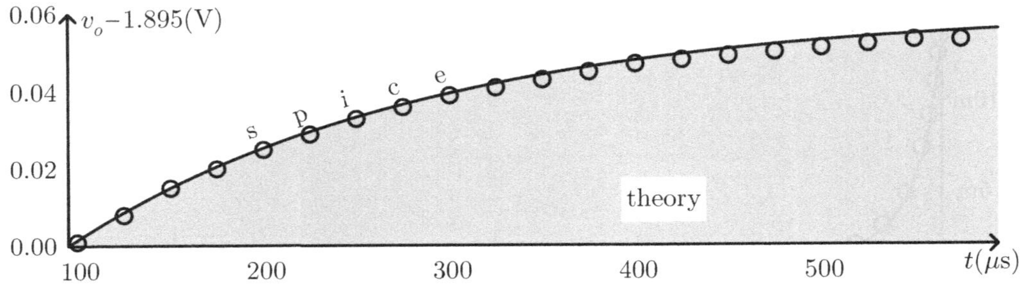

Fig. 40.31 Sample solution to Problem 5

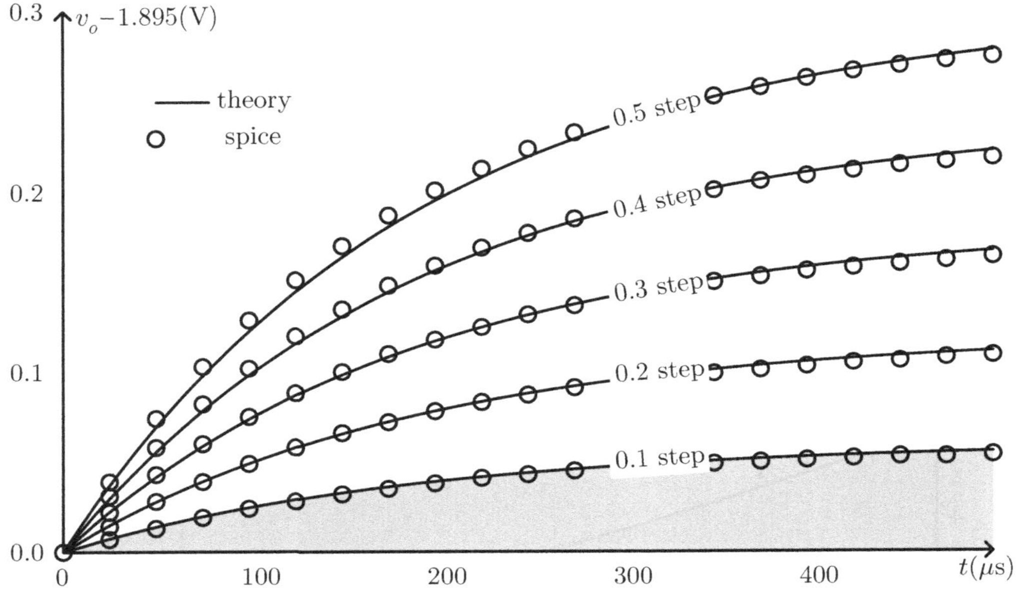

Fig. 40.32 Sample solution to Problem 6

Fig. 40.33 Statement to Problem 7

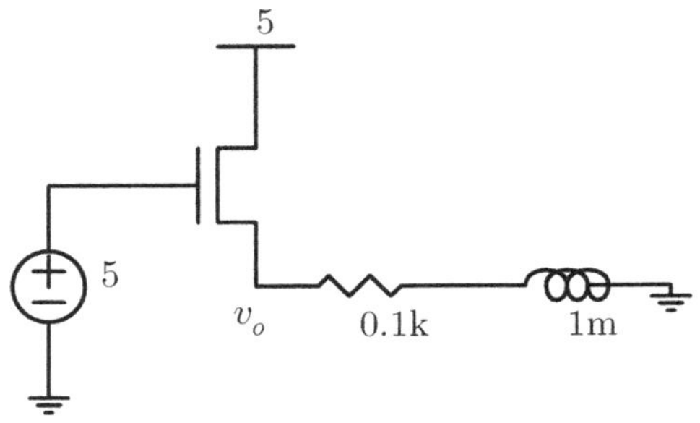

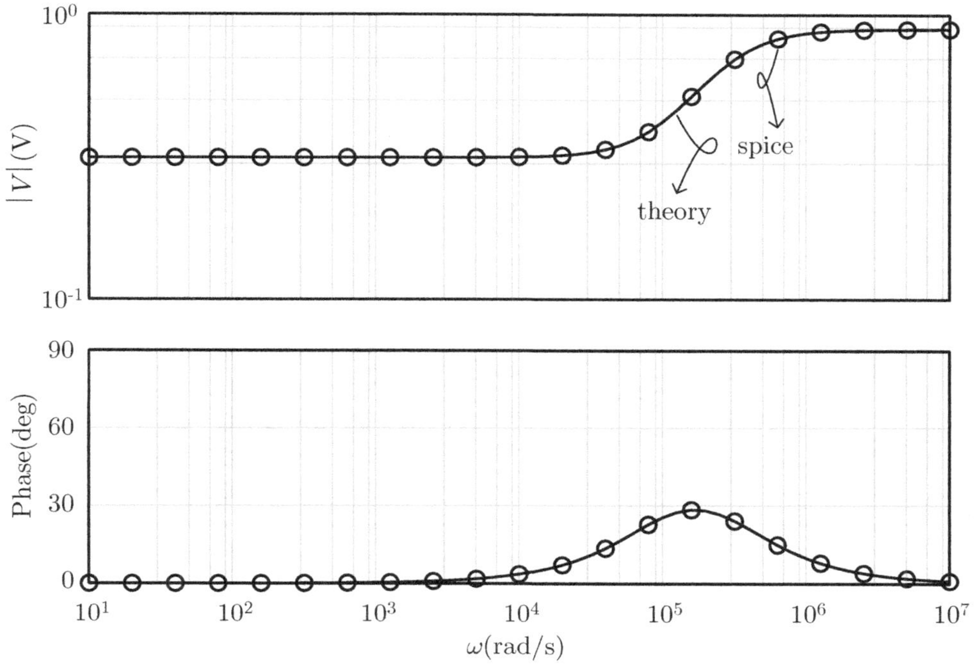

Fig. 40.34 Sample solution to Problem 7

is an AC source applied on top of the DC gate level. Plot results and compare to SPICE; see sample solution in Fig. 40.34.
Answer:

$$V_S{=}0.845,\quad g_m{=}4.891\,\text{mA/V},\quad r_o{=}1.675\,\text{k}\Omega$$

8. Take the transfer function derived in Problem 7 and use it to figure impulse response. Then figure step response; that is, if a step is applied to the gate (on top of the preexisting 5 V DC level), what would be the output voltage change (from the DC operating point)? In order to ease things on the operating point DC levels, we don't want to apply a full step; instead apply a 0.1 step, and observe results. Then try 0.2 and so forth till 0.5. For each case plot output voltage and compare to SPICE; see sample solution in Fig. 40.35.

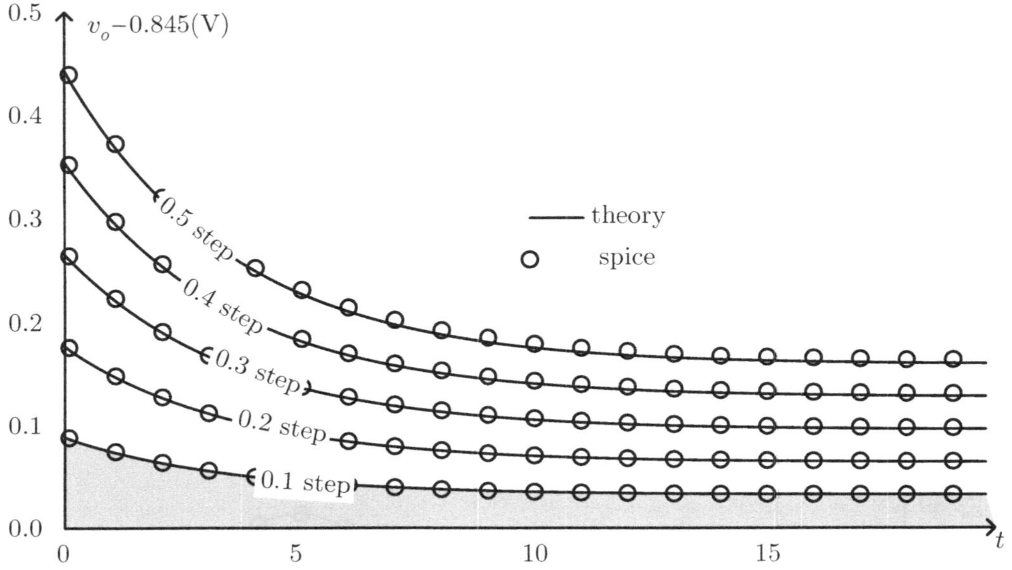

Fig. 40.35 Sample solution to Problem 8

41 Op-Amp Filters

41.1 Introduction

The operational amplifier filters differ from the *RLC* filters in the sense that the former is an "active" while latter is a "passive" filter. Intrinsically the op-amp is an involved multi-transistor circuit with nonlinear effects and with certain frequency dependence. But if the amp can be abstracted as a box and if the nonlinearities are small then the resulting setup lends itself nicely to spectral and convolution analysis. The amp is typically characterized by multiple variables, such as gain, bandwidth, and input and output impedance. The exact dependence of gain on frequency is also of special importance. In addition to the amp, there typically is a feedback network that samples the output and feeds it back to the amp, and the characteristics of this feedback network are of much relevance. Combined together the treatment of amplifiers and feedback networks greatly expands our capabilities and opens interesting and challenging applications. For reference this chapter relies heavily on Chap. 36 (feedback) and Chap. 40 (transistor modeling).

41.2 The Ideal Amp

In essence an amplifier takes a difference in input voltage and produces a magnified version thereof. That is, if input is v_{i1} and v_{i2} then output is

$$v_o = A(v_{i1} - v_{i2}) \tag{41.1}$$

This is shown on the left side of Fig. 41.1. For the special case where one input is grounded we get

$$\boxed{v_o = Av_i} \tag{41.2}$$

which is shown on the right side of Fig. 41.1. The op-amp, in the end, derives power from a set of finite rails; so, above equation is valid only for a finite (limited) set of input voltage.

The ideal amp is characterized by two requirements:

1. The amp gain A is large (infinite).
2. Input currents are zero.

Both requirements are relatively reasonable; first is achieved by large gain, and second is a statement that no DC current flows into the input of the amp, since input terminals are typically tied to gates of MOSFETs. How close a real amp from an ideal one is debatable, but for a starter and for reference at least we want to be able to analyze the ideal one.

41.3 Complete Feedback

The above analysis has the output decoupled from the input; i.e., *output does not impact input.* Now, if we were to tie the output node to the input one, as shown in Fig. 41.2, we would get a differ-

F. Badrieh, *Spectral, Convolution and Numerical Techniques in Circuit Theory*,
https://doi.org/10.1007/978-3-319-71437-0_41

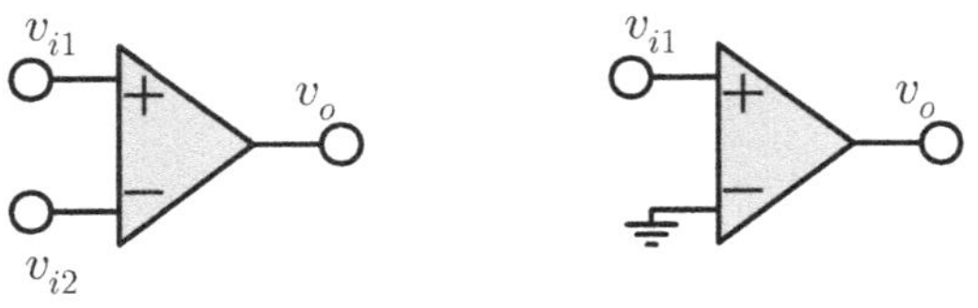

Fig. 41.1 Ideal amp with two inputs (left) and one input grounded (right)

Fig. 41.2 Complete feedback from output to input

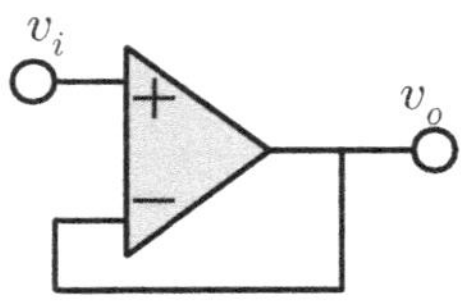

ent scenario. We start with the basic requirement that

$$v_o = A(v_{i+} - v_{i-}) \tag{41.3}$$

But v_{i-} is nothing more than v_o; hence

$$v_o = A(v_i - v_o) \tag{41.4}$$

Rearrange:

$$v_o(1 + A) = Av_i \tag{41.5}$$

from which we finally get

$$\boxed{v_o = \frac{A}{1 + A} v_i} \tag{41.6}$$

For the case where $A \gg 1$ we get

$$v_o = v_i \tag{41.7}$$

That is, output equals input.

41.4 Partial Feedback

The prior section had the whole output voltage fed back to input; but we could also feed only a portion of output, as shown in Fig. 41.3.

Fig. 41.3 Partial feedback from output to input

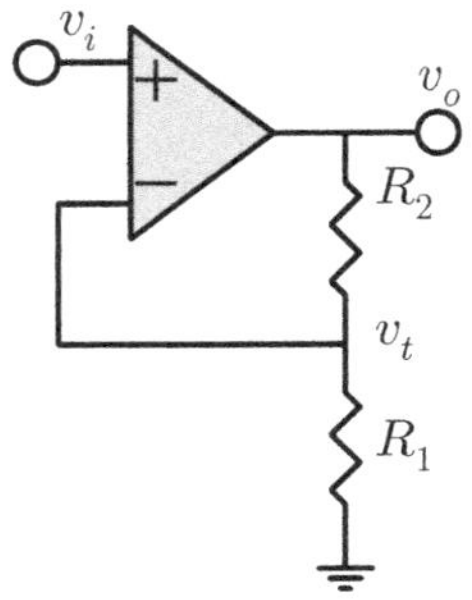

Again we start with basic amp equation

$$v_o = A(v_i - v_t) \tag{41.8}$$

But v_t is nothing more than

$$v_t = \frac{v_o}{R_1 + R_2} R_1 \tag{41.9}$$

(Remember, no current is allowed into the input of the amp, which means current through R_1 is the same as that through R_2.) If we define

$$\beta = \frac{R_1}{R_1 + R_2} \tag{41.10}$$

then we get

$$v_o = A(v_i - \beta v_o) \tag{41.11}$$

Rearrange and solve for v_o to get

$$\boxed{v_o = \frac{A}{1 + \beta A} v_i} \tag{41.12}$$

Notice the two interesting limits:

1. $\beta = 0$: in this case above equation reduces to

$$v_o = Av_i \tag{41.13}$$

which is nothing more than the open loop operation; that is, if we feed nothing back to input, then we regain the open loop scenario.

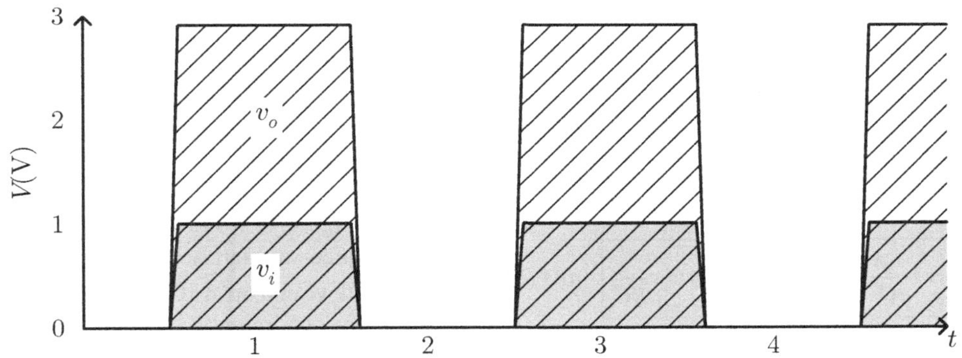

Fig. 41.4 Simulation results for schematics in Fig. 41.3

2. $\beta = 1$: in this case above equation reduces to

$$v_o = \frac{A}{1+A} v_i \tag{41.14}$$

which is nothing more than the close-loop case with full feedback, in agreement with Eq. (41.6)!

For the case when $A \gg 1$, then our reference Eq. (41.12) reduces to

$$\boxed{v_o = \frac{1}{\beta} v_i, \qquad A \gg 1} \tag{41.15}$$

If we plug back for β we get

$$v_o = \left[1 + \frac{R_2}{R_1}\right] v_i \tag{41.16}$$

That is, the output is a scaled up version of the input. A sample run with $R_1 = 1$ and $R_2 = 2$ is shown in Fig. 41.4. Notice that as expected, output is 3 times that of input.

41.5 Low-Pass Feedback

So far we have dealt with three kinds of feedback: none, total, or partial, but in all cases the feedback did not depend on frequency. Consider the setup shown in Fig. 41.5 where a low-pass filter is used as the feedback circuit. It is called alow-pass

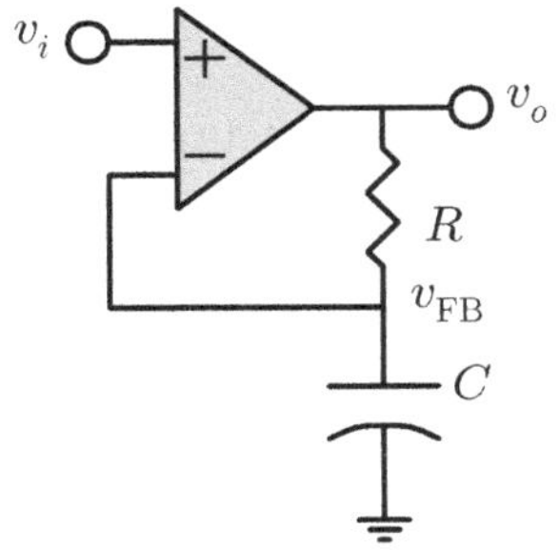

Fig. 41.5 Low-pass feedback from output to input

because if the operating frequency is low, then the cap acts as an open and all of output voltage is fed back to the amp; i.e., feedback signal passes at low frequency. At high frequency, on the other hand, the cap acts like short and the inverting terminal of the amp is effectively tied to ground; i.e., no feedback takes place.

The feedback voltage is the output current times the cap impedance:

$$v_{FB} = \frac{v_o}{R + \frac{1}{sC}} \frac{1}{sC} = \frac{v_o}{RC} \frac{1}{s + \frac{1}{RC}} \tag{41.17}$$

Notice when figuring output current we assumed that resistor current equals capacitor current; that is, no current enters the inverting terminal of the amp! The output voltage is then

$$\begin{aligned} v_o &= A(v_i - v_{FB}) \\ &= A\left(v_i - a v_o \frac{1}{s+a}\right), \quad a = \frac{1}{RC} \end{aligned} \tag{41.18}$$

Rearrange

$$v_o\left(1+A\frac{a}{s+a}\right)=Av_i \Rightarrow v_o\frac{s+a+Aa}{s+a}=Av_i \Rightarrow v_o\frac{s+a(1+A)}{s+a}=Av_i \qquad (41.19)$$

The voltage transfer function is then

$$\boxed{\frac{v_o}{v_i}=\frac{A(s+a)}{a(1+A)+s}, \qquad a=\frac{1}{RC}} \qquad (41.20)$$

Notice that at low frequency we get the limit

$$\left.\frac{v_o}{v_i}\right|_{s=0}=\frac{Aa}{a(1+A)}=\frac{A}{1+A} \qquad (41.21)$$

in agreement with Eq. (41.14), since the cap there is open and we have full feedback. On the other hand the high frequency limit is

$$\left.\frac{v_o}{v_i}\right|_{s\to\infty}=\frac{As}{s}=A \qquad (41.22)$$

in agreement with the open loop case, since the cap there is short and we don't have any feedback. Figure 41.6 shows sample simulations results. Notice we start around 1, since $\frac{A}{A+1}=\frac{100}{101}\sim 1$ and we end up at 100 since $A=100$. The phase starts at 0°, ramps to 90°, and then back to 0°. Whenever the magnitude is flat we get zero phase; on the other hand when the magnitude is ramping up at 20 dB/dec we get positive 90°.

Let's reflect on what we have accomplished. We started with an abstract concept of gain and then added in a frequency dependent network. Now we have a system with a frequency dependent transfer function which fits very nicely with what we have done throughout most of the text. That is, once we boil down the problem into a transfer function, then we can apply spectral techniques, do manipulations in the frequency domain, and then use inverse transform to go back to the time domain. We could also figure impulse response (step one, ...) and use convolution

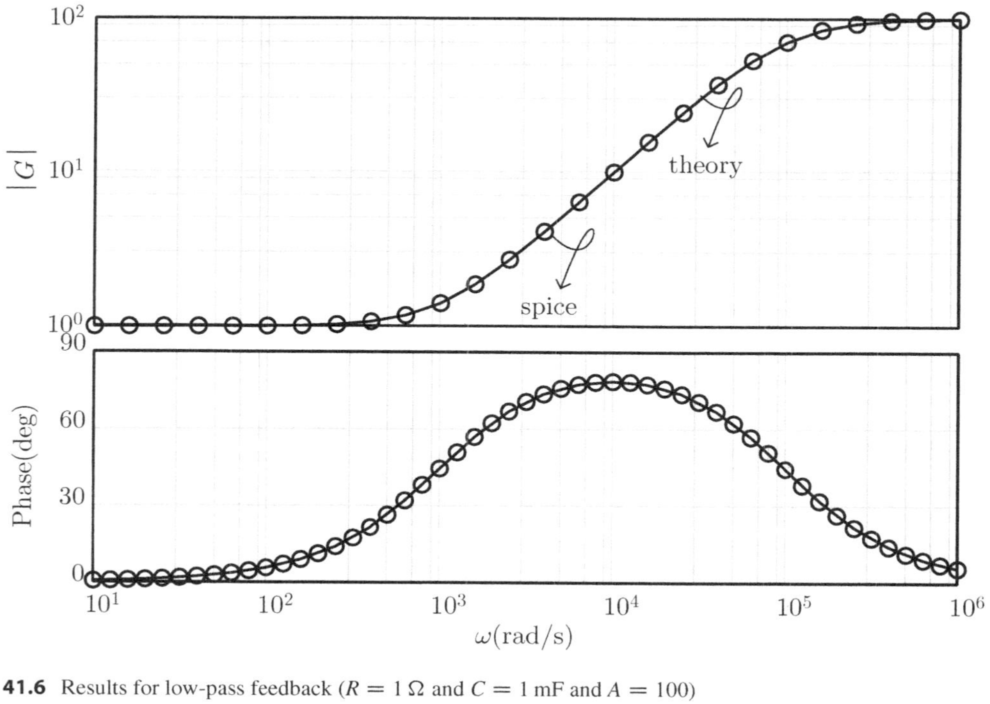

Fig. 41.6 Results for low-pass feedback ($R = 1\,\Omega$ and $C = 1$ mF and $A = 100$)

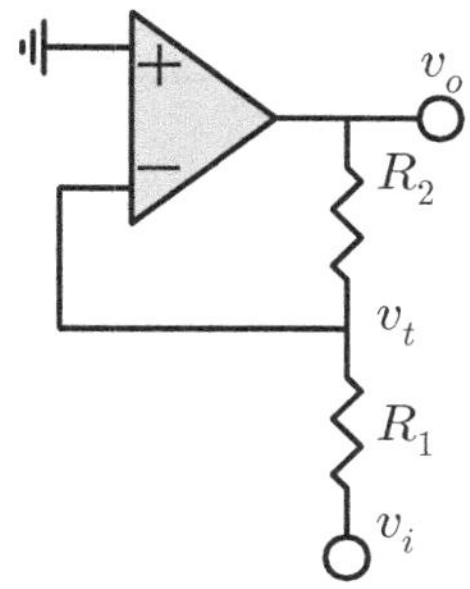

Fig. 41.7 Inverting input configuration

as an alternative. Either way we have a solution to the problem.

41.6 Inverting Amp

A special feedback case is shown in Fig. 41.7. Notice that the positive input side has been grounded. Starting with the basic amp equation we get

$$v_o = A(0 - v_t) \tag{41.23}$$

But v_t is given by voltage division as

$$\begin{aligned} v_t &= v_i + \frac{v_o - v_i}{R_1 + R_2} R_1 \\ &= v_i \left[1 - \frac{R_1}{R_1 + R_2}\right] + v_o \frac{R_1}{R_1 + R_2} \\ &= v_i \frac{R_2}{R_1 + R_2} + v_o \frac{R_1}{R_1 + R_2} \end{aligned} \tag{41.24}$$

Plugging back into Eq. (41.23) we get

$$v_o = -A \left[v_i \frac{R_2}{R_1 + R_2} + v_o \frac{R_1}{R_1 + R_2}\right] \tag{41.25}$$

Rearranging we get

$$v_o \left[1 + \frac{AR_1}{R_1 + R_2}\right] = -v_i \frac{AR_2}{R_1 + R_2} \tag{41.26}$$

$$v_o \frac{R_1(A + 1) + R_2}{R_1 + R_2} = -v_i \frac{AR_2}{R_1 + R_2} \tag{41.27}$$

or

$$\boxed{v_o = -\frac{AR_2}{R_1(A + 1) + R_2} v_i} \tag{41.28}$$

For the case $A \gg 1$ we get

$$\boxed{v_o = -\frac{R_2}{R_1} v_i, \qquad A \gg 1} \tag{41.29}$$

That is, output is a *negative* scaled version of input. A sample result with $R_1 = 1\,\Omega$, $R_2 = 2\,\Omega$, and $A = 100$ is shown in Fig. 41.8. Notice that as expected, output is the negative of (twice) the input.

41.7 Simplified Method to Deal with Ideal Op-Amps

For the case of very large gain, it can be shown easily that all the above derived equations can be figured easily if we assume that the input terminals of the amp are at the same potential. In other words, with large amp gain, the difference between the two input terminals of the amp must by very small (ideally zero). If we use this assumption, then we can greatly simplify our analysis. For example let's reconsider Sect. 41.4 which dealt with partial feedback. If we assume $v_t = v_i$, then output current would be

$$i = \frac{v_i}{R_1} \tag{41.30}$$

Accordingly output voltage would be v_i plus this current times R_2:

$$v_o = v_i \left[1 + \frac{R_2}{R_1}\right] \tag{41.31}$$

which is exactly Eq. (41.16).

41.8 Integrator Amp

Going back to our inverting amp (Fig. 41.7) if we change R_2 to a cap we get the configuration shown in Fig. 41.9. Following the idealized assumptions in the last section we arrive at conclusion that

$$v_t = 0 \tag{41.32}$$

Then current across the resistor (and the cap) is simply

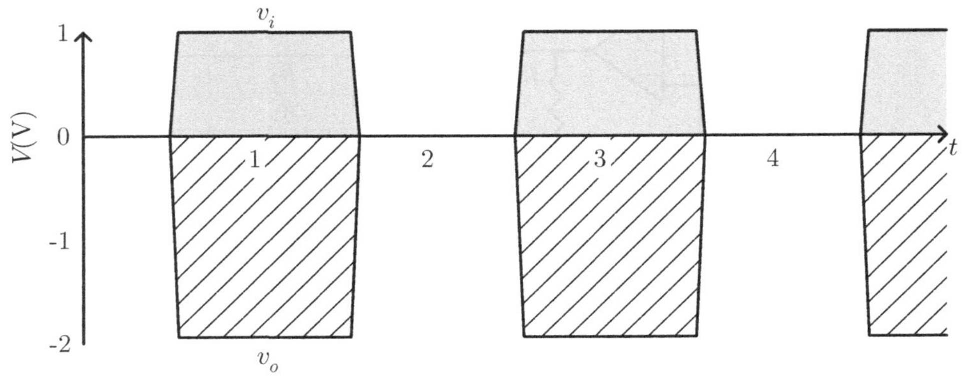

Fig. 41.8 Simulation results for schematics in Fig. 41.7

Fig. 41.9 Amp configured as an integrator

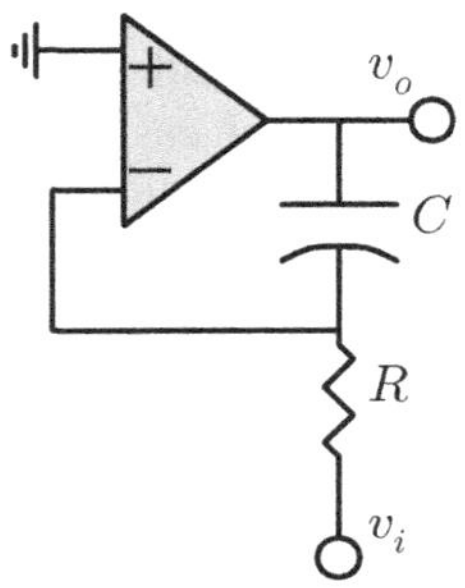

$$i = -\frac{v_i}{R} \tag{41.33}$$

The output voltage would then simply be this current times impedance of cap:

$$\boxed{V_o(s) = -\frac{V_i(s)}{R}\frac{1}{sC}} \tag{41.34}$$

In the time domain this translates to simple integration

$$\boxed{v_o(t) = -\frac{1}{RC}\int_0^t v_i(\tau)d\tau} \tag{41.35}$$

That is, the output voltage is (negative) the time integral of input voltage! A first sample application is shown in Fig. 41.10 with $R = 1\,\Omega$ and $C = 1\,\text{F}$ where input is a step function. Notice that output is simply a (negative) ramp function, which is the integral of the step; the slope is $1/RC$ which is 1 here. Another sample application is for a pulse input, as shown in Fig. 41.11. Again, output is simply (negative) the time integral of input. If we can do integration then shouldn't we be able to do differentiation?

41.9 Differentiator Amp

If we swap the R and C in Fig. 41.9 we get the configuration shown in Fig. 41.12. Again following the idealized assumptions in Sect. 41.7 we arrive at conclusion that

$$v_t = 0 \tag{41.36}$$

The current across the cap is

$$i_C(t) = -C\frac{dv_i}{dt} \tag{41.37}$$

In the frequency domain we get (in accordance with Eq. (16.23) and assuming initial voltage is zero)

$$I_C(s) = -sCV_i(s) \tag{41.38}$$

This same current goes through the resistor and output voltage is simply

$$\boxed{V_o(s) = -sRCV_i(s)} \tag{41.39}$$

In the time domain we get

$$\boxed{v_o(t) = -RC\frac{dv_i}{dt}} \tag{41.40}$$

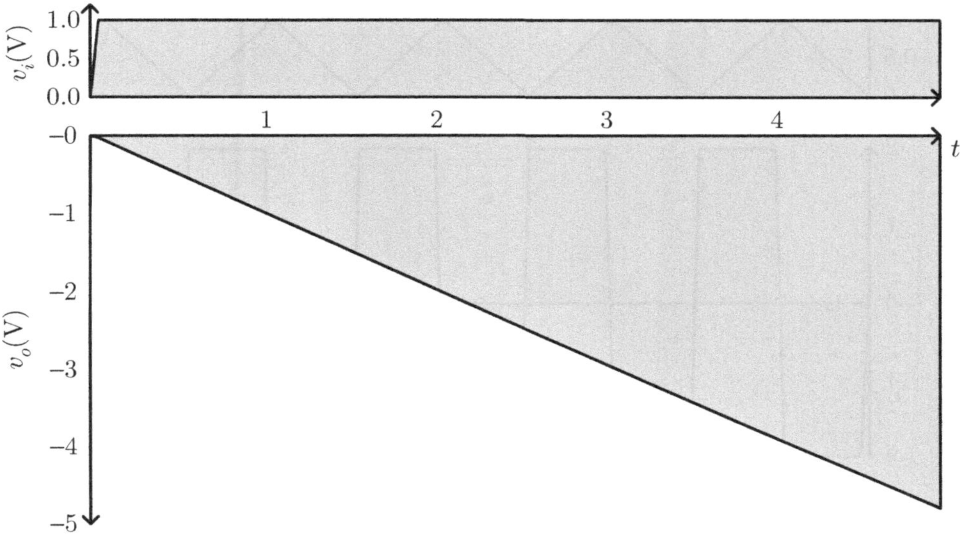

Fig. 41.10 Integration of step input

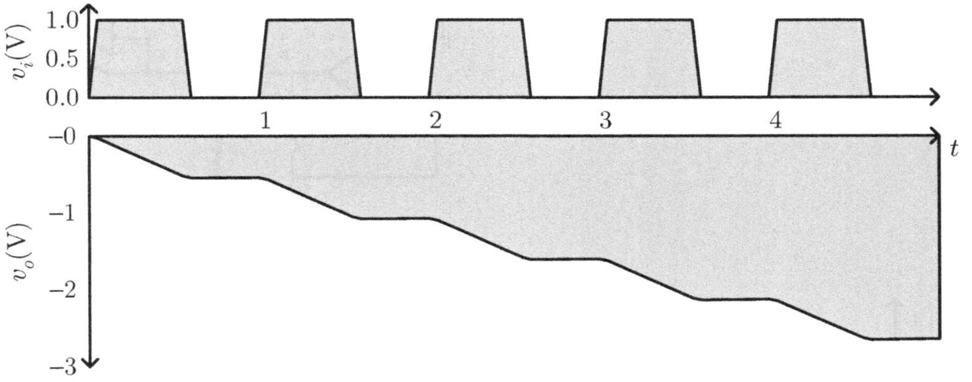

Fig. 41.11 Integration of pulse input

As a sample application, consider the input voltage of zigzag shape, with ramp rate 2. Assuming $R = 1\,\Omega$ and $C = 1\,\text{F}$ we get the results in Fig. 41.13. Clearly the derivative of a triangle comes out a pulse. Notice the inversion in sign!

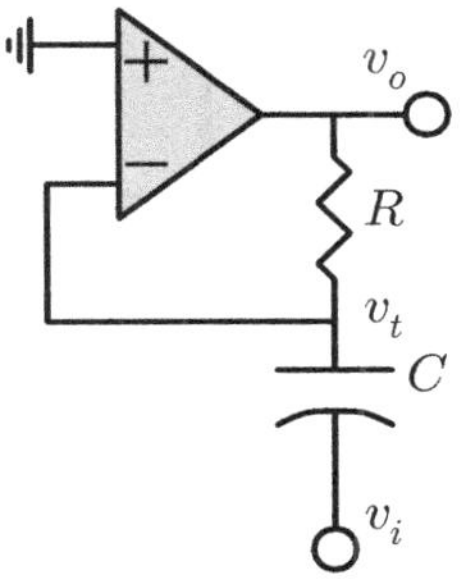

Fig. 41.12 Amp configured as a differentiator

41.10 Second Order Differentiator Amp

By same logic, if we cascade two differentiating amps as shown in Fig. 41.14 we ought to get the second derivative of the input. As a sample application we apply an input pattern as shown at the top of Fig. 41.15 and get second derivative as shown at the bottom of the same figure. Looking

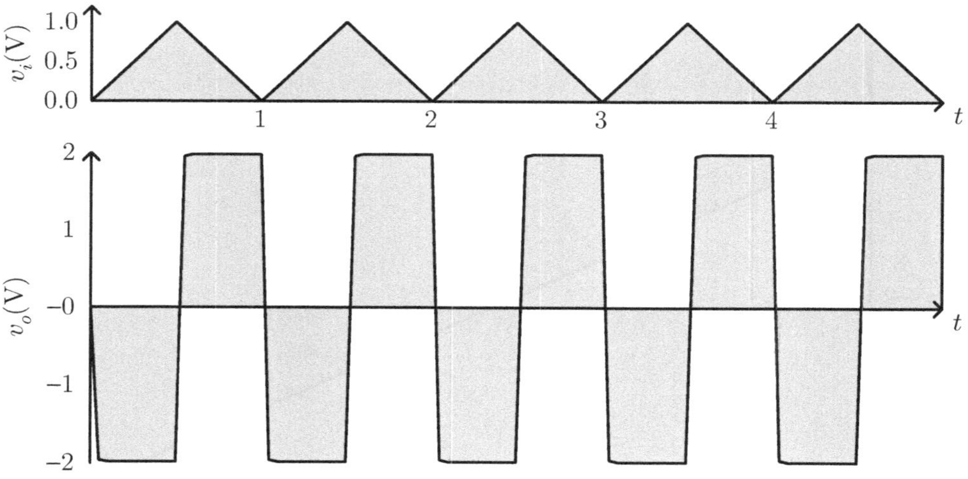

Fig. 41.13 Differentiation of zigzag input

Fig. 41.14 Amp configured as second order differentiator

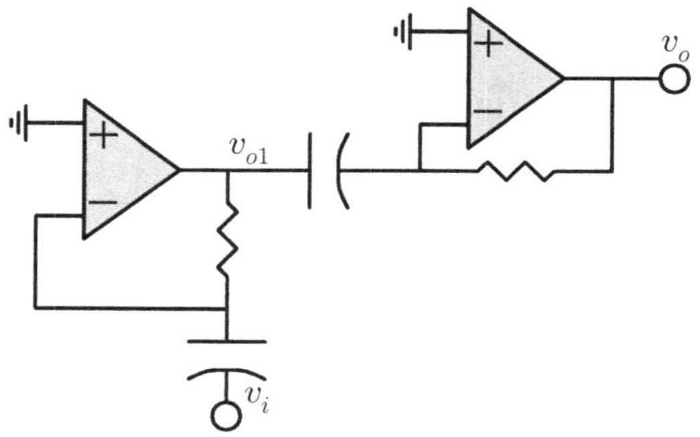

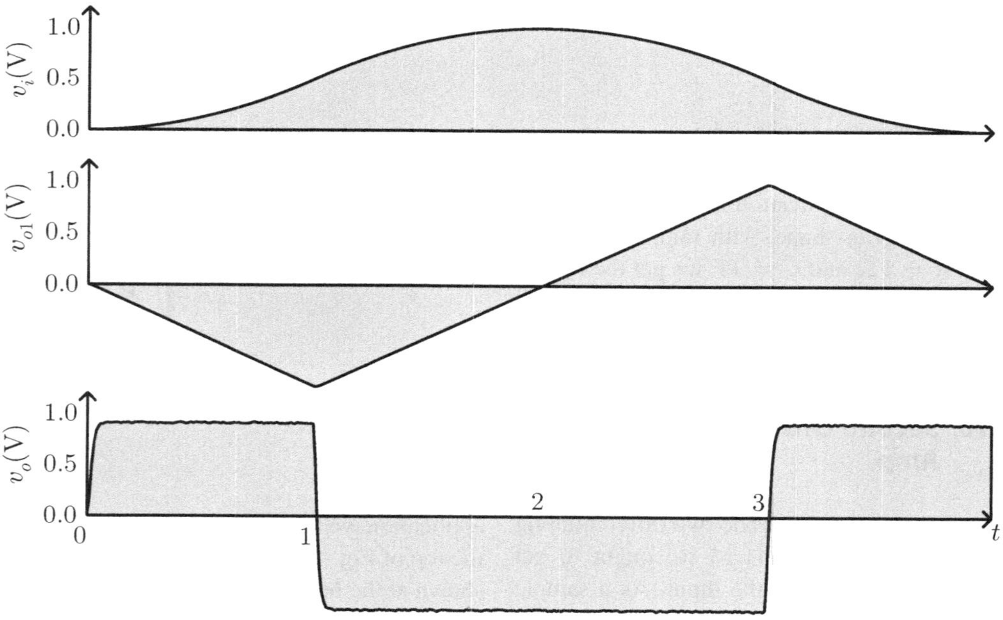

Fig. 41.15 Second order differentiation of input signal

closer at the figure we conclude that $v_o(t)$ is in fact the second derivative of $v_i(t)$. Notice that in this case there is *no* sign inversion because two negatives give a positive! By the same token we can cascade more amps to get even higher derivatives.

41.11 Differential Amplifier Small Signal Model

The amplifiers presented so far have been assumed ideal in the sense we did not worry how they were actually built. In reality amplifiers are typically built from devices (MOSFETs or BJTs). And then the amplification is really the small signal amplification, around an operating point. Without delving too much in details, a typical amplifier may have two stages: a differential input one, and an output one. The small signal model of these two stages may look like that in Fig. 41.16. The overall setup should look familiar having studied small signal models of MOSFETs in Chap. 40. Our interest for now is not how to derive this topology, but rather how to solve it. In particular we want to find output voltage as a function of input one, which we define as the frequency dependent gain (of this two stage amplifier). The open loop transfer function can be obtained as follows. The first stage gain is

$$\begin{aligned} v_1(s) &= -g_{m1}v_iR_1||C_1 = -g_{m1}v_i\frac{R_1\frac{1}{sC_1}}{R_1+\frac{1}{sC_1}} = -g_{m1}v_i\frac{R_1}{1+sR_1C_1} \\ &= -g_{m1}v_i\frac{1}{C_1}\frac{1}{s+\frac{1}{R_1C_1}} = -g_{m1}v_i\frac{1}{C_1}\frac{1}{s+a} \end{aligned} \tag{41.41}$$

so that

$$\frac{v_1(s)}{v_i(s)} = -g_{m1}\frac{1}{C_1}\frac{1}{s+a}, \qquad a = \frac{1}{R_1C_1} \tag{41.42}$$

Similarly the second stage gain is given by

$$\frac{v_o(s)}{v_1(s)} = -g_{m2}\frac{1}{C_2}\frac{1}{s+b}, \qquad b = \frac{1}{R_2C_2} \tag{41.43}$$

Total gain is then

$$\boxed{\frac{v_o(s)}{v_i(s)} = g_{m1}g_{m2}\frac{1}{C_1C_2}\frac{1}{s+a}\frac{1}{s+b}} \tag{41.44}$$

We can expand using partial fractions and the result is

$$\frac{v_o(s)}{v_i(s)} = \frac{g_{m1}g_{m2}}{C_1C_2(b-a)}\left[\frac{1}{s+a} - \frac{1}{s+b}\right] \tag{41.45}$$

We expand $b-a$ as

$$b-a = \frac{1}{R_2C_2} - \frac{1}{R_1C_1} = \frac{R_1C_1 - R_2C_2}{R_1R_2C_1C_2} \tag{41.46}$$

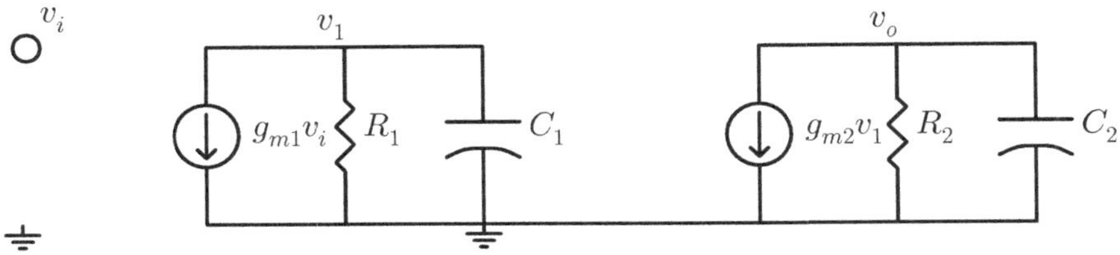

Fig. 41.16 Small signal model for op-amp

Plugging this back into Eq. (41.45) we get

$$\frac{v_o(s)}{v_i(s)} = g_{m1}g_{m2}\frac{R_1R_2}{R_1C_1 - R_2C_2}\left[\frac{1}{s+\frac{1}{R_1C_1}} - \frac{1}{s+\frac{1}{R_2C_2}}\right] \tag{41.47}$$

Notice that the DC gain is

$$\left.\frac{v_o(s)}{v_i(s)}\right|_{s=0} = g_{m1}g_{m2}R_1R_2 \tag{41.48}$$

Notice also that we have two poles:

$$\text{First pole at } \omega = \frac{1}{R_1C_1}, \text{ second pole at } \omega = \frac{1}{R_2C_2} \tag{41.49}$$

Initially the gain drops 20 dB/decade; once the second pole is reached the gain drops at 40 dB/decade. Recall

$$\mathrm{dB}(x) = 20\log(x) \tag{41.50}$$

The open loop transfer function of the amplifier is shown in Fig. 41.17. This particular plot had

$$g_{m1} = 0.1, g_{m2} = 0.1, R_1 = 1\,\mathrm{k}, C_1 = 1\,\mu, R_2 = 2\,\mathrm{k}, C_2 = 5\,\mathrm{n} \tag{41.51}$$

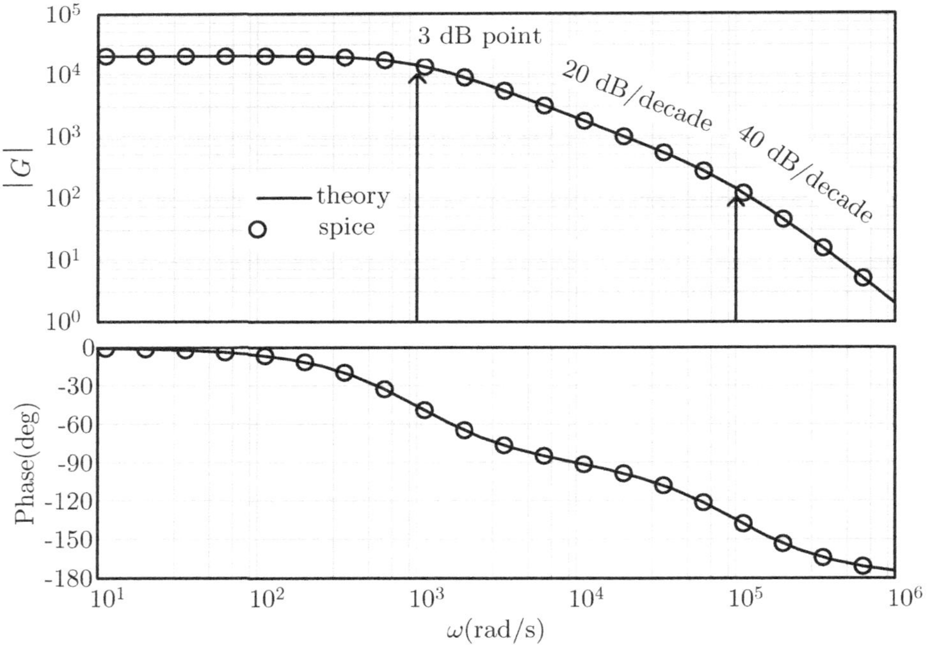

Fig. 41.17 Open loop response of op-amp: top—magnitude, bottom—phase

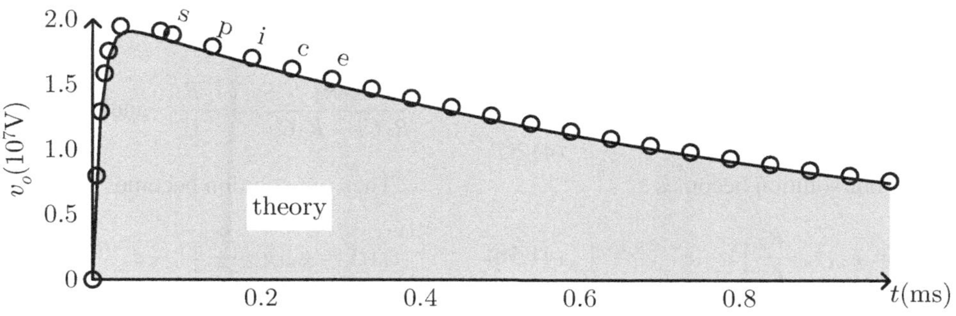

Fig. 41.18 Impulse response of amplifier

Impulse Response Once the transfer function is known, we can find the impulse response of this amplifier, by taking the inverse Laplace transform of Eq. (41.47). We get

$$\boxed{v_o(t) = g_{m1}g_{m2}\frac{R_1R_2}{R_1C_1 - R_2C_2}\left[e^{-t/(R_1C_1)} - e^{-t/(R_2C_2)}\right]} \tag{41.52}$$

The impulse response of the amplifier and comparison to SPICE simulation are shown in Fig. 41.18.

Some Limiting Cases We examine some limits to the impulse response as shown in Eq. (41.52).

- Small R_1: In this case we have the following limit

$$\frac{R_1R_2}{R_1C_1 - R_2C_2} \to -\frac{R_1}{C_2}, \quad \text{and} \quad e^{-t/(R_1C_1)} \to 0 \tag{41.53}$$

Then our solution becomes

$$v_o(t) = g_{m1}g_{m2}\frac{R_1}{C_2}e^{-t/(R_2C_2)} \tag{41.54}$$

That is, the first stage simply takes the impulse and scales it by $g_{m1}R_1$ and the second stage poses as a simple parallel RC with its $1/C_2e^{-t/(R_2C_2)}$ impulse response (scaled by g_{m2} and the scaled impulse response of the first stage).

- Small R_2: In this case we have the following limits:

$$\frac{R_1R_2}{R_1C_1 - R_2C_2} \to \frac{R_2}{C_1}, \quad \text{and} \quad e^{-t/(R_2C_2)} \to 0 \tag{41.55}$$

$$v_o(t) = g_{m1}g_{m2}\frac{R_2}{C_1}e^{-t/(R_1C_1)} \tag{41.56}$$

That is, the first stage would have its typical parallel RC response with value $g_{m1}\frac{1}{C_1}e^{-t/(R_1C_1)}$ and this simply scales with the second resistor R_2 (with multiplication of g_{m2}) since all the current in the second stage would go only through R_2 (it being the smaller impedance path).

- Large R_1: In this case we have the following limits:

$$\frac{R_1 R_2}{R_1 C_1 - R_2 C_2} \to \frac{R_2}{C_1}, \text{ and } e^{-t/(R_1 C_1)} \to 1 \quad (41.57)$$

Then our solution becomes

$$v_o(t) = g_{m1} g_{m2} \frac{R_2}{C_1} \left[1 - e^{-t/(R_2 C_2)}\right] \quad (41.58)$$

That is, with R_1 big, all the impulse current would charge the first cap; so the first voltage becomes $g_{m1}\frac{1}{C_1}$. This would form a step function input to the second stage. We know the step response of a parallel RC circuit is $R(1 - e^{-t/RC})$; hence the product of the two gives the above limit.

- Large R_2: In this case we have the following limits:

$$\frac{R_1 R_2}{R_1 C_1 - R_2 C_2} \to -\frac{R_1}{C_2}, \text{ and } e^{-t/(R_2 C_2)} \to 1 \quad (41.59)$$

Then our solution becomes

$$v_o(t) = g_{m1} g_{m2} \frac{R_1}{C_2} \left[1 - e^{-t/(R_1 C_1)}\right] \quad (41.60)$$

That is, the impulse response to the first stage would be $-g_{m1}\frac{1}{C_1}e^{-t/(R_1 C_1)}$. This gets scaled by g_{m2} and integrated (and divided by C_2) to form the response of the second stage, since the second stage now is simply a cap (with R_2 large). The integration steps are as follows:

$$\begin{aligned} -g_{m1} g_{m2} \frac{1}{C_1 C_2} \int_0^t e^{-\tau/R_1 C_1} d\tau &= g_{m1} g_{m2} \frac{1}{C_1 C_2} R_1 C_1 e^{-\tau/R_1 C_1} \Big|_0^t \\ &= g_{m1} g_{m2} \frac{R_1}{C_2} \left[1 - e^{-t/R_1 C_1}\right] \quad (41.61) \end{aligned}$$

in agreement with Eq. (41.60).

So it appears that our derived answer does reduce to expected ones under some simplifying limits. To recap, we started with the small signal model of the differential amplifier; derived the transfer function tying input voltage to output one; plotted the transfer function and observed the impact of the two poles; went back to the time domain via the impulse response; and finally took some limits of this latter one and rationalized them.

41.12 Output Impedance of Ideal Amp

In anticipation of the nonideal case, let's first derive the ideal amp output impedance. Consider the circuit in Fig. 41.19. The output voltage is simply

$$v_o = A v_i \quad (41.62)$$

We can convey this by using the voltage-dependence voltage source, shown at the right of the figure. Now suppose we increase the output current i_o by some amount—what would the output voltage do? Since output voltage is strictly dependent on input voltage, output voltage will *not* change; in other words

$$\boxed{r_o = \frac{dv_o}{di_o} = \frac{0}{di_o} = 0} \quad (41.63)$$

That is, the output impedance of an ideal amp is zero.

41.13 Output Impedance of Nonideal Amp

Now suppose the amp has nonzero output impedance denoted by r_o. Then we can convey this by schematics shown in Fig. 41.20.

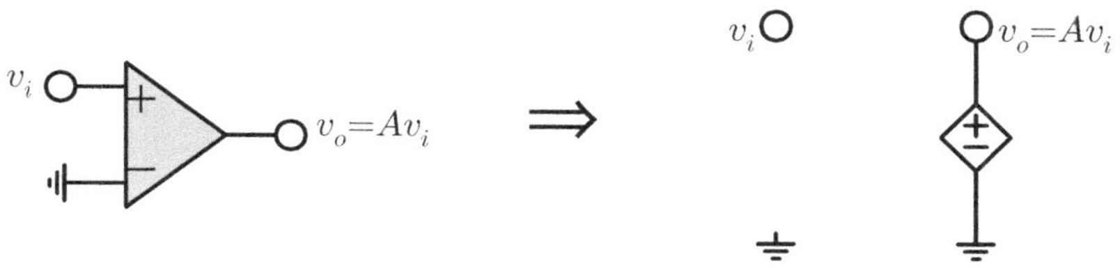

Fig. 41.19 Amplifier with zero output resistance

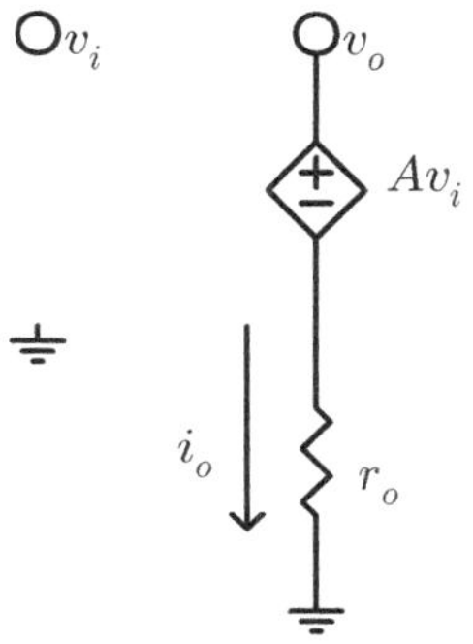

Fig. 41.20 Amplifier with nonzero output resistance

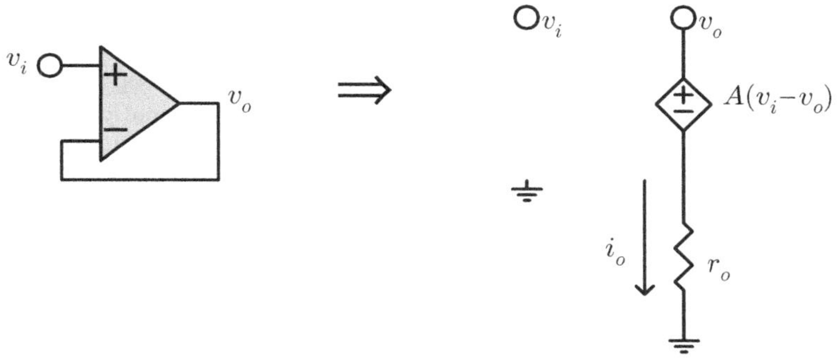

Fig. 41.21 Amp with negative feedback and its output impedance

Let's verify that this schematics does convey our intention. From the schematics we read

$$v_o = Av_i + i_o r_o \tag{41.64}$$

Let's vary the output current and derive output impedance:

$$R = \frac{dv_o}{di_o} = \frac{d}{di_o}(Av_i + i_o r_o) = r_o \tag{41.65}$$

So in fact output impedance is r_o. Notice that we used the fact that $dv_i/di_o = 0$ since input voltage does not depend on output current. With the open-loop amp output impedance understood we next mode to the closed-loop case.

41.14 Output Impedance of Nonideal Amp with Full Feedback

Let's consider an amp with full feedback as shown in Fig. 41.21. The output voltage is given by

$$v_o = A(v_i - v_o) + i_o r_o \tag{41.66}$$

It is tempting to assume that output impedance remains the same (r_o), but let's see the proper derivation. Rewrite above equation as

$$v_o(1 + A) = Av_i + i_o r_o, \quad \text{or} \tag{41.67}$$

$$v_o = \frac{Av_i}{1+A} + i_o \frac{r_o}{1+A} \tag{41.68}$$

Now find output resistance as derivative of voltage to current:

$$R_o = \frac{dv_o}{di_o} = \frac{d}{di_o}\left[\frac{Av_i}{1+A} + i_o\frac{r_o}{1+A}\right], \tag{41.69}$$

such that

$$\boxed{R_o = \frac{r_o}{1+A}} \tag{41.70}$$

where again we used the fact that v_i does not depend on i_o. We can verbally summarize our results in multiple ways:

1. The effective output impedance of an amplifier, with negative feedback, and with loop gain A is the open loop output impedance divided by $(1+A)$.
2. For large gain, the effective output impedance of a feedback amp is greatly diminished.
3. A feedback amp can provide current without noticeably changing its output voltage.
4. An amp with negative feedback, even being nonideal, appears almost as an ideal voltage source.

41.15 Output Impedance of Nonideal Amp with Partial Feedback

Now consider the amp with partial negative feedback as shown in Fig. 41.22. The output voltage is given by

$$v_o = A(v_i - \beta v_o) + i_o r_o, \quad \text{where } \beta = \frac{R_2}{R_1+R_2} \tag{41.71}$$

That is, the fed back portion of v_o is βv_o. Rewrite above equation as

$$v_o(1+\beta A) = Av_i + i_o r_o, \qquad \text{or} \tag{41.72}$$

$$\boxed{v_o = \frac{Av_i}{1+\beta A} + i_o\frac{r_o}{1+\beta A}} \tag{41.73}$$

To find output impedance again we differentiate output voltage with respect to output current

$$\boxed{R_o = \frac{dv_o}{di_o} = \frac{r_o}{1+\beta A}} \tag{41.74}$$

Again output impedance will be smaller than open loop one. Notice that when we feedback all of output voltage, $\beta = 1$, we regain Eq. (41.70). On the other hand, if we feedback nothing ($\beta = 0$), then output impedance collapses to that of open loop case (r_o).

41.16 Output Impedance of Nonideal Amp with Remote Sensing

Consider next the case of an amp whose output goes through a series resistor (R_1), and then is fed back to the inverting terminal, as shown in Fig. 41.23. Doing KVL gives

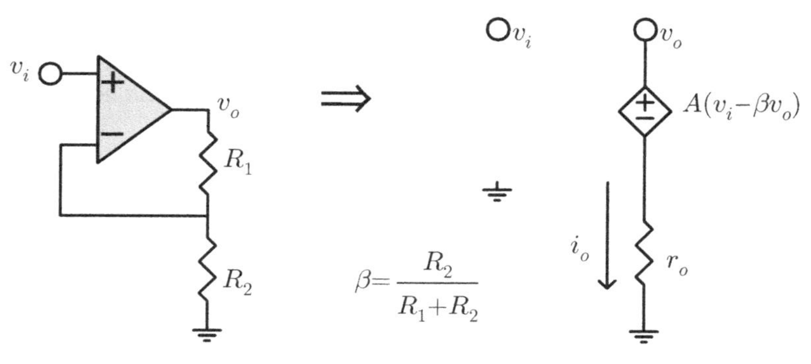

Fig. 41.22 Amp with partial negative feedback and its output impedance

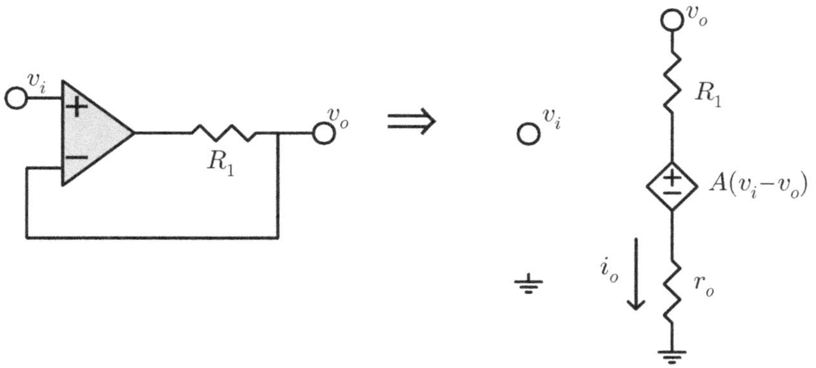

Fig. 41.23 Amp with remote sensing

$$v_o = A(v_i - v_o) + i_o(r_o + R_1) \quad (41.75)$$

Rearranging we get

$$v_o(1 + A) = Av_i + i_o(r_o + R_1), \quad \text{or} \quad (41.76)$$

$$\boxed{v_o = v_i\frac{A}{1 + A} + i_o\frac{r_o + R_1}{1 + A}} \quad (41.77)$$

Output impedance, looking from remote sense point, is then

$$\boxed{R_o = \frac{r_o + R_1}{1 + A}} \quad (41.78)$$

If the amp intrinsic output impedance r_o is zero then system output impedance is

$$R_o = \frac{R_1}{1 + A}, \quad (\text{case } r_o = 0) \quad (41.79)$$

That is, as seen from load, the impedance R_1 connecting load to amplifier output appears much smaller, specifically by ratio $1 + A$.

41.17 Output Impedance of Amp with Remote Sensing and Low-Pass Feedback Filter

Let's revise our assumption of full feedback and instead put a low-pass filter as shown in

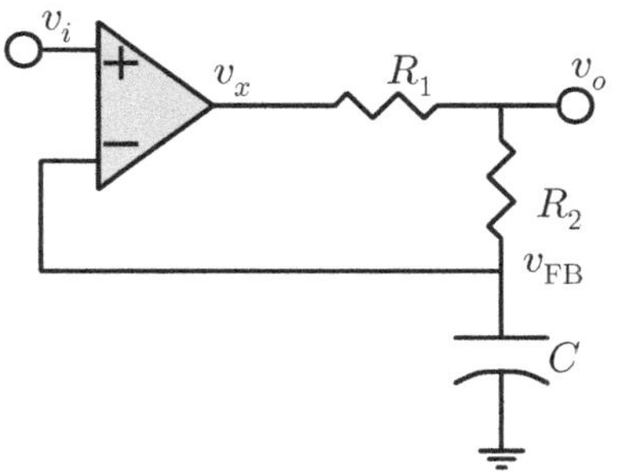

Fig. 41.24 Amp with remote sensing and low-pass filter feedback

Fig. 41.24. Let's assume that the intrinsic output impedance of the amp r_o is zero (if not, we can simply lump it into R_1). The feedback voltage is

$$v_{FB} = av_o\frac{1}{s + a}, \qquad a = \frac{1}{R_2C} \quad (41.80)$$

The output voltage of the amp v_x is given by

$$v_x = A(v_i - v_{FB}) = -Aav_o\frac{1}{s + a} \quad (41.81)$$

where we set input voltage as zero AC (we are after impedance). Applying KCL at the output node we get

$$i_o = \frac{v_o}{R_2 + \frac{1}{sC}} + \frac{v_o - v_x}{R_1} = \frac{sCv_o}{1 + sCR_2} + \frac{v_o}{R_1} + \frac{Aav_o}{R_1}\frac{1}{s+a} \quad (41.82)$$

$$i_o = \frac{v_o}{R_2}\frac{s}{s+a} + \frac{v_o}{R_1} + \frac{Aav_o}{R_1}\frac{1}{s+a} \quad (41.83)$$

$$i_o = \frac{v_o}{R_1} + \frac{v_o s/R_2 + Aav_o/R_1}{s+a} \quad (41.84)$$

$$i_o = v_o\left[\frac{1}{R_1} + \frac{s/R_2 + Aa/R_1}{s+a}\right] = v_o\frac{s + a + sR_1/R_2 + aA}{R_1(s+a)} \quad (41.85)$$

$$i_o = v_o\frac{a(1+A) + s(1 + R_1/R_2)}{R_1(s+a)} = v_o\frac{a(1+A) + s\frac{R1+R_2}{R_2}}{R_1(s+a)} \quad (41.86)$$

$$i_o = v_o\frac{R_1 + R_2}{R_2}\frac{s + \frac{a(1+A)R_2}{R_1+R_2}}{R_1(s+a)} = v_o\frac{R_1 + R_2}{R_1R_2}\frac{s + \frac{1+A}{C(R_1+R_2)}}{s + \frac{1}{R_2C}} \quad (41.87)$$

Solve for v_o to get

$$v_o = i_o\frac{R_1R_2}{R_1 + R_2}\frac{s + \frac{1}{R_2C}}{s + \frac{1+A}{C(R_1+R_2)}} \quad (41.88)$$

so that output impedance is

$$\boxed{Z_o(s) = \frac{R_1R_2}{R_1 + R_2}\frac{s + \frac{1}{R_2C}}{s + \frac{1+A}{C(R_1+R_2)}}} \quad (41.89)$$

Let's take some limits. At DC we get

$$Z_o(0) = \frac{R_1R_2}{R_1 + R_2}\frac{1}{R_2C}\frac{C(R_1 + R_2)}{1+A} = \frac{R_1}{1+A} \quad (41.90)$$

in agreement with Eq. (41.79). This means that at DC and since the cap is open we have full feedback and output impedance is simply R_1 divided by gain (plus 1). On the other hand, at high frequency we get the limit

$$Z_o(\infty) = \frac{R_1R_2}{R_1 + R_2} \quad (41.91)$$

which is nothing other than the parallel impedance of R_1 and R_2. That is, at high frequency the cap shorts and we have *no feedback*. Then R_1 shows its full value, and that goes in parallel with R_2. (Remember in this case we assumed $r_o = 0\,\Omega$, but if not we can simply lump it into R_1.) A sample run case is shown in Fig. 41.25; in this case we have

$$R_1 = R_2 = 10\,\Omega, C = 1\,\text{mF} \quad \text{and} \quad A = 100 \quad (41.92)$$

We see that at DC we get $0.1\,\Omega$, which is $R_1/(A+1)$. At high frequency we get $5\,\Omega$ which is $R_1||R_2$. The zero happens at

$$\text{zero at}\omega = \frac{1}{R_2C} = 100\text{ rad/s} \quad (41.93)$$

The pole happens at

$$\text{pole at } \omega = \frac{1+A}{C(R_1+R_2)} = \frac{101}{1\text{mF}(20\,\Omega)} \sim 5\text{ k rad/s} \quad (41.94)$$

All of these are confirmed in Fig. 41.25.

41.18 Main Idea Behind Feedback, Revisited

In wrapping up this chapter let's reflect a little bit on the underlying mechanism of feedback. Consider the feedback system as shown in Fig. 41.26. Assume the following

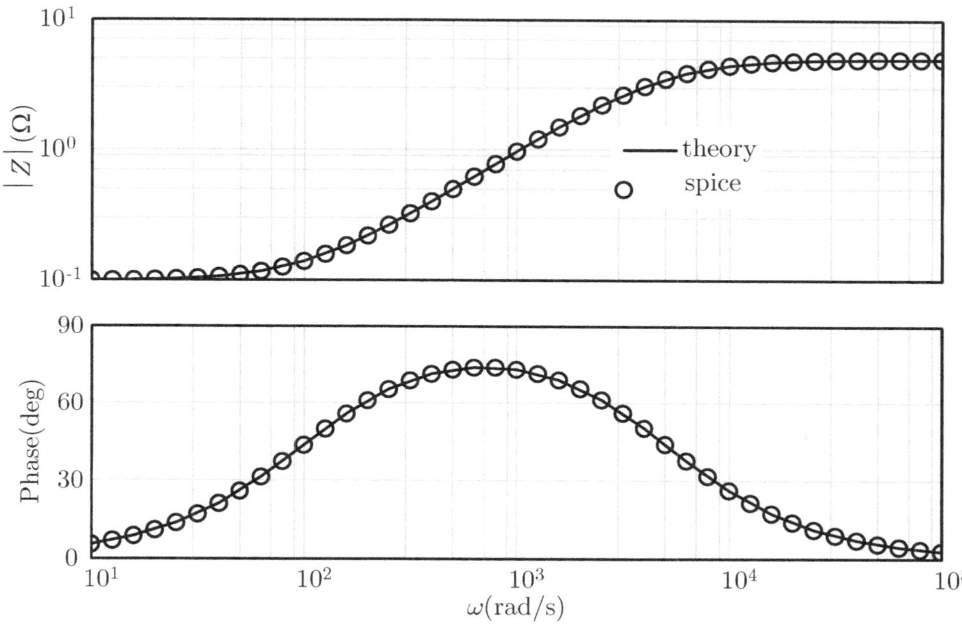

Fig. 41.25 Output impedance of Fig. 41.24

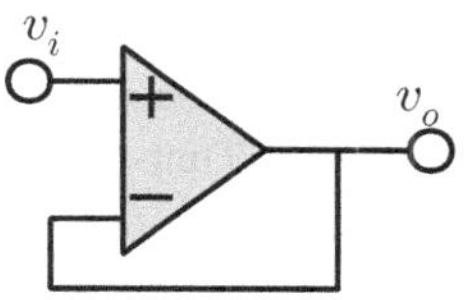

Fig. 41.26 Amplifier with feedback

1. Input voltage is 1 V.
2. Gain is 10.
3. Min max rail is 10 ± 10 V.

Assume initially output voltage drifted to max rail which is 10 V. If $v_o = 10$ and $v_i = 1$, then output voltage wants to be at

$$v_o = 10(1 - 10) = -90\,\text{V} \tag{41.95}$$

That is, there is a push for v_o to drop (ideally to −90 V). Assume that v_o drops from 10 to 8 V. Now we apply the amp equation and get

$$v_o = 10(1 - 8) = -70\,\text{V} \tag{41.96}$$

Again there is a push for v_o to drop (ideally to −70 V). Assume that v_o drops from 8 to 4 V. Now we apply the amp equation and get

$$v_o = 10(1 - 4) = -30\,\text{V} \tag{41.97}$$

Push to reduce more. Assume v_o drops from 4 to 1 V. Then the amplifier equation gives

$$v_o = 10(1 - 1) = 0 \tag{41.98}$$

That is, the output is at 1 V but it needing to get to 0 V. This is much better than what we started with, which is the amp at 10 V and it needs to get to −90 V. That is, looks like we are converging to a satisfactory point for the current voltage and what it needs to be. Now assume v_o drops from 1 V to 0.91 V; then we get

$$v_o = 10(1 - 0.91) = 0.9\,\text{V} \tag{41.99}$$

Getting real close—output at 0.91 V and it needing only to get to 0.9 V. If output drops from 0.91 V to 0.909 V then

$$v_o = 10(1 - 0.909) = 0.91\,\text{V} \tag{41.100}$$

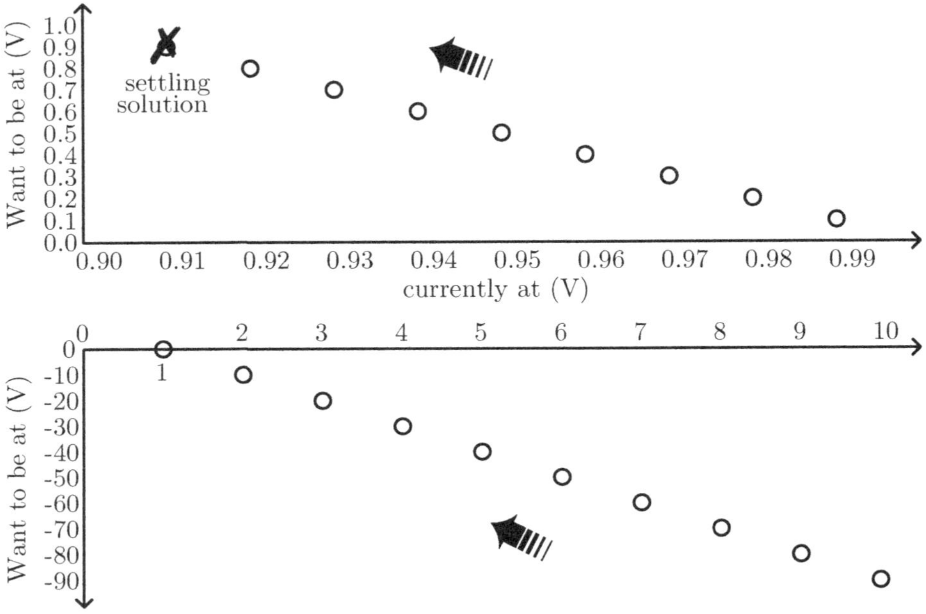

Fig. 41.27 Process of feedback. x- axis is where output voltage currently resides; y-axis is where it wants to be. Bottom figure—large voltage scale; top figure—small voltage scale

which is for all practical purposes equal to 0.909 V. That is, we converged to a settling value. Of course we already know what this value should be, and that is

$$v_o = \frac{A}{1+A} = \frac{10}{1+10} = 0.9091\,\text{V} \quad (41.101)$$

This iterative process is shown in Fig. 41.27 which shows on the x-axis the current output voltage, and on the y-axis where it wants to be. We can see that as the output drops in voltage, the difference between where it is and where it wants to be diminishes, and in the limit the difference goes to zero, at the settling point.

41.19 Summary

The op-amp is a 3-port unique device in that around an operating point it amplifies the voltage difference between its two input terminals. Typically the amp is used in conjunction with feedback tying its output terminal to one of its input terminals (typically the inverting one). Depending on the exact port connectivity, and the kind of feedback network the op-amp can serve as a voltage scaler, an inverting one, an integrator, differentiator, and other functionalities. What matters for our study is the op-amp transfer function, its behavior in the frequency domain in terms of magnitude, phase, poles and zeroes, and its behavior in the time domain in terms of impulse response, unit step one, and so forth. An amp is typically characterized in terms of a gain and a bandwidth. Around an operating point the amp acts as a linear device and as such lends itself to spectral and convolution methods. Other than the input/output transfer function we also studied the output impedance of the amp and the effective output impedance with feedback.

41.20 Problems

1. Starting with Eq. (41.20), identify the location of the zero and the pole, and then confirm from Fig. 41.6.
2. Consider the *RL* circuit in Fig. 41.28; figure the input/output transfer function and compare

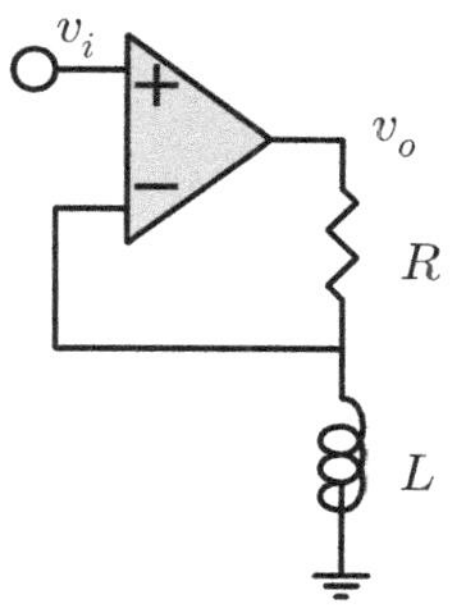

Fig. 41.28 Statement to Problem 2

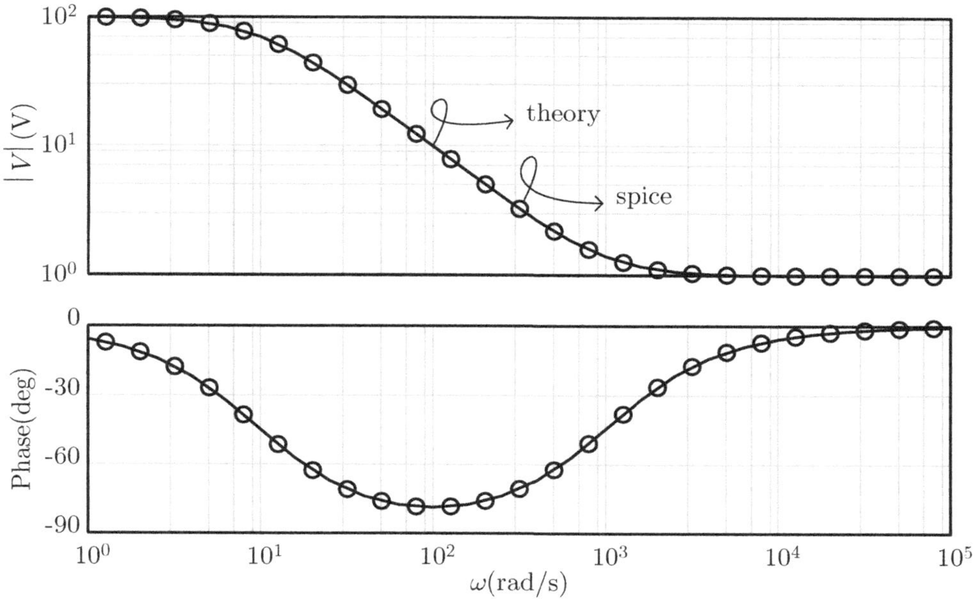

Fig. 41.29 Sample solution to Problem 2

to SPICE for the case $R = 1$, $L = $mH and $A = 100$; see sample solution in Fig. 41.29. Identify the location of the zero and the pole. Also identify the low- and high-frequency limits of the transfer function.
Answer:

$$H(s) = \frac{A}{1+A}\frac{s+\frac{R}{L}}{s+\frac{R}{L(1+A)}}$$

3. The integrating amp in Fig. 41.30 has an input sine function $v_i(t) = \sin t$; find the output voltage and compare to SPICE; see sample solution in Fig. 41.31.
4. Redo the example in Sect. 41.6 using the assumptions in Sect. 41.7 and show agreement in the limit as the gain $A \to \infty$.
5. Consider the filter in Fig. 41.32. What does it do? If the input voltage is a unit step one, what is the output voltage both in the frequency and time domain? See sample solution in Fig. 41.33.
6. Consider the LC circuit in Fig. 41.34. It is stimulated by an input voltage of the form $v_i(t) = \sin t$. Figure output voltage both in the frequency domain and in the time one; compare the latter to SPICE. See sample solution in Figs. 41.35 and 41.36. Note: don't

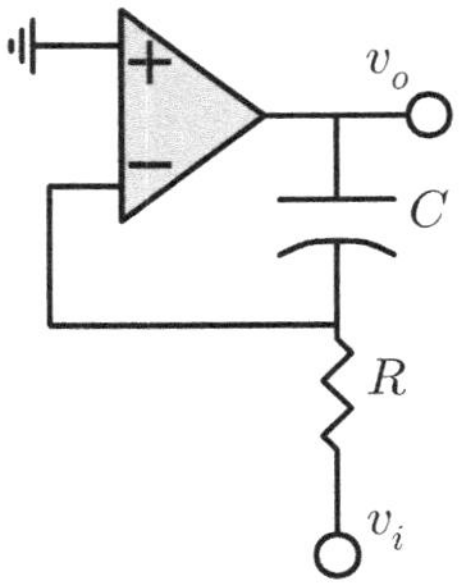

Fig. 41.30 Statement to Problem 3

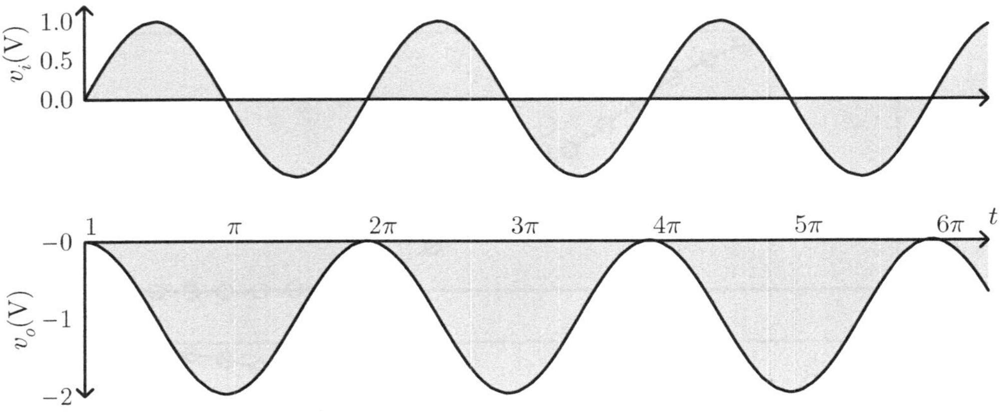

Fig. 41.31 Sample solution to Problem 3

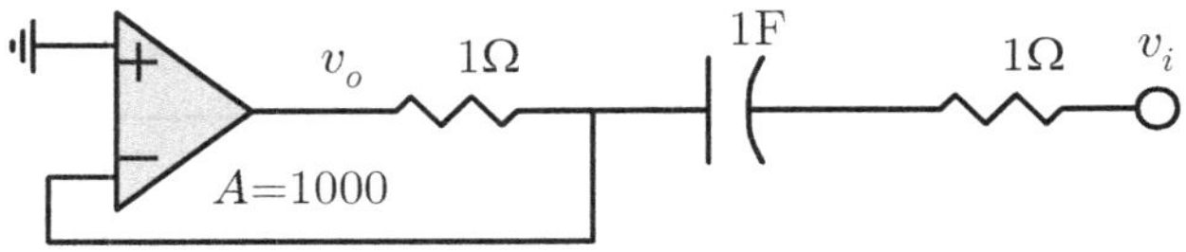

Fig. 41.32 Specifications to Problem 5

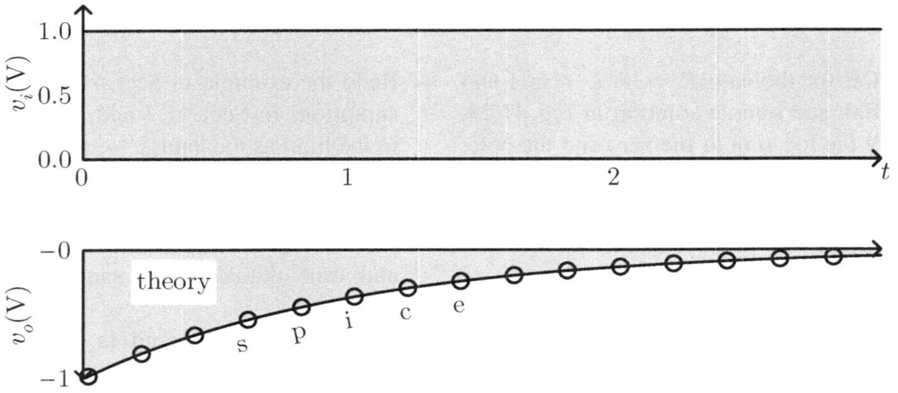

Fig. 41.33 Sample solution to Problem 5

Fig. 41.34 Specifications to Problem 6

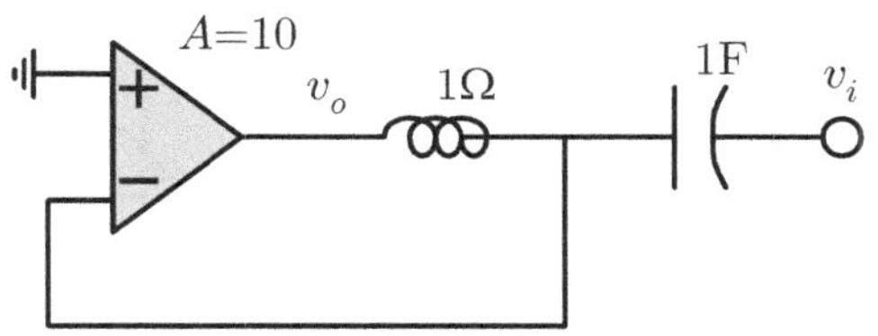

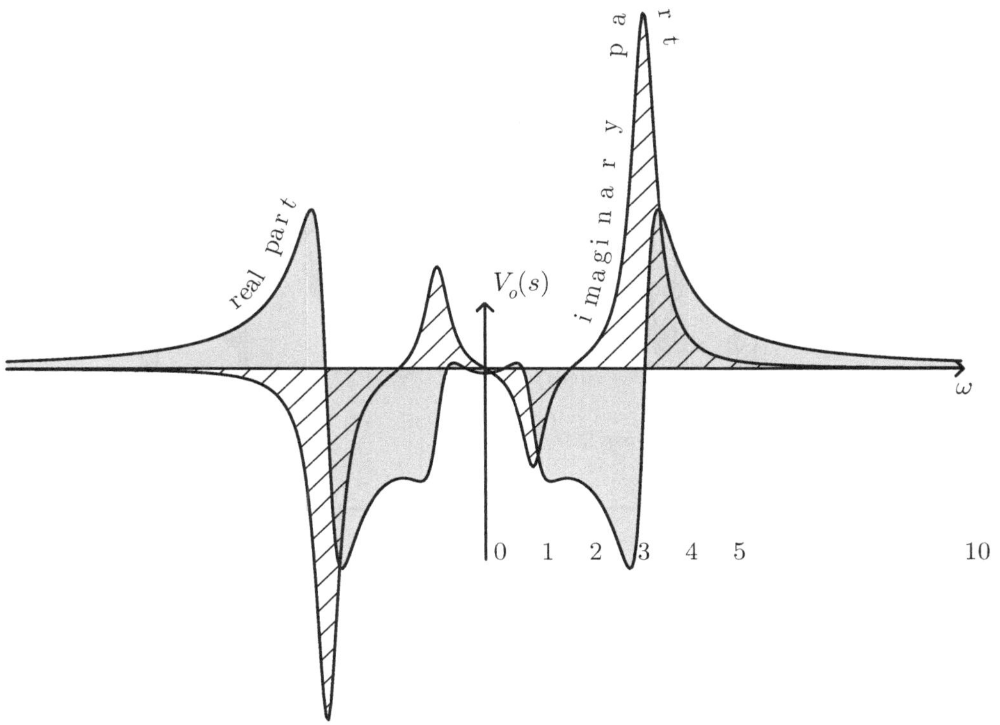

Fig. 41.35 Sample solution to Problem 6 (part 1/2); sample case of $\sigma = 0.3$

use the simple approximation of Sect. 41.7; instead use full KVL/KCL analysis.
Answer:

$$V_o(s) = -A\frac{s^2}{s^2 + 1 + A}\frac{1}{s^2 + 1}$$

7. Consider the two-stage circuit in Fig. 41.37. Derive the input/output transfer function $H(s) = V_o(s)/V_i(s)$ and plot it alongside comparison to SPICE. Then assuming input is $v_i(t) = \sin 2\pi t$ plot the output voltage in the frequency domain. See sample solution in Fig. 41.38.
8. Starting with Problem 7 and knowing output voltage in the frequency domain, find it in the time domain, and compare to SPICE; see sample solution in Fig. 41.39.
9. What is the output impedance of the circuit in Fig. 41.40? What are the DC and high frequency limits? Explain. Where are the locations of the zero and the pole? Plot and compare to SPICE; see sample solution in Fig. 41.41.
Answer:

$$Z_o(s) = \frac{R_1}{1 + A}\frac{s + a}{s + \frac{a+b}{1+A}}, \; a = \frac{R_2}{L}, \; b = \frac{R_1}{L}$$

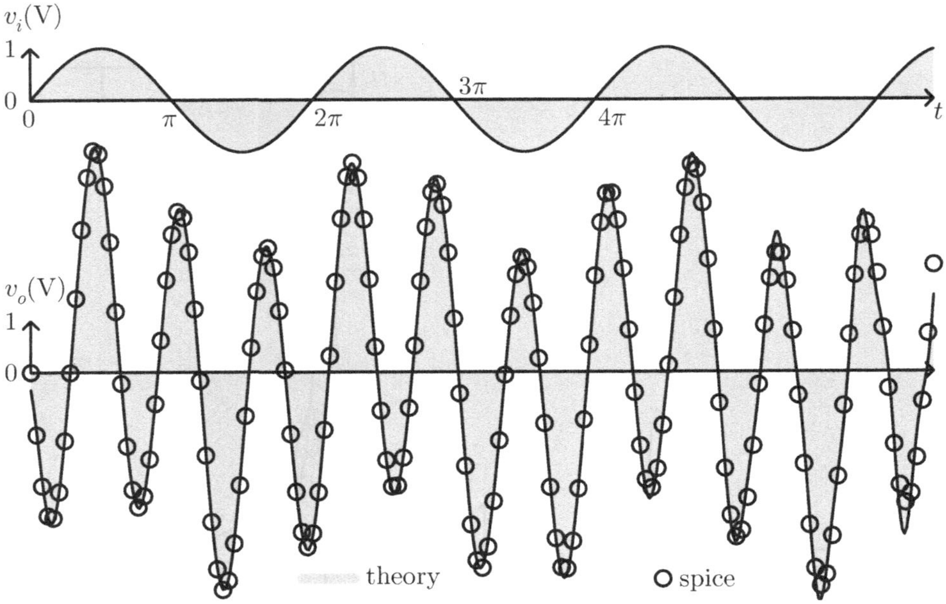

Fig. 41.36 Sample solution to Problem 6 (part 2/2)

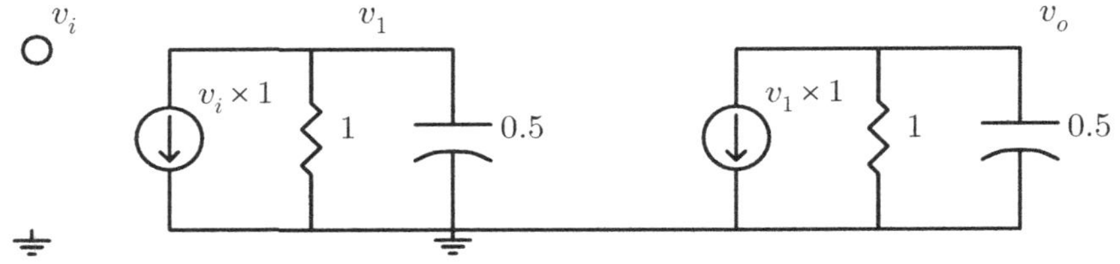

Fig. 41.37 Specifications to Problem 7

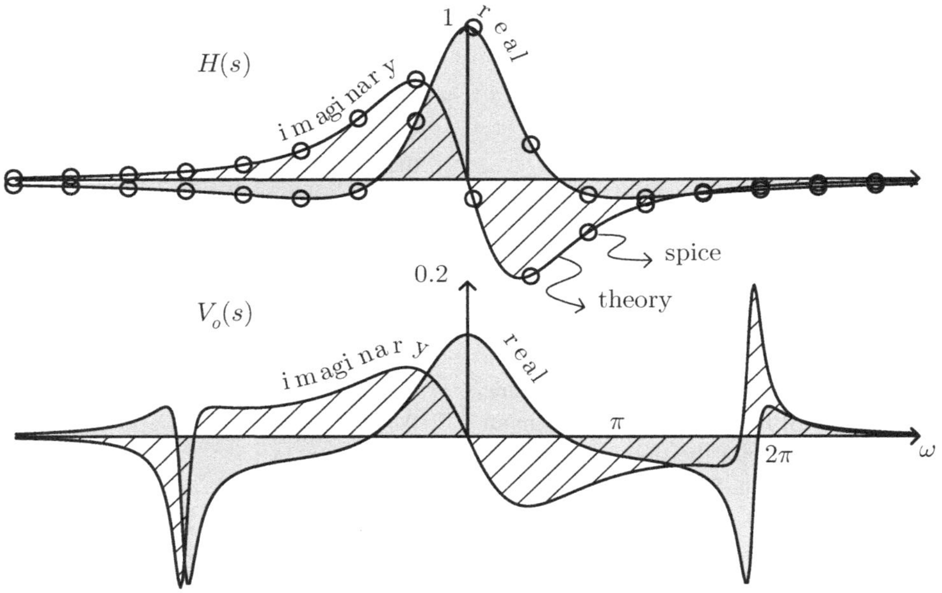

Fig. 41.38 Sample solution to Problem 7; top—$\sigma = 0.0$, bottom—$\sigma = 0.2$

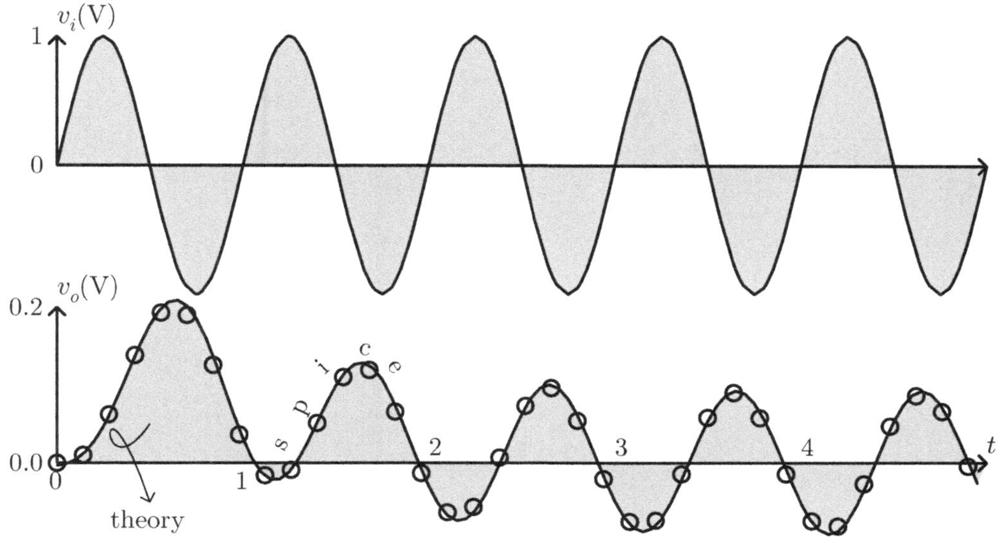

Fig. 41.39 Sample solution to Problem 8

Fig. 41.40 Specifications to Problem 9

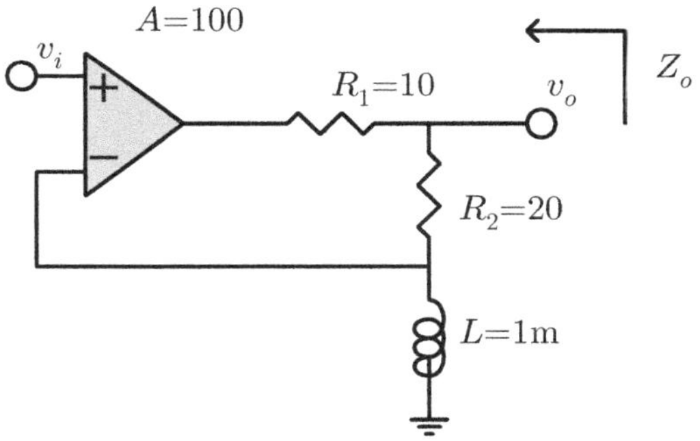

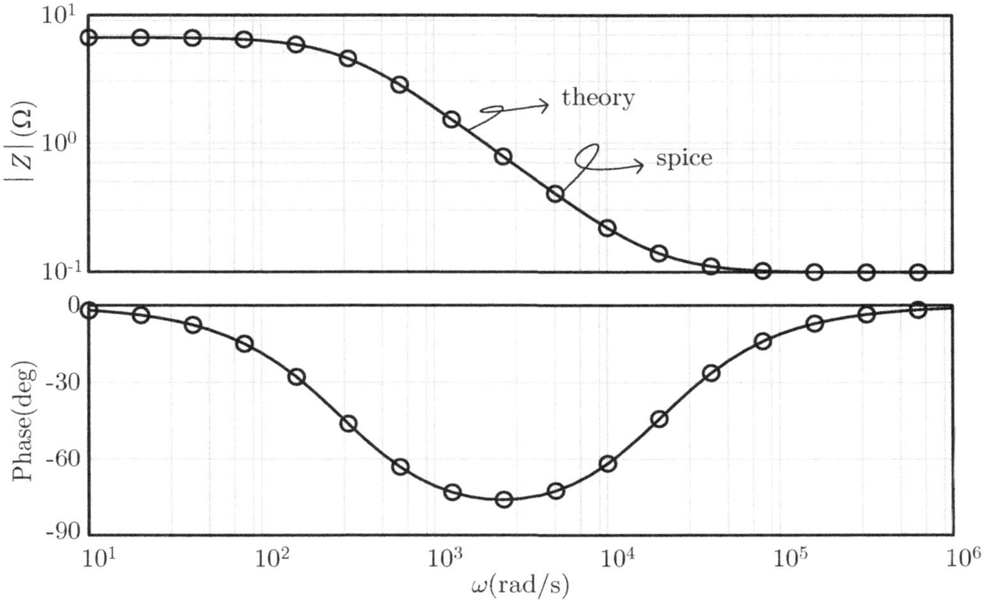

Fig. 41.41 Sample solution to Problem 9

Multi-Port Network: Z- and Y-Parameters

42

42.1 Introduction

Just about any physical block can be modeled as a multi-port network. In almost all cases the block has some spatial distribution effects, meaning currents, voltages, and other variables that depend both on the x- and y-axis. The block would have a number of ports, depending on input and output, and the inners of the block modeled via some mechanism. There are a number of such mechanisms including flat RLC grids, Z-parameters, Y-parameters, S-parameters, behavior models, or some other methods. There maybe advantages of using each method, but the premise in all is to model the block accurately, in the time and frequency domain. In the following sections we will focus on some of the most common block models encountered, including the flat grid, Z- and Y-parameters, and defer S-parameters treatment for a separate chapter.

42.2 Flat Grid Models

In this method, the block of concern is modeled using a flat grid of RLC elements. Depending on the case, one may end up using only a subset of the RLC set. For example, at DC one may simply use a resistor grid; or at real high frequency one may drop the R and keep the L since inductive impedance would dominate resistive impedance.

Resistive Grid Example While using a resistive grid is limiting, it illustrates the basic idea behind the flat grid. Even more, we know that each of inductance and capacitance at a given frequency has a unique impedance (resistance); so from that point of view, what applies for the resistor grid should apply for the LC one. Figure 42.1 shows a slab of an electrical block with a simplified R model representation. The resistor grid is built by simply dividing the slab into a number of small segments (i.e., rectangles) and modeling the resistance of each segment. Then when connecting all segments one should have captured the current and voltage behavior throughout the slab.

In particular one can either force a set of voltages at the ports and measure the corresponding currents, or the other way (or a mix thereof). Either way, once the stimulus at the ports is applied, solving KCL equations should solve for all unknown variables.

Advantages of Flat Grid There are several advantages of the flat grid approach:

- **Simplicity**: This method is simple and relies on characterizing a small number of elementary blocks, such as a square, and then simply tiling the big block with a mosaic of the small blocks.
- **Access to internal nodes**: This method solves for everything, every node, and every current. This means we can get a topographical map

F. Badrieh, *Spectral, Convolution and Numerical Techniques in Circuit Theory*,
https://doi.org/10.1007/978-3-319-71437-0_42

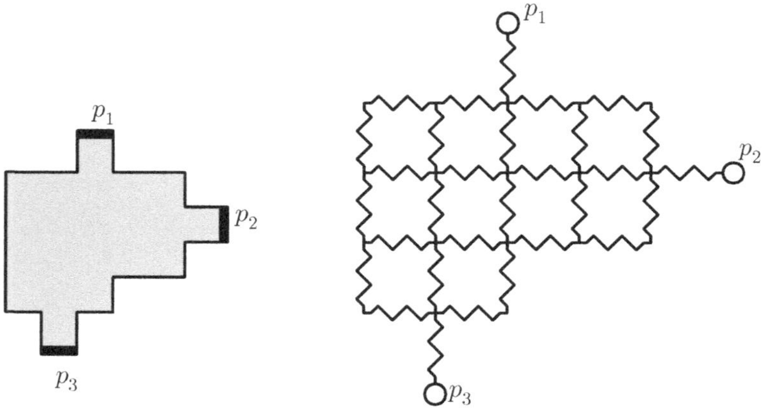

Fig. 42.1 Physical layout and R-grid model

of the voltage across the whole structure, and look at the spatial distribution of all currents.

- **Ideal for compute-intensive machines**: This method becomes more viable for computer processing, with even more powerful processors.

Disadvantages of Flat Grid The disadvantages of this method are listed below:

- **Brute force**: This method can be characterized as a brute force method, as it does not lend itself to analytic techniques.
- **More difficult with** LC **elements**: When combining LC elements this method becomes more challenging, especially in capturing loop effects.
- **Slow**: Since the number of unknowns is large, it is expected that this method to be slow as compared to others.
- **Has to be solved every time**: For each different stimulus, the set of KVL/KCL equations needs to be solved all over again.

Sample Run A sample run case is shown in Fig. 42.2. In the figure each resistor is 1 Ω. A unit voltage is applied to the bottom terminal, while the other three terminals are grounded. The figure shows the corresponding current. The current is denoted via arrows the size of which denotes current magnitude while the direction denotes current direction. We see very clearly how the current enters the structure, redistributes both horizontally and vertically, and then pours (sinks) into the top three terminals. We are assured, via KCL, that the sum of current (both horizontal and vertical) at a given node is zero; we can almost tell this by visual inspection. Once the branch currents and node voltages are known, the problem is solved; we know everything! Of course if the terminal voltages were to ever change, the whole process of solving KVL/KCL (for every node and every branch) has to be redone all over again! There is not pretty much anything that can be reused between one stimulus case and another. Nonetheless this method still works and can be resorted to when needed.

42.3 *Z*-Parameters

Z-parameters define a set of self and mutual *impedances* between the network ports. For each port we have a self and mutual impedance to the other ports. What happens inside the modeled block is inaccessible to the outside world, but the external behavior is fully captured.

Set of Governing Equations Given a number of ports (one of them being a reference one), the set of governing equations ties port voltages to port currents as follows:

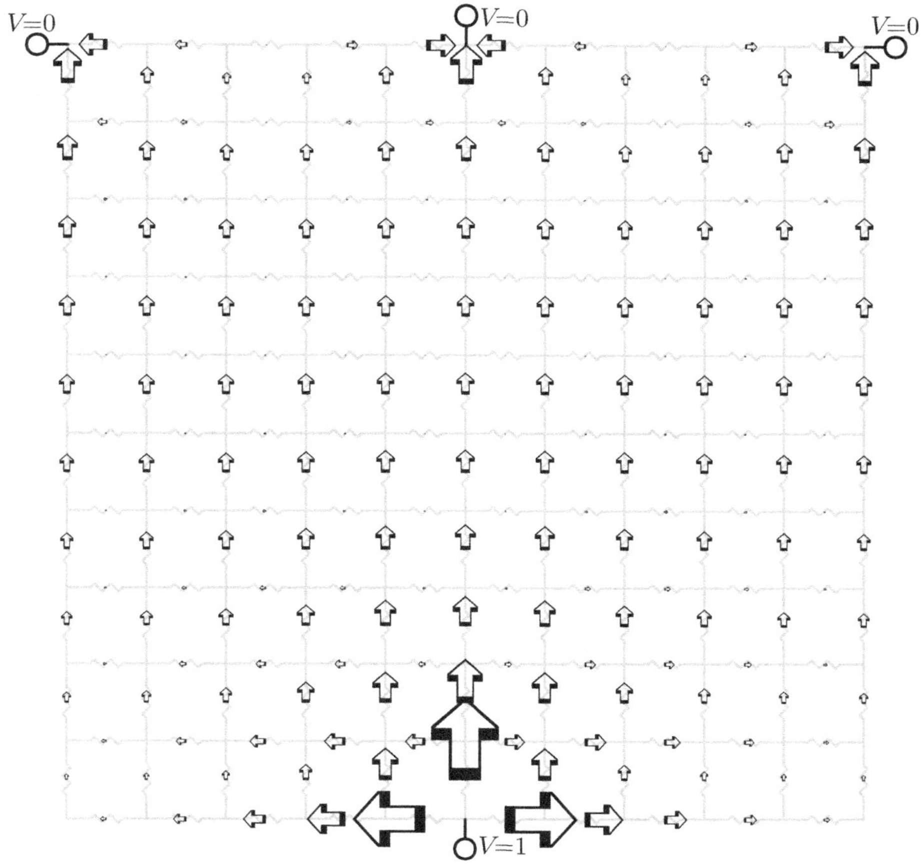

Fig. 42.2 Sample run of distributed grid showing input and output current subject to prescribed terminal voltages

$$
\begin{aligned}
V_1(s) &= Z_{11}(s)I_1(s)+Z_{12}(s)I_2(s)+Z_{13}(s)I_3(s)+\cdots \\
V_2(s) &= Z_{21}(s)I_1(s)+Z_{22}(s)I_2(s)+Z_{23}(s)I_3(s)+\cdots \\
V_3(s) &= Z_{31}(s)I_1(s)+Z_{32}(s)I_2(s)+Z_{33}(s)I_3(s)+\cdots
\end{aligned}
\tag{42.1}
$$

and so forth. In matrix notation we have

$$
\begin{bmatrix} V_1(s) \\ V_2(s) \\ V_3(s) \\ \cdots \end{bmatrix} = \begin{bmatrix} Z_{11}(s) & Z_{12}(s) & Z_{13}(s) & \cdots \\ Z_{21}(s) & Z_{22}(s) & Z_{23}(s) & \cdots \\ Z_{31}(s) & Z_{32}(s) & Z_{33}(s) & \cdots \\ \cdots & \cdots & \cdots & \cdots \end{bmatrix} \begin{bmatrix} I_1(s) \\ I_2(s) \\ I_3(s) \\ \cdots \end{bmatrix}
\tag{42.2}
$$

Notice that the reference port does not have its equation; its purpose is to serve as a reference for defining the various port voltages. To figure the Z- elements we selectively open all ports, and excite only a single one. Reading the voltage at that port gives the self impedance, while reading the voltages at all other ports gives the mutual impedance to those ports. Repeat the process for the next port until all ports have been exhausted! Let's demonstrate this on a few examples.

DC Resistor Example As a first demonstration of the Z-parameter network, consider the circuit in Fig. 42.3. Notice that while the network has three ports, we label only two of them; the third (bottom) one serves as a reference, or ground.

We define the terminal voltages as follows:

$$
\begin{aligned}
v_1(t) &= Z_{11}i_1(t) + Z_{12}i_2(t) \\
v_2(t) &= Z_{21}i_1(t) + Z_{22}i_2(t)
\end{aligned}
\tag{42.3}
$$

That is, the voltage at terminal 1 is related to the current at that terminal times the self

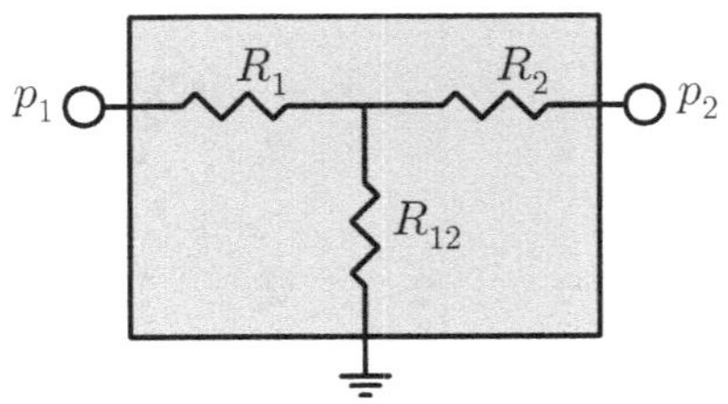

Fig. 42.3 Simple resistive circuit illustrating Z-parameters

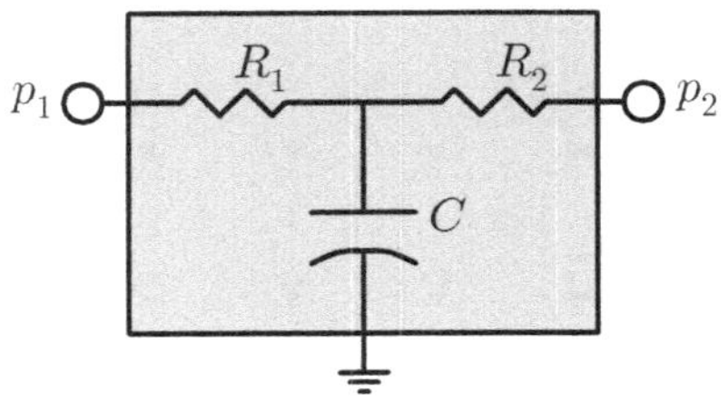

Fig. 42.4 Simple *RC* circuit illustrating *Z*-parameters

impedance plus the current at the other terminal times the mutual impedance. To find the self impedance of terminal 1 we simply set the current at the other terminal to zero. That is, we open terminal 2:

$$Z_{11} = \frac{v_1}{i_1}\Big|_{i_2=0} \tag{42.4}$$

For the simple test case at hand, it is clear that if i_2 is set to zero, and i_1 injected, then ratio of voltage to current would be

$$Z_{11} = R_1 + R_{12} \tag{42.5}$$

While at the same setup we can also find Z_{21} which relates the voltage at terminal 2 as a result of current at terminal 1:

$$Z_{21} = \frac{v_2}{i_1}\Big|_{i_2=0} \tag{42.6}$$

For the case at hand, and by inspection we conclude that mutual impedance is simply

$$Z_{21} = R_{12} \tag{42.7}$$

Now we open port 1 and inject current through port 2. Similar to above we get

$$\begin{aligned} Z_{22} &= R_2 + R_{12} \\ Z_{12} &= R_{12} \end{aligned} \tag{42.8}$$

Now supposedly we have fully captured the behavior of this network. That is, we don't need to know the details of the internal connectivity or operation of the circuit. We have captured the network behavior at the macroscopic level—at the port level. So far as port 1 and 2 currents and voltages are concerned, and with respect to the reference port, and under any voltage/current excitations they are a priori known. Notice, though, that once we migrate to the *Z*-parameter description of the network we have lost access to the internal node(s)—in this case the node connecting the three resistors. But viewed to the outside world, most of the time this is not a problem; the outside world only cares about the connection points to the block—by definition the block ports.

AC Cap-Resistor Example As a second demonstration of *Z*-parameters consider the network in Fig. 42.4.

Following the same procedure as that in the prior section we derive the self and mutual impedances as

$$\begin{aligned} Z_{11} &= R_1 + \frac{1}{sC} \\ Z_{22} &= R_2 + \frac{1}{sC} \\ Z_{12} &= \frac{1}{sC} \\ Z_{21} &= \frac{1}{sC} \end{aligned} \tag{42.9}$$

Notice that now all of the impedances are frequency dependent. Specifically, both self impedances start at open (at DC) and flatten to R_1 and R_2, respectively, at high frequency. On the other hand, mutual impedance starts open and shorts at high frequency. That is, at high frequency this circuit exhibits little mutual effects. A sample run with $R_1 = 1$, $R_2 = 2$, and $C = 1$ mF is shown in Fig. 42.5.

Again once we know the *Z*-parameters we don't need to keep track of the internals of the circuit; we don't even need to know that it has

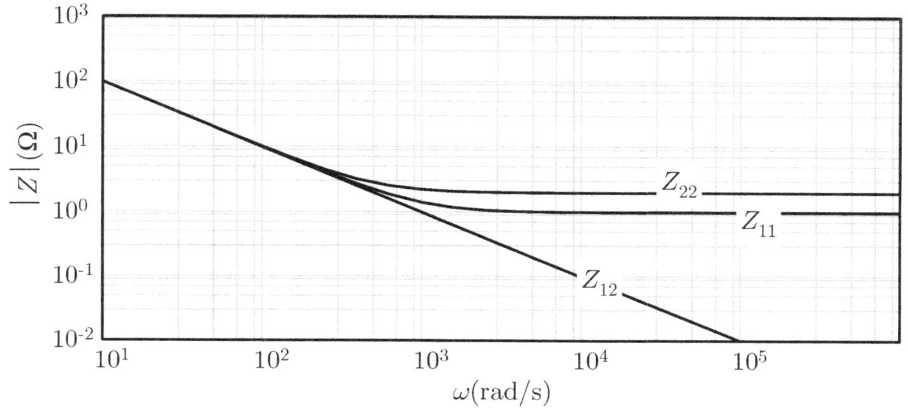

Fig. 42.5 Impedance versus frequency for circuit in Fig. 42.4

resistors or capacitors, or how they are connected. Everything has been captured by the Z-parameters, defined in Eq. (42.9). For example if we know the set of current stimulus $I_1(s)$ and $I_2(s)$ we can figure the corresponding voltages as

$$\begin{bmatrix} V_1(s) \\ V_2(s) \end{bmatrix} = \begin{bmatrix} Z_{11}(s) \; Z_{12}(s) \\ Z_{21}(s) \; Z_{22}(s) \end{bmatrix} \begin{bmatrix} I_1(s) \\ I_2(s) \end{bmatrix} \quad (42.10)$$

Once we know $V(s)$ we can go back to the time domain using inverse Laplace transform.

Three Terminal Example In this example we consider that case of three ports, as shown in Fig. 42.6. Notice again that the implicit fourth port serves as a reference port, and does not count in the matrix size. To find Z_{11} we open ports 2 and 3, force unit current through port 1, and measure voltage. By inspection we get

$$Z_{11} = R_1 + R_4 \quad (42.11)$$

Similarly we get the other two self impedances

$$Z_{22} = R_2 + R_4$$

$$Z_{33} = R_3 + R_4 \quad (42.12)$$

To find mutual impedance between port 1 and 2 we again open all ports other than port 1, force current and measure voltage at port 2. By inspection we get

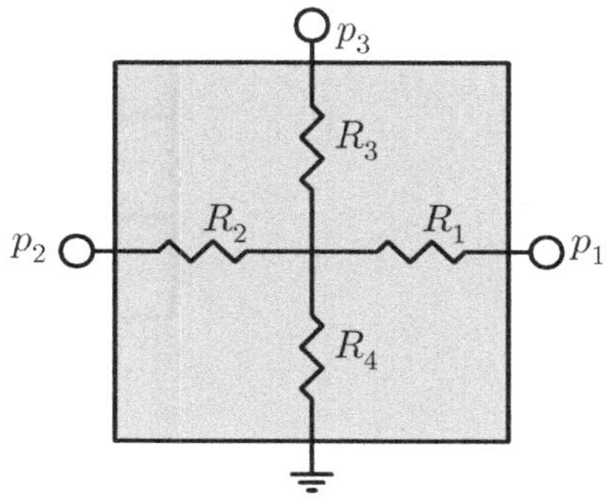

Fig. 42.6 Three terminal network for Z-parameter calculations

$$Z_{12} = R_4 \quad (42.13)$$

Similarly we get mutual impedance between port 1 and 3

$$Z_{13} = R_4 \quad (42.14)$$

Even more, mutual impedance between 2 and 3 comes out the same

$$Z_{23} = R_4 \quad (42.15)$$

It is clear from this example that for this kind of topology, the common path to ground comes out as the mutual impedance between the various ports. Our final Z-parameter matrix is then

$$Z(s) = \begin{bmatrix} R_1 + R_4 & R_4 & R_4 \\ R_4 & R_2 + R_4 & R_4 \\ R_4 & R_4 & R_3 + R_4 \end{bmatrix} \quad (42.16)$$

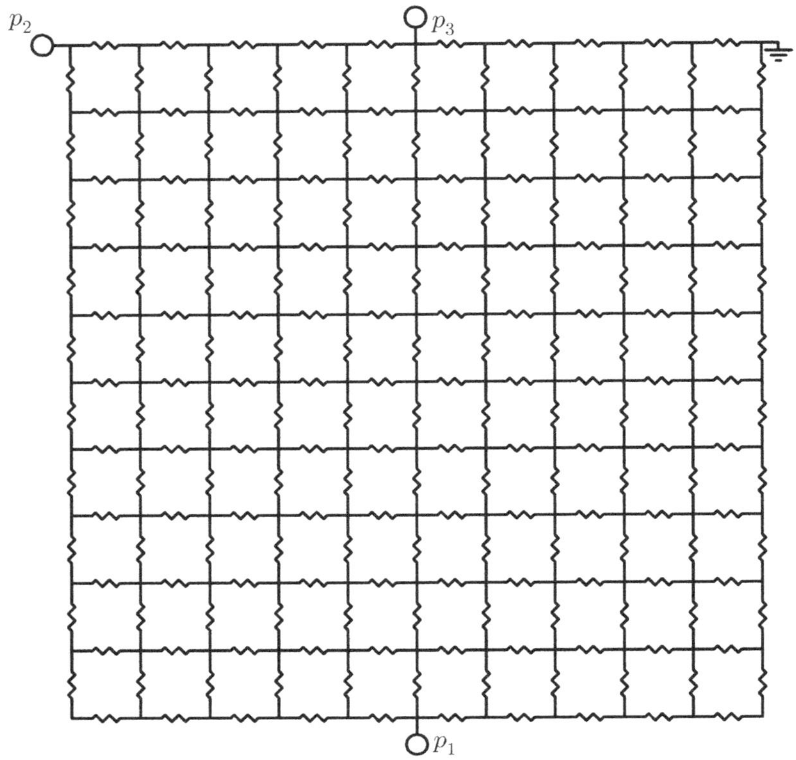

Fig. 42.7 220-resistor power plane with three ports

220-Resistor Power Plane Example Consider the 220-resistor power plane shown in Fig. 42.7. Each resistor is set to 1 Ω. The upper right corner is designated as a reference port (i.e., it is not considered in port count). The three ports are labeled $p_1 - p_3$. Assume the following bias conditions:

$$V_1 = 1; \qquad V_2 = V_3 = 0V \tag{42.17}$$

We would like to find out the port currents. First we would need to figure the Z-matrix. This can be measured or simulated. Either way we have it as

$$Z = \begin{bmatrix} 2.456 & 1.454 & 1.236 \\ 1.454 & 2.908 & 1.454 \\ 1.236 & 1.454 & 1.896 \end{bmatrix} \tag{42.18}$$

To solve for current we write down the matrix equation

$$V = ZI, \qquad \text{or} \tag{42.19}$$

$$\begin{bmatrix} 1.0 \\ 0.0 \\ 0.0 \end{bmatrix} = \begin{bmatrix} 2.456 & 1.454 & 1.236 \\ 1.454 & 2.908 & 1.454 \\ 1.236 & 1.454 & 1.896 \end{bmatrix} \times \begin{bmatrix} i_1 \\ i_2 \\ i_3 \end{bmatrix} \tag{42.20}$$

If we do the inversion we get

$$I = [0.6634 - 0.1873 - 0.2889] \tag{42.21}$$

We also ran SPICE on the raw resistive grid and also got $I = [0.6635 - 0.1873 - 0.2889]$. Let's reflect for a second what was accomplished. The whole 220-resistor grid in Fig. 42.7 has been abstracted to a mere 3 × 3 matrix as defined in Eq. (42.18)! That is an enormous simplification! Even more the claim is that for any current stimulus applied to the three ports we are guaranteed to figure the corresponding port voltages. The same is true if voltages were applied and the unknowns were the port currents. Sure we have lost connection with the internal nodes, but if this block was a device, do we really need

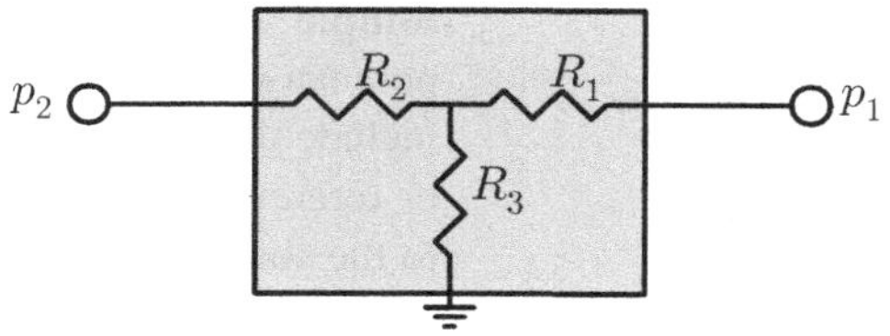

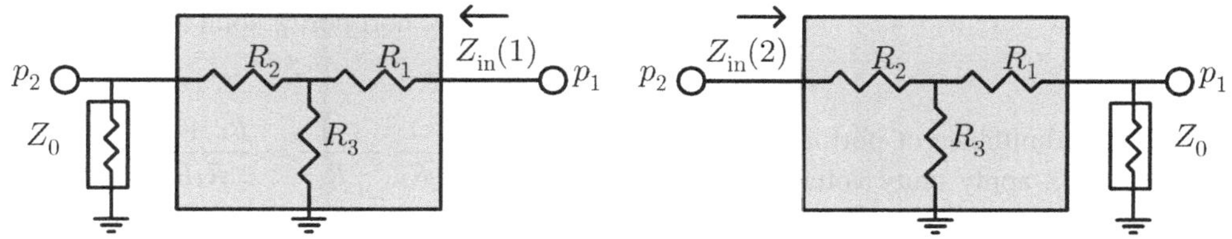

Fig. 42.8 Setup to measure input impedance

to keep track of all internal node voltages and branch currents? Most likely not! And thence the enormous savings become apparent and the justification and need for Z and other network parameters become apparent.

Input Impedance In the above sections we defined the self and mutual impedances between various ports. Another common impedance of interest is what is referred to as input impedance. Unlike the setup we used before to find self and mutual impedance, where one selectively opened ports, the setup used here is to terminate all other ports by their characteristics impedances. This is shown in Fig. 42.8. By inspection we can tell that input impedance at port 1 is

$$Z_{\text{in}}(1) = (Z_0 + R_2)||R_3 + R1 \tag{42.22}$$

Similarly input impedance at port 2 is

$$Z_{\text{in}}(2) = (Z_0 + R_1)||R_3 + R2 \tag{42.23}$$

For example, if $R_1 = 1$, $R_2 = 2$, $R_3 = 3$, and $Z_0 = 50$ then we would get

$$Z_{\text{in}}(1) = (50{+}2)||3{+}1 = \frac{52 \cdot 3}{52 + 3}{+}1 = 3.836\,\Omega \tag{42.24}$$

and

$$Z_{\text{in}}(2) = (50{+}1)||3{+}2 = \frac{51 \cdot 3}{51 + 3}{+}2 = 4.833\,\Omega \tag{42.25}$$

as has been confirmed by running SPICE.

42.4 *Y*-Parameters

Y or *admittance* parameters tie voltage to current, in a way similar to Z-parameters. The difference is that here the input variables are voltages and output ones are currents.

Governing Equations We know for a single impedance element, the relation between current and voltage is

$$I = GV \tag{42.26}$$

where G is the admittance. For multi-port networks, the governing equations become

$$\begin{aligned} I_1 &= G_{11}V_1 + G_{12}V_2 + G_{13}V_3 + \cdots \\ I_2 &= G_{21}V_1 + G_{22}V_2 + G_{23}V_3 + \cdots \\ I_3 &= G_{31}V_1 + G_{32}V_2 + G_{33}V_3 + \cdots \end{aligned} \tag{42.27}$$

In matrix format we have

$$\begin{bmatrix} I_1 \\ I_2 \\ I_3 \\ \cdots \end{bmatrix} = \begin{bmatrix} G_{11} & G_{12} & G_{13} & \cdots \\ G_{21} & G_{22} & G_{23} & \cdots \\ G_{31} & G_{32} & G_{33} & \cdots \\ \cdots & \cdots & \cdots & \cdots \end{bmatrix} \begin{bmatrix} V_1 \\ V_2 \\ V_3 \\ \cdots \end{bmatrix} \tag{42.28}$$

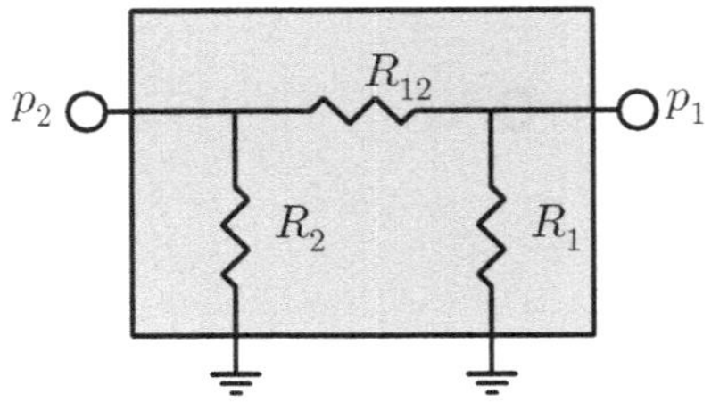

Fig. 42.9 Two-port resistive network for Y-parameter calculations

To find self admittance of port n we short all ports (to ground), apply unity voltage at port n, and measure port n current. To find mutual to port m we repeat and measure current through port m. Notice that in all cases we measure current *into* the port as being positive. Let's run a few examples to illustrate the meaning and extraction of Y-parameters.

Simple 2-Port DC Example Consider the 2-port network in Fig. 42.9. Notice that we don't include the ground port(s) in port count; ground or reference ports are needed but they don't add to the size of the Y-parameter count. To find self admittance of port 1 we short port two (apply zero voltage), apply unity voltage at port 1, and measure port 1 current.

When we short port 2, total admittance at port 1 is

$$Y_{11} = \frac{1}{R_1} + \frac{1}{R_{12}} = \frac{R_1 + R_{12}}{R_1 R_{12}} \tag{42.29}$$

To find mutual admittance between port 1 and 2 we repeat above but now measure current through port 2. We know, by current division, that port 2 current is equal to the above current times R_1 divided by $R_1 + R_{12}$; that is

$$Y_{12} = -Y_{11}\frac{R_1}{R_1 + R_{12}} = -\frac{R_1 + R_{12}}{R_1 R_{12}}\frac{R_1}{R_1 + R_{12}} = -\frac{1}{R_{12}} \tag{42.30}$$

Similarly we get

$$Y_{22} = \frac{1}{R_2} + \frac{1}{R_{12}} = \frac{R_2 + R_{12}}{R_2 R_{12}}, \quad \text{and} \quad Y_{21} = Y_{12} \tag{42.31}$$

Notice that mutual elements are *negative*; for example, to find Y_{12} we force unit voltage at p_1, ground p_2, and measure current at p_2. Since current is flowing *out* of the circuit, it would be negative; in other words, since p_1 is higher potential than p_2, current in latter would be negative. With the Y-matrix at hand we are able to figure port current for any applied port voltage.

Simple 2-Port AC Example Let's use the same topology as used in prior section, but change the coupling term to a capacitor one, as shown in Fig. 42.10. Similar to the prior section we get the following self and mutual admittances:

$$\begin{aligned} Y_{11} &= \frac{1}{R_1} + sC \\ Y_{22} &= \frac{1}{R_2} + sC \\ Y_{12} &= -sC \quad (= Y_{21}) \end{aligned} \tag{42.32}$$

For the case $R_1 = 1\,\Omega$, $R_2 = 2\,\Omega$, and $C = 1\,\text{mF}$ we get the plot in Fig. 42.11. Notice how the mutual admittance is negative (purely imaginary, with $-90°$ phase).

Again the premise is that since we know the Y-matrix, we can figure port current for any port voltage stimulus, including transient ones. First we find the Laplace transform of the input voltage

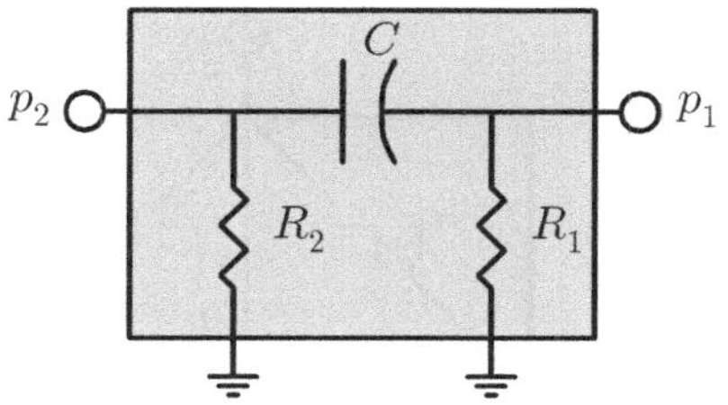

Fig. 42.10 Two-port AC network for Y-parameter calculations

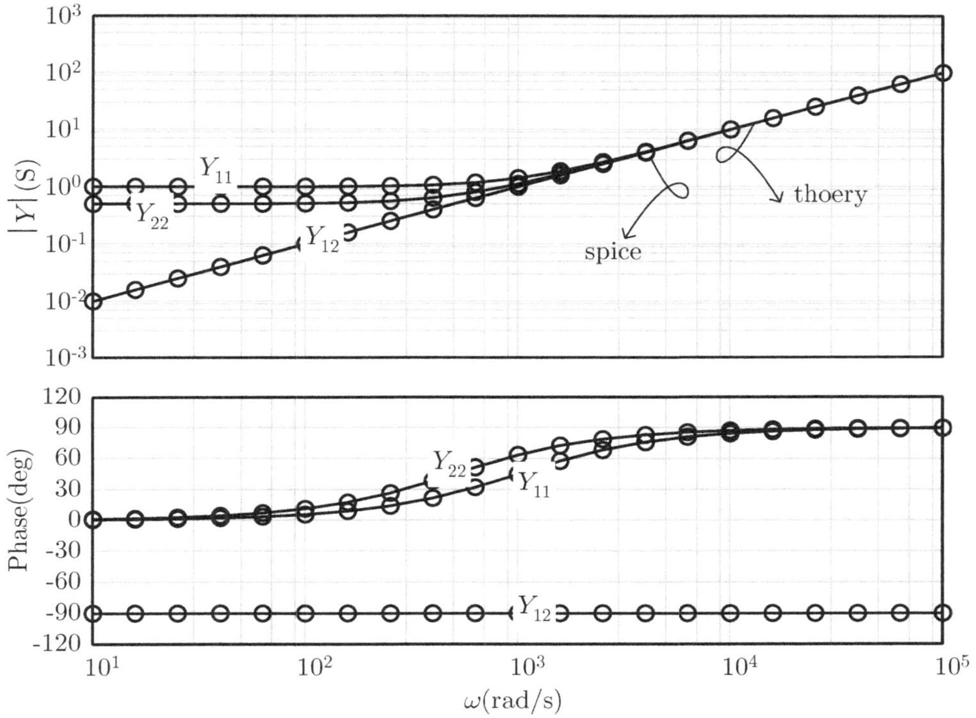

Fig. 42.11 Admittance versus frequency for circuit in Fig. 42.10

and map it to the frequency domain. Next we solve for the port current using $I(s) = G(s)V(s)$. And lastly we convert current back into the time domain.

Three-Port DC Example As a more complicated example, consider the network in Fig. 42.12. To find self admittance of port 1 we short the other two ports to ground, apply unity volt at port 1, and measure current. It is evident that the admittance would be

$$Y_{11} = \frac{1}{R_1} + \frac{1}{R_{12}} + \frac{1}{R_{13}} \tag{42.33}$$

The mutuals between 1 and 2 and 3 are simply

$$Y_{12} = -\frac{1}{R_{12}}$$

$$Y_{13} = -\frac{1}{R_{13}} \tag{42.34}$$

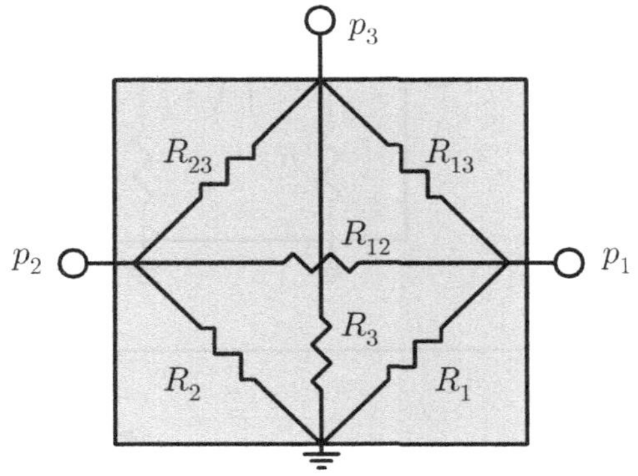

Fig. 42.12 Three-port DC network for Y-parameter calculations

Next we move to port 2. Its self admittance is given by

$$Y_{22} = \frac{1}{R_2} + \frac{1}{R_{12}} + \frac{1}{R_{23}} \tag{42.35}$$

Similarly for port 3

$$Y_{33} = \frac{1}{R_3} + \frac{1}{R_{13}} + \frac{1}{R_{23}} \tag{42.36}$$

The mutual admittance between 2 and 1 is simply that between 1 and 2

$$Y_{21} = Y_{12} \tag{42.37}$$

and the same between 3 and 1

$$Y_{31} = Y_{13} \tag{42.38}$$

Finally the mutual admittance between 2 and 3

$$Y_{23} = Y_{32} = -\frac{1}{R_{23}} \tag{42.39}$$

For a numerical example, consider the case with

$$R_1 = 1, \quad R_2 = 2, \quad R_3 = 3, \quad R_{12} = 4, \quad R_{13} = 5, \quad R_{23} = 6 \tag{42.40}$$

The admittance matrix would then come out to be

$$Y = \begin{bmatrix} 1.45 & -0.25 & -0.2 \\ -0.25 & 0.92 & -0.17 \\ -0.2 & -0.17 & 0.7 \end{bmatrix} \tag{42.41}$$

which has been verified by SPICE simulations.

42.5 Relation Between *Z*- and *Y*-Parameters

Both Z- and Y-parameters relate voltages to currents across the network ports. They differ in which set of variables is the *input* and which is the *output*. But since in the end the network is the same we should expect some relation between the two sets of parameters. For simplicity let us assume that our network is comprised of three ports; having more should not change our conclusions. We know that Z-parameters relate input current to output voltage as follows:

$$\begin{bmatrix} V_1 \\ V_2 \\ V_3 \end{bmatrix} = \begin{bmatrix} Z_{11} & Z_{12} & Z_{13} \\ Z_{21} & Z_{22} & Z_{23} \\ Z_{31} & Z_{32} & Z_{33} \end{bmatrix} \begin{bmatrix} I_1 \\ I_2 \\ I_3 \end{bmatrix} \tag{42.42}$$

Similarly we know that Y-parameters relate input voltage to output current as follows:

$$\begin{bmatrix} I_1 \\ I_2 \\ I_3 \end{bmatrix} = \begin{bmatrix} Y_{11} & Y_{12} & Y_{13} \\ Y_{21} & Y_{22} & Y_{23} \\ Y_{31} & Y_{32} & Y_{33} \end{bmatrix} \begin{bmatrix} V_1 \\ V_2 \\ V_3 \end{bmatrix} \tag{42.43}$$

Let us use this last equation for current and put it in Eq. (42.42) for voltage

$$\begin{bmatrix} V_1 \\ V_2 \\ V_3 \end{bmatrix} = \begin{bmatrix} Z_{11} & Z_{12} & Z_{13} \\ Z_{21} & Z_{22} & Z_{23} \\ Z_{31} & Z_{32} & Z_{33} \end{bmatrix} \begin{bmatrix} Y_{11} & Y_{12} & Y_{13} \\ Y_{21} & Y_{22} & Y_{23} \\ Y_{31} & Y_{32} & Y_{33} \end{bmatrix} \begin{bmatrix} V_1 \\ V_2 \\ V_3 \end{bmatrix} \tag{42.44}$$

Examining this equation closely we see that this can hold true only if

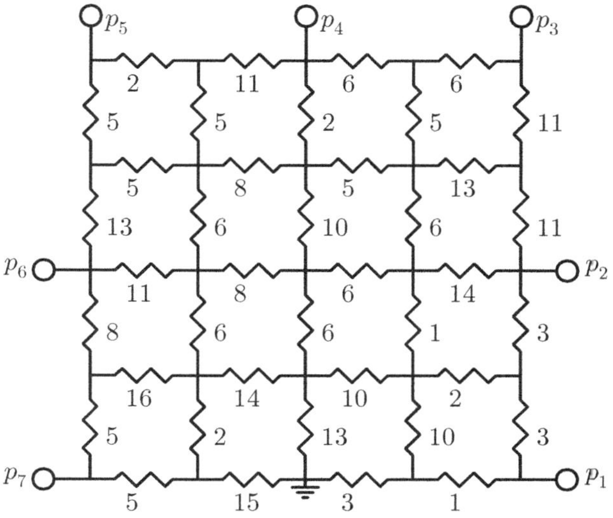

Fig. 42.13 Seven-port test case to validate relation between Z- and Y-matrix

$$\begin{bmatrix} Z_{11} & Z_{12} & Z_{13} \\ Z_{21} & Z_{22} & Z_{23} \\ Z_{31} & Z_{32} & Z_{33} \end{bmatrix} \begin{bmatrix} Y_{11} & Y_{12} & Y_{13} \\ Y_{21} & Y_{22} & Y_{23} \\ Y_{31} & Y_{32} & Y_{33} \end{bmatrix} = \begin{bmatrix} 1 & 0 & 0 \\ 0 & 1 & 0 \\ 0 & 0 & 1 \end{bmatrix} \tag{42.45}$$

Or in symbolic format, and in terms of the identity matrix I

$$\boxed{[Z][Y] = [I]} \tag{42.46}$$

Another way of stating this is to say that Z- and Y-parameters form inverse matrix relations to each other; i.e., they are reciprocal:

$$[Z] = [Y]^{-1}, \qquad [Y] = [Z]^{-1} \tag{42.47}$$

Sample Demonstration of Reciprocity (DC) Consider the 7-port test case in Fig. 42.13. The 40 resistors have been assigned randomly.

If we pass this network to SPICE and extract the Z matrix we get

$$Z = \begin{bmatrix} 3.212 & 2.512 & 2.134 & 1.956 & 1.775 & 1.548 & 1.298 \\ 2.512 & 6.132 & 4.135 & 3.503 & 3.120 & 2.671 & 2.218 \\ 2.134 & 4.135 & 11.440 & 6.013 & 4.915 & 3.894 & 3.112 \\ 1.956 & 3.503 & 6.013 & 8.019 & 6.115 & 4.633 & 3.631 \\ 1.775 & 3.120 & 4.915 & 6.115 & 10.430 & 6.099 & 4.495 \\ 1.548 & 2.671 & 3.894 & 4.633 & 6.099 & 10.110 & 6.131 \\ 1.298 & 2.218 & 3.112 & 3.631 & 4.495 & 6.131 & 9.586 \end{bmatrix} \tag{42.48}$$

On the other hand if we extract the Y matrix we get

$$Y = \begin{bmatrix} 0.470 & -0.164 & -0.009 & -0.026 & -0.005 & -0.006 & -0.007 \\ -0.164 & 0.295 & -0.047 & -0.041 & -0.007 & -0.007 & -0.007 \\ -0.009 & -0.047 & 0.156 & -0.090 & -0.003 & -0.002 & -0.002 \\ -0.026 & -0.041 & -0.090 & 0.318 & -0.108 & -0.023 & -0.013 \\ -0.005 & -0.007 & -0.003 & -0.108 & 0.206 & -0.064 & -0.011 \\ -0.006 & -0.007 & -0.002 & -0.023 & -0.064 & 0.206 & -0.090 \\ -0.007 & -0.007 & -0.002 & -0.013 & -0.011 & -0.090 & 0.175 \end{bmatrix} \tag{42.49}$$

If we form the product sure enough we get the identity matrix:

$$ZY = \begin{bmatrix} 1\,0\,0\,0\,0\,0\,0 \\ 0\,1\,0\,0\,0\,0\,0 \\ 0\,0\,1\,0\,0\,0\,0 \\ 0\,0\,0\,1\,0\,0\,0 \\ 0\,0\,0\,0\,1\,0\,0 \\ 0\,0\,0\,0\,0\,1\,0 \\ 0\,0\,0\,0\,0\,0\,1 \end{bmatrix} \qquad (42.50)$$

So we have in fact validated that the Y matrix is the inverse of the Z matrix and vice versa.

Sample Demonstration of Reciprocity (AC) While the above example demonstrated the reciprocity relation between Z and Y for a resistive DC grid, here we show a more complicated case—that shown in Fig. 42.14. We already derived Y-parameters for this network, and we can derive Z- ones here. After that we can form the ZY product. But since our goal is to show the ZY product, we might as well use a SPICE simulator to figure both Z- and Y-matrices, and leave to us only the task of multiplying the matrices. As the matrices would vary versus frequency we will need to do the product for at least a few frequencies, and if all come out unity, then we have confirmed one more time the reciprocity relation between Z and Y. Assume for now that

$$R_1 = 1\,\Omega, \qquad R_2 = 2\,\Omega, \qquad C = 1\,\text{mF} \qquad (42.51)$$

Notice that we should expect both Z and Y matrices to be complex! Listing below shows Z- and Y-matrices for three frequencies: 1 Hz, 1 kHz, and 1 MHz. For each case we form the product ZY. As shown below, sure enough, for each case we get the identity matrix!

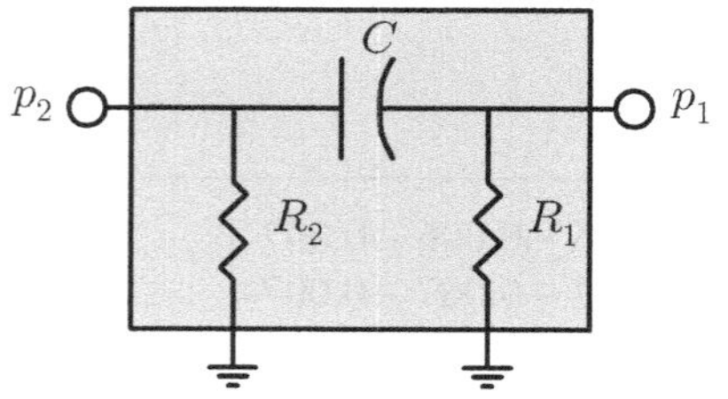

Fig. 42.14 Two-port AC network for ZY product validation

```
[1] "f=1Hz"
[1] "Z = "
                [,1]              [,2]
[1,] 0.9999-0.0063i 0.0002+0.0126i
[2,] 0.0002+0.0126i 1.9990-0.0251i
[1] "Y = "
           [,1]        [,2]
[1,] 1+0.0063i 0.0-0.0063i
[2,] 0-0.0063i 0.5+0.0063i
[1] "Z*Y = "
      [,1] [,2]
[1,] 1+0i 0+0i
[2,] 0+0i 1+0i


[1] "f=1kHz"
[1] "Z = "
                [,1]              [,2]
[1,] 0.6676-0.0176i 0.6648+0.0353i
[2,] 0.6648+0.0353i 0.6704-0.0705i
[1] "Y = "
          [,1]       [,2]
[1,] 1+6.283i 0.0-6.283i
[2,] 0-6.283i 0.5+6.283i
[1] "Z*Y = "
      [,1] [,2]
[1,] 1+0i 0+0i
[2,] 0+0i 1+0i


[1] "f=1MHz"
[1] "Z = "
           [,1]            [,2]
[1,] 0.6667+0i 0.6667+0.0000i
[2,] 0.6667+0i 0.6667-0.0001i
[1] "Y = "
         [,1]      [,2]
[1,] 1+6283i 0.0-6283i
[2,] 0-6283i 0.5+6283i
[1] "Z*Y = "
      [,1] [,2]
[1,] 1+0i 0+0i
[2,] 0+0i 1+0i
```

Again, for any frequency we have the relation

$$Z(s)Y(s) = I \qquad \text{(identity matrix)} \qquad (42.52)$$

42.6 Summary

A physical network is unique; whether we describe it with a flat grid, Z- or Y-parameters, given a set of excitations the network response (in the form of port voltages and currents) must be the same. In this chapter we showed the basic idea behind spatial distribution and distinguished between two classes of network representation: the flat one and the port parameter one. In the flat one we have access to all internal nodes (as well as port ones), but the computation complexity is expensive. For every new set of voltage/current excitation the whole grid needs to be resolved for. In the port parameter class (Z- or Y-parameters), on the other hand, we don't have access to internal nodes, but we form a complete description of the network as seen through the input/output ports. We can now relate port currents to voltages without having to resolve the Z- or the Y-matrices; those are solved for only once. The advantage of the network parameter method is that network characterization is done once as opposed to the flat grid method which needs to characterize the grid every time. Our conclusion is that linear port parameters are sufficient and efficient to characterize the network and its interactions to the outside world. This was demonstrated on a few cases, including DC and AC, and for system sizes as large as 220 elements. It is worth noting that there are yet other kinds of network parameters, such as hybrid ones or transmission ones; but the most relevant for circuit design are the scattering parameter ones, and those will be covered next.

42.7 Problems

1. Consider the R- grid in Fig. 42.15. Four DC current sources are injected across the four ports; the current amount is denoted in the figure. Solve the system and figure voltage at all nodes, using either SPICE or via solving a linear system. For each port ensure that the three outgoing branch currents (through the resistors) are equal to the injected current. See sample solution in the same figure (node voltages marked numerically at each node).
2. Starting with the R-grid in Fig. 42.15 open all current sources and selectively excite one of them, and then measure voltage at all ports; then repeat exciting the other ports and the measurement process until all ports have been exhausted. So in summary, four simulations need to be done and 16 measurements need to be done in total. What is the impedance matrix?
   Answer:
$$Z = \begin{bmatrix} 1.917 & 1.233 & 1.050 & 1.200 \\ 1.233 & 1.855 & 1.294 & 0.997 \\ 1.050 & 1.294 & 1.770 & 0.892 \\ 1.200 & 0.997 & 0.892 & 1.770 \end{bmatrix}$$
3. Starting with Problem 2 and using the impedance matrix derived therein, assume the four current sources are the same as those in Problem 1: what are the four port voltages using the equation $[V] = [Z][I]$? How do those voltage, using Z-parameters, compare to the flat-grid voltages obtained in Problem 1?
4. Consider the RC-grid in Fig. 42.16. Stimulate it with an AC current source as shown in the figure and measure (using SPICE) v_1 which would be the self impedance (Z_{11}) and v_2 which would be the mutual impedance (Z_{12}). Plot the self and mutual impedance as a function of frequency; see sample solution in Fig. 42.17. Explain why both impedances equate at low frequency, and why one is larger than the other at high frequency.
5. Show that we can estimate the self and mutual impedances in Problem 4 with
$$Z_{11}(s) \sim \frac{1+0.7s}{s}, \quad \text{and} \quad Z_{12}(s) \sim \frac{1+0.085s}{s}$$
6. Find the impulse response of the two transfer functions of Problem 5 and then the unit step response. Knowing the latter derive the periodic pulse response, of pulse width 1 and period 2. Plot both $v_1(t)$ and $v_2(t)$ and

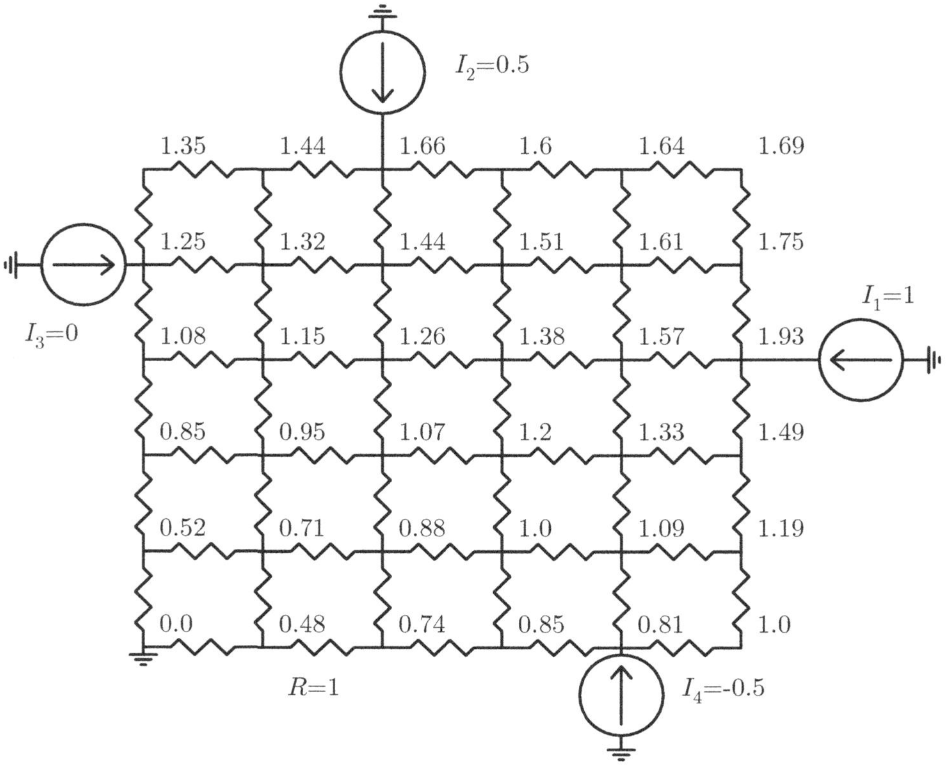

Fig. 42.15 Specification and sample solution to Problem 1

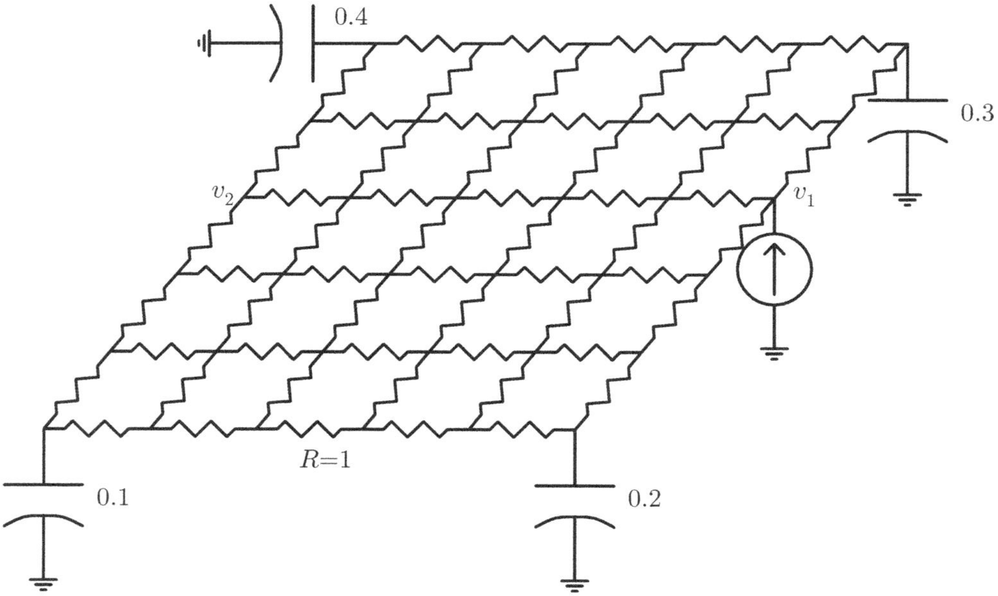

Fig. 42.16 Specification to Problem 4

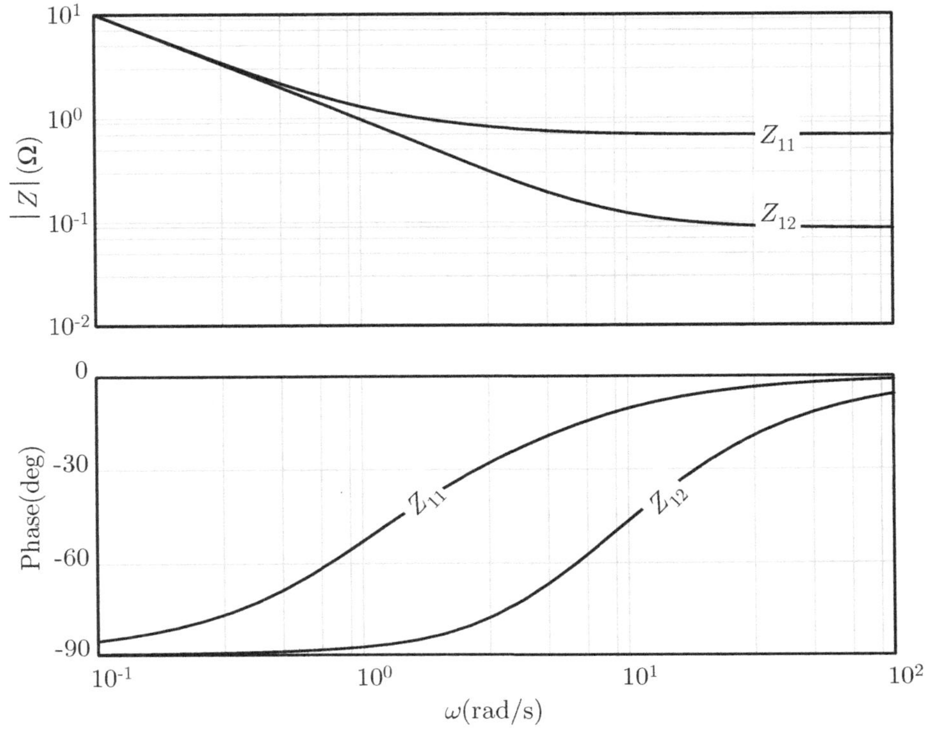

Fig. 42.17 Sample solution to Problem 4

compare to SPICE (running transient simulations on the latter). Keep in mind we won't get 100% match since we had already approximated the transfer functions in Problem 5. See sample solution in Fig. 42.18.

7. Consider again the grid from Problem 1 but now stimulated by voltage sources as shown in Fig. 42.19. Solve for all branch currents (for example using SPICE) as shown in sample solution on the same figure. See note in figure about current polarity. Also, sum currents into each port and compare to those imposed in Problem 1.
8. Starting with the R-grid in Problem 7, zero out all 4 voltage sources and selectively enable one at a time (setting it to 1V), each time measuring all 4 port currents; this by definition is the Y matrix. What is it? Also, assuming Z is the impedance matrix from Problem 2 and Y admittance matrix from this problem, evaluate the ZY and YZ products—explain!
Answer:

$$Y = \begin{bmatrix} 1.175 & -0.416 & -0.148 & -0.488 \\ -0.416 & 1.367 & -0.680 & -0.146 \\ -0.148 & -0.680 & 1.214 & -0.129 \\ -0.488 & -0.146 & -0.129 & 1.043 \end{bmatrix}$$

9. Consider the RC grid in Fig. 42.20. Two voltage sources are applied to the network: one (V_1) with unity magnitude, and the other (V_2) is grounded. Solve for current into the grid points which tie to the two sources using SPICE or a linear solver. The current into the point connecting to V_1 is named Y_{11}, while the current into the point connecting to V_2 is named Y_{12}. Plot both and explain the overall behavior and relative magnitudes. See sample solution in Fig. 42.21.

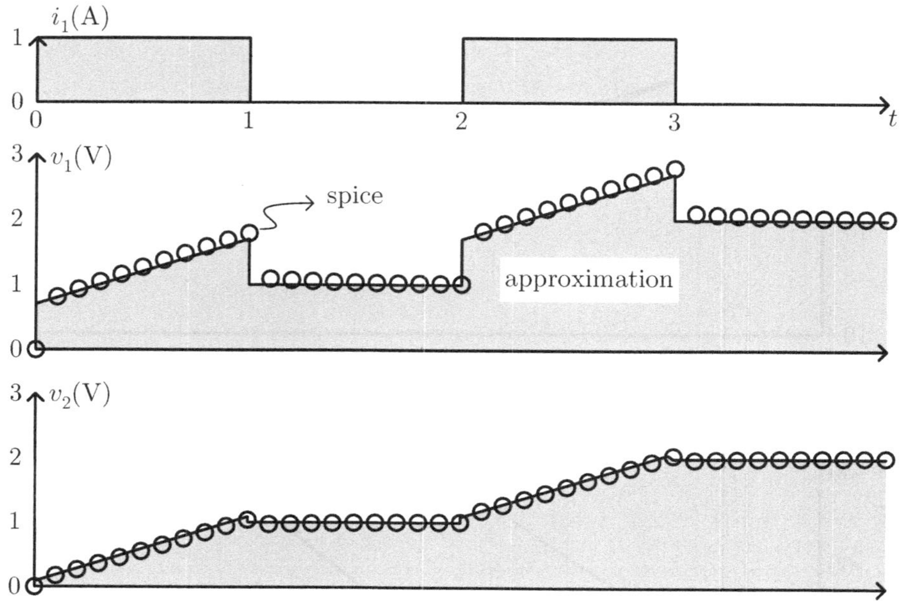

Fig. 42.18 Sample solution to Problem 6

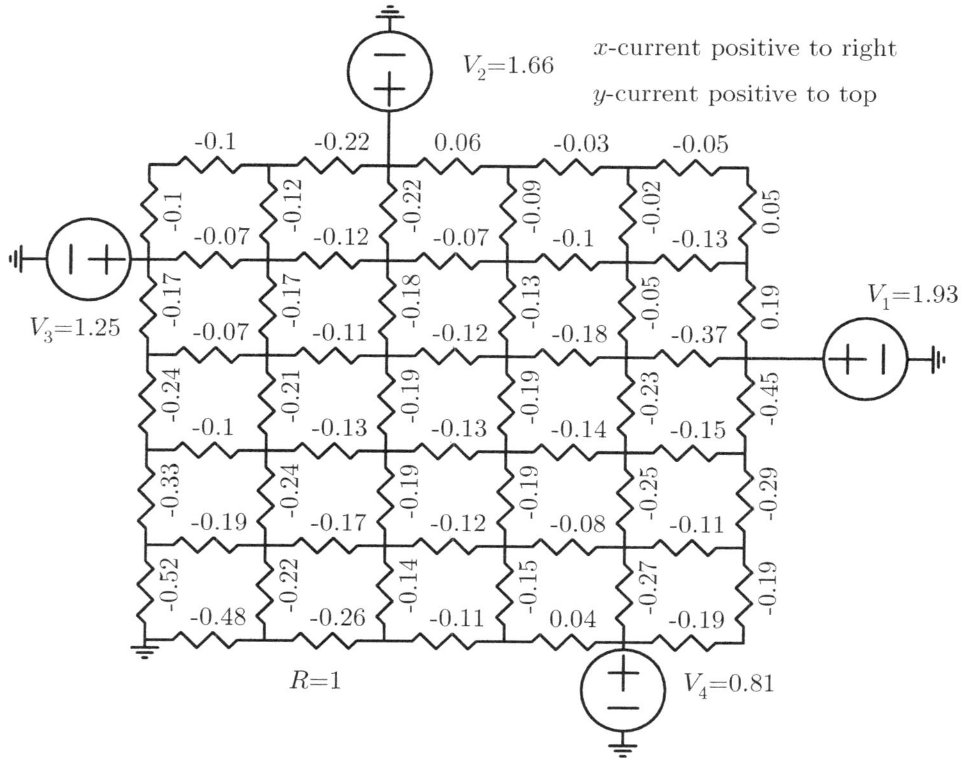

Fig. 42.19 Specification and sample solution to Problem 7

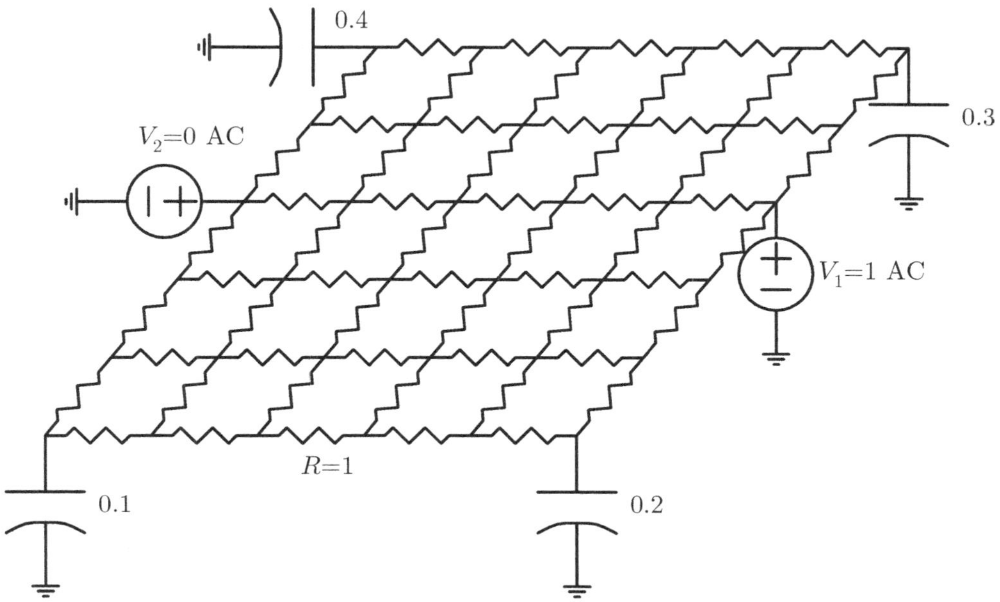

Fig. 42.20 Specification to Problem 9

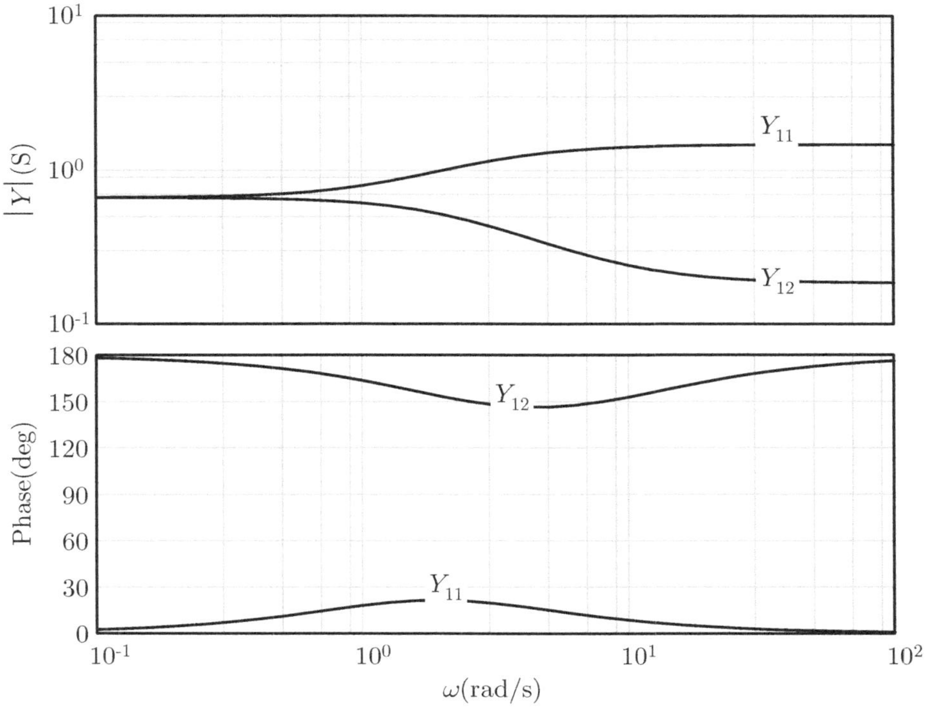

Fig. 42.21 Sample solution to Problem 9

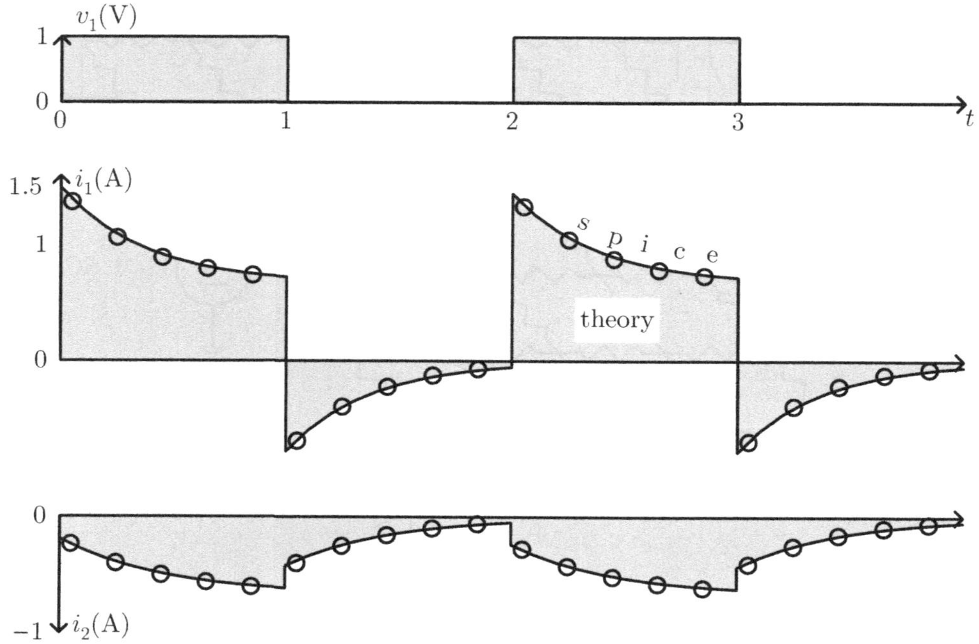

Fig. 42.22 Sample solution to Problem 11

10. Show that the self and mutual admittances in Problem 9 can be approximated by

$$Y_{11} = 1.5\frac{s+1.2}{s+2.7}, \quad \text{and} \quad Y_{12} = -0.19\frac{s+7.7}{s+2.2}$$

11. Knowing the transfer function from Problem 10, figure the current response (at both injection nodes on the grid) in Problem 9 given that $v_1(t)$ is a periodic pulse with period 2 and pulse width 1 and compare to SPICE; see sample solution in Fig. 42.22.

Scattering ($S-$) Parameters

43

43.1 Introduction

Why another set of network parameters? Aren't the Z- and Y-parameters sufficient? First, even if Z- and Y-parameters work fine, that alone does not automatically preclude the introduction of yet another set of linear parameters! On a more practical side, the introduction and use of S-parameters are really rooted in the limitations of the former two. As we recall from the last chapter, the way we measure self and mutual YZ-parameters is via the selective opening or shorting of ports. In other words, we demand certain ports to have zero impedance while other ones to be open. This works just fine for lumped elements and low frequency. But for distributed elements and at high frequency the notion of "short" and "open" becomes blurred! As such a more general method of treating ports is necessary.

When we dealt with YZ-networks our main variables were port currents and voltages. We were able to selectively set some of those to zero; that is port n is open and hence its current is zero, while port m is shorted and hence its terminal voltage is zero. In dealing with S-parameters, on the other hand, the things that we set to zero are the *waves*! We can at will select a port to have zero incident or reflected scattering waves! This is accomplished by *terminating* the port with some *characteristic impedance*. When a port is properly terminated, a wave incident on it will not reflect; it would be completely absorbed. So the analogy here is that we can selectively set some scattering waves to zero, as compared to setting some port voltages/currents to zero in the YZ-counterpart.

43.2 Scattering Parameters in the Emag, Microwaves, and RF Worlds

S-parameters have picked up a lot of momentum in diverse fields, such as Microwave and RF design. There are dedicated tools to deal with them, and those techniques made their way full circle back to the SPICE world. We could argue whether they are really mandatory, but perhaps we are better off simply learning them and utilizing the tools that use them, especially if they happen to be the only medium of communication for certain design problems. For example, package and board models almost universally come in S-par format. The reason there is that quasi-static models (RLC ones) have limited bandwidth, while S-par one don't. As such it is important to have some working knowledge with S-par models and format, but always remembering they are just network description parameters, and what really matters in the end is the network response itself.

F. Badrieh, *Spectral, Convolution and Numerical Techniques in Circuit Theory*,
https://doi.org/10.1007/978-3-319-71437-0_43

43.3 Power Waves

In the frame of S-parameters we deal with incident and reflected waves. But these are not voltage or current waves; instead they are power wave. To be exact, square root of power waves. So we have the following convention:

$$[\text{incident/reflected waves}]^2 \rightarrow \text{power} \quad (43.1)$$

To get voltage out of these waves we use the following unit conversion:

$$\text{incident/reflected waves} \rightarrow \frac{V}{\sqrt{Z_0}} \quad (43.2)$$

where Z_0 is some reference impedance. Notice that when squaring both sides of above equation we get power units, since V^2/R is power.

43.4 Setup for S-Parameters

A sample 2-port network is shown in Fig. 43.1. We designate two incident waves a_1 and a_2 and a byproduct two reflected waves b_1 and b_2. Again these waves have units of $\sqrt{\text{power}}$.

S-parameters come into play by defining a relation between incident and reflected waves: specifically, and for a 2-port network we have

$$\begin{bmatrix} b_1 \\ b_2 \end{bmatrix} = \begin{bmatrix} S_{11} & S_{12} \\ S_{21} & S_{22} \end{bmatrix} \begin{bmatrix} a_1 \\ a_2 \end{bmatrix} \quad (43.3)$$

To reiterate, ***S*-parameters relate reflected waves (output) to incident waves (input).**

43.5 Relation Between Waves and Terminal Currents and Voltages

In the last section we tied input and output waves via the scattering matrix. Here we tie the waves to voltages and currents. With relation to the terminal currents and voltages as shown in Fig. 43.2 the incoming and reflected waves are defined as

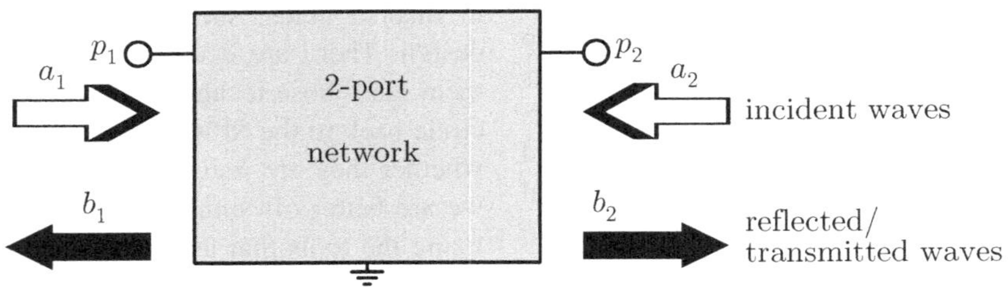

Fig. 43.1 S-parameter network

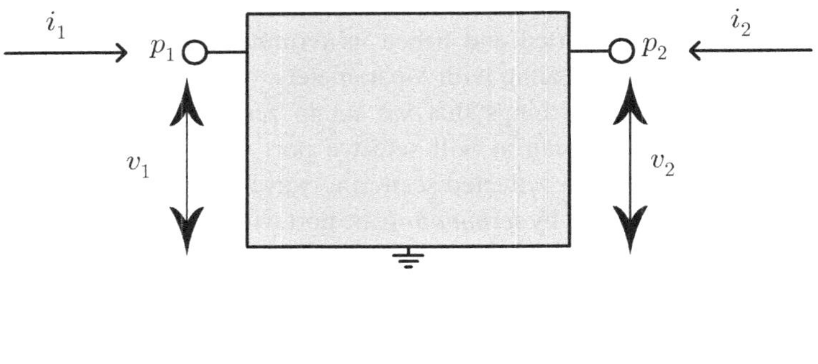

Fig. 43.2 S-parameter network and terminal voltages and currents

$$[\text{incident waves}] \qquad a_1 = \frac{v_1 + Z_0 i_1}{2\sqrt{Z_0}}, \qquad a_2 = \frac{v_2 + Z_0 i_2}{2\sqrt{Z_0}} \tag{43.4}$$

and

$$[\text{reflected/transmitted waves}] \qquad b_1 = \frac{v_1 - Z_0 i_1}{2\sqrt{Z_0}}, \qquad b_2 = \frac{v_2 - Z_0 i_2}{2\sqrt{Z_0}} \tag{43.5}$$

Notice again that each of incident, reflected, and transmitted waves has functional behavior of $\frac{V}{\sqrt{Z}}$; that is voltage divided by square root of impedance, such that waves^2 $\rightarrow$ power. We can rearrange the above equations to isolate terminal voltages and currents such that

$$v_1 = \sqrt{Z_0}\,[a_1 + b_1], \qquad v_2 = \sqrt{Z_0}\,[a_2 + b_2], \tag{43.6}$$

and

$$i_1 = \frac{1}{Z_0}\,[a_1 - b_1], \qquad i_2 = \frac{1}{Z_0}\,[a_2 - b_2] \tag{43.7}$$

So if we know terminal currents and voltages we can figure incident/reflected waves, and conversely if we know port incident/reflected waves we can figure port currents and voltages. The above relations are critically important! We always need to remember how to tie back scattering parameters to real terminal currents and voltages. Finally, if we know incident/reflected waves, we will be able to figure the scattering parameter matrix, as shown next.

43.6 Calculations of *S*-Parameters

We start with Eq. (43.3) defining the *S*-parameter matrix. We expand the matrix and pick the first term:

$$b_1 = S_{11}a_1 + S_{12}a_2 \tag{43.8}$$

To isolate S_{11} we can simply set a_2 to zero; what this means is setting the incident wave at port 2 to zero. Under that condition we have

$$\begin{aligned} S_{11} &= \left.\frac{b_1}{a_1}\right|_{a_2=0} \\ &= \text{ratio of reflected wave at port 1} \\ &\quad \text{to incident wave at same port} \end{aligned} \tag{43.9}$$

Similarly

$$\begin{aligned} S_{22} &= \left.\frac{b_2}{a_2}\right|_{a_1=0} \\ &= \text{ratio of reflected wave at port 2} \\ &\quad \text{to incident wave at same port} \end{aligned} \tag{43.10}$$

To find the pass-through *S*-parameter we again start with

$$b_1 = S_{11}a_1 + S_{12}a_2 \tag{43.11}$$

but this time set a_1 to zero. Then we get

$$\begin{aligned} S_{12} &= \left.\frac{b_1}{a_2}\right|_{a_1=0} \\ &= \text{ratio of transmitted wave at port 1} \\ &\quad \text{to incident wave at port 2} \end{aligned} \tag{43.12}$$

To get S_{21} we start with

$$b_2 = S_{21}a_1 + S_{22}a_2 \tag{43.13}$$

and set $a_2 = 0$. Then

$$\begin{aligned} S_{21} &= \left.\frac{b_2}{a_1}\right|_{a_2=0} \\ &= \text{ratio of transmitted wave at port 2} \\ &\quad \text{to incident wave at port 1} \end{aligned} \tag{43.14}$$

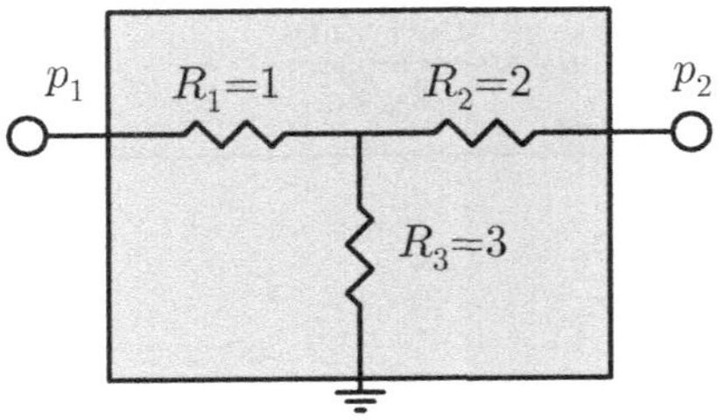

Fig. 43.3 Application of S-parameters on 2-port resistor network

Notice that in all four cases above the S-parameter value is defined in terms of a reflected/transmitted wave divided by an incident wave. Notice also that the b terms can be either reflected or transmitted waves. For example if a_2 is set to zero and a_1 is the only incident wave (at port 1), then b_1 is thought of as the *reflected* wave (at port 1, due to an incident wave at port 1). But if we reverse things such that a_1 is zero and a_2 is the only incident wave, then the same b_1 is now thought of as being the *transmitted* wave (again at port 1, but now due to an incident wave at port 2).

43.7 Sample Application of S-Parameter on a 2-Port Resistor Network

As a simple application of S-parameters consider the network shown in Fig. 43.3. To find S_{11} we apply a voltage source to terminal 1, with its Z_0 characteristic impedance, terminate port 2 with Z_0, figure terminal voltages, break them in terms of incident and reflected waves, and apply S-parameter definitions. In particular assume $Z_0 = 50$ and $V_s = 1$; then we get Fig. 43.4.

To find input current and voltage we need to find total impedance as seen from the applied voltage source. It is

$$Z_{t1} = Z_0 + Z_1 \tag{43.15}$$

Z_1 in turn is given by

$$Z_1 = R_1 + \frac{(R_2 + Z_0)R_3}{R_2 + R_3 + Z_0} = 3.836 \tag{43.16}$$

Hence

$$Z_{t1} = 50 + 3.836 = 53.836 \tag{43.17}$$

Input current is then

$$i_1 = \frac{1}{Z_{t1}} = 0.0186 \tag{43.18}$$

Input voltage is

$$v_1 = i_1 Z_1 = 0.0186 \times 3.836 = 0.071 \tag{43.19}$$

Incident wave is simply

$$a_1 = \frac{v_1 + i_1 Z_0}{2\sqrt{Z_0}} = \frac{0.071 + 0.0186 \times 50}{2\sqrt{Z_0}} = \frac{1}{2\sqrt{Z_0}} \tag{43.20}$$

Input reflected wave is

$$b_1 = \frac{v_1 - i_1 Z_0}{2\sqrt{Z_0}} = \frac{0.071 - 0.0186 \times 50}{2\sqrt{Z_0}} = \frac{-0.859}{2\sqrt{Z_0}} \tag{43.21}$$

We now have all that is needed to figure s_{11}

$$s_{11} = \frac{b_1}{a_1} = -0.859 \tag{43.22}$$

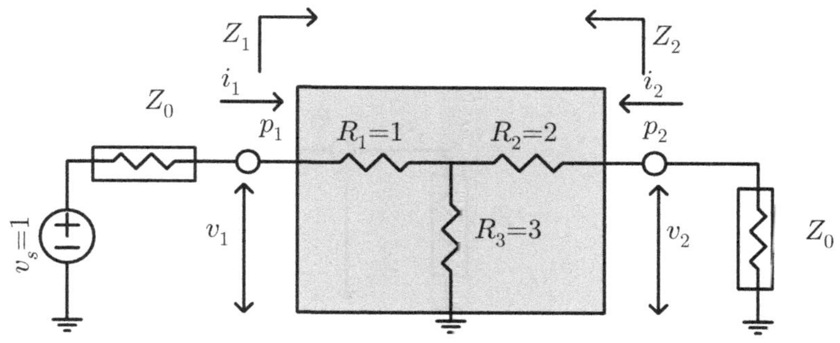

Fig. 43.4 Setup for solving for S_{11} and S_{21}

In closed form, and for future reference, we can summarize as

$$s_{11} = i_1(Z_1 - Z_0) \tag{43.23}$$

or

$$\boxed{s_{11} = \frac{Z_1 - Z_0}{Z_1 + Z_0}} \tag{43.24}$$

To find s_{21} we need output current and voltage. Output current is figured by current division:

$$i_2 = -i_1 \frac{R_3}{R_2 + R_3 + Z_0} = -1.015 \times 10^{-3} \tag{43.25}$$

and output voltage is

$$v_2 = -i_2 Z_0 = 0.051 \tag{43.26}$$

It follows that

$$b_2 = \frac{v_2 - i_2 Z_0}{2\sqrt{Z_0}} = \frac{1}{2\sqrt{Z_0}} 0.101 \tag{43.27}$$

Then

$$s_{21} = \frac{b_2}{a_1} = 0.101 \tag{43.28}$$

In closed form, and for future reference, we can summarize as

$$s_{21} = v_2 - i_2 Z_0 = -i_2 Z_0 - i_2 Z_0 = -2 i_2 Z_0 \tag{43.29}$$

or

$$\boxed{s_{21} = 2Z_0 \frac{1}{Z_1 + Z_0} \frac{R_3}{R_2 + R_3 + Z_0}} \tag{43.30}$$

To find the other two S-parameters we use the setup in Fig. 43.5. We can use the above equations provided we swap R_1 and R_2. For example

$$Z_2 = R_2 + \frac{(R_1 + Z_0)R_3}{R_1 + R_3 + Z_0} = 4.833 \tag{43.31}$$

such that total impedance seen from the voltage source is

$$Z_{t2} = Z_2 + Z_0 = 50 + 4.833 = 54.833 \tag{43.32}$$

Input current is then

$$i_2 = \frac{1}{Z_{t2}} = 0.0182 \tag{43.33}$$

Input voltage is

$$v_2 = i_2 Z_2 = 0.0182 \times 4.833 = 0.088 \tag{43.34}$$

Incident wave is simply

$$a_2 = \frac{1}{2\sqrt{Z_0}} \tag{43.35}$$

and input reflected wave is

$$b_2 = \frac{v_2 - i_2 Z_0}{2\sqrt{Z_0}} = \frac{0.088 - 0.0182 \times 50}{2\sqrt{Z_0}} = -\frac{0.822}{2Z_0} \tag{43.36}$$

It follows:

$$s_{22} = \frac{b_2}{a_2} = -0.822 \tag{43.37}$$

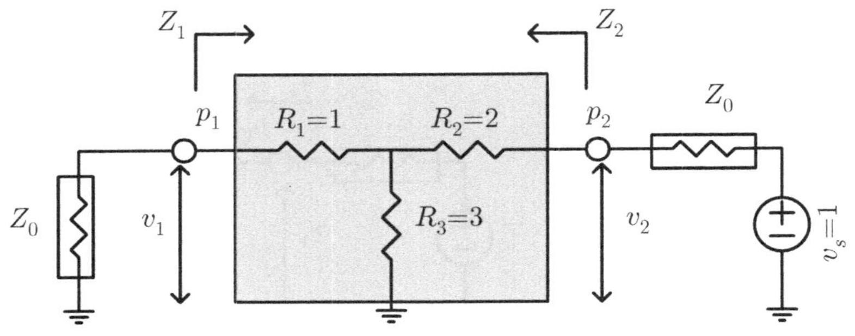

Fig. 43.5 Setup for solving for S_{22} and S_{12}

In closed form we have

$$\boxed{s_{22} = \frac{Z_2 - Z_0}{Z_2 + Z_0}} \tag{43.38}$$

By symmetry we get

$$s_{12} = 0.101 \tag{43.39}$$

Hence our s-matrix is

$$S = \begin{bmatrix} -0.859 & 0.101 \\ 0.101 & -0.822 \end{bmatrix} \tag{43.40}$$

All these numbers have been verified by SPICE simulations.

43.8 Pass-Through Resistor

Let's find the s-matrix of a simple pass through resistor as shown in Fig. 43.6. Our starting point is Eq. (43.24), repeated here for convenience:

$$s_{11} = \frac{Z_1 - Z_0}{Z_1 + Z_0} \tag{43.41}$$

where Z_1 would be $10 + Z_0 = 60$; hence

$$s_{11} = \frac{60 - 50}{60 + 50} = 0.091 \tag{43.42}$$

Also from the prior section we use Eq. (43.30), again repeated here for convenience

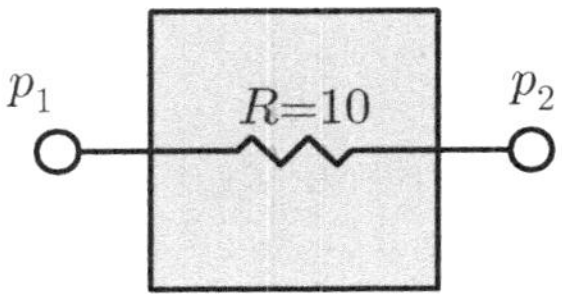

Fig. 43.6 Application of S-parameter pass-through resistor

$$s_{21} = 2Z_0 \frac{1}{Z_1 + Z_0} \frac{R_3}{R_2 + R_3 + Z_0} \tag{43.43}$$

Notice here we have $R_3 = \infty$ and hence above equation collapses to

$$s_{21} = 2Z_0 \frac{1}{Z_1 + Z_0} = 2 \times 50 \frac{1}{60 + 50} = 0.909 \tag{43.44}$$

The other two elements are calculated similarly; our final s-matrix is then

$$S = \begin{bmatrix} 0.091 & 0.909 \\ 0.909 & 0.091 \end{bmatrix} \tag{43.45}$$

This was verified in SPICE!

43.9 Shunt Resistor

Let's find the s-matrix of the shunt resistor shown in Fig. 43.7. Our starting point is Eq. (43.24)

$$s_{11} = \frac{Z_1 - Z_0}{Z_1 + Z_0} \tag{43.46}$$

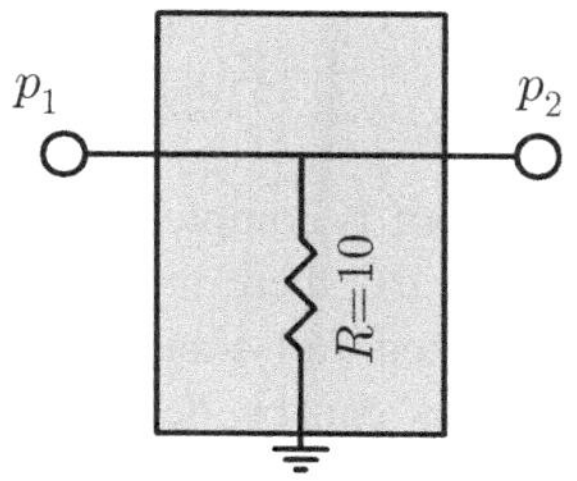

Fig. 43.7 Application of S-parameter to shunt resistor

Here Z_1 is the parallel combination of R and the termination resistance $Z_0 = 50$; hence we have

$$Z_1 = 10||50 = 8.333\,\Omega \tag{43.47}$$

It follows then that

$$s_{11} = \frac{8.333 - 50}{8.333 + 50} = -0.714 \tag{43.48}$$

For s_{21} we reuse Eq. (43.30)

$$s_{21} = 2Z_0 \frac{1}{Z_1 + Z_0} \frac{R_3}{R_2 + R_3 + Z_0} \tag{43.49}$$

In our case, we have $R = R_3$ and $R_2 = 0$; hence we have

$$s_{21} = 2Z_0 \frac{1}{Z_1 + Z_0} \frac{R}{R + Z_0} = 2 \times 50 \frac{1}{8.333 + 50} \frac{10}{10 + 50} = 0.286 \tag{43.50}$$

Hence our s-matrix is

$$s = \begin{bmatrix} -0.714 & 0.286 \\ 0.286 & -0.714 \end{bmatrix} \tag{43.51}$$

43.10 Shunt Capacitor

Let's find the s-matrix of the shunt cap shown in Fig. 43.8. Similar to last section we start by finding Z_1

$$Z_1 = \frac{1}{sC}||Z_0 = \frac{1}{C} \frac{1}{s + \frac{1}{Z_0C}} \tag{43.52}$$

Next we plug into the s_{11} formula

$$s_{11} = \frac{\frac{1}{C}\frac{1}{s+\frac{1}{Z_0C}} - Z_0}{\frac{1}{C}\frac{1}{s+\frac{1}{Z_0C}} + Z_0} = \frac{\frac{1}{C} - sZ_0 - \frac{1}{C}}{\frac{1}{C} + sZ_0 + \frac{1}{C}} = \frac{-sZ_0}{sZ_0 + \frac{2}{C}} \tag{43.53}$$

$$\boxed{s_{11} = -\frac{s}{s + \frac{2}{Z_0C}}} \tag{43.54}$$

For s_{21} we have

$$s_{21} = 2Z_0 \frac{1}{\frac{Z_0}{1+sZ_0C} + Z_0} \frac{\frac{1}{sC}}{Z_0 + \frac{1}{sC}} = 2 \frac{1}{\frac{1}{1+sZ_0C} + 1} \frac{1}{1 + sZ_0C} \tag{43.55}$$

$$s_{21} = \frac{2}{1 + 1 + sZ_0C} = \frac{2}{2 + sZ_0C} = \boxed{\frac{1}{Z_0C} \frac{2}{s + \frac{2}{Z_0C}}} \tag{43.56}$$

Results and comparison to SPICE are shown in Fig. 43.9. By symmetry we can get s_{22} and s_{12}. Notice that unlike prior section, the S-parameters are now frequency dependent.

In passing let us rationalize for example why s_{11} is zero at low frequency? Recall the definitions of $s_{11} = \frac{b_1}{a_1}$ (when p_2 is terminated). If s_{11} is zero this would imply the b_1 is zero too. So let's prove that to be the case. Recall from Eq. (43.5) that

$$b_1 = \frac{v_1 - Z_0 i_1}{2\sqrt{Z_0}} \quad (43.57)$$

So what we want to prove is that v_1 comes out equal to $Z_0 i_1$. At DC the cap is open and the network collapses to that of an input voltage source applied across two Z_0 impedances (due to source and termination impedances). Hence, by voltage division, voltage at terminal 1 comes out $v_1 = \frac{1}{2}$. By the same token input current comes out $i_1 = \frac{1}{2Z_0}$. Plugging back we get

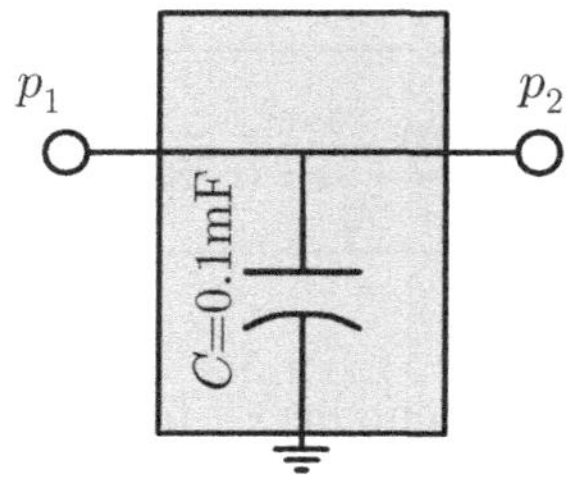

Fig. 43.8 Application of S-parameter to shunt capacitor

$$b_1 \sim v_1 - Z_0 i_1 = \frac{1}{2} - Z_0 \frac{1}{2Z_0} = \frac{1}{2} - \frac{1}{2} = 0 \quad (43.58)$$

Hence the proof is complete! Essentially this is saying that at DC there is no reflection, and by the same token full transmission. At high frequency, and as can be deciphered from the graph, roles flip and we get full reflection and zero transmission.

Looking at the phase it mostly makes sense other than that of s_{11}. It would appear that since s_{11} is ramping up, it should have had a phase of 90° and when it settles down at high frequency it should have had a zero phase! But looking closer at Eq. (43.54) we observe a negative sign! Nonetheless let's do a quick verification. At small frequency s_{11} approaches $-s$ (times a constant) and when using $s = -j\omega$ we get on the complex plane a point that lies at $-90°$ phase; so that limit is sound. On the other hand at high frequency s_{11} approaches -1 and that simply has the phase $-180°$ which also matches the plot. So by quickly validating the two frequency extremes we are assured that the phase plot is sound too.

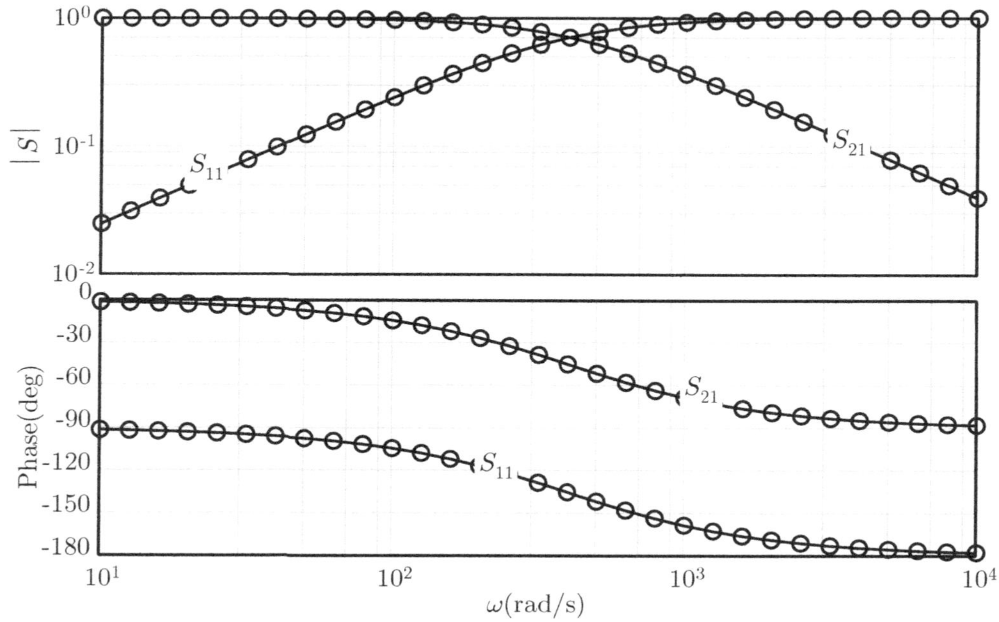

Fig. 43.9 S-parameters of shunt capacitor

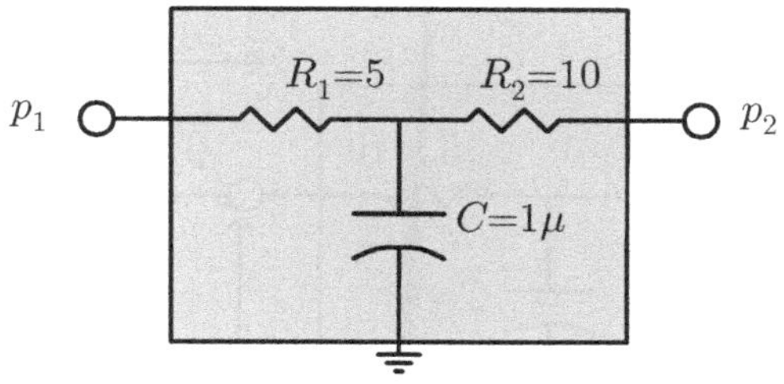

Fig. 43.10 Application of S-parameter on 2-port RC network

43.11 Sample Application of S-Parameter on a 2-Port RC Network

Let's find the s-matrix of the network shown in Fig. 43.10. The setup for finding s_{11} and s_{12} is shown in Fig. 43.11. From Eq. (43.24) we know that

$$s_{11} = \frac{Z_1 - Z_0}{Z_1 + Z_0} \tag{43.59}$$

where

$$Z_1 = R_1 + \frac{R_2 + Z_0}{1 + (R_2 + Z_0)sC} \tag{43.60}$$

A comparison between SPICE and our results is shown in Fig. 43.12. Next is s_{12} which is given by (based on Eq. (43.30))

$$s_{12} = 2Z_0 \frac{1}{Z_1 + Z_0} \frac{\frac{1}{sC}}{R_2 + Z_0 + \frac{1}{sC}} \tag{43.61}$$

$$s_{12} = 2Z_0 \frac{1}{Z_1 + Z_0} \frac{1}{1 + sC(R_2 + Z_0)} \tag{43.62}$$

A comparison between SPICE and our results is shown in Fig. 43.13. Let us compare the magnitude of S_{11} and S_{12}; this is shown in Fig. 43.14. Notice that at low frequency, the cap is open and we have

$$Z_1(0) \sim R_1 + R_2 + Z_0 = 65 \tag{43.63}$$

Hence

$$s_{11}(0) \sim \frac{65 - 50}{65 + 50} = 0.13 \tag{43.64}$$

as shown in the figure. On the other hand, at high frequency the cap shorts and we have the approximation

$$Z_1(\infty) \sim 5 \tag{43.65}$$

Hence

$$s_{11}(\infty) \sim \frac{5 - 50}{5 + 50} = 0.82 \tag{43.66}$$

again as shown in the figure. Moving to s_{12} and using Eq. (43.62) at DC we have the limit

$$s_{12}(0) \sim 2Z_0 \frac{1}{Z_1 + Z_0} = 2Z_0 \frac{1}{R_1 + R_2 + Z_0 + Z_0} = 2 \times 50 \frac{1}{5 + 10 + 50 + 50} = 0.87 \tag{43.67}$$

as can be confirmed in the figure. Finally at high frequency and again using Eq. (43.62) we see the limit

$$s_{12}(\infty) \sim 0 \tag{43.68}$$

as can be confirmed in the figure. The exact same procedure can be done to figure s_{22} and s_{21}.

43.12 Relationship Between S- and Z-Matrix

Recall (Eq. (43.4)) the definition of the incident waves (for a two-port case here)

$$a_1 = \frac{v_1 + i_1 Z_0}{2\sqrt{Z_0}}, \quad \text{and} \quad a_2 = \frac{v_2 + i_2 Z_0}{2\sqrt{Z_0}} \tag{43.69}$$

Fig. 43.11 Setup to find s_{11} for network in Fig. 43.10

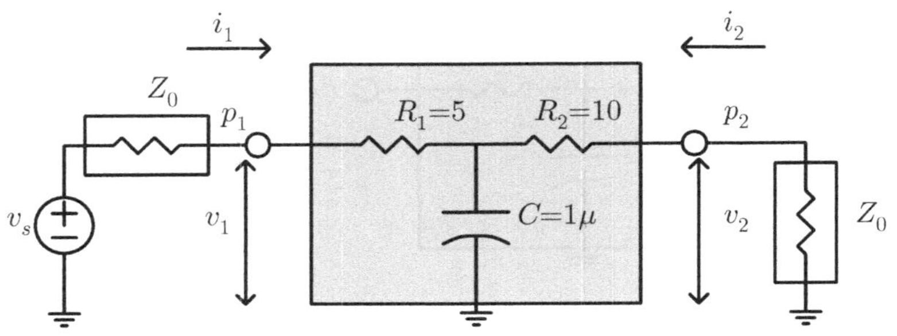

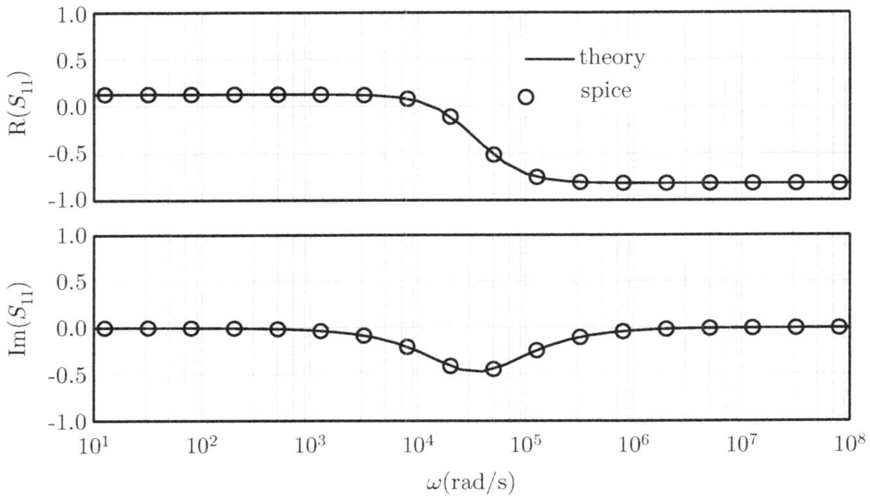

Fig. 43.12 S_{11} results for Fig. 43.10

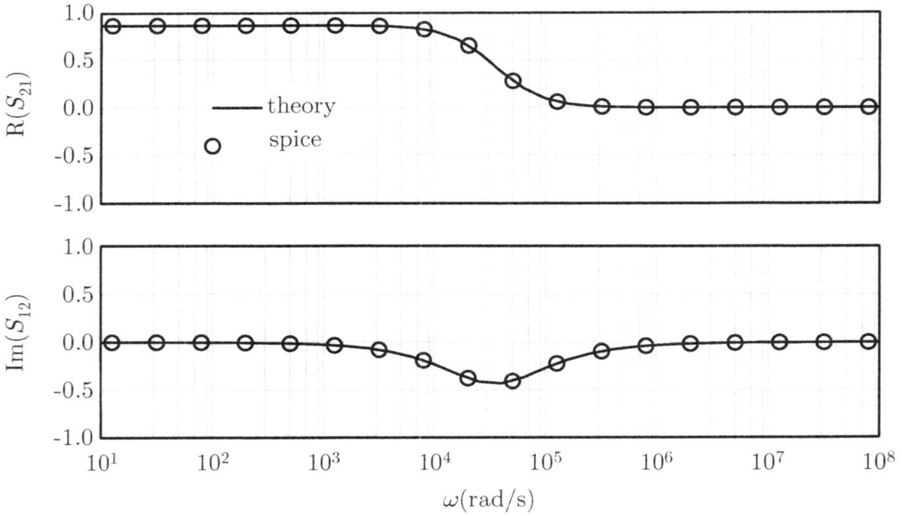

Fig. 43.13 S_{12} results for Fig. 43.10

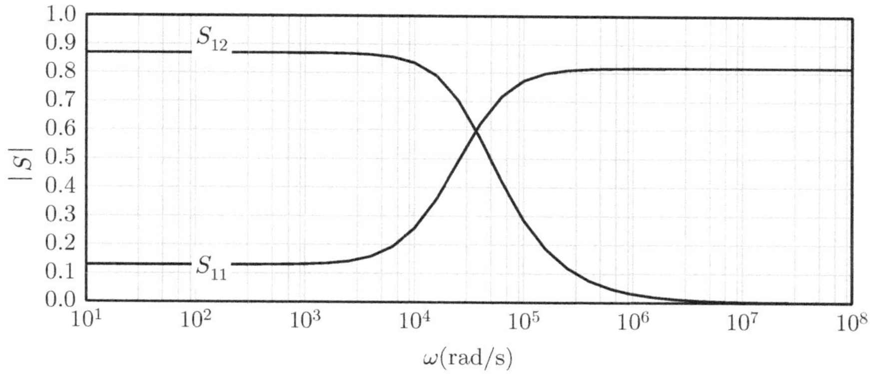

Fig. 43.14 Magnitude comparison between S_{11} and S_{12} results for Fig. 43.10

Similarly (Eq. (43.5)) we get for the reflected waves

$$b_1 = \frac{v_1 - i_1 Z_0}{2\sqrt{Z_0}}, \quad \text{and} \quad b_2 = \frac{v_2 - i_2 Z_0}{2\sqrt{Z_0}} \tag{43.70}$$

Let us define

$$\mathbf{a}=\begin{bmatrix} a_1 \\ a_2 \end{bmatrix}, \quad \mathbf{b}=\begin{bmatrix} b_1 \\ b_2 \end{bmatrix}, \quad \mathbf{v}=\begin{bmatrix} v_1 \\ v_2 \end{bmatrix}, \quad \mathbf{i}=\begin{bmatrix} i_1 \\ i_2 \end{bmatrix} \tag{43.71}$$

Then we can recast the above equations as

$$\mathbf{a} = \frac{\mathbf{v} + \mathbf{i}Z_0}{2\sqrt{Z_0}}, \quad \text{and} \quad \mathbf{b} = \frac{\mathbf{v} - \mathbf{i}Z_0}{2\sqrt{Z_0}} \tag{43.72}$$

With the new definitions, Eq. (43.6)—which tied incident and reflected waves to terminal voltages—then becomes

$$\mathbf{v} = \sqrt{Z_0}(\mathbf{a} + \mathbf{b}) \tag{43.73}$$

Let us replace

$$\mathbf{b} = S\mathbf{a} \tag{43.74}$$

to get

$$\mathbf{v} = \sqrt{Z_0}(I + S)\mathbf{a} \tag{43.75}$$

where I is the identity matrix. Now we plug back for $\mathbf{a}$ to get

$$\mathbf{v} = \sqrt{Z_0}(I + S)\frac{\mathbf{v} + Z_0\mathbf{i}}{2\sqrt{Z_0}} = \frac{1}{2}(I + S)(\mathbf{v} + Z_0\mathbf{i}) \tag{43.76}$$

Collect terms

$$(I - S)\,\mathbf{v} = Z_0\,(I + S)\,\mathbf{i} \tag{43.77}$$

Multiply both sides by the matrix inverse of $(I - S)$ to get

$$\mathbf{v} = Z_0(I - S)^{-1}(I + S)\mathbf{i} \tag{43.78}$$

If we compare this to the definition of the Z matrix

$$\mathbf{v} = Z\mathbf{i} \tag{43.79}$$

we conclude that

$$\boxed{Z = Z_0(I - S)^{-1}(I + S)} \tag{43.80}$$

This relation gives Z knowing S. To find S out of Z we rewrite as

$$(1 - S)Z = Z_0(I + S) \tag{43.81}$$

Collect terms

$$S(Z_0 I + Z) = Z - Z_0 I \tag{43.82}$$

then solve for S

$$\boxed{S = (Z - Z_0 I)(Z + Z_0 I)^{-1}} \tag{43.83}$$

At last we made the connection between S- and Z-parameters. We were so accustomed to Z-parameters (and Y-ones) that introducing S-ones left us with a bit of unease! But now we know knowing one gives the other; there is nothing mystical about S-parameters. It goes unsaid that we now also know the relation between S-parameters and Y-ones as well.

To test the above relations let's reconsider the simple RC network shown in Fig. 43.15. We already extracted the S-matrix for this network in Sect. 43.11. Now if we take this S-matrix, at each frequency, then apply Eq. (43.80) we would get the Z-matrix. If we then plot the Z-matrix and compare it to the real one (extracted separately) we get Fig. 43.16 for Z_{11} and Fig. 43.17 for Z_{12}. Notice the exact match and also that *results are independent of the characteristics impedance* Z_0 *choice*.

Notice that as expected Z_{11} starts open due to the cap and settles to $R_1 = 5\,\Omega$. Its phase starts at $-90°$ (capacitive) and settles to zero (resistive). On the other hand Z_{12} is simply $\frac{1}{sC}$ which is open at DC and short at high frequency; it has a consistent $-90°$ for all frequencies. Finally for more details about scattering parameters see Orfanidis, Chap. 13.

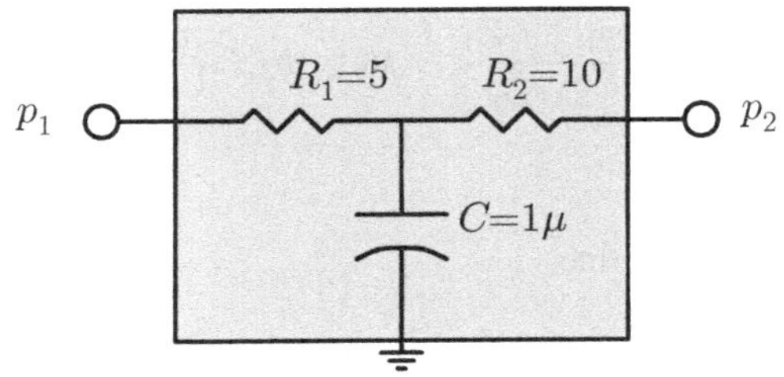

Fig. 43.15 Simple RC network to test relation between S- and Z-matrices

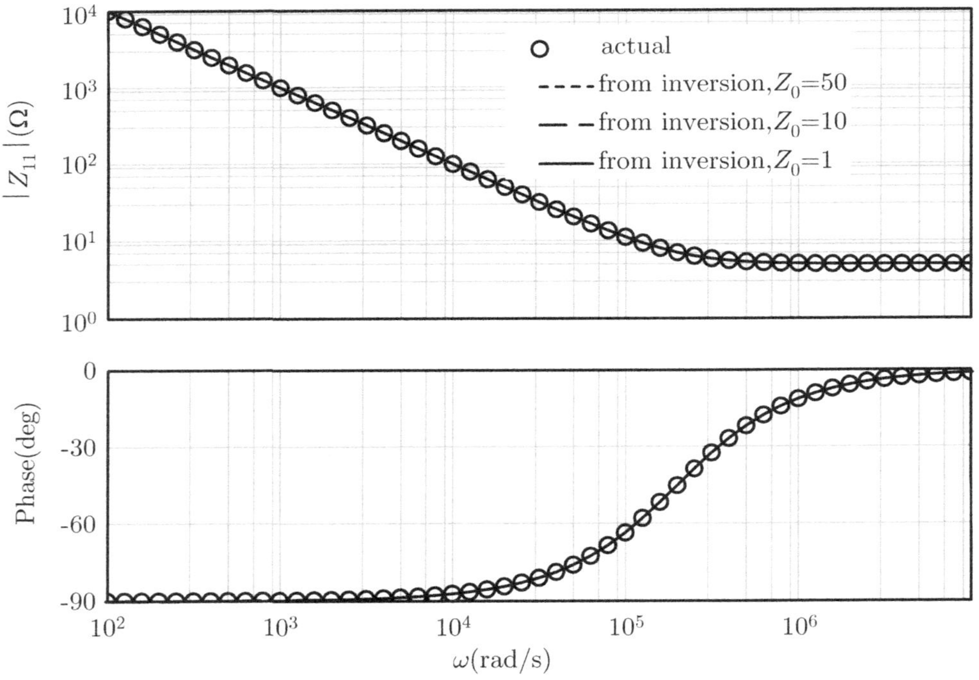

Fig. 43.16 Z_{11} of circuit in Fig. 43.15 and comparison to that extracted from S matrix inversion

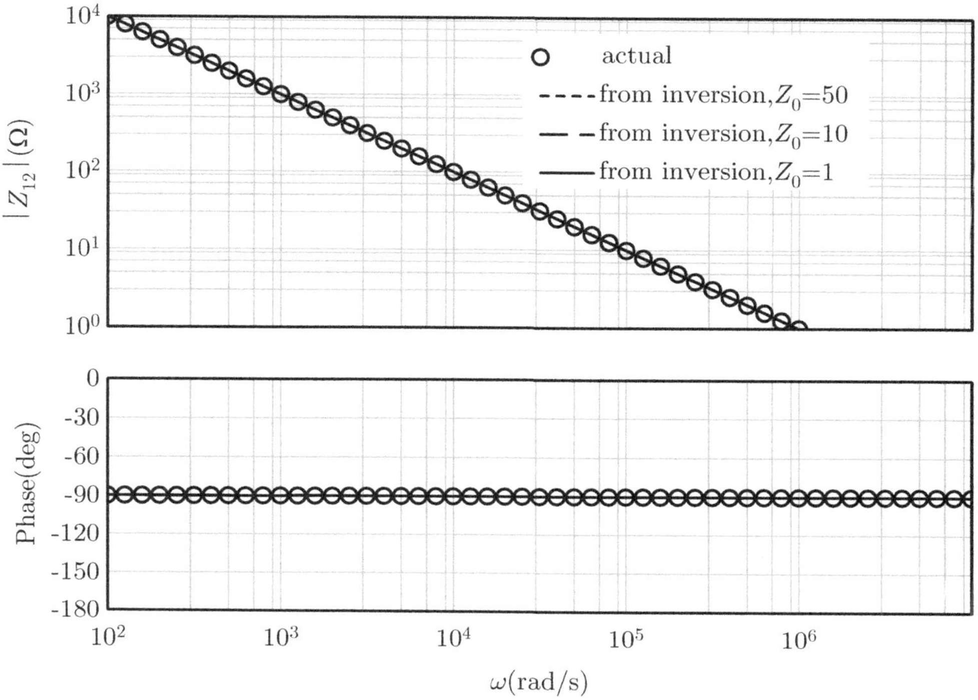

Fig. 43.17 Z_{12} of circuit in Fig. 43.15 and comparison to that extracted from S matrix inversion

43.13 Summary

Scattering parameters are extremely important in various fields where signal speed is high, and of particular importance for us in the field of circuits, package, and board design. At the core level S-parameters behave the same as Z- or Y-ones in the sense of describing multi-port networks in the frequency domain. They hold an advantage at high frequency where the other two tend to suffer due to the lack of clear definition of open and short. Scattering parameters deal with incident, reflected, and transmitted waves at the various network ports. The waves utilized are power waves—more specifically square root of power waves. Scattering parameters are calculated by terminating each port with a characteristic impedance Z_0 to ensure the selective elimination of some S-parameters while enabling the calculation of the rest. Similar to Z- and Y-parameters, S-parameters have magnitude and phase, or real and imaginary parts. Once in that form they are merely a transfer function with certain numerical dependence on frequency. There is a direct conversion relation between Z-matrices and S-ones and vice versa; and the same holds for S-to-Y conversion. In this chapter, and after deriving the S-matrix and its relation to port voltages and currents we extracted the S-matrix for a number of circuits and validated with SPICE simulations. We wrapped the chapter with a confirmation of the S-to-Z conversion formula.

43.14 Problems

1. Consider the two-port RL network in Fig. 43.18a. Calculate and plot the scattering matrix, and compare to SPICE. Use $Z_0 = 50\,\Omega$ in the calculations; see sample solution in Fig. 43.19.

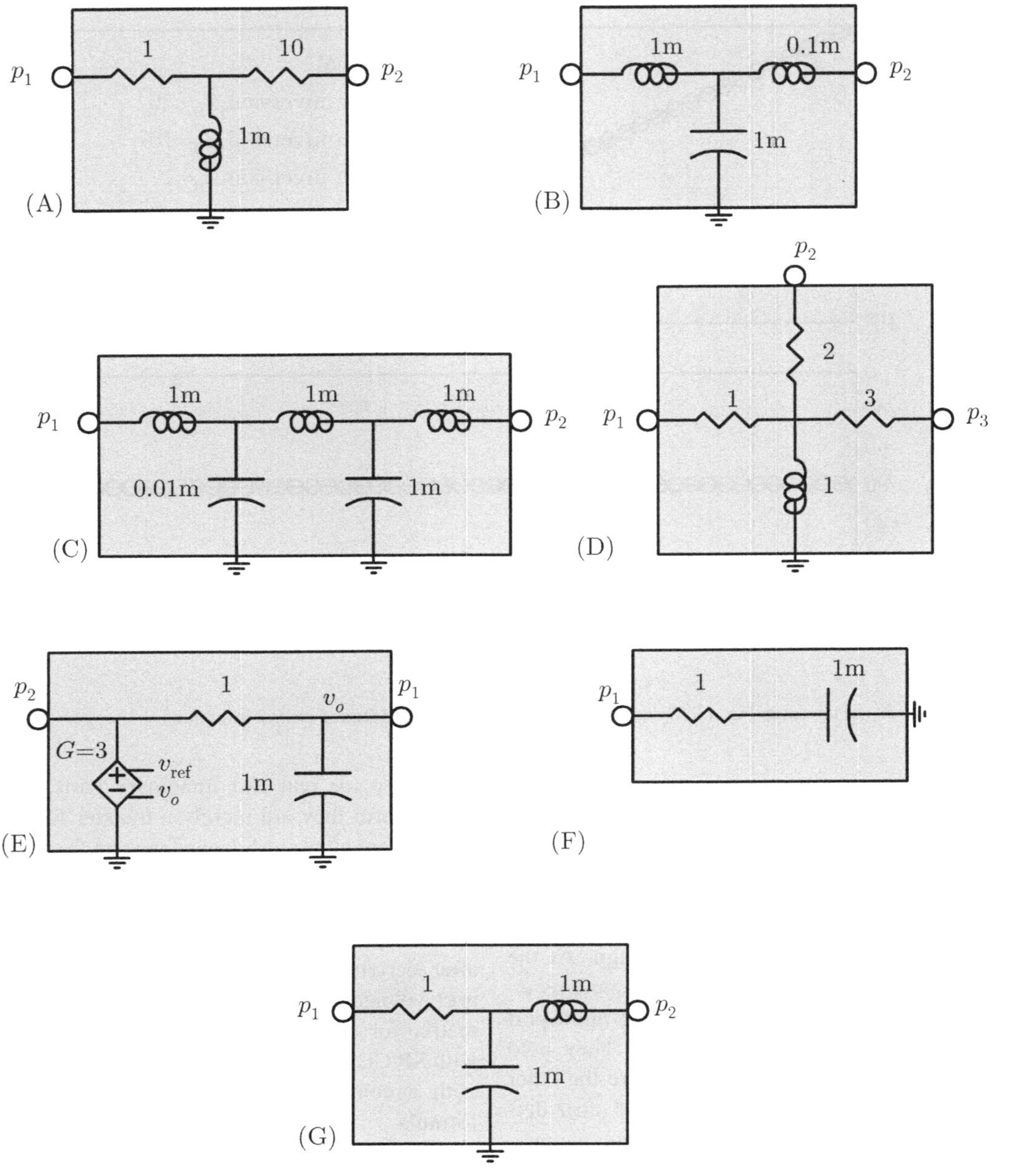

Fig. 43.18 Various circuits used in Problems section

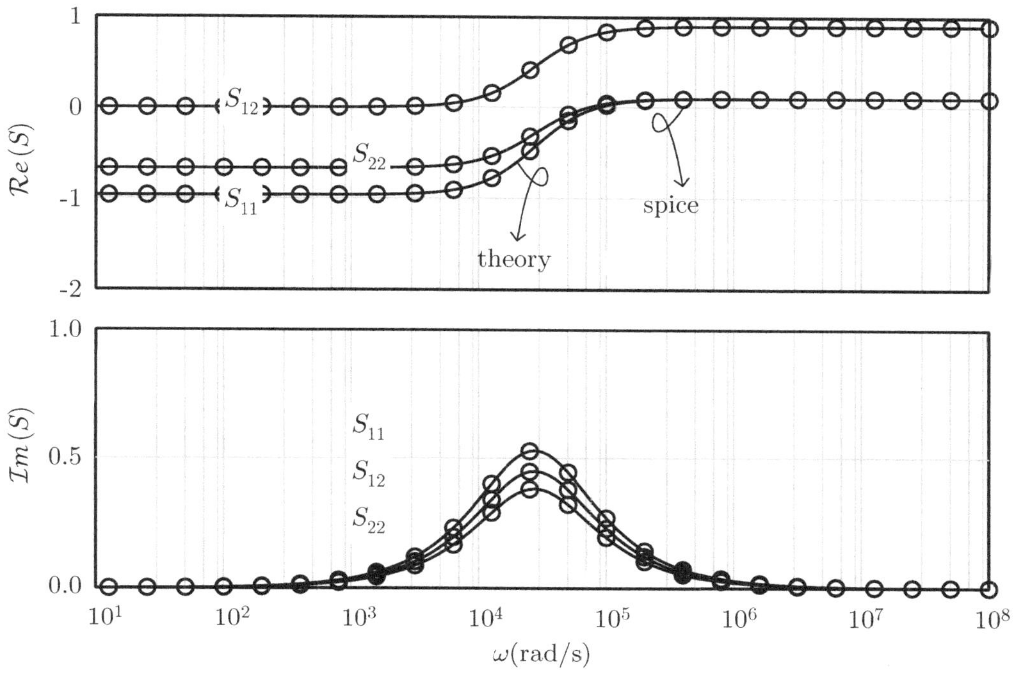

Fig. 43.19 Sample solution to Problem 1

2. Consider the two-port *LC* network in Fig. 43.18b. Calculate and plot the scattering matrix, and compare to SPICE. Use $Z_0 = 50\,\Omega$ in the calculations; see sample solution in Fig. 43.20.
3. Consider the two-port *LC* network in Fig. 43.18c. Calculate and plot S_{11} and S_{12}, and compare to SPICE. Use $Z_0 = 50\,\Omega$ in the calculations; see sample solution in Fig. 43.21.
4. Consider the 3-port *RL* network in Fig. 43.18d. Extract the scattering parameter matrix at frequency $f = 1$ Hz and compare to SPICE. Use $Z_0 = 10$ in the calculations.
Answer:

$$S = \begin{bmatrix} -0.349 + 0.298i & 0.430 + 0.273i & 0.397 + 0.252i \\ 0.430 + 0.273i & -0.272 + 0.250i & 0.364 + 0.231i \\ 0.397 + 0.252i & 0.364 + 0.231i & -0.202 + 0.213i \end{bmatrix}$$

5. Consider the two-port *RC* network in Fig. 43.18e. Calculate and plot S_{11} and S_{12}, and compare to SPICE. Use $Z_0 = 1\,\Omega$ in the calculations; see sample solution in Fig. 43.22.
6. Consider the single-port *RC* network in Fig. 43.18f. It could for example represent a nonideal cap model. Calculate and plot S_{11}, and compare to SPICE. Use $Z_0 = 1\,\Omega$ in the calculations; see sample solution in Fig. 43.23.
7. Consider again the single-port *RC* network in Fig. 43.18f. Recalculate and plot S_{11} for 4 different Z_0 values—1, 2, 5, and 10; and compare to SPICE. See sample solution in Fig. 43.24.

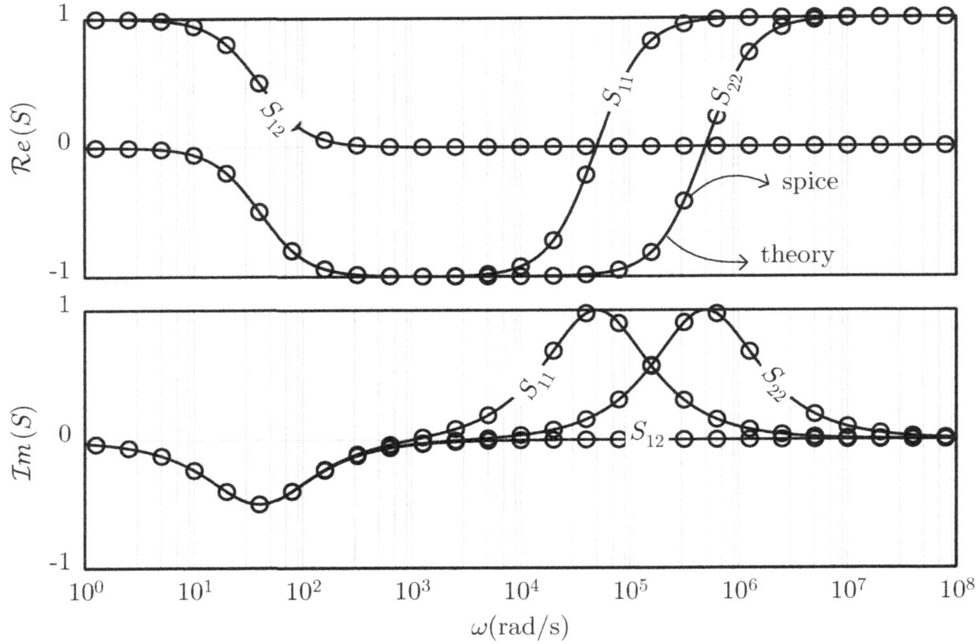

Fig. 43.20 Sample solution to Problem 2

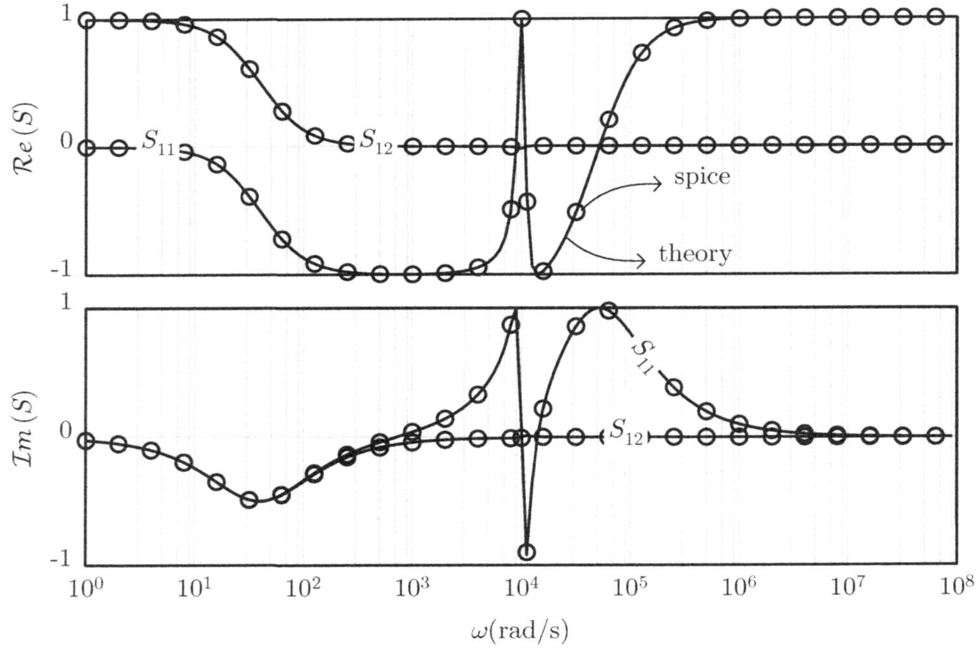

Fig. 43.21 Sample solution to Problem 3

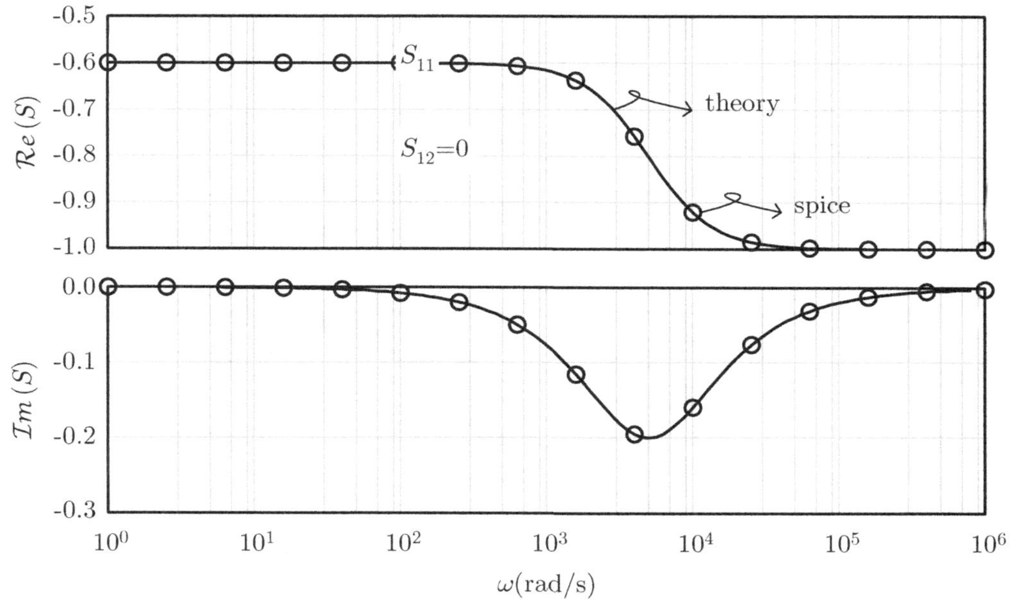

Fig. 43.22 Sample solution to Problem 5

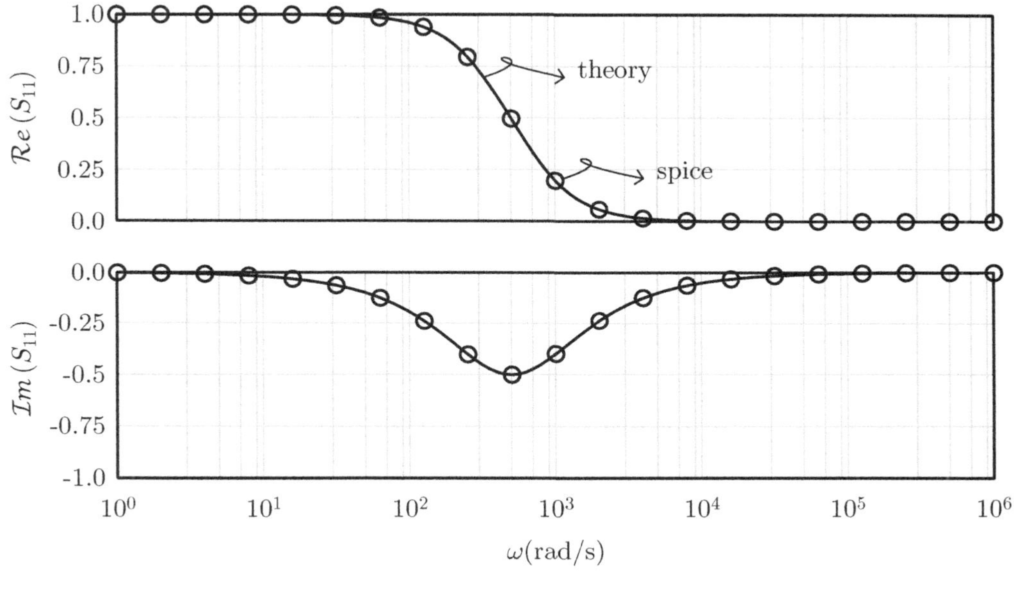

Fig. 43.23 Sample solution to Problem 6

8. Having extracted in the prior problem the S-par model for the single-port RC network in Fig. 43.18f, for the 4 different Z_0 values, use now Eq. (43.80) for each extraction to figure input impedance $Z_{11}(s)$ and plot results. Do the numbers make sense in terms of slope and saturation values? See sample solution in Fig. 43.25.

9. Consider the 2-port RLC network in Fig. 43.18g. Find the impedance transfer function and then use Eq. (43.83) to figure the scattering parameter matrix. Plot results (using $Z_0 = 10\,\Omega$) and compare to SPICE; see sample solution in Fig. 43.26.

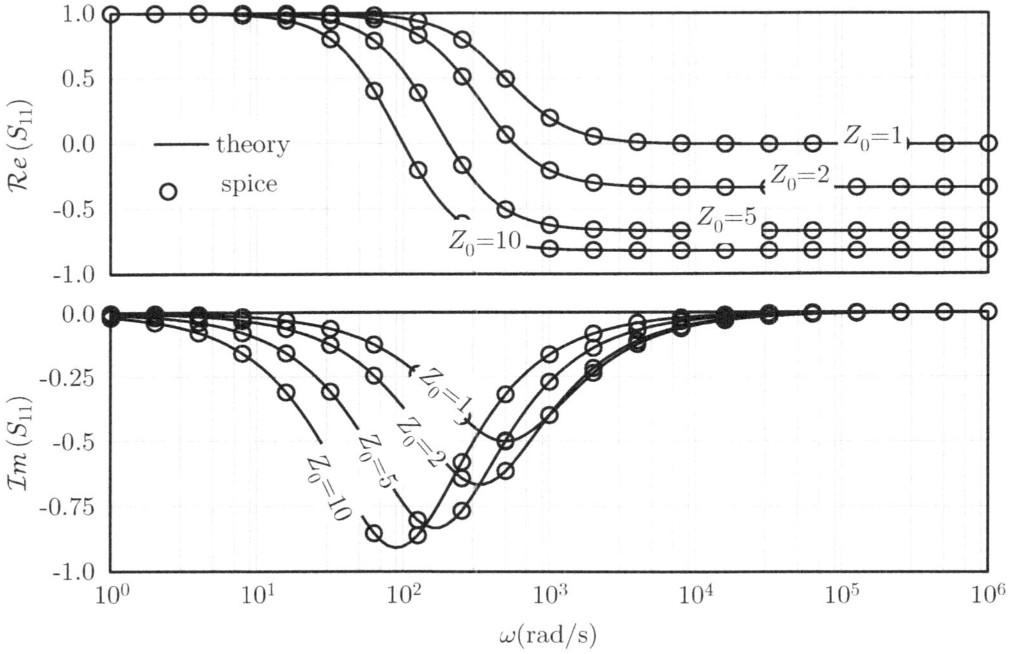

Fig. 43.24 Sample solution to Problem 7

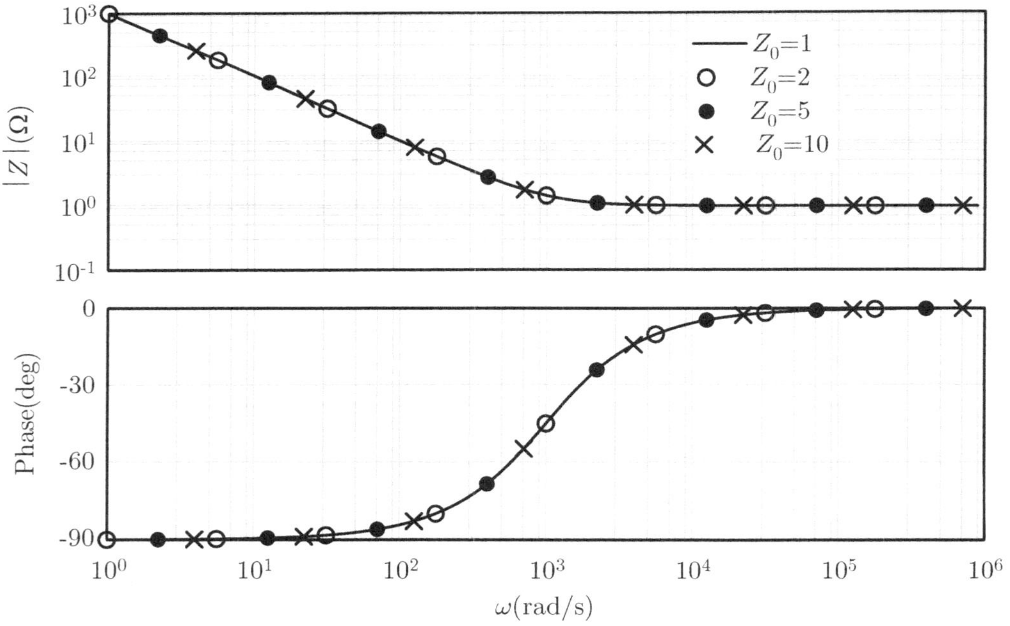

Fig. 43.25 Sample solution to Problem 8

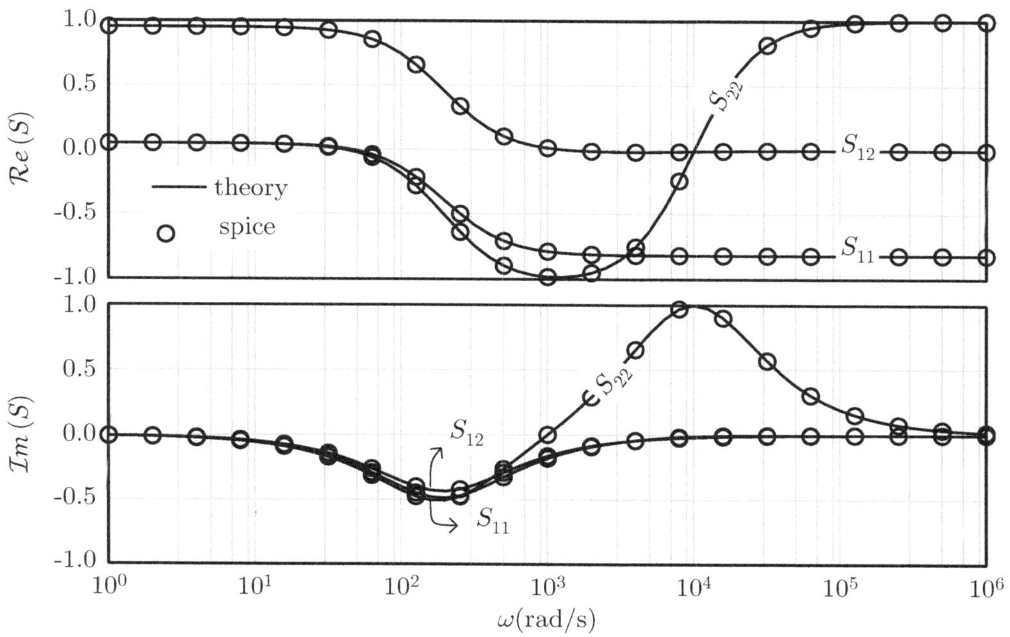

Fig. 43.26 Sample solution to Problem 9

44 Application of Spectral Techniques to Solving 2D Electrostatic Problems

44.1 Introduction

Other than applications in circuit theory, Fourier series, transform, and the Laplace transform find applications in numerous other fields. In this chapter we apply spectral techniques to solving *electrostatic* problems. Yes we are departing the world of ideal circuits and visiting the world of electromagnetics! Ultimately and at the deepest level the world of circuits and that of electromagnetics of course coalesce; but for our purpose and for the sake of simplicity we are treating them as separate.

44.2 Main Idea

Just like we did in the *time* domain where we split the solution in terms of time harmonics, we can do that same but now in the *space* domain. A signal, be it voltage or electric field, in space can be decomposed in terms of sinusoids in space. The sinusoids form our *basis functions* and are easier to deal with. Combining weighted versions of those, as will be shown in this chapter, we are able to figure total solution to meet a rather arbitrary set of boundary conditions. Notice the mention of the new term *boundary conditions* as opposed to what we've been accustomed to so far which are *initial conditions*. We will illustrate the use of spectral techniques in solving electrostatic problems via a handful of examples.

44.3 Fourier Series in Solving 2D Electrostatic Problems

Consider the 2D setup shown in Fig. 44.1. It is comprised of four walls, each perfect conductor, and each holding a certain voltage. The left, right, and bottom walls have zero voltage (grounded). The top wall is split into three smaller walls, with outer two having zero voltage, and inner one having unity voltage. We are interested in finding the voltage everywhere inside the square. We start with *Coulomb's law*

$$\nabla \cdot E = \frac{\rho}{\epsilon} \tag{44.1}$$

where E is the electric field and ρ the charge density. Using the following equation for voltage V

$$V = -\nabla E \tag{44.2}$$

and knowing there is no charge inside the square we end up with *Laplace's equation*

$$\frac{\partial^2 v(x, y)}{\partial x^2} + \frac{\partial^2 v(x, y)}{\partial y^2} = 0 \tag{44.3}$$

F. Badrieh, *Spectral, Convolution and Numerical Techniques in Circuit Theory*,
https://doi.org/10.1007/978-3-319-71437-0_44

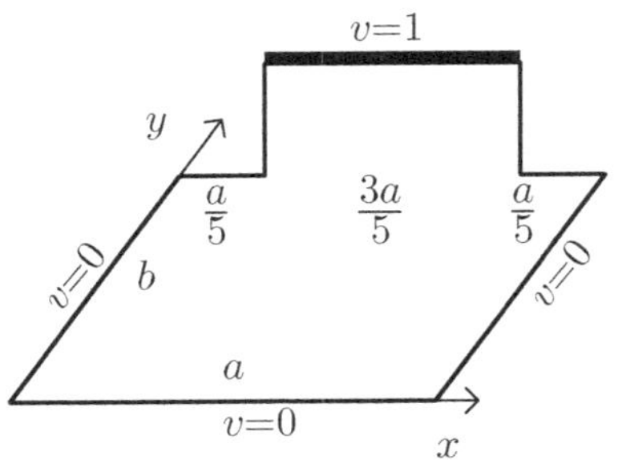

Fig. 44.1 2D electrostatic problem with prescribed boundary conditions

This is a partial differential equation governing the dependence of the scalar $v(x, y)$ on space. This partial differential equation is subject to the following boundary conditions

$$
\begin{aligned}
v(0, y) &= 0 \quad \text{(left wall)} \\
v(a, y) &= 0 \quad \text{(right wall)} \\
v(x, 0) &= 0 \quad \text{(bottom wall)} \\
v(x, b) &= \begin{cases} 0 & 0 < x < 0.2a \\ 1 & 0.2a < x < 0.8a \\ 0 & 0.8a < x < a \end{cases} \quad \text{(top wall(s))}
\end{aligned} \tag{44.4}
$$

We use the method of *separation of variables* in which we assume

$$v(x, y) = X(x)Y(y) \tag{44.5}$$

That is we assume we can split the voltage in terms of two components: one that exclusively depends on x while the other on y. We cannot be sure we will get somewhere with this assumption, but until we are proven wrong there is nothing that can stop us from trying! (Additionally people have done this for a very long time, so we know we are on the right track!) Plugging back into Laplace's equation we arrive at

$$\frac{d^2X(x)}{dx^2}Y(y) + \frac{d^2Y(y)}{dy^2}X(x) = 0 \tag{44.6}$$

Divide both sides with $X(x)Y(y)$ and get

$$\frac{X''(x)}{X(x)} + \frac{Y''(y)}{Y(y)} = 0 \tag{44.7}$$

The only way this could be true, for all x and y, is if *both* terms are constant such that

$$\frac{X''(x)}{X(x)} = -\lambda^2 \tag{44.8}$$

$$\frac{Y''(y)}{Y(y)} = +\lambda^2 \tag{44.9}$$

The solution to the X equation is either a sine or a cosine, but the cosine does not vanish at the boundaries; so we end up with:

$$X(x) = \sin \lambda x \tag{44.10}$$

To meet the zero boundary conditions on the right wall ($x = a$) we arrive at allowed values of λ

$$\lambda = \frac{n\pi}{a} \tag{44.11}$$

Hence

$$\boxed{X(x) = \sin \frac{n\pi x}{a}} \tag{44.12}$$

Let's double check quickly. The second derivative of this function is

$$\frac{d^2}{dx^2} \sin \frac{n\pi x}{a} = -\left(\frac{n\pi}{a}\right)^2 \sin \frac{n\pi x}{a} = -\lambda^2 \sin \frac{n\pi x}{a} \tag{44.13}$$

which satisfies Eq. (44.8). Additionally the boundary conditions are such that $X(0) = X(a) = 0$ and those also satisfy the requirements in Eq. (44.4). The solution to the Y equation, on the other hand, is

$$Y(y) = Be^{\lambda y} + Ce^{-\lambda y} \tag{44.14}$$

which when using the prefigured λ gives

$$Y(y) = Be^{\frac{n\pi y}{a}} + Ce^{-\frac{n\pi y}{a}} \tag{44.15}$$

To satisfy the bottom zero boundary condition we relate B to C and end up with

$$\boxed{Y(y) = A \sinh \frac{n\pi y}{a}} \tag{44.16}$$

Our generic solution so far is then

$$\boxed{v(x, y) = A \sin \frac{n\pi x}{a} \sinh \frac{n\pi y}{a}} \tag{44.17}$$

In fact any linear combination of the above would also be a solution:

$$v(x,y) = \sum_n A_n \sin\frac{n\pi x}{a} \sinh\frac{n\pi y}{a} \tag{44.18}$$

This solution automatically satisfies the three zero boundary conditions (left, right, and bottom walls). But it does not automatically satisfy the top boundary condition. At the top wall, we have to select the A_n to satisfy the top boundary conditions. Specifically, at $y = b$ we have

$$v(x,b) = \sum_n A_n \sin\frac{n\pi x}{a} \sinh\frac{n\pi b}{a} \tag{44.19}$$

Multiply both sides by $\sin\frac{m\pi x}{a}$ and integrate, and use the orthogonality of the sine function to get

$$\frac{a}{2} A_n \sinh\frac{n\pi b}{a} = \int_0^a v(x,b) \sin\frac{n\pi x}{a} dx \tag{44.20}$$

which solves for

$$A_n = \frac{2}{a} \frac{1}{\sinh\frac{n\pi b}{a}} \int_0^a v(x,b) \sin\frac{n\pi x}{a} dx \tag{44.21}$$

Performing the x integration we get

$$A_n = -\frac{2}{a} \frac{1}{\sinh\frac{n\pi b}{a}} \frac{a}{n\pi} [\cos(0.8n\pi) - \cos(0.2n\pi)] \tag{44.22}$$

Now that we know the various constants A_n, we know the total solution as per Eq. (44.18). The resolution of the solution would depend on the number of included harmonics. Figure 44.2 shows sample runs versus n. Figure 44.3 shows a comparison between our results and a field solver one ($n = 20$ case). We can go further and calculate the electric field

$$\mathbf{E} = -\nabla v \tag{44.23}$$

Knowing that

$$v(x,y) = \sum_n A_n \sin\frac{n\pi x}{a} \sinh\frac{n\pi y}{a} \tag{44.24}$$

we get

$$\mathbf{E}(x,y) = -\frac{n\pi}{a}\left[\mathbf{i}\sum_n A_n \cos\frac{n\pi x}{a} \sinh\frac{n\pi y}{a} + \mathbf{j}\sum_n A_n \sin\frac{n\pi x}{a} \cosh\frac{n\pi y}{a}\right] \tag{44.25}$$

Figure 44.4 shows results and comparison to field solver ones. Notice excellent agreement!

44.4 Wall with Odd Boundary Condition

Consider next the case where top wall has the following boundary conditions as shown in Fig. 44.5.

$$v(x,b) = \begin{cases} 0 & x < \frac{1}{2} \\ 1 & x > \frac{1}{2} \end{cases} \tag{44.26}$$

The following steps are exactly as those in the prior section. The thing that changes is A_n; recall Eq. (44.21) repeated here for convenience

$$A_n = \frac{2}{a} \frac{1}{\sinh\frac{n\pi b}{a}} \int_0^a v(x,b) \sin\frac{n\pi x}{a} dx$$

This gives

$$A_n = \frac{2}{a}\frac{1}{\sinh\frac{n\pi b}{a}}\int_{a/2}^{a}\sin\frac{n\pi x}{a}dx = \frac{2}{a}\frac{1}{\sinh\frac{n\pi b}{a}}\frac{a}{n\pi}\left[\cos\frac{n\pi}{2} - \cos n\pi\right] \tag{44.27}$$

Results and comparison to field solver ones are shown in Fig. 44.6. Electric field results are shown in Fig. 44.7.

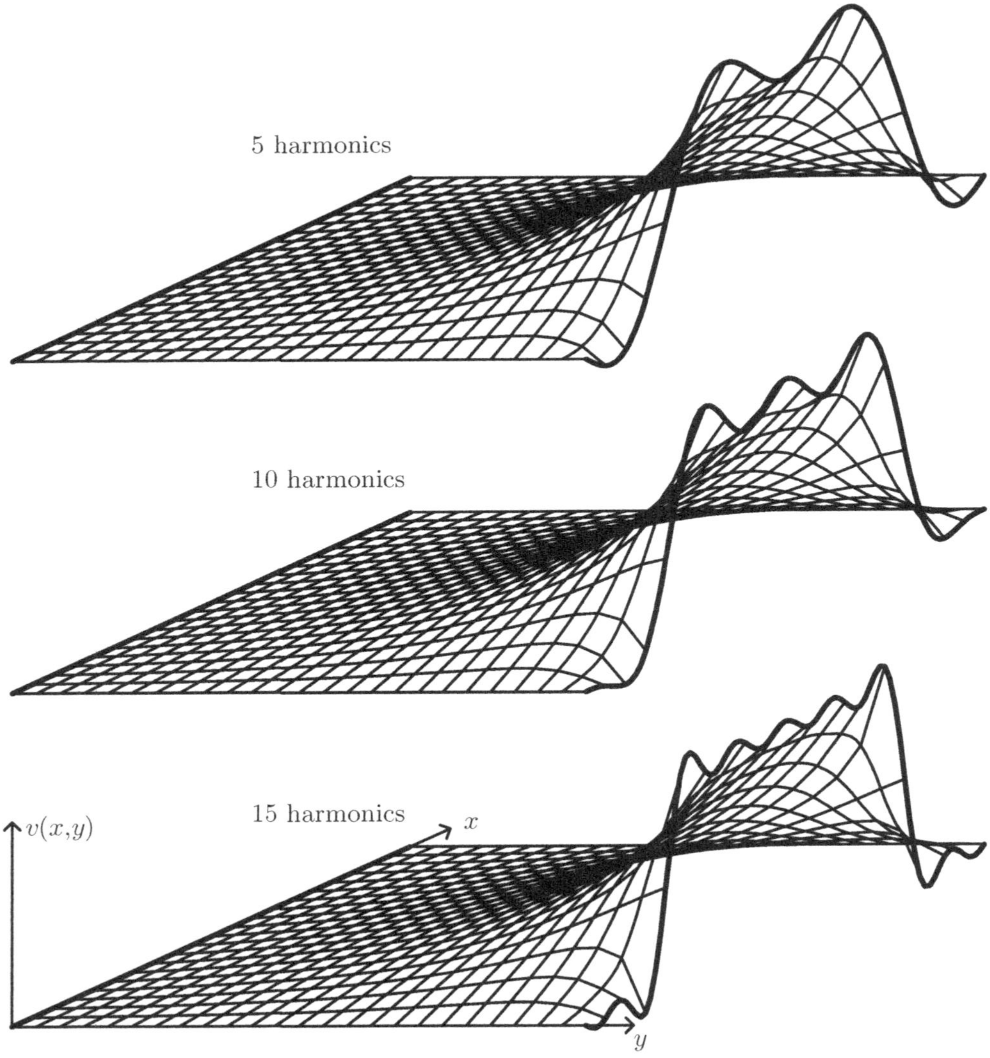

Fig. 44.2 Fourier series solution for setup in Fig. 44.1, vs harmonic count

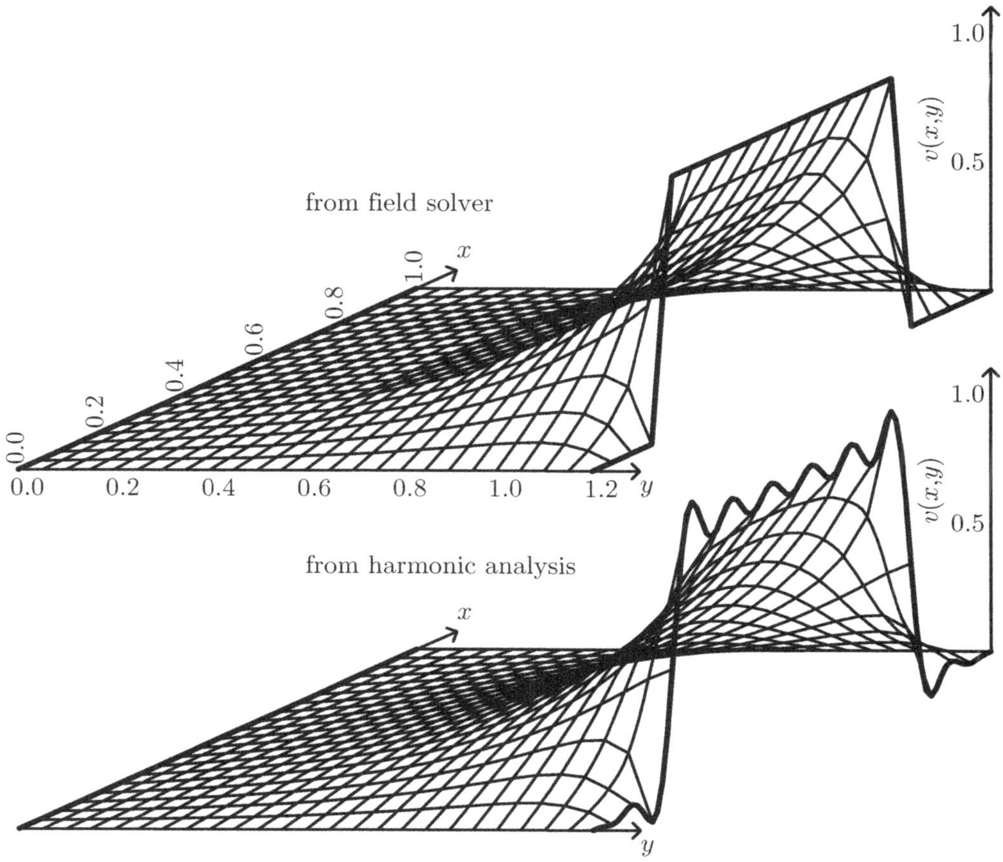

Fig. 44.3 Field solver (top) and analytic answer (bottom) solution to the 2D setup shown in Fig. 44.1

44.5 Case with Two Wall Boundary Conditions

The above examples had all three walls set to zero, while last wall at some potential (potentially varying in x or y). Now we allow for two of the walls to have nonzero potential. For example, consider the case shown in Fig. 44.8. It has zero potential on left and bottom sides, but nonzero ones on top and right sides. To tackle this we use superposition. We know the solution for the case of three zero-potential sides and one nonzero side (*top* side), as shown in the prior section (Eq. (44.27)). We call this

$$v_1(x, y) = \sum_m A_n \sin\frac{n\pi x}{a} \sinh\frac{n\pi y}{a} \quad (44.28)$$

$$A_n = \frac{2}{a}\frac{1}{\sinh\frac{n\pi b}{a}}\frac{a}{n\pi}\left[\cos\frac{n\pi}{2} - \cos n\pi\right] \quad (44.29)$$

We do the same thing for the case of three zero potentials and one nonzero on the *right* side and get

$$v_2(x, y) = \sum_m B_n \sin\frac{n\pi y}{b} \sinh\frac{n\pi x}{b} \quad (44.30)$$

$$B_n = \frac{2}{b}\frac{1}{\sinh\frac{n\pi a}{b}}\frac{b}{n\pi}\left[\cos\frac{n\pi 7}{12} - \cos n\pi\right] \quad (44.31)$$

By super position, then, total solution would be the sum of both solutions

$$v(x, y) = v_1(x, y) + v_2(x, y) \quad (44.32)$$

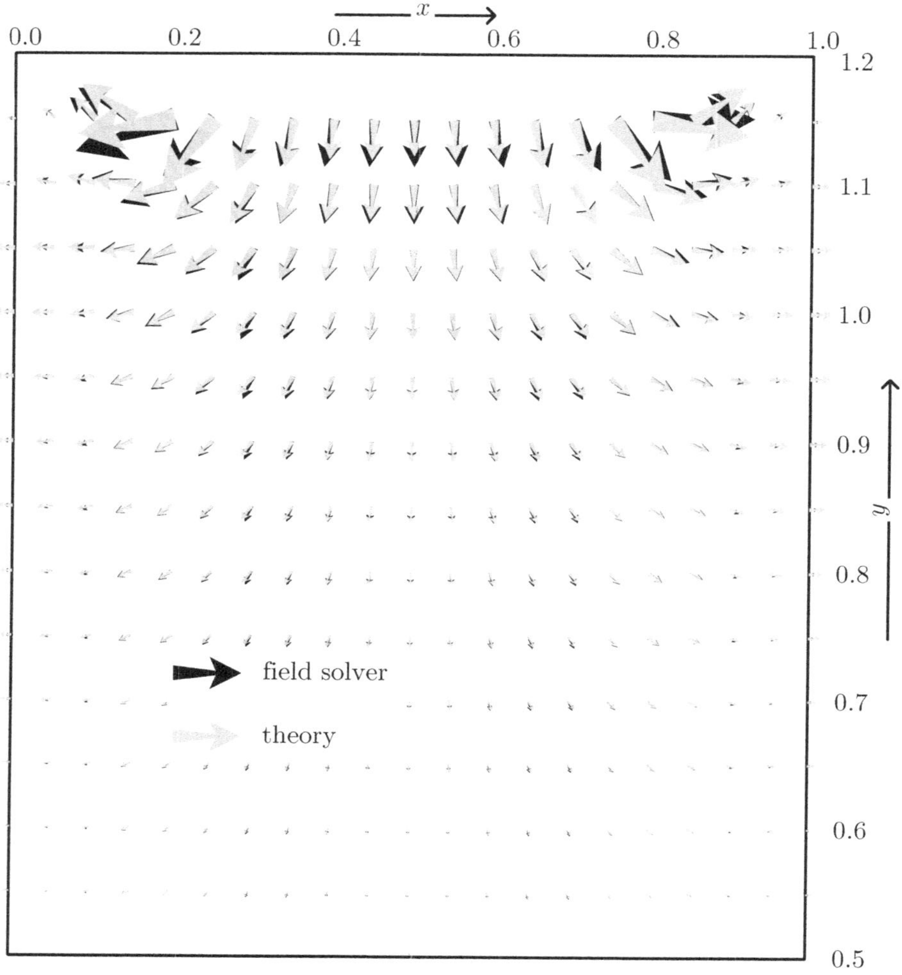

Fig. 44.4 Electric field corresponding to potential in Fig. 44.3

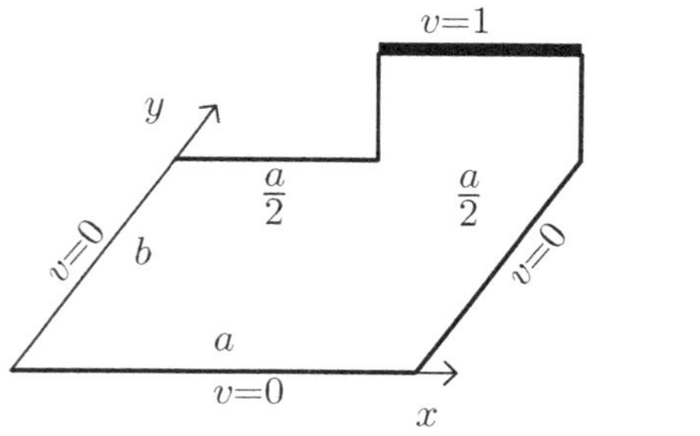

Fig. 44.5 2D electrostatic problem with three ground walls and odd boundary condition on last wall

Results and comparison to field solver ones are shown in Fig. 44.9. The same process can be applied for the case of three nonzero potential sides.

44.6 Notes on Superposition

In wrapping up this chapter lets (re-)reflect a little on the principle of superposition. As was demonstrated in the last section we can solve a

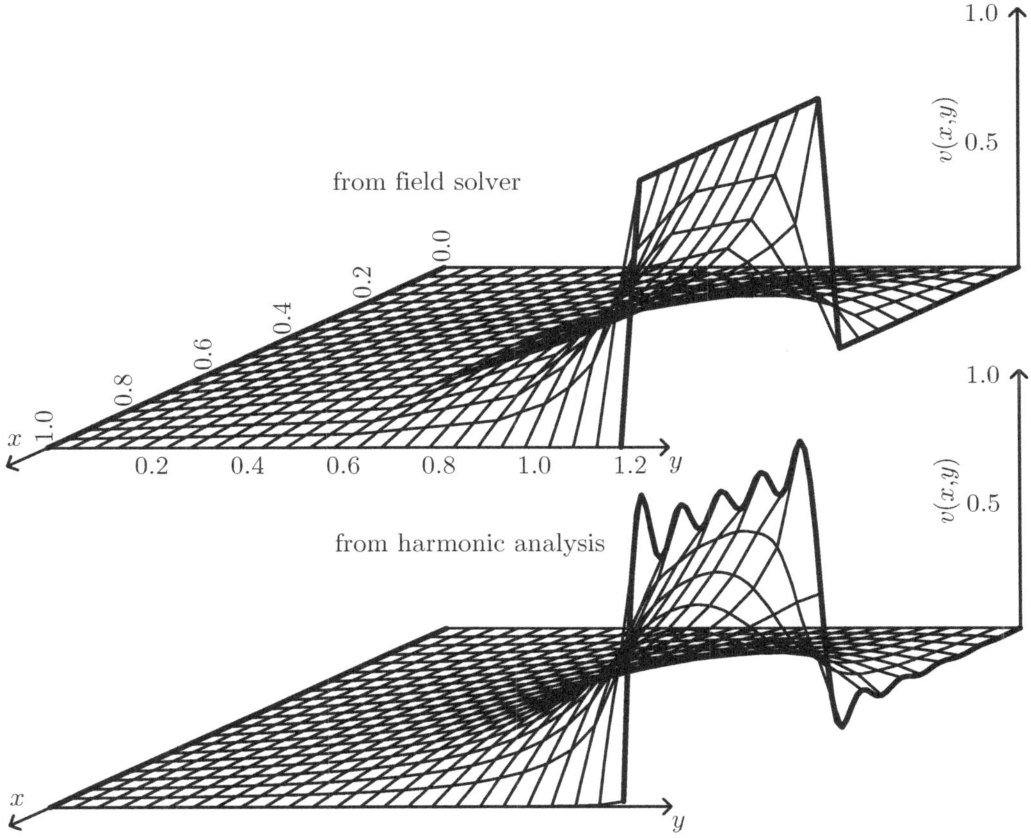

Fig. 44.6 Field solver (top) and analytic answer (bottom) solution to the 2D setup shown in Fig. 44.5

multi-boundary condition problem, with different sets of prescribed boundary conditions on different walls by selectively enabling one of the boundary condition sets at time (while enforcing the others to zero), getting the corresponding solution, repeating for the next boundary condition (till all cases have been exhausted), and finally adding all intermediate solutions. This reminds us of Chap. 38 which dealt with networks with multi-sources. There too we selectively enabled one source at a time (while disabling the others), obtained the corresponding solution, moved to the next source, did the same, and finally added all solutions. This is a very nice bonus showing the generality and power of the principle of superposition. We see it disguised in time (such as in Fourier decomposition, and in multi-source networks) and in space (such as in electrostatic problems); in fact this principle can be used in many other cases, and in other fields. It is a good technique to keep in ones arsenal when tackling problems; sort of divide-and-conquer strategy!

44.7 Summary

In this chapter we expanded the use of spectral techniques from the time (frequency) domain to the space domain. Now instead of talking about waves in the time domain, we talk about waves in the space domain. Different application, but

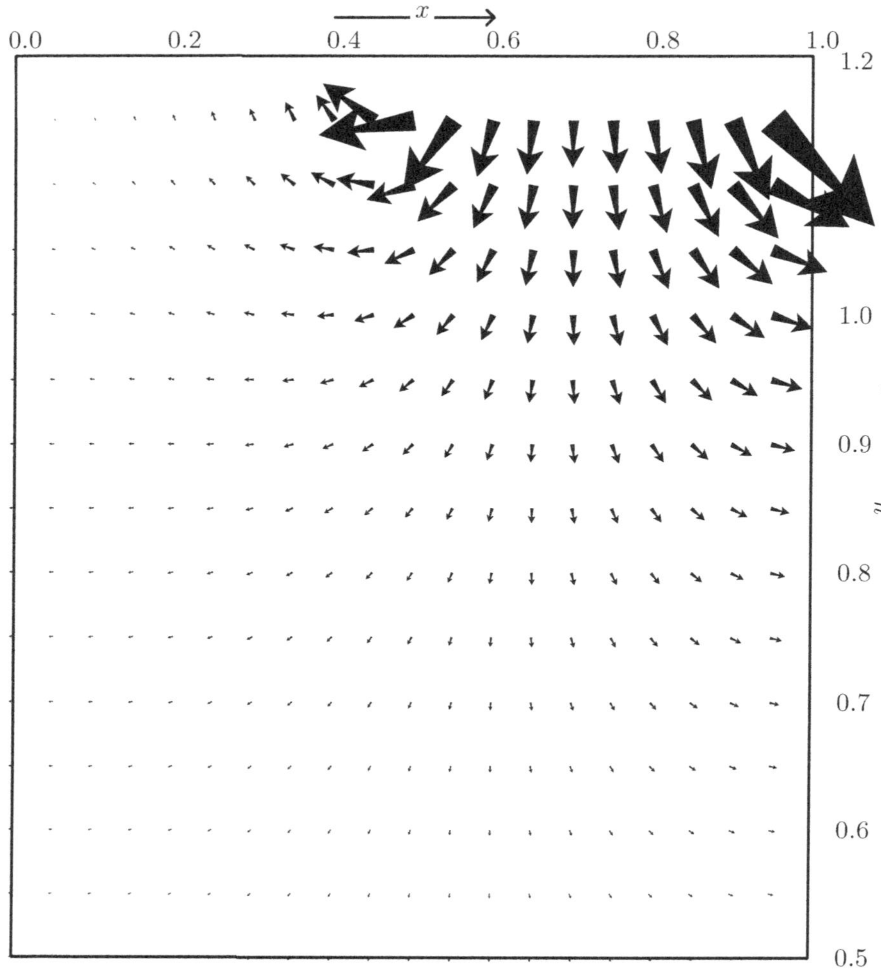

Fig. 44.7 Electric field corresponding to potential in Fig. 44.6

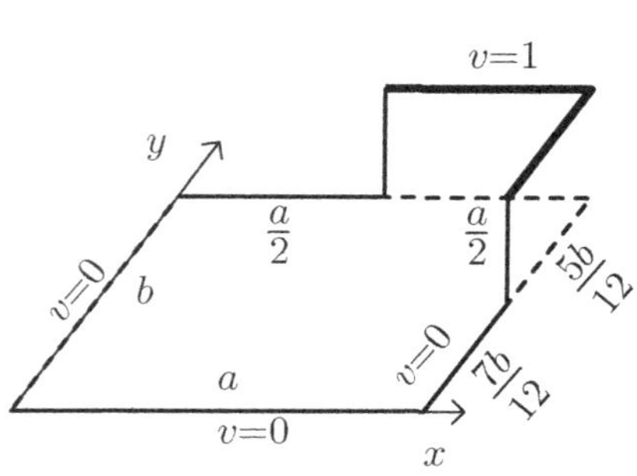

Fig. 44.8 2D electrostatic problem with nonzero prescribed boundary conditions on two walls

same idea: expand solution in terms of basis functions. More specifically we tackled the class of electrostatic problems dealing with 2D space with prescribed boundary conditions. Quite often a number of the walls is grounded while the remaining walls have nonzero potential. The nonzero boundary conditions could be constant or space dependent. After using the method of separation of variables where we expressed the solution in terms of an x-dependent function and a y-dependent one, we used the Fourier series expansion to ensure we satisfy the boundary

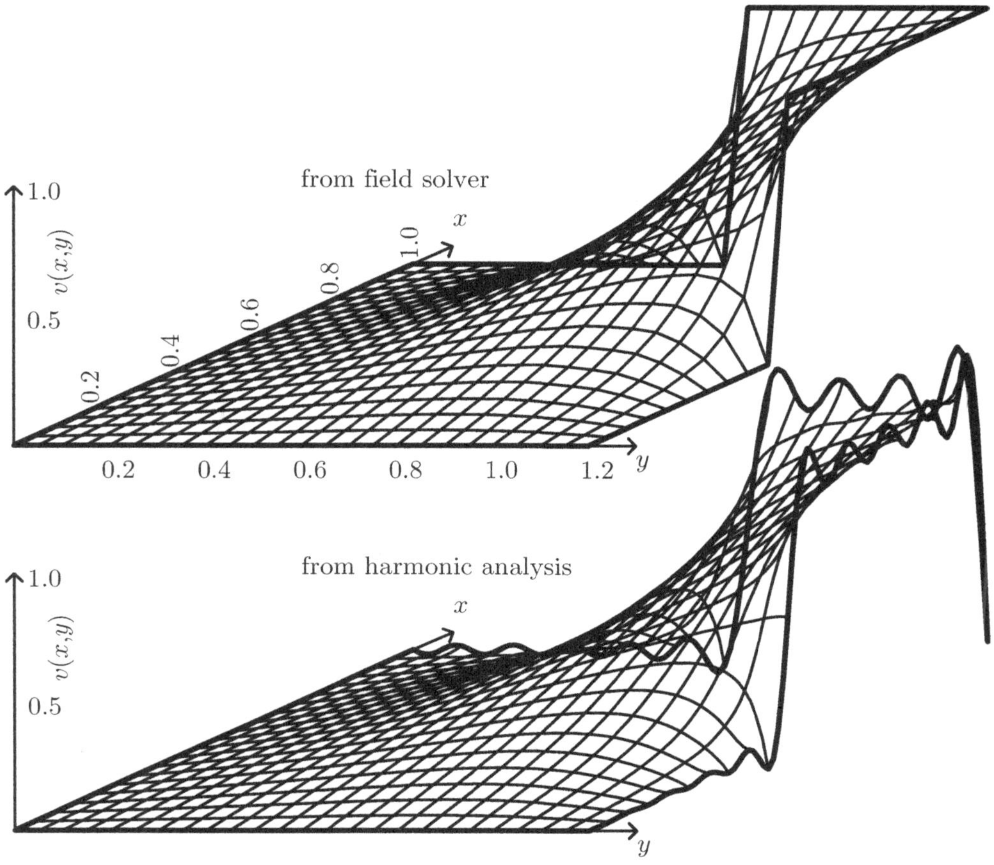

Fig. 44.9 Field solver (top) and analytic answer (bottom) solution to the 2D setup shown in Fig. 44.8

conditions. We were able to extract both potential and electric field and get excellent match compared to field solver results. We wrapped the chapter with some notes on the generality and application of the principle of superposition. While our expansion from the time domain to the space one may appear simple, this is in fact a huge step in generalizing our analytics and methods and demonstrating the power of mathematical analysis. Once we've converted the problem into equations and numbers, the Fourier analysis methods (and friends) are not discriminant in lending a hand!

44.8 Problems

1. A rectangle of width 1 and height 1.2 has zero boundary conditions on all three walls, except the $y = 1.2$ one, where it is set to $v(x, 1.2) = x$. The setup is shown in Fig. 44.10a. Find the potential everywhere and plot it using 20 harmonics; see sample solution in Fig. 44.11. Answer:

$$a_n = -\frac{2}{a}\frac{1}{\sinh(n\pi b/a)}\frac{a}{n\pi}\cos(n\pi a)$$

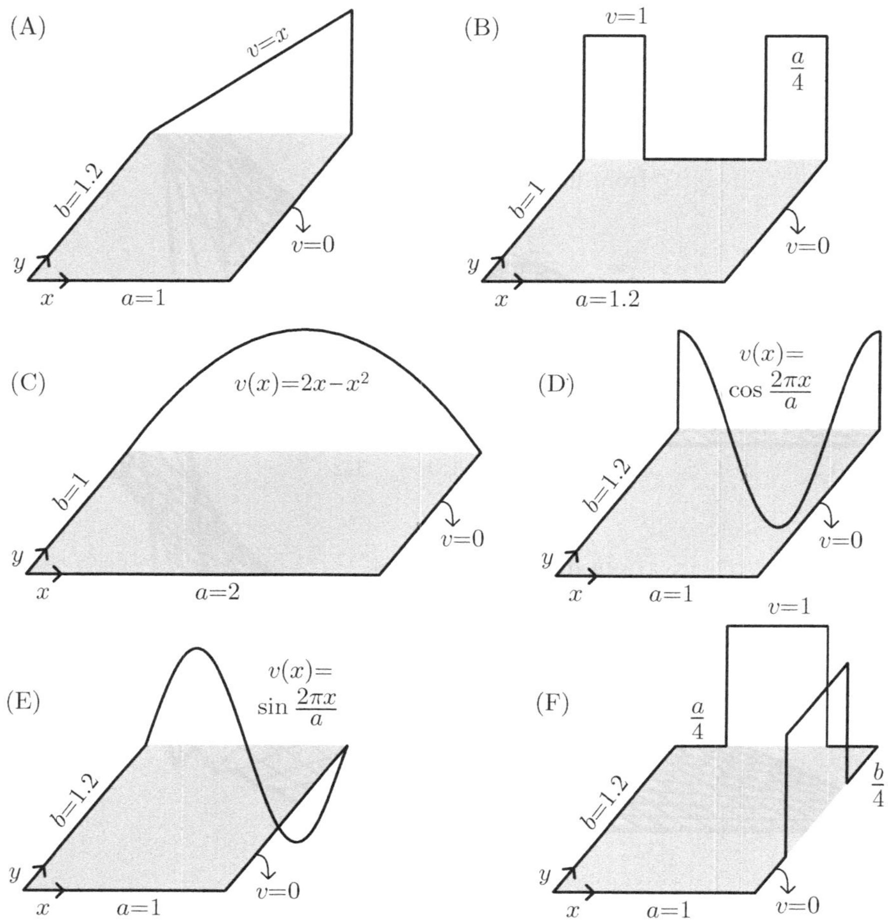

Fig. 44.10 Specifications for various end-of-chapter problems

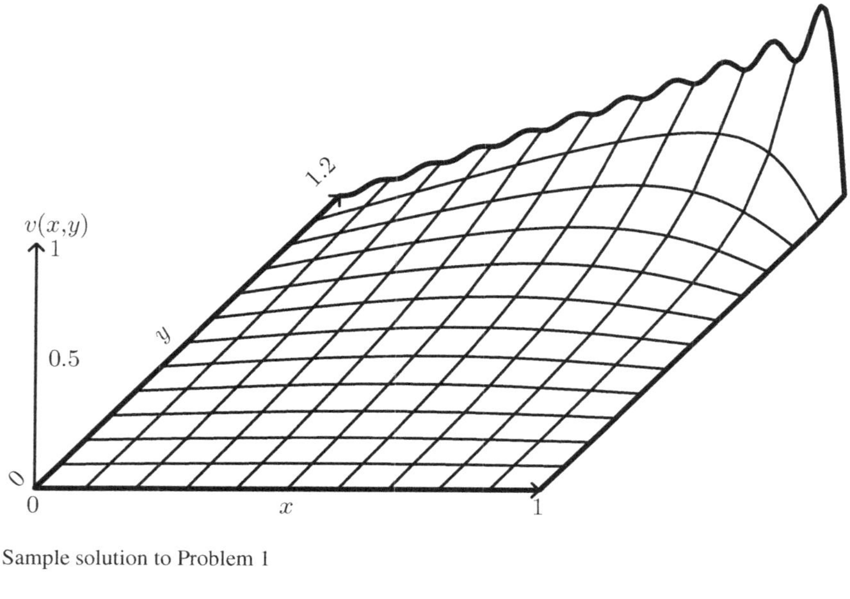

Fig. 44.11 Sample solution to Problem 1

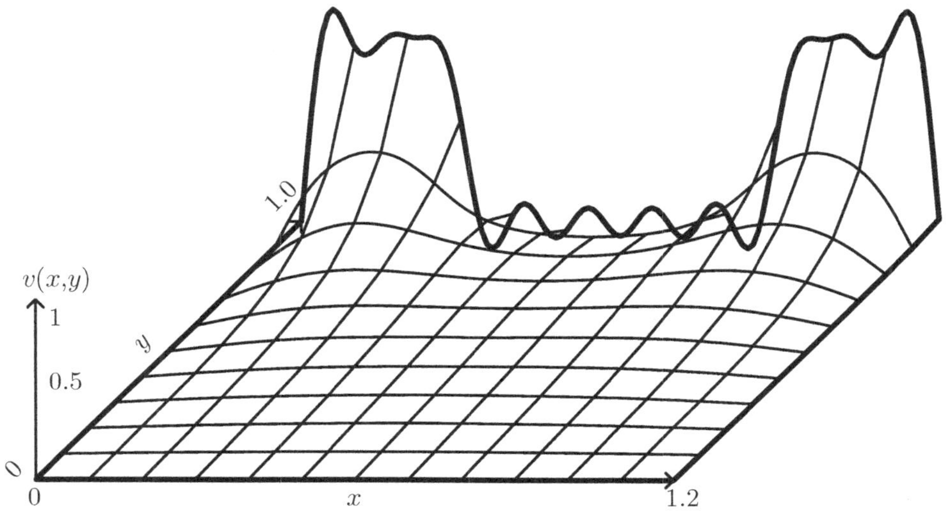

Fig. 44.12 Sample solution to Problem 2

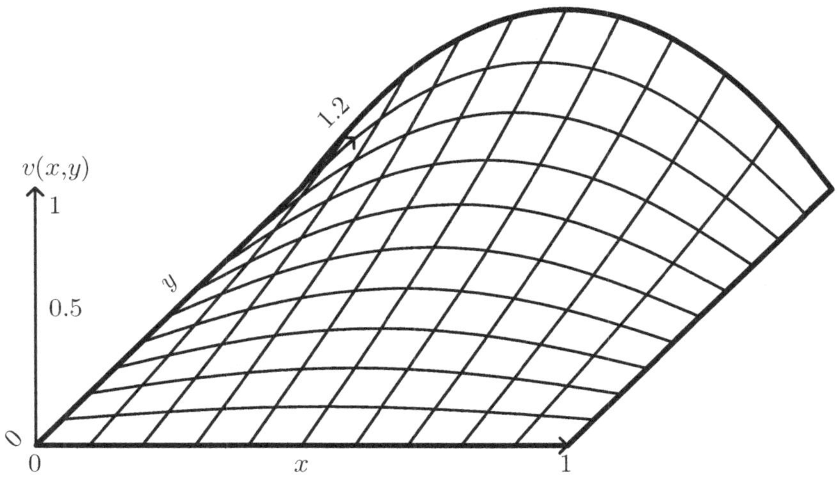

Fig. 44.13 Sample solution to Problem 3

2. Consider the setup and boundary conditions in Fig. 44.10b. Find the potential everywhere and plot it using 20 harmonics; see sample solution in Fig. 44.12.
   Answer:

$$a_n = -\frac{2}{a}\frac{1}{\sinh(n\pi b/a)}\frac{a}{n\pi}\left[\cos(n\pi/4) - 1\right.$$
$$\left. + \cos(n\pi) - \cos(n\pi \times 3/4)\right]$$

3. Consider the setup and boundary conditions in Fig. 44.10c where three of the boundaries are set to zero voltage while the other has a quadratic shape. Find the potential everywhere and plot it using 20 harmonics; see sample solution in Fig. 44.13.
   Answer:

$$a_n = \frac{2}{a}\frac{1}{\sinh(n\pi b/a)}\times 2\left(\frac{a}{n\pi}\right)^3 [1 - \cos(n\pi)]$$

4. Consider the setup and boundary conditions in Fig. 44.10d where the nonzero boundary is a cosine one: $v(x, b) = \cos 2\pi x/a$. Find the potential everywhere and plot it using 20 harmonics; see sample solution in Fig. 44.14. Answer:

$$a_n = \frac{2}{a}\frac{1}{\sinh(n\pi b/a)}\frac{1}{2}\left[\frac{a}{n\pi + 2\pi}(1 - \cos(n\pi + 2\pi)) + \frac{a}{n\pi - 2\pi}(1 - \cos(n\pi - 2\pi))\right]$$

5. Consider the setup and boundary conditions in Fig. 44.10 (E) where the nonzero boundary is a sine one: $v(x, b) = \sin 2\pi x/a$. Find the potential everywhere and plot it using 20 harmonics; see sample solution in Fig. 44.15.

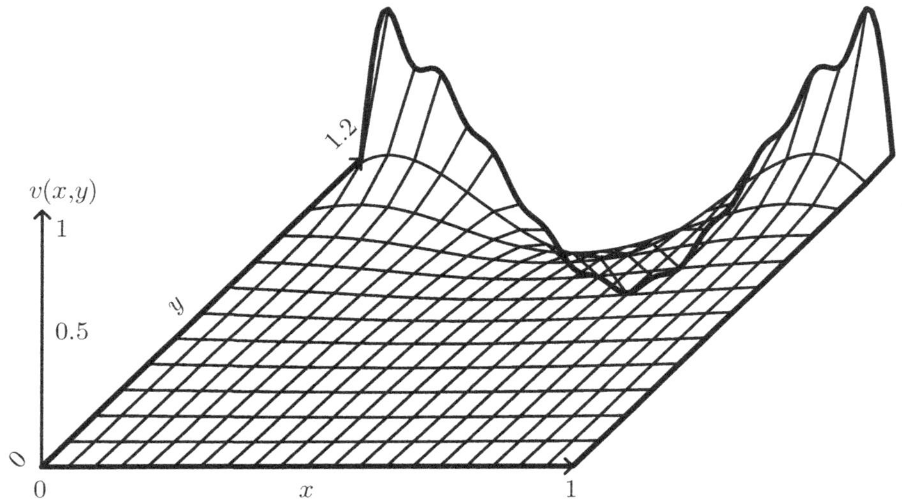

Fig. 44.14 Sample solution to Problem 4

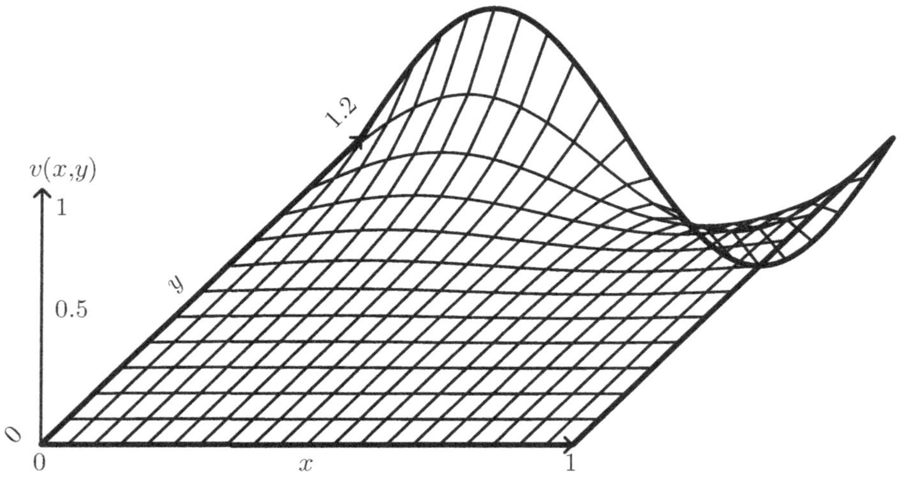

Fig. 44.15 Sample solution to Problem 5

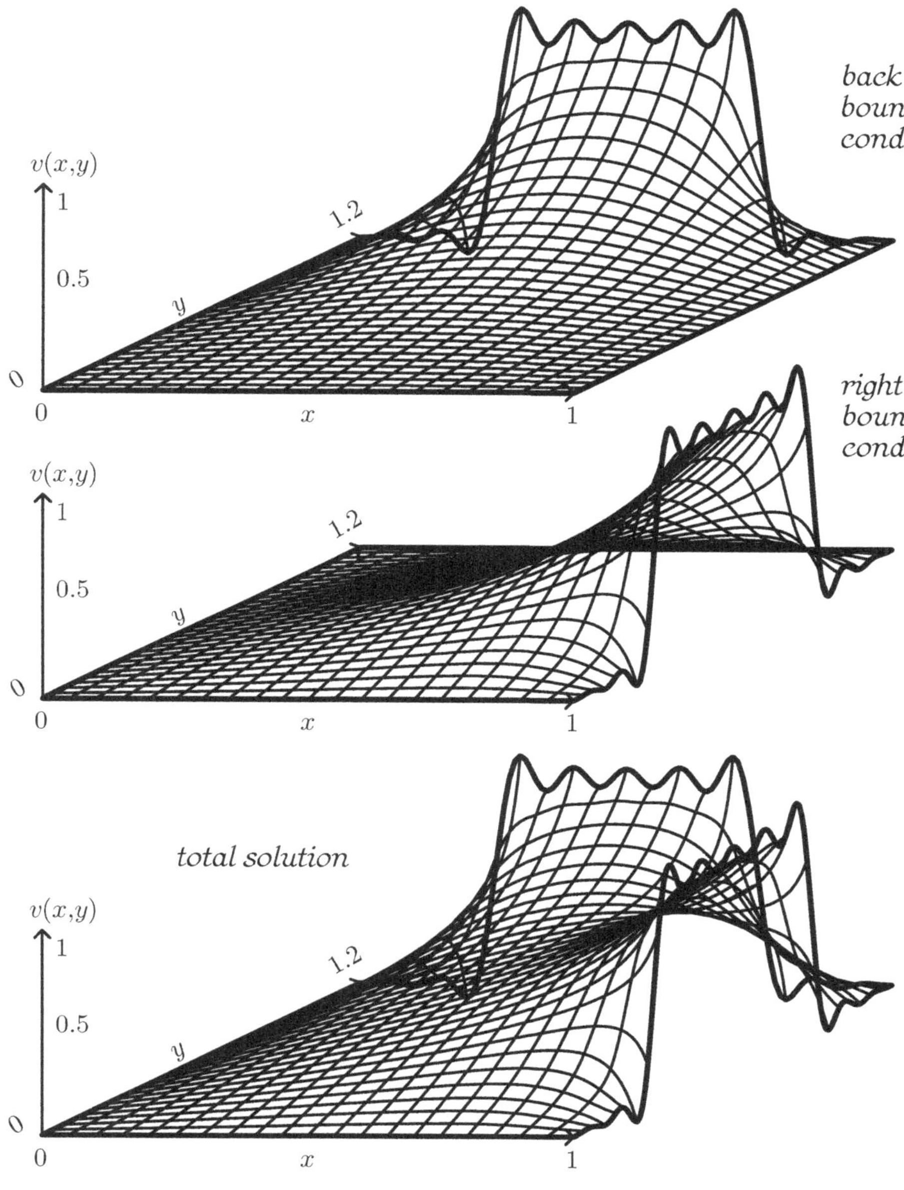

Fig. 44.16 Sample solution to Problem 6

Answer:

$$a_2 = \frac{1}{\sinh(2\pi b/a)}$$

6. Consider the setup and boundary conditions in Fig. 44.10 (F) where two walls have nonzero boundary conditions. Solve this problem by superposition. First find solution satisfying three zero BCs and one nonzero BC at the *top* wall; then find solution satisfying three zero BCs and one nonzero BC at the *right* wall. Finally add both solutions; see sample solution in Fig. 44.16.

Answer:

$$a_n = -\frac{2}{a}\frac{1}{\sinh(n\pi b/a)}\frac{a}{n\pi}\left[\cos\frac{3n\pi}{4} - \cos\frac{n\pi}{4}\right]$$
$$b_n = -\frac{2}{b}\frac{1}{\sinh(n\pi a/b)}\frac{b}{n\pi}\left[\cos\frac{3n\pi}{4} - \cos\frac{n\pi}{4}\right]$$

$$v(x, y) = \sum_n a_n \sin\frac{n\pi x}{a}\sinh\frac{n\pi y}{a} + \sum_n b_n \sinh\frac{n\pi x}{b}\sin\frac{n\pi y}{b}$$

45 Application of Spectral Techniques in Solving Diffusion Problems

45.1 Introduction

Another area where we can use spectral techniques is the distributed *RC* line, whose governing equation obeys the diffusion equation. The distributed *RC* line can have initial conditions and boundary conditions, or combination thereof. Similar to the last chapter this chapter deals with *distributed* systems as opposed to prior chapters which dealt with *lumped* (or discretized) systems (such as resistors, inductors, and so forth). Of course we can try discretize a distributed system in terms of a lumped one, but if we can solve it as is why not? In what follows we will first derive the diffusion equation and then apply spectral techniques on a few sample cases.

45.2 Derivation of *RC* Diffusion Equation

Consider an electrical line above a ground plane as shown in Fig. 45.1. Assume line length is L, resistance per unit length R_0, and capacitance per unit length C_0. Assume further that the end points are tied to ground (not necessary, but simple for now). We can then model the line as shown in Fig. 45.2. The difference in voltage between two cap nodes is given by the iR product

$$\frac{dv(x,t)}{dx} = -i(x,t)R_0 \tag{45.1}$$

Notice the negative sign which states that the voltage at the right side would be lower than that at the left if current was flowing left-to-right! That is, using the conventional definition of the derivative (which is right minus left) and using the conventional positive direction flow of current, along the x-axis, we need the negative sign in Eq. (45.1). If we take the spatial derivative of this we get

$$\frac{d^2v(x,t)}{dx^2} = -R_0\frac{di(x,t)}{dx} \tag{45.2}$$

The difference in current between two resistors is proportional to (again negative) the rate of change of the cap voltage:

$$\frac{di(x,t)}{dx} = -C_0\frac{dv(x,t)}{dt} \tag{45.3}$$

That is, if the intermediate cap is charging then current at right must be *less* than current at left, and hence the negative sign! Plugging this back into Eq. (45.2) we arrive at the *diffusion equation*:

$$\boxed{\frac{d^2v(x,t)}{dx^2} = R_0C_0\frac{dv(x,t)}{dt}} \tag{45.4}$$

F. Badrieh, *Spectral, Convolution and Numerical Techniques in Circuit Theory*,
https://doi.org/10.1007/978-3-319-71437-0_45

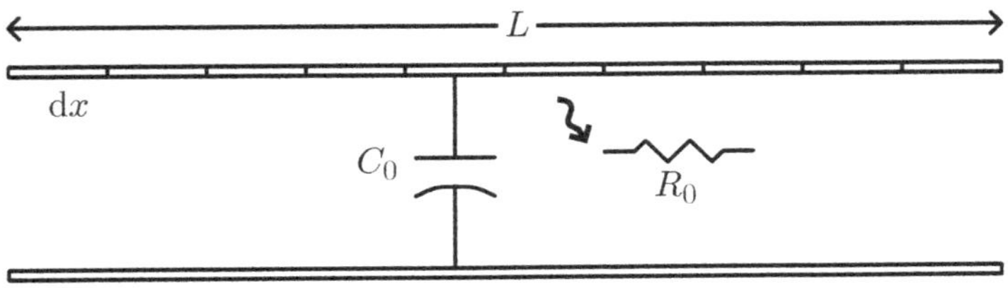

Fig. 45.1 *RC* ladder

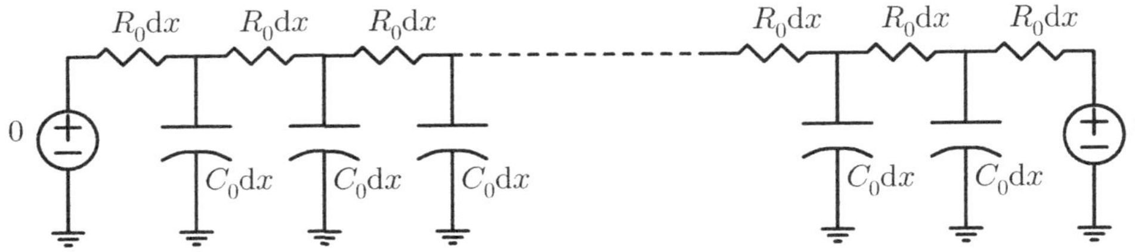

Fig. 45.2 Model of *RC* ladder

This partial differential equation with its corresponding *initial* conditions and *boundary* conditions, and the corresponding solution (both in space and time) forms the basis for the remainder of the current chapter.

45.3 Method of Separation of Variables

Equation (45.4) is a partial differential equation in two variables—space (x) and time (t). Just like was done in the last chapter we can use the method of separation of variables to solve this problem. Assume that we can split the total solution in terms of a spatial one and a temporal one:

$$v(x,t) = X(x)T(t) \tag{45.5}$$

where again $X(x)$ depends exclusively on x and $Y(y)$ exclusively on y. Plug back into the diffusion equation and get

$$\frac{d^2X(x)}{dx^2}T(t) = R_0C_0\frac{dT(t)}{dt}X(x) \tag{45.6}$$

Divide by $X(x)T(t)$ and get

$$\frac{X''(x)}{X(x)} = R_0C_0\frac{T'(t)}{T(t)} \tag{45.7}$$

Since the left side is a function of x only and the right side is a function of t only, the equality (for all time and space) implies that each side is a constant. Furthermore assume this constant to be $-\lambda^2$; then we get

$$\frac{X''(x)}{X(x)} = -\lambda^2 \tag{45.8}$$

$$\frac{T'(t)}{T(t)} = -\frac{1}{R_0C_0}\lambda^2 \tag{45.9}$$

The solution to the spatial equation is given as

$$X(x) = A\sin\lambda x + B\cos\lambda x \tag{45.10}$$

On the other hand the solution to the temporal equation is given by

$$T(t) = Ce^{-\lambda^2 t/R_0C_0} \tag{45.11}$$

To find the various constants we need to satisfy the boundary and initial conditions. Let us try some cases.

45.4 Case with Zero Boundary Conditions and Nonzero Initial Conditions

Consider the case of zero boundary conditions (zero Dirichlet conditions) and a nonzero initial condition comprised of a pulse as shown in Fig. 45.3. If we apply the left boundary condition (zero) we arrive at the requirement that $B = 0$; this implies that

$$X(x) = A \sin \lambda x \tag{45.12}$$

When we apply the right boundary condition (zero again) we arrive at the requirement that

$$\boxed{\lambda = \frac{n\pi}{L}} \tag{45.13}$$

where L is total length. The x solution comes out

$$\boxed{X(x) = A \sin \frac{n\pi x}{L}} \tag{45.14}$$

The time solution has the form

$$T(t) = De^{-t\lambda^2/(R_0C_0)}, \qquad D \quad \text{constant} \tag{45.15}$$

When we plug in for λ we get

$$\boxed{T(t) = D \exp\left[-\frac{n^2\pi^2}{L^2}\frac{t}{R_0C_0}\right]} \tag{45.16}$$

Our total solution is the product of space and time ones:

$$\boxed{v(x,t) = A \sin \frac{n\pi x}{L} e^{-\frac{n^2\pi^2}{L^2}\frac{t}{R_0C_0}}} \tag{45.17}$$

where we lumped the D into A. In fact any linear combination of the above solution also works

$$\boxed{\boxed{v(x,t) = \sum_n A_n \sin \frac{n\pi x}{L} e^{-\frac{n^2\pi^2}{L^2}\frac{t}{R_0C_0}}}} \tag{45.18}$$

Notice that the above solution satisfies the boundary conditions (zero both at 0 and at L) but does not necessarily satisfy the *initial conditions*! The various coefficient A_n are figured dependent on the initial conditions; i.e., at time zero. Specifically at time zero we have

$$v(x,0) = \sum_n A_n \sin \frac{n\pi x}{L} \tag{45.19}$$

Similar to how we found the Fourier coefficients and similar to last chapter if we multiply both sides by a sine and integrate we can single out A_n which comes out

$$\boxed{A_n = \frac{2}{L}\int_0^L v(x,0) \sin \frac{n\pi x}{L} dx} \tag{45.20}$$

Assume for example that total length is $L = 50$ mm. Furthermore assume that all caps between 21 and 29 mm are pre-charged to 1 V; the rest to zero. Then A_n evaluates to

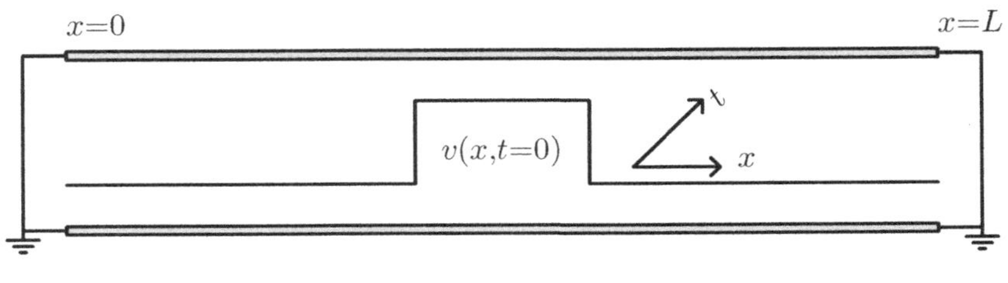

Fig. 45.3 Zero Dirichlet boundary conditions and nonzero initial conditions

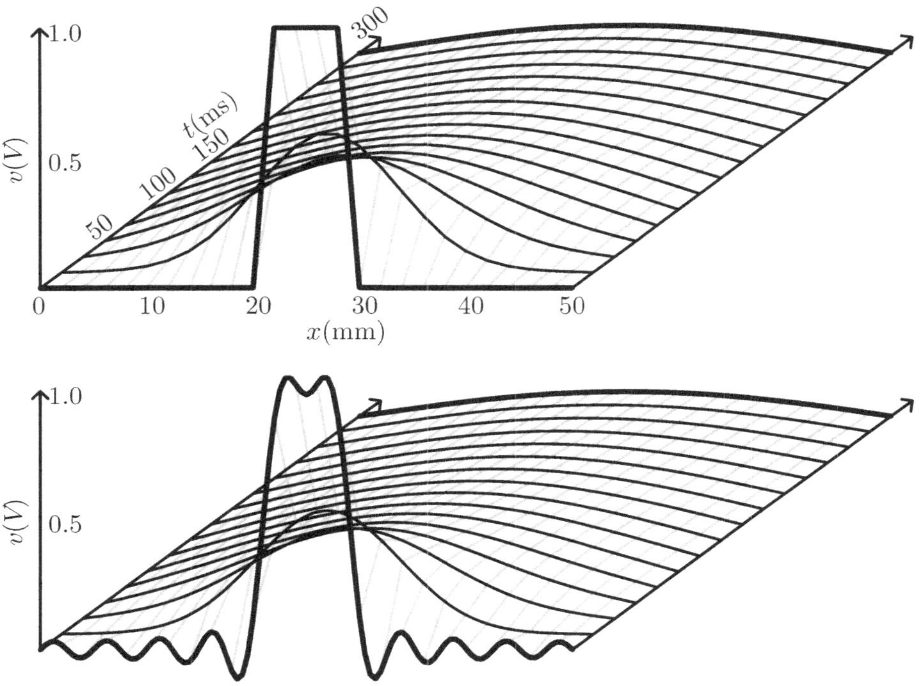

Fig. 45.4 Space/time response of RC ladder shown in Fig. 45.3. Assumed values are $R_0 = 1\,\Omega/\text{mm}$, $C_0 = 1\,\text{mF/mm}$, and $L = 50\,\text{mm}$

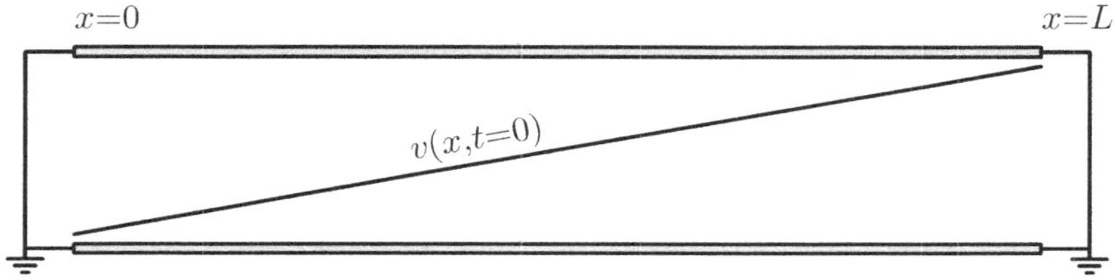

Fig. 45.5 Another example of zero Dirichlet boundary conditions but nonzero initial conditions

$$\boxed{A_n = \frac{-2}{L}\frac{L}{n\pi}\left[\cos\frac{n\pi 29}{50} - \cos\frac{n\pi 21}{50}\right]} \tag{45.21}$$

Now that we know A_n, we know the total solution by using Eq. (45.18). Figure 45.4 shows SPICE results (top) and analytic results (bottom) for a sample run. We observe excellent agreement. Notice how the charge discharges, and eventually all nodes would go to zero voltage, as expected (since both ends are tied to ground)

Another Example Consider another example, still with zero boundary conditions, but with a different set of initial conditions as shown in Fig. 45.5. That is, before we apply the boundary conditions, we pre-charge the various caps such that the left-most cap is at zero voltage, while right-most cap is at unity voltage. Immediately after time zero, we apply zero voltage sources to both left and right sides. No change at the left side, but the right side will experience immediate discharge. The other intermediate caps would now react in accordance with the diffusion equation. We still have the general solution in Eq. (45.18) but with a different set of A_n coefficients. To find A_n we carry on the integral

$$A_n = \frac{2}{L}\int_0^L v(x,0)\sin\frac{n\pi x}{L}dx = \frac{2}{L^2}\int_0^L x\sin\frac{n\pi x}{L}dx \tag{45.22}$$

Using integration by parts we get

$$A_n = -\frac{2}{L^2}\frac{L}{n\pi}x\cos\frac{n\pi x}{L}\Big|_0^L + \frac{2}{L^2}\frac{L}{n\pi}\int_0^L \cos\frac{n\pi x}{L}dx \tag{45.23}$$

(the last term is struck through and marked 0)

$$\boxed{A_n = -\frac{2}{n\pi}\cos(n\pi)} \tag{45.24}$$

and again

$$\boxed{v(x,t) = \sum_n A_n \sin\frac{n\pi x}{L}e^{-\frac{n^2\pi^2}{L^2}\frac{t}{R_0C_0}}} \tag{45.25}$$

By merely finding the new set of A_n we are able to accommodate an entirely new set of initial conditions! The corresponding results and a comparison to SPICE ones are shown in Fig. 45.6. Notice the excellent match which would be the better with more harmonics. The line starts with the prescribed initial conditions, and in time will completely discharge since both left and right sides were forced to zero potential.

45.5 Case with Floating Boundary Conditions and Nonzero Initial Conditions

Rather than forcing the boundary conditions at the edges of the *RC* line, in this case we *float* them! That is, we don't apply any voltage sources at the ends of the line. Since we don't have *RC* segments to the left or to the right of the line, and since we don't have voltage sources there, we are forced to acknowledge that the *current* at the two edges must vanish:

$$i(0,t) = i(L,t) = 0 \tag{45.26}$$

But we know that current is related to voltage by

$$R_0 i(x) = \frac{dv}{dx} \tag{45.27}$$

Hence we conclude that the implicit boundary conditions are such that the *derivative* (rather than the value itself) must vanish at the ends; this is normally referred to as the Neumann boundary conditions.

$$\frac{dv(x,t)}{dx}\Big|_{x=0} = \frac{dv(x,t)}{dx}\Big|_{x=L} = 0 \tag{45.28}$$

We established before when doing the method of separation of variables that the most generic x solution is

$$X(x) = A\sin\lambda x + B\cos\lambda x \tag{45.29}$$

Taking the derivative we get

$$\frac{dX(x)}{dx} = \lambda\left[A\cos\lambda x - B\sin\lambda x\right] \tag{45.30}$$

Applying the Neumann boundary condition at $x = 0$ we are forced to set

$$A = 0 \tag{45.31}$$

such that

$$X(x) = B\cos\lambda x \tag{45.32}$$

Applying next the Neumann boundary condition at $x = L$ we are forced to set

$$\lambda_n = \frac{n\pi}{L} \tag{45.33}$$

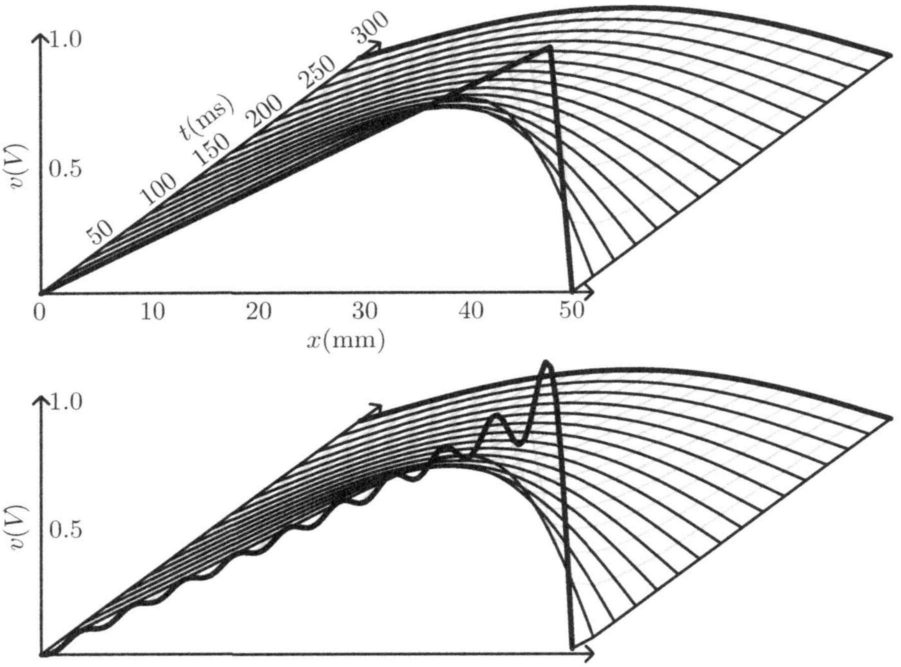

Fig. 45.6 Space/time response of RC ladder with BCs and ICs as shown in Fig. 45.5. Assumed values are $R_0 = 1\,\Omega/\text{mm}$ and $C_0 = 1\,\text{mF/mm}$ and $L = 50\,\text{mm}$

such that

$$X(x) = B\cos\frac{n\pi x}{L} \tag{45.34}$$

Let us do a quick sanity check. First we know Eq. (45.34) satisfies the x component of the diffusion equation (Eq. (45.8))

$$\frac{d^2}{dx^2}X(x) = -\left(\frac{n\pi}{L}\right)^2 X(x) \tag{45.35}$$

Next we know that the derivative of this is 0 both at $x = 0$ and $x = L$:

$$-B\frac{n\pi}{L}\sin\frac{n\pi x}{L}\bigg|_{x=0,x=L} = 0 \tag{45.36}$$

Hence the x solution is sound. Plugging back into $v(x,t)$ and using superposition we end up with

$$\boxed{v(x,t) = \sum_n A_n \cos\frac{n\pi x}{L} e^{-\frac{n^2\pi^2}{L^2}\frac{t}{R_0C_0}}} \tag{45.37}$$

The various A_n coefficients are figured depending on the initial conditions. Suppose for example we have the same ICs as in Sect. 45.3, and reshown in Fig. 45.7. Carrying on the integration we arrive at

$$\boxed{A_n = \frac{2}{L}\frac{L}{n\pi}\left[\sin\frac{n\pi 29}{50} - \sin\frac{n\pi 21}{50}\right]} \tag{45.38}$$

Resulting solution and comparison to SPICE one are shown in Fig. 45.8. Notice that unlike before, the resulting solution does *not* settle down to zero at large time; instead the whole line is charged to a value equal to the average value of the initial charging voltage. In this example, initially 9 points out of 51 points were charged to 1 V while the rest to zero. The average initial voltage is then 9/51 = 0.18 V; this is exactly the settling value observed for all points at larger time. In other words since charge has nowhere to go, but to redistribute along the line, and due

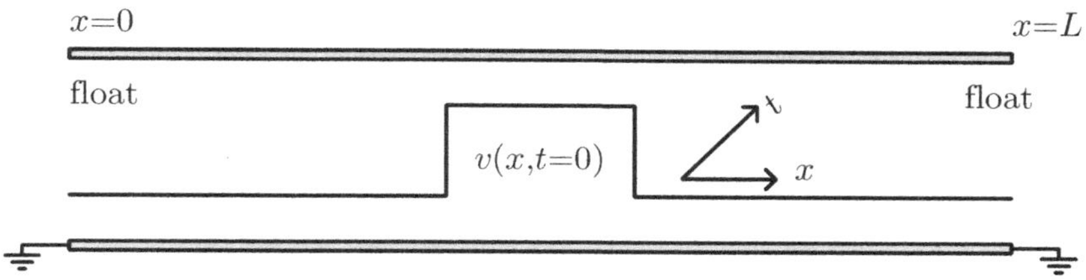

Fig. 45.7 Zero Neumann boundary conditions and nonzero initial conditions

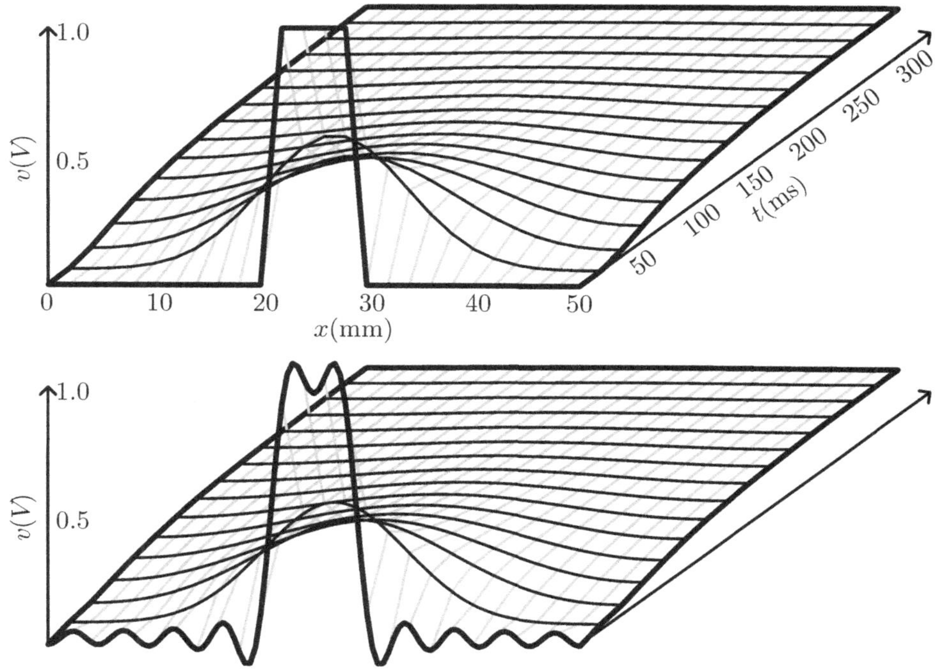

Fig. 45.8 Space/time response of RC ladder shown in Fig. 45.7. Assumed values are $R_0 = 1\,\Omega/\text{mm}$, $C_0 = 1\,\text{mF/mm}$, and $L = 50\,\text{mm}$. Top SPICE, bottom theory

to symmetry, all caps settle to a nonzero voltage value, and to an equal one.

45.6 Mixed Zero and Floating Boundary Conditions

Now consider the case where one boundary (left) is fixed (to zero, for example) while the other (right) is left floating; something like the setup shown in Fig. 45.9 Again via separation of variables we get the X solution

$$X(x) = A\sin\lambda x + B\cos\lambda x \tag{45.39}$$

Applying the zero left boundary conditions forces to set

$$B = 0 \tag{45.40}$$

such that

$$X(x) = A\sin\lambda x \tag{45.41}$$

At the right boundary we set the derivative to zero

$$A\lambda\cos\lambda x\big|_{x=L} = 0 \tag{45.42}$$

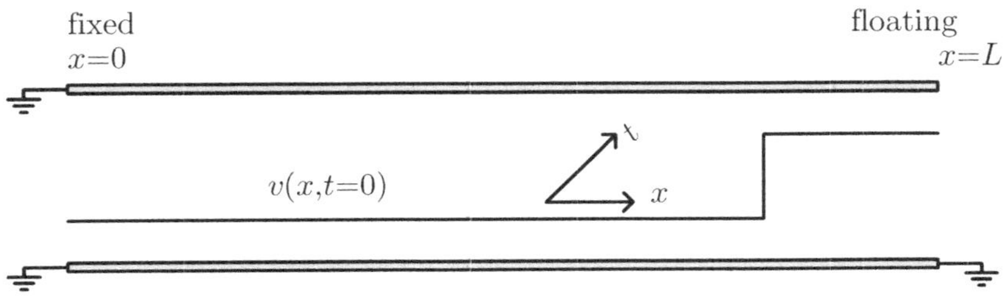

Fig. 45.9 Setup with fixed left boundary condition, and floating right BC; nonzero initial conditions

This would happen when

$$\lambda_n = \frac{(2n+1)\pi}{2L} \tag{45.43}$$

Hence we have

$$v(x,t) = \sum_n A_n \sin(\lambda_n x) e^{-t\lambda_n^2/(R_0 C_0)} \tag{45.44}$$

To figure A_n we need the initial conditions; suppose those to be zero except for $40 < x < 50$ mm, again as shown in Fig. 45.9. Then

$$A_n = \frac{2}{L\lambda_n} \cos\left[\lambda_n \frac{40}{50} L\right] \tag{45.45}$$

With that ready we are able to plot final solution and compare to SPICE as shown in Fig. 45.10. Notice excellent agreement. Notice also that the left side voltage does not change in time; after all it is held by a zero voltage source! But the right side discharges and does so only to the left side. That is, charge from right segments have nowhere else to go but to the left (since there is no metal to the right of the line), and in doing so a discharged right segment charges a left segment, and that is why we see middle line segments charge in time. But if we wait long enough even those will eventually discharge and in the end the zero-voltage forcing source on the left will have its way!

45.7 Case with Zero Initial Conditions and Constant Boundary Conditions

So far we have dealt with problems which had either zero boundary conditions or floating ones. When BCs were zeroes we used sine basis functions, while for floating BCs we used cosine basis functions. For the special case of mix of zero and floating BCs we resorted to sine basis function of "quarter wave length." For nonzero, forced BCs using any of the above basis functions won't do! We will need to expand our flow. For example consider an RC line whose left side tied to ground, while the right side held at unity, as shown in Fig. 45.11. To solve this problem assume we can decompose the solution into two parts: one with zero BCs and one which strictly satisfies the BCs

$$v(x,t) = w(x,t) + v_b(x) \tag{45.46}$$

Notice that the boundary satisfying function $v_b(x)$ depends only on x since the right boundary does not depend on time. For example we could have

$$v_b(x,t) = \frac{x}{L} \tag{45.47}$$

This function is zero at the left boundary, while 1 at right one. So it takes care of the BCs. Moving to $w(x,t)$ we may enquire, what differential equation does it satisfy? To that end we start with

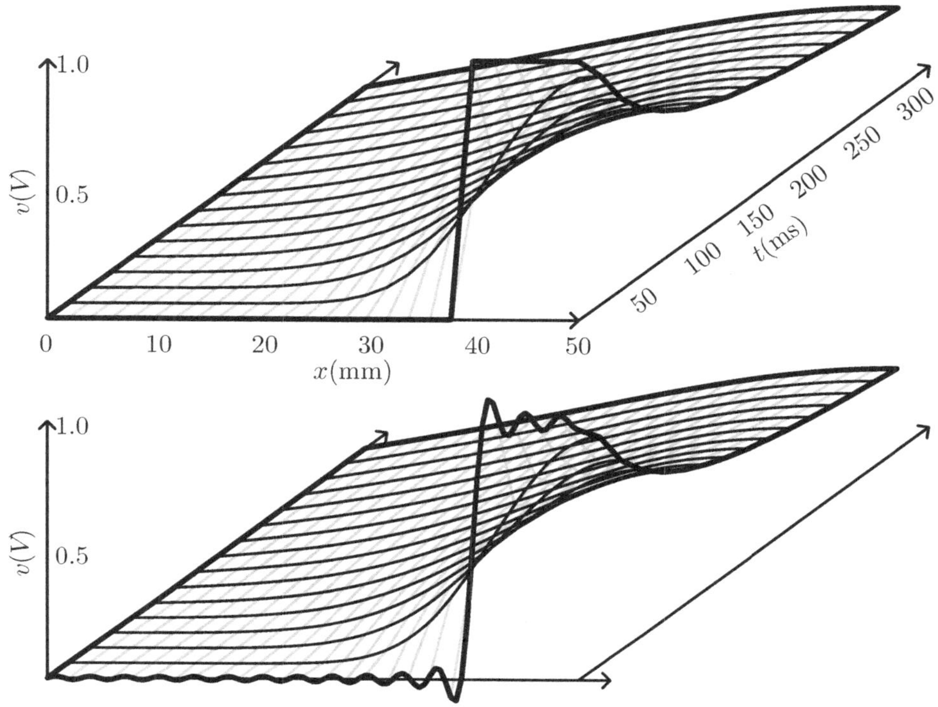

Fig. 45.10 Results corresponding to setup in Fig. 45.9. Assumed values are $R_0 = 1\,\Omega/\text{mm}$, $C_0 = 1\,\text{mF/mm}$, and $L = 50\,\text{mm}$. Top SPICE; bottom theory

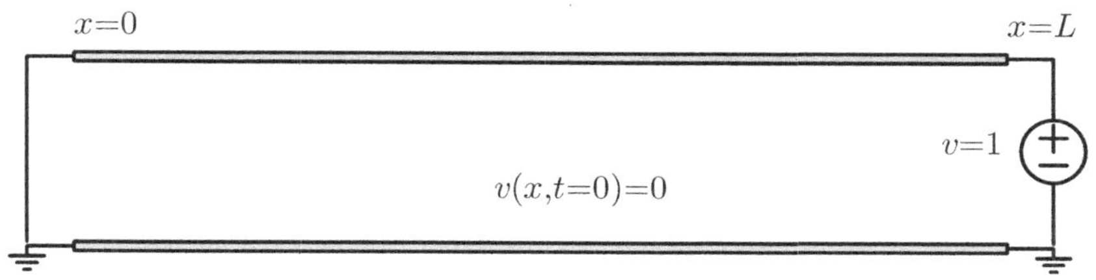

Fig. 45.11 Nonzero Dirichlet boundary conditions and zero initial conditions

$$v_{xx}(x,t) = \frac{d^2}{dx^2}\left[w(x,t) + \frac{x}{L}\right] = w_{xx}(x,t) \tag{45.48}$$

Next

$$v_t(x,t) = \frac{d}{dt}\left[w(x,t) + \frac{x}{L}\right] = w_t(x,t) \tag{45.49}$$

Plugging back into the diffusion equation we get

$$w_{xx}(x,t) = R_0 C_0 w_t(x,t) \tag{45.50}$$

And what boundary conditions does this satisfy? We already set it up to satisfy zero BCs; that is

$$w(0,t) = w(L,t) = 0 \tag{45.51}$$

Perfect! So now we have the diffusion equation with zero boundary conditions! We already know how to solve this; that is

$$w(x,t) = \sum_m A_n \sin\frac{n\pi x}{L} e^{-\frac{t}{R_0 C_0}\frac{n^2\pi^2}{L^2}} \tag{45.52}$$

The only thing remaining is to determine the expansion coefficients A_n. This is done via initial conditions. And this leads to the second question: what initial conditions should $w(x,t)$ satisfy? Recall

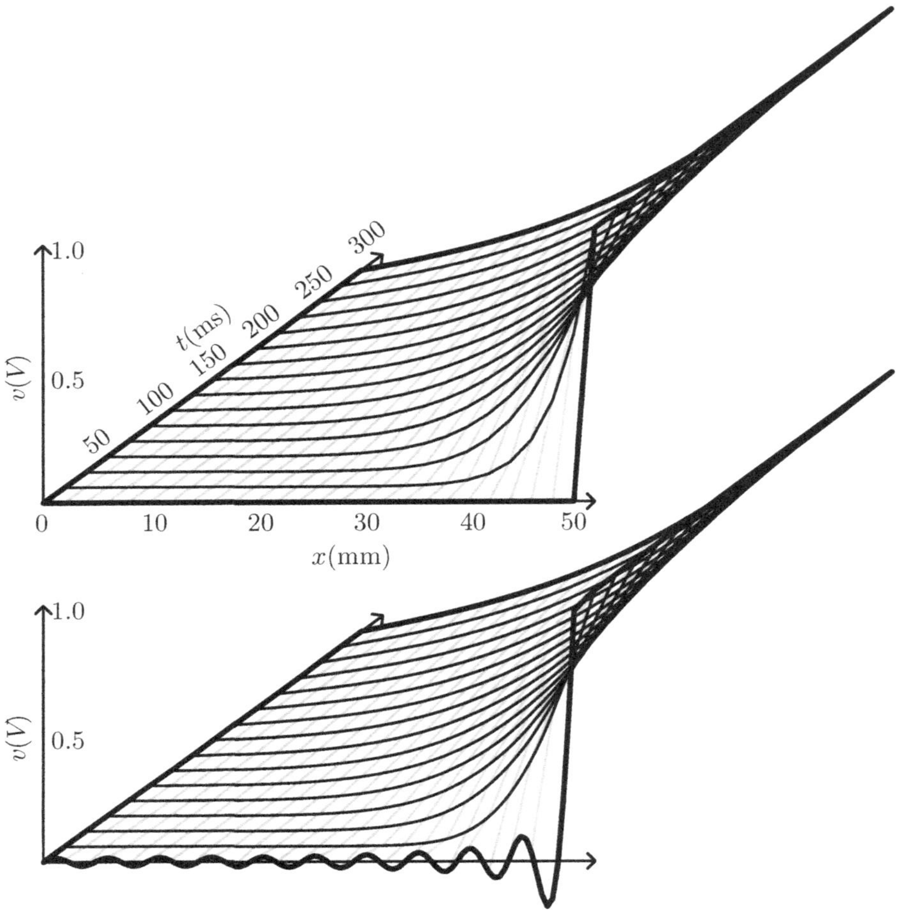

Fig. 45.12 Solution to diffusion problem with initial and boundary conditions as prescribed in Fig. 45.11. Assumed values are $R_0 = 1\,\Omega/\text{mm}$, $C_0 = 1\,\text{mF/mm}$, and $L = 50\,\text{mm}$. Top SPICE; bottom theory

$$v(x, t) = w(x, t) + \frac{x}{L} \tag{45.53}$$

and we know that $v(x, 0) = 0$; this implies that

$$0 = w(x, 0) + \frac{x}{L} \tag{45.54}$$

or

$$\sum A_n \sin \frac{n\pi x}{L} = -\frac{x}{L} \tag{45.55}$$

But this is nothing other than the Fourier series of the function x. We already know this from Eq. (45.24), which gives

$$A_n = \frac{2}{n\pi} \cos(n\pi) \tag{45.56}$$

Putting now all the pieces together we finally get

$$v(x, t) = \frac{x}{L} + \sum_n A_n \sin \frac{n\pi x}{L} e^{-\frac{n^2\pi^2}{L^2}\frac{t}{R_0C_0}} \tag{45.57}$$

Corresponding results are shown in Fig. 45.12. Notice that sure enough at time zero voltage is zero everywhere. Immediately after time zero, the right side jumps to 1, by the forcing voltage function at the right. Only afterwards, and gradually does the rest of the wire incrementally rises in voltage, while maintaining at all times the zero BCs at the far left. In the end we would expect

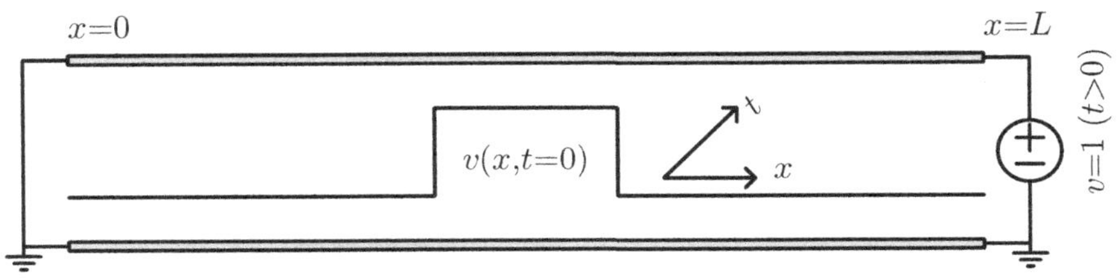

Fig. 45.13 Nonzero initial conditions and nonzero boundary conditions

the voltage to simply be zero at left, 1 at right, and linearly increasing in between!

45.8 Case with Nonzero Initial Conditions and Constant Boundary Conditions

This case has nonzero initial conditions and constant boundary conditions. Consider for example the setup shown in Fig. 45.13. Similar to the last section, we will stitch together various solutions such that in the end the differential equation is satisfied, and so would the initial and boundary conditions. In particular we will use the solution in Sect. 45.3 in conjunction with the last section's (Sect. 45.7) solution. The former gave us the solution to a zero-boundary condition setup but with nonzero initial conditions. The latter, on the other hand, gave us the solution to a nonzero boundary conditions but with zero initial conditions.

Specifically for the zero boundary condition case, but nonzero initial conditions we had

$$A_n = \frac{-2}{n\pi}\left[\cos\frac{n\pi 29}{50} - \cos\frac{n\pi 21}{50}\right] \quad (45.58)$$

And, on the other hand, for the nonzero boundary condition case but with zero initial conditions we had

$$B_n = \frac{2}{n\pi}\cos(n\pi) \quad (45.59)$$

as well as the $\frac{x}{L}$ particular solution. If we add all together we finally get

$$\boxed{v(x,t) = \frac{x}{L} + \sum_n (A_n + B_n)\sin\frac{n\pi x}{L}e^{-\frac{n^2\pi^2}{L^2}\frac{t}{R_0C_0}}} \quad (45.60)$$

That is, using superposition in time, we are claiming that the total solution is the sum of two cases:

1. Case of zero boundary conditions, but nonzero initial conditions; and
2. Case of zero initial conditions, but nonzero boundary conditions.

In other words, the response due to a system with initial conditions and a forced voltage is the sum of an unforced system with initial conditions plus a driven system with no initial conditions!

Taking the risk of repeating oneself but just in case it still does not make absolute sense, the claim is that the solution is the sum of two solutions: $v_1(x,t)$ and $v_2(x,t)$. Each solution satisfies the diffusion equation, so the sum automatically satisfies the diffusion equation; good! Next, we need to ensure that the sum satisfies the boundary conditions. Since the former has zero BCs and the latter is built to satisfy the BCs, then the sum automatically satisfies the BCs, which is also good. Lastly we need to ensure that the sum automatically satisfies the initial conditions. Since the former satisfies the initial conditions while the latter has zero initial conditions, then the sum indeed satisfies the ICs. Hence we have proven that the sum (a) satisfies the *RC* diffusion equation, (b) satisfies the boundary conditions, and (c) satisfies the initial conditions. By virtue of this we arrive at the inevitable conclusion that this solution is a valid one and in fact the only one! Figure 45.14 shows our results and comparison to SPICE; clearly our assumptions were valid all along!

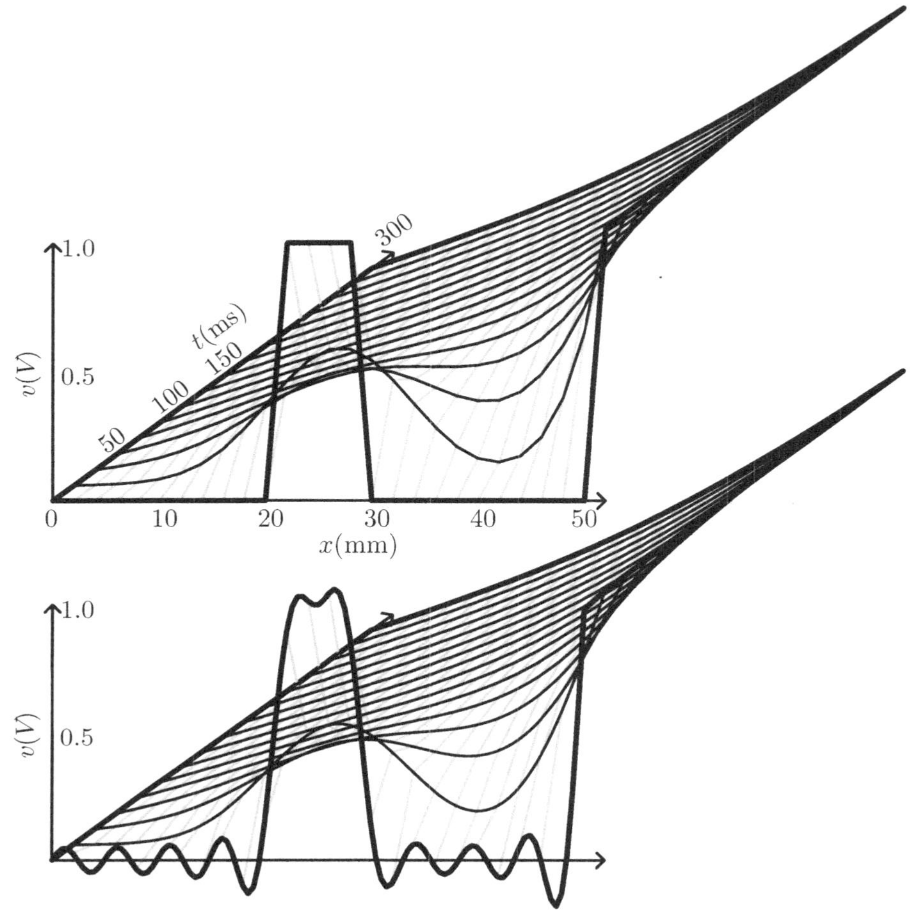

Fig. 45.14 Solution to diffusion problem with initial and boundary conditions as prescribed in Fig. 45.13. Assumed values are $R_0 = 1\ \Omega/\text{mm}$, $C_0 = 1\ \text{mF/mm}$, and $L = 50\ \text{mm}$. Top SPICE; bottom theory

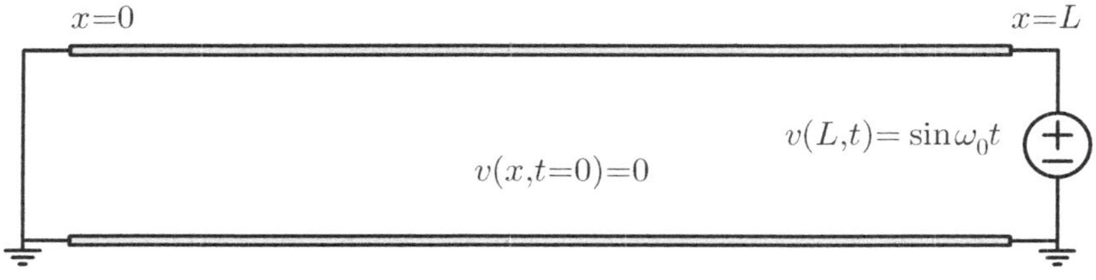

Fig. 45.15 Zero initial conditions and time-dependent boundary conditions

45.9 Case with Zero Initial Conditions and Harmonic Time-Varying Boundary Conditions

In the last section we forced a nonzero boundary condition, but it was constant (i.e., time independent). In this section we remove that restriction. Consider for example the case where the right boundary is driven by a sine function as shown in Fig. 45.15:

$$v(L,t) = \sin\omega_0 t \qquad \text{and} \qquad v(0,t) = 0 \tag{45.61}$$

Since the voltage does not die off at the boundary, again we cannot assume the solution to be expandable in terms of spatial eigenfunctions! As before, instead of dealing with the function $v(x,t)$ directly we introduce another function defined as

$$w(x,t) = v(x,t) - v_b(x,t) \tag{45.62}$$

where $v_b(x,t)$ satisfies the original boundary conditions. One simple such function is

$$v_b(x,t) = \frac{x}{L}\sin\omega_0 t \tag{45.63}$$

Notice that this function satisfies both left and right boundary conditions. Then

$$w(x,t) = v(x,t) - \frac{x}{L}\sin\omega_0 t \tag{45.64}$$

$$v(x,t) = w(x,t) + \frac{x}{L}\sin\omega_0 t \tag{45.65}$$

To find the differential equation governing $w(x,t)$ we start again with the diffusion equation:

$$\frac{d^2 v(x,t)}{dx^2} = R_0 C_0 \frac{dv(x,t)}{dt} \tag{45.66}$$

Plugging in for $v(x,t)$ we get

$$\frac{d^2 w(x,t)}{dx^2} = R_0 C_0 \frac{dw(x,t)}{dt} + R_0 C_0 \omega_0 \frac{x}{L}\cos\omega_0 t \tag{45.67}$$

Now this equation is subject to homogeneous boundary conditions

$$w(0,t) = w(L,t) = 0 \tag{45.68}$$

That is, the function $w(x,t)$ satisfies zero boundary conditions. As such we can now expand this one in terms of spatial eigenfunctions.

$$w(x,t) = \sum_n a_n(t)\sin\frac{n\pi x}{L} \tag{45.69}$$

Plugging back into Eq. (45.67) we get

$$-\frac{n^2\pi^2}{L^2}\sum_n a_n(t)\sin\frac{n\pi x}{L} = R_0 C_0 \sum_n \frac{da_n(t)}{dt}\sin\frac{n\pi x}{L} + R_0 C_0 \omega_0 \frac{x}{L}\cos\omega_0 t \tag{45.70}$$

Recall that on the axis $0 \to L$ we can represent the function x as

$$x = \sum_n b_n \sin\frac{n\pi x}{L} \tag{45.71}$$

where

$$b_n = \frac{2}{L}\int_0^L x\sin\frac{n\pi x}{L}dx = \frac{2}{L}\frac{-L}{n\pi}x\cos\frac{n\pi x}{L}\Big|_0^L = -\frac{2L}{n\pi}\cos n\pi = -\frac{2L}{n\pi}(-1)^n \tag{45.72}$$

Plugging back into Eq. (45.70) we get

$$-\frac{n^2\pi^2}{L^2}\sum_n a_n(t)\sin\frac{n\pi x}{L} = R_0 C_0 \sum_n \frac{da_n(t)}{dt}\sin\frac{n\pi x}{L} + \frac{R_0 C_0 \omega_0}{L}\cos\omega_0 t \sum_n \frac{-2L(-1)^n}{n\pi}\sin\frac{n\pi x}{L} \tag{45.73}$$

or

$$-\frac{n^2\pi^2}{L^2}\sum_n a_n(t)\sin\frac{n\pi x}{L} = R_0C_0\sum_n \frac{da_n(t)}{dt}\sin\frac{n\pi x}{L} - 2R_0C_0\omega_0\cos\omega_0 t\sum_n \frac{(-1)^n}{n\pi}\sin\frac{n\pi x}{L} \quad (45.74)$$

This can be true only if

$$-\frac{n^2\pi^2}{L^2}a_n(t) = R_0C_0\frac{da_n(t)}{dt} - 2R_0C_0\omega_0\frac{(-1)^n}{n\pi}\cos\omega_0 t \quad (45.75)$$

or

$$\boxed{\frac{n^2\pi^2}{L^2}a_n(t) + R_0C_0\frac{da_n(t)}{dt} = 2R_0C_0\omega_0\frac{(-1)^n}{n\pi}\cos\omega_0 t} \quad (45.76)$$

Hence we need to solve for $a_n(t)$. For a particular solution we assume

$$a_{np}(t) = A\sin\omega_0 t + B\cos\omega_0 t \quad (45.77)$$

Plugging in we get

$$\frac{n^2\pi^2}{L^2}[A\sin\omega_0 t + B\cos\omega_0 t] + R_0C_0\omega_0[A\cos\omega_0 t - B\sin\omega_0 t] = 2R_0C_0\omega_0\frac{(-1)^2}{n\pi}\cos\omega_0 t \quad (45.78)$$

Equating sine terms we get

$$\frac{n^2\pi^2}{L^2}A - R_0C_0\omega_0 B = 0 \quad (45.79)$$

which gives

$$\boxed{B = \frac{1}{R_0C_0\omega_0}\frac{n^2\pi^2}{L^2}A} \quad (45.80)$$

Equating cosine terms we get

$$\frac{n^2\pi^2}{L^2}\frac{1}{R_0C_0\omega_0}\frac{n^2\pi^2}{L^2}A + R_0C_0\omega_0 A = 2R_0C_0\omega_0\frac{(-1)^2}{n\pi} \quad (45.81)$$

which gives

$$\boxed{A = \frac{2R_0C_0\omega_0\frac{(-1)^n}{n\pi}}{\frac{n^4\pi^4}{L^4}\frac{1}{R_0C_0\omega_0} + R_0C_0\omega_0}} \quad (45.82)$$

So now we know both A and B for the particular solution. The homogeneous solution is simply

$$a_{nh}(t) = D\exp\left(-t\frac{n^2\pi^2}{L^2}\frac{1}{R_0C_0}\right) \quad (45.83)$$

The initial conditions on $a_n(t)$ are $a_n(0) = 0$ since the RC line is presumed uncharged. This then sets

$$D = -B \tag{45.84}$$

So finally the $a_n(t)$ solution is

$$\boxed{a_n(t) = A\sin\omega_0 t + B\cos\omega_0 t - B\exp\left(-t\frac{n^2\pi^2}{L^2}\frac{1}{R_0C_0}\right)} \tag{45.85}$$

where A was given in Eq. (45.82) and B in Eq. (45.80). Knowing $a_n(t)$ we plug back into

$$\boxed{w(x,t) = \sum_n a_n(t)\sin\frac{n\pi x}{L}} \tag{45.86}$$

and from this we finally get

$$\boxed{v(x,t) = w(x,t) + \frac{x}{L}\sin\omega_0 t} \tag{45.87}$$

Results and comparison to SPICE are shown in Fig. 45.16. Notice the excellent match. Sure enough, at time zero the whole line is grounded. After that, the left side remains grounded while the right side is driven by the sine input voltage. The internal nodes are governed by the diffusion equation whose solution we just derived!

45.10 Case with Zero Initial Conditions and Arbitrary Time-Varying Boundary Conditions

In the last section we derived the solution due to a sinusoidal boundary conditions. Namely if the boundary condition at $x = L$ is $v(L,t) = \sin\omega_0 t$ then we go the solution in Eq. (45.87). Let us label this solution as $v_{\omega_0}(x,t)$ implying it belongs to the special case of forced boundary conditions of the form $v(L,t) = \sin\omega_0 t$. That is

$$\text{case of } v(L,t) = \sin\omega_0 t \overset{\text{gives solution}}{\longrightarrow} v_{\omega_0}(x,t) \tag{45.88}$$

Now we tackle the more general case of an *arbitrary*, time varying boundary condition; the only restriction for now is that the boundary condition is assumed to be periodic in $\frac{2\pi}{\omega_0}$; i.e., it has the same angular frequency as the sinusoidal one. That is any periodic signal (with period $\frac{2\pi}{\omega_0}$) applied at $x = L$. Assume then we can represent this signal as a Fourier series in $\sin m\omega_0 t$; that is

$$v(L,t) = \sum_m b_m \sin m\omega_0 t \tag{45.89}$$

Then by superposition (in time) we are led to conclude that the corresponding solution would be

$$v(x,t) = \sum_m b_m v_{m\omega_0}(x,t) \tag{45.90}$$

where $v_{\omega_0}(x,t)$ was given in Eq. (45.88). To reiterate the solution $v_{m\omega_0}(x,t)$ corresponds to an excitation of the form

$$\text{case of } v(L,t) = \sin(m\omega_0 t) \overset{\text{gives solution}}{\longrightarrow} v_{m\omega_0}(x,t) \tag{45.91}$$

Keeping in mind that $v_{\omega_0}(x,t)$ itself was a Fourier series expansion in *space* we see now that the general solution encompasses another Fourier series, but this time in *time*! So now the solution

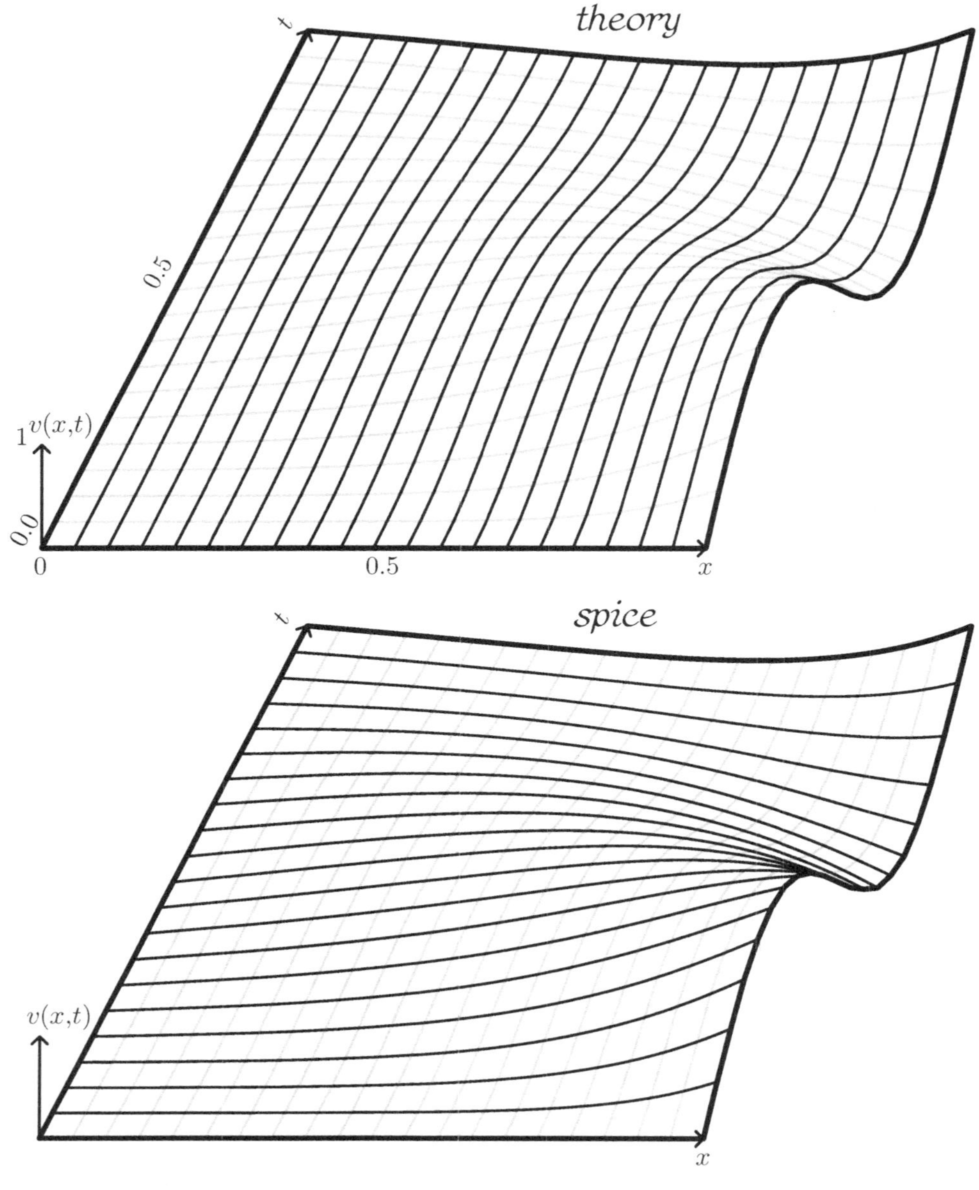

Fig. 45.16 Solution to diffusion problem with zero initial conditions and forced sinusoid on right boundary; assumed values are $R = 1\,\Omega/\text{m}$, $C = 2\,\text{F/m}$, and $L = 1\,\text{m}$

is a doubly Fourier series! Though it is a lot of book keeping, it cannot be emphasized enough the importance of capturing the exact solution analytically. A sample application of this case is shown in Problem 6.

To recap, we know the solution to the case of a sinusoidal boundary condition of angular frequency ω_0; call it the harmonic solution. Now if the boundary condition is another signal, which is still periodic in $\frac{2\pi}{\omega_0}$, such is the case the signal can be constructed in terms of a Fourier series in time, then the solution to that case would be a Fourier series of the harmonic solution with the same Fourier coefficients used to generate the arbitrary signal in terms of the time harmonics with angular frequency ω_0.

45.11 Current Profile Through *RC* Diffusive Networks

In the various prior sections we tackled figuring out voltage, as a function of space and time, for various initial and boundary conditions. End result was a 3D plot showing voltage vs. space/time. In this quick section we show a sample current profile, again vs position and time. This case is a bit more difficult to visualize since current is a *vector*, and in fact we have two current profiles: that across the resistors, and that across the capacitors. So we would have to get a bit creative in visualizing the current. In the prior sections, and for better visualization we discretize the x-axis into some 50 segments; in order not to blur things, however, in this section we discretize only into 10 sections, as shown in Fig. 45.17. In conjunction with the 10 resistors, we pick up 9 capacitors. We also impose zero boundary conditions and nonzero initial conditions, again as shown in the figure.

We know after time zero, the charge would smear out (diffuse) and we know steady state solution is such that the whole line discharges (into the two voltage sources at the extreme ends). As the various caps discharge/charge, we ought to observe current. If we were to plot this current in the shape of an arrow whose *direction* points in the direction of current flow, and whose *size* mimics the magnitude of the current; and if we further plot such currents both along the resistors and along the caps we end up with Fig. 45.18. This figure takes a bit of a time to get adjusted to, so patience is advised! The background of the figure shows the actual *RC* grid, and each row corresponds to a different time step. Notice how initially current gushes out of the middle cap, both to the left and to the right. By all measures the initial current is huge. At the same time the end current is zero; that is, no charge has traveled down the line (yet). At any point along the line, the sum of input current at a node must equal that out of the node. Current along the resistors is not uniform, as can be seen clearly; reason being the presence of the caps. At any point if the cap is charging/discharging, this would imply that current across the resistor to the left of the cap would *not* equal that across the resistor to the right, and hence the asymmetry. As time progresses, all currents decay in magnitude, as can be observed towards the bottom of the figure. In the end all dies and we are left with an uncharged line, with zero voltage all across! Again it is advised to spend some time absorbing the figure and tracing the various arrows. The reader may even try reading the plot backwards in the sense of starting at the bottom and reading upwards, back in time; this way the initial arrows are small in size and don't overlap much.

One goal of this sort of plot is to get a feel or appreciation how current flows. It is critical to remember that current is the flow of charge. In the end we ought to be able to account for charge at any point in time or position in space. Current across a resistor is literally charge flow through

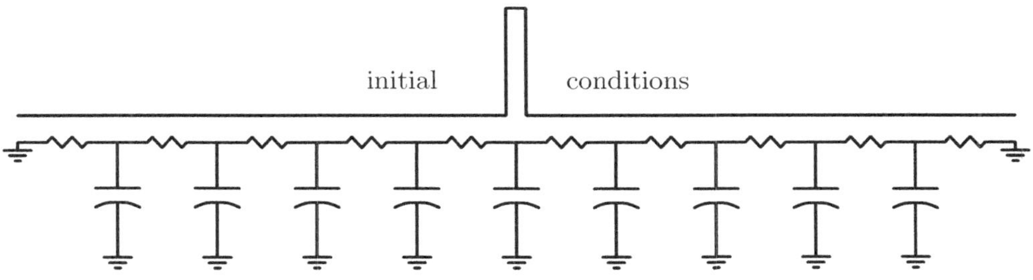

Fig. 45.17 Diffusion line with zero boundary conditions and nonzero initial ones

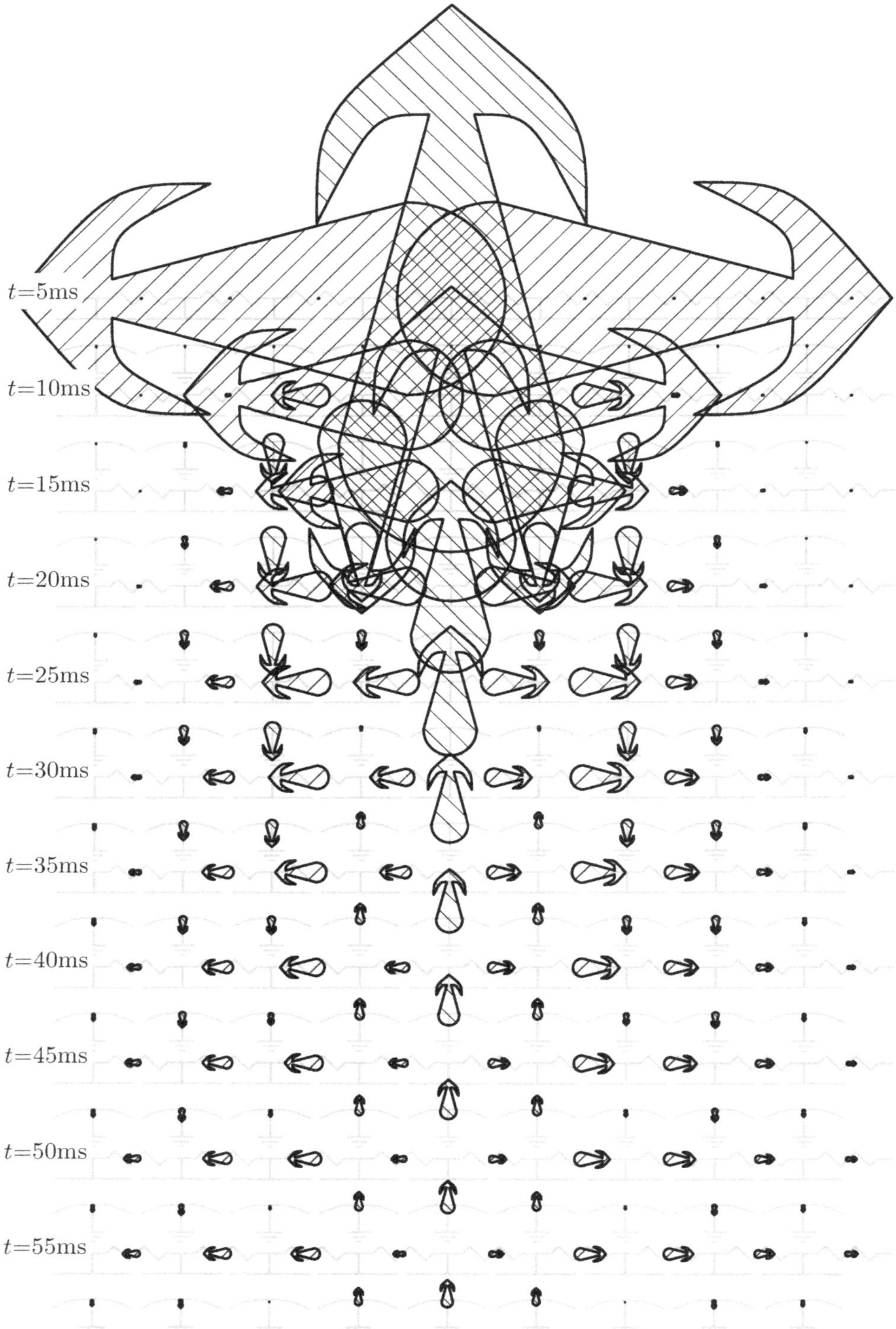

Fig. 45.18 Current flow corresponding to Fig. 45.17. In this case line length is 50 mm, and resistance/cap per segment is 5 Ω and 5 mF

it. But current across the cap is a bit different! Just in case the reader was not sure, charge does *not* cross the cap! It does not penetrate through the dielectric (or air) between the cap terminals. (Might be surprised how many believe so!) Charge only *accumulates* at the cap terminal, and this has the side effect of altering current through adjacent resistors; and hence we say that the current across the left and right resistors differs because of the cap current. What we really mean by this is that not all charge passing from left resistor makes it to right resistor, because some accumulated (stayed) at the intermediate cap terminal; hence we have input current and output current across the left and right resistors, as well as what looks like a current across the intermediate cap. Of course cap current takes place only in the transient world; at DC or when things settle down only resistor current remains (if any) and caps no longer change voltage.

45.12 Summary

Continuing on with distributed media this chapter dealt with the *RC* diffusion lines. The voltage/current across the line is characterized by the diffusion equation. This equation is second order in space and first order in time. The typical setup is an *RC* line of certain length, certain resistance/capacitance (R_0/C_0) per unit length, an initial condition describing voltage across the line, and a set of boundary conditions describing voltage at both ends. Initial conditions range from zero ones, meaning the line is pre-charged to zero; to space dependent ones describing the initial voltage across the line. The boundary conditions can be either fixed (Dirichlet) or floating (Neumann), or a mix thereof. The fixed boundary conditions can be either time independent or time dependent ones. In the chapter we worked many examples ranging from the simplest of zero boundary conditions and some prescribed initial condition to the most complex case which is a line driven by a time dependent signal. Our strategy all along was to decompose the solution in terms of spatial harmonic ones. But for the more complex case of time dependent boundary conditions we had to resort to also utilizing time dependent harmonics; so the most generic solution of the distributed *RC* line is a doubly Fourier series, both in space and time. This just shows the potential and power of spectral methods. No doubt the whole thing can be cast and reworked in terms of convolution methods as well, but for lack of space we stop here. For each of the many tackled examples we plotted voltage versus position and monitored it versus time. We also compared all cases against SPICE simulations and obtained excellent agreement. Finally we did a visualization exercise in demonstrating current flow (which is a vector phenomena) again versus space and time to illustrate the overall phenomena of resistive current and capacitive (charging) current and to put Kirchoff's current law in perspective.

45.13 Problems

1. Consider the distributed *RC* line with zero boundary conditions, and initial conditions as shown in Fig. 45.19a. The line has length 1 m, $R_0 = 1\,\Omega/\text{m}$, and $C_0 = 2\,\text{F/m}$. Solve for the potential for all space and time, and plot it using 20 harmonics. Compare to SPICE; see sample solution in Fig. 45.20.
   Answer:

$$a_n = \frac{2}{L}\left[-\frac{L}{n\pi}\frac{1}{2}\cos\frac{n\pi 3}{4} + \left(\frac{L}{n\pi}\right)^2\left(\sin\frac{n\pi 3}{4} - \sin\frac{n\pi}{2}\right)\right]$$

2. Consider the distributed *RC* line with zero boundary conditions, and initial conditions as shown in Fig. 45.19b. Use line specifications in Problem 1. Solve for the potential for all space and time, and plot it using 20 harmonics. Compare to SPICE; see sample solution in Fig. 45.21.
   Answer:

$$a_n = \frac{2}{L}\frac{L}{n\pi}\left[-\cos\frac{n\pi 3}{4} - \cos\frac{n\pi 1}{4} + 2\cos\frac{n\pi 2}{4}\right]$$

3. Consider the distributed *RC* line with *floating* boundary conditions, and initial conditions as shown in Fig. 45.19c. Use line specifications in Problem 1. Solve for the potential for all space and time, and plot it using 20 harmonics. Compare to SPICE; see sample solution in Fig. 45.22.
   Answer:

$$a_n = -\frac{2}{L}\frac{L}{n\pi}2\sin\frac{n\pi}{2}$$

4. Consider the distributed *RC* line again with *floating* boundary conditions, and initial conditions as shown in Fig. 45.19d. Use line specifications in Problem 1. Solve for the potential for all space and time, and plot it using 20 harmonics. Compare to SPICE; see sample solution in Fig. 45.23.
   Answer:

$$a_n = \frac{2}{L}\frac{L}{n\pi}[\sin(n\pi \times 0.8) - \sin(n\pi \times 0.6) + \sin(n\pi \times 0.4) - \sin(n\pi \times 0.2)]$$

5. Consider the distributed *RC* line with zero initial conditions, but with one side driven by a voltage source $v(L, t) = t$ as shown in Fig. 45.24a. Use line specifications in Problem 1. Solve for the potential for all space and time, and plot it using 20 harmonics. Compare to SPICE; see sample solution in Fig. 45.25.
   Answer:

$$A = 2R_0C_0\frac{(-1)^n}{n\pi}\frac{1}{n^2\pi^2/L^2}$$

$$a_n(t) = A - A\exp\left(-\frac{t}{R_0C_0}\frac{n^2\pi^2}{L^2}\right)$$

$$v(x,t) = \sum_n a_n(t)\sin\frac{n\pi x}{L} + \frac{x}{L}t$$

6. Consider the distributed *RC* line with zero initial conditions, but with one side driven by a voltage source which is a periodic pulse of period 1, and 0 average, and as shown in Fig. 45.24b. Use line specifications in Problem 1. Solve for the potential for all space and time, and plot it using 20 harmonics. Compare to SPICE; see sample solution in Fig. 45.26.

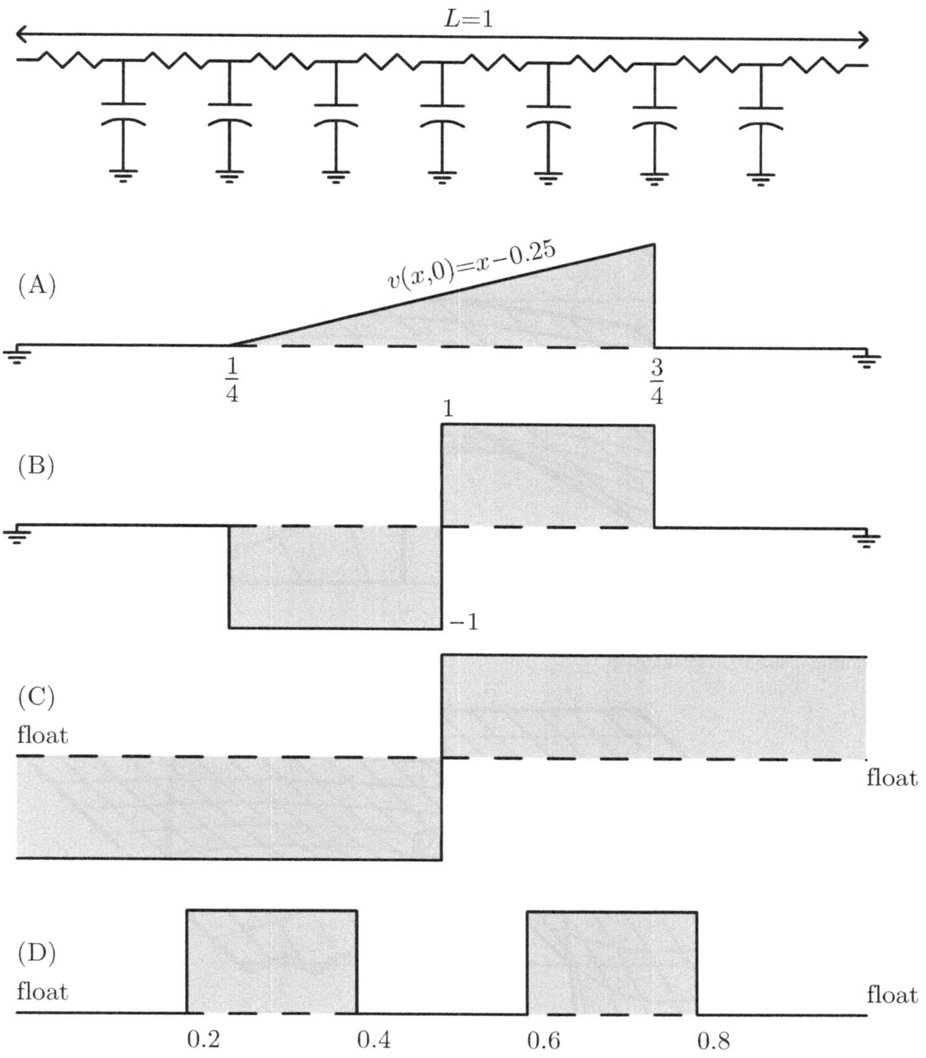

Fig. 45.19 Initial conditions used in Problems section

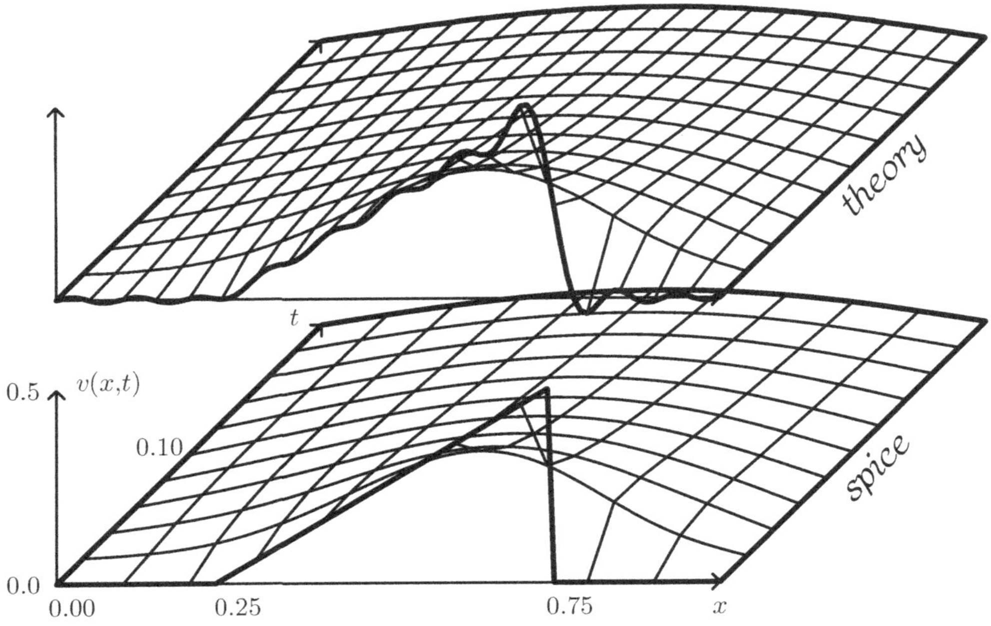

Fig. 45.20 Sample solution to Problem 1

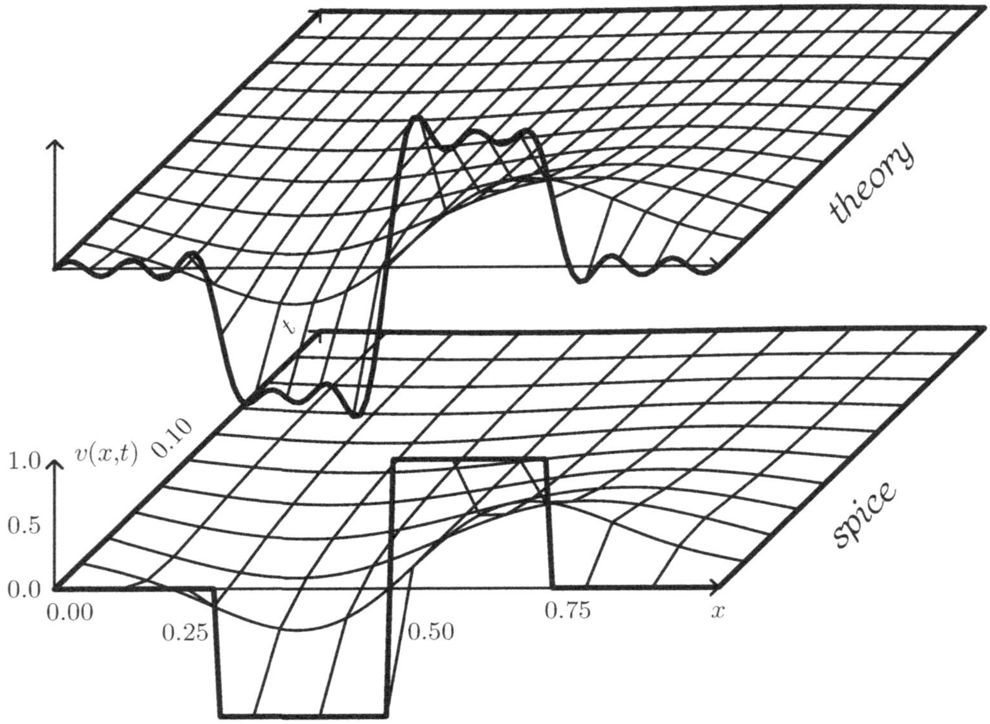

Fig. 45.21 Sample solution to Problem 2

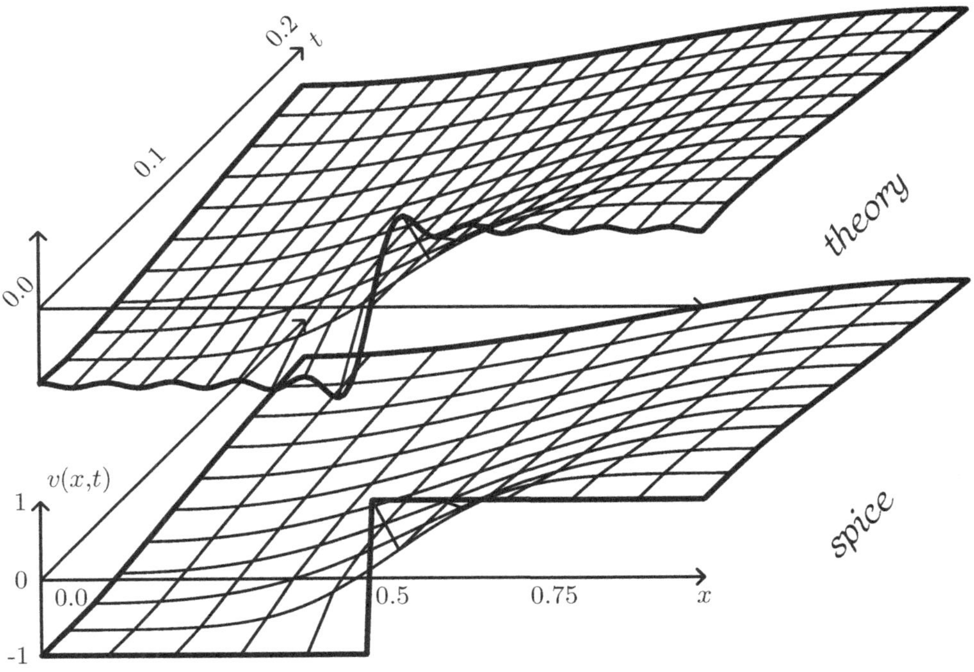

Fig. 45.22 Sample solution to Problem 3

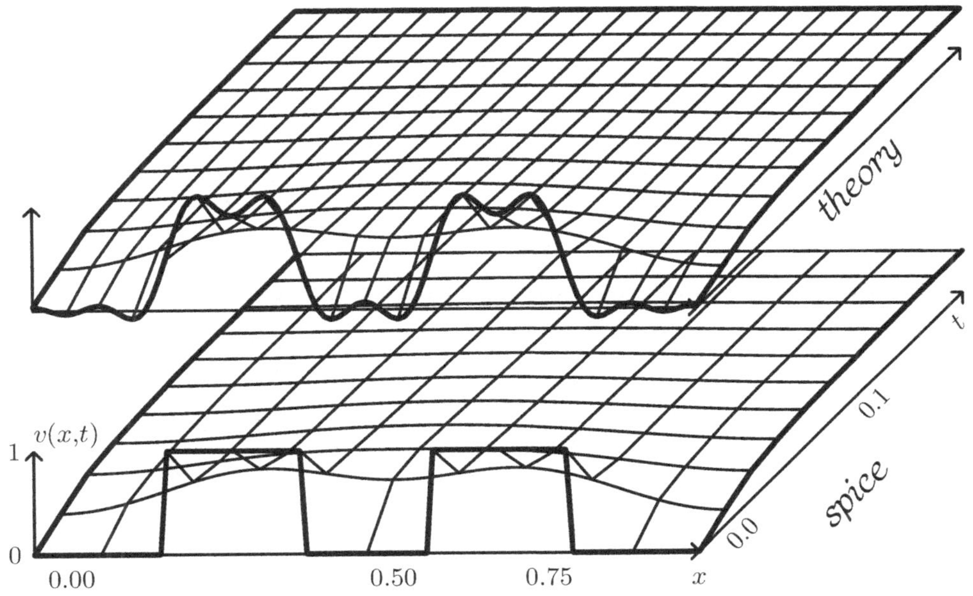

Fig. 45.23 Sample solution to Problem 4

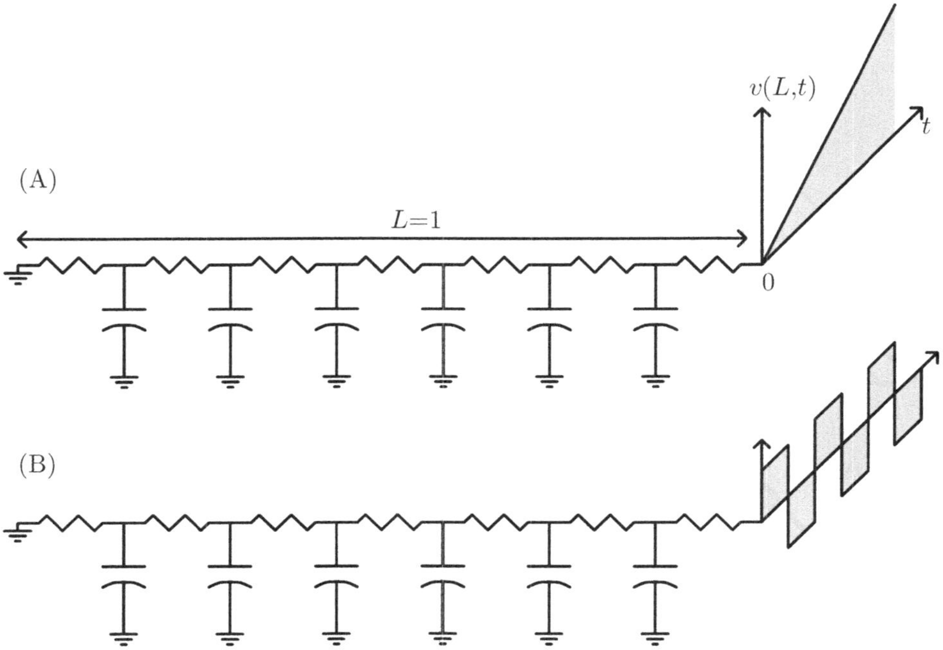

Fig. 45.24 Specification to Problems 5 and 6

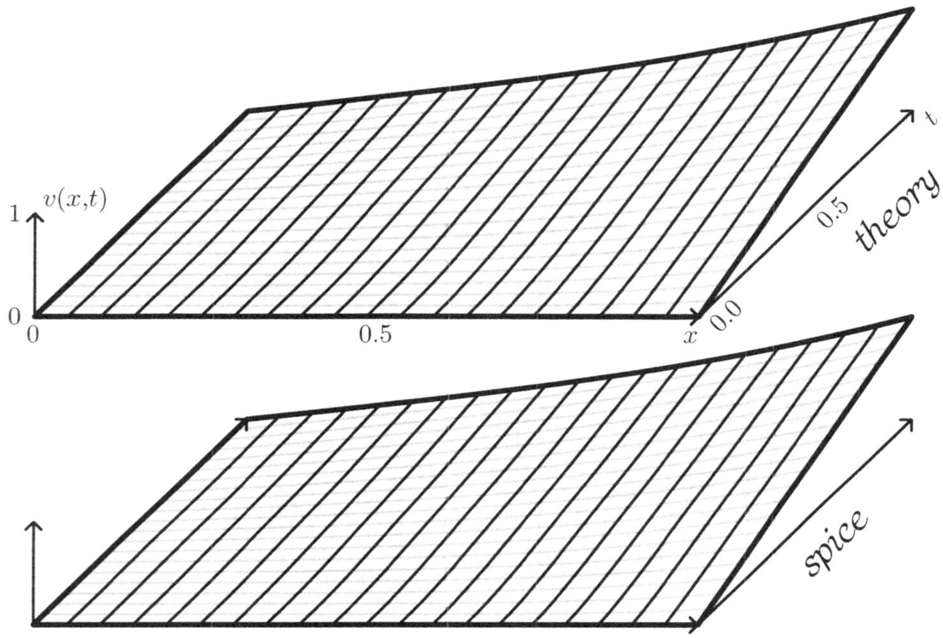

Fig. 45.25 Sample solution to Problem 5

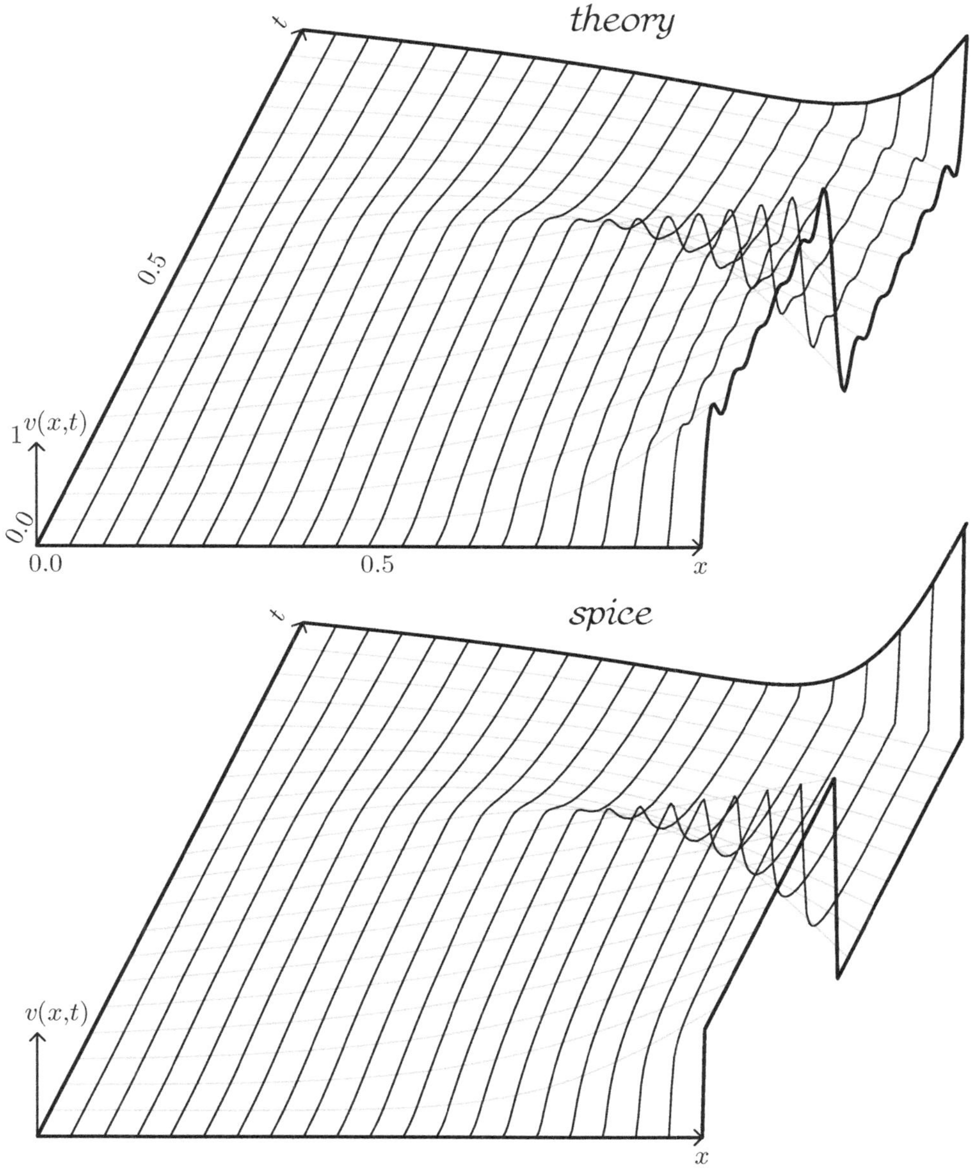

Fig. 45.26 Sample solution to Problem 6

46 Application of Spectral Techniques in Solving the Wave Equation

46.1 Introduction

Continuing on with distributed media, and after having covered Laplace's equation in solving electrostatic problems and the diffusion equation in solving distributed RC lines we wrap up this series of chapters, and in fact the text with the application of spectral techniques in solving the wave equation across transmission lines. Transmission lines are of great interest and have been studied for decades if not centuries. Their analytic solution in the form of a Fourier series and Laplace transform is always a beautiful testimony of the harmony between physics, engineering, and mathematics!

46.2 Wave Equation Derivation

For ease and simplicity we will assume that the transmission line is lossless; that is energy is conserved and not dissipated (as in heat for example); as such we can represent the electrical transmission line as a sequence of inductors and capacitors as shown in Fig. 46.1.

If we discretize the line into x segments then the voltage difference between two segments would simply be the Ldi/dt across the connecting inductor:

$$\frac{dv}{dx} = -L\frac{di}{dt} \tag{46.1}$$

where L is inductance per unit length. Notice the sign, because if current is flowing to the right (positive) and increasing in time then the right end would have *less* voltage than the left end! Next, at the intersection between two inductors, and a cap, if we measure the current through the right inductor and compare to that in the left one, the difference (if any) must be attributed to any charge accumulation across the terminals of the cap:

$$\frac{di}{dx} = -C\frac{dv}{dt} \tag{46.2}$$

where C is capacitance per unit length. Again notice the negative sign. That is, if the cap is charging, output current must be *less* than input one (the difference being the current used to charge the cap). If we differentiate Eq. (46.1) again with respect to x we get

$$\frac{d^2v}{dx^2} = -L\frac{d}{dx}\frac{di}{dt} \tag{46.3}$$

Plugging in for di/dt (Eq. (46.2)) we get

$$\boxed{\frac{d^2v}{dx^2} = LC\frac{d^2v}{dt^2}} \tag{46.4}$$

This is the partial differential equation that governs the voltage along the transmission line, in space and time. Let's define a new variable, which we call the velocity c:

F. Badrieh, *Spectral, Convolution and Numerical Techniques in Circuit Theory*, https://doi.org/10.1007/978-3-319-71437-0_46

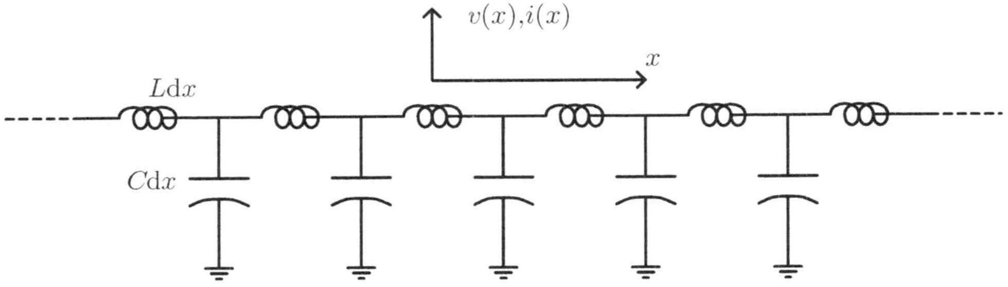

Fig. 46.1 Lossless transmission line supporting wave transmission

$$c^2 = \frac{1}{LC}, \Rightarrow c = \frac{1}{\sqrt{LC}} \quad (46.5)$$

Then the *wave* equation can be written as

$$\boxed{\frac{d^2v}{dx^2} = \frac{1}{c^2}\frac{d^2v}{dt^2}} \quad (46.6)$$

46.3 Method of Separation of Variables

Similar to last chapter, let's assume that the solution can be written in the form

$$v(x,t) = X(x)T(t) \quad (46.7)$$

where $X(x)$ depends exclusively on x and $T(t)$ on t. Plugging in the wave equation we get

$$T(t)\frac{d^2X(x)}{dx^2} = \frac{1}{c^2}X(x)\frac{d^2T(t)}{dt^2} \quad (46.8)$$

Divide both sides by $X(x)T(t)$ to get

$$\frac{X''(x)}{X(x)} = \frac{1}{c^2}\frac{T''(t)}{T(t)} \quad (46.9)$$

Since the left side is purely a function of x and the right side purely a function of t, the only way they equate (for all x and t) is if both sides are equal to a constant:

$$\frac{X''(x)}{X(x)} = -\lambda^2, \qquad \frac{1}{c^2}\frac{T''(t)}{T(t)} = -\lambda^2 \quad (46.10)$$

The choice of λ would depend on the boundary conditions. Let's try a few cases.

46.4 Case of Zero Boundary Conditions and Nonzero Initial Conditions

If the end boundary conditions are forced to zero, and the line is initialized to some IC, then the derivation continues as follows. The solution of the space equation is

$$X(x) = A\sin\lambda x + B\cos\lambda x \quad (46.11)$$

If we apply the left boundary condition of $X(0) = 0$ we are forced to set B to zero such that

$$X(x) = A\sin\lambda x \quad (46.12)$$

Next if we apply the right condition such that $X(l) = 0$, where l is line length, we are forced to set

$$\lambda l = n\pi; \qquad \Rightarrow \lambda = \frac{n\pi}{l} \quad (46.13)$$

such that

$$X(x) = A\sin\frac{n\pi x}{l} \quad (46.14)$$

(Notice that we forewent the use of L for length since that is now used for inductance!) Similarly we get for the time solution

$$T(t) = D\cos\frac{cn\pi t}{l} + E\sin\frac{cn\pi t}{l} \quad (46.15)$$

(Notice that we did not use C as a constant because that is reserved for capacitance!) In order to determine the two time constants we need the initial condition and time derivative thereof. Assume that the initial condition is set, and that the initial time derivative is zero. For this last condition we get

$$T'(0) = \frac{cn\pi}{l}\left[-D\sin(0) + E\cos(0)\right] = 0 \qquad (46.16)$$

which would imply that $E = 0$; hence

$$T(t) = D\cos\frac{cn\pi t}{l} \qquad (46.17)$$

Total solution (after variable change) then follows

$$v(x,t) = A\sin\frac{n\pi x}{l}\cos\frac{cn\pi t}{l} \qquad (46.18)$$

This satisfies the partial differential equation and the initial time and initial time derivative conditions. To find A we need to satisfy the initial conditions; assume

$$v(x,0) = f(x) \qquad (46.19)$$

If we try $t = 0$ in Eq. (46.18) we don't get $f(x)$. However, and using superposition, let's assume

$$\boxed{v(x,t) = \sum_n A_n \sin\frac{n\pi x}{l}\cos\frac{cn\pi t}{l}} \qquad (46.20)$$

Then at time zero we get

$$v(x,0) = \sum_n A_n \sin\frac{n\pi x}{l} = f(x) \qquad (46.21)$$

This is nothing other than the typical Fourier series. We can find the coefficients A_n as follows:

$$\boxed{A_n = \frac{2}{l}\int_0^l f(x)\sin\frac{n\pi x}{l}dx} \qquad (46.22)$$

For example assume initial condition of a line of length 50 mm such that between 22 and 28 mm voltage is pre-charged to 1; let's find voltage for all position and time! First we evaluate the A_n coefficients in Eq. (46.22) as

$$A_n = -\frac{2}{n\pi}\left[\cos\frac{n\pi 28}{50} - \cos\frac{n\pi 22}{50}\right] \qquad (46.23)$$

Then we put back into the series expansion Eq. (46.20). Results and comparison to SPICE are shown in Fig. 46.2. Notice how the initial pulse splits into two and each of the two propagates outwards. Let's rotate the view for a better visualization—see Fig. 46.3. We can see that the initial pulse splits into two, and each sets out propagating along the transmission line. To convey this even clearer, let's plot the voltage across the T-line for different time steps; see Fig. 46.4. Clearly now we see the wave nature of propagation, all the while preserving the shape of the pulse (since we assumed lossless case).

46.5 Wave Reflection at Boundaries

In the prior section we saw how the pulse propagates from the center of the line outwards, towards left and right ends. What happens when the pulse hits the edges? It depends on the "termination"! If the edges are terminated by the "characteristic" impedance of the wire (Z_0), then nothing happens—the pulse simply crosses the edges and never comes back. But if the edges are terminated by something other than Z_0 then we should expect to see *reflection* effects. The details of this would depend on the exact termination; for example, if the end is open circuited, then we should expect complete reflection; if short circuited then complete reflection and sign reversal. For example, for the case studied in last section, with shorted circuited ends, the voltage reflection is shown in Fig. 46.5. Notice when the pulse hits the edges, it flips sign and comes back (with full magnitude), and so forth.

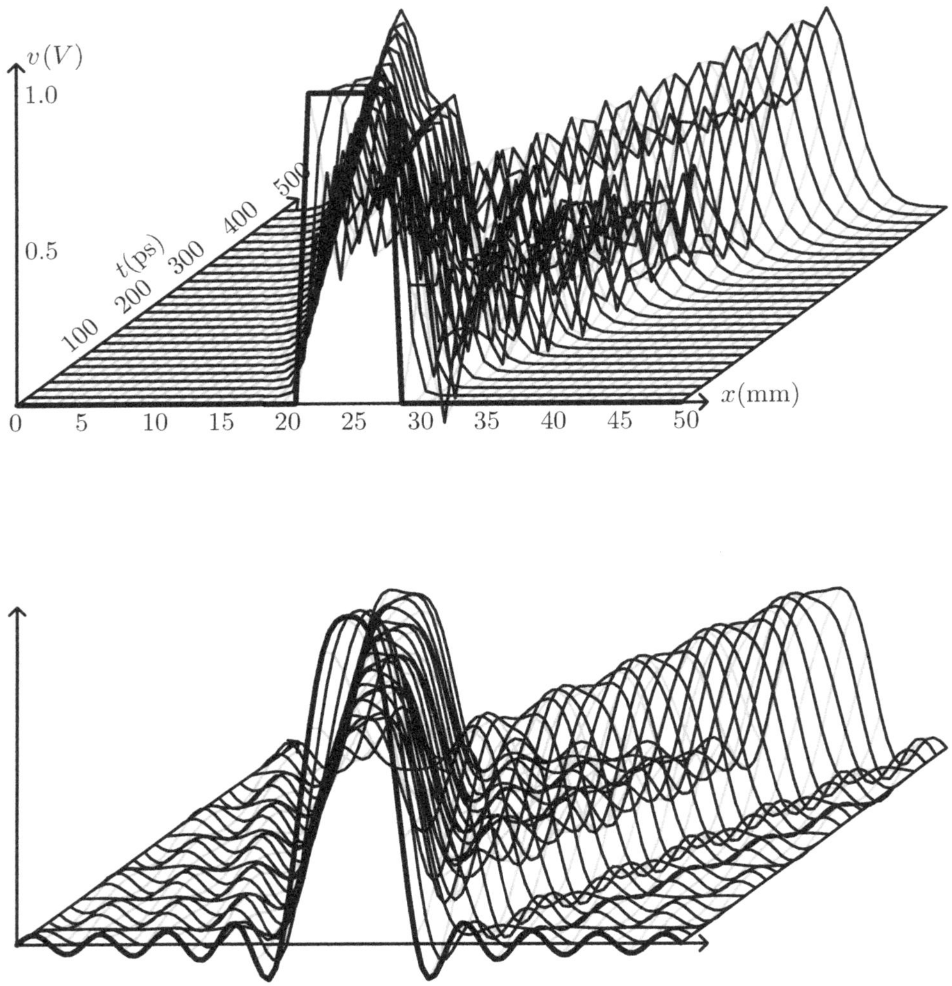

Fig. 46.2 Voltage across transmission line with zero boundary conditions and nonzero initial conditions. Assumed values are $L = 1\,\mu\text{H/m}$, $C = 1\,\text{nF/m}$, and $l = 50\,\text{mm}$. Top – SPICE; bottom – theory

46.6 Case of Zero Initial Conditions and Nonzero Boundary Conditions

This more relevant case is shown in Fig. 46.6. Essentially the wire is driven by a source and terminated by a load. Following the conventional treatment we will depart slightly from the derivation of the last two chapters and treat the problem in the frequency domain from the start. The strategy is as follows: Assume driving signal is of the form

$$V_s(t) = e^{st} \tag{46.24}$$

Find the current and voltage along the wire, then figure the transfer function. Once that is known, we are assured that we can derive the response for any input voltage (rather than limiting ourselves to the assumed e^{st} form.)

If the stimulus is of the form e^{st} then it is guaranteed that the various branch currents and

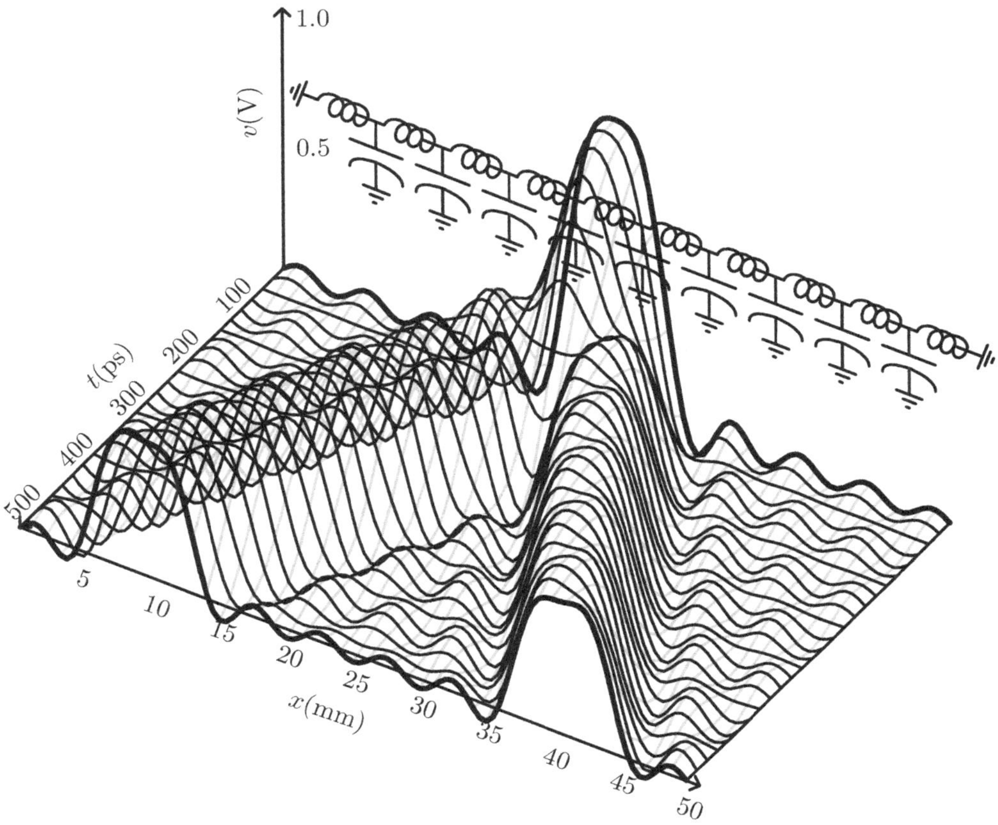

Fig. 46.3 Different view of Fig. 46.2

voltage *would also have the form* e^{st}, though they would be scaled, depending on the particular frequency. That is,

$$v(x,t) = V(x,s)e^{st}, \qquad i(x,t) = I(x,s)e^{st} \tag{46.25}$$

Taking out the implicit time dependence we end up with the two unknowns

$$V(x,s), \qquad I(x,s) \tag{46.26}$$

If we look at the difference in voltage between two nodes we get

$$\frac{dV(x,s)}{dx} = -I(x,s)(R+sL) \tag{46.27}$$

That is the voltage drop between two points along the line is due to current times the impedance between the two points which is $(R+sL)dx$; notice the negative sign because we are looking at right voltage minus left one! Similarly if we look at the difference in current we arrive at

$$\frac{dI(x,s)}{dx} = -V(x,s)(G+sC) \tag{46.28}$$

That is the difference in current between two points along the line is due to capacitive current plus loss current across the dielectric; again notice the negative sign. Differentiating Eq. (46.27) and using last equation we get

$$\frac{d^2V(x,s)}{dx^2} = (R+sL)(G+sC)V(x,s) \tag{46.29}$$

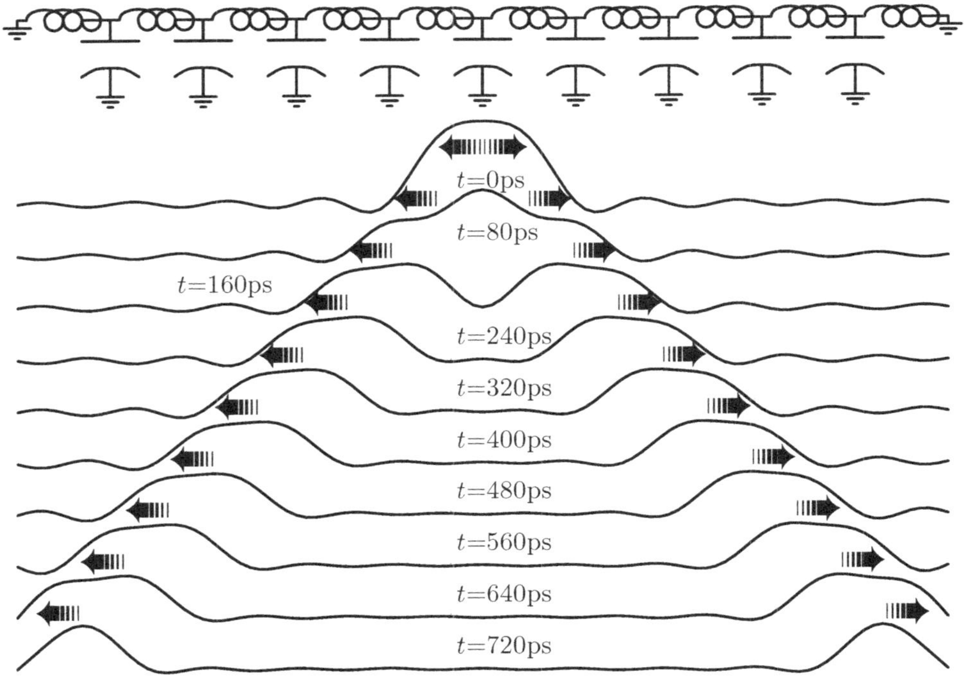

Fig. 46.4 2D slices of Fig. 46.3

Introducing the *propagation* constant γ

$$\gamma^2 = (R + sL)(G + sC) \tag{46.30}$$

we get

$$\frac{d^2V(x,s)}{dx^2} = \gamma^2 V(x,s) \tag{46.31}$$

The solution to this is

$$V(x,s) = Ae^{-\gamma x} + Be^{\gamma x} \tag{46.32}$$

To get current we reuse Eq. (46.27)

$$-I(x,s)(R+sL) = \gamma\,[-Ae^{-\gamma x} + Be^{\gamma x}] \tag{46.33}$$

or

$$I(x,s)(R+sL) = \sqrt{(R+sL)(G+sC)}\,[Ae^{-\gamma x} - Be^{\gamma x}] \tag{46.34}$$

$$I(x,s) = \sqrt{\frac{G+sC}{R+sL}}\left[Ae^{-\gamma x} - Be^{+\gamma x}\right] \tag{46.35}$$

If we define the *characteristic impedance*

$$\boxed{Z_0(s) = \sqrt{\frac{R+sL}{G+sC}}} \tag{46.36}$$

then we get

$$I(x,s) = \frac{1}{Z_0}\,[Ae^{-\gamma x} - Be^{\gamma x}] \tag{46.37}$$

Digression: why $\sqrt{L/C}$ named an impedance? Let's at least check the units. We know L has units $\frac{\Omega}{\text{rad/s}}$; we also know C has units of $\frac{1}{\Omega\text{rad/s}}$; then

$$\sqrt{\frac{L}{C}} \rightarrow \sqrt{\frac{\Omega}{\text{rad/s}}\Omega\text{rad/s}} = \sqrt{\Omega^2} = \Omega \tag{46.38}$$

which makes sense! To recap we have

$$\boxed{V(x,s) = Ae^{-\gamma x} + Be^{+\gamma x}} \tag{46.39}$$

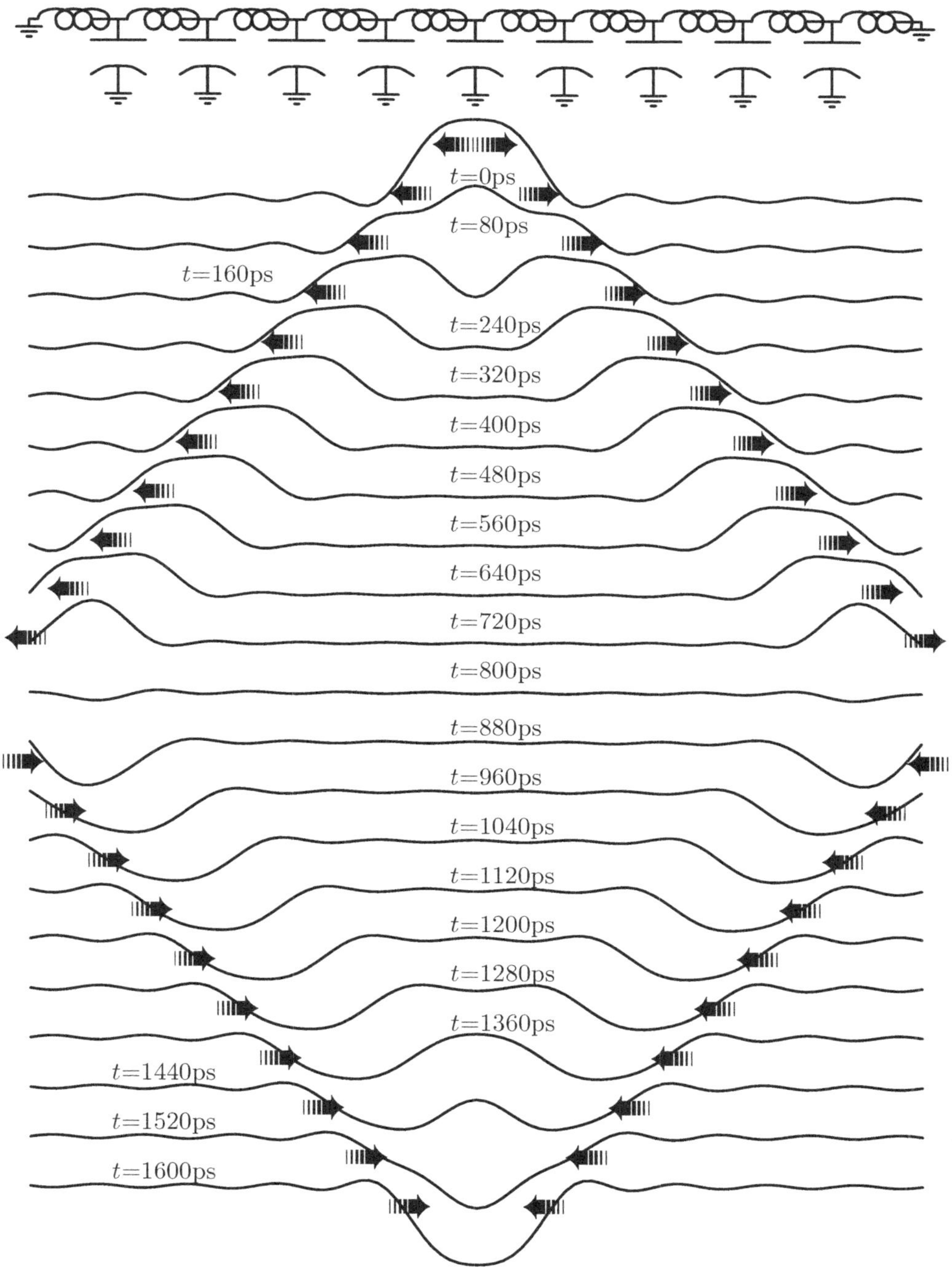

Fig. 46.5 Wave propagation and complete (negative) reflection

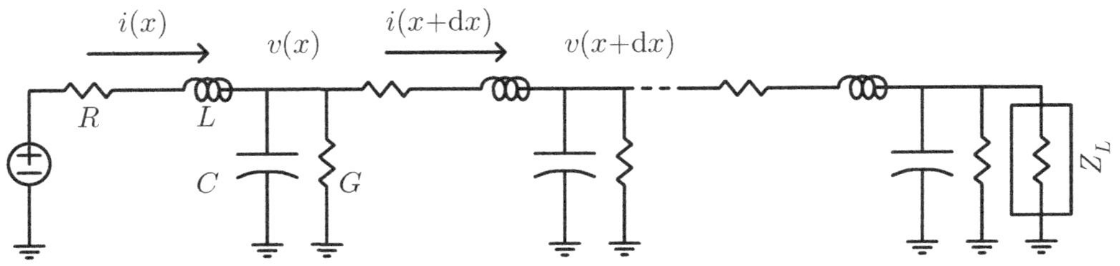

Fig. 46.6 Transmission line with zero initial conditions and nonzero boundary conditions

$$\boxed{I(x,s) = \frac{1}{Z_0}\left[Ae^{-\gamma x} - Be^{\gamma x}\right]} \tag{46.40}$$

Recall these are the transfer functions in response to a stimulus input of the form e^{st}. So putting back the implicit time dependence we get

$$\boxed{v(x,t) = \left[Ae^{-\gamma x} + Be^{\gamma x}\right]e^{st}} \tag{46.41}$$

$$\boxed{i(x,t) = \frac{1}{Z_0}\left[Ae^{-\gamma x} - Be^{\gamma x}\right]e^{st}} \tag{46.42}$$

To find the constants A and B we would need the boundary conditions. Right at the left of the wire $z = 0$, we have

$$e^{st} = [A + B]\,e^{st} \tag{46.43}$$

which implies

$$1 = A + B \tag{46.44}$$

That's the first equation. The second equation derives from the boundary condition at the right edge; there the ratio of voltage to current should match the load impedance $Z_L(s)$

$$Z_0(s)\frac{Ae^{-\gamma l} + Be^{\gamma l}}{Ae^{-\gamma l} - Be^{\gamma l}} = Z_L(s) \tag{46.45}$$

Rearrange

$$Z_0\left[Ae^{-\gamma l} + Be^{\gamma l}\right] = Z_L\left[Ae^{-\gamma l} - Be^{\gamma l}\right] \tag{46.46}$$

Collect terms

$$Ae^{-\gamma l}\left[Z_0 - Z_L\right] + Be^{\gamma l}\left[Z_0 + Z_L\right] = 0 \tag{46.47}$$

Use Eq. (46.44) for B and get

$$Ae^{-\gamma l}\left[Z_0 - Z_L\right] + [1 - A]e^{\gamma l}\left[Z_0 + Z_L\right] = 0 \tag{46.48}$$

Define *reflection coefficient*

$$\boxed{\rho(s) = \frac{Z_L(s) - Z_0(s)}{Z_L(s) + Z_0(s)}} \tag{46.49}$$

to get

$$-Ae^{-\gamma l}\rho + [1 - A]e^{\gamma l} = 0 \tag{46.50}$$

Multiply by $e^{\gamma l}$

$$-A\rho + [1 - A]e^{2\gamma l} = 0 \tag{46.51}$$

Collect terms

$$A\left[\rho + e^{2\gamma l}\right] = e^{2\gamma l} \tag{46.52}$$

So we finally get

$$\boxed{A = \frac{1}{1 + \rho e^{-2\gamma l}}} \qquad (46.53)$$

This automatically gives B

$$B = 1 - A = 1 - \frac{1}{1 + \rho e^{-2\gamma l}}$$
$$= \frac{1 + \rho e^{-2\gamma l} - 1}{1 + \rho e^{-2\gamma l}} = \boxed{\frac{\rho e^{-2\gamma l}}{1 + \rho e^{-2\gamma l}}} \qquad (46.54)$$

Plugging back for voltage (Eq. (46.41)) we get

$$v(x,t) = \frac{1}{1 + \rho e^{-2\gamma l}} \left[e^{-\gamma x} + \rho e^{-2\gamma l} e^{\gamma x} \right] e^{st} \qquad (46.55)$$

Multiplying both numerator and denominator by $e^{\gamma l}$ we get

$$\boxed{v(x,t) = \frac{1}{e^{\gamma l} + \rho e^{-\gamma l}} \left[e^{-\gamma (x-l)} + \rho e^{\gamma (x-l)} \right] e^{st}} \qquad (46.56)$$

And similarly for current

$$\boxed{i(x,t) = \frac{1}{Z_0} \frac{1}{e^{\gamma l} + \rho e^{-\gamma l}} \left[e^{-\gamma (x-l)} - \rho e^{\gamma (x-l)} \right] e^{st}} \qquad (46.57)$$

Sanity Check of Voltage and Current Equations Let's evaluate the voltage at $x = 0$:

$$v(0,t) = \frac{e^{\gamma l} + \rho e^{-\gamma l}}{e^{\gamma l} + \rho e^{-\gamma l}} e^{st} = e^{st} \qquad (46.58)$$

as ought to be the case. Let's next evaluate the ratio of voltage to current at the end of the wire

$$\frac{v(l,t)}{i(l,t)} = Z_0 \frac{1+\rho}{1-\rho} = Z_0 \frac{1 + \dfrac{Z_L - Z_0}{Z_L + Z_0}}{1 - \dfrac{Z_L - Z_0}{Z_L + Z_0}} = Z_0 \frac{Z_L + Z_0 + Z_L - Z_0}{Z_L + Z_0 - Z_L + Z_0}$$
$$= Z_0 \frac{2Z_L}{2Z_0} = Z_0 \frac{Z_L}{Z_0} = Z_L \qquad (46.59)$$

again as ought to be the case. So we have confirmed that both voltage and current behave correctly at the boundaries; they also obey the wave differential equation; hence we have confidence in the solution!

46.7 Recap of Case of Forced Input

We've done quite a few acrobatic moves in the last section, introduced quite a few new variables, and arrived at the most succinct expressions both for voltage and current as a function of time and voltage! The wave equation and its solution for transmission line is one of the cornerstones of electromagnetics and needless to say took a long time to get to that stage. We are lucky this was done for us, and for our purpose we can even take the solution as an axiom. Again we are interested in applying spectral techniques on distributed media and our main concern is the generalization of our tools and testing thereof. So let's unwind, gather our findings so far, and prepare them for use throughout the rest of the chapter. Given a transmission line of length l and of resistance, inductance, capacitance, and conductance (all per unit length)

$$R, L, C, G \tag{46.60}$$

we first find the characteristic impedance

$$Z_0(s) = \sqrt{\frac{R + sL}{G + sC}} \tag{46.61}$$

This line impedance is frequency dependent. We next find the propagation constant

$$\gamma^2(s) = (R + sL)(G + sC) \tag{46.62}$$

Again this also is frequency dependent. Next take note of the load impedance which in the most generic case could be frequency dependent too

$$\text{termination impedance} = Z_L(s) \tag{46.63}$$

Next we figure the reflection coefficient

$$\rho(s) = \frac{Z_L - Z_0}{Z_L + Z_0} \tag{46.64}$$

which again is frequency dependent! Now the response due to an input voltage of the form e^{st} at any position x along the line and at any time t is given by

$$v(x,t) = \frac{1}{e^{\gamma l} + \rho e^{-\gamma l}} \left[e^{-\gamma(x-l)} + \rho e^{\gamma(x-l)} \right] e^{st} \tag{46.65}$$

and the current at the same position and time is given by

$$i(x,t) = \frac{1}{Z_0} \frac{1}{e^{\gamma l} + \rho e^{-\gamma l}} \left[e^{-\gamma(x-l)} - \rho e^{\gamma(x-l)} \right] e^{st} \tag{46.66}$$

That is for any transmission line, with any termination—be it open, short, and any *RLC* combination—we are able to figure both voltage and time at any position along the line and at any point in time! Of course at this time our input is restricted to being exclusively of the form e^{st}, but at least we have the leeway of setting the frequency s to anything. Well if we know the solution to any stimulus of the form e^{st}, for any frequency s, and if superposition still holds (which of course it does), then can we not apply a bunch of inputs each of the form of e^{st}, but with different frequency s? Sure we can, but why? Because we can scale the different input signals each differently such that when combined together they give us any desired effective input, such as a unit step, pulse, periodic pulse, and so forth. That, after all, is the whole idea behind the Fourier series/transform and the Laplace transform! Let's try some examples.

46.8 Forced Pulse Through Ideal T-Line with Open Termination

Let's apply results from last section to the special case of ideal T-line with open termination. In other words,

$$R = 0; \qquad G = 0; \qquad Z_L = \infty \tag{46.67}$$

We know from the prior section the voltage along the line, at any time, given the input is the exponential function, as was given in Eq. (46.65). Our input, however, is not an exponential function! But we know by now how to represent this input in terms of the exponential function; namely using the Laplace transform

$$\text{pulse of width } t_0 = \frac{1}{2\pi} \int F(s) e^{st} ds, \qquad \text{where} \qquad F(s) = \frac{1 - e^{-st_0}}{s} \tag{46.68}$$

Hence the response due to the pulse function is

$$v(x,t) = \frac{1}{2\pi}\int \left\{ \frac{1}{e^{\gamma l} + \rho e^{-\gamma l}} \left[e^{-\gamma(x-l)} + \rho e^{\gamma(x-l)} \right] \right\} \frac{1-e^{-st_0}}{s} e^{st} ds \tag{46.69}$$

How magnificent is this? A closed-form solution for the response of a transmission line (ideal for now) for a pulse of arbitrary width with abrupt edges, which gives voltage (and later current) for all position and time, including total reflection both at right edge and left one (with sign reversal)! Sure we have an integration to be carried out, but even without a closed-form solution for the integral itself we can simply convert the integral to a summation and if we judiciously include a few dominant terms we are assured at least a first order approximation to the answer. Of course if we add more terms in the summation we should get better results, and so forth.

Results and comparison are shown in Fig. 46.7. This sample case has $L = C = 0.5$ (per meter) and length 10 m (gigantic, but to keep numbers simple!). The pulse width is set to 1 s. Since the termination is open ended, we would expect full reflection. Since the source has zero impedance, then we would expect full, negative reflection at the source side (after the signal bounced back from the load side). It is expected that as we add more harmonics to the above expression (and we add more LC segments to the SPICE case), both theoretical and SPICE results would match better, and the signal would better approximate a rectangular pulse. Figure 46.8 shows a 3D view of the same results.

In both cases, observe signal propagation and reflection (both positive at right edge and negative at left one), and observe signal shape. Since the line is ideal it is expected that the signal will maintain an exact square shape for all position and time; i.e., the channel is distortionless. While we almost see it, the pulse does not actually keep a perfect shape in actual results—neither in SPICE nor in our simulations—but that is only because of limited resolution. In the SPICE case we have limited resolution in space discretization (i.e., we used a finite set of Ls and Cs.) In the theoretical results, on the other hand, and on top of limitation in space discretization limitation, we have a finite set of harmonics and that can only reproduce an exact pulse to a certain resolution—the larger the harmonic count, the finer the resolution.

46.9 Impact of *L* and *C* (or Wave Speed)

The propagation of the signal along the transmission line is dictated by many variables, including line resistance, inductance, capacitance, and conductance. Also, source and termination impedance play a role so far as reflection/transmission at the ends. Here we specifically analyze the impact of L and C which directly impact the speed of the wave. Namely

$$c = \frac{1}{\sqrt{LC}} \tag{46.70}$$

For example, if we half both L and C, so that the product gets reduced to a quarter, and the inverse thereof goes up by a factor of 4, and finally the square root thereof goes up by a factor of 2, we should expect the wave speed to double!

$$\text{Halving } L \text{ and } C \text{ increases speed by 2} \tag{46.71}$$

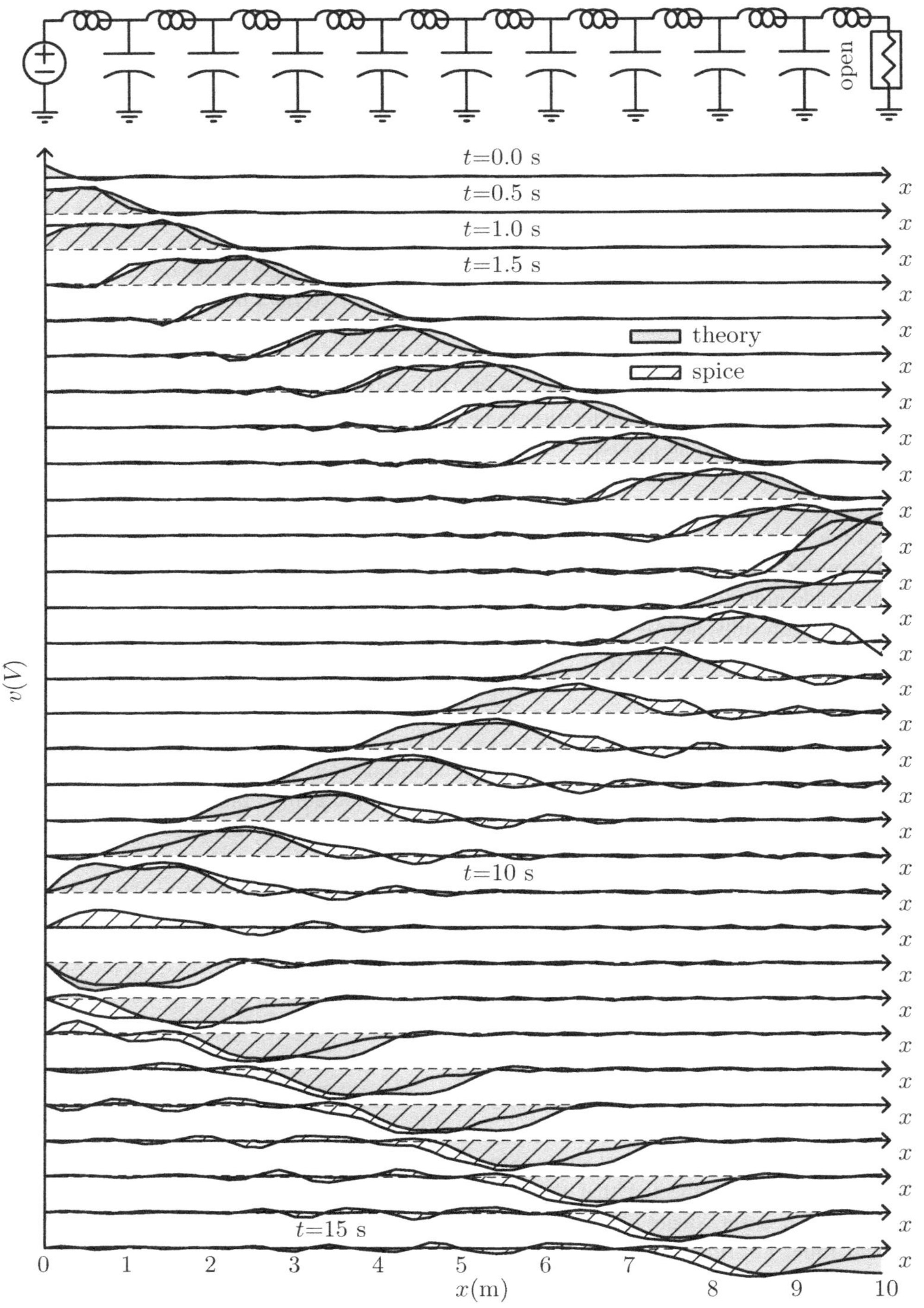

Fig. 46.7 Forced pulse through ideal T-line with open termination; case of $C = 0.5$ and $L = 0.5$ per meter, and length $= 10\,\text{m}$

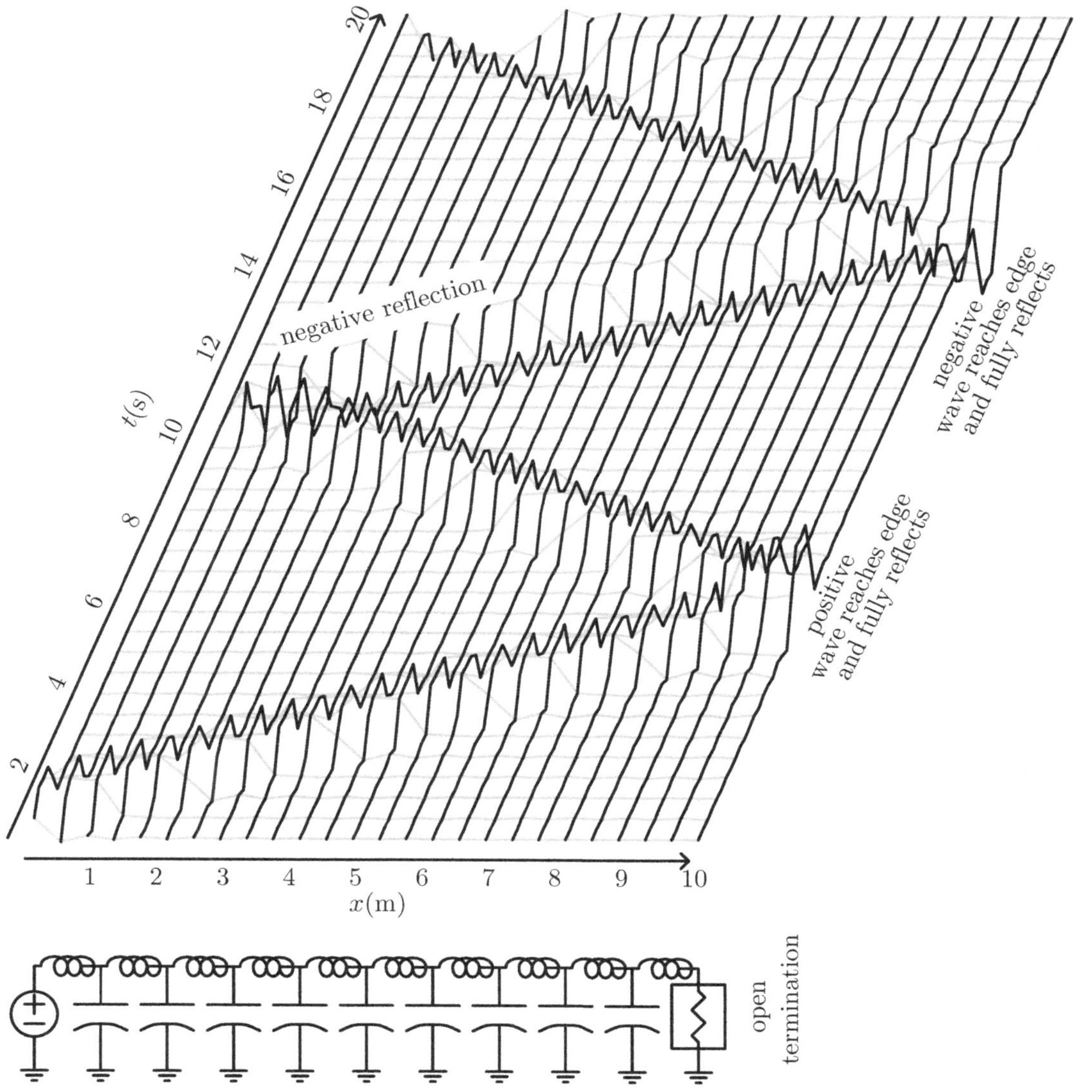

Fig. 46.8 3D view of Fig. 46.7

There is also another side effect of this, and that is the pulse width (in space) increases in two as well!

$$\text{Halving } L \text{ and } C \text{ increases pulse spatial width by 2} \tag{46.72}$$

Recall

$$\text{distance} = \text{velocity} \times \text{time} \tag{46.73}$$

So for the same time duration, and given twice the speed, we should expect the distance (or pulse width) to double. Put another way: say the pulse has a width in space of 1 m. Assume at some time the right edge of the pulse is at $x = 5\,\text{m}$ while the left edge is at $x = 4\,\text{m}$. If the new pulse has twice the speed this means that an edge on the faster pulse would have traveled twice the distance as compared to the same edge on the reference pulse. This would apply both for the left and the right edges of the pulse. For our specific case the right edge which was at $x = 5$ m would now be at $x = 10$ m; similarly the left edge which was at $x = 4$ m would now be at $x = 8$ m. The spatial difference between the left and right edges of the faster pulse, which defines the pulse width in space, is $w = 10\,\text{m} - 8\,\text{m} = 2\,\text{m}$ which is twice the spatial width of the reference case! Hence we have showed qualitatively why a faster propagation prolongs the spatial extent of a signal. All of these are confirmed in simulation results as shown in Fig. 46.9. Notice that the 2× case reaches the right end and fully reflects before the reference case does so.

46.10 Impact of *R* (or Series Loss)

As we saw above, the ideal case of lossless case results in pulse propagation without distortion; that is the pulse enters a square and *remains* a square for all time. It moves and reflects around, but at any time it still looks like a square. The more generic case, of no zero-loss, has a somehow different behavior, depending on how large R is. Let's take our reference case of

$$L = C = 0.5/\text{m}, \qquad G = 0 \tag{46.74}$$

Let's set R to 0.5/m so that it is comparable to the ideal characteristic impedance

$$Z_0(\text{lossless}) = \sqrt{L/C} = 1\,\Omega \tag{46.75}$$

Results are shown in Fig. 46.10. Notice that as the signal propagates to the right, the amplitude is reduced! And the reduction is more so towards the (right) end of the signal. That is, the signal is being dissipated. As time proceeds the "left over" of the signal is reduced; and after a long time the signal will be no more! It has been dissipated through due to line resistance. Notice, at least for this case, the nonzero resistance had no apparent impact on the speed of the signal. It looks like both intact and lossy signals propagate with about the same speed. We will next follow a more systematic treatment for the nonzero resistance case.

46.11 Case of Small Series Loss

Recall the expression for the voltage along the T-line as a function of position and time, Eq. (46.65) repeated here for convenience:

$$v(x,t) = \frac{1}{e^{\gamma l} + \rho e^{-\gamma l}} \left[e^{-\gamma(x-l)} + \rho e^{\gamma(x-l)}\right] e^{st} \tag{46.76}$$

Notice that for a given $RLCG$, termination impedance Z_L, and line length l the factor on the left at a given frequency gives a simple constant; as such and after taking out the time dependence we have

$$v(x) \sim e^{-\gamma(x-l)} + \rho e^{\gamma(x-l)} \tag{46.77}$$

As seen above, what determines the shape of the signal along the line is the propagation constant

$$\gamma = \sqrt{(R + sL)(G + sC)} \tag{46.78}$$

Assume for now zero shunt conductance

$$\gamma = \sqrt{(R + sL)(sC)}, \qquad \text{case } G = 0 \tag{46.79}$$

Factor out sL

$$\gamma = \sqrt{s^2 LC \left(1 + \frac{R}{sL}\right)}, \qquad \text{case } G = 0 \tag{46.80}$$

Factor out s^2LC

$$\gamma = s\sqrt{LC}\sqrt{1 + R/sL} \tag{46.81}$$

Assume for simplicity $s = j\omega$; then

$$\gamma = j\omega\sqrt{LC}\sqrt{1 + R/(j\omega L)} \tag{46.82}$$

Now for the case of small R, and by small we mean

$$R \ll sL \tag{46.83}$$

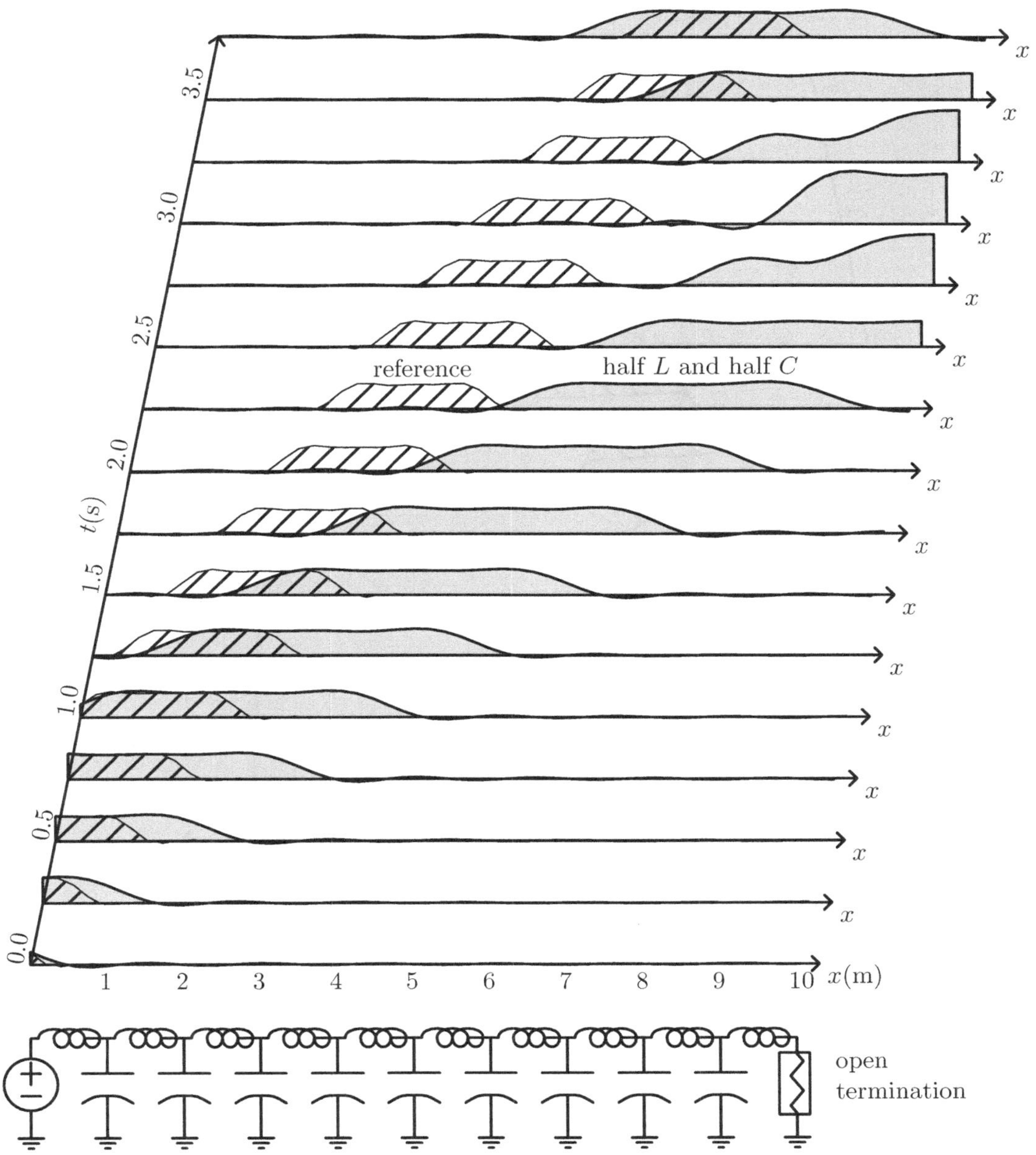

Fig. 46.9 Theoretical results showing impact of L and C on signal propagation; case of $C = 0.5$ and $L = 0.5$ per meter, and length = 10 m

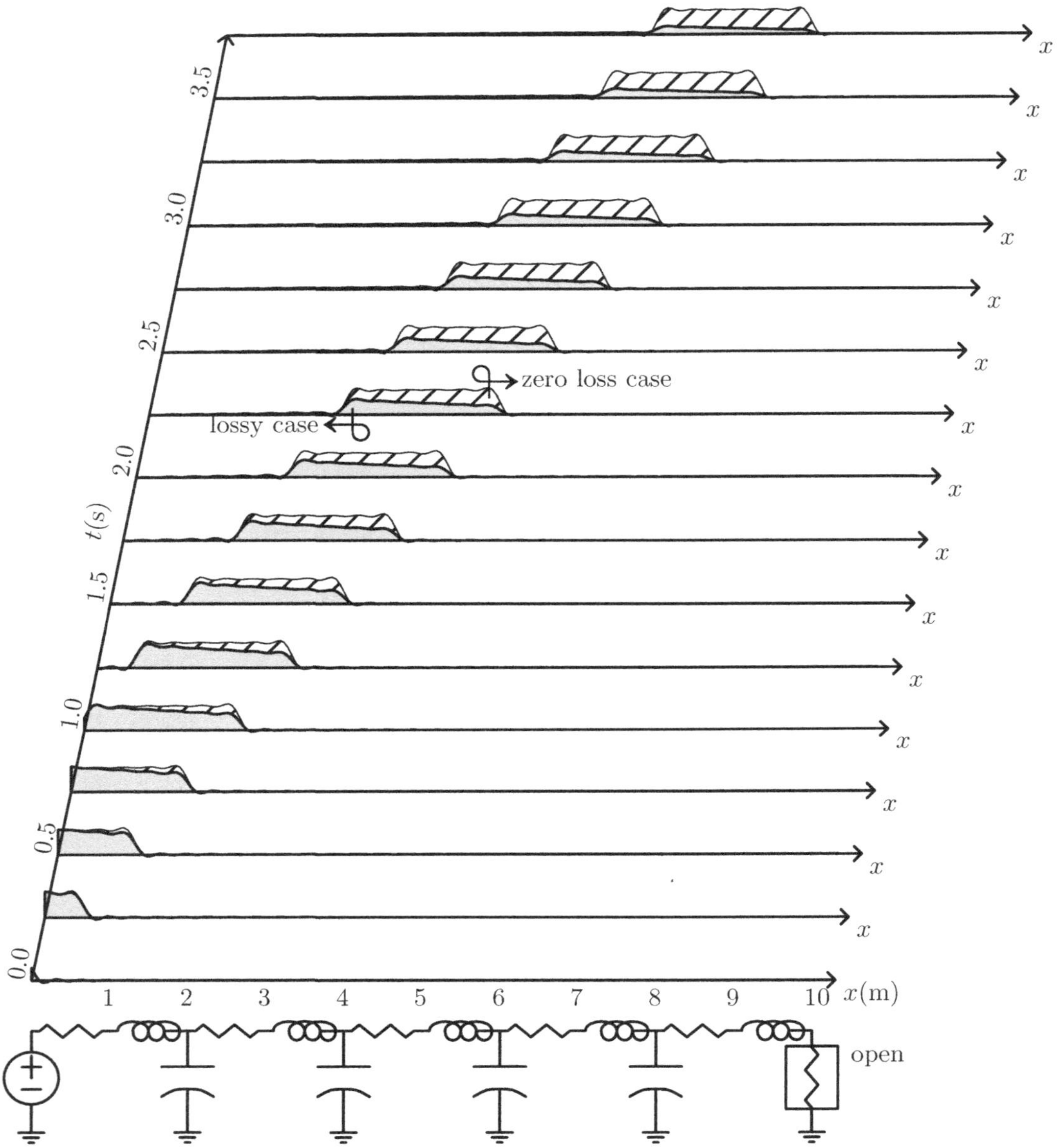

Fig. 46.10 Pulse propagation along T-line with nonzero R; open termination case of $L = C = 0.5[H, F]/\mathrm{m}$, and $R = 0.5\,\Omega/\mathrm{m}$

we can use the following approximation

$$\sqrt{1+x} \sim 1 + \frac{x}{2} \tag{46.84}$$

For example

$$\sqrt{1.1} = 1.049 \sim 1 + \frac{0.1}{2} \tag{46.85}$$

Then the propagation constant approximates

$$\begin{aligned}\gamma &\sim j\omega\sqrt{LC}\left[1 + \frac{R}{2j\omega L}\right] \\ &= j\omega\sqrt{LC} + \frac{\sqrt{LC}R}{2L}\end{aligned} \tag{46.86}$$

So finally

$$\boxed{\gamma \sim j\omega\sqrt{LC} + \frac{R}{2\sqrt{L/C}}}, \quad \text{case of small } R \tag{46.87}$$

So we have an oscillation (in x) with angular frequency

$$\text{angular spatial frequency } = \omega\sqrt{LC} \tag{46.88}$$

and a dissipation (along x) with magnitude

$$\text{spatial dissipation } = \frac{R}{2\sqrt{L/C}} \tag{46.89}$$

Notice that for the ideal, lossless T-line the propagation constant reduces to

$$\gamma = j\omega\sqrt{LC} \text{ (lossless case)} \tag{46.90}$$

So the harmonic part of the spatial solution (i.e., the complex exponential) retains the same angular frequency, which is $\omega\sqrt{LC}$; it's just that the harmonic of the lossy case now gets multiplied by a decaying function. Again approximation Eq. (46.87) applies when R is much smaller than the product sL. For single-tone input, s is well defined; but for a random signal, such as a pulse or a unit step, we have a collection of frequencies. Clearly the above smallness relation would hold for the higher harmonics, but not so for the lower ones. So this approximation is clearer to apply for single-tone inputs, but more difficult for aggregate tone case. Nonetheless, the main premise holds, which is the approximation holds better for larger harmonics, but not so for lower ones. As an example, consider the pulse case, discussed so far. Assume again that $L = C = 0.5$[H,F/m]; and assume R as large as $0.5\,\Omega$. Notice that this value is non-negligible taking into account the ideal characteristic impedance in this case is

$$Z_0(\text{lossless}) = \sqrt{\frac{L}{C}} = \sqrt{\frac{0.5}{0.5}} = 1\,\Omega \tag{46.91}$$

Even with such large R value, the approximation, as shown in Fig. 46.11, shows very good accuracy; that is, using Eq. (46.87) as an approximation for the propagation constant, where we split γ into real (dissipative) and imaginary components, we are able to get a good feel of the dissipation effects of series resistance, analytically and numerically!

46.12 Unit Step Propagation

Another important signal to consider for wave propagation along a transmission line is of course the unit step. Assume input source is a unit step function, and assume for simplicity (and without loss of generality) that the line is lossless. Assume further that we have open-end termination. How does the voltage across the line look like? Since the LT of the unit step function is simply $1/s$, the general solution as compared to the pulse solution in Eq. (46.69) comes out

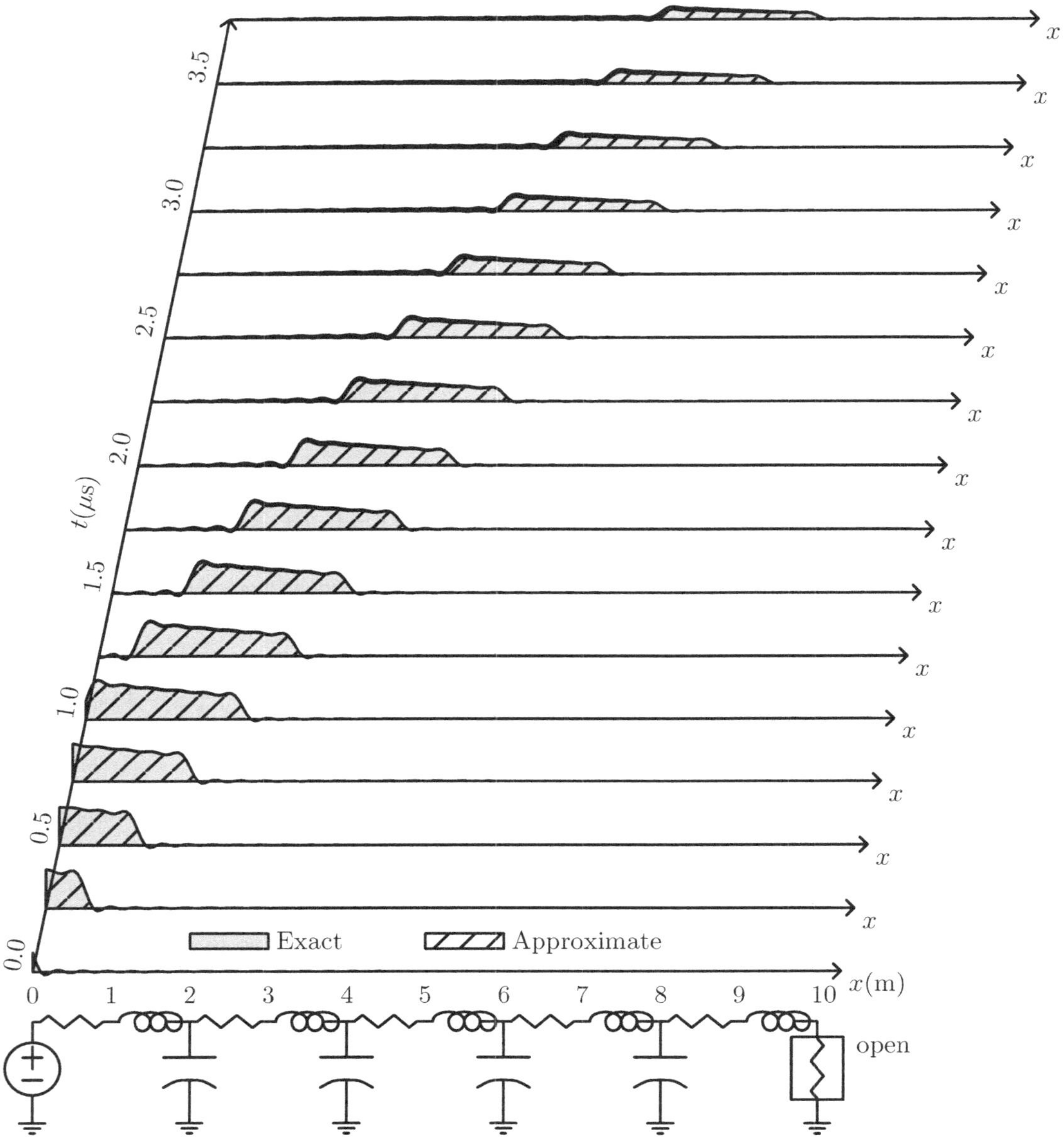

Fig. 46.11 Pulse propagation along open-ended T-line with nonzero R and comparison to approximation solution; case of $C = 0.5\,\mathrm{F/m}$, $L = 0.5\,\mathrm{H/m}$, $R = 0.5\,\Omega/\mathrm{m}$, and $l = 10\,\mathrm{m}$

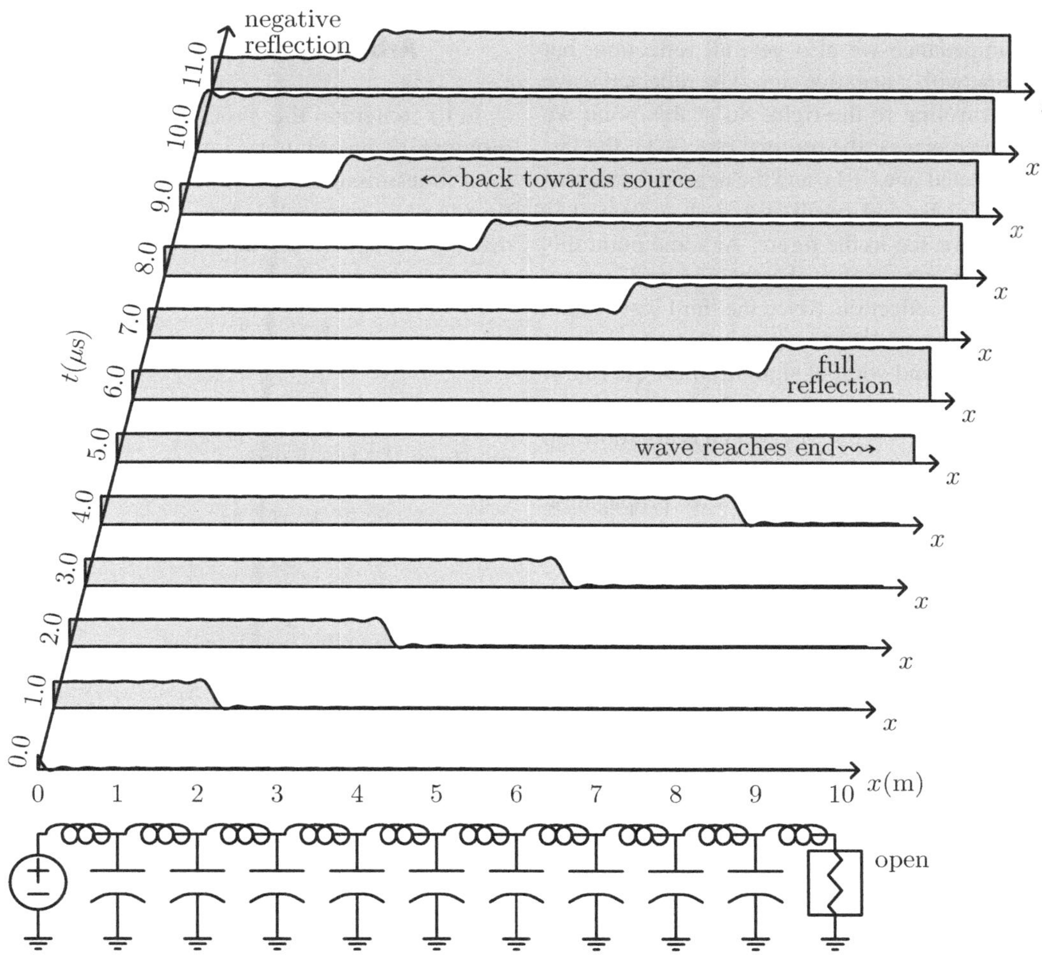

Fig. 46.12 Unit step propagation along open-ended loss-less T-line; case of $C = 0.5$ F/m, $L = 0.5$ H/m and $l = 10$ m

$$\boxed{v(x,t) = \frac{1}{2\pi}\int\left\{\frac{1}{e^{\gamma l}+\rho e^{-\gamma l}}\left[e^{-\gamma(x-l)}+\rho e^{\gamma(x-l)}\right]\right\}\frac{1}{s}e^{st}ds} \tag{46.92}$$

Figure 46.12 shows theoretical results. Before the step is applied the voltage across the whole T-line is zero. When the step is applied it does not show up along the T-line immediately; it makes its way gradually and with a finite speed. In this particular case the speed of propagation down the line is simply $c = \frac{1}{\sqrt{LC}}$. As the because step makes its way down the line, it does not get distorted (deformed in shape) the line is assumed lossless here. After some time the front edge of the step reaches the far end. Since the T-line is open terminated we get full positive reflection. This reflection simply rides on top of the continuously "broadcasted" original unit step for a total sum of 2 V, when applicable. This means that the reflected wave also travels with a finite speed, and as such it would alter the overall voltage along the line only for those relevant times where the front edge of the reflected wave has covered. At some point the reflected wave hits the source. Since the source is short terminated, i.e., assumed to have zero impedance we also get full reflection, but this time with a negative sign. The reflected wave starts traveling to the right. So at this point we have three waves: the original one (+1), the far-end reflected one (+1), and the near-end reflected one (−1); the end result is $1 + 1 - 1 = 1$ V, and that we see in the figure. At some point this negative wave reaches the far end and we get yet another reflection. Since the third wave had a negative sign, and since it gets reflected totally at the far end (and without sign flip) now the fourth wave has a magnitude of −1 V. So the sum of all 4 waves is $1+1-1-1 = 0$ V, which is not shown in the figure because we have not covered that long of a time. The back-and-forth wave propagation will continue (forever) and if we were to look at any point along the T-line we will see its voltage jumps between 0, 1, 2, 1 and then back to 0 V. On average, though the voltage would be 1 V since that is what we are injecting at the source!

46.13 Recap of Process of Figuring T-Line Response Due to an Arbitrary Stimulus

Let's reiterate the process of finding the response to an arbitrary input voltage applied to a transmission line. First we know from Eq. (46.65), repeated below for convenience, that the voltage versus time and position due to an input source of the form e^{st} is

$$v(x,t) = \frac{1}{e^{\gamma l} + \rho e^{-\gamma l}} \left[e^{-\gamma(x-l)} + \rho e^{\gamma(x-l)} \right] e^{st} \tag{46.93}$$

Next assume our new input has the Laplace transform $V_i(s)$ such that

$$v_i(t) = \frac{1}{2\pi} \int_s V_i(s) e^{st} ds \tag{46.94}$$

Then by sheer superposition we now get the response due to this new input as

$$\boxed{v(x,t) = \frac{1}{2\pi} \int_s \frac{1}{e^{\gamma l} + \rho e^{-\gamma l}} \left[e^{-\gamma(x-l)} + \rho e^{\gamma(x-l)} \right] V_i(s) e^{st} ds} \tag{46.95}$$

For the unit step case, Sect. 46.12 $V_i(s)$ was simply $\frac{1}{s}$. For the pulse case, Sect. 46.8 $V_i(s)$ was simply $\frac{1-e^{-st_0}}{s}$. For any other input, such as a causal sine or cosine, or a parabola, and so forth, all that has to be done is plug in the corresponding Laplace transform and carry on the integration (most likely numerically).

46.14 Impact of Source Impedance

So far in this chapter we have assumed that the voltage source is *ideal*; that is it has zero impedance. In other words, looking from the transmission line into the input voltage source we see zero impedance. This resulted in any waves reflected back from the far-end to the source to be completely, and negatively reflected. The more general case is a finite source impedance as shown in Fig. 46.13. Even in this case we still have the general solution

$$V(x) = Ae^{-\gamma x} + Be^{\gamma x} \tag{46.96}$$

And similar to before, we need two equations to evaluate A and B. The first equation stems from the fact that the voltage at the start of the T-line must equal the source voltage minus any current (at $x = 0$) times the source impedance Z_s; that is

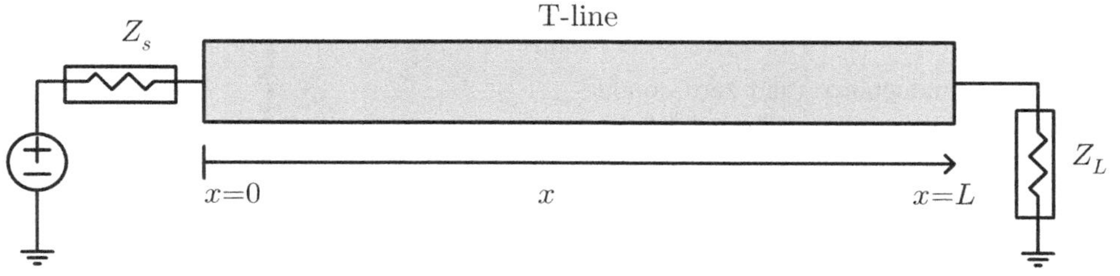

Fig. 46.13 Transmission line with nonideal source characterized by an input impedance Z_s

$$A + B = 1 - i(0)Z_s \tag{46.97}$$

Notice we used "1" for the source, implying it is e^{st}. Plugging in for current at $x = 0$ we get

$$A + B = 1 - [A - B]\frac{Z_s}{Z_0} \tag{46.98}$$

This is the first equation. The second one is again at the load and it states that the ratio of voltage to current there is simply Z_L:

$$\frac{Ae^{-\gamma l} + Be^{\gamma l}}{Ae^{-\gamma l} - Be^{\gamma l}} = \frac{Z_L}{Z_0} \tag{46.99}$$

Between these two equations and two unknowns we ought to figure A and B, and as such voltage and current for all space and time! A simple case happens when the source is matched; that is it has the same impedance as the characteristic of the line:

$$Z_s = Z_0 \qquad \text{(matched source)} \tag{46.100}$$

For this case Eq. (46.98) becomes

$$A + B = 1 - [A - B] \tag{46.101}$$

and this gives

$$A = \frac{1}{2} \tag{46.102}$$

Plugging this into the second Eq. (46.99) we get

$$\frac{\frac{1}{2}e^{-\gamma l} + Be^{\gamma l}}{\frac{1}{2}e^{-\gamma l} - Be^{\gamma l}} = \frac{Z_L}{Z_0} \tag{46.103}$$

Rearrange and collect terms

$$Be^{\gamma l}[Z_0 + Z_L] = \frac{1}{2}e^{-\gamma l}[Z_L - Z_0] \tag{46.104}$$

which gives

$$B = \frac{1}{2}e^{-2\gamma l}\frac{Z_L - Z_0}{Z_L + Z_0} = \frac{1}{2}e^{-2\gamma l}\rho \tag{46.105}$$

So our voltage in space is

$$\boxed{v(x) = \frac{1}{2}e^{-\gamma x} + \frac{1}{2}\rho e^{-2\gamma l}e^{\gamma x}} \tag{46.106}$$

or in space and time as

$$\boxed{v(x,t) = \frac{1}{2}\left[e^{-\gamma x} + \rho e^{-2\gamma l}e^{\gamma x}\right]e^{st}} \tag{46.107}$$

And finally for an arbitrary input with Laplace transform $F(s)$

$$\boxed{v(x,t) = \frac{1}{2\pi}\int_s \left\{\frac{1}{2}\left[e^{-\gamma x} + \rho e^{-2\gamma l}e^{\gamma x}\right]e^{st}\right\}F(s)ds} \tag{46.108}$$

An example demonstration applied for the case of a unit step input can be seen in Problem 10 at the end of the chapter. For the rest of the chapter, however, and throughout the Problems (unless explicitly stated) we will use the simplifying assumption that the driving source is ideal (knowing we can always treat the nonideal version if needed).

46.15 Case of Zero Inductance

The case of zero inductance (and zero conductance) puts us back into the diffusive RC line. Even though all of the above derivations apply for the RC case, it is instructive to re-derive and take advantage of the simplifications of $L = G = 0$. Again assume source is of the form e^{st} and get the spatial dependence in terms of the propagation constant γ; however, this comes out in a simplified format

$$\gamma = \sqrt{(R+sL)(G+sC)} = \sqrt{RCs} = \sqrt{RC}\sqrt{s} \tag{46.109}$$

Notice we cannot simply pull s out of the square root, since s is complex. Applying the same boundary conditions and utilizing the reflection coefficient we arrive at the spatial dependence as derived from the x–component of Eq. (46.55):

$$v(x) = \frac{1}{1+\rho e^{-2\sqrt{RC}\sqrt{s}l}}\left[e^{-[\sqrt{RC}\sqrt{s}]x} + \rho e^{-2\sqrt{RC}\sqrt{s}l}e^{[\sqrt{RC}\sqrt{s}]x}\right] \tag{46.110}$$

To reiterate we do the exact analysis we did for the general transmission line with the only exception of now using the new γ as shown above. Putting back the time dependence we get

$$\boxed{v(x,t) = \frac{1}{e^{\sqrt{sRC}l} + \rho e^{-\sqrt{sRC}l}}\left[e^{-\sqrt{RCs}[x-l]} + \rho e^{\sqrt{RCs}[x-l]}\right]e^{st} \quad \text{(case of) } L = G = 0} \tag{46.111}$$

Notice that this is essentially Eq. (46.65), but again with a new γ. For the generic input with Laplace transform $F(s)$, and using superposition we get

$$\boxed{v(x,t) = \frac{1}{2\pi}\int_s F(s)\frac{1}{e^{\sqrt{sRC}l} + \rho e^{-\sqrt{sRC}l}}\left[e^{-\sqrt{RCs}[x-l]} + \rho e^{\sqrt{RCs}[x-l]}\right]e^{st}ds \quad \text{(case of) } L = G = 0} \tag{46.112}$$

Let us test this on a simple case.

Open-End Termination and Response to Unit Step Input Assume open-end termination such that

$$\rho = \frac{Z_L - Z_0}{Z_L + Z_0} = \frac{\infty - Z_0}{\infty + Z_0} = \frac{\infty}{\infty} = 1 \tag{46.113}$$

Furthermore, assume unit step input such that

$$F(s) = \frac{1}{s} \tag{46.114}$$

Then our response would be

$$v(x,t) = \frac{1}{2\pi}\int_s \frac{1}{s}\frac{1}{e^{\sqrt{sRC}l} + e^{-\sqrt{sRC}l}}\left[e^{-\sqrt{RCs}[x-l]} + e^{\sqrt{RCs}[x-l]}\right]e^{st}ds \tag{46.115}$$

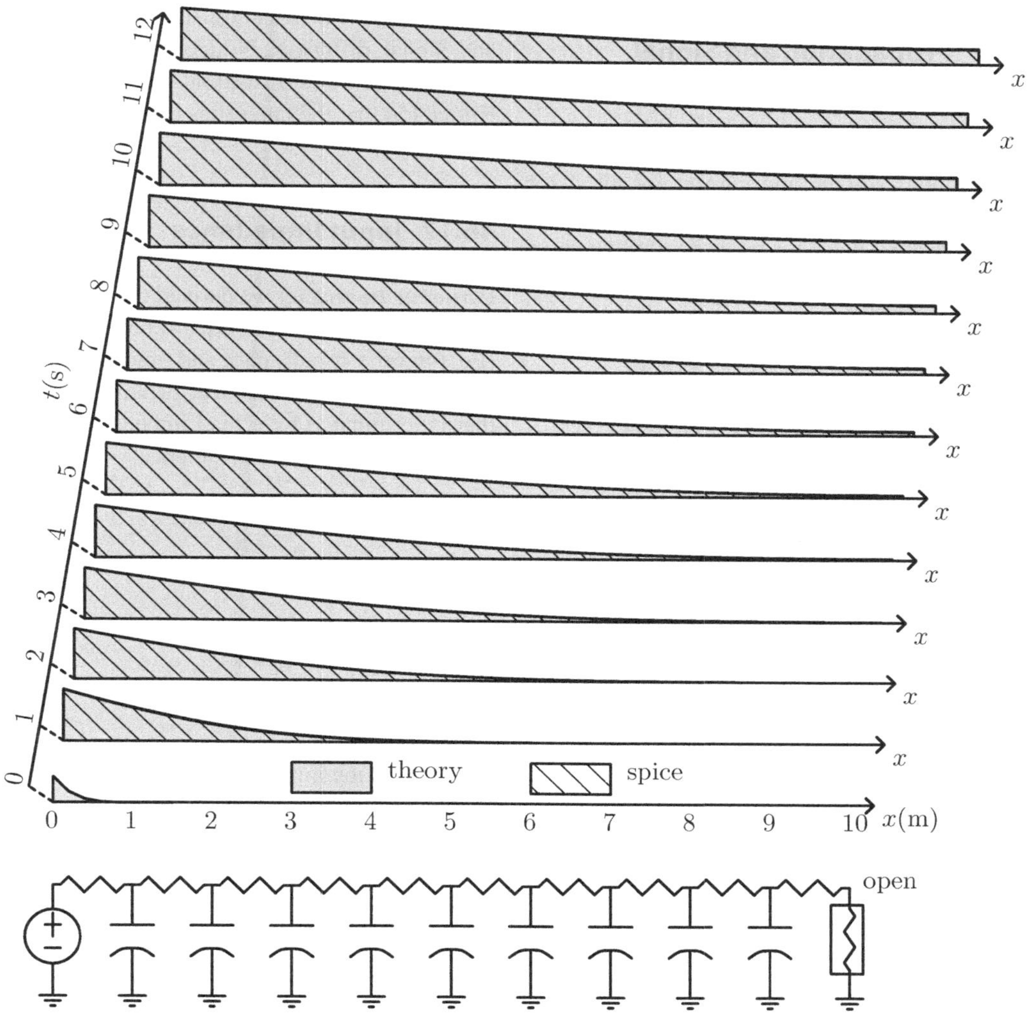

Fig. 46.14 Diffusive *RC* line (open ended) response to unit step input; case of $R = 1\ \Omega/\mathrm{m}$, $C = 0.5\ \mathrm{F/m}$ and $l = 10\ \mathrm{m}$

Figure 46.14 shows our results and comparison to SPICE. Notice that how the line slowly charges, and eventually would have unity voltage all along. Notice too that the overall behavior is different than the wave propagation one.

It is important to stress that we were successful in predicting the behavior of a diffusive *RC* line even though our original derivation steps belonged to the wave equation class. That is, the wave equation steps and derivation correctly predict the simpler case of diffusion for the special case of zero inductance. Let's next look closer how diffusion phenomena differs from wave propagation one.

46.16 Difference Between Wave Propagation and Signal Diffusion

The general wave equation derived in this chapter showing both voltage and current for all space and time applies equally to LC and RC networks. Yet the physical behavior between the two cases is quite different. Of particular importance are the two limiting cases:

- Wave propagation through ideal LC line: In this case both R and G are set to zero, and the signal is carried faithfully down the line without distortion. That is, while the signal (spatial) width is shrunk or prolonged (depending on speed), the overall shape (again measured in space) remains the same, as compared to that in time. In other words, if we send out a pulse in time, it will still look like a pulse in space. The signal is originated at the source, travels along the line, and either gets absorbed, transmitted, or reflected.
- Signal propagation through RC line: In this case, the inductance L (as well as the conductance G) are set to zero; that is we fall back onto the classical RC diffusion line. While the signal is still originated at the source, the way it proceeds down the line is quite different than that of wave propagation. For one thing, the signal is distorted—that is, if we apply a pulse in time, it will *not* look like a pulse in space! Also, for a finite signal energy, for example as in a pulse, the long-term solution is zero signal. This means that if the signal is stopped at the source, after long enough time, the signal along the wire will die. This is in contrast to the LC case, where the signal does not have to die; in particular it could bounce back and forth between the source and the termination.

To best illustrate the difference between the two cases, let's take the case of a pulse in time applied to both the LC and to the RC ladder, where in both cases we use the same C value. Figure 46.15 shows sample results clarifying the different movement nature of wave propagation and signal diffusion. Notice the distortion-less of the former, and distortion(-full!) of the latter. Notice too that the latter dies off in time.

46.17 Input Impedance

Earlier in the chapter we derived the most generic expression for voltage (Eq. (46.65)) and current (Eq. (46.66)), for all space and time, and for a line of length l that has an ideal driving source and that is terminated by a load impedance Z_L; they are repeated here for convenience:

$$v(x,t) = \frac{1}{e^{\gamma l} + \rho e^{-\gamma l}} \left[e^{-\gamma(x-l)} + \rho e^{\gamma(x-l)}\right] e^{st} \tag{46.116}$$

$$i(x,t) = \frac{1}{Z_0} \frac{1}{e^{\gamma l} + \rho e^{-\gamma l}} \left[e^{-\gamma(x-l)} - \rho e^{\gamma(x-l)}\right] e^{st} \tag{46.117}$$

The ratio of voltage to current at a given space point yields the impedance at that point:

$$z(x) = \frac{v(x,t)}{i(x,t)} = Z_0 \frac{e^{-\gamma(x-l)} + \rho e^{\gamma(x-l)}}{e^{-\gamma(x-l)} - \rho e^{\gamma(x-l)}} \tag{46.118}$$

Of particular interest is impedance as seen by the input source; that is impedance at $x = 0$; for this we get

$$\boxed{Z_{\text{in}} = z(0) = Z_0 \frac{e^{\gamma l} + \rho e^{-\gamma l}}{e^{\gamma l} - \rho e^{-\gamma l}}} \tag{46.119}$$

This gives the impedance as seen at the source, looking into the transmission line, with an arbitrary termination Z_L. Recall that ρ depends on Z_L. We will next consider two special cases.

Input Impedance with Open Termination For the special case of open termination we have full reflection

$$\rho = 1 \tag{46.120}$$

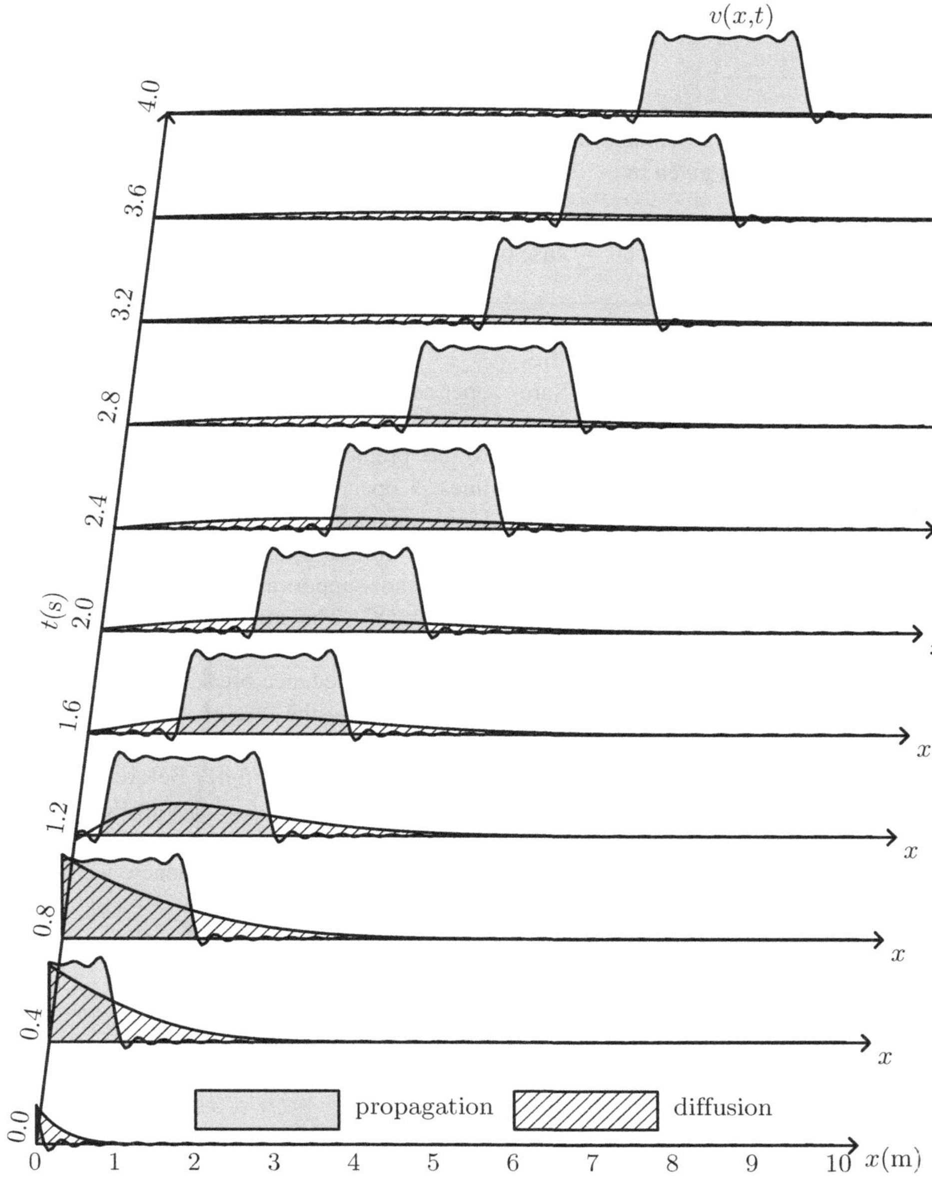

Fig. 46.15 Difference between wave propagation and diffusion (analytic results). Both cases have $C = 0.5\,\text{F/m}$; LC case has $L = 0.5\,\text{H/m}$; RC case has $R = 1.0\,\Omega/\text{m}$

and hence

$$Z_{in} = Z_0 \frac{e^{\gamma l} + e^{-\gamma l}}{e^{\gamma l} - e^{-\gamma l}} \tag{46.121}$$

For the special case of zero loss, the propagation constant is imaginary and given by

$$\gamma = j\omega\sqrt{LC} \tag{46.122}$$

and we end up with

$$\boxed{Z_{in} = -jZ_0 \cot\left(\omega l\sqrt{LC}\right)}, \quad \text{case of lossless and open termination} \tag{46.123}$$

Let's plot this equation as a function of frequency; results are shown in Fig. 46.16. Sure enough we start with infinite impedance at DC, since the line is *open*. But after that, impedance goes down and in fact something special happens at the following frequency:

$$\omega l\sqrt{LC} = \frac{\pi}{2} \tag{46.124}$$

or

$$\omega = \frac{\pi}{2l\sqrt{LC}} \tag{46.125}$$

At that frequency the input impedance is zero!! We'll look at the meaning of this in the next section. After that frequency the impedance goes negative, and again blows up. Then switch sign and repeat. Clearly the effective impedance of the T-line is pretty elaborate and needs careful attention. No longer can we say the line is open and hence it simply has open (infinite) impedance. Notice that viewed naively one would predict that if the end of two parallel lines is open, then sure enough looking at the loop impedance from the other side should register infinite one; after all, the loop is open and cannot support any current! This is true, but only at DC. The moment we talk about time varying phenomena, the moment our conclusion of infinite impedance breaks. Let's next take the other extreme—the case of short termination.

Why the Negative Sign? Before moving on, though, why the negative sign in Eq. (46.123)—wouldn't be prettier without it? Let's do a quick sanity check. At low frequency we have the approximation

$$\cot x \sim \frac{1}{x}, \qquad x \ll 1 \tag{46.126}$$

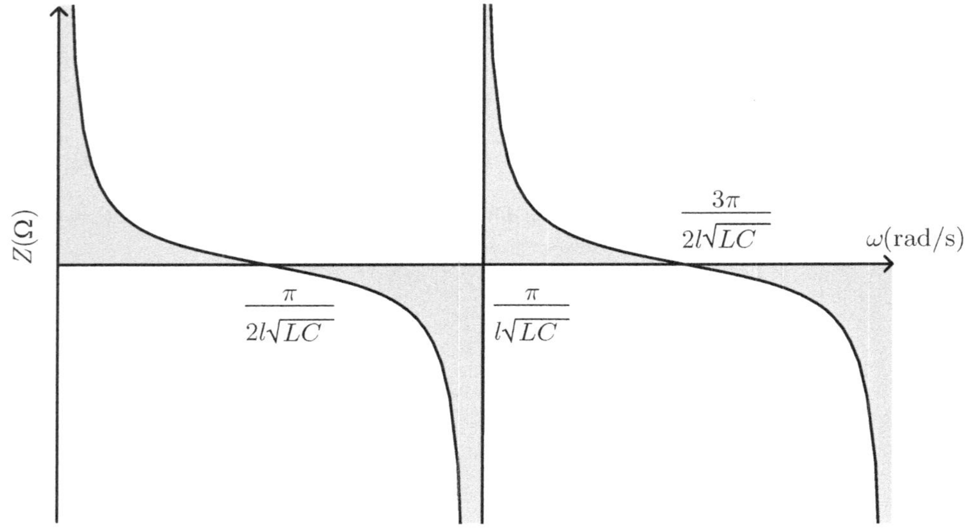

Fig. 46.16 Input impedance of lossless T-line with open termination. Case of $L = C = 0.5\,[\text{H,F}]/\text{m}$ and $l = 10\,\text{m}$

and Eq. (46.123) approaches

$$Z_{\text{in}}(0) \sim -jZ_0\frac{1}{\omega l\sqrt{LC}} = -j\sqrt{\frac{L}{C}}\frac{1}{\omega l\sqrt{LC}} = -j\frac{1}{\omega lC} = \frac{1}{j\omega lC} \tag{46.127}$$

which is nothing more than the typical $\frac{1}{j\omega C}$ impedance of a cap, but this time a cap of total capacitance Cl since C here is per unit length and l is the total length. So we have confirmed that the negative sign is in fact needed, at least so far as corroborating expected DC impedance!

Input Impedance with Short Termination Similar to above we get for the short termination an impedance that oscillates in frequency in accordance to

$$\boxed{Z_{\text{in}} = jZ_0\tan\left(\omega l\sqrt{LC}\right)}, \text{ case of lossless and short termination} \tag{46.128}$$

Again the impedance at DC makes sense (zero) but with larger frequency the impedance goes up and in fact when the argument of the tan function is $\pi/2$ impedance blows up. That is, even though the line is short terminated, at some frequencies it appears open! So the golden concepts of simple open and simple short are being challenged when input source is time dependent. Let's next take a close look at what exactly does it mean to have an infinite or zero impedance, for a given stimulus frequency.

46.18 Meaning of Zero and Infinite Transmission Line Input Impedance

As shown in the last section, an open line may appear as a short, and as viewed from source side, if the length of the line is odd multiples of quarter spatial wave lengths! Mathematically there is not much about reading a zero number for impedance; but physically how can that happen and what is its meaning? Best way to answer this is to setup a similar case, and probe voltage and current at the input of the line. All that is needed is a sinusoid voltage at the input, and a line that is odd multiples of quarter spatial wavelength—say 3×. Notice we mentioned spatial wave length and not temporal one; recall that while the temporal frequency is s, the spatial frequency is s/c, where c is the wave speed. Our demo vehicle is shown in Fig. 46.17. The line has $C = 0.5$ F/m and $L = 0.5$ H/m and the input source has an angular frequency of $\omega_0 = 2\pi$. Hence the spatial angular frequency would be

$$\frac{\omega_0}{c} = \omega_0\sqrt{LC} = 2\pi\sqrt{0.25} = \pi \tag{46.129}$$

and the spatial wave length would be

$$\text{spatial wave length} = \frac{2\pi}{\pi} = 2 \tag{46.130}$$

and finally the quarter wave length would be

$$\text{quarter wave length} = \frac{1}{2} \tag{46.131}$$

such that 3× quarter wave length equals 1.5 m which again is our assumed length for the transmission line. The figure shows *current* along the wire for different time points. Initially the current wave is initiated at the source (left) and is traveling towards the right. Since we have full termination we get full reflection. Notice that because current is directional the reflected wave subtracts from the incident wave (as opposed to voltage which would have both incident and reflected add up). As such right at the right edge, current is zero! (And this makes sense since the line is open right there!) This reflected wave travels back to the sources and is again fully

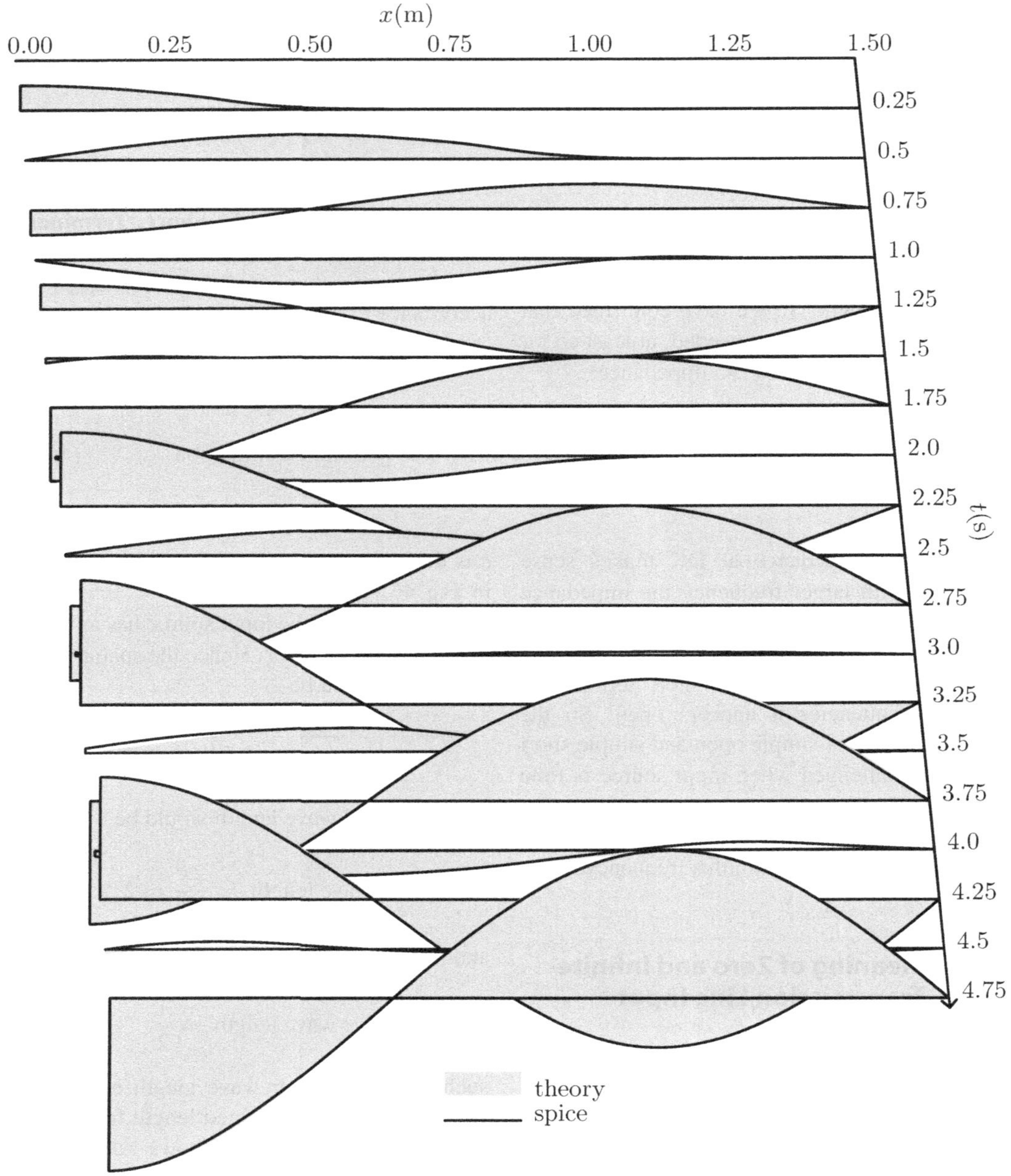

Fig. 46.17 Current through lossless transmission with ideal source and open termination and with line that is 3 quarter wave lengths in length (1.5 m). Case of $L = 0.5$ H/m, $C = 0.5$ F/m, and applied frequency $\omega_0 = 2\pi$

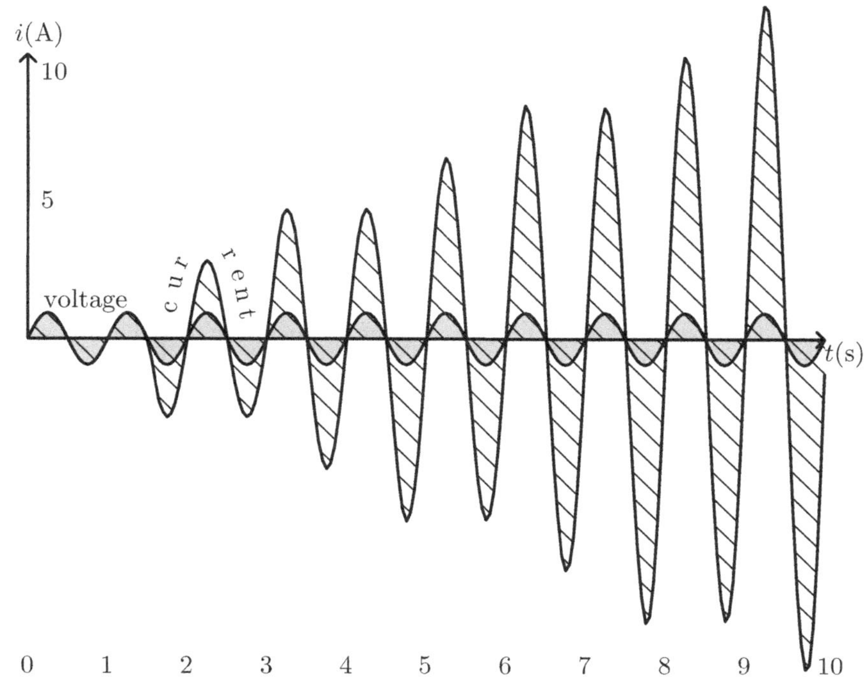

Fig. 46.18 Replot of Fig. 46.17 showing current as a function of time right at the left edge of line

reflected, with a sign reversal (since $\rho = -1$ at the source). As the source continues sending out waves, and as the waves bounce back and forth, it just happens that the sum of theses waves *adds constructively*. As seen in the figure, current is growing in magnitude; it continues to be a sine, but the peak values are increasing. As more time elapses, the current grows without bound. Hence we say the current is infinite! So looking right at the left edge of the line, we see a sine current, of infinite value. How about the voltage right there? By construction we are putting a unity sine voltage, hence it is one. Now let's form the ratio of voltage divided by current

$$Z_{\text{in}} = \frac{v(0,t)}{i(0,t)} = \frac{1}{\infty} = 0 \qquad (46.132)$$

which is exactly what we are after! Keep in mind that the effective impedance *grows* into being zero. It is not zero immediately—it takes time and it tends to zero as time proceeds. The equation in the prior section predicting zero identically has an implicit built-in assumption of steady state. Figure 46.18 shows a different view of Fig. 46.17 such that current at left edge of line is plotted versus time; notice again how it blows up, indicating short effective impedance.

Similar arguments can be made for the case of a *short-ended* T-line. At odd multiples of quarter (spatial) waves, the input impedance to the line appears open. To demonstrate this, one forces a unit AC current source at the input, and ought to observe how the voltage bounces back and forth, increasing in magnitude each round trip, till it blows up! Dividing voltage (infinite here) by input current (unity here) we arrive at infinite impedance, even though the line is actually shorted at the far end!

46.19 Transmission Line Discretization

When simulating transmission lines in SPICE, and short of using a dedicated T-line model, quite often we can approximate the T-line by a discretized version built from discrete *RLGC* elements. The assumption is implicit that if we make element count large, or segment size small, then we ought to get results closer to those of the continuous case. As a demonstration of this process let's consider the input impedance of a T-line. Figure 46.19 shows input impedance to a lossless T-line with open termination, and comparison to that of the finite-segment implementation using SPICE. As shown, the more segments we add, the more accurate the finite segment implementation becomes, in comparison to the exact case. Notice too that each additional *LC* segment can capture an addition resonance peak.

The main take from here is that we can approximate transmission line effects, such as wave propagation, reflection, and input impedance (the case here) by using a train of simple discrete *RLGC* elements. There does not appear to be any "magical" thing about wave phenomena that forces us to resort to full wave solutions and that is an encouraging thing which we can take advantage of it. Of course we are assuming that things are linear all along, such that for example line resistance (per unit length) is not frequency dependent and such.

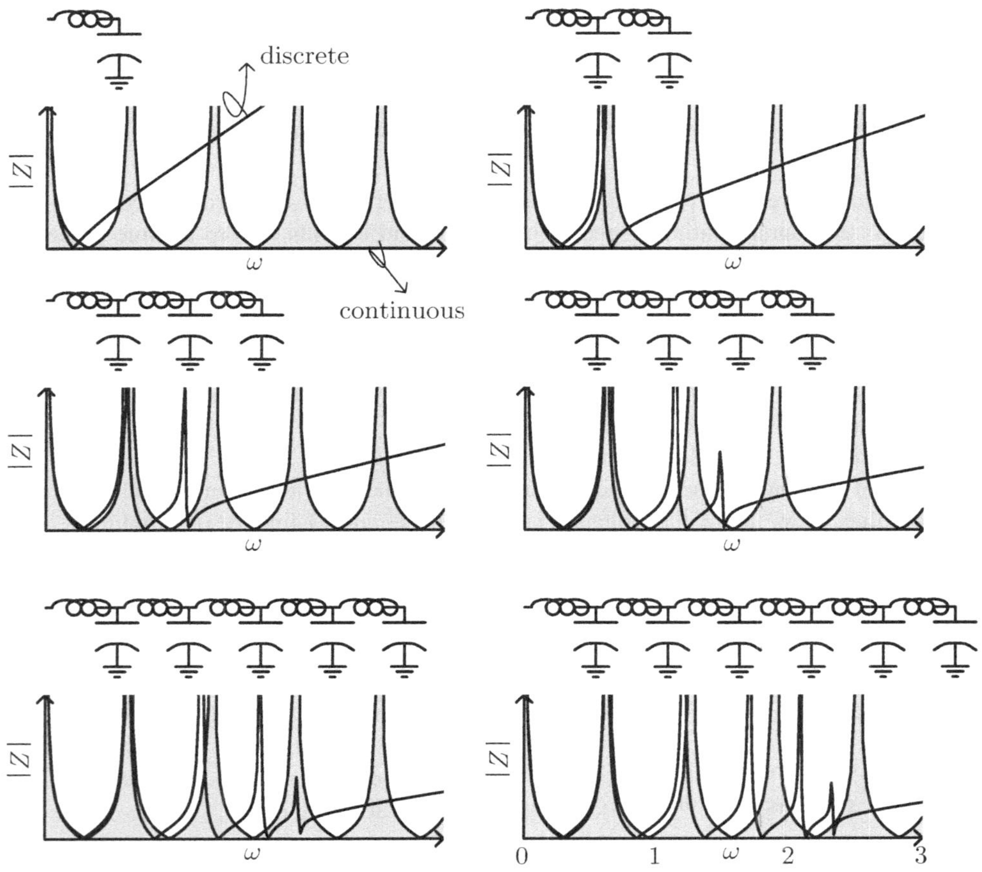

Fig. 46.19 Input impedance to ideal transmission line (length 10 m, and L/C of 0.5/0.5 [H/m,F/m]) with open termination, and comparison to SPICE with finite segment count

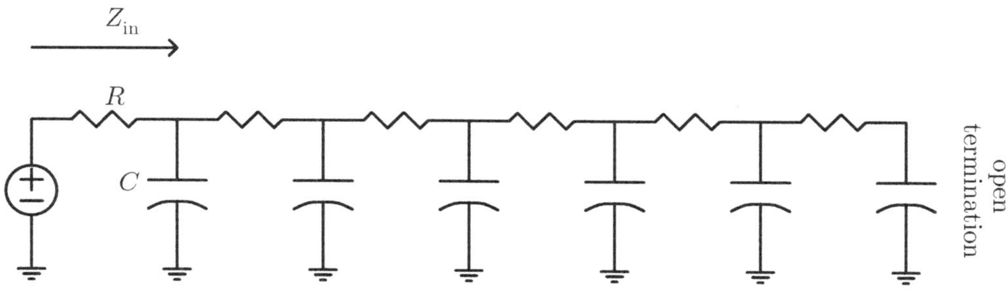

Fig. 46.20 RC transmission line input impedance

46.20 Application to *RC* Diffusion Line: Input Impedance

We wrap this chapter with a quick section on finding input impedance of an RC transmission line, shown in Fig. 46.20. We derived already in Eq. (46.121) the input impedance of a transmission line with open termination as

$$Z_{\text{in}} = Z_0 \frac{e^{\gamma l} + e^{-\gamma l}}{e^{\gamma l} - e^{-\gamma l}} \tag{46.133}$$

Plugging in for Z_0 and γ we finally arrive at

$$\boxed{Z_{\text{in}} = \sqrt{\frac{R}{sC}} \coth\left[\sqrt{RsC} \times l\right]} \tag{46.134}$$

Figure 46.21 shows theoretical results for case of $l = 10\,\text{m}$, $R = 1\,\Omega/\text{m}$, and $C = 1\,\text{mF/m}$. Also shown are SPICE results with finite number of segments. We see from the figure that as we increase segment count, we do in fact converge to the exact answer given above! Notice too from the plot the presence of almost two distinct regions—low and high frequency. Let's get an approximation to each.

Low-Frequency Approximation Recall that the hyperbolic cotangent is given by

$$\coth(x) = \frac{e^x + e^{-x}}{e^x - e^{-x}} \tag{46.135}$$

Expanding in terms of Taylor series we get the approximation for small x

$$\coth(x) \sim \frac{1+1}{1 + x - (1 - x)} = \frac{2}{2x} = \frac{1}{x} \tag{46.136}$$

With $x = \sqrt{RsC} \times l$ we get for the low-frequency approximation of input impedance

$$Z_{\text{in}} = \sqrt{\frac{R}{sC}} \frac{1}{\sqrt{RsC} \times l} = \boxed{\frac{1}{sCl}}, \quad \text{low frequency} \tag{46.137}$$

which is nothing other than the impedance of cap with total capacitance Cl. (Recall C is cap per length.) That is, at low-frequency we simply see the capacitance of the line, as that is much larger than the series resistance.

High-Frequency Approximation The high-frequency limit can be figured if we note the coth approaches unity for large argument

$$\lim_{x \to \infty} \coth(x) \sim \frac{e^x}{e^x} = 1 \tag{46.138}$$

Then our impedance becomes

$$\boxed{Z_{\text{in}} = \sqrt{\frac{R}{sC}}}, \quad \text{high frequency} \tag{46.139}$$

Notice that the high-frequency impedance does depend on R, and sC, but with a *square root* relation. This impedance cannot be simply expressed in terms of resistors and capacitors, since neither have square root. In other words, the only way to capture this is using a transmission line. Of course putting many series segments would approach the exact solution, but never

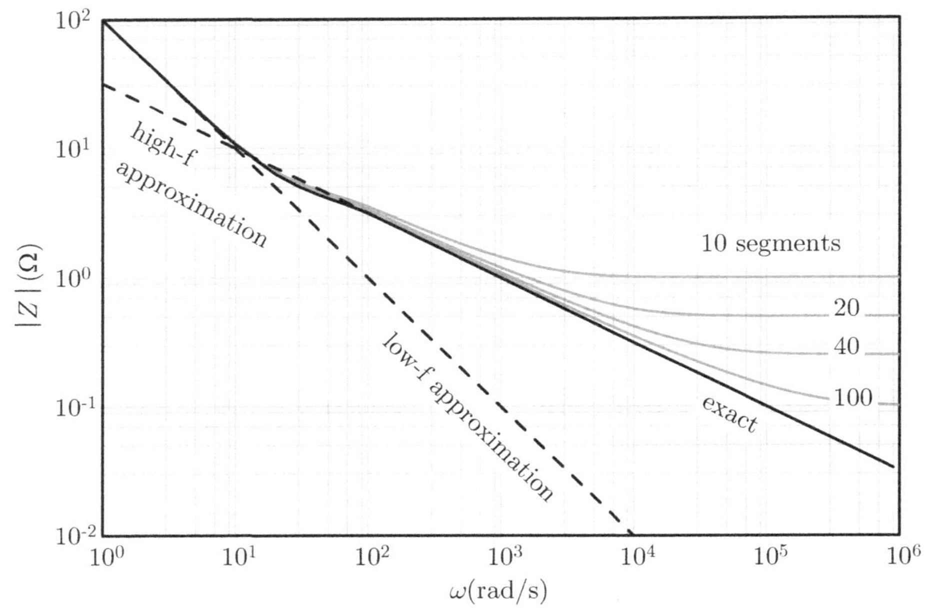

Fig. 46.21 *RC* transmission line input impedance and comparison to SPICE with finite segment count; case of $l = 10\,\text{m}$, $R = 1\,\Omega/\text{m}$, and $C = 1\,\text{mF/m}$

exactly! Both of these limits have been annotated in the figure as well.

This is but an example of how we can use analytics and theory in practice. It is almost granted that many problems call for a distributed *RC* line or the likes. Having an exact solution and corresponding plot that shows the clear dependence of impedance on frequency, the low- and high-frequency limits, the number of poles and zeroes, and perhaps a simple approximation function—is deemed for sure to be helpful in analyzing and better understanding the problem, and even checking the simulator itself.

46.21 Summary

The field of transmission lines is truly enormous and in fact transmission lines can be considered a cornerstone of electrical engineering! The amount of physics it contains from voltages to currents, electric and magnetic fields, the concept of the lumped elements *RLGC*, wave propagation, transmission, reflection, space, and time, and even the speed of light (and relativity) are all tied to this topic. Also the amount of variation in frequency and space dimensions is astounding; for example conventional transmission lines (for power distribution) run at 60 Hz and span hundreds of miles, while high-speed computing channels run in the GHz and span inches to millimeters! Either way, when boiled down to the essentials we arrive at the wave equation with spatial and temporal dependence. The aim of this chapter was to capture this dependence using spectral techniques. The plan was simple: if an input of single temporal frequency stimulated a line and gave us a particular solution (also harmonic in space) then we can figure the response due to any input by simply representing the new input as a Fourier series or as a Fourier/Laplace transform and by adding the corresponding solutions. We illustrated the process on many cases with various source/load terminations and for each case we plotted the corresponding solution and got excellent match to that of SPICE. We also dealt with the special case of un-driven line, with initial conditions and with either hard or floating boundary conditions. We see it again and again—the final solution disguised as a Fourier series/transform of harmonics in space and in time. Finally we touched on the important topic of input impedance and illustrated it for both the *LC* line and the *RC* one.

46.22 Problems

1. An ideal transmission line with length 50 mm and *LC* values of 1 μH/m and 1 nF/m is grounded on both ends as shown in Fig. 46.22a. The line is precharged such that the right half is at unity potential while the left one at zero potential. Calculate the potential across the line, as a function of time and plot it. Compare to SPICE; see sample solution in Fig. 46.23.
   Answer:
   $$a_n = -\frac{2}{n\pi}\left[\cos(n\pi) - \cos(0.5n\pi)\right]$$
2. An ideal transmission line has the same specifications as that in Problem 1. Instead of it being grounded, however, both sides are floating as shown in Fig. 46.22b. The line is precharged such that the middle fifth segment is at unity, and the rest grounded. Calculate the potential across the line, as a function of time and plot it. Compare to SPICE; see sample solution in Fig. 46.24.
   Answer:
   $$a_n = \frac{2}{n\pi}\left[\sin(n\pi\frac{3}{5}) - \sin(n\pi\frac{2}{5})\right], \quad a_0 = 0.2$$

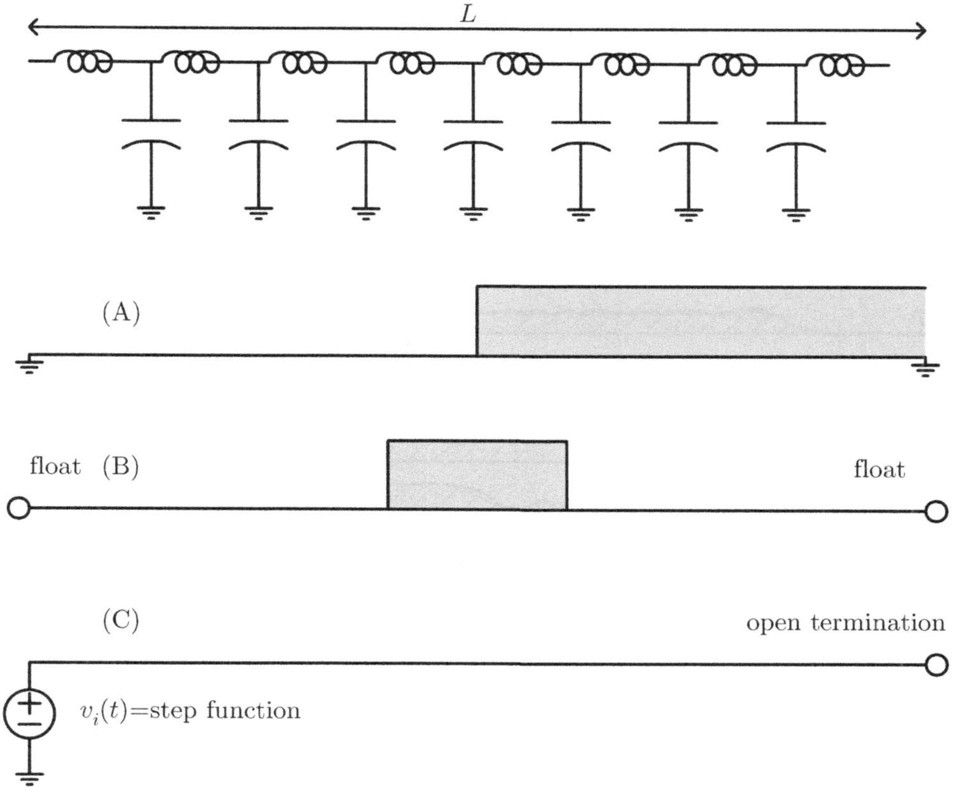

Fig. 46.22 Specifications for various problems

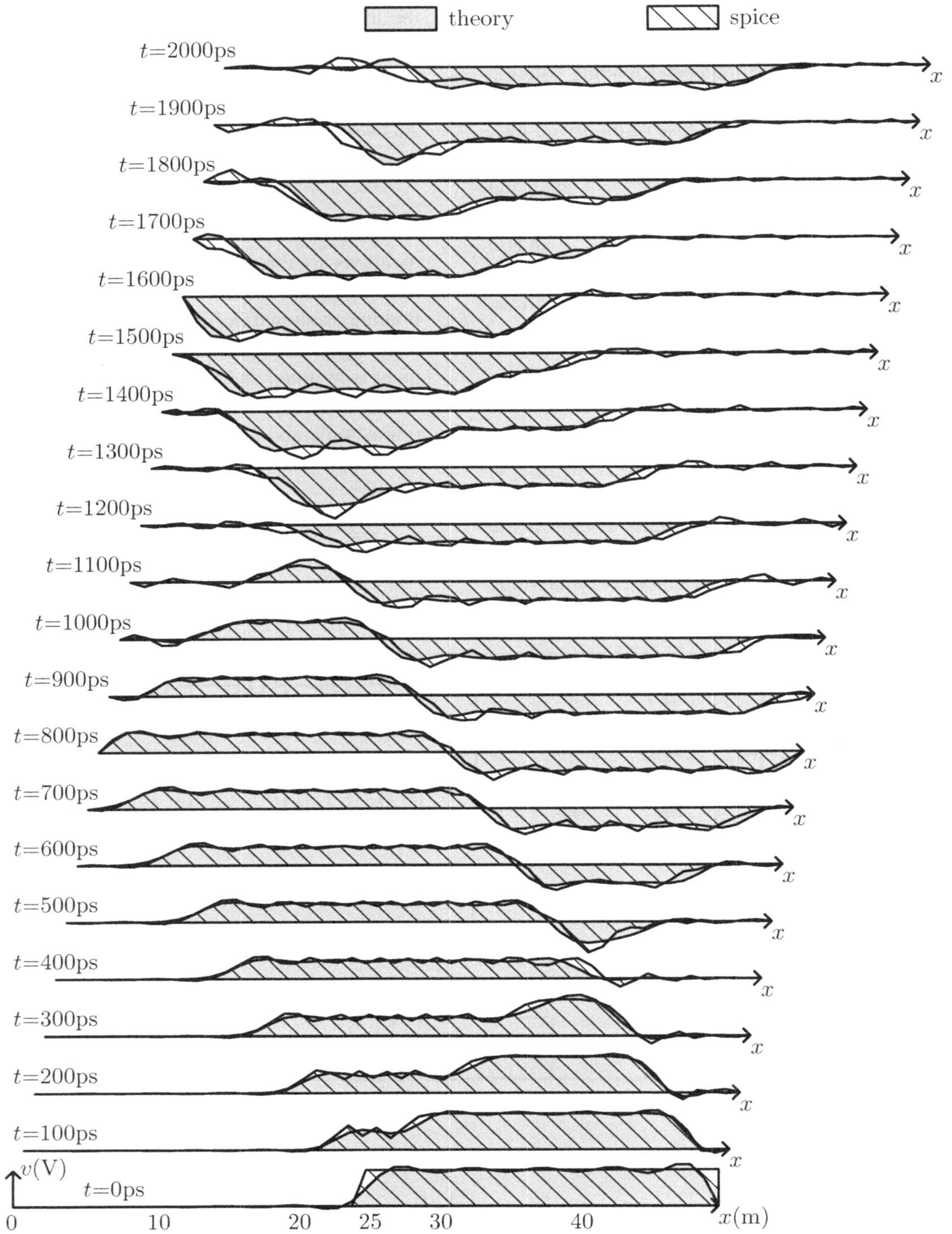

Fig. 46.23 Sample solution to Problem 1

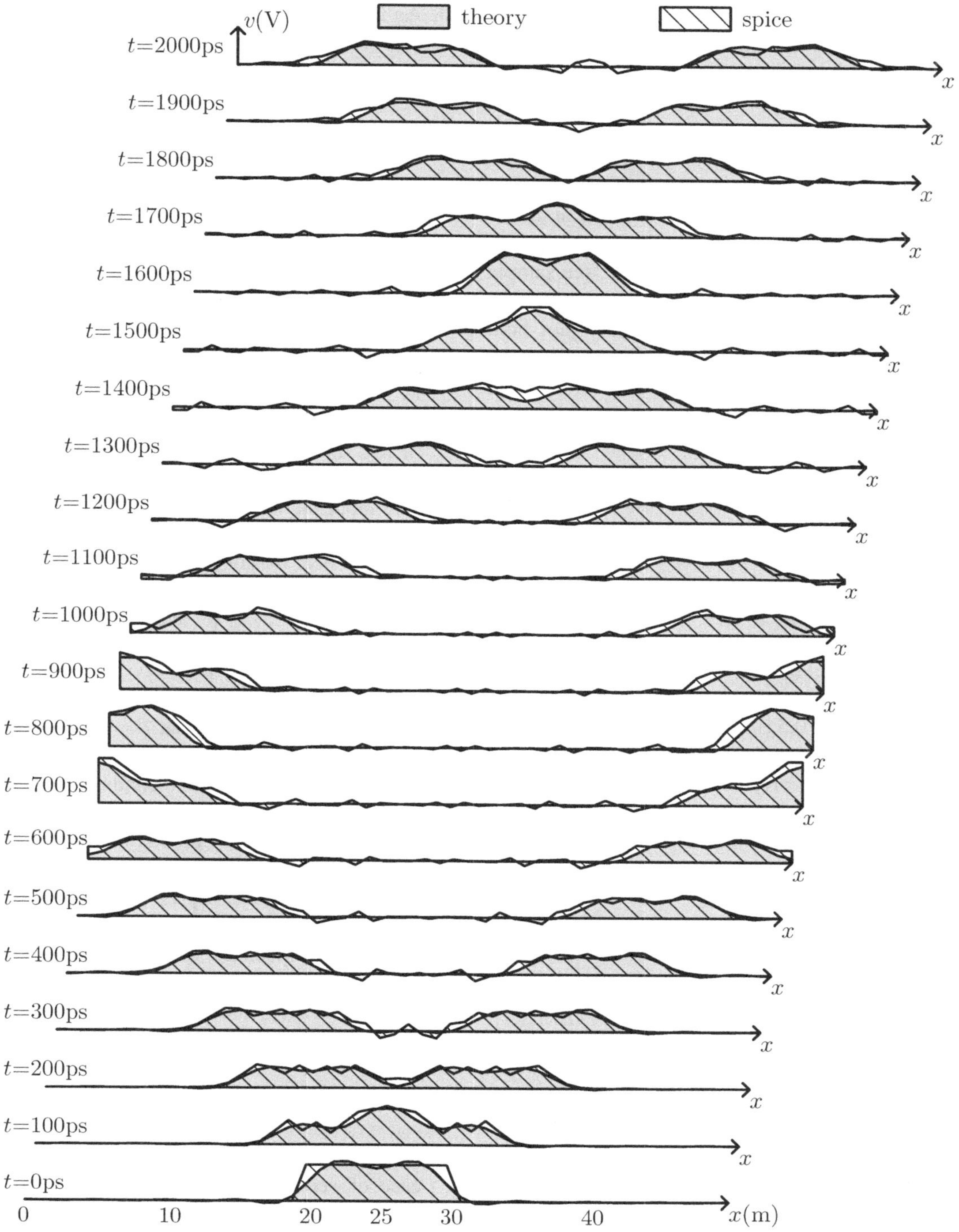

Fig. 46.24 Sample solution to Problem 2

3. An ideal transmission line with length 10 m and LC values of 0.5 H/m and 0.5 F/m is driven on the left by an ideal unit step function (meaning with zero impedance), and open terminated on the right; see Fig. 46.22c. Calculate the potential across the line, as a function of time and plot it. How long does it take for the initial signal to arrive at the far end? How long does it take to come back to the source? Compare to SPICE; see sample solution in Fig. 46.25.
4. Repeat Problem 3 but this time plot both voltage and current. Explain results; see sample solution in Fig. 46.26. What is the min, max, and average of voltage? How about current? Explain!
5. It was shown in the chapter that halving both L and C resulted in 2-times larger speed; what would happen if L was upped by 2 while C downed by 2, such that the LC product remains the same? What is the impact on the voltage profile (if any) for the case of open termination. If no impact, where would the impact be?
6. A transmission line with length 10 m and LC values of 0.5 H/m and 0.5 F/m has a finite resistance of 0.5 Ω/m and is open terminated. The line is driven by an ideal voltage source that is given by $v_i(t) = \sin \pi t$, and as shown in Fig. 46.27a. Find the voltage for all space and time and plot it; compare to SPICE. See sample solution in Fig. 46.28.
7. What is the velocity of propagation of the wave in Problem 6. Measure it from the plot in Fig. 46.28 by dividing distance by time change. How does that compare to $\frac{1}{LC}$?
8. If a T-line is driven by a sine function of angular frequency ω_0 such that the *temporal* period is $T = \frac{2\pi}{\omega_0}$, what is the *spatial* period? Assume the line is ideal. Compare your prediction to the lossy case in Problem 6 and identify how good of an approximation that is even for the lossy case.
9. Repeat Problem 6 but this time use the low-resistance approximation in Eq. 46.87. Compare both solutions.
10. An ideal transmission line with length 10 m and LC values of 0.5 H/m and 0.5 F/m is driven on the left by a unit step voltage, and is open terminated as shown in Fig. 46.27b. The source is matched to the line, such that source impedance is $Z_s = \sqrt{\frac{L}{C}} = 1$. Figure voltage for all space and time and plot it; compare to SPICE as shown in Fig. 46.29.
11. A transmission line with length 10 m has zero inductance, $R = 0.5\,\Omega$/m and $C = 0.5$ F/m. The right-end side is terminated by a short; i.e., $Z_L = 0$. The left-end side is driven by an ideal step function, and as shown in Fig. 46.27c. Compute voltage across the line versus time, and compare to SPICE. Do the results make sense? Explain! See sample solution in Fig. 46.30.
12. The input impedance of a transmission line of length l was derived in Eq. 46.119, and repeated here for convenience

$$Z_{\text{in}} = Z_0 \frac{e^{\gamma l} + \rho e^{-\gamma l}}{e^{\gamma l} - \rho e^{-\gamma l}}$$

For the case of a short-ended line (i.e., zero termination), such that $\rho = -1$, derive the input impedance for the special case of zero cap (and zero shunt conductance) as shown in Fig. 46.31a. Show that it comes out $Z(s) = (R_0 + sL_0)l$ which is nothing other than the series impedance of the line, with resistance/inductance R_0/L_0 per unit length. Use the approximation $e^x \sim 1+x$ for small x.
13. The input impedance of an ideal LC line which is short terminated was derived in Eq. 46.128 and repeated here for convenience

$$Z_{\text{in}} = jZ_0 \tan\left(\omega l \sqrt{L_0 C_0}\right)$$

Derive the input impedance of the special case of zero cap as shown in Fig. 46.31b. Show input impedance comes out $Z \sim j\omega L_0 l$ which is nothing other than the total inductance of the line. Use the approximation $\tan x \sim x$ for small x.

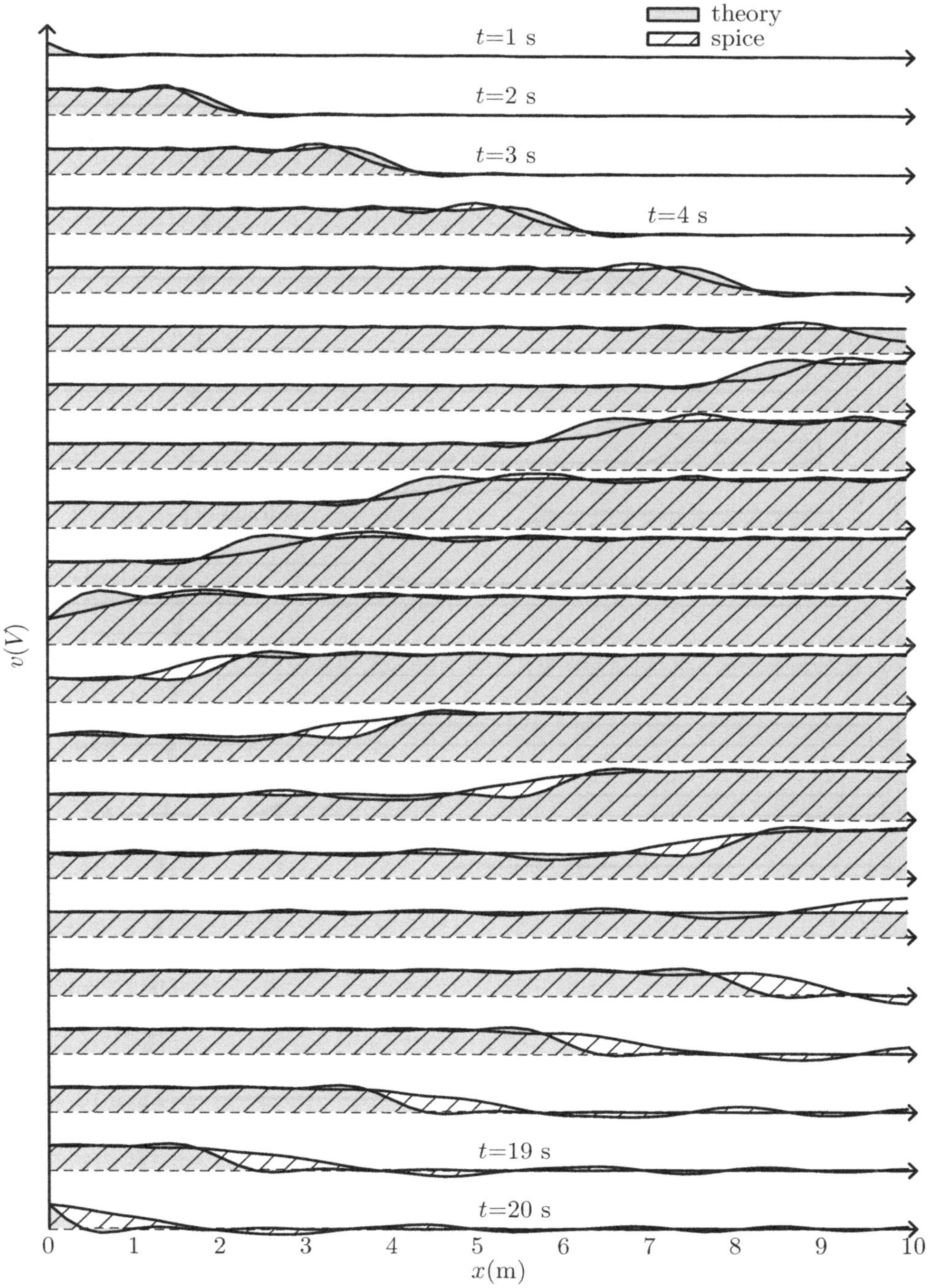

Fig. 46.25 Sample solution to Problem 3

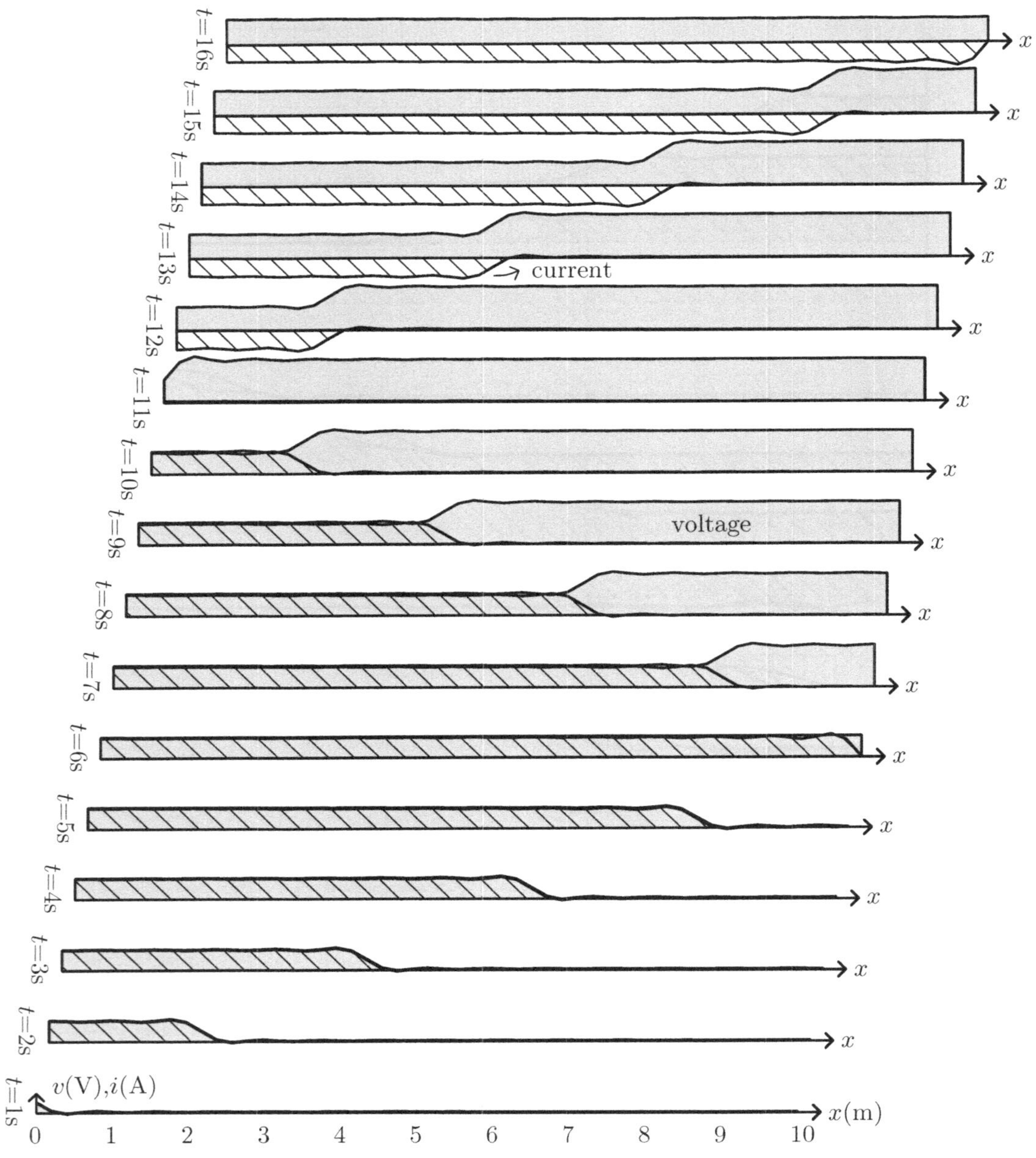

Fig. 46.26 Sample solution to Problem 4

14. The input impedance of an ideal *LC* line which is *open* terminated was derived in Eq. 46.123 and repeated here for convenience

$$Z_{\text{in}} = -jZ_0 \cot\left(\omega l \sqrt{L_0 C_0}\right)$$

Derive the input impedance of the special case of zero *inductance* as shown in Fig. 46.31c. Show it comes out $Z \sim \frac{1}{j\omega C_0 l}$ which is nothing other than the total capacitance of the line. Use the approximation $\cot x \sim \frac{1}{x}$ for small x.

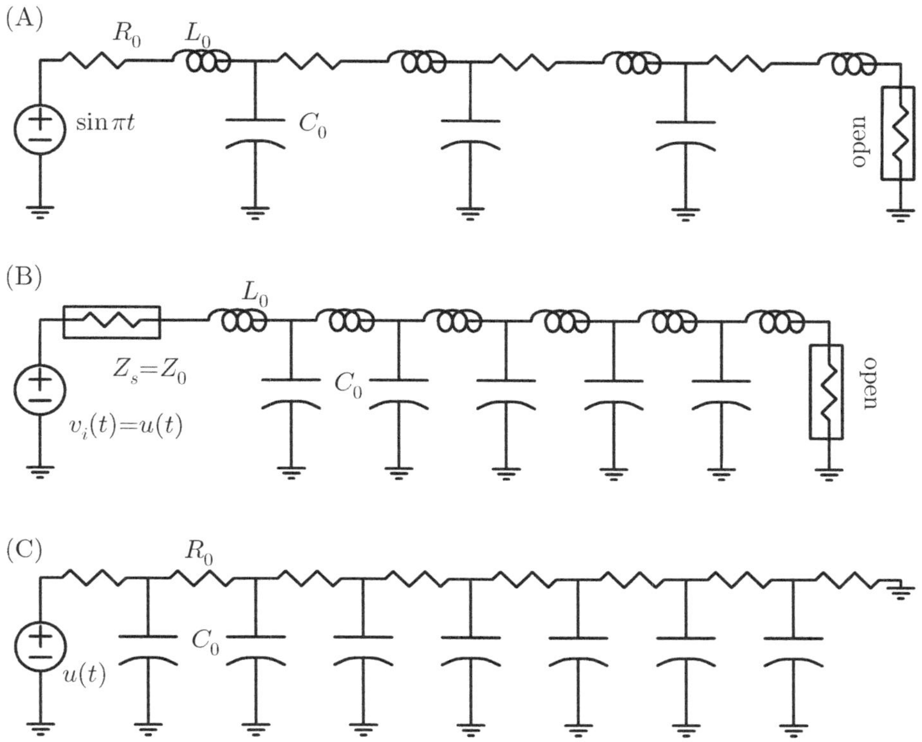

Fig. 46.27 Specifications for various problems

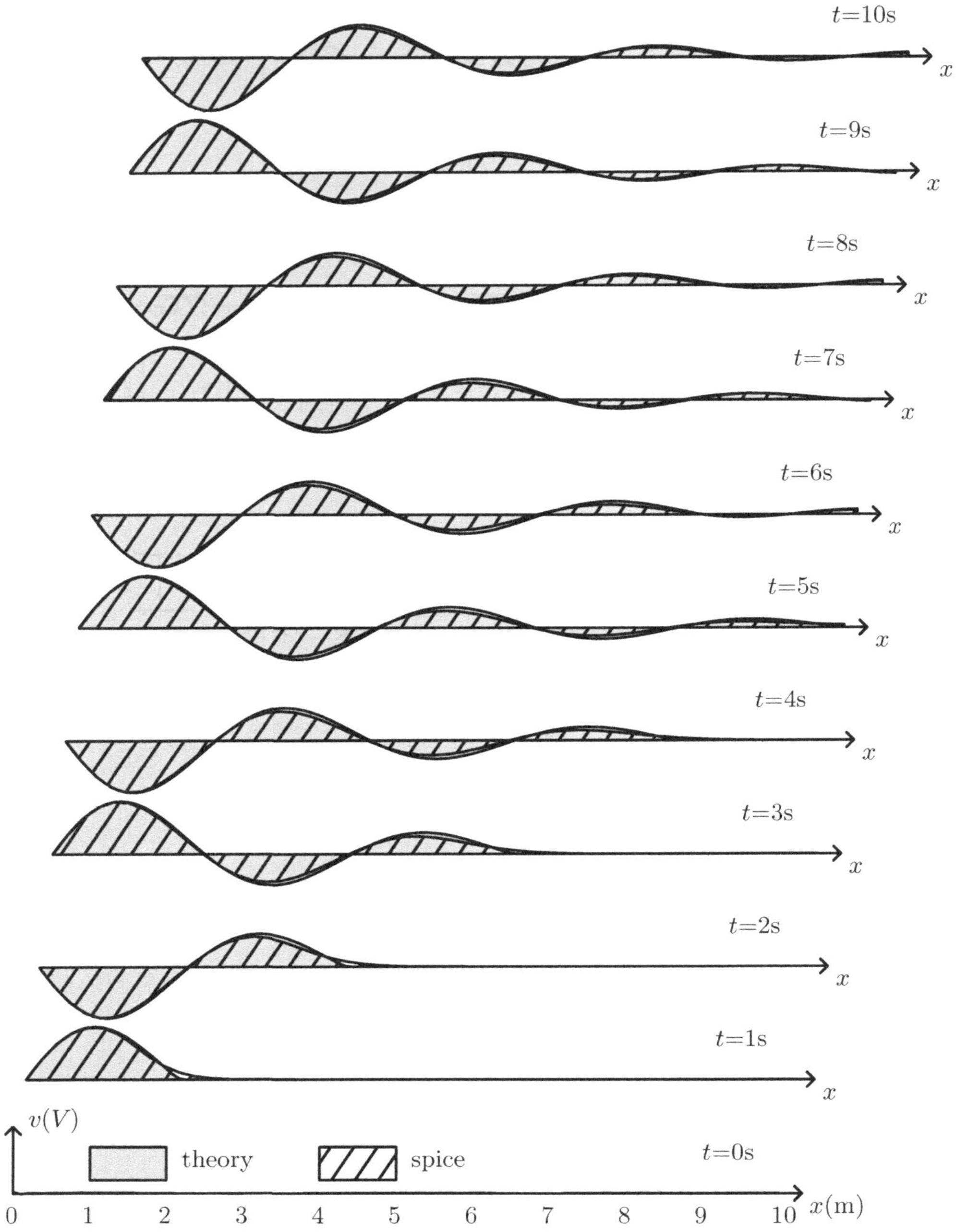

Fig. 46.28 Sample solution to Problem 6

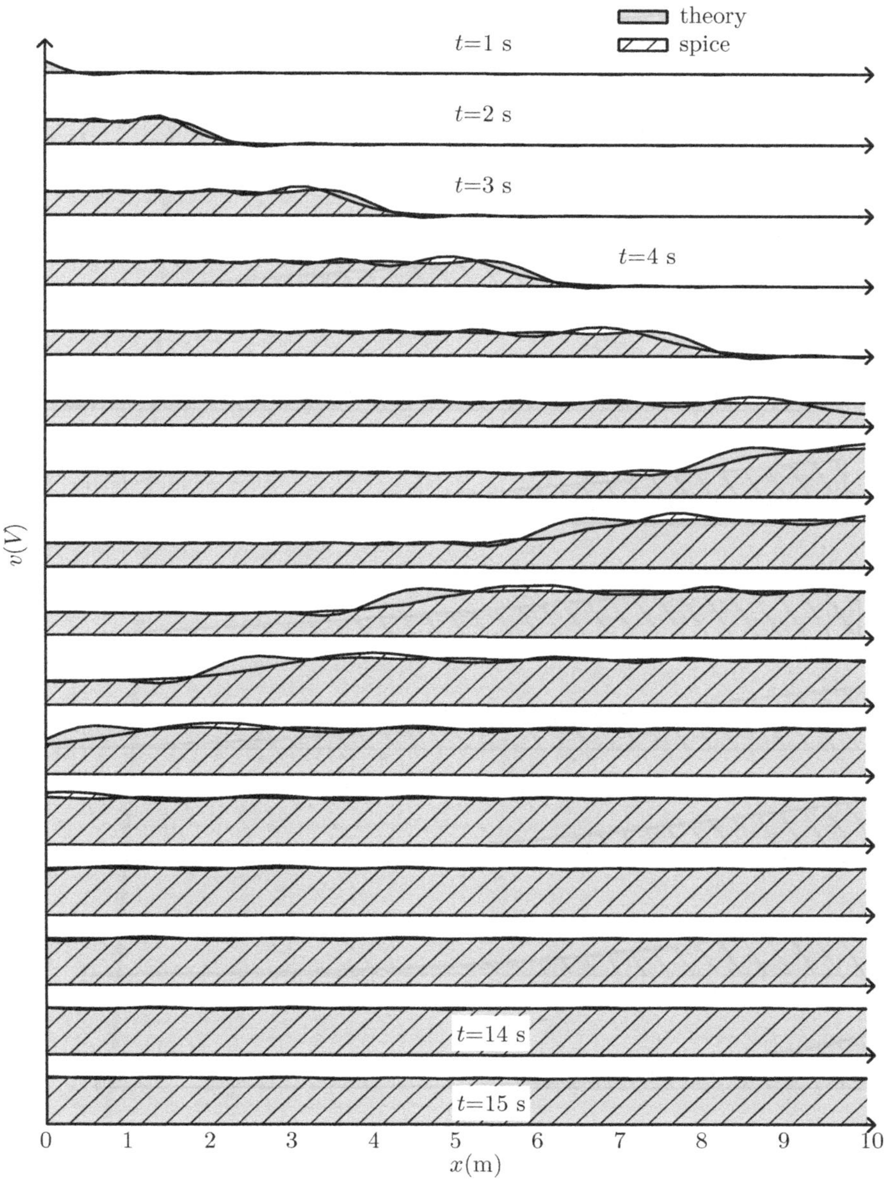

Fig. 46.29 Sample solution to Problem 10

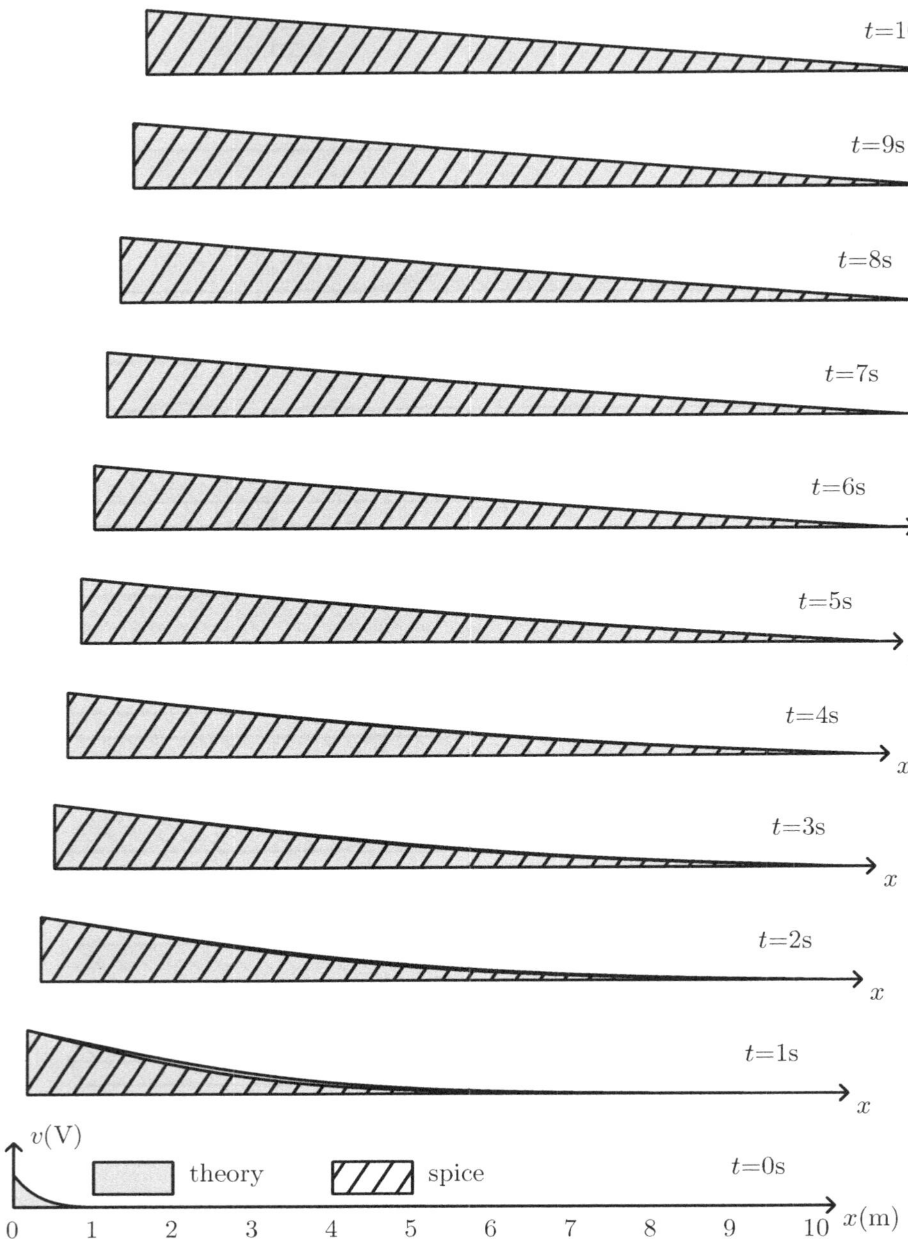

Fig. 46.30 Sample solution to Problem 11

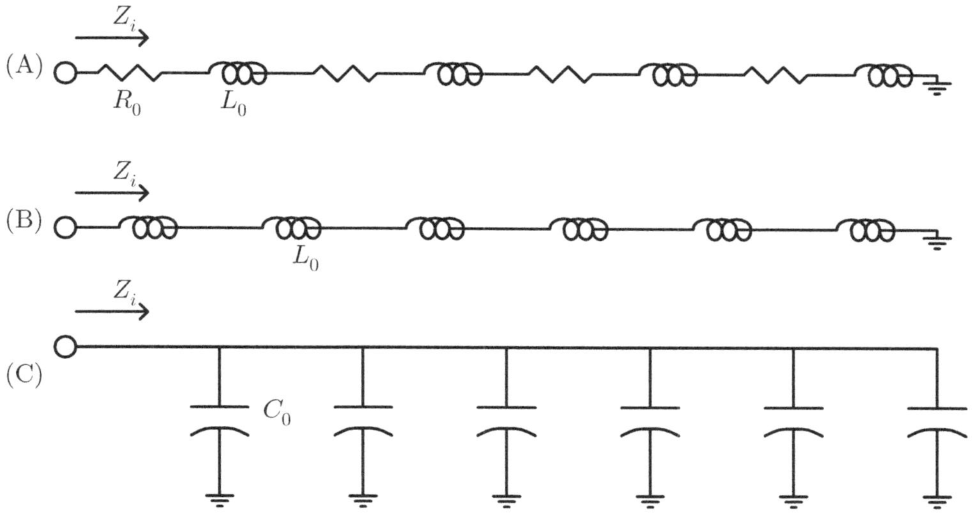

Fig. 46.31 Specifications to various problems

Mathematics of Complex Analysis and Inverse Transforms

A

A.1 Introduction to Complex Analysis

Complex analysis deals with complex numbers. Complex numbers are in the form of a real number plus an imaginary one:

$$z = x + jy \tag{A.1}$$

The evolution of the complex number is long, but in essence it follows the evolution of real numbers. First there were the positive integers; then came negative ones; then came fractions; then came irrational numbers, such as e and π. Finally came the imaginary number j. The applications of complex numbers are numerous. For our subject at hand, who would not have seen the following formulas?

$$\text{Impedance of capacitor} = \frac{1}{j\omega C} \tag{A.2}$$

$$\text{Impedance of inductor} = j\omega L \tag{A.3}$$

$$\text{Wave equation} = f(x,t) = e^{j[\omega t - kx]} \tag{A.4}$$

$$\text{Fourier Transform } F(\omega) = \int_{-\infty}^{\infty} f(t)e^{-j\omega t}dt \tag{A.5}$$

$$\text{Laplace Transform } F(s) = \int_{-\infty}^{\infty} f(t)e^{-st}dt \tag{A.6}$$

$$\text{where} \quad s = \sigma + j\omega \tag{A.7}$$

Clearly the use of complex numbers and analysis is ubiquitous and time invested becoming familiar with complex analysis is time well spent!

The Imaginary Number Complex numbers combine real numbers and imaginary ones. It all stems from the definition of the imaginary number

$$j^2 = -1, \quad \text{or} \quad j = \sqrt{-1} \tag{A.8}$$

By allowing for imaginary numbers, we expand the real axis to a 2D plane. While this seems to be adding complications, it comes with associated benefits as has been shown throughout the text.

Polar Representation of Complex Numbers Any complex number in the form of

$$z = x + jy \tag{A.9}$$

can be represented in polar coordinates as

$$z = Re^{j\theta} \tag{A.10}$$

if the following two conditions are met

$$R = \sqrt{x^2 + y^2} \quad \text{and}$$
$$\theta = \text{atan}\frac{y}{x} \tag{A.11}$$

We can verify this by referring to Fig. A.1 and recognizing

$$\text{if } \theta = \text{atan}\frac{y}{x} \quad \text{then } \cos\theta = \frac{x}{\sqrt{x^2 + y^2}}$$

F. Badrieh, *Spectral, Convolution and Numerical Techniques in Circuit Theory*,
https://doi.org/10.1007/978-3-319-71437-0

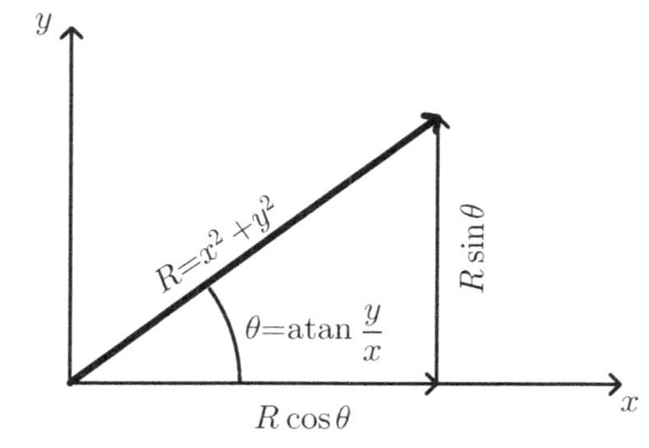

Fig. A.1 Polar representation of complex numbers

$$\text{and } \sin\theta = \frac{y}{\sqrt{x^2+y^2}} \tag{A.12}$$

We start with the polar representation and reduce

$$\begin{aligned} z &= Re^{j\theta} \\ &= \sqrt{x^2+y^2}\,[\cos\theta + j\sin\theta] \\ &= \sqrt{x^2+y^2}\left[\frac{x}{\sqrt{x^2+y^2}} + j\frac{y}{\sqrt{x^2+y^2}}\right] \\ &= x + jy \end{aligned} \tag{A.13}$$

which is what we started with.

Complex Number Multiplication Quite often there arises the need to multiply two (or more) complex numbers. There are (at least) two ways about this. First consider two complex numbers defined by

$$z_1 = x_1 + jy_1, \quad z_2 = x_2 + jy_2 \tag{A.14}$$

The first method of multiplication uses the distributive law as follows:

$$\begin{aligned} z_1 z_2 &= (x_1 + jy_1)(x_2 + jy_2) \\ &= (x_1x_2 - y_1y_2) + j(x_1y_2 + y_1x_2) \end{aligned} \tag{A.15}$$

That is, the product gives a real part which picks up the product of the real terms and product of imaginary terms (with a negative sign); and an imaginary term which picks up the cross coupling terms between real and imaginary. The second method relies on the polar representation of both numbers. First rewrite each number as

$$\begin{aligned} z_1 &= R_1 e^{j\theta_1}, \quad \text{where} \quad R_1^2 = (x_1^2 + y_1^2), \quad \text{and} \\ \theta_1 &= \operatorname{atan}\frac{y_1}{x_1} \\ z_2 &= R_2 e^{j\theta_2}, \quad \text{where} \quad R_2^2 = (x_2^2 + y_2^2), \quad \text{and} \\ \theta_2 &= \operatorname{atan}\frac{y_2}{x_2} \end{aligned} \tag{A.16}$$

Now simply form the product

$$z_1 z_2 = R_1 R_2 e^{j(\theta_1+\theta_2)} \tag{A.17}$$

We can verify that both results are similar by taking the latter one and expanding

$$\begin{aligned} z_1 z_2 &= R_1 R_2 [\cos\theta_1 + j\sin\theta_1][\cos\theta_2 + j\sin\theta_2] \\ &= R_1 R_2 \frac{x_1 + jy_1}{\sqrt{x_1^2 + y_1^2}} \frac{x_2 + jy_2}{\sqrt{x_2^2 + y_2^2}} \\ &= \sqrt{(x_1^2 + y_1^2)(x_2^2 + y_2^2)}\,\frac{(x_1 + jy_1)(x_2 + jy_2)}{\sqrt{(x_1^2 + y_1^2)(x_2^2 + y_2^2)}} \\ &= (x_1 + jy_1)(x_2 + jy_2) \\ &= (x_1x_2 - y_1y_2) + j(x_1y_2 + y_1x_2) \end{aligned} \tag{A.18}$$

in agreement with Eq. (A.15).

Complex Number Manipulations In theory any complex number can be represented in the form of a real number plus an imaginary one

$$z = x + jy \tag{A.19}$$

In some cases things may not be as clear.

First Example Consider for example the number

$$z = Re^{j\theta} \tag{A.20}$$

We can easily recast this in the form of $x + jy$ by using Euler's formula

$$z = R\cos\theta + jR\sin\theta \tag{A.21}$$

which is in the form of $x + jy$.

Second Example Consider another example

$$z = \frac{1}{1+j} \tag{A.22}$$

Multiply both sides by $1 - j$ and get

$$\frac{1}{1+j} = \frac{1}{1+j}\frac{1-j}{1-j} = \frac{1-j}{1^2 - j^2} = \frac{1-j}{2} = \frac{1}{2} - j\frac{1}{2} \tag{A.23}$$

which is in the form of $x + jy$.

Third Example Another example is

$$z = (1+j)^2 \tag{A.24}$$

Again we simplify as

$$(1+j)^2 = (1+j)(1+j) = 1 - 1 + 2j = 0 + j2 \tag{A.25}$$

which is in the form of $x + jy$.

Fourth Example Another example is

$$z = \cos(1+j) \tag{A.26}$$

We use Euler's formula

$$\begin{aligned}\cos(1+j) &= \frac{e^{j(1+j)} + e^{-j(1+j)}}{2} = \frac{e^{(-1+j)} + e^{(1-j)}}{2} \\ &= \frac{e^{-1}\left[\cos(1) + j\sin(1)\right] + e^{1}\left[\cos(1) - j\sin(1)\right]}{2} \\ &= \frac{\cos(1)\left[e^{-1} + e^{1}\right]}{2} + j\frac{\sin(1)\left[e^{-1} - e^{1}\right]}{2}\end{aligned} \tag{A.27}$$

which again is in the form of $x + jy$.

Complex Functions If x is used to denote the real axis, such that a typical real function is represented as $f(x)$, then a complex function utilizes both x and jy such that our domain variable is

$$z = x + jy \tag{A.28}$$

Some examples of complex functions are

$$f(z) = z = x + jy \tag{A.29}$$

$$f(z) = z + 1 = (x+1) + jy \tag{A.30}$$

and

$$f(z) = z^2 = (x+jy)(x+jy) = (x^2 - y^2) + j(2xy) \tag{A.31}$$

Notice that in all of the above cases we are able to write the function as

$$f(z) = u(x,y) + jv(x,y) \tag{A.32}$$

In other words, a complex function takes as input a complex argument and produces as output another complex quantity. The process of mapping $z = x + jy$ into $f(z) = u(x,y) + jv(x,y)$ is shown in Fig. A.2.

A.2 Analytic Functions

Analytic functions play a very important role in complex analysis. One of the most relevant consequences of being analytic or not thereof is when doing contour integration. As will be

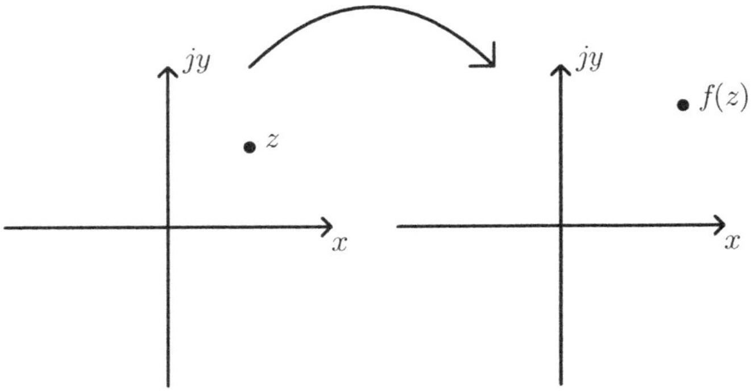

Fig. A.2 Process of mapping between $z = x + jy$ and $f(z) = u(x, y) + jv(x, y)$

shown, analytic functions have the property that a contour integration in the complex domain yields zero results. Nonanalytic functions, on the other hand, have contour integrals which yield nonzero results. So, when utilizing contour integration, knowing whether the function is analytic or not is very important.

Analytic Function Criterion An analytic function is a differentiable one, and specifically if

$$f(z) = u(x, y) + jv(x, y) \tag{A.33}$$

then $f(z)$ is analytic (differentiable) if

$$\frac{\partial u}{\partial x} = \frac{\partial v}{\partial y} \quad \text{and} \tag{A.34}$$

$$\frac{\partial u}{\partial y} = -\frac{\partial v}{\partial x} \tag{A.35}$$

Let's examine a few examples.

First Example of Analytic Function: $f(z) = z$ As a first example consider the function

$$f(z) = z \tag{A.36}$$

We can rewrite as

$$f(z) = z = x + jy \tag{A.37}$$

Here we have

$$u(x, y) = x \quad \text{and} \quad v(x, y) = y \tag{A.38}$$

Furthermore,

$$\frac{\partial u}{\partial x} = \frac{\partial v}{\partial y} = 1 \tag{A.39}$$

and

$$\frac{\partial u}{\partial y} = -\frac{\partial v}{\partial x} = 0 \tag{A.40}$$

Hence the function $f(z) = z$ is analytic.

Second Example of Analytic Function: $f(z) = z^2$ As a second example of analytic functions consider the function

$$f(z) = z^2 \tag{A.41}$$

We rewrite the function as

$$z^2 = (x + jy)(x + jy) = x^2 - y^2 + j2xy \tag{A.42}$$

which implies that

$$u(x, y) = x^2 - y^2, \quad \text{and} \quad v(x, y) = 2xy \tag{A.43}$$

We proceed to finding the partial derivatives

$$\frac{\partial u(x, y)}{\partial x} = \frac{d\left(x^2 - y^2\right)}{dx} = 2x \tag{A.44}$$

Next

$$\frac{\partial v(x, y)}{\partial y} = \frac{d(2xy)}{dy} = 2x \tag{A.45}$$

Since that last two are equal, the first condition of analyticity is satisfied. Next

$$\frac{\partial u(x,y)}{\partial y} = \frac{d\left(x^2 - y^2\right)}{dy} = -2y \qquad (A.46)$$

and

$$\frac{\partial v(x,y)}{\partial x} = \frac{d(2xy)}{dx} = 2y \qquad (A.47)$$

Since the last two equations are equal, with a negative sign, then the second condition of analyticity is also satisfied. Hence, the function z^2 is analytic.

Third Example of Analytic Function: $f(z) = 1$ Here is the trivial case of

$$f(z) = 1 \qquad (A.48)$$

In terms of real and imaginary components we have

$$u(x,y) = 1, \quad \text{and} \quad v(x,y) = 0 \qquad (A.49)$$

Next we evaluate the partial derivatives

$$\frac{\partial u(x,y)}{\partial x} = \frac{\partial v(x,y)}{\partial y} = 0 \qquad (A.50)$$

and

$$\frac{\partial u(x,y)}{\partial y} = -\frac{\partial v(x,y)}{\partial x} = 0 \qquad (A.51)$$

Hence, the function is analytic, and so are all constant functions.

Fourth Example of Analytic Function: $f(z) = e^z$ Consider the function

$$f(z) = e^z \qquad (A.52)$$

We rewrite this function in terms of u and v as follows:

$$e^z = e^{x+jy} = e^x e^{jy} = e^x\left[\cos y + j \sin y\right] \quad (A.53)$$

so that

$$u(x,y) = e^x \cos y, \quad \text{and} \quad v(x,y) = e^x \sin y \qquad (A.54)$$

We proceed to evaluate the partial derivatives:

$$\frac{\partial u(x,y)}{\partial x} = \frac{\partial\left(e^x \cos y\right)}{\partial x} = e^x \cos y \qquad (A.55)$$

Next

$$\frac{\partial v(x,y)}{\partial y} = \frac{\partial\left(e^x \sin y\right)}{\partial y} = e^x \cos y \qquad (A.56)$$

Since these last two are equal, then the first condition of analyticity is satisfied. Next we evaluate

$$\frac{\partial u(x,y)}{\partial y} = \frac{\partial\left(e^x \cos y\right)}{\partial y} = -e^x \sin y \qquad (A.57)$$

and

$$\frac{\partial v(x,y)}{\partial x} = \frac{\partial\left(e^x \sin y\right)}{\partial x} = e^x \sin y \qquad (A.58)$$

Since these last two are equal, with a negative sign, then the second condition of analyticity is satisfied, and the function is deemed analytic.

Example of Function that is *not* Analytic: $\frac{1}{z}$ Consider the function

$$f(z) = \frac{1}{z} \qquad (A.59)$$

We can rewrite in terms of u and v

$$f(z) = \frac{1}{x + jy} = \frac{x}{x^2 + y^2} + j\frac{-y}{x^2 + y^2} \qquad (A.60)$$

Hence

$$u(x,y) = \frac{x}{x^2 + y^2} \quad \text{and}$$
$$v(x,y) = \frac{-y}{x^2 + y^2} \qquad (A.61)$$

Furthermore

$$\frac{\partial u}{\partial x} = \frac{1}{x^2+y^2} - \frac{2x^2}{(x^2+y^2)^2}$$
$$= \frac{x^2+y^2-2x^2}{(x^2+y^2)^2} = \frac{-x^2+y^2}{(x^2+y^2)^2} \tag{A.62}$$

which equals

$$\frac{\partial v}{\partial y} = -\frac{1}{x^2+y^2} + \frac{2y^2}{(x^2+y^2)^2}$$
$$= \frac{-x^2-y^2+2y^2}{(x^2+y^2)^2} = \frac{-x^2+y^2}{(x^2+y^2)^2} \tag{A.63}$$

Also

$$\frac{\partial u}{\partial y} = -\frac{2xy}{(x^2+y^2)^2} \tag{A.64}$$

which equals the negative of

$$\frac{\partial v}{\partial x} = \frac{2xy}{(x^2+y^2)^2} \tag{A.65}$$

This function is then analytic everywhere, except at the point $(x, jy) = 0$. To show this, consider for example the following partial derivative

$$\frac{\partial u(x,y)}{\partial x} = \frac{-x^2+y^2}{(x^2+y^2)^2} \tag{A.66}$$

Let us approach the origin first using the following sequence. Set $x = 0$ and get

$$\left.\frac{\partial u(x,y)}{\partial x}\right|_{x=0} = \left.\frac{-x^2+y^2}{(x^2+y^2)^2}\right|_{x=0} = \frac{1}{y^2} \tag{A.67}$$

So as we approach $(0, 0)$ along the y-axis, we get a large positive number. Now let us approach the origin using the following modified sequence. Set $y = 0$ and get

$$\left.\frac{\partial u(x,y)}{\partial x}\right|_{y=0} = \left.\frac{-x^2+y^2}{(x^2+y^2)^2}\right|_{y=0} \tag{A.68}$$
$$= -\frac{1}{x^2}$$

So as we approach $(0, 0)$ along the x-axis, we get a large *negative* number. Clearly the above two results are not consistent! Hence, the derivative is not well defined at the origin, and the function is said to be *not* analytic at that point.

Last Example of Analytic Function: $\frac{1}{z^2}$ Consider the function

$$f(z) = \frac{1}{z^2} \tag{A.69}$$

Using the notation $z = x + jy$ we rewrite this as follows:

$$\frac{1}{z^2} = \frac{1}{(x+jy)(x+jy)} = \frac{1}{(x^2-y^2)+j(2xy)}$$
$$= \frac{(x^2-y^2)-j(2xy)}{(x^2-y^2)^2+4x^2y^2} \tag{A.70}$$

so that

$$u(x,y) = \frac{x^2-y^2}{(x^2-y^2)^2+4x^2y^2}$$

and

$$v(x,y) = \frac{-2xy}{(x^2-y^2)^2+4x^2y^2} \tag{A.71}$$

We proceed in finding the various partial derivatives; buckle up! First,

$$\begin{aligned}
\frac{u(x,y)}{\partial x} &= \frac{\partial}{\partial x}\frac{x^2-y^2}{(x^2-y^2)^2+4x^2y^2} \\
&= \frac{2x}{(x^2-y^2)^2+4x^2y^2} - (x^2-y^2)\frac{2(x^2-y^2)2x+8xy^2}{[(x^2-y^2)^2+4x^2y^2]^2} \\
&= \frac{2x}{(x^2-y^2)^2+4x^2y^2} - (x^2-y^2)\frac{4x^3+4xy^2}{[(x^2-y^2)^2+4x^2y^2]^2} \\
&= \frac{2x}{(x^2-y^2)^2+4x^2y^2} - \frac{4x^5-4x^3y^2+4x^3y^2-4xy^4}{[(x^2-y^2)^2+4x^2y^2]^2} \\
&= \frac{2x}{(x^2-y^2)^2+4x^2y^2} - \frac{4x^5-4xy^4}{[(x^2-y^2)^2+4x^2y^2]^2} \\
&= \frac{2x[(x^2-y^2)^2+4x^2y^2]-4x^5+4xy^4}{[(x^2-y^2)^2+4x^2y^2]^2} \\
&= \frac{2x[(x^4+y^4-2x^2y^2)+4x^2y^2]-4x^5+4xy^4}{[(x^2-y^2)^2+4x^2y^2]^2} \\
&= \frac{2x^5+2xy^4-4x^3y^2+8x^3y^2-4x^5+4xy^4}{[(x^2-y^2)^2+4x^2y^2]^2} \\
&= \frac{-2x^5+6xy^4+4x^3y^2}{[(x^2-y^2)^2+4x^2y^2]^2}
\end{aligned} \tag{A.72}$$

Next

$$\begin{aligned}
\frac{v(x,y)}{\partial y} &= \frac{\partial}{\partial y}\frac{-2xy}{(x^2-y^2)^2+4x^2y^2} \\
&= \frac{-2x}{(x^2-y^2)^2+4x^2y^2} + 2xy\frac{-2(x^2-y^2)2y+8x^2y}{[(x^2-y^2)^2+4x^2y^2]^2} \\
&= \frac{-2x}{(x^2-y^2)^2+4x^2y^2} + 2xy\frac{4y^3+4x^2y}{[(x^2-y^2)^2+4x^2y^2]^2} \\
&= \frac{-2x}{(x^2-y^2)^2+4x^2y^2} + \frac{8xy^4+8x^3y^2}{[(x^2-y^2)^2+4x^2y^2]^2} \\
&= \frac{-2x[(x^2-y^2)^2+4x^2y^2]+8xy^4+8x^3y^2}{[(x^2-y^2)^2+4x^2y^2]^2} \\
&= \frac{-2x[x^4+y^4-2x^2y^2+4x^2y^2]+8xy^4+8x^3y^2}{[(x^2-y^2)^2+4x^2y^2]^2} \\
&= \frac{-2x[x^4+y^4+2x^2y^2]+8xy^4+8x^3y^2}{[(x^2-y^2)^2+4x^2y^2]^2} \\
&= \frac{-2x^5-2xy^4-4x^3y^2+8xy^4+8x^3y^2}{[(x^2-y^2)^2+4x^2y^2]^2} \\
&= \frac{-2x^5+6xy^4+4x^3y^2}{[(x^2-y^2)^2+4x^2y^2]^2}
\end{aligned} \tag{A.73}$$

Though it took a while, we do observe that the last two results are the same. Next

$$
\begin{aligned}
\frac{u(x,y)}{\partial y} &= \frac{\partial}{\partial y}\frac{x^2-y^2}{(x^2-y^2)^2+4x^2y^2}\\
&= \frac{-2y}{(x^2-y^2)^2+4x^2y^2} - (x^2-y^2)\frac{-2(x^2-y^2)2y+8x^2y}{[(x^2-y^2)^2+4x^2y^2]^2}\\
&= \frac{-2y}{(x^2-y^2)^2+4x^2y^2} - (x^2-y^2)\frac{4y^3+4x^2y}{[(x^2-y^2)^2+4x^2y^2]^2}\\
&= \frac{-2y}{(x^2-y^2)^2+4x^2y^2} - \frac{4x^2y^3-4y^5+4x^4y-4x^2y^3}{[(x^2-y^2)^2+4x^2y^2]^2}\\
&= \frac{-2y}{(x^2-y^2)^2+4x^2y^2} - \frac{-4y^5+4x^4y}{[(x^2-y^2)^2+4x^2y^2]^2}\\
&= \frac{-2y[x^4+y^4-2x^2y^2+4x^2y^2]+4y^5-4x^4y}{[(x^2-y^2)^2+4x^2y^2]^2}\\
&= \frac{-2x^4y-2y^5+4x^2y^3-8x^2y^3+4y^5-4x^4y}{[(x^2-y^2)^2+4x^2y^2]^2}\\
&= \frac{-6x^4y+2y^5-4x^2y^3}{[(x^2-y^2)^2+4x^2y^2]^2}
\end{aligned}
\tag{A.74}
$$

Finally,

$$
\begin{aligned}
\frac{v(x,y)}{\partial x} &= \frac{\partial}{\partial x}\frac{-2xy}{(x^2-y^2)^2+4x^2y^2}\\
&= \frac{-2y}{(x^2-y^2)^2+4x^2y^2} + 2xy\frac{2(x^2-y^2)2x+8xy^2}{[(x^2-y^2)^2+4x^2y^2]^2}\\
&= \frac{-2y}{(x^2-y^2)^2+4x^2y^2} + 2xy\frac{4x^3+4xy^2}{[(x^2-y^2)^2+4x^2y^2]^2}\\
&= \frac{-2y}{(x^2-y^2)^2+4x^2y^2} + \frac{8x^4y+8x^2y^3}{[(x^2-y^2)^2+4x^2y^2]^2}\\
&= \frac{-2y[x^4+y^4-2x^2y^2+4x^2y^2]+8x^4y+8x^2y^3}{[(x^2-y^2)^2+4x^2y^2]^2}\\
&= \frac{-2x^4y-2y^5+4x^2y^3-8x^2y^3+8x^4y+8x^2y^3}{[(x^2-y^2)^2+4x^2y^2]^2}\\
&= \frac{6x^4y-2y^5+4x^2y^3}{[(x^2-y^2)^2+4x^2y^2]^2}
\end{aligned}
\tag{A.75}
$$

Again while it took a bit of algebra, we notice that the last two results are identical, with a negative sign. Hence we have established that the function $f(z) = \frac{1}{z^2}$ is analytic. Notice that this function—unlike the $\frac{1}{z}$ function—does not have the sign reversal of the derivative at zero, and hence it is analytic at zero as well.

A.3 Cauchy's Integral Theorem

Complex integration, also known as contour integration or the calculus of residues, is a very powerful tool in dealing with inverse Fourier and Laplace transforms. Recall in both cases we took a real time function and found its transform in the complex frequency domain; in the Fourier transform we had the $j\omega$ domain while in the Laplace transform we had the $s = \sigma + j\omega$ domain. In many cases, and for a given transform, a sort of look-up table is used to figure out the inverse transform, which is the time version of the signal. However, in some cases we may have to resort into actual integration, and it is towards this goal that contour integration comes in very handy.

Cauchy's Integral Theorem If a function is *analytic* in a neighborhood, then the contour integration of the function is zero

$$\boxed{\int_C f(z) = 0 \quad \text{analytic function}} \tag{A.76}$$

Let's demonstrate this via an example.

Example Find the integral of the function $f(z) = z$ around the unity circle.

$$\int_{R=1} z dz \tag{A.77}$$

We know from before that this function is analytic. Around the unity circle we can represent the function as

$$f(z) = e^{j\theta} \quad \text{and} \quad dz = je^{j\theta} d\theta \tag{A.78}$$

Then the integral becomes

$$\int_{\theta=0}^{\theta=2\pi} e^{j\theta} je^{j\theta} d\theta = j \int_0^{2\pi} e^{j2\theta} d\theta$$
$$= \frac{1}{2} e^{j2\theta} \Big|_0^{2\pi} = \frac{1}{2} [\cos 4\pi - \cos 0] = 0 \tag{A.79}$$

Hence we have shown that since the function is analytic, its contour integration is zero.

Functions with Singularities and Their Contour Integration If a function is analytic, but has a finite number of singularities (poles) then the contour integration around the singularity will not in general be zero.

Example Take the function

$$f(z) = \frac{1}{z} \tag{A.80}$$

We know from before that this function is analytic, except at the point $z = 0$. To integrate this function we use

$$\int_{R=1} \frac{dz}{z} = \int_{\theta=0}^{\theta=2\pi} \frac{je^{j\theta} d\theta}{e^{j\theta}} = \int_0^{2\pi} jd\theta = 2\pi j \tag{A.81}$$

Hence we have shown that the integral of $\frac{1}{z}$ around the unity circle, which encloses a singularity, is not equal to zero, even though the function is analytic everywhere (except at the singularity).

Cauchy's Residue Theorem If a function is analytic and has a number of singularities, then the contour integration of the function around a path enclosing the singularities equals the sum of the residues

$$\boxed{\oint_C f(z)dz = 2\pi j \sum \text{residues}} \tag{A.82}$$

where the residue is the $\frac{1}{z}$ coefficient in the Laurent expansion of the function:

$$\boxed{\text{residue} = \text{coefficient of} \frac{1}{z}} \tag{A.83}$$

Finding Residues The following examples illustrate how to find the residues.

Example of Simple Pole suppose the function has a single singularity around the point z_0; then the function can be written as

$$f(z) = \frac{a_{-1}}{z - z_0} + g(z) \quad (A.84)$$

where $g(z)$ is analytic at the point z_0. To isolate the a_{-1} term we multiply both sides by $(z - z_0)$ and evaluate at $z = z_0$

$$(z - z_0)f(z) = (z - z_0)\frac{a_{-1}}{z - z_0} + (z - z_0)g(z)\Big|_{z=z_0} \quad (A.85)$$

Since $g(z)$ is analytic at z_0 then

$$(z - z_0)g(z)\Big|_{z=z_0} = 0 \quad (A.86)$$

and we end up with

$$\boxed{\text{residue} = a_{-1} = (z - z_0)f(z)\Big|_{z=z_0}}, \quad \text{case of simple pole} \quad (A.87)$$

Example of Pole of Order 2 Suppose now that the function has a pole of order 2 at z_0; then the function can be written as

$$f(z) = \frac{a_{-2}}{(z - z_0)^2} + \frac{a_{-1}}{z - z_0} + g(z) \quad (A.88)$$

where $g(z)$ is analytic at z_0. Now multiply both sides by $(z - z_0)^2$

$$(z - z_0)^2 f(z) = (z - z_0)^2 \frac{a_{-2}}{(z - z_0)^2} + (z - z_0)^2 \frac{a_{-1}}{z - z_0} + (z - z_0)^2 g(z) \quad (A.89)$$

or

$$(z - z_0)^2 f(z) = a_{-2} + (z - z_0)a_{-1} + (z - z_0)^2 g(z) \quad (A.90)$$

If we evaluate now at $z = z_0$ we would pick a_{-2} which is not what we want; recall we are after the $\frac{1}{z}$ term (or $\frac{1}{z-z_0}$ term in this case). Let's instead differentiate both sides with respect to z

$$\frac{d}{dz}(z - z_0)^2 f(z) = \frac{da_{-2}}{dz} + \frac{d}{dz}(z - z_0)a_{-1} + \frac{d}{dz}(z - z_0)^2 g(z) \quad (A.91)$$

Now the a_{-2} terms vanishes and the desired term a_{-1} remain. We have then

$$\frac{d}{dz}(z - z_0)^2 f(z) = a_{-1} + \frac{d}{dz}(z - z_0)^2 g(z) \quad (A.92)$$

Since $g(z)$ is analytic at z_0, then it can be shown that

$$\frac{d}{dz}(z - z_0)^2 g(z)\Big|_{z=z_0} = 0 \quad (A.93)$$

(just use the chain derivative rule) and we finally end up with

$$\boxed{\text{residue} = a_{-1} = \frac{d}{dz}\left[(z - z_0)^2 f(z)\right]_{z=z_0}}, \quad \text{case of pole of order 2} \quad (A.94)$$

Case of Higher Order Poles For the more generic case of a pole of order n the residue is

$$\boxed{\text{residue} = \frac{1}{(n-1)!}\frac{d^{n-1}}{dz^{n-1}}\left[(z-z_0)^n f(z)\right]_{z=z_0}}, \quad \text{case of pole of order } n \tag{A.95}$$

Examples of Cauchy's Residue Theorem Here we show a few applications of Cauchy's residue theorem.

Example Find the following integral inside the unity circle

$$\oint \frac{1}{z}dz \tag{A.96}$$

Clearly the residue here is 1, and hence the integral is

$$\oint \frac{1}{z}dz = 2\pi j \tag{A.97}$$

which agrees with results in Eq. (A.81).

Example Find the following integral inside the unity circle

$$\oint \frac{1}{z^2}dz \tag{A.98}$$

Here we have a pole of order 2 at zero. The residue is

$$\text{residue} = \frac{d}{dz}\frac{z^2}{z^2} = \frac{d}{dz}1 = 0 \tag{A.99}$$

which makes sense, since the function $1/z^2$ is analytic everywhere inside the unit circle; remember, the contour integral of an analytic function is zero!

Example Find the following integral inside the unity circle

$$\oint \frac{1}{z^3}dz \tag{A.100}$$

Here we have a pole of order 3 at zero. The residue is

$$\text{residue} = \frac{1}{2}\frac{d^2}{dz^2}\frac{z^3}{z^3} = \frac{1}{2}\frac{d^2}{dz^2}1 = 0 \tag{A.101}$$

Clearly other than $\frac{1}{z}$, all functions of the form $\frac{1}{z^n}$ have zero integral around zero.

A.4 Cartesian Complex Integration

We had shown above how to integrate functions such as $f(z) = z$ and $f(z) = \frac{1}{z}$ around a unity circle. Namely

$$\oint z dz = 0 \tag{A.102}$$

and

$$\oint \frac{dz}{z} = 2\pi j \tag{A.103}$$

We were able to do this by using the following substitution

$$z = Re^{j\theta} \quad \text{and} \quad dz = jRe^{j\theta} \quad R = 1 \tag{A.104}$$

Here we show an alternate method, mostly to convince the reader that the above results are valid. Recall that

$$z = j + jy \tag{A.105}$$

Then

$$dz = dx + jdy \tag{A.106}$$

Now rather than integrating around the unit circle, we will integrate around a square, of side length 2 and center at the origin, as shown in Fig. A.3. We claim that the contour integration is path independent so long as it encompasses any singularities. If we label the four sides of the square such that

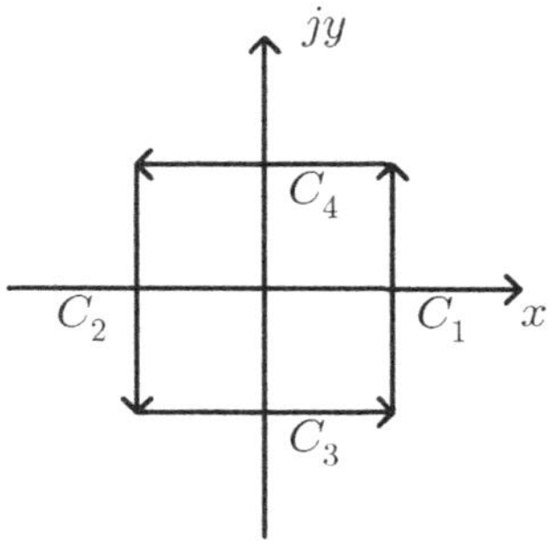

Fig. A.3 Complex integration using rectangular geometry

- Right side C_1 with $x = 1, y = y, z = 1 + jy$, and $dz = jdy$,
- Left side C_2 with $x = -1, y = y, z = -1 + jy$, and $dz = jdy$,
- Bottom side C_3 with $x = x, y = -1, z = x - j$, and $dz = dx$,
- Top side C_4 with $x = x, y = 1, z = x + j$, and $dz = dx$,

then we'd need to do four separate integrals.

Example Where Integral Evaluates to Zero For the case $f(z) = z$ the first integral becomes

$$I_1 = \int_{-1}^{1} (1 + jy)j\,dy = j\left[y + j\frac{y^2}{2}\right]_{-1}^{1}$$
$$= j\left[1 + \frac{j}{2} - \left(-1 + \frac{j}{2}\right)\right] = 2j \tag{A.107}$$

The second integral becomes

$$I_2 = \int_{1}^{-1} (-1 + jy)j\,dy = j\left[-y + j\frac{y^2}{2}\right]_{1}^{-1}$$
$$= j\left[1 + \frac{j}{2} - \left(-1 + \frac{j}{2}\right)\right] = 2j \tag{A.108}$$

The third integral becomes

$$I_3 = \int_{-1}^{1} (x - j)dx = \left[\frac{x^2}{2} - jx\right]_{-1}^{1}$$
$$= \frac{1}{2} - j - \left(\frac{1}{2} + j\right) = -2j \tag{A.109}$$

Finally the fourth integral becomes

$$I_4 = \int_{1}^{-1} (x + j)dx = \left[\frac{x^2}{2} + jx\right]_{1}^{-1}$$
$$= \left[\frac{1}{2} - j - \left(\frac{1}{2} + j\right)\right] = -2j \tag{A.110}$$

The whole integral then becomes

$$\oint z dz = I_1 + I_2 + I_3 + I_4$$
$$= 2j + 2j - 2j - 2j = 0 \tag{A.111}$$

as expected! That is, since the function $f(z) = z$ is analytic, its contour integration is zero!

Example Where Integral Evaluates to Nonzero This second example is more complicated and deals with $f(z) = \frac{1}{z}$, and again we are interested in evaluating this function around the origin. Similar to the prior case we take path of integration the square with 2 side length and center at the origin. Along the four sides we have the following

- Right side C_1 with $x = 1, y = y, z = \frac{1}{1+jy}$, and $dz = jdy$,
- Left side C_2 with $x = -1, y = y, z = \frac{1}{-1+jy}$, and $dz = jdy$,
- Bottom side C_3 with $x = x, y = -1, z = \frac{1}{x-j}$, and $dz = dx$,
- Top side C_4 with $x = x, y = 1, z = \frac{1}{x+j}$, and $dz = dx$.

The first integral becomes

$$I_1 = \int_{-1}^{1} \frac{1}{1 + jy} jdy = \ln(1 + jy)\Big|_{-1}^{1} = \ln\frac{1 + j}{1 - j} \tag{A.112}$$

The second integral becomes

$$I_2 = \int_1^{-1} \frac{1}{-1+jy} jdy = \ln(-1+jy)\Big|_1^{-1} = \ln\frac{-1-j}{-1+j} = \ln\frac{1+j}{1-j} \tag{A.113}$$

The third integral becomes

$$I_3 = \int_{-1}^{1} \frac{1}{x-j} dx = \ln(x-j)\Big|_{-1}^{1} = \ln\frac{1-j}{-1-j} = \ln\frac{-1+j}{1+j} \tag{A.114}$$

And the fourth integral becomes

$$I_4 = \int_1^{-1} \frac{1}{x+j} dx = \ln(x+j)\Big|_1^{-1} = \ln\frac{-1+j}{1+j} \tag{A.115}$$

The whole integral becomes

$$\begin{aligned} I &= I_1 + I_2 + I_3 + I_4 \\ &= \ln\frac{1+j}{1-j} + \ln\frac{1+j}{1-j} \\ &\quad + \ln\frac{-1+j}{1+j} + \ln\frac{-1+j}{1+j} \\ &= 2\left[\ln\frac{1+j}{1-j} + \ln\frac{-1+j}{1+j}\right] \\ &= 2\ln\left[\frac{1+j}{1-j}\cdot\frac{-1+j}{1+j}\right] \\ &= 2\ln\left[\frac{-1+j-j-1}{1+j-j+1}\right] = 2\ln\left[\frac{-2}{2}\right] \\ &= 2\ln(-1) \end{aligned} \tag{A.116}$$

How do we go about finding the log of a negative number? We can represent -1 as

$$-1 = e^{j\pi} \tag{A.117}$$

Then

$$I = 2\ln\left[e^{j\pi}\right] = 2j\pi \tag{A.118}$$

where we have used the identity $\ln e^x = x$. So our final result is

$$\oint \frac{dz}{z} = 2\pi j \tag{A.119}$$

in exact agreement with polar integration and using the Residue theorem (see Eq. (A.97))!

A.5 Application of Contour Integration in Finding Indefinite Integrals

Having established the theory of complex integration we turn now to real applications. Our ultimate goal is to be able to do the inverse Fourier and Laplace transforms, but as an intermediate goal we now show how contour integration can be used for real integration.

Evaluating Indefinite Integrals Here the goal is to evaluate an indefinite integral of the form

$$\int_{-\infty}^{\infty} f(x)dx \tag{A.120}$$

where $f(x)$ has the form of

$$f(z) = \frac{g(x)}{h(x)} \tag{A.121}$$

and order of $h(x)$ is at least 2 more than $g(x)$.

Example Evaluating $\int \frac{dx}{x^2+1}$ As a first example, consider evaluating the integral

$$\int_{-\infty}^{\infty} \frac{dx}{x^2+1} \tag{A.122}$$

Notice that here $g(x) = 1$ and $h(x) = x^2 + 1$, and that latter is at least 2 orders more than former. For reference we will first do this integral using the standard trig substitution method, then later using contour integration. Recall

$$\sin^2\theta + \cos^2\theta = 1 \tag{A.123}$$

Dividing both sides by $\cos^2\theta$ we get

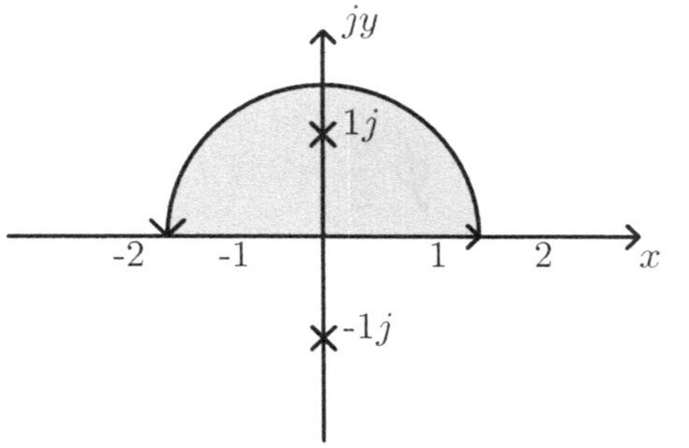

Fig. A.4 Contour integration for $f(z) = \frac{1}{z^2+1}$

$$\tan^2 + 1 = \sec^2\theta \tag{A.124}$$

This suggests the substitution

$$x = \tan\theta \quad \text{and} \quad dx = \sec^2\theta d\theta \tag{A.125}$$

Then

$$x^2 + 1 = \tan^2 + 1 = \sec^2\theta \tag{A.126}$$

and the integral becomes

$$\int \frac{\sec^2\theta d\theta}{\sec^2\theta} = \int d\theta = \theta \tag{A.127}$$

Now reverse substitute and get

$$\int_{-\infty}^{\infty} \frac{dx}{x^2+1} = \text{atan} x\Big|_{-\infty}^{\infty} = \frac{\pi}{2} + \frac{\pi}{2} = \pi \tag{A.128}$$

This is our reference answer. Next let's use contour integration. We map the x function into z such that

$$f(z) = \frac{1}{z^2+1} \tag{A.129}$$

Instead of evaluating along the x axis we evaluate along the contour formed by the x axis and a circle covering the upper complex plane as shown in Fig. A.4. The premise here is that we are able to calculate the contour integral. If the part of integration around the arc goes to zero, as the circle radius goes to infinity, then the remaining part (which is the sought real integral) would be the answer.

Since the denominator goes as R^2 and the arc as R we are guaranteed that the contour integration around the arc would go to zero. Then all that is left is evaluating the contour integral in terms of residues. As shown in the figure the function $f(z)$ has poles at $+j$ and $-j$, both of which are simple. Let us integrate around the top half of the complex plane. There, the residue is

$$\text{residue} = (z-j)\frac{1}{(z-j)(z+j)}\Big|_{z=j} = \frac{1}{z+j}\Big|_{z=j} = \frac{1}{2j} \tag{A.130}$$

The contour integration is then

$$\oint \frac{dz}{z^2+1} = 2\pi j \sum \text{residue} = 2\pi j \frac{1}{2j} = \pi \tag{A.131}$$

in agreement with our trig substitution method (Eq. (A.128))! Hence we have shown that using complex integration enables us to correctly figure the value of an indefinite real integral! Let's take another, more complicated example.

Example Evaluating $\int \frac{dx}{x^4+1}$ As another example take the indefinite integral

$$\int_{-\infty}^{\infty} \frac{dx}{x^4+1} \tag{A.132}$$

Unlike the prior integral, this one may not easily be integrated using conventional techniques. To do this using complex integration first we change in terms of z

$$\oint \frac{dz}{z^4+1} \tag{A.133}$$

Notice that since the denominator degree is at least twice that of numerator, we are guaranteed that the branch integral over the arc extending to infinity will vanish, and as such then the desired integral over the real axis would equal the sum of residues. So the only needed work is in evaluating the residues. First rewrite the denominator as

$$z^4 + 1 = (z^2 - j)(z^2 + j) \tag{A.134}$$

The first quadratic $(z^2 - j) = 0$ gives us two roots

$$z_1 = \sqrt{j}, \quad z_2 = -\sqrt{j} \tag{A.135}$$

While the second quadratic $(z^2 + j) = 0$ gives us two other roots

$$z_3 = j\sqrt{j}, \quad z_4 = -j\sqrt{j} \tag{A.136}$$

In order to locate those poles in the complex plane, we'd need to express them in terms of real and imaginary numbers. The difficulty lies in evaluating $\sqrt{j}$. First notice that

$$j = e^{j\pi/2} \tag{A.137}$$

Then

$$\begin{aligned}\sqrt{j} &= \left[e^{j\pi/2}\right]^{1/2} = e^{j\pi/4}\\ &= \cos\frac{\pi}{4} + \sin\frac{\pi}{4} = \frac{1}{\sqrt{2}}(1+j)\end{aligned} \tag{A.138}$$

Then the four poles become

$$\begin{aligned}z_1 &= \frac{1}{\sqrt{2}}(1+j), z_2 = \frac{1}{\sqrt{2}}(-1-j),\\ z_3 &= \frac{1}{\sqrt{2}}(-1+j), z_4 = \frac{1}{\sqrt{2}}(1-j)\end{aligned} \tag{A.139}$$

and they are shown in Fig. A.5.

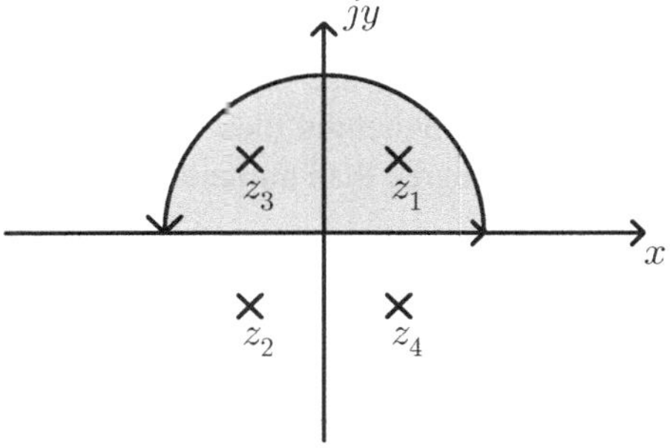

Fig. A.5 Contour integration for $f(z) = \frac{1}{z^4+1}$

If we are to integrate in the upper complex half plane, then only z_1 and z_3 count. The first residue is

$$\begin{aligned}\text{1st residue} &= \left.\frac{(z-z_1)}{(z-z_1)(z-z_2)(z^2+j)}\right|_{z=z_1=\frac{1}{\sqrt{2}}(1+j)}\\ &= \frac{1}{\frac{1}{\sqrt{2}}(1+j)+\frac{1}{\sqrt{2}}(+1+j)}\frac{1}{2j}\\ &= \frac{1}{2\sqrt{2}j}\frac{1}{1+j} = \frac{1}{2\sqrt{2}}\frac{1}{-1+j}\end{aligned} \tag{A.140}$$

The second residue is

$$\begin{aligned}\text{2nd residue} &= \left.\frac{(z-z_3)}{(z^2-j)(z-z_3)(z-z_4)}\right|_{z=z_3=\frac{1}{\sqrt{2}}(-1+j)}\\ &= \frac{1}{-2j}\frac{1}{\frac{1}{\sqrt{2}}(-1+j)+\frac{1}{\sqrt{2}}(-1+j)}\\ &= \frac{1}{-2\sqrt{2}j}\frac{1}{-1+j} = \frac{1}{2\sqrt{2}}\frac{1}{1+j}\end{aligned} \tag{A.141}$$

The sum of the residues is then

$$\begin{aligned}\sum \text{residues} &= \frac{1}{2\sqrt{2}}\left[\frac{1}{1+j} - \frac{1}{1-j}\right]\\ &= \frac{1}{2\sqrt{2}}\frac{1-j-(1+j)}{(1+j)(1-j)}\\ &= \frac{1}{2\sqrt{2}}\frac{1-j-1-j}{2}\\ &= \frac{1}{2\sqrt{2}}\frac{-2j}{2} = \frac{-j}{2\sqrt{2}}\end{aligned} \tag{A.142}$$

The integral is then

$$\begin{aligned}\oint \frac{dz}{z^4+1} &= 2\pi j \sum \text{residues}\\ &= 2\pi j\frac{-j}{2\sqrt{2}} = \frac{\pi}{\sqrt{2}}\end{aligned} \tag{A.143}$$

The author is unaware of any other method to get this integral, but a numerical evaluation thereof did indeed yield the provided answer. Between this and the prior example, we have demonstrated that in fact using contour integration we are able to figure normal, real integration provided the arc-part of the integration goes to zero. Next we extend this idea of using contour integration but this time to figure inverse Fourier and Laplace transforms.

A.6 Application of Contour Integration in Finding Inverse Fourier Transforms

Having developed the basic theory of contour integration and residue theorem, and applications to indefinite integrals, now we turn to our main goal behind this whole effort, and that is finding inverse Fourier and Laplace transforms. We demonstrate this via a few examples.

Contour Integration of Transform of Signum Function Recall the signum function is given by

$$\mathrm{sig}(t) = -\frac{1}{2} \quad t < 0 \tag{A.144}$$

$$= +\frac{1}{2} \quad t > 0 \tag{A.145}$$

We had found earlier (Eq. (8.49)) that the Fourier transform of this function is

$$\mathcal{F}[\mathrm{sig}(t)] = \frac{1}{j\omega} \tag{A.146}$$

This then implies that

$$\mathrm{sig}(t) = \frac{1}{2\pi} \int_{-\infty}^{\infty} \frac{e^{j\omega t}}{j\omega} d\omega \tag{A.147}$$

How can this integral be validated? Using contour integration! In the original integral we integrate over ω from $-\infty$ to ∞. To cast this in terms of a complex integral we need to migrate to the complex plane. If we set up the real axis as having symbol σ and imaginary axis as having $j\omega$, and now construct a contour which has the shape of half a circle, with the straight side being the imaginary axis, then we can relate the integral over the ω domain to that around the contour. In order to succeed with this we'd need to ensure that the branch of the integral around the arc goes to zero as the arc radius tends to infinity. First we recast the integral in terms of complex variable z

$$I = \frac{1}{2\pi j} \oint \frac{e^{zt} dz}{z} \tag{A.148}$$

Notice that we used the fact that

$$z = \sigma + j\omega \tag{A.149}$$

and that on the $\sigma = 0$ axis

$$z = j\omega, \quad dz = jd\omega \Rightarrow d\omega = \frac{1}{j} dz \tag{A.150}$$

If $t < 0$ we'd want to use an arc on the right side of the complex plane to ensure that the exponential dies of at large radius, as shown in Fig. A.6 (left). As shown in the figure, the contour path (which is clockwise) passes directly over a singularity (at 0). To overcome this, we bend the path around that singularity, but in doing so we pick up half a residue, namely πj. (Notice this all works if pole is simple; higher order poles cannot use this trick.) If we label the three integration paths as

- I_{arc} as the integral around the large arc,
- I_{desired} as the desired integral over ω,
- I_{origin} as the small bend around the origin,

we have

$$I_{\mathrm{arc}} + I_{\mathrm{desired}} + I_{\mathrm{origin}} = 2\pi j \sum \text{residues} \tag{A.151}$$

We know that the arc integral will be zero, since both the exponential dies off and denominator is larger degree than numerator; that is

$$I_{\mathrm{arc}} = 0 \tag{A.152}$$

(Notice that in this case the denominator order is only 1 larger than that of numerator, but that is still good enough for the arc integral to vanish, utilizing Jordan's lemma.) We also know that the integration around the origin (half circle) gave half a residue; that is

$$I_{\text{origin}} = \pi j \tag{A.153}$$

Finally we know that there are no residues *inside* the whole contour; that is

$$2\pi j \sum \text{residues} = 0 \tag{A.154}$$

Then

$$0 + I_{\text{desired}} + \pi j = 0 \tag{A.155}$$

which implies that

$$I_{\text{desired}} = -\pi j \tag{A.156}$$

If we scale down by the $2\pi j$ in front of the inverse Fourier transform we finally get

$$\frac{1}{2\pi}\int_{-\infty}^{\infty}\frac{e^{j\omega t}d\omega}{j\omega} = \frac{1}{2\pi j}\times(-\pi j) = -\frac{1}{2} \quad (t<0) \tag{A.157}$$

Similarly, if $t > 0$ we would choose the contour integration as shown in Fig. A.6 (right). Now, however, the integration around the singularity around zero gives *negative* half a residue, since we are going there in a clockwise direction. Our three integration paths then give

$$0 + I_{\text{desired}} - \pi j = 0 \tag{A.158}$$

which implies that

$$I_{\text{desired}} = +\pi j \tag{A.159}$$

If we scale down by the $2\pi j$ in front of the inverse Fourier transform we finally get

$$\frac{1}{2\pi}\int_{-\infty}^{\infty}\frac{e^{j\omega t}d\omega}{j\omega} = +\frac{1}{2} \quad (t>0) \tag{A.160}$$

Our final result is then

$$\frac{1}{2\pi}\int_{-\infty}^{\infty}\frac{e^{j\omega t}d\omega}{j\omega} = \begin{cases} -\frac{1}{2} & (t<0) \\ +\frac{1}{2} & (t>0) \end{cases} \tag{A.161}$$

which is exactly the signum function, and exactly what we set out to do using contour integration! Let us reflect quickly on this before moving on. We carried the full blown complex integration, in the frequency domain and arrived at a time function—a unique one, to say the least—which is $-\frac{1}{2}$ for negative time and $\frac{1}{2}$ for positive time. This transient function by all means is a "digital" function which has an enormous discontinuity at time zero, yet it can be simply represented as an integral in the frequency domain, an integral which we just evaluated analytically! The main take is the potential and power of using

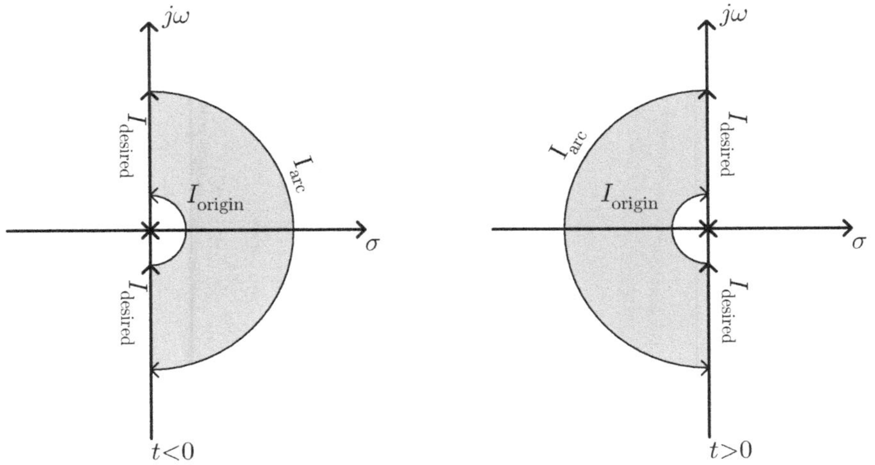

Fig. A.6 Contour integration of $\frac{e^{zt}}{z}$

the frequency domain—albeit complex one—in representing nonconventional functions, such as the signum one.

Contour Integration of Transform of $u(t)e^{-t}$
We know (Eq. (8.19)) that the Fourier transform of the function $u(t)e^{-t}$ is

$$u(t)e^{-t} \rightarrow \frac{1}{1+j\omega} \tag{A.162}$$

This means that using the inverse Fourier transform we should get

$$u(t)e^{-t} = \frac{1}{2\pi}\int_{-\infty}^{\infty}\frac{e^{j\omega t}}{1+j\omega}d\omega \tag{A.163}$$

We'd like to validate this integral. Again let's migrate into the complex domain with independent variable z:

$$\begin{aligned} z &= \sigma + j\omega \\ dz &= d\sigma + jd\omega \end{aligned} \tag{A.164}$$

so that on the $j\omega$ axis (since $d\sigma = 0$)

$$d\omega = \frac{dz}{j} \quad \text{(on } j\omega \text{ axis)} \tag{A.165}$$

We define this integral around a contour and ensure that part of this contour encompasses the frequency integration. The contour integral becomes

$$I = \frac{1}{2\pi j}\oint\frac{e^{zt}}{1+z}dz \tag{A.166}$$

Notice that the integrand has a simple pole at $z = -1$. Now we specify the contour to go on the $j\omega$ axis and either wrap to positive infinity or negative infinity, as shown in Fig. A.7.

- For negative time we'd want to take the right-hand half plane since we are assured that $e^{zt}/(1+z)$ would decay to zero around the arc. Since there are no residues in the right-hand half plane, we conclude that the integral is zero

$$I_1 = \frac{1}{2\pi j}\oint\frac{e^{zt}}{1+z}dz = 0 \quad (t<0) \tag{A.167}$$

- For positive negative time we'd want to rake the left-hand half plane, again since we are assured that $e^{zt}/(1+z)$ would decay to zero around the arc. We know that there is a pole in that plane, so we need to find the residues there. Since this is a simple residue we use the formula

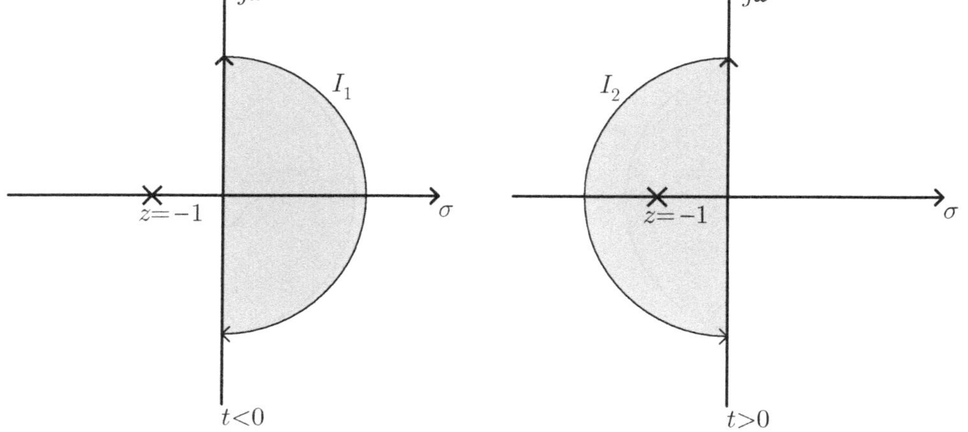

Fig. A.7 Contour integration of $\frac{e^{zt}}{z+1}$

$$\text{Residue} = (z - z_0)f(z)\big|_{z=z_0} = (z+1)\frac{e^{zt}}{1+z}\bigg|_{z=-1} = e^{-t} \quad \text{(A.168)}$$

Our integral then becomes

$$I_2 = \frac{1}{2\pi j}\oint \frac{e^{zt}}{1+z}dz = \frac{1}{2\pi j}2\pi j\sum \text{Residues} = e^{-t} \quad (t > 0) \quad \text{(A.169)}$$

In summary we have shown that

$$\frac{1}{2\pi}\int_{-\infty}^{\infty}\frac{e^{j\omega t}}{1+j\omega}d\omega = \begin{cases} 0 & t < 0 \\ e^{-t} & t > 0 \end{cases} \quad \text{(A.170)}$$

which is what we set out to prove. Again the transient function is not a conventional one! It is not simply the negative exponential! It is the *causal* negative exponential in the sense of it being zero for negative time and being the normal negative exponential for positive time. Amongst other things this means that at time zero there is an abrupt discontinuity. Yet despite this discontinuity and all the associated difficulties we can still represent this analog/digital function in terms of smooth integral in the complex frequency domain.

Contour Integration of Transform of $e^{-|t|}$ We know that the Fourier transform of the function

$$f(t) = e^{-|t|} \quad \text{(A.171)}$$

is

$$F(\omega) = \frac{1}{1+j\omega} + \frac{1}{1-j\omega} = \frac{2}{1+\omega^2} \quad \text{(A.172)}$$

(see Eq. (8.25)). Using the inverse Fourier transform this would imply that

$$e^{-|t|} = \frac{1}{2\pi}\int_{-\infty}^{\infty}\frac{2e^{j\omega t}}{1+\omega^2}d\omega \quad \text{(A.173)}$$

Again we'd like to verify this integral. When converting to the complex variable z the above integral transforms to

$$I = \frac{1}{2\pi j}\oint \frac{2e^{zt}}{1-z^2}dz \quad \text{(A.174)}$$

Notice that the integrand has two poles: one at $z = 1$ and the other at $z = -1$; both poles are simple. Again in defining the contour path we ensure it covers the $j\omega$ axis and either wraps up at positive or negative infinity, as shown in Fig. A.8.

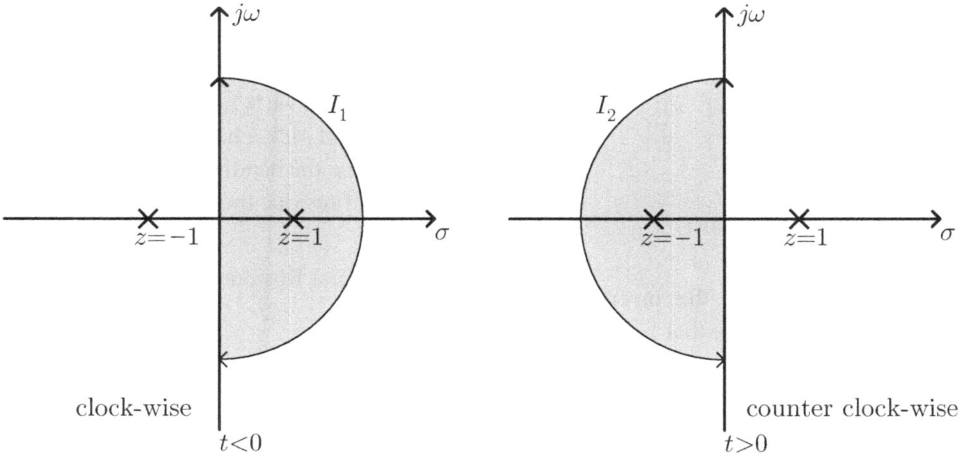

Fig. A.8 Contour integration of $\frac{2e^{zt}}{1-z^2}$

- If time is negative, then we take the right half plane contour. The residue there is

$$\text{First Residue} = [-](z-1)\frac{2e^{zt}}{(1-z)(1+z)}\bigg|_{z=1} = e^t \quad (A.175)$$

Notice that we put a negative sign since we are going clockwise. The integral then becomes

$$I = \frac{1}{2\pi j}2\pi j \sum \text{Resides} = e^t \quad (A.176)$$

- If time is positive, then we take the left half plane contour. The residue there is

$$\text{Second Residue} = (z+1)\frac{2e^{zt}}{(1-z)(1+z)}\bigg|_{z=-1} = e^{-t} \quad (A.177)$$

Notice here we did not put a negative sign, since we are going counterclockwise.

In summary we have shown that

$$\frac{1}{2\pi}\int_{-\infty}^{\infty}\frac{2e^{j\omega t}}{1+\omega^2}d\omega = \begin{cases} e^t & t < 0 \\ e^{-t} & t > 0 \end{cases} \quad (A.178)$$

which is what we wanted to show. Again a fabulous frequency representation of this symmetric by abrupt transient function.

Contour Integration of Transform of Pulse Function We know that the Fourier transform of the pulse function defined by

$$p(t) = \begin{cases} 1 & |t| < 1 \\ 0 & |t| > 1 \end{cases} \quad (A.179)$$

is

$$F(\omega) = 2\frac{\sin\omega}{\omega} \quad (A.180)$$

(see Eq. (8.13)). Using the inverse Fourier transform we ought to prove that

$$p(t) = \frac{1}{2\pi}\int_{-\infty}^{\infty}\frac{2\sin(\omega)e^{j\omega t}}{\omega}d\omega \quad (A.181)$$

First we rewrite the sine in terms of complex exponentials

$$\sin\omega = \frac{e^{j\omega} - e^{-j\omega}}{2j} \quad (A.182)$$

Then the desired integral becomes

$$p(t) = \frac{1}{2\pi j}\int_{-\infty}^{\infty}\frac{e^{j\omega(t+1)} - e^{j\omega(t-1)}}{\omega}d\omega \quad (A.183)$$

Again defining the complex variable z

$$z = \sigma + j\omega \quad (A.184)$$

the above integral gets transformed to a contour one

$$p(t) = \frac{1}{2\pi j}\oint\frac{e^{z(t+1)} - e^{z(t-1)}}{z}dz \quad (A.185)$$

Let's split this into two integrals

$$p_1(t) = \frac{1}{2\pi j}\oint\frac{e^{z(t+1)}}{z}dz \quad (A.186)$$

$$p_2(t) = \frac{1}{2\pi j}\oint\frac{e^{z(t-1)}}{z}dz \quad (A.187)$$

First Integral $p_1(t)$ The first integral has a simple pole at $z = 0$. The contour path would depend on time as shown in Fig. A.9.

- If $t < -1$, the exponential would tend to vanish around the arc extending in the right plane, so we take the right contour as shown in the left side of Fig. A.9. As we cross the zero z point, we bend around it as shown in the figure, and pick a half a residue (with positive sign since the bending was counterclockwise). The half residue there is

$$\text{Half First Residue} = \frac{1}{2}z\frac{e^{z(t+1)}}{z}\bigg|_{z=0} = \frac{1}{2} \quad (A.188)$$

Since the total path encompasses zero residues, we are left with the conclusion that the total integral equals negative half the residue at zero

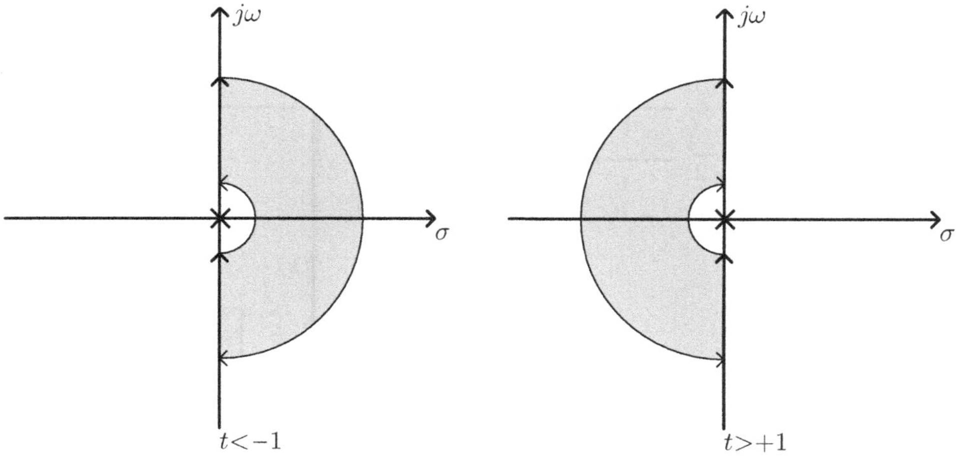

Fig. A.9 Contour integration of $\frac{e^{z(t+1)}}{z}$

$$p_{1a}(t) = -\frac{1}{2\pi j} 2\pi j \times \text{Half First Residue} = -\frac{1}{2} \tag{A.189}$$

- If $t > -1$, the exponential would tend to vanish around the arc extending in the left plane, so we take the left contour as shown in the right part of Fig. A.9. As we cross the zero z point we bend around it as shown in the figure, and pick a half a residue (with negative sign since the bending was clockwise)

$$\text{Half First Residue} = -\frac{1}{2} z \frac{e^{z(t+1)}}{z}\bigg|_{z=0} = -\frac{1}{2} \tag{A.190}$$

Since the total path encompasses zero residues, we are left with the conclusion that the total integral equals negative half the residue at zero

$$p_{1b}(t) = -\frac{1}{2\pi j} 2\pi j \times \text{Half First Residue} = \frac{1}{2} \tag{A.191}$$

In summary we have

$$p_1(t) = \begin{cases} -\frac{1}{2} & t < -1 \\ +\frac{1}{2} & t > -1 \end{cases} \tag{A.192}$$

as shown in Fig. A.10

Second Integral $p_2(t)$ The second integral $p_2(t)$ also has a simple pole at $z = 0$. The contour path would depend on time:

- If $t < +1$, the exponential would tend to vanish around the arc extending in the right plane, so we take the right contour. Similar to the first integral, as we cross the zero z point, we bend around it, and pick a half a residue which is

$$\text{Half Second Residue} = \frac{1}{2} z \frac{e^{z(t-1)}}{z}\bigg|_{z=0} = \frac{1}{2} \tag{A.193}$$

Since the total path encompasses zero residues, we are left with the conclusion that the total integral equals negative half the residue at zero

$$p_{2a}(t) = -\frac{1}{2\pi j} 2\pi j \times \text{Half Second Residue} = -\frac{1}{2} \tag{A.194}$$

- If $t > +1$, the exponential would tend to vanish around the arc extending in the left plane, so we take the left contour. Again similar to

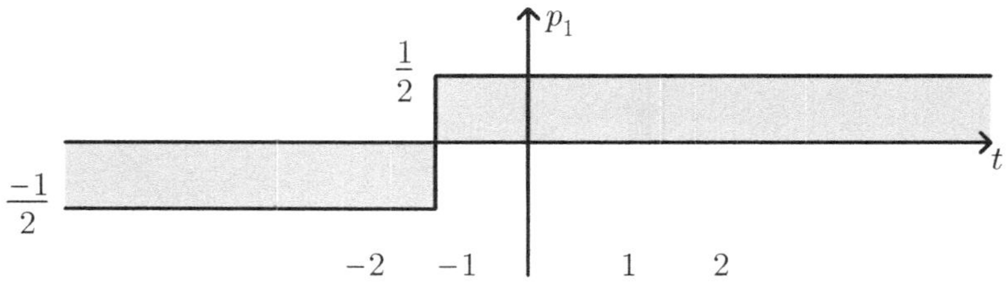

Fig. A.10 Result of first half contour integration

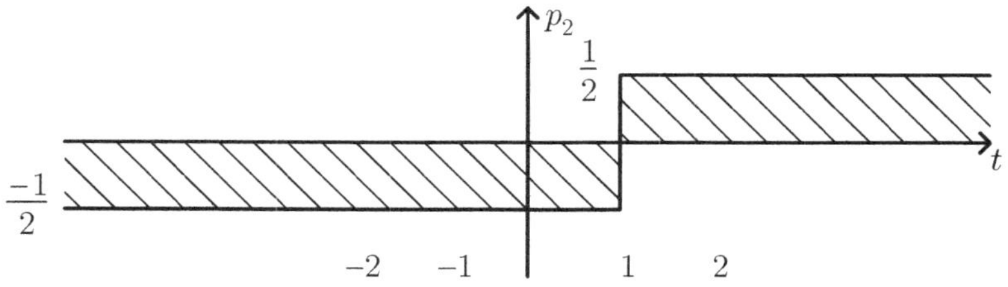

Fig. A.11 Result of second half contour integration

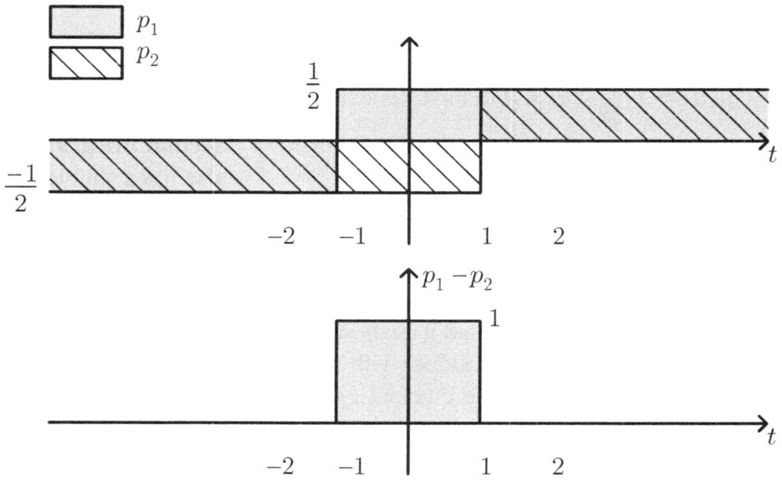

Fig. A.12 Various components of pulse function as extracted from contour integration

the first integral, as we cross the zero point we pick a half a residue (with negative sign since we are going clockwise)

$$\text{Half Second Residue} = -\frac{1}{2} z \frac{e^{z(t-1)}}{z}\bigg|_{z=0} = -\frac{1}{2} \tag{A.195}$$

Since the total path encompasses zero residues, we are left with the conclusion that the total integral equals negative half the residue at zero

$$p_{2b}(t) = -\frac{1}{2\pi j} 2\pi j \times \text{Half Second Residue} = \frac{1}{2} \tag{A.196}$$

In summary we have

$$p_2(t) = \begin{cases} -\frac{1}{2} & t < +1 \\ +\frac{1}{2} & t > +1 \end{cases} \tag{A.197}$$

as shown in Fig. A.11

Sum of Both Integrals $p_1(t) - p_2(t)$ Now that we have both integrals, we add them and get results in Fig. A.12, which in equation form comes out as shown below:

$$p(t) = p_1(t) - p_2(t) \tag{A.198}$$

$$= \begin{cases} 0 & t < -1 \\ 1 & -1 < t < 1 \\ 0 & t > 1 \end{cases} \tag{A.199}$$

This is exactly the pulse function!

Contour Integration of Transform of $u(t)te^{-t}$ We know the FT of the function $f(t) = u(t)te^{-t}$ is

$$F(\omega) = \frac{1}{(1 + j\omega)^2} \tag{A.200}$$

(see Eq. (9.112)). Using inverse FT we should get

$$u(t)te^{-t} = \frac{1}{2\pi}\int_{-\infty}^{\infty} \frac{e^{j\omega t}}{(1 + j\omega)^2} d\omega \tag{A.201}$$

We'd like to verify this complex integral. Again we introduce the complex variable

$$z = \sigma + j\omega \tag{A.202}$$

and convert the linear integration into a contour one

$$I = \frac{1}{2\pi j}\oint \frac{e^{zt}}{(1 + z)^2} dz \tag{A.203}$$

Notice that the integrand has a pole at $z = -1$, and it is of double order. The path of integration, as shown in Fig. A.13, would depend on time. If time is less than zero, then we'd want to take the right half plane, as shown in the left side of the figure. Because there are no residues there, the integral would be zero.

$$I = 0 \quad (t < 0) \tag{A.204}$$

If time is larger than zero, then we'd want to take the left half plane. Now the contour integral encircles a residue, and hence the overall integral would not be zero. The arc part of the integral would be zero, though, and this would imply that the whole integral would equal the integral along the imaginary axis (which is the desired part). To find the residue, we'd need to multiply the integrand by $(1 + z)^2$, differentiate once with respect to z and evaluate at $z = -1$:

$$\begin{aligned} \text{Residue} &= \frac{d}{dz}\left[(1 + z)^2 \frac{e^{zt}}{(1 + z)^2}\right]_{z=-1} \\ &= \frac{d}{dz}e^{zt}\Big|_{z=-1} = te^{zt}\Big|_{z=-1} = te^{-t} \end{aligned} \tag{A.205}$$

Then our integral for this case becomes

$$I = \frac{1}{2\pi j}2\pi j \sum \text{Residues} = te^{-t} \quad (t > 0) \tag{A.206}$$

So in summary we have

$$I = \begin{cases} 0 & t < 0 \\ te^{-t} & t > 0 \end{cases} \tag{A.207}$$

which is what we set to prove.

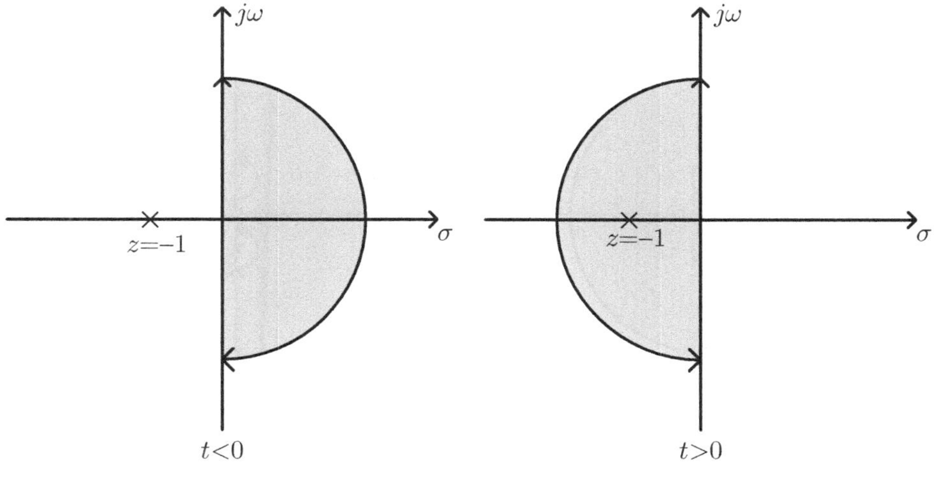

Fig. A.13 Contour integration of $\frac{e^{zt}}{(1+z)^2}$

Contour Integration of Transform of sin(t) · signum(t) Our starting function is

$$f(t) = \begin{cases} -\frac{1}{2}\sin(t) & t < 0 \\ +\frac{1}{2}\sin(t) & t > 0 \end{cases} \tag{A.208}$$

From before (see Eq. (8.73), second half and Fig. 8.17) we know that the Fourier transform of this function is

$$F(\omega) = \frac{1}{1-\omega^2} \tag{A.209}$$

This would imply that

$$\sin(t)\cdot \text{signum}(t) = \frac{1}{2\pi}\int_{-\infty}^{\infty} \frac{1}{1-\omega^2} e^{j\omega t} d\omega \tag{A.210}$$

We'd like to prove this. First let's use the complex variable z defined as

$$z = \sigma + j\omega \tag{A.211}$$

Along the $j\omega$-axis we have

$$z = j\omega, \quad z^2 = -\omega^2, \quad \text{and} \quad d\omega = \frac{1}{j}dz \tag{A.212}$$

Our integral then becomes

$$I = \frac{1}{2\pi j}\int_{-\infty}^{\infty} \frac{1}{z^2+1} e^{zt} dz \tag{A.213}$$

We will expand this integral in the form of a contour one, and assuming that the arc-part of the integral evaluates to zero, then the contour integral is our sought integral:

$$I = \frac{1}{2\pi j}\oint \frac{1}{z^2+1} e^{zt} dz \tag{A.214}$$

Since the denominator is at least two orders higher than the numerator, our arc-vanishing integral is valid. The integral has two poles: one at $z = j$ and the other at $z = -j$, so that the FT can be written as

$$\frac{1}{z^2+1} = \frac{1}{(z-j)(z+j)} \tag{A.215}$$

So far as the details of the arc would depend on time:

- Time less than zero: for this case we'd want to take the right half plane as shown in Fig. A.14. Inside the contour there are no residues; however, going along the imaginary axis we do pass by two poles, and each pole would contribute half a residue. If we define I_1 as the arc

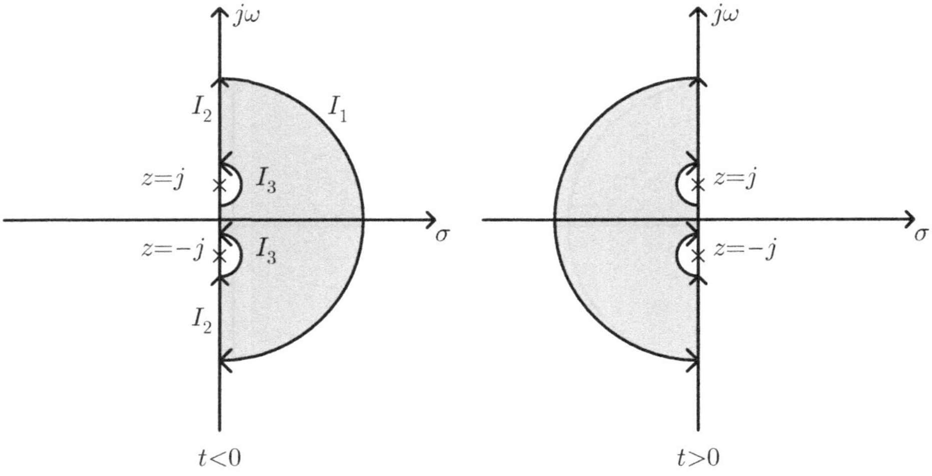

Fig. A.14 Contour integration of $\frac{e^{zt}}{z^2+1}$

integral, I_2 our desired integral and I_3 as that part of integral going above the two poles we then have

$$I_1 + I_2 + I_3 = 0 \tag{A.216}$$

which implies that

$$I_2 = -I_3 \tag{A.217}$$

So far as I_3 is concerned, it is given by

$$I_3 = \frac{1}{2\pi j} 2\pi j\,[\text{Residue at}(z = j) + \text{Residue at}(z = -j)] \tag{A.218}$$

The first residue is

$$\text{First Half Residue at } (z = j) = \frac{1}{2}\frac{(z-j)e^{zt}}{(z-j)(z+j)}\bigg|_{z=j} = \frac{1}{2}\frac{e^{jt}}{(j+j)} = \frac{1}{2}\frac{e^{jt}}{2j} \tag{A.219}$$

The second residue is

$$\text{Second Half Residue at } (z = -j) = \frac{1}{2}\frac{(z+j)e^{zt}}{(z-j)(z+j)}\bigg|_{z=-j} = \frac{1}{2}\frac{e^{-jt}}{(-j-j)} = -\frac{1}{2}\frac{e^{-jt}}{2j} \tag{A.220}$$

Then I_3 becomes

$$I_3 = \frac{1}{2}\left[\frac{e^{jt}}{2j} - \frac{e^{-jt}}{2j}\right] = \frac{1}{2}\sin(t) \tag{A.221}$$

Our sought integral is then

$$I_2 = -I_3 = -\frac{1}{2}\sin(t) \quad (t < 0) \tag{A.222}$$

- Time larger than zero: for this case we'd want to take the left half plane as shown in Fig. A.14. Again inside the contour there are no residues; however, going along the imaginary axis we do pass by two poles, and each pole would contribute half a residue. Since we pass clockwise, we pick a negative sign for the residues. Following exactly the same steps as the $t < 0$ case we arrive at

$$I_2 = \frac{1}{2}\sin(t) \quad (t > 0) \tag{A.223}$$

Combining the above two results we finally get

$$f(t) = \text{signum}(t) \cdot \sin(t) \tag{A.224}$$

which is what we want to prove!

Inverse Transform of $\frac{1}{j\omega(a+j\omega)}$ Let's find the inverse transform of the transfer function

$$F(\omega) = \frac{1}{j\omega(a + j\omega)} \tag{A.225}$$

$$f(t) = \frac{1}{2\pi}\int_{-\infty}^{\infty} \frac{e^{j\omega t}}{j\omega(a + j\omega)} d\omega \tag{A.226}$$

Replace $j\omega$ by z as follows:

$$j\omega = z, \quad d\omega = \frac{1}{j}dz \tag{A.227}$$

and the integral becomes

$$\frac{1}{2\pi j}\int_{-j\infty}^{j\infty} \frac{e^{zt}}{z(z + a)} dz \tag{A.228}$$

We have two poles for the integration, located at $z = 0$ and $z = -a$. We turn the integral into a contour one as shown in Fig. A.15.

$$I = \frac{1}{2\pi j}\oint \frac{e^{zt}}{z(z + a)} dz \tag{A.229}$$

We split the integral into three parts, namely I_1 which is the path around the arc, which would go

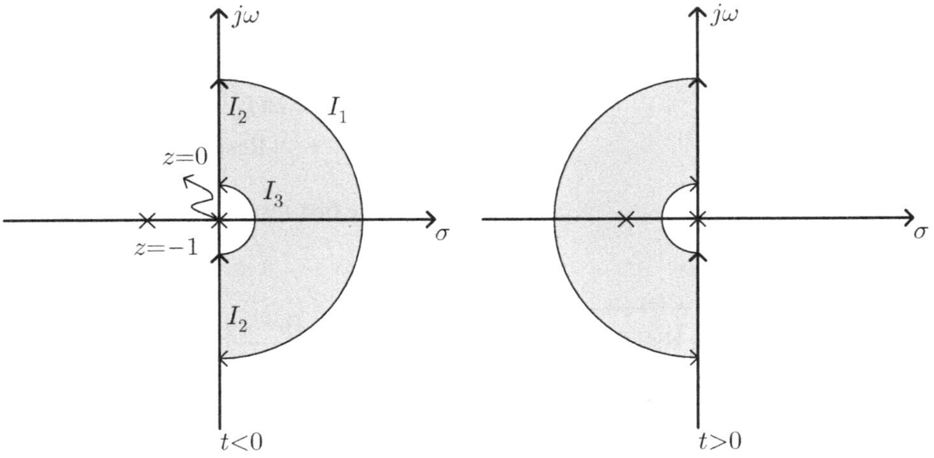

Fig. A.15 Contour integration of $\frac{e^{zt}}{z(z+a)}$

to zero. The second one I_2 is the desired integral; and the third one I_3 is the path around the poles. Since $I_1 = 0$ we would have

$$I_2 + I_3 = \sum \text{Residues} \tag{A.230}$$

- Negative time: here we would have to take the contour on the right-hand side. There are no residues inside so total integral would be zero, which would imply that

$$I_2 = -I_3 \tag{A.231}$$

  When evaluating I_3 we pass by $z = 0$ and we would pick half a residue.

$$\begin{aligned} \text{First residue/2} &= \frac{1}{2} z \frac{e^{zt}}{z(z+a)}\bigg|_{z=0} \\ &= \frac{1}{2a} \end{aligned} \tag{A.232}$$

  Then we would have

$$I_2 = -I_3 = -\frac{1}{2a} \tag{A.233}$$

- Positive time: here we would have to take the contour on the left-hand side. We would pick the same half residue, but with a sign flip (since we are going clockwise).

$$I_3 = -\frac{1}{2a} \tag{A.234}$$

  Also, we now have a residue inside the contour. The total integral would give

$$I_2 = -I_3 + \text{Second residue} \tag{A.235}$$

  The second residue at $z = -a$ is

$$\text{Second residue} = (z+a)\frac{e^{zt}}{z(z+a)}\bigg|_{z=-a} = -\frac{e^{-at}}{a} \tag{A.236}$$

  Our target integral then becomes

$$\begin{aligned} I_2 &= -I_3 + \text{Second residue} \\ &= \frac{1}{2a} - \frac{e^{-at}}{a} \end{aligned} \tag{A.237}$$

In summary we now have

$$f(t) = \begin{cases} -\frac{1}{2a} & t < 0 \\ +\frac{1}{2a} - \frac{e^{-at}}{a} & t > 0 \end{cases} \tag{A.238}$$

Or we can utilize the signum function

$$f(t) = \frac{1}{a}\left[\text{signum}(t) - u(t)e^{-at}\right] \tag{A.239}$$

Notice that we could have arrived at the same answer by using partial fraction:

$$\frac{1}{j\omega(a+j\omega)} = \frac{1}{a}\left[\frac{1}{j\omega} - \frac{1}{a+j\omega}\right] \quad (A.240)$$

Taking inverse transform, the first terms give the signum function, while the second the negative exponential!

Contour Integration of Transform of Stair Signum Function As was shown before (Eq. (10.6)), we had established that the Fourier transform of the stair signum function was

$$F(\omega, \tau) = \frac{\cos\omega\tau}{j\omega} \quad (A.241)$$

This would imply that the stair signum function can be extracted from the inverse FT

$$f(t,\tau) = \frac{1}{2\pi}\int_{-\infty}^{\infty} \frac{\cos\omega\tau}{j\omega} e^{j\omega t} d\omega \quad (A.242)$$

Let's rewrite using Euler's identity

$$f(t,\tau) = \frac{1}{2\pi}\int_{-\infty}^{\infty} \frac{e^{j\omega\tau} + e^{-j\omega\tau}}{2j\omega} e^{j\omega t} d\omega \quad (A.243)$$

Now replace $j\omega$ by z and convert into contour integration

$$\begin{aligned} f(t,\tau) &= \frac{1}{4\pi j}\oint \frac{e^{z\tau} + e^{-z\tau}}{z} e^{zt} dz \\ &= \frac{1}{4\pi j}\oint \frac{e^{z(t-\tau)} + e^{z(t+\tau)}}{z} dz \\ &= \frac{1}{4\pi j}[I_1 + I_2] \end{aligned} \quad (A.244)$$

where

$$I_1 = \oint \frac{e^{z(t-\tau)}}{z} dz \quad \text{and} \quad I_2 = \oint \frac{e^{z(t+\tau)}}{z} dz \quad (A.245)$$

Both integrals have a pole at $z = 0$ and both give half a residue there. For the first integral, if $t < \tau$ we would integrate around the right half plane and pick a $-j\pi$. For $t > \tau$ we would use the left half plane and pick $j\pi$. This translates to

$$I_1 = j2\pi \cdot \text{signum}(t-\tau) \quad (A.246)$$

For the second integral, if $t < -\tau$ we would integrate around the right half plane and pick a $-j\pi$. For $t > -\tau$ we would use the left half plane and pick $j\pi$. This translates to

$$I_2 = j2\pi \cdot \text{signum}(t+\tau) \quad (A.247)$$

Our desired integral is then

$$\begin{aligned} f(t,\tau) &= \frac{1}{4\pi j}[I_1 + I_2] \\ &= \frac{1}{4\pi j}[j2\pi \cdot \text{signum}(t-\tau) \\ &\quad + j2\pi \cdot \text{signum}(t+\tau)] \\ &= \frac{1}{2}[\text{signum}(t-\tau) + \text{signum}(t+\tau)] \end{aligned} \quad (A.248)$$

Results are shown in Fig. A.16.

A.7 Application of Contour Integration in Fourier Transforms

Just like we used contour integration to figure inverse Fourier transform, we can use contour integration to find the Fourier transform itself. We will illustrate this via an example.

Fourier Transform of $\frac{1}{1+t^4}$ Let's find the Fourier transform of the function

$$f(t) = \frac{1}{1+t^4} \quad (A.249)$$

$$F(\omega) = \int_{-\infty}^{\infty} \frac{e^{-j\omega t}}{1+t^4} dt \quad (A.250)$$

Change variables from t to x in anticipation of going to the $z = x + jy$ complex plane:

$$F(\omega) = \int_{-\infty}^{\infty} \frac{e^{-j\omega x}}{1+x^4} dx \quad (A.251)$$

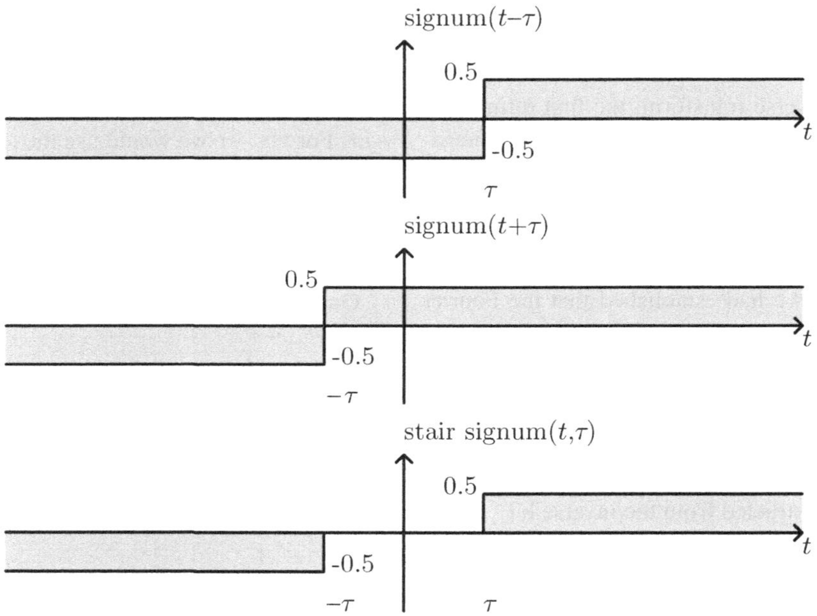

Fig. A.16 Stair signum function as a combination of regular signum functions

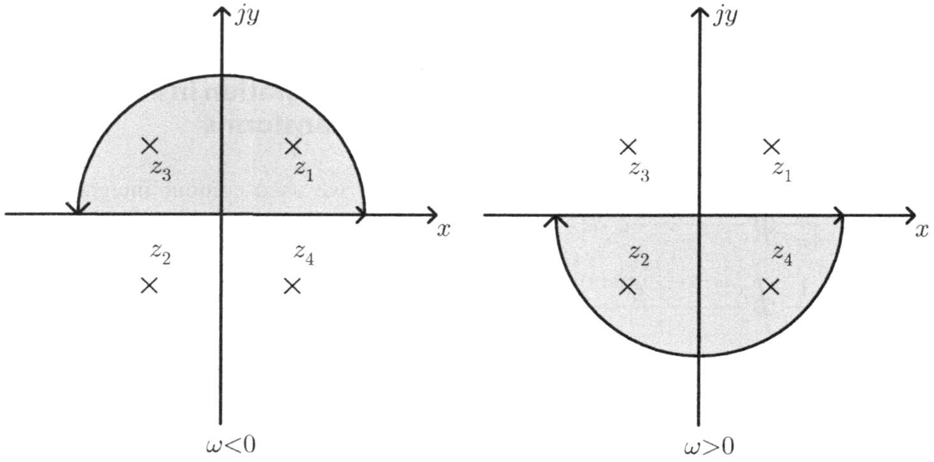

Fig. A.17 Contour integral of $\frac{e^{-j\omega z}}{1+z^4}$

Now set the complex variable z to

$$z = x + jy \quad \text{such that} \quad dz = dx + jdy \tag{A.252}$$

Along the real axis, we can set $z = x$ and $dz = dx$; then we would have

$$F(\omega) = \int_{-\infty}^{\infty} \frac{e^{-j\omega z}}{1+z^4} dz \tag{A.253}$$

Change the linear integration to a contour one

$$F(\omega) = \oint \frac{e^{-j\omega z}}{1+z^4} dz \tag{A.254}$$

on the premise that the arc part of the integral would vanish. The integrand denominator has four roots, as shown in Fig. A.17.

$$z_1 = \frac{1}{\sqrt{2}}(1+j), z_2 = \frac{1}{\sqrt{2}}(-1-j), z_3 = \frac{1}{\sqrt{2}}(-1+j), z_4 = \frac{1}{\sqrt{2}}(1-j) \tag{A.255}$$

Our integrand becomes

$$\text{integrand} = \frac{e^{-j\omega z}}{\left[z - \frac{1}{\sqrt{2}}(1+j)\right]\left[z - \frac{1}{\sqrt{2}}(-1-j)\right]\left[z - \frac{1}{\sqrt{2}}(-1+j)\right]\left[z - \frac{1}{\sqrt{2}}(1-j)\right]} \tag{A.256}$$

When $\omega < 0$ we want to pick the upper half of the complex plane to ensure thearc part of the integral dies out; there the contour would encircle z_1 and z_3. The residue at z_1 is given by

$$\begin{aligned}
\text{first residue} &= \left.\frac{e^{-j\omega z}}{\left[z - \frac{1}{\sqrt{2}}(-1-j)\right]\left[z - \frac{1}{\sqrt{2}}(-1+j)\right]\left[z - \frac{1}{\sqrt{2}}(1-j)\right]}\right|_{z=\frac{1}{\sqrt{2}}(1+j)} \\
&= \frac{e^{-j\omega \frac{1}{\sqrt{2}}(1+j)}}{\left[\frac{2}{\sqrt{2}}(1+j)\right]\left[\frac{1}{\sqrt{2}}(1+j) - \frac{1}{\sqrt{2}}(-1+j)\right]\left[\frac{1}{\sqrt{2}}(1+j) - \frac{1}{\sqrt{2}}(1-j)\right]} \\
&= \frac{e^{\omega/\sqrt{2}}e^{-j\omega/\sqrt{2}}}{\left[\frac{2}{\sqrt{2}}(1+j)\right]\left[\frac{2}{\sqrt{2}}\right]\left[\frac{2j}{\sqrt{2}}\right]} = \frac{e^{\omega/\sqrt{2}}e^{-j\omega/\sqrt{2}}}{2\sqrt{2}(-1+j)} \\
&= \frac{e^{\omega/\sqrt{2}}\left[\cos\omega/\sqrt{2} - j\sin\omega/\sqrt{2}\right]}{2\sqrt{2}(-1+j)}
\end{aligned} \tag{A.257}$$

$$\text{first residue} = \frac{e^{\omega/\sqrt{2}}\left[\cos\omega/\sqrt{2} - j\sin\omega/\sqrt{2}\right][-1-j]}{4\sqrt{2}} \tag{A.258}$$

The second residue at z_3 is

$$\begin{aligned}
\text{second residue} &= \left.\frac{e^{-j\omega z}}{\left[z - \frac{1}{\sqrt{2}}(1+j)\right]\left[z - \frac{1}{\sqrt{2}}(-1-j)\right]\left[z - \frac{1}{\sqrt{2}}(1-j)\right]}\right|_{z=\frac{1}{\sqrt{2}}(-1+j)} \\
&= \frac{e^{-j\omega \frac{1}{\sqrt{2}}(-1+j)}}{\left[-\frac{2}{\sqrt{2}}\right]\left[\frac{1}{\sqrt{2}}(-1+j) - \frac{1}{\sqrt{2}}(-1-j)\right]\left[\frac{1}{\sqrt{2}}(-1+j) - \frac{1}{\sqrt{2}}(1-j)\right]} \\
&= \frac{e^{-j\omega \frac{1}{\sqrt{2}}(-1+j)}}{\left[-\frac{2}{\sqrt{2}}\right]\left[\frac{2j}{\sqrt{2}}\right]\left[\frac{2}{\sqrt{2}}(-1+j)\right]} = \frac{e^{\omega/\sqrt{2}}e^{j\omega/\sqrt{2}}}{2\sqrt{2}(1+j)} \\
&= \frac{e^{\omega/\sqrt{2}}\left[\cos\omega/\sqrt{2} + j\sin\omega/\sqrt{2}\right][1-j]}{4\sqrt{2}}
\end{aligned} \tag{A.259}$$

The sum of the residues is then

$$\text{sum of res} = \frac{e^{\omega/\sqrt{2}}\left[\cos\omega/\sqrt{2} - j\sin\omega/\sqrt{2}\right][-1-j]}{4\sqrt{2}} + \frac{e^{\omega/\sqrt{2}}\left[\cos\omega/\sqrt{2} + j\sin\omega/\sqrt{2}\right][1-j]}{4\sqrt{2}} \tag{A.260}$$

$$\begin{aligned}\text{sum of res} &= \frac{e^{\omega/\sqrt{2}}}{4\sqrt{2}}\Big[-\cos\omega/\sqrt{2} - \sin\omega/\sqrt{2} - j\cos\omega/\sqrt{2} + j\sin\omega/\sqrt{2} \\ &\quad + \cos\omega/\sqrt{2} + \sin\omega/\sqrt{2} - j\cos\omega/\sqrt{2} + j\sin\omega/\sqrt{2}\Big] \\ &= \frac{e^{\omega/\sqrt{2}}}{2\sqrt{2}}\left[-j\cos\frac{\omega}{\sqrt{2}} + j\sin\frac{\omega}{\sqrt{2}}\right]\end{aligned} \tag{A.261}$$

The integral then becomes

$$\begin{aligned} I &= 2\pi j \times \text{sum of res} \\ &= \boxed{\frac{\pi}{\sqrt{2}} e^{\omega/\sqrt{2}}\left[\cos\frac{\omega}{\sqrt{2}} - \sin\frac{\omega}{\sqrt{2}}\right]}, \quad \omega < 0 \end{aligned} \tag{A.262}$$

Similarly for $\omega > 0$ we get

$$I = \boxed{\frac{\pi}{\sqrt{2}} e^{-\omega/\sqrt{2}}\left[\cos\frac{\omega}{\sqrt{2}} + \sin\frac{\omega}{\sqrt{2}}\right]}, \quad \omega > 0 \tag{A.263}$$

A plot of this Fourier transform is shown in Fig. A.18. Notice that since the time function is both real and symmetric, the Fourier transform is also real and symmetric! How do we know if our results are valid? We plot the inverse transform of the transform! This is shown in Fig. A.19; as can be seen our results are good.

A.8 Complex Identities

We wrap up this chapter with a few useful identities. Start with

$$e^{jA} = \cos A + j\sin A \tag{A.264}$$

Then on the one hand

$$e^{jA}e^{jB} = e^{j(A+B)} \tag{A.265}$$

$$= \cos(A+B) + j\sin(A+B) \tag{A.266}$$

and

$$e^{jA}e^{-jB} = e^{j(A-B)} \tag{A.267}$$

$$= \cos(A-B) + j\sin(A-B) \tag{A.268}$$

On the other hand

$$\begin{aligned} e^{jA}e^{jB} &= [\cos A + j\sin A][\cos B + j\sin B] \\ &= \cos A\cos B - \sin A\sin B \\ &\quad + j[\cos A\sin B + \sin A\cos B] \end{aligned} \tag{A.269}$$

and

$$\begin{aligned} e^{jA}e^{-jB} &= [\cos A + j\sin A][\cos B - j\sin B] \\ &= \cos A\cos B + \sin A\sin B \\ &\quad + j[-\cos A\sin B + \sin A\cos B] \end{aligned} \tag{A.270}$$

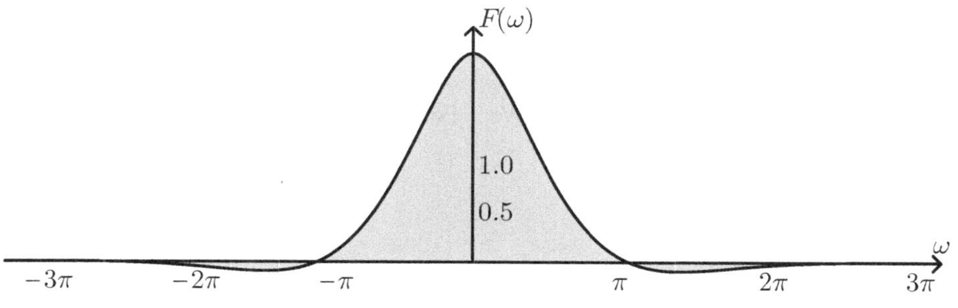

Fig. A.18 Fourier transform of $f(t) = \frac{1}{1+t^4}$

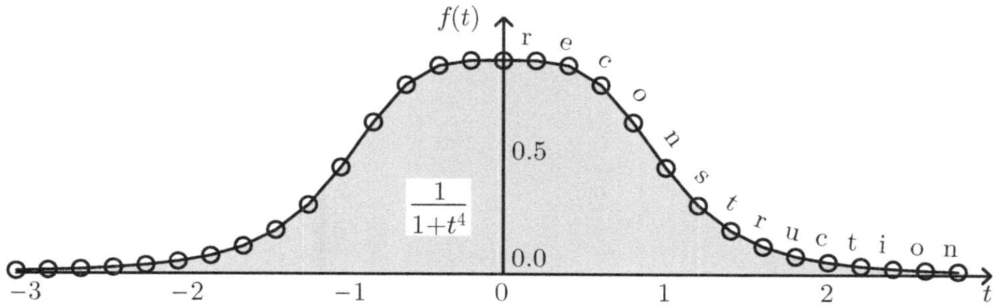

Fig. A.19 Inverse Fourier transform of $\mathcal{F}\left[\frac{1}{1+t^4}\right]$ and comparison to original function

If we next add

$$e^{jA}e^{jB} + e^{jA}e^{-jB} \tag{A.271}$$

in both cases and equate real parts we get

$$\cos A \cos B = \frac{\cos(A+B) + \cos(A-B)}{2} \tag{A.272}$$

Equating imaginary parts we get

$$\sin A \cos B = \frac{\sin(A-B) + \sin(A+B)}{2} \tag{A.273}$$

Similarly if we subtract

$$e^{jA}e^{-jB} - e^{jA}e^{jB} \tag{A.274}$$

and equate real parts we get

$$\sin A \sin B = \frac{\cos(A-B) - \cos(A+B)}{2} \tag{A.275}$$

A.9 Summary

In this appendix we got a first exposition to complex analysis. This topic, which is a cornerstone of mathematical analysis, is very elaborate and vast and there are whole books and treatises covering it. We could only scratch the surface and hopefully convince the reader to take it seriously, see first hand how it can be used in real life applications and perhaps potentially pursue some further learning in it if needed. After covering some introductory material on complex analysis and the complex plane, including analytic functions and the Cauchy theorem, we showed how complex integration can be used to figure indefinite integrals. Then we dived into our main goal of the chapter which is figuring inverse Fourier transforms. The trick is to map the linear integration (over frequency) to a contour one (in the complex plane), and ensure that arc-part of the integral goes to zero. Then the contour integral can be evaluated by simply adding the residues inside the integration semi-circle. We demonstrated the process with multiple examples and we wrapped the chapter with the parallel path of using contour integration to figure the actual Fourier transform (as opposed to its inverse).

References

Books on Fourier Series and Transform

1. B.P. Lathi, *Signals, Systems and Communications* (Wiley, New York, 1965)
2. B.P. Lathi, *Linear Systems and Signals* (Oxford University Press, New York, 2005)
3. A.V. Oppenheim, A.S. Willsky, *Signals and Systems* (Prentice-Hall, New York, 1982)
4. D.K. Cheng, *Analysis of Linear Systems* (Addison-Wesley, Reading, MA, 1959)
5. B. Boulet, *Fundamentals of Signals and Systems* (Da Vinci Engineering Press, Hingham, MA, 2006)
6. E. Kreyszig, *Advanced Engineering Mathematics*, 4th edn. (Wiley, New York, 1999)
7. K. Morita, *Applied Fourier Transform* (IOS Press, Amsterdam, 1995)

Books on Laplace Transform

8. B.P. Lathi, *Signals, Systems and Communications* (Wiley, New York, 1965)
9. B.P. Lathi, *Linear Systems and Signals* (Oxford University Press, New York, 2005)
10. D.K. Cheng, *Analysis of Linear Systems* (Addison-Wesley, Reading, MA, 1959)
11. B. Boulet, *Fundamentals of Signals and Systems* (Da Vinci Engineering Press, Hingham, MA, 2006)
12. J.L. Schiff, *The Laplace Transform: Theory and Applications* (Springer, New York, 1991)
13. E. Kreyszig, *Advanced Engineering Mathematics*, 4th edn. (Wiley, New York, 1999)

Books on Impulse Response and Convolution

14. B.P. Lathi, *Signals, Systems and Communications* (Wiley, New York, 1965)
15. D.K. Cheng, *Analysis of Linear Systems* (Addison-Wesley, Reading, MA, 1959)

Books on Basic Circuit Theory

16. D.E. Johnson, J.L. Hilburn, J.R. Johnson, *Basic Electric Circuit Analysis* (Prentice-Hall, New Jersey, 1978)

Books on Feedback Systems

17. S.M. Shinners, *Modern Control System Theory and Application* (Addison-Wesley, Reading, MA, 1978)

Books on MOSFET Circuits

18. R.J. Baker, *CMOS Circuit Design, Layout and Simulations* (Wiley, New York, 2005)
19. P.R. Gray, R.G. Meyer, *Analysis and Design of Analog Integrated Circuits* (Wiley, New York, 1990)
20. P.E. Allen, D.R. Holberg, *CMOS Analog Circuit Design* (Oxford University Press, New York, 1987)

F. Badrieh, *Spectral, Convolution and Numerical Techniques in Circuit Theory*,
https://doi.org/10.1007/978-3-319-71437-0

Books on Electromagnetics

21. H.A. Haus, J.R. Melcher, *Electromagnetic Fields and Energy* (Prentice-Hall, New Jersey, 1989)

Books on Transmission Lines

22. R.E. Matick, *Transmission Lines for Digital And Communication Networks* (IEEE Press, New York, 1995)
23. H.A. Haus, J.R. Melcher, *Electromagnetic Fields and Energy* (Prentice-Hall, New Jersey, 1989)

Books on Complex Analysis

24. W.W. Chen, *Introduction to Complex Analysis (lecture notes), Imperial College* (University of London, London, 2008)
25. J.L. Schiff, *The Laplace Transform: Theory and Applications* (Springer, New York, 1991)
26. T.W. Gamelin, *Complex Analysis* (Springer, New York, 2001)
27. E. Kreyszig, *Advanced Engineering Mathematics*, 4th edn. (Wiley, New York, 1999)

Books on Scattering Parameters

28. S.J. Orfanidis, *Electromagnetic Waves and Antennas*, available at www.ece.rutgers.edu/~orfanidi/ewa (2016)

Books on Partial Differential Equations

29. H.F. Weinberger, *A First Course In Partial Differential Equations* (Dover, New York, 1995)
30. J.D. Logan, *Applied Partial Differential Equations*, 2nd edn. (Springer, New York, 2004)
31. R. Haberman, *Applied Partial Differential Equations with Fourier Series and Boundary Value Problems*, 5th edn. (Pearson, New York, 2012)
32. S.J. Farlow, *Partial Differential Equations for Scientists and Engineers* (Dover, New York, 1993)
33. I.G. Petrovsky, *Lectures On Partial Differential Equations* (Dover, New York, 1992)

Books on Numerical Analysis

34. A. Ralston, P. Rabinowitz, *A First Course in Numerical Analysis*, 2nd edn. (Dover, New York, 2001)
35. E. Kreyszig, *Advanced Engineering Mathematics*, 4th edn. (Wiley, New York, 1999)

C-Programming Manual

36. S. Loosemore et al., *The GNU C Library Reference Manual* (Free Software Foundation, Boston, 2016)

SPICE Simulator

37. M. Engelhardt, *LTSPICE Getting Started Guide* (Linear Technology, Milpitas, CA, 2011)

Index

F. Badrieh, *Spectral, Convolution and Numerical Techniques in Circuit Theory*,
https://doi.org/10.1007/978-3-319-71437-0

D

E

F

T

U

V

W

The manufacturer's authorised representative in the EU is Springer Nature Customer Service Centre GmbH, Europaplatz 3, 69115 Heidelberg, Germany. If you have any concerns regarding our products, please contact ProductSafety@springernature.com

Printed and bound by CPI Group (UK) Ltd, Croydon, CR0 4YY

30/09/2026

02225433-0005